W0037601

Invasive Species in Forests and Rangelands of the United States

Therese M. Poland • Toral Patel-Weynand
Deborah M. Finch • Chelcy Ford Miniat
Deborah C. Hayes • Vanessa M. Lopez
Editors

Invasive Species in Forests and Rangelands of the United States

A Comprehensive Science Synthesis for the United States Forest Sector

Editors
Therese M. Poland
U.S. Department of Agriculture
Forest Service, Northern Research Station
Lansing, MI, USA

Toral Patel-Weynand
U.S. Department of Agriculture
Forest Service, Sustainable Forest
Management Research
Washington, DC, USA

Deborah M. Finch
U.S. Department of Agriculture
Forest Service, Rocky Mountain Research Station
Albuquerque, NM, USA

Chelcy Ford Miniat
U.S. Department of Agriculture
Forest Service, Southern Research Station,
Coweeta Hydrologic Lab
Otto, NC, USA

Deborah C. Hayes
U.S. Department of Agriculture
Forest Service, Sustainable Forest
Management Research
Washington, DC, USA

Vanessa M. Lopez
U.S. Department of Agriculture
Forest Service, State and Private Forestry
Washington, DC, USA

ISBN 978-3-030-45366-4 ISBN 978-3-030-45367-1 (eBook)
https://doi.org/10.1007/978-3-030-45367-1

Photo Captions and Credits for the Front Cover: From left to right: Yellow floating heart creates dense mats and outcompetes native aquatic plants, Bruce Hansen, Aquatic Ecologist, USDA Forest Service, Pacific Northwest Research Station. Cheatgrass-invaded landscapes increase the risk of wildfires that threaten adjacent forests, Nolan Preece, Freelance Artist and Photographer. The emerald ash borer is the most devastating invasive forest insect pest in North America and threatens ash trees, Deborah L. Miller, Entomologist, USDA Forest Service, Northern Research Station. The hemlock woolly adelgid causes widespread death and destruction to hemlock trees across the landscape, Chelcy Ford Miniat, Research Ecologist, USDA Forest Service, Southern Research Station.

This Springer imprint is published by the registered company Springer Nature Switzerland AG
The registered company address is: Gewerbestrasse 11, 6330 Cham, Switzerland

In memory of Steven J. Seybold, October 14, 1959 – November 15, 2019, for his valuable contributions to this book and the science of forest entomology and invasive species.

Acknowledgments

The book editors and chapter authors thank the many reviewers who helped improve the chapters and the overall volume. Reviews consisted of multiple stages including informal chapter reviews requested by the authors, formal chapter peer reviews arranged by the editors, content and grammatical reviews across the entire volume, and high-level reviews for publishing approval. We also thank Becky Kerns and Robert Progar for their valuable contributions to compiling, synthesizing, reviewing, and editing the document.

Executive Summary

Invasive Species in Forests and Rangelands of the United States is a sector-wide scientific assessment of the current state of invasive species science and research in the United States. Leading experts on invasive pests, climate change, social sciences, and forest and rangeland management contributed to highlighting the science and identifying knowledge gaps on a diverse array of topics related to invasive species. Stakeholders from nongovernmental organizations, academic institutions, professional organizations, private corporations, and state and federal agencies representing public, private, and tribal interests also provided input to the assessment. Input from these stakeholders helped to frame the subject matter content and management options presented in this report, ensuring relevance for decision-makers and resource managers.

Invasive species can be serious threats to native ecosystems. With ever-increasing world trade, they have caused, and will continue to cause, enormous ecological and economic damage. Understanding and managing invasive species are critical for protecting and restoring resilient forest and grassland ecosystems.

This report is divided into the following chapters that address the most significant aspects and issues related to invasive species:

- Impacts of Invasive Species in Terrestrial and Aquatic Systems
- Impacts of Invasive Species on Forest and Grassland Ecosystem Processes
- Effects of Climate Change on Invasive Species
- Invasive Species Response to Natural and Anthropogenic Disturbance
- Early Intervention Strategies for Invasive Species Management
- Management of Landscapes for Established Invasive Species
- Rehabilitation and Restoration of Landscapes and Habitats Affected by Established Invasive Species
- Sectoral Impacts of Invasive Species and Approaches to Management
- Inventory and Monitoring of Invasive Species
- Tools and Technologies for Quantifying Spread and Impacts of Invasive Species
- Social and Cultural Dynamics of Non-native Invasive Species
- The Role of International Cooperation in Invasive Species Research
- Economics of Invasive Species
- Legislation and Policy
- Future Invasive Species Research Challenges and Opportunities

The sections below discuss the most important findings of this report, including summaries of key issues in each major region of the United States. These sections take a comprehensive look at ecological and economic impacts of invasive species, interactions of invasive species with changing environmental conditions, social dynamics related to invasive species, early intervention and management of invasive species, managing for resilience, restoring ecosystems impacted by invasive species, regional summaries, and an imperative for action.

Ecological and Economic Impacts of Invasive Species

The introduction and establishment of invasive species continue to increase with world trade, and there appears to be no sign of saturation for most taxonomic groups. Invasive species have significant impacts on ecosystems, ecological processes, and both local and national economies. They can substantially alter the composition, structure, or function of native terrestrial and aquatic systems. Their impacts may occur directly or indirectly at the genetic, organism, population, community, and ecosystem levels. A short synopsis of the ecological impacts of invasive organisms in both terrestrial and aquatic environments below, illustrates the breadth and depth of coverage in this area of the assessment. The assessment also addresses management strategies and potential economic implications of invasive species and their spread.

Invasive terrestrial plants can alter key system processes such as productivity and nutrient cycling, which can impact native plants and animals at higher trophic levels through the food web. Invasive plants frequently accelerate carbon cycling, potentially making more energy available to other species, but these effects depend on traits of the invader, resident species in the invaded ecosystem, and environmental conditions. Invasive species that accelerate disturbance regimes can lead to significant declines in carbon storage. In general, invasive plants in grasslands and forests increase the amount of nitrogen in aboveground plant tissues and the availability of inorganic nitrogen in invaded soils and can stimulate microbial activity. Invasion of riparian areas can also affect water quality by altering stream channel morphology, leading to altered timing of hydrologic cycles, such as flood frequency and severity.

Freshwater aquatic habitats are even more vulnerable than terrestrial habitats to invasive plants, which can reduce sunlight penetration and dissolved oxygen content and can alter water temperature, pH, and nutrient concentrations. There is considerable diversity of invasive aquatic animals. Their impacts range in type, severity, and level of biological organization, from genes to organisms to populations to communities to ecosystems. Invasive aquatic plants can greatly impair human uses of water bodies by restricting navigation; impeding water movement important for flood control, irrigation, or hydropower; disrupting recreational activities; and decreasing aesthetics, property values, and tourism.

Invasive insects and plant pathogens (or complexes involving both) cause tree mortality, resulting in canopy gaps, stand thinning, or overstory removals that, in turn, alter microenvironments and hydrologic or biogeochemical cycling regimes. These changes can shift the overall species composition and structure of the plant community, with associated effects on terrestrial and aquatic fauna. In the short term, invasive insects and diseases can generally reduce productivity of desired species in forests. Tree mortality or defoliation can affect leaf-level transpiration rates, affecting watershed hydrology. Tree mortality caused by invasive insects and pathogens also leads to enormously high costs for tree removal, other management responses, and reduced property values in urban and residential landscapes.

Invasive terrestrial vertebrates impact native systems through a wide variety of mechanisms, including overgrazing, seed consumption, predation and related indirect trophic effects, resource competition, and hybridization with native species. They can alter soil nutrient cycling processes and food webs by consuming and redistributing litter and soil, and altering soil physical structure. Invasive vertebrates can also introduce or serve as vectors of harmful pathogens that cause disease in wildlife, livestock, and humans.

Control of invasive species with chemicals or barriers can have unintended negative consequences for native species. Aside from the harm they may cause to native species, control efforts are often unsuccessful, reinforcing the importance of prevention, early detection, and rapid response in managing biological invasions.

Invasive species cause **damage to economically valuable host resources** and negatively affect the state of native ecosystems and economically important crops. They also impact non-market values and ecosystem services, including social infrastructure, recreational use, aesthetics, and factors associated with human health. In general, the value of the damages and impacts on these ecosystem services is difficult to measure because these services are not traded in markets and do not, therefore, have observable prices.

Interactions of Invasive Species with Changing Environmental Conditions

Invasive species tend to have high dispersal abilities, rapid growth with short generation times, and high tolerance of broad environmental conditions. Collectively, these traits are associated with successional species regardless of their geographic origin. These traits greatly enhance the ability of species to adapt to new habitats and cope with rapid changes in environmental conditions, such as those associated with climatic variability and change. Climate-related changes will almost certainly lead to changes in the distribution of invasive species, as their populations respond to variability and changes in temperature, moisture, and biotic interactions. Predicting how invasive species will respond under potential climate change scenarios is difficult but essential to developing effective prevention, control, and restoration strategies.

Disturbances can also have a major influence on ecosystem vulnerability. Increased levels of disturbances generally increase the probability of invasive plant species becoming established and spreading, while increased native biodiversity, in the absence of disturbances, reduces the probability of invasion. Invasive plants that were deliberately introduced for horticulture are possibly more likely to invade adjacent undisturbed forests than are weedy species that are accidentally introduced. The probability of insect pest and plant pathogen invasion is indirectly tied to disturbances and native biodiversity through their effects on the availability and spatial distribution of suitable hosts.

Social Dynamics Related to Invasive Species

Social-ecological issues related to invasive species and their management are complex. Humans can both cause and address invasive species problems through their complex interactions with ecosystems. Public awareness of invasive species and support for management and control are highly variable. Engaging the public is essential to promoting cultural changes in the understanding of invasive species and to the implementation of widespread management actions at landscape levels. Public participation in invasive species monitoring and management can both complement and amplify the work of natural resource professionals. Innovative techniques often simultaneously address various human-dimension issues, including attitude change, stakeholder engagement, a landscape-level perspective, a common vision leading to cross-boundary cooperation, behavior changes, and aiding and participation in public agency work. Public engagement efforts that incorporate input from the public and stakeholders about their priorities and values are often more successful than efforts to simply educate and persuade the public. Citizen science, geospatial participatory modeling, and social marketing offer promise for enhancing public science and engagement for invasive species.

Early Intervention and Management of Invasive Species

Early intervention is the most cost-effective approach to invasive species management. Potential damage from invasive species can be avoided if invasion is prevented through regulatory or technical approaches. Pest risk assessments evaluate the likelihood that individual species will invade and cause economic, ecological, or social harm and may include pathway analyses, which attempt to characterize how arrays of pests might be moved into areas of concern. Risk assessment results can be used at the landscape scale to distinguish and prioritize species that are already present and require management from those that have naturalized but are unlikely to have significant negative impacts. In addition, assessments and risk mapping can help identify species of high invasion risk that should be the focus of early detection/rapid response programs. New tools are being developed to improve early detection, including improvements in inventory and monitoring, remote sensing, environmental DNA, decision-support systems, and improved models to predict invasion and spread.

For many non-native species established in the United States, when discovered, little is initially known about their biology, ecology, host interactions, ecological and economic impacts, or how to manage them. Considerable research has been conducted to understand the biology and ecology of invasive species, as well as to develop and evaluate management approaches for established populations. Federal and state quarantines regulate the movement of many significant invasive pests. Recent research addresses the efficacy of current regulations and has led to new regulations for the treatment of solid wood packing and shipping material. Public outreach and education promote awareness and support for regulations and control actions toward invasive vertebrates and aquatic animals, including trapping and shooting.

The efficacy of physical control is being evaluated for invasive vertebrates, including fencing, sound devices to frighten animals, and lethal control in the form of trapping and shooting, as well as for invasive aquatic organisms, including angling, netting, water skimmers, and hand removal of egg clutches. Current control practices which have been developed and implemented for the management of invasive plant species involve mulching, irrigation, mechanical root cutting, sanitation, harvesting, prescribed fire, and silvicultural manipulations. Research on pesticides includes evaluating efficacy of new rodenticides, insecticides, and chemical controls for swine feral (*Sus scrofa*) and identifying safe and effective piscicides for invasive fish. Biological control research has led to identifying natural enemies of hemlock woolly adelgid (*Adelges tsugae*) and emerald ash borer (*Agrilus planipennis*) in their native ranges; developing and evaluating rearing, release, and recovery methods for natural enemies; using insects for biological control of invasive weeds; using soil bacteria for biological control of diseases such as the white-nose syndrome (*Pseudogymnoascus destructans*) in bats; and evaluating predators to control invasive aquatic organisms. Vaccination and immunomodulation are also being evaluated to control vertebrate diseases, including sylvatic plague (*Yersinia pestis*), white-nose syndrome, and amphibian chytridiomycosis.

Considerable research also has been conducted on developing host resistance to tree diseases and insect pests, including chestnut blight (*Cryphonectria parasitica*), Dutch elm disease (*Ophiostoma* spp.), beech bark disease (*Neonectria* spp.), emerald ash borer, and hemlock woolly adelgid. Research on reproduction control has led to mating disruption for managing the gypsy moth (*Lymantria dispar*) and immunocontraceptives for feral horses (*Equus ferus caballus*) and swine.

Integrated pest management programs consolidate multiple techniques—such as biological control, habitat manipulation, and modification of cultural practices and use of pesticides and resistant varieties—into a unified program. Integrated ecosystem- or landscape-level programs are being developed and evaluated for hemlock woolly adelgid, emerald ash borer, gypsy moth, invasive plants, and aquatic organisms.

Managing for Resilience and Restoring Ecosystems Impacted by Invasive Species

Recovery of ecosystems impacted by invasive species falls along a gradient between being restored to a known historic state and being rehabilitated to a defined desired state. Generally, invasive species must be controlled to some threshold level either prior to or in conjunction with restoration activities. To ensure long-term success, restoration and rehabilitation efforts need to emphasize building ecosystem resilience and resistance to future invasions. If there is an adequate native seedbank, the restoration of invaded plant communities may occur passively. Active restoration can be approached either by introducing the desired final species or by using a predictable successional trajectory to eventually reach the desired final species composition. Establishing a rich native plant community can increase a system's resistance to reinvasion.

Restoration in forests severely impacted by invasive insects and disease is often focused on identifying and enriching for genetic resistance in native populations of the impacted host tree species. Many invasive insects and diseases leave behind such a small number of widely dispersed surviving individuals that they are unable to recover naturally. Thus, restoration often requires selection and breeding of the infrequently occurring resistant individuals. Examples of breeding programs for resistance to tree diseases include white pine blister rust (*Cronartium ribicola*), chestnut blight, Dutch elm disease, butternut canker (*Ophiognomonia clavigignenti-juglandacearum*), Port-Orford-cedar root disease (*Phytophthora lateralis*), and sudden oak death disease (*Phytophthora ramorum*). Breeding programs have also been established for developing resistance to invasive insects including hemlock woolly adelgid and emerald ash borer. Deployment strategies for establishing resistant hosts also need to be developed, including methods for seed production, mass propagation, site preparation, planting, and maintenance.

Regional Effects of Invasive Species

Alaska

- Alaska has fewer invasive species and is less impacted than most places on Earth due to its cold climate and its comparative lack of roads and other development.
- Warming climate trends and more extensive wildland fires, combined with increasing activity in mining, oil and gas extraction, and wilderness tourism, are extending the network of travel corridors and altered landscapes increase vulnerability to establishment and spread of invasive species.
- Fox, cattle, deer, and elk have been intentionally introduced in Alaska, as have several plants for agriculture or erosion control, including white sweetclover (*Melilotus albus*).
- Introductions have increased in recent years due to world trade, including weed seeds in hay and straw, wood-boring beetles in bundles of firewood, and European and Asian gypsy moths (*Lymantria* spp.) transported on vehicles or on cargo ships.
- Some invasive species have been introduced due to recreational activities, including nonnative plants along hiking trails and earthworms (suborder *Lumbricina*) originally brought in by anglers.
- Releases from aquaria have resulted in the aquatic plant Elodea (*Elodea* spp.) and red-legged frogs (*Rana aurora*) becoming established.
- Recently, the larch sawfly (*Pristiphora erichsonii*) killed approximately 80% of the larch (*Larix laricina*) trees in 240 000 ha in interior Alaska, and the green alder sawfly (*Monsoma pulveratum*) was found to be widespread in southcentral Alaska, where it completely defoliated large patches of alder (*Alnus* spp.).
- Invasive species regulations are broadening in Alaska, as are public outreach and education to increase awareness among land managers and the public.

Hawaii and the US-Affiliated Pacific Islands

- The Hawaiian Islands are the most isolated archipelago on Earth. Extreme gradients of elevation, productivity, and climate mean they have very high native biodiversity as well as extreme habitat complexity.
- The historical and continued degradation of Hawaiian and Pacific Island ecosystems has opened the door to invasive species that have transformed entire ecosystems.
- Approximately 22 taxa per year are introduced, with 869 alien species becoming established in the last 200 years. As a result, the native flora and its ecosystems are considered depauperate because they have been replaced by invasive species.
- Over the past century, wildfire frequency and size have increased dramatically as a result of invasion by fire-promoting alien grasses.

- There are more endangered species per square mile on these islands than in any other place on Earth, including one-third of federally listed threatened and endangered species and almost half of all listed plants. Many of the endangered species and their ecosystems are not found anywhere else in the world.

Northwest

- The Northwest region contains major ports, waterways, and highway arteries that provide pathways for invasive plants, pathogens, insects, and vertebrates.
- The region has a major horticultural industry, extensive areas of mesic and dryland agriculture, and abundant urban and native forests that provide hosts or alternate hosts for invasive species.
- Forested lands are regionally vital to the forest industry and as a recreational base for millions in the Northwest.
- Nearly 190 species and species groups have been identified as regional invasive or nuisance species of key concern.
- Invasive pathogens of concern include sudden oak death disease, Port-Orford-cedar root disease, and white pine blister rust.
- European and Asian gypsy moths are constant and recurring threats that are continually detected in the region. Established invasive insects of concern include balsam woolly adelgid (*Adelges piceae*), larch casebearer (*Coleophora laricella*), spotted wing drosophila (*Drosophila suzukii*), and brown marmorated stink bug (*Halyomorpha halys* Stål).
- The most significant invasive aquatic and terrestrial animals include Asian clams (*Corbicula fluminea*), New Zealand mud snails (*Potamopyrgus antipodarum*), zebra mussels and quagga mussels, American bullfrogs (*Lithobates catesbeianus*), red-legged frogs, Atlantic salmon (*Salmo salar*), Amur goby (*Rhinogobius brunneus*), golden shiners (*Notemigonus crysoleucas*), mute swans (*Cygnus olor*), feral swine, and nutria (*Myocastor coypus*).

Southwest

- The Southwest region has an extensive border with Mexico.
- Changing climate and water stress may facilitate the northern movement of invasive species across the border.
- This region also features a wide range of non-native ornamental plants in urban and rural areas that may favor the establishment and spread of non-native species.
- Terrestrial invasive plants in the Southwest region include annual, biennial, and perennial species of grasses, forbs, shrubs, and trees. Buffel grass (*Pennisetum ciliare*), musk thistle (*Carduus nutans*), salt cedar (*Tamarix ramosissima*), and Russian olive (*Elaeagnus angustifolia*) are among the most significant invasive plant species.
- Invasive pathogens are a significant problem in both urban and rural forests, including white pine blister rust, sudden oak death disease, pitch canker (*Fusarium circinatum*), and Port-Orford-cedar root disease.
- Much of the vast Southwest region has not been invaded by non-native forest insects. However, 22 species of invasive bark and ambrosia beetles have recently been found, including polyphagous shot-hole borer (*Scolytus rugulosus*) and banded elm bark beetle (*Scolytus schevyrewi*).
- A number of ungulate species introduced into the Southwest region have become problematic, including Barbary sheep (*Ammotragus lervia*) and African oryx (*Oryx gazella*). Feral swine are of particular concern.

Great Plains

- The Great Plains region is a diverse landscape consisting of a complex matrix of native, seminative, and non-native grasslands intermixed with riparian and prairie woodlands, shrublands, forests, and intensively cultivated agricultural lands.

- Increasing pressure for intensive urban, agricultural, and energy development, coupled with climate change, is threatening the maintenance of goods and services in the region.
- Some of the most significant invasive species in the region include Russian olive, non-native perennial grass assemblages, buffel grass, absinth wormwood (*Artemisia absinthium*), Johnson grass (*Sorghum halepense*), tumble mustard (*Sisymbrium altissimum*), whitetop (*Lepidium draba*), field (Japanese) brome (*Bromus arvensis*), feral swine, feral horses and burros (*Equus africanus asinus*), Dutch elm disease, and emerald ash borer.

Midwest

- Forest ecosystems of the Midwest are diverse and complex, dominated by northern and central hardwood forests and bordered by northern boreal forest to the north and prairie ecosystems to the south and west. Five states in the region border the Great Lakes, and this region contains 40% of all the surface water in the continental United States.
- The region has many large cities as well as a very high agriculture and industry presence. The actions of humans and their interactions with the environment exacerbate the movement and impacts of invasive species.
- Invasive insects of the most significance in the region include the gypsy moth, hemlock woolly adelgid, Asian long-horned beetle (*Anoplophora glabripennis*), and emerald ash borer.
- Invasive pathogens cause serious ecological and economic impacts to Midwestern forests. A few of the more significant current problems include white pine blister rust, Dutch elm disease, oak wilt (*Bretziella fagacearum*), butternut canker, beech bark disease, and *Phytophthora* diseases.
- There are many invasive terrestrial and aquatic plants distributed throughout the Midwest region. Many are causing considerable damage in woodlands, including garlic mustard (*Alliaria petiolata*), Japanese barberry (*Berberis thunbergii*), common buckthorn (*Rhamnus cathartica*), invasive honeysuckles (*Lonicera* spp.), tree of heaven (*Ailanthus altissima*), and reed canary grass (*Phalaris arundinacea*).
- The most significant invasive animals in the Midwest region are feral swine and invasive earthworms.
- Invasive aquatic organisms causing significant damage include fish such as sea lamprey (*Petromyzon marinus*), bighead carp (*Hypophthalmichthys nobilis*), and silver carp (*Hypophthalmichthys molitrix*); mollusks such as zebra mussels and quagga mussels; crustaceans such as rusty crayfish (*Orconectes rusticus*) and spiny water flea (*Bythotrephes longimanus*); and pathogens such as viral hemorrhagic septicemia (VHS).

Northeast

- The Northeast region is heavily forested, with a high diversity of hardwood and conifer forest tree species, and is water rich, with 10% of the surface area covered by water.
- The Northeast region is the most densely populated region of the country, with many opportunities for the human-mediated introduction of pests, including international shipping ports, large urban/rural interfaces, highly industrialized areas, and high recreational use of forests. This region was colonized by Europeans earlier than most of the rest of the country and, coincidentally, has the highest concentrations of invasive forest insects and pathogens in the country.
- The most damaging invasive forest insects in the region include gypsy moth, hemlock woolly adelgid, Asian long-horned beetle, and emerald ash borer.
- There are many significant invasive diseases of trees, including chestnut blight, white pine blister rust, beech bark disease, Dutch elm disease, butternut canker, and oak wilt.
- White-nose syndrome in bats is now widespread in the Northeast region.
- The aquatic animals that have had the greatest impacts in the Northeast region are sea lamprey, zebra mussels, and quagga mussels.

- State committees and working groups in the Northeast region have ranked the significance of invasive plant species. Species with very high significance rankings include 22 species of terrestrial plants, such as Norway maple (*Acer platanoides*), garlic mustard, Japanese barberry, invasive honeysuckles, Japanese stilt grass (*Microstegium vimineum*), and multiflora rose (*Rosa multiflora*), and 10 species of aquatic plants, including water thyme (*Hydrilla verticillata*), purple loosestrife (*Lythrum salicaria*), broadleaf and Eurasian watermilfoils (*Myriophyllum* spp.), common reed grass (*Phragmites australis*), and water chestnut (*Trapa natans*).
- Education has led to action. Many states in the region now have prohibited plant lists, as well as resources directed at controlling and protecting the most threatened natural resources.

Southeast and Caribbean

- The Southeastern and Caribbean region is characterized by wide climatic variation, including tropical, subtropical, warm-temperate, and temperate environments, as well as diverse ecosystems from coastal wetlands and dunes to piedmont savannahs and montane forests.
- The region is experiencing rapid population growth and landscape fragmentation, which, along with a changing climate, are likely to put stressors on ecosystems that may increase their invasion by, or decrease their resilience to, invasive species.
- Some of the most significant invasive species in the Southeast include cogongrass (*Imperata cylindrica*), Chinese privet (*Ligustrum sinense*), common water hyacinth (*Eichhornia crassipes*), chestnut blight, beech bark disease, Dutch elm disease, thousand cankers disease (*Geosmithia morbida*), hemlock woolly adelgid, emerald ash borer, Burmese python (*Python bivittatus*), feral swine, and nutria.
- Introduction of species has enriched the flora and fauna of Puerto Rico and the Virgin Islands, and many species have become naturalized. The invasive nature of many introduced species allowed lands that had been deforested for agriculture to be restored to forest conditions.
- Some introduced species are considered invasive or problematic, while others coexist with native species in novel forest types.
- Invasive pest species include some mammals, such as rats (*Cricetomys gambianus*), mongoose (*Herpestes javanicus*), feral swine, goats (*Capra aegagrus hircus*), burros, red deer (*Cervus elaphus*), and monkeys (*Macaca mulatta*); amphibians and reptiles, such as the green iguana (*Iguana iguana*); and birds, such as parakeets (*Psittacula krameri* and *Myiopsitta monachus*) and shiny cowbirds (*Molothrus bonariensis*).

An Imperative for Action

Scientific research has improved our understanding of invasion ecology and the impacts of invasive species on ecosystems and ecological processes. Understanding the biology and ecology of invasive species is necessary for effective management. Research has also led to the development of management strategies and tools. Prioritizing invasive species management and decision-making depends on scientific risk assessments and economic analyses. The hallmark of successful invasive species management efforts is strong collaborative partnerships between scientists and managers within public agencies and with various external stakeholders. State and federal regulatory research and resource management agencies must continue to cooperate to ensure that the best science-based information is incorporated in import and quarantine regulations, detection surveys, early response, and long-term management strategies.

Management of invasive species requires work across institutional and ownership boundaries by developing, sharing, and implementing effective regulatory, survey, and management approaches. This assessment incorporates information from multiple partners and stakeholders, summarizes recent research on significant invasive species issues in forests and grasslands, and identifies major research needs and gaps. Our challenge will be to build on this assessment, continue to work together to conduct new research to further our understanding, and develop and implement new and improved tools and management programs for existing and future invasive species that threaten our nation's forests and grasslands.

Contents

Author Biographies

Amanda M. Countryman is an associate professor, Department of Agricultural and Resource Economics, Colorado State University, Fort Collins, CO 80523.

Janice Alexander is a forest health program coordinator, University of California Cooperative Extension, Marin County, Novato, CA 94947.

Sybill K. Amelon is a research wildlife biologist, U.S. Department of Agriculture, Forest Service, Northern Research Station, Columbia, MO 65211.

Jenny L. Apriesnig is an assistant professor, College of Business, Michigan Technological University, Houghton, MI 49931.

Dana Blumenthal is a research ecologist, U.S. Department of Agriculture, Agricultural Research Service, Rangeland Resources and Systems Research Unit, 1701 Center Ave., Fort Collins, CO 80526.

Stephen Boyte is a research geographer, U.S. Department of the Interior, U.S. Geological Survey, Earth Resources and Observation Science Center, 47914 252nd Street, Sioux Falls, SD 57198.

Steven T. Brantley is an associate scientist, Jones Center at Ichauway, 3988 Jones Center Drive, Newton, Georgia 39870.

Mark Brunson is a professor of environment and society, Utah State University, Logan, UT 84322.

Douglas A. Burkett is an environmental biologist/entomologist, U.S. Department of Defense, Armed Forces Pest Management Board, 2460 Linden Lane, Silver Spring, MD 20910.

Jack L. Butler is a research ecologist (retired), U.S. Department of Agriculture, Forest Service, Rocky Mountain Research Station, Rapid City, SD 57702.

Tommy Cabe is a forest resource specialist, Natural Resource Department, Eastern Band of Cherokee Indians, Cherokee, NC 28719.

Mac A. Callaham Jr is a research ecologist, U.S. Department of Agriculture, Forest Service, Southern Research Station, Center for Forest Disturbance Science, 320 Green Street, Athens, GA 30602.

Faith Campbell is an invasive species analyst, Center for Invasive Species Prevention, 8208 Dabney Avenue, Springfield, VA 22152.

Gang Chen is an associate professor, Department of Geography and Earth Sciences, University of North Carolina at Charlotte, Charlotte, NC 28223.

Richard C. Cobb is an assistant professor of forest health, Department of Natural Resources Management and Environmental Sciences, California State Polytechnic University, San Luis Obispo, CA 93407.

Susan Cordell is a research forest ecologist, U.S. Department of Agriculture, Forest Service, Pacific Southwest Research Station, Institute of Pacific Islands Forestry, 60 Nowelo Street, Hilo HI 96720.

Alycia Crall is a research scientist, National Ecological Observatory Network, Boulder, CO 80301.

Samuel A. Cushman is a research ecologist, U.S. Department of Agriculture, Forest Service, Rocky Mountain Research Station, Flagstaff, AZ 86011.

Alison Dalsimer was the former natural resources program manager, U.S. Department of Defense, Mark Center, Alexandria, VA 22350.

Jamie L. Davidson is a forest supervisor, U.S. Department of Agriculture, Forest Service, Allegheny National Forest, 4 Farm Colony Drive, Warren, PA 16365.

Hilda Diaz-Soltero is Caribbean Advisor to the APHIS Administrator, U.S. Department of Agriculture, Animal and Plant Health Inspection Service, 1201 Ceiba Street, Jardín Botánico Sur, San Juan, PR 00926.

Michael J. Dockry is an assistant professor of tribal natural resource management, Department of Forest Resources, University of Minnesota, Saint Paul, MN 55108.

Jeffrey S. Dukes is a professor of forestry and natural resources and biological sciences, Purdue University, 715 West State Street, West Lafayette, IN 47907.

Marla R. Emery is a research geographer, U.S. Department of Agriculture, Forest Service, Northern Research Station, 81 Carrigan Drive, Burlington, VT 05405.

Christopher J. Fettig is a research entomologist, U.S. Department of Agriculture, Forest Service, Pacific Southwest Research Station, 1731 Research Park Drive, Davis, CA 95618.

Deborah M. Finch is a biological scientist and program manager, U.S. Department of Agriculture, Forest Service, Rocky Mountain Research Station, 333 Broadway, SE, Suite 115, Albuquerque, NM 87102.

Rebecca Flitcroft is a research fish biologist, U.S. Department of Agriculture, Forest Service, Pacific Northwest Research Station, 3200 SW Jefferson Way, Corvallis, OR 97331.

Susan J. Frankel is a plant pathologist, U.S. Department of Agriculture, Forest Service, Pacific Southwest Research Station, 800 Buchanan Street, Albany, CA 94710.

Jennifer M. Fraterrigo is an associate professor in the Department of Natural Resources and Environmental Sciences and in the Program in Ecology, Evolution and Conservation Biology, University of Illinois, Urbana, IL 61801.

Christian P. Giardina is a research ecologist, U.S. Department of Agriculture, Forest Service, Pacific Southwest Research Station, 60 Nowelo Street, Hilo, Hawaii 96720.

Doria R. Gordon is a lead senior scientist, Environmental Defense Fund, 1875 Connecticut Ave. NW, Washington, DC 20036.

Andrew D. Graves is a forest entomologist, U.S. Department of Agriculture, Forest Service, Forest Health Protection, New Mexico Zone, Southwestern Region, 333 Broadway Blvd, SE Albuquerque, NM 87101.

Qinfeng Guo is a research ecologist, U.S. Department of Agriculture, Forest Service, Southern Research Station, Eastern Forest Environmental Threat Assessment Center, 3041 Cornwallis Road, Research Triangle Park, NC 27709.

Wendell R. Haag is a research fisheries biologist, U.S. Department of Agriculture, Forest Service, Southern Research Station, Center for Bottomland Hardwoods Research, 3761 Georgetown Road, Frankfort, KY 40601.

Christopher Haberland is an ORISE fellow, U.S. Department of Agriculture, Forest Service, National Center for Natural Resource Economics, 1400 Independence Ave., SW, Washington, DC 20024.

Robert G. Haight is a research forester, U.S. Department of Agriculture, Forest Service, Northern Research Station, 1992 Folwell Ave, St Paul, MN 55108.

Nina Hapner is the director of the Kashia Department of Environmental Planning, Kashia Band of Pomo Indians, Santa Rosa, CA 95403.

Bonnie L. Harper-Lore is a restoration ecologist (retired), U.S. Department of Transportation, Federal Highway Administration, Washington, DC 20590.

Linda Haugen is a plant pathologist, U.S. Department of Agriculture, Forest Service, Northeastern Area State and Private Forestry, Forest Health Protection, 1992 Folwell Avenue, St. Paul, MN 55108.

Nathan Havill is a research entomologist, U.S. Department of Agriculture, Forest Service, Northern Research Station, 51 Mill Pond Rd., Hamden, CT 06514.

Deborah C. Hayes is a research ecologist and invasive species specialist (retired), U.S. Department of Agriculture, Forest Service, Sustainable Forest Management Research, Washington, DC 20024.

Jeffrey A. Hicke is a professor, Department of Geography, University of Idaho, Moscow, ID 83844.

Caleb R. Hickman is a supervisory fish and wildlife biologist, Natural Resource Department, Eastern Band of Cherokee Indians, Cherokee, NC 28719.

Thomas P. Holmes is an emeritus scientist, U.S. Department of Agriculture, Forest Service, Southern Research Station, 3041 Cornwallis Road, Research Triangle Park, NC 27709.

Cynthia D. Huebner is a research botanist, U.S. Department of Agriculture, Forest Service, Northern Research Station, 180 Canfield St., Morgantown, WV 26505.

Inés Ibáñez is an associate professor in ecology, School of Environment and Sustainability, University of Michigan, Ann Arbor, MI 48109.

Mike Ielmini is a National Invasive Species program manager, U.S. Department of Agriculture, Forest Service, 201 14th St., Washington, DC 20024.

Catherine Jarnevich is a research ecologist, U.S. Deparatment of Interior, U.S. Geological Survey, Fort Collins Science Center, 2150 Centre Ave Bldg C, Fort Collins, CO 80526.

M. Tracy Johnson is a research entomologist, USDA Forest Service, Pacific Southwest Research Station, Institute of Pacific Islands Forestry, PO Box 236, Volcano, HI 96785.

Rebecca Jordan is professor and chair, Department of Community Sustainability, Michigan State University, East Lansing, MI 48824.

Shibu Jose Associate Dean for Research, College of Agriculture, Food and Natural Resources, University of Missouri, 2-44 Agriculture Building, Columbia, MO 65211.

Jennifer Juzwik is a research plant pathologist, U.S. Department of Agriculture, Forest Service, Northern Research Station, 1561 Lindig Street, St. Paul, MN 55108.

Becky K. Kerns is a research ecologist, U.S. Department of Agriculture, Forest Service, Pacific Northwest Research Station, Fire and Environmental Applications Research Team, 3200 SW Jefferson Way, Corvallis, OR 97331.

John C. Kilgo is a research wildlife biologist, U.S. Department of Agriculture, Forest Service, Southern Research Station, Center for Forest Watershed Research, P.O. Box 700, New Ellenton, SC 29809.

Francis F. Kilkenny is a research biologist, U.S. Department of Agriculture, Forest Service, Rocky Mountain Research Station, Boise, ID 83702.

Bradley Kinder is a public engagement and collaboration specialist, U.S. Department of Agriculture, Forest Service, Washington Office, 201 14th Street, SW, Washington, DC 20024.

Patrice N. Klein is a senior veterinary medical officer and national program lead for fish and wildlife health, U.S. Department of Agriculture, Forest Service, Research and Development, Landscape and Ecosystem Services Research, Washington, DC 20250.

Frank H. Koch is a research ecologist, U.S. Department of Agriculture, Forest Service, Southern Research Station, Eastern Forest Environmental Threat Assessment Center, 3041 East Cornwallis Road, Research Triangle Park, NC 27709.

Jennifer Koch is a research biologist, U.S. Department of Agriculture, Forest Service, Northern Research Station, 359 Main Rd., Delaware, OH 43015.

John Kyhl is a forest entomologist, USDA Forest Service, Northeastern Area, State and Private Forestry, 1992 Folwell Avenue, St. Paul, MN 55108.

David R. Lance laboratory director (retired), U.S. Department of Agriculture, Animal and Plant Health Inspection Service, Plant Protection and Quarantine, Center for Plant Health Science and Technology, Buzzards Bay, MA 02542.

Daniel J. Larkin is an assistant professor and extension specialist, Department of Fisheries, Wildlife and Conservation Biology and the Minnesota Aquatic Invasive Species Research Center, University of Minnesota, St. Paul, MN 55108.

Michael J. LaVoie is a program manager, Natural Resource Department, Eastern Band of Cherokee Indians, Cherokee, NC 28719.

Jana C. Lee is a research entomologist, U.S. Department of Agriculture, Agricultural Research Service, Corvallis, OR 97330.

Andrew M. Liebhold is a research entomologist, U.S. Department of Agriculture, Forest Service, Northern Research Station, 180 Canfield St., Morgantown, WV 26505.

Vanessa M. Lopez is the invasive plants and biological control national program manager, U.S. Department of Agriculture, Forest Service, State and Private Forestry, 201 14th St., Washington, DC 20024.

Gary Lovett is a senior scientist, Cary Institute of Ecosystem Studies, Box AB, 2801 Sharon Turnpike, Millbrook, NY 12545.

Rima D. Lucardi is a research ecologist, U.S. Department of Agriculture, Forest Service, Southern Research Station, 320 Green Street, Athens, GA 30606.

Ariel E. Lugo is an ecologist and director, U.S. Department of Agriculture, Forest Service, International Institute of Tropical Forestry, 1201 Ceiba Street, Jardín Botánico Sur, Río Piedras, Puerto Rico 00926-1115.

Zhao Ma is an associate professor of natural resource social science, Department of Forestry and Natural Resources, Purdue University, West Lafayette, IN 47907.

Richard MacKenzie is an aquatic ecologist, U.S. Department of Agriculture, Forest Service, Pacific Southwest Research Station, Institute of Pacific Islands Forestry, 60 Nowelo St., Hilo, HI 96720.

Humfredo Marcano-Vega is a research biologist, U.S. Department of Agriculture, Forest Service, Southern Research Station, Forest Inventory and Analysis Program, International Institute of Tropical Forestry, Jardín Botánico Sur, 1201 Calle Ceiba, San Juan, PR 00926.

Anne S. Marsh is the program lead for bioclimatology and climate change, Sustainable Forest Management Research, U.S. Department of Agriculture, Forest Service, 201 14th St., Washington, DC 20024.

Albert E. Mayfield III is a research entomologist, U.S. Department of Agriculture, Forest Service, Southern Research Station, Insects, Diseases and Invasive Plants of Southern Forests, 200 W.T. Weaver Blvd., Asheville, NC 28804.

Ross K. Meentemeyer is the Goodnight distinguished professor of geospatial analytics and director of the Center for Geospatial Analytics, North Carolina State University, Raleigh, NC 27695.

Susan E. Meyer is a research ecologist, U.S. Department of Agriculture, Forest Service, Rocky Mountain Research Station, Shrub Sciences Laboratory, Provo, UT 84606.

Laura A. Meyerson is a professor of invasion and restoration ecology, University of Rhode Island, Kingston, RI 02881.

Chelcy Ford Miniat is a research ecologist, U.S. Department of Agriculture, Forest Service, Southern Research Station, Coweeta Hydrologic Lab, 3160 Coweeta Lab Road, Otto, NC 28763.

Bruce D. Moltzan is a national forest pathologist, U.S. Department of Agriculture, Forest Service, State and Private Forestry, Forest Health Protection, Washington Office, 201 14th St. SW, Washington, DC 20250.

A. Steve Munson is group leader, Forest Health Protection (retired), U.S. Department of Agriculture, Forest Service, Intermountain Region, 1900 E. 4746 S., Ogden, Utah 84403.

Deanna H. Olson is a research ecologist, U.S. Department of Agriculture, Forest Service, Pacific Northwest Research Station, 3200 SW Jefferson Way, Corvallis, OR 97331.

Chris Oswalt is a research forester, U.S. Department of Agriculture, Forest Service, Forest Inventory and Analysis Program, and adjunct assistant professor of forestry, Department of Forestry, Wildlife, and Fisheries, University of Tennessee in Knoxville, TN 37996.

Sonja Oswalt is a forester, U.S. Department of Agriculture, Forest Service, Southern Research Station, Forest Inventory and Analysis Program, 4700 Old Kingston Pike, Knoxville, TN 37919.

Toral Patel-Weynand is the director of Sustainable Forest Management Research, Research and Development, U.S. Department of Agriculture, Forest Service, 201 14th St., Washington, DC 20024.

Dean E. Pearson is a research ecologist, U.S. Department of Agriculture, Forest Service, Wildlife and Terrestrial Ecosystems Program Rocky Mountain Research Station, 800 East Beckwith Avenue, Missoula, MT 59801 and adjunct faculty Ecology and Evolution, Division of Biological Sciences, University of Montana, 32 Campus Drive, Missoula MT 59812.

Robert K. D. Peterson is a professor of entomology, Department of Land Resources and Environmental Sciences, Montana State University, Bozeman, MT 59717.

Therese M. Poland is a research entomologist, U.S. Department of Agriculture, Forest Service, Northern Research Station, 3101 Technology Blvd., Ste. F, Lansing, MI 48910.

Robert Progar is a national program lead of entomology, pathology, and invasive species research, Research and Development, U.S. Department of Agriculture, Forest Service, 201 14th St., Washington, DC 20024.

Robert Rabaglia is a national entomologist, U.S. Department of Agriculture, Forest Service, State and Private Forestry, Forest Health Protection, Washington Office, 201 14th St, SW, FHP 3CE, Washington, DC 20250.

Thomas J. Rawinski is a botanist, U.S. Department of Agriculture, Forest Service, State and Private Forestry, Durham Field Office, 271 Mast Rd., Durham, NH 03824.

Matt Reeves is a research ecologist, U.S. Department of Agriculture, Forest Service, Rocky Mountain Research Station, Human Dimensions Program, Department of Forestry and Environmental Resources Campus, 800 E. Beckwith, Missoula, MT 59801.

Karen L. Ripley is a forest entomologist and forest health monitoring coordinator, U.S. Department of Agriculture, Forest Service, Pacific Northwest Regional Office, 1220 SW Third Avenue, Portland, OR, 97204.

John D. Rothlisberger is a research program manager, U.S. Department of Agriculture, Forest Service, Washington Office, 1400 Independence Ave. SW, Washington, DC 20250.

Allen Rowley is Associate Deputy Chief, National Forest System, U.S. Department of Agriculture, Forest Service, Washington Office, 201 14th Street, SW, Suite 3SE, Washington, DC 20250.

Justin B. Runyon is a research entomologist, U.S. Department of Agriculture, Forest Service, Rocky Mountain Research Station, 1648 S 7th Ave, Bozeman, MT 59717.

Frank Sapio is assistant director (retired), Forest Health Assessment and Applied Sciences Team, Forest Health Protection, U.S. Department of Agriculture, Forest Service, Fort Collins, CO 80526.

Jeffrey D. Schardt was the invasive aquatic plant management program coordinator (retired), Florida Fish and Wildlife Conservation Commission, 810 Remington Ave., Thomasville, GA 31792.

John Schelhas is a research forester, U.S. Department of Agriculture, Forest Service, Southern Research Station, Athens, GA 30602.

Noel F. Schneeberger is an area entomologist (retired), U.S. Department of Agriculture, Forest Service, State and Private Forestry, Forest Health Protection, 11 Campus Blvd., suite 200, Newtown Square, PA 19073.

Michele Schoeneberger is research program lead and soil scientist (retired), U.S. Department of Agriculture, Forest Service, Research and Development, National Agroforestry Center, 1945 N 38th Street, Lincoln, NE 68583.

Bethany Schulz is a research ecologist, U.S. Department of Agriculture, Forest Service, Pacific Northwest Research Station, Resource Monitoring and Assessment Program, 161 E 1st Ave Door 8, Anchorage, AK 99501.

Michael K. Schwartz is director, U.S. Department of Agriculture, Forest Service, National Genomics Center for Wildlife and Fish Conservation, 800 E. Beckwith Ave., Missoula MT 59801.

Steven J. Seybold formerly a research entomologist (deceased), U.S. Department of Agriculture, Forest Service, Pacific Southwest Research Station, Invasives and Threats Team in the Ecosystem Function and Health Program, 1731 Research Park Drive, Davis, CA 95618.

Sharlene E. Sing is a research entomologist, U.S. Department of Agriculture, Forest Service, Rocky Mountain Research Station, 1648 South 7th Avenu, Montana State University, Bozeman, MT 59717.

Richard A. Sniezko is center geneticist, U.S. Department of Agriculture, Forest Service, Dorena Genetic Resource Center, 34963 Shoreview Drive, Cottage Grove, OR 97424.

Joe Starinchak is an applied social scientist, U.S. Department of the Interior, U.S. Fish and Wildlife Service – Headquarters, Fish and Aquatic Conservation Program, 5275 Leesburg Pike, Falls Church, VA 22041.

Jane E. Stewart is an assistant professor of forest pathology, Department of Bioagricultural Sciences and Pest Management, Colorado State University, Fort Collins, CO 80523.

Carol A. Sutherland is extension entomologist, New Mexico State University, and state entomologist, New Mexico Department of Agriculture, Las Cruces, NM 88003.

Robert C. Venette is a research biologist, U.S. Department of Agriculture, Forest Service, Northern Research Station, 1561 Lindig Street, St. Paul, MN 55108.

Jelena Vukomanovic is an assistant professor, Department of Parks, Recreation, and Tourism Management, and a faculty fellow, Center for Geospatial Analytics, North Carolina State University, Raleigh, NC 27695.

Steven D. Warren is a research disturbance ecologist, U.S. Department of Agriculture, Forest Service, Rocky Mountain Research Station, Shrub Sciences Laboratory, Grassland, Shrubland, and Desert Ecosystems program, 735 North 500 East, Provo, UT 84606.

Travis Warziniack is a research economist, U.S. Department of Agriculture, Forest Service, Rocky Mountain Research Station, 240 W Prospect Road, Fort Collins, CO 80525.

Allen White is a biologist, U.S. Department of Agriculture, Forest Service, 333 Broadway Blvd SE, Albuquerque NM 87111.

Tricia L. Wurtz is an ecologist, U.S. Department of Agriculture, Forest Service, Alaska Region, 3700 Airport Way, Fairbanks, AK 99709.

Bruce K. Wylie is a research physical scientist, U.S. Department of Interior, U.S. Geological Survey, EROS Science Division, 47914 252nd St., Sioux Falls, SD 57198.

Denys Yemshanov is a research scientist, Natural Resources Canada, Canadian Forest Service, Great Lakes Forestry Centre, 1219 Queen Street East, Sault Ste. Marie, ON, Canada, P6A2E5.

Michael K. Young is a research fisheries biologist, U.S. Department of Agriculture, Forest Service, Rocky Mountain Research Station and National Genomics Center for Wildlife and Fish Conservation, 800 E. Beckwith Avenue, Missoula, MT 59801.

Nicole Zimmerman served as an Oak Ridge Institute for Science Education (ORISE) fellow (2015–2017), U.S. Department of Agriculture, U.S. Forest Service, Research and Development, Landscape and Ecosystem Services Research, and National Forest System – Watershed, Fish, Wildlife, Air, and Rare Plants, 201 14th St SW, Washington, DC 20250 and is now the public affairs manager, Ecological Society of America, 1990 M St NW, Suite 700, Washington, DC 20036.

List of Boxes

List of Figures

List of Tables

The original version of this book was revised: Author/Editor has provided belated corrections which has been updated throughout the book. The correction to this book is available at https://doi.org/10.1007/978-3-030-45367-1_17

Introduction

1

Deborah C. Hayes, Becky K. Kerns, Toral Patel-Weynand, and Deborah M. Finch

Invasive species are a historical, long-term, and continually growing threat to the ecology, economy, and infrastructure of the United States. Widely recognized as one of the most serious threats to the health, sustainability, and productivity of native ecosystems, invasive species issues have commonly been viewed as problems specific to Federal, State, and private landowners. However, it is increasingly apparent that the impacts from these species are all encompassing, affecting ecosystem processes in addition to the economics of land management, public and private infrastructure, the energy sector, international trade, cultural practices, and many other sectors in the United States.

In the United States, the President issued Executive Order (EO) 13112 in 1999, providing a common definition of the term "invasive species" across the Federal agencies for the first time (see below). The definitions within the EO provided Federal agencies with consistent terminology that greatly facilitates discussion across taxa. Up until this time, invasive species issues were considered for individual taxa such as weeds, forest pests and diseases, aquatic animal species, and terrestrial animals. The common language provided by the EO also facilitated discussion at the international scale regardless of differing species of concern. Executive Order 13112 prompted basic research within the Federal scientific community concerning foreign exploration, ecological functioning, pest detection, prevention of pest invasion, and control and management. In the time since the EO was signed, human and animal diseases have increasingly been included in the discussion of invasive species. Biosecurity has also become an issue as impacts on military readiness have become apparent.

Definitions within Executive Order 13112 Section 1:
(e) "Introduction" means the intentional or unintentional escape, release, dissemination, or placement of a species into an ecosystem as a result of human activity.
(f) "Invasive species" means an alien species whose introduction does or is likely to cause economic or environmental harm or harm to human health.

Executive Order 13112 also specifically defined invasive species as relating only to the introduction and movement "as a result of human activity." This anthropogenic definition excluded natural background rates of species introductions by natural or nonhuman pathways such as natural disasters and animal movement. There was a great deal of discussion on this specificity in the definition from numerous viewpoints. However, many argued against an unrestricted definition because it would overlap with natural succession and other processes that are not driven by human activities, such as movement of species onto volcanic areas, new islands, or fire- or flood-ravaged areas.

The EO was intended to establish Federal Government definitions of key terms, with the hope that other public and private entities would adopt the definitions therein. However, many sectors use other terminology and definitions, and terminology continues to be debated in the literature. For example, terms such as "forest pests and diseases" are still used, including both native and non-native organisms. Other terms include "alien" or "exotic" and "nonindigenous" or "non-native." Typically, these terms refer to non-native species that may or may not be invasive. Unfortunately, the need remains for an agreed-upon definition within the larger community concerned about invasive species. This assessment

D. C. Hayes · T. Patel-Weynand (✉)
U.S. Department of Agriculture, Forest Service, Sustainable Forest Management Research, Washington, DC, USA
e-mail: toral.patel-weynand@usda.gov

B. K. Kerns
U.S. Department of Agriculture, Forest Service, Pacific Northwest Research Station, Corvallis, OR, USA

D. M. Finch
U.S. Department of Agriculture, Forest Service, Rocky Mountain Research Station, Albuquerque, NM, USA

T. M. Poland et al. (eds.), *Invasive Species in Forests and Rangelands of the United States*,
https://doi.org/10.1007/978-3-030-45367-1_1

uses the EO 13112 definitions, and we avoid using other terms unless the cited source specifically uses them.

The issues surrounding invasive species are complex and diverse, interdisciplinary, and multi-jurisdictional and occur at local, landscape, national, and international scales. Issues also vary across temporal scales or invasion stage, such as species introduction, establishment, and spread, which impact options for control. Invasive plants, animals, diseases, and pathogens can affect ecosystems through changes in nutrient cycling, biodiversity, tree mortality, forest dynamics, fire regimes, and hydrology. Population dynamics and competitive interactions are affected as highly competitive invasive plants or insects can outcompete native species, creating monocultures and threatening or endangering native species. Current research examining the effect of climate change on invasive species' establishment and spread suggests that some invasive species (e.g., plants) may have more adaptive plasticity, with potential fitness advantages in the invaded range. Historical approaches to prevention, control, and management all need to be conducted with follow-up evaluation to determine the actual effectiveness of the measures taken or their effect upon the environment. As invasive species research has developed, more emphasis has been placed upon evaluating the effectiveness and economic viability of these measures and resources needed to reduce the introduction and spread of invasive species.

The purpose of this assessment is to present the state of science for invasive species studies, from impacts at the ecosystem and population levels to knowledge about international impacts and restricting vectors of introduction. Increased science that focuses on quantifying invasive species' biology, impacts, and interactions, along with managing invasive species and altered ecosystems, were identified as priorities in a USDA Forest Service Technical Report, *A Dynamic Invasive Species Research Vision: Opportunities and Priorities 2009–2029*. This assessment is largely restricted to exploration of topics associated with nonagricultural lands and does not extensively cover issues related to croplands, orchards, or vineyards, although some public forests and rangelands (e.g., national forests) are administered within the U.S. Department of Agriculture, and some may consider them agricultural lands due to the production of fiber and food. However, Chap. 9 provides a brief review of many invasive species impacts related to these other sectors. Key information and knowledge gaps are included for each chapter, and discussion is highlighted by region as well as in the appendices.

The first part of the assessment explores the ecological impacts of invasive species from aquatic species to insects and diseases. In Chap. 4, the authors discuss the potential influence of climate change on invasive species, and in Chap. 5, invasive species' response to natural and anthropogenic disturbance is reviewed. Research applications for early intervention, prevention, and rapid response, and the challenges surrounding controlling established populations are presented in Chaps. 6 and 7. Chapter 8 reviews the relatively new topic of restoration and rehabilitation of lands after control or eradication. In some cases, the landscape needs rehabilitation before restoration can occur. Research has also shown that often the first plants coming into an area after a control action against an invasive plant is another invasive plant species. The national inventories presented in Chap. 10 provide information for conducting risk assessments to evaluate vulnerable ecosystems and enable models predicting spread, but not all species have reliable national data sets, and many local data sets remain isolated. Tools for quantifying spread and impacts of invasive species are discussed in Chap. 11, including newer technologies such as remote sensing and eDNA.

The authors of Chap. 12 present a valuable survey of social science research across a diversity of ecosystems and stakeholders that provides a foundation for understanding social and cultural dimensions of invasive species, including impacts on Native American tribes. The cultural impact of native species no longer available for Native American tribal rituals can be devastating. In addition, the perceptions and attitudes of affected human populations can have an enormous effect on the ability to direct resources for both State and Federal Government. Yet new social science approaches exist that are increasingly good at engaging these human populations through citizen science programs, bringing in the layperson, amateur, and paraprofessional into the discussion of what action an individual can take to influence the status of an invasive species on the ground.

Chapter 13 points out that international collaboration plays a key role in research on prevention of future invasions. In addition, understanding species in their native ranges and how they enter the invasion pathway is critical to the analysis of risk used to guide quarantine measures. Invasive species can also have a major impact on the economics of an area, as noted in Chap. 14. For example, the forest products and ranching industries have been greatly affected. Exploring the economics of invasive species also highlights the human dimensions of the topic, an important aspect regarding invasive species issues. The Federal legislation that provides standing for Federal actions on invasive species is highlighted in Chap. 15. Some of the most important Federal laws are aimed at preventing introduction and interstate spread of known or potential invaders, while other laws regulate modes of transport as well as the organisms themselves. Rarely has the impact of invasive species upon the sectoral parts of the U.S. economy, including military preparedness, human health, the energy and utility industry, and transportation infrastructure been examined. A general but not exhaustive review of sectoral values and impacts is included in Chap. 9. These issues typically are not well known by the invasive species research and management community, and only recently have more rigorous and in-depth impact assessments become available.

This assessment synthesizes local to international research and information on a comprehensive array of topics pertinent to invasive species and identifies future needs and gaps for research. The information provided is intended to be useful for a range of stakeholders, including researchers, managers, and decision makers working on a variety of invasive species issues. Much progress in research has been made on understanding the major aspects of invasive species, yet considerable challenges remain. Advances in these challenge areas are critical to improve prevention and management of invasive species and reduce impacts to humans and ecosystem services.

Disclaimer Text The findings and conclusions in this publication are those of the authors and should not be construed to represent any official USDA or U.S. Government determination or policy.

Impacts of Invasive Species in Terrestrial and Aquatic Systems in the United States

2

Albert E. Mayfield III, Steven J. Seybold, Wendell R. Haag, M. Tracy Johnson, Becky K. Kerns, John C. Kilgo, Daniel J. Larkin, Rima D. Lucardi, Bruce D. Moltzan, Dean E. Pearson, John D. Rothlisberger, Jeffrey D. Schardt, Michael K. Schwartz, and Michael K. Young

2.1 Introduction

The introduction, establishment, and spread of invasive species in terrestrial and aquatic environments is widely recognized as one of the most serious threats to the health, sustainability, and productivity of native ecosystems (Holmes et al. 2009; Mack et al. 2000; Pyšek et al. 2012; USDA Forest Service 2013). In the United States, invasive species are the second leading cause of native species endangerment and extinction, and their costs to society have been estimated at $120 billion annually (Crowl et al. 2008; Pimentel et al. 2000, 2005). These costs include lost production and revenue from agricultural and forest products, compromised use of waterways and terrestrial habitats, harm to human and animal health, reduced property values and recreational opportunities, and diverse costs associated with managing (e.g., monitoring, preventing, controlling, and regulating) invasive species (Aukema et al. 2011; Pimentel et al. 2005). The national significance of these economic, ecological, and social impacts in the

A. E. Mayfield III (✉)
U.S. Department of Agriculture, Forest Service, Southern Research Station, Insects, Diseases and Invasive Plants of Southern Forests, Asheville, NC, USA
e-mail: albert.e.mayfield@usda.gov

S. J. Seybold
U.S. Department of Agriculture, Forest Service, Pacific Southwest Research Station, Davis, CA, USA

W. R. Haag
U.S. Department of Agriculture, Forest Service, Southern Research Station, Center for Bottomland Hardwoods Research, Frankfort, KY, USA

M. T. Johnson
U.S. Department of Agriculture, Forest Service, Pacific Southwest Research Station, Institute of Pacific Islands Forestry, Volcano, HI, USA

B. K. Kerns
U.S. Department of Agriculture, Forest Service, Pacific Northwest Research Station, Corvallis, OR, USA

J. C. Kilgo
U.S. Department of Agriculture, Forest Service, Southern Research Station, Center for Forest Watershed Research, New Ellenton, SC, USA

D. J. Larkin
Department of Fisheries, Wildlife and Conservation Biology and the Minnesota Aquatic Invasive Species Research Center, University of Minnesota, St. Paul, MN, USA

R. D. Lucardi
U.S. Department of Agriculture, Forest Service, Southern Research Station, Athens, GA, USA

B. D. Moltzan
U.S. Department of Agriculture, Forest Service, State and Private Forestry, Forest Health Protection, Washington Office, Washington, DC, USA

D. E. Pearson
U.S. Department of Agriculture, Forest Service, Rocky Mountain Research Station, Missoula, MT, USA

Ecology and Evolution, Division of Biological Sciences, University of Montana, Missoula, MT, USA

J. D. Rothlisberger
U.S. Department of Agriculture, Forest Service, Washington Office, Washington, DC, USA

J. D. Schardt
Florida Fish and Wildlife Conservation Commission, Thomasville, GA, USA

M. K. Schwartz
U.S. Department of Agriculture, Forest Service, National Genomics Center for Wildlife and Fish Conservation, Missoula, MT, USA

M. K. Young
U.S. Department of Agriculture, Forest Service, Rocky Mountain Research Station and National Genomics Center for Wildlife and Fish Conservation, Missoula, MT, USA

T. M. Poland et al. (eds.), *Invasive Species in Forests and Rangelands of the United States*, https://doi.org/10.1007/978-3-030-45367-1_2

United States has prompted various actions by both legislative and executive branches of the Federal Government (e.g., the Nonindigenous Aquatic Nuisance Prevention and Control Act of 1990; the Noxious Weed Control and Eradication Act of 2002; Executive Order 13112 of 1999, amended in 2016).

Because the concept of impact is inevitably influenced by human perceptions and biases, the scientific literature often characterizes the environmental effects of invasive species in terms of "ecological impacts." Ecological impacts refer to measurable changes to the properties of an ecosystem, which may be considered positive or negative depending on context (Ricciardi et al. 2013). In this chapter, we consider invasive species as a subset of non-native species that substantially affect the composition, structure, or function of native populations, communities, or ecosystems, and particularly impact these systems in a manner that decreases the ecosystem values and services (e.g., economic, aesthetic, and/or social benefits) that they provide to humans (NISC 2005; Pejchar and Mooney 2009; Walsh et al. 2016). To more effectively manage invasive species and mitigate their negative impacts on native systems, a better understanding of the nature of their ecological impacts and the mechanisms that underlie the impacts is needed.

2.1.1 Conceptualizing Mechanisms of Invasive Species Ecological Impacts

The classic model developed by Parker et al. (1999), in which the impact of an invasive species is a function of its range, abundance, and per capita effect, has provided a basic conceptual framework for quantifying impacts, but it has rarely been applied to specific systems (Pearson et al. 2016a; Thiele et al. 2010). A more highly synthetic framework for understanding the full range of invader impacts remains elusive (Ricciardi et al. 2013), though approaches for broad categorization of impacts have been proposed (Blackburn et al. 2014). Here, we draw from basic community ecology concepts to characterize the mechanisms and processes by which invasive species impact native systems (Fig. 2.1). Because invasive predators, pathogens, consumers, decomposers, and primary producers impact native species through different mechanisms (Fig. 2.1), and impacts can manifest differently in aquatic and terrestrial systems (Cox and Lima 2006; Moorhouse and Macdonald 2015), in this chapter, we review invasive species ecological impacts taxonomically by invasive plants, pathogens, invertebrates, and vertebrates in terrestrial and aquatic systems in the United States. Examples of the consequences of several specific invasive species are highlighted in boxes accompanied by figures.

2.1.2 Direct and Indirect Impacts: Density-Mediated vs Trait-Mediated

Invasive species may affect native organisms positively or negatively through direct and indirect interactions that are transmitted through density- or trait-mediated mechanisms (Fig. 2.1). These interactions may ultimately affect native organisms at multiple ecological levels, i.e., organisms, populations, communities, and ecosystems (e.g., Wootton and Emmerson 2005). Direct negative impacts of invasive species include effects on the abundance, distribution, or function of native species through predation (including, for our purposes, infection, herbivory, or parasitism) or competition for resources with potentially lasting and profound changes to native biodiversity (Allen et al. 2004, 2015; Blackburn et al. 2014; Crowl et al. 2008; Wagner and Van Driesche 2010).

Direct impacts of invaders are often obvious, such as when they damage, kill, consume, or overgrow native species, but cryptic indirect effects can be similarly powerful (White et al. 2006). Density-mediated indirect interactions arise when the invader alters the abundance of a native species, and this, in turn, alters the abundance of other native species via interaction chains (e.g., Ortega et al. 2006). Trait-mediated indirect interactions (also referred to as interaction modifications (Wootton 1994)) arise when the invader alters interactions between species (two natives or a native and a non-native) in ways that change how strongly those species interact, i.e., the per capita interaction strength (e.g., Pearson 2010). Such changes arise because the invader changes the traits (phenology, morphology, or behavior) rather than the abundance of the intermediate species in ways that alter the interaction strength between the invader and the receiver species. For example, an invasive predator may kill some individuals in a native herbivore's population (a density-mediated direct effect), which could reduce foraging on its preferred forage plant (a density-mediated indirect effect), but it could also alter the herbivore's behavior (trait effect) due to predator avoidance such that the whole herbivore population dramatically reduces its impacts on its preferred plant by shifting its foraging to other plant species (trait-mediated indirect interaction; e.g., Schmitz et al. 1997). Although trait-mediated indirect interactions are more cryptic than density-mediated pathways, studies in native systems indicate that they are ubiquitous and frequently as strong as or stronger than density-mediated interactions (Schmitz et al. 2004; Trussell et al. 2006; Werner and Peacor 2003).

Other mechanisms by which invasive species impact native systems further illustrate the cryptic or complex nature of their effects. In some systems, impacts can be transmitted genetically as non-native species alter the gene pools of native species via introgression (Lockwood et al. 2007), or initial invasions may facilitate additional invaders with

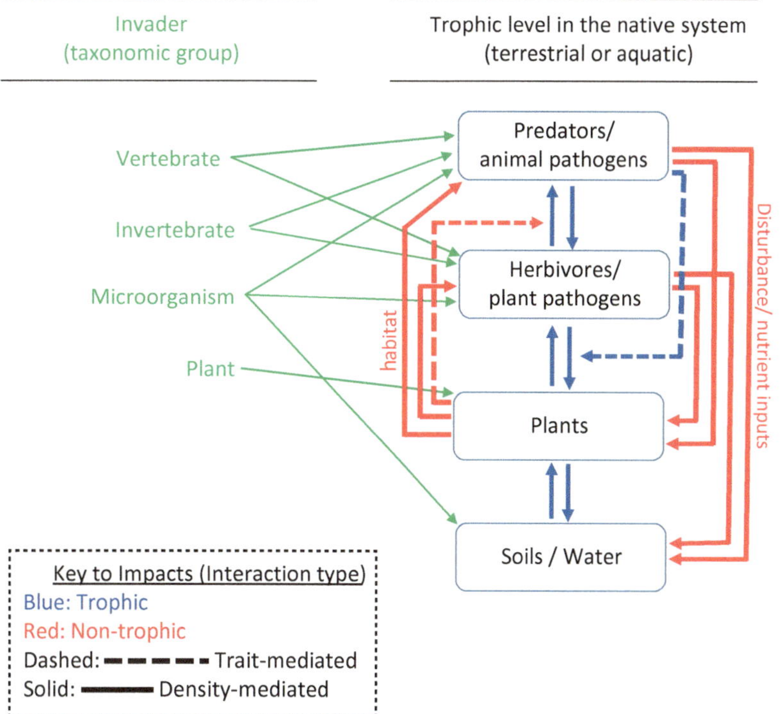

Fig. 2.1 Conceptual diagram illustrating how differences among invader taxa or functional roles determine how invasive species impact native systems. Trophic interactions are depicted as simplified food webs (*blue arrows*) integrated with non-trophic interactions (*red arrows*). *Solid arrows* indicate density-mediated effects, whereas *dotted arrows* indicate trait-mediated effects (interaction modifications). Only a subset of possible interactions are highlighted for simplicity. Invaders from various taxonomic groups (*green*) enter the functional network from the left. For example, in this framework, an invasive terrestrial vertebrate that enters the system as a predator can impact native herbivores directly through food web interactions, but it can only affect native plants indirectly via either density-mediated or trait-mediated indirect interactions. The predator could, however, impact native plants and soils directly through non-trophic interactions such as disturbance or nutrient inputs. The predator can also impact other predators on the same trophic level (interactions not highlighted here) via interference (direct) or resource competition (indirect), thereby initiating indirect effects on native species at lower trophic levels

cascading ecological effects, referred to as invasional meltdown (Gandhi and Herms 2010a; Simberloff and Von Holle 1999). Because of their complexity, attempting to mitigate invader impacts without sufficient understanding of the impact mechanism and community context can result in serious unintended consequences (Bergstrom et al. 2009; Boettner et al. 2000; Pearson et al. 2016b; Zavaleta et al. 2001).

2.1.3 Understanding Invaders in a Community Context

A single invasive species can exhibit very different behaviors in different communities (Zenni and Nuñez 2013). Hence, it is critical to understand invader traits in the context of the invaded community in order to characterize its impacts (Kolar and Lodge 2001; Pyšek et al. 2012). In general, the novelty of a species relative to other community members influences whether it contributes uniquely to community function and productivity, or perhaps serves more redundantly as a buffer against disruptive forces (e.g., Duffy 2002). For example, the introduction of predators, consumers, or plants to islands where they represent novel functions frequently results in dramatic impacts (Pyšek et al. 2012; Simberloff 1995; Vitousek 1990). Research indicates that evaluating the novelty of an invader's traits relative to the recipient community can explain which native species will be affected and how (Ortega et al. 2014; Pearson 2009). Not surprisingly, invaders that act as ecosystem engineers (i.e., organisms that modify the availability of resources to other species by causing physical state changes in biotic or abiotic materials; Jones et al. 1996) or impact native ecosystem engineers can alter community context with substantial impacts across many species and ecological levels (e.g., Blackburn et al. 2014; Crooks 2002; Pearson 2010; Rodriguez 2006; White et al. 2006).

While an invader's novelty has clear ramifications for its ability to enter a community and impact natives when the invader represents a separate trophic level (e.g., an established invasive predator can readily impact native prey where no predators previously existed), novelty of invaders has presented a conundrum as it relates to understanding impacts within trophic levels (MacDougall et al. 2009). On one hand, an invading species that is unique should have an advantage in entering a community as it should experience minimal niche overlap with and resistance from native species, but if it has minimal niche overlap, how does it impact natives within the same trophic level? On the other hand, if it exhibits high niche overlap with natives, how does it enter the community? Application of coexistence theory to invasions demonstrates how invaders with high or low niche overlap with natives can invade and impact native communities as a function of the interplay between their fitness and niche differences with the natives (MacDougall et al. 2009).

2.1.4 Research Gaps for Understanding Ecological Impact

There are several research needs or gaps in our understanding of the impacts of invasive species on terrestrial and aquatic systems. Invader impacts can be quantified as $I_{impact} = R_{range} \times A_{abundance} \times E_{per\ capita\ effect}$ (Parker et al. 1999), but this approach has rarely been applied quantitatively due to the challenges associated with estimating per capita effects (i.e., the effect per individual or per biomass unit) of invaders within native communities (Barney et al. 2013; Pearson et al. 2016a; Ricciardi et al. 2013). Although this is a model that largely assumes a linear impact (see below) that is invariant over space and time, there may be conceptual value in attempting to apply this approach to other invasive taxa besides plants, for which it was developed. Evaluating impacts that involve multiple invaders at regional scales can be critical for prioritizing invasive species management (Chap. 7). However, most work quantifying invader impacts has been invader-specific and focused on local scales (Hulme et al. 2013). Since most systems experience multiple invaders (Kuebbing et al. 2013), understanding the additive and interactive effects of invaders, including invasional meltdown, will be a key research need for understanding overall invader impacts.

Furthermore, very few studies have examined regional impacts, especially for multiple invaders, and these have usually (but not exclusively) examined the impacts of invasive plants (Morin and Liebhold 2015; Pearson et al. 2016a; Thiele et al. 2010). These studies indicate that invader impacts are highly variable, and they may be linear or nonlinear as a function of invader abundance, and depending in part on the selection of a response metric, e.g., native abundance versus diversity (Barney et al. 2013; Pearson et al. 2016a; Thiele et al. 2010). Nonlinear impacts can result in thresholds that complicate quantification and, ultimately, management of invader impacts, but nonlinearities can be addressed (Thiele et al. 2010; Yokomizo et al. 2009). Hence, understanding how invader abundance relates to impact for different response metrics is another important research gap.

Building on the impact models of Parker et al. (1999) and Ricciardi (2003), Lockwood et al. (2007) suggested that accounting for variable success of the invader at different stages in the invasion process (e.g., transport, introduction, establishment, or spread) is important for determining overall impact. Thus, time since introduction is a significant factor when assessing impact of an invasive species. In addition to the complex mix of contributing factors such as species characteristics, environmental site characteristics, ecological interactions, and invasion history, our interpretations of impact are also shaped inevitably by human biases and limits on scientific perception and detection (Lockwood et al. 2007).

Community interactions are context dependent as are invasion outcomes (Cox and Lima 2006; Kolar and Lodge 2001). Hence, anthropogenic changes such as nitrification, increasing temperature, increasing carbon dioxide (CO_2), and increased or altered disturbance regimes can have large ramifications for invasion outcomes and invader impacts (Walther et al. 2009). Invader impacts may also change over time as a function of increasing invader abundance, cumulative effects, or changing soil feedbacks (e.g., Lankau et al. 2009). Additionally, management strategies intended to mitigate the effects of invasive species (Chap. 7) may themselves result in undesired effects by way of complex interactions or simple side effects of management tools (Bergstrom et al. 2009; Boettner et al. 2000; Pearson et al. 2016b; Zavaleta et al. 2001), resulting in a need to better understand and mitigate against these unintended effects. In order to move beyond speculation to more predictive science regarding invasive species impacts, we need to advance research in all of the areas addressed above. In the following sections, we outline current understandings of the state of the science regarding mechanisms of invader impacts by taxonomic group, highlighting key information needs for invasive plants, pathogens, invertebrates, and vertebrates in terrestrial and aquatic systems.

2.1.5 Key Findings

- Invasive species are a subset of non-native organisms that substantially alter composition, structure, or function of native terrestrial and aquatic systems. Their ecological impacts can include direct and indirect effects at multiple levels (organisms, populations, communities, and ecosystems). Invasive species can also impact the genetic makeup of native species populations.
- Invasive species impacts may be considered positive or negative depending on the environmental context, the stage of the invasion process, and human biases and perceptions. Nonetheless, invasive species can cause large negative ecological impacts in ways that decrease the

economic, aesthetic, and social benefits of native ecosystems and cost taxpayers billions of dollars annually.

- The distribution, abundance, and per capita effects of invasive species are primary determinants of their ecosystem impacts. However, predicting or quantifying these impacts is difficult due to differences among organisms, their environments, and the numerous complex interactions among organisms and their environments.
- Attempts to manage invasive species can have unintended negative consequences when those complex interactions are poorly understood. Therefore, the potential consequences of applicable management options (including decisions to refrain from treatment efforts) should be considered when developing any pest management plans.

2.1.6 Key Information Needs

- Quantitative assessments of the ecological effects of a much wider variety of plants, animals, and pathogens would help to provide more comprehensive and accurate estimates of invasive species impact.
- Better characterization of large-scale, regional impacts of invasive species, including co-occurring impacts of multiple invaders, is needed for prioritizing invasive species management.
- When comparing impact of different invasive species, scientists should account for potential differences in invader abundance, the stage of the invasion, the type of ecosystem, and the type of impact measurements used.
- Research aimed at an understanding of how control of an invasive species affects the whole ecosystem will help avoid undesired, nontarget impacts of management.

2.2 Impacts of Invasive Plants in Terrestrial Systems

2.2.1 Invasive Plant Impacts on Community Structure and Function

Invasive plants impact native terrestrial systems by altering species abundances and distributions, fire regimes, belowground biotic and abiotic processes, and resource availability to other taxa. Impacts caused by invasive plants may differ from those caused by other invasive taxa in several ways. As autotrophs, invasive plants alter the base of the food chain and thereby key processes like primary productivity and nutrient cycling. By disrupting these basic processes, invasive plants can restructure extant ecological interactions and alter future trajectories of the community (Didham et al. 2007). Furthermore, all plants act as ecosystem engineers to varying degrees because plants provide habitats for animals and arenas for their interactions, thus impacting animals through various non-trophic and trophic pathways (Crooks 2002).

As with other invaders, key components of impact by invasive plants include the strength of their interaction with native species and their novelty in the system (i.e., the degree to which the invasive species brings new traits or functional roles). While invasive plants that are similar in form and function to native species can impact systems, those with novel traits frequently change the way systems function (Crooks 2002; Li et al. 2015; MacDougall et al. 2009) (Fig. 2.2). Recent studies that have quantified invasive plant impacts (Pearson et al. 2016a; Thiele et al. 2010) based on Parker et al.'s (1999) framework (see Sect. 2.1) demonstrate that the local abundance of the invader is a critical factor in determining the strength of plant invader impacts on native plant abundance. Accordingly, plant traits that favor increased local abundance are key to driving local impacts such as clonality, resource reallocation to larger body size, and/or release from natural enemies (Blossey and Nötzold 1995; Pyšek and Richardson 2008; Rejmánek 1996; Suda et al. 2015). Furthermore, traits linked to spread, such as increased fecundity and dispersal, facilitate the dissemination of those impacts over larger spatial scales only for species that can achieve high local abundance (Pearson et al. 2016a). In this regard, plants are unique in that polyploidy events (the nuclear accumulation of multiple sets of chromosomes) are not always fatal (as they are in animals) and can be associated with the development of traits such as larger body size or increased seed production. Historically, analyses attempting to predict invader impacts based on plant traits alone have met with limited success (Pyšek et al. 2012). However, distinguishing between traits associated with invasiveness (the effectiveness of the invader at establishing populations over wide areas) versus impact (the actual effect of the

Fig. 2.2 Schematic of the grass-fire cycle

invader on native species or systems) may improve prediction of both invasiveness and impact, particularly if invader traits are evaluated relative to the context of the recipient community (see Pearson et al. 2016a).

2.2.2 Invasive Plants: Competition and System Engineering

One of the most direct ways in which invasive plants impact native plants is through resource competition. According to competition theory, the plant predicted to win in head-to-head resource competition will be the species that can utilize a limiting resource at lower resource levels than its competitor (Tilman 1982). Different life history strategies and associated trait sets will generate cost–benefit tradeoffs that favor different individuals or species under different resource and environmental conditions (Grime 1988). Research indicates that increased availability of a limiting resource tends to favor invaders, whereas decreased resource availability can favor natives (Daehler 2003; Seabloom et al. 2015). Accordingly, directional anthropogenic changes to the environment (e.g., nitrification, elevated CO_2, and associated climate change, altering disturbance regimes) can favor invasive plants when they increase availability of limiting resources but favor natives when they reduce limiting resources. We need to better understand how limiting resources affect invasive plant impacts on natives particularly in the context of human-caused environmental change to understand and predict the effects of invasions in the context of directional anthropogenic change.

Disturbances, whether natural or anthropogenic, are often important facilitators of invasion (Davis et al. 2000). For example, when a large, mature tree falls in an uninvaded forest (a natural disturbance), the light availability to the ground increases and mineral soil may be moved or exposed, providing an opportunity for an invading plant to establish (Colautti et al. 2006; Hierro et al. 2005). In such a scenario, the success of the plant invader is often attributed more to ruderal traits associated with establishment as compared to competitive traits (ideal weed hypothesis, Baker and Stebbins 1965; Rejmánek and Richardson 1996). Alternatively, some forests and prairies require regular intervals of disturbance to flourish and maintain their biodiversity, and thereby the community's ability to resist or minimize the establishment and impact of invasion by terrestrial plants (biotic resistance hypothesis, Elton 1958). For example, in the Southeastern United States, the land area in longleaf pine (*Pinus palustris*) savannahs, characterized by widely spaced trees and sparse grassy understory, has been critically reduced due to human land-use changes and fire suppression (Landers et al. 1995). The resulting lack of low- to moderate-intensity fires has led to population reductions for a federally endangered keystone species, the gopher tortoise (*Gopherus polyphemus*), whose burrows serve as shelter for more than 330 other animal species (Van Lear et al. 2005). Plant invasions in this system contribute to thick understories, outcompeting fire-adapted native plant species, displacing or extirpating species like the gopher tortoise, and altering the vertical and trophic structure, negatively impacting the system.

Allelopathy, the chemical inhibition of one species by another, is another means by which invasive plants can directly impact native plants (Callaway and Aschehoug 2000). Many high-impact invaders are purported to produce allelopathic compounds that potentially impact native plants (Hierro and Callaway 2003). However, studies demonstrating allelopathic effects of invasive plants in natural conditions are uncommon (Hierro and Callaway 2003), and more definitive work is required to understand how this mechanism functions and the degree to which it can explain invasive plant impacts (Blair et al. 2006).

Ecosystem engineering is an extremely important means by which invasive plants can impact native plants, animals, and system processes (Crooks 2002). Plant litter deposition and turnover rates can substantially influence abiotic conditions and biotic interactions (Xiong and Nilsson 1999). Invasive plants can differ substantially in litter production and decomposition rates from natives due to differences in growth rates and tissue composition (Allison and Vitousek 2004; Holly et al. 2009; Liao et al. 2008), and these differences likely contribute to invader impacts at multiple ecological scales (Ashton et al. 2005). Alterations in litter production and decomposition rates can also modify fire regimes in plant communities by increasing the frequency or intensity of fire events, or suppressing fire events in fire-adapted communities. For example, cogongrass (*Imperata cylindrica*) is a well-established and widespread invasive grass across the southern Gulf Coast, and now Atlantic States, due to multiple introductions with a proportionally high degree of introduced genetic variation and intrinsic phenotypic plasticity (Lucardi et al. 2014). Cogongrass infestations increase leaf-litter production (Holly et al. 2009; Terry et al. 1997), promoting more frequent, intense fire events that result in significant timber loss and monotypic stands (Lippincott 2000; MacDonald 2004). Serious modifications in fire frequency impact both ecological and human values and tend to form positive feedback cycles with large-scale negative effects and long-term consequences (Brooks et al. 2004; D'Antonio and Vitousek 1992).

Invaders that alter fire regimes are recognized globally as some of the most important ecosystem-altering species on the planet (Box 2.1) (Balch et al. 2013; Brooks et al. 2004; D'Antonio and Vitousek 1992; Estrada and Flory 2015). Positive feedback cycles related to invasive grasses and disturbances are noted for invasive grasses such as cheatgrass (*Bromus tectorum*) and other annual grass species in the

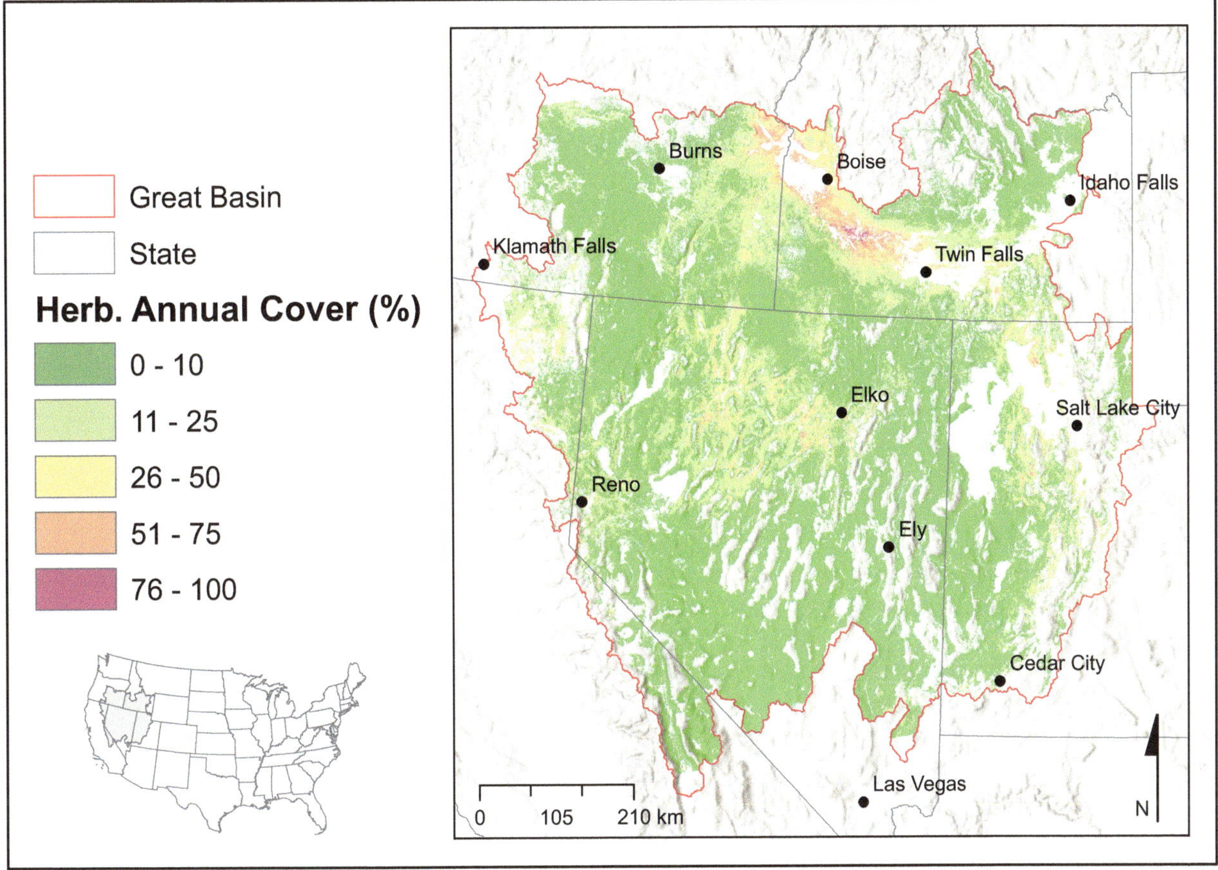

Fig. 2.3 Early 2017 season (May 1) annual herbaceous cover in the Great Basin. (Boyte and Wylie 2017)

Western United States (Fig. 2.3), buffelgrass (*Pennisetum ciliare*) in the Southwestern United States, and cogongrass in the Southeastern United States (see also the Northwest, Southwest, and Southeastern and Caribbean Regional Summaries in the appendix). These grass invaders initially fill the interspaces between native plants and provide horizontal continuity that can fuel fires that would typically not carry well due to the natural fuel limitations of many of these systems. These species can also dry out earlier in the season than native plants, creating a dangerous fire hazard. After fire, annual grasses may invade and exclude native species. At fine spatial scales, cheatgrass establishment in areas with cheatgrass in the vicinity is correlated with burn extent (Kerns and Day 2017). The "grass-fire" cycle can drive an ecosystem further from its original state and may eventually lead to a novel ecosystem that has no historical analog (Box 2.1). Increased fire occurrence, intensity, and severity have been observed in association with these types of grass invasions across the globe (Balch et al. 2013). For example, one fire history study in Idaho estimated a fire return interval of 3–5 years in cheatgrass-dominated rangelands, compared with 60–100 years in native sagebrush (*Artemisia* spp.)-dominated rangelands (Whisenant 1990).

Plant–soil feedbacks are an important indirect interaction by which invasive plants impact native plants. Plant–soil interactions can result in positive or negative feedbacks between plants and soil microbial communities (Wolfe and Klironomos 2005). Pathogenic soil microbes can negatively affect plants by attacking them directly. Soil microbes can positively affect plants by increasing nutrient availability or uptake, usually via a symbiotic mycorrhizal relationship wherein the plant reciprocates by providing carbon to the microbes. Plant–soil feedbacks occur when plants or soil microbial communities influence these interactions, for example, when plants generate litter, carbon, or secondary compounds that influence the relative abundance of pathogenic versus symbiotic microbes in ways that affect their own populations or those of neighboring plant species (Wolfe and Klironomos 2005). Different species of invasive plants associate with different species or functional groups of soil microorganisms that alter plant–soil feedback after invasion. For instance, garlic mustard (*Alliaria petiolate*) is nonmycorrhizal and may impact natives by depleting mycorrhizal inoculum to the detriment of native host plants (Stinson et al. 2006). Also, mycorrhizal fungi that are introduced with invasive plants may enhance those plants' abilities to spread (Dickie et al. 2010; Schwartz et al. 2006; Urcelay et al.

Box 2.1 Invasive Grasses and the Grass-Fire Cycle: Saving the Sagebrush Biome

Invasion by grass species following fire and other disturbances can promote strong feedbacks. In the case of fire, this process is frequently referred to as the "grass-fire cycle" (Fig. 2.2). Once a system is dominated by invasive grasses, e.g., cheatgrass (*Bromus tectorum*) or buffelgrass (*Pennisetum ciliare*), restoration or rehabilitation may be difficult or prohibitively expensive to accomplish, especially across large spatial scales. Emergent risks of habitat degradation due to invasive-dominated grasslands that readily burn are now widely recognized. Cheatgrass invasion and the grass-fire cycle are now known to be one of the primary mechanisms altering contemporary sagebrush (*Artemisia* spp.) ecosystems of the Great Basin and the sagebrush biome (Chambers et al. 2013).

The Sagebrush Biome of the Western United States: An Imperiled Ecosystem

The Great Basin of Western North America is a large (541,727 km^2; Coates et al. 2016), cold desert ecosystem dominated by sagebrush shrubs (Fig. 2.3). This sagebrush steppe ecosystem has long been included among the most imperiled in North America (Noss et al. 1995), and it provides a case study of how altered wildfire regimes driven by invasive grass can rapidly change a fragile regional ecosystem and threaten native habitats and sensitive species. Much of the Great Basin has been invaded to some extent by annual grasses such as species in the genus *Bromus*, medusahead (*Taeniatherum caput-medusae*), and ventenata (*Ventenata dubia*) (Fig. 2.3). While other factors such as climate change, conifer expansion, land-use change, and development have been important, the spread of non-native invasive plant species is one of the major and persistent threats in these ecosystems (Chambers et al. 2017; Coates et al. 2016; Davies et al. 2011).

Much of the sagebrush biome is home to the greater sage-grouse (*Centrocercus urophasianus*), a large gallinaceous bird that requires distinct sagebrush habitats to survive. Populations of sage-grouse have declined in concert with the overall loss and fragmentation of the sagebrush biome following Euro-American settlement of the Western United States. The species is estimated to occupy about half of its historic distribution (Schroeder et al. 2004).

Owing to the population decline of the greater sage-grouse, several evaluations for protection under the Endangered Species Act have been conducted. While the species has not been listed, population declines have motivated unprecedented multi-agency coordinated Federal land management and An Integrated Rangeland Fire Management Strategy (IRFMS) (U.S. Department of Interior 2015). The IRFMS outlines longer term actions needed to implement policies and strategies for preventing and suppressing rangeland fire and restoring rangeland landscapes affected by fire in the Western United States. As part of this strategy, a science framework for conservation and restoration of the sagebrush biome (Part 1) was recently released (Chambers et al. 2017). The forthcoming Part 2 will focus on management considerations.

Restoration of cheatgrass-dominated landscapes in the sagebrush biome emphasizes resilience to wildfire and resistance to cheatgrass invasion. Research suggests that resilience and resistance are strongly associated with soil moisture and temperature regimes in these semiarid ecosystems (Chambers et al. 2013; Maestas et al. 2016). Management focused on resilience and resistance is critical for local communities that depend on ecosystems services from rangelands such as water for consumption, forage, and recreational opportunities.

Because of the strong feedbacks due to the grass-fire cycle in the Great Basin, Coates et al. (2016) suggested that areas mapped with low resilience and resistance to cheatgrass could be targeted for wildfire suppression efforts to protect vulnerable sage-grouse habitat. The IRFMS used a mid-scale approach to prioritize areas for management and treatment focused on six steps: (1) identifying focal species, resources, or habitats; (2) mapping soil temperature and moisture regimes; (3) development of a decision-support matrix; (4) threat assessment; (5) prioritization; and (6) appropriate management strategies (Chambers et al. 2017).

Integrated approaches such as the IRFMS that consider multiple factors may provide the most likely approach to restore the sagebrush biome. However, many challenges for restoration remain. Davies et al. (2011) noted that research is needed to develop either long-term control or reduction in invasive annual grasses, and that there is a lack of knowledge regarding native seedling establishment ecology and variability (in time and space) to provide opportunities to successfully restore these plant communities (see additional details on restoration in Chap. 8). Despite these hurdles and the significant threats to the sagebrush biome, with sufficient resources, multi-agency cooperation, and use of integrated management approaches, restoration for many sites can be successful.

2017). Theoretically, invasive plants could generally benefit over natives if they experience reduced negative soil feedbacks or increased positive soil feedbacks relative to natives (Reinhart and Callaway 2006). However, recent studies suggest that invasive plant responses to plant–soil feedbacks may be more idiosyncratic (Suding et al. 2013).

2.2.3 Invasive Plants and Pollinators

Insect-based pollination in US wildland and agroecosystems is strongly impacted by invasive species of plants and insects (Box 2.2). Invasive plants directly impact native pollinators (Moroń et al. 2009, in Europe) and indirectly impact native plants via interactions with pollinators. Invasive flowering plants may enhance or reduce pollinator services to native flowering plants by increasing or decreasing pollinator visitation rates or by increasing heterospecific pollen transfer, which is essentially interference competition (Brown et al. 2002; Morales and Traveset 2009). However, in some instances non-native plants may provide benefits to native pollinators (Goodell 2008; Russo et al. 2016; Stout and Morales 2009; Tepedino et al. 2008). Ironically, removal of invasive legumes during tallgrass prairie restoration may negatively impact pollinator conservation efforts (Harmon-Threatt and Chin 2016). Invasive plants may be relatively successful reproductively in new habitats either because they are either visited by pollinators at similar or greater rates to native plants or because they are self-fertilizing (i.e., autogamous) (Baker 1974; Harmon-Threatt et al. 2009; Powell et al. 2011).

Two of the primary managed pollinators in the United States, the western honeybee (*Apis mellifera*) (Fig. 2.4a) and the alfalfa leafcutter bee (*Megachile rotundata*), are non-native insects. The latter is a key pollinator for alfalfa seed production, which forms the basis for hay production for livestock (Pitts-Singer and Cane 2011). Unmanaged invaders

Box 2.2 Interactions Among Invasive and Native Plants and Pollinators in the United States

In the strict sense, the workhorse of pollinators, the western honeybee (*Apis mellifera*) (Fig. 2.4a), is a managed, non-native species originating from Western Europe (Franck et al. 1998). The pollination services that it provides in the United States have come under increasing scrutiny as the number of pollination-dependent crops and their planting acreages have grown (Aizen and Harder 2009; Williams et al. 2010) and the general health of the commercial colonies has declined (Cox-Foster et al. 2007; Ellis et al. 2010; vanEngelsdorp et al. 2009). An issue of concern associated with Colony Collapse Disorder is the spread of an invasive natural enemy of the western honeybee, the Varroa mite (*Varroa destructor*) (Fig. 2.4b), which was introduced into the United States in 1987 (NRC 2007). Originating from Southeast Asia, this invasive pest is parasitic and transmits viruses to the western honeybee (NRC 2007). Another invasive organism originating from Asia is the pathogenic microsporidian *Nosema ceranae* (not pictured), which also appears to have played a relatively major role in the declining health of the western honeybee in the United States (Chen et al. 2008; Higes et al. 2008; Klee et al. 2007). These are issues of major concern to the U.S. beekeeping industry (vanEngelsdorp and Meixner 2010), which have led to a growing research effort to understand the role of native pollinators (Fig. 2.4d) in wildland and agroecosystems (Artz et al. 2013; Koh et al. 2015).

When invasive plants that do not provide floral resources (e.g., cheatgrass, Fig. 2.4c) replace native flowering plants, native pollinators lose access to pollen and nectar. Conversely, in some instances, invasive flowering plants (e.g., *Melilotus* spp. and *Sonchus* spp.) have provided the bulk of pollen collected by managed populations of western honeybees in foraging areas in the Prairie Pothole Region of North Dakota (Smart 2015). Other potential sources of impact on native pollinators are unintentionally introduced species such as the wool carder bee (*Anthidium manicatum*) (Fig. 2.4e), a solitary bee introduced to North America in the late 1960s from Europe (Gibbs and Sheffield 2009; Miller et al. 2002; Strange et al. 2011; Zavortink and Shanks 2008). The fortuitous management of another accidentally introduced species (from the Near East), the alfalfa leafcutter bee (*Megachile rotundata*), has had dramatically positive impacts on the production of alfalfa seed in North America (Cane 2003; Pitts-Singer and Cane 2011). However, several other species of invasive megachilid bees in California may negatively impact native bees in the same family by excluding them from suitable nesting cavities (Cane 2003). The intentionally introduced Japanese bee *Osmia cornifrons* appears to have been accompanied or joined by a Japanese mite, a parasitic wasp, and the look-alike bee *O. taurus*. Throughout the Eastern United States, these two Japanese *Osmia* are displacing the native bee *O. lignaria*, partly through aggressive competition for nesting sites (Cane 2018). Little is known about interactions between the recently detected Asian giant hornet, *Vespa mandarinia* Smith, and native plants or pollinators. It is the world's largest hornet and a very aggressive predator that specializes in mass attacking nests of other species and is a major pest in its native range. Thus, it potentially threatens honey bees and native pollinators throughout North America (Matsuura and Yamane 1990, Tripodi and Hardin 2020).

Fig. 2.4 A nexus of invasive and native plants, pollinators, and pathogens: (**a**) The western honeybee (*Apis mellifera*) with a full pollen basket on its hind leg. (Photo by David Cappaert, Bugwood.org); (**b**-1 and **b**-2) a western honeybee infected by a phoretic Varroa mite (*Varroa destructor*). (Photo by Kathy Keatley Garvey, University of California, Davis); (**c**) Cheatgrass (Bromus tectorum). (Photo by John M. Randall, The Nature Conservancy, Bugwood.org); (**d**) native bees (Stephen Buchmann, Pollinator Partnership, www.pollinator.org); and (**e**) the invasive wool carder bee (*Anthidium manicatum*). (Photo by Kimberly Steinmann, University of California, Bugwood.org)

(e.g., the wool carder bee (*Anthidium manicatum*), Fig. 2.4e) may also impact pollination in various systems. Including these three species, there are at least 25 non-native bees established in the United States and Canada (Cane 2003; NRC 2007), comprising less than 1% of the continental native bee fauna. Cane (2003) reported that most of the non-native bee species have come from Europe, most nest in stems or wood, and most use many different floral hosts as pollen sources (polylecty). Non-native bees in the United States (NRC 2007) may also impact native pollinators (Cane and Tepedino 2017; Fürst et al. 2014 (in the United Kingdom); Goulson 2003; Stout and Morales 2009) and native plants (Goodell 2008; Goulson 2003; Stout and Morales 2009).

Arguably, the strongest effect of invasive plants on native plant–pollinator interactions may be the large-scale species transformation of plant communities, such as the transition from diverse forb communities to wind-pollinated grasses such as cheatgrass (Fig. 2.4c) or crested wheatgrass (*Agropyron cristatum*) (Gunnell et al. 2010), across the Great Basin (Mack 1981). Much in the way that invasive grasses have transformed fire regimes over large regions (D'Antonio and Vitousek 1992 (Box 2.1), these wind-pollinated grasses have transformed pollinator systems.

Although there is strong evidence that invasive flowering plants can negatively impact native plants by competing for pollination services (Morales and Traveset 2009), it is unclear how strong these indirect effects are on native plants relative to direct impacts via competition (Brown et al. 2002). In general, the effects of invasive plants on pollinators and pollination services are complex and not fully understood. For example, Russo et al. (2016) demonstrated in Pennsylvania that the invasive spiny plumeless thistle (*Carduus acanthoides*) was both highly visited and strongly preferred by bees relative to other flowering species. Indeed, greater than four times more *Bombus* species, a group of native bumblebees recently found to be in decline, were attracted to patches that included the invasive thistle. However, in many cases greater visitation of invasive plants by bees may primarily be about their greater abundance rather than pollinator preference (Williams et al. 2011). Research in California has shown that protein and amino acid compositions of pollen were comparable among invasive and native flowering plants in a plant community visited by bumblebees (Harmon-Threatt and Kremen 2015). Conversely, removal of the invasive shrub Chinese privet (*Ligustrum sinense*) from the riparian forests of the Southeastern United States dramatically increases the abundance and diversity of pollinator communities, but the potentially intricate mechanisms behind these changes and their relationship to native flora require further study (Hanula and Horn 2011a, b; Hudson et al. 2013). More work is also needed to understand invasive plant effects on pollination, particularly regarding large-scale transitions from forbs to wind-pollinated grasses.

2.2.4 Invasive Plant Impacts on Trophic Interactions: A Tangled Web

Apparent competition, an indirect interaction of two prey species via differential impacts from a shared predator (Holt 1977), is a potential mechanism by which invasive plants impact native plants (Noonburg and Byers 2005). Instances of apparent competition between invasive and native plants are not well demonstrated empirically, but variations on this theme are known. For example, "second-order apparent competition" has been shown to significantly impact native plant recruitment in a weed biocontrol system in which the biocontrol insect is fed upon by a native rodent that also acts as a seed predator on native plants (Pearson and Callaway 2008). "Refuge-mediated apparent competition" (Orrock et al. 2010), wherein an invasive plant provides habitat instead of food to native rodent seed predators, also has been shown to facilitate increased seed predation on native plants (Dangremond et al. 2010; Orrock et al. 2008). Apparent competition and its variations may be an important but understudied means by which invasive plants impact native plants via seed predators and other herbivores.

Invasive plant impacts on native plants can produce negative or positive effects on native animals by altering the quantity or quality of food resources (Lockwood and Burkhalter 2015; Waring et al. 1993), an effect that can transmit indirectly to higher trophic levels (Ortega et al. 2006). Invasive plants can also directly and indirectly affect native animals by altering vegetation architecture (Box 2.3, Fig. 2.5), which can improve or degrade habitat for nesting, hiding, and foraging (Lockwood and Burkhalter 2015). For example, cheatgrass invasion can create such dense stands in the Great Basin relative to native vegetation that it inhibits movement of native rodents (Rodentia) and lizards (Squamata) (Rieder et al. 2010). In contrast, changes in plant architecture have dramatically increased the abundance of native spiders (Araneae) in some grassland systems by altering the quality and quantity of web substrates, which results in indirect negative impacts on spider prey species through both density- and trait-mediated interaction pathways (Pearson 2009, 2010). These examples demonstrate how invasive plants can have either positive or negative direct effects on native animals.

2.2.5 Invasive Plant Impacts and Evolution

Plant invasions that shift the structure and function of native communities can alter the evolutionary trajectories of both native and invasive plant species. Plant invaders may adapt to new environments through contemporary evolution, i.e., observable evolutionary change occurring over tens of generations or fewer (Buswell et al. 2011; Carroll et al. 2007) and hybridization (Vellend et al. 2007), and in some cases native plants and animals can adapt to invaders (Lau 2006; Strauss et al. 2006; Vellend et al. 2007). The study of invasions has increased our recognition of contemporary evolution among a wide array of taxa and how evolution of invaders and invaded systems interacts, modifying perceivable impact (Colautti and Lau 2015; Cox 2004; Whitney and Gabler 2008). Examining evolutionary adaptations of invaders suggests that traits associated with impact can differ from those associated with spread, with traits linked to spread being associated more with invasion fronts (Phillips et al. 2010; Sakai et al.

Fig. 2.5 Stand of strawberry guava (*Psidium cattleianum*) in Glenwood, HI. (Photo courtesy of Jack Jeffrey, taken 2009)

Box 2.3 Strawberry Guava Invasion and Impacts in Tropical Island Ecosystems

As an ornamental fruit tree, strawberry guava (*Psidium cattleianum*) has been moved by humans from its native Brazil to new habitats worldwide (Ellshoff et al. 1995). With seeds readily dispersed by birds and pigs, it has become a dominant invader in wet forests of tropical islands (Lorence and Sussman 1986; Space 2013), for example, forming dense thickets and displacing native species across tens of thousands of hectares in Hawaii (Fig. 2.5). In addition to reducing habitat for many endangered species (State of Hawaii 2011), strawberry guava increases water loss from forested watersheds (Takahashi et al. 2011), impedes sustainable native hardwood forestry (Baker et al. 2009), and serves as the primary reservoir host for a major fruit fly pest of agriculture (Vargas et al. 1990). The USDA Forest Service developed a leaf-galling scale insect (*Tectococcus ovatus*) as a biological control agent, with the intention of substantially reducing vegetative growth and fruit production of strawberry guava (State of Hawaii 2011). This insect was released in Hawaii in 2012, and monitoring is now underway to measure the benefit of biocontrol to agricultural and native forest ecosystems.

2001). Invader plant impacts may also decline over time as the invader and community settle into a new equilibrium such that coexistence with native taxa may occur due to accumulation of natural enemies (Lankau et al. 2009).

2.2.6 Invasive Plant Impacts: Looking Forward

Invasive plant impacts vary among communities as a function of differences in the recipient community's susceptibility to invasion (Guo et al. 2015) and environmental context. The biotic resistance hypothesis postulates that higher local species richness increases a community's intrinsic resistance to invasion (Elton 1958). However, research examining this relationship has generated conflicting results depending on spatial scales that are likely linked to underlying resource gradients (Iannone et al. 2015; Stohlgren et al. 2003). This indicates the need to better understand how environmental context interacts with diversity to influence community invasibility and susceptibility to invasive plants impacts. Over large spatial and temporal scales, invasive plant impacts can result in biotic homogenization (a global mixing of highly successful and typically ruderal organisms) and the extirpation or mass extinction of unique, constrained, rare, and/or endemic taxa (McKinney and Lockwood 1999; Olden et al. 2004).

2.2.7 Key Findings

- Invasive plants can alter key system processes such as productivity and nutrient cycling, affecting not only native plants but also animals that feed on them.
- There are numerous mechanisms by which terrestrial invasive plants impact native systems, including resource competition, allelopathy, ecosystem engineering, plant–soil feedbacks, effects on pollinators, and apparent competition.
- Competitiveness of invasive plants may be favored when there is increased availability of limiting resources, which in turn may be influenced by natural or human-caused disturbances.
- Invasive plant impacts can result in homogenization of ecosystems and the loss of unique native species.

2.2.8 Key Information Needs

- There is a need to better quantify the impacts of multiple invaders on native communities across spatial and temporal scales.
- A better understanding of how resource limitation influences invasive plant impacts on native plants and ecosystems is needed, particularly in relation to human-caused disturbances.
- Conservation of pollinator species is of increasing importance in the management of natural and agricultural systems, and additional research on the degree to which invasive plants affect pollinator populations and networks is needed to inform management strategies.
- Although diverse communities are often thought to be more resistant to plant invasion, a better understanding of how environmental context interacts with diversity to influence community susceptibility to invasive plant impacts is needed (e.g., evaluating the biotic resistance hypothesis).
- Certain mechanisms of invasive plant impact such as allelopathy, plant-soil feedbacks, and apparent competition have received relatively little attention but may be important to understanding effects on ecosystems.

2.3 Impacts of Invasive Phytophagous Insects and Plant Pathogens in Terrestrial Systems

Non-native invasive phytophagous (plant-feeding) insects and plant pathogens have impacted forests and other terrestrial systems throughout the United States for nearly 150 years (Liebhold et al. 1995; Niemelä and Mattson 1996). Many of these non-native organisms have had only minor, localized, or regional impacts on forest or shade trees, whereas a small proportion has killed millions of trees (Anagnostakis 1987; Herms and McCullough 2014) (Box 2.4, Fig. 2.6) or pushed

Fig. 2.6 The emerald ash borer (*Agrilus planipennis*) and its impact. Adult beetle (**a**) and larva (**b**). The adult disperses through flight; the larva damages the phloem (inner bark) of ash (*Fraxinus* spp.) trees. (Photos courtesy of David Cappaert, Michigan State University, Bugwood.org). Ornamental ash tree killed by emerald ash borer (**c**). (Photo courtesy of Daniel Herms, The Ohio State University, Bugwood.org). Landscape-level impacts of emerald ash borer (**d**) (photo courtesy of Troy Kimoto, Canadian Food Inspection Agency, Bugwood.org). Surveying for damage by emerald ash borer (**e**). (Photo courtesy of Erin Morris, Michigan State University)

ecologically foundational species toward functional extinction (Ellison et al. 2005). In a recent assessment of more than 450 invasive forest insect species that have established populations in the United States, Aukema et al. (2010) considered 14% (62 species), plus an additional 16 invasive pathogens, as "high-impact" species, i.e., of regulatory significance or having caused notable damage to forests or urban forest trees. Although these impacts can frequently be attributed primarily to the action of a single invasive organism, a number of important historic and emerging forest diseases are caused by insect–pathogen complexes in which one or more of the organisms are not native to the ecosystem (Houston 1994; Hulcr and Dunn 2011; Sinclair and Campana 1978). In some cases, these complexes of multiple invaders (see Sect. 2.1) provide important examples of how invasive species can function symbiotically to exert strong impact in terrestrial systems. We have highlighted 15 invasive forest insects, pathogens, or insect–pathogen complexes with historic, current, or emerging importance for terrestrial systems in the United States (Table 2.1).

In this subsection, we focus on impacts of invasive insect and pathogen pests of trees not only because of the economic and aesthetic importance of trees to human society but also because trees frequently shape the structure and function of the terrestrial ecosystems in which they occur. Nonetheless, it should be recognized that invasive insects, other invertebrates, and pathogens of other plant or even non-plant hosts cause a variety of impacts in terrestrial systems (e.g., Bohlen et al. 2004; Daszak et al. 2000; Snyder and Evans 2006).

Invasive phytophagous insects and plant pathogens impact forest ecosystems directly by causing damage to, or mortality of, host trees through herbivory or disease development. Structural or physiological damage may result via a number of different mechanistic or phytochemical pathways. For example, high levels of defoliation by larval gypsy moth (*Lymantria dispar*) reduce carbohydrate allocation to roots and shoots, cause nitrogen (N) deficiency due to reduced N uptake by roots, and result in growth loss and top dieback (Kosola et al. 2001). Phloem-feeding by larval emerald ash borer (*Agrilus planipennis*) hinders transport of photosynthates and nutrients, decreases essential foliar amino acids, and culminates in rapid tree mortality (Chen et al. 2011). In addition to depleting carbohydrates, feeding in the vascular system of firs (*Abies* spp.) and hemlocks (*Tsuga* spp.) by invasive adelgids (*Adelges* spp.) induces abnormal xylem formation, reducing water use and producing drought-like symptoms (Domec et al. 2013). Even minor damage associated with feeding or host colonization by certain invasive insects can predispose (e.g., beech scale (*Cryptococcus fagisuga*)) or inoculate (e.g., redbay ambrosia beetle (*Xyleborus glabratus*) or smaller European elm bark beetle (*Scolytus multistriatus*)) trees with virulent pathogens, resulting in host mortality (Houston 1994; Fraedrich et al. 2008). Several damaging forest pathogens extract carbohydrates and cause cankers or necrosis of the cambium (e.g., chestnut blight caused by *Cryphonectria parasitica*, Anagnostakis 1987), roots (e.g., root disease caused by *Phytophthora lateralis*, Hansen et al. 2000), or shoots, branches, or stems (e.g., white pine blister rust caused by *Cronartium ribicola*, Maloy 2001). Other pathogens induce extreme hypersensitive responses in their hosts, such as systemic tyloses triggered by presence of the laurel wilt pathogen (*Raffaelea lauricola*) in North American Lauraceae (Inch et al. 2012).

Although several authors have proposed empirical models to describe the impact of invasive species (see Sect. 2.1)

Box 2.4 Profound and Widespread Impact of an Invasive Beetle on Naive Ash Trees in the United States

The emerald ash borer (*Agrilus planipennis*) is a flat-headed woodborer from Asia that was first discovered in North America in 2002 (Herms and McCullough 2014). Adult beetles in this group are also known as metallic woodborers, as emphasized in this case by the brilliant green color of emerald ash borer adults (Fig. 2.6a). The beetle has spread from the original site of detection in southeastern Michigan (where it was likely established in the early 1990s) to 31 US states (Emerald Ash Borer Information Network 2018) and the Canadian provinces of Manitoba, Ontario, and Quebec as of January 2018 (Canadian Food Inspection Agency 2018; Kimoto 2018). The spread has been described as "stratified," consisting of natural flight dispersal of the adults and longer distance human-assisted spread, largely through infested ash (*Fraxinus* spp.), firewood, nursery stock, and logs (Herms and McCullough 2014).

The damaging life stage is the larva (Fig. 2.6b), which mines at the phloem–xylem interface of ash trees, eventually causing tree mortality (Fig. 2.6c). Adults feed incidentally on foliage. North American impacts of emerald ash borer have been characterized as ecological, economic, and cultural (Herms and McCullough 2014). Ecological effects include altered understory environment, nutrient cycles, and successional trajectories; facilitation of the spread of light-limited invasive plants; and increased coarse woody debris. Elimination of ash as a consequence of the feeding activity of larval emerald ash borers threatens nearly 100 species of ash-dependent native invertebrate herbivores (Wagner and Todd 2015). Economically, emerald ash borer is the most destructive and costly forest insect to have invaded the United States (Aukema et al. 2011). The multi-billion dollar cost projections are based largely on removal costs for infested and dead ash trees in communities throughout the Midwest. Cultural impacts are centered on the Native American basket weaving tradition that utilizes black ash (*Fraxinus nigra*) as the source material (Poland et al. 2015b). The impact of emerald ash borer on ash (Fig. 2.6d) has been monitored in Indiana, Michigan, and Ohio through a series of survey plots (Fig. 2.6e) established in 2002 and 2003 (Marshall et al. 2013; Mercader et al. 2016).

(Lockwood et al. 2007; Parker et al. 1999; Ricciardi 2003), to our knowledge there has been little attempt to apply these quantitative models to predict impacts from specific invasive terrestrial insects or pathogens of trees. Using a more qualitative conceptual framework, Lovett et al. (2006) proposed that the magnitude of short-term (weeks to years) and long-term (decades to centuries) impacts is at least partially a function of three key features of an invasive insect or pathogen—mode of action, host specificity, and virulence—and three key features of its host plant(s)—importance or dominance in the stand, ecological uniqueness, and phytosociology. Noting that these six features vary continuously and independently, Lovett et al. (2006) suggest that exceptionally severe long-term impacts should be expected for a highly virulent, host-specific agent that kills dominant, abundant, and ecologically unique hosts. Although abundance of their primary hosts varies geographically, this scenario is closely representative of impacts caused by hemlock woolly adelgid (*Adelges tsugae*) (Vose et al. 2013), emerald ash borer (Gandhi and Herms 2010a, 2010b), chestnut blight, and white pine blister rust (Loo 2009). It should be noted that even when host plant species comprise only a very small percentage of the regional biota, impacts may be considered severe in systems where the host is locally abundant. For example, the highly virulent, host-specific pathogen of laurel wilt disease kills nearly all mature stems of native *Persea* spp. that are substantial canopy components of unique Everglades tree islands, southeastern bayhead swamps, and coastal mixed maritime forests, where their importance to biodiversity and ecological function is high (Hughes et al. 2015; Rodgers et al. 2014; Snyder 2015).

The success of invasive species in their non-native ranges has been viewed traditionally as a function of release from natural enemies, and/or a lack of co-evolved defenses in naive hosts (Elton 1958), frequently resulting in greater negative impacts than that occur in interactions among co-evolved species (Paolucci et al. 2013). A number of studies and reviews have evaluated the mixed evidence for the enemy release hypothesis, primarily as it relates to the success of invasive plants (Colautti et al. 2004; Keane and Crawley 2002; Liu and Stiling 2006), but to a much lesser extent for invasive herbivores (Blossey 2011). For invasive insects, evidence for release from enemies has been derived primarily from the successful biological control of several species in natural systems including the European spruce sawfly (*Gilpinia hercyniae*), larch casebearer (*Coleophora laricella*), and winter moth (*Operophtera brumata*) (see Van Driesche et al. 2010; Van Driesche and Reardon 2014; and references therein). Of course, not all invasive insects or pathogens succeed primarily due to enemy release or can be effectively managed via biological control. For example, a robust biological control program is being pursued for the emerald ash borer, and there is recent evidence of released and native parasitoids reducing population growth (Duan et al. 2015). However, North American ash (*Fraxinus* spp.)

Table 2.1 Selected high-impact invasive insects, pathogens, or insect-pathogen complexes established in the United States

Organism	Impact			Selected references
	Disease or complex name	Hosts	US region	
Insects				
Agrilus planipennis (emerald ash borer)	N/A	Ash	Eastern and western	Herms and McCullough (2014)
Anoplophora glabripennis (Asian longhorned beetle)	N/A	Maple, poplar, willow, other	Eastern	Dodds and Orwig (2011) and Hu et al. (2009)
Adelges tsugae (hemlock woolly adelgid)	N/A	Hemlock	Eastern	Havill et al. (2014) and Vose et al. (2013)
Adelges piceae (balsam woolly adelgid)	N/A	Fir	Eastern and western	Smith and Nicholas (1998) and Davis et al. (2020)
Lymantria dispar (gypsy moth)	N/A	Oak, numerous other	Eastern	Davidson et al. (2001)
Pathogens				
Cryphonectria parasitica	Chestnut blight	Chestnut	Eastern	Anagnostakis (1987)
Phytophthora ramorum	Sudden oak death	Oak, tanoak	Western	Grünwald et al. (2008, 2012) and Rizzo et al. (2002)
Cronartium ribicola	White pine blister rust	Five-needle pines	Eastern and western	Maloy (2001)
Phytophthora lateralis	Port-Orford-cedar root disease	Port-Orford cedar	Western	Jules et al. (2002)
Sirococcus clavigignenti-juglandacearum	Butternut canker	Butternut	Eastern	Broders et al. (2015)
Insect-pathogen complexes				
Insect: *Cryptococcus fagisuga* (beech scale) Pathogens: *Nectria coccinea* var. *faginata*, *Nectria galligena*	Beech bark disease	Beech	Eastern	Houston (1994)
Insects: *Scolytus multistriatus* (smaller European elm bark beetle), *Scolytus schevyrewi* (banded elm bark beetle), *Hylurgopinus rufipes* (native elm bark beetle) Pathogens: *Ophiostoma ulmi*, *Ophiostoma novo-ulmi*	Dutch elm disease	Elm	Eastern and western	Brasier and Buck (2001), Jacobi et al. (2007, 2013), Negrón et al. (2005), and Sinclair and Campana (1978)
Insect: *Xyleborus glabratus* (redbay ambrosia beetle) Pathogen: *Raffealea lauricola*	Laurel wilt	Redbay, sassafras, others	Eastern	Fraedrich et al. (2008) and Hughes et al. (2015)
Insect: *Pityophthorus juglandis* Pathogen: *Geosmithia morbida*	Thousand cankers disease	Walnut, butternut, wingnut	Eastern and western	Seybold et al. (2016) and Tisserat et al. (2009)
Insect: *Euwallacea* spp. (polyphagous shot hole borer) Pathogens: *Fusarium* spp., *Acremonium* spp., *Graphium* spp.	Fusarium wilt	Alder, boxelder, cottonwood, sycamore, willow, others	Western (California)	Eskalen et al. (2013), Lynch et al. (2016), and Umeda et al. (2016)

planted in the native range of emerald ash borer in Asia (i.e., ostensibly with the full complement of native natural enemies) are heavily attacked and killed (Liu et al. 2003), suggesting that top-down pressure by natural enemies alone does not preclude substantial impact. With invasive insects, in addition to release from natural enemies as an explanation for invasion success, there is the potential for non-native insects to utilize a niche marked by "pheromone-free space" from native insects in the same guild or feeding group. This concept, based on a reduction in competition from reduced overlap of behavioral chemical "channels" of communication, is being explored with roundheaded woodborers (also known as longhorned beetles) (Millar and Hanks 2017; Mitchell et al. 2015).

Lack of co-evolutionary history may result in defensive mismatches between native plants and their invasive herbivores or pathogens, such as when elicitor-receptor-based defenses are not recognized or only weakly induced in a plant upon attack, or when toxin-based constitutive defenses have not been selected naturally for the specific invader (Desurmont et al. 2011; Verhoeven et al. 2009). A growing body of research on ash indicates that differences in both constitutive and induced defenses (especially in phloem chemistry) confer resistance of Asian ash species to the emerald ash borer relative to susceptible North American congeners (Poland et al. 2015a). Evidence for greater susceptibility or vulnerability of naive host plants compared with co-evolved hosts has been presented for numerous other invasive forest insects and pathogens including the hemlock woolly adelgid (Havill et al. 2011; Montgomery et al. 2009), viburnum leaf beetle (*Pyrrhalta viburni*) (Desurmont et al. 2011), the thousand cankers disease vector *Pityophthorus juglandis* (Hefty et al. 2018), pathogen *Geosmithia morbida* (Utley et al. 2013), the laurel wilt disease pathogen *R. lauricola* (Fraedrich et al. 2015), and the sudden oak death pathogen *Phytophthora ramorum* (Rizzo et al. 2005). The escape of pathogens from their usual selection pressures when introduced into new environments and onto new hosts can provide opportunities for rapid evolution and hybridization that may also influence the magnitude of their impact (Brasier 2001; Hansen 2008; Parker and Gilbert 2004).

In the last decade, a number of authors have reviewed the varied, interacting, and sometimes cascading ecological effects of outbreaks of invasive insects and/or pathogens in forests or other natural systems (Gandhi and Herms 2010a; Kenis et al. 2009; Loo 2009; Lovett et al. 2006; Moser et al. 2009). Tree mortality caused by invasive insects or pathogens thins or creates gaps in the forest canopy to varying degrees depending on key characteristics of invader and host summarized above (Lovett et al. 2006) (Box 2.5, Fig. 2.7). Over the short term, this canopy thinning can reduce host species' stem density and basal area; alter understory microenvironmental factors such as light availability, temperature, and moisture regimes; and increase organic inputs to the forest floor in the form of downed coarse woody debris, leaf fragments, frass (insect excrement and feeding debris), or insect biomass (Cobb et al. 2012; Gandhi and Herms 2010a; Vose et al. 2013). These changes in turn can alter hydrologic and biogeochemical cycling regimes (see Chap. 3) (Brantley et al. 2013, 2015; Clark et al. 2010; Lovett et al. 2010); the establishment, relative abundance, and growth of native and non-native understory plants; and the overall species composition and structure of the plant community (Gandhi and Herms 2010a; Morin and Liebhold 2015). Associated impacts on terrestrial and aquatic fauna may occur due to short- or long-term shifts in the availability and quality of food (including the invasive organism itself) (Barber et al. 2008; Koenig et al. 2011, 2013), host plants (Chupp and Battaglia 2014; Gandhi and Herms 2010b; Wagner 2007), habitat (Rabenold et al. 1998; Tingley et al. 2002), competitors (Work and McCullough 2000), or natural enemies (Gandhi and Herms 2010a; Redman and Scriber 2000). These impacts are particularly pronounced when the host plant of the invasive insect or pathogen is a "foundation species" that defines the structure of the community and has a stabilizing effect on ecosystem processes (Ellison et al. 2005).

Box 2.5 Responding to Sudden Oak Death Through Collaborative Management

Sudden oak death (SOD) caused by the invasive pathogen *Phytophthora ramorum* threatens oak woodlands, urban forests, and horticultural industries. Currently, the disease is established and regulated in forests in California and Oregon. Because numerous eastern oak species and certain associated understory forest plants have shown susceptibility to this disease, there is a potential risk to oak forests beyond the regulated areas. Once established, SOD may continue to have significant negative impact beyond the known infestations, since the loss of oak would adversely affect ecosystem functions such as water quality, biodiversity, and forest structure.

The USDA Forest Service has responsibility in developing, implementing, and promoting innovative management strategies in response to threats to the nation's forests. Likewise, USDA Animal and Plant Health Inspection Service (APHIS) has the lead regulatory role to prevent further spread of damaging agents and protect natural resources. State forestry and agriculture agencies have corresponding roles within their respective States, especially the State Plant Regulatory Official (SPRO), who must determine the extent of any regulatory action needed once SOD has been positively confirmed. It is clear that once established in a forest, complete removal of *P. ramorum* has a low likelihood of success given the broad host range of the pathogen, available pathways for spread, and complex biological life history. It takes continued diligence, collaboration, and constant monitoring, especially given many invasive species have long lag periods prior to fully expressing their disease potential. Collaborative management in practice has many challenges, but with lessons learned in California and Oregon, it may be possible to successfully mitigate the negative impacts of *P. ramorum*. To this end, the USDA agencies and their partners continue to address the ongoing threat of SOD.

From the perspective of human populations, the impact of invasive insects and pathogens on species, communities, and ecosystems is most acutely experienced via changes in the

Fig. 2.7 Tree mortality caused by sudden oak death in (**a**) Southern Oregon. (Photo courtesy of Bruce Moltzan, USDA Forest Service, Washington Office) and (**b**) Monterey County, CA. (Photo courtesy of Tom Coleman, USDA Forest Service, suddenoakdeath.org)

services the affected urban and wildland ecosystems provide to human welfare. Ecological impacts of major forest pests naturally translate into effects on various types of services, i.e., provisioning (e.g., timber and non-timber products); regulating (e.g., air and water quality/quantity, climate regulation); and cultural (e.g., recreation, aesthetics, shade) services (Charles and Dukes 2008; Pejchar and Mooney 2009). However, relatively few studies have quantified impacts on these services by invasive forest insects and pathogens (Holmes et al. 2009). Efforts to quantify direct costs associated with timber losses (Houston 1994), tree removals and replacements (Kovacs et al. 2010; Sydnor et al. 2007), and reduced property values (Holmes et al. 2005) have been made for major invasive pests such as the emerald ash borer, beech scale, and hemlock woolly adelgid, but assessments of direct economic impacts for many other invasive pests are lacking, and studies of indirect or non-market impacts are largely absent. Aukema et al. (2011) conservatively estimated an economic cost of nearly $5 billion per year on governments, landowners, and households due to timber losses, reduced property values, and other expenditures associated with invasive forest insect infestations in the United States, but did not include impacts to non-market ecosystem services due to the scarcity of available data. Useful impact metrics for non-market values and ecosystem services may better inform policy decisions that could help mitigate against the negative effects of invasive insect and pathogen introduction, establishment, and spread (Boyd et al. 2013; Chornesky et al. 2005).

2.3.1 Key Findings

- Invasive plant-feeding insects and plant pathogens (or combinations of both) have large ecological impacts in forests. As they kill the dominant trees, they alter the sunlight, temperature, water and nutrient cycles, and plant composition of the forest, causing potential shifts in animal communities as well.
- Tree mortality caused by invasive insects and pathogens leads to enormously high costs of tree removal, other management responses, and reduced property values in urban and residential landscapes.
- The severity of impact caused by invasive insects or pathogens is at least partially a function of key traits of the invader (e.g., mode of action, host specificity, and virulence), as well as key characteristics of its host plant(s) (e.g., dominance, uniqueness, phytosociology).
- Impacts are particularly large when the affected native organism is a "foundation species" that defines and stabilizes ecosystem processes.
- Factors influencing the success of invasive insects and pathogens include their release from natural enemies, "pheromone-free space," defensive mismatches between plants and invaders, and/or rapid change when released from usual selection pressures.

2.3.2 Key Information Needs

- Models for characterizing the impact of invasive insects and pathogens have been mostly qualitative, and application of more quantitative approaches could help improve impact predictions and allow for better comparisons among invaders.
- In order to adequately assess the potential value of biological control strategies, more experimental research is needed to determine the degree to which release from natural enemies contributes to the success of specific invasive herbivores (e.g., evaluating the enemy release hypothesis).
- Accurate assessments of both economic and non-market impacts are needed for most invasive insects and pathogens, including development of impact metrics for eco-

system values and services that can be used to better inform policy and management decisions.

2.4 Impacts of Invasive Vertebrates in Terrestrial Systems

Invasive terrestrial vertebrates (i.e., amphibians, reptiles, birds, and mammals) in the United States include species that have been relatively recently introduced (since European colonization of North America) and species that persist due to close association with human habitation (i.e., peridomestics), such as several rat and avian species. Invasive vertebrates are estimated to cause more than $46 billion in damage and control costs in the United States annually, including more than $37 billion for control and damage from mammals, $1.9 billion for birds, and $5.6 million for reptiles (Pimentel 2011; Pimentel et al. 2000). This is a conservative estimate in that it accounts for only a subset of all 81 mammal, 99 bird, 69 reptile, and 11 amphibian species that could potentially be considered (Fall et al. 2011). Additionally, the estimates do not consider all costs. For example, damage estimates for feral swine (*Sus scrofa*) ($1.5 billion U.S.) (Pimentel 2011) include only crop damage and control costs, but not damage to other property such as landscaping and vehicles (via collision) or to native plant and animal species, soil, or water. Traditionally excluded from discussions of invasive terrestrial vertebrates are a group of species that have become naturalized and culturally accepted, such as feral horses (*Equs caballus*). In this section, we will also exclude species undergoing natural range expansion into previously unoccupied areas, even when facilitated by a human modified landscape, such as coyotes (*Canis latrans*) spreading to the Eastern United States. Here we cover economic and ecological impacts of vertebrates that alter the biological and physical composition, structure, or function of native populations, communities, or ecosystems in ways that decrease the ecosystem services or have other undesired ecological effects (Ricciardi et al. 2013) while also noting some examples of positive effects on particular species.

Native flora are often affected directly by the actions of invasive vertebrate grazers such as goats (Caprinae), rabbits (Leporidae), and horses and burros (Equidae). In Olympic National Park in Washington State, mountain goats (*Oreamnos* spp.) were introduced in the 1920s (Houston and Schreiner 1995) and are known to impact vegetation near rocky outcrops, as goats preferentially graze on alpine plants such as Idaho fescue (*Festuca idahoensis*) and showy sedge (*Carex spectabilis*) (Pfitsch and Bliss 1985). They also damage other native vegetation through the creation of dirt wallows (Pfitsch and Bliss 1985). Similarly, wild burros (*Equus africanus asinus*) impacted sensitive vegetation through grazing in Grand Canyon National Park in Arizona and were ultimately removed (Houston and Schreiner 1995). Removal of grazers is often a management goal, but ecosystems are complex and removal of grazers can lead to novel issues. For example, on Santa Cruz Island in California, the removal of introduced grazers created an increase in the abundance of invasive weeds (Myers et al. 2000).

Direct agricultural costs are associated with invasive wildlife. Starlings (*Sturnus vulgaris*) occur at high densities in agricultural areas and are estimated to cause more than $800 million in damage by eating grain (Pimentel et al. 2000). Sixty individual starlings were introduced in the United States in 1890 when an attempt was made to bring in every bird species described in the plays of William Shakespeare. This frivolous action has led to a population explosion, where today starling numbers are greater than 150 million individuals and span the entire continent (Homan et al. 2017). Similarly, domestic hogs (*Sus scrofa*) were introduced into mainland North America in the sixteenth century by Spanish explorers and later by settlers looking to have a constant and familiar food supply. The native range of Eurasian wild boar (also *Sus scrofa*, from which domestic hogs are derived) is North Africa and Eurasia, but feral swine (including feral hogs, wild boar, and admixtures of the two) have expanded in North America to at least 38 states (Bevins et al. 2014). Massive agricultural damage (Jay et al. 2007; McClure et al. 2015) and contamination of agricultural crops and potable water with *E. coli* and other disease pathogens (Jay et al. 2007; Kaller and Kelso 2006) have accompanied this population and range expansion.

Invasive terrestrial vertebrates can also directly impact native fauna. Invasive black rats (*Rattus rattus*) are voracious nest predators on many Hawaiian bird species, including the ‘Elepaio (*Chasiempis* species complex), monarch flycatchers rated as “vulnerable” by the International Union for the Conservation of Nature (Amarasekare 1993; Loope et al. 1988; Vanderwerf and Smith 2002). On O’ahu, after rat control was instituted via snap traps and poison baits, researchers noted a 112% increase in ‘Elepaio reproduction; a 66% increase in survival of female ‘Elepaio; restoration of site fidelity, female age structure, and female recruitment; and an increase in the population growth rate (Vanderwerf and Smith 2002). Globally, island ecosystems, especially the avifauna, have been strongly impacted by invasive terrestrial vertebrates (Harper and Bunbury 2015; Jones et al. 2008, 2016).

Although most direct effects of invasive vertebrate species on native flora and fauna are detrimental to the native populations, there are cases where invasive species improve conditions for native populations (Rodriguez 2006). Invasive vertebrates can act as a food subsidy to native predator populations. In the United Kingdom, invasive rabbits (*Oryctolagus cuniculus*) have increased populations of European wild cats (*Felis silvestris*), polecats (*Mustela putorius*), red kites (*Milvus milvus*), and common buzzards (*Buteo buteo*) (Lees and Bell 2008). In the United States, there is evidence that invasive terrestrial vertebrates can make up substantial portions of the diet of native carnivores, including endangered species. For example, the dominant prey item in the diet of the Florida panther (*Puma concolor*) is feral swine (42% of scats contained this species) (Maehr et al. 1990).

Terrestrial invasive species can have direct impacts not only on other species but on the geomorphology of the landscape itself. These "ecosystem engineers" (Jones et al. 1996) can change the ecology of an entire system by disrupting the hydrology or changing the vegetative community. One terrestrial invasive vertebrate species acting as an ecosystem engineer in the United States is nutria (*Myocastor coypus*), a semiaquatic rodent native to South America. The species was introduced at the end of the nineteenth century into California and later into the Southeastern United States for fur farming (Evans 1970). Nutria are known as major consumers of vegetation and can completely denude an area surrounding a waterbody (Fall et al. 2011). They have aquatic and terrestrial dispersal capabilities and are spreading throughout the United States (Guichón and Cassini 1999). Nutria can change the course of a waterway and are blamed for turning vegetated marshes into open water ponds, impacting natural tidal flood controls (Swank and Petrides 1954). Nutria also use a burrow for rearing young, which has been shown to weaken levees and other irrigation structures. They are also major crop pests in the Southeastern United States where they damage rice, sugarcane, cereal, grain, beets, peanuts, melons, and alfalfa (Fall et al. 2011). Overall, the species has the ability through its burrowing and food habits to morphologically change an area, which in turn impacts entire aquatic communities.

Invasive feral swine are also ecosystem engineers. The damage they cause is due to their digging, rooting, and plowing activities, which damage crops, lead to soil erosion, and have been shown to disrupt wetland ecosystems (Engeman et al. 2007; Fall et al. 2011). This damage is greatest in wet environments but can also impact terrestrial wildlands and agricultural areas. From an ecological perspective, feral swine rooting has been shown to influence plant succession and species composition, which in turn has trophic effects on other species that utilize those plant communities (Campbell and Long 2009; Engeman et al. 2007). Feral swine also depredate many native vertebrates including birds, reptiles, and, reportedly, domestic livestock (Seward et al. 2004).

Ecosystem engineers are a type of keystone species, which is a species whose impact is large relative to its proportional biomass in the community (Mills et al. 1993). Burmese pythons (*Python bivittatus*) in the Florida Everglades could be considered a keystone invasive terrestrial vertebrate. Burmese pythons were imported to the United States from Southeast Asia through the pet trade and now are a top predator in the Everglades, where they were first sighted in the 1980s. They have spread throughout southern Florida including all of Everglades National Park (Dorcas et al. 2011). Road surveys from 2002 to 2011 have documented declines in sightings of native vertebrates: raccoons (*Procyon lotor*) (99.3%), opossum (*Didelphis virginiana*) (98.9%), bobcats (*Lynx rufus*) (87.5%), and rabbits (*Sylvilagus* spp.) (100%) (Dorcas et al. 2012). All of these species are prey of Burmese pythons.

Indirect effects of an invasive vertebrate species on native populations can take several forms, including negative impacts on the health of humans and native fauna. Invasive vertebrates can harbor pathogens that cause disease, act as new vectors for disease pathways, reduce crop yield or harm livestock that in turn leads to increased human disease, and contaminate potable water (Davis 2009; Fall et al. 2011; Hall et al. 2008; Meng et al. 2009). A simple example is the ability of feral swine to spread diseases such as brucellosis, leptospirosis, influenza, and pseudorabies, which impact humans either through direct pathogen transmission or through indirect transmission through livestock (Hall et al. 2008; Witmer et al. 2003). In a review of emerging infectious diseases in humans, Jones et al. (2008) found that 60.3% were zoonoses, with 71.8% originating in wildlife; it is not surprising that introductions of terrestrial vertebrates are linked to increases in human disease. Similarly, non-native arctic fox (*Alopex lagopus*) on the Pribilof Islands, a critical northern fur seal (*Callorhinus ursinus*) breeding ground, may be responsible for the spread of *Salmonella enteritidis* in seals, as foxes are often seen feeding on placentae and dead pups and are a known reservoir of this disease in other locations (Stroud and Roelke 1980).

Invasive vertebrates can compete with native fauna for resources. Feral swine consume oak (*Quercus* spp.) and hickory (*Carya* spp.) mast, thereby competing directly with native wildlife for this important food source (Elston and Hewitt 2010; Henry and Conley 1972). Competition can also be for the same resource used in different ways. For example, non-native grazers, such as European rabbits (*O. cuniculus*) or goats (*Capra hircus*), can reduce the abundance and distribution of native vegetation, which may be used by other species for cover and food. Between 1903 and 1923, rabbit grazing was responsible for the local extinction of 26 species of plants on Laysan in Hawaii (Atkinson 1989; Courchamp et al. 1999). Transformation of this native ecosystem by rabbits has been blamed for the loss of several species of land birds on Laysan and the reduction of several reptile populations (Atkinson 1989). These indirect effects can also impact multiple trophic levels. On islands where there are both non-native domestic cats and rats, rats can serve as an important overwinter food source for cats, which in turn affect predation rates of native birds (Courchamp et al. 1999; Zavaleta et al. 2001).

Trophic level effects can be difficult to demonstrate as the strength of interaction can attenuate throughout a food web or chain; alternatively, a trophic cascade can occur if a top predator's abundance changes, thereby altering prey density, which releases the lower trophic level. Some of the best trophic interaction studies have been conducted on oceanic islands. Pigs were introduced to California's Channel Islands, became feral, and rapidly increased in abundance. Golden eagles (*Aquila chrysaetos*) were drawn to the islands and ultimately colonized due to the constant food subsidy of the feral pigs. Golden eagles then began preying on native Channel Island foxes (*Urocyon littoralis*), whose numbers declined precipitously, which released the competitively inferior native spotted skunk (*Spilogale gracilis amphiala*), whose numbers boomed (Roemer et al. 2002). Similarly, Croll et al. (2005) show that released foxes on the Aleutian Islands severely reduce seabird numbers (see below), which then ultimately impacted nutrient transport from the ocean to the land. The change in nutrient transport influenced soil fertility and changed native grasslands to forb- and shrub-dominated ecosystems (Croll et al. 2005). These oceanic island studies consistently show that a terrestrial invasive apex predator can affect food webs in complex ways involving top-down forcing and bottom-up nutrient exchanges (Maron et al. 2006).

A plethora of indirect economic costs are associated with managing invasive terrestrial vertebrate species. Costs of removing predators such as arctic fox, cats (*Felis catus*), and rats (*Rattus* spp.) on islands are substantial, but necessary to protect threatened seabirds (Ebbert and Byrd 2002). Foxes were first brought to the Aleutian Islands in the mid-1700s, and a fox farming boom continued there through the early part of the 1900s. However, escaped foxes that established invasive populations caused much ecological damage to seabird colonies requiring active management remediation. Removal of foxes on 39 islands, while expensive, was immediately successful with a fivefold increase in nesting birds in 10 years (Ebbert and Byrd 2002). The native Aleutian goose (*Branta hutchinsii leucopareia*) population in the Aleutian Islands went from 1000 birds in 1975 to over 35,000 by 2000 once invasive mammal predators were removed. Similarly, restoration by the removal of wild pigs has been necessary for the recovery of native mammals on California's Channel Islands (Roemer et al. 2002). Likewise, in Hawaii, almost all of the native birds are threatened by invasive rodents and other mammalian predators (Amarasekare 1993; Hammond et al. 2015; Harper and Bunbury 2015). In addition to eradication costs are the costs of threatened species recovery (e.g., captive breeding programs) for those endemic Hawaiian birds impacted by invasive predators.

2.4.1 Key Findings

- Invasive terrestrial vertebrates cause impacts to native systems through a wide variety of mechanisms, including overgrazing, seed consumption, predation and related indirect trophic effects, resource competition, and hybridization with native species.
- Invasive vertebrates (e.g., feral swine or nutria) that act as ecosystem engineers through rooting, burrowing, or altering hydrologic patterns can cause substantial ecological and economic impacts.
- Invasive vertebrates can also introduce or serve as vectors of harmful pathogens that cause disease in wildlife, livestock, and humans.
- Economic impacts associated with invasive vertebrates include not only damage to crops and ecosystems but costs of their control or eradication and the cost of recovering or conserving native species threatened by the invasion.

2.4.2 Key Information Needs

- Economic impacts of invasive vertebrates are likely underestimated, and better accounting of the full costs associated with damage to native flora and fauna, soil, water, property, and human health are needed, in addition to costs associated with crop damage and control efforts.
- Effects of introducing invasive vertebrates can attenuate through food chains when the species functions as either prey or predator, resulting in numerous direct or indirect interactions. Additional research should focus on identifying and quantifying complex impacts that cascade across trophic levels.

2.5 Impacts of Invasive Plants in Aquatic Systems

Invasive plants are found throughout the United States across a wide variety of aquatic habitats, including lakes, ponds, rivers, streams, estuaries, and wetlands. Species of concern represent diverse taxonomic groups and include not only angiosperms (monocotyledonous and dicotyledonous flowering plants) but also macroalgae with plant-like growth forms. There are three main growth forms of aquatic invasive plants: floating (on the water's surface), submersed (rooted underwater but potentially topping out to form surface mats), and emergent (erect stems above or on the surface of the water/saturated soils).

Regardless of growth form, aquatic invasive plants can have severe effects, altering environmental conditions, ecosystem processes, plant and animal communities, and human uses of water bodies. Freshwater aquatic habitats appear to be disproportionately vulnerable to and negatively affected by invasive species compared to terrestrial habitats (Moorhouse and Macdonald 2015). This is because of both the wide range of vectors available for spread of live organisms, such as boats, ballast water, and the aquarium trade, and the susceptibility of aquatic systems to hydrologic, nutrient, and other disturbances (Lodge et al. 1998; Zedler and Kercher 2004). Aquatic invasive plants are not only "drivers" of change that directly alter habitats but also "passengers" of change that have become more abundant in response to anthropogenic stressors and disturbances (sensu MacDougall and Turkington 2005). The extent to which invasive plants impair aquatic habitats, and to which they can be effectively controlled, depends on a variety of factors, including site conditions, detection and response times, and management decisions.

As invasive plants become more abundant in an aquatic habitat, their potential impacts on environmental conditions increase. These environmental changes can then have cascading effects on biodiversity, ecosystem functioning, and human uses. A fundamental attribute altered by invasive plants is light availability, particularly when canopy-producing floating (e.g., water hyacinth (*Eichhornia crassipes*)) or submersed species (e.g., hydrilla (*Hydrilla verticillata*) and Eurasian watermilfoil (*Myriophyllum spicatum*)) reduce underwater penetration of sunlight (Smith and Barko 1990; Villamagna and Murphy 2010) or tall emergent species (e.g., non-native common reed (*Phragmites australis* haplotype M) and hybrid cattail (*Typha* × *glauca*)) overtop shorter native plants (Larkin et al. 2012a). Aquatic invasive plants can also alter microclimates, for example, reducing water and substrate temperatures and variability (Larkin et al. 2012a; Schmitz et al. 1993), and water chemistry parameters (e.g., pH, redox potential, and nutrient concentrations) (Carpenter and Lodge 1986; Hummel and Kiviat 2004; Posey et al. 1993). Dissolved oxygen is of particular concern, as it can become depleted when high biomass of senesced invasive plants undergoes microbial decomposition or at night or during prolonged cloud cover when oxygen-producing photosynthesis gives way to oxygen-consuming respiration in large mats of vegetation (Caraco and Cole 2002; Pennington 2014; Sousa 2011).

Changes to key ecosystem processes accompany these shifts in environmental conditions. For example, primary productivity often increases with aquatic plant invasions (Nichols and Shaw 1986; Zedler and Kercher 2004). Movement of water may be disrupted by dense growth of non-native vegetation and sediment can be trapped at higher rates (Petticrew and Kalff 1992; Rooth et al. 2003). Rates of litter decomposition, sediment accumulation, and cycling of carbon, nitrogen, and other nutrients can be altered (Joyce et al. 1992). Changes in dominant vegetation following invasions influence the food webs and flow of energy through aquatic systems (Gratton and Denno 2006; Kelly and Hawes 2005).

In evaluating these and other impacts of invasive species, it is important to bear in mind that both native and non-native species can act as ecosystem engineers in aquatic habitats (Crain and Bertness 2005; Duffy 2006; Posey et al. 1993). It is when non-native species have traits that are novel in the habitats they invade, i.e., when they occupy distinct ecological niches, that their impacts are potentially greatest (Moles et al. 2008; Ricciardi and Atkinson 2004). For example, invasive aquatic plants may occupy areas of habitat that would otherwise be unvegetated, produce substantially more biomass, or differ from native species with respect to growth form, phenology, tissue chemistry, position in the habitat, or other functional traits (Bolduan et al. 1994; Nichols and Shaw 1986; Posey et al. 1993). Under these circumstances, there is high potential for harm to aquatic habitats and associated human uses of these systems.

One of the major impacts of invasive aquatic plants is loss of plant diversity. Native macrophytes can be displaced by invasive species that are superior competitors for space, light, or nutrients (Gettys et al. 2014). Earlier phenology, more rapid growth, multiple reproductive mechanisms, or other "weedy" traits enable invasive macrophytes to outcompete native species for light (Woolf and Madsen 2003; Zedler and Kercher 2004). High stem densities and thick floating mats monopolize space, depriving native plants of suitable habitat (Hummel and Kiviat 2004; Schooler et al. 2006). Higher rates of nutrient uptake and utilization can also confer competitive advantages to invasive aquatic species over native species (Larkin et al. 2012b; Van et al. 1999).

The effects of aquatic invasive plant species on fish and other wildlife are of great concern. By reducing dissolved oxygen, dense growth of invasive plants can drive off or cause mortality of invertebrates and fish (Madsen 1997). Invasive plants can also have less acute effects that nonetheless deprive invertebrates, fish, and birds of suitable habitat, foraging resources, or nursery/breeding sites (Able and Hagan 2000; Glisson et al. 2015). However, because invasive plants may provide shelter and primary production that benefit certain animals, aquatic invasive plants do not necessarily decrease total invertebrate or fish diversity but may instead alter the community composition of these groups by favoring species adapted to higher stem densities or other differences in structure (Chick and Mlvor 1997; Engel 1987; Theel et al. 2007). Despite the clear value of macrophytes in general for fish and other organisms, the importance of particular plant species to particular animal species is generally not well resolved (Kovalenko et al. 2010). This makes it difficult to predict how aquatic plant invasions will ultimately affect animal communities. For example, canopy-producing invasive plants may enhance fish and wildlife habitat at early invasion stages by providing food, shelter, or substrate for macroinvertebrates but can eventually decrease these ecological services if the invader becomes overly abundant.

In addition to their broad environmental and ecological impacts, aquatic invasive plants can impair economic and recreational uses of aquatic systems. Dense mats of floating and submersed plants restrict navigation and impede water movement important for flood control, irrigation, and hydropower (Eiswerth and Johnson 2002; Penfound and Earle 1948). Recreational uses of lakes and rivers are disrupted when nuisance vegetation fouls boat motors or interferes with waterskiing, fishing, or wildlife viewing (Gettys et al. 2014). Decreased aesthetic value and interference with recreation reduce property values and tourism income (Charles and Dukes 2008; Horsch and Lewis 2009). While a dollar value is not easy to assign to the impacts of aquatic invasive plants, they are an important component of the estimated $120 billion per year that invasive species of all types impose on the US economy (Pimentel et al. 2000).

The risk posed by invasive plants to a given aquatic habitat depends on geography, climate, propagule pressure from source populations, habitat characteristics that influence susceptibility to invasion, and management effort. Regional priorities for early detection and management are a moving target given that new species continue to be introduced to the United States or expand their ranges into new regions within the United States (Essl et al. 2011; Maki and Galatowitsch 2004). This uncertainty is compounded by climate change, which will enable certain aquatic invasive species to invade new areas and habitat types (Hellmann et al. 2008; Rahel and Olden 2008).

The degree to which a particular water body is susceptible to invasion depends on landscape and local factors. Locations closer to, or hydrologically connected to, source populations of aquatic invasive plants or those that are heavily used by boaters are more likely to receive seeds, fragments, or other propagules that can establish new populations (Buchan and Padilla 1999; Jackson and Pringle 2010). High levels of development activities by humans along shorelines or within watersheds act as disturbances that benefit opportunistic invasive species, for example, by altering hydrology, sedimentation, and nutrient loading (Vander Zanden and Olden 2008; Zedler and Kercher 2004).

The physical and chemical characteristics of a water body have a large influence on invasion risk. A fundamental characteristic of aquatic systems is their trophic state. In shallow lakes, differences in trophic status are associated with regime shifts between clear, macrophyte-dominated and turbid, phytoplankton-dominated alternative states (Scheffer 2004). Turbid states can support harmful algal blooms, but high light penetration in the desirable clear state can promote high productivity of submersed plants, including invasive species (Gettys et al. 2014; Scheffer 2004). Water depth affects light penetration and temperature, which are often limiting factors for growth of submersed invasive species (Barko and Smart 1981; Madsen et al. 1991). Thus, water drawdowns during droughts can lead to new infestations or accelerate expansion of existing infestations (Barrat-Segretain and Cellot 2007). Bathymetry and extent of the littoral zone determine potential habitat area available to submersed and emergent invasive species (Vis et al. 2003). Whether water flow is relatively static, moderate, or fast dictates which species are able to establish in a given water body (Madsen et al. 2001). Once established, sediment and water column characteristics—e.g., sediment texture and bulk density, nutrient availability, pH, and oxidation—influence aquatic plants' fitness and productivity, potentially making the difference between an invasive species simply being present versus occurring at nuisance levels (Barko and Smart 1986; Fleming and Dibble 2015; Squires and Lesack 2003).

Ultimately, the impacts of aquatic invasive plants depend greatly on the diligence of aquatic and wetland monitoring and the degree of management responses directed against the invasive plants. Proactive efforts to prevent new invasions can be cost-effective relative to post-invasion, long-term management (Keller et al. 2008). Early detection and rapid response efforts can identify new infestations at stages when eradication may still be feasible (Anderson 2005). The use of existing invasive or prohibited plant lists and predictive models can help guide search efforts (Tamayo and Olden 2014), and advances in citizen science offer opportunities to develop large-scale detection networks for invasive plants (Crall et al. 2015). When invasive populations are already well established, sustained active management efforts can achieve effective control and conserve ecological structure and function (Kovalenko et al. 2010). However, large-scale or long-term management efforts may be complex and expensive. Thus, it is important to proceed with clearly defined goals, consult the most current control technologies, and implement a robust monitoring program to enable course corrections (Blossey 1999; Zedler 2005).

2.5.1 Key Findings

- In aquatic environments, invasive plants of various growth forms (floating, submersed, or emergent) can have large negative effects that can alter environmental conditions, ecosystem processes, plant and animal communities, and biological diversity. Compared to terrestrial habitats, freshwater aquatic habitats are disproportionately vulnerable to plant invasions.
- Aquatic invasive plant impacts generally increase as their abundance increases and when novel traits allow them to exploit distinct ecological niches. Mechanisms by which invasive plants affect aquatic systems include reducing sunlight penetration; altering water temperature, pH, and nutrient concentrations; and reducing dissolved oxygen content.
- Invasive aquatic plants can greatly impair human uses of water bodies by restricting navigation; impeding water movement important for flood control, irrigation, or hydropower; disrupting recreational activities; and decreasing aesthetics, property values, and tourist income.

- The physical and chemical characteristics of a water body, such as trophic state, depth, flow rate, and sediment characteristics greatly influence its vulnerability to invasive aquatic plants.

2.5.2 Key Information Needs

- Research to clarify the importance of particular plant species to particular animal species in aquatic systems is needed in order to better predict how aquatic plant invasions will affect animal communities.
- Understanding how climate change is likely to affect the ability of invasive aquatic species to invade new areas and habitat types would help improve regional efforts at early detection and management prioritization.

2.6 Impacts of Invasive Animals in Aquatic Systems

The introduction and establishment of invasive animals in aquatic environments is a key threat to the conservation of aquatic biodiversity and the provisioning of aquatic ecosystem services. The distribution of most aquatic animals is naturally constrained by drainage divides, but these barriers are frequently overcome by human-mediated movement of organisms among watersheds at many scales. Over 300 non-native aquatic animals are established in North America, and these species represent a broad taxonomic spectrum, including mollusks, crustaceans, insects, fish, amphibians, reptiles, mammals, and microorganisms (Strayer 2010). These include species translocated from other continents or from one part of North America to another.

Not all non-native aquatic animals are invasive. Those non-native species recognized as having net negative effects on native biodiversity, ecosystem functioning, human health, or economic conditions are labeled as invasive (Sala et al. 2000), and the focus of this science synthesis is on species in this category. However, the effects of non-native species and human perceptions of these effects vary widely according to context. For example, brook trout (*Salvelinus fontinalis*) are the focus of conservation efforts as they decline in their native range in Eastern North America, but they negatively impact native fish populations when released as non-natives into the Rocky Mountains (Fausch et al. 2009) (Box 2.6, Fig. 2.8d). Non-native species, particularly sport fishes, may be esteemed by some segments of society for recreation, but considered invasive by others because of their negative ecological impacts. Some non-native species have been present in their non-native range for so long, and it is difficult to evaluate their effects. The Asian clam (*Corbicula fluminea*) was introduced into North America in the 1930s and is now the dominant bivalve there, but evidence of its negative effects is equivocal (Haag 2012; Strayer 1999). Still other non-native aquatic species appear to have no obvious negative effects, but in most cases these have not been studied well enough to determine what, if any, more subtle effects they may have on populations, communities, and ecosystems. Although our focus here is on aquatic invasive species with well-recognized impacts, it is essential to note that for

Fig. 2.8 Aquatic invasive species. (**a**) Zebra mussels (*Dreissena polymorpha*). (Photo courtesy of Randy Westbrooks, Invasive Plant Control, Inc., Bugwood.org); (**b**) silver carp (*Hypophthalmichthys molitrix*). (Photo courtesy of Kate Gardiner, https://www.flickr.com/photos/ennuiislife/4120213381); (**c**) rusty crayfish (*Orconectes rusticus*). (Photo courtesy of U.S. Geological Survey, Bugwood.org); (**d**) brook trout (*Salvelinus fontinalis*). (Photo courtesy of Michael Young, USDA Forest Service)

Box 2.6 Aquatic Invasive Animal Species in Varying Contexts

Aquatic invasive species include a wide range of taxa that have a variety of impacts. Taxa and impacts range from invasive bivalves, such as zebra mussels (*Dreissena polymorpha*) (Fig. 2.8a), which encrust native mollusks, interfering with their feeding and reproduction, to fish like silver carp (*Hypophthalmichthys molitrix*) (Fig. 2.8b), which have become the dominant species by biomass in some sections of North American rivers and can outcompete juvenile native fish for planktonic food resources. Zebra mussels and silver carp are not native to North America, originating from Northcentral Europe and Asia, respectively. In contrast, rusty crayfish (*Orconectes rusticus*) (Fig. 2.8c) and brook trout (*Salvelinus fontinalis*) (Fig. 2.8d, lower fish) are native to North America, but humans have moved them beyond their native ranges to other regions of the continent, where they have net negative impacts on the species native to those regions and on the aquatic ecosystems upon which native species depend. Rusty crayfish, native to the Southcentral United States, prey on, compete with, and hybridize with native crayfish in the Upper Midwest. Brook trout, native to the Eastern United States, compete with native cutthroat trout (*Oncorhynchus clarkii*) (Fig. 2.8d, upper fish) in the Western United States and prey on their offspring, contributing to concerns about native species' population viability.

many non-native aquatic species, additional research is needed to determine whether or not they are invasive.

The breadth of ecosystem effects of non-native species, from indeterminate to extreme, is illustrated in the Laurentian Great Lakes, where there are established populations of at least 90 non-native aquatic animals, most of which appear to have been introduced via commercial shipping (Holeck et al. 2004; Ricciardi 2001). Of the dozens of zooplankton species introduced into the Great Lakes, dramatic effects are demonstrated for only a few, and these effects vary spatially (e.g., Strecker and Arnott 2008). In contrast, zebra mussels and quagga mussels (*Dreissena* spp.) have radically transformed aquatic ecosystems in the Great Lakes region and throughout the rest of North America. Their effects include dramatic declines in phytoplankton and zooplankton and the near-elimination of native bivalves (Strayer et al. 1999) (Fig. 2.8a).

The distribution, abundance, and per capita effects (i.e., the effect per individual or per biomass unit, see Sect. 2.1.4) of non-native species in recipient habitats are the primary determinants of their ecosystem impacts (Parker et al. 1999; Ricciardi et al. 2013). When non-native species exploit previously unoccupied ecological niches, they may have particularly large effects. For example, epibenthic bivalves were absent in the Great Lakes prior to the introduction of *Dreissena*, which may have been a factor in their success. However, ecological novelty is not a prerequisite for becoming invasive. Rusty crayfish (*Orconectes rusticus*) displace native crayfish species to which they appear ecologically similar (Wilson et al. 2004) (Fig. 2.8c). Similarly, filter-feeding Asian carps can compose up to 80% of the overall fish biomass in midwestern rivers despite the historical presence of a variety of native filter-feeding fishes (e.g., gizzard shad (*Dorosoma cepedianum*) and paddlefish (*Polyodon spathula*)) (Sass et al. 2010) (Fig. 2.8b). Regardless of the factors that influence invasiveness (i.e., propensity of a species to cause net negative ecological impacts), the effects of aquatic invasive animal species can be manifested in several ways, including alteration of food webs or trophic relationships, habitat modification, and genetic hybridization.

Food web or trophic impacts can take a variety of forms. Many aquatic invasive animal species exert ecosystem effects by direct predation on native species. For example, non-native lake trout (*Salvelinus namaycush*) consume juvenile native cutthroat trout (*Oncorhynchus clarkii*) in Yellowstone Lake (Ruzycki et al. 2003), and rusty crayfish consume eggs of native fish in the family Centrarchidae (Wilson et al. 2004). Invasive species also may parasitize native species (e.g., sea lamprey (*Petromyzon marinus*) parasitism of native lake trout in the Great Lakes (Lawrie 1970)) or as pathogens cause disease in native species (e.g., whirling disease, caused by *Myxobolus cerebralis*, infecting native salmonids (Gilbert and Granath 2003)). Invasive species can compete directly or indirectly for food with native organisms. Adult silver carp (*Hypophthalmichthys molitrix*) compete directly with juvenile yellow perch (*Perca flavescens*) for plankton, and reduction of phytoplankton abundance by *Dreissena* may be a major factor in the decline of native bivalves (Strayer 1999). The feeding behavior of invasive species may also have indirect, cascading effects on other organisms at progressively higher or lower trophic levels. An example of a trophic cascade occurred with the introduction of the opossum shrimp (*Mysis diluviana*) in Flathead Lake, MT (Ellis et al. 2011). Opossum shrimp preyed heavily on native zooplankton, causing significant declines in their abundance. In response to declining zooplankton, the native planktivorous fish that fed on them also declined, reducing the food available for piscivorous kokanee salmon (*Oncorhynchus nerka*), resulting in their decline. Ironically, kokanee are also introduced in this ecosystem, but they sustain a popular sport fishery and attract a concentration of migrating bald eagles (*Haliaeetus leucocephalus*), which feed on kokanee. After introduction of opossum shrimp, the crash in the kokanee population effectively ended the sport fishery and forced the eagles to relocate. A similarly complex outcome is invasional meltdown, in which invasions by one species facilitate invasions by others and, as a group, these non-native species can completely restructure faunal communities and trophic linkages. One example of this type of invasional meltdown is the establishment of round goby (*Neogobius melanostomus*) in the Great Lakes, which was facilitated by the presence of one of its preferred prey items from its native Ponto-Caspian range, zebra mussels (*Dreissena polymorpha*), which had previously invaded and become abundant in the ecosystem (Ricciardi 2001).

Invasive aquatic animals can modify the physical and chemical environment, which affects native species and ecosystems in various ways. Common carp (*Cyprinus carpio*) feed by rooting in soft substrates, and this behavior can increase water column turbidity and nutrient levels and result in a decrease in aquatic plants and macroinvertebrates (Parkos et al. 2003). Accumulations of *Dreissena* can transform areas of soft sediments into reefs of dead and living shells, which provide habitat for a different array of macroinvertebrates than would be present otherwise (Radziejewska et al. 2009). Waste excretion by dense aggregations of New Zealand mud snails (*Potamopyrgus antipodarum*) can dramatically raise nitrogen levels in otherwise low-productivity habitats (Hall et al. 2003). These changes can produce environmental conditions more conducive to non-native species than to native species. For example, common carp are adapted to turbid conditions that result from their feeding behavior, but bluegill (*Lepomis macrochirus*), smallmouth bass (*Micropterus dolomieu*), and other native centrarchids in these systems are visual feeders that may be affected negatively by increases in turbidity (Wolfe et al. 2009).

Hybridization between native and invasive species is also a widespread and growing concern (Perry et al. 2002). Hybridization poses little risk for species that are sufficiently divergent and have strong reproductive barriers, but closely related taxa are vulnerable to hybridization. Non-introgressive hybridization, such as between invasive brook trout and federally listed bull trout (*Salvelinus confluentus*) in the Northwest (DeHaan et al. 2010), results in offspring that are largely inviable or infertile, causing reduced recruitment of the native species. Introgressive hybridization produces viable offspring that can spread non-native genes in later generations. Native Pecos pupfish (*Cyprinodon pecosensis*) in the Southwestern United States have been rapidly assimilated by invasive sheepshead minnow (*Cyprinodon variegatus*), resulting in loss of pure pupfish populations (Rosenfield et al. 2004). In some fish species, introgressive hybridization has led to range reductions of native species, with formerly occupied areas hosting parental forms of the non-native taxon or admixed individuals (McKelvey et al. 2016; Ward et al. 2012).

Humans experience a range of direct and indirect impacts of invasive aquatic animals. For example, *Dreissena* can clog municipal and industrial water intakes, and cyanobacterial blooms associated with high *Dreissena* densities cause taste and odor problems in water supplies (Vanderploeg et al. 2001). *Dreissena* can also interfere with recreational opportunities, including angling, boating, and swimming (Lovell et al. 2006; Rothlisberger et al. 2012). Reduced commercial fish stocks and increased costs of power generation have also been identified as impacts of *Dreissena* (Rothlisberger et al. 2012).

Attempts to control invasive species can have unintended negative consequences for native species and, as such, are an indirect ecological impact of aquatic invasive species. Mechanical or chemical control of invasive species, such as rotenone treatments for fish or chelated copper for *Dreissena*, may inadvertently harm populations of non-target organisms, such as amphibians and macroinvertebrates (Billman et al. 2011; Hamilton et al. 2009; Montz et al. 2010). Unnatural barriers designed to prevent the spread of invasive aquatic animals may interfere with the life cycle and habitat requirements of native fish species (Fausch et al. 2009). Aside from the harm they may cause to native species, control efforts are often unsuccessful, reinforcing the importance of prevention, early detection, and rapid response in managing biological invasions (Vander Zanden and Olden 2008).

The taxonomic diversity of invasive aquatic animals and the wide range of mechanisms by which they affect native ecosystems make it impossible to generalize what types of impacts are the most, or least, important, common, or severe. The effects of invasive aquatic animals are not only wide ranging as to type and severity; they are also manifest at multiple levels of biological organization, from genes to organisms to populations to communities to ecosystems. Human interests, including recreation, navigation, and water quality, are also affected negatively by the ecological changes caused by invasive aquatic animals. Additional research is needed to understand and quantify these ecological changes to better inform societal responses to aquatic invasive animals (Walsh et al. 2016).

2.6.1 Key Findings

- Invasive aquatic animals cause wide-ranging ecological and economic impacts. For example, invasive dreissenid mussels have transformed food webs in the Laurentian Great Lakes and elsewhere in North America. Their effects include dramatic declines in phytoplankton and zooplankton and the near-elimination of native bivalves.

- Types of impacts of invasive aquatic animals include alteration of food webs and trophic relationships, habitat modification, and genetic hybridization with native species.
- Many non-native aquatic animals have not been studied well enough to fully determine their ecological and economic impacts.

2.6.2 Key Information Needs

- Additional study is needed to clarify the role and effects of non-native animals in aquatic systems.
- For the vast majority of invasive aquatic animals, accurate assessments of ecological and economic impacts are needed.
- Studies of per capita effects of invasive aquatic animals have been more frequent than studies of population-level impacts. Additional work is needed for more invasive aquatic animals to scale information about per capita effects to the population and community levels.

Acknowledgments We greatly appreciate preliminary reviews of this chapter by M.A. Barnes (Natural Resources Management, Texas Tech University), Y. Ortega (USDA FS Rocky Mountain Research Station), and R.C. Venette (USDA FS Northern Research Station and Director, Minnesota Invasive Terrestrial Plants and Pests Center, University of Minnesota, St. Paul). We sincerely thank K.D. Klepzig (formerly USDA FS Southern Research Station), J. Lockwood (Department of Ecology, Evolution, and Natural Resources, Rutgers University), and D.A. Herms (Department of Entomology, The Ohio State University), and four anonymous reviewers for formal reviews of the chapter. J.H. Cane (USDA ARS, Logan, Utah), N.M. Williams (UC Davis Department of Entomology and Nematology), and K. Ullmann (formerly Xerces Society for Invertebrate Conservation, Davis, CA) provided excellent reviews of the content on insect pollinators, whereas L.S. Bauer and T.M. Poland (both USDA FS Northern Research Station, Lansing, MI) provided reviews of content and images for the call-out example of the emerald ash borer.

Disclaimer Text The findings and conclusions in this publication are those of the authors and should not be construed to represent any official USDA or U.S. Government determination or policy.

Literature Cited

Able KW, Hagan SM (2000) Effects of common reed (*Phragmites australis*) invasion on marsh surface macrofauna: response of fishes and decapod crustaceans. Estuaries 23:633–646

Aizen MA, Harder LD (2009) The global stock of domesticated honey bees is growing slower than agricultural demand for pollination. Curr Biol 19:915–918

Allen CR, Epperson D, Garmestani A (2004) The impacts of fire ants on wildlife: a decade of research. Am Midl Nat 152:88–103

Allen CR, Uden DR, Johnson AR, Angeler DG (2015) Spatial modelling approaches for understanding and predicting the impacts of invasive alien species on native species and ecosystems. In: Venette RC (ed) Pest risk modelling and mapping for invasive alien species. CAB International, New York, pp 162–170

Allison SD, Vitousek PM (2004) Rapid nutrient cycling in leaf litter from invasive plants in Hawai'i. Oecologia 141:612–619

Amarasekare P (1993) Potential impact of mammalian nest predators on endemic forest birds of western Mauna Kea, Hawaii. Conserv Biol 7:316–324

Anagnostakis SL (1987) Chestnut blight: the classical problem of an introduced pathogen. Mycologia 79:23–37

Anderson LW (2005) California's reaction to *Caulerpa taxifolia*: a model for invasive species rapid response. Biol Invasions 7:1003–1016

Artz DR, Allan MJ, Wardell GI, Pitts-Singer TL (2013) Nesting site density and distribution affect *Osmia lignaria* (Hymenoptera: Megachilidae) reproductive success and almond yield in a commercial orchard. Insect Conserv Divers 6:715–724

Ashton IW, Hyatt LA, Howe KM et al (2005) Invasive species accelerate decomposition and litter nitrogen loss in a mixed deciduous forest. Ecol Appl 15:1263–1272

Atkinson IAE (1989) Introduced animals and extinctions. In: Western D, Pearl MC (eds) Conservation for the twenty-first century. Oxford University Press, New York, pp 54–75

Aukema JE, McCullough DG, Von Holle B et al (2010) Historical accumulation of nonindigenous forest pests in the continental US. Bioscience 60:886–897

Aukema JE, Leung B, Kovacs K et al (2011) Economic impacts of non-native forest insects in the continental United States. PLoS ONE 6(9):e24587

Baker HG (1974) The evolution of weeds. Annu Rev Ecol Syst 5:1–24

Baker HG, Stebbins GL (1965) The genetics of colonizing species. Academic, New York. 588 p

Baker PJ, Scowcroft PG, Ewel JJ (2009) Koa (*Acacia koa*) ecology and silviculture. General Technical Report PSW-GTR-211. Pacific Southwest Research Station. USDA Forest Service, Washington, DC

Balch JK, Bradley BA, D'Antonio CM, Gomez-Dans J (2013) Introduced annual grass increases regional fire activity across the arid western USA (1980–2009). Glob Chang Biol 19:173–183

Barber NA, Marquis RJ, Tori WP (2008) Invasive prey impacts the abundance and distribution of native predators. Ecology 89:2678–2683

Barko JW, Smart RM (1981) Comparative influences of light and temperature on the growth and metabolism of selected submersed freshwater macrophytes. Ecol Monogr 51:219–235

Barko JW, Smart RM (1986) Sediment-related mechanisms of growth limitation in submersed macrophytes. Ecology 67:1328–1340

Barney JN, Tekiela DR, Dollete ES, Tomasek BJ (2013) What is the "real" impact of invasive plant species? Front Ecol Environ 11:322–329

Barrat-Segretain M-H, Cellot B (2007) Response of invasive macrophyte species to drawdown: the case of *Elodea* sp. Aquat Bot 87:255–261

Bergstrom DM, Lucieer A, Kiefer K et al (2009) Indirect effects of invasive species removal devastate World Heritage Island. J Appl Ecol 46:73–81

Bevins SN, Pedersen K, Lutman MW et al (2014) Consequences associated with the recent range expansion of nonnative feral swine. Bioscience 64:291–299

Billman HG, St-Hilaire S, Kruse CG et al (2011) Toxicity of the piscicide rotenone to Columbia spotted frog and boreal toad tadpoles. Trans Am Fish Soc 140:919–927

Blackburn TM, Essl F, Evans T et al (2014) A unified classification of alien species based on the magnitude of their environmental impacts. PLoS Biol 12(5):e1001850

Blair AC, Nissen SJ, Brunk GR, Hufbauer RA (2006) A lack of evidence for an ecological role of the putative allelochemical (±)-catechin in spotted knapweed invasion success. J Chem Ecol 32:2327–2331

Blossey B (1999) Before, during and after: the need for long-term monitoring in invasive plant species management. Biol Invasions 1:301–311

Blossey B (2011) Enemy release hypothesis. In: Simberloff D, Rejmanek M (eds) Encyclopedia of biological invasions. University of California Press, Berkeley, pp 193–196

Blossey B, Notzold R (1995) Evolution of increased competitive ability in invasive nonindigenous plants: a hypothesis. J Ecol 83:887–889

Boettner GH, Elkinton JS, Boettner CJ (2000) Effects of a biological control introduction on three nontarget native species of saturniid moths. Conserv Biol 14:1798–1806

Bohlen PJ, Scheu S, Hale CM et al (2004) Non-native invasive earthworms as agents of change in northern temperate forests. Front Ecol Environ 2:427–435

Bolduan BR, Van Eeckhout GC, Quade HW, Gannon JE (1994) *Potamogeton crispus* – the other invader. Lake Reservoir Manag 10:113–125

Boyd IL, Freer-Smith PH, Gilligan CA, Godfray HCJ (2013) The consequence of tree pests and diseases for ecosystem services. Science 342:1235773. https://doi.org/10.1126/science.1235773

Boyte SP, Wylie BK (2017) Estimates of herbaceous annual cover in the sagebrush ecosystem. (May 1, 2017). U.S. Geological Survey data release. https://doi.org/10.5066/F7445JZ9

Brantley S, Ford CR, Vose JM (2013) Future species composition will affect forest water use after loss of eastern hemlock from southern Appalachian forests. Ecol Appl 23:777–790

Brantley ST, Miniat CF, Elliott KJ et al (2015) Changes to southern Appalachian water yield and stormflow after loss of a foundation species. Ecohydrology 8:518–528

Brasier CM (2001) Rapid evolution of introduced plant pathogens via interspecific hybridization. Bioscience 51:123–133

Brasier CM, Buck KW (2001) Rapid evolutionary changes in a globally invading fungal pathogen (Dutch elm disease). Biol Invasions 3:223–233

Broders K, Boraks A, Barbison L et al (2015) Recent insights into the pandemic disease butternut canker caused by the invasive pathogen *Ophiognomonia clavigignenti-juglandacearum*. For Pathol 45:1–8

Brooks ML, D'Antonio CM, Richardson DM et al (2004) Effects of invasive alien plants on fire regimes. Bioscience 54:677–688

Brown BJ, Mitchell RJ, Graham SA (2002) Competition for pollination between invasive species (purple loosestrife) and a native congener. Ecology 83:2328–2336

Buchan LA, Padilla DK (1999) Estimating the probability of long-distance overland dispersal of invading aquatic species. Ecol Appl 9:254–265

Buswell JM, Moles AT, Hartley S (2011) Is rapid evolution common in introduced plant species? J Ecol 99:214–224

Callaway RM, Aschehoug ET (2000) Invasive plants versus their new and old neighbors: a mechanism for exotic invasion. Science 290:521–523

Campbell TA, Long DB (2009) Feral swine damage and damage management in forested ecosystems. For Ecol Manag 257:2319–2326

Canadian Food Inspection Agency (2018) Areas regulated for the emerald ash borer. http://www.inspection.gc.ca/plants/plant-pests-invasive-species/insects/emerald-ash-borer/areas-regulated/eng/1347625322705/1367860339942

Cane JH (2003) Exotic non-social bees (Hymenoptera: Apoidea) in North America: ecological implications. In: Strickler KV, Cane JH (eds) For non-native crops, whence pollinators of the future? Thomas Say Publications in Entomology, Entomological Society of America, Lanham, pp 113–126

Cane JH (2018) Personal correspondence, email letter dated January 8, 2018 to Steven Seybold. On file with Seybold at U.S. Department of Agriculture, Forest Service, Pacific Southwest Research Station, 1731 Research Park Drive, Davis, CA 95618.

Cane JH, Tepedino VJ (2017) Gauging the effect of honey bee pollen collection on native bee communities. Conserv Lett 10:205–210

Caraco NF, Cole JJ (2002) Contrasting impacts of a native and alien macrophyte on dissolved oxygen in a large river. Ecol Appl 12:1496–1509

Carpenter SR, Lodge DM (1986) Effects of submersed macrophytes on ecosystem processes. Aquat Bot 26:341–370

Carroll SP, Hendry AP, Reznick DN, Fox CW (2007) Evolution on ecological time-scales. Funct Ecol 21:387–393

Chambers JC, Bradley BA, Brown CA et al (2013) Resilience to stress and disturbance, and resistance to *Bromus tectorum* L. invasion in cold desert shrublands of western North America. Ecosystems 17:360–375

Chambers JC, Beck JL, Bradford JB et al (2017) Science framework for conservation and restoration of the sagebrush biome: linking the Department of Interior's integrated rangeland fire management strategy to long-term strategic conservation actions. Part 1. Science basis and applications. RMRS-GTR-360. U.S. Department of Agriculture, Forest Service, Rocky Mountain Research Station, Fort Collins. 213 p

Charles H, Dukes JS (2008) Impacts of invasive species on ecosystem services. In: Nentwig W (ed) Biological invasions. Chapter 13, 1st edn. Springer, Berlin, pp 217–237

Chen Y, Evans JD, Smith IB, Pettis JS (2008) *Nosema ceranae* is a long-present and wide-spread microsporidian infection of the European honey bee (*Apis mellifera*) in the United States. J Invertebr Pathol 97:186–188

Chen Y, Whitehill JG, Bonello P, Poland TM (2011) Feeding by emerald ash borer larvae induces systemic changes in black ash foliar chemistry. Phytochemistry 72:1990–1998

Chick JH, Mlvor CC (1997) Habitat selection by three littoral zone fishes: effects of predation pressure, plant density and macrophyte type. Ecol Freshw Fish 6:27–35

Chornesky EA, Bartuska AM, Aplet GH et al (2005) Science priorities for reducing the threat of invasive species to sustainable forestry. Bioscience 55:335–348

Chupp AD, Battaglia LL (2014) Potential for host shifting in *Papilio palamedes* following invasion of laurel wilt disease. Biol Invasions 16:2639–2651

Clark KL, Skowronski N, Hom J (2010) Invasive insects impact forest carbon dynamics. Glob Chang Biol 16:88–101

Coates PS, Riccaa MA, Prochazkaa BG et al (2016) Wildfire, climate, and invasive grass interactions negatively impact an indicator species by reshaping sagebrush ecosystems. Proc Natl Acad Sci USA 113:12745–12,750

Cobb RC, Chan MN, Meentemeyer RK, Rizzo DM (2012) Common factors drive disease and coarse woody debris dynamics in forests impacted by sudden oak death. Ecosystems 15:242–255

Colautti RI, Lau JA (2015) Contemporary evolution during invasion: evidence for differentiation, natural selection, and local adaptation. Mol Ecol 24:1999–2017

Colautti RI, Ricciardi A, Grigorovich IA, MacIsaac HJ (2004) Is invasion success explained by the enemy release hypothesis? Ecol Lett 7:721–733

Colautti RI, Grigorovich IA, MacIsaac HJ (2006) Propagule pressure: a null model for biological invasions. Biol Invasions 8:1023–1037

Courchamp F, Langlais M, Sugihara G (1999) Control of rabbits to protect island birds from cat predation. Biol Conserv 89:219–225

Cox GW (2004) Alien species and evolution: the evolutionary ecology of exotic plants, animals, microbes, and interacting native species. Island Press, Washington, DC. 379 p

Cox JG, Lima SL (2006) Naiveté and an aquatic–terrestrial dichotomy in the effects of introduced predators. Trends Ecol Evol 21:674–680

Cox-Foster DL, Conlan S, Holmes EC (2007) A metagenomics survey of microbes in honey bee colony collapse disorder. Science 318:283–287

Crain CM, Bertness ND (2005) Community impacts of a tussock sedge: is ecosystem engineering important in benign habitats? Ecology 86:2695–2704

Crall A, Jarnevich C, Young N et al (2015) Citizen science contributes to our knowledge of invasive plant species distributions. Biol Invasions 17:2415–2427

Croll DA, Maron JL, Estes JA et al (2005) Introduced predators transform subantarctic islands from grassland to tundra. Science 307:1959–1961

Crooks JA (2002) Characterizing ecosystem-level consequences of biological invasions: the role of ecosystem engineers. Oikos 97:153–166

Crowl TA, Cris TO, Parmenter RR et al (2008) The spread of invasive species and infectious disease as drivers of ecosystem change. Front Ecol Environ 6:238–246

D'Antonio CM, Vitousek PM (1992) Biological invasions by exotic grasses, the grass/fire cycle, and global change. Annu Rev Ecol Syst 23:63–87

Daehler CC (2003) Performance comparisons of co-occurring native and alien invasive plants: implications for conservation and restoration. Annu Rev Ecol Evol Syst 34:183–211

Dangremond EM, Pardini EA, Knight TM (2010) Apparent competition with an invasive plant hastens the extinction of an endangered lupine. Ecology 91:2261–2271

Daszak P, Cunningham AA, Hyatt AD (2000) Emerging infectious diseases of wildlife – threats to biodiversity and human health. Science 287:443–449

Davidson CB, Gottschalk KW, Johnson JE (2001) European gypsy moth (*Lymantria dispar* L.) outbreaks: a review of the literature. General Technical Report NE-278. USDA Forest Service, Northeastern Research Station, Newtown Square. 15 p

Davies KW, Boyd C, Beck JL et al (2011) Saving the sagebrush sea: an ecosystem conservation plan for big sagebrush plant communities. Biol Conserv 144:2573–2584

Davis MA (2009) Invasion biology. Oxford University Press on Demand

Davis MA, Grime JP, Thompson K (2000) Fluctuating resources in plant communities: a general theory of invasibility. J Ecol 88:528–534

Davis G, Lowrey L, Eckberg T, et al (2020) Notes on balsam woolly adelgid, *Adelgespiceae* (Ratzeburg, 1844) (Hemiptera: Adelgidae), range expansion in Idaho, Montana and Utah. Pan-Pacific Entomol 96:129–133

DeHaan PW, Schwabe LT, Ardren WR (2010) Spatial patterns of hybridization between bull trout, *Salvelinus confluentus*, and brook trout, *Salvelinus fontinalis*, in an Oregon stream network. Conserv Genet 11:935–949

Desurmont GA, Donoghue MJ, Clement WL, Agrawal AA (2011) Evolutionary history predicts plant defense against an invasive pest. Proc Natl Acad Sci USA 108:7070–7074

Dickie IA, Bolstridge N, Cooper JA, Peltzer DA (2010) Co-invasion by *Pinus* and its mycorrhizal fungi. New Phytol 187:475–484

Didham RK, Tylianakis JM, Gemmell NJ et al (2007) Interactive effects of habitat modification and species invasion on native species decline. Trends Ecol Evol 22:489–496

Dodds KJ, Orwig DA (2011) An invasive urban forest pest invades natural environments – Asian longhorned beetle in northeastern US hardwood forests. Can J For Res 41:1729–1742

Domec JC, Rivera LN, King JS et al (2013) Hemlock woolly adelgid (*Adelges tsugae*) infestation affects water and carbon relations of eastern hemlock (*Tsuga canadensis*) and Carolina hemlock (*Tsuga caroliniana*). New Phytol 199:452–463

Dorcas ME, Willson JD, Gibbons JW (2011) Can invasive Burmese pythons inhabit temperate regions of the southeastern United States? Biol Invasions 13:793–802

Dorcas ME, Willson JD, Reed RN et al (2012) Severe mammal declines coincide with proliferation of invasive Burmese pythons in Everglades National Park. Proc Natl Acad Sci USA 109:2418–2422

Duan JJ, Bauer LS, Abell KJ et al (2015) Population dynamics of an invasive forest insect and associated natural enemies in the aftermath of invasion: implications for biological control. J Appl Ecol 52:1246–1254

Duffy JE (2002) Biodiversity and ecosystem function: the consumer connection. Oikos 99:201–219

Duffy JE (2006) Biodiversity and the functioning of seagrass ecosystems. Mar Ecol Prog Ser 311:233–250

Ebbert SE, Byrd GV (2002) Eradications of invasive species to restore natural biological diversity on Alaska Maritime National Wildlife Refuge. Turning the tide: the eradication of invasive species. In: Veitch CR, Clout MN (eds) IUCN Invasive Species Specialist Group. Gland/Cambridge, pp 102–109

Eiswerth ME, Johnson WS (2002) Managing nonindigenous invasive species: insights from dynamic analysis. Environ Resour Econ 23:319–342

Ellis JD, Evans JD, Pettis J (2010) Colony losses, managed colony population decline, and Colony Collapse Disorder in the United States. J Agric Res 49:134–136

Ellis BK, Stanford JA, Goodman D et al (2011) Long-term effects of a trophic cascade in a large lake ecosystem. Proc Natl Acad Sci USA 108:1070–1075

Ellison AM, Bank MS, Clinton BD et al (2005) Loss of foundation species: consequences for the structure and dynamics of forested ecosystems. Front Ecol Environ 3:479–486

Ellshoff ZE, Gardner DE, Wikler C, Smith CW (1995) Annotated bibliography of the genus *Psidium*, with emphasis on *P. cattleianum* (strawberry guava) and *P. guajava* (common guava), forest weeds in Hawai'i. Technical Report 95. Cooperative National Park Resources Studies Unit. University of Hawai'i, Manoa. 102 p

Elston JJ, Hewitt DG (2010) Intake of mast by wildlife in Texas and the potential for competition with wild boars. Southwest Nat 55:57–66

Elton CS (1958) The ecology of invasions by animals and plants. Methuen, London. 181 p

Emerald Ash Borer Information Network (2018) Map of EAB confirmations by state. http://www.emeraldashborer.info

Engel S (1987) The impact of submerged macrophytes on largemouth bass and bluegills. Lake Reservoir Manag 3:227–234

Engeman RM, Stevens A, Allen J et al (2007) Feral swine management for conservation of an imperiled wetland habitat: Florida's vanishing seepage slopes. Biol Conserv 134:440–446

Eskalen A, Stouthamer R, Lynch SC et al (2013) Host range of Fusarium dieback and its ambrosia beetle (Coleoptera: Scolytinae) vector in southern California. Plant Dis 97:938–951

Essl F, Dullinger S, Rabitsch W et al (2011) Socioeconomic legacy yields an invasion debt. Proc Natl Acad Sci USA 108:203–207

Estrada JA, Flory LS (2015) Cogongrass (*Imperata cylindrica*) invasions in the US: mechanisms, impacts, and threats to biodiversity. Glob Ecol Conserv 3:1–10

Evans J (1970) About nutria and their control. Resource Publication No. 86. U.S. Department of the Interior, Bureau of Sport Fisheries and Wildlife, Denver. 65 p

Fall MW, Avery ML, Campbell TA et al (2011) Rodents and other vertebrate invaders in the United States. In: Pimentel D (ed) Biological invasions: economic and environmental costs of alien plant, animal, and microbe species. CRC Press, Boca Raton, pp 381–410

Fausch KD, Rieman BE, Dunham JB et al (2009) Invasion versus isolation: trade-offs in managing native salmonids with barriers to upstream movement. Conserv Biol 23:859–870

Fleming JP, Dibble ED (2015) Ecological mechanisms of invasion success in aquatic macrophytes. Hydrobiologia 746:23–37

Fraedrich SW, Harrington TC, Rabaglia RJ et al (2008) A fungal symbiont of the redbay ambrosia beetle causes a lethal wilt in redbay and other Lauraceae in the southeastern United States. Plant Dis 92:215–224

Fraedrich SW, Harrington TC, Best GS (2015) *Xyleborus glabratus* attacks and systemic colonization by *Raffaelea lauricola* associated with dieback of *Cinnamomum camphora* in the southeastern United States. For Pathol 45:60–70

Franck P, Garnery L, Solignac M, Cornuet J-M (1998) The origin of west European subspecies of honeybees (*Apis mellifera*): new insights from microsatellite and mitochondrial data. Evolution 52:1119–1134

Fürst MA, McMahon DP, Osborne JL et al (2014) Disease associations between honeybees and bumblebees as a threat to wild pollinators. Nature 506:364–366

Gandhi KJK, Herms DA (2010a) Direct and indirect effects of alien insect herbivores on ecological processes and interactions in forests of eastern North America. Biol Invasions 12:389–405

Gandhi KJK, Herms DA (2010b) North American arthropods at risk due to widespread *Fraxinus* mortality caused by the alien emerald ash borer. Biol Invasions 12:1839–1846

Gettys LA, Haller WT, Petty DG (2014) Biology and control of aquatic plants. A best management practices handbook, 3rd edn. Aquatic Ecosystem Restoration Foundation, Marietta. 200 p

Gibbs J, Sheffield CS (2009) Rapid range expansion of the wool-carder bee, *Anthidium manicatum* (Linnaeus) (Hymenoptera: Megachilidae), in North America. J Kansas Entomol Soc 82:21–29

Gilbert MA, Granath WO (2003) Whirling disease of salmonid fish: life cycle, biology and disease. J Parasitol 89:658–667

Glisson WJ, Brady RS, Paulios AT et al (2015) Sensitivity of secretive marsh birds to vegetation condition in natural and restored wetlands in Wisconsin. J Wildl Manag 79:1101–1116

Goodell K (2008) Invasive exotic plant-bee interactions. In: James RR, Pitts-Singer TL (eds) Bees in agricultural ecosystems. Oxford University Press, New York, pp 166–183

Goulson D (2003) Effects of introduced bees on native ecosystems. Annu Rev Ecol Evol Syst 34:1–26

Gratton C, Denno RF (2006) Arthropod food web restoration following removal of an invasive wetland plant. Ecol Appl 16:622–631

Grime JP (1988) The C-S-R model of primary plant strategies—origins, implications and tests. In: Gottlieb LD, Jain SK (eds) Plant evolutionary biology. Springer, Dordrecht, pp 371–393

Grünwald NJ, Goss EM, Press CM (2008) *Phytophthora ramorum*: a pathogen with a remarkably wide host range causing sudden oak death on oaks and ramorum blight on woody ornamentals. Mol Plant Pathol 9:729–740

Grünwald NJ, Garbelotto M, Goss EM et al (2012) Emergence of the sudden oak death pathogen *Phytophthora ramorum*. Trends Microbiol 20:131–138

Guichón ML, Cassini MH (1999) Local determinants of coypu distribution along the Luján River, eastcentral Argentina. J Wildl Manag 63:895–900

Gunnell KL, Monaco TA, Call CA, Ransom CV (2010) Seedling interference and niche differentiation between crested wheatgrass and contrasting native Great Basin species. Rangel Ecol Manag 63:443–449

Guo Q, Fei S, Dukes JS et al (2015) A unified approach for quantifying invasibility and degree of invasion. Ecology 96:2613–2621

Haag WR (2012) North American freshwater mussels. Natural history, ecology and conservation. Cambridge University Press, New York, 540 p

Hall RO, Tank JL, Dybdahl MF (2003) Exotic snails dominate nitrogen and carbon cycling in a highly productive stream. Front Ecol Environ 1:407–411

Hall JS, Minnis RB, Campbell TA et al (2008) Influenza exposure in United States feral swine populations. J Wildl Dis 44:362–368

Hamilton BT, Moore SE, Williams TB et al (2009) Comparative effects of rotenone and antimycin on macroinvertebrate diversity in two streams in Great Basin National Park, Nevada. N Am J Fish Manag 29:1620–1635

Hammond RL, Crampton LH, Foster JT (2015) Breeding biology of two endangered forest birds on the island of Kauai, Hawaii. Condor 117:31–40

Hansen EM (2008) Alien forest pathogens: *Phytophthora* species are changing world forests. Boreal Environ Res 13:33–41

Hansen EM, Goheen DJ, Jules ES, Ullian B (2000) Managing Port-Orford-cedar and the introduced pathogen *Phytophthora lateralis*. Plant Dis 84:4–14

Hanula JL, Horn S (2011a) Removing an exotic shrub from riparian forests increases butterfly abundance and diversity. For Ecol Manag 262:674–680

Hanula JL, Horn S (2011b) Removing an invasive shrub (Chinese privet) increases native bee diversity and abundance in riparian forests of the southeastern United States. Insect Conserv Divers 4:275–283

Harmon-Threatt A, Chin K (2016) Common methods for tallgrass prairie restoration and their potential effects on bee diversity. Nat Areas J 36:400–411

Harmon-Threatt A, Kremen C (2015) Bumble bees selectively use native and exotic species to maintain nutritional intake across highly variable and invaded local floral resource pools. Ecol Entomol 40:471–478

Harmon-Threatt A, Burns JH, Shemyakina LA, Knight TM (2009) Breeding system and pollination ecology of introduced plants compared to their native relatives. Am J Bot 96:1544–1550

Harper GA, Bunbury N (2015) Invasive rats on tropical islands: their population biology and impacts on native species. Glob Ecol Conserv 3:607–627

Havill NP, Montgomery ME, Keena M (2011) Hemlock woolly adelgid and its hemlock hosts: a global perspective. In: Onken B, Reardon R (eds) Implementation and status of biological control of the hemlock woolly adelgid. FHTET-2011-04. USDA Forest Service Forest Health Technology Enterprise Team, Morgantown, pp 3–14

Havill NP, Vieira LC, Salom SM (2014) Biology and control of hemlock woolly adelgid. FHTET-2014-05. USDA Forest Service, Forest Health Technology Enterprise Team, Morgantown. 21 p

Hefty AR, Aukema BH, Venette RC et al (2018) Reproduction of walnut twig beetle across the Juglandaceae. Biol Invasions 20:2141–2155

Hellmann JJ, Byers JE, Bierwagen BG, Dukes JS (2008) Five potential consequences of climate change for invasive species. Conserv Biol 22:534–543

Henry VG, Conley RH (1972) Fall foods of European wild hogs in the southern Appalachians. J Wildl Manag 36:854–860

Herms DA, McCullough DG (2014) Emerald ash borer invasion of North America: history, biology, ecology, impacts, and management. Annu Rev Entomol 59:13–30

Hierro JL, Callaway RM (2003) Allelopathy and exotic plant invasion. Plant Soil 256:29–39

Hierro JL, Maron JL, Callaway RM (2005) A biogeographical approach to plant invasions: the importance of studying exotics in their introduced and native range. J Ecol 93:5–15

Higes M, Martín-Hernández R, Botías C et al (2008) How natural infection by *Nosema ceranae* causes honeybee colony collapse. Environ Microbiol 10:2659–2669

Holeck KT, Mills EL, MacIsaac HJ et al (2004) Bridging troubled waters: biological invasions, transoceanic shipping, and the Laurentian Great Lakes. Bioscience 54:919–929

Holly DC, Ervin GN, Jackson CR et al (2009) Effect of an invasive grass on ambient rates of decomposition and microbial community structure: a search for causality. Biol Invasions 11:1855–1868

Holmes TP, Murphy EA, Royle DD (2005) The economic impacts of hemlock woolly adelgid on residential landscape values: Sparta, New Jersey case study. In: Onken B, Reardon R (eds) Third symposium on hemlock woolly adelgid in the eastern United States, 1–3 February 2005, Asheville, North Carolina 2005. U.S. Dep. Agric.–Forest Service, Morgantown, pp 15–24

Holmes TP, Aukema JE, Von Holle B et al (2009) Economic impacts of invasive species in forests. Ann N Y Acad Sci 1162:18–38

Holt RD (1977) Predation, apparent competition, and the structure of prey communities. Theor Popul Biol 12:197–229

Homan HJ, Johnson RJ, Thiele JR, Linz GM (2017) European starlings. Wildlife Damage Management Technical Series 13. USDA Animal and Plant Health Inspection Service, Fort Collins, 26 p

Horsch EJ, Lewis DJ (2009) The effects of aquatic invasive species on property values: evidence from a quasi-experiment. Land Econ 85:391–409

Houston DR (1994) Major new tree disease epidemics: beech bark disease. Annu Rev Phytopathol 32:75–87

Houston DB, Schreiner EG (1995) Alien species in national parks: drawing lines in space and time. Conserv Biol 9:204–209

Hu J, Angeli S, Schuetz S et al (2009) Ecology and management of exotic and endemic Asian longhorned beetle *Anoplophora glabripennis*. Agric For Entomol 11:359–375

Hudson JR, Hanula JL, Horn S (2013) Removing Chinese privet from riparian forests still benefits pollinators five years later. Biol Conserv 167:355–362

Hughes MA, Smith JA, Ploetz RC et al (2015) Recovery plan for laurel wilt on redbay and other forest species caused by *Raffaelea lauricola* and disseminated by *Xyleborus glabratus*. Plant Health Progr. https://doi.org/10.1094/PHP-RP-15-0017

Hulcr J, Dunn RR (2011) The sudden emergence of pathogenicity in insect–fungus symbioses threatens naive forest ecosystems. Proc R Soc Lond B Biol Sci 278:2866–2873

Hulme PE, Pyšek P, Jarošík V et al (2013) Bias and error in understanding plant invasion impacts. Trends Ecol Evol 28:212–218

Hummel M, Kiviat E (2004) Review of world literature on water chestnut with implications for management in North America. J Aquat Plant Manag 42:17–27

Iannone BV, Oswalt CM, Liebhold AM et al (2015) Region-specific patterns and drivers of macroscale forest plant invasions. Divers Distrib 21:1181–1192

Inch S, Ploetz R, Held B, Blanchette R (2012) Histological and anatomical responses in avocado, *Persea americana*, induced by the vascular wilt pathogen, *Raffaelea lauricola*. Botany 90:627–635

Jackson CR, Pringle CM (2010) Ecological benefits of reduced hydrologic connectivity in intensively developed landscapes. Bioscience 60:37–46

Jacobi WR, Koski RD, Harrington TC et al (2007) Association of *Ophiostoma novo-ulmi* with *Scolytus schevyrewi* (Scolytidae) in Colorado. Plant Dis 91:245–247

Jacobi WR, Koski RD, Negrón JF (2013) Dutch elm disease pathogen transmission by the banded elm bark beetle *Scolytus schevyrewi* (Curculionidae subfamily Scolytinae). For Pathol 43:232–237

Jay MT, Cooley M, Carychao D et al (2007) *Escherichia coli* O157: H7 in feral swine near spinach fields and cattle, central California coast. Emerg Infect Dis 13:1908–1911

Jones CG, Lawton JH, Shachak M (1996) Organisms as ecosystem engineers. In: Samson FB, Knopf FL (eds) Ecosystem management. Springer, New York, pp 130–147

Jones HP, Tershy BR, Zavaleta ES et al (2008) Severity of the effects of invasive rats on seabirds: a global review. Conserv Biol 22:16–26

Jones HP, Holmes ND, Butchard SHM et al (2016) Invasive mammal eradication on islands results in substantial conservation gains. Proc Natl Acad Sci USA 113:4033–4038

Joyce J, Langeland K, Van T, Vandiver V (1992) Organic sedimentation associated with hydrilla management. J Aquat Plant Manag 30:20–23

Jules ES, Kauffman MJ, Ritts WD, Carroll AL (2002) Spread of an invasive pathogen over a variable landscape: a nonnative root rot on Port Orford cedar. Ecology 83:3167–3181

Kaller MD, Kelso WE (2006) Swine activity alters invertebrates and microbial communities in a coastal plain watershed. Am Midl Nat 156:163–177

Keane RM, Crawley MJ (2002) Exotic plant invasions and the enemy release hypothesis. Trends Ecol Evol 17:164–170

Keller RP, Frang K, Lodge DM (2008) Preventing the spread of invasive species: economic benefits of intervention guided by ecological predictions. Conserv Biol 22:80–88

Kelly DJ, Hawes I (2005) Effects of invasive macrophytes on littoral-zone productivity and foodweb dynamics in a New Zealand high-country lake. J N Am Benthol Soc 24:300–320

Kenis M, Auger-Rozenberg MA, Roques A et al (2009) Ecological effects of invasive alien insects. Biol Invasions 11:21–45

Kerns BK, Day MA (2017) The importance of disturbance by fire and other abiotic and biotic factors in driving cheatgrass invasion varies based on invasion stage. Biol Invasions 19:1853–1862

Kimoto T (2018) Personal correspondence, email letter dated January 10, 2018 to Steven Seybold. On file with Seybold at U.S. Department of Agriculture, Forest Service, Pacific Southwest Research Station, 1731 Research Park Drive, Davis, CA 95618

Klee J, Besana AM, Genersch E et al (2007) Widespread dispersal of the microsporidian *Nosema ceranae*, an emergent pathogen of the western honey bee, *Apis mellifera*. J Invertebr Pathol 96:1–10

Koenig WD, Walters EL, Liebhold AM (2011) Effects of gypsy moth outbreaks on North American woodpeckers. Condor 113:352–361

Koenig WD, Liebhold AM, Bonter DN et al (2013) Effects of the emerald ash borer invasion on four species of birds. Biol Invasions 15:2095–2103

Koh I, Lonsdorf EV, Williams NM (2015) Modeling the status, trends, and impacts of wild bee abundance in the United States. Proc Natl Acad Sci USA 113:140–145

Kolar CS, Lodge DM (2001) Progress in invasion biology: predicting invaders. Trends Ecol Evol 16:199–204

Kosola KR, Dickmann DI, Paul EA, Parry D (2001) Repeated insect defoliation effects on growth, nitrogen acquisition, carbohydrates, and root demography of poplars. Oecologia 129:65–74

Kovacs KF, Haight RG, McCullough DG et al (2010) Cost of potential emerald ash borer damage in US communities, 2009–2019. Ecol Econ 69:569–578

Kovalenko KE, Dibble ED, Slade JG (2010) Community effects of invasive macrophyte control: role of invasive plant abundance and habitat complexity. J Appl Ecol 47:318–328

Kuebbing SE, Simberloff D, Nuñez MA (2013) Current mismatch between research and conservation efforts: the need to study co-occurring invasive plant species. Biol Conserv 160:121–129

Landers JL, Van Lear DH, Boyer WD (1995) The longleaf pine forests of the Southeast: requiem or renaissance? J For 93:39–44

Lankau RA, Nuzzo V, Spyreas G, Davis AS (2009) Evolutionary limits ameliorate the negative impact of an invasive plant. Proc Natl Acad Sci USA 106:15362–15367

Larkin DJ, Freyman MJ, Lishawa SC et al (2012a) Mechanisms of dominance by the invasive hybrid cattail *Typha* × *glauca*. Biol Invasions 14:65–77

Larkin DJ, Lishawa SC, Tuchman NC (2012b) Appropriation of nitrogen by the invasive cattail *Typha* × *glauca*. Aquat Bot 100:62–66

Lau JA (2006) Evolutionary responses of native plants to novel community members. Evolution 60:56–63

Lawrie AH (1970) The sea lamprey in the Great Lakes. Trans Am Fish Soc 99:766–775

Lees AC, Bell DJ (2008) A conservation paradox for the 21st century: the European wild rabbit *Oryctolagus cuniculus*, an invasive alien and an endangered native species. Mammal Rev 38:304–320

Li S-P, Cadotte MW, Meiners SJ et al (2015) The effects of phylogenetic relatedness on invasion success and impact: deconstructing Darwin's naturalisation conundrum. Ecol Lett 18:1285–1292

Liao C, Peng R, Luo Y et al (2008) Altered ecosystem carbon and nitrogen cycles by plant invasion: a meta-analysis. New Phytol 177:706–714

Liebhold AM, MacDonald WL, Bergdahl D, Mastro VC (1995) Invasion by exotic forest pests: a threat to forest ecosystems. For Sci 41:1–49

Lippincott CL (2000) Effects of *Imperata cylindrica* (L.) Beauv. (Cogongrass) invasion on fire regime in Florida sandhill (USA). Nat Areas J 20:140–149

Liu H, Stiling P (2006) Testing the enemy release hypothesis: a review and meta-analysis. Biol Invasions 8:1535–1545

Liu H, Bauer LS, Gao R et al (2003) Exploratory survey for the emerald ash borer, *Agrilus planipennis* (Coleoptera: Buprestidae), and its natural enemies in China. Great Lakes Entomol 36:191–204

Lockwood JL, Burkhalter JC (2015) The impact of invasive species on wildlife habitat. In: Morrison ML, Mathewson HA (eds) Wildlife habitat conservation. Johns Hopkins University Press in association with The Wildlife Society, Baltimore, pp 102–114

Lockwood JL, Hoopes MF, Marchetti MP (2007) Invasion ecology. Wiley, New York. 304 p

Lodge DM, Stein RA, Brown KM et al (1998) Predicting impact of freshwater exotic species on native biodiversity: challenges in spatial scaling. Aust Ecol 23:53–67

Loo JA (2009) Ecological impacts of non-indigenous invasive fungi as forest pathogens. Biol Invasions 11:81–96

Loope LL, Hamann O, Stone CP (1988) Comparative conservation biology of oceanic archipelagoes: Hawaii and the Galapagos. Bioscience 38:272–282

Lorence DH, Sussman RW (1986) Exotic species invasion into Mauritius wet forest remnants. J Trop Ecol 2:147–162

Lovell SJ, Stone SF, Fernandez L (2006) The economic impacts of aquatic invasive species: a review of the literature. Agric Resour Econ Rev 35:195–208

Lovett GM, Canham CD, Arthur MA et al (2006) Forest ecosystem responses to exotic pests and pathogens in eastern North America. Bioscience 56:395–405

Lovett GM, Arthur MA, Weathers KC, Griffin JM (2010) Long-term changes in forest carbon and nitrogen cycling caused by an introduced pest/pathogen complex. Ecosystems 13:1188–1200

Lucardi RD, Wallace LE, Ervin GN (2014) Invasion success in Cogongrass (*Imperata cylindrica*): a population genetic approach exploring genetic diversity and historical introductions. Invasive Plant Sci Manag 7:59–75

Lynch SC, Twizeyimana M, Mayorquin JS et al (2016) Identification, pathogenicity and abundance of *Acremonium pembeum* sp. nov. and *Graphium euwallaceae* sp. nov.- two newly discovered mycangial associates of the polyphagous shot hole borer (*Euwallacea* sp.) in California. Mycologia 108:313–329

MacDonald GE (2004) Cogongrass (*Imperata cylindrica*) – biology, ecology, and management. Crit Rev Plant Sci 23:367–380

MacDougall AS, Turkington R (2005) Are invasive species the drivers or passengers of change in degraded ecosystems? Ecology 86:42–55

MacDougall AS, Gilbert B, Levine JM (2009) Plant invasions and the niche. J Ecol 97:609–615

Mack RN (1981) Invasion of *Bromus tectorum* L. into western North America: an ecological chronicle. Agro-Ecosystems 7:145–165

Mack RN, Simberloff D, Lonsdale WM et al (2000) Biotic invasions: causes, epidemiology, global consequences, and control. Ecol Appl 10:689–710

Madsen JD (1997) Methods for management of nonindigenous aquatic plants. In: Luken JO, Thieret JW (eds) Assessment and management of plant invasions, 1st edn. Springer, New York, pp 145–171

Madsen JD, Hartleb CF, Boylen CW (1991) Photosynthetic characteristics of *Myriophyllum spicatum* and six submersed aquatic macrophyte species native to Lake George, New York. Freshw Biol 26:233–240

Madsen JD, Chambers PA, James WF et al (2001) The interaction between water movement, sediment dynamics and submersed macrophytes. Hydrobiologia 444:71–84

Maehr DS, Belden RC, Land ED, Wilkins L (1990) Food habits of panthers in southwest Florida. J Wildl Manag 54:420–423

Maestas JD, Campbell SB, Chambers JC et al (2016) Tapping soil survey information for rapid assessment of sagebrush ecosystem resilience and resistance. Rangelands 38:120–128

Maki K, Galatowitsch S (2004) Movement of invasive aquatic plants into Minnesota (USA) through horticultural trade. Biol Conserv 118:389–396

Maloy OC (2001) White pine blister rust. Plant Health Progr. https://doi.org/10.1094/PHP-2001-0924-01-HM

Maron JL, Estes JA, Croll DA et al (2006) An introduced predator alters Aleutian Island plant communities by thwarting nutrient subsidies. Ecol Monogr 76:3–24

Marshall JM, Smith EL, Mech R, Storer AJ (2013) Estimates of *Agrilus planipennis* infestation rates and potential survival of ash. Am Midl Nat 169:179–193

Matsuura M, Yamane S (1990) Biology of the Vespine wasps. Springer-Verlag, Berlin, pp 1–323

McClure ML, Burdett CL, Farnsworth ML et al (2015) Modeling and mapping the probability of occurrence of invasive wild pigs across the contiguous United States. PLoS ONE 10(8):e0133771

McKelvey KS, Young MK, Wilcox TM et al (2016) Patterns of hybridization among cutthroat trout and rainbow trout in northern Rocky Mountain streams. Ecol Evol 6:688–706

McKinney ML, Lockwood JL (1999) Biotic homogenization: a few winners replacing many losers in the next mass extinction. Trends Ecol Evol 14:450–453

Meng XJ, Lindsay DS, Sriranganathan N (2009) Wild boars as sources for infectious diseases in livestock and humans. Philos Trans R Soc Lond B Biol Sci 364:2697–2707

Mercader RJ, McCullough DG, Storer AJ et al (2016) Estimating local spread of recently established emerald ash borer, *Agrilus planipennis,* infestations and the potential to influence it with a systemic insecticide and girdled ash trees. For Ecol Manag 366:87–97

Millar JG, Hanks LM (2017) Chemical ecology of cerambycids. In: Wang Q (ed) Cerambycidae of the world: biology and pest management. CRC Press/Taylor & Francis Group, Boca Raton, pp 161–208

Miller SR, Gaebel R, Mitchell RJ, Arduser M (2002) Occurrence of two species of old world bees, *Anthidium manicatum* and *A. oblongatum* (Apoidea: Megachilidae), in northern Ohio and southern Michigan. Great Lakes Entomol 35:65–69

Mills LS, Soulé ME, Doak DF (1993) The keystone species concept in ecology and conservation. Bioscience 43:219–224

Mitchell RF, Reagel PF, Wong JCH et al (2015) Cerambycid beetle species with similar pheromones are segregated by phenology and minor pheromone components. J Chem Ecol 41:431–440

Moles AT, Gruber MAM, Bonser SP (2008) A new framework for predicting invasive plant species. J Ecol 96:13–17

Montgomery ME, Bentz SE, Olsen RT (2009) Evaluation of hemlock (*Tsuga*) species and hybrids for resistance to *Adelges tsugae* (Hemiptera: Adelgidae) using artificial infestation. J Econ Entomol 102:1247–1254

Montz GR, Hirsch J, Rezanka R, Staples DF (2010) Impacts of copper on a lotic benthic invertebrate community: response and recovery. J Freshw Ecol 25:575–587

Moorhouse TP, Macdonald DW (2015) Are invasives worse in freshwater than terrestrial ecosystems? Wiley Interdiscip Rev Water 2:1–8

Morales CL, Traveset A (2009) A meta-analysis of impacts of alien vs. native plants on pollinator visitation and reproductive success of co-flowering native plants. Ecol Lett 12:716–728

Morin RS, Liebhold AM (2015) Invasive forest defoliator contributes to the impending downward trend of oak dominance in eastern North America. Forestry. https://doi.org/10.1093/forestry/cpv053

Moroń D, Lenda M, Skorka P et al (2009) Wild pollinator communities are negatively affected by invasion of alien goldenrods in grassland landscapes. Biol Conserv 142:1322–1332

Moser WK, Barnard EL, Billings RF et al (2009) Impacts of nonnative invasive species on US forests and recommendations for policy and management. J For 107:320–327

Myers JH, Simberloff D, Kuris AM, Carey JR (2000) Eradication revisited: dealing with exotic species. Trends Ecol Evol 15:316–320

National Invasive Species Council [NISC] (2005) Five-year review of Executive Order 13112 on Invasive Species. National Invasive Species Council, Department of the Interior, Office of the Secretary (OS/SIO/NISC), Washington, DC. 44 p. https://www.invasivespeciesinfo.gov/laws/execorder.shtml#sec1

National Research Council [NRC] (2007) Chapter 3: Causes of pollinator declines and potential threats. In: Committee on the Status of Pollinators in North America. Status of pollinators in North America. The National Academies Press, Washington, DC, pp 75–103. http://www.nap.edu/read/11761/chapter/5#90. Accessed 10 Jan 2018

Negrón JF, Witcosky JJ, Cain RJ et al (2005) The banded elm bark beetle: a new threat to elms in North America. Am Entomol 51:84–94

Nichols SA, Shaw BH (1986) Ecological life histories of the three aquatic nuisance plants, *Myriophyllum spicatum*, *Potamogeton crispus* and *Elodea canadensis*. Hydrobiologia 131:3–21

Niemelä P, Mattson WJ (1996) Invasion of North American forests by European phytophagous insects. Bioscience 46:741–753

Noonburg EG, Byers JE (2005) More harm than good: when invader vulnerability to predators enhances impact on native species. Ecology 86:2555–2560

Noss RF, LaRoe ET, III, Scott JM (1995) Endangered ecosystems of the United States: a preliminary assessment of loss and degradation, vol. 28. US Department of the Interior, National Biological Service, Washington, DC. 95 p. http://sciences.ucf.edu/biology/king/wp-content/uploads/sites/106/2011/08/Noss-et-al-1995.pdf

Olden JD, Poff NL, Douglas MR et al (2004) Ecological and evolutionary consequences of biotic homogenization. Trends Ecol Evol 19:18–24

Orrock JL, Witter MS, Reichman OJ (2008) Apparent competition with an exotic plant reduces native plant establishment. Ecology 89:1168–1174

Orrock JL, Holt RD, Baskett ML (2010) Refuge-mediated apparent competition in plant–consumer interactions. Ecol Lett 13:11–20

Ortega YK, McKelvey KS, Six DL (2006) Invasion of an exotic forb impacts reproductive success and site fidelity of a migratory songbird. Oecologia 149:340–351

Ortega YK, Greenwood L, Callaway RM, Pearson DE (2014) Differential response of congeneric consumers to an exotic food resource: who gets the novel resource prize? Biol Invasions 16:1757–1767

Paolucci EM, MacIsaac HJ, Ricciardi A (2013) Origin matters: alien consumers inflict greater damage on prey populations than do native consumers. Divers Distrib 19:988–995

Parker IM, Gilbert GS (2004) The evolutionary ecology of novel plant-pathogen interactions. Annu Rev Ecol Evol Syst 35:675–700

Parker IM, Simberloff D, Lonsdale WM et al (1999) Impact: toward a framework for understanding the ecological effects of invaders. Biol Invasions 1:3–19

Parkos JJ III, Santucci VJ Jr, Wahl DH (2003) Effects of adult common carp (*Cyprinus carpio*) on multiple trophic levels in shallow mesocosms. Can J Fish Aquat Sci 60:182–192

Pearson DE (2009) Invasive plant architecture alters trophic interactions by changing predator abundance and behavior. Oecologia 159:549–558

Pearson DE (2010) Trait- and density- mediated indirect interactions initiated by an exotic plant autogenic ecosystem engineer. Am Nat 176:394–403

Pearson DE, Callaway RM (2008) Weed biocontrol insects reduce native plant recruitment through second-order apparent competition. Ecol Appl 18:1489–1500

Pearson DE, Ortega YK, Eren O, Hierro JL (2016a) Quantifying "apparent" impact and distinguishing impact from invasiveness in multispecies plant invasions. Ecol Appl 26:162–173

Pearson DE, Ortega YK, Runyon J, Butler J (2016b) Secondary invasion: the bane of weed management. Biol Conserv 197:8–17

Pejchar L, Mooney HA (2009) Invasive species, ecosystem services and human well-being. Trends Ecol Evol 24:497–504

Penfound WT, Earle TT (1948) The biology of the water hyacinth. Ecol Monogr 18:447–472

Pennington T (2014) Egeria. In: Gettys LA, Haller WT, Bellaud M (eds) Biology and control of aquatic plants. A best management practices handbook, 3rd edn. Aquatic Ecosystem Restoration Foundation, Marietta, pp 129–134

Perry WL, Lodge DM, Feder JL (2002) Importance of hybridization between indigenous and nonindigenous freshwater species: an overlooked threat to North American biodiversity. Syst Biol 51:255–275

Petticrew EL, Kalff J (1992) Water flow and clay retention in submerged macrophyte beds. Can J Fish Aquat Sci 49:2483–2489

Pfitsch WA, Bliss LC (1985) Seasonal forage availability and potential vegetation limitations to a mountain goat population, Olympic National Park. Am Midl Nat 113:109–121

Phillips BL, Brown GP, Shine R (2010) Evolutionarily accelerated invasions: the rate of dispersal evolves upwards during the range advance of cane toads. J Evol Biol 23:2595–2601

Pimentel D (2011) Environmental and economic costs associated with alien invasive species in the United States. In: Pimentel D (ed) Biological invasions: economic and environmental costs of alien plant, animal, and microbe species. CRC Press, Boca Raton, pp 411–430

Pimentel D, Lach L, Zuniga R, Morrison D (2000) Environmental and economic costs of nonindigenous species in the United States. Bioscience 50:53–65

Pimentel D, Zuniga R, Morrison D (2005) Update on the environmental and economic costs associated with alien-invasive species in the United States. Ecol Econ 52:273–288

Pitts-Singer TL, Cane JH (2011) The alfalfa leafcutting bee, *Megachile rotundata*: the world's most intensively managed solitary bee. Annu Rev Entomol 56:221–237

Poland TM, Chen Y, Koch J, Pureswaran D (2015a) Review of the emerald ash borer (Coleoptera: Buprestidae), life history, mating behaviours, host plant selection, and host resistance. Can Entomol 147:252–262

Poland TM, Ciaramitaro TM, Emery M et al (2015b) Submergence of black ash logs to control emerald ash borer and preserve wood for American Indian basketmaking. Agric For Entomol 17:412–420

Posey MH, Wigand C, Stevenson JC (1993) Effects of an introduced aquatic plant, *Hydrilla verticillata*, on benthic communities in the upper Chesapeake Bay. Estuar Coast Shelf Sci 37:539–555

Powell KI, Krakos KN, Knight TM (2011) Comparing the reproductive success and pollination biology of an invasive plant to its rare and common native congeners: a case study in the genus *Cirsium* (Asteraceae). Biol Invasions 13:905–917

Pyšek P, Richardson DM (2008) Traits associated with invasiveness in alien plants: where do we stand? In: Nentwig W (ed) Biological invasions. Ecological studies (analysis and synthesis), vol 193. Springer, Berlin/Heidelberg, pp 97–125

Pyšek P, Jarošík V, Hulme PE et al (2012) A global assessment of invasive plant impacts on resident species, communities and ecosystems: the interaction of impact measures, invading species' traits and environment. Glob Chang Biol 18:1725–1737

Rabenold KN, Fauth PT, Goodner BW et al (1998) Response of avian communities to disturbance by an exotic insect in spruce-fir forests of the southern Appalachians. Conserv Biol 12:177–189

Radziejewska T, Fenske C, Wawrzyniak-Wydrowska B et al (2009) The zebra mussel (*Dreissena polymorpha*) and the benthic community in a coastal Baltic lagoon: another example of enhancement? Mar Ecol 30(Suppl. 1):138–150

Rahel FJ, Olden JD (2008) Assessing the effects of climate change on aquatic invasive species. Conserv Biol 22:521–533

Redman AM, Scriber JM (2000) Competition between the gypsy moth, *Lymantria dispar,* and the northern tiger swallowtail, *Papilio canadensis*: interactions mediated by host plant chemistry, pathogens, and parasitoids. Oecologia 125:218–228

Reinhart KO, Callaway RM (2006) Soil biota and invasive plants. New Phytol 170:445–457

Rejmánek M (1996) A theory of seed plant invasiveness: the first sketch. Biol Conserv 78:171–181

Rejmánek M, Richardson DM (1996) What attributes make some plant species more invasive? Ecology 77:1655–1661

Ricciardi A (2001) Facilitative interactions among aquatic invaders: is an "invasional meltdown" occurring in the Great Lakes? Can J Fish Aquat Sci 58:2513–2525

Ricciardi A (2003) Predicting the impacts of an introduced species from its invasion history: an empirical approach applied to zebra mussel invasions. Freshw Biol 48:972–981

Ricciardi A, Atkinson SK (2004) Distinctiveness magnifies the impact of biological invaders in aquatic ecosystems. Ecol Lett 7:781–784

Ricciardi A, Hoopes MF, Marchetti MP, Lockwood JL (2013) Progress toward understanding the ecological impacts of nonnative species. Ecol Monogr 83:263–282

Rieder JP, Newbold TAS, Ostoja SM (2010) Structural changes in vegetation coincident with annual grass invasion negatively impacts sprint velocity of small vertebrates. Biol Invasions 12:2429–2439

Rizzo DM, Garbelotto M, Davidson JM et al (2002) *Phytophthora ramorum* as the cause of extensive mortality of *Quercus* spp. and *Lithocarpus densiflorus* in California. Plant Dis 86:205–214

Rizzo DM, Garbelotto M, Hansen EM (2005) *Phytophthora ramorum*: Integrative research and management of an emerging pathogen in California and Oregon forests. Annu Rev Phytopathol 43:309–335

Rodgers L, Derksen A, Pernas T (2014) Expansion and impact of laurel wilt in the Florida Everglades. Fla Entomol 97:1247–1250

Rodriguez LF (2006) Can invasive species facilitate native species? Evidence of how, when, and why these impacts occur. Biol Invasions 8:927–939

Roemer GW, Donlan CJ, Courchamp F (2002) Golden eagles, feral pigs, and insular carnivores: how exotic species turn native predators into prey. Proc Natl Acad Sci USA 99:791–796

Rooth JE, Stevenson JC, Cornwall JC (2003) Increased sediment accretion rates following invasion by *Phragmites australis*: the role of litter. Estuaries 26(2B):475–483

Rosenfield JA, Nolasco S, Lindauer S et al (2004) The role of hybrid vigor in the replacement of Pecos pupfish by its hybrids with sheepshead minnow: genetic homogenization of endemic pupfish. Conserv Biol 18:1589–1598

Rothlisberger JD, Finnoff DC, Cooke RM, Lodge DM (2012) Shipborne nonindigenous species diminish Great Lakes ecosystem services. Ecosystems 15:1–15

Russo L, Nichol C, Shea K (2016) Pollinator floral provisioning by a plant invader: quantifying beneficial effects of detrimental species. Divers Distrib 22:189–198

Ruzycki JR, Beauchamp DA, Yule DL (2003) Effects of introduced lake trout on native cutthroat trout in Yellowstone Lake. Ecol Appl 13:23–37

Sakai AK, Allendorf FW, Holt JS et al (2001) The population biology of invasive species. Annu Rev Ecol Syst 32:305–332

Sala OE, Chapin FS, Armesto JJ et al (2000) Global biodiversity scenarios for the year 2100. Science 287:1770–1774

Sass GG, Cook TR, Irons KS et al (2010) A mark-recapture population estimate for invasive silver carp (*Hypophthalmichthys molitrix*) in the La Grange Reach, Illinois River. Biol Invasions 12:433–436

Scheffer M (2004) Ecology of shallow lakes. Kluwer Academic Publishers, Dortdrecht. 357 p

Schmitz DC, Schardt JD, Leslie AJ et al (1993) The ecological impact and management history of three invasive alien aquatic plant species in Florida. In: McKnight BN (ed) Biological pollution: The control and impact of invasive exotic species. Proceedings of a symposium held at Indianapolis, Indiana, USA, 25–26 October 1991. Indiana Academy of Science, Indianapolis, pp 173–194

Schmitz OJ, Beckerman AP, O'Brien KM (1997) Behaviorally mediated trophic cascades: effects of predation risk on food web interactions. Ecology 78:1388–1399

Schmitz OJ, Krivan V, Ovadia O (2004) Trophic cascades: the primacy of trait-mediated indirect interactions. Ecol Lett 7:153–163

Schooler SS, McEvoy PB, Coombs EM (2006) Negative per capita effects of purple loosestrife and reed canary grass on plant diversity of wetland communities. Divers Distrib 12:351–363

Schroeder MA, Aldridge CL, Apa AD et al (2004) Distribution of sage-grouse in North America. Condor 106:363–376

Schwartz MW, Hoeksema JD, Gehring CA et al (2006) The promise and the potential consequences of the global transport of mycorrhizal fungal inoculum. Ecol Lett 9:501–515

Seabloom EW, Borer ET, Buckley YM et al (2015) Plant species' origin predicts dominance and response to nutrient enrichment and herbivores in global grasslands. Nat Commun 6:7710

Seward NW, VerCauteren KC, Witmer GW, Engeman RM (2004) Feral swine impacts on agriculture and the environment. Sheep Goat Res J 19:34–40

Seybold SJ, Penrose RL, Graves AD (2016) Invasive bark and ambrosia beetles in California Mediterranean forest ecosystems. In: Paine TD, Lieutier F (eds) Insects and diseases of Mediterranean forest systems. Springer, Cham, pp 583–662

Simberloff D (1995) Why do introduced species appear to devastate islands more than mainland areas? Pac Sci 49:87–97

Simberloff D, Von Holle B (1999) Positive interactions of nonindigenous species: Invasional meltdown? Biol Invasions 1:21–32

Sinclair WA, Campana RJ (eds) (1978) Dutch elm disease: perspectives after 60 years. Northeast Regional Research Publication, Agriculture 8(5). Cornell University Agricultural Experiment Station, Ithaca. 52 p

Smart MD (2015) The influence of mid-continent agricultural land use on the health and survival of commercially managed honey bee (*Apis mellifera* L.) colonies. Ph.D. Dissertation, University of Minnesota, St. Paul. 245 p

Smith CS, Barko J (1990) Ecology of Eurasian watermilfoil. J Aquat Plant Manag 28:55–64

Smith GF, Nicholas NS (1998) Patterns of overstory composition in the fir and fir-spruce forests of the Great Smoky Mountains after balsam woolly adelgid infestation. Am Midl Nat 139:340–352

Snyder JR (2015) Ecological implications of laurel wilt infestation on Everglades tree islands, southern Florida. U.S. Geological Survey Open-File Report 2014-1225. 18 p. https://doi.org/10.3133/ofr20141225. (10 January 2018)

Snyder WE, Evans EW (2006) Ecological effects of invasive arthropod generalist predators. Annu Rev Ecol Evol Syst 37:95–122

Sousa WTZ (2011) *Hydrilla verticillata* (Hydrocharitaceae), a recent invader threatening Brazil's freshwater environments: a review of the extent of the problem. Hydrobiologia 669:1–20

Space JC (2013) USDA Forest Service, Pacific Island Ecosystems at Risk (PIER). Online resource at http://www.hear.org/pier/

Squires MM, Lesack LFW (2003) The relation between sediment nutrient content and macrophyte biomass and community structure along a water transparency gradient among lakes of the Mackenzie Delta. Can J Fish Aquat Sci 60:333–343

State of Hawaii (2011) Final Environmental Assessment: Biological control of strawberry guava by its natural control agent for preservation of native forests in the Hawaiian Islands. 712 p. http://oeqc2.doh.hawaii.gov/EA_EIS_Library/2011-11-08-ST-FEA-Biocontrol-Strawberry-Guava.pdf

Stinson KA, Campbell SA, Powell JR et al (2006) Invasive plant suppresses the growth of native tree seedlings by disrupting belowground mutualisms. PLoS Biol 4(5):e0040140

Stohlgren TJ, Barnett DT, Kartesz JT (2003) The rich get richer: patterns of plant invasions in the United States. Front Ecol Environ 1:11–14

Stout JC, Morales CL (2009) Ecological impacts of invasive alien species on bees. Apidologie 40:388–409

Strange JP, Koch JB, Gonzalea VH et al (2011) Global invasion by *Anthidium manicatum* (Linnaeus) (Hymenoptera: Megachilidae): assessing potential distribution in North America and beyond. Biol Invasions 13:2115–2133

Strauss SY, Lau JA, Carroll SP (2006) Evolutionary responses of natives to introduced species: what do introductions tell us about natural communities? Ecol Lett 9:357–374

Strayer DL (1999) Effects of alien species on freshwater mollusks in North America. J N Am Benthol Soc 18:74–98

Strayer DL (2010) Alien species in fresh waters: ecological effects, interactions with other stressors, and prospects for the future. Freshw Biol 55:152–174

Strayer DL, Caraco NF, Cole JJ et al (1999) Transformation of freshwater ecosystems by bivalves: a case study of zebra mussels in the Hudson River. Bioscience 49:19–27

Strecker AL, Arnott SE (2008) Invasive predator, *Bythotrephes*, has varied effects on ecosystem function in freshwater lakes. Ecosystems 11:490–503

Stroud RK, Roelke ME (1980) *Salmonella meningoencephalomyleitis* in a northern fur seal (*Callorhinus ursinus*). J Wildl Dis 16:15–18

Suda J, Meyerson LA, Leitch IJ, Pyšek P (2015) The hidden side of plant invasions: the role of genome size. New Phytol 205:994–1007

Suding KN, Stanley Harpole W, Fukami T et al (2013) Consequences of plant–soil feedbacks in invasion. J Ecol 101:298–308

Swank WG, Petrides GA (1954) Establishment and food habits of the nutria in Texas. Ecology 35:172–176

Sydnor TD, Bumgardner M, Todd A (2007) The potential economic impacts of emerald ash borer (*Agrilus planipennis*) on Ohio, US, communities. Arboricult Urban For 33:48–54

Takahashi M, Giambelluca TW, Mudd RG et al (2011) Rainfall partitioning and cloud water interception in native forest and invaded forest in Hawai'i Volcanoes National Park. Hydrol Process 25:448–464

Tamayo M, Olden JD (2014) Forecasting the vulnerability of lakes to aquatic plant invasions. Invasive Plant Sci Manag 7:32–45

Tepedino VJ, Bradley BA, Griswold TL (2008) Might flowers of invasive plants increase native bee carrying capacity? Intimations from Capitol Reef National Park. Utah Nat Areas J 28:44–50

Terry PJ, Adjers G, Akobundo IO et al (1997) Herbicides and mechanical control of *Imperata cylindrica* as a first step in grassland rehabilitation. Agrofor Syst 36:151–179

Theel HJ, Dibble ED, Madsen JD (2007) Differential influence of a monotypic and diverse native aquatic plant bed on a macroinvertebrate assemblage; an experimental implication of exotic plant induced habitat. Hydrobiologia 600:77–87

Thiele J, Kollmann J, Markussen B (2010) Impact assessment revisited: improving the theoretical basis for management of invasive alien species. Biol Invasions 12:2025–2035

Tilman D (1982) Resource competition and community structure. Princeton University Press, Princeton. 296 p

Tingley MW, Orwig DA, Field R, Motzkin G (2002) Avian response to removal of a forest dominant: consequences of hemlock woolly adelgid infestations. J Biogeogr 29:1505–1516

Tisserat N, Cranshaw W, Leatherman D et al (2009) Black walnut mortality in Colorado caused by the walnut twig beetle and thousand cankers disease. Plant Health Progr. https://doi.org/10.1094/PHP-2009-0811-01-RS

Tripodi A, Hardin T (2020) New Pest Response Guidelines. *Vespa mandarinia* Asian Giant Hornet, United States Department of Agriculture Animal and Plant Health Inspection Service. https://cms.agr.wa.gov/WSDAKentico/Documents/PP/PestProgram/Vespa_mandarinia_NRPG-10Feb2020-(002).pdf

Trussell GC, Ewanchuk PJ, Matassa CM (2006) Habitat effects on the relative importance of trait-and density-mediated indirect interactions. Ecol Lett 9:1245–1252

U.S. Department of Interior (2015) An Integrated Rangeland Fire Management Strategy. Final Report to the Secretary of the Interior. https://www.forestsandrangelands.gov/rangeland/documents/IntegratedRangelandFireManagementStrategy_FinalReportMay2015.pdf

Umeda C, Esaklen A, Paine TD (2016) Polyphagous shot hole borer and *Fusarium* dieback in California. In: Paine TD, Lieutier F (eds) Insects and diseases of Mediterranean forest systems. Springer, Cham, pp 757–767

Urcelay C, Longo S, Geml J et al (2017) Co-invasive exotic pines and their ectomycorrhizal symbionts show capabilities for wide distance and altitudinal range expansion. Fungal Ecol 25:50–58

USDA Forest Service (2013) Forest Service national strategic framework for invasive species management. FS-1017. United States Department of Agriculture, Forest Service, Washington, DC, 35 p. http://www.fs.fed.us/foresthealth/publications/Framework_for_Invasive_Species_FS-1017.pdf

Utley C, Nguyen T, Roubtsova T et al (2013) Susceptibility of walnut and hickory species to *Geosmithia morbida*. Plant Dis 97:601–607

Van Driesche R, Reardon R (2014) The use of classical biological control to preserve forests in North America. Publication FHTET-2013-2. USDA Forest Service, Forest Health Technology Enterprise Team, Morgantown, 414 p

Van Driesche RG, Carruthers RI, Center T et al (2010) Classical biological control for the protection of natural ecosystems. Biol Control 54:S2–S33

Van Lear DH, Carroll WD, Kapeluck PR, Johnson R (2005) History and restoration of the longleaf pine-grassland ecosystem: implications for species at risk. For Ecol Manag 211:150–165

Van TK, Wheeler GS, Center TD (1999) Competition between *Hydrilla verticillata* and *Vallisneria americana* as influenced by soil fertility. Aquat Bot 62:225–233

Vander Zanden MJ, Olden JD (2008) A management framework for preventing the secondary spread of aquatic invasive species. Can J Fish Aquat Sci 65:1512–1522

Vanderploeg HA, Liebig JR, Carmichael WW et al (2001) Zebra mussel (*Dreissena polymorpha*) selective filtration promoted toxic *Microcystis* blooms in Saginaw Bay (Lake Huron) and Lake Erie. Can J Fish Aquat Sci 58:1208–1221

Vanderwerf EA, Smith DG (2002) Effects of alien rodent control on demography of the O'ahu'Elepaio, an endangered Hawaiian forest bird. Pac Conserv Biol 8:73–81

vanEngelsdorp D, Meixner MD (2010) A historical review of managed honey bee populations in Europe and the United States and the factors that may affect them. J Invertebr Pathol 103:S80–S95

vanEngelsdorp D, Evans JD, Saegerman C et al (2009) Colony collapse disorder: a descriptive study. PLoS ONE 4(8):e6481

Vargas RI, Stark JD, Nishida T (1990) Population dynamics, habitat preference, and seasonal distribution patterns of oriental fruit fly and melon fly in an agricultural area. Environ Entomol 19:1820–1828

Vellend M, Harmon LJ, Lockwood JL et al (2007) Effects of exotic species on evolutionary diversification. Trends Ecol Evol 22:481–488

Verhoeven KJ, Biere A, Harvey JA, Van Der Putten WH (2009) Plant invaders and their novel natural enemies: who is naïve? Ecol Lett 12:107–117

Villamagna AM, Murphy BR (2010) Ecological and socio-economic impacts of invasive water hyacinth (*Eichhornia crassipes*): a review. Freshw Biol 55:282–298

Vis C, Hudon C, Carignan R (2003) An evaluation of approaches used to determine the distribution and biomass of emergent and submerged aquatic macrophytes over large spatial scales. Aquat Bot 77:187–201

Vitousek PM (1990) Biological invasions and ecosystem processes: towards an integration of population biology and ecosystem studies. Oikos 57:7–13

Vose JM, Wear DN, Mayfield AE, Nelson CD (2013) Hemlock woolly adelgid in the southern Appalachians: control strategies, ecological impacts, and potential management responses. For Ecol Manag 291:209–219

Wagner DL (2007) Emerald ash borer threatens ash-feeding Lepidoptera. News Lepidopterists' Soc 49:10–11

Wagner DL, Todd KJ (2015) Ecological impacts of emerald ash borer. In: Van Driesche RG, Reardon RC (eds) Biology and control of the emerald ash borer. FHTET-2014-09. USDA Forest Service, Forest Health Technology Enterprise Team, Morgantown, pp 15–63

Wagner DL, Van Driesche RG (2010) Threats posed to rare or endangered insects by invasions of nonnative species. Annu Rev Entomol 55:547–568

Walsh JR, Carpenter SR, Vander Zanden MJ (2016) Invasive species triggers a massive loss of ecosystem services through a trophic cascade. Proc Natl Acad Sci USA 113:4081–4085

Walther GR, Roques A, Hulme PE et al (2009) Alien species in a warmer world: risks and opportunities. Trends Ecol Evol 24:686–693

Ward JL, Blum MJ, Walters DM et al (2012) Discordant introgression in a rapidly expanding hybrid swarm: hybridization and biological invasion in shiners. Evol Appl 5:380–392

Waring GH, Loope LL, Medeiros AC (1993) Study on use of alien versus native plants by nectarivorous forest birds on Maui, Hawaii. Auk 110:917–920

Werner EE, Peacor SD (2003) A review of trait-mediated indirect interactions in ecological communities. Ecology 84:1083–1100

Whisenant SG (1990) Changing fire frequencies on Idaho's Snake River plains: ecological and management implications. In: McArthur ED, Romney EM, Smith SD, Tueller PT (eds) Proceedings from the symposium on cheatgrass invasion, shrub dieoff and other aspects of shrub biology and management, General Technical Report INT-276. USDA Forest Service, Ogden, pp 5–7

White EM, Wilson JC, Clarke AR (2006) Biotic indirect effects: a neglected concept in invasion biology. Divers Distrib 12:443–455

Whitney KD, Gabler CA (2008) Rapid evolution in introduced species, 'invasive traits' and recipient communities: challenges for predicting invasive potential. Divers Distrib 14:569–580

Williams GR, Tarpy DR, vanEngelsdorp D et al (2010) Colony collapse disorder in context. Bioessays 32:845–846

Williams NM, Cariveau D, Winfree R, Kremen C (2011) Bees in disturbed habitats use, but do not prefer, alien plants. Basic Appl Ecol 12:332–341

Wilson KA, Magnuson JJ, Lodge DM et al (2004) A long-term rusty crayfish (*Orconectes rusticus*) invasion: dispersal patterns and community change in a north temperate lake. Can J Fish Aquat Sci 61:2255–2266

Witmer GW, Sanders RB, Taft AC (2003) Feral swine – are they a disease threat to livestock in the United States? USDA National Wildlife Research Center – Staff Publications. Paper 292. http://digitalcommons.unl.edu/icwdm_usdanwrc/292

Wolfe BE, Klironomos JN (2005) Breaking new ground: soil communities and exotic plant invasion. Bioscience 55:477–487

Wolfe MD, Santucci VJ, Einfalt LM, Wahl DH (2009) Effects of common carp on reproduction, growth, and survival of largemouth bass and bluegills. Trans Am Fish Soc 138:975–983

Woolf TE, Madsen JD (2003) Seasonal biomass and carbohydrate allocation patterns in southern Minnesota curlyleaf pondweed populations. J Aquat Plant Manag 41:113–118

Wootton JT (1994) The nature and consequences of indirect effects in ecological communities. Annu Rev Ecol Evol Syst 25:443–466

Wootton JT, Emmerson M (2005) Measurement of interaction strength in nature. Annu Rev Ecol Evol Syst 36:419–444

Work TT, McCullough DG (2000) Lepidopteran communities in two forest ecosystems during the first gypsy moth outbreaks in northern Michigan. Environ Entomol 29:884–900

Xiong S, Nilsson C (1999) The effects of plant litter on vegetation: a meta-analysis. J Ecol 87:984–994

Yokomizo H, Possingham HP, Thomas MB, Buckley YM (2009) Managing the impact of invasive species: the value of knowing the density-impact curve. Ecol Appl 19:376–386

Zavaleta ES, Hobbs RJ, Mooney HA (2001) Viewing invasive species removal in a whole-ecosystem context. Trends Ecol Evol 16:454–459

Zavortink TJ, Shanks SS (2008) *Anthidium manicatum* (Linnaeus) (Hymenoptera: Megachilidae) in California. Pan-Pac Entomol 84:238–241

Zedler JB (2005) Ecological restoration: guidance from theory. San Franc Estuary Watershed Sci 3(2):31 pp. http://repositories.cdlib.org/jmie/sfews/vol3/iss2/art4

Zedler JB, Kercher S (2004) Causes and consequences of invasive plants in wetlands: opportunities, opportunists, and outcomes. Crit Rev Plant Sci 23:431–452

Zenni RD, Nuñez MA (2013) The elephant in the room: the role of failed invasions in understanding invasion biology. Oikos 122:801–815

Impacts of Invasive Species on Forest and Grassland Ecosystem Processes in the United States

3

Chelcy Ford Miniat, Jennifer M. Fraterrigo, Steven T. Brantley, Mac A. Callaham Jr, Susan Cordell, Jeffrey S. Dukes, Christian P. Giardina, Shibu Jose, and Gary Lovett

3.1 Introduction

In this chapter, we describe current understanding of and identify research gaps on how invasive species directly, and indirectly, affect ecosystem processes. Specifically, we focus on how invasive species can alter the terrestrial carbon, nitrogen, and hydrologic cycles and how changes to these terrestrial cycles cascade to affect water quantity and quality. While invasive species may alter other ecosystem processes, we focus on these due to their importance to policy, to the public, and to their likely interaction with climate change effects. For example, carbon sequestration and surface water supply originating from forests and grasslands (Caldwell et al. 2014) are important policy and public concerns, and drought frequency and intensity will likely increase with climate change (Vose et al. 2016a). Our goal is to draw generalizations rather than provide details on invasive species effects on a case-by-case basis. We do, however, provide case studies for illustration and draw linkages with other chapters that provide detailed coverage to disturbance regimes (Chap. 5) and types and mechanisms of ecological impact caused by invasive insects (Chap. 2).

Ecosystem processes are hypothesized to change when invading species: (1) acquire resources differently from native species, including differences in space or time, (2) use acquired resources with efficiencies that differ from native species, (3) alter trophic linkages and resulting food webs, and (4) alter the frequency, duration, extent, and/or intensity of disturbances (Vitousek 1990). Most studies agree that invasive plant species increase ecosystem productivity by enhancing carbon sequestration, storage, and cycling (see review by Liao et al. 2008), although results can vary with the invaded ecosystem type and associated climate (Qiu 2015). Invasive species also increase rates of nitrogen cycling in forests and grasslands, but as with carbon, increased cycling rates are not necessarily sustainable and may lead to longer-term shifts in species composition, disturbance regimes, indirect effects of pathogens, invasive animals, and harm to the environment (Lovett et al. 2010b; Qiu 2015). Changes to the terrestrial carbon and nitrogen cycles directly affect the hydrologic balance, with invasive species generally using more water than natives, resulting in lower soil moisture (Cavaleri and Sack 2010; Pysek et al. 2012). Indirect effects to the hydrologic cycle are especially obvious when invasive species differ phenologically or physiologically from natives (Baer et al. 2006; Brantley et al. 2015;

C. F. Miniat (✉)
U.S. Department of Agriculture, Forest Service, Southern Research Station, Coweeta Hydrologic Lab, Otto, NC, USA
e-mail: chelcy.f.miniat@usda.gov

J. M. Fraterrigo
Department of Natural Resources and Environmental Sciences, and the Program in Ecology, Evolution and Conservation Biology, University of Illinois, Urbana, IL, USA

S. T. Brantley
Jones Center at Ichauway, Newton, GA, USA

M. A. Callaham Jr.
U.S. Department of Agriculture, Forest Service, Southern Research Station, Center for Forest Disturbance Science, Athens, GA, USA

S. Cordell
U.S. Department of Agriculture, Forest Service, Pacific Southwest Research Station, Institute of Pacific Islands Forestry, Hilo, HI, USA

J. S. Dukes
Forestry and Natural Resources and Biological Sciences, Purdue University, West Lafayette, IN, USA

C. P. Giardina
U.S. Department of Agriculture, Forest Service, Pacific Southwest Research Station, Hilo, HI, USA

S. Jose
College of Agriculture, Food and Natural Resources, University of Missouri, Columbia, MO, USA

G. Lovett
Cary Institute of Ecosystem Studies, Millbrook, NY, USA

T. M. Poland et al. (eds.), *Invasive Species in Forests and Rangelands of the United States*,
https://doi.org/10.1007/978-3-030-45367-1_3

Cordell and Sandquist 2008). And as with carbon and nitrogen cycling, changes to water cycling become more complex over time as interactive and cascading impacts are fully realized (see Chap. 2).

3.2 Impacts on Carbon Cycling in Forests and Grasslands

The impacts of invasive species on the carbon (C) cycle have societal importance for two main reasons. First, carbon inputs are a measure of the amount of energy that plants make available to other species in the ecosystem (e.g., as fiber and forage). Second, carbon stored in the ecosystem is not in the atmosphere, and thus does not contribute to climate change. If invasive species reduce the stores of carbon in the soil or plants, this will tend to warm the planet through a net release of carbon dioxide (CO_2) to the atmosphere; invasive species that bolster ecosystem carbon storage will tend to reduce atmospheric CO_2. Carbon stocks in the soil typically turn over very slowly (Schmidt et al. 2011), so ecosystems in which most of the carbon is belowground are relatively resistant to declines in carbon stocks following invasion (Liao et al. 2008). Those ecosystems with the largest fraction of carbon stored aboveground are more sensitive, as turnover is faster and shifts in plant species composition or mortality can affect the bulk of the carbon stock. Unless a given invasive species covers vast areas and dramatically changes carbon stores of the invaded areas, its overall impact on the global carbon cycle, or even that of the invaded region, will be quite small (see Chap. 2). In aggregate, if invasive species were to have consistent and strong effects on carbon cycling, a perceptible climate feedback is possible.

Research to date shows that invasive species affect carbon cycling in terrestrial ecosystems through a variety of mechanisms (Peltzer et al. 2010). Carbon enters ecosystems through plant production and senescence and is lost via decomposition of senesced or exuded plant material or via disturbance events such as fire. Rate of carbon input, the quality of those inputs, or where in the ecosystem those inputs are released regulates ecosystem carbon balance and system capacity to retain carbon. Because this balance of inputs and losses drives ecosystem carbon storage, invasive plant species that differ meaningfully from natives with respect to inputs or losses will alter terrestrial carbon balance. Such difference can emerge when invasive species have different rates of primary productivity, produce litter or exudates with different chemistry, or allocate carbon differently than natives. Many studies have documented how invasive plant species affect primary production and litter stocks of forests, grasslands, and other ecosystems (e.g., Bradley et al. 2006; Litton et al. 2008). A meta-analysis by Liao et al. (2008) found that across all invasive plant species and ecosystem types, aboveground net primary production and litter carbon mass increased by 83% and 49%, respectively, in invaded ecosystems as compared with native ecosystems. However, it is not clear what fraction of this carbon is sequestered in the ecosystem. Of the studies examining litter decomposition in plant communities with and without invasive species, most have found an acceleration of decomposition after invasion (but see Jo et al. 2016 and Pysek et al. 2012). As a result, invasive species do not always enhance soil carbon stocks and may cause existing soil carbon stocks to decline (e.g., Tamura and Tharayil 2014).

Differences in productivity, litter chemistry, and carbon allocation between native and invasive species can be linked to differences in plant functional traits (van Kleunen et al. 2010), including those associated with resource acquisition, use efficiency, and retention. For example, invasive species often have greater light and nutrient use efficiencies and relative growth rates than native species (Funk and Vitousek 2007; Heberling and Fridley 2013), even among phylogenetically related species (Matzek 2012). Effects on carbon cycling increase with greater differences in traits between the invasive plant species and the resident species they replace (Dukes 2002; Ehrenfeld 2010) and with greater dominance of the invasive plant (Craig et al. 2015; Kramer et al. 2012). However, knowledge of which traits directly affect ecosystem processes, specifically carbon cycling, and how invasion alters allocation patterns remains limited (Drenovsky et al. 2012).

Plant functional types (e.g., deciduous shrubs, annual grasses) represent clusters of plant functional traits and provide a simplified construct for generalizing effects of invasive species on carbon cycling within the context of the ecosystem (e.g., grasslands or forests). In grasslands, invasive deciduous shrubs generally cause carbon stocks to increase in aboveground biomass, coarse roots, and, following senescence and decomposition, soil organic matter (Qiu 2015; Vila et al. 2011). In forests and shrublands, invasive annual grasses generally cause soil carbon stocks to decrease (Kramer et al. 2012; Strickland et al. 2010; Wheeler et al. 2016); however, environmental conditions can alter these effects. In the Southwestern United States, for example, losses of soil carbon exceeded gains in plant biomass carbon following invasion of woody plants on wet but not dry grasslands. This resulted in a net loss in ecosystem carbon in wetter areas and a net gain in ecosystem carbon in drier areas (Jackson et al. 2002).

Biotic conditions can also alter the effects of invasive species. In a replicated common garden experiment, Ammondt and Litton (2012) found that while the invasive guinea grass (*Megathyrsus maximus*) had 39–94% higher maximum photosynthetic rate than three native grass species, when it was planted with native species, its aboveground, belowground, and total biomass and tiller production were all lower than when it was planted alone, suggesting competition for resources reduced this invasive species' impact. Indeed, resource competition has been proposed to explain why the same invasive species can have differing effects on native communities in different geographical locations (Fraterrigo et al. 2014). In support of this hypothesis, experimental evidence shows that changing the availability of limiting resources (e.g., light, nutrients) can shift the competitive balance between invasive and native species (Concilio et al. 2016; Pearson et al. 2017; Prevey and Seastedt 2014). This could lead to variation in ecosystem impacts if ecosystem processes scale with the invasive species' abundance (Parker et al. 1999). Thus, although there are some general patterns with respect to the effects of invasive species on carbon cycling, the magnitude of these effects remains uncertain. Additionally, human activities or disturbances that alter the availability of limiting resources can influence the effects of invasive species. In many cases, an increase in the availability of a limiting resource will enhance invasive species performance (Daehler 2003), leading to more severe effects. For example, human activities or disturbances that result in increased light availability to forest understories commonly lead to increases in the frequency (Belote et al. 2008) and productivity (Eschtruth and Battles 2009) of invasive species with less shade tolerance, which could enhance invasion effects on resident communities. However, other outcomes are possible. For instance, use of fossil fuels and fertilizers over the past century has increased the amount of biologically reactive nitrogen in the atmosphere, leading to large increases in nitrogen deposition in terrestrial and aquatic ecosystems (Galloway et al. 1995). Elevated nitrogen deposition can impede organic matter decomposition by suppressing extracellular enzyme activities (Janssens et al. 2010). This may lessen the negative effects of invasive species that mine the soil organic matter for nitrogen. For example, the invasive Japanese stiltgrass (*Microstegium vimineum* (Trin.) A. Camus) was associated with a decline in forest soil carbon stocks when ambient nitrogen availability was low but an increase in forest soil carbon stocks when ambient nitrogen availability was high (Craig et al. 2015). Additional research is needed to determine whether broadscale processes such as atmospheric nitrogen deposition commonly modify carbon effects of other invasive species.

Invasive species on other trophic levels can also alter carbon cycling. Herbivores (including insects) and plant pathogens can alter carbon input rates by affecting plant growth and, over longer periods of time, causing a shift in selective pressures and recruitment success, and thus plant species composition. When these trophic effects result in only minor or short-term changes to plant species composition, ecosystem effects are likely to be restricted to changes in the annual carbon budgets of the ecosystem. However, more profound effects would be expected if the changes result in plants being replaced by other species that possess different resource acquisition and use strategies. While most introduced herbivores and pathogens likely have little effect on carbon inputs and outputs, herbivorous species that reach high densities and pathogens that cause high rates of plant mortality can lead to dramatic changes in rates of carbon cycling (Lovett et al. 2006; Peltzer et al. 2010). In general, widespread defoliation or tree mortality reduces net primary productivity, and increased litterfall and dead plant material enhances decomposition and soil respiration. Over decadal scales, reduction in soil carbon inputs causes a decline in soil and ecosystem respiration (Hicke et al. 2012; Moore et al. 2013). Longer-term effects on carbon cycling are mediated by changes in tree species composition and the resulting alterations of productivity, litter quality, and soil organic matter production and turnover (Lovett et al. 2006).

Invasive earthworms (in North America belonging primarily to two families, the Lumbricidae and the Megascolecidae) can profoundly influence soil physical, chemical, and biological characteristics (Edwards et al. 2013). In agroecosystems, earthworm activity can increase the total amount of carbon protected in slower cycling pools (e.g., Bossuyt et al. 2005). When invasive earthworms establish in forest soils, they can facilitate a redistribution of carbon through the soil profile, with dramatic reductions in the carbon stored in the litter layer (Bohlen et al. 2004; Eisenhauer et al. 2007; Hale et al. 2008). As European earthworm invasions progress in boreal forests, this may have important positive implications for the amount of carbon lost following fires (Cameron et al. 2015). This redistribution of soil carbon also results in changes in microbial community structure (Dempsey et al. 2011), decreases in native arthropod populations (Snyder et al. 2011), and reduced microbial respiration (C mineralization) (Eisenhauer et al. 2007), ultimately affecting the total ecosystem carbon cycle. It is important to note that the effects of earthworms on the carbon cycle of a particular system will depend upon the species involved, with some species having greater relative impacts than others (Chang et al. 2016).

Invasive species can also affect carbon cycling by altering disturbance regimes. Perhaps the most dramatic examples entail invaders changing or introducing fire regimes (Brooks et al. 2004; D'Antonio and Vitousek 1992). For example, in a wide range of seasonally dry tropical forests, invasion by fire-prone grasses can result in accentuated or in some cases novel fire regimes (D'Antonio and Vitousek 1992; Hoffmann et al. 2004). In the case of forests or shrublands, heat-related impacts to the roots and stems of overstory vegetation can result in dramatic conversions from forest ecosystems that store large amounts of carbon to low storage grasslands (Litton et al. 2006). Cheatgrass (*Bromus tectorum* L.), an aggressive and widespread invader in the Western United States, provides an important example of how an invader in a temperate climate can alter carbon and nutrient process rates and storage (Jones et al. 2015). Bradley et al. (2006) suggested that cheatgrass has turned the Western United States from a carbon sink to a source; the frequent fires in cheatgrass-invaded regions of the Western United States have released 8 ± 3 Tg C to the atmosphere and over the next decades will put another 50 ± 20 Tg C at risk. In deciduous forests of the Eastern United States, where fire regimes are characterized by low-intensity fires, the invasive Japanese stiltgrass increases fire intensity, which enhances its own recruitment and growth in subsequent years (Wagner and Fraterrigo 2015), and can also suppress tree regeneration (Flory et al. 2015), thereby altering future carbon storage potential.

In summary, invasive species frequently accelerate carbon cycling, potentially making more energy available to other species; however, these effects and their consequences for carbon storage depend on the context—the traits of the invader, the resident species in the invaded ecosystem, and the environmental conditions. Where accelerated carbon uptake is paired with accelerated decomposition, the consequences for carbon storage are not clear. On the other hand, invasive species that accelerate disturbance regimes can lead to significant declines in carbon storage. This pattern has been found for invasive grasses in more arid regions of the Western United States where carbon stocks in soil are relatively low and suggests that this may be a general pattern in other arid areas with low belowground carbon stocks.

3.3 Impacts on Nutrient Cycling in Forests and Grasslands

Invasive plant, insect, pathogen, and animal species can drastically alter nutrient cycling in forests and grasslands. Changes in soil nutrient stocks and fluxes can have important implications for ecosystem productivity, atmospheric concentrations of greenhouse gasses, and water quality. Consequently, predicting invasion effects on nutrient cycling is an imperative.

Effects of plant invasion on nutrient cycling have been widely documented and vary considerably among invaders, but several consistent patterns have emerged through meta-analysis of this body of work. Specifically, invasion of grasslands and forests generally increases the amount of nitrogen in aboveground plant tissues (Castro-Diez et al. 2014; Liao et al. 2008; Sardans et al. 2016), stimulates microbial activity, and increases the availability of inorganic nitrogen in invaded soils (Castro-Diez et al. 2014; Lee et al. 2017; Liao et al. 2008). Ehrenfeld et al. (2005) suggested that soil nitrogen availability is enhanced by plant invasion and may promote the fitness of invasive plants, thereby creating a positive feedback. Lee et al. (2012) provided experimental evidence of this positive feedback in soils dominated by Japanese stiltgrass; however, the links between such biogeochemical changes and fitness remain poorly understood for most invasive species. More recently, the results of a global meta-analysis linked the magnitude of invasion effects on nitrogen cycling to trait dissimilarities (Lee et al. 2017). Specifically, effects of plant invasion on soil inorganic nitrogen content were more severe when invaded communities had higher litter nitrogen contents and lower litter carbon-to-nitrogen ratios than native communities. Plant invasion also increased nitrogen mineralization rates and decreased soil carbon-to-nitrogen ratios to a greater degree when invaded communities had much lower leaf carbon-to-nitrogen ratios than reference communities. These findings reinforce earlier conclusions regarding the importance of context for understanding and predicting the consequences of invasive species.

Although plant invasion can increase litter nitrogen, the rate of litter decomposition does not consistently increase (Castro-Diez et al. 2014; Jo et al. 2016) and may actually decrease (Vila et al. 2011). This pattern is consistent with the finding that many invasive plant species have higher tissue carbon-to-nitrogen ratios than native species in the communities they invade and can shift nitrogen pools from aboveground to belowground plant tissues (Daneshgar and Jose 2009). Compared to native species, invasive plant species retain a higher proportion of nutrients in photosynthetic tissues (Sardans et al. 2016), potentially reducing the rate at which nitrogen and phosphorus are released to other plant species (Laungani and Knops 2009). This hypothesis is supported by evidence that fluxes of inorganic nitrogen from soil to plants and of stabilized nitrogen from soil organic matter (SOM) to plants consistently increase with invasion, whereas invasion effects on the nitrogen flux from plants to soil are less consistent (Castro-Diez et al. 2014).

Invasion of woodlands by grasses in particular has been shown to accelerate nitrogen-cycling rates. If grass invasion increases fire frequency, then nutrient loss may also increase through burning and volatilization of nutrient-containing compounds (Mack and D'Antonio 2003). Grass invasions can also alter soil pH. For example, cogongrass (*Imperata cylindrica* (L.) P. Beauv) caused a decrease in soil pH and in the availability of soil potassium (K) in invaded pine flatwoods in the Southern United States (Collins and Jose 2008), whereas Japanese stiltgrass caused an increase in soil pH in invaded mixed deciduous forests in the Eastern United States (Craig et al. 2015; Ehrenfeld et al. 2001). In general, soil acidification can be expected to occur when invasive species take up and sequester a large proportion of the cations available in a system or promote nitrate leaching. In contrast, soil alkalinization can be expected to occur when invasive species sequester a large proportion of available anions, such as nitrate.

Invasive species that can symbiotically fix atmospheric nitrogen (N_2) have distinct effects on nitrogen cycling. Results of meta-analyses repeatedly show that nitrogen-fixing invasive species increase ecosystem nitrogen pool sizes and rates of nitrification to a greater extent and more consistently than nonnitrogen-fixing invasive species (Lee et al. 2017; Liao et al. 2008; Vila et al. 2011). There is limited evidence that invasion by nitrogen-fixing species also has a large, positive effect on the emission of nitrous oxide (N_2O), a potent greenhouse gas (Qiu 2015). Invasive nitrogen fixers generally reduce tissue carbon-to-nitrogen ratio; however, responses are highly variable (Liao et al. 2008; Vila et al. 2011).

In Hawaii, where invasion dynamics may differ because the native flora evolved on young, nitrogen-poor volcanic substrates, nitrogen-fixing trees and nutrient-acquisitive species such as grasses can quickly outcompete native species by altering nutrient-cycling regimes. Hughes and Denslow (2005) showed that nitrogen mass from litterfall in invaded forests was 55 times higher than in native forests. Allison and Vitousek (2004) found a 50-fold increase in leaf litter decay rates for invasive plants. It is likely that this magnitude of change not only allows for displacement of the slower-growing native flora but also potentially facilitates further invasion by nonnitrogen-fixing species, resulting in cascading impacts (see Chap. 2).

Long-term studies are needed to understand fully the ecosystem-level effects of shifts in nutrient dynamics induced by invasion. For example, in a seasonally dry submontane tropical forest, grass invasion can lead to a grass-fire cycle partly because postfire nitrogen-rich soils favor grasses (D'Antonio and Vitousek 1992). Over time, these feedbacks can weaken, and elevated nitrogen mineralization rates can return to preinvasion levels. Yet, instead of facilitating native species, this shift benefits invasive nitrogen-fixing woody species that are competitively dominant when nitrogen availability is low (Yelenik and D'Antonio 2013).

Invasive insects and pathogens can cause both short- and long-term changes in nutrient cycling (Lovett et al. 2006). Short-term increases in soil nitrogen availability and leaching can result from the pulse of litter and the reduction in plant nitrogen uptake that may follow defoliation or tree mortality (Orwig et al. 2008; Webb et al. 1995). Productivity in forests is generally reduced in the short term by invasive insects and diseases. For instance, defoliation of eastern forests by the invasive gypsy moth (*Lymantria dispar* L.) has been shown to reduce net ecosystem production (Clark et al. 2010), host tree growth (Fajvan et al. 2008), and seed production (Gottschalk 1990) in the years immediately following the defoliation. Long-term changes in nutrient cycling arise from indirect effects of invasive species, specifically when the replacement species differs from the host species with respect to their patterns of nutrient uptake, growth, and litter quality. For example, in eastern forests, a complex of an invasive scale insect and an invasive fungus, together known as beech bark disease, is causing American beech (*Fagus grandifolia*) decline which results in subsequent replacement by sugar maple (*Acer saccharum*). As a consequence, the ecosystem shows increases in litter decomposition, nitrification, and nitrate leaching from soils (Lovett et al. 2010a) and decreases in soil CO_2 efflux (Hancock et al. 2008). As New England stands of eastern hemlock (*Tsuga canadensis* (L.) Carr.) decline after infestations of the invasive hemlock woolly adelgid (*Adelges tsugae*) (HWA), their species composition often shifts to dominance by black birch (*Betula lenta* L.) which results in increased aboveground production and rates of nitrogen uptake but no significant effect on soil respiration (Finzi et al. 2014). Complex tree-soil interactions can also mediate potential responses. For example, in contrast to the responses in New England, hemlock stands in the Southern Appalachians experienced increased soil CO_2 efflux and no changes to the nitrogen cycle following eastern hemlock mortality (Knoepp et al. 2011; Nuckolls et al. 2009). This was due primarily to the co-occurrence of rosebay rhododendron (*Rhododendron maximum* (L.)) and its plant-soil-fungal feedbacks (Hoover and Crossley Jr. 1995; Wurzburger and Hendrick 2007; Wurzburger and Hendrick 2009). When rhododendron was absent from declining hemlock stands, soil nitrogen availability increased after hemlock mortality (Block et al. 2013).

Invasive animals substantially affect forest ecosystem nutrient cycles, and responses depend primarily on the animal's foraging and sheltering behaviors. Perhaps the best studied examples in the continental US forests are feral swine (*Sus scrofa*)—generalist omnivores that disturb soil and eat numerous types of seeds, invertebrates, and herpetofauna (Bratton 1975; Jolley et al. 2010)—and earthworms (see Chap. 5) that can alter soil nutrient-cycling processes and food webs by consuming and redistributing litter and soil, creating soil pores, and altering soil physical structure due to their burrows and casts (Bohlen et al. 2004; Migge-Kleian et al. 2006). Invasive earthworms can nearly eliminate the forest floor if they feed on surface litter (Bohlen et al. 2004). In a study of earthworm-invaded forests in New York State, loss of forest floor reduced the soil carbon pool but not the nitrogen pool and consequently decreased the soil carbon-to-nitrogen ratio (see Sect. 3.2) (Bohlen et al. 2004).

In summary, invasive species frequently accelerate rates of soil and ecosystem nutrient cycling. Emerging evidence suggests that the degree of invasion effects will be larger when the traits of invasive species are distinct from those of the native community. In accordance with this hypothesis, nitrogen-fixing invasive species have among the most pronounced effects on nutrient cycling. Invasive species that cause increased fire severity may also enhance the loss of ecosystem nutrients, which can contribute to weakened plant-soil feedbacks over time. As a result, impacts of invasive species may change over time. Additional research is needed to fully understand the long-term effects of invasive plants on nutrient cycling and their role in plant-soil-disturbance feedbacks. In the case of invasive insects and pathogens, long-term changes in nutrient cycling can be expected when the replacement species differs from the host species with respect to their patterns of nutrient uptake, growth, and litter quality.

3.4 Effects of Invasive Species on Water Quantity and Quality

3.4.1 Direct Effects of Invasive Plants on Water Quantity

One of the most important ecosystem processes affected by invasive species is evapotranspiration (ET). Changes in ET can affect multiple components of watershed function including water yield, runoff timing (e.g., stormwater mitigation and maintenance of baseflow), groundwater recharge, and dilution capacity. Water yield and stormflow mitigation are particularly important ecosystem services that support municipal, industrial, and agricultural water supplies and protect human systems from flooding. While scarce, studies that have quantified the effects of invasive species on streamflow at catchment scales show that species conversion from native to invasive species causes substantial (>10%) declines in streamflow (Jain et al. 2015; reviewed by Salemi et al. 2012). Invasion of riparian areas can alter stream channel morphology, leading to altered timing of hydrologic cycles, such as flood frequency and severity (reviewed by Zavaleta 2000), or can alter groundwater dynamics attributed to greater plant water use (Gordon 1998; Saha et al. 2015).

Because direct measurements of changes in hydrology at catchment scales are scarce (Miller et al. 2013), effects often are inferred from comparative measurements of ET or soil moisture at smaller spatial scales. Many studies that examine water use by invasive plant species have found that ET, or water use per unit ground area, is higher for invasive species than for native species. In a meta-analysis, Calaveri and Sack (2010) showed that across 15 invasive/native paired stands, water use per unit ground area was almost 50% higher in invaded stands, and some invaders have at least a twofold greater sap-flux density compared to the native tree species (Cavaleri et al. 2014). As a consequence, soil moisture in invaded stands is consistently lower than in uninvaded stands (Pysek et al. 2012), and removing the invasive species generally increases soil moisture (Hata et al. 2015; Michaud et al. 2015). While these studies do not conclusively show that stream runoff and timing are altered (Owens and Moore 2007), they do provide evidence that the overall water balance is altered, with possible changes to baseflow and water yield.

Stand-level changes in transpiration large enough to affect runoff can occur from major changes in either stand structure (e.g., leaf area) or physiology, phenology, or morphology of the dominant species (e.g., transpiration rates, water use efficiency, leaf phenology, xylem or rooting characteristics) (Le Maitre et al. 2015). Invaders have been shown to have functional morphological and/or phenological traits that allow them to maintain rapid gas exchange rates and thus high productivity in the invaded range at times when native species cannot. Analyzing over 80 species in North Carolina, Wolkovich and Cleland (2010) found that invasive species leafed out earlier than native species. And in a common garden experiment involving 43 native, noninvasive and 30 invasive eastern US forest understory species, Fridley (2012) demonstrated that invasive species prolonged the window of carbon gain later into the fall as compared to native species. Both studies suggest that the timing of carbon gain, and thus water use, may be altered in invaded systems. Invasive plants may exploit water resources that native plants do not. For example, morphological traits that allow greater water access and storage (e.g., larger root systems, sapwood cross-sectional area, xylem-leaf area ratios) also facilitate greater water use by invaders compared to native species (Caplan and Yeakley 2013; Glenn et al. 2012; Nippert et al. 2010). Xylem anatomy can also confer greater water use by invasive species compared to native species in vernal and autumnal windows in deciduous forests. Among 82 native and non-native understory, deciduous, woody species common to eastern US deciduous forests, invasive species had xylem traits that conferred higher freezing resistance or drought-induced cavitation resistance in autumn, thus promoting delayed autumn leaf fall and continued carbon gain when native species were dormant (Yin et al. 2016) (see Box 3.1).

Characteristics such as higher leaf area, dimorphic rooting morphology, or longer leaf-on may have little or no effect in energy-limited systems (see Jones et al. 2012) but substantial effects in water-limited systems as increases in ET tend to be manifested in streamflow (Oishi et al. 2010). Thus, ecosystems characterized by periods of chronic or episodic water limitation may be particularly vulnerable to invasions. Invasive plant species in water-limited ecosystems are not necessarily more tolerant of water stress than native species (Pratt and Black 2006); instead, their invasiveness is due in part to a myriad of physiological and life history traits that allow them to avoid water stress (detailed above). For example, invasive species may possess dimorphic root systems (e.g., saltcedar, *Tamarix ramosissima* Ledeb.) or adaptive drought dormancy, which allow them to avoid drought conditions and maintain high rates of water use (Ammondt et al. 2013; Germino et al. 2016; Nagler et al. 2003). Many inva-

Box 3.1 Traits of the Invasive Shrub Amur Honeysuckle

The invasive shrub Amur honeysuckle (*Lonicera maackii* (Rupr.) Maxim) is one of the most widespread invaders of the Ohio River Valley. Xylem anatomical traits allow this species to leaf out earlier, withstand early spring freezes without leaf mortality, and senesce later in autumn than native shrub species. These traits suggest that the extended leaf-on period and freeze tolerance allow greater water use and carbon gain (McEwan et al. 2009). Further, Amur honeysuckle's transpiration rate is roughly proportional to its basal area, which suggests that it has the potential to reduce streamflow by 10% or more and affect hydroperiod of ephemeral ponds and streams (Boyce et al. 2012).

Leaves and flowers of Amur honeysuckle (upper), *and efforts to remove honeysuckle from the understory of a deciduous forest stand, before and after* (middle). *Note early leaf-out of honeysuckle in* lower photos *in spring when other deciduous species have yet to leaf out. Photos courtesy of Bugwood.org, taken by L. Mehrhoff and T. Evans. Distribution of number of reported cases by county of Amur honeysuckle as of December 2016 (*lower*)(EDDMapS* 2016*).*

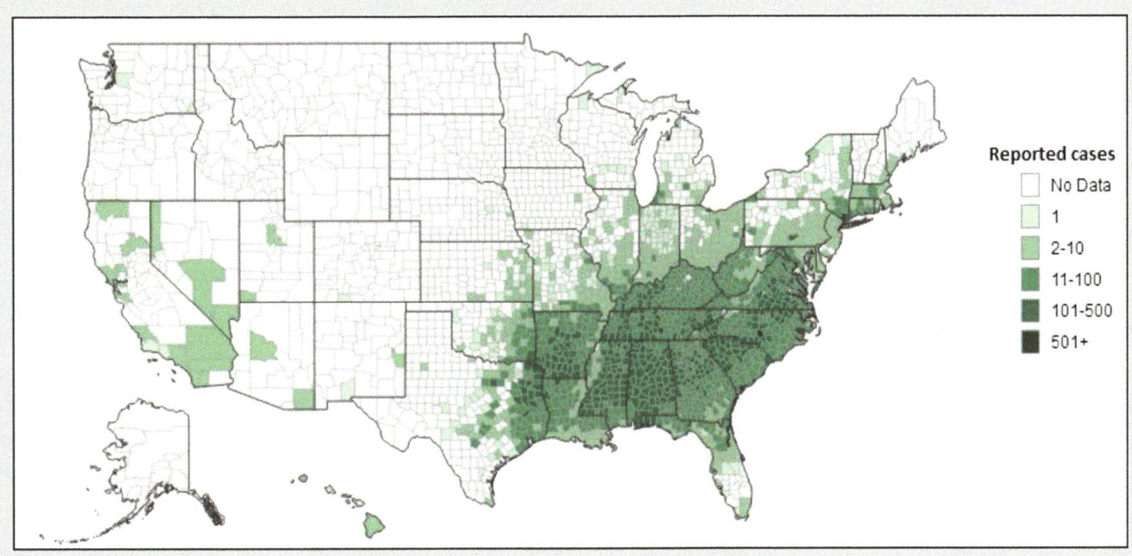

sive species tend to have low leaf construction costs, which facilitate rapid growth and high assimilation and transpiration rates, but these traits may render them more prone to water stress in low-precipitation years if they can't avoid drought (Cordell et al. 2002). Rapid growth and liberal water use may also allow invasive species to take advantage of both carbon fertilization and water savings as climate changes (Peterson et al. 2014). Elevated concentrations of CO_2 typically cause a decrease in stomatal conductance, and as a result, stand water use can decline, and soil moisture can increase. One example of this is Dalmatian toadflax (*Linaria dalmatica* ssp. *dalmatica*), an invasive forb with C3 metabolism. This invader has an order of magnitude higher biomass and seed production as a result of both enhanced carbon supply and increased soil water compared to the native C3 and C4 plant community (Blumenthal et al. 2013). Despite the lack of information in the literature, effects of invasion on stream runoff and timing may be highly variable and transient in duration and may interact with other factors such as climate and spatial dynamics (Le Maitre et al. 2015).

3.4.2 Indirect Effects of Invasive Insects and Pathogens on Water Quantity

Effects of invasive species aren't limited to the direct effects of invasive plants on ecosystem ET. Invasive insects and pathogens may also affect ecosystem structure and species composition resulting in cascading effects on ecosystem function. Effects from short-term defoliation are moderate or undetectable, but in extreme cases, invasive insects and pathogens may extirpate dominant plant species that have few, if any, defenses against them (Flower et al. 2013; Ford et al. 2012; McManamay et al. 2011; Riscassi and Scanlon 2009). If native species that replace the affected host species have different leaf area, leaf phenology, and/or functional traits than the extirpated species, dramatic alterations in hydrology can occur. For example, when the evergreen eastern hemlock was nearly extirpated from Southern Appalachian riparian forests, the deciduous species that replaced hemlock had lower leaf area, higher leaf-level transpiration rates, and a more pronounced seasonal water use pattern, all affecting watershed hydrology (Brantley et al. 2013; Ford and Vose 2007). While initial annual and winter stand transpiration declined by 22% and 74%, respectively, following loss of hemlock, positive growth responses of deciduous tree species resulted in 12% higher stand annual transpiration, 9% lower annual water yield, and 20% higher dormant-season peak flows after occurrence of the most extreme storm events (Brantley et al. 2013; Brantley et al. 2015).

3.4.3 Direct Effects of Invasive Plants on Water Quality

Invasive plant species can affect water quality through a number of mechanisms including reduced dilution capacity of streams from lower runoff (see Sects. 3.4.1 and 3.4.2), higher suspended sediment loads from increased soil erosion, increased nutrient leaching from altered biogeochemical cycles, and addition of novel plant exudates (Chamier et al. 2012; Ehrenfeld 2003; Nagler et al. 2008). Specific changes to terrestrial biogeochemical cycles that increase nutrient leaching include altered foliar chemistry (e.g., higher foliar N content), faster N mineralization rates, and shifts in soil pH which can cause displacement of cations (see Sect. 3.3). These changes often lead to increased concentrations of water-soluble nutrients, increased rates of nutrient leaching, and ultimately elevated nutrient concentrations in both surface water and groundwater. For example, in the Southwestern United States, invasion of riparian corridors by saltcedar, a halophytic shrub, can elevate soil salinity over time due to existence of salt secretions on the leaf surfaces (Merritt and Shafroth 2012), thus potentially affecting water quality during pulse flooding in this arid region. Increased riparian leaf area after saltcedar invasion also decreases incident light, decreases aquatic macrophyte biomass and chlorophyll in streams, and increases carbon input into streams from leaf litter (Kennedy and Hobbie 2004). Also, in the Southwestern United States, the invasive cheatgrass has displaced native grass species such as black grama (*Bouteloua eriopoda*). Compared to other native and nonnative grasses, black grama root morphology and resilience to drought stabilize the easily erodible soil; thus, when it disappears, soil erosion increases (Germino et al. 2016) which increases the potential for subsequent loss of water quality.

Invasive plant species that fix atmospheric N_2 can cause some of the greatest observed effects on water quality. Symbiotic N_2-fixing invasive plants increase rates of N inputs into ecosystems and increase soil water N (Baer et al. 2006; Goldstein et al. 2010; Vitousek and Walker 1989). Although direct impacts to groundwater and surface water are rare, Jovanovic et al. (2009) showed that groundwater nitrate (NO_3) concentrations increased after invasion of shrublands with nitrogen-fixing orange wattle (*Acacia saligna* (Labill.) Wendl. f.). Increasing N inputs may also exacerbate soil acidification leading to greater leaching of cations into groundwater and surface water (Matson et al. 1999). With increasing frequency and intensity of droughts (Vose et al. 2016b), the impact of N_2-fixing invasive plants on water quality may be even greater than anticipated. Drought can decrease soil inorganic N supply (He and Dijkstra 2014; Rennenberg et al. 2009), and in response, N_2-fixing plants have been shown to have a competitive advantage over nonfixing species when both water and N were limiting, due to their ability to upregulate N_2 fixation (Wurzburger and Miniat 2014).

Other effects of invasive species on water quality and quantity are more complex and/or may only manifest themselves over time. For example, invasions that alter fire regimes may have substantial effects on coupled hydrologic and biogeochemical cycles. Plant invasion often increases standing biomass, changes plant flammability characteristics, alters fuel continuity, and may prolong fire intervals, any or all of which can result in increased wildfire intensity and severity (Chamier et al. 2012; Le Maitre et al. 2014; Smith et al. 2011). A common effect of increasing fire intensity is the generation of water-repellent soil layers, which increase overland flow and soil erosion during storms, resulting in reduced water quality (Smith et al. 2011). For example, areas of grassland that have been invaded by cheatgrass, which increases fuel flammability and the fire return interval by as much as fourfold (Balch et al. 2013; Brooks et al. 2004), have experienced higher rates of soil erosion following fire (Germino et al. 2016), with the potential for reducing water quality.

3.4.4 Indirect Effects of Invasive Insects and Pathogens on Water Quality

Invasive insects and pathogens can also have significant impacts on water quality when they induce changes in vegetation structure and composition. Like the indirect effects of invasive insects and pathogens on water quantity, changes tend to be caused by either defoliation or mortality which initiates a cascade of changes in ecosystem structure and function. The most extreme impacts occur when dominant native species are extirpated. Similar to the direct effects of plant invasions, these impacts can be expressed through changes in water chemistry (N, P, sulfur (S)), clarity (total suspended solids (TSS)), temperature, and other parameters. For example, in the Eastern United States, small forested watersheds tend to retain the vast majority of N (90%+) under undisturbed or unsaturated conditions (Swank and Vose 1997), e.g., if they receive low N deposition (Adams et al. 2014; Lovett et al. 2000). One of the most apparent effects of invasive species that cause defoliation is an increase in export of watershed NO_3. Large-scale defoliation and/or mortality of dominant tree species caused by invasive species can result in immediate and dramatic increases in NO_3 flux to streams (Adams et al. 2014; Lovett et al. 2000; Swank and Vose 1997), with potential negative consequences to water quality downstream. Reasons for increased N leakage may include less uptake by affected trees, increased litter inputs into streams during defoliation and/or after mortality, increased inputs of insect frass, and a general loosening of the normally conservative N cycle in these nutrient-limited forests.

The best examples of these effects can be attributed to two invasive insects that occur in the forests of the Central and Southern Appalachian Mountains. In the Central Appalachians, severe gypsy moth defoliation of native hardwoods between 1987 and 1992 resulted in increased groundwater NO_3 levels for several years after disturbance (Riscassi and Scanlon 2009). In the Southern Appalachians, HWA-caused mortality of eastern hemlock in the Coweeta Basin that began in 2004 (Ford et al. 2012) resulted in significant increases in NO_3 exports in four different forested headwater catchments (Fig. 3.1). Increases in annual N exports (up to 300% higher) lasted for 4 years after infestation; however, changes in NO_3 exports varied seasonally with greater relative changes in summer. Summer N exports remained significantly higher 7 years after infestation. Other effects of HWA-induced hemlock mortality include increased inputs of litter and wood into streams from dying hemlock and changes in stream temperature due to altered light regimes (Webster et al. 2012). These collective changes may have detrimental impacts on downstream ecosystems that rely on cool, clean water from forested headwaters (Ross et al. 2003). These impacts associated with invasion aren't isolated to the Southern Appalachians. In New England where HWA can decimate entire stands of hemlock, mobile soil N and stream NO_3 are also elevated in areas experiencing higher mortality compared to healthy hemlock stands (Cessna and Nielsen 2012).

Although many forested headwater watersheds are characterized by high water quality, the cumulative effects of invasive species on receiving water bodies could include substantial reduction of water quality with the potential for eutrophication in extreme circumstances. However, more research is needed to directly assess the effects of invasive insects and pathogens on water quality in forests and grasslands.

3.5 Key Findings

In general, invasive species effects on carbon and nutrient cycling depend on severity of the invasion, differences in structure (e.g., growth form or leaf area) and function (e.g., phenology) between the invasive species and the species it replaces, and where in the cycle the impact occurs (Table 3.1). For instance, an invasive species could affect the carbon budget by altering productivity, allocation, litter production, decomposition, herbivory, disturbance regimes, or food web structure, and each of these will have different consequences for the ecosystem (Peltzer et al. 2010). We have an emerging understanding of why and under what conditions plant invasions will have the largest effects on

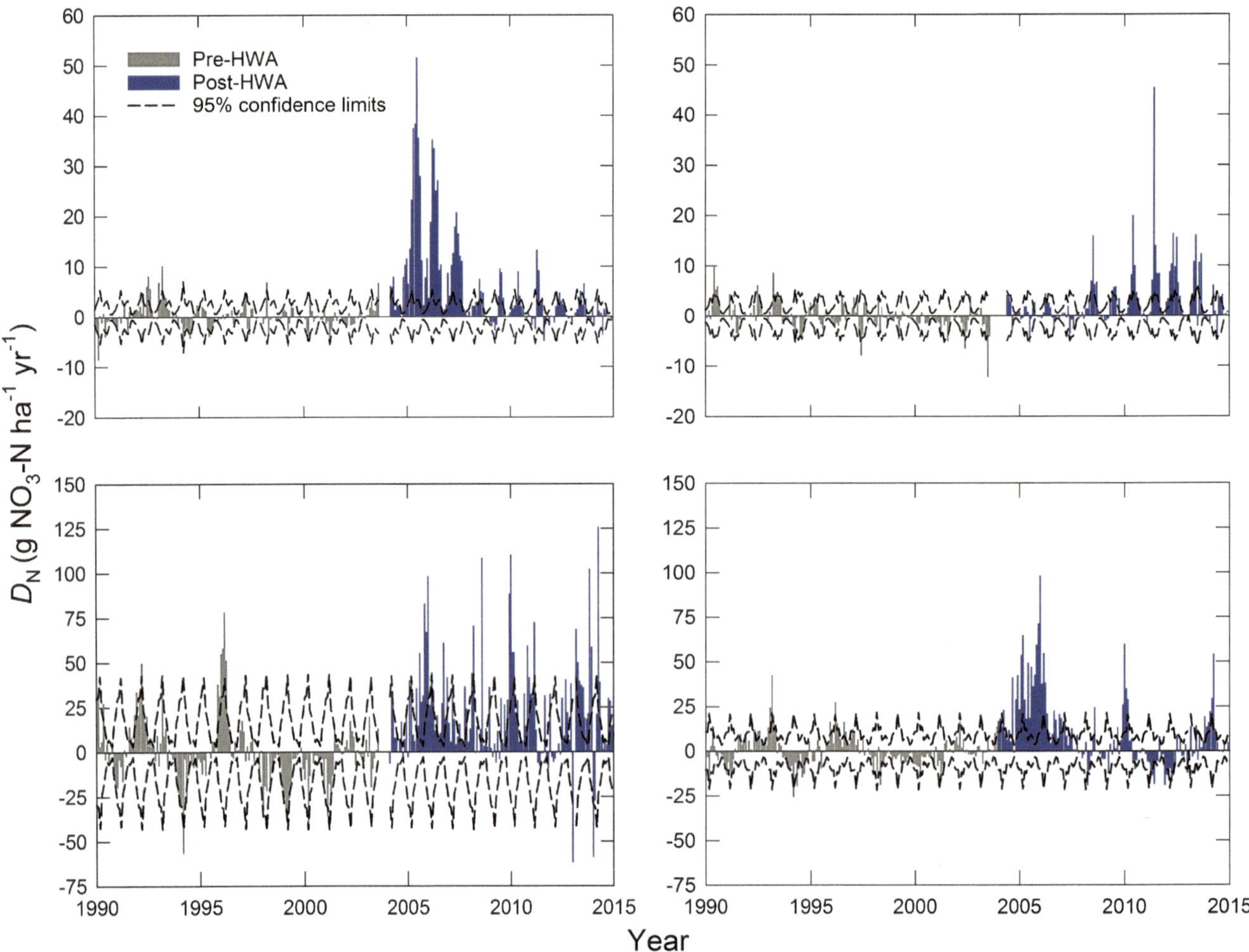

Fig. 3.1 Monthly deviations between observed and expected NO_3-N flux (D_N) for four headwater catchments experiencing severe infestation by hemlock woolly adelgid (*Adelges tsugae*), an invasive insect (see Chap. 2), in the Coweeta Basin. *Top two panels* are low-elevation (cove hardwood and mixed oak communities) watersheds (WS14 and WS18), while *bottom two panels* are high-elevation (northern hardwood, cove hardwood, and mixed oak communities) watersheds (WS27 and WS36). Note scale difference between upper and lower panels. *Gray bars* represent D_N during the calibration period; *blue bars* represent D_N during and after infestation. *Lines* are 95% confidence intervals, bars lying within which are not statistically significantly different than expected (S. Brantley and C. Miniat, *unpublished data*). Briefly, we used the paired watershed approach, pairing reference watersheds with relatively high proportions of eastern hemlock basal area in the riparian zones with a reference watershed with relatively low eastern hemlock basal area in the riparian zone (WS2, see Brantley et al. 2015). Site descriptions and water chemistry methods are detailed by Webster et al. (2016)

biogeochemical processes (Pysek et al. 2012). Effects often vary substantially across space for the same species, suggesting that environmental conditions play an important but overlooked role in determining invasive species impacts (Block et al. 2012; Block et al. 2013; Craig et al. 2015). Recent studies indicate that many invasive plants possess functional traits associated with high capacity for nutrient acquisition and carbon fixation (Heberling and Fridley 2013; Jo et al. 2015). Because such traits are linked to biogeochemical cycling (Diaz and Cabido 2001), assessment of these attributes will lead to new insights about how invasion may alter biogeochemical processes. In addition, there is growing appreciation that biotic interactions with species of the recipient community may determine invasion effects (Ehrenfeld 2003; Kumschick et al. 2015). For invasive insects and pathogens, the key factors that determine the impact on ecosystem processes are the lethality and host specificity of the insect or pathogen and the dominance and uniqueness of the host tree (Lovett et al. 2006). With regard to ecosystem carbon and nutrient cycling, a key question is whether the invasion produces a long-term shift in the dominant plant species in the ecosystem. If so, enhanced knowledge of the community dynamics and the functional traits of the declining and increasing species will be needed in order to predict the future functioning of the ecosystem. To predict how biogeochemical processes will change after invasions, it's imperative that we develop a mechanistically based framework that merges our understanding of how the

Table 3.1 Summary of observed changes in community structure, corresponding changes in ecosystem function, and the effects on ecosystem services from invasive species

	Mechanisms		Commonly observed effects
	Frequently observed changes in structure	*Examples of changes in function*	
<u>*Carbon cycling*</u>	Graduated increase/decrease in plant functional type for the ecosystem context (e.g., increase, change from grasses in grasslands to shrubs, change from shrubs in shrublands to trees; decrease, forests or shrublands invaded by grasses)	Increases/decreases in aboveground and belowground biomass and soil organic matter	Increases/decreases in carbon stocks but depends on N and water availability and competition
	Decrease in leaf area from herbivory	Short-term decrease in primary productivity and increased litterfall and dead plant material	Altered disturbance regime (e.g., fire)
	Altered plant species composition (mortality of dominant species)	Decreased litter layer carbon, increased soil organic matter	Decreased annual carbon budget
	Introduction of earthworms		Protection of carbon in slower cycling pools
<u>*Nutrient cycling*</u>	Increase in biomass of species associated with high rates of resource acquisition	Short- to long-term increases in soil nitrogen availability and nitrogen-cycling rates	Increased litter nitrogen content, lower tissue carbon-to-nitrogen ratios, increased fluxes of inorganic nitrogen from soil to plants, and stabilized nitrogen from soil organic matter to plants
	Grass invasion of woodlands or shrublands	Loss of nutrients through burning or volatilization of nutrient-containing compounds	Increase in fire severity
	Introduction of nitrogen-fixing invasive species	Replacement by fast-growing species with low tissue construction costs	Increased ecosystem nitrogen pools and rates of nitrification
	Altered species composition (mortality of dominant species)		Increased litter decomposition, nitrification, and nitrate leaching from soils
<u>*Water quantity*</u>	Altered species composition	Altered canopy phenology (e.g., longer leaf-on, lower winter leaf area)	Decline in water yield (e.g., reduced stream runoff)
	Changes in dominant plant functional group (i.e., tree, shrub, vine, or grass)	1.1.Increased water use at tree, stand, and watershed scales	1.1.Altered stormflow dynamics (e.g., higher peak flows and/or lower minimum flows)
	Changes in leaf habit (deciduous vs. evergreen)		
	Increases in sapwood area and leaf area		
<u>*Water quality*</u>	Altered species composition	Shift to N_2 fixation	Increased stream nutrient concentrations (especially N and cations)
	Changes in dominant plant functional group (i.e., tree, shrub, vine, or grass)	Altered soil pH	Increased stream salinity
	Increases in sapwood area and leaf area	Reduced watershed nutrient retention	Higher sedimentation
	Altered leaf chemistry	Changes in fire frequency	1.1.Changes in stream community composition and/or productivity
	Changes in root morphology	Increased fire severity and intensity	
		Loss of soil stability	

functional traits of invaders and other species in the recipient community, and their biotic interactions, create context dependency in invasion effects.

In conclusion, many of our ecosystems have reached a point where healthy functions that effectively store carbon and promote sustainable nutrient and water balance are in a more tenuous balance owing to the effects of invasive species. Sustaining ecosystems that store more carbon than they release and that regulate nutrient and water cycles will become more challenging in the future and will require using a creative blend of old and new land management tools.

3.6 Key Information Needs

1. While invasive species increase ecosystem productivity in many cases, it is not clear what possible and potential tradeoffs associated with increased productivity are. Several questions need to be answered. Over time, are the levels of carbon inputs due to invasive species sustainable? What are the direct and indirect impacts of other trophic level invaders on carbon cycling? For example, what are the long-term carbon cycle impacts attributed to invasive species-induced shifts in species composition? How will the indirect effects of invasive pathogens alter

carbon cycling if tree mortality and defoliation episodes increase?
2. Additional research is needed to determine whether broadscale processes such as atmospheric nitrogen deposition can explain context dependence in carbon effects of other invasive species.
3. In order to predict the impact of invasions on biogeochemical processes, we need to develop a mechanistically based framework that merges our understanding of how the functional traits of invaders and species in the recipient community, and their biotic interactions, create context dependency in invasion effects.
4. Does enhanced soil N availability induced by plant invasion promote the fitness of invasive plants, thereby creating a positive feedback? The links between such biogeochemical changes and fitness remain poorly understood for most invasive species.
5. Long-term studies, or expanded use of experimental addition and removal studies, are needed to understand fully the ecosystem-level effects of shifts in nutrient dynamics. Specifically, before-and-after invasion measurements on ecosystem processes are needed over the long periods required to characterize the full range of variability of ecosystem processes (Stricker et al. 2015).
6. Increased efforts to document the extent and severity of invasions are recommended to establish broader-scale impacts.
7. Few studies have assessed the direct and indirect effects of invasive species on water quality and quantity. Impacts have been inferred from measurements at smaller spatial scales that may not be appropriate for larger-scale processes (Owens and Moore 2007). More whole-watershed studies are needed to directly assess water quality effects of invasive insects and pathogens in forests and grasslands.

Disclaimer Text The findings and conclusions in this publication are those of the authors and should not be construed to represent any official USDA or U.S. Government determination or policy.

Literature Cited

Adams MB, Knoepp JD, Webster JR (2014) Inorganic nitrogen retention by watersheds at Fernow Experimental Forest and Coweeta Hydrologic Laboratory. Soil Sci Soc Am J 78:S84–S94

Allison SD, Vitousek PM (2004) Rapid nutrient cycling in leaf litter from invasive plants in Hawai'i. Oecologia 141:612–619

Ammondt SA, Litton CM (2012) Competition between native Hawaiian plants and the invasive grass *Megathyrsus maximus*: implications of functional diversity for ecological restoration. Restor Ecol 20:638–646

Ammondt SA, Litton CM, Ellsworth LM, Leary JK (2013) Restoration of native plant communities in a Hawaiian dry lowland ecosystem dominated by the invasive grass *Megathyrsus maximus*. Appl Veg Sci 16:29–39

Baer SG, Church JM, Williard KWJ, Groninger JW (2006) Changes in intrasystem N cycling from N-2-fixing shrub encroachment in grassland: multiple positive feedbacks. Agric Ecosyst Environ 115:174–182

Balch JK, Bradley BA, D'Antonio CM, Gomez-Dans J (2013) Introduced annual grass increases regional fire activity across the arid western USA (1980–2009). Glob Chang Biol 19:173–183

Belote RT, Jones RH, Hood SM, Wender BW (2008) Diversity-invasibility across an experimental disturbance gradient in Appalachian forests. Ecology 89:183–192

Block CE, Knoepp JD, Elliott KJ, Fraterrigo JM (2012) Impacts of hemlock loss on nitrogen retention vary with soil nitrogen availability in the southern Appalachian Mountains. Ecosystems 15:1108–1120

Block CE, Knoepp JD, Fraterrigo JM (2013) Interactive effects of disturbance and nitrogen availability on phosphorus dynamics of southern Appalachian forests. Biogeochemistry 112:329–342

Blumenthal DM, Resco V, Morgan JA et al (2013) Invasive forb benefits from water savings by native plants and carbon fertilization under elevated CO_2 and warming. New Phytol 200:1156–1165

Bohlen PJ, Groffman PM, Fahey TJ et al (2004) Ecosystem consequences of exotic earthworm invasion of north temperate forests. Ecosystems 7:1–12

Bossuyt H, Six J, Hendrix PF (2005) Protection of soil carbon by microaggregates within earthworm casts. Soil Biol Biochem 37:251–258

Boyce RL, Durtsche RD, Fugal SL (2012) Impact of the invasive shrub *Lonicera maackii* on stand transpiration and ecosystem hydrology in a wetland forest. Biol Invasions 14:671–680

Bradley BA, Houghton RA, Mustard JF, Hamburg SP (2006) Invasive grass reduces aboveground carbon stocks in shrublands of the Western US. Glob Chang Biol 12:1815–1822

Brantley S, Ford CR, Vose JM (2013) Future species composition will affect forest water use after loss of eastern hemlock from southern Appalachian forests. Ecol Appl 23:777–790

Brantley ST, Miniat CF, Elliott KJ et al (2015) Changes to southern Appalachian water yield and stormflow after loss of a foundation species. Ecohydrology 8:518–528

Bratton SP (1975) Effect of European wild boar, *Sus scrofa*, on gray beech forest in Great Smoky Mountains. Ecology 56:1356–1366

Brooks ML, D'Antonio CM, Richardson DM et al (2004) Effects of invasive alien plants on fire regimes. Bioscience 54:677–688

Caldwell P, Muldoon C, Miniat CF et al (2014) Quantifying the role of National Forest System Lands in providing surface drinking water supply for the Southern U.S. USDA-Forest Service, Southern Research Station, Asheville, 144 p

Cameron EK, Shaw CH, Bayne EM et al (2015) Modelling interacting effects of invasive earthworms and wildfire on forest floor carbon storage in the boreal forest. Soil Biol Biochem 88:189–196

Caplan JS, Yeakley JA (2013) Functional morphology underlies performance differences among invasive and non-invasive ruderal *Rubus* species. Oecologia 173:363–374

Castro-Diez P, Godoy O, Alonso A et al (2014) What explains variation in the impacts of exotic plant invasions on the nitrogen cycle? A meta-analysis. Ecol Lett 17:1–12

Cavaleri MA, Sack L (2010) Comparative water use of native and invasive plants at multiple scales: a global meta-analysis. Ecology 91:2705–2715

Cavaleri MA, Ostertag R, Cordell S, Sack L (2014) Native trees show conservative water use relative to invasive trees: results from a removal experiment in a Hawaiian wet forest. Conser Physiol 2:1–14

Cessna JF, Nielsen C (2012) Influences of hemlock woolly adelgid-induced stand-level mortality on nitrogen cycling and stream water nitrogen concentrations in southern Pennsylvania. Castanea 77:127–135

Chamier J, Schachtschneider K, le Maitre DC et al (2012) Impacts of invasive alien plants on water quality, with particular emphasis on South Africa. Water SA 38:345–356

Chang C-H, Szlavecz K, Buyer JS (2016) Species-specific effects of earthworms on microbial communities and the fate of litter-derived carbon. Soil Biol Biochem 100:129–139

Clark KL, Skowronski N, Hom J (2010) Invasive insects impact forest carbon dynamics. Glob Chang Biol 16:88–101

Collins AR, Jose S (2008) *Imperata cylindrica*, an exotic invasive grass, changes soil chemical properties of forest ecosystems in the southeastern United States. In: Kohli RK, Jose S, Singh HP, Batish DR (eds) Invasive plants and Forest ecosystems. CRC Press, Boca Raton, 454 p

Concilio AL, Nippert JB, Ehrenfeucht S et al (2016) Imposing antecedent global change conditions rapidly alter plant community composition in a mixed-grass prairie. Oecologia 182:899–911

Cordell S, Sandquist DR (2008) The impact of an invasive African bunchgrass (*Pennisetum setaceum*) on water availability and productivity of canopy trees within a tropical dry forest in Hawai'i. Funct Ecol 22:1008–1017

Cordell S, Cabin R, Hadway L (2002) Physiological ecology of native and alien dry forest shrubs in Hawai'i. Biol Invasions 4:387–396

Craig ME, Pearson SM, Fraterrigo JM (2015) Grass invasion effects on forest soil carbon depend on landscape-level land use patterns. Ecology 96:2265–2279

Daehler CC (2003) Performance comparisons of co-occurring native and alien invasive plants: implications for conservation and restoration. Ann Rev Ecol Evol Syst 34:183–211

Daneshgar P, Jose S (2009) *Imperata cylindrica*, an alien invasive grass, maintains control over nitrogen availability in an establishing pine forest. Plant Soil 320:209–218

D'Antonio CM, Vitousek PM (1992) Biological invasions by exotic grasses, the grass/fire cycle, and global change. Annu Rev. Ecol Syst 23:63–87

Dempsey MA, Fisk MC, Fahey TJ (2011) Earthworms increase the ratio of bacteria to fungi in northern hardwood forest soils, primarily by eliminating the organic horizon. Soil Biol Biochem 43:2135–2141

Diaz S, Cabido M (2001) Vive la difference: plant functional diversity matters to ecosystem processes. Trends Ecol Evol 16:646–655

Drenovsky RE, Grewell BJ, D'Antonio CM et al (2012) A functional trait perspective on plant invasion. Ann Bot 110:141–153

Dukes JS (2002) Species composition and diversity affect grassland susceptibility and response to invasion. Ecol Appl 12:602–617

EDDMapS (2016) Early detection & distribution mapping system. The University of Georgia - Center for Invasive Species and Ecosystem Health. Available online at http://www.eddmaps.org/. Last accessed 19 Dec, 2016.

Edwards CA, Hendrix PF, Arancon N (2013) Biology and ecology of Earthworms, 4th edn. Springer US, New York, 422 p

Ehrenfeld JG (2003) Effects of exotic plant invasions on soil nutrient cycling processes. Ecosystems 6:503–523

Ehrenfeld JG (2010) Ecosystem consequences of biological invasions. Annu Rev. Ecol Evol Syst 41:59–80

Ehrenfeld JG, Kourtev P, Huang WZ (2001) Changes in soil functions following invasions of exotic understory plants in deciduous forests. Ecol Appl 11:1287–1300

Ehrenfeld JG, Ravit B, Elgersma K (2005) Feedback in the plant-soil system. Annu Rev. Environ Resour 30:75–115

Eisenhauer N, Partsch S, Parkinson D, Scheu S (2007) Invasion of a deciduous forest by earthworms: changes in soil chemistry, microflora, microarthropods and vegetation. Soil Biol Biochem 39:1099–1110

Eschtruth AK, Battles JJ (2009) Assessing the relative importance of disturbance, herbivory, diversity, and propagule pressure in exotic plant invasion. Ecol Monogr 79:265–280

Fajvan MA, Rentch J, Gottschalk K (2008) The effects of thinning and gypsy moth defoliation on wood volume growth in oaks. Trees-Structure Funct 22:257–268

Finzi AC, Raymer PCL, Giasson M-A, Orwig DA (2014) Net primary production and soil respiration in New England hemlock forests affected by the hemlock woolly adelgid. Ecosphere 5:1–16

Flory SL, Clay K, Emery SM et al (2015) Fire and non-native grass invasion interact to suppress tree regeneration in temperate deciduous forests. J Appl Ecol 52:992–1000

Flower CE, Knight KS, Gonzalez-Meler MA (2013) Impacts of the emerald ash borer (*Agrilus planipennis* Fairmaire) induced ash (*Fraxinus* spp.) mortality on forest carbon cycling and successional dynamics in the eastern United States. Biol Invasions 15:931–944

Ford CR, Vose JM (2007) *Tsuga canadensis* (L.) Carr. mortality will impact hydrologic processes in southern Appalachian forest ecosystems. Ecol Appl 17:1156–1167

Ford CR, Elliott KJ, Clinton BD et al (2012) Forest dynamics following eastern hemlock mortality in the southern Appalachians. Oikos 121:523–536

Fraterrigo JM, Wagner S, Warren RJ (2014) Local-scale biotic interactions embedded in macroscale climate drivers suggest Eltonian noise hypothesis distribution patterns for an invasive grass. Ecol Lett 17:1447–1454

Fridley JD (2012) Extended leaf phenology and the autumn niche in deciduous forest invasions. Nature 485:359–362

Funk JL, Vitousek PM (2007) Resource-use efficiency and plant invasion in low-resource systems. Nature 446:1079–1081

Galloway JN, Schlesinger WH, Levy H et al (1995) Nitrogen fixation: atmospheric enhancement - environmental response. Glob Biogeochem Cycles 9:235–252

Germino MJ, Belnap J, Stark JM et al (2016) Ecosystem impacts of exotic annual invaders in the genus Bromus. In: Germino M, Chambers J, Brown C (eds) Exotic Brome-Grasses in arid and semiarid ecosystems of the Western US. Springer Cham, Heidelberg, pp 61–95

Glenn EP, Morino K, Nagler PL et al (2012) Roles of saltcedar (*Tamarix* spp.) and capillary rise in salinizing a non-flooding terrace on a flow-regulated desert river. J Arid Environ 79:56–65

Goldstein CL, Williard KWJ, Schoonover JE et al (2010) Soil and groundwater nitrogen response to invasion by an exotic nitrogen-fixing shrub. J Environ Qual 39:1077–1084

Gordon DR (1998) Effects of invasive, non-indigenous plant species on ecosystem processes: lessons from Florida. Ecol Appl 8:975–989

Gottschalk KW (1990) Gypsy moth effects on mast production. In: McGee CE (ed) Proceedings of the workshop: Southern Appalachian mast management; 1989 August 14–16. University of Tennessee, Knoxville, pp 42–50

Hale CM, Frelich LE, Reich PB, Pastor J (2008) Exotic earthworm effects on hardwood forest floor, nutrient availability and native plants: a mesocosm study. Oecologia 155:509–518

Hancock JE, Arthur MA, Weathers KC, Lovett GM (2008) Carbon cycling along a gradient of beech bark disease impact in the Catskill Mountains, New York. Can J For Res 38:1267–1274

Hata K, Kawakami K, Kachi N (2015) Higher soil water availability after removal of a dominant, nonnative tree (*Casuarina equisetifolia* Forst.) from a subtropical forest. Pac Sci 69:445–460

He MZ, Dijkstra FA (2014) Drought effect on plant nitrogen and phosphorus: a metaanalysis. New Phytol 204:924–931

Heberling JM, Fridley JD (2013) Resource-use strategies of native and invasive plants in Eastern North American forests. New Phytol 200:523–533

Hicke JA, Allen CD, Desai AR et al (2012) Effects of biotic disturbances on forest carbon cycling in the United States and Canada. Glob Chang Biol 18:7–34

Hoffmann WA, Lucatelli VMPC, Silva FJ et al (2004) Impact of the invasive alien grass *Melinis minutiflora* at the savanna-forest ecotone in the Brazilian Cerrado. Divers Distrib 10:99–103

Hoover CM, Crossley DA Jr (1995) Leaf litter decomposition and microarthropod abundance along an altitudinal gradient. In: Collins HP, Robertson GP, Klug MJ (eds) The significance and regulation of soil biodiversity, Developments in Plant and Soil Sciences, vol 63. Springer, Dordrecht, pp 287–292

Hughes RF, Denslow JS (2005) Invasion by a N_2-fixing tree alters function and structure in wet lowland forests of Hawai'i. Ecol Appl 15:1615–1628

Jackson RB, Banner JL, Jobbagy EG et al (2002) Ecosystem carbon loss with woody plant invasion of grasslands. Nature 418:623–626

Jain S, Ale S, Munster CL et al (2015) Simulating the hydrologic impact of *Arundo donax* invasion on the headwaters of the Nueces River in Texas. Hydrology 2:134–147

Janssens IA, Dieleman W, Luyssaert S et al (2010) Reduction of forest soil respiration in response to nitrogen deposition. Nat Geosci 3:315–322

Jo I, Fridley JD, Frank DA (2015) Linking above- and belowground resource use strategies for native and invasive species of temperate deciduous forests. Biol Invasions 17:1545–1554

Jo I, Fridley JD, Frank DA (2016) More of the same? In situ leaf and root decomposition rates do not vary between 80 native and nonnative deciduous forest species. New Phytol 209:115–122

Jolley DB, Ditchkoff SS, Sparklin BD et al (2010) Estimate of herpetofauna depredation by a population of wild pigs. J Mammal 91:519–524

Jones JA, Creed IF, Hatcher KL et al (2012) Ecosystem processes and human influences regulate streamflow response to climate change at long-term ecological research sites. Bioscience 62:390–404

Jones R, Chambers JC, Johnson DW et al (2015) Effect of repeated burning on plant and soil carbon and nitrogen in cheatgrass (*Bromus tectorum*) dominated ecosystems. Plant Soil 386:47–64

Jovanovic NZ, Israel S, Tredoux G et al (2009) Nitrogen dynamics in land cleared of alien vegetation (*Acacia saligna*) and impacts on groundwater at Riverlands Nature Reserve (Western Cape, South Africa). Water SA 35:37–44

Kennedy TA, Hobbie SE (2004) Saltcedar (*Tamarix ramosissima*) invasion alters organic matter dynamics in a desert stream. Freshw Biol 49:65–76

Knoepp JD, Vose JM, Clinton BD, Hunter MD (2011) Hemlock infestation and mortality: impacts on nutrient pools and cycling in Appalachian forests. Soil Sci Soc Am J 75:1935–1945

Kramer TD, Warren RJ II, Tang Y, Bradford MA (2012) Grass invasions across a regional gradient are associated with declines in belowground carbon pools. Ecosystems 15:1271–1282

Kumschick S, Gaertner M, Vila M et al (2015) Ecological impacts of alien species: quantification, scope, caveats, and recommendations. Bioscience 65:55–63

Laungani R, Knops JMH (2009) Species-driven changes in nitrogen cycling can provide a mechanism for plant invasions. Proc Natl Acad Sci U S A 106:12400–12405

Le Maitre DC, Kotzee IM, O'Farrell PJ (2014) Impacts of land-cover change on the water flow regulation ecosystem service: invasive alien plants, fire and their policy implications. Land Use Policy 36:171–181

Le Maitre DC, Gush MB, Dzikiti S (2015) Impacts of invading alien plant species on water flows at stand and catchment scales. AoB Plants 7

Lee MR, Flory SL, Phillips RP (2012) Positive feedbacks to growth of an invasive grass through alteration of nitrogen cycling. Oecologia 170:457–465. https://doi.org/10.1007/s00442-012-2309-9

Lee MR, Bernhardt ES, van Bodegom PM et al (2017) Invasive species' leaf traits and dissimilarity from natives shape their impact on nitrogen cycling: a meta-analysis. New Phytol 213:128–139

Liao CZ, Peng RH, Luo YQ et al (2008) Altered ecosystem carbon and nitrogen cycles by plant invasion: a meta-analysis. New Phytol 177:706–714

Litton CM, Sandquist DR, Cordell S (2006) Effects of non-native grass invasion on aboveground carbon pools and tree population structure in a tropical dry forest of Hawai'i. For Ecol Manag 231:105–113

Litton CM, Sandquist DR, Cordell S (2008) A non-native invasive grass increases soil carbon flux in a Hawai'ian tropical dry forest. Glob Chang Biol 14:726–739

Lovett GM, Weathers KC, Sobczak WV (2000) Nitrogen saturation and retention in forested watersheds of the Catskill Mountains, New York. Ecol Appl 10:73–84

Lovett GM, Canham CD, Arthur MA et al (2006) Forest ecosystem responses to exotic pests and pathogens in eastern North America. Bioscience 56:395–405

Lovett GM, Arthur M, Weathers K, Griffin J (2010a) Long-term changes in forest carbon and nitrogen cycling caused by an introduced pest/pathogen complex. Ecosystems 13:1188–1200

Lovett GM, Arthur MA, Weathers KC, Griffin JM (2010b) Long-term changes in forest carbon and nitrogen cycling caused by an introduced pest/pathogen complex. Ecosystems 13:1188–1200

Mack MC, D'Antonio CM (2003) Exotic grasses alter controls over soil nitrogen dynamics in a Hawai'ian woodland. Ecol Appl 13:154–166

Matson PA, McDowell WH, Townsend AR, Vitousek PM (1999) The globalization of N deposition: ecosystem consequences in tropical environments. Biogeochemistry 46:67–83

Matzek V (2012) Trait values, not trait plasticity, best explain invasive species' performance in a changing environment. PLoS One 7:e48821

McEwan RW, Birchfield MK, Schoergendorfer A, Arthur MA (2009) Leaf phenology and freeze tolerance of the invasive shrub Amur honeysuckle and potential native competitors. J Torrey Botanical Soc 136:212–220

McManamay RH, Resler LM, Campbell JB, McManamay RA (2011) Assessing the impacts of balsam woolly adelgid (*Adelges piceae* Ratz.) and anthropogenic disturbance on the stand structure and mortality of fraser fir [*Abies fraseri* (Pursh) Poir.] in the Black Mountains, North Carolina. Castanea 76:1–19

Merritt DM, Shafroth PB (2012) Edaphic, salinity, and stand structural trends in chronosequences of native and non-native dominated riparian forests along the Colorado River, USA. Biological Invasions 14:2665–2685

Michaud J, Cordell S, Cole TC, Ostertag R (2015) Drought in an invaded Hawai'ian lowland wet forest. Pac Sci 69:367–383

Migge-Kleian S, McLean MA, Maerz JC, Heneghan L (2006) The influence of invasive earthworms on indigenous fauna in ecosystems previously uninhabited by earthworms. Biol Invasions 8:1275–1285

Miller JH, Lemke D, Coulston J (2013) The invasion of southern forests by nonnative plants: current and future occupation, with impacts, management strategies, and mitigation approaches. In: Wear DN, Greis JG (eds) Southern Forest Futures Project: technical report, Gen. Tech. Rep. SRS-GTR-178. USDA-Forest Service, Southern Research Station, Asheville, pp 397–456

Moore DJP, Trahan NA, Wilkes P et al (2013) Persistent reduced ecosystem respiration after insect disturbance in high elevation forests. Ecol Lett 16:731–737

Nagler PL, Glenn EP, Thompson TL (2003) Comparison of transpiration rates among saltcedar, cottonwood and willow trees by sap flow and canopy temperature methods. Agric For Meteorol 116:73–89

Nagler PL, Glenn EP, Didan K et al (2008) Wide-area estimates of stand structure and water use of *Tamarix* spp. on the Lower Colorado River: implications for restoration and water management projects. Restor Ecol 16:136–145

Nippert JB, Butler JJ, Kluitenberg GJ et al (2010) Patterns of *Tamarix* water use during a record drought. Oecologia 162:283–292

Nuckolls AE, Wurzburger N, Ford CR et al (2009) Hemlock declines rapidly with hemlock woolly adelgid infestation: impacts on the carbon cycle of southern Appalachian forests. Ecosystems 12:179–190

Oishi AC, Oren R, Novick KA et al (2010) Interannual invariability of forest evapotranspiration and its consequence to water flow downstream. Ecosystems 13:421–436

Orwig DA, Cobb RC, D'Amato AW et al (2008) Multi-year ecosystem response to hemlock woolly adelgid infestation in southern New England forests. Can J For Res 38:834–843

Owens MK, Moore GW (2007) Saltcedar water use: realistic and unrealistic expectations. Rangel Ecol Manag 60:553–557

Parker IM, Simberloff D, Lonsdale WM et al (1999) Impact: toward a framework for understanding the ecological effects of invaders. Biol Invasions 1:3–19

Pearson DE, Ortega YK, Maron JL (2017) The tortoise and the hare: reducing resource availability shifts competitive balance between plant species. J Ecol 105:999–1009

Peltzer DA, Allen RB, Lovett GM et al (2010) Effects of biological invasions on forest carbon sequestration. Glob Chang Biol 16:732–746
Peterson DL, Vose JM, Patel-Weynand T (2014) Climate change and United States Forests. Springer, Heidelberg, 261 p
Pratt RB, Black RA (2006) Do invasive trees have a hydraulic advantage over native trees? Biol Invasions 8:1331–1341
Prevey JS, Seastedt TR (2014) Seasonality of precipitation interacts with exotic species to alter composition and phenology of a semi-arid grassland. J Ecol 102:1549–1561
Pysek P, Jarosik V, Hulme PE et al (2012) A global assessment of invasive plant impacts on resident species, communities and ecosystems: the interaction of impact measures, invading species' traits and environment. Glob Chang Biol 18:1725–1737
Qiu JX (2015) A global synthesis of the effects of biological invasions on greenhouse gas emissions. Glob Ecol Biogeogr 24:1351–1362
Rennenberg H, Dannenmann M, Gessler A et al (2009) Nitrogen balance in forest soils: nutritional limitation of plants under climate change stresses. Plant Biol 11:4–23
Riscassi AL, Scanlon TM (2009) Nitrate variability in hydrological flow paths for three mid-Appalachian forested watersheds following a large-scale defoliation. J Geophys Res Biogeosci 114:G2
Ross RM, Bennett RM, Snyder CD et al (2003) Influence of eastern hemlock (*Tsuga canadensis* L.) on fish community structure and function in headwater streams of the Delaware River basin. Ecol Freshw Fish 12:60–65
Saha S, Sadle J, van der Heiden C, Sternberg L (2015) Salinity, groundwater, and water uptake depth of plants in coastal uplands of Everglades National Park (Florida, USA). Ecohydrology 8:128–136
Salemi LF, Groppo JD, Trevisan R et al (2012) Riparian vegetation and water yield: a synthesis. J Hydrol 454–455:195–202
Sardans J, Bartrons M, Margalef O et al (2016) Plant invasion is associated with higher plant-soil nutrient concentrations in nutrient poor-environments. Glob Chang Biol 23:1282–1291
Schmidt MW, Torn MS, Abiven S et al (2011) Persistence of soil organic matter as an ecosystem property. Nature 478:49–56
Smith HG, Sheridan GJ, Lane PNJ et al (2011) Wildfire effects on water quality in forest catchments: a review with implications for water supply. J Hydrol 396:170–192
Snyder BA, Callaham MA, Hendrix PF (2011) Spatial variability of an invasive earthworm (*Amynthas agrestis*) population and potential impacts on soil characteristics and millipedes in the Great Smoky Mountains National Park, USA. Biol Invasions 13:349–358
Stricker KB, Hagan D, Flory SL (2015) Improving methods to evaluate the impacts of plant invasions: lessons from 40 years of research. AoB Plants. plv028.
Strickland MS, Devore JL, Maerz JC, Bradford MA (2010) Grass invasion of a hardwood forest is associated with declines in below-ground carbon pools. Glob Chang Biol 16:1338–1350
Swank WT, Vose JM (1997) Long-term nitrogen dynamics of Coweeta forested watersheds in the southeastern United States of America. Glob Biogeochem Cycles 11:657–671
Tamura M, Tharayil N (2014) Plant litter chemistry and microbial priming regulate the accrual, composition and stability of soil carbon in invaded ecosystems. New Phytol 203:110–124
van Kleunen M, Weber E, Fischer M (2010) A meta-analysis of trait differences between invasive and non-invasive plant species. Ecol Lett 13:235–245
Vila M, Espinar JL, Hejda M et al (2011) Ecological impacts of invasive alien plants: a meta-analysis of their effects on species, communities and ecosystems. Ecol Lett 14:702–708
Vitousek PM (1990) Biological invasions and ecosystem processes: towards an integration of population biology and ecosystem studies. Oikos 57:7–13
Vitousek PM, Walker LR (1989) Biological invasion by *Myrica faya* in Hawai'i: plant demography, nitrogen-fixation, ecosystem effects. Ecol Monogr 59:247–265
Vose JM, Clark JS, Luce C, Patel-Weynand T (eds) (2016a) Effects of drought on forests and rangelands in the United States: a comprehensive science synthesis, General Technical Report WO-93b. USDA Forest Service, Washington Office, Washington, DC, 289 p
Vose JM, Miniat CF, Luce CH et al (2016b) Ecohydrological implications of drought. In: Vose JM, Clark JS, Luce C, Patel-Weynand T (eds) Effects of drought on forests and rangelands in the United States: a comprehensive science synthesis, General Technical Report WO-93b. USDA Forest Service, Washington Office, Washington, DC, pp 231–251
Wagner SA, Fraterrigo JM (2015) Positive feedbacks between fire and non-native grass invasion in temperate deciduous forests. For Ecol Manag 354:170–176
Webb JR, Cosby BJ, Deviney FA et al (1995) Change in the acid-base status of an Appalachian mountain catchment following forest defoliation by the gypsy moth. Water Air Soil Pollut 85:535–540
Webster JR, Morkeski K, Wojculewski CA et al (2012) Effects of hemlock mortality on streams in the southern Appalachian Mountains. Am Midl Nat 168:112–131
Webster J, Knoepp J, Swank W, Miniat C (2016) Evidence for a regime shift in nitrogen export from a forested watershed. Ecosystems 19:881–895
Wheeler MM, Dipman MM, Adams TA et al (2016) Carbon and nitrogen storage in California sage scrub and non-native grassland habitats. J Arid Environ 129:119–125
Wolkovich EM, Cleland EE (2010) The phenology of plant invasions: a community ecology perspective. Front Ecol Environ 9:287–294
Wurzburger N, Hendrick RL (2007) Rhododendron thickets alter N cycling and soil extracellular enzyme activities in southern Appalachian hardwood forests. Pedobiologia 50:563–576
Wurzburger N, Hendrick RL (2009) Plant litter chemistry and mycorrhizal roots promote a nitrogen feedback in a temperate forest. J Ecol 97:528–536
Wurzburger N, Miniat CF (2014) Drought enhances symbiotic dinitrogen fixation and competitive ability of a temperate forest tree. Oecologia 174:1117–1126
Yelenik SG, D'Antonio CM (2013) Self-reinforcing impacts of plant invasions change over time. Nature 503:517–520
Yin J, Fridley JD, Smith MS, Bauerle TL (2016) Xylem vessel traits predict the leaf phenology of native and non-native understorey species of temperate deciduous forests. Funct Ecol 30:206–214
Zavaleta E (2000) The economic value of controlling an invasive shrub. AMBIO J Hum Environ 29:462–467

Effects of Climate Change on Invasive Species

4

Deborah M. Finch, Jack L. Butler, Justin B. Runyon, Christopher J. Fettig, Francis F. Kilkenny, Shibu Jose, Susan J. Frankel, Samuel A. Cushman, Richard C. Cobb, Jeffrey S. Dukes, Jeffrey A. Hicke, and Sybill K. Amelon

4.1 Introduction

Mean surface temperatures have increased globally by ~0.7 °C per century since 1900 and 0.16 °C per decade since 1970 (Levinson and Fettig 2014). Most of this warming is believed to result from increases in atmospheric concentrations of greenhouse gases produced by human activity. Temperature increases have been greater in winter than in summer, and there is a tendency for these increases to be manifested mainly by changes in minimum (nighttime low) temperatures (Kukla and Karl 1993). Changes in precipitation patterns have also been observed, but are more variable than those of temperature. Even under conservative emission scenarios, future climatic changes are likely to include further increases in temperature with significant drying (drought) in some regions and increases in the frequency and severity of extreme weather events (IPCC 2007). For example, multimodel means of annual temperature from climate projections predict an increase of 3–9 °C in the United States over the next century combined with reductions in summer precipitation in certain areas (Walsh et al. 2014). These changes will affect invasive species in several ways. Furthermore, climate change may challenge the way we perceive and consider nonnative invasive species, as impacts to some will change and others will remain unaffected; other nonnative species are likely to become invasive; and native species are likely to shift their geographic ranges into novel habitats.

The ability to predict accurately how invasive species distributions and their impacts will change under projected climate scenarios is essential for developing effective preventive, control, and restoration strategies. Climate variables are known to influence the presence, absence, distribution, reproductive success, and survival of both native and nonnative species. Environmental selection for traits that enhance reproduction in warming climates will enable range expansion of some invasive species. Also, the availability of "empty" niches in the naturalized range, an escape from natural enemies, and a capacity to adapt to new habitats can

D. M. Finch (✉)
U.S. Department of Agriculture, Forest Service, Rocky Mountain Research Station, Albuquerque, NM, USA
e-mail: deborah.m.finch@usda.gov

J. L. Butler
U.S. Department of Agriculture, Forest Service, Rocky Mountain Research Station, Rapid City, SD, USA

J. B. Runyon
U.S. Department of Agriculture, Forest Service, Rocky Mountain Research Station, Bozeman, MT, USA

C. J. Fettig
U.S. Department of Agriculture, Forest Service, Pacific Southwest Research Station, Davis, CA, USA

F. F. Kilkenny
U.S. Department of Agriculture, Forest Service, Rocky Mountain Research Station, Boise, ID, USA

S. Jose
College of Agriculture, Food and Natural Resources, University of Missouri, Columbia, MO, USA

S. J. Frankel
U.S. Department of Agriculture, Forest Service, Pacific Southwest Research Station, Albany, CA, USA

S. A. Cushman
U.S. Department of Agriculture, Forest Service, Rocky Mountain Research Station, Flagstaff, AZ, USA

R. C. Cobb
Department of Natural Resources Management and Environmental Sciences, California State Polytechnic University, San Luis Obispo, CA, USA

J. S. Dukes
Forestry and Natural Resources and Biological Sciences, Purdue University, West Lafayette, IN, USA

J. A. Hicke
Department of Geography, University of Idaho, Moscow, ID, USA

S. K. Amelon
U.S. Department of Agriculture, Forest Service, Northern Research Station, Columbia, MO, USA

T. M. Poland et al. (eds.), *Invasive Species in Forests and Rangelands of the United States*,
https://doi.org/10.1007/978-3-030-45367-1_4

enhance an invader's ability to respond positively to climate change (Jarnevich et al. 2014).

In this chapter, we explain how the adaptive traits, genetic variability, and physiology of certain invasive species provide them with the competitive ability to grow, reproduce, and spread successfully under conditions of climate change. Our chapter offers examples of biological responses, distributional changes, and impacts of invasive species in relation to climate change and describes how these vary among plants, insects, and pathogens, as well as by species, and by type and extent of change. We also review attributes of plants, insects, and pathogens that enhance their ability to adapt to changes in hosts, native species, and environments affected by climate change.

Our assessment of the literature reveals that, for a given invasive species at a given location, the consequences of climate change depend on (1) direct effects of altered climate on individuals, (2) indirect effects that alter resource availability and interactions with other species, and (3) other factors such as human influences that may alter the environment for an invasive species. Manipulative experiments on invasive species, while uncommon, have shown that some species respond strongly to elevated carbon dioxide (CO_2) (e.g., Dukes et al. 2011) but less so to temperature and precipitation (Dukes et al. 2011). Insects are not directly affected by elevated CO_2, but they can be affected indirectly by responses of plants to CO_2. However, increasing temperatures can positively affect invasive insects by influencing their movements, growth rates, phenology, dispersal, and survival. Conversely, elevated temperatures also have the potential to affect invasive insects negatively by disrupting their synchrony with their hosts and altering their overwintering environments. Climate change can directly affect invasive pathogens through effects on formation of spores, host infection success, or selection pressures. For example, some invasive pathogens are sensitive to changes in timing and amount of precipitation and to changes in ambient temperature or humidity, whereas others are more responsive to changes in host stress. Briefly, effects of climate change on pathogens vary depending on how the change is expressed and how hosts are affected.

We describe and provide examples of how indirect effects of climate change are mediated through changes in habitats, hosts, other disturbances, trophic interactions, and land use or management. Our chapter provides information on how host-invasive species relationships and trophic interactions can be modified by climate change while recognizing that important knowledge gaps remain and need to be addressed. Our review revealed that disturbances (e.g., fire, storms) associated with, or exacerbated by, climate change can result in large releases of CO_2, an increase in bare ground available for invasions, and mortality of native species, all of which can potentially enhance invasive species performance. Similarly, management practices implemented in response to effects of disturbances and climate can alter the susceptibility to invasions in positive or negative directions (Chapter 7). For example, reseeding a disturbed area after a climate-related event with seed contaminated with an aggressive invasive plant like cheatgrass (*Bromus tectorum*) can unintentionally promote its spread.

This chapter covers the genetic basis of, and environmental selection on, several factors including (1) adaptive traits of invasive species, (2) evolutionary trends of invasive species in changing climates, and (3) interacting drivers and evolutionary responses of ecological communities to invasion. Climate change and invasive species are drivers of global environmental change that interact across biological communities in ways that have eco-evolutionary consequences. Successful invasions are dependent on the genetic makeup of a species. High levels of additive genetic variation tend to be linked to successful invasions (Crawford and Whitney 2010) and the ability of an invader to evolve in response to novel environments or changing conditions. Rapid adaptation to local climates can facilitate range expansions of invasive species (Colautti and Barrett 2013), even beyond the climatic distributions in their native ranges (Petitpierre et al. 2012).

We discuss and provide examples of how and why carbon cycling and carbon storage change, their relationship with insect outbreaks, and how climate can influence those changes. Insect and disease outbreaks can affect ecosystem-level carbon cycling and storage by reducing growth, survival, or distribution of trees. Under climate change, invasive organisms are likely to vary in their impact and rate of spread, depending on their sensitivities to climate variation and on the extent and type of climate change.

In order to manage invasive species under a changing climate, it is important to anticipate which species will spread to new habitats and when, and to understand how the characteristics of specific invaders may disrupt or have the potential to disrupt invaded ecosystems. Of utmost importance in containing the spread of invasive species, managers must have the ability to (1) predict which species will positively respond to climate change, (2) predict and detect sites likely to be invaded, and (3) deter incipient invasions before they are beyond control. We outline methods for developing the capability to predict and monitor invasive species in order to forecast their spread and increase their detection. Key findings and key research needs are included for each section.

4.2 The Influence of Climate Change on Invasive Species Distributions

At large spatial scales, climate variables are considered to be the dominant factors influencing species presence or absence (Thuiller et al. 2007). Relating climatic conditions to occurrence data is a widely used biogeographic approach to describe contemporary species distributions (Pearman et al. 2007; Peterson 2011) and for predicting how climate change may impact distributions (Guisan et al. 2014; Jeschke and Strayer 2008). The basic approach to predicting the potential geographic distribution of invasive species in their naturalized range involves developing statistical models that describe their native range in relation to climatic variables (their climatic niche) and then applying the models to the naturalized range (Broennimann and Guisan 2008; Early and Sax 2014; Jeschke and Strayer 2008). It is generally assumed that climatic models of native range distributions reflect interactions with nonclimatic factors (competition, predation, parasites, dispersability, edaphic factors, etc.) (Pearman et al. 2007).

The ability to predict the future distribution of invasive species in response to climate change is a complicated task, considering that numerous factors influence local and short-term patterns of invasion (Mainali et al. 2015), and because invasive species and concurrent climate and land-use changes are dynamically linked (Bellard et al. 2013; Smith et al. 2012). This is a linkage that already represents a significant component of global change (Vitousek 1994). At the broadest level, climate change may create conditions that favor the introduction of new invasive species into habitats where suitability was improved while altering local distribution and abundance of existing invasive species (Hellmann et al. 2008; Walther et al. 2009). Climate change is also likely to modify competitive interactions, resulting in native communities that are more or less susceptible to colonization by new invaders or expansion by established invaders. If the competitive ability of primary invaders is lessened by climate change, the ecological and economic impact of the invader may be reduced to the point where it would no longer be considered invasive (Bellard et al. 2013; Bradley et al. 2010; Pyke et al. 2008). Conversely, climate change-induced interchange of biotic interactions may also expedite the conversion of benign, resident nonindigenous species to invaders (Richardson et al. 2000). Climate change could also facilitate the increased abundance of secondary invaders by reducing the competitive ability of the primary invader or by altering the effectiveness of management strategies (Pearson et al. 2016). The significance of secondary invasions is increasingly being recognized, and it may arise either from invasive species subordinate to primary invaders (Pearson et al. 2016) or from the pool of nonindigenous species that often co-occur with the primary or secondary invaders (Kuebbing et al. 2013). Collectively, if climate change increases the abundance and distribution of some invasive species while decreasing or converting others, the net result may be no change in species richness of either invasive or nonnative species (Hellmann et al. 2008).

4.2.1 Altered Pathways for Invasive Species Introductions

Climate change will almost certainly alter pathways for the movement of invasive species on a global scale (Walther et al. 2009). During the invasion process, potential invasive species must successfully pass through a variety of environmental filters (Theoharides and Dukes 2007). However, invasive species must first overcome major geographic barriers to their spread, which is currently facilitated largely by human activities (Lehan et al. 2013) and likely will be enhanced under climate change (Pyke et al. 2008; Seebens et al. 2015). Many existing and potential invasive species spread into new areas as stowaways in and on cargo ships (in cargo holds, containers, or ballast water; as contaminants in agricultural crops; or on ships hulls) (Hulme 2009). In the United States, current inspection of cargo ships for invasive species involves examining a small percentage of cargo imports for a small subset of federally listed species while leaving the vast majority unchecked; some of these overlooked species could potentially become invaders under a scenario facilitated by climate change (Lehan et al. 2013).

Global warming is reducing the extent and thickness of sea ice, resulting annually in more open water for longer periods of time (Liu et al. 2013; Stroeve et al. 2012). For invasive species that may arrive as stowaways, reductions in ice pack affecting both oceanic and freshwater shipping routes have globally extended shipping seasons while reducing travel time for cargo ships; this will likely increase survival rates of propagules and potentially enhance the probability of establishments in the new range (Pyke et al. 2008). Loss of sea ice has also increased the frequency and ease of movement of marine birds and mammals between the Pacific and Atlantic Ocean basins (McKeon et al. 2016). This could enhance long-distance dispersal of many sessile organisms (Viana et al. 2016) that potentially could become invasive in newly created suitable climates in the invaded range. Additionally, migration times of birds have been shown to be impacted by climate change (see review by Miller-Rushing et al. 2008), which may play a role in expanding the distribution of aquatic invasive species carried by avian vectors (Coughlan et al. 2015; Reynolds et al. 2015).

Another mechanism for short- and long-term dispersal of invasive species is the increased frequency, intensity, and duration of extreme weather events that are correlated with climate change (IPCC 2007; Melillo et al. 2014). Hurricanes and other strong winds can carry invading propagules, insects, marine larvae, and birds; reduce existing vegetation; and create bare soil, enhancing opportunities for colonization (Michener et al. 1997; Richardson and Nemeth 1991; Schneider et al. 2005; Walther et al. 2009). Frequent and more expansive flooding associated with climate change would likely increase connectivity among different habitats for invasive species, although this pathway has received little attention.

There is a 400-year history of importing and cultivating introduced grasses, forbs, shrubs, and trees into the United States for ornamental or agricultural purposes (Mack and Erneberg 2002; Reichard and White 2001). Although the vast majority of the plants intentionally introduced are not invasive (Reichard and White 2001), deliberate introductions are the primary source of invasive plants, especially for trees and shrubs, in the Eastern United States (Lehan et al. 2013). The risk of ornamentals escaping and becoming invasive escalates among an increasingly urbanized and affluent population that has an aversion for ornamental lawns and gardens (Marco et al. 2010). This is compounded by minimal legal restrictions to oversee the introduction of plants into the United States (Reichard and White 2001), and a complete disconnect between the financial benefit realized by the horticultural industry from selling imported plants and the economic and ecological costs attributed to escaped ornamentals that become invasive (Barbier et al. 2011). Typically, plants selected for introduction for ornamental or agricultural purposes have broad climatic tolerances and phylogenetic traits that favor their rapid establishment and growth, thus enhancing their potential for invasiveness in response to climate and land-use changes (Bradley et al. 2010). Likewise, under climate change, the demand for introduced plants that can better tolerate drought and high temperatures is likely to increase (Bradley et al. 2012), which will dramatically increase propagule pressure of potential invasive species (Lockwood et al. 2005). Meanwhile, native plants may experience "migration lag" to climate change (sensu Corlett and Westcott 2013), which is likely to put them at a competitive disadvantage, thereby creating vegetation gaps potentially filled by introduced species. Increased introductions of phenotypically plastic, preadapted, nonnative species (Turner et al. 2015), coupled with a lag in native species migration (Corlett and Westcott 2013), are likely to lead to the creation of novel communities that possess unknown ecological characteristics (Bernard-Verdier and Hulme 2015).

Even without considering the effects of global change on invasive species distributions, humans have deliberately or accidentally moved thousands of species beyond their native ranges.

Published estimates of the number of introduced species in the United States range from 4000 (Stein and Flack 1996) to 5000 species (Morse et al. 1995). Currently, only a small fraction of the pool of introduced species is classified as invasive, defined as adversely impacting native species, communities, and ecosystems (Hiebert 1997; Skinner et al. 2000). However, it is proposed that climate change will enhance and accelerate pathways for new introductions and, thus, dramatically increase the risk of invasion by potentially damaging species (Bradley et al. 2012; Hellmann et al. 2008). Developing the proficiency to predict which species will successfully emerge among the next wave of invaders is the subject of increased experimental research and species distribution modeling. Published literature on the impact of climate change on invasive species distribution has increased substantially since 2000, and most is focused on North America (see review by Smith et al. 2012).

4.2.2 Distribution Changes During Invasion

The tendency of invaders to inhabit similar climatic niches in both the native and introduced ranges was confirmed in a large-scale survey of 50 terrestrial plant invaders (Petitpierre et al. 2012). However, other studies have shown that invasive species are able to successfully establish and reproduce in climates different from those found in their native range (Beaumont et al. 2009; Bradley et al. 2015; Broennimann et al. 2007). Incorporating multiple-scale climatic, biotic, and land-use variables into distribution models can improve the models' performance in predicting changes in the distribution of invasive species in response to future climates (Jarnevich et al. 2014; Pearson and Dawson 2003), especially if monitoring data were available to validate prediction models (Jones 2012; Sheppard et al. 2014; Smith et al. 2012).

While many species distribution models can successfully predict areas of potential introduction, the lack of nonclimatic data in these models often inhibits their ability to predict the total extent of invasion in the naturalized range (Bradley et al. 2015). For example, Broennimann et al. (2007) reported a dramatic climatic niche shift for spotted knapweed (*Centaurea maculosa*), an aggressive plant invader in Western North America. However, nonclimatic factors, especially lack of natural enemies in the native range, may play a major role in the climatic expansion of invasive plants, as observed with spotted knapweed (Corn et al. 2006; Maines et al. 2013; Seastedt et al. 2007; Story et al. 2006).

Once established in a community, invasive species, along with native and nonnative constituents, must track future climate change in order to survive (Corlett and Westcott 2013). The consequences of not doing so may be less severe for invasive species than for native species because of partial or total release of abiotic and biotic constraints. The availability of empty niches in the naturalized range, coupled with adaptive plasticity and evolutionary changes, can enhance the ability of invasive species to shift into new habitats and climates (Higgins and Richardson 2014; Jarnevich et al. 2014; Kumschick et al. 2013; Turner et al. 2015). However, documenting evolutionary niche shifts is likely only possible when invasive species have occupied their naturalized ranges for relatively long periods of time. Broennimann et al. (2007) suggested that for spotted knapweed, this period is likely greater than 120 years, although the timeframe is probably strongly influenced by propagule pressure driven by multiple introductions (Lockwood et al. 2005).

4.2.3 Interacting Drivers of Global Environmental Change and Evolutionary Responses of Native Ecological Communities to Invasion

Interactions between species play an important role in structuring ecological communities, and these interactions are often climate dependent (Dunson and Travis 1991; Norberg 2004). A review of 688 published studies showed that drivers of global environmental change, including climate change and species invasions, are likely to influence virtually every type of species interaction across all terrestrial ecosystems (Tylianakis et al. 2008). Due to the immense complexity of global ecological communities, it is extremely difficult to decipher the higher order effects of interactions between drivers of environmental change, which can both mitigate and exacerbate one another. However, Tylianakis et al. (2008) were able to glean some generalities across terrestrial ecosystems, indicating that these drivers will generally (1) reduce the strength of positive mutualisms (i.e., where two species benefit in a relationship) involving plants, such as seed dispersal and pollination; (2) have effects on soil food webs that will shift ecosystem process rates, including for beneficial ecosystem services; (3) change the relative abundance and dominance of all taxa across ecological networks; (4) increase herbivory rates from all animal taxa, which will be mitigated only by variable levels of attack by predators or parasites; and (5) lead to a higher frequency and severity of pathogen effects on plants and animals across ecosystems.

Theoretical and applied models are being developed to help disentangle and better predict the effects of multiple drivers of global change on ecological communities (Gilman et al. 2010). However, few of these models take into account the effect of evolution on ecological interactions. Invasion by nonnative species can act as a novel and strong selective pressure for native species that compete with invaders for resources or interact with them in other ways (Leger and Espeland 2010). Native species that survive initially aggressive invasions are likely to compete with the invaders and enter into coevolutionary feedback loops. For example, some populations of the red-bellied black snake (*Pseudechis porphyriacus*) in Australia, a predator that has suffered population declines due to invasion of the cane toad (*Bufo marinus*), and therefore faces strong selection pressure, have evolved a physiological tolerance to cane toad toxin, which is normally lethal (Phillips and Shine 2006). In another example, invasive giant hogweed (*Heracleum mantegazzianum*) became less dominant in Eastern European sites that had been invaded the longest (~30 years); common garden soil inoculation experiments indicated that negative plant–soil feedbacks may be responsible, possibly through the evolution of soil microbiota (Dostál et al. 2013). Knowledge of the evolutionary responses of native species to invaders can help us better understand how ecological communities might resist invasion. To gain insight, novel models of resistance and resilience will need to be constructed that take into account the potential impact of climate change on eco-evolutionary processes.

As invasive species populations expand in space (overcoming geographic barriers) and time (under climate change), they are exposed to novel environments and, therefore, subject to different selection pressures as compared to those that occur in their native ranges (Mooney and Cleland 2001). Rapid adaptation to novel environments can provide both native and introduced species opportunities for expansion under a changing climate (Barrett 2000); however, introduced species that have escaped their natural enemies may have a competitive advantage (Blossey and Nötzold 1995), thus increasing their likelihood for spread. Furthermore, evidence indicates that populations of some invasive species may arrive in their introduced range preadapted to conditions extant in the new range. For example, spotted knapweed occurs as both diploid and tetraploid cytotypes in its native range in Europe, but currently, only the tetraploid cytotype has been reported in its introduced range in North America (Mráz et al. 2014). The tetraploid cytotype has higher drought tolerance than the diploid cytotype (Mráz et al. 2014), which might contribute to its successful invasion success in Western North America where severe drought frequently occurs.

Preadaptation, genetic variation, hybridization, and multiple introductions provide the raw material that allows introduced populations to adapt rapidly to broad-scale and dynamic environmental conditions. Kumschick et al. (2013) reported that invasive common mullein (*Verbascum thapsus*) has evolved into a fast-growing phenotype with a strong response to a wide array of abiotic conditions, with and without competition. This may provide a competitive advantage in environments where precipitation is variable. Similarly, Turner et al. (2015) found that populations of diffuse knapweed (*Centaurea diffusa*) that occurred in its introduced range were genetically and phenotypically different from populations existing in its native range. The collective evidence indicates that environmental selection for genotypes that can grow and reproduce under broad climatic conditions will provide opportunities for existing invasive species to expand eventually into a greater diversity of environments.

4.2.4 Key Findings

Invasive species tend to have high dispersal rates, rapid growth rates with short generation times, and high capacity to tolerate broad environmental conditions. Collectively, these traits greatly enhance their ability to cope with rapid changes in abiotic and biotic conditions such as those associated with climate change. When these traits are compounded by the existence of ineffective international trade regulations, climate change-induced alterations in geographic barriers, temperature and moisture constraints, and biotic interactions, considerable evidence suggests that climate change will almost certainly lead to changes in the distribution of invasive species. Predicting how invasive species will respond under predicted climate change scenarios is a serious challenge but essential to developing effective strategies for preventing and controlling invasive species and for restoring invaded habitats.

An extensive review showed that drivers of global environmental change will influence every type of species interaction across all terrestrial ecosystems. Theoretical and applied models are being developed to help disentangle the effects of multiple drivers of global change on ecological communities, but few of these models take into account the effect of evolution on ecological interactions. Native species that survive initially aggressive invasions are likely to compete with the invaders and enter into coevolutionary feedback loops.

4.2.5 Key Information Needs

Current guidelines for detecting and managing new and existing invasive species that have the potential to spread have limited applicability under scenarios that include climate and land-use changes. Research is needed that evaluates population- and landscape-level responses of invasive species to multiple spatial and temporal stressors and disturbances when they are operating simultaneously (e.g., extreme climatic events in relation to increased global commerce and changes in fire regimes). The ability to project accurately how future invasive species distributions respond to climate change is usually enhanced when ensembles of climate envelope models are used in conjunction with multiple climate change scenarios. Field evaluation of predictions is essential for improving model performance. Understanding the evolutionary responses of native species and species assemblages to invaders can help us understand how ecological communities might respond to invasion.

Theoretical models on interacting drivers of global change, including changing climates and invasive species, need to incorporate explicitly the influence of evolutionary processes. Extensive research will be necessary to explain coevolutionary feedback loops between native communities and invaders, specifically in the context of climate change.

4.3 Adaptive Responses of Invasive Species to a Changing Climate

One of the lessons learned from evolutionary biology is that evolutionary processes, such as natural selection and genetic drift, often happen at ecologically relevant time scales (Carrol et al. 2007) and that these evolutionary processes can drive change over only a few generations, often well within a human lifetime (e.g., Kilkenny and Galloway 2013; Phillips et al. 2006). When environmental fluctuations occur within a narrow time frame, as with normal year-to-year variation, directional evolutionary shifts tend to counteract one another so that characteristics of species and populations remain relatively stable for long periods (Grant and Grant 2002). However, when the biotic or abiotic environment that a species experiences moves beyond stable boundaries, either through prolonged directional change or increased variability, then long-term evolutionary changes are likely to occur (Carrol et al. 2007). Climate change and invasive species are drivers of global environmental change that are likely to interact across ecological communities in ways that will have long-lasting eco-evolutionary impacts.

4.3.1 Selection on Adaptive Traits and Evolutionary Trends in Changing Climates

Phenotypic variation in climate adaptation can arise across a species' invaded range during the invasion process. As individuals move into new areas, selection is expected to operate on traits by favoring individuals that can thrive under local biotic and abiotic conditions (Parker et al. 2003). For example, a number of studies have shown that invasive plant species can adapt rapidly to local conditions along environmental clines that can be regional or range wide (e.g., Alexander et al. 2009; Kooyers and Olsen 2012; Maron et al. 2004). Rapid adaptation to local climates can facilitate range expansion and overcome early environmental constraints on propagule production (Colautti and Barrett 2013). Indeed, adaptation to local environments can facilitate invasive species expanding into areas that possess novel environmental conditions that they did not experience in their native habitats (Broennimann et al. 2007). A comprehensive meta-analysis showed that 15% of invasive species show evidence of range shifts that are significantly outside the climatic distributions in their native ranges (Petitpierre et al. 2012). While the authors of this meta-analysis considered 15% to be "rare," it nevertheless suggests that a large number of invasive species are adapting to novel climate conditions and may continue to do so. For example, ongoing adaptation of populations of the invasive vine Japanese honeysuckle (*Lonicera japonica*) at the margin of its expanding northern range in Eastern North America suggests that this species will continue to spread under changing climatic conditions (Fig. 4.1; Kilkenny and Galloway 2016).

Similar selection pressures may be operating for plant species such as Lehmann lovegrass (*Eragrostis lehmanniana*), which was introduced for agricultural purposes.

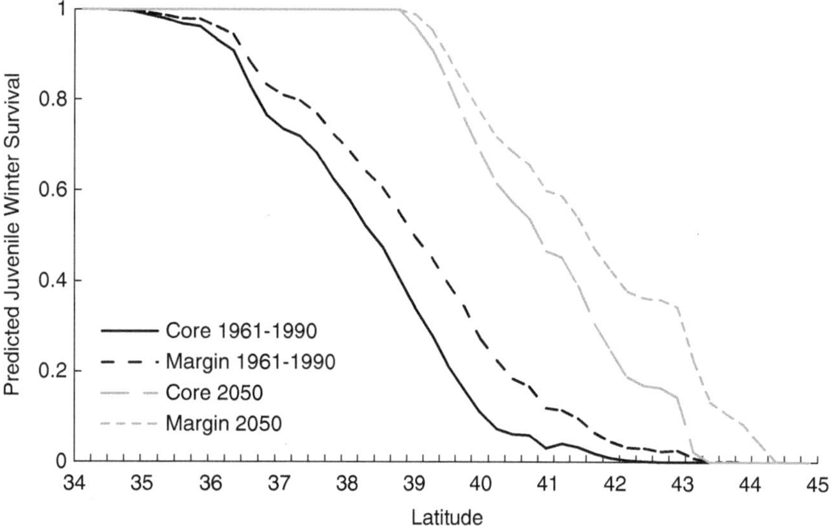

Fig. 4.1 Predicted juvenile winter survival of Japanese honeysuckle (*Lonicera japonica*) by latitude under recent climate normals (1961–1990) and a consensus of future climate scenarios for 2050 (scenario A2 for CCCMA, CSIRO, and HADCM3) for populations sourced from the core and margin of the invaded range (From Kilkenny and Galloway 2016)

Lehmann lovegrass was selected and introduced specifically for its drought tolerance, which likely will favor its success in drier conditions that are projected for the Southwestern United States (Archer and Predick 2008). The primary expansion of buffelgrass (*Pennisetum ciliare*), another neotropical species introduced in the Southwestern United States, began during the warmer winters in the 1980s, and buffelgrass continues to expand upslope with increasing winter temperatures (Archer and Predick 2008). Similar expansions of buffelgrass are predicted to occur in Australia under a climate change scenario (Martin et al. 2015).

The process of invasion and range expansion can select for traits that promote colonization itself, such as those that enhance dispersal (e.g., Lombaert et al. 2014), enhance inter specific competition (e.g., Lankau et al. 2009), or promote more rapid reproduction (e.g., Kilkenny and Galloway 2013). Dispersal traits may be particularly important, because individuals with the greatest dispersal abilities are more likely to initiate new populations, which can lead to "spatial sorting" (Travis and Dytham 2002) and widespread selection on dispersal traits across the expansion front (Chuang and Peterson 2016).

Theoretical studies indicate that selection based on dispersal ability and adaptation to local climates will likely interact in complex ways under changing climates. Hargreaves et al. (2015) demonstrated that dispersal ability is generally only favored in situations where dispersal has little to no direct fitness costs, such as low survival of long-distance dispersers, regardless of whether the climate is stable or changing. However, when local adaptation is taken into account, the situation becomes more complex. In a stable climate, local adaptation will limit selection based on dispersal ability, even if there are no other direct fitness costs. But in a changing climate, local adaptation may enhance selection for dispersal ability even if it does result in some fitness costs because populations or ecotypes are likely to be at an advantage if they can spatially track the climates they are most adapted to.

While the occurrence of adaptive processes is widespread during invasions, repeated founder events can drive nonadaptive evolution and mimic some of the patterns characteristic of climatic adaptation (Keller and Taylor 2008). This is because founding propagules are more likely to establish in climates similar to those that match their source environment. Therefore, what appears during an invasion to be adaptive evolution may simply reflect sampling processes that have led to the foundation of invading populations that contain preadapted genotypes (Keller et al. 2009). Understanding how adaptive and nonadaptive processes contribute to the distribution of an invasive species may be critical in predicting how that species might spread under climate change, because this balance can affect future evolutionary potential.

4.3.2 Key Findings

Evolutionary processes can drive rapid change in species but tend to cancel out over time under stable environmental conditions. When the environments are no longer stable, long-term evolutionary change is more likely. Drivers of global environmental change are likely to interact in ways that will produce long-lasting eco-evolutionary impacts.

As species invade new areas, natural selection will favor individuals that can thrive under biotic and abiotic conditions common to the new habitat. As much as 15% of all invasive species are adapting to climate conditions that they never experienced in their native ranges. The process of invasion can select for traits that promote colonization, including traits that enhance dispersal and rapid establishment.

Theoretical studies indicate that selection on dispersal ability and adaptation to local climates will likely interact in complex ways under changing climates. Repeated founder events can drive nonadaptive evolution during invasions and can mimic some of the patterns of climatic adaptation.

4.3.3 Key Information Needs

Research is needed to determine how and to what extent climate change acts as a selection process on invasive species. Acquiring knowledge on the interaction between climate change and genetic processes such as hybridization and polyploidization will be essential to our ability to predict how invasive species adapt to climate change. Understanding how climate change influences natural selection on invasive species undergoing range expansion, including selection on dispersal traits, will improve our ability to manage spreading populations effectively.

4.4 Impacts of Climate Change on Physiology, Survival, Productivity, Phenology, and Behavior of Invasive Plants, Insects, and Pathogens

4.4.1 Impacts on Invasive Plants

Changes in the climate and atmosphere are provoking a wide variety of responses from invasive plants (for in-depth reviews, see Blumenthal and Kray 2014; Bradley et al. 2010; Dukes 2011; Leishman and Gallagher 2016). The consequences of climate change for a given plant species at a particular location depend on three types of mechanisms: direct effects of climate change that alter physiology, growth, and survival; indirect biological effects that alter resource availability, competition, herbivory, disease, and resistance to human management; and indirect societal effects that may alter the value of resources affected by the invasive species, and thus the degree to which the species is subjected to human management. In a given location, any one of these mechanisms may have the greatest influence. These local-scale concerns, though, occur in the context of the larger landscape; propagules of species (including the invasive species) move around, and the composition of communities can change with time. These changes also have consequences for invasive plant species.

At a basic level, many studies have examined the direct effects of climate and atmospheric change on a variety of invasive plant species grown in isolation (Dukes 2000; Leishman and Gallagher 2016; Sorte et al. 2013; Verlinden and Nijs 2010; Ziska and Dukes 2011). Elevated CO_2 generally favors invasive plants, though not necessarily much more than natives, and changes in warming and precipitation can favor or disfavor them, depending on the magnitude of change (and in the case of precipitation, the direction of change as well). However, in natural and managed ecosystems, these direct effects do not occur in isolation. Field manipulations in which both the invasive plants and their surrounding communities experience simulated future conditions unavoidably include at least some of the indirect biotic effects that will influence the success of invasive species. Relatively few of these studies have been conducted, but some show strong responses, such as those of the invasive forbs yellow star-thistle (*Centaurea solstitialis* L.) (Dukes et al. 2011) and Dalmatian toadflax (*Linaria dalmatica* (L.) Mill.) (Blumenthal et al. 2013), to elevated CO_2 in annual grassland and mixed-grass prairie, respectively. No realistic studies of this type have been conducted on invasive trees, but an invasive shrub responded positively to elevated CO_2 (Belote et al. 2003) in a forest plantation in Tennessee. Such responses may depend on other conditions, such as soil moisture (Smith et al. 2000). The response of invasive species to warming and precipitation manipulations have generally been less dramatic than their responses to CO_2 (e.g., Blumenthal et al. 2013; Dukes et al. 2011; Maron and Marler 2008), but precipitation changes can sometimes have important consequences in concert with other environmental or biological factors (Blumenthal et al. 2008; Suttle et al. 2007).

Potential responses of species' distributions to changes in mean climatic conditions can be predicted using a variety of habitat suitability models; these models have been applied to a growing number of invasive plant species (Bradley 2014). Distributions of invasive plant species are generally projected to expand outward from the current colder edges of their habitat and shift away from the warmer edges, leaving potential opportunities for preemptive restoration (Bradley and Wilcove 2009). However, these models make a variety of assumptions and only provide predictions of potential range—the expansion of species into these ranges would depend on numerous other factors.

Recent evidence indicates that many invasive species occupy unique phenological niches and track climate change more closely than native species (Willis et al. 2010; Wolkovich and Cleland 2011, 2014; Wolkovich et al. 2013). Willis et al. (2010) discussed how phenological flexibility and the existence of vacant niches may contribute to the success of nonnative species under conditions of climate change, because those species with the most flexible phenologies also flowered earlier than native plants that had not responded to earlier warming. In a study across five North American sites, Wolkovich et al. (2013) found that nonnative species shifted flowering in relation to climate change while native species, on average, did not. They also reported that in mesic systems, invasive species exhibited greater tracking of interannual variation in temperature than did native species, while in grasslands, invasive species differed from native species primarily in their responses to precipitation and soil moisture but not to temperature. Their findings provide cross-site support for explaining the role of phenology and climate change and possibly for predicting species invasions. In addition, because climate change may alter the timing and severity of ecosystem stress and disturbance, it could provide unique opportunities for invasion (Fig. 4.2) (Wolkovich and Cleland 2014). Based on experimental manipulations of climate change, Wainwright et al. (2012) suggest that managers might be able to trick phenotypically plastic invasive plants into germinating earlier than appropriate for the local climate, resulting in reduced survivorship, but this is yet to be demonstrated at large scales.

Recipient plant communities will likely become more susceptible to climate change (through a reduction of the biotic resistance of the communities) as the climate becomes less optimal for resident species (Dukes and Mooney 1999) and as extreme events become more frequent and more disruptive to the resident community (Diez et al. 2012). Because many invasive species are able to disperse rapidly over long distances, they may be able to rapidly colonize areas with lowered resistance, such as those disturbed by extreme events. The relatively broad climatic tolerances found in many invasive species (Bradley et al. 2015) may confer greater tolerance to changing climatic conditions than what is typically observed in native species.

The effectiveness of techniques for managing some invasive plant species may be affected by climate and atmospheric changes (Ziska and Dukes 2011). When exposed to enriched CO_2, some invasive plants became more tolerant to the widely used herbicide glyphosate (Manea et al. 2011; Ziska et al. 2004). Biocontrol species may also be affected by climate change, thus potentially altering their effectiveness (Hellmann et al. 2008).

Finally, as climatic disruption progresses, the management of invasive plant species may change (Dukes 2011), either because they are considered to be more harmful under climate change, for example, if they are perceived to deplete a resource such as water that increases in value, or because they are perceived to have greater value, for example, if they are selected to be grown for bioenergy.

4.4.2 Impacts on Invasive Insects

The physiology of insects is highly sensitive to temperature and climate warming, and thus climate change is predicted to be largely beneficial to invasive insects, as least directly (Bale et al. 2002; Deutsch et al. 2008). Warming tempera-

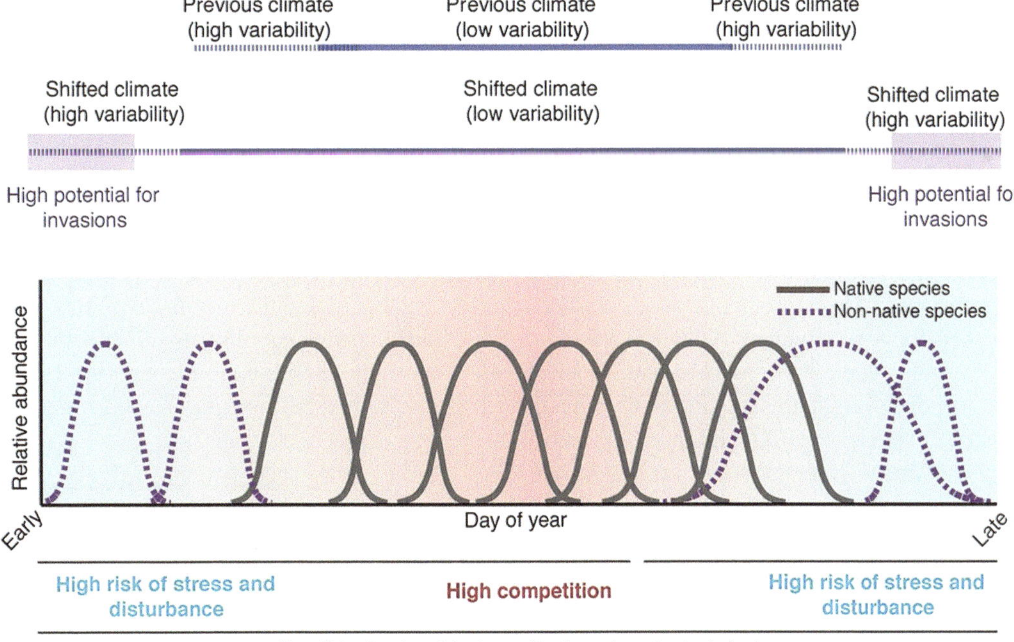

Fig. 4.2 Conceptual model of a hypothesized mesic temperate system showing idealized niche diagrams for four nonnative species (*dashed-line distributions*) and seven native species (*gray distributions*) where temperature limits viable periods for plant growth. Variation in stress, disturbance, and competition may dictate optimal phenological strategies, with benefits for early- and late-flowering invasive species. With climate change extending viable periods for plant growth (*dark blue lines*), nonnative species with highly plastic phenologies may have an increased opportunity for invasion at the start and end of the growing seasons (From Wolkovich and Cleland 2014)

tures tend to increase insect movement, feeding rate, growth rate, dispersal, and survival (Bale et al. 2002), but very high temperatures can sharply reduce fitness of insects (Deutsch et al. 2008). The distribution and abundance of insects respond quickly to climatic change owing to insects' high mobility, short generation time, physiological sensitivity to temperature, and high reproductive potential (Weed et al. 2013). For example, the distribution of the invasive hemlock woolly adelgid (*Adelges tsugae*) in the Eastern United States is currently limited by cold winter temperatures; but predicted future warming could allow hemlock woolly adelgid to spread unchecked throughout the range of hemlock (*Tsuga* spp.) in North America (Dukes et al. 2009). Additionally, climate affects tree defenses, tree tolerance, and community interactions involving enemies, competitors, and mutualists of insects and diseases (Weed et al. 2013).

Warming could negatively affect invasive insects, on the other hand, by disrupting developmental synchrony with their host plants (Bale et al. 2002; see biocontrol discussion in Sect. 4.7.2). Moreover, negative effects can occur if warming leads to a reduction of insulating snow cover, thus exposing overwintering life stages to freezing conditions (Bale and Hayward 2010). There is evidence that periods of extreme winter warm snaps followed by extreme cold can kill emerald ash borer (*Agrilus planipennis*) larvae (Sobek-Swant et al. 2012). Despite the great influence of temperature on insects, we still cannot confidently predict how climate warming will affect most invasive species.

Unlike plants, insects are not directly affected by elevated CO_2 (Guerenstein and Hildebrand 2008); but insects, especially herbivores, can be indirectly affected through responses of plants to CO_2. For example, most plants grown under a regime of elevated CO_2 have a higher carbon-to-nitrogen ratio in their tissues which dilutes nutrient content, most notably nitrogen which is a factor that usually limits growth of insects (Coviella and Trumble 1999). As a result, insects must consume more plant tissue to obtain enough nutrients for growth and development (Cannon 1998; Coviella and Trumble 1999; Dermody et al. 2008; Johnson and McNicol 2010). However, in some forests, the amount of insect feeding can be reduced under elevated CO_2 (Knepp et al. 2005). Plant secondary chemistry—a central factor regulating herbivore growth and survival—can be affected by most climate change factors, including elevated CO_2, temperature, ozone (O_3), drought, and ultraviolet (UV) light (Bidart-Bouzat and Imeh-Nathaniel 2008; Burkle and Runyon 2016; Jamieson et al. 2017; Kolb et al. 2016; Yuan et al. 2009). A plant's ability to enact chemical defenses in response to feeding by invasive herbivores can also be altered by climate change (Zavala et al. 2008). However, information available on climate-induced changes in plant chemistry is limited, and the response (increase, decrease, or no effect) is dependent on the plant species involved as well as the class of chemicals examined (Bidart-Bouzat and Imeh-Nathaniel 2008; Lindroth 2010). Additionally, the response to climate change varies with herbivore species. A recent meta-analysis found that in response to elevated CO_2, the abundance of some arthropod herbivore groups increased (e.g., mites and thrips) whereas others decreased (e.g., Lepidoptera and leaf miners) (Robinson et al. 2012). These highly context-dependent and species-specific findings have hindered our attempts to identify general patterns.

4.4.3 Impacts on Invasive Pathogens

There are numerous microbes that are considered to be serious pathogens of ecological communities, and several factors, that is, their small physical size, cryptic symptoms, and poor detection methods, have contributed to rapid and extensive invasions. Invasive microbes are among some of our most destructive forest pathogens, and most of them were initially introduced or their spread was facilitated by outplanting of infected nursery stock, international plant trade, or poor nursery cultural practices (Anagnostakis 1982; Maloy 1997; Rizzo et al. 2005). While there is a clear need and strong desire to improve cultural practices in the nursery industry to reduce pathogen invasions (Brasier 2008), there is also an inescapable need to understand why some of the many established pathogens eventually emerge as disease agents or suddenly expand into new locations and hosts. In our effort to understand the interaction of invasive pathogens with climate change, it is important to distinguish the condition (disease) from the biological agents (pathogens) involved. The emergence of disease can almost always be framed as a three-way interaction among pathogens, their hosts, and the environment (Fig. 4.3). This disease triangle, sometimes expanded to include the effects of management, can be used to demonstrate how environmental change can affect pathogens and thereby disease (Alexander 2010).

Climate change can directly affect pathogen populations by controlling sporulation, affecting the likelihood of successful infection, or imposing selection on pathogen popula-

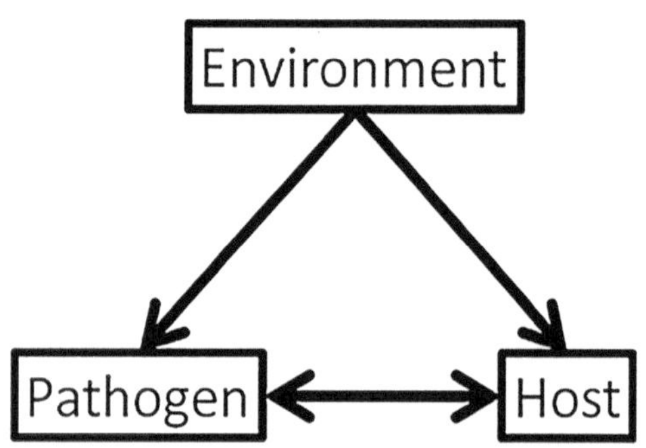

Fig. 4.3 The three-way interactions of the disease triangle

tions for certain phenotypic characteristics (Davidson et al. 2008; Eyre et al. 2013; Woods et al. 2005). Over the last decade, extensive and severe drought has been a major focus of research and management in forests of Western North America (Vose et al. 2016); while our understanding of drought–pathogen interactions is notable for many uncertainties (Desprez-Loustau et al. 2006; Kolb et al. 2016; Sturrock et al. 2011), broader climate change such as changes in temperature or the timing and type of precipitation (rain vs. snow) can also alter host–pathogen relationships in ways that foster disease emergence (Sturrock et al. 2011). Fungi are without doubt the most important group of forest pathogens and are very sensitive to the timing and quantity of precipitation, ambient temperature, relative humidity, and other factors that influence leaf surface or soil moisture (Davidson et al. 2008; Meentemeyer et al. 2011; Woods et al. 2005). Because fungi are such a diverse group of microbes, it is essential to evaluate each fungus-caused disease problem on a case-by-case basis. For any analysis of risk or threat posed by invasive pathogens, the specific biology of each organism is critical in determining how climate change will interact with each respective organism. Unfortunately, that biological understanding is weak for many important groups of damaging fungal microbes (Hansen 2015). This has led to a number of ecological surprises where a pathogen was thought to be insignificant in one environment but was very destructive in another. This can be due to congeneric hosts in the new environment but also due to direct effects of environmental change. Where climate change increases the sporulation, growth rate, or survival of individual pathogens, it is possible that unanticipated disease epizootics can occur. For example, when shifts in precipitation forms (greater precipitation in rain vs. snow) occur, foliar pathogen outbreaks can be unprecedented (Woods et al. 2016). Unplanned, natural experiments using nonnative timber species planted in environments warmer or wetter than their native range can also result in unexpected pathogen outbreak, sometimes with devastating impacts to timber resources (Brasier and Webber 2010).

The effects of climate change on pathogen-caused diseases can also be facilitated indirectly by their effect on host plants. Plant stress resulting from decreased carbon reserves or loss of hydraulic function (Adams et al. 2009; Anderegg et al. 2012; Hartmann et al. 2013) can alter plant defensive responses in ways that can increase plant susceptibility to infection and possibly enhance subsequent spread (Bostock et al. 2014). Different plant pathogens impact different plant parts and utilize varying modes of infection (Oliva et al. 2014). Some plants may become more susceptible to initial infection during periods of environmental stress, such as drought, and, thus, facilitate an increase in spread rates of invasive pathogens. Alternatively, plants may become more likely to be damaged by previously established pathogens or by unremarkable, but possibly nonnative, endophytic microbes (Stergiopoulos and Gordon 2014). Our superficial understanding of existing microbial communities in wild plants and in wildland ecosystems is a significant barrier to our ability to predict the emergence of diseases, because many widespread invasive pathogens will be detected only after a plant health problem emerges. Thus, climate change, through its influence on host physiology (McDowell et al. 2011), is likely to divulge the presence of potentially damaging, invasive, microbial pathogens which are widespread within populations, but only after eradication is no longer feasible (Filipe et al. 2012).

Our success in managing disease-causing organisms will depend on our ability to predict their occurrence under changing climate conditions and to attack vulnerable points in the disease cycle (i.e., weak links in the infection chain). Thorough understanding of the disease cycle, including the climatic and other environmental factors that influence the cycle, is essential to effective management.

4.4.4 Key Findings

All aspects of climate change have the potential to directly and indirectly affect invasive species in important ways. Direct effects vary depending on the climate change component and species involved. For example, elevated CO_2 can have a dramatic, positive effect on growth of invasive plants but little or no direct effect on invasive insects or pathogens, which are more affected by changes in temperature and precipitation. Indirect effects of climate change on invasive species can be equally important and occur through changes in interactions with, and/or status of, competitors and hosts. The landscape to global occurrence and distribution of invasive species can also be altered by climate change.

4.4.5 Key Information Needs

Most studies have examined the impacts of climate change on invasive species occurring in isolation or in simplified systems; in order to better understand impacts, more realistic studies need to be conducted in natural settings and over larger landscapes. Similarly, most studies have examined impacts of only one component of climate change on invasive species (e.g., drought or elevated CO_2 but not both), even though multiple climate variables are changing simultaneously. Consequently, studies should examine the combined effects of variables on invasive species. In order to manage biological communities that are more resistant to invasions, managers need a practical understanding of how these climate variables influence invasive plants, insects, and pathogens.

4.5 Ecosystem Responses to Climate Change That Affect Invasive Species

The distribution and abundance of a species are governed by natality, growth, mortality, and dispersal of individuals comprising a population. These variables are influenced by environmental factors such as climate, among others, as mediated through fluctuations in resource availability, fecundity, fitness, and survivorship. The direct effects of climate change on invasive taxa are discussed above. Here we focus on the indirect effects as mediated through changes in habitat, hosts, disturbance, trophic interactions, and land management.

4.5.1 Habitat and Host Range

Climate is a primary factor regulating the geographic distributions of plants. For example, the current distribution of coniferous vegetation across Western North America resulted from climatic shifts dating back millions of years (Brunsfeld et al. 2001), along with more recent recolonization of deglaciated lands (Godbout et al. 2008). Plants tend to be adapted to a range of climatic conditions (niches), and climate change may cause shifts in the geographic distribution of these niches (Parmesan 2006) with broad implications for other species (e.g., invasive herbivores) that rely on these plants for food and/or shelter. Substantial shifts in the geographic distributions of bioclimatic envelopes (climatic niches) have been projected for grass species, shrub species, tree species, and entire communities in North America (e.g., Bradley 2009; McKenney et al. 2007; Rehfeldt et al. 2012; Wang et al. 2012). To the extent that dispersal and resource availability allow, these species and communities are expected to track associated shifts in bioclimatic envelopes over time (Pearson and Dawson 2003). For example, Parmesan and Yohe (2003) conducted a meta-analysis indicating that climate change caused an average boundary shift of 6.1 km per decade northward (or 6.1 m in elevation upward) for 99 species of birds, plants, and insects. However, climate-induced downhill shifts of plant communities can also occur (Crimmins et al. 2011), thus illustrating the diversity and complexity of plant responses to climate change. In general, shifts are expected to be most noticeable along present-day ecotones, but the fate of any individual, species, or community will depend on genetic variation, phenotypic variation, fecundity, and dispersal mechanisms. Furthermore, the resilience of plants to a multitude of stressors may be affected by climate change (Fettig et al. 2013).

4.5.2 Host Physiology and Phenology

For the 1000 years prior to the Industrial Revolution, concentrations of atmospheric CO_2 remained stable at ~270 ppm. Atmospheric CO_2 is ~407 ppm (December 2017, www.esrl.noaa.gov) and is projected to reach 550 ppm by the middle of this century and to surpass 700 ppm by the end of the century (IPCC 2007). While elevated CO_2 has the potential to affect many metabolic processes in terrestrial plants with C3 photosynthetic pathways, impacts on Rubisco (the enzyme by which atmospheric CO_2 is converted to energy in plants) and stomatal movement have been consistently demonstrated to occur within the range of CO_2 concentrations associated with climate change (Long et al. 2004). Elevated CO_2 increases net photosynthesis and decreases transpiration through reduced stomatal conductance and increased water-use efficiency (Wand et al. 1999), thus influencing plant growth and competition.

Smith et al. (2000) studied the effects of CO_2 enrichment on growth in creosote bush (*Larrea tridentata*) and two deciduous shrub species in Nevada. Significant increases in shoot production were observed with a 50% increase in atmospheric CO_2 in a high rainfall year but not during a low rainfall year. Similar results were observed for several annual plants including red brome (*Bromus* spp.), a nonnative annual grass that has invaded portions of the Southwestern United States (Hunter 1991). The density of red brome increased as a result of CO_2 enrichment (Smith et al. 2000), demonstrating that increased atmospheric CO_2 can influence an invasive plant through modification of its physiology and competitive interactions. Similarly, the growth of cheatgrass, a notable invasive grass in the Western United States, is also enhanced by elevated CO_2 (Smith et al. 1987; Ziska et al. 2005) and increased temperature (Zelikova et al. 2013), specifically during periods of high soil moisture. While desert plants are likely to be among the most responsive to elevated CO_2 (i.e., due to increases in water-use efficiency), similar relationships have been observed in many plant species. In general, elevated CO_2 results in increased plant growth manifested as increased leaf area, increased leaf thickness, and larger shoots, stems, and branches (Pritchard et al. 1999).

Drought affects many components of plant nutritional quality and morphology of importance to invasive species. Most research has focused on indirect effects of drought on folivores as mediated through changes in host quality, primarily leaf chemistry, and palatability (Kolb et al. 2016). Drought often increases plant tissue concentrations of nitrogen compounds such as amino acids and nitrate, osmolytes such as sugars and inorganic ions, and allelochemicals such as cyanogenic glycosides, terpenoids, and alkaloids. These compounds were observed to increase in tissue concentration during periods of mild or moderate drought, when water stress constrains growth more than photosynthesis and root uptake of nutrients, and decrease during periods of long and severe drought, when intense water stress constrains growth, photosynthesis, and root uptake (Kolb et al. 2016). In particular, increases in the concentration of nitrogen compounds may affect performance of insect folivores, as nitrogen is often a limiting factor in their growth (Mattson 1980). For example, Rouault et al. (2006) commented that some defoliating insects benefited from increased nitrogen in plant tissues associated with moderate water stress during the drought and heat waves that occurred in Europe in 2003. Drought also affects certain morphological characteristics of plants, causing a decrease in leaf toughness and an increase in dry matter content, which typically reduces folivore feeding as leaf water content decreases. Interestingly, drought-stressed plants are consistently warmer than unstressed plants because reduced transpiration limits plant cooling, with differences as great as 15 °C being observed (Mattson and Haack 1987). This has obvious implications to invasive insects due to the positive responses of most insect herbivores to increasing temperature (Bale et al. 2002). Overall, there is likely to be considerable variation in the magnitude and direction of responses to drought by invasive insects and pathogens, similar to that observed in other groups. Droughts are expected to accelerate the pace of invasion by some nonindigenous plants (Finch et al. 2016). For example, saltcedar (*Tamarix ramosissima*) is more drought-tolerant than co-occurring native species, and its capacity to invade is thought to increase with drought (Cleverly et al. 1997).

Plants and animals exhibit seasonality in the timing of life history events associated with temporal variation in habitat suitability. In particular, plants and insects are finely tuned to the seasonality of their environment, and shifts in phenology provide some of the most compelling evidence that species and ecosystems are being influenced by climate change (Cleland et al. 2007). Climate change has the capacity to cause phenological shifts that may result in asynchrony between different trophic levels. The potential consequences of phenological asynchrony have been demonstrated in several terrestrial and aquatic systems (Winder and Schindler 2004) and have been well documented in insect folivores of forest trees, where it has been demonstrated that timing of bud burst and shoot development can have marked impacts on insect growth and survival (Watt and McFarlane 2002). Such climate-induced developmental asynchrony has implications for both native and invasive species.

4.5.3 Disturbances

Disturbances (e.g., storms, wildfire, and herbivory) are relatively discrete events that affect the structure, composition, and function of ecosystems through alterations of the physical environment (White and Pickett 1985). Some disturbances result in the release of large amounts of CO_2, thereby further contributing to climate change. Climate change is expected to exacerbate the frequency and severity of many disturbances (Fettig et al. 2013; Westerling et al. 2006), which in turn influence the distribution, abundance, and impact of invasive species. For example, bark beetles feed on the phloem of trees and are important disturbances in conifer forests worldwide (Raffa et al. 2015). In Western North America, recent outbreaks of the native mountain pine beetle (*Dendroctonus ponderosae*) have been severe, long lasting, and well documented and have been linked to climate change (Bentz et al. 2010) and other factors (Fettig et al. 2007). Mountain pine beetle outbreaks increase host mortality rates and can result in subsequent replacement by other plant associations, including invasive species (Fettig et al. 2015; Fig. 4.4). Furthermore, outbreaks alter forest fuels with consequences to the frequency, severity, and intensity of wildfires (Jenkins et al. 2014).

A recent global meta-analysis of relevant literature concluded that wildfires, which are increasing due to climate change in many systems (Westerling et al. 2006), enhance the composition and performance of invasive plants, while having no effect on the composition and reducing performance of native plants (Alba et al. 2015). Additionally, responses appear to vary by habitat type. Invasive species groups respond most positively to wildfire in arid shrublands, temperate forests, and heathlands (Alba et al. 2015). Fire likely promotes invasion due to increased resource availability and nutrient inputs. Some invasive species (e.g., cheatgrass) create a feedback loop in which fire-promoted nonnative species further alter the fire regime to the detriment of native species (Brooks et al. 2004; D'Antonio and Vitousek 1992).

4.5.4 Trophic Interactions

Trophic interactions will undoubtedly be influenced by climate change, although little is known about these relationships. Some fungal pathogens of insects are important in regulating insect populations and are likely to be impacted by climate change. For example, *Entomophaga maimaiga*, which causes extensive epizootic in populations of the invasive gypsy moth (*Lymantria dispar*) in the Eastern United States, requires high levels of moisture for conidial production and discharge (Hajek 1999). Consequently, drought is expected to reduce this pathogen's impact on gypsy moth populations (Kolb et al. 2016). Studies indicate that climate change could alter the phenology of insect and plant pathogens, modify host resistance, and result in changes in the physiology of host–pathogen interactions (Coakley et al. 1999), likely with differential effects to invasive species.

Fig. 4.4 In recent decades, billions of conifers across millions of hectares have been killed by native bark beetles in forests ranging from Alaska to Mexico, and several recent outbreaks are considered among the largest and most severe in recorded history. Temperature influences several important life history traits of bark beetles, and recent outbreaks have been linked to climate change (Bentz et al. 2010). Mountain pine beetle (*Dendroctonus ponderosae*) outbreaks have been particularly severe, long lasting, and well documented, with over 27 million ha impacted. One potential consequence, particularly in areas of high (>50%) tree mortality, is subsequent invasion by nonnative plants, in this case by Canada thistle (*Cirsium arvense*) and bull thistle (*C. vulgare*) in a lodgepole pine (*Pinus contorta*) forest in Colorado (Fettig et al. 2015). Any activity that increases resource availability (e.g., water, nutrients, and light), increases disturbance (e.g., when trees fall), and/or decreases plant competition may promote plant invasions (Photo by Justin Runyon, USDA Forest Service)

Most species possess a large array of ecto- and endosymbiotic organisms that exhibit highly complex interactions that are often poorly understood but which may be influenced by climate change. For example, changes in temperature have been demonstrated to cause shifts in the composition of two native symbiotic (bluestain) fungi associated with conifer bark beetles (Six and Bentz 2007). *Grosmannia clavigera* predominates during cool periods but decreases in prevalence as daily maximum temperatures approach 25 °C, becoming extremely rare when temperatures reach or exceed 32 °C. In contrast, *Ophiostoma montium* increases as temperatures approach 25 °C and becomes the predominant symbiont when temperatures reach or exceed 32 °C (Six and Bentz 2007). While this may be important in brood development, it is unknown if one fungus is more beneficial than the other or if effects vary by temperature. Similar relationships have been demonstrated in other systems. For example, Prado et al. (2010) showed decreases in stink bug (*Acrosternum hilare* and *Murgantia histrionica*) fitness associated with loss of gut symbionts within two generations when insects were reared at 30 °C as compared to 25 °C.

4.5.5 Influence of Land Management on Invasive Species in a Changing Climate

Land management to benefit native species impacted by climate change may influence shifts in the geographic distributions of invasive species through effects on their dispersal routes and mechanisms. For example, in response to climate change, managers may consider assisted migration (e.g., the practice of planting tree species outside of their current distribution due to anticipated changes in the climatic niche) (Andalo et al. 2005; Rehfeldt et al. 1999). While most efforts involving assisted migration are still experimental, large-scale plantings could result in unintended introductions of other plant and animal species and/or provide new dispersal routes for established invasive species, both with unintended impacts to recipient communities.

Land management practices often influence susceptibility to disturbances exacerbated by climate change. For example, wildfires have sculpted seasonally dry forests in the Western United States for millennia. Such events reduced the quantity and continuity of forest fuels and discouraged establishment of fire-intolerant species. However, during the last century, fire suppression and preferential harvest of certain trees, among other factors, have increased fuels and changed forest conditions over extensive areas. As a result, wildfires tend to be larger and more severe, a trend likely to only intensify as a result of climate change (Westerling et al. 2006). Accidental introduction of invasive species, specifically plants, is frequently of concern in areas that are rehabilitated after fire (Keeley 2006). For example, following the 2000 Cerro Grande Fire in New Mexico, contamination of aerial seeding sources was responsible for inadvertently broadcasting cheatgrass seeds across recently burned areas (Keeley et al. 2006). Relatedly, prescribed fire and thinning of small-diameter trees are used to reduce fuels in order to increase the resilience of forests to high-intensity wildfire (Stephens et al. 2012), but some studies have shown that these treatments promote an increase in invasive species richness (Schwilk et al. 2009). Climate change may also affect the efficacy of tools used to manage invasive species (Sect. 4.7).

4.5.6 Key Findings

The indirect effects of climate change on invasive species are primarily mediated through changes in habitats and hosts, the frequency and severity of other disturbances, trophic interactions, and land use. Climate is a primary factor regulating plants, and consequently, climate change can have an important influence on the abundance and distribution of suitable habitats and hosts and on the phenology, physiology, and morphology of hosts. In particular, plants and insects are finely tuned to the seasonality of their environment, and changes in phenology may result in asynchrony between different trophic levels affecting performance at one or more levels. Furthermore, climate change exacerbates the frequency and severity of many disturbances (e.g., wildfire), which affects the distribution, spread, abundance, and impact of invasive species. Trophic interactions will undoubtedly be influenced by climate change, although little is known about these relationships. Changes in land-use patterns and management practices in response to climate change may alter susceptibility to invasions in a variety of ways, but primarily through alterations of dispersal routes and mechanisms and accidental transport. For example, accidental introduction of invasive plants is a major concern in forests and grasslands being rehabilitated after wildfire.

4.5.7 Key Information Needs

Bioclimatic models being used to project changes in the distribution of invasive species, hosts, habitats, and communities would be more helpful if they were further refined and downscaled. Further studies on the effects of elevated CO_2 on plant growth and invasiveness are fundamental to our understanding of how plants respond to climate change. More information is needed on the response of invasive species to other disturbances exacerbated by climate change. There is a critical need to develop adaptation strategies to manage native and invasive species (and their many interactions) effectively in the face of climate change.

4.6 Influence of Invasive Species on Climate Change and Carbon Sequestration

Insect and disease outbreaks can have substantial impacts on ecosystem-level carbon cycling and storage (Cobb et al. 2012a; Morehouse et al. 2008; Nuckolls et al. 2009). The consequences to carbon storage of outbreaks of invasive insects and pathogens are poorly understood as compared to other disturbances such as forest harvesting or wildfire (Nave et al. 2010, 2011). However, it is becoming increasingly clear that these events can alter forest-level carbon (C) storage. A better understanding of outbreaks will improve the understanding of such threats to the important atmospheric greenhouse gas (GHG) sink associated with forests (Hicke et al. 2012). Invasive insects and pathogens can alter the process of CO_2 sequestration as biomass by reducing tree growth, killing trees, and altering the distribution of carbon within forests (Albani et al. 2010; Cobb et al. 2013; Kurz et al. 2008a). Increases in dead woody biomass (fuels or coarse woody debris) are commonly associated with outbreaks (Cobb et al. 2012a; Hoffman et al. 2012; Valachovic et al. 2011). Although dead wood is an important component of forest-level carbon storage (Harmon 2009), these increases can also lead to changes in fire behavior or impacts in complex ways (Buma 2015; Jenkins et al. 2014; Meigs et al. 2016; Metz et al. 2011; Simard et al. 2010).

Carbon storage in trees is an important component of local and regional policy aimed at capping or ameliorating GHG emissions. The potential for episodes of tree mortality to interfere seriously with these goals has been recognized for over a decade (Breshears and Allen 2002), but there has been less effort expended to predict these impacts and address them in formal forest management policies (Hicke et al. 2012; Kurz et al. 2008b). Furthermore, in terms of GHGs, forests are not solely CO_2 sinks or sources. Soil microbial communities also emit methane (CH_4) and N_2O, gases, which have far higher radiative heating capacity on a per molecule basis than CO_2 (Smeets et al. 2009). Ecosystem studies have focused mainly on CO_2 because it is the major component of GHG sources within forest ecosystems, and, compared to CH_4 and N_2O, changes in CO_2 storage and exchange are better understood. In general, ecosystem C storage in the living plant biomass pool is second only to the soil C pool (Domke et al. 2018). Although plant biomass is a heterogeneous collection of compounds with different decomposition rates, those constituent compounds are relatively short-lived compared to many of the organic compounds, such as humic and fulvic acids, that form soil organic matter (Harmon 2009; Lewis et al. 2014). Reasonable estimates of changes in C storage during invasive pest outbreaks can be made when tree biomass can be mapped accurately, tree mortality rates and spatial patterns are known, and decomposition rates of woody debris can be measured or estimated from reliable data (Albani et al. 2010; Hicke et al. 2012; Kurz et al. 2008a).

Many invasive insects and pathogens are likely to influence ecosystem C storage by reducing growth and killing trees; consequently, explosive episodes of population increases and spread of these organisms are certain to have significant net impacts on this important ecosystem resource. Knowledge gained from evaluating impacts of invasive and analogous native insects and pathogens at the ecosystem level suggests that several broad categories of outbreaks will have different impacts on ecosystem C cycling (Table 4.1). The most subtle changes are likely to occur due to organisms that reduce growth rather than those that kill individual trees (Eviner and Likens 2008). Changes in C cycling can be expected to intensify depending on the interplay between stand composition and invasive insect or pathogen host range. As the range of hosts killed by these invasive organisms increases, the rate of infection/attack and the prevalence of hosts affected at stand or landscape scales will also increase, thus intensifying changes in C cycling. For example, the invasive hemlock woolly adelgid (HWA) alters litter decomposition, litterfall, and soil respiration rates in ways that reduce C storage at the stand level (Cobb 2010; Finzi et al. 2014; Orwig et al. 2013). HWA attacks on eastern hemlock (*Tsuga canadensis*) and Carolina hemlock (*T. caroliniana*) in the Eastern United States have altered C cycling by reducing C storage, a process that escalates with the amount of preoutbreak C stored in at-risk forests (Albani et al. 2010). *Phytophthora ramorum*, the invasive species that causes sudden oak death, is a broad-host-range pathogen that decreases litterfall and aboveground biomass (Cobb et al. 2012b, 2013); in contrast to the HWA, *P. ramorum* has differential spread and impacts among host species and causes different amounts of mortality across forest types (Metz et al. 2012).

Integrating the influence of climate change on both invasive insects and pathogens, as well as on their hosts, is critical to understanding and predicting when and how much C will be lost during an outbreak (Alexander 2010; Sturrock et al. 2011). Because spread rates of microbes are difficult to measure directly, proxy measurements such as repeated aerial surveys (of tree mortality) are often used to estimate pathogen dispersal and lags between exposure and mortality

Table 4.1 Types of insect or pathogen effects on host trees that affect carbon storage, with examples of both invasive and native species

Type of effect	Invasive species causing effect	Native species causing effect
Growth reduction	1. Many *Phytophthora* species such as *P. nemorosa*	1. Forest tent caterpillar (*Malacosoma disstria*)
	2. Elongate hemlock scale (*Fiorinia externa*)	2. *Pseudomonas syringae*
	3. Gypsy moth (*Lymantria dispar*)	3. Swiss needle cast (*Phaeocryptopus gaeumannii*)
Scattered individual tree mortality	1. Gypsy moth	1. Many heart rot fungi such as *Ganoderma* spp.
	2. Asian long-horned beetle (*Anoplophora glabripennis*)	2. Engraver beetles (*Ips* spp.)
	3. *P. cactorum*	
Extensive mortality of one species (in low-diversity stands)	1. Hemlock woolly adelgid (*Adelges tsugae*) (Northeastern United States)	1. *Heterobasidion irregulare*
	2. White pine blister rust (*Cronartium ribicola*)	2. Fusiform rust (*Cronartium fusiforme*)
	3. *P. ramorum* (in Japanese larch (*Larix kaempferi*) plantations)	3. Dothistroma needle blight (*Dothistroma septosporum*)
Extensive selective mortality of one species (in multi-species stands)	1. *Cryphonectria parasitica*	1. Spruce budworm (*Choristoneura occidentalis*) (some stand compositions)
	2. *P. austrocedri*	2. *H. occidentale*
	3. Emerald ash borer (*Agrilus planipennis*)	
Mortality of many trees of multiple species (in diverse stands)	1. *P. ramorum*	1. Armillaria root disease (*Armillaria* spp.)
	2. *P. cinnamomi*	2. *H. irregulare* (some stand compositions)
	3. Gypsy moth	

(Filipe et al. 2012; Fitzpatrick et al. 2011; Meentemeyer et al. 2011). Detailed biological understanding of emergent invasive insects and pathogens is a prerequisite to mapping their potential areas of outbreak risk (Meentemeyer et al. 2004, Orwig et al. 2012). However, invasive insects and pathogens are a diverse group of organisms with equally variable evolutionary histories and sensitivities to environmental variation. This implies that some invasive organisms will have increased impacts or greater spread under one climate change scenario but not under another. For example, greater winter minimum temperatures are likely to increase the potential range of HWA, which should exacerbate the loss of C storage at landscape and regional scales (Orwig et al. 2012, 2013). A reduction in snow cover, attributed to either an increase in the proportion of precipitation occurring as rain or lower overall precipitation, may decrease insulation and result in greater HWA overwinter mortality (Fitzpatrick et al. 2011; Orwig et al. 2012). Similarly, spread of *P. ramorum* is likely to be slowed by regional drought, thus affecting the distribution of this pathogen in California (Meentemeyer et al. 2011). However, changes in *P. ramorum* population levels may facilitate rapid reinvasion of areas once drought abates (Eyre et al. 2013).

4.6.1 Key Findings

Infestations of invasive insects, pathogens, and plants can disrupt forest carbon storage and rates of sequestration because they influence tree and plant growth, mortality, decomposition rates, and ecosystem processes. A better understanding of invasive species outbreaks will accelerate our understanding of the threats to the important atmospheric greenhouse gas (GHG) sink that forests represent.

Generally, ecosystem C storage is greatest in the soil and belowground pools followed by the live plant biomass pool. Soil microbial communities emit CH_4 and N_2O gases, which have greater radiative heating capacity relative to CO_2. Therefore, invasive organisms that alter ecosystem processes in the litter layer may have significant effects on atmospheric greenhouse gases. The changes in carbon storage attributed to outbreaks of invasive insects and pathogens can be reasonably estimated when biomass levels are accurately mapped, mortality rates and spatial patterns are known, and decomposition rates are measured or estimated from reliable data.

4.6.2 Key Information Needs

Given the inherent risks associated with managing forests under a changing environment, increased monitoring of invasive species distribution and impacts is needed in order to provide a baseline for comparing and understanding changes in their spread and behavior in the future. Improved techniques for detecting invasive species and an expansion of surveys are needed. A more thorough biological understanding of emergent invasive insects and pathogens is essential to improve our ability to project patterns of risk. Field experiments and modeling to understand invasive species effects on carbon cycles and to identify techniques to sustain or restore carbon sequestration affected by such species are needed to develop strategies to manage forest carbon sequestration given invasive insect and pathogen activity.

4.7 Predicting, Monitoring, and Managing Invasions Under a Changing Climate

4.7.1 Modeling Future Scenarios to Predict Effects of Climate Change on Species Invasions

As previously discussed, climate change will dramatically alter the rates and patterns of species invasion, and in many cases, it is expected to facilitate expansion of invasive species into areas that previously were climatically inaccessible through synergistic interactions (Walther et al. 2009). As such, an effective management approach would involve anticipating which species will spread to what locations at what times, monitoring to detect incipient invasions rapidly, and applying treatments to prevent or deter establishment. Once established, invasive species are usually extremely difficult to contain and almost impossible to eradicate.

A number of novel approaches for monitoring and modeling invasive species can facilitate our ability to predict their future spread under a range of climate change scenarios and disturbance regimes. These approaches are synergistic with ongoing broad-scale efforts to delineate the current distributions and patterns of spread of invasive species and to incorporate experimental data on the ecological and evolutionary characteristics of invasive species (Chown et al. 2014; Chuang and Peterson 2016). In addition, these approaches can be combined into a comprehensive monitoring, experimental, and modeling framework that includes the following components: (1) collection of georeferenced information on the extent, pattern, and genetic characteristics of current invasions; (2) genomic analysis of relatedness and genetic variation among invading populations; (3) assessment of functional and adaptive traits related to survival and spread into novel locations; and (4) development of models needed to predict rates and patterns of spread of each species as functions of disturbance history, landscape context, climate, biotic interactions, ecologically relevant species traits, and pathways of spread (such as human transportation networks). Specific methodologies utilized for each of these components can vary, but the conceptual approaches are broadly applicable.

Developing a comprehensive predictive framework involves obtaining reliable information about the current distribution and pattern of spread of invasive species to provide a baseline for monitoring future spread and acquiring data to develop models to predict the drivers of past spread and forecast future spread (Hulme 2006). Development of high-quality datasets will require a concerted effort to collect geographic coordinates and obtain tissue samples for genetic analysis across an invasion and with sufficient sampling density, frequency, and design optimization to obtain a representative sample with a high probability of detecting occurrence. In order to model future spread effectively, reliable information is needed about how landscape and ecological features affect spread probability. Sampling along climatic gradients to document relationships between invading populations and environmental variables such as temperature and precipitation is critical for identifying climatic drivers of spread and predicting effects of climate change on future spread (Shirk et al. 2018). In addition, biological composition, disturbances, land use, and probable invasion pathways should be identified in order to build comprehensive predictive models (Cushman 2015). While making robust data collections is ideal, alternate data sources such as herbarium or museum collections for evaluating distribution and spread (Elith and Leathwick 2009; Peterson 2003) may be available and useful.

Genomic analysis of relatedness and knowledge of the genetic variation among invading populations are important inputs to the development of robust models of spread in a changing climate. Knowing the genetic structure of invading populations is necessary in order to identify the factors that drive or inhibit their spread reliably (Cushman 2015). Genomic analysis can shed light on past spread such as founder events, vicariance, and secondary contact, as well as the degree to which these events have affected the gene pools of invading populations. Genomic analysis can also help determine evolutionary potential under changing climates, by separating out adaptive and nonadaptive processes (Keller and Taylor 2008), and by determining levels of genetic variation available for natural selection (Xu et al. 2015). To develop robust models of genetic structure across an invasion, DNA samples need to be obtained from multiple individuals from as many invasive populations as possible. Current technology allows for the rapid sequencing of a large number of loci for a large number of individuals, which provides rich information on the genetic structure of invading populations and supports the development of robust models on the pattern and timing of their past spread, as well as provides support for models of future spread (Chown et al. 2014).

Traits that facilitate the spread of invasive species into, and favor their persistence in, novel environments are expected to undergo natural selection during the invasion process. This can result in evolutionary shifts that can affect predictions of future spread and may be particularly important in understanding the effects of climate change on species invasions (Hargreaves et al. 2015). Data from experimental investigations, such as controlled common garden, greenhouse and growth chamber studies of plants, and quarantine laboratory tests for invasive taxa such as vertebrates, insects, and disease-causing organisms, can be incorporated into predictive models to determine how evolutionary dynamics and climate might interact to drive future spread (Kilkenny and Galloway 2016). Trait data can also be taken from measurements of existing invading populations and can be especially useful for model parameterization if studies involve experimental treatments such as additions of CO_2 or manipulations of temperature and moisture. Trait means and variance components, as well as trait × environment reaction curves, can be incorporated into a large variety of statistical and process-based predictive models.

Ultimately, construction of a comprehensive framework for prediction requires the development and application of spatially explicit models of spread, and usually, it involves incorporating species distribution models (SDM) of some type. SDMs predict the expected distribution of species based on species relationships to landscape characteristics (Elith and Leathwick 2009), and they have been used successfully to predict species invasions (Fernández and Hamilton 2015; Peterson 2003) and species range shifts under climate change (Hijmans and Graham 2006; Pearson and Dawson 2003). SDMs can vary widely in method, from statistical to process based. Research is progressing on ways to improve SDMs, including (1) developing better ways to characterize and reduce uncertainty in model predictions and to assess its impact on management decisions; (2) developing better strategies for model selection and evaluation, including integrating tools from other fields, such as machine learning (a method of data analysis that automates analytical model building); and (3) developing better procedures to address the complexities of both spatial and temporal scales (Elith and Leathwick 2009).

Increasingly, novel data types and modeling procedures are being incorporated into SDMs. For example, Kilkenny and Galloway (2016) used trait data from a large-scale reciprocal transplant study of the invasive vine Japanese honeysuckle to develop response functions (Wang et al. 2006) that modeled the relationship between juvenile winter survival and winter temperature for populations from two different regions (core and northern margin). Core and margin populations showed adaptive differentiation in their ability to survive, with margin populations having a significant survival advantage under controlled conditions in gardens planted beyond the current range edge. When the response functions were used to model spread under future climate scenarios, the survival advantage of margin populations persisted.

In another example, Cushman (2015) developed a spatially explicit spread model that combined both local spread as a function of landscape resistance and long-distance colonization through transportation networks to improve predictions of the distribution and spread of invasive species. This type of spread modeling, when combined with advanced individual-based genetic simulations (CDPOP; Landguth and Cushman 2010), landscape dynamic simulation modeling (Cushman et al. 2011), and sophisticated multimodel optimization of patterns of gene flow (Cushman et al. 2013), can enable researchers and managers to: (1) identify the climatic, disturbance, management, and landscape factors driving spread of different invasive species; (2) predict patterns of future spread under a range of scenarios involving altered climate, disturbance, and management regimes; and (3) utilize this information to optimize monitoring in areas predicted to be most vulnerable to future invasion, or where biocontrol will be most effective.

4.7.2 Management and Restoration Techniques

Lovett et al. (2016) explored a broad range of management and policy applications for managing invasive species. Critical among these was the need to strengthen defenses against pest arrival and establishment, including taking measures to inspect and ensure clean shipments of plant and wood products prior to and after shipment and implementing postentry measures such as quarantines, surveillance, and eradication programs. Improving such pre- and postentry defenses will help reduce rates of initial introductions of invasive species whose spread may be favored by climate variation.

Impacts of Climate Change on Mechanical Control Strategies Mechanical control strategies (e.g., cutting, girdling, and tilling) are useful to combat some invasive species, especially plants. The effectiveness of this technique may vary as a result of climate change. In areas where cold temperatures and hard freezes make mechanical control feasible, warmer winter temperatures may make it more expensive. If warmer winter temperatures allow these species to overwinter, greater survival rates and an increased number of generations can be expected. While fire may be helpful in controlling insects and pests, it may facilitate dominance by several fire-adapted invasive plants. For example, under hotter temperatures and reduced moisture associated with climate change, cheatgrass tends to enhance the size of wildfires and cause the wildfire season to begin earlier and continue later into the fall. Cheatgrass and other invasive species can recover faster following fires and thus suppress many native species. Consequently, more frequent fires lead to irreversible losses of native shrubs and grasses, threatening not only the habitat but the survival of species such as greater sage grouse (*Centrocercus urophasianus*), which is dependent on the shrub–steppe environment. These interactions complicate decision-making on how best to manage and restore infested areas.

Impacts of Climate Change on Effectiveness of Herbicides Herbicides are commonly used to manage invasive plants, and there is evidence suggesting that climate change could alter their effectiveness. For example, increases in CO_2 can increase the tolerance of some weeds to the herbicide glyphosate, though the underlying mechanism is not fully understood (Ziska et al. 1999). Moreover, herbicides applied during periods of drought are generally less effective than those applied when moisture is adequate (Bussan and Dyer 1999; Kogan and Bayer 1996). For example, Morrison et al. (1995) found that drought stress reduced translocation of herbicides and thus adversely affected control of Russian knapweed (*Acroptilon repens*). Kogan and Bayer (1996) determined that plants responded to drought stress by reducing uptake and translocation of herbicides. When applied during high temperatures, herbicides, including the active ingredients therein, are much more likely to volatilize (Behrens and Lueschen 1979; Tabernero et al. 2000), resulting in drift to nontarget areas (Jordan et al. 2009) and inadequate control of target invasive plants. Active ingredients may also be heat labile, causing them to break down before they can be translocated. However, we still lack a good understanding of the potential impacts of climate change on the effective use of herbicides for managing invasive plants.

Impacts of Climate Change on Biological Control of Invasive Species The efficacy of certain biocontrol agents (e.g., pathogenic fungi, insect predators, and plant herbivores) may be impacted by climate change. Many studies have highlighted the importance of environmental temperature in mediating the outcome of host–pathogen and host–parasite interactions (Thomas and Blanford 2003) and are relevant to managing invasive species under a changing climate. For example, climate change can exert important direct and indirect effects on insect herbivores commonly used in managing invasive plants (Runyon et al. 2012). Increasing temperatures generally lead to shorter development times and greater survival of insect herbivores (Bale et al. 2002) and, in some species, can increase the number of generations per year (Tobin et al. 2008; Altermatt 2010). Drought can also have either a positive or negative effect on herbivores (Finch et al. 2016). Climate warming could also shift the geographic distribution of invasive plants and biocontrol agents.

There is evidence that range shifts caused by warming can increase the impact of biocontrol agents on nontarget plants (Lu et al. 2015). However, Lu et al. (2016) reported that warming can also shift plant communities from invader dominated to native dominated in the presence of a biocontrol beetle. Perhaps the most significant impact of climate change on biocontrol is through disruption in plant–herbivore synchrony. Rising temperatures could result in the herbivore being present when the targeted plant or plant stage is absent; this can occur because temperature and photoperiod can have a variable effect on the development of plant and herbivore species (Bale et al. 2002). Although there are no reported cases of such asynchrony in weed biocontrol, there are examples of climate-induced asynchrony occurring in some plant/herbivore species (van Asch et al. 2007); therefore, biocontrol practitioners should be aware of this potential response to climate change.

Climate change can also indirectly affect biocontrol by altering the basic nutritional value of plants. For example, elevated CO_2 increases plant growth (the "fertilizer effect") and the ratio of carbon-to-nitrogen (C:N) in plant tissues, which reduces the nutritional quality for nitrogen-limited insects (Coviella and Trumble 1999). This causes insects to consume more foliage to compensate for reduced nitrogen content (Coviella and Trumble 1999; Dermody et al. 2008; Johnson and McNicol 2010). Elevated CO_2 can also increase leaf sugar content of plants and spur herbivores to consume more foliage from plants growing under high CO_2 conditions (Hamilton et al. 2005). These findings suggest that elevated CO_2 could enhance biocontrol if it induces herbivores to increase the rate and volume or biomass of invasive plants consumed. Climate change may also affect plant nutritional value by altering chemical defenses against herbivores. Climatic factors including drought, elevated CO_2, temperature, ozone (O_3), and UV light, singly and in combination, can affect levels of plant secondary chemicals (Bidart-Bouzat and Imeh-Nathaniel 2008; Huberty and Denno 2004; Percy et al. 2002; Yuan et al. 2009). Interestingly, flavonoid and cyanide concentrations decreased in foliage of invasive garlic mustard (*Alliaria petiolata*) growing under elevated CO_2 and temperatures (Anderson and Cipollini 2013), which could enhance the efficacy of biocontrol agents. However, we have a poor understanding of how climate-induced changes influence secondary plant chemistry, and available knowledge indicates that the response is dependent on the plant and insect species involved and the class of chemicals examined (Bidart-Bouzat and Imeh-Nathaniel 2008). A better understanding of how climate change will impact the interactions between invasive organisms and their biocontrol agents is needed.

Impacts of Climate Change on Restoration Strategies Climate change may alter the success of restoration strategies used to combat invasive species and may also affect the utility of restoring natural disturbance regimes as a strategy to control invasive species (Hellmann et al. 2008). Revegetation is often necessary following the control of invasive plant species in natural environments; otherwise, the empty niche may be occupied by other undesirable species (Pearson et al. 2016). The success of such revegetation attempts can be hindered by extreme weather events such as frequent floods and droughts. As mentioned previously, prescribed fire can be used either alone or when combined with other control tactics as a strategy to control several invasive species, including plants, insects, and pathogens. However, altered fire regimes that are associated with climate change can adversely affect the use of prescribed fire as a restoration tool.

4.7.3 Key Findings

A number of novel monitoring and modeling approaches can be used to facilitate prediction of the future spread of invasive species under a range of climate change scenarios and disturbance regimes. Effective modeling of future spread requires acquiring reliable information not only about how landscape and ecological features affect spread probability but also how climatic gradients influence invading populations and their interactions with other environmental variables. Genomic analysis can shed light on past spread such as founder events, vicariance, and secondary contact, as well as the degree to which these events have affected the gene pools of invading populations.

Mechanical control strategies and herbicides are commonly used to combat some invasive species, but their effectiveness and cost may change as a result of climate change. Climate change and drought can also exert important direct and indirect effects on insect herbivores used to manage invasive plants. Increasing temperatures generally result in shorter development times and greater survival of insect herbivores and can increase the number of generations per year in some species. Warming climates could conceivably improve effectiveness of insect biocontrol agents applied in invaded areas, provided the ranges of host and biocontrol insects are matched rather than altered by climate change and plant chemical defenses are not intensified. Success of restoration strategies, such as those involving revegetation and prescribed fire, can be impacted by climate change.

4.7.4 Key Information Needs

Sampling along climatic gradients is needed to document relationships between invading populations and environmental variables, such as temperature and precipitation, and is critical for identifying climatic drivers of spread and predicting effects of climate change on future spread. Also needed are better range models for predicting invasive species occurrence that incorporate dispersal and demographic processes. Genomic analysis of relatedness and genetic variation among invading populations is also needed to develop robust models of spread under a changing climate. Knowledge of the genetic structure of invading populations is required to reliably identify how climate change can drive or inhibit their spread. With respect to methods for controlling invasive species, we still lack a good understanding of the potential impacts of climate change on the effective use of herbicides or mechanical methods utilized to manage invasive plants. We need a better understanding of how climate change will impact the interactions among invasive organisms and their biocontrol agents, as well as the effectiveness of biocontrol agents. Finally, new knowledge is needed on improving success of restoration strategies under varying climates following control and/or removal of invasive species.

Acknowledgments We thank David Hawksworth for assistance in eliminating duplicate references and integrating them and Yancey Ranspot for help in adjusting format. We thank Sharlene Sing, Matt Ayres, Eric Holzmueller, and Chris Weber for their manuscript reviews. We also thank the editors for their instrumental work on the volume.

Disclaimer Text The findings and conclusions in this publication are those of the authors and should not be construed to represent any official USDA or U.S. Government determination or policy.

Literature Cited

Adams HD, Guardiola-Claramonte M, Barron-Gafford GA et al (2009) Temperature sensitivity of drought-induced tree mortality portends increased regional die-off under global-change-type drought. Proc Natl Acad Sci 106:7063–7066

Alba C, Skálová H, McGregor KF et al (2015) Native and exotic plant species respond differently to wildfire and prescribed fire as revealed by meta-analysis. J Veg Sci 26:102–113

Albani M, Moorcroft PR, Ellison AM et al (2010) Predicting the impact of hemlock woolly adelgid on carbon dynamics of eastern United States forests. Can J For Res 40:119–133

Alexander HM (2010) Disease in natural plant populations, communities, and ecosystems: insights into ecological and evolutionary processes. Plant Dis 94:492–503

Alexander JM, Edwards PJ, Poll M et al (2009) Establishment of parallel altitudinal clines in traits of native and introduced forbs. Ecology 90:612–622

Altermatt F (2010) Climatic warming increases voltinism in European butterflies and moths. Proc R Soc Lond B Biol Sci 277:1281–1287

Anagnostakis SL (1982) Biological control of chestnut blight. Science 215:466–471

Andalo C, Beaulieu J, Bousquet J (2005) The impact of climate change on growth of local white spruce populations in Quebec. Canada Forest Ecol Manag 205:169–182

Anderegg WRL, Berry JA, Smith DD et al (2012) The roles of hydraulic and carbon stress in a widespread climate-induced forest die-off. Proc Natl Acad Sci 109:233–237

Anderson LJ, Cipollini D (2013) Gas exchange, growth, and defense responses of invasive *Alliaria petiolata* (Brassicaceae) and native *Geum vernum* (Rosaceae) to elevated atmospheric CO_2 and warm spring temperatures. Am J Bot 100:1544–1554

Archer SR, Predick KI (2008) Climate change and ecosystems of the southwestern United States. Rangelands 30:23–28

Bale JS, Hayward SAL (2010) Insect overwintering in a changing climate. J Exp Biol 213:980–994

Bale JS, Masters GJ, Hodkinson ID et al (2002) Herbivory in global change research: direct effects of rising temperature on insect herbivores. Glob Chang Biol 8:1–16

Barbier EB, Gwaitpedza J, Knowlder D et al (2011) The North American horticultural industry and the risk of plant invasion. Agric Econ 42:113–129

Barrett SCH (2000) Microevolutionary influences of global changes on plant invasions. In: Mooney HA, Hobbs RJ (eds) Invasive species in a changing world. Island Press, Washington, DC, pp 115–139

Beaumont LJ, Gallagher RV, Thuiller W et al (2009) Different climatic envelopes among invasive populations may lead to underestimations of current and future biological invasions. Divers Distrib 15:409–420

Behrens R, Lueschen WE (1979) Dicamba volatility. Weed Sci 27:486–493

Bellard C, Thuiller W, Leroy B et al (2013) Will climate change promote future invasions? Glob Chang Biol 19:3740–3748

Belote RT, Weltzin JF, Norby RJ (2003) Response of an understory plant community to elevated [CO_2] depends on differential responses of dominant invasive species and is mediated by soil water availability. New Phytol 161:827–835

Bentz BJ, Régnière J, Fettig CJ et al (2010) Climate change and bark beetles of the western United States and Canada: direct and indirect effects. Bioscience 60:602–613

Bernard-Verdier M, Hulme PE (2015) Alien and native plant species play different roles in plant community structure. J Ecol 103:143–145

Bidart-Bouzat MG, Imeh-Nathaniel A (2008) Global change effects on plant chemical defenses against insect herbivores. J Integr Plant Biol 50:1339–1354

Blossey B, Nötzold R (1995) Evolution of increased competitive ability in invasive nonindigenous plants: a hypothesis. J Ecol 83:887–889

Blumenthal DM, Kray JA (2014) Climate change, plant traits and invasion in natural and agricultural ecosystems. In: Ziska LH, Dukes JS (eds) Invasive species and global climate change. CAB International, Oxfordshire, pp 62–78

Blumenthal D, Chimner RA, Welker JM, Morgan JA (2008) Increased snow facilitates plant invasion in mixedgrass prairie. New Phytol 179:440–448

Blumenthal DM, Resco V, Morgan JA et al (2013) Invasive forb benefits from water savings by native plants and carbon fertilization under elevated CO_2 and warming. New Phytol 200:1156–1165

Bostock RM, Pye MF, Roubtsova TV (2014) Predisposition in plant disease: exploiting the nexus in abiotic and biotic stress perception and response. Annu Rev. Phytopathol 52:517–549

Bradley BA (2009) Regional analysis of the impacts of climate change on cheatgrass invasion shows potential risk and opportunity. Glob Chang Biol 15:196–208

Bradley BA (2014) The biogeography of invasive plants – projecting range shifts with climate change. In: Ziska LH, Dukes JS (eds) Invasive species and global climate change. CAB International, Oxfordshire, pp 240–252

Bradley BA, Wilcove DS (2009) When invasive plants disappear: transformative restoration possibilities in the Western United States resulting from climate change. Restor Ecol 17:715–721

Bradley BA, Blumenthal DM, Wilcove DS, Ziska LH (2010) Predicting plant invasions in an era of global change. Trends Ecol Evol 25:310–318

Bradley BA, Bluementhal DM, Early R et al (2012) Global change, global trade, and the next wave of plant invasions. Front Ecol Environ 10:20–28

Bradley BA, Early R, Sorte CJB (2015) Space to invade? Comparative range infilling and potential range of invasive and native plants. Glob Ecol Biogeogr 24:348–359

Brasier CM (2008) The biosecurity threat to the UK and global environment from international trade in plants. Plant Pathol 57:792–808

Brasier CM, Webber J (2010) Plant pathology: sudden larch death. Nature 466:824–825

Breshears DD, Allen CD (2002) The importance of rapid, disturbance-induced losses in carbon management and sequestration. Glob Ecol Biogeogr 11:1–5

Broennimann O, Guisan A (2008) Predicting current and future biological invasions: both native and invaded ranges matter. Biol Lett 4:585–589

Broennimann O, Treier UA, Müller-Schärer H et al (2007) Evidence of climatic shifts during biological invasion. Ecol Lett 10:701–709

Brooks ML, D'Antonio CM, Richardson DM et al (2004) Effects of invasive alien plants on fire regimes. Bioscience 54:677–688

Brunsfeld SJ, Sullivan J, Soltis DE, Soltis PS (2001) Comparative phylogeography of northwestern North America: A synthesis. In: Silvertown J, Antonovics J (eds) Integrating ecology and evolution in a spatial context. Blackwell Science, Oxford, pp 319–339

Buma B (2015) Disturbance interactions: characterization, prediction, and the potential for cascading effects. Ecosphere 6:70

Burkle LA, Runyon JB (2016) Drought and leaf herbivory influence floral volatiles and pollinator attraction. Glob Chang Biol 22:1644–1654

Bussan AJ, Dyer WE (1999) Herbicides and rangeland. In: Sheley R, Petroff J (eds) Biology and management of noxious rangeland weeds. Oregon State University Press, Corvallis, pp 116–132

Cannon RJ (1998) The implications of predicted climate change for insect pests in the UK, with emphasis on non-indigenous species. Glob Chang Biol 4:785–796

Carroll SP, Hendry AP, Reznick DN, Fox CW (2007) Evolution on ecological time-scales. Funct Ecol 21:387–393

Chown SL, Hodgins KA, Griffin PC, Oakeshott JG, Byrne M, Hoffmann AA (2014) Biological invasions, climate change and genomics. Evol Appl 8:23–46

Chuang A, Peterson CR (2016) Expanding population edges: theories, traits, and trade-offs. Glob Chang Biol 22:494–512

Cleland EE, Chuine I, Menzel A et al (2007) Shifting plant phenology in response to global change. Trends Ecol Evol 22:357–365

Cleverly JR, Smith SD, Sala A, Devitt DA (1997) Invasive capacity of *Tamarix ramosissima* in a Mojave Desert floodplain: the role of drought. Oecologia 111:12–18

Coakley SM, Scherm H, Chakraborty S (1999) Climate change and plant disease management. Annu Rev. Phytopathol 37:399–426

Cobb RC (2010) Species shift drives decomposition rates following invasion by hemlock woolly adelgid. Oikos 119:1291–1298

Cobb RC, Chan MN, Meentemeyer RK, Rizzo DM (2012a) Common factors drive disease and coarse woody debris dynamics in forests impacted by sudden oak death. Ecosystems 15:242–255

Cobb RC, Filipe JAN, Meentemeyer RK, Gilligan CA, Rizzo DM (2012b) Ecosystem transformation by emerging infectious disease: loss of large tanoak from California forests. J Ecol 100:712–722

Cobb RC, Eviner VT, Rizzo DM (2013) Mortality and community changes drive sudden oak death impacts on litterfall and soil nitrogen cycling. New Phytol 200:422–431

Colautti RI, Barrett SCH (2013) Rapid adaptation to climate facilitates range expansion of an invasive plant. Science 342:364–366

Corlett RT, Westcott DA (2013) Will plant movements keep up with climate change? Trends Ecol Evol 28:482–488

Corn JG, Story JM, White LJ (2006) Impacts of the biological control agent *Cyphocleonus achates* on spotted knapweed, *Centaurea maculosa*, in experimental plots. Biol Control 37:75–81

Coughlan NE, Kelly TC, Jansen MAK (2015) Mallard duck (*Anas platyrhynchos*)-mediated dispersal of Lemnacae: a contributing factor in the spread of the invasive *Lemna minuta*? Plant Biol 17:108–114

Coviella CE, Trumble JT (1999) Effects of elevated atmospheric carbon dioxide on insect-plant interactions. Conserv Biol 13:700–712

Crawford KM, Whitney KD (2010) Population genetic diversity influences colonization success. Mol Ecol 19:1253–1263

Crimmins SM, Dobrowski SZ, Greenberg JA et al (2011) Changes in climatic water balance drive downhill shifts in plant species' optimum elevations. Science 331:324–327

Cushman SA (2015) Pushing the envelope in genetic analysis of species invasion. Mol Ecol 24:259–262

Cushman SA, Wasserman TN, McGarigal K (2011) Modeling landscape fire and wildlife habitat. In: McKenzie D, Miller C, Falk DA (eds) The landscape ecology of fire. Ecological studies (analysis and synthesis). Springer, Dordrecht, pp 223–245

Cushman SA, Wasserman TN, Landguth EL, Shirk AJ (2013) Re-evaluating causal modeling with Mantel tests in landscape genetics. Diversity 5:51–72

D'Antonio CM, Vitousek PM (1992) Biological invasions by exotic grasses, the grass/fire cycle, and global change. Annu Rev. Ecol Syst 23:63–87

Davidson JM, Patterson HA, Rizzo DM (2008) Sources of inoculum for *Phytophthora ramorum* in a redwood forest. Phytopathology 98:860–866

Dermody O, O'Neill BF, Zangerl AR et al (2008) Effects of elevated CO_2 and O_3 on leaf damage and insect abundance in a soybean agro-ecosystem. Arthropod Plant Interact 2:125–135

Desprez-Loustau ML, Marçais B, Nageleisen LM et al (2006) Interactive effects of drought and pathogens in forest trees. Ann For Sci 63:597–612

Deutsch CA, Tewksbury JJ, Huey RB et al (2008) Impacts of climate warming on terrestrial ectotherms across latitude. Proc Natl Acad Sci 105:6668–6672

Diez JM, D'Antonio CM, Dukes JS et al (2012) Will extreme climatic events facilitate biological invasions? Front Ecol Environ 10:249–257

Domke G, Williams CA, Birdsey R et al (2018) Chap. 9: Forests. In: Cavallaro N, Shrestha G, Birdsey R, Mayes MA, Najjar RG et al (eds) Second state of the carbon cycle report (SOCCR2): a sustained assessment report. Washington, DC: U.S. Global Change Research Program. Pages: 365–398

Dostál P, Müllerová J, Pyšek P et al (2013) The impact of an invasive plant changes over time. Ecol Lett 16:1277–1284

Dukes JS (2000) Will the increasing atmospheric CO_2 concentration affect the success of invasive species? In: Mooney HA, Hobbs RJ (eds) Invasive species in a changing world. Island Press, Washington, DC, pp 95–113

Dukes JS (2011) Responses of invasive species to a changing climate and atmosphere. In: Richardson DM (ed) Fifty years of invasion ecology: the legacy of Charles Elton. Oxford, UK, Wiley-Blackwell, pp 345–357

Dukes JS, Mooney HA (1999) Does global change increase the success of biological invaders? Trends Ecol Evol 14:135–139

Dukes JS, Pontius J, Orwig D et al (2009) Responses of insect pests, pathogens, and invasive plant species to climate change in the forests of northeastern North America: what can we predict? Can J For Res 39:231–248

Dukes JS, Chiariello NR, Loarie SR, Field CB (2011) Strong response of an invasive plant species (*Centaurea solstitialis* L.) to global environmental changes. Ecol Appl 21:1887–1894

Dunson WA, Travis J (1991) The role of abiotic factors in community organization. Am Nat 138:1067–1091

Early R, Sax DF (2014) Climatic niche shifts between species' native and naturalized range raise concern for ecological forecasts during invasion and climate change. Glob Ecol Biogeogr 23:1356–1365

Elith J, Leathwick JR (2009) Species distribution models: ecological explanation and prediction across time and space. Annu Rev. Ecol Evol Syst 40:677–697

Eviner VT, Likens GE (2008) Effects of pathogens on terrestrial ecosystem function. In: Ostfeld RS, Keesing F, Eviner VT (eds) Infectious disease ecology. Effects of ecosystems on disease and disease on ecosystems. Princeton University Press, Princeton, pp 260–283

Eyre CA, Kozanitas M, Garbelotto M (2013) Population dynamics of aerial and terrestrial populations of *Phytophthora ramorum* in a California forest under different climatic conditions. Phytopathology 103:1141–1152

Fernández M, Hamilton H (2015) Ecological niche transferability using invasive species as a case study. PLoS One 10:e0119891

Fettig CJ, Klepzig KD, Billings RF et al (2007) The effectiveness of vegetation management practices for prevention and control of bark beetle infestations in coniferous forests of the western and southern United States. For Ecol Manag 238:24–53

Fettig CJ, Reid ML, Bentz BJ et al (2013) Changing climates, changing forests: a western North American perspective. J For 111:214–228

Fettig CJ, Gibson KE, Jørgensen CL et al (2015) The impacts of mountain pine beetle (*Dendroctonus ponderosae*) outbreaks on forest conditions in the Intermountain West. In: Potter KM, Conkling BL (eds) Forest health monitoring: national status, trends, and analysis, 2014. Forest health monitoring national technical report. U.S. Department of Agriculture, Forest Service, Southern Research Station, Asheville, pp 117–183

Filipe JAN, Cobb RC, Meentemeyer RK et al (2012) Landscape epidemiology and control of pathogens with cryptic and long-distance dispersal: sudden oak death in northern Californian forests. PLoS Comput Biol 8:e1002328

Finch DM, Pendleton RL, Reeves MC et al (2016) Rangeland drought: Effects, restoration, and adaptation [Chap. 8]. In: Vose JM, Clark JS, Luce CH, Patel-Weynand T (eds) Effects of drought on forests and rangelands in the United States: a comprehensive science synthesis. General technical report WO-93b. Washington, DC: U.S. Department of Agriculture, Forest Service, Washington Office:155–194

Finzi AC, Raymer PCL, Giasson M-A, Orwig DA (2014) Net primary production and soil respiration in New England hemlock forests affected by the hemlock woolly adelgid. Ecosphere 5:98

Fitzpatrick MC, Preisser EL, Porter A et al (2011) Modeling range dynamics in heterogeneous landscapes: invasion of the hemlock woolly adelgid in eastern North America. Ecol Appl 22:472–486

Gilman SE, Urban MC, Tewksbury J et al (2010) A framework for community interactions under climate change. Trends Ecol Evol 25:325–331

Godbout J, Fazekas A, Newton CH, Yeh FC (2008) Glacial vicariance in the Pacific Northwest: evidence from a lodgepole pine mitochondrial DNA minisatellite for multiple genetically distinct and widely separated refugia. Mol Ecol 17:2463–2475

Grant PR, Grant BR (2002) Unpredictable evolution in a 30-year study of Darwin's finches. Science 296:707–711

Guerenstein PG, Hildebrand JG (2008) Roles and effects of environmental carbon dioxide in insect life. Annu Rev Entomol 53:161–178

Guisan A, Petitpierre B, Broennimann O et al (2014) Unifying niche shift studies: insights from biological invasions. Trends Ecol Evol 29:260–269

Hajek AE (1999) Pathology and epizootiology of *Entomophaga maimaiga* infections in forest Lepidoptera. Microbiol Mol Biol Rev 63:814–835

Hamilton JG, Dermody O, Aldea M et al (2005) Anthropogenic changes in tropospheric composition increase susceptibility of soybean to insect herbivory. Environ Entomol 34:479–485

Hansen EM (2015) *Phytophthora* species emerging as pathogens of forest trees. Curr For Rep 1:16–24

Hargreaves AL, Bailey SF, Laird RA (2015) Fitness declines towards range limits and local adaptation to climate affect dispersal evolution during climate-induced range shifts. J Evol Biol 28:1489–1501

Harmon ME (2009) Woody detritus mass and its contribution to carbon dynamics of old-growth forests: The temporal context. In: Wirth C, Gleixner G, Heimann M (eds) Old-growth forests. Ecological studies (analysis and synthesis), vol 207. Springer, Berlin/Heidelberg, pp 159–190

Hartmann H, Ziegler W, Kolle O et al (2013) Thirst beats hunger – declining hydration during drought prevents carbon starvation in Norway spruce saplings. New Phytol 200:340–349

Hellmann JJ, Byers JE, Bierwagen BG, Dukes JS (2008) Five potential consequences of climate change for invasive species. Conserv Biol 22:534–543

Hicke JA, Allen CD, Desai AR et al (2012) Effects of biotic disturbances on forest carbon cycling in the United States and Canada. Glob Chang Biol 18:7–34

Hiebert RD (1997) Prioritizing invasive plants and planning for management. In: Luken JO, Thieret JW (eds) Assessment and management of plant invasion. New York: Springer-Verlag, Inc. Pages 195–212

Higgins SI, Richardson DM (2014) Invasive plants have broader physiological niches. Proc Natl Acad Sci 111:10610–10614

Hijmans RJ, Graham CH (2006) The ability of climate envelope models to predict the effect of climate change on species distributions. Glob Chang Biol 12:2272–2281

Hoffman C, Morgan P, Mell W et al (2012) Numerical simulation of crown fire hazard immediately after bark beetle-caused mortality in lodgepole pine forests. For Sci 58:178–188

Huberty AF, Denno RF (2004) Plant water stress and its consequences for herbivorous insects: a new synthesis. Ecology 85:1383–1398

Hulme PE (2006) Beyond control: wider implications for the management of biological invasions. J Appl Ecol 43:835–847

Hulme PE (2009) Trade, transport and trouble: managing invasive species pathways in an era of globalization. J Appl Ecol 46:10–18

Hunter R (1991) *Bromus* invasions on the Nevada Test Site: present status of *B. rubens* and *B. tectorum* with notes on their relationship to disturbance and altitude. Great Basin Naturalist 51:176–182

Intergovernmental Panel on Climate Change (IPCC) (2007) Climate change 2007: The physical science basis. In: Solomon S, Qin D, Manning M et al (eds) Contribution of Working Group I to the Fourth Assessment Report of the Intergovernmental Panel on Climate Change. Cambridge University Press, Cambridge, UK/New York. 996 p

Jamieson MA, Burkle LA, Manson JS et al (2017) Global change effects on plant insect interactions: the role of phytochemistry. Curr Opin Insect Sci 23:70–80

Jarnevich CS, Holcombe TR, Bella EM et al (2014) Cross-scale assessment of potential habitat shifts in a rapidly changing climate. Invasive Plant Sci Manag 7:491–502

Jenkins MJ, Runyon JB, Fettig CJ et al (2014) Interactions among the mountain pine beetle, fires, and fuels. For Sci 60:489–501

Jeschke JM, Strayer DL (2008) Usefulness of bioclimatic models for studying climate change and invasive species. Ann N Y Acad Sci 1134:1–24

Johnson SN, McNicol JW (2010) Elevated CO_2 and aboveground-belowground herbivory by the clover root weevil. Oecologia 162:209–216

Jones CC (2012) Challenges in predicting the future distribution of invasive plant species. For Ecol Manag 284:69–77

Jordan T, Nice G, Johnson B, Bauman T (2009) Reducing spray drift from glyphosate and growth regulator herbicide drift caution. Purdue University Extension Weed Science. https://ag.purdue.edu/btny/weedscience/Documents/ReducingDrift09.pdf

Keeley JE (2006) Fire management impacts on invasive plants in the western United States. Conserv Biol 20:375–384

Keeley JE, Allen CD, Betancourt J et al (2006) A 21st century perspective on postfire seeding. J For 104:1–2

Keller SR, Taylor DR (2008) History, chance and adaptation during biological invasion: separating stochastic phenotypic evolution from response to selection. Ecol Lett 11:852–866

Keller SR, Sowell DR, Neiman M, Wolfe LM, Taylor DR (2009) Adaptation and colonization history affect the evolution of clines in two introduced species. New Phytol 183:678–690

Kilkenny FF, Galloway LF (2013) Adaptive divergence at the margin of an invaded range. Evolution 67:722–731

Kilkenny FF, Galloway LF (2016) Evolution of marginal populations of an invasive vine increases the likelihood of future spread. New Phytol 209:1773–1780

Knepp RG, Hamilton JG, Mohan JE et al (2005) Elevated CO_2 reduces leaf damage by insect herbivores in a forest community. New Phytol 167:207–218

Kogan M, Bayer DE (1996) Herbicide uptake as influenced by plant water status. Pestic Biochem Physiol 56:174–182

Kolb TE, Fettig CJ, Bentz BJ et al (2016) Insects and pathogens. In: Vose J, Clark J, Luce C, Patel-Weynand T (eds) Effects of drought on forest ecosystems: a comprehensive science synthesis. WO-GTR-93b. Washington, DC: U.S. Department of Agriculture, Forest Service, Washington Office: 113–133

Kooyers NJ, Olsen KM (2012) Rapid evolution of an adaptive cyanogenesis cline in introduced North American white clover (*Trifolium repens* L.). Mol Ecol 21:2455–2468

Kuebbing SE, Nuñez MA, Simberloff D (2013) Current mismatch between research and conservation efforts: the need to study co-occurring invasive plant species. Biol Conserv 160:121–129

Kukla G, Karl TR (1993) Nighttime warming and the greenhouse effect. Environ Sci Technol 27:1468–1474

Kumschick S, Hufbauer RA, Alba C et al (2013) Evolution of fast-growing and more resistant phenotypes in introduced common mullein (*Verbascum thapsus*). J Ecol 101:378–387

Kurz WA, Dymond CC, Stinson G et al (2008a) Mountain pine beetle and forest carbon feedback to climate change. Nature 452:987–990

Kurz WA, Stinson G, Rampley GJ et al (2008b) Risk of natural disturbances makes future contribution of Canada's forests to the global carbon cycle highly uncertain. Proc Natl Acad Sci 105:1551–1555

Landguth EL, Cushman SA (2010) CDPOP: An individual-based, cost-distance spatial population genetics model. Mol Ecol Resour 10:156–161

Lankau RA, Nuzzo V, Spyreas G, Davis AS (2009) Evolutionary limits ameliorate the negative impact of an invasive plant. Proc Natl Acad Sci U.S.A 106:15362–15367

Leger EA, Espeland EK (2010) Coevolution between native and invasive plant competitors: implications for invasive species management. Evol Appl 3:169–178

Lehan NE, Murphy JR, Thorburn LP, Bradley BA (2013) Accidental introductions are important source of invasive plants in the Continental United States. Am J Bot 100:1287–1293

Leishman MR, Gallagher RV (2016) Will alien plant invaders be advantaged under future climates? In: Canning-Clode J (ed) Biological invasions in changing ecosystems. De Gruyter Open Ltd, Warsaw/Berlin, pp 368–388

Levinson DH, Fettig CJ (2014) Climate change: Overview of data sources, observed and predicted temperature changes, and impacts on public and environmental health. In: Pinkerton KE, Rom W (eds) Climate change and global public health. Springer-Verlag, New York, pp 31–49

Lewis DB, Castellano MJ, Kaye JP (2014) Forest succession, soil carbon accumulation, and rapid nitrogen storage in poorly remineralized soil organic matter. Ecology 95:2687–2693

Lindroth RL (2010) Impacts of elevated atmospheric CO_2 and O_3 on forests: phytochemistry, trophic interactions, and ecosystem dynamics. J Chem Ecol 36:2–21

Liu JP, Song MR, Horton RM, Hu YY (2013) Reducing spread in climate model projections of a September ice-free Arctic. Proc Natl Acad Sci 110:12571–12576

Lockwood JL, Cassey P, Blackburn T (2005) The role of propagule pressure in explaining species invasion. Trends Ecol Evol 20:223–228

Lombaert E, Estoup A, Facon B et al (2014) Rapid increase in dispersal during range expansion in the invasive ladybird *Harmonia axyridis*. J Evol Biol 27:508–517

Long SP, Ainsworth EA, Rogers A, Ort DR (2004) Rising atmospheric carbon dioxide: plants FACE the future. Annu Rev Plant Biol 55:591–628

Lovett GM, Weiss M, Leibhold AM et al (2016) Nonnative forest insects and pathogens in the United States: impacts and policy options. Ecol Appl 26:1437–1455

Lu X, Siemann E, He M et al (2015) Climate warming increases biological control agent impact on a non-target species. Ecol Lett 18:48–56

Lu X, Siemann E, He M et al (2016) Warming benefits a native species competing with an invasive congener in the presence of a biocontrol beetle. New Phytol 211:1371–1381

Mack RN, Erneberg M (2002) The United States naturalized flora: largely the product of deliberate introductions. Ann Mo Bot Gard 89:176–189

Mainali KP, Warren DL, Dhileepan K et al (2015) Projecting future expansion of invasive species: comparing and improving methodologies for species distribution modeling. Glob Chang Biol 21:4464–4480

Maines A, Knochel D, Seastedt T (2013) Biological control and precipitation effects on spotted knapweed (*Centaurea stoebe*): empirical and modeling results. Ecosphere 4:1–14

Maloy OC (1997) White pine blister rust control in North America: a case history. Annu Rev. Phytopathol 35:87–109

Manea A, Leishman MR, Downey PO (2011) Exotic C4 grasses have increased tolerance to glyphosate under elevated carbon dioxide. Weed Sci 59:28–36

Marco A, Lavergne S, Dutoit T et al (2010) From the backyard to the backcountry: how ecological and biological traits explain the escape of garden plants into Mediterranean old fields. Biol Invasions 12:761–779

Maron JL, Marler M (2008) Field-based competitive impacts between invaders and natives at varying resource supply. J Ecol 96:1187–1197

Maron JL, Vilà M, Bommarco R et al (2004) Rapid evolution of an invasive plant. Ecol Monogr 74:261–280

Martin TG, Murphy H, Liedloff A et al (2015) Buffel grass and climate change: a framework for projecting invasive species distributions when data are scarce. Biol Invasions 17:3197–3210

Mattson WJ Jr (1980) Herbivory in relation to plant nitrogen content. Annu Rev Ecol Syst 11:119–161

Mattson WJ Jr, Haack RA (1987) The role of drought in outbreaks of plant-eating insects. Bioscience 37:110–118

McDowell NG, Beerling DJ, Breshears DD et al (2011) The interdependence of mechanisms underlying climate-driven vegetation mortality. Trends Ecol Evol 26:523–532

McKenney DW, Pedlar JH, Lawrence K et al (2007) Potential impacts of climate change on the distribution of North American trees. Bioscience 57:939–948

McKeon CS, Weber MX, Alter SE et al (2016) Melting barriers to faunal exchange across ocean barriers. Glob Chang Biol 22:465–473

Meentemeyer R, Rizzo D, Mark W, Lotz E (2004) Mapping the risk of establishment and spread of sudden oak death in California. For Ecol Manag 200:195–214

Meentemeyer RK, Cunniffe NJ, Cook AR et al (2011) Epidemiological modeling of invasion in heterogeneous landscapes: spread of sudden oak death in California (1990–2030). Ecosphere 2:1–24

Meigs GW, Zald HSJ, Campbell JL et al (2016) Do insect outbreaks reduce the severity of subsequent forest fires? Environ Res Lett 11:045008

Melillo JM, Richmond TC, Yohe GW (eds) (2014) Climate change impacts in the United States: the third national climate assessment. U.S. Global Change Research Program, 841 p. https://doi.org/10.7930/J0Z31WJ2

Metz MR, Frangioso KM, Meentemeyer RK, Rizzo DM (2011) Interacting disturbances: wildfire severity affected by stage of forest disease invasion. Ecol Appl 21:313–320

Metz MR, Frangioso KM, Wickland AC et al (2012) An emergent disease causes directional changes in forest species composition in coastal California. Ecosphere 3:1–23

Michener WK, Blood ER, Bildstein KL, Brinson MM, Gardner L (1997) Climate change, hurricanes and tropical storms, and rising sea level in coastal wetlands. Ecol Appl 7:770–801

Miller-Rushing AJ, Lloyd-Evans TL, Primack RB et al (2008) Bird migration times, climate change, and changing population size. Glob Chang Biol 14:1959–1972

Mooney HA, Cleland EE (2001) The evolutionary impact of invasive species. Proc Natl Acad Sci 98:5446–5451

Morehouse K, Johns T, Kaye J, Kaye M (2008) Carbon and nitrogen cycling immediately following bark beetle outbreaks in Southwestern ponderosa pine forests. For Ecol Manag 255:2698–2708

Morrison RG, Lownds NK, Sterling TM (1995) Picloram uptake, translocation, and efficacy in relation to water status of Russian knapweed (*Acroptilon repens*). Weed Sci 43:34–39

Morse LE, Kartesz JT, Kutner LS (1995) Native vascular plants. In: LaRoe ET, Farris GS, Puckett CE, Doran PD, Mac MJ (eds) Our living resources: a report to the nation on the distribution, abundance, and health of U.S. plants, animals, and ecosystems. U.S. Department of the Interior, National Biological Service, Washington, DC, pp 205–209

Mráz P, Tarbush E, Müller-Schärer H (2014) Drought tolerance and plasticity in the invasive knapweed *Centaurea stoebe* s.l. (Asteraceae): effect of populations stronger than those of cytotype and range. Ann Bot 114:289–299

Nave LE, Vance ED, Swanston CW, Curtis PS (2010) Harvest impacts on soil carbon storage in temperate forests. For Ecol Manag 259:857–866

Nave LE, Vance ED, Swanston CW, Curtis PS (2011) Fire effects on temperate forest soil C and N storage. Ecol Appl 21:1189–1201

Norberg J (2004) Biodiversity and ecosystem functioning: a complex adaptive systems approach. Limnol Oceanogr 49:1269–1277

Nuckolls AE, Wurzburger N, Ford CR et al (2009) Hemlock declines rapidly with hemlock woolly adelgid infestation: impacts on the carbon cycle of southern Appalachian forests. Ecosystems 12:179–190

Oliva J, Stenlid J, Martínez-Vilalta J (2014) The effect of fungal pathogens on the water and carbon economy of trees: implications for drought-induced mortality. New Phytol 203:1028–1035

Orwig DA, Thompson JR, Povak NA et al (2012) A foundation tree at the precipice: *Tsuga canadensis* health after the arrival of *Adelges tsugae* in central New England. Ecosphere 3:10

Orwig DA, Plotkin AAB, Davidson EA et al (2013) Foundation species loss affects vegetation structure more than ecosystem function in a northeastern USA forest. Peer J 1:e41

Parker IM, Rodriguez J, Loik M (2003) An evolutionary approach to understanding the biology of invasions: local adaptation and general purpose genotypes in the weed *Verbascum thapsus*. Conserv Biol 17:59–72

Parmesan C (2006) Ecological and evolutionary responses to recent climate change. Annu Rev. Ecol Evol Syst 37:637–669

Parmesan C, Yohe G (2003) A globally coherent fingerprint of climate change impacts across natural systems. Nature 421:37–42

Pearman PB, Guisan A, Broennimann O et al (2007) Niche dynamics in space and time. Trends Ecol Evol 23:149–158

Pearson RG, Dawson TP (2003) Predicting the impacts of climate change on the distribution of species: are bioclimate envelope models useful? Glob Ecol Biogeogr 12:361–371

Pearson DE, Ortega YK, Runyon JB et al (2016) Secondary invasion: the bane of weed management. Biol Conserv 197:8–17

Percy KE, Awmack CS, Lindroth RL et al (2002) Altered performance of forest pests under atmospheres enriched by CO_2 and O_3. Nature 420:403–407

Peterson AT (2003) Predicting the geography of species' invasions via ecological niche modeling. Q Rev. Biol 78:419–433

Peterson AT (2011) Ecological niche conservatism: a time structured review of evidence. J Biogeogr 38:817–827

Petitpierre B, Kueffer C, Broennimann O et al (2012) Climatic niche shifts are rare among terrestrial plant invaders. Science 335:1344–1347

Phillips BL, Shine R (2006) An invasive species induces rapid adaptive change in a native predator: cane toads and black snakes in Australia. Proc R Soc B 273:1545–1550

Phillips BL, Brown GP, Webb JK, Shine R (2006) Invasion and the evolution of speed in toads. Nature 439:803–803

Prado SS, Hung KY, Daugherty MP, Almeida RPP (2010) Indirect effects of temperature on stink bug fitness, via maintenance of gut-associated symbionts. Appl Environ Microbiol 76:1261–1266

Pritchard SG, Rogers HH, Prior SA, Peterson CM (1999) Elevated CO_2 and plant structure: a review. Glob Chang Biol 5:807–837

Pyke CR, Thomas R, Porter RD et al (2008) Current practices and future opportunities for policy on climate change and invasive species. Conserv Biol 22:585–592

Raffa KF, Gregoire J-C, Lindgren BS (2015) Natural history and ecology of bark beetles. In: Vega FE, Hofstetter RW (eds) Bark beetles: biology and ecology of native and invasive species. Springer, London, 40 p

Rehfeldt GE, Tchebakova NM, Barnhardt LK (1999) Efficacy of climate transfer functions: introduction of Eurasian populations of *Larix* into Alberta. Can J For Res 29:1660–1668

Rehfeldt GE, Crookston NL, Sáenz-Romero C, Campbell EM (2012) North American vegetation model for land-use planning in a changing climate: a solution to large classification problems. Ecol Appl 22:119–141

Reichard SH, White P (2001) Horticulture as a pathway of invasive plant introductions in the United States. Bioscience 51:103–113

Reynolds C, Miranda NAF, Cumming GS (2015) The role of waterbirds in the dispersal of aquatic alien and invasive species. Divers Distrib 21:744–754

Richardson H, Nemeth DJ (1991) Hurricane-borne African locusts (*Schistocerca gregaria*) on the Windward Islands. Geo J 23:349–357

Richardson DM, Pysek P, Rejmanek M et al (2000) Naturalization and invasion of invasive plants: concepts and definitions. Divers Distrib 6:93–107

Rizzo DM, Garbelotto M, Hansen EM (2005) *Phytophthora ramorum*: Integrative research and management of an emerging pathogen in California and Oregon forests. Annu Rev. Phytopathol 43:309–335

Robinson EA, Ryan GD, Newman JA (2012) A meta-analytical review of the effects of elevated CO_2 on plant–arthropod interactions highlights the importance of interacting environmental and biological variables. New Phytol 194:321–336

Rouault G, Candau J-N, Lieutier F et al (2006) Effects of drought and heat on forest insect populations in relation to the 2003 drought in Western Europe. Ann For Sci 63:613–624

Runyon JB, Butler JL, Friggens MM et al (2012) Invasive species and climate change (Chapter 7). In: Finch DM (ed) Climate change in grasslands, shrublands, and deserts of the interior American West: a review and needs assessment. General technical report RMRS-GTR-285. U.S. Department of Agriculture, Forest Service, Rocky Mountain Research Station, Fort Collins, pp 97–115

Schneider RW, Hollier CA, Whitam HK et al (2005) First report of soybean rust caused by *Phakopsora pachyrhizi* in the continental United States. Plant Dis 89:774

Schwilk DW, Keeley JE, Knapp EE et al (2009) The National Fire and Fire Surrogate Study: effects of alternative fuel reduction methods on forest structure. Ecol Appl 19:285–304

Seastedt TR, Knochel DG, Garmoe M et al (2007) Interactions and effects of multiple biological control insects on diffuse and spotted knapweed in the Front Range of Colorado. Biol Control 42:345–354

Seebens H, Essl F, Dawson W et al (2015) Global trade will accelerate plant invasions in emerging economies under climate change. Glob Chang Biol 21:4128–4140

Sheppard CS, Burns BR, Stanley MC (2014) Predicting plant invasions under climate change: are species distribution models validated by field trials. Glob Chang Biol 20:2800–2814

Shirk, AJ, Cushman SA, Waring KM et al (2018) Southwestern white pine (Pinus strobiformis) species distribution models project a large range shift and contraction due to regional climatic changes. Forest Ecol Manag 411:127–186

Simard M, Romme WH, Griffin JM, Turner MG (2010) Do mountain pine beetle outbreaks change the probability of active crown fire in lodgepole pine forests? Ecol Monogr 81:3–24

Six DL, Bentz BJ (2007) Temperature determines symbiont abundance in a multipartite bark beetle-fungus ectosymbiosis. Microb Ecol 54:112–118

Skinner K, Smith L, Rice P (2000) Using noxious weeds to prioritize targets for developing weed management strategies. Weed Sci 48:640–644

Smeets CJPP, Holzinger R, Vigano I et al (2009) Eddy covariance methane measurements at a Ponderosa pine plantation in California. Atmos Chem Phys 9:8365–8375

Smith SD, Strain BR, Sharkey TD (1987) Effects of CO_2 enrichment on four Great Basin grasses. Funct Ecol 1:139–143

Smith SD, Huxman TE, Zitzer SF et al (2000) CO_2 increases productivity and invasive species success in an arid ecosystem. Nature 408: 79–82

Smith AL, Hewitt N, Klenk N, et al (2012) Effects of climate change on the distribution of invasive alien species in Canada: a knowledge synthesis of range change projections in a warming world. Environ Rev 20:1–16

Sobek-Swant S, Crosthwaite JC, Lyons DB et al (2012) Could phenotypic plasticity limit an invasive species? Incomplete reversibility of mid-winter deacclimation in emerald ash borer Biol Invasions 14:115–125

Sorte CJ, Ibanez I, Blumenthal DM et al (2013) Poised to prosper? A cross-system comparison of climate change effects on native and non-native species performance. Ecol Lett 16:261–270

Stein BA, Flack SR (1996) America's least wanted: alien species invasions of U.S. ecosystems. The Nature Conservancy, Arlington, 31 p

Stephens SL, McIver JD, Boerner REJ et al (2012) Effects of forest fuel-reduction treatments in the United States. Bioscience 62: 549–560

Stergiopoulos I, Gordon TR (2014) Cryptic fungal infections: the hidden agenda of plant pathogens. Plant-Microbe Interact 5:506

Story JM, Callan NW, Corn JG et al (2006) Decline of spotted knapweed density at two sites in western Montana with large populations of the introduced root weevil, *Cyphocleonus achates* (Fahraeus). Biol Control 38:227–232

Stroeve JC, Serreze MC, Holland MM et al (2012) The Arctic's rapidly shrinking sea ice cover: a research synthesis. Clim Chang 110:1005–1027

Sturrock RN, Frankel SJ, Brown AV et al (2011) Climate change and forest diseases. Plant Pathol 60:133–149

Suttle KB, Thomsen MA, Power ME (2007) Species interactions reverse grassland responses to changing climate. Science 315:640–642

Tabernero MT, Álvarez-Benedí J, Atienza J, Herguedas A (2000) Influence of temperature on the volatilization of triallate and terbutryn from two soils. Pest Manag Sci 56:175–180

Theoharides KA, Dukes JS (2007) Plant invasion across space and time: factors affecting non-indigenous species success during four stages of invasion. New Phytol 176:256–273

Thomas MB, Blanford S (2003) Thermal biology in insect-parasite interactions. Trends Ecol Evol 18:344–350

Thuiller W, Albert C, Araujo MB et al (2007) Predicting global change impacts on plant species' distributions: future challenges. Perspect Plant Ecol Evol Syst 9:137–157

Tobin PC, Nagarkatti S, Loeb G, Saunders MC (2008) Historical and projected interactions between climate change and insect voltinism in a multivoltine species. Glob Chang Biol 14:951–957

Travis JMJ, Dytham C (2002) Dispersal evolution during invasions. Evol Ecol Res 4:1119–1129

Turner KG, Fréville H, Riesberg LH (2015) Adaptive plasticity and niche expansion in an invasive thistle. Ecol Evol 5:3183–3197

Tylianakis JM, Didham RK, Bascompte J, Wardle DA (2008) Global change and species interactions in terrestrial ecosystems. Ecol Lett 11:1351–1363

Valachovic YS, Lee CA, Scanlon H et al (2011) Sudden oak death-caused changes to surface fuel loading and potential fire behavior in Douglas-fir-tanoak forests. For Ecol Manag 261:1973–1986

van Asch M, Tienderen V, Peter H et al (2007) Predicting adaptation of phenology in response to climate change, an insect herbivore example. Glob Chang Biol 13:1596–1604

Verlinden M, Nijs I (2010) Alien plant species favoured over congeneric natives under experimental climate warming in temperate Belgian climate. Biol Invasions 12:2777–2787

Viana DS, Gangoso L, Bouten W, Figuerola J (2016) Overseas seed dispersal by migratory birds. Proc R Soc B 283:2015–2046

Vitousek PM (1994) Beyond global warming – ecology and global change. Ecology 75:1861–1876

Vose J, Clark J, Luce C, Patel-Weynand T (eds) (2016) Effects of drought on forest ecosystems: a comprehensive science synthesis. WO-GTR-93b. U.S. Department of Agriculture, Forest Service, Washington Office, Washington, DC, 289 p

Wainwright CE, Wolkovich EC, Cleland EE (2012) Seasonal priority effects: implications for invasion and restoration in semi-arid system. J Appl Ecol 49:234–241

Walsh J, Wuebbles D, Hayhoe K et al (2014) Appendix 3: climate science supplement. In: Melillo JM, Richmond TC, Yohe GW (eds) Climate change impacts in the United States: the third national climate assessment. Washington, DC: U.S. Global Change Research Program. Pages: 735–789

Walther GR, Roques A, Hulme PE et al (2009) Alien species in a warmer world: risks and opportunities. Trends Ecol Evol 24:686–693

Wand SJE, Midgley GF, Jones MH, Curtis PS (1999) Responses of wild C_4 and C_3 grass (Poaceae) species to elevated atmospheric CO_2 concentration: a meta-analytic test of current theories and perceptions. Glob Chang Biol 5:723–741

Wang T, Hamann A, Yanchuck A, O'Neill GA, Aitkin SN (2006) Use of response functions in selecting lodgepole pine populations for future climates. Glob Chang Biol 12:2404–2416

Wang T, Campbell EM, O'Neill A, Aitken SN (2012) Projecting future distributions of ecosystem climate niches: uncertainties and management applications. For Ecol Manag 279:128–140

Watt AD, McFarlane AM (2002) Will climate change have a different impact on different trophic levels? Phenological development of winter moth *Operophtera brumata* and its host plants. Ecol Entomol 27:254–256

Weed AS, Ayres MP, Hicke JA (2013) Consequences of climate change for biotic disturbances in North American forests. Ecol Monogr 83:441–470

Westerling AL, Hidalgo HG, Cayan DR, Swetnam TW (2006) Warming and earlier spring increases western U.S. forest wildfire activity. Science 313:940–943

White PS, Pickett STA (1985) The ecology of natural disturbance and patch dynamics. Academic Press, Orlando, 472 p

Willis CG, Ruhfel BR, Primack RB et al (2010) Favorable climate change response explains non-native species' success in Thoreau's woods. PLoS One 5:e8878

Winder M, Schindler DE (2004) Climate change uncouples trophic interactions in an aquatic ecosystem. Ecology 85:2100–2106
Wolkovich EM, Cleland EE (2011) The phenology of plant invasions: a community ecology perspective. Front Ecol Environ 9:287–229
Wolkovich EM, Cleland EE (2014) Phenological niches and the future of invaded ecosystems with climate change. AoB Plants 6:plu013
Wolkovich EM, Davis TJ, Schaefer H, Cleland EE, Cook BI, Travers SE, Willis CG, Davis CC (2013) Temperature-dependent shifts in phenology contribute to the success of exotic species with climate change. Am J Bot 100:1407–1421
Woods A, Coates KD, Hamann A (2005) Is an unprecedented Dothistroma needle blight epidemic related to climate change? Bioscience 55:761–769
Woods AJ, Martín-García J, Bulman L et al (2016) Dothistroma needle blight, weather and possible climatic triggers for the disease's recent emergence. For Pathol 46:443–452
Xu C-Y, Tang S, Fatemi M et al (2015) Population structure and genetic diversity of invasive *Phyla canescens*: implications for the evolutionary potential. Ecosphere 6:162
Yuan JS, Himanen SJ, Holopainen JK et al (2009) Smelling global climate change: mitigation of function for plant volatile organic compounds. Trends Ecol Evol 24:323–331
Zavala JA, Casteel CL, DeLucia EH et al (2008) Anthropogenic increase in carbon dioxide compromises plant defense against invasive insects. Proc Natl Acad Sci 105:5129–5133
Zelikova TJ, Hufbauer RA, Reed SL et al (2013) Eco-evolutionary responses of *Bromus tectorum* to climate change: implications for biological invasions. Ecol Evol 3:1374–1387
Ziska LH, Dukes JS (2011) Weed biology and climate change. Wiley-Blackwell, Ames, 235 pages
Ziska LH, Teasdale JR, Bunce JA (1999) Future atmospheric carbon dioxide may increase tolerance to glyphosate. Weed Sci 47:608–615
Ziska LH, Faulkner S, Lydon J (2004) Changes in biomass and root: shoot ratio of field-grown Canada thistle (*Cirsium arvense*), a noxious, invasive weed, with elevated CO_2: implications for control with glyphosate. Weed Sci 52:584–588
Ziska LH, Reeves JB III, Blank B (2005) The impact of recent increases in atmospheric CO_2 on biomass production and vegetative retention of cheatgrass (*Bromus tectorum*): implications for fire disturbance. Glob Chang Biol 11:1325–1332

Invasive Species Response to Natural and Anthropogenic Disturbance

5

Susan E. Meyer, Mac A. Callaham Jr, Jane E. Stewart, and Steven D. Warren

5.1 Introduction

Much of the literature dealing with the biology and management of invasive species has focused on the damaging ecological and economic consequences of invasions (see Chaps. 2, 3, and 14 of this volume for review). In this chapter, we shift the focus to the causes of invasion, with the goal of proactively limiting or preventing invasions rather than reacting to them once they have occurred. Preventing the introduction of invasive species is one key element in this proactive approach (Chap. 6, this volume). Here, we specifically focus on ecosystem attributes that affect whether or not an ecosystem is vulnerable to invasion, that is, the features that affect its invasibility (Lonsdale 1999), with particular emphasis on the role of natural and anthropogenic disturbance.

The question of what makes an ecosystem susceptible to invasion has been the subject of intense scientific investigation for several decades, and the literature on this topic is very extensive. Unfortunately, scrutiny of the many thousands of studies on invasion has not resulted in the emergence of a clear paradigm that can explain the invasion process. Several research groups have attempted to use this large body of knowledge to create a conceptual framework or synthesis for invasion biology that can at least provide a basis for the design of studies that can fill knowledge gaps and critically address long-held assumptions based on limited information (Catford et al. 2009; Davis et al. 2000; Funk et al. 2008; Sher and Hyatt 1999; Theoharides and Dukes 2007). These syntheses often have the specific intent of describing the concepts of invasion biology under the more general rubric of ecology, with the idea that the underlying processes regulating community assembly in native ecosystems should be the same as those that mediate invasions (Davis et al. 2005; Facon et al. 2006; Gurevitch et al. 2011; MacDougall et al. 2009; Moles et al. 2012; Shea and Chesson 2002). More recently, there has been recognition of the complexity of the invasion process and the low likelihood that broad generalizations applicable across classes of organisms (e.g., plants, pathogens, insects, earthworms, vertebrates) will emerge (Catford et al. 2012a; Heger et al. 2013; Jeschke et al. 2012; Jeschke 2014; Kueffer et al. 2013). Nonetheless, these syntheses provide a good starting point for discussing the factors that predispose an ecosystem to invasion. A principal tool of more recent syntheses is meta-analysis, where the results of numerous published studies are analyzed together to look for underlying patterns.

Historically, the concepts of ecosystem invasibility and species invasiveness have been considered largely separately. Many have attempted to define what attributes make an ecosystem prone to invasion (e.g., Catford et al. 2012a; Davis et al. 2000; Fridley et al. 2007; Lonsdale 1999) and also what attributes make a species a potentially serious invader (e.g., Colautti et al. 2014a; Dick et al. 2014; Hayes and Barry 2008; Moles et al. 2008; Rejmanek and Richardson 1996; Shea et al. 2004). In practice, however, invasibility is not a static property of an ecosystem but instead results from an interplay between current ecosystem condition and ecological properties of the potential invader. The probability of a significant invasion is a result of the interaction of these two sets of factors, along with factors involved with the current population status of the

S. E. Meyer (✉)
U.S. Department of Agriculture, Forest Service, Rocky Mountain Research Station, Shrub Sciences Laboratory, Provo, UT, USA
e-mail: susan.meyer@usda.gov

M. A. Callaham Jr
U.S. Department of Agriculture, Forest Service, Southern Research Station, Center for Forest Disturbance Science, Athens, GA, USA

J. E. Stewart
Department of Bio-agricultural Sciences and Pest Management, Colorado State University, Fort Collins, CO, USA

S. D. Warren
U.S. Department of Agriculture, Forest Service, Rocky Mountain Research Station, Shrub Sciences Laboratory, Provo, UT, USA

T. M. Poland et al. (eds.), *Invasive Species in Forests and Rangelands of the United States*,
https://doi.org/10.1007/978-3-030-45367-1_5

potential invader, particularly as it affects local propagule pressure, that is, the number of invader individuals in proximity to an uninvaded area of interest (Simberloff 2009; Strayer et al. 2006).

5.2 Natural Versus Anthropogenic Disturbance

Disturbance regime is a key ecosystem attribute that can play a major role in mediating invasibility. Ecosystem response to disturbance depends on ecological resilience, that is, the capacity to return to a relatively undisturbed condition following disturbance (Gunderson 2000; Holling 1973). A resilient ecosystem can rebound from disturbance rather than shifting to a different structural and functional condition that is likely to represent a permanent change, that is, a transition to an alternative stable state (Briske et al. 2008). Disturbance regimes that are within the historic range of variation, within the bounds that existed prior to the initiation of human influence, have persisted through long periods of time. This implies that the ecosystem has generally been resilient to this range of disturbance variation and can return to its previous condition following a disturbance event (Landres et al. 1999).

Anthropogenic (human-caused) disturbance, on the other hand, can quickly take an ecosystem outside the bounds of the natural range of disturbance variation and potentially into another stable state. Ecological resilience is not a constant property of an ecosystem, but is itself subject to change. Processes that erode ecological resilience increase the chances of apparently catastrophic changes in state that are often interpreted by managers as "ecological crises" (Gunderson 2000). A decrease in resilience may result from narrowly focused management for a single goal such as timber harvest or livestock production, which can itself impose severe disturbance or chronic stress. For example, reduced resilience, largely as a consequence of the chronic stress of livestock grazing, resulted in a transition from savanna to wooded thickets as an alternative stable state in parts of the American Southwest (Gunderson 2000).

Ecological resilience to disturbance is directly tied to resistance to invasion and therefore to invasibility. This is because disturbance usually decreases resistance to invasion, particularly for invasive plants (Davis et al. 2000; Hobbs and Huenneke 1992). When an invasive species enters the scene after ecosystem resilience and resistance to invasion have been reduced, its arrival represents a disturbance event that can lead to expanding invader monocultures that replace native vegetation as an alternative stable state (e.g., Chambers et al. 2014).

We define natural disturbance very broadly to include episodic events such as earth movement (e.g., landslides), fire, flooding, weather-related events such as wind throw and ice-storm breakage, and biological phenomena such as native insect and disease outbreaks. Land-use changes for agricultural and urban development are examples of drastic anthropogenic disturbance, but road construction, mining, energy development, hunting, recreational use, and harvest of timber, firewood, and other natural products are also important sources of anthropogenic disturbance. We also include chronic disturbance or stressors such as livestock grazing, nutrient pollution, changes in hydrologic regime, and prolonged drought. Many sources of chronic stress are anthropogenic in origin.

We also recognize that disturbance occurs on many temporal and spatial scales and that there is a continuum between processes, such as the activities of native herbivores, that may not disturb the system in the short run, but which, when they occur at abnormally high levels, can qualify as sources of disturbance (e.g., Kalisz et al. 2014). Other examples of this continuum of disturbance include the effects of endemic versus epidemic levels of native insects and pathogens (e.g., bark beetles (*Ips* spp.; Raffa et al. 2008)).

5.3 Resistance to Invasion

The invasion process can be considered in three stages: introduction, establishment, and spread. The introduction stage is considered in detail in Chap. 6 of this volume. Here we consider factors that affect the likelihood of establishment and spread. Much of the discussion in this section pertains to plant invasions. The broader concepts of invasibility and resistance to invasion can meaningfully be applied to consumer invasions, however, as discussed later in this chapter.

5.3.1 Abiotic Resistance

In order to succeed, a potential invader must first establish locally, a process that is regulated largely by the "match" of its ecological tolerance to climatic and other environmental conditions in the ecosystem potentially invaded. If the match is poor, the potential invader has arrived in an environment that is outside its range of environmental tolerance. This means that the invasion will not take place regardless of the potential invasibility of the ecosystem. This type of resistance to invasion is called abiotic resistance and is probably the most common reason that invaders fail to establish. Abiotic resistance to invasion by a newly arrived species is a direct consequence of the species' lack of ecological match to the environment. An ecosystem could have high abiotic resistance to one invader, yet present low abiotic resistance to another invader with contrasting ecological tolerance that is a better match.

If the potential invader is only marginally adapted to the abiotic environment but able to establish a founder population, it may undergo microevolution in situ, either through selection on standing genetic variation present in the founder population or through selection on novel forms that arise by chance, thereby increasing its degree of adaptation to the abiotic environment. This microevolutionary process has been documented for several invasive plants and is thought to be one explanation for the commonly observed "lag phase" of plant invasion (Aikio et al. 2010; Crooks 2005). It is important to realize that invader populations are not necessarily static in terms of genetic composition, and that such evolution can take place over short time scales and play an important part in increasing the magnitude and severity of particular invasions (Bossdorf et al. 2005; Lee 2002). Another explanation for the "lag phase" is the trend for exponential increase, that is, even for a well-adapted invader, propagule pressure takes time to build to levels that permit large-scale invasion (Crooks 2005).

5.3.2 Biotic Resistance

Another form of resistance to invasion that is presented by natural ecosystems is termed biotic resistance. Biotic resistance can be defined as the sum of all negative ecological impacts that the native organisms (and any other organisms) in an ecosystem can have on a particular invasive species and that can reduce or prevent its success (Levine et al. 2004). For plant invaders, these impacts can include competition, herbivory, impacts from pathogens and parasites, and negative soil microbial effects. In a meta-analysis of the sources of biotic resistance to plant invasion, Levine et al. (2004) found that both competition from resident native plants and herbivory from resident native herbivores had a large negative effect on invasibility, while the few studies that examined the effects of the soil microbial community produced contradictory evidence. They found that these sources of biotic resistance could not prevent initial colonization but instead operated to slow the rate of population increase and spread. However, biotic resistance may only constrain the extent of an invasion when propagule pressure is low (Kerns and Day 2017).

Biotic resistance to population increase and spread is probably another reason why most non-native species do not become invasive even in ecosystems where they are adapted to the abiotic regime. Most remain minor players in the ecosystems they colonize, and only a few become problematic as major invaders. The distinction between "weak" and "strong" invaders is an important one (Ortega and Pearson 2005). Weak invaders may behave much like native species in the regional species pool, able to establish and persist under conditions of moderate disturbance but not likely to experience major population growth. Although they are non-native, they do not meet the definition of an invasive species. This may be due to intrinsic features of life history that make the introduced species unable to experience explosive growth even under favorable conditions, or it may be due to a poor ecological fit with a particular ecosystem. The distinction between weak and strong invaders can therefore sometimes be ecosystem-dependent.

Disturbance and Resource Availability

On a community level, processes that increase resource availability (e.g., water, light, nutrients) have been found to make a community more open to invasion by both natives in the regional species pool and non-native species, while processes that reduce resource availability make it less open to invasion. Resource availability can be increased through physical disturbance or other processes that reduce plant biomass and therefore resource demand, or through resource pulses, which temporarily flood the system with resources that the resident community cannot efficiently use (Davis et al. 2000). The unpredictability of invasion hinges on the necessity of propagule arrival that coincides temporally with such a period of increased resource availability, which itself fluctuates unpredictably through time.

Disturbance can occur on multiple spatial scales, and the scale of disturbance has a major effect on both resource availability and on the likelihood that newly arriving invader propagules will "find" a resource-rich disturbance. Wildfire probably represents the broadest scale natural disturbance in many ecosystems, and its effects on invasibility for disturbance-responsive invaders are well documented (Zouhar et al. 2008). Kerns and Day (2017) found that at fine scales, cheatgrass (*Bromus tectorum*) establishment in areas with cheatgrass in the vicinity was correlated with burn extent even for low-intensity prescribed burning in forested stands. But even the most stable ecosystems have fine-scale disturbances that can provide an initial toe-hold for invaders, for example, rodent mounds, new soil deposition from localized flooding events, or gaps formed by the death of individual trees.

The distinction between invasibility and invadedness is crucial in the interpretation of correlative studies of factors such as disturbance that mediate species invasions. Level of invadedness results from the interaction of ecosystem invasibility with the invader species pool and the ecological attributes of both invaders and recipient ecosystems, as well as with propagule pressure and the timeline of invasion (Guo et al. 2015; Simberloff 2009; Strayer et al. 2006). Post hoc level of invadedness is therefore not a reliable indicator of ecosystem invasibility. For example, Moles et al. (2012) found in a meta-analysis that current and past disturbance regimes per se had low ability to predict the degree of plant invasion, accounting for <10% of the variation in invadedness. They used a correlative approach to relate the degree of invadedness in a large number of different ecosystems to reported levels of disturbance in those ecosystems. This approach conflates invasibility with observed levels of invadedness, which casts doubt on its conclusion that disturbance is a relatively unimportant factor in invasibility.

It has been pointed out that most studies of invasion deal with species that are disturbance-responsive even in their native range, and this may have led to an overemphasis on disturbance as a mediator of invasion (Martin et al. 2009). The first serious plant invaders to arrive in North America were likely inadvertently introduced by people, often in the context of agriculture or animal husbandry. They originated in human-disturbed landscapes and were likely favored by such disturbance in both the native and introduced ranges. More recently, however, the majority of new invaders have been introduced deliberately, often as horticultural selections, and species attractive to horticulturalists can originate from any ecosystem in any condition of disturbance (Reichard and White 2001). Undisturbed forests have been thought to be relatively resistant to invasion, but this could have been an artifact of the poor match of most of these early invaders to the deep shade environment of mature forests. The list of invasive plant species that have succeeded in mature forests is becoming longer through time, and virtually all of these invaders are horticultural introductions (Martin et al. 2009). This emphasizes the importance of the ecological match of the new invader to the ecosystem and implies that protecting an ecosystem from disturbance is no guarantee that it will remain uninvaded, particularly by this new invader class, which presents a much wider array of ecological tolerances than invaders introduced as accidents of agriculture.

Species Composition

Another community-level factor that has been found in many small plot studies to reduce invasibility on a local scale is biodiversity or species richness. Ecological niche theory predicts that the more species included in a community, the more completely the niche space will be filled, and therefore the more completely resources will be utilized (Shea and Chesson 2002). Experimental studies of community assembly almost always yield this negative effect of species richness on probability of subsequent successful invasion by novel species (e.g., Fargione and Tilman 2005; Kennedy et al. 2002; Maron and Marler 2007). Theoretical work has also supported this hypothesis (Case 1990).

Sometimes the negative effect of increased diversity on invasibility can be credited to the increased presence of species that are functionally similar to a newly invading species or in some other way better able to fill its niche (Fargione and Tilman 2005). Evidence for the importance of the presence of functionally similar species in biotic resistance to plant invasion is equivocal, however. For example, Price and Partel (2013) showed in a meta-analysis that functionally similar species could limit herbaceous dicot invasion but not grass invasion. Similarly, in a manipulative removal study, Pokorney et al. (2005) showed that removal of a functionally similar group (herbaceous dicots) increased invasion of spotted knapweed (*Centaurea stoebe*) more than removal of dominant grasses. Even uncommon species can add to the overall biotic resistance of a plant community to a potential invader (Lyons and Swartz 2001). Experimental changes in biodiversity through random removal of species may not reflect the real impact of species loss on invasibility (Zavaleta and Hulvey 2007). Selments et al. (2012) showed in experiments in serpentine grassland that realistic species removal, i.e., removal of species most likely to be lost during extended drought, increased invasibility more than random species removal.

Biodiversity studies at regional scales have often yielded results that seem to contradict the results of local-scale studies, in that native biodiversity is commonly positively associated with non-native biodiversity (e.g., Stohlgren et al. 1999; Stohlgren et al. 2003). This alarming result seems to suggest that native biodiversity "hot spots" are more prone to invasion by non-native species than less diverse ecosystems. The reasons for this "invasion paradox" appear to be complex (Fridley et al. 2007). One possibility is that, on a regional scale, high environmental heterogeneity creates a multiplicity of niches that can potentially be filled by any species from the regional species pool, whether native or introduced. As mentioned earlier, high non-native biodiversity can result from colonization of weak invaders from the regional species pool and may not indicate imminent invasion by a strong invader. This regional-scale relationship does not negate the finding that high native biodiversity at a local scale can decrease community invasibility.

Recent work by Iannone et al. (2016) on the relationship between non-native species richness and community attributes in eastern hardwood forests showed that native and introduced species richness were positively correlated at the landscape scale as in earlier large-scale studies. However, a measure of native diversity that incorporated degree of phylogenetic relatedness ("evolutionary diversity") was negatively correlated with introduced species diversity, indicating that species richness per se may not be the best measure for predicting biotic resistance. Their study would predict, for example, that a forest dominated by multiple species of oak (*Quercus* spp.) would be more easily invaded than a forest dominated by the same number of tree species that are not so closely related and perhaps better able to fill the available niche space.

Natural Enemies

One of the principal hypotheses explaining why a species that is not invasive in its native range becomes invasive in the introduced range is the idea of escape from natural enemies that keep it in check, particularly host-specific natural enemies that are unlikely to have homologues in the introduced range (Heger and Jeschke 2014; Keane and Crawley 2002). This hypothesis is the basis of classical biocontrol, which seeks to reunite invasive species with their specialist natural enemies from the native range (Clewley et al. 2012; see Chap. 7 this volume). Generalist natural enemies, on the other hand, are often represented by similar species in the introduced range, and these can contribute substantially to biotic resistance to invasion if they feed on the invasive species. Native generalist herbivores often increase resistance to invasion by introduced plant species (Parker et al. 2006), though there are exceptions (Maron and Vila 2001; Vavra et al. 2007). For example, excessive native deer (*Odocoileus virginianus* Boddaert) herbivory on native vegetation caused by overstocking has been shown to facilitate plant invasion (Eschtruth and Battles 2008; Kalisz et al. 2014).

Facilitation of plant invasion by non-native herbivores is a much more common pattern than facilitation by native herbivores. It represents the type of synergistic interaction among invaders that has been termed "invasional meltdown" (Simberloff and Von Holle 1999). An example of this involves the facilitation of plant invasion by invasive defoliating insects that open the forest canopy and increase light levels in the understory (Eschtruth and Battles 2009). Sometimes a single defoliation event can set a long-term trajectory of plant invasion in motion so that years later, it would not be possible to detect the cause of the population increase that initiated the successful plant invasion (Eschtruth and Battles 2014).

In a meta-analysis of generalist herbivore effects on plant invasion, Parker et al. (2006) found that native generalist herbivores often added to biotic resistance to invasion, whereas introduced generalist herbivores, including domestic ungulates, facilitated non-native plant invasion in an "invasional meltdown" effect. They suggested that removal of introduced generalist herbivores, e.g., domestic livestock, might be as effective as introduction of specialist herbivores in control of invasive plants, and that this would not involve the perceived risks associated with introducing non-native specialist herbivores. This idea has never been explicitly tested, however.

5.4 Stress and Invasibility

Ecosystems vary widely in their general level of abiotic stress, but this stress level is part of the natural regime to which resident species are already adapted. Many ecologists believe that invasibility is negatively correlated with abiotic stress, largely because stressful abiotic environments often impose severe resource restrictions that limit productivity and niche space, whereas less stressful, more productive environments are usually less resource-limited and therefore more likely to present unused resources and niche space (Albert et al. 2000). Some of the least-invaded ecosystems, including alpine tundra and more arid deserts, are also among the most abiotically stressful, whereas highly productive ecosystems such as grasslands are often highly invaded. On the other hand, productive ecosystems with high standing biomass, such as closed forests, are also often relatively uninvaded, for reasons that are discussed later in this chapter.

In contrast to episodic disturbance such as fire, geomorphic movement, or insect or disease outbreaks, chronic stress as defined here represents disturbance that operates over a longer time scale and generally at a lower grade (Albert et al. 2000). Chronic stress may not open a system to invasion immediately, but it tends to erode ecological resilience, so that resistance to invasion and to other forms of irreversible change is decreased over the long term. Many sources of chronic stress that can eventually push the disturbance regime beyond the natural range of variation are anthropogenic, but some, such as decadal drought, are of natural origin.

One good example of a chronic anthropogenic stressor is nutrient pollution. Many natural ecosystems have evolved with very low levels of plant macronutrients, particularly nitrogen and phosphorus. The influx of high levels of one of these nutrients into such an oligotrophic (nutrient-limited) system over time can have a strong destabilizing effect. One of the best-documented cases involves phosphorus pollution from agriculture and its impacts on the oligotrophic wetland system of the Everglades (Childers et al. 2003). Excess phosphorus strongly favored the increase of cattails (*Typha domingensis* Pers.), which rapidly became dominant in the system, causing loss of the structure of sawgrass (*Cladium jamaicense* Crantz) wetland and tree hummocks that had been the long-term ecosystem configuration. Eutrophication is also implicated in the spread of many aquatic invaders, e.g., water hyacinth (*Eichornia crassipes*; Coetzee and Hill 2012). Invasive plants that are nitrogen fixers can also be agents of nutrient enrichment, dramatically increasing nitrogen availability in naturally nitrogen-poor ecosystems, thereby increasing resource availability that facilitates invasion by additional species (e.g., fire tree (*Myrica faya*) in Hawai'i; Vitousek and Walker 1989).

In another example of nutrient pollution, the serpentine grasslands of California represent an oligotrophic plant community high in endemism and biodiversity. Experimental work showed that low nitrogen was a main factor preventing the invasion and domination of this plant community by introduced annual grasses from surrounding areas with higher soil fertility (Huenneke et al. 1990). Subsequently, it was shown that nitrogen deposition originating from automobile catalytic converters was causing the breakdown of this invasion barrier, so that serpentine grasslands are now readily invaded by these nitrogen-loving annual grasses, which have the capacity to outcompete the endemic species and potentially cause the extinction of a rare butterfly that depends on them (Weiss 1999).

5.5 Propagule Pressure and Dispersal Corridors

The importance of propagule pressure in mediating invasion has recently received more emphasis (Simberloff 2009), and experimental studies have shown that propagule pressure can frequently overwhelm biotic resistance (Von Holle and Simberloff 2005). Several methods are now available to measure plant propagule pressure independently (e.g., Eschtruth and Battles 2011; Miller et al. 2014). In one study, it was found that the higher levels of invadedness of riparian versus upland forest sites were due, not to higher invasibility as had been thought, but to a higher rate of arrival of invader propagules, that is, higher propagule pressure (Eschtruth and Battles 2011). Mechanistic studies that tease apart the factors responsible for a given level of invadedness across invasion stages (establishment, spread, and impact) are essential for understanding the invasion process.

Another reason that plant invasions are often associated with disturbance at least initially is that the dispersal corridors that introduce propagules into natural ecosystems are almost always associated with humans and therefore with anthropogenic disturbances such as roads, trails, and pipeline and powerline corridors (Gelbard and Belnap 2003; Wilson et al. 2009).

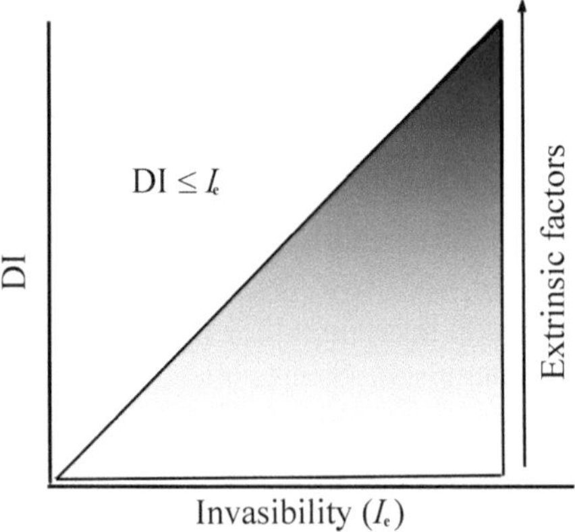

Fig. 5.1a The relationship between invasibility (I_e), extrinsic factors, and degree of invasion (*DI*). Invasibility sets the upper limit to degree of invasion, whereas extrinsic factors (e.g., propagule pressure, time, species-specific abiotic resistance) determine the realized degree of invasion below this upper limit (From Guo et al. 2015)

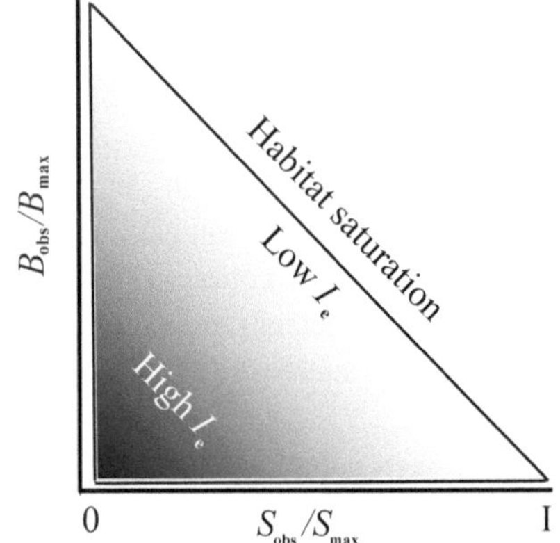

Fig. 5.1b Conceptual model for invasibility (I_e) of a plant community in terms of relative biomass (B_{obs}/B_{max}) and relative species richness (S_{obs}/S_{max}). Relative biomass increases with lower disturbance and later successional stage, whereas relative species richness is highest at intermediate disturbance (see Fig. 5.2). Where both relative biomass and relative species richness are low (high disturbance, low competition, early successional stage), invasibility is highest. Increase in either relative biomass or relative species richness reduces invasibility, which is very low at habitat saturation (From Guo et al. 2015)

5.6 Modeling Invasibility and Invadedness

Guo et al. (2015) recently proposed a methodology for quantifying invasibility and degree of invadedness independently. Their conceptual model captures the idea that the degree of invasion is ultimately constrained by community invasibility, but that at a given level of invasibility, extrinsic factors determine invasion success. These factors could include ecological match, invader ecological attributes (e.g., weak or strong invader), propagule pressure, and invasion timeline (Fig. 5.1a). Their scheme for quantifying invasibility accounts for the effects of both disturbance (expressed in terms of biomass as a proportion of maximum, i.e., biomass in the undisturbed condition) and species richness (expressed as a proportion of maximum species richness) on invasibility (Fig. 5.1b). They used a very large long-term data set generated as part of the Forest Inventory and Analysis program for three eastern deciduous forest types to demonstrate the usefulness of this approach.

The method of Guo et al. (2015) for quantifying invasibility depends on biomass and species richness data from reference areas in undisturbed condition to estimate relative values that can then be compared across ecosystems. This need for reference or historical data is a potential limitation, but even approximations of these reference values can be useful for management of invasive species. To calculate degree of invadedness, in contrast, requires only current data (Guo and Symstad 2008). The two-dimensional surface for quantifying degree of invadedness is similar to Fig. 5.1b, except that introduced species biomass and species richness relative values are calculated as a fraction of total current biomass and species richness. This has the advantages of incorporating both diversity and dominance measures of invadedness and of using unitless measures that are useful for community cross-comparisons.

In support of the model of Guo et al. (2015), Iannone et al. (2016) found that non-native species richness was negatively correlated with both tree biomass and evolutionary diversity.

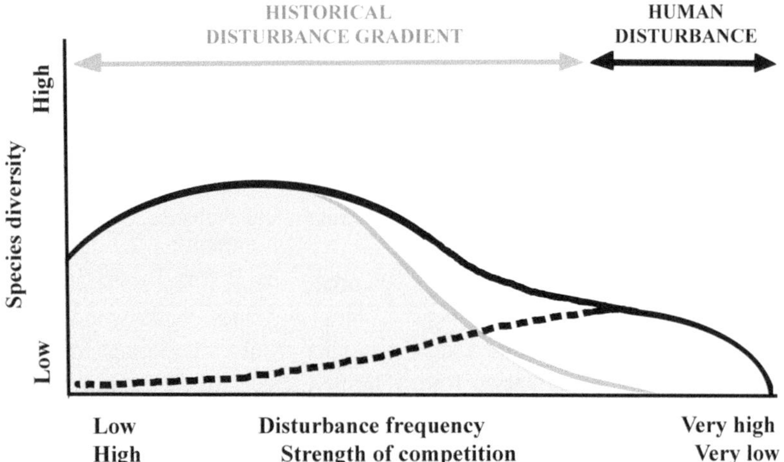

Fig. 5.2 The relationship between disturbance frequency over the natural (historical) and anthropogenic (human-caused) ranges and post-invasion diversity of native plant species (*gray line*), exotic plant species (*dashed line*), and total plant species (*black line*). Strength of competition is shown as inversely proportional to disturbance frequency. *Shaded curve* represents species diversity under the historical disturbance regime prior to any invasion (From Catford et al. 2012b)

This supports the idea that the combined effects of increased competition and increased biodiversity could reduce invasion and therefore be used to predict potential levels of invasion. Overall, this study provided evidence for biotic resistance to invasion at the landscape scale in eastern hardwood forests but also revealed some regional differences in effect.

Relative biomass is likely to impact directly the level of competition encountered by an invader and is also a reasonable surrogate for disturbance history in that reduced biomass relative to the maximum likely reflects a system in recovery from disturbance. Native species diversity, on the other hand, usually peaks at intermediate disturbance frequency. This is because after long periods with low disturbance, more competitive species become dominant and species diversity is reduced, while in the period following high disturbance, species diversity is colonization-limited. This relationship has been codified as the Intermediate Disturbance Hypothesis (IDH; Shea et al. 2004).

Catford et al. (2012b) examined how the IDH pattern of species diversity as a function of disturbance frequency could impact plant invasion (Fig. 5.2). In this scheme, disturbance frequency increases through the range of natural disturbance regimes from left to right, then enters the anthropogenic zone of more frequent (and/or more intense and/or novel) disturbance. Native species richness increases with disturbance as natural disturbance frequency increases to an intermediate level, then decreases over the remainder of the natural range. It is reduced to very low levels under anthropogenic disturbance regimes. Introduced species richness remains at low levels until intermediate levels of natural disturbance are reached, then increases through most of the range of anthropogenic disturbance, with only the most recently disturbed sites still colonization-limited. This difference between native and introduced species is predicated on the assumption that invaders are faster at colonization and better adapted to recent disturbance than native species, life-history traits that characterize many if not most invasive plant species (Rejmanek and Richardson 1996). This model predicts that maximum invasibility will occur under anthropogenically amplified disturbance regimes, but that some degree of invasion is also possible under most natural disturbance regimes. High native species diversity at intermediate levels of disturbance can compensate for reduced biomass and thus reduced competition from community dominants at intermediate disturbance levels, but moderately disturbed communities may still be more invasible than closed communities with very low levels of disturbance.

5.7 Disturbance and Plant Invasion in Different Vegetation Types

Each of the four major vegetation types in the continental United States is characterized by a suite of natural and anthropogenically modified disturbance regimes. Natural and anthropogenic disturbances in each of these vegetation types have functioned to regulate the level of invasion by introduced plant species. We use a few key examples involving major invaders to explain how disturbance and invasion interact in each of these broad vegetation types.

5.7.1 Forest Vegetation

The natural range of variability in disturbance regime in North American forests is driven primarily by fire (Attiwill

1994; Zouhar et al. 2008). All but the wettest coastal rainforests are characterized by fire return intervals that have shaped community structure, and forests typically exist in mosaics or patches with different fire histories, creating characteristic patterns of environmental heterogeneity (Churchill et al. 2013). Other natural sources of sometimes large-scale disturbance include weather-related phenomena such as long-term drought, extreme wind events, and ice storms (Lafon 2015; Millward and Kraft 2004), and outbreaks of native insects (Raffa et al. 2008). At mid-scale, geomorphic disturbances such as flooding, avalanches, and landslides can have dramatic effects, while at a smaller scale, gap formation from the death of individual trees as well as the activities of ungulates and rodents and local-scale flooding and frost-heaving can create openings in otherwise undisturbed vegetation.

The anthropogenic impacts on forest disturbance regimes have been great, beginning with the clearing of large areas for agriculture and resulting forest fragmentation. This has been exacerbated by the ever-increasing network of roads, which greatly increases the effects of forest edges on ecological processes, including invasion by non-native species. Even after abandonment and forest regrowth, the impact of human disturbance can have long-term effects. A second major impact has been through fire suppression, which has major consequences for successional trajectories in all forest types (Zouhar et al. 2008). Timber harvest and associated silvicultural activities, including control of understory vegetation, represent other major sources of anthropogenic change in disturbance regimes (Cyr et al. 2009).

Norway Maple

As mentioned earlier, intact forests are generally thought to offer substantial biotic resistance to plant invasion, and the primary limiting resource is usually light. Most forest invaders are more light-limited than the dominant tree species, and invasion is almost always initiated at forest edges where human-assisted dispersal is also more likely. A notable exception to this rule is Norway maple (*Acer platanoides*), which invades old-growth beech-maple (*Fagus-Acer*) forests in the Northeastern United States (Martin 1999; Martin and Marks 2006). This species combines exceptionally high growth rates with a remarkable degree of shade tolerance, rather than exhibiting the usual tradeoff between these two traits (Martin et al. 2010). It produces a canopy that creates conditions too shady for recruitment by the native sugar maple (*A. saccharum*), but it can easily recruit under a sugar maple canopy. It also appears to have been released from herbivores in its introduced range, experiencing only one-third the damage experienced in the native range in an experimental study (Adams et al. 2009). It also experiences significantly less herbivory than sugar maple where the two species grow together in North American forests (Cincotta et al. 2009). These studies support the enemy release hypothesis and demonstrate weak biotic resistance to invasion in these North American forests. Invasion by Norway maple has a long initial lag time because of a prolonged juvenile period under deep shade conditions, and it may be punctuated by multiple lag periods as successive generations mature, but its invasion, though slow, may be inexorable (Wangen and Webster 2006).

Kudzu

Vines or lianas, both woody and herbaceous, represent another group of important forest invaders. These may have limited shade tolerance, but they have the advantage of a climbing habit, which permits them to reach more favorable light conditions by overtopping trees. Kudzu (*Pueraria montana*) is a major invasive species in forests of the Southeastern United States, where its negative impacts are well documented (Forseth and Innis 2004). It has limited ability to establish in forest interiors, but because it can shade out and kill the trees it overtops, it can improve light conditions as it migrates further into the forest, thus reducing the effectiveness of biotic resistance through light limitation. Experimental evidence suggests that kudzu has benefitted from escape from natural enemies in its home range, enabling it to divert resources from defense to growth and thereby evolve increased growth rate and competitive ability (Yang et al. 2014). Climatic niche modeling shows that there are areas in North America where kudzu could potentially become invasive, including parts of the Western United States. (Cullen and Miller 2015). The invasion of kudzu in the Southeastern United States was greatly accelerated by a very extensive planting effort in the first half of the twentieth century (Forseth and Innis 2004). When many of these plantings were abandoned and the areas underwent natural reforestation, kudzu was already a dominant component of the resulting vegetation.

Japanese Honeysuckle

Japanese honeysuckle (*Lonicera japonica*) is another non-native vine that is a successful invader of mesic eastern US forests (Schierenbeck 2004). It has high shade tolerance and is favored by fire suppression, although it is a root sprouter that is also quite fire-tolerant. Closed-canopy forests offer some biotic resistance to invasion by this species, but it can penetrate forest interiors at low rates of spread, then take advantage of the improved light conditions in small-scale disturbances such as treefall gaps to increase its reproductive success (Horvitz et al. 1998). The fruits are bird-dispersed, which provides an avenue of introduction into forest interiors, but the species is reported to be pollinator-limited in at least some of the invaded range (Larson et al. 2002). Japanese honeysuckle was less impacted by herbivory than a native congener in an observational study and also exhibited compensatory growth in response to herbivory, which suggests that native herbivores are not an effective component of biotic resistance (Schierenbeck et al. 1994).

Garlic Mustard

Garlic mustard (*Alliaria petiolata*) is a deliberately introduced herbaceous forest invader that has undergone explosive range expansion (Kurtz and Hansen 2014; Rodgers et al. 2008). A great deal of research has been carried out on this species, and it has been proposed as a model species for the study of plant invasion (Colautti et al. 2014b). This biennial plant is quite tolerant of low light conditions and can invade forest interiors, albeit slowly. It is most successful in deciduous forests where it can complete much of its life cycle in spring before canopy closure. Once established, it benefits from natural and anthropogenic disturbances that increase light availability (Eschtruth and Battles 2014). It is not palatable to deer, and several studies have shown that its spread is accelerated by deer overstocking, which impacts more palatable natives and gives garlic mustard a competitive advantage (Eschtruth and Battles 2008; Kalisz et al. 2014). There is some indirect evidence that forest communities differ in their biotic resistance to this species, with some studies reporting large negative effects on the native understory that would be expected from a strong invader (Stinson et al. 2007) and others reporting no effect and a positive association between garlic mustard abundance and overall species richness, as would be expected from a weak invader (Davis et al. 2014, 2015). This difference could reflect intrinsic differences in the structure of the invaded communities or possibly differences in modern or historic disturbance regimes (Nuzzo 1999). It is clear, however, that this species has become a permanent member of the forests that have been invaded.

A common thread that runs through the narrative for almost all invaders of relatively undisturbed forests is that these species were introduced deliberately for horticultural purposes and were not accidental arrivals. As mentioned earlier, this makes it more likely that they will not be constrained by the light limitation in forests that would prevent invasion by the great majority of accidental introductions (Martin et al. 2009).

5.7.2 Wetland and Riparian Vegetation

Wetland and riparian ecosystems occupy the transition zone between aquatic and upland terrestrial ecosystems. They are characterized by highly dynamic disturbance regimes, with frequent, intense disturbance and broad natural ranges of variability as the norm (Middleton 1999). Consequently, the plant communities in wetland and riparian ecosystems are rarely in an equilibrium condition and are often far from equilibrium. This increases the likelihood of state changes as a result of natural disturbances (Richardson et al. 2007). These natural disturbances often involve dramatic changes in hydrologic regime, including seasonal flooding or episodic flooding associated with extreme weather events like hurricanes. Water level fluctuations due to drought or changes in flow caused by erosion and aggradation processes as well as modification of woody debris structure of stream channels and beaver (*Castor canadensis*) activity are additional sources of frequent natural disturbance. The complex community structure and high native biodiversity characteristic of these ecosystems are maintained by these natural disturbance regimes, and efforts to restore wetland or riparian communities to static, ideal states cannot be successful in the long term (Middleton 1999; Zedler and Kercher 2004). Long-term stress from drought and episodic weather events such as out-of-season frosts and windstorms are additional sources of disturbance. Finally, periodic fire is also an essential component of the disturbance regime of many wetland and riparian ecosystems, which changes surface litter and nutrient cycling and also plays a role in slowing or preventing succession from wetland to swamp forest, thus maintaining a mosaic of forested hummocks and wetland vegetation dominated by grasses and other graminoids such as sawgrass (Childers et al. 2003).

Wetlands and riparian areas have been heavily impacted by anthropogenic disturbance, much of it associated with land use of surrounding upland areas or with modifications of the hydrologic regime designed to meet human goals. Wetlands have been much reduced in areal extent in many parts of the country due to draining for agriculture or urban development. Development on adjacent uplands can also lead to sedimentation and eutrophication, often accompanied by drastic changes in community structure, as discussed earlier for conversion from sawgrass to cattail vegetation in the Everglades. Because wetlands tend to be sinks in terms of their position on the landscape, they are more dramatically affected by waterborne nutrient pollution than other ecosystems. Fragmentation reduces the connectivity of wetlands and thus their resilience in the face of disturbance, and disruption of natural fire regimes through fire suppression can lead to state changes that are difficult to reverse.

Riparian systems have been dramatically impacted by human efforts to capture water, generate electricity, and reduce the perceived negative effects of flooding. These anthropogenic disturbance factors include dam construction, stream diversion and inter-basin water transfer, and canalization, all of which have the effect of dewatering riparian areas and generating major vegetation changes in response to changed hydrologic regimes (Stromberg et al. 2007). Dam construction has had enormous impacts on almost all US rivers, with very few that are still free-flowing. Other anthropogenic disturbances in riparian areas include livestock grazing and recreational use.

Wetlands and riparian areas are often highly invaded by introduced plant species. One reason is that these areas act either as conduits of seed dispersal in the case of riparian systems or seed collectors in the case of wetlands. They are thus characterized by high propagule pressure from both native and introduced species (Stohlgren et al. 1998). And because these ecosystems are essentially subject to perpetual natural disturbance and far from equilibrium, the niche space is rarely filled, creating continuous opportunities for new species to colonize and establish (Catford et al. 2011; Chipps et al. 2006).

Saltcedar

Invasive species management in wetlands and riparian areas is complicated by the often tight linkage between anthropogenic disturbance and invasion, making it difficult to know where control efforts are best applied. Saltcedar (*Tamarix* spp.) invasion in the river systems of the desert Southwest, for example, is largely driven by changes in hydrologic regime caused by dam construction (Stromberg et al. 2007). Its rapid rise to dominance along many southwestern rivers has elicited a major control effort, but some ecologists argue that this species is in fact dominant only in a novel niche that cannot be filled by native riparian species under the current hydrologic regime (Richardson et al. 2007). Its invasion into the riparian communities of free-flowing rivers has apparently not resulted in the rapid rise to dominance seen along dammed rivers with dewatered floodplains, but instead seems to result in its integration into the native riparian vegetation with little loss of native biodiversity or cover (Stromberg 1998). In environments where light is the main limiting factor, native trees like box elder (*Acer negundo*) can effectively compete with saltcedar (Dewine and Cooper 2008). The case for saltcedar as an invasive species is therefore context-dependent and subject to considerable debate (Stromberg et al. 2009).

Paperbark

Paperbark (*Melaleuca quinquenervia*) invasion of large areas of sawgrass marsh in the Everglades is less subject to the kind of debate applied to saltcedar invasion, as it is apparently capable of invading under the natural range of variability in disturbance, though some might argue that anthropogenic influence is pervasive throughout the Everglades. This Australian native was intentionally introduced to Florida and was even seeded into parts of the Everglades, setting the stage for massive invasion and consequent state change to a wooded swamp vegetation type (Serbesoff-King 2003; Turner et al. 1998). It is a prodigious seed producer that can release seeds continuously but that also reserves a large seed bank in canopy fruits for release after fire. Seedlings can reach reproductive maturity in as little as one year, and mature trees are highly fire-tolerant. Sawgrass marshes are maintained free of less fire-tolerant woody vegetation by the natural fire regime, but this does nothing to stop paperbark invasion. In addition, this species is tolerant to drought, flooding, and moderate freezes, further expanding its niche. It is limited to subtropical climates, but could further expand its range across wetlands along the Gulf Coast States under current climate scenarios (Watt et al. 2009).

Common Reed

Another major invader throughout the wetlands of the United States, particularly on the eastern seaboard, is common reed (*Phragmites australis*). There is little doubt that this species is a strong invader that can interact synergistically with anthropogenic disturbance to create a state change to near-monoculture of common reed (Chambers et al. 1999; Kettering et al. 2012). Management is complicated by the fact that both noninvasive native genotypes and invasive Eurasian genotypes are widespread in North America. In coastal wetlands, there is good evidence that invasion and subsequent dominance by introduced genotypes of common reed are exacerbated by land-use patterns in adjacent terrestrial vegetation (Minchinton and Bertness 2003; Silliman and Bertness 2004). Removal of woody vegetation as part of shoreline development results in increased nitrogen availability and decreased salinity in coastal salt marsh, both factors that favor common reed invasion. Anthropogenic disturbance within the marsh, as simulated in removal experiments, also greatly accelerated the invasion rate. Minimizing within-marsh disturbance and maintaining buffer zones of woody vegetation between salt marshes and adjacent developed areas can be used as a means to slow the rate of common reed invasion. In a study in a riparian system, it was found that high native functional group diversity, and the presence of native pioneer species that could provide immediate competition following disturbance, could greatly reduce seedling establishment of common reed in previously uninvaded habitat (Byun et al. 2013).

5.7.3 Grassland Vegetation

Grassland was historically the dominant vegetation type across much of the Great Plains, as well as in many other parts of the country, for example, the Palouse prairie of the Pacific Northwest. The natural or historic range of variability in disturbance regime for grasslands includes periodic fire, small-scale physical disturbance, and nutrient pulses. Physical disturbance resulted from the burrowing activity of rodents, from the actions of native ungulates (e.g., pronghorn (*Antilocapra americana*), bison (*Bison bison*)), and from soil movement due to frost heaving and hydrological effects. Native ungulates can also create an uneven distribution of soil nutrients with temporary, locally enriched patches. Intensive grazing by large herds of bison probably created short-term impacts on prairie vegetation, but their wide-ranging habits kept these at relatively low frequency, so that the vegetation had time to recover. Large herding ungulates were largely absent from grasslands of the Interior West, so that intensive grazing rarely if ever took place there. Most grasslands in the United States historically experienced fire, although natural fire regimes varied widely depending on climate and soils. Grasslands also experience invasion of native woody species from adjacent vegetation types during periods of low fire frequency, especially in more mesic grasslands such as tallgrass prairie. Drought of varying length and severity is another natural chronic stress that probably caused shifts in species composition.

The arrival of European humans drastically changed the disturbance regimes of grassland ecosystems. Large-scale conversion for agriculture nearly eliminated tallgrass prairie and Palouse prairie grasslands, while other grassland ecosystems, such as the Mediterranean grasslands of California, the mountain meadow ecosystems of the Interior West, and the high desert grasslands of New Mexico and Texas, were heavily exploited for livestock production. Excessive grazing pressure pushed some of these systems beyond their ability to recover. For example, excessive grazing on montane ranges in Utah caused denuded mountainsides and severe soil erosion and flooding. The management response was to seed fast-growing introduced forage grasses for soil stabilization, primarily smooth brome (*Bromus inermis*). This has apparently caused a permanent state change in these ecosystems, which are still dominated by this introduced perennial grass nearly a century after treatment. In central California, the ecosystem response to anthropogenic disturbance was a type conversion to a relatively high-diversity mixed introduced-native system dominated by introduced winter annual grasses from the Mediterranean region, mainly wild oat (*Avena fatua*), annual ryegrass (*Lolium multiflorum*), and soft brome (*Bromus hordeaceus*), with native perennial grassland present only in remnant areas. European humans also impacted natural fire regimes in many grasslands, in most cases through fire suppression.

Anthropogenically altered disturbance regimes in grasslands opened the way for the invasion of a large suite of non-native species that can exploit these new regimes. Increased nutrient availability has emerged as a key variable regulating invasibility in many grassland studies. For example, when fire retardants are applied to intermountain grasslands during control activities, they cause an intense pulse of both nitrogen and phosphorus (Besaw et al. 2011). Experimental work showed that this resource pulse created a competitive advantage for the winter annual invaders cheatgrass and tumble mustard (*Sisymbrium altissimum*) over both native grasses and herbaceous perennial dicots and the perennial invader spotted knapweed (*Centaurea stoebe*). This effect was increased by burning. Knapweed showed a positive response to this resource pulse only in the absence of competition from the more resource-responsive annuals.

The knapweeds (*Centaurea* spp.) are among the most serious invaders of grassland and associated ecosystems in the Western United States (LeJeune and Seastedt 2001). Spotted knapweed invades and increases in response to "improper grazing" of bunchgrass communities but also apparently has the capacity to invade "well-managed rangeland" (Sheley et al. 1998). Ortega and Pearson (2005) showed that dominance by spotted knapweed in an invaded bunchgrass community in western Montana likely was the cause of reduced native species diversity rather than the consequence. In contrast, Maron and Marler (2007) found that native biodiversity in experimental bunchgrass communities was the apparent cause of reduced spotted knapweed success in post-seedling stages. This negative effect was mediated through competition for resources but persisted even at enhanced resource levels.

A somewhat similar scenario has emerged with studies of yellow starthistle (*Centaurea solstitialis*). This species is a late-season annual that is able to use deeper soil moisture than most co-occurring species, enabling it to complete its life cycle in the summer even in summer-dry environments (Roche and Thill 2001). Kyser and DiTomaso (2002) studied the effect of late-season prescribed burning on yellow starthistle populations in a remnant perennial grassland in California. Repeated burning before seed dispersal effectively eliminated yellow starthistle and triggered a strong positive response in the fire-adapted native plant community. However, when burning was discontinued, the competitive advantage shifted away from the native community, and the site was reinvaded.

Frequent fire was an especially important part of the historic fire regime in tallgrass prairie in terms of its effect on invasibility. Smith and Knapp (1999) found that invasion by introduced cool-season species at Konza Prairie in Kansas, though never a serious problem, increased with fire suppression. Burning increased the dominance of the warm-season grasses that provide strong competition to invaders. Bison grazing at moderate levels, on the other hand, increased invasion through reduction of competitive abilities of the warm-season dominants. Species richness of both native and introduced species was increased in the grazing treatment, however, suggesting that bison grazing historically was a force for the maintenance of native species diversity. In contrast, Milchunas et al. (1989) found that heavy cattle grazing in shortgrass prairie in central Colorado increased the biomass production of the dominant grass, blue grama (*Bouteloua gracilis*), a compensatory response that is probably the result of long evolution with bison. Invasibility may not be strongly impacted by grazing in shortgrass prairie, though this question was not examined. In experimental work in tallgrass prairie, both Kennedy et al. (2002) and Fargione and Tilman (2005) found that increased plant diversity decreased invasibility.

5.7.4 Shrubland Vegetation

Shrub-dominated vegetation occurs at some scale under most climate regimes in the United States, but shrubs achieve regional dominance primarily in the arid and semiarid Interior West. We focus on these shrublands, which occur over a wide productivity gradient that is related to precipitation. Most precipitation is received in winter; shrublands tend to give way to desert grasslands where monsoonal moisture dominates the precipitation regime (Brooks and Chambers 2011). The natural disturbance regime in these shrublands also varies systematically according to precipitation, which is itself usually correlated with elevation. At the low-precipitation and high-temperature end of the gradient, in the low-elevation Mojave Desert of the Southwestern United States, the historic disturbance regime was likely dominated by the periodicity and severity of long-term drought (Hereford et al. 2006; McAuliffe and Hamerlynk 2010). Fire caused by lightning strikes burned only small areas because of low productivity and the large proportion of bare ground. Even in high-precipitation years, the native desert annuals that make up much of the biodiversity in the Mojave Desert as well as any perennials that were present tended to be concentrated beneath shrubs because of the "fertile island" effect, with very little fuel production in the interspaces (Thompson et al. 2005). A similar pattern of historical disturbance is characteristic of the salt desert shrublands of the Great Basin, with long-term drought a major force for vegetation change but with fire playing a very minor role.

Other disturbance factors on a more local scale in these low-productivity deserts included the effects of overland flow from severe rainfall events on erosional and depositional processes and also more large-scale geomorphic events such as debris flows at longer intervals (Webb et al. 1987). Rodent workings are another important source of disturbance in deserts and can have a large impact on vegetation dynamics on a local scale (Kitchen and Jorgensen 1999).

In the more productive cold desert vegetation dominated by Wyoming big sagebrush (*Artemisia tridentata* ssp. *wyomingensis*) at mid-elevation on the valley floors and bajadas in the Great Basin, decadal drought was probably still important as a disturbance factor. Fire was more important relative to the warm and salt deserts because of increased fuel loads, the presence of perennial bunchgrasses and perennial dicots in the shrub interspaces, and the lack of fire tolerance in the dominant shrub species. The fires were likely still relatively small and patchy, with return times measured in decades, resulting in a shifting mosaic of shrub-dominated and perennial grass-dominated vegetation (Whisenant 1990). On more productive and cooler upland sites, Wyoming big sagebrush gives way to mountain big sagebrush (*Artemisia tridentata* ssp. *vaseyana*) steppe, with a concomitant decrease in the importance of drought as a historic disturbance factor and an increase in the importance of, and resilience to, fire.

One historic disturbance factor that was conspicuously lacking in the arid and semiarid shrublands of the Interior West was the effect of large herding ungulates such as bison that played such an important role in grasslands to the east of the Rocky Mountains. When Europeans arrived on the scene, they brought domestic livestock in large numbers to these ecosystems, and the results were devastating (Mack and Thompson 1982). The bunchgrasses characteristic of desert and steppe vegetation were largely intolerant to grazing and were quickly eliminated, while the biological soil crusts that formed a continuous cover in the interspaces were readily destroyed by trampling. The stress of unregulated livestock grazing continued for decades in these ecosystems and caused a reduction in resilience that has resulted in dramatic vegetation changes. Another source of anthropogenic disturbance was clearing for rainfed agriculture that ultimately failed, leaving a legacy of huge tracts of abandoned farmland (Morris et al. 2011). Mining and energy development have created major disturbances and have also left a web-like network of roads over much of the desert, providing corridors of disturbance for plant invasion and also an invitation to off-road vehicle recreational use (Gelbard and Belnap 2003).

The introduction of livestock grazing into the shrublands of the Interior West created the "perfect storm" for plant invasion and the subsequent shift to an alternative stable state. The winter annual grasses that became the primary invaders were pre-adapted to livestock grazing and other forms of intense disturbance in their native Eurasian range. They were largely introduced incidental to agriculture, for example, as contaminants in grain seed, and they could readily disperse with livestock as the animals were allowed to range freely across the landscape (Mack 1981). They encountered environmental conditions similar to the steppe and desert habitats of Eurasia, and experienced little biotic resistance due to the destruction of both the perennial understory and the biological soil crust that could deter recruitment from seed. The depleted understory quickly became dominated by these grasses, chiefly cheatgrass and medusahead rye (*Taeniatherum caput-medusae*).

Once these disturbance-responsive annual grasses became dominant in the understory, the shrub overstory could not suppress them effectively, especially in years of above-average precipitation, when they formed a dense, continuous layer of highly flammable fine fuel. As a result, wildfire, always part of the disturbance regime in these systems, became the ascendant disturbance, resulting in the loss of shrub cover and conversion to communities that are hardly more than annual grass monocultures that burn repeatedly (Brooks et al. 2004; D'Antonio and Vitousek 1992). In this case of the interplay of chronic livestock grazing and resulting lowered resilience, annual grass invasion, and a subsequently altered fire regime were necessary to push the sagebrush steppe ecosystem over to an alternative stable state over millions of acres in the Intermountain West (Chambers et al. 2014). In an experimental study, cheatgrass increased in response to both perennial biomass reduction and burning, but the synergistic effect of both of these factors together increased its seed production by an order of magnitude relative to each factor alone (Chambers et al. 2007).

A similar process has played out with red brome (*Bromus rubens*) as the invader over large areas in the Mojave Desert, where wildfire has also become the ascendant disturbance factor in a system that basically evolved without fire and therefore has little fire resilience (Brooks and Chambers 2011). Increased fuel loads in the interspaces in high-precipitation years are the main drivers of these fires, which kill many desert shrub species outright and reduce the competitive ability of those that do manage to resprout. Even in highly fire-adapted shrubland ecosystems such as the California chaparral, disturbance of the understory and subsequent invasion of non-native winter annual grasses have increased the frequency of fire beyond the capacity of chaparral species to regenerate post-burn, so that these ecosystems too are undergoing conversion to annual grass-dominated systems (Keeley 2001; Zedler et al. 1983). Another anthropogenic disturbance potentially driving these changes to alien annual dominance is nitrogen deposition (Brooks 2003).

Most invasive plants in deserts have been accidentally introduced and are disturbance-responsive species, but there is also a potential for deliberately introduced species to become invasive in desert ecosystems, as they could be pre-adapted for success even in relatively undisturbed xeric environments. Invasive forage grasses such as buffelgrass (*Pennisetum ciliare*) are good examples (Stevens and Falk 2009).

5.8 Disturbance and Invasions of Consumer Organisms

Much of the literature on the invasion process has focused on invasive plants, and these studies are emphasized in the preceding discussion. It would be a mistake, however, to discount the importance of invasive species that are consumers rather than producers, including vertebrate animals, specialist insects, plant pathogens, and generalists such as earthworms. It is well known that specialist invaders, for example, the white pine blister rust pathogen (*Cronartium ribicola*; Kinloch 2003) and the hemlock woolly adelgid (*Adelges tsugae;* Eschtruth et al. 2006), can have profound and far-reaching negative effects on ecosystem structure and function that can rise to the level of an irreversible change of ecological state. What is less obvious is how invasive consumer organisms are impacted by disturbance regimes.

The concept of invasibility has been applied mainly to plant invaders (Lonsdale 1999), but the same principles can potentially be applied to the factors regulating the colonization and spread of invasive consumer organisms. Disturbance is less likely to be a direct regulator of consumer invasion than of plant invasion because plant establishment depends directly on the physical space or resource that is made more available by disturbance. In a comprehensive literature review, Lozon and MacIsaac (1997) concluded that disturbance was far less likely to be implicated in consumer invasions than in plant invasions. Notwithstanding, the broader concept of invasibility, which includes disturbance as one factor, can be useful in understanding invasions of consumer organisms.

5.8.1 Invasive Plant Pathogens

Non-native pathogen invasion is a widespread and global problem that is continually increasing (Fisher et al. 2012; see Chap. 2 of this volume for discussion). Though many introductions likely occur, there fortunately are still relatively few examples of invasive plant pathogens that have caused widespread ecosystem change in the United States (Aukema et al. 2010). These include white pine blister rust on five-needle pines (*Pinus* spp.; Richardson et al. 2010), chestnut blight (caused by *Cryphonectria parasitica*) on American chestnut (*Castanea dentata*; Anagnostakis 1987), and sudden oak death (caused by *Phytophthora ramorum*) on a variety of broadleaf trees in California (Grünwald et al. 2012; Rizzo and Garbelotto 2003).

The many ecological roles that fungi play within ecosystems, including nutrient acquisition, nitrogen cycling, and soil formation, make their connections to biological invasion and disturbance complex (Van Der Heijden et al. 2008). Little research exists on factors that drive successful introductions of plant pathogens, though studies have examined how native plant pathogens successfully infect their native hosts (Ennos 2014). Further, only a handful of studies examine how native pathogens fare under disturbance. Much of the research on biological invasion of fungi is focused on how, through anthropogenic introductions, fungi can be the drivers of disturbance as invasive or emerging plant pathogens or as fungal mutualists helping invasive plants with co-establishment in a new environment by improving their fitness and success (Desprez-Loustau et al. 2007). These topics are beyond the scope of this chapter.

Establishment or persistence of invasive plant pathogens depends on several basic principles for transmission to occur between the initial infected host and subsequent infections (Gilligan and van den Bosch 2008). Duration of the infection period is critical because infection must be sustained to ensure high propagule pressure to infect additional susceptible hosts; otherwise, the invasion will fail (Ennos 2014). Important to invasibility as well is the ability of the pathogen to persist during unfavorable environmental and various host demographic conditions (Gilligan and van den Bosch 2008). These can directly affect the ecology of the pathogen, as there are specific abiotic and biotic conditions that are required for success of a particular pathogen (Ennos 2014). Woods et al. (2005) showed that *Dothistroma septosporum*, the causal agent of the red band needle blight, required sustained wetness at 15–20 °C for successful spore dispersal and germination. In sites where these conditions are not met, the pathogen cannot thrive. Spatial and genetic heterogeneity of the host population can also be directly linked to the ability of a pathogen to become invasive (Burdon et al. 2006), as invasive pathogens can spread more quickly in single-species plantings such as poplar or pine plantations than in diverse, species-rich natural forest. Additional factors that can influence invasibility and rate of spread include transmission rate, host density, host susceptibility, and weather conditions (Gilligan and van den Bosch 2008).

Success by invasive pathogens is also thought to be driven by a lack of resistance in native hosts, the ability of fungal pathogens to evolve quickly, and their ability to produce many reproductive propagules (Gladieux et al. 2015). Host jumps occur through anthropogenic invasion events where pathogens are moved to new environments and are in contact with a susceptible host for the first time (Burdon et al. 2006). Native hosts have not evolved genetic resistance against new, invasive pathogens. Environment plays an important role as well. When ecosystems are subject to chronic stress due to drought or other factors, plants become more susceptible to invasive pathogens because their defense mechanisms are weaker. This could allow for host jumps that might not occur if plants were not under chronic stress conditions (Burdon et al. 2006).

Host jumps can also occur through pathogen evolutionary change resulting in the ability to be pathogenic on a new host present in the invaded range. One evolutionary mechanism is hybridization between two pathogens, allowing for a broader host range. Hybrids of *Melampsora medusae* (North American origin) and *M. larici-populina* (European origin) in New Zealand have resulted in offspring with a broader host range within *Populus* species than either parent (Burdon et al. 2006; Spiers and Hopcroft 1994), thus facilitating the spread of a new invasive pathogen.

5.8.2 Invasive Insects

As with plants and plant pathogens, most insect introductions likely fail to establish due to abiotic factors or other forms of environmental mismatch (e.g., lack of suitable host species) and also to Allee effects, i.e., processes that tend to cause extinction from initially low population size (Liebhold and Tobin 2008). Presumably, biotic resistance operates to keep most non-native insects at levels that have little impact, but cases where this biotic resistance fails can result in invasions with major ecological effects (see Chap. 2 of this volume for discussion).

Hemlock Woolly Adelgid

Biotic resistance to herbivorous insect invasion involves at least three basic mechanisms: host resistance, native natural enemies, and vegetation patterns, including the distribution of host plants relative to non-host plants on the landscape. The effectiveness of these mechanisms can vary even for the same insect invader. For example, the host-specific hemlock woolly adelgid from Japan has become a high-impact invader in eastern forests, where the hemlock species have little or no resistance, and natural parasitoid populations are inadequate to keep the invader in check (McClure and Cheah 1999; Wallace and Hain 2000). A related but genetically distinct species native in Western North American hemlock forests has only a minor effect, largely because western hemlock species have co-evolved with this insect and have some resistance (Havill et al. 2011). Native parasitoids are also more abundant and effective in western hemlock forests (Kohler et al. 2008). Predation by parasitoids has the potential to be a major component of biotic resistance to herbivorous insect invasion (Owen and Lewis 2001). This concept is the basis for the development of biocontrols for insect pests of plants, including hemlock woolly adelgid (Van Driesche et al. 2008).

Locally, the hemlock woolly adelgid is dispersed passively by wind. Experimental work has shown that even this passive mode of dispersal can result in dispersal distances of >1200 m in spring before deciduous trees leaf out and reduce wind speeds; even in summer with closed canopies, dispersal distances can exceed 400 m (Turner et al. 2011). The role of vegetation pattern in limiting rate of spread has been established for another passively dispersed insect, maritime pine bast scale (*Matsucoccus feytaudii*), a host-specific pest of *Pinus pinaster* in Europe (Rigot et al. 2014). These researchers found that the rate of spread was slowed by patches of non-host deciduous trees surrounding pine patches, an example of associational resistance on a landscape scale (Barbosa et al. 2009). The rate of spread was increased by high connectivity between patches of pine but also by the presence of non-forested open areas. This effect of openings could be interpreted as a result of disturbance that lowered biotic resistance to invasion.

Emerald Ash Borer

Insects that have an actively flying dispersal stage are much less likely to be slowed in rate of spread by heterogeneous landscapes and can achieve much larger dispersal distances. Emerald ash borer (*Agrilus planipennis*) has had a devastating impact on highly susceptible North American ash (*Fraxinus*) species (Herms and McCullough 2014). Mated females can disperse up to 10 km in their lifetime, enabling them to locate even widely dispersed host trees, and human-aided "host jump" dispersal through long-distance transport further accelerates the rate of spread (Muirhead et al. 2006). Few negative impacts of native predators or parasitoids on emerald ash borer have been reported (Duan et al. 2012). Trees that are stressed by other factors may be attacked first, and in the native range of emerald ash borer in China and Russia where the ash species have evolved resistance, only stressed trees are strongly impacted. In the Eastern United States, even otherwise healthy trees are frequently killed. Open-grown trees in landscape plantings are more likely to be killed than trees growing in shaded conditions, but this associational resistance is weak. The rapid spread and high ecological impact of emerald ash borer represents an invasion into "defense-free space" (Gandhi and Herms 2010), which represents the worst-case scenario for impacts of an insect invader—a case where biotic resistance does not seem to operate at all. Introduction of emerald ash borer natural enemies from the native range is one of the few remaining options for control in wildland settings (Duan et al. 2013). Habitats that support *Fraxinus* species are also likely to support emerald ash borer, meaning that abiotic constraints are not likely to limit its spread.

Gypsy Moth

Biotic resistance to generalist herbivore insect invaders presents a different scenario. Undoubtedly, the best studied of these is the gypsy moth (*Lymantria dispar*), a species that is known to feed on hundreds of species of woody plants, though it has a feeding preference for oaks (Elkinton and Liebhold 1990). Natural enemies and habitat heterogeneity both play a role in limiting spread of this species, or at least in regulating its patterns of outbreak and decline. Different classes of natural enemies operate to regulate gypsy moth abundance when it is at high versus low abundance. Host-specific pathogens play a major role at high densities, causing population crashes following epidemic outbreaks (Hajek et al. 2015). One of these, the Nuclear Polyhedrosis Virus (NPV), appears to have traveled with gypsy moth from its native range (Hajek et al. 2015), while the other, a fungal pathogen (*Entomophthora maimaiga*), reappeared as a natural control many decades after a biocontrol introduction from the native range in Japan (Andreadis and Weseloh 1990). Thus, these pathogens do not represent biotic resistance in the sense that they are native to the invaded ecosystem. Population regulation at low levels has been found to involve native rodent predation on the immature stages (Liebhold et al. 2000). The deer mice (*Peromyscus* spp.) that eat caterpillars and pupae are generalists that also rely heavily on oak mast (acorn production) and that prefer acorns to caterpillars as a food source. Caterpillar mortality levels are a function of rodent abundance (which in turn is correlated with weather patterns) and exhibit high spatial synchrony. This could be an explanation for the observed spatial synchrony of gypsy moth outbreaks (Liebhold et al. 2000).

Rates of spread for gypsy moth are uneven, resulting in patchy distributions that are not necessarily associated with underlying environmental heterogeneity. Jankovic and Petrovskii (2013) showed that Allee effects at low densities and disease (NPV) at high densities coupled with a diffusion dispersal model can generate these patterns. Because gypsy moth females are flightless and attract males through pheromones, the difficulty of mate-finding at low densities can cause local establishment failures, limiting the effectiveness of human-assisted long-distance dispersal. The moths disperse locally only as ballooning first instar larvae, which limits their primary dispersal distance, and epidemic disease at gypsy moth outbreak levels as described earlier also affects spread dynamics.

Habitat heterogeneity has also been shown to affect rates of gypsy moth spread. Topographic relief increases reproductive asynchrony and thus further increases the difficulty of finding mates, increasing the Allee effect (Walter et al. 2015). On the other hand, forest fragmentation that increases the relative importance of forest edges on the landscape increased gypsy moth mating success via its effects on male flight behavior and the aerodynamics of pheromone plumes (Thompson et al. 2016). This could be interpreted to mean that disturbance that creates fragmentation and forest edges could be a factor accelerating gypsy moth spread.

Generalist Insect Predators

Generalist predators are another class of insect invaders. These have a propensity for occupying areas disturbed by human activity, which makes them more noticeable to humans (Snyder and Evans 2006). Important generalist predators are invasive ants and to a lesser extent vespid wasps, mantids, and ladybird beetles. Many species are almost completely restricted to anthropogenic disturbances; these include the "tramp ants" that are rarely found far from human habitation (Holway et al. 2002) and mantids that can occupy old fields but cannot move into nearby forested areas (Snyder and Evans 2006). Even species like fire ants (*Solenopsis invicta*) that can invade natural systems are favored by natural disturbances that resemble the floodplain disturbances that are characteristic of their habitat in the native range (Snyder and Evans 2006). Therefore, the biotic resistance presented by undisturbed native vegetation is probably a factor in reducing the importance and spread of these generalist predators in North America. However, there are exceptions to the generalization that biotic resistance can reduce ant invasion and subsequent negative impacts, particularly on oceanic islands in the tropics (Hoffman and Saul 2010; O'Dowd et al. 2003; Wetterer 2005).

Native ant species can also present biotic resistance to introduced ant species through interference competition, though the outcome is often the opposite due to the aggressive nature of many of the introduced species. But while short-term negative impacts on native ants have been reported, in one long-term study, native ants returned to pre-introduction levels in spite of the continued presence of fire ants, a clear case of biotic resistance (Morrison 2002).

5.8.3 Invasive Earthworms

Earthworms are among the most abundant soil macroinvertebrates in temperate and tropical ecosystems. They are ecologically and taxonomically diverse, exhibiting the capacity to occupy different habitats ranging from truly arboreal to deep mineral soil horizons and have approximately 3,000 described species worldwide (Edwards et al. 2013). Earthworms are often termed ecosystem engineers (*sensu* Jones et al. 1994) because of their ability to physically alter the environments they inhabit. They are responsible for the processing of surface and mineral-associated soil organic matter, and through their burrowing and feeding activity, they are known to increase soil porosity, water infiltration, gas diffusion, and soil aggregate formation and to contribute to increased plant productivity (Edwards et al. 2013). See Chap. 3 of this volume for a more complete description of earthworm effects on biogeochemical processes.

Earthworms also occupy a critical position in terrestrial food webs and represent an important channel through which nutrients and energy flow from detrital pools to higher trophic levels, as they are consumed by other invertebrates, birds, herptiles, and mammals (Blackman et al. 2012; Catania 2008; Gorsuch and Owen 2014; Maerz et al. 2005; Richardson et al. 2015). There are numerous "peregrine" earthworm species that have been transported and introduced around the globe, and of these, there are perhaps 45 species introduced into North American soils (Hendrix et al. 2008). When non-native earthworms become established, they have been shown to significantly alter ecosystem properties ranging from plant population dynamics (e.g., Gundale 2002) to biogeochemical cycling (e.g., Cameron et al. 2015).

In North American soils, there are two distinct scenarios that must be considered when discussing earthworm invasions. These have to do with the fact that much of the continent was covered in glacial ice sheets as recently as 10,000 years ago. In the first case, following the retreat of glaciers, the underlying soils developed in the absence of earthworm influences, whereas in the second case, soils of mesic environments that did not experience glaciations were continuously inhabited by native earthworm species (James 2004). These different starting conditions have had a strong influence on the impacts and invasion patterns that have been observed for invasive earthworm species introduced primarily from Europe and Asia (Frelich et al. 2006; Hendrix et al. 2008).

Disturbances and Invasive Earthworms in Glaciated Soils

The earliest records of European and Asian earthworms in the glaciated portion of North America come primarily from agricultural soils, from extensive surveys conducted along roadsides, and generally from areas that otherwise experienced some form of human disturbance (Reynolds 1995). These records suggest that human-mediated transport of earthworms has been responsible for establishment of invasive populations, and the dumping of ship ballast, movement of horticultural materials, and use of non-native earthworms as fish bait have all been implicated as important vectors of the invasive species (Callaham et al. 2006a; Hendrix and Bohlen 2002). More recently, however, it has become clear that invasive earthworms are now actively colonizing less disturbed or even pristine forested habitats of the glaciated portions of the continent (Frelich et al. 2006).

The introduction of invasive earthworms into less disturbed habitats is still thought to be driven primarily by human activities (i.e., disturbances), and in one study the age of roads constructed in northern boreal forests was a strong predictor of the extent of earthworm invasion (Cameron and Bayne 2009). These authors also showed that the distance of a particular forested habitat from active agriculture was related to earthworm invasion, and they further developed a predictive model showing the relationship between road density and the likely extent of new earthworm invasions. Another study found that road building and harvesting in a small portion of a forested watershed were enough disturbance to allow earthworms to become established throughout the watershed, and that the earthworms were passively dispersed along stream corridors into areas otherwise unaffected by the harvesting activity (Costello et al. 2011). In addition to this general pattern of association with roads, earthworm invasions can also occur in essentially roadless areas such as the Boundary Waters Canoe Area Wilderness where recreational use is high.

Disturbances and Earthworm Invasions in Non-glaciated Soils

The relationship between soil profile-disrupting disturbances and the advent of non-native-dominated earthworm communities have long been recognized (Stebbings 1962), as early observers noted the loss of native species from and the incursion of European species into agricultural soils of the Midwest. Indeed, in soils where native earthworm species might be found (i.e., in soils south of the most recent glacial maximum, including the Southeastern States and some soils along the Pacific coast (James 1995, 2004)), it is extremely rare to find pure assemblages of native earthworm species in soils that experience regular tillage or other profile disruption (Hendrix et al. 2008). However, these profile-disrupting disturbances need not affect large areas of land, and other researchers have found that roads built through otherwise undisturbed habitats can serve as corridors for the introduction of non-native earthworms into less disturbed habitats (Kalisz and Dotson 1989; Kalisz 1993). In other parts of non-glaciated North America, workers have observed a relationship between land-use history and the prevalence of non-native earthworms in soil invertebrate communities, with a clear indication that perturbations to vegetation and soil result in a dominance of the earthworm community by European species (e.g., Callaham et al. 2006b; Sanchez de Leon and Johnson-Maynard 2009; Winsome et al. 2006). This and other evidence has led to the development of ideas around the influence of disturbance gradients and the resulting likelihood that a particular community will include non-native invasive earthworms (Fig. 5.3). Included in this line of reasoning is the observation that native earthworms and/or

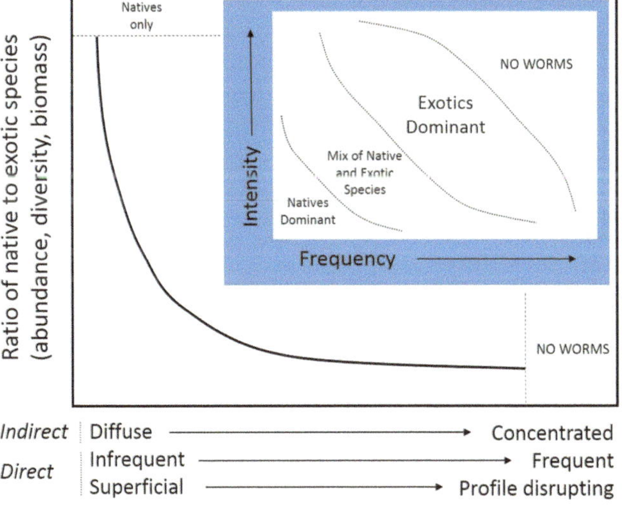

Fig. 5.3 Conceptual diagram showing the relationship between the composition of the earthworm community (ratio of native to non-native species) and disturbances. Observations from the field support the hypothesis that in soils where both native and non-native earthworms occur, the ratio of native to non-native earthworm abundance is related to disturbance level. The *inset* provides a more nuanced view of the nature of these relationships and acknowledges the interplay of disturbance frequency and intensity. Disturbances (*x-axis of main plot*) can be indirect or direct. An example of an indirect disturbance is the establishment and spread of an invasive plant, which modifies the environment to be favorable for invasive earthworms. An example of a direct disturbance would be annual tillage of an agricultural field (Modified from Hendrix et al. 2008)

other macroinvertebrates in the forest floor may offer some biotic resistance to the invasion of non-native earthworms. Snyder et al. (2011) found that earthworm invasions in the field were associated with lower diversity and abundance of millipedes. In laboratory microcosms, Snyder et al. (2013) found that millipedes prevented the earthworms from producing cocoons, and this may limit or slow invasive spread into forests.

Invasive Earthworm Interactions with Other Invasive Species

There is growing evidence that introduced earthworms and invasive plants may be involved in a synergistic "invasional meltdown" relationship (*sensu* Simberloff and Von Holle 1999). This has been suggested in shrub invasions of ecosystems both in the Great Lakes region as well as in the riparian forests of the Southeastern United States. In deciduous forests of the Great Lakes, European buckthorn (*Rhamnus cathartica*) is an invasive shrub that is currently expanding its range, and there have been observations of greater densities of non-native European earthworms beneath the canopy of this species in some cases (Heneghan et al. 2007; Wykoff et al. 2014). Furthermore, in other studies where European buckthorn was removed through management activities, the density of European earthworms declined (Madritch and Lindroth 2009). Similarly, in southern riparian forests where Chinese privet (*Ligustrum sinense*) has invaded extensive areas, the abundance of non-native invasive European earthworms increased, but these densities were reduced five years after removal of the shrub (Lobe et al. 2014). Removal of the shrub was also associated with the recovery of native earthworm populations, and the authors hypothesized that the plant's influence on soil pH may be the underlying mechanism driving these earthworm species responses (Lobe et al. 2014).

Invasive Earthworms and Fire

Studies on the potential interactions between fire and invasive earthworms have been few, but due to the sometimes profound effects that fires can have on soils and vegetation, there is the expectation that there should be at least indirect effects on earthworms. In North American tallgrass prairies, fire is considered a necessary component of the ecosystem, and when fires are suppressed, the vegetation will shift away from grasses and toward woody species (Knapp et al. 1998). In one study explicitly examining the effects of prescribed fire on prairie invertebrates, Callaham et al. (2003) found that native earthworms were favored by fire, but that when fire was excluded for 20+ years, European earthworms dominated the community. In a study of earthworm/fire interactions in boreal forests of North America, Cameron et al. (2015) suggested that introduced earthworms interact with soil carbon in such a way that losses of forest floor carbon attributable to wildfires should decline as European earthworm invasions progress in these forests. In still another earthworm/fire interaction study, Ikeda et al. (2015) explored the use of prescribed fire as a way to manage invasion of southern forests by litter-dwelling Asian earthworm species, and found that experimental fires did not have significant direct effects on adult worms, but that the fires had strong negative impacts on the cocoons of the invasive species, potentially limiting their population levels in the following year. There have also been documented cases of completely neutral effects of fire on invasive earthworms (Scharenbroch et al. 2012).

5.8.4 Invasive Terrestrial Vertebrates

Introduced terrestrial vertebrates in the United States are well documented to cause major ecological harm (see Chap. 2 of this volume). The conditions upon which successful vertebrate invasions are predicated were examined in a meta-analysis that included invasions in both Europe and North America (Jeschke and Strayer 2006). Frequency of introduction and "human affiliation" were identified as the most important factors. In contrast to the case with invasive plants, traits associated with life history were relatively unimportant in invasion success; instead, the direct and indirect effects of human activities were much more likely to predispose a vertebrate species to invasion success than any specific set of biological attributes.

The close association of many invasive vertebrates, for example, rats (*Rattus* spp.) and English sparrows (*Passer domesticus*), with humans implies that anthropogenic disturbance favors their increase and that relatively undisturbed natural systems must possess some biotic resistance to invasion. But there is little evidence to suggest that disturbance of a natural system reduces its resistance to invasion by these "human commensal" species, unless this disturbance results in a major land-use change to a human-dominated landscape (Jeschke and Strayer 2006). Invasive vertebrates that pose a more direct threat to natural ecosystems are more often species that were introduced for hunting or that represent escapes from the exotic pet trade. The invasibility of a natural system to an introduced vertebrate that is preadapted to the environment does not seem to be contingent upon level of disturbance. This conclusion is supported by Lozon and MacIsaac (1997), who reported that only 11% of examined studies on terrestrial vertebrate invasions mentioned disturbance as a predisposing factor.

European Starling

The European starling (*Sturnus vulgaris*) is a classic example of a human commensal invasive species that does tremendous damage to US agriculture (Linz et al. 2007). Starlings also have negative impacts on natural systems, specifically in competition with native cavity-nesting birds (Purcell 2015). In a study in the steppe of central Washington, starlings visited sagebrush steppe habitats but did not nest there (Brandt and Rickard 1994). Purcell (2015) showed that starlings forage in mowed areas in preference to areas with taller vegetation, which could explain why they preferred abandoned agricultural land to sagebrush steppe in the Washington study. She suggested that managing vegetation height might be a way to protect native cavity-nesting birds from the negative impacts of starlings.

Wild Boar

An example of an invasive vertebrate that was introduced for hunting is the wild boar (*Sus scrofa*). This species can thrive over an extremely wide range of environmental conditions, as evidenced by both its large native range in Eurasia and its ability to invade both tropical and temperate ecosystems worldwide (Barrios-Garcia and Ballari 2012). The spread of wild boars is favored by the expansion of agriculture because of the rich food source represented by agricultural fields, and wild boar abundance in natural systems depends on the availability of preferred foods rather than on the level of disturbance. The main effect of wild boars on natural ecosystems is through their rooting activities, which can have a major negative impact on herbaceous vegetation. There is evidence that plant communities that are adapted to frequent surface disturbance are more resilient to wild boar rooting disturbance, with even some positive effects reported (Baron 1982; Kotanen 1995). This represents a form of biotic resistance in that it potentially reduces the negative impacts of this invader, but there is no reported mechanism of biotic resistance that prevents its establishment and spread.

Burmese Python

The Burmese python (*Python bivittatus*), one of the largest snake species in the world, represents a new category of vertebrate invader that has resulted from the rapidly expanding market for exotic pets. These kinds of animals are not human commensals and do not depend on anthropogenic disturbance for their success in either the native or the introduced range. Consequently, when they are released deliberately or accidentally into the wild in ecologically suitable areas, they are able to establish and spread rapidly across a broad spectrum of disturbance regimes. Burmese pythons apparently have few or no natural enemies in their introduced range in south Florida, and their dramatic negative impact on native mammals has manifested itself very rapidly (Dorcas et al. 2012). The spread of Burmese pythons in the United States is limited by cold temperature and dry conditions, but climate modeling, based on the occupied native range, indicates that they could spread across the South and even along the Pacific Coast under the current climate regime and could spread even farther north in response to climate change (Rodda et al. 2009).

5.9 Disturbance and Invasive Species Management

The strong link between anthropogenic disturbance and invasion, especially for plants, suggests that managing to reduce anthropogenic disturbance in vegetation not yet impacted by strong invaders is the most straightforward and cost-effective tool for reducing the probability of such an invasion (Firn et al. 2008; MacDougall and Turkington 2005; Marvier et al. 2004). This generalization applies primarily to species that are disturbance-dependent, but as anthropogenic disturbance also increases the probability of propagule arrival, decreasing disturbance can also reduce the probability of arrival of an invader that does not depend strongly on disturbance to establish. These authors emphasize the difficulty and expense of attempting to manage the ever-growing cadre of invasive organisms one species at a time through active control. Extraordinary investment of time, resources, and funding in active control of invasive species has often yielded equivocal results (e.g., Martin and Blossey 2013 for common reed). It should be more cost-effective to address the underlying facilitator of most plant invasions, namely anthropogenic disturbance, more directly. While many human-related activities that create disturbance in natural vegetation are probably difficult to regulate or eliminate, this tactic should be considered and used to the maximum extent possible. It may become more of a priority to limit human impacts if a known strong invader has already established along disturbance corridors in the area.

Weed-Shaped Holes

An often-overlooked source of disturbance in invasion biology is the disturbance created by weed control activities themselves. These activities create "weed-shaped holes" that are prime sites for invasion by secondary weeds and also for reinvasion (Buckley et al. 2007). Pearson et al. (2016) in a meta-analysis found that a large majority of weed control efforts that are successful in terms of controlling the target weed are subsequently beset by the problem of secondary weed invasion and experience very limited recovery of native vegetation. The importance of minimizing anthropogenic disturbance especially in areas adjacent to weed-infested patches has also been emphasized (Firn et al. 2008; Marvier et al. 2004), along with the obvious need to target secondary weeds in the initial control effort, to fill the "weed-shaped hole" with invasion-resistant native vegetation as quickly and efficiently as possible, and to protect it from subsequent disturbance (Pearson et al. 2016).

Identifying Problematic Invaders

Another difficult problem for managers is determining which introduced species are likely to be strong invaders with potentially major impacts on invaded plant communities (Ortega and Pearson 2005). As discussed earlier, the majority of non-native species that arrive and establish along disturbance corridors in a particular vegetation type are unlikely to have a major impact due to a combination of abiotic and biotic resistance and intrinsic species traits (e.g., growth rate, seed production). Given the globalization of world trade and the range of broad-scale anthropogenic disturbances such as global warming and nutrient pollution that are largely beyond management control, it seems inevitable that the species composition of natural vegetation will include increasing numbers of introduced species (Davis et al. 2008). Ecologists predict the emergence of "novel ecosystems" that will include species combinations that have never before coexisted (Hobbs et al. 2009).

The inevitability of colonization by non-native species does not mean, however, that efforts to prevent invasive species from pushing ecosystems into less desirable alternative stable states should be abandoned. Monitoring the abundance of individual species through time, perhaps in plots already established for some other purpose, can provide this kind of information (Strayer et al. 2006). The lag-time phenomenon in biological invasions adds to the difficulty of distinguishing the weak from the strong, however, because it may take a long time for a potentially strong invader to manifest itself with respect to a particular ecosystem. However, a potentially strong invader has probably shown its true colors in similar ecosystems elsewhere, and this is one of the best predictors of subsequent invasion behavior.

5.10 Knowledge Gaps and Research Needs

Many of the review papers cited earlier that address the history and impacts of even important and apparently well-studied invaders remark on the surprising lack of experimental studies or even rigorously collected quantitative field data that address the question of the underlying causes of invasion into a particular ecosystem. Ecologists have tended to focus on theoretical questions and to use whatever native or non-native species were tractable for addressing those questions, and even when these studies produce results that are useful to managers, they often are not readily accessible. Recently, the focus on "model invasive species" (e.g., Colautti et al. 2014b) attempts to address this somewhat piecemeal approach, but it hinges on the assumption that knowing everything there is to know about a few species will cast light on the invasion process for many others. One alternative, namely in-depth experimental and field studies of each invasive species of interest, is clearly not practical, given the urgency of the problems and the decisions that managers must make to address them. A more useful approach for basic research might be to classify environmental factors that drive species invasion and perform studies designed to elucidate the relationship between these drivers and consequent invasion for multiple invasive species.

The work of Guo et al. (2015) and Iannone et al. (2016) suggests another approach that might provide information on invasibility and invasiveness more quickly than experimental studies, and this approach could be applied to specific ecosystems, as well as to specific invasive species. As discussed earlier, long-term data sets available from the Forest Inventory and Analysis Program (www.fia.fs.fed.us/) were used to address the relationship between vegetation characteristics and both invasibility and invasion by non-native plants in general. The conceptual model of Guo et al. (2015; Fig. 5.1a, 5.1b) could also be used as a framework for addressing questions about the distribution and abundance of individual species relative to these vegetation characteristics. There may be other long-term data sets, for example, from LTER (Long Term Ecological Research Network) sites and USDA Forest Service Experimental Forest and Range sites that could be used for this purpose in non-forested systems. With careful matching, a suitable reference data set for the undisturbed stage could be selected for a particular site and then compared with current biomass and species richness data as an estimate of invasibility (Fig. 5.1a). Relative cover of individual non-native species on multiple dates could be used as a measure of the progress of the invasion through time. Although this approach is strictly correlative and cannot establish causal relationships, it could give managers some valuable information about the ecological behavior of both weak and strong invaders. The practicality of this approach depends on the availability of suitable data to address the management question at hand. A first step would be to determine whether existing data sets could be used in this way.

5.11 Key Findings

- The probability of establishment and spread of invasive plant species generally increases with increased disturbance and decreases with increased native biodiversity.
- Invasive plants that were deliberately introduced for horticulture are more likely to invade undisturbed forests than accidentally introduced weedy species.
- The probability of insect pest and plant pathogen invasion is not closely tied to either disturbance or native biodiversity but instead depends more on availability and spatial distribution of suitable hosts.

5.12 Key Information Needs

- Better methodology for assessing invasibility (vulnerability to invasion) and degree of invadedness for different ecosystem types.
- Improved understanding of the role of dispersal corridors and propagule pressure in mediating the pattern and speed of invasion.
- Experimental studies that address specific environmental drivers that affect groups of invasive species in a similar way, enabling generalization beyond individual species.
- Improved understanding of the ecological differences between invasive species and non-native species that do not become invasive, including development of models predicting the likelihood of invasion into specific ecosystems based on species functional traits.

Literature Cited

Adams JN, Fong W, Callaway R et al (2009) A cross-continental test of the enemy release hypothesis: leaf herbivory on *Acer platanoides* is three times lower in North America than in its native Europe. Biol Invasions 11:1005–1016

Aikio S, Duncan RP, Hulme PE (2010) Lag-phases in alien plant invasions: separating the facts from the artefacts. Oikos 119:370–378

Albert P, Bone E, Holzapfel C (2000) Invasiveness, invasibility, and the role of environmental stress in the spread of non-native plants. Perspect Plant Ecol Evol Syst 3:52–66

Anagnostakis SL (1987) Chestnut blight: the classical problem of an introduced pathogen. Mycologia 79:23–37

Andreadis TG, Weseloh RM (1990) Discovery of *Entomophaga maimaiga* in North American gypsy moth (*Lymantria dispar*). Proc Natl Acad Sci 87:2461–2465

Attiwill PM (1994) The disturbance of forest ecosystems: the ecological basis for conservation management. For Ecol Manag 63:247–300

Aukema J, McCullough DG, Von Holle B et al (2010) Historical accumulation of non-indigenous forest pests in the continental United States. Bioscience 60:886–897

Baron J (1982) Effects of feral hogs (*Sus scrofa*) on the vegetation of Horn Island, Mississippi. Am Midl Nat 107:202–205

Barbosa P, Hines J, Kaplan I (2009) Associational resistance and associational susceptibility: having right or wrong neighbors. Annu Rev Ecol Syst 40:1–20

Barrios-Garcia MN, Ballari SA (2012) Impact of wild boar (*Sus scrofa*) in its introduced and native range: a review. Biol Invasions 14:2283–2300

Besaw LM, Thelen GE, Sutherland S et al (2011) Disturbance, resource pulses and invasion: short-term shifts in competitive effects, not growth responses, favor exotic annuals. J Appl Ecol 48:998–1006

Blackman EB, Deperno CS, Heiniger RW et al (2012) Effects of crop field characteristics on nocturnal winter use by American woodcock. J Wildl Manag 76:528–533

Bossdorf FO, Auge H, Lafuma L et al (2005) Phenotypic and genetic differentiation between native and introduced plant populations. Oecologia 144:1–11

Brandt CA, Rickard WH (1994) Alien taxa in the North American shrub-steppe four decades after cessation of livestock grazing and cultivation agriculture. Biol Conserv 68:95–105

Briske DD, Bestelmeyer BT, Stringham TK, Shaver PL (2008) Recommendations for development of resilience-based state and transition models. Rangel Ecol Manag 61:359–367

Brooks ML (2003) Effects of increased soil nitrogen on the dominance of alien annual plants in the Mojave Desert. J Appl Ecol 40:344–353

Brooks ML, Chambers JC (2011) Resistance to invasion and resilience to fire in desert shrublands of North America. Rangel Ecol Manag 64:431–438

Brooks ML, D'Antonio CM, Richardson DM et al (2004) Effects of alien annual plants on fire regimes. Bioscience 54:677–688

Buckley YM, Bolker BM, Rees M (2007) Disturbance, invasion, and re-invasion: managing the weed-shaped hole in disturbed ecosystems. Ecol Lett 10:809–817

Burdon JJ, Thrall PH, Ericson L (2006) The current and future dynamics of disease in plant communities. Annu Rev Phytopathol 44:19–39

Byun C, deBlois S, Brisson J (2013) Plant functional group identity and diversity determine biotic resistance to an invasive grass. J Ecol 101:128–139

Callaham MA Jr, Blair JM, Todd TC, Kitchen DJ, Whiles MR (2003) Macroinvertebrates in North American tallgrass prairie soils: effects of fire, mowing, and fertilization on density and biomass. Soil Biol Biochem 35:1079–1093

Callaham MA Jr, González G, Hale CM et al (2006b) Policy and management responses to earthworm invasions. Biol Invasions 8:1317–1329

Callaham MA Jr, Richter DD, Coleman DC, Hofmockel M (2006a) Long-term land use effects on soil invertebrate communities in Southern Piedmont soils. Eur J Soil Biol 42:S150–S156

Cameron EK, Bayne EM (2009) Road age and its importance in earthworm invasions of northern boreal forests. J Appl Ecol 46:28–36

Cameron EK, Shaw CH, Bayne EM et al (2015) Modelling interacting effects of invasive earthworms and wildfire on forest floor carbon storage in the boreal forest. Soil Biol Biochem 88:189–196

Case TJ (1990) Invasion resistance arises in strongly interacting species-rich model competition communities. Proc Natl Acad Sci 87:9610–9614

Catania KC (2008) Worm grunting, fiddling, and charming – humans unknowingly mimic a predator to harvest bait. PLoS One 3(e):3472

Catford J, Jansson R, Nilsson C (2009) Reducing redundancy in invasion ecology by integrating hypotheses into a single theoretical framework. Divers Distrib 15:22–40

Catford J, Daehler CC, Murphy HT et al (2012b) The intermediate disturbance hypothesis and plant invasions: implications for species richness and management. Perspect Plant Ecol Evol Syst 14:231–241

Catford JA, Downes BJ, Gippel CJ, Vesk PA (2011) Flow regulation reduces native plant cover and facilitates exotic plant invasion in riparian wetlands. J Appl Ecol 48:432–442

Catford JA, Vesk PA, Richarson DM, Pysek P (2012a) Quantifying levels of biological invasion: towards the objective classification of invaded and invasible ecosystems. Glob Chang Biol 18:44–62

Chambers JC, Bradley BA, Brown CS et al (2014) Resilience to stress and disturbance, and resistance to *Bromus tectorum* L. invasion in cold desert shrublands of western North America. Ecosystems 17:360–375

Chambers JC, Roundy BA, Blank RR et al (2007) What makes Great Basin sagebrush ecosystems invasible by *Bromus tectorum*? Ecol Monogr 77:117–145

Chambers RM, Meyerson LA, Saltonstall K (1999) Expansion of *Phragmites australis* into tidal wetlands of North America. Aquat Bot 64:261–273

Childers DL, Doren RF, Jones R et al (2003) Decadal change in vegetation and soil phosphorus patterns across the Everglades landscape. J Environ Qual 32:344–362

Chipps SR, Hubbard DE, Werlink B et al (2006) Association between wetland disturbance and biological attributes on floodplain wetlands. Wetlands 26:497–508

Churchill DJ, Larson AJ, Dahlgreen MC et al (2013) Restoring forest resilience: from reference spatial patterns to silvicultural prescriptions and monitoring. For Ecol Manag 291:442–457

Cincotta CH, Adams JM, Holzapfel C (2009) Testing the enemy release hypothesis: a comparison of foliar herbivory of the exotic Norway maple (*Acer platanoides* L.) and the native sugar maple (*A. saccharum* L.). Biol Invasions 11:379–388

Clewley GD, Eschen R, Shaw RH, Wright DJ (2012) The effectiveness of biological control of invasive plants. J Appl Ecol 49:1287–1295

Coetzee JA, Hill MP (2012) The role of eutrophication in the biological control of water hyacinth, *Eichhornia crassipes*, in South Africa. BioControl 57:247–261

Colautti RI, Franks SJ, Hufbauer RA (2014b) The global garlic mustard field survey (GGMFS): challenges and opportunities of a unique, large-scale collaboration for invasion biology. NeoBiota 21:29–47

Colautti RI, Parker JD, Cadotte MW et al (2014a) Quantifying the invasiveness of species. NeoBiota 21:7–27

Costello DM, Teigs SD, Lamberti GA (2011) Do non-native earthworms in Southeast Alaska use streams as invasional corridors in watersheds harvested for timber? Biol Invasions 13:177–187

Crooks JA (2005) Lag times and exotic species: the ecology and management of biological invasions in slow motion. Ecoscience 12:316–329

Cullen ST, Miller AJ (2015) Signatures of niche conservation and niche shift in the North American (*Pueraria montana*) invasion. Divers Distrib 21:853–863

Cyr D, Gaulthier S, Bergeron Y, Carcaillet C (2009) Forest management is driving the eastern North American boreal forest outside its natural range of variability. Forest Ecol Environ 7:519–524

D'Antonio CM, Vitousek PM (1992) Biological invasions by exotic grasses, the grass/fire cycle, and global change. Annu Rev Ecol Syst 23:63–87

Davis MA, Anderson MD, Brock-Brownstein L et al (2015) Little evidence of native and non-native species influencing one another's abundance in an herb layer of an oak woodland. J Veg Sci 26:1005–1012

Davis MA, Chew MK, Hobbs RJ (2008) Don't judge species on their origins. Nature 474:153–154

Davis MA, Grime JP, Thompson K (2000) Fluctuating resources in plant communities: a general theory of invasibility. J Ecol 88:528–534

Davis MA, Thompson K, Grime JP (2005) Invasibility: the local mechanism driving community assembly and species diversity. Ecography 28:696–704

Davis MA, MacMillen C, LeFevre-Levy M et al (2014) Population and plant community dynamics involving garlic mustard (*Alliaria petiolata*) in a Minnesota oak woodland: a four-year study. J Torrey Bot Soc 141:205–216

Desprez-Loustau ML, Robin C, Buee M et al (2007) The fungal dimension of biological invasions. Trends Ecol Evol 22:472–480

Dewine JM, Cooper DJ (2008) Canopy shade and the successional replacement of tamarisk by native box elder. J Appl Ecol 45:505–514

Dick JTA, Alexander ME, Jeschke JM et al (2014) Advancing impact prediction and hypothesis testing in invasion ecology using a comparative functional response approach. Biol Invasions 16:735–753

Dorcas ME, Willson JD, Reed RN et al (2012) Severe mammal declines coincide with proliferation of invasive Burmese pythons in Everglades National Park. Proc Natl Acad Sci 109:2418–2422

Duan JJ, Bauer LS, Abell KJ et al (2013) Establishment and abundance of *Tetrastichus plenipennisi* (Hymenoptera: Eulophidae) in Michigan: potential for success in classical biocontrol of the invasive emerald ash borer (Coleoptera: Buprestidae). J Econ Entomol 106:1145–1154

Duan JJ, Bauer LS, Abell KJ, Van Driesche R (2012) Population responses of hymenopteran parasitoids to the emerald ash borer (Coleoptera: Buprestidae) in recently invaded areas in north central United States. BioControl 57:199–209

Edwards CA, Hendrix PF, Arancon N (2013) Biology and ecology of earthworms, 4th edn. Springer US, New York, 422 p

Elkinton JS, Liebhold AM (1990) Population dynamics of gypsy moth in North America. Annu Rev Entomol 35:571–596

Ennos RA (2014) Resilience of forests to pathogens; an evolutionary ecology perspective. Forestry 88:41–52

Eschtruth AK, Battles JJ (2008) Acceleration of exotic plant invasion in a forested ecosystem by a generalist herbivore. Conserv Biol 23:388–399

Eschtruth AK, Battles JJ (2009) Assessing the relative importance of disturbance, herbivory, diversity, and propagule pressure on exotic plant invasion. Ecol Monogr 79:265–280

Eschtruth AK, Battles JJ (2011) The importance of quantifying propagule pressure to understand invasion: an examination of riparian forest invasibility. Ecology 92:1314–1322

Eschtruth AK, Battles JJ (2014) Ephemeral disturbances have long-lasting impacts on forest invasion dynamics. Ecology 95:1770–1779

Eschtruth AK, Cleavitt NL, Battles JJ et al (2006) Vegetation dynamics in declining eastern hemlock stands: 9 years of forest response to hemlock woolly adelgid infestation. Can J For Res 36:1435–1450

Facon B, Genton BJ, Shykoff J et al (2006) A general eco-evolutionary framework for understanding bioinvasions. Trends Ecol Evol 21:130–135

Fargione JE, Tilman D (2005) Diversity decreases invasion by both complementarity and sampling effects. Ecol Lett 8:604–611

Firn J, Rout T, Possingham H, Buckley Y (2008) Managing beyond the invader: manipulating disturbance of natives simplifies control efforts. J Appl Ecol 45:1143–1151

Fisher MC, Henk DA, Briggs CJ et al (2012) Emerging fungal threats to animal, plant and ecosystem health. Nature 484:186–194

Forseth JN Jr, Innis AF (2004) Kudzu (*Puereria montana*): history, physiology, and ecology combine to make a major ecosystem threat. Crit Rev Plant Sci 23:401–413

Frelich LE, Hale CM, Scheu S et al (2006) Earthworm invasions into previously earthworm-free temperate and boreal forests. Biol Invasions 8:1235–1245

Fridley JD, Stachowicz JJ, Naeem S et al (2007) The invasion paradox: reconciling pattern and process in species invasions. Ecology 88:3–17

Funk JL, Cleland EE, Suding KN, Zavaleta ES (2008) Restoration through reassembly: plant traits and invasion resistance. Trends Ecol Evol 23:695–703

Gandhi KJK, Herms DA (2010) Direct and indirect effects of insect herbivores on ecological processes and interactions in forests of eastern North America. Biol Invasions 12:389–405

Gelbard JI, Belnap J (2003) Roads as conduits for exotic plant invasions in a semiarid landscape. Conserv Biol 17:420–432

Gilligan CA, van den Bosch F (2008) Epidemiological models for invasion and persistence of pathogens. Annu Rev Phytopathol 46:385–418

Gladieux P, Feurtey A, Hood ME et al (2015) The population biology of fungal invasions. Mol Ecol 24:1969–1986

Gorsuch JP, Owen PC (2014) Potential edaphic and aquatic predators of a nonindigenous Asian earthworm (*Amynthas agrestis*) in the eastern United States. Northeast Nat 21:652–661

Grünwald NJ, Garbelotto M, Goss EM et al (2012) Emergence of the sudden oak death pathogen *Phytophthora ramorum*. Trends Ecol Evol 20:131–138

Gundale MJ (2002) Influence of exotic earthworms on the soil organic horizon and the rare fern *Botrychium mormo*. Conserv Biol 16:1555–1561

Gunderson LH (2000) Ecological resistance - in theory and application. Annu Rev Ecol Syst 31:425–439

Guo Q, Symstad A (2008) A two-part measure of degree of invasion for cross-community comparisons. Conserv Biol 22:666–672

Guo Q, Fei S, Dukes JS et al (2015) A unified approach for quantifying invasibility and degree of invasion. Ecology 96:2613–2621

Gurevitch J, Fox GA, Wardle GM et al (2011) Emergent insights from the synthesis of conceptual frameworks for biological invasions. Ecol Lett 14:407–418

Hajek AE, Tobin PC, Haynes KJ (2015) Replacement of a dominant viral pathogen by a fungal pathogen does not alter the collapse of a regional forest insect outbreak. Oecologia 177:785–797

Havill N, Montgomery ME, Keena MA (2011) Hemlock woolly adelgid and its hemlock hosts: a global perspective. In: Onkin B, Reardon R, tech. coord (eds) Implementation and status of biocontrol of the hemlock woolly adelgid. U.S Forest Service, Forest Health Technology Enterprise Team. FHTET_2011-04, pp 3–14

Hayes KR, Barry SC (2008) Are there any consistent predictors of invasion success? Biol Invasions 10:483–506

Heger T, Jeschke JM (2014) The enemy release hypothesis as a hierarchy of hypotheses. Oikos 123:741–750

Heger T, Pahl AT, Botta-Dukat Z et al (2013) Conceptual frameworks and methods for advancing invasion ecology. Ambio 42:527–540

Hendrix PF, Bohlen PJ (2002) Exotic earthworm invasions in North America: ecological and policy implications. Bioscience 52:801–811

Hendrix PF, Callaham MA Jr, Drake JM et al (2008) Pandora's Box contained bait: the global problem of introduced earthworms. Annu Rev Ecol Evol Syst 39:593–613

Heneghan L, Steffen J, Fagen K (2007) Interactions of an introduced shrub and introduced earthworms in an Illinois urban woodland: Impact on leaf litter decomposition. Pedobiologia 50:543–551

Hereford R, Webb RH, Longpre CI (2006) Precipitation history and ecosystem response to multidecadal precipitation variability in the Mojave Desert region, 1893–2001. J Arid Environ 67:13–34

Herms DA, McCullough DE (2014) Emerald ash borer invasion of North America: history, biology, ecology, impacts, and management. Annu Rev Entomol 59:13–30

Hobbs RJ, Higgs E, Harris JA (2009) Novel ecosystems: implications for conservation and restoration. Trends Ecol Evol 24:599–605

Hobbs RJ, Huenneke LF (1992) Disturbance, diversity, and invasion: implications for conservation. Conserv Biol 6:324–337

Hoffmann BD, Saul WC (2010) Yellow crazy ant (*Anoplolepis gracilipes*) invasions within undisturbed mainland Australian habitats: no support for biotic resistance hypothesis. Biol Invasions 12:3093–3108

Holling CS (1973) Resilience and stability of ecological systems. Annu Rev Ecol Syst 4:1–23

Holway DA, Lach L, Suarez AV et al (2002) The causes and consequences of ant invasions. Annu Rev Ecol Syst 33:181–233

Horvitz CC, Pascarella JB, McMann S et al (1998) Functional roles of invasive non-indigenous plants in hurricane-affected subtropical hardwood forests. Ecol Appl 8:947–974

Huenneke LF, Hamburg SP, Koide R et al (1990) Effects of soil resources on plant invasion and community structure in Californian serpentine grasslands. Ecology 71:478–491

Iannone BV III, Potter KM, Dixon-Hamil K et al (2016) Evidence of biotic resistance to invasions in forests of the eastern USA. Landsc Ecol 31:85–99

Ikeda H, Callaham MA Jr, O'Brien JJ et al (2015) Can the invasive earthworm, *Amythas agrestis*, be controlled with prescribed fire? Soil Biol Biochem 82:21–27

James SW (1995) Chapter 2: Systematics, biogeography, and ecology of Nearctic earthworms from eastern, central, southern, and southwestern United States. In: Hendrix PF (ed) Earthworm ecology and biogeography in North America. CRC Press, Boca Raton, pp 29–51

James SW (2004) Planetary processes and their interactions with earthworm distributions and ecology. In: Edwards CA (ed) Earthworm ecology, 2nd edn. CRC Press, Boca Raton, pp 53–62. Chapter 3

Jankovic M, Petrovskii S (2013) Gypsy moth invasions in North America: a simulation study of the spatial pattern and rate of spread. Ecol Complex 14:132–144

Jeschke JM (2014) General hypotheses in invasion ecology. Divers Distrib 20:1229–1234

Jeschke JM, Gomez-Aparicio L, Haider S et al (2012) Support for major hypotheses in invasion biology is uneven and declining. NeoBiota 14:1–20

Jeschke JM, Strayer DL (2006) Determinants of vertebrate invasion success in Europe and North America. Glob Chang Biol 12:1608–1619

Jones CG, Lawton JH, Shachak M (1994) Organisms as ecosystem engineers. Oikos 69:373–386

Kalisz PJ (1993) Chapter 7: Native and exotic earthworms in deciduous forest soils of eastern North America. In: McKnight BN (ed) Biological pollution: the control and impact of invasive exotic species. Indiana Academy of Sciences, Indianapolis, pp 93–100

Kalisz PJ, Dotson DB (1989) Land use history and the occurrence of exotic earthworms in the mountains of eastern Kentucky. Am Midl Nat 122:288–297

Kalisz S, Spigler RB, Horvitz CC (2014) In a long-term demographic study, excluding ungulates reversed invader's explosive population growth and restored natives. Proc Natl Acad Sci 111:4501–4506

Keane RM, Crawley MJ (2002) Exotic plant invasions and the enemy release hypothesis. Trends Ecol Evol 17:164–170

Keeley JE (2001) Fire and invasive species in Mediterranean-climate ecosystems of California. In: Galley KEM, Wilson TP (eds) Proceedings of the invasive species workshop: the role of fire in the control and spread of invasive species. fire conference 2000: the first national congress on fire ecology, prevention, and management. Miscellaneous Publication No. 11, Tall Timbers Research Station, Tallahassee, FL, pp 81–94

Kennedy TA, Naeem S, Howe KM et al (2002) Biodiversity as a barrier to biological invasion. Nature 417:636–638

Kerns BK, Day MA (2017) The importance of disturbance by fire and other abiotic and biotic factors in driving cheatgrass invasion varies based on invasion stage. Biol Invasions 19:1853–1862

Kettering KM, De Blois S, Hauber DP (2012) Moving from a regional to a continental perspective of *Phragmites australis* invasion in North America. AoB Plants:pls040

Kinloch BB Jr (2003) White pine blister rust in North America: past and prognosis. Phytopathology 93:1044–1047

Kitchen SG, Jorgensen GL (1999) Annualization of rodent burrow clusters and winterfat decline in a salt-desert community. In: McArthur ED, Ostler WK, Wambolt CL, comps (eds) Proceedings: shrubland ecotones; 1998 August 12–14; Ephraim, UT. Proc. RMRS-P-11. U.S. Department of Agriculture, Forest Service, Rocky Mountain Research Station, Ogden, pp 175–180

Knapp AK, Briggs JM, Hartnett DC, Collins SL (1998) Grassland dynamics: long-term ecological research in Tallgrass Prairie. Oxford University Press, New York, 364 p

Kohler GR, Stiefel VL, Wallin KF, Ross DW (2008) Predators associated with the hemlock woolly adelgid (Hemiptera: Adelgidae) in the Pacific Northwest. Environ Entomol 37:494–504

Kotanen PM (1995) Responses of vegetation to a changing regime of disturbance: effects of feral pigs in a Californian coastal prairie. Ecography 18:190–199

Kueffer C, Pysek P, Richardson DM (2013) Integrative invasion science: model systems, multi-site studies, focused meta-analysis, and invasion syndromes. New Phytol 200:615–633

Kurtz CM, Hansen MH (2014) An assessment of garlic mustard in northern US forests. Research Note NRS-199. Newtown Square PA: USDA Forest Service, Northern Research Station, 5 p

Kyser GB, DiTomaso JM (2002) Instability in a grassland community after the control of yellow starthistle (*Centaurea solstitialis*) with prescribed burning. Weed Sci 50:648–657

Lafon CW (2015) Ice storms in central hardwood forests: the disturbance regime, spatial patterns, and vegetation influences. In: Greenberg CH, Collins BS (eds) Natural disturbance and historic range of variation; type, frequency, severity, and post-disturbance structure in Central Hardwood Forests, USA, Managing Forest Ecosystems, vol 32. Springer, New York, pp 147–166

Landres PB, Morgan P, Swanson FJ (1999) Overview of the use of natural variability concepts in the management of natural systems. Ecol Appl 9:1179–1188

Larson KL, Fowler SP, Walker JC (2002) Lack of pollinators limits fruit set in the exotic *Lonicera japonica*. Am Midl Nat 148:54–60

Lee CE (2002) Evolutionary genetics of invasive species. Trends Ecol Evol 17:386–391

LeJeune KD, Seastedt TR (2001) *Centaurea* species: the forb that won the West. Conserv Biol 15:1568–1574

Levine JM, Adler PB, Yelenik SG (2004) A meta-analysis of biotic resistance to exotic plant invaders. Ecol Lett 7:975–989

Liebhold A, Elkinton J, Williams D, Muzika R (2000) What causes outbreaks of gypsy moth in North America? Popul Ecol 42:257–266

Liebhold AM, Tobin PC (2008) Population ecology of insect invasions and their management. Annu Rev Entomol 53:387–408

Linz GM, Homan JL, Gaulker SM et al (2007) European starlings: a review of an invasive species with far-reaching impacts. In: Witmer GW, Pitt WC, Fagerstone KA (eds) Managing vertebrate invasive species. USDA/APHIS/WS, National Wildlife Research Center, Ft. Collins CO, pp 378–386

Lobe JW, Callaham MA Jr, Hendrix PF, Hanula JL (2014) Removal of an invasive shrub (Chinese privet: *Ligustrum sinense* Lour) reduces exotic earthworm abundance and promotes recovery of native North American earthworms. Appl Soil Ecol 83:133–139

Lonsdale WM (1999) Global patterns of plant invasion and the concept of invasibility. Ecology 80:1522–1536

Lozon JD, MacIsaac HJ (1997) Biological invasions: are they dependent on disturbance? Environ Rev 5:131–144

Lyons KG, Swartz MW (2001) Rare species loss alters ecosystem function – invasion resistance. Ecol Lett 4:358–365

MacDougall AS, Gilbert B, Levine JM (2009) Plant invasions and the niche. J Ecol 97:609–615

MacDougall AS, Turkington R (2005) Are invasive species the drivers or passengers of change in degraded ecosystems? Ecology 86:42–55

Mack RN (1981) Invasion of *Bromus tectorum* L into western North America: an ecological chronicle. Agro-Ecosystems 7:145–165

Mack RN, Thompson JN (1982) Evolution in steppe with few large hooved mammals. Am Nat 119:757–773

Madritch MD, Lindroth RL (2009) Removal of invasive shrubs reduces exotic earthworm populations. Biol Invasions 11:663–671

Maerz JC, Karuzas JM, Madison DM, Blossey B (2005) Introduced invertebrates are important prey for a generalist predator. Biodiversity Res 11:83–90

Maron JM, Marler M (2007) Native plant diversity resists invasion at both low and high resource levels. Ecology 88:2651–2661

Maron JM, Vila M (2001) When do herbivores affect plant invasion? Evidence for the natural enemies and biotic resistance hypotheses. Oikos 95:361–373

Martin PH (1999) Norway maple (*Acer platanoides*) invasion of a natural forest stnd: understory consequences and regeneration pattern. Biol Invasions 1:215–222

Martin LJ, Blossey B (2013) The runaway weed: costs and failures of *Phragmites australis* management in the USA. Estuar Coasts 36:626–632

Martin PH, Canham CD, Kobe RK (2010) Divergence from the growth-survival trade-off and extreme high growth rates drive patterns of exotic tree invasions in closed-canopy forests. J Ecol 98:778–789

Martin PH, Canham CD, Marks PL (2009) Why forests appear resistant to exotic plant invasions: intentional introductions, stand dynamics, and the role of shade tolerance. Front Ecol Environ 7:142–149

Martin PH, Marks PL (2006) Intact forests provide only weak resistance to a shade tolerant invasive Norway maple (*Acer platanoides* L.). J Ecol 94:1070–1079

Marvier M, Kareiva P, Neubert MG (2004) Habitat destruction, fragmentation, and disturbance promote invasion by habitat generalists in a multi-species population. Risk Anal 24:869–878

McAuliffe JR, Hamerlynck EP (2010) Perennial plant mortality in the Sonoran and Mojave deserts in response to severe, multi-year drought. J Arid Environ 74:885–896

McClure MS, Cheah CASJ (1999) Reshaping the ecology of invading populations of hemlock woolly adelgid, *Adelges tsugae* (Homoptera: Adelgidae), in eastern North America. Biol Invasions 9:247–254

Middleton B (1999) Wetland restoration: flood pulsing and disturbance dynamics. Wiley and Sons, New York, 389 p

Milchunas DG, Lauenroth WK, Chapman PL, Kazempour MK (1989) Effect of grazing, topography, and precipitation on the structure of a semiarid grassland. Vegetatio 80:11–23

Miller AL, Diez JM, Sullivan JJ et al (2014) Quantifying invasion resistance: the use of recruitment functions to control for propagule pressure. Ecology 95:920–929

Millward AA, Kraft KE (2004) Physical influences of landscape on a large-extent ecological disturbance: the northeastern North American ice storm of 1998. Landsc Ecol 19:99–111

Minchinton TE, Bertness MD (2003) Disturbance-mediated competition and the spread of *Phragmites australis* in a coastal marsh. Ecol Appl 13:1400–1416

Moles AT, Gruber MAM, Bonser SP (2008) A new framework for predicting invasive species. J Ecol 96:13–17

Moles AT, Flores-Moreno H, Bonser SP et al (2012) Invasions: the trail behind, the path ahead, and a test of a disturbing idea. J Ecol 100:116–127

Morris LR, Monaco TA, Sheley RL (2011) Land-use legacies and vegetation recovery 90 years after cultivation in Great Basin sagebrush ecosystems. Rangel Ecol Manag 64:488–497

Morrison LW (2002) Long term impacts of an arthropod community invasion by the imported fire ant, *Solenopsis invicta*. Ecology 83:2337–2345

Muirhead JR, Leung B, Van Overdijk C et al (2006) Modelling local and long-distance dispersal of invasive emerald ash borer *Agrilus plenipennis* (Coleoptera) in North America. Divers Distrib 12:71–79

Nuzzo Y (1999) Invasion pattern of the herb garlic mustard (*Alliaria petiolata*) in high quality forests. Biol Invasions 1:169–179

O'Dowd DJ, Green PT, Lake PS (2003) Invasional 'meltdown' on an oceanic island. Ecol Lett 6:812–817

Ortega YK, Pearson DE (2005) Weak vs. strong invaders of natural plant communities: assessing invasibility and impact. Ecol Appl 15:651–661

Owen MR, Lewis MA (2001) How predation can slow, stop, or reverse a prey invasion. Bull Math Biol 63:655–684

Parker JD, Burkepile DE, Hay ME (2006) Opposing effects of native and exotic herbivores on plant invasions. Science 311:1459–1461

Pearson DE, Ortega YK, Runyon JB, Butler JL (2016) Secondary invasion: the bane of weed management. Biol Conserv 197:8–17

Pokorney M, Sheley RL, Zabinski CL et al (2005) Plant functional group diversity as a mechanism for invasion resistance. Restor Ecol 13:448–459

Price JN, Partel M (2013) Can limiting similarity increase invasion resistance: a meta-analysis of experimental studies. Oikos 122:649–656

Purcell KL (2015) Foraging behavior of European starlings: implications for mitigating their impact on native species. In: Standiford RB, Purcell KL, tech. cords (eds) Proceedings of the seventh California oak symposium: managing oak woodlands in a dynamic world, Gen. Tech. Rep. PSW-GTR-251. U.S. Department of Agriculture, Forest Service, Pacific Southwest Research Station, Berkeley, CA, pp 73–81

Raffa KF, Aukema BH, Bentz BJ et al (2008) Cross-scale drivers of natural disturbances prone to anthropogenic amplification: the dynamics of bark beetle eruptions. Bioscience 58:501–517

Reichard SH, White P (2001) Horticulture as a pathway of invasive plant introductions in the United States. Bioscience 51:103–113

Rejmanek M, Richardson DM (1996) What attributes make some plant species more invasive? Ecology 77:1655–1661

Reynolds JW (1995) Status of exotic earthworm systematics and biogeography in North America. In: Hendrix PF (ed) Earthworm ecology and biogeography in North America. CRC Press, Boca Raton, pp 1–27. Chapter 1

Richardson BA, Ekramoddoulah AKM, Liu JJ et al (2010) Current and future molecular approaches to investigate the white pine blister rust pathosystem. For Pathol 40:314–331

Richardson DM, Holmes PM, Esler KJ et al (2007) Riparian vegetation: degradation, alien invasion, and restoration prospects. Divers Distrib 13:126–139

Richardson JB, Görres JH, Jackson BP, Friedland AJ (2015) Trace metals and metalloids in forest soils and exotic earthworms in northern New England. USA Soil Biol Biochem 85:190–198

Rigot T, VanHalder I, Jactel H (2014) Landscape diversity slows the spread of an invasive forest pest species. Ecography 37:1–11

Rizzo DM, Garbelotto M (2003) Sudden oak death: endangering California and Oregon forest ecosystems. Front Ecol Environ 1:197–204

Roche CT, Thill DC (2001) Biology of common crupina and yellow starthistle, two Mediterranean winter annual invaders in western North America. Weed Sci 49:439–447

Rodda GR, Jarnevich CS, Reed RN (2009) What parts of the US mainland are climatically suitable for invasive alien pythons spreading from Everglades National Park? Biol Invasions 11:241–252

Rodgers VL, Stinson KA, Finzi AC (2008) Ready or not, garlic mustard is moving in: *Alliaria petiolata* as a member of eastern North American forests. Bioscience 58:426–436

Sanchez-deLeon Y, Johnson-Maynard J (2009) Dominance of an invasive earthworm in native and non-native grassland ecosystems. Biol Invasions 11:1393–1401

Scharenbroch BC, Nix B, Jacobs KA, Bowles ML (2012) Two decades of low-severity prescribed fire increases soil nutrient availability in a Midwestern, USA oak (*Quercus*) forest. Geoderma 183:80–91

Schierenbeck KA (2004) Japanese honeysuckle (*Lonicera japonica*) as an invasive species: history, ecology, and context. Crit Rev Plant Sci 23:391–400

Schierenbeck KA, Mack RN, Sharitz RR (1994) Effects of herbivory on growth and biomass allocation in native and introduced species of honeysuckle. Ecology 75:1661–1672

Selments PC, Zavaleta ES, Pasari JR, Hernandez DL (2012) Realistic plant species losses reduce invasion resistance in a California serpentine grassland. J Ecol 100:723–731

Serbesoff-King K (2003) Melaleuca in Florida: A literature review on the taxonomy, distribution, biology, ecology, economic importance and control measures. J Aquat Plant Manag 41:98–112

Shea K, Chesson P (2002) Community ecology theory as a framework for biological invasions. Trends Ecol Evol 17:170–176

Shea K, Roxburgh SH, Rauchert ESJ (2004) Moving from pattern to process: coexistence mechanisms under intermediate disturbance regimes. Ecol Lett 7:491–508

Sheley RL, Jacobs JS, Carpinelli MF (1998) Distribution, biology, and management of diffuse knapweed (*Centaurea diffusa*) and spotted knapweed (*Centaurea maculosa*). Weed Technol 12:353–362

Sher AA, Hyatt LA (1999) The disturbed resource-flux invasion matrix: a new framework for patterns of plant invasion. Biol Invasions 1:107–114

Silliman BR, Bertness MD (2004) Shoreline development drives invasion of *Phragmites australis* and loss of plant diversity on New England salt marshes. Conserv Biol 18:1424–1434

Simberloff D (2009) The role of propagule pressure in biological invasions. Annu Rev Ecol Evol Syst 40:81–102

Simberloff D, Von Holle B (1999) Positive interactions of nonindigenous species: invasional meltdown? Biol Invasions 1:21–32

Smith MD, Knapp AK (1999) Exotic plant species in a C4-dominated grassland; invasibility, disturbance, and community structure. Oecologia 120:605–612

Snyder BA, Callaham MA Jr, Hendrix PF (2011) Spatial variability of an invasive earthworm (*Amynthas agrestis*) population and potential impacts on soil characteristics and millipedes in the Great Smoky Mountains National Park, USA. Biol Invasions 13:349–358

Snyder BA, Callaham MA Jr, Lowe CN, Hendrix PF (2013) Earthworm invasion in North America: Food resource competition affects native millipede survival and invasive earthworm reproduction. Soil Biol Biochem 57:212–216

Snyder WE, Evans EW (2006) Ecological effects of invasive arthropod generalist predators. Annu Rev Ecol Evol Syst 37:95–122

Spiers AG, Hopcroft DH (1994) Comparative studies of the poplar rusts *Melampsora medusae, M. larici-populina* and their interspecific hybrid *M. medusae-populina*. Mycol Res 98:889–903

Stebbings JH (1962) Endemic-exotic earthworm competition in the American Midwest. Nature 196:905–906

Stevens J, Falk DA (2009) Can buffelgrass invasions be controlled in the American Southwest? Using invasion ecology theory to understand buffelgrass success and develop comprehensive restoration and management. Ecol Restor 27:417–427

Stinson KA, Kaufman S, Durbin L, Lowenstein F (2007) Impacts of garlic mustard invasion on a forest understory community. Northeast Nat 14:73–88

Stohlgren TJ, Barnett DT, Kartesz JT (2003) The rich get richer: patterns of plant invasions in the United States. Front Ecol Environ 1:11–14

Stohlgren TJ, Binkley D, Chong GW et al (1999) Exotic plant species invade hotspots of native plant diversity. Ecol Monogr 69:25–46

Stohlgren TJ, Bull KA, Otsuka Y et al (1998) Riparian zones as havens for exotic plant species in the central grasslands. Plant Ecol 138:113–125

Strayer DL, Eviner VT, Jaschke JM, Pace ML (2006) Understanding the long-term effects of species invasions. Trends Ecol Evol 21:645–651

Stromberg JC (1998) Functional equivalency of saltcedar (*Tamarix chinensis*) and Fremont cottonwood (*Populus fremontii*) along a free-flowing river. Wetlands 18:675–686

Stromberg JC, Lite SJ, Macher R et al (2007) Altered stream flow regimes and invasive plant species: the *Tamarix* case. Glob Ecol Biogeogr 16:381–393

Stromberg JC, Chew MK, Nagler PL, Glenn EP (2009) Changing perceptions of change: the role of scientist in *Tamarix* and river management. Restor Ecol 17:177–186

Theoharides KA, Dukes JS (2007) Plant invasion across space and time: factors affecting non-indigenous species success during four stages of invasion. New Phytol 176:256–273

Thompson DB, Walker LR, Landau FH, Stark LR (2005) The influence of elevation, shrub species, and biological soil crust on fertile islands in the Mojave Desert, USA. J Arid Environ 61:609–629

Thompson LM, Grayson KL, Johnson DM (2016) Forest edges enhance mate finding in the invasive European gypsy moth (*Lymantria dispar*). Entomol Exp Appl 158:295–303

Turner CE, Center TD, Burrows DW, Buckingham GR (1998) Ecology and management of *Melaleuca quinquenervia*, an invader of wetlands in Florida, USA. Wetl Ecol Manag 5:165–178

Turner JL, Fitzpatrick MC, Preisser EL (2011) Simulating the dispersal of hemlock woolly adelgid in the temperate forest understory. Entomol Exp Appl 141:216–223

Van Der Heijden MGA, Bardgett R, Van Straalen NM (2008) The unseen majority: soil microbes as drivers of plant diversity and productivity in terrestrial ecosystems. Ecol Lett 11:296–310

Van Driesche R, Hoddle M, Center T (2008) Control of pests and weeds by natural enemies: an introduction to biological control. Wiley, Malden, Massachusetts. 488 p

Vavra M, Parks CG, Wisdom M (2007) Biodiversity, exotic plant species, and herbivory: the good, the bad, and the ungulate. For Ecol Manag 246:66–72

Vitousek PM, Walker LR (1989) Biological invasion by *Myrica faya* in Hawai'i: plant demography, nitrogen fixation, ecosystem effects. Ecol Monogr 59:247–265

Von Holle B, Simberloff D (2005) Ecological resistance to biological invasion overwhelmed by propagule pressure. Ecology 86:3212–3218

Wallace MS, Hain P (2000) Field surveys and evaluation of native and established predators of the hemlock woolly adelgid (Homoptera: Adelgidae) in the southeastern United States. Environ Entomol 29:638–644

Walter JA, Meixler MS, Mueller T et al (2015) How topography induces reproductive asynchrony and alters gypsy moth invasion dynamics. J Anim Ecol 84:188–198

Wangen SR, Webster CR (2006) Potential for multiple lag phases during biotic invasions: reconstructing an invasion of the exotic tree *Acer platanoides*. J Appl Ecol 43:258–268

Watt MS, Kriticos DJ, Manning LK (2009) The current and potential future distribution of *Melaleuca quinqenervia*. Weed Res 49:381–390

Webb RH, Steiger JW, Turner RM (1987) Dynamics of Mojave desert shrub assemblages in the Panamint Mountains, California. Ecology 68:478–490

Weiss SB (1999) Cows, cars, and checkerspot butterflies: nitrogen deposition and management of nutrient-poor grasslands for a threatened species. Conserv Biol 13:1476–1486

Wetterer JK (2005) Worldwide distribution and potential spread of the long-legged ant, *Anoplolepis gracilipes* (Hymenoptera: Formicidae). Sociobiology 45:77–97

Whisenant SG (1990) Changing fire frequencies on Idaho's Snake River Plains: ecological and management implications. In: McArthur ED, Romney EM, Smith SD, Tueller PT (eds) Proceedings: symposium on cheatgrass invasion, shrub die-off and other aspects of shrub biology and management 1989, Las Vegas, NV, USA. USDA General Technical Report INT-276. Intermountain Research Station, Ogden, pp 4–10

Wilson JRU, Dormontt EE, Prentis PJ et al (2009) Something in the way you move: dispersal pathways affect invasion success. Trends Ecol Evol 24:136–144

Winsome T, Epstein L, Hendrix PF, Horwath WR (2006) Competitive interactions between native and exotic earthworm species as influenced by habitat quality in a California grassland. Appl Soil Ecol 32:38–53

Woods A, Coates KD, Hamann A (2005) Is an unprecedented *Dothistroma* needle blight epidemic related to climate change? Bioscience 55:761–769

Wykoff PH, Shaffer A, Hucka B et al (2014) No evidence of facilitation between invasives *Rhamnus cathartica* (European buckthorn) and invasive earthworms in west central Minnesota. Pedobiologia 57:311–317

Yang X, Huang W, Ding J (2014) Differences in growth and herbivore damage of native and invasive kudzu (*Pueraria montana* var. *lobata*) populations grown in the native range. Plant Ecol 215:339–346

Zavaleta ES, Hulvey KB (2007) Realistic variation in species composition affects grassland production, resource use, and invasion resistance. Plant Ecol 188:39–51

Zedler JB, Kercher S (2004) Causes and consequences of invasive plants in wetlands: opportunities, opportunists, and outcomes. Crit Rev Plant Sci 23:431–452

Zedler PH, Gautier CR, McMaster CS (1983) Vegetation change in response to extreme events: the effect of a short interval between fires in California chaparral and coastal scrub. Ecology 64:809–818

Zouhar K, Smith JK, Sutherland S Brooks ML eds. (2008) Wildfire in ecosystems: fire and nonnative invasive plants. USDA Forest Service Rocky Mountain Research Station General Technical Report RMRS-GTR-42-Volume 6.

6 Early Intervention Strategies for Invasive Species Management: Connections Between Risk Assessment, Prevention Efforts, Eradication, and Other Rapid Responses

Robert C. Venette, Doria R. Gordon, Jennifer Juzwik, Frank H. Koch, Andrew M. Liebhold, Robert K. D. Peterson, Sharlene E. Sing, and Denys Yemshanov

6.1 Introduction

Managing invasive species becomes increasingly difficult and expensive as populations of new pathogens, plants, insects, and other animals (i.e., pests) spread and reach high densities. Research over the past decade confirms the value of early intervention strategies intended to (1) prevent invasive species from arriving within an endangered area or (2) detect and respond quickly to new species incursions (Baker et al. 2009; Ewel et al. 1999; Holden et al. 2016; Leung et al. 2014). The goal of such biosecurity approaches is to keep or return the density of invasive species to zero so that damages from those pests might be prevented or to confine populations to localized areas so that damage from those species might be limited (Magarey et al. 2009). Prediction, prevention, early detection, eradication, and other rapid responses, all components of proactive management, are less costly and more effective than reactive tactics (Epanchin-Niell and Liebhold 2015; Leung et al. 2002; Lodge et al. 2006; Rout et al. 2014) (Fig. 6.1). Prediction is achieved through *risk assessment* (a process to forecast the likelihood and consequence of an invasion) and *pathway analysis* (a process to evaluate the means by which invasive species might be brought into an area of concern). Prevention is achieved through a variety of measures including regulations and quarantine treatments. Indeed, pathway analyses and subsequent regulation of those pathways are considered "the frontline in the prevention of biological invasions" (Hulme 2009) and cost-effective approaches (Essl et al. 2015; Keller et al. 2007; Leung et al. 2002; Tidbury et al. 2016). Surveillance is fundamental to early detection, and if a target species is detected, the primary rapid responses are eradication, containment, or suppression (reviewed in Beric and MacIsaac 2015). Early intervention strategies often operate at spatial scales that are much greater than the scale at which most land managers operate. Success thus requires effective coordination among researchers, regulators, and managers at international, national, sub-national, and local levels.

Early intervention strategies for invasive species share many elements with integrated pest management (IPM) approaches that are used against well-established pests (Venette and Koch 2009). In broad terms, IPM requires (1) clear articulation of a goal for the system; (2) background

R. C. Venette (✉) · J. Juzwik
U.S. Department of Agriculture, Forest Service, Northern Research Station, St. Paul, MN, USA
e-mail: robert.c.venette@usda.gov

D. R. Gordon
Environmental Defense Fund, Washington, DC, USA

F. H. Koch
U.S. Department of Agriculture, Forest Service, Southern Research Station, Eastern Forest Environmental Threat Assessment Center, Research Triangle Park, NC, USA

A. M. Liebhold
U.S. Department of Agriculture, Forest Service, Northern Research Station, Morgantown, WV, USA

R. K. D. Peterson
Department of Land Resources and Environmental Sciences, Montana State University, Bozeman, MT, USA

S. E. Sing
U.S. Department of Agriculture, Forest Service, Rocky Mountain Research Station, Bozeman, MT, USA

D. Yemshanov
Natural Resources Canada, Canadian Forest Service, Great Lakes Forestry Centre, Sault Ste. Marie, ON, Canada

T. M. Poland et al. (eds.), *Invasive Species in Forests and Rangelands of the United States*,
https://doi.org/10.1007/978-3-030-45367-1_6

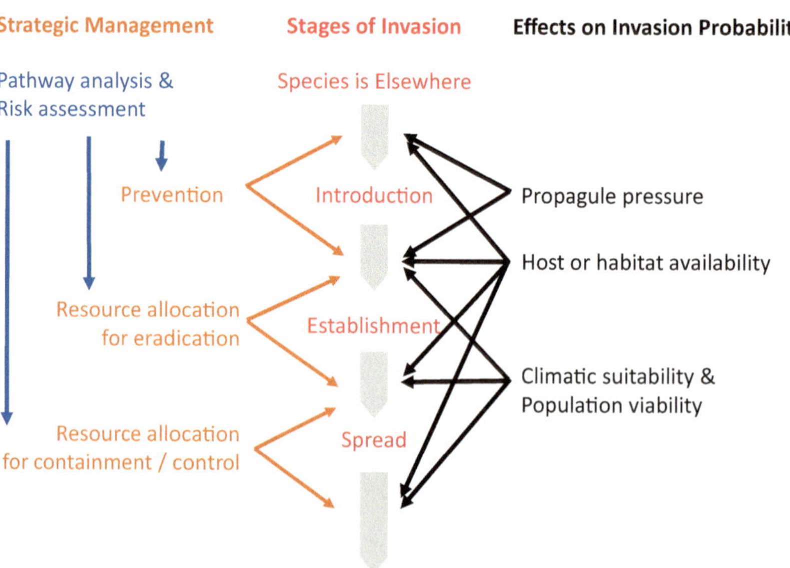

Fig. 6.1 Relationship between the stages of pest invasion, opportunities for strategic management, and factors that affect invasion probability (*red* depicts stages of invasion; *black* depicts factors influencing invasion probability; *blue* depicts analyses that facilitate when and where management should be targeted; *orange* depicts management options)

knowledge of the complex of pestiferous species (species with ability to cause harm) that might affect a system (i.e., prior experience); (3) systems to monitor for the presence and abundance of those species (i.e., sampling tools and plans); (4) guidance on when management is worthwhile (e.g., economic thresholds); (5) a suite of complementary tools and tactics to affect the abundance or impact of unwanted species (e.g., resistant plants, pesticides, and biological control agents); and (6) follow-up methods to ensure that interventions are successful. Current IPM programs have evolved through years of intensive research on the biology and management of single species in a range of systems and environments.

Early intervention strategies for invasive species expand on principles derived from IPM. For example, prior experience is supplemented with information about the suite of pestiferous species that affect similar ecosystems globally. Pest risk assessments attempt to help distinguish those non-native species with a high probability of causing harm from those that might not be harmful. Likewise, both general and specific tools and techniques are needed to find newly invading species and quickly and accurately identify them. Many responses to invasive species are similar to those for well-established pests, but early intervention strategies for invasive species may also involve quarantines, regulations, or more intensive approaches to ensure pest elimination or containment. These measures may be imposed and paid for by governments and immediately affect producers and other stakeholders. Because early intervention efforts have the potential to conflict with other social values (e.g., limits to freedom of personal movement or trade), a reliable, scientifically credible assessment of the likelihood that an alien species will cause harm is needed to determine whether the benefits of a preventative measure outweigh its costs. The design and implementation of early intervention strategies often do not have the benefit of years of research and must contend with significant uncertainties about the biology of threatening alien species, how those species might affect different ecosystems, and the effectiveness of management responses, especially under budget constraints. Research is underway to more accurately measure these uncertainties, reduce them, and provide tools to address uncertainty in decision-making (e.g., Koch et al. 2009; Yemshanov et al. 2015).

This chapter summarizes major research accomplishments on early intervention strategies, with a special emphasis on risk assessment, for invasive species. We emphasize results that apply to multiple alien taxa. References to particular invasive pathogens, plants, or other pests are provided to illustrate general concepts. The unique interplay between science and regulation needed to devise early intervention strategies may be unfamiliar to some researchers, so we provide overviews of regulatory procedures to illustrate how research results may inform regulatory decisions. Space constraints prevent us from addressing the diverse research projects that provide a basic understanding of the biology of threatening invasive species, even though such knowledge is imperative for conducting rigorous pest risk assessments and effective early intervention strategies.

6.2 Risk Assessment

6.2.1 Definitions of Risk

The word "risk" has many definitions across disciplines. "Risk" is used colloquially to describe an undesired consequence of an event (e.g., cancer as a consequence of smoking), but is technically defined as the product of the probability that an undesired event will occur, sometimes described as 'exposure,' and the consequences of the event, sometimes described as 'effect' (Kaplan and Garrick 1981). This definition underpins the definition of pest risk assessment for invasive species (IPPC 2016b). The unwanted event is typically the entry (i.e., introduction or arrival), establishment, and spread of a particular alien species into an uninvaded area (all related to 'exposure' in broader risk assessment parlance), and the consequences are the economic, ecological, or social impacts of invasion (all related to 'effect').

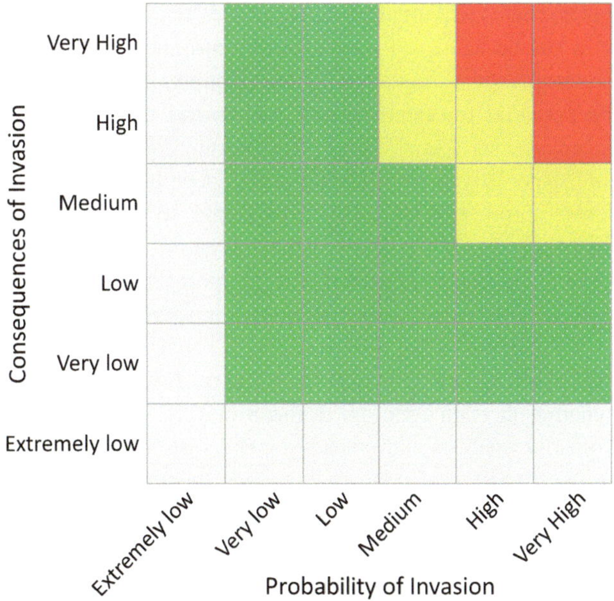

Fig. 6.2 Pest risk matrix depicting how the likelihood and consequences of invasion affect overall risk (*red* = high risk; *yellow* = moderate risk; *green* = low risk; *grey* = effectively no risk)

6.2.2 Introduction to Risk Assessment

Risk assessment is broadly defined as a process to determine the probability that a specified negative event will occur and the magnitude of its effect. While the process sounds simple and general guidance is available (e.g., Baker et al. 2009; Venette 2015), no requisite standards or techniques exist to quantify risks for invasive species (Hulme 2003). Typical approaches for assessing risks associated with invasive species often focus on identifying pathways and processes of introduction and movement, characterizing susceptible hosts and suitable environments, and evaluating the potential consequences of spread and establishment in previously uninvaded areas (Andersen et al. 2004; Pheloung et al. 1999; Venette 2015). However, data on the behavior of alien species and their biology in novel landscapes are often scarce or nonexistent, which leads to coarse representations of risk that are based extensively on expert judgment or simple analytical approaches (Andersen et al. 2004; Gray et al. 1998; Landis 2003; Landis and Wiegers 1997; Rafoss 2003). The results of such analyses are largely qualitative and usually are assigned an ordinal risk rating (e.g., high, moderate, or low risk). Qualitative assessments (Fig. 6.2) may be adequate to assist managers or policymakers in making decisions, such as whether to allow importation of certain commodities or to prioritize particular pests for survey. Baker et al. (2015) provide a decision-support system to determine when qualitative or quantitative analyses may be needed for decision-making.

Quantitative estimates of risk may help to focus discussions on complex policy issues (Gray et al. 1998). Such advanced models require numerical models capable of representing invasion processes in realistic environments and processing large geographical data sets (Yemshanov et al. 2009b). In fact, Andersen et al. (2004) identified multiscale decision-support systems as one of the urgent research needs for better risk assessments of invasive species.

Ideally, risk assessment is conducted within a preventative approach to screen species before the species arrives in a new country or region (i.e., "pre-border"). Because a history of harmful invasion elsewhere is a consistently accurate predictor of invasion in a new region (e.g., Gordon et al. 2008), cost-effective risk management could start with this single question (as implemented for the USDA Animal and Plant Health Inspection Service Plant Protection and Quarantine (APHIS PPQ) Not Authorized Pending Pest Risk Analysis list; USDA 2015). However, as more species are moved with global trade (Bain et al. 2010; Kaluza et al. 2010; Yemshanov et al. 2012), a history of previous invasion may be unavailable to use as a guide. Therefore, risk assessment is also conducted "post-border" after a damaging alien species has been detected within a country to prioritize management efforts. Frequently, assessments must be performed rapidly, incorporating any available information, especially in response to new pest incursions.

Methods for risk assessment depend on both the mode(s) of potential entry into the region of interest and the type of species involved. Unintentional introductions are most common among those species that are inadvertently moved with the transport of people, goods, or commodities (e.g., marine organisms in ballast water, forest insects in solid wood packing, or crop pests on imported plants) and often involve alien species that have caused harm elsewhere (i.e., proven to be invasive outside the area of concern). Conversely, plants, pets, livestock, and biological control agents are often deliberately introduced. These alien species may or may not have a history of causing harm and will generally have perceived benefits, for which they are being imported. While pre-border pathway analysis is required to assess the probability that a species will unintentionally arrive, that probability can be assumed to be 1.0 for deliberate introductions. Though environmental variables (climate, substrate, etc.) may limit the potential for establishment and spread of many introduced species (e.g., Kearney and Porter 2009), deliberate care, especially of intentionally introduced plants, may overcome initial environmental constraints to establishment (Mack et al. 2000). Assessments of entry, establishment, spread, and impact are required to support biosecurity decisions for intentional and unintentional introductions.

Pest Risk Assessment and Commerce Entry potential is dynamic through time, so decision-makers need guidance from risk analysts and other researchers on the potential of an alien species of concern to invade locations of interest (Lodge et al. 2006; Muirhead et al. 2008). Recent research has demonstrated that the entry (and often the subsequent spread) of invasive organisms has been facilitated by humans and their various economic activities (Hulme 2009; Hulme et al. 2008; Kaluza et al. 2010; Lounibos 2002; Westphal et al. 2008). The long-distance spread of alien species has been linked to patterns of historical settlement (Brawley et al. 2009), marine and terrestrial trade and transportation (Bain et al. 2010; Blakeslee et al. 2010; Kaluza et al. 2010; Yemshanov et al. 2013), and human population density, and national wealth benchmarks (Pyšek et al. 2010). Most markedly, increases in the number of new invasive species that have invaded the United States have corresponded with the expansion of international trade, which now regularly features long-distance, rapid transport of raw commodities and finished goods (Bain et al. 2010; Bradley et al. 2012; Pyšek et al. 2010).

In North America and elsewhere, the rate of growth of trade volumes is expected to exceed the rate of economic growth (UNCTAD 2007; WTO 2008). The transportation corridors that facilitate this trade also have become critical avenues for introducing alien species (Tatem et al. 2006). The complexity of modern transportation networks and the range of socioeconomic factors that influence trade flows (and the potential spread of alien species) are also projected to increase (Pyšek et al. 2010). Under these circumstances, rapid assessments of the potential origins of new (or anticipated) species introductions are a critical starting point in identifying possible pest outbreaks and strategizing measures for immediate response and screening. General biosecurity concerns are not grounds to impede trade, so the challenge becomes to identify specific threats and take appropriate actions to mitigate those threats based on the best available science following International Plant Protection Convention (IPPC) standards (Devorshak 2012). Research in this area has benefitted significantly from international cooperation (Chap. 13).

Assessments of the Potential Entry of Alien Species Assessments of entry potential can be undertaken with modeling tools that trace the movement pathways of an alien organism from its suspected region(s) of origin to locations of interest (e.g., Carey 1996; Muirhead et al. 2006; Pitt et al. 2009; Wang and Wang 2006; Yemshanov et al. 2013). In general, data on the gross trade volume may serve as a crude proxy to estimate the anticipated number of new pest incursions in a region of interest. Several studies have modeled the entry potential of pests as a function of changing climate (e.g., Magarey et al. 2007), socio-political and economic events (such as the recent global financial crisis; e.g., Koch et al. 2011), or the impact of new trade rules (e.g., Costello et al. 2007). A dynamic representation of the pest entry process also provides a more reliable depiction of multiple reintroductions over time (Koch et al. 2009; Rafoss 2003; Yemshanov et al. 2009a). However, determining the finer scale geographic distribution of these new incursions usually requires a more detailed analysis of the movement of specific commodities and cargoes that may have a high probability to carry invasive species through a region's network of trade routes and transportation corridors (Hulme 2009; Hulme et al. 2008; Kenis et al. 2009). Where comprehensive data on commodity movement and species incursions are available, data-driven models of invasion risks can be produced. For example, Koch et al. (2011) outlined procedures to combine broad- and fine-scale data on trade and commodity movement with historical pest records to estimate establishment rates for alien forest insect species in urban areas across the United States. Increasingly, direct-to-consumer import via internet sales overcomes earlier pathway constraints and poses an additional threat for purposeful imports (Humair et al. 2015).

Pathway analyses provide keen insights on propagule pressure, now recognized as a key determinant of invasion success (e.g., Lockwood et al. 2005; Simberloff 2009; Wilson et al. 2009). Propagule pressure describes the composite number of individuals of an alien species that are introduced to an area and is a reflection of the number of introduction events (i.e., propagule number) and the number of individuals introduced per event (i.e., propagule size). As propagule pressure increases, the probability of establishment in otherwise suitable environments is likely to increase, but propagule size and number may affect the nature of this relationship differently (Lockwood et al. 2005). Propagule number can affect the likelihood that a species arrives during climatically suitable periods while propagule size can affect the level of genetic diversity in a given introduction (Novak 2007). Propagule size also affects the ability of the nascent population to overcome random demographic effects, like chance variation in the number of females born to a population, or Allee effects, processes that disproportionately affect small populations and can lead to negative population growth rates, such as the challenge of finding a mate (Drake and Lodge 2006; Leung et al. 2004). Policy analysts have suggested that placing a greater emphasis on devising methods to reduce propagule pressure may provide substantial gains in efforts to prevent future invasions (Hulme et al. 2008; Meyerson and Pyšek 2013; Reaser et al. 2008).

Assessment of Areas Suitable for Establishment of Alien Species Assessments of the potential for establishment typically focus on a single pest and require extensive information about the threatening or invading species and the endangered area. Frequently, analyses begin with listing the environmental factors and resources (e.g., soils or hosts) that might support or limit a pest's distribution. As the development of many pathogens, plants, arthropods, and some vertebrates is dictated by temperature and moisture, an evaluation of climate suitability can be particularly informative. Climate suitability for pest establishment can be assessed by analyzing the climatic conditions of regions where the species is known to exist and using the resulting models (alternatively known as ecological-niche, species-distribution, or climate-envelope models) to forecast the quality of the environment for establishment in endangered areas (e.g., Jarvis and Baker 2001; Peterson et al. 2011a; Venette et al. 2010). Alternatively, data from properly designed experiments to ascertain how population growth or decline is governed by temperature or moisture can be used to develop mechanistic models of the suitability of climates for the persistence of an invading population through time (e.g., Pattison and Mack 2009).

Assessments of Potential Spread by Alien Species The study of the ecology and mathematics of spread by alien species is a long-standing, rich, active area of research (e.g., Phillips 2015; Shigesada and Kawasaki 1997; Skellam 1951). The potential for an invasive species to spread from points of introduction into climatically and ecologically suitable areas largely depends on that species' biological capacities, specifically its population growth rate and dispersal ability, and other means by which the species may be moved. Spread can be facilitated by humans (such as by transportation or movement of goods and commodities), hence the assessment of spread risk often involves characterization of patterns and modes of human movement, but not always in the specific context of trade; for instance, Tatem (2009) investigated the spread of invasive species via airline passenger travel. In any case, when knowledge about the factors that control the behavior and microevolution of a species in a novel environment is lacking, estimates of a species' survival and spread are ambiguous. In this case, comparing historical spread records of the species in similar climatic regions, or in other areas where it is known to exist, can help to estimate an approximate range of spread rates in the area of concern.

Assessments of Potential Impacts from Alien Species Risk assessments also depend on forecasts of the potential extent of ecological (Chaps. 2 and 3), social (Chap. 12), or economic impacts (Chap. 14). Alien organisms can damage economically valuable host resources and negatively affect the state of economically important agricultural systems and native ecosystems (e.g., estuaries). Assessing economic risks implies a valuation of economic consequences and impacts from an introduction and spread of an alien organism. The potential extent of economic damages may justify enacting quarantines or other regulatory actions aimed to eradicate or contain the spreading populations or, if containment is no longer feasible, to slow the rate of its spread (Epanchin-Niell and Wilen 2012).

Pest risk assessments can also focus on indirect economic effects, such as impacts on trade (Arthur 2006; Breukers et al. 2008; Surkov et al. 2009), anticipated changes for exports and access to markets (Cook 2008; Elliston et al. 2005; Juliá et al. 2007), changes to the production costs in domestic markets (Macleod et al. 2003; Soliman et al. 2010), or large-scale impacts at the macroeconomic level (Wittwer et al. 2005). Some other harder-to-assess risks include potential impacts on ecosystem structure or function, social infrastructure, recreational activities (e.g., fishing or use of firewood), and factors associated with human health (e.g., water quality or productivity of important agricultural crops). The estimation of non market impacts caused by alien invasive species requires application of special techniques, such as hedonic analysis (Holmes et al. 2010), contingent valuation (Mohammed 2014), stated preference (Morse-Jones et al. 2014), and benefit transfer methods (Loomis et al. 2014).

Impacts from invasion have proven difficult to forecast reliably, and methods to more accurately forecast impacts over space and time are an active area of research (Kumschick et al. 2015; Venette et al. 2010). The framework to assess impact as proposed by Parker et al. (1999) and reviewed in Chap. 2 is extremely useful conceptually. The framework asks (1) where an alien species is, now or in the future; (2) how abundant is it or might it be; and (3) what impact it is having or might have on a per capita basis. The ecological impact of each alien species is not expected to be constant in space or time but will depend on the response of interest (e.g., species losses or changed abundance), an outcome of complex interactions between the invading species and biotic and abiotic components of the recipient ecosystem. As a result, some previous efforts to measure impact have met with mixed results. For example, assessments of impacts from wetland invasions by purple loosestrife (*Lythrum salicaria*) have ranged from no clear impact or insufficient evidence (Farnsworth and Ellis 2001; Hager and McCoy 1998; Lavoie 2010) to clear negative effects (Blossey et al. 2001; Schooler et al. 2006).

6.2.3 Assessments for Intentional Introductions

Here, we focus on risk assessments for two types of intentional introductions: alien plants for consumption or planting and classical biological control agents for alien plants or arthropods.

Assessments for the Intentional Introduction of Alien Plants The "Weed Risk Assessment" (WRA) system developed in Australia (Pheloung et al. 1999) is widely used, either in its original form or with slight modifications, to assess intentional introductions of plants. Research has demonstrated that this tool accurately identifies over 90% of harmful plant invaders, misidentifies fewer than 10% of non-invaders as invasive, and requires further evaluation (biased toward non-invaders) for fewer than 15% of species; this accuracy is consistent across temperate, tropical, island, and continental applications (Gordon et al. 2008). This primarily trait-based tool was originally designed for pre-border use. The weed risk assessment system used by the USDA APHIS PPQ Plant Epidemiology and Risk Analysis Laboratory, hereafter referred to as PPQ WRA (Box 6.1), is based on the Australian approach (Koop et al. 2012). The PPQ WRA framework draws from international standards for phytosanitary measures (IPPC 2016a, 2016c).

Assessments for the Intentional Introduction of Classical Biological Control Agents for Invasive Plants or Arthropods The enemy release hypothesis contends that invasive species are problematic because they have escaped the effects of natural enemies (e.g., herbivores, predators, parasitoids, or pathogens) that kept the invader at a low density in its native range (reviewed in Liu and Stiling 2006). So, the premise of classical biological control is that reintroducing those natural enemies to established invading pests should lower the densities of those invading pests, an approach that is more sustainable and less disruptive than many chemical or physical approaches to pest management. For classical biological control agents of plants, the challenge is to ensure that agents, typically pathogens or insect herbivores, only affect the targeted weed, not other valued plants such as crops, ecologically important plants, or federally listed threatened and endangered species (reviewed in Schaffner 2001). These efforts are meant to guard against unintended outcomes. For example, the weevil *Rhinocyllus conicus* was introduced from Europe into North America in 1968 to control invasive thistles (primarily in the genus *Carduus*) but has now been recovered from at least four *Cirsium* spp., including Platte thistle (*C. canescens*), a close relative of the federally listed Pitcher's thistle (*C. pitcheri*) (reviewed in Louda 2000). Similarly, assessments for classi-

Box 6.1: Overview of the Weed Risk Assessment (WRA) Framework Used by the US Department of Agriculture, Animal, and Plant Health Inspection Service, Plant Protection and Quarantine (USDA APHIS PPQ)

The USDA APHIS PPQ uses the WRA when an applicant seeks a permit to import or export a new, as-yet-not-approved alien plant species for planting into the United States. The agency conducts its own analyses with the best scientific information available, some of which may be provided by the applicant, but typically it would not be conducting primary research in support of the application. The assessments are conducted to evaluate the likelihood of a plant taxon becoming weedy or invasive, and to determine where it might become established in the United States. Analyses are based on a logistic regression model that is used to quantify a plant taxon's ability to escape, establish, and spread outside of intentional cultivation, and thereby cause harm to U.S. plant resources (Koop et al. 2012). The PPQ WRA relies on a series of questions to generate risk scores for the plant taxon's entry, establishment, spread, and impact potential. Decision or risk thresholds (1) maximize the model's ability to correctly identify the likelihood that a plant taxon will become a non-, minor, or major invader; (2) minimize predictive errors; and (3) translate risk scores into final risk ratings: low, moderate, or high. Taxa rated as a moderate risk undergo further screening of life history and behavioral traits associated with invasiveness, as expressed throughout the taxon's geographical distribution. The global distribution of a plant taxon is used to infer which plant hardiness zones, Köppen-Geiger climate classes, and mean annual precipitation bands might be needed for establishment and to identify areas in the United States that meet those criteria. Entry potential is assessed only if the taxon is not already present in the United States and is based on the likelihood of intentional or accidental entry. Risk scores are generally higher for taxa valued by society or cultivated outside the United States. Uncertainty in the risk score is assessed by using Monte Carlo simulations to generate 5000 simulated risk scores and analyzing the distribution of outcomes.

cal biological control agents of invasive arthropods are intended to ensure that the proposed agent affects only the targeted species (reviewed in van Lenteren et al. 2006). The released tachinid fly *Compsilura concinnata* in 1906 to control the gypsy moth (*Lymantria dispar*) and browntail moth (*Euproctis chrysorrhoea*) causes significant mortality in the cecropia moth (*Hyalophora cecropia*) (Elkinton and Boettner 2012) and exemplifies the undesired outcome. These examples of unintended consequences of biological control are relatively limited, and modern pre-release screening procedures and safety reviews minimize potential impacts to non-target species (Hajek et al. 2016).

In the United States, screening of alien natural enemies to assess their safety and suitability for environmental release as biological control agents of invasive plants or arthropods does not involve a formal quantitative risk assessment process, and decision-making does not rely exclusively on data acquired, and analyses generated, by the APHIS PPQ. Instead, researchers (also known as 'petitioners') may submit petitions to the APHIS PPQ, ultimately to gain approval to release classical biological control agents into the environment. Petitions summarize taxonomy, geographic distribution, life history, and ecology of the target species and candidate biological control agent(s). Frequently, the petition includes the results of pre-release host range tests to determine the suite of species upon which the agent might feed or infect. Host range testing often follows a centrifugal phylogenetic approach, with extensive testing of the target and closely related taxa and less emphasis on more distantly related taxa (e.g., Evans and Tomley 1994). The petition includes a description of experimental methodology, results, and analyses used to assess host specificity and impact of the candidate agent. Known and potential environmental impacts associated with the target plant and candidate biological control agent(s) are described. A general description of procedures to obtain the approval to release a new biological control agent in the United States is given for invasive plants in Box 6.2 and for arthropods in Box 6.3.

6.2.4 Assessments for Unintentional Introductions of Alien Species

Assessments of unintentional introductions typically focus on alien species that are likely to cause harm, often with an emphasis on a single species or a suite of species associated with an imported good. A complete assessment would evaluate both the likelihood that a species would invade and the consequences of that invasion (Venette et al. 2010). Clear

Box 6.2: Overview of Procedures to Obtain Approval to Release Classical Biological Control Agents for Weeds in the United States

All relevant information that must be included in a petition to release a new, non-native biological control agent of weeds is described in the USDA (2016). This guide is also used by the Technical Advisory Group for Biological Control Agents of Weeds (TAG-BCAW, or TAG) to appropriately review petitions. TAG is a scientifically independent, voluntary committee comprising members appointed by Federal agencies, such as the USDA, U.S. Department of the Interior, Environmental Protection Agency, and U.S. Department of Defense, along with representatives of the National Plant Board, the Weed Science Society of America, and Canadian and Mexican governments. TAG advises petitioners and provides USDA Animal and Plant Health Inspection Service Plant Protection and Quarantine (APHIS PPQ) with recommendations about the safety of releasing candidate agents based on the completeness and robustness of information presented in petition documents. TAG serves in a purely advisory role.

If TAG recommends that the APHIS PPQ approve the release of a specific petitioned agent and APHIS PPQ concurs with that recommendation, then the ensuing issuance of a permit by APHIS for the environmental release of the agent is considered a Federal action, requiring compliance with the Endangered Species Act (ESA) and the National Environmental Policy Act (NEPA). To address NEPA requirements, the APHIS PPQ develops an environmental assessment (EA) providing a concise summary of the material presented in the petition and potential effects on the quality of the human environment that may be associated with the release of the candidate agent, and compares these to potential effects of alternative actions, including a no action option. Evidence and analysis provided by the EA determines if a Finding of No Significant Impact (FONSI) can be reached; if not, then a more detailed environmental impact statement (EIS) must be produced. The EA's 30-day public comment period is publicized in the Federal Register. The EA is one of the relevant reports included in the biological assessment (BA) submitted to the U.S. Fish and Wildlife Service (USFWS) for a so-called Sect. 7 consultation, to satisfy ESA compliance. The EA provides descriptions of the action to be considered (i.e., release of the agent); specific areas that may be affected by that action; listed species (i.e., threatened, endangered, or species of interest), or their critical habitats that may

(continued)

Box 6.2 (continued)

be affected by the action; the manner in which the action may affect listed species or critical habitats, and an analysis of any cumulative effects; relevant reports; and any other relevant information on the action, the affected species, or critical habitat. If the USFWS concurs with "may affect, not likely to adversely affect" determinations regarding listed species and critical habitats included in the BA, they then send the APHIS PPQ a concurrence letter, which completes the ESA consultation; the concurrence from the USFWS is then incorporated into the EA. Although many groups comment on the safety and host specificity of candidate classical biological control agents, none of the stages in the review process leading to the issue of a permit involves a formal quantitative risk assessment.

Box 6.3: Overview of Procedures to Obtain Approval to Release Classical Biological Control Agents for Arthropod Pests in the United States

Petitions must be submitted to the USDA Animal and Plant Health Inspection Service Plant Protection and Quarantine (APHIS PPQ) in a format that follows North American Plant Protection Organization (NAPPO 2015). The APHIS PPQ issues permits required for interstate movement of non-native entomophagous biocontrol organisms for the purpose of environmental release or for research or releases that will occur outside of containment facilities. Regardless of the number of scenarios requiring a permit, issuing a permit triggers the same requirements for the APHIS PPQ's compliance with the National Environmental Policy Act (NEPA) and the Endangered Species Act (ESA) as described in Box 6.2 for phytophagous biocontrol agents. Environmental assessments (EAs) are produced by the APHIS PPQ and then publicized in the Federal Register with notification of a 30-day public comment period; if no additional, credible adverse effects stemming from the release of the agent are identified, then a Finding of No Significant Impact (FONSI) is issued. The EAs for entomophagous biological control agents contain generic statements about the potential impact a candidate agent might have on threatened or endangered species. However, these statements likely have little bearing on the outcome of the actual Sect. 7 consultations with the U.S. Fish and Wildlife Service.

standards govern the conduct of a risk assessment when the assessment will factor into decisions on international trade (IPPC 2016a, 2016c). In the United States, for example, those assessments are typically prepared by the APHIS PPQ when an importer/exporter makes a request to begin shipping a plant or plant product to this country that has not been approved previously. However, other organizations within the United States may conduct their own risk assessments to prioritize their own management activities or support local biosecurity needs. Those assessments would not necessarily follow international standards.

While pest risk assessment strives to assess the joint probability of introduction (or entry), establishment, spread, and impact, typically for an individual species or a suite of related taxa, pathway analysis focuses on introduction events, often for multiple alien species that might be moved into an area of concern on a common conveyance. Pathway analysis can be a component of pest risk assessment (Box 6.4), or it can be conducted on its own, for example, to identify introduction hotspots. Assessing the risk of introduction may require attention to multiple pathways of introduction, including the identification of potential vectors (such as wood packaging materials for wood-boring alien insects; IPPC 2016c) and regions from which the species is most likely to arrive.

Pathway analysis has several interpretations. Perhaps the most common interpretation is that it examines, in a broad sense, the kinds of species, their relative rates of arrival and, in some cases, their most prominent destinations associated with a commodity type or group (e.g., avocados, Hennessey 2004; live plant imports, Liebhold et al. 2012). Alternatively, pathway analyses may target an industry that depends on a particular commodity (e.g., the horticulture industry, which relies on the global trade of live plants and seeds; Reichard and White 2001). This particular interpretation also extends to categories unrelated to trade (e.g., airline passenger baggage; Liebhold et al. 2006). A somewhat different interpretation of "pathway" involves a more geographically explicit perspective, which focuses on the primary routes between origin and destination locations. This latter type of pathway analysis has been applied, for example, to examine the trade and transport of goods that may carry wood-boring insects (Colunga-Garcia et al. 2009; Koch et al. 2011; Yemshanov et al. 2012), and can have a domestic (e.g., recreational travel and firewood movement in the United States; Koch et al. 2012, 2014) or international focus. Such analyses also may include quantitative modeling of the links between origins and destinations using geospatially depicted networks (e.g., Koch et al. 2014; Paini and Yemshanov 2012; Yemshanov et al. 2012, 2013). No matter the interpretation of the pathway analysis concept, a potential outcome of such analyses is an assessment of the likelihood of pest introduction or spread that can feed into more comprehensive pest risk assessment efforts.

Box 6.4: Overview of Procedures for Pest Risk Assessment

General guidance for the preparation of pest risk assessments that may affect international trade is described in the International Standards for Phytosanitary Measures from the International Plant Protection Convention (IPPC). For many alien species, pest risk assessment starts with identifying the potential pest species of concern, the area for which information is needed, possible locations of pest origin, and likely pathways by which the species could enter an uninvaded area (IPPC 2016a). At this stage, qualitative and descriptive information is collected to help understand the species' present distribution and to identify susceptible hosts and possible vectors of spread. Some of this qualitative and quantitative information is intended to provide insight on how the species might enter, spread, and establish viable populations in the uninvaded area.

The next stage of the assessment may include more sophisticated analyses of the likelihoods of the organism's introduction and spread, as well as analyses of potential economic consequences and environmental impacts (IPPC 2016c). Key information collected for these later stages may include details on dispersal mechanisms (e.g., rates and patterns of movement), relative susceptibility of known host species, reconstruction of the history and timing of the invasion, and identification of the critical vectors of entry that must be controlled to prevent new arrivals of the species. Other relevant information that may affect the likelihood of establishment includes an invader's life cycle, survival rates, and natural enemies in the uninvaded area. Such knowledge helps assessors to understand whether the organism under consideration can be expected to establish and cause recurring harm in a newly invaded area or might be present for a short time and have transient effects. Ultimately, the level of complexity that is incorporated into the risk assessment may depend on decision-making goals (e.g., possible imposition of trade restrictions may necessitate a detailed assessment) or the nature of the species of concern (e.g., a well-known species that is expected to be low-impact may only require a basic assessment).

6.2.5 Assessments of Management Tactics

While pest risk assessments focus on biological invasions as the unwanted event, the same general risk assessment framework can be used to prioritize management efforts and estimate the risks posed by strategies and tactics used to manage the invasive species (Sing et al. 2005). The latter approach can help identify whether "the cure is worse than the disease." In this context, the additional unwanted event is harm from efforts to manage invasive species. For example, the unwanted event could involve reduced density of native plants as a consequence of herbicide applications. The risk assessment framework here, termed "comparative risk assessment," provides researchers and policymakers with guidance to estimate and compare risks from the invasive species and its potential management strategies.

The purpose of comparative risk assessment is to qualitatively and quantitatively compare different environmental risks for the purpose of improved decision-making. Despite the need to systematically compare risks to make more effective policy decisions, there are relatively few examples of this activity in the literature (Peterson 2010; Peterson and Shama 2005). In some cases, the necessary risk assessments have been conducted, but the outcomes have not been directly compared (Antwi et al. 2008; Davis et al. 2007; Peterson et al. 2006, 2011b; Schleier et al. 2008). Sing and Peterson (2011) argued that the decision to initiate control programs for invasive pests often occurs without first considering the ecological or economic evidence to support that decision. Frequently, risks from associated management tactics are not formally part of the decision matrix.

The comparative risk assessment approach often is limited by a lack of quantitative effect and exposure data (Drake et al. 2006; Drake and Lodge 2006). In addition, the data that are available may be highly uncertain, especially when the proposed management strategy is biological control (Schleier et al. 2008; Sing et al. 2005). However, effect and exposure data for other management tactics, such as pesticides, may be more certain and readily available. The problem then becomes one of comparing risks among stressors in which the accuracy and uncertainty of individual risk assessments vary appreciably.

When risks are difficult to compare quantitatively because of challenges in identifying common endpoints and the existence of large differences in uncertainties associated with estimating effect and exposure, the use of comparative qualitative or semi-quantitative risk assessments may be a solution. Although quantitative risk assessments are almost always preferred over qualitative risk assessments (Cox et al. 2005; Schleier et al. 2008), employing comparative qualitative assessments has been proposed to unify seemingly disparate assessments and establish a common frame of reference for subsequent decisions (Gentile and Harwell 2001; Landis and Wiegers 1997). For example, Sing et al. (2005) retrospectively evaluated risks associated with insects that feed on invasive toadflax (*Linaria* spp.), and Sing and Peterson (2011) assessed risks for Dalmatian toadflax (*L. dalmatica*) and yellow toadflax (*L. vulgaris*) in North

Fig. 6.3 Dalmatian toadflax (*Linaria dalmatica*) and two biological control agents, *Rhinusa* spp. (*top right*) and *Mecinus janthinus* (*lower right*)

America (Fig. 6.3). A third environmental risk assessment could be conducted on the herbicides used on the two toadflax species. Based on the three risk assessments, a unified comparative risk assessment could be conducted, possibly using simple yet quantitative risk metrics such as risk quotients (Peterson 2006). This type of comprehensive assessment would at least provide a starting point for evaluating multiple risks of the invasive species and the management tactics being proposed.

6.2.6 Key Findings for Risk Assessment

- Early intervention is the most cost-effective approach to manage invasive species. By keeping invasive species out of an area of concern through regulatory or technical approaches, the potential damages from those species are avoided.
- Global trade has provided several pathways for new pest introductions. The number of countries engaged in trade and the diversity and volume of products moved in trade create significant opportunities for the movement of a pest species outside its native range.
- Risk assessment provides a useful framework conceptually and analytically to evaluate the potential for future adverse impacts from unwanted events. The outcome of pest risk assessment typically provides a clear strategic direction for biosecurity decisions and a foundation for tactical actions. Pest risk assessments attempt to forecast the likelihood that individual species will invade and cause economic, ecological, or social harm. Pathway analyses, which may be part of pest risk assessments, attempt to characterize how suites of pests might be moved into areas of concern.
- Effective risk assessment requires close collaboration between scientists (i.e., risk assessors) and decision-makers (i.e., risk managers). The challenge for scientists is to balance rigor and timeliness to obtain an acceptable degree of accuracy in their assessments, while the challenge for risk managers is to clearly articulate information needs to support time-critical decision-making (Venette 2015).

6.3 Prediction and Prevention

Risk assessments provide the backbone of prediction and prevention, often viewed as the first lines of defense in proactive, pre-border, biosecurity strategies (Venette 2015). Prediction is fundamentally the outcome of the pathway analysis or pest risk assessment. Prevention refers to the integrated suite of tools and strategies that are intended to lower risks from those pathways or species to acceptable levels. Quarantine regulations (e.g., prohibiting species or items from entering an area of concern because they may harbor threatening species) are a prominent component of prevention, as are quarantine treatments designed to disinfest pathways of threatening species. Several analyses have indicated that prevention is one of the most efficient strategies for managing invasions; by preventing propagules from arriving, all of the costs associated with impacts and management can be prevented (Leung et al. 2002; Lodge et al. 2006).

6.3.1 Prediction

Researchers have determined that the importance of different introduction pathways varies considerably among invasive species. For example, the dominant pathway responsible for the transport of most invasive plants has been intentional imports for ornamental, agricultural, soil-stabilization, or other uses (Hulme et al. 2008). Several species of plants had been introduced to arboreta or other cultivated settings where they subsequently spread into surrounding regions. Intentional introduction is also considered the most common pathway for invasions by birds, mammals, and fish. Some insect species that were introduced as biological control agents at a time when assessment standards were less rigorous also have spread into unintended environments (e.g., Louda 2000).

In contrast to such intentional introductions for the above groups, most invasive insects and plant pathogens have entered either with plants, wood, or as "hitchhikers" on other material (Kenis et al. 2007; Kiritani and Yamamura 2003). Analysis of pest interception data from the APHIS PPQ revealed that the pathway responsible for the entry of most forest insects and diseases into North America has been importation of plants (Liebhold et al. 2012). Plants are the perfect medium for moving herbivorous pests because they provide food and shelter during transportation. Historically, large numbers of sap-feeding and foliage-feeding insects accidentally entered the United States when unregulated imports of plants allowed infested plants to freely enter the United States in large numbers (Liebhold and Griffin 2016). Enactment of the Plant Quarantine Act of 1912 led to quarantine restrictions on plant imports by the USDA and subsequently decreased the establishment rate of new plant pests.

However, following World War II, the movement toward free trade led to enormous increases in import rates. These trends plus the advent of more efficient trans-oceanic shipping technologies (e.g., containerized cargo) led to massive movement of solid wood packing material (SWPM). Though not fully recognized until the last two decades, SWPM provides a very effective pathway to move pests, particularly bark- and wood-boring insects (Haack 2001, 2006). Examples of pests that have likely entered North America with SWPM include the emerald ash borer (*Agrilus planipennis*), the Asian longhorned beetle (*Anoplophora glabripennis*), and laurel wilt disease (caused by the fungus *Raffaelea lauricola* and vectored by the beetle *Xyleborus glabratus*). Current increases in the online trade of plants (Humair et al. 2015) have created new opportunities to import potentially problematic plants and pests (Keller and Lodge 2007).

6.3.2 Prevention

The search for quarantine treatments for goods and commodities is a broad and active area of research. For example, new solutions are being sought to disinfest ballast water of aquatic alien invasive species (e.g., Tsolaki and Diamadopoulos 2010) or commodities of pests that might affect forests or rangelands (e.g., compression of imported hay to control insects; Yokoyama 2011). Many approaches focus on specific technologies, while systems approaches rely on integrating several techniques to rid a commodity of invasive species when any one technique may be insufficient to achieve a desired biosecurity standard (Follett and Neven 2006). The Forest Service is actively conducting research to identify quarantine treatments capable of eliminating invasive pathogens or insects from wood or wood products.

In 2002, the IPPC, recognizing the potential for damage from invasive pests, adopted a harmonized international standard for phytosanitary measure (ISPM) for treating SWPM (IPPC 2016d). The standard, called ISPM 15, requires the treatment of SWPM with heat or methyl bromide fumigation to eliminate wood- and bark-boring insects (Box 6.5). Specifications for these treatments were developed, in part, from investigations conducted by the Forest Service Research and Development scientists (Haack and Petrice 2009). The addition of a bark standard that requires nearly all bark to be removed from SWPM has contributed to a large reduction in risks (IPPC 2016d). The IPPC requires that exporting countries use a stamp on each piece of SWPM to certify that ISPM 15 treatments were conducted (Fig. 6.4).

A series of studies, organized by the National Center for Ecological Analysis and Synthesis, sought to quantify the potential economic benefit of ISPM 15. Part of this work included quantifying the rate at which wood-boring insects have entered the United States and their economic impacts (Aukema et al. 2010, 2011). Other research quantified the effectiveness of ISPM 15 in reducing woodborer approach rates (Haack et al. 2014) and the costs of ISPM 15 to trade (Strutt et al. 2013). Finally, Leung et al. (2014) used all of this information in a cost/benefit analysis to show that, while ISPM 15 had a negative economic effect in the initial decade

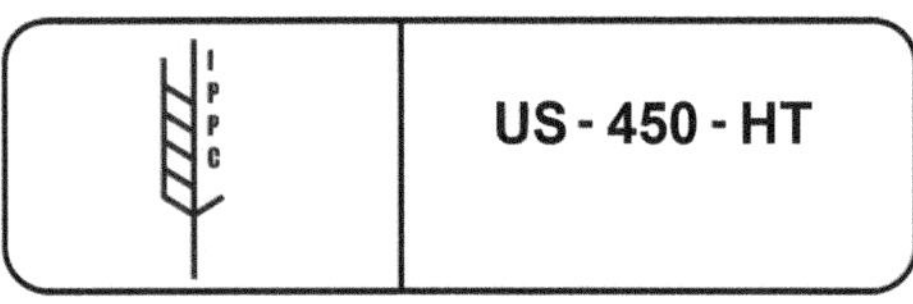

Fig. 6.4 Example of an approved stamp for solid wood packing materials. The stamp, denoting the country of origin, the treatment facility, and the treatment type, signifies that a piece of wood has been treated in compliance with International Standards for Phytosanitary Management (ISPM 15) from the International Plant Protection Convention

Box 6.5: Preventing the Movement of Forest Pests in Wood Packing and Lumber: Research to Demonstrate the Value and Achieve the Goals of International Standards for Phytosanitary Measures No. 15 (ISPM 15)

Investigations into treatments for ISPM 15 have benefitted from, and contributed to, research into developing quarantine treatments for export of lumber or whole logs. The most common treatment mandated for international movement of dried lumber or green lumber is heat applied until the core temperature reaches 56 °C for 30 min, often called the "56/30 standard." The deleterious effects of methyl bromide on the environment and stratosphere have led to global efforts to drastically reduce the production and use of this fumigant. Concomitantly, research on promising alternatives to methyl bromide has been ongoing since the early 1990s. Although log schedules have been devised and set between countries engaged in log trade, no comprehensive, international convention has been established for treatment of whole logs in international trade.

Past testing of heat treatment focused on insect and nematode pests, but this has recently shifted to evaluating the 56/30 standard for its utility to kill fungal pathogens in wood. Heat treatment is not suitable for wood or wood products where quality (e.g., color change or drying effects) is a concern; however, steam treatment was found to be effective in heating large timbers (Simpson 2001). Vacuum plus steam thermal treatment is currently being evaluated as an alternative to heat treatment and fumigation for eliminating invasive insects and tree pathogens in logs. Log degrade was minor, and product (veneer) quality was unaffected in a vacuum steam trial with logs from five hardwood species (Chen et al. 2016). Time to reach 56 °C for 30 min (to core) required 17–29 h of treatment under 200 mm Hg vacuum.

Dielectric heating with microwaves or radio frequencies simultaneously heats throughout the wood profile as compared with kiln and oven treatments that rely on thermal conduction from outer wood to the core. Industrial-sized wood blocks that were subjected to microwave energy to reach 56 °C for 1 min resulted in 100% mortality of high numbers of the pinewood nematode (*Bursaphelenchus xylophilus*) (Hoover et al. 2010). Microwave treatment is more rapid and similar in efficacy to previously tested treatments for this pest. Further investigations are needed to ensure that minimum lethal temperatures for target pests are reached and that the desired internal temperature is reached based on predictions from surface temperatures. Radio frequency heating was found to reach or exceed 56 °C for 1-min hold time in trials with large wood blocks infested with high numbers of pinewood nematodes (Uzunovic et al. 2013). One hundred-percent mortality of the nematodes was achieved. Evaluation of dielectric heating for ability to deliver 56 °C throughout a commercial wood profile in industrial-scale operations is needed.

Whole-log fumigation with methyl bromide for export from the United States is currently one of the largest Quarantine and Pre-Shipment (QPS) use exemptions for that chemical. The best available fumigant alternative options for quarantine-level disinfestation of logs and other wood products are sulfuryl fluoride and phosphine as their use at the commercial scale would require few or no changes to current industry practices and infrastructure. Data on pest eradication efficacy and economic viability have been the focus of recent and ongoing research on these alternatives (e.g., Barak et al. 2006, 2010). Because high doses of methyl bromide over a significant time period are required to kill pinewood nematodes and the oak wilt fungus (*Bretziella fagacearum*), much of the ongoing fumigant research has been focused on these organisms (Schmidt et al. 1997; Tubajika and Barak 2011).

after implementation, it ultimately had a positive net benefit via reduced rates of forest pest establishment.

6.3.3 Key Findings for Prediction and Prevention

- Pest risk assessments and related pathway analyses provide a clear, scientific basis to identify future invasive species (not yet present) that may affect forests, grasslands, wetlands, and water bodies. Those analyses support strategic biosecurity decisions (e.g., uses of quarantine treatments or other regulations) to prevent real threats from arriving into an area of concern.
- The use of risk assessment to support complex decision-making can reduce the likelihood of unintended consequences of intentional introductions, such as with plants for planting or potential biological control agents. The goal is to prevent the introduction of seemingly beneficial species from causing unintended environmental, economic, or social harm.
- A potentially daunting aspect of prediction and prevention is the enormity of the number of species or pathways that could be evaluated. Pest risk assessment is not a pan-

acea and is unlikely to uncover *all* potential threats; however, the approach provides a clear, systematic, rational basis for making strategic decisions in light of the significant number of pest threats that must be addressed.
- Development of phytosanitary measures to preclude the movement of pests in pathways (e.g., fumigation of wood containers and pallets) can cost-effectively reduce the risk of unintentional introductions.

6.4 Early Detection and Rapid Response

Early-detection (i.e., biosecurity surveillance) and rapid-response strategies for invasive species provide a biosecurity safety net should prevention efforts fail. These strategies hinge on effective surveillance of the landscape to locate and recognize new species incursions while those populations are localized (Venette et al. 2010). Often the landscape is enormous relative to the resources that are available to conduct surveys. Research has addressed this problem in three general ways. Firstly, a variety of spatial analyses have been developed to support program planning and implementation. These analyses are useful to stratify the landscape into areas where invasions are more or less likely and determine the appropriate amount and allocation of resources in those areas to achieve programmatic goals (e.g., Koch et al. 2011). Secondly, researchers have made technological advancements to find invasive species, such as with environmental DNA (eDNA) (e.g., Jerde et al. 2011) or remote sensing (e.g., Hestir et al. 2008), and improved our understanding of chemical and behavioral ecology to produce better attractants and traps (e.g., Allison et al. 2004). Thirdly, advancements with computer-aided identification, genomic testing, and other molecular diagnostics support the rapid, reliable confirmation of species' identity (McCartney et al. 2003). Broad lines of research address the appropriate response to incursions. Eradication may be difficult, but achievable, particularly if populations can be driven to densities (i.e., Allee thresholds) that are too low for population growth, so that populations go extinct (Liebhold et al. 2016).

Program managers face a difficult challenge in implementing surveillance and response strategies. Often, the overall budget is fixed, forcing a difficult tradeoff between surveillance and response (Bogich et al. 2008; Cacho et al. 2007; Mehta et al. 2007; Chap. 14). The response cannot occur until the pest is detected. The decision to allocate more funds to response potentially allows large populations to build and extensive damage to accrue before detection occurs. Depending on how budgets are allocated, more funds for surveillance may limit response options once the pest is found. Epanchin-Niell and Hastings (2010) note the complexity of the allocation decision as being dependent on program goals, attributes of the invading species, extent and timing of damages, and the effectiveness of the response. Adjusting surveillance efforts to account for spatial variation in the likelihood of pest establishment can substantially reduce overall management costs (Epanchin-Niell et al. 2012).

6.4.1 Spatial Analysis for Program Planning (aka Pest Risk Maps)

For some alien organisms, the amount of available information enables risks to be assessed with a finer grain, spatially explicit approach (Koch and Smith 2008; Venette et al. 2010; Volin et al. 2004). Pest risk maps can integrate several models (i.e., pathway analyses, species distribution models, spread models, and/or impact models) to describe how the probabilities of invasion by a non-native species, and the magnitude of its impacts, might vary spatially within an area of concern (Venette 2015). Pest risk maps are based upon fundamental ecological concepts that address factors governing species' distribution and abundance. The construction of these maps helps to reveal a species' potential distribution, hotspots of entry and establishment, and those areas that are most vulnerable. Maps provide a powerful means to communicate spatial variation in the risk that species will establish and cause damage, and have therefore become a common decision-support tool for managing invasive species outbreaks (Venette et al. 2010).

For decision-makers, risk maps essentially represent a prioritization surface that guides them in allocating tactics aimed to detect and control the spread of invasive species (e.g., Volin et al. 2004). Risk maps are extremely useful to determine whether quarantine restrictions might be warranted if the alien species is not known to be present in the area of concern, to structure an early detection survey if the species might be present, or to describe the potential extent of impact if the species is not managed effectively. For example, Fig. 6.5 describes the potential spread of the redbay ambrosia beetle (*Xyleborus glabratus*) through areas of the Southeastern United States with its preferred hosts: redbay (*Persea borbonia*) and sassafras (*Sassafras albidium*) (Koch and Smith 2008). The value of an individual risk map for decision-making is subject to the constraints of available knowledge about the biology of the species of interest and conditions within the area of concern, as well as the economic and logistical constraints on map production.

6.4.2 Implementation of Early Detection

The extent of biosecurity surveillance depends on budgets and other technical support. Unfortunately, time, infrastructure, and funding constraints seldom if ever meet the con-

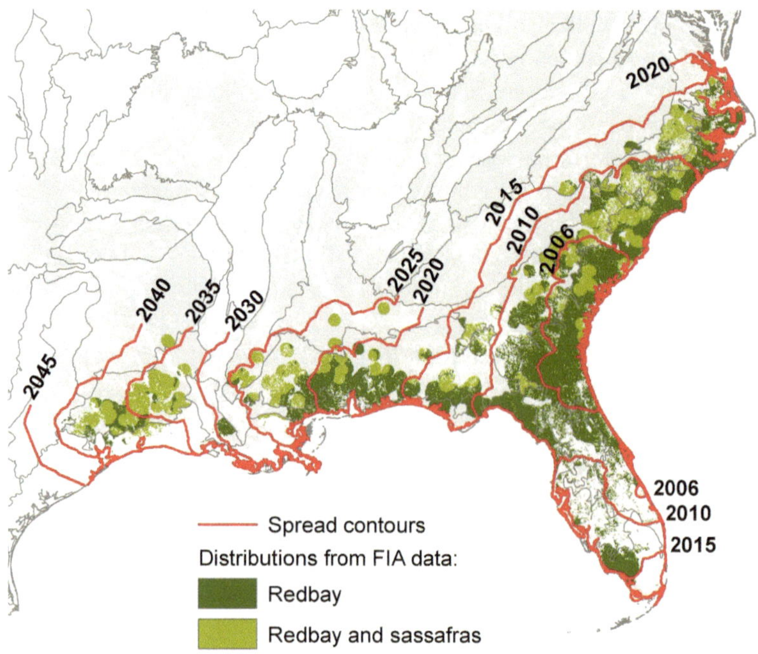

Fig. 6.5 Pest risk map for redbay ambrosia beetle (*Xyleborus glabratus*) spread through the Southeastern United States. (Reproduced from Koch and Smith 2008)

tinuous demand for detection, identification, and response (Saccaggi et al. 2016). Westbrooks (2004) recommends six actions to improve early detections of, and rapid responses to, invasive plants species: (1) public and private partnerships for "early detection and reporting of suspected new plants to appropriate officials; (2) identification and vouchering of submitted specimens by designated botanists; (3) verification of suspected new State, regional, and national plant records; (4) archival of new records in designated regional and plant databases; (5) rapid assessment of confirmed new records; and (6) rapid response to new records that are determined to be invasive." Similar principles were embodied in a national program for early detection and rapid response to invasive bark and ambrosia beetles (Rabaglia et al. 2008); identification of all submitted specimens led to the detection of several invasive species.

APHIS PPQ's Cooperative Agricultural Pest Survey (CAPS) Program funds a network of cooperators to conduct surveys for the early detection of plant pests that are threats to U.S. agriculture or the environment. CAPS targets specific alien invasive pests, diseases, and plants that are not yet established in the conterminous United States. A science-based pest prioritization model is used to determine which pests will be included on annual CAPS Priority Pest lists. Subject matter experts in biology and economics evaluate pest species individually against a weighted set of criteria that address environmental and economic impacts. The Analytic Hierarchy Process (Golden et al. 2012) is used to produce a prioritized pest list.

Detecting invasions of alien species not previously or widely reported in the United States relies on surveillance and reporting by regulatory and research communities, with significant contributions from knowledgeable citizen scientists. Environmental DNA (eDNA)-based detection has improved the accuracy, price, and efficiency for confirming the presence of non-native species, particularly for invasive fish at low population densities within large bodies of water (Handley 2015; Rees et al. 2014). The pivotal challenge for lay contributors to early detection and rapid response is the accurate identification of specimens; this has been somewhat offset by continually improving online identification resources. Currently, documentation of invasive plant and insect identification and distribution can be accessed and records of sightings can be added online through an early detection and distribution mapping system (EDDMapS) website (https://www.eddmaps.org/).

Sentinel sites for invaders can be established outside the known infested area to provide early warning of spread. One example of a collaborative, private–public partnership for early detection is for northward spread of Old World climbing fern (*Lygodium microphyllum*) in central Florida. A similar approach has been used to detect incipient tree pathogens in Europe (Vettraino et al. 2015). Surveillance for alien insects is typically semiochemically based, using strategically arranged traps baited with either pheromones or host attractants (Berec et al. 2015).

Methodologies for the detection of cryptic pathogens in plant tissues and on insect associates have greatly evolved over the past decade (see Chap. 7). Molecular tools are available for screening large numbers of samples collected during detection surveys using high-throughput methods. Detection of multiple invasive pathogens is possible using specific TaqMan® real-time PCR detection assays (Lamarche et al. 2015). The same PCR conditions, utilizing the same thermocycling parameters and chemistry, allowed for high-throughput assay for 10 high-priority and unwanted alien pathogens of trees in Canada (Lamarche et al. 2015).

Other major scientific advances have been made in developing accurate, sensitive, species-specific, rapid, and "suitable for field use" technologies for invasive tree pathogens. For example, such an assay was recently developed for *Phytophthora ramorum* (sudden oak death pathogen) using recombinase polymerase amplification that does not require DNA extraction or extensive training to complete (Miles et al. 2015). Most recently, DNA hybridization assays utilizing specific capture probes and complementary DNA target sequences have been developed with hybridization signaled by fluorescent dyes, chemically induced color changes, radioactivity, or surface-enhanced Raman spectroscopy (SERS) consisting of silver nanoparticles (Yuksel et al. 2015).

6.4.3 Options for Rapid Response

Upon the detection of an incursion by an invasive species, managers generally have four options: (1) eradication, (2) containment, (3) continued monitoring, or (4) do nothing. Options (1) and (2) qualify as rapid responses. Eradication refers to the total elimination (i.e., intentionally driven to extinction) of a species from a specific area. While the concept is not necessarily new, it is only in the last few decades that it has been widely applied to successfully prevent the establishment of invading species, with several hundreds of examples of successful eradication of insects (Liebhold et al. 2016; Mack and Foster 2009; Simberloff 2009; Tobin et al. 2013). Among pathogens, bacteria and viruses are more likely to be eradicated than fungi (Pluess et al. 2012). The most important determinant of a successful eradication is the availability of sensitive tools for detecting the target species, thus allowing for early detection and accurate spatial delimitation. While eradication does not preclude reintroduction, it can reduce the extent of invasion and propagule pressure. Eradications are most successful when infestations are small, for plants, generally <1 ha (Rejmánek and Pitcairn 2002) to ~5000 ha or within 4 years of first detection (Pluess et al. 2012). Simberloff (2009) identified sufficient funding, including for follow-up surveys and treatment, coordination, and enforcement, and an understanding of the biology and ecology of the target organism as components of effective eradication efforts. Additional components for successful eradication include a sustained effort, initial focus on outlying infestations, prohibited reintroduction, and public cooperation (Mack and Foster 2009).

Many types of treatment are used in eradication. For plants, eradication is typically carried out either through physical removal or herbicide treatments. Methods used for eradicating insects include synthetic or microbial pesticides, mating disruption, male annihilation (e.g., trap-out), and the sterile male technique. For vertebrates, newly established, isolated populations may be eradicated with an intensive effort that combines multiple approaches; one example is the eradication of feral swine (*Sus scrofa*) from Santa Cruz Island in California (Parkes et al. 2010). After fencing the island into five zones, pigs were systematically removed from each zone first by trapping, then aerial shooting, followed by ground-based shooting, trailing with dogs, and finally the use of Judas pigs. Over 411 days, 5036 pigs were removed. Genetic engineering technologies are providing new tools such as gene drives, which have been proposed to eradicate alien insects like non-native mosquitoes carrying dengue and Zika virus (NAS 2016).

The effectiveness of an eradication treatment may depend on the extent to which the treatment creates or enhances an existing Allee effect (Liebhold et al. 2016). Because low-density populations, such as those encountered during the early stages of invasion, are prone to extinction as a result of Allee effects, treatments that enhance Allee effects may be particularly efficient (Liebhold and Tobin 2008; Tobin et al. 2011). For example, in sexually reproducing species, mate-location failure may cause a strong Allee effect, resulting in a threshold below which populations decline towards extinction. Tactics such as mating disruption may strengthen such an Allee effect and thus facilitate eradication. Bio-economic models can be used to identify the optimal allocation among multiple treatments, exploiting synergistic influences on Allee effects (Blackwood et al. 2012).

More invading species are likely to arrive in urban/suburban areas (Colunga-Garcia et al. 2010) than in rural areas, suggesting that eradication projects will increasingly occur in residential areas. In these areas, some residents may object to aerial spraying of pesticides or other proposed treatments. This situation presents several challenges: treatment technologies are needed that are widely acceptable to the general public, and new approaches to public outreach and engagement are needed to avoid conflict (Gamble et al. 2010; Liebhold et al. 2016).

Containment is meant to prevent or slow the spread of an invading species and is usually attempted through treatments of delimited populations and imposition of quarantines and other regulations (Pasquali et al. 2015). Many of the same tools for eradication are used for containment, but for containment, the goal is to limit the extent of damages, not eliminate the target pest. Withrow et al. (2015) demonstrated the value of pre-emptive domestic quarantines as a component of rapid response plans, especially if the target species, like the emerald ash borer, is difficult to detect. Technologies for containment are often not specific, which can lead to "scorched-earth" responses (Britton et al. 2011).

6.4.4 Key Findings for Early Detection and Rapid Response

- Risk assessments for unintentionally and deliberately introduced species can be used productively at the landscape scale to distinguish or prioritize species that are already present and require management from those that have naturalized but are unlikely to have significant negative impacts. Where species are already present, information on field invasiveness and impacts should inform those prioritization efforts.
- Management may be warranted for any high-risk species that are: (1) not yet spreading but have been introduced recently; (2) present at low levels but not yet prioritized for management; or (3) not present in the area of interest but where a probable introduction pathway exists. "High-risk" status must be determined at a spatial and temporal scale that matches operational management decision-making. Assessments and risk mapping further can help identify and prioritize species of high invasion risk that should be the focus of early detection/rapid response programs.
- More research is needed to determine an appropriate balance of generalized prevention strategies that exclude many, but perhaps not all, alien species of concern versus specialized prevention strategies that are highly effective at excluding a specific species of concern but may miss an array of other alien pests.

6.5 Information Gaps and Future Directions

Ecological risk assessment emerged as a discipline in the 1970s; however, formal applications of ecological risk assessments to invasive species did not begin until the 1990s and early 2000s (Yoe 2012). In the last two decades, the number of research ideas to improve pest risk assessments has expanded rapidly, especially with respect to species-distribution and spread models (Venette et al. 2010). For example, incorporating effects of climate change (Chap. 4) and human behaviors into the assessments could provide valuable new insights (Venette et al. 2010); few of those ideas to improve pest risk assessment as yet can be considered fully mature.

The greatest barrier to the development of pest risk assessments has been the lack of information about pathways of pest introduction, the distribution and ecology of invading species, the biotic and abiotic conditions within geographic areas of concern, and resultant impacts of invasion. To be useful for many applications to invasive species, spatially explicit data must be collected globally, consistently, and repeatedly, similar to what has been done with the acquisition of meteorological data. Historical presence/absence records for species' distributions are useful, but current information on the phenology and dynamics of a species at several locations may be much more valuable to risk assessment.

This lack of information fundamentally interferes with the development of pest risk assessment as a science. In essence, forecasts from pest risk assessment are hypotheses about the state of future conditions. Those forecasts are grounded in current knowledge but inherently require extrapolations beyond what it is known. How will a species behave if it arrives in an area where it has never occurred? Research, by its nature, cannot prove that a forecast is correct, only that it is wrong. The true test of a pest risk assessment occurs when an alien species begins to invade forecasted areas where it has historically never occurred, an event many organizations and individuals are actively seeking to prevent. Extensive empirical observations are needed of invasive species in their native ranges and in areas where they are invading to rigorously test new theories and models and identify opportunities for substantive improvements.

Some have argued that pest risk assessments have limited value because they are so severely encumbered by associated uncertainties (e.g., Simberloff 2005). Future research is needed to provide ways to meaningfully characterize that uncertainty and formally incorporate it into risk management decisions (Koch et al. 2009; Yemshanov et al. 2013, 2015). This transition may require new thinking about the nature of risk itself.

One important, but sometimes overlooked, aspect of risk is that it can be described in many dimensions. This need should be addressed during future research on pest risk assessment. Although most definitions of risk follow a two-dimensional interpretation (i.e., risk as the product of probability and severity), Yellman (2000) presented a more complex, three-faceted view of risk which includes expected loss, variability of loss values, and uncertainty arising from how risk perception (i.e., the uncertainty of how risk is perceived by decision-makers) is modeled. The best (i.e., the most rigorous) risk assessments extend beyond simple estimates of risk values and attempt to narrow the bounds of uncertainty associated with the phenomenon of interest, so that the decision-making options for responding to risk can be reduced to a manageable size. For industrial applications, the International Organization for Standardization defines risk as an "effect of uncertainty on objectives," where an effect is a positive or negative deviation from what is expected.[1] This definition recognizes that a decision-maker operates in an uncertain environment, so there is always a chance

[1] ISO 31000 is a generic risk management standard that is not specific to any sector or industry and could be applied in a wide range of disciplines.

that the decision-making objectives will not be achieved. Similarly, in engineering disciplines, technical risk denotes the odds that a project will fail to meet the performance criteria (Pennock and Haimes 2002). For project management, risk is often defined as an undesirable situation that has both a likelihood of occurring and a potentially negative consequence for the project (ESA 2000).

The common rationale behind the notion of risk in these diverse contexts is that decisions, and subsequent actions predicated on those decisions, must be undertaken under the assumption that the outcome of those actions is uncertain. Uncertainty is assumed to always be present as a component of risk. This uncertainty can stem from a lack of information about the process of interest or poor understanding of the consequences of decision-making actions based on incomplete information. With respect to invasive species, uncertainty arises when knowledge about the biology, ecology, impact, or management of an alien organism is limited. This uncertainty is exacerbated by the unknown state of future conditions, such as climate, land use, nitrogen deposition, and species composition.

Protection of natural resources from the seeming onslaught of new invading species requires robust management plans that emphasize early intervention strategies. Successful early intervention strategies will require close collaborations between biologists, modelers, resource managers, and policymakers. Researchers will need to work diligently to measure, describe, and reduce sources of uncertainty in their assessments. Policymakers are likely to need more sophisticated tools to understand how scientific uncertainties might affect their decisions. Success in reducing uncertainty will be aided by international collaborations and future interactions with citizen scientists to provide useful real-time information on the extent and impact of invasions as they occur.

Disclaimer Text The findings and conclusions in this publication are those of the authors and should not be construed to represent any official USDA or U.S. Government determination or policy.

Literature Cited

Allison JD, Borden JH, Seybold SJ (2004) A review of the chemical ecology of the Cerambycidae (Coleoptera). Chemoecology 14(3–4):123–150

Andersen MC, Adams H, Hope B, Powell M (2004) Risk analysis for invasive species: general framework and research needs. Risk Anal 24(4):893–900

Antwi F, Shama LM, Peterson RKD (2008) Risk assessments for the insect repellents DEET and picaridin. Regul Toxicol Pharmacol 51:31–36

Arthur M (2006) An economic analysis of quarantine: the economics of Australia's ban on New Zealand apple imports. In: 2006 Conference of the New Zealand Agricultural and Resource Economics Society; August 24–25, 2006; Nelson, New Zealand. http://purl.umn.edu/31959

Aukema JE, McCullough DG, Von Holle B et al (2010) Historical accumulation of nonindigenous forest pests in the continental US. Bioscience 60:886–897

Aukema JE, Leung B, Kovacs K et al (2011) Economic impacts of non-native forest insects in the continental United States. PLoS One 6(9):e24587

Bain MB, Cornwell ER, Hope KM et al (2010) Distribution of an invasive aquatic pathogen (viral hemorrhagic septicemia virus) in the Great Lakes and its relationship to shipping. PLoS One 5:e10156

Baker RHA, Battisti A, Bremmer J et al (2009) PRATIQUE: a research project to enhance pest risk analysis techniques in the European Union. EPPO Bull 39:87–93

Baker R, Eyre D, Brunel S et al (2015) Mapping endangered areas for pest risk analysis. In: Pest risk modelling and mapping for invasive alien species, vol 7, pp 18–34

Barak AV, Wang Y, Zhan G et al (2006) Sulfuryl fluoride as a quarantine treatment for *Anoplophora glabripennis* (Coleoptera: Cerambycidae) in regulated wood packing material. J Econ Entomol 99:1628–1635

Barak AV, Messenger M, Neese P et al (2010) Sulfuryl fluoride as a quarantine treatment for emerald ash borer (Coleoptera: Buprestidae) in ash logs. J Econ Entomol 103:603–611

Berec L, Kean JM, Epanchin-Niell R et al (2015) Designing efficient surveys: spatial arrangement of sample points for detection of invasive species. Biol Invasions 17:445–459

Beric B, MacIsaac HJ (2015) Determinants of rapid response success for alien invasive species in aquatic ecosystems. Biol Invasions 17(11):3327–3335

Blackwood J, Berec L, Yamanaka T et al (2012) Bioeconomic synergism between tactics for insect eradication in the presence of Allee effects. Proc R Soc B 279:2807–2815

Blakeslee AMH, McKenzie CH, Darling JA et al (2010) A hitchhiker's guide to the Maritimes: anthropogenic transport facilitates long-distance dispersal of an invasive marine crab to Newfoundland. Divers Distrib 16:879–891

Blossey B, Skinner LC, Taylor J (2001) Impact and management of purple loosestrife (*Lythrum salicaria*) in North America. Biodivers Conserv 10(10):1787–1807

Bogich TL, Liebhold AM, Shea K (2008) To sample or eradicate? A cost minimization model for monitoring and managing an invasive species. J Appl Ecol 45(4):1134–1142

Bradley BA, Blumenthal DM, Early R et al (2012) Global change, global trade, and the next wave of plant invasions. Front Ecol Environ 10(1):20–28

Brawley SH, Coyer JA, Blakeslee AMH et al (2009) Historical invasions of the intertidal zone of Atlantic North America associated with distinctive patterns of trade and emigration. Proc Natl Acad Sci 106:8239–8244

Breukers A, Mourits M, van der Werf W, Oude Lansink A (2008) Costs and benefits of controlling quarantine diseases: a bio-economic modeling approach. Agric Econ 38:137–149

Britton JR, Gozlan RE, Copp GH (2011) Managing non-native fish in the environment. Fish Fish 12(3):256–274

Cacho OJ, Hester S, Spring D (2007) Applying search theory to determine the feasibility of eradicating an invasive population in natural environments. Aust J Agric Resour Econ 51(4):425–443

Carey JR (1996) The future of the Mediterranean fruit fly *Ceratitis capitata* invasion of California: a predictive framework. Biol Conserv 78:35–50

Chen Z, White MS, Mack R (2016) Evaluating vacuum and steam process on hardwood veneer logs for export. Eur J Wood Wood Prod 75:1–8

Colunga-Garcia M, Haack RA, Adelaja AO (2009) Freight transportation and the potential for invasions of exotic insects in urban and periurban forests of the United States. J Econ Entomol 102(1):237–246

Colunga-Garcia M, Haack RA, Magarey RA, Margosian ML (2010) Modeling spatial establishment patterns of exotic forest insects in urban areas in relation to tree cover and propagule pressure. J Econ Entomol 103(1):108–118

Cook DC (2008) Benefit cost analysis of an import access request. Food Policy 33(3):277–285

Costello C, Springborn M, McAusland C, Solow A (2007) Unintended biological invasions: does risk vary by trading partner? J Environ Econ Manag 54:262–276

Cox LA, Babayev D, Huber W (2005) Some limitations of qualitative risk rating systems. Risk Anal 25:651–662

Davis RS, Peterson RKD, Macedo PA (2007) An ecological risk assessment for insecticides used in adult mosquito management. Integr Environ Assess Manag 3:373–382

Devorshak C (2012) Plant pest risk analysis: concepts and applications. CAB International, Wallingford, 296 p

Drake JM, Lodge DM (2006) Allee effects, propagule pressure and the probability of establishment: risk analysis for biological invasions. Biol Invasions 8:365–375

Drake JM, Drury KLS, Lodge DM et al (2006) Demographic stochasticity, environmental variability, and windows of invasion risk for *Bythotrephes longimanus* in North America. Biol Invasions 8:843–861

Elkinton JS, Boettner GH (2012) Benefits and harm caused by the introduced generalist tachinid, *Compsilura concinnata*, in North America. BioControl 57:277–288

Elliston L, Hinde R, Yainshet A (2005) Plant disease incursion management. Lect Notes Comput Sci 3415:225–235

Epanchin-Niell RS, Hastings A (2010) Controlling established invaders: integrating economics and spread dynamics to determine optimal management. Ecol Lett 13(4):528–541

Epanchin-Niell RS, Liebhold AM (2015) Benefits of invasion prevention: effect of time lags, spread rates, and damage persistence. Ecol Econ 116:146–153

Epanchin-Niell RS, Wilen JE (2012) Optimal spatial control of biological invasions. J Environ Econ Manag 63(2):260–270

Epanchin-Niell RS, Haight RG, Berec L et al (2012) Optimal surveillance and eradication of invasive species in heterogeneous landscapes. Ecol Lett 15(8):803–812

ESA, European Space Agency (2000) European Space Project Management: risk assessment (ECSS-M-00-03A). European Cooperation for Space Standardization, Noordwijk, p 40. http://everyspec.com/ESA/download.php?spec=ecss-m-00-03a.002569.pdf

Essl F, Bacher S, Blackburn TM et al (2015) Crossing frontiers in tackling pathways of biological invasions. Bioscience 65(8):769–782

Evans HC, Tomley AJ (1994) Studies on the rust, *Maravalia cryptostegiae*, a potential biological control agent of rubber vine weed, *Cryptostegia grandiflora* (Asclepiadaceae, Periplocoideae), in Australia, III: Host range. Mycopathologia 126(2):93–108

Ewel JJ, O'Dowd DJ, Bergelson J et al (1999) Deliberate introductions of species: research needs – benefits can be reaped, but risks are high. Bioscience 49:619–630

Farnsworth EJ, Ellis DR (2001) Is purple loosestrife (*Lythrum salicaria*) an invasive threat to freshwater wetlands? Conflicting evidence from several ecological metrics. Wetlands 21(2):199–209

Follett PA, Neven LG (2006) Current trends in quarantine entomology. Annu Rev Entomol 51:359–385

Gamble JC, Payne T, Small B (2010) Interviews with New Zealand community stakeholders regarding acceptability of current or potential pest eradication technologies. N Z J Crop Hortic Sci 38:57–68

Gentile JH, Harwell MA (2001) Strategies for assessing cumulative ecological risks. Hum Ecol Risk Assess 7:239–246

Golden BL, Wasil EA, Harker PT (2012) The analytic hierarchy process: applications and studies. Springer, Berlin, 265 p

Gordon DR, Onderdonk DA, Fox AM, Stocker RK (2008) Consistent accuracy of the Australian weed risk assessment system across varied geographies. Divers Distrib 14:234–242

Gray GM, Allen JC, Burmaster DE et al (1998) Principles for conduct of pest risk analyses: report of an expert workshop. Risk Anal 18(6):773–780

Haack RA (2001) Intercepted Scolytidae (Coleoptera) at US ports of entry: 1985–2000. Integr Pest Manag Rev 6(3–4):253–282

Haack RA (2006) Exotic bark-and wood-boring Coleoptera in the United States: recent establishments and interceptions. Can J For Res 36(2):269–288

Haack RA, Petrice TR (2009) Bark-and wood-borer colonization of logs and lumber after heat treatment to ISPM 15 specifications: the role of residual bark. J Econ Entomol 102(3):1075–1084

Haack RA, Britton KO, Brockerhoff EG et al (2014) Effectiveness of the International Phytosanitary Standard ISPM no. 15 on reducing wood borer infestation rates in wood packaging material entering the United States. PLoS One. 9(5):e96611

Hager HA, McCoy KD (1998) The implications of accepting untested hypotheses: a review of the effects of purple loosestrife (*Lythrum salicaria*) in North America. Biodivers Conserv 7(8):1069–1079

Hajek AE, Hurley BP, Kenis M et al (2016) Exotic biological control agents: a solution or contribution to arthropod invasions? Biol Invasions 18(4):953–969

Handley LL (2015) How will the 'molecular revolution' contribute to biological recording? Biol J Linn Soc 115:750–766

Hennessey MK (2004) Quarantine pathway pest risk analysis at the APHIS Plant epidemiology and risk analysis laboratory. Weed Technol 18(1):1484–1485

Hestir EL, Khanna S, Andrew ME et al (2008) Identification of invasive vegetation using hyperspectral remote sensing in the California Delta ecosystem. Remote Sens Environ 112(11):4034–4047

Holden MH, Nyrop JP, Ellner SP (2016) The economic benefit of time-varying surveillance effort for invasive species management. J Appl Ecol 53(3):712–721

Holmes TP, Murphy EA, Bell KP, Royle DD (2010) Property value impacts of hemlock woolly adelgid in residential forests. For Sci 56(6):529–540

Hoover K, Uzunovic A, Gething B et al (2010) Lethal temperature for pinewood nematode, *Bursaphelenchus xylophilus*, in infested wood using microwave energy. J Nematol 42:101–110

Hulme PE (2003) Biological invasions: winning the science battles but losing the conservation war? Oryx 37(2):178–193

Hulme PE (2009) Trade, transport and trouble: managing invasive species pathways in an era of globalization. J Appl Ecol 46:10–18

Hulme PE, Bacher S, Kenis M et al (2008) Grasping at the routes of biological invasions: a framework for integrating pathways into policy. J Appl Ecol 45:403–414

Humair F, Humair L, Kuhn F, Kueffer C (2015) E-commerce trade in invasive plants. Conserv Biol 29:1658–1665

IPPC, International Plant Protection Convention (2016a) International Standards for Phytosanitary Management (ISPM) 2: framework for Pest risk analysis. Food and Agriculture Organization of the United Nations, Rome, p 16. https://www.ippc.int/static/media/files/publication/en/2016/01/ISPM_02_2007_En_2015-12-22_PostCPM10_InkAmReformatted.pdf

IPPC, International Plant Protection Convention (2016b) International Standards for Phytosanitary Management (ISPM) 5: Glossay of Phytosanitary terms. Food and Agriculture Organization of the United Nations, Rome, p 34. https://www.ippc.int/static/media/files/publication/en/2016/05/ISPM_05_2016_En_2016-05-20.pdf

IPPC, International Plant Protection Convention (2016c) International Standards for Phytosanitary Management (ISPM) 11: Pest risk analysis for quarantine pests. Food and Agriculture Organization of the United Nations, Rome, p 39. https://www.ippc.int/static/media/files/publication/en/2016/01/ISPM_11_2013_En_2015-12-22_PostCPM10_InkAmReformatted.pdf

IPPC, International Plant Protection Convention (2016d) International Standards for Phytosanitary Management (ISPM) 15: regulation of wood packing material in international trade. Food and Agriculture Organization of the United Nations, Rome, p 19. https://www.ippc.int/static/media/files/publication/en/2016/01/ISPM_15_2013_En_2015-12-22_PostCPM10_InkAmReformatted.pdf

Jarvis CH, Baker RHA (2001) Risk assessment for nonindigenous pests: 2. Accounting for interyear climate variability. Divers Distrib 7:237–248

Jerde CL, Mahon AR, Chadderton WL, Lodge DM (2011) "Sight-unseen" detection of rare aquatic species using environmental DNA. Conserv Lett 4(2):150–157

Juliá R, Holland DW, Guenthner J (2007) Assessing the economic impact of invasive species: the case of yellow starthistle (*Centaurea solsitialis* L.) in the rangelands of Idaho, USA. J Environ Manag 85:876–882

Kaluza P, Kolzsch A, Gastner MT, Blasius B (2010) The complex network of global cargo ship movements. J R Soc Interface 7:1093–1103

Kaplan S, Garrick BJ (1981) On the quantitative definition of risk. Risk Anal 1:11–27

Kearney M, Porter W (2009) Mechanistic niche modelling: combining physiological and spatial data to predict species' ranges. Ecol Lett 12(4):334–350

Keller RP, Lodge DM (2007) Species invasions from commerce in live aquatic organisms: problems and possible solutions. Bioscience 57:428–436

Keller RP, Lodge DM, Finnoff DC (2007) Risk assessment for invasive species produces net bioeconomic benefits. Proc Natl Acad Sci U S A 104(1):203–207

Kenis M, Rabitsch W, Auger-Rozenberg M-A, Roques A (2007) How can alien species inventories and interception data help us prevent insect invasions? Bull Entomol Res 97:489–502

Kenis M, Auger-Rozenberg M, Roques A et al (2009) Ecological effects of invasive alien insects. Biol Invasions 11:21–45

Kiritani K, Yamamura K (2003) Exotic insects and their pathways for invasion. In: Ruiz GM, Carlton JT (eds) Invasive species: vectors and management strategies. Island Press, Washington, pp 44–67

Koch FH, Smith WD (2008) Spatio-temporal analysis of *Xyleborus glabratus* (Coleoptera : Circulionidae : Scolytinae) invasion in eastern US forests. Environ Entomol 37(2):442–452

Koch FH, Yemshanov D, McKenney DW, Smith WD (2009) Evaluating critical uncertainty thresholds in a spatial model of forest pest invasion risk. Risk Anal 29(9):1227–1241

Koch FH, Yemshanov D, Colunga-Garcia M et al (2011) Establishment of alien-invasive forest insect species in the United States: where and how many? Biol Invasions 13:969–985

Koch FH, Yemshanov D, Magarey RD, Smith WD (2012) Dispersal of invasive forest insects via recreational firewood: a quantitative analysis. J Econ Entomol 105(2):438–450

Koch FH, Yemshanov D, Haack RA, Magarey RD (2014) Using a network model to assess risk of forest pest spread via recreational travel. PLoS One 9(7):e102105

Koop A, Fowler L, Newton L, Caton B (2012) Development and validation of a weed screening tool for the United States. Biol Invasions 14:273–294

Kumschick S, Gaertner M, Vila M et al (2015) Ecological impacts of alien species: quantification, scope, caveats, and recommendations. Bioscience 65(1):55–63

Lamarche J, Potvin A, Pelletier G et al (2015) Molecular detection of 10 of the most unwanted alien forest pathogens in Canada using real-time PCR. PLoS One 10(8):e0134265

Landis WG (2003) Ecological risk assessment conceptual model formulation for nonindigenous species. Risk Anal 24(4):847–858

Landis WG, Wiegers JA (1997) Design considerations and a suggested approach for regional and comparative ecological risk assessment. Hum Ecol Risk Assess 3:287–297

Lavoie C (2010) Should we care about purple loosestrife? The history of an invasive plant in North America. Biol Invasions 12(7):1967–1999

Leung B, Lodge DM, Finnoff D et al (2002) An ounce of prevention or a pound of cure: bioeconomic risk analysis of invasive species. Proc R Soc Lond B Biol Sci 269(1508):2407–2413

Leung B, Drake JM, Lodge DM (2004) Predicting invasions: propagule pressure and the gravity of allee effects. Ecology 85(6):1651–1660

Leung B, Springborn MR, Turner JA, Brockerhoff EG (2014) Pathway-level risk analysis: the net present value of an invasive species policy in the US. Front Ecol Environ 12(5):273–279

Liebhold AM, Griffin R (2016) The legacy of Charles Marlatt and efforts to limit plant pest invasions. Am Entomol 62(4):218–227

Liebhold AM, Tobin PC (2008) Population ecology of insect invasions and their management. Annu Rev Entomol 53:387–408

Liebhold AM, Work TT, McCullough DG, Cavey JF (2006) Airline baggage as a pathway for alien insect species invading the United States. Am Entomol 52(1):48–54

Liebhold AM, Brockerhoff EG, Garrett LJ et al (2012) Live plant imports: the major pathway for forest insect and pathogen invasions of the United States. Front Ecol Environ 10:135–143

Liebhold AM, Berec L, Brockeroff EG et al (2016) Eradication of invading insect populations: from concepts to applications. Annu Rev Entomol 61:335–352

Liu H, Stiling P (2006) Testing the enemy release hypothesis: a review and meta-analysis. Biol Invasions 8(7):1535–1545

Lockwood JL, Cassey P, Blackburn T (2005) The role of propagule pressure in explaining species invasions. Trends Ecol Evol 20(5):223–228

Lodge DM, Williams S, MacIsaac HJ et al (2006) Biological invasions: recommendations for US policy and management. Ecol Appl 16(6):2035–2054

Loomis J, Richardson L, Kroeger T, Casey F (2014) Valuing ecosystem services using benefit transfer: separating credible and incredible approaches. In: Ninan KN (ed) Valuing ecosystem services: methodological issues and case studies. Edward Elgar, Cheltenham, pp 78–89

Louda SM (2000) Negative ecological effets of the musk thistle biological control agent, *Rhinocyllus conicus*. In: Follett PA, Duan JJ (eds) Nontarget effects of biological control. Springer, New York, pp 215–243

Lounibos LP (2002) Invasions by insect vectors of human disease. Annu Rev Entomol 47:233–266

Mack RN, Foster SK (2009) Eradicating plant invaders: combining ecologically based tactics and broad-sense strategy. In: Inderjit (ed) Management of Invasive Weeds. Springer, Heidelberg, pp 35–60

Mack RN, Simberloff D, Lonsdale WM et al (2000) Biotic invasions: causes, epidemiology, global consequences, and control. Ecol Appl 10(3):689–710

Macleod A, Head J, Gaunt A (2003) The assessment of the potential economic impact of *Thrips palmi* on horticulture in England and the significance of a successful eradication campaign. Crop Prot 23:601–610

Magarey RD, Borchert DM, Fowler GL et al (2007) NAPPFAST, an internet system for the weather-based mapping of plant pathogens. Plant Dis 91:336–345

Magarey RD, Colunga-Garcia M, Fieselmann DA (2009) Plant biosecurity in the United States: roles, responsibilities, and information needs. Bioscience 59(10):875–884

McCartney HA, Foster SJ, Fraaije BA, Ward E (2003) Molecular diagnostics for fungal plant pathogens. Pest Manag Sci 59(2):129–142

Mehta SV, Haight RG, Homans FR et al (2007) Optimal detection and control strategies for invasive species management. Ecol Econ 61(2–3):237–245

Meyerson LA, Pyšek P (2013) Manipulating alien plant species propagule pressure as a prevention strategy for protected areas. In: Foxcroft LC, Pyšek P, Richardson DM, Genovesi P (eds) Plant invasions in protected areas: patterns, problems and challenges.

Invading nature-springer series in invasion ecology, vol 7. Springer, Dordrecht, pp 473–486
Miles TD, Martin FN, Coffey MD (2015) Development of rapid isothermal amplification assays for detection of *Phytophthora* spp. in plant tissue. Phytopathology 105(2):265–278
Mohammed EY (2014) Contingent valuation responses and hypothetical bias. In: Ninan KN (ed) Valuing ecosystem services: methodological issues and case studies. Edward Elgar, Cheltenham, pp 90–108
Morse-Jones S, Bateman IJ, Kontoleon A et al (2014) Stated preferences for tropical wildlife conservation amongst distant beneficiaries: charisma, endemism, scope and substitution effects. In: Ninan KN (ed) Valuing ecosystem services: methodological issues and case studies. Edward Elgar, Cheltenham, pp 109–131
Muirhead JR, Leung B, van Overdijk C et al (2006) Modelling local and long-distance dispersal of invasive emerald ash borer *Agrilus planipennis* (Coleoptera) in North America. Divers Distrib 12:71–79
Muirhead JR, Gray DK, Kelly DW et al (2008) Identifying the source of species invasions: sampling intensity vs. genetic diversity. Mol Ecol 17:1020–1035
NAPPO, North American Plant Protection Organization (2015) Regional standards for Phytosanitary management 12: guidelines for petition for first release of non-indigenous Entomophagous biological control agents Ottawa. Secretariat of the North American Plant Protection Organization, Ontario, p 14. http://www.nappo.org/files/1814/4065/2949/RSPM12_30-07-2015-e.pdf
NAS (2016) Gene drives on the horizon: advancing science, navigating uncertainty, and aligning research with public values. The National Academies of Sciences, Engineering, and Medicine. The National Academies Press, Washington, DC
Novak SJ (2007) The role of evolution in the invasion process. Proc Natl Acad Sci U S A 104(10):3671–3672
Paini DR, Yemshanov D (2012) Modeling the arrival of invasive organisms via the international marine shipping network: a Khapra beetle study. PLoS One 7(9):e44589
Parker IM, Simberloff D, Lonsdale WM et al (1999) Impact: toward a framework for understanding the ecological effects of invaders. Biol Invasions 1:3–19
Parkes JP, Ramsey DSL, Macdonald N et al (2010) Rapid eradication of feral pigs (*Sus scrofa*) from Santa Cruz Island, California. Biol Conserv 143(3):634–641
Pasquali S, Gilioli G, Janssen D, Winter S (2015) Optimal strategies for interception, detection, and eradication in plant biosecurity. Risk Anal 35(9):1663–1673
Pattison RR, Mack RN (2009) Environmental constraints on the invasion of *Triadica sebifera* in the eastern United States: an experimental field assessment. Oecologia 158(4):591–602
Pennock MJ, Haimes YY (2002) Principles and guidelines for project risk management. Syst Eng 5(2):98–108
Peterson RKD (2006) Comparing ecological risks of pesticides: the utility of a risk quotient ranking approach across refinements of exposure. Pest Manag Sci 62:46–56
Peterson RKD (2010) Mosquito management and risk. Wing Beats 21:28–31
Peterson RKD, Shama LM (2005) Comparative risk assessment of genetically engineered, mutagenic, and conventional wheat production systems. Transgenic Res 14:859–875
Peterson RKD, Macedo PA, Davis RS (2006) A human-health risk assessment for West Nile virus and insecticides used in mosquito management. Environ Health Perspect 114:366–372
Peterson AT, Soberon J, Pearson RG et al (2011a) Ecological niches and geographic distributions. Monographs in population biology 49. Princeton University Press, i–x, 1–314 p
Peterson RKD, Barber LM, Schleier JJ III (2011b) Net risk: a risk assessment of long-lasting insecticide bed nets used for malaria management. Am J Trop Med Hyg 84:951–956
Pheloung PC, Williams PA, Halloy SR (1999) A weed risk assessment model for use as a biosecurity tool evaluating plant introductions. J Environ Manag 57:239–251
Phillips BL (2015) Evolutionary processes make invasion speed difficult to predict. Biol Invasions 17(7):1949–1960
Pitt JPW, Worner SP, Suarez AV (2009) Predicting Argentine ant spread over the heterogeneous landscape using a spatially explicit stochastic model. Ecol Appl 19:1176–1186
Pluess T, Jarošik V, Pyšek P et al (2012) Which factors affect the success or failure of eradication campaigns against alien species? PLoS One 7(10):11
Pyšek P, Jarošik V, Hulme PE et al (2010) Disentangling the role of environmental and human pressures on biological invasions across Europe. Proc Natl Acad Sci 107:12157–12162
Rabaglia R, Duerr D, Acciavatti R, Ragenovich I (2008) Early detection and rapid response for non-native bark and Ambrosia beetles. U.S. Deptartment of Agriculture Forest Service, Forest Health Protection, Washington, DC, p 12. http://www.fs.fed.us/foresthealth/publications/EDRRProjectReport.pdf
Rafoss T (2003) Spatial stochastic simulation offers potential as a quantitative method for pest risk analysis. Risk Anal 23(4):651–661
Reaser JK, Meyerson LA, Von Holle B (2008) Saving camels from straws: how propagule pressure-based prevention policies can reduce the risk of biological invasion. Biol Invasions 10(7):1085–1098
Rees HC, Maddison BC, Middleditch DJ et al (2014) Review: the detection of aquatic animal species using environmental DNA – a review of eDNA as a survey tool in ecology. J Appl Ecol 51:1450–1459
Reichard SH, White P (2001) Horticulture as a pathway of invasive plant introductions in the United States. Bioscience 51(2):103–113
Rejmánek M, Pitcairn MJ (2002) When is eradication of exotic pest plants a realistic goal? In: Veitch D, Clout M (eds) Turning the tide: the eradication of invasive species. Invasive Species Specialty Group of the World Conservation Union (IUCN), Auckland, pp 249–253
Rout TM, Moore JL, McCarthy MA (2014) Prevent, search or destroy? A partially observable model for invasive species management. J Appl Ecol 51(3):804–813
Saccaggi DL, Karsten M, Robertson MP et al (2016) Methods and approaches for the management of arthropod border incursions. Biol Invasions 18:1057–1075
Schaffner U (2001) Host range testing of insects for biological weed control: how can it be better interpreted? Bioscience 51(11):951–959
Schleier JJ III, Sing SE, Peterson RKD (2008) Regional ecological risk assessment for the introduction of *Gambusia affinis* (western mosquitofish) into Montana watersheds. Biol Invasions 10:1277–1287
Schmidt E, Juzwik J, Schneider B (1997) Sulfuryl fluoride fumigation of red oak logs eradicates the oak wilt fungus. Holz Roh Werkst 55:315–318
Schooler SS, McEvoy PB, Coombs EM (2006) Negative per capita effects of purple loosestrife and reed canary grass on plant diversity of wetland communities. Divers Distrib 12(4):351–363
Shigesada N, Kawasaki K (1997) Biological invasions: theory and practice. Oxford University Press, Oxford
Simberloff D (2005) The politics of assessing risk for biological invasions: the USA as a case study. Trends Ecol Evol 20(5):216–222
Simberloff D (2009) We can eliminate invasions or live with them. Successful management projects. Biol Invasions 11:149–157
Simpson WT (2001) Heating times for round and rectangular cross sections of wood in steam. U.S. Deptartment of Agriculture Forest Service, Forest Products Laboratory, Madison, 103 p
Sing SE, Peterson RKD (2011) Assessing environmental risks for established invasive weeds: Dalmatian (*Linaria dalmatica*) and yellow (*L. vulgaris*) toadflax in North America. Int J Environ Res Public Health 8:2828–2853

Sing SE, Peterson RKD, Weaver DK et al (2005) A retrospective analysis of known and potential risks associated with exotic toadflax-feeding insects. Biol Control 35:276–287

Skellam JG (1951) Random dispersal in theoretical populations. Biometrika 38:196–218

Soliman T, Mourits MCM, Oude Lansink AGJM, van der Werf W (2010) Economic impact assessment in pest risk analysis. Crop Prot 29:517–524

Strutt A, Turner JA, Haack RA, Olson LJ (2013) Evaluating the impacts of an international phytosanitary standard for wood packaging material: global and United States trade implications. Forest Policy Econ 27:54–64

Surkov IV, Oude Lansink AGJM, van der Werf W (2009) The optimal amount and allocation of sampling effort for plant health inspection. Eur Rev Agric Econ 36:295–320

Tatem AJ (2009) The worldwide airline network and the dispersal of exotic species: 2007–2010. Ecography 32(1):94–102

Tatem AJ, Rogers DJ, Hay SI (2006) Global transport networks and infectious disease spread. Adv Parasitol 62:293–343

Tidbury HJ, Taylor NGH, Copp GH et al (2016) Predicting and mapping the risk of introduction of marine non-indigenous species into Great Britain and Ireland. Biol Invasions 18(11):3277–3292

Tobin PC, Berec L, Liebhold AM (2011) Exploiting Allee effects for managing biological invasions. Ecol Lett 14:615–624

Tobin PC, Blackburn LM, Gray RH et al (2013) Using delimiting surveys to characterize the spatiotemporal dynamics facilitates the management of an invasive non-native insect. Popul Ecol 55(4):545–555

Tsolaki E, Diamadopoulos E (2010) Technologies for ballast water treatment: a review. J Chem Technol Biotechnol 85(1):19–32

Tubajika KM, Barak AV (2011) Fungitoxicity of methiyl iodide, sulfuryl fluoride, and methyl bromide to *Ceratocystis fagacearum* in red oak, maple, poplar, birch and pine wood. Am J Plant Sci 2:268–275

UNCTAD, United Nations Conference on Trade and Development (2007) Review of maritime transport 2007. United Nations, Geneva, p 153. http://unctad.org/en/docs/rmt2007_en.pdf

USDA, U.S. Department of Agriculture (2016) Technical advisory Group for Biological Control Agents of weeds manual, Interim edn. USDA, Washington, DC, p 156. https://www.aphis.usda.gov/import_export/plants/manuals/domestic/downloads/tag-bcaw_manual.pdf

USDA, U.S.D.o.A (2015) Not Authorized Pending Pest Risk Analysis (NAPPRA). https://www.aphis.usda.gov/aphis/ourfocus/planthealth/import-information/permits/plants-and-plant-products-permits/plants-for-planting/ct_nappra

Uzunovic A, Gething B, Coelho A et al (2013) Lethal temperature for pinewood nematode, *Bursaphelenchus xylophilus*, in infested wood using radio frequency (RF) energy. J Wood Sci 59:160–170

van Lenteren JC, Bale J, Bigler E et al (2006) Assessing risks of releasing exotic biological control agents of arthropod pests. Annu Rev Entomol 51:609–634

Venette RC (2015) Pest risk modelling and mapping for invasive alien species. CAB International, Wallingford, 268 p

Venette RC, Koch RL (2009) IPM for invasive species. In: Radcliffe EB, Hutchison WD, Cancelado RE (eds) Integrated Pest management. Cambridge University Press, Cambridge, pp 424–436

Venette RC, Kriticos DJ, Magarey R et al (2010) Pest risk maps for invasive alien species: a roadmap for improvement. Bioscience 60:349–362

Vettraino A, Roques A, Yart A et al (2015) Sentinel trees as a tool to forecast invasions of alien plant pathogens. PLoS One 10(3):e0120571

Volin JC, Lott MS, Muss JD, Owen D (2004) Predicting rapid invasion of the Florida Everglades by Old World climbing fern (*Lygodium microphyllum*). Divers Distrib 10:439–446

Wang R, Wang YZ (2006) Invasion dynamics and potential spread of the invasive alien plant species *Ageratina adenophora* (Asteraceae) in China. Divers Distrib 12(4):397–408

Westbrooks RG (2004) New approaches for early detection and rapid response to invasive plants in the United States. Weed Technol 18(1):1468–1471

Westphal MI, Browne M, MacKinnon K, Noble I (2008) The link between international trade and the global distribution of invasive alien species. Biol Invasions 10:391–398

Wilson JRU, Dormontt EE, Prentis PJ et al (2009) Something in the way you move: dispersal pathways affect invasion success. Trends Ecol Evol 24(3):136–144

Withrow JR, Smith EL, Koch FH, Yemshanov D (2015) Managing outbreaks of invasive species – a new method to prioritize preemptive quarantine efforts across large geographic regions. J Environ Manag 150:367–377

Wittwer G, McKirdy S, Wilson R (2005) Regional economic impacts of a plant disease incursion using a general equilibrium approach. Aust J Agric Resour Econ 49:75–89

WTO, World Trade Organization (2008) World trade report 2008: trade in a globalizing world. WTO Secretariat, Geneva, p 178. http://www.wto.org/english/res_e/booksp_e/anrep_e/world_trade_report08_e.pdf

Yellman T (2000) The three facets of risk. SAE Trans 109(1):1244–1257

Yemshanov D, Koch FH, McKenney DW et al (2009a) Mapping invasive species risks with stochastic models: a cross-border United States-Canada application for *Sirex noctilio* Fabricius. Risk Anal 29:868–884

Yemshanov D, McKenney DW, Pedlar JH et al (2009b) Towards an integrated approach to modelling the risk and impacts of invasive forest species. Environ Rev 17:163–178

Yemshanov D, Koch FH, Ducey M, Koehler K (2012) Trade-associated pathways of alien forest insect entries in Canada. Biol Invasions 14:797–812

Yemshanov D, Koch FH, Ducey MJ et al (2013) Exploring critical uncertainties in pathway assessment of human-assisted introductions of alien forest species in Canada. J Environ Manag 129:173–182

Yemshanov D, Koch FH, Ducey M (2015) Making invasion models useful for decision makers: incorporating uncertainty, knowledge gaps and decision-making preferences. In: Pest risk modelling and mapping for invasive alien species, vol 7, pp 206–222

Yoe C (2012) Risk analysis: decision making under uncertainty. CRC Press, Boca Raton, 553 p

Yokoyama VY (2011) Approved quarantine treatment for hessian fly (Diptera: Cecidomyiidae) in large-size hay bales and hessian fly and cereal leaf beetle (Coleoptera: Chrysomelidae) control by bale compression. J Econ Entomol 104(3):792–798

Yuksel S, Schwenkbier L, Pollok S et al (2015) Label-free detection of *Phytophthora ramorum* using surface-enhanced Raman spectroscopy. Analyst 140:7254–7262

Management of Landscapes for Established Invasive Species

7

Therese M. Poland, Jennifer Juzwik, Allen Rowley, Cynthia D. Huebner, John C. Kilgo, Vanessa M. Lopez, Deanna H. Olson, Dean E. Pearson, Robert Progar, Robert Rabaglia, John D. Rothlisberger, Justin B. Runyon, and Sharlene E. Sing

7.1 Introduction

Long-term management strategies are invoked once an invasive species has become established and spread beyond feasible limits for eradication or containment. Although an invasive species may be well-established in small to large geographical areas, prevention of its spread to non-affected areas (e.g., sites, regions, and cross-continent) through early detection and monitoring is an important management activity. The level for management of established invasive species in the United States has increasingly shifted to larger geographical scales in the past several decades. Management of an invasive fish may occur at the watershed level in the western States, with watershed levels defined by their hydrologic unit codes (HUC) ranging from 2 digits at the coarsest level to 8 digits at the finest level (USGS 2018). Invasive plant management within national forests, grasslands, and rangelands can be implemented at the landscape level (e.g., Chambers et al. 2014), although management can still occur at the stand or base level. Landscapes in this chapter refer to areas of land bounded by large-scale physiographic features integrated with natural or man-made features that govern weather and disturbance patterns and limit frequencies of species movement (Urban et al. 1987). These are often at a large physical scale, such as the Great Basin.

This chapter considers the continuum from application of broad-scale invasive species management to implementation of specific local tactics (Fig. 7.1). Several foundational principles are discussed in Sect. 7.2. Considerations for natural resource managers faced with invasive species issues within the context of an ecosystem (Pickett and Candenasso 2002) or ecological community of variable size, landscape, or watershed management are then presented in Sect. 7.3. In Sect. 7.4, we address strategies, approaches, and tactics but in the context of recent advances made in (1) the sciences (e.g., biology, ecology, and epidemiology) involved and (2) strategies, approaches, and tools for invasive species man-

T. M. Poland (✉)
U.S. Department of Agriculture, Forest Service, Northern Research Station, Lansing, MI, USA
e-mail: therese.poland@usda.gov

J. Juzwik
U.S. Department of Agriculture, Forest Service, Northern Research Station, St. Paul, MN, USA

A. Rowley
U.S. Department of Agriculture, Forest Service, National Forest System, Washington Office, Washington, DC, USA

C. D. Huebner
U.S. Department of Agriculture, Forest Service, Northern Research Station, Morgantown, WV, USA

J. C. Kilgo
U.S. Department of Agriculture, Forest Service, Southern Research Station, Center for Forest Watershed Research, New Ellenton, SC, USA

V. M. Lopez · R. Progar
U.S. Department of Agriculture, Forest Service, State and Private Forestry, Washington, DC, USA

D. H. Olson
U.S. Department of Agriculture, Forest Service, Pacific Northwest Research Station, Corvallis, OR, USA

D. E. Pearson
U.S. Department of Agriculture, Forest Service, Rocky Mountain Research Station, Missoula, MT, USA

Ecology and Evolution, Division of Biological Sciences, University of Montana, Missoula, MT, USA

R. Rabaglia
U.S. Department of Agriculture, Forest Service, State & Private Forestry, Forest Health Protection, Washington Office, Washington, DC, USA

J. D. Rothlisberger
U.S. Department of Agriculture, Forest Service, Washington Office, Washington, DC, USA

J. B. Runyon · S. E. Sing
U.S. Department of Agriculture, Forest Service, Rocky Mountain Research Station, Bozeman, MT, USA

T. M. Poland et al. (eds.), *Invasive Species in Forests and Rangelands of the United States*,
https://doi.org/10.1007/978-3-030-45367-1_7

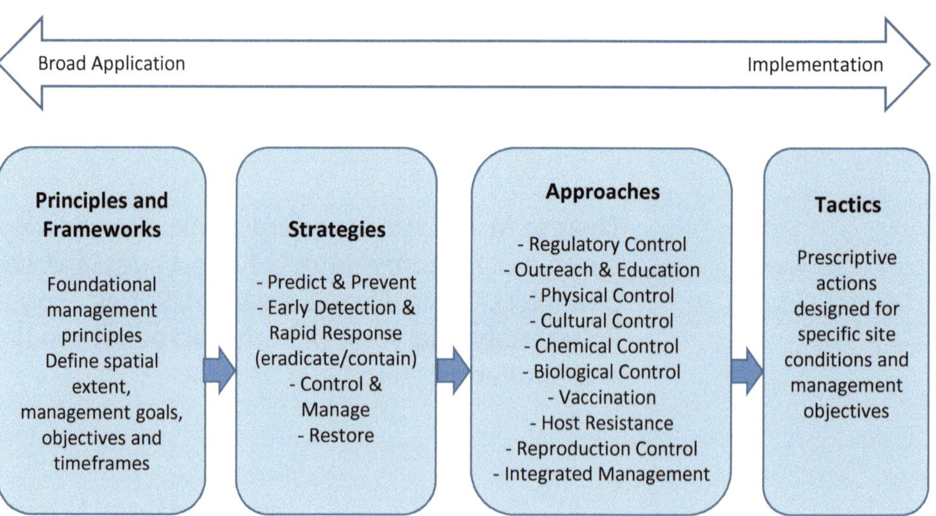

Fig. 7.1 A continuum of management responses to address management objectives at appropriate scales. (Adapted from Fig. 4.2 in Millar et al. 2012)

agement; key findings, knowledge gaps, and needs are included. This chapter considers invasive species that affect features in landscapes containing the Nation's forests, grasslands, and rangelands. Types of invaders included in this synthesis include insect pests of trees and disease vectors, pathogens of trees and wildlife, terrestrial and aquatic plants, and terrestrial and aquatic wildlife or other animals.

7.2 Invasive Species Management Principles

Organizations and agencies have utilized various concepts or principles in written invasive species management plans such as the training module on managing invasive plants by the US Fish and Wildlife Service (US FWS 2017a). The USDA Forest Service uses the Forest Service National Strategic Framework for Invasive Species Management to prioritize and guide management of all invasive species using the Invasive Species Systems Approach (ISSA). For the purposes of this chapter, four general principles are considered relevant to natural resource managers and land managers faced with the invasive species issues in the context of this volume. The following principles apply broadly to decision-making and implementation of management strategies, approaches, and tactics (Fig. 7.1).

7.2.1 Understanding Impacts of Invasive Species Is Essential for Effective Management

Impacts of invasive species can be highly complex deriving from a variety of direct and indirect interaction pathways (see Chap. 3). Mitigating impacts of invasive species and restoring affected systems therefore require an understanding of impacts and ecological pathways leading to those impacts at all levels in the affected community (see example, Box 7.1). Invasive species management in natu-

Box 7.1 Understanding Impacts of Invasive Species for Effective Management: Example, Spotted Knapweed (*Centaurea stoebe*)

Spotted knapweed infestations impact the food chain in pine savannas in western Montana and can delay reproduction of chipping sparrows (*Spizella passerina*), increase their dispersal, and reduce return rates of resident birds to breeding sites (Ortega et al. 2006). Spotted knapweed can outcompete native plants (Maron and Marler 2008) and consequently reduce populations of native insects that serve as important food sources required by chipping sparrows while nesting and rearing their young. Hence, suppressing spotted knapweed populations and restoring some native plant species but failing to restore those plant species that support these insect foods might mitigate some but not all of the impacts caused by spotted knapweed in this system. For example, the use of broadleaf herbicides can favor native grasses over native forbs (Ortega and Pearson 2010). Whether restoring the system to native grasses at the expense of native forbs would be considered successful would depend on management objectives. If the primary objective was to restore the displaced knapweed with native grasses that are needed to increase winter forage for elk (*Cervus elaphus*), this outcome could be deemed successful (e.g., Thompson 1996). However, further restoration efforts might be required to restore these sites for chipping sparrow breeding areas. This example illustrates both the importance of understanding ecological pathways leading to impacts at all levels and the value of designing management strategies essential to address specific objectives.

ral areas can be far more complex than in managed agricultural systems, and many factors can impede management objectives (see Pearson et al. 2016a). In extreme cases, mitigation may fail even if invasive species populations have been locally extirpated. For example, invasive plants may alter soil properties in ways that persist after the invasive species is eliminated (Magnoli et al. 2013), chemical control measures may suppress nontarget natives (Ortega and Pearson 2010), and secondary invasion may result in replacement of the target invasive species with another pest (Pearson et al. 2016b). Hence, it is necessary to anticipate and understand the full range of impacts, any side effects of management actions, and complicating factors, prior to applying adaptive management to meet management objectives.

7.2.2 Effective Management Is Specific to Ecosystem, Landscape, and Forest Management Objectives

Invasive species management is an important component of a comprehensive management plan for any ecosystem, landscape, or forest. With increasing global trade (among other factors), it is likely that the number of potential invasions will continue to increase (Chornesky et al. 2005). Furthermore, the full effect of already established invasive species on US forests has not been realized. Thus, management success as defined by management objectives must include specific desired outcomes related to existing or likely invasive species. However, management objectives must be formulated in accordance with the governing processes established (land management planning process for Federal lands or other appropriate planning and decision-making process for other ownerships) for the affected system or land area and the ecological processes extant. A partial list of important factors to consider might include system resilience, or the capacity of an ecosystem to respond to a disturbance by recovering quickly (Beisner et al. 2003), susceptibility to invasion, and directional changes in abiotic conditions, e.g., anthropogenic eutrophication, climate change, and shifts in disturbance regimes (see Chaps. 4 and 5).

Much of our knowledge about resilience theory and thresholds as they relate to plant invasions has been gained through research conducted in the Great Basin (e.g., Chambers et al. 2014). For example, studies in the Great Basin that consider system resilience in relation to disturbance, susceptibility to invasion, climate change, and shifts in disturbance regimes illustrate how focusing invasive species management on sites exhibiting higher resilience to disturbance and greater resistance to invasion by cheatgrass (*Bromus tectorum*) can favor success (Chambers et al. 2014). On the other hand, in systems where humans have altered disturbance regimes to the point that they are pushed beyond historic equilibrium states, non-natives may be better adapted than native species to thrive under the new ecosystem conditions (e.g., MacDougall and Turkington 2005), thus creating a situation where management efforts are likely to fail unless the disturbance regime itself is restored. It is important to recognize that systems are changing over global scales in ways that may favor non-native invasive species over native species that are no longer able to adapt to changing local conditions. In some cases, systems have already transitioned into novel ecosystems, with biotic, abiotic, and social components that have been altered by human influence and comprise a combination of introduced species which either have attained or are well on their way toward becoming new stable equilibrium states (Hobbs et al. 2006). In such cases, restoration to the original state is likely infeasible, and the best approach may be managing for form and function that best serve to generate the ecosystem services we desire (Hobbs 2007). Hence, management objectives must account for system processes to be successful, and management and restoration goals will tend to range along the full gradient from restoration to pre-invasion conditions to management of novel ecosystems.

7.2.3 Threshold Concept Aids Decision-Making for Invasive Species Management

In general terms, threshold is considered a defined level (e.g., magnitude or intensity) that, if exceeded, will lead to a change in condition or result in an action. Three types of thresholds (ecological, utility, and decision) are relevant to decision-making in natural resource management (Nichols et al. 2014) and, by extension, invasive species management within forest and rangelands. Ecological thresholds have been defined in various ways. Common terms to those definitions include "a point or zone at which there is a sudden change in the condition or dynamics of a biological system" (Nichols et al. 2014; see p. 10). Utility thresholds are "values of state or performance variables at which small changes yield substantial changes in the value of management outcomes" (Nichols et al. 2014; see p. 12). Human values are the "drivers" of utility thresholds, although there may be an ecological basis as well. Decision thresholds are "values of system state variables that should prompt specific management activities" (Nichols et al. 2014; see p. 13). Generally speaking, management objectives, available control actions, and predictive models of an invasive species population or other measures in system dynamics are the basis of decision thresholds. Examples include action thresholds for management of invasive plants (US FWS 2017a) and thresholds of the delimiting and priority indices within the gypsy moth (*Lymantria dispar*) Slow the Spread program which results in a recommended course of action (Tobin and Blackburn 2007).

7.2.4 Prioritization of Limited Resources for Effective Management

Invasions of non-native pests cost the US economy an estimated $120 billion US annually (Pimentel et al. 2005). Current and future mitigations may require difficult tradeoffs about which species in which locations to address and how to allocate resources between detection, treatment, and monitoring. As a result, effective management depends on careful prioritization to ensure that the limited resources are implemented for maximal benefit. Approaches for natural resource managers to prioritize invasive species management issues for purposes of effort or resource allocation are discussed in the following section.

7.3 Framework for Management and the Prioritization or Allocation of Limited Resources

In a general sense, management is the act or skill of controlling and making decisions about a business or enterprise. The heart of decision-making comes down to the allocation of limited resources necessary to stabilize, expand, or ensure the longevity of some specific part of a, or an entire, business. The business or enterprise in this effort is the management of natural resources for beneficial use, ecosystem services, intrinsic societal value, observation, and preservation for scientific study. Invasive species are clearly capable of negatively impacting natural resources (Chap. 2) and preventing us from reaching one or more of our goals in achieving natural resource management.

For the purposes of this chapter, management of invasive species refers to any activity that is used to minimize the spread and address adverse effects of an invader. Activities used to accomplish these goals include (1) preventing, surveying, detecting, identifying, monitoring, inventorying, eradicating, containing, and controlling invasive species; (2) rehabilitating and restoring affected lands (see Chap. 8); and (3) providing technical outreach and educational activities to various audiences in support of these activities as a means to achieve the specified goals.

For the forest, grassland, or rangeland manager, prioritization of invasive species and the ecosystems they threaten is needed for wise use of resources available for their management. Multifaceted inputs are required for this prioritization exercise (Box 7.2).

At the simplest level, the outcome of this process is to allocate resources to the highest priority work. Numerous decision-support tools are available to help identify the highest priority work, such as linear optimization programs and cost-benefit analysis programs available from business. We often lack complete information about the new invaders or their impacts needed to support implementing these kinds of models. In this situation, other methods can be implemented for identifying relative priorities for invasive species management. These are discussed in the following section.

Box 7.2 Inputs Required for Prioritization of Invasive Species Management

1. Level of threat or potential impact from the invading species
2. The ecological, economic, and/or societal value of the recipient ecosystem or community, its susceptibility to invasion, and capacity for restoration
3. Spatial extent and temporal stage of the invasion within the ecosystem under consideration
4. Goals and objectives of potential invasive species management effort(s)
5. Available tools and their relative effectiveness in managing invasive species under the conditions existing in the threatened or already affected ecosystem (see Sect. 7.4)

7.3.1 The Threat or Impact from Invading Species

Species with the greatest negative impact, such as wildfire threat, rate of spread, or ecosystem impact, would be given the highest priority if only one factor is considered. For example, an aggressive tree-killing insect or pathogen that has the potential to threaten the survival of a single genus of trees, such as emerald ash borer (*Agrilus planipennis*) or Dutch elm disease (*Ophiostoma novo-ulmi*), would have a high priority. Similarly, annual invasive grasses increase wildfire threat and degrade habitat quality of greater sage-grouse (*Centrocercus urophasianus*) in the bird's western (Washington, Oregon, Idaho) and southwestern ranges (California, Nevada, Arizona, New Mexico), thus posing a significant threat to the ecosystem. With new invasive species, for which we may not have a significant level of knowledge, the threat can be estimated by examining historical data on past impacts of the species elsewhere. For example, we know the potential impacts of buffelgrass (*Pennisetum ciliare*) because Forest Service noxious weed managers have observed its expansion in the Sonoran Desert, and thus we assume that it will behave in a similar manner in other desert ecosystems. Or, in contrast, there may be a species that has displayed a very narrow habitat preference which may be listed as a lower priority because of its more limited threat of spread. Plant invader impacts can now be estimated and ranked from empirical surveys that provide managers with critical information for prioritizing invaders for management action according to their relative impacts on the system (Pearson et al. 2016b). Federal noxious weed lists and State lists can be used to assess priority (USDA NRCS 2016a). In all cases, impacts can and should be measured or considered at multiple levels, with ecosystem transformation being listed as the most severe (see Chap. 2; Barney et al. 2013; Ricciardi et al. 2013).

7.3.2 Prioritizing Communities or Ecosystems for Invasive Species Management Based on Their Value

Unique, highly specialized ecosystems, communities, or even specific sites that provide ecological goods and services used by people or rare wildlife species may rank "high" in a single-factor priority system. For example, the North American Committee on Cooperation for Wilderness and Protected Area Conservation (NAWPA) determined that only 2% of our native grassland ecosystems remain in North America (Davidson 2009). Thus, a high priority may be assigned for invasive species management action in a native grassland ecosystem due to its rarity. Other examples of ecosystems with high ecological or societal value include the Florida Everglades, forests that produce highly valued fungi like morels (*Morchella esculenta*) and white truffles (*Tuber magnatum*) or vegetation such as western huckleberry (*Vaccinium membranaceum*) on the West Coast, or forests that produce ginseng (*Panax quinquefolius*) in the East. An ecosystem may also be prioritized based on its known susceptibility to negative impacts from invaders. A floristically simple ecosystem may be prioritized if the invader is projected to have negative impacts on key species. In the case of some high-elevation subalpine western forests that contain only two or three tree species, the negative effect of invasive species can be magnified. For example, whitebark pine (*Pinus albicaulis*), which is the only five-needle pine in some subalpine forest communities, is a significant mast-producing tree for wildlife forage. However, whitebark pine is susceptible to the invasive white pine blister rust (*Cronartium ribicola*), and thus its negative effect on whitebark pine in this system is significant because no other tree species can compensate for the amount of lost forage for wildlife, including the threatened mainland grizzly or brown bear (*Ursus arctos horribilis*). Therefore, the management of invasive species that threaten whitebark pine may be a high management priority.

7.3.3 Spatial Extent and Temporal Stage of the Invasion Within the Ecosystem Under Consideration

The spatial scale and stage of infestation can affect the outcome of mitigation efforts. Early detection and monitoring have been highlighted as key activities to discover non-native species at the initial stage of invasion, providing an opportunity for rapidly initiating eradication measures and implementing responses to prevent spread and permanent establishment, reducing costs and damage. Based on a review of 53 invasive plant eradication projects in California, Rejmanek and Pitcairn (2002) found that attempts to eradicate invasive weed infestations smaller than 2.5 acres (1 ha) were generally successful, while infestations over 2500 acres had almost no chance of success. For other invader types that are inherently more mobile than plants (e.g., insects, aquatic organisms), the size of the infested area above which eradication may not be possible is likely considerably smaller. Tobin et al. (2013) published a review of over 600 different arthropod eradication programs encompassing 130 species in 91 countries to examine the effect of different factors on success or failure on eradication. They concluded that factors that most strongly influenced success included the size of the infested area, relative detectability of the target species, method of detection, and the primary feeding guild of the target species. The probability of success may be even lower for taxa that are also difficult to detect. Wood-boring beetles, for example, are notoriously difficult to detect since they spend the majority of their life cycle inside their tree host. Typically, these species are not identified until negative impacts on the landscape become widespread and are apparent. In some cases, this awareness can be years following their initial introduction and establishment, thus making eradication attempts challenging. Emerald ash borer was established in southeast Michigan in the early 1990s but was not detected and identified as the cause of extensive ash (*Fraxinus* spp.) mortality until 2002. By 2003, eradication efforts were initiated, but this management strategy was eventually terminated due to the amplified magnitude of the infestation and economic and technological constraints (Herms and McCullough 2014). However, eradication of Asian longhorned beetle (*Anoplophora glabripennis*) was successful from sites in Islip, Manhattan, and Staten Island, NY; Carteret and Jersey City, NJ; Chicago, IL; and Boston, MA, even though establishment of this woodborer likely occurred several years prior to its first US detection in 1996 (Meng et al. 2015). The rapid and coordinated detection and removal of infested trees and effective community outreach and engagement likely influenced the successful eradication of Asian longhorned beetle from these urban areas. However, the more recently detected infestations in Worcester, MA, in 2008, and Bethel, OH, in 2011 may present additional challenges to current eradication efforts since these infestations were likely established for a longer period of time prior to their initial detection. In addition, these infestations are also located within heavily wooded suburban/rural landscapes that are connected to contiguous tracts of eastern deciduous hardwood forests, thereby providing Asian longhorned beetle populations with an abundance of preferred hosts and enhancing the potential for spread (Lopez et al. 2017; Trotter and Hull-Sanders 2015). Spotted lanternfly (*Lycorma deli-*

catula), an exotic species native to Asia, was found infesting tree of heaven (*Ailanthus altissima*) on three residential properties and one commercial property within a 2-mile radius in Boyertown, Berks County, PA, in September 2014 (Parra et al. 2017). The likelihood of eradication is considered to be low pending availability of improved detection methods and availability of new control methods that do not rely as heavily on the use of trap trees and host removal (Parra et al. 2017). Aquatic species that are small in size at some point in their life cycle and may occur anywhere in an aquatic system are also particularly difficult to detect at low population levels. If the establishment of an aquatic invasive species is not detected and acted upon almost immediately, eradication is extremely unlikely (Simberloff 2003). For example, the non-native marine alga *Caulerpa taxifolia* was left untreated when first detected in the Mediterranean Sea near the coast of Monaco. It subsequently spread and now blankets thousands of hectares of coastal substrate in the region, rendering futile any hope of eradication (Meinesz et al. 2001). In contrast, *C. taxifolia* was effectively eradicated in California due to timely identification and rapid implementation of containment and treatment (Anderson 2005). Even when natural resource managers detected populations of aquatic invasive species at low population levels and acted decisively, eradication was successful in relatively few cases and usually at great expense. For instance, the polychaete *Terebrasabella heterouncinata*, a parasite of South African abalones, was detected in California and successfully eradicated by removing 1.6 million of its most preferred and susceptible hosts in the intertidal area (Culver and Kuris 2000). Although zebra mussels (*Dreissena polymorpha*) have been present in the United States for more than 30 years and have continued to spread to new waterways during that time, the only sites from which they have been eradicated are a handful of isolated, abandoned quarries, and only after heavy applications of molluscicide (Strayer 2009). In general, the smaller the infestation and the earlier the stage of invasion, the more likely eradication and mitigation efforts will have a successful outcome.

7.3.4 Goals and Objectives of Potential Invasive Species Management Efforts

It is important to clearly state and establish appropriate goals and objectives for management actions against the invasive species under consideration. For example, eradication of a species or the restoration of the ecosystem or community to both its pre-invasion species composition and functional state may not be possible. If so, decision-making and priority setting would need to incorporate integration of the negative impacts of the invasive species, importance of the invaded community, efficacy of any proposed actions, the actions chosen, and the desired goal or "end state." The goal may then be to build/manage/repair the affected area to a functional resilient state. For example, cheatgrass may always be present in a plant community at some reduced level, such as 10% cover. However, presence of desirable native bunch grasses such as bluebunch wheatgrass (*Pseudoroegneria spicata*) in such a community provides good wildlife habitat, domestic livestock forage, and long-term soil protection. The occurrence of these native grasses with a mix of native forbs and shrubs that existed prior to cheatgrass invasion will yield a resilient landscape that provides multiple benefits in spite of the low-level occurrence of cheatgrass. Management plans aimed at building landscape resilience, decreasing negative impacts, and preventing or slowing establishment into uninfested areas can also be adaptable, especially when developed for well-established species known to have periodically fluctuating population densities. For example, depending on current biotic (e.g., population levels and stand densities) and abiotic (e.g., climate) conditions, management of gypsy moth populations can vary yearly to encompass a variety of techniques including stand thinning, mass trapping, microbial or chemical controls, and detection and monitoring surveys (Schweitzer et al. 2014; Sharov et al. 2002).

7.3.5 Effectiveness of Available Tools or Their Potential for Success

Experiential knowledge, published reports on the effectiveness of available tools or tactics, and online maps are also useful in setting priorities. For example, the USDA Natural Resources Conservation Service has published an interactive map of ecosystem resilience and resistance for the Great Basin ecoregion (Chambers et al. 2014). The map provides an index of relative ecosystem resilience to disturbance and resistance to cheatgrass invasion based on underlying soil, temperature, and moisture regimes. Thus, the most resilient and resistant areas would have the highest potential for successful management and would receive the highest priority in a single-factor system. An overview of categories of tactics or tools and synthesis of recently developed tools are provided in the next section of this chapter.

7.3.6 Integration of Input

In reality, resource managers will seldom be operating in a single-factor priority system. Instead, they will need to integrate all five of the prioritized factors discussed above, with emphasis on (1) the potential tools or techniques that may be used and (2) the ultimate management goal. Potential tools or techniques might be implemented singly or in combinations. Integrated pest management is a site-specific, multi-tactic, decision-making process that optimizes pest control in an economically and ecologically sound manner. Approaches such as regulatory control; education and outreach; physical, cultural, chemical, and biological control; vaccination; host resistance; and control of reproduction may be integrated and consolidated into a unified program. A discussion of these approaches is found in the next section.

7.3.7 Key Findings

- Prioritization of invasive species and the ecosystems at risk is needed in order to allocate limited resources available for their management.
- Input from many factors including the degree of threat of the invasive species, the value of the ecosystem, the spatial extent of the invasion, management goals, and available management tools must be integrated to set priorities and make management decisions.
- Early detection, inventory, and monitoring effectiveness provide the base data needed to analyze threats and define treatments.

7.3.8 Key Information Needs

- Models for analyzing risk and uncertainty to better prioritize management decision-making
- Mechanisms for feedback on the efficacy of management actions for evaluating management decisions and incorporating new information into future actions
- Large scale predictive models on the impact of invasive species on ecosystem changes to estimate if the new invader will dominate the invaded ecosystem, be restricted to microsites, or persist at a lower population level and thereby allow the components of the pre-invasion community to also persist
- Priority-setting models that integrate inputs including ecosystem uniqueness and value, potential invader impacts, management goals, available tools, and probability of success
- Improved tools for early detection and rapid response and guidelines for optimal implementation in time and space to enhance their efficacy

7.4 Recent Advances in Understanding the Biology and Ecology of Invasive Species

More than 450 non-native forest insects (some of which are invasive) (Aukema et al. 2011), at least 197 invasive pathogens of plants and animals (CISEH 2016), over 1600 invasive plants (CISEH 2016), at least 261 species of non-native terrestrial vertebrates (some of which are invasive) (Witmer et al. 2007), and more than 186 species of invasive aquatic organisms (CISEH 2016) are established in the United States. Several invasive species have caused impacts severe enough to inflict heavy damage both economically and ecologically and thus warrant management attention.

Many established non-native species are not economic pests in their native range where they coevolved with natural enemies and, along with host resistance, they typically coexist in equilibrium with native populations. When an invasive species is detected in the United States, little is generally known about its biology in its native range, and even less is known about its ecology, dispersal, and interactions with hosts and the environment, knowledge which is critical to guide management in the introduced range. Fortunately, significant advances have been achieved in understanding the biology and ecology of some of our most damaging invasive insects, pathogens, plants, vertebrates, and aquatic organisms. Additional information about the biology and impacts of damaging invasive species is given in Chap. 2.

7.4.1 Invasive Insects

Some examples of the most threatening invasive forest insects that have become established in North America and either impact trees directly or vector tree diseases are listed in Table 7.1 along with management approaches that will be discussed in Sect. 7.4.2. Native insects such as mosquitoes (Diptera: Culicidae) may also vector invasive pathogens including West Nile virus and Zika virus that impact animals and humans (Fauci and Morens 2016; Reisen 2013).

Understanding the life history of invasive insects is critical for predicting and modeling population growth and spread, timing the application of control tactics to target vulnerable life stages, and directing the location and implementation of survey and management strategies. Basic biology and life cycles have been studied for many invasive insects: environmental factors that influence 1-year or 2-year development and thus affect population growth and spread rates of emerald ash borer (Tluczek et al. 2011); factors that influence development, longevity, and fecundity of Asian longhorned beetle (Keena 2002) and goldspotted oak borer (*Agrilus auroguttatus*) (Lopez and Hoddle 2014); phenology and seasonal flight of redbay ambrosia beetle (*Xyleborus glabratus*) (Hanula et al. 2008); and the complex life cycle of sirex woodwasp (*Sirex noctilio*) and its relationships with and horizontal transmission of different species of mutualistic fungi (Morris and Hajek 2014). Information on the biology, economic impacts (from damage and control), and pest management of spotted lanternfly is currently incomplete for fully informing the feasibility of eradication.

Table 7.1 Examples of significant invasive insects of forest trees and management approaches under development or in operational use[a]

Insect	Scientific name	Major forest hosts	Year of introduction or detection	Management approaches[b]
Emerald ash borer	*Agrilus planipennis* Fairmaire (Coleoptera: Buprestidae)	*Fraxinus* spp.	2002	RC, PC, CuC, CC, BC, HR, IPM
Gypsy moth	*Lymantria dispar* L. (Lepidoptera: Lymantriidae)	Wide host range; preferred genera include *Alnus*, *Fagus*, *Betula*, *Quercus*, *Populus*, and *Salix*	1869	RC, PC, CuC, CC, BC, HR, R, IPM
Hemlock woolly adelgid	*Adelges tsugae* Annand (Hemiptera: Adelgidae)	*Tsuga* spp. (eastern and Carolina hemlocks are more susceptible than western and Asian species)	1951	RC, PC, CuC, CC, BC, HR, IPM
Asian longhorned beetle	*Anoplophora glabripennis* Motschulsky (Coleoptera: Cerambycidae)	Wide host range; preferred genera include *Acer*, *Populus*, *Salix*, and *Ulmus*	1996	RC, PC, CC
Sirex woodwasp	*Sirex noctilio* F. (Hymenoptera: Siricidae)	*Pinus* spp.	2004	RC, PC, CuC, CC, BC
Winter moth	*Operophtera brumata* L. (Lepidoptera: Geometridae)	Wide host range; preferred genera include *Quercus*, *Acer*, *Prunus*, *Tilia*, *Fraxinus*, and *Ulmus*	1930s	RC, PC, CuC, CC, BC
Goldspotted oak borer	*Agrilus auroguttatus* Schaeffer (Coleoptera: Buprestidae)	*Quercus* spp.	1990s	PC, CuC, CC
Balsam woolly adelgid	*Adelges piceae* Ratz. (Hemiptera: Adelgidae)	*Abies* spp.	Around 1900	RC, PC, CuC, CC
Polyphagous shot hole borer	*Euwallacea* spp. (Coleoptera: Curculionidae: Scolytinae)	Wide host range including *Quercus* spp., *Salix* spp., *Platanus* spp., and *Populus* spp.	2003	RC, PC, CuC
Spotted lanternfly	*Lycorma delicatula* (Hemiptera: Fulgoridae)	*Ailanthus altissima* is preferred but feeds on hosts from 20 plant families	2014	RC, PC, CC, IPM

[a]Management approaches listed are not "recommended"; rather, they are a summary of approaches that have been studied and may also be used in some operational invasive species management programs

[b]*RC* regulatory control, *PC* physical control, *CuC* cultural control, *CC* chemical control, *BC* biological control, *HR* host resistance, *R* reproductive, *IPM* integrated pest management

Recent advances in molecular techniques and DNA analysis have been used to identify populations and country of origin for several species, including hemlock woolly adelgid (*Adelges tsugae*) (Havill et al. 2006), emerald ash borer (Bray et al. 2011), gypsy moth (Keena et al. 2008), Asian longhorned beetle (Carter et al. 2010), and sirex woodwasp (Boissin et al. 2012). Identification of the country of origin facilitates exploration for natural enemies, location of genetic material for developing host resistance, and evaluation of control strategies in the native range with well-established populations.

Sophisticated techniques have been developed and used to measure insect flight capacity and spread: harmonic radar for Asian longhorned beetle (Williams et al. 2004); computer-monitored flight mills for sirex woodwasp (Bruzzone et al. 2009), emerald ash borer (Taylor et al. 2010), and Asian longhorned beetle (Lopez et al. 2017); dendrochronology-based models for emerald ash borer (Siegert et al. 2014); geographic-, host-, and environment-based models for spread of hemlock woolly adelgid (Morin et al. 2009) and emerald ash borer (Prasad et al. 2010); and trap-based monitoring for the spread of gypsy moth (Sharov et al. 2002). Understanding dispersal by investigating a species' behavior and physiological limits is critical for establishing quarantine boundaries and determining zones for implementation of control measures. This information is also useful for predicting the spread and subsequent distribution of new populations, thereby improving rapid detection and eradication efforts.

Determining the range of host species and host interactions of invasive insects in the introduced ecosystem is essential for effective management. Invasive insects that reach high densities, which then encounter different tree species within their host genera as well as other genera in their new environment, may respond by colonizing a range of new hosts that are not infested when densities are lower. The emerald ash borer has recently been found to infest a novel host, the native white fringetree (*Chionanthus virginicus*) in North America (Cipollini 2015). The Asian longhorned beetle attacks >100 species of trees but prefers maples (*Acer* spp.), poplars (*Populus* spp.), willows (*Salix* spp.), and elms (*Ulmus* spp.) (Meng et al. 2015); however, susceptibility among poplar species and hybrids varies considerably (Hu et al. 2009). Although all North American species of ash encountered by emerald ash borer to date are susceptible, preferences differ among species and are related to differences in host volatiles, nutrition, and defense compounds (Chen and Poland 2010; Chen et al. 2011). Sirex woodwasp infests a wide range of pine (*Pinus*) species across its global distribution; however, preferences among species are poorly understood because attacked trees are often growing in monocultures (Slippers et al. 2015). Goldspotted oak borer colonizes several species of oaks in California, including coast live oak (*Quercus agrifolia*), California black oak (*Q. kelloggii*), and canyon live oak (*Q. chrysolepis*), as well as other red oak species, but rarely infests white oaks (Coleman and Seybold 2011). Redbay ambrosia beetle attacks redbay (*Persea borbonia*) and several other tree species in the family Lauraceae, including sassafras (*Sassafras albidum*) and avocado (*Persea americana*) (Mayfield et al. 2013). The polyphagous shot hole borer (*Euwallacea* spp.) attacks over 200 species of trees in California including oaks, sycamore (*Platanus occidentalis*), cottonwood (*Populus* spp.), willow, and avocados (Eskalen et al. 2013). The spotted lanternfly is known to feed on plants in more than 20 families; however, the relationship between it and tree of heaven provides an opportunity to reduce tree of heaven populations using a combination of pest population reduction and host removal (e.g., using systemic insecticide treatments and removal or herbicide treatment of some tree of heaven, when appropriate) (Parra et al. 2017).

Significant advances have been made in analytical chemistry techniques for identifying semiochemical attractants including insect-produced pheromones and host kairomones. Semiochemical attractants are used for detecting and monitoring many species of insects which rely heavily on olfactory cues for mate and host selection. For example, the semiochemical lure quercivorol was recently found to be highly attractive to polyphagous shot hole borer and Kuroshio shot hole borer (*Euwallacea* spp.) and is being used to monitor shot hole borer invasions and dispersal (Dodge et al. 2017). Coupled gas-chromatographic-electroantennographic detection has been used to identify male-produced aggregation pheromones that attract both sexes of sirex woodwasp (Cooperband et al. 2012). Identification of insect-produced pheromones has been more challenging for Asian longhorned beetle and emerald ash borer. In these species, host volatiles are considered to be more effective for long-distance attraction and for synergizing attraction to close range or contact pheromones (Crook and Mastro 2010; Nehme et al. 2010, 2014; Ryall et al. 2012). Volatiles from the symbiotic laurel wilt fungus (*Raffaelea lauricola*) synergize host volatiles present in manuka (*Leptospermum scoparium*) oil in facilitating attraction of redbay ambrosia beetle (Kuhns et al. 2014).

7.4.2 Invasive Pathogens of Trees

There have been recent advances in knowledge and understanding of the basic biology, ecology, dispersal, and host interactions of invasive tree pathogens. Examples of some of the most significant diseases caused by invasive pathogens infecting trees and wildlife in North America are summarized in Table 7.2.

Understanding genetics of invasive pathogens aids in accurate identification of causal agents of disease. Multilocus microsatellite genotyping of *Phytophthora ramorum*, the causative agent of sudden oak death (Garnica et al. 2006; Ivors et al. 2006; Prospero et al. 2007), has led to characterization of clonal lineages and their distributions (COMTF 2016). Results are organized in a searchable database that is categorized by three lineages (PRMGP 2016). A previously undescribed and presumably non-native pathogen, *R. lauricola*, the causative agent of laurel wilt that is also a fungal symbiont of the invasive redbay ambrosia beetle, was recently discovered (Fraedrich et al. 2008; Harrington et al. 2008). In addition, other related fungal symbionts of the same insect have been identified and described (Harrington et al. 2010).

Confidence in detection methods used to assess expanding disease distributions and an understanding of dominant modes of pathogen spread also are important in the management of invasive tree pathogens. For example, study of *P. ramorum*-caused disease of tanoaks (*Notholithocarpus densiflorus*) and documentation of disease patterns in the forest landscapes of Oregon led to understanding the correlation between aerial dispersal of inoculum and disease pattern and the lack of correlation with dispersal of inoculum in streams and soil (Hansen et al. 2014). These findings support the continued use of aerial surveys for *P. ramorum* in Oregon's forested landscapes. Investigation of the relative importance of multiple putative insect vectors of *Ceratocystis fagacearum* (the oak wilt fungus), a long-established pathogen, led to the conclusion that two nitidulid beetle species are the principal vectors species of oak wilt (Ambourn et al. 2005; Juzwik et al. 2004), whereas the smaller oak bark beetle (*Pseudopityophthorus minutissimus*) is minimally important in pathogen transmission in Minnesota (Ambourn et al. 2006). Frequencies of pathogen-contaminated nitidulid beetles (*Colopterus truncatus* and *Carpophilus sayi*) present in freshly made wounds, and the nearly immediate arrival of *C. truncatus* to such wounds, have fostered greater adherence for following guidelines for removing recently wilted red oaks (sanitation), disposing of diseased material, and developing harvesting guidelines to reduce the potential for new infections via insect transmission.

Better understanding of spatial patterns of trees that survive disease may indicate environmental differences that affect the pathogen. For example, a recent analysis of surviv-

Table 7.2 Examples of some of the most significant invasive pathogens and associated diseases of trees and wildlife of forests, grasslands, and rangelands and management approaches[a]

Disease	Pathogen/parasite and key insect associates	Major forest/grassland/rangeland hosts	Year of introduction or detection	Management approaches[b]
Tree diseases				
Rapid \`ōhi'a death	*Ceratocystis* A and *Ceratocystis* B	*Metrosideros polymorpha*	2014	RC
Laurel wilt disease	*Raffaelea lauricola* (pathogen); *Xyleborus glabratus* (insect vector)	Lauraceae, e.g., *Persea borbonia*, *Sassafras albidum*, *Litsea aestivalis*, *Lindera melissaefolia*	2003	PC, CuC, HR
Sudden oak death	*Phytophthora ramorum*	*Quercus* spp., *Lithocarpus* spp.	2002	RC, PC, CuC, CC, IPM
Butternut canker	*Ophiognomonia clavigignenti-juglandacearum*	*Juglans cinerea*	1967	PC, CuC, HR
Oak wilt	*Ceratocystis fagacearum* (pathogen); sap beetle vectors (*Colopterus* spp.; *Carpophilus sayi*) and bark beetle vectors (*Pseudopityophthorus* spp.)	*Quercus* spp.	1942	RC, PC, CuC, CC, IPM
Beech bark disease	*Neonectria* spp. (pathogen); *Cryptococcus fagisuga*, *Xylococculus betulae* (scale insects provide entry wound)	*Fagus grandifolia*	~1890	PC, CuC, HR
Dutch elm disease	*Ophiostoma novo-ulmi*, *O. ulmi* (pathogen); *Scolytus multistriatus*, *S. schevyrewi*, *Hylurgopinus rufipes* (insect vectors)	*Ulmus* spp.	~1928 (*O. ulmi*); ~1940 (*O. novo-ulmi*)	PC, CuC, CC, HR
Chestnut blight	*Cryphonectria parasitica*	*Castanea dentata*	1905	PC, HR
White pine blister rust	*Cronartium ribicola*	Five-needle pines, e.g., *Pinus strobus*, *P. albicaulis*, *P. lambertiana*, *P. monticola*	~1900	RC, PC, CuC, HR
Wildlife diseases				
Sylvatic plague	Gram-negative bacterium *Yersinia pestis* (pathogen); fleas on rodents (vectors)	Prairie dogs (*Cynomys* spp.); black-footed ferret (*Mustela nigripes*)	1900	PC, CuC, V
West Nile virus	Arbovirus (*Flavivirus* spp.) (*Flaviviridae*) (pathogen); mosquitos (*Culex* spp.) (vectors)	Wide range of bird species, e.g., bald eagle (*Haliaeetus leucocephalus*), greater sage-grouse (*Centrocercus urophasianus*), western scrub-jay (*Aphelocoma californica*), red-tailed hawk (*Buteo jamaicensis*), great horned owl (*Bubo virginianus*)	1999	PC, CuC, V
White nose syndrome	*Pseudogymnoascus destructans*	Many species of bats, e.g., little brown bat (*Myotis lucifugus*), northern long-eared bat (*M. septentrionalis*)	2006	RC, CuC, PC

[a]Management approaches listed are not "recommended"; rather, they are a summary of approaches that have been studied and may also be used in some operational invasive species management programs

[b]*RC* regulatory control, *PC* physical control, *CuC* cultural control, *CC* chemical control, *BC* biological control, *HR* host resistance, *V* vaccination, *IPM* integrated pest management

ing butternut (*Juglans cinerea*) trees indicated that drier, upland sites were correlated with increased likelihood of butternut survival (LaBonte et al. 2015). These findings suggest the need for further disease assessment of butternut plantings on open, well-drained sites.

7.4.3 Invasive Pathogens of Animals

Research on the biology of invasive pathogens of animals increases our understanding of how they are vectored and might be managed through preventing transmission. West Nile virus is an arbovirus typically vectored by non-native mosquitos (*Culex* spp.). Successful transmission of the virus to uninfected birds depends on the engorged female mosquitos living long enough for virus in the blood to replicate to transmissible levels in their salivary glands. West Nile virus may persist in mosquito hosts through facultative diapause or localized adaptation to overwintering of infected mosquito adults, or through vertical transmission of the virus to F1 progeny (Reisen 2013). Overwintering persistence of West Nile virus in vertebrate hosts has not been confirmed; however, recent research suggests that the house finch (*Carpodacus mexicanus*), house sparrow (*Passer domesti-

cus), and western scrub-jay (*Aphelocoma californica*) could serve as overwintering reservoirs (VanDalen et al. 2013; Wheeler et al. 2012). The source of initial viral infection in uninfected mosquitos in spring has not been conclusively attributed to relapse and recurrence of viral activity and symptoms in persistently infected birds; however, viral transmission to avian predators feeding on West Nile virus-infected live or dead avian prey is possible (Nemeth et al. 2009; Pérez-Ramírez et al. 2014; Reisen 2013).

White nose syndrome is a cutaneous infection of bats caused by the invasive (Leopardi et al. 2015) psychrophilic fungus *Pseudogymnoascus* (formerly *Geomyces*) *destructans* (Gargas et al. 2009; Lorch et al. 2011) (Table 7.2). The disease is named for the white fungus that grows on the muzzles, ears, and wing membranes of infected bats (Blehert et al. 2009). White nose syndrome has led to the local extirpation of bat populations and may eventually cause the extinction of the endangered little brown bat (*Myotis lucifugus*). The six native bat species currently known to be susceptible to infection by *P. destructans* are insectivorous and hibernate under thermally stable, cool, and moist conditions in caves and mines where they congregate to overwinter and effectively reduce their metabolic function during the seasonal absence of food (Blehert and Meteyer 2011). These behavioral and physiological adaptations for overwintering survival, coupled with geographic features and environmental factors (Maher et al. 2012), may explain why white nose syndrome has spread so rapidly. Deleterious physiological (Cryan et al. 2010; Verant et al. 2014; Warnecke et al. 2013; Willis and Wilcox 2014) and behavioral (Brownlee-Bouboulis and Reeder 2013; Wilcox et al. 2014) changes during hibernation have been linked to *P. destructans* colonization of bat dermis and epidermis.

Amphibian chytridiomycosis likely originated from regions of Asia, Africa, and/or Brazil and is caused by the amphibian generalist fungal pathogen *Batrachochytrium dendrobatidis* (*Bd*), which releases aquatic flagellated zoospores (Berger et al. 1998). Infection by *Bd* disrupts cutaneous osmoregulatory function among phylogenetically distant amphibian taxa (Voyles et al. 2009) by inhibiting electrolyte transport across the epidermis, thus causing significant reductions in plasma sodium and potassium concentrations, leading to asystolic cardiac arrest.

A morphologically, genetically, and functionally distinct congeneric species, *B. salamandrivorans* (*Bsal*), likely originating from Asia and first detected in the Netherlands (Martel et al. 2013), has not yet been confirmed as present in the United States (Grant et al. 2016). Since the Eastern United States has the highest diversity of salamanders (Salamandridae) in the world, high-risk areas have been identified (Richgels et al. 2016) and a national response strategy has been developed (Grant et al. 2016).

A greater understanding of the distribution of *Bd* and areas with amphibians at risk of infection with chytridiomycosis is required to effectively manage this disease. Recent analyses of database information accessed from the Global *Bd* Mapping Project indicate that *Bd* was prevalent in many countries, amphibian families, and species and that sites in the montane grasslands and shrublands biome had the highest probability of *Bd* occurrence (Olson et al. 2013). A comprehensive species distribution model for *Bd* in the Americas based on habitat parameters projected higher suitability of *Bd* infection in Western North America (James et al. 2015) than earlier models (Liu et al. 2013; Rödder et al. 2009). The optimal (17–25 °C) and physiological (4–28 °C) temperature ranges for growth of *Bd* ultimately constrain its distribution (Piotrowski et al. 2004), and the probability of *Bd* infection was found to shift between seasons along a latitudinal/precipitation gradient. Prevalence of early-season infections was associated with higher latitudes receiving decreased (recent) precipitation, while late-season prevalence was higher at low elevations receiving increased (recent) precipitation (Petersen et al. 2016).

Recent research has provided insights into mechanisms of amphibian resistance to *Bd* infection. MCH (major histocompatibility complex) alleles encode receptors at the cell surface that are responsible for induction and regulation of acquired immune responses to pathogens. *Bd*-resistant amphibians across four continents share common amino acids in three peptide binding pockets of the MCH class II antigen binding groove (Bataille et al. 2015). Characterizing MCH class II-based resistance in the North American native lowland leopard frog (*Lithobates yavapaiensis*) serves as an important initial step in developing genetically informed breeding programs for amphibian species recovery (Savage and Zamudio 2011).

7.4.4 Invasive Plants

A summary of the most common invasive plants is presented in Table 7.3.

The successful invasion and persistent establishment of non-native plants can frequently be attributed to their mating system (Pannell 2015). Further, mating system plasticity, as exemplified by purple viper's bugloss (*Echium plantagineum*) and yellow star-thistle (*Centaurea solstitialis*), has allowed for some obligate native-range outcrossers to be self-compatible in the invaded range (Petanidou et al. 2012). Mixed-mating systems also exist in invasive plants such as Japanese stiltgrass (*Microstegium vimineum*), with both cleistogamous (obligate selfing) and chasmogamous flowers (may outcross) (Cheplick 2005). It may be possible to manipulate rates of inbreeding depression to reduce the invasion potential of selfing species. Indeed, Japanese stiltgrass

Table 7.3 The most common significant invasive plants in forests and management approaches[a]

Plant	Common name	State and/or Federal regulation (if any)	States in which found	Approx. year of introduction or detection	Management approaches[b]
Acer platanoides	Norway maple	CT; MA	MT; ID; WA; OR; MN; WI; MI; IL; IN; OH; KY; TN; WV; VA; NC; VA; MD; PA; NY; NJ; CT; MA; RI; VT; NH; ME	1756	PC, CC, IPM
Ailanthus altissima	Tree of heaven	CT; MA; VT; NH	All but AK; MT; WY; ND; SD; MN; VT; NH	1784	PC, CC, IPM
Akebia quinata	Chocolate vine	None	LA; MO; IL; MI; IN; OH; KY; WV; PA; GA; SC; NC; VA; MD; DE; NJ; NY; MA; CT; RI	1845	PC, CC
Albizia julibrissin	Mimosa	None	CA; AZ; UT; TX; NM; OK; LA; AR; MO; IL; IN; OH; KY; TN; NC; SC; GA; MS; FL; WV; VA; MD; DE; NY; PA; RI; MA; CT; AL; NJ	1745	PC, CC
Alliaria petiolata	Garlic mustard	AL; CT; MA; MN; VT; NH; OR; WA	WA; OR; ID; UT; CO; KA; NE; OK; AR; MO; IA; MN; WI; IL; IN; KY; TN; OH; MI; WV; PA; GA; SC; NC; VA; MD; DE; NJ; NY; CT; MA RI; NH; VT; ME; AK	1868	PC, CuC, CC
Berberis thunbergii	Japanese barberry	CT; MA; MI	WA; WY; ND; SD; NE; KA; MO; IA; MN; WI; IL; MI; IN; KY; TN; OH; WV; PA; VA; NC; SC; GA; MD; DE; NJ; NY; CT; MA; RI; VT; NH; ME	1875	PC, CuC, CC, IPM
Bromus tectorum	Cheatgrass	CO; CT	All 50 States; not PR or VI	1890s	PC, CC
Celastrus orbiculatus	Oriental bittersweet	CT; MA; VT; NH; NC	AR; GA; SC; NC; TN; VA; WV; KY; IL; IN; OH; WV; MD; DE; PA; NJ; NY; VT; NH; MA; CT; RI; ME	1736	PC, CC
Centaurea diffusa	Diffuse knapweed	AZ; CA; CO; ID; MT; NE; NV; NM; ND; SD; UT; WA; WY; OR	WA; OR; CA; NV; AZ; NM; UT; CO; WY; MT; ID; NE; IA; MO; MI; IL; IN; KY; TN; OH; NY; MA; CT; NJ	1907	PC, CC, BC
Centaurea solstitialis	Yellow star-thistle	AZ; CA; CO; ID; MT; NV; NM; ND; OR; SD; UT; WA	All but AR; LA; MI; AL; GA; HI; AK; PR; VI	1852	BC, PC, CC
Elaeagnus umbellata	Autumn olive	CT; MA; NH; WV	WA; OR; MT; NE; KS; IA; MO; AR; LA; MI; WI; IL; IN; OH; KY; TN; AL; MS; GA; FL; SC; NC; WV; VA; MD; DE; NJ; NY; PA; CT; MA; RI; VT; NH; ME; HI	1830	PC, CuC, CC
Euonymus alatus	Winged burning bush	CT; MA	MT; IA; MO; IL; WI; MI; IN; OH; KY; WV; PA; VA; NC; SC; GA; NY; NJ; DE; MD; VT; NH; MA; CT; RI	1860	PC, CC
Euonymus fortunei	Wintercreeper euonymus	None	TX; KS; MO; WI; IL; IN; MI; OH; KY; TN; MI; AL; GA; SC; NC; VA; MD; PA; NJ; DE; RI; NY; CT; MA; RI	1907	PC, CC
Euphorbia esula	Leafy spurge	AK; AZ; CA; CO; CT; HI; ID; IA; KS; MA; MN; MT; NE; NM; ND; SD; UT; WA; WI; WY	All but TX; OK; AR; LA; MI; AL; KY; TN; NC; SC; GA; FL; HI; PR; VI	1827	PC, CC, BC
Falcataria moluccana	*Albizia*; peacocks plume	None	HI	1917	PC, CuC, CC, IPM

(continued)

Table 7.3 (continued)

Plant	Common name	State and/or Federal regulation (if any)	States in which found	Approx. year of introduction or detection	Management approaches[b]
Fallopia japonica	Japanese knotweed	AL; CA; OR; WA; CT; MA; VT; NH	All but ND; WY; NV; AZ; NM; TX; AL; FL; HI; PR; VI	Late 1800s	PC, CC, IPM
Fallopia sachalinense	Giant knotweed	CA; OR; WA; CT	AK; WA; OR; CA; MT; ID; MN; WI; IL; MI; KY; TN; LA; OH; WV; VA; NC; PA; MD; DE; NJ; NY; CT; VT; MA; RI; ME	Late 1800s	PC, CC, IPM
Ficaria verna	Lesser celandine	CT; MA	WA; OR; TX; MO; WI; IL; IN; MI; OH; KY; TN; WV; PA; VA; MD; DE; NY; NJ; CT; MA; RI; NH	1867	PC, CC
Frangula alnus	European buckthorn	MN; CT; MA; VT; NH	ID; WY; CO; NE; IA; MN; IL; IN; MI; KY; TN; OH; NC; WV; PA; MD; NJ; NY; CT; RI; MA; VT; NH; ME	1864	PC, CC, IPM
Hedera helix	English ivy	OR; WA	HI; WA; OR; CA; ID; UT; AZ; TX; LA; AR; MO; IL; IN; MI; AL; GA; FL; OH; KY; VA; WV; NC; SC; MD; DE; PA; NJ; NY; CT; MA	1800	PC, CC
Hedychium gardnerianum	Himalayan ginger	None	HI	Mid-1900s	PC, CC
Heracleum mantegazzianum	Giant hogweed	Federal noxious weed; AL; CA; FL; MN; NC; OR; SC; WA; CT; MA;VT; NH; PA	WA; OR; IL; WI; NC; PA; NY; CT; MA; ME; MI	1917	PC, CC
Hydrilla verticillata	Hydrilla	Federal noxious weed; AL; AZ; CA; CO; FL; ME; MS; NV; NM; NC; OR; SC; TX; WA; CT; MA; VT	WA; CA; AZ; TX; LA; AR; IA; MS; AL; FL; GA; SC; NC; TN; KY; IN; VA; MD; PA; DE; NJ; NY; CT; MA; ME	1960	PC, CC
Imperata cylindrica	Cogongrass	Federal noxious weed; AL; CA; FL; HI; MN; MS; NC; OR; SC; VT	OR; TX; LA; MS; AL; GA; FL; SC; VA	1912	PC, CC, IPM
Lespedeza bicolor	Shrubby lespedeza	None	TX; KS; IA; MO; AR; LA; IL; IN; KY; TN; MS; AL; GA; FL; SC; NC; VA; WV; OH; PA; MD; DE; NJ; NY; CT; MA; MI	1856	PC, CC
Lespedeza cuneata	Sericea lespedeza	CO; KS	HI; NE; KS; OK; TX; LA; AR; MO; IA; WI; IL; IN; MI; IN; KY; TN; MS; AL; FL; GA; SC; NC; VA; WV; OH; PA; MD; DE; NJ; NY; CT; MA	1896	PC, CuC, CC
Ligustrum obtusifolium	Privet	CT; MA; NH	WA; IA; MO; IL; IN; MI; OH; KY; TN; NC; VA; MD; PA; NJ; NY; CT; MA; RI; NH; VT	1860	PC, CC
Ligustrum sinense	Chinese privet	None	TX; OK; MO; AR; LA; MS; AL; FL; GA; TN; KY; VA; NC; SC; MD; NJ; CT; MA; RI; HI	1852	PC, CC
Linaria dalmatica	Dalmatian toadflax	CO; ID; MT; NV; OR; SD; WY	All but TX; MO; AR; LA; MS; AL; TN; KY; WV; GA; FL; VA; MD; DE; HI; PR; VI	Late 1800s	PC, CC, BC, IPM
Linaria vulgaris	Yellow toadflax	ID; MT; NV; OR; SD; WA; WY; NM	All but HI; PR; VI	Late 1600s	PC, CC, BC, IPM

(continued)

Table 7.3 (continued)

Plant	Common name	State and/or Federal regulation (if any)	States in which found	Approx. year of introduction or detection	Management approaches[b]
Lonicera × bella	Bell's honeysuckle	CT; MA; VT; NH	WA; WY; NM; MN; IA; IL; IN; MI; KY; OH; PA; VA; NC; SC; NJ; MD; NY; CT; RI; MA; NH; VT; ME	Hybrid of *L. tatarica* and *L. morrowii*	PC, CC
Lonicera japonica	Japanese honeysuckle	CT; MA; VT; NH	All but OR; ID; MT; WY; CO; ND; SD; MN; IA; VT; VI	1806	PC, CuC, CC, IPM
Lonicera maackii	Amur honeysuckle	CT; MA; VT; NH	OR; ND; NE; KS; TX; IA; MO; AR; WI; IL; IN; MI; KY; TN; OH; WV; VA; NC; SC; GA; MS; PA; NY; MD; NJ; DE; CT; MA	1855–1860	PC, CC
Lonicera morrowii	Morrow honeysuckle	CT; MA; VT; NH	WY; CO; NM; MN; IA; MO; AR; WI; IL; MI; OH; KY; TN; WV; PA; NY; VA; NC; SC; NJ; MD; DE; CT; RI; VT; NH; MA; ME	1975	PC, CC
Lonicera tatarica	Tatarian honeysuckle	CT; MA; NH; VT	All but NV; AZ; OK; MO; AR; LA; AL; MS; TN; GA; FL; SC; NC; HI; VI	1752	PC, CC
Lygodium japonicum	Japanese walking fern	AL; FL	TX; AR; LA; MS; AL; GA; FL; SC; NC; PA; HI; PR	1930s	CC
Lythrum salicaria	Purple loosestrife	AL; AZ; AR; CA; CO; FL; ID; IN; IA; MI; MN; MO; NV; NM; NC; ND; OH; OR; PA; SC; SD; TN; TX; UT; VA; WA; WI; WY; CT; MA; VT; NH	All but AZ; LA; GA; FL; SC; HI; AK; PR; VI	1800	PC, CC, BC, IPM
Melaleuca quinquenervia	Cayeput; melaleuca	Federal noxious weed; AL; CA; FL; MA; NC; OR; SC; TX; VT	LA; FL; HI; PR	Early 1900s	PC, CC, BC, IPM
Mesembryanthemum crystallinum	Iceplant	None	CA; AZ; PA	Early 1800s	CC
Miconia calvescens	Miconia	HI	HI	1960s	PC, CC, BC, IPM
Microstegium vimineum	Japanese stiltgrass	AL; CT; MA	TX; LA; AR; MO; IL; MS; AL; IN; KY; TN; GA; FL; OH; WV; VA; NC; SC; MD; DE; PA; NJ; NY; CT; MA; PR	1919	PC, CC
Miscanthus sinensis	Chinese silvergrass	CT	CA; CO; LA; MO; IL; KY; MI; MS; AL; GA; FL; SC; NC; TN; WV; OH; PA; MD; DE; NJ; NY; CT; RI; MA	Early 1940s	PC, CuC, CC
Morella faya	Fire tree	HI	HI	1800s	PC, CC, BC
Oeceoclades maculata	Monk orchid	None	FL; PR; VI	1960s	PC
Paulownia tomentosa	Princess tree	CT	WA; TX; OK; LA; MO; AR; IL; IN; KY; TN; MS; AL; GA; FL; WV; VA; NC; SC; PA; MD; MD; DE; NJ; PA; NY; CT; RI; MA	1834	PC, CC
Pennisetum setaceum	Fountain grass	HI	OR; CA; AZ; NM; CO; LA; KY; TN; FL; HI	Early 1900s	PC, CC
Persicaria perfoliata	Mile-a-minute weed	AL; CT; MA; OH; NC; SC	OR; OH; KY; WV; VA; NC; PA; MD; DE; NJ; NY; CT; MA	1930s	PC, CC, BC, IPM

(continued)

Table 7.3 (continued)

Plant	Common name	State and/or Federal regulation (if any)	States in which found	Approx. year of introduction or detection	Management approaches[b]
Phragmites australis ssp. *australis*	Common reed	AL; SC; WA; CT; MA; VT	WA; OR; CA; NV; UT; ID; WY; NE; TX; LA; MN; IA; WI; IL; IN; OH; PA; VA; NC; SC; MD; DE; NY; NJ; CT; VT; NH; MA; RI; ME	1800s	PC, CC
Psidium cattleianum	Strawberry guava	None	FL; HI; PR	1800s	PC, CC, BC
Pueraria montana var. *lobata*	Kudzu vine	FL; KS; KY; MS; OR; WA; MO; TX; CT; MA; IL; PA; WV	WA; OR; NE; KS; OK; TX; MO; AR; LA; IL; IN; KY; TN; MS; AL; GA; FL; WV; OH; VA; NC; SC; MD; DE; PA; NJ; NY; CT; MA	Late 1800s	PC, CuC, CC, IPM
Pyrus calleryana	Callery pear	None	TX; OK; AR; LA; MS; IL; IN; OH; KY; TN; AL; GA; FL; SC; NC; VA; WV; PA; MD; DE; NJ; NY; CT; MA	1908	PC, CC
Rhamnus cathartica	Common buckthorn	IA; MN; MA; VT; NH	CA; ID; UT; MT; WY; CO; ND; SD; NE; KS; MN; IA; MO; WI; IL; IN; MI; KY; TN; OH; WV; VA; NC; MD; PA; DE; NY; CT; RI; VT; NH; MA	Early 1800s	PC, CC, IPM
Rhodotypos scandens	Black jetbead	None	WI; MO; IL; IN; OH; KY; WV; VA; PA; NJ; DE; NY; CT; VT; NH; MA; SC; AL	1866	PC, CC
Rosa multiflora	Multiflora rose	AL; KY; MO; SD; CT; MA; IA; IN; NH; PA; WI; WV	WA; OR; CA; NM; TX; NE; KS; OK; MN; IA; MO; AR; LA; WI; IL; IN; MI; KY; TN; MS; AL; GA; FL; SC; NC; VA; WV; OH; PA; MD; DE; NJ; NY; CT; RI; VT; NH; MA; ME	1868	PC, CuC, CC, IPM
Rubus armeniacus	Himalayan blackberry	OR	WA; OR; CA; NV; AZ; UT; ID; MT; CO; NM; MO; AR; IL; KY; TN; AL; OH; VA; PA; DE; NJ; MA; HI	1885	PC, CuC, CC
Rubus ellipticus	Himalayan raspberry	None	HI	Mid-1900s	PC, CC
Rubus phoenicolasius	Wineberry	CT; MA	MI; IL; IN; AR; OH; KY; TN; GA; SC; NC; VA; MD; WV; DE; NJ; PA; NY; CT; RI; VT	1890	PC, CuC, CC, IPM
Rumex acetosella	Sheep sorrel	AR; CT; IA	All 50 States	1700s or earlier	PC, CC
Schinus terebinthifolius	Christmas berry, Brazilian peppertree	FL; TX	CA; TX; AL; FL; HI; PR; VI	1891	PC, CC, BC, IPM
Senecio jacobaea	Tansy ragwort	AZ; CA; CO; CT; ID; MT; OR; WA	WA; OR; CA; ID; MT; WY; IL; IN; MI; PA NY; NJ; MA; ME	1922	CC, BC, IPM
Senecio madagascariensis	Fireweed	HI	HI	Early 1980s	PC, CC, BC
Syzygium jambos	Rose apple	None	FL; PR; VI	1800s	PC, CC
Taeniatherum caput-medusae	Medusahead	CA; CO; NV; OR; UT	WA; OR; CA; NV; ID; UT; MT; PA; NY; CT	1887	CC
Tamarix ssp. (*T. ramosissima* is one of most common)	Saltcedar, tamarisk	CO; MT; NE; NM; ND; OR; SD; TX; WA; WY	CA; NV; UT; AZ; CO; NM; TX; ND; SD; NE; KS; OK; AR; LA; MS; GA; SC; NC; VA	Early 1800s	PC, CC, BC, IPM

(continued)

Table 7.3 (continued)

Plant	Common name	State and/or Federal regulation (if any)	States in which found	Approx. year of introduction or detection	Management approaches[b]
Triadica sebifera	Chinese tallow tree	FL; LA; MS; TX	CA; TX; AR; LA; MS; AL; GA; FL; SC; NC	Late 1700s	PC, CuC, CC
Triphasia trifolia	Sweet lime	None	TX; FL; PR; VI	Possibly 1950s	CC
Ulex europaeus	Gorse	CA; HI; OR; WA	WA; OR; CA; NY; PA; WV; VA; MA; HI	Late 1800s	PC, CuC, CC, IPM
Vinca minor	Common periwinkle	WI	WA; OR; MT; UT; AZ; NE; KS; TX; MT; IA; MO; AR; LA; WI; IL; IN; MI; KY; TN; MS; AL; GA; FL; SC; NC; WV; VA; MD; OH; PA; DE; NJ; NY; CT; RI; VT; NH; MA; ME	1700s	PC, CC
Wisteria sinensis	Wisteria	None	TX; MO; AR; LA; IL; KY; MI; TN; MS; AL; GA; FL; SC; NC; WV; MD; DE; PA; NJ; NY; CT; MA; VT; HI	1816	PC, CC

[a]Management approaches listed are not "recommended"; rather, they are a summary of approaches that have been studied and may also be used in some operational invasive species management programs

[b]*PC* physical control, *CuC* cultural control, *CC* chemical control, *BC* biological control, *IPM* integrated pest management

does vary its chasmogamous and cleistogamous flower ratios in response to the environment (Cheplick 2005).

Outcrossing breeding systems, including dioecy, may facilitate high genetic variation in invasive plant populations, which in turn increases the likelihood of their successful adaptation to the wide range of novel environmental conditions that may be encountered in the invaded range (Guggisberg et al. 2012). Though only ~7% of all flowering plants are dioecious (obligate outcrossing between individuals that are separate sexes) (Renner 2014), a disproportionate percentage of common US woody invaders are dioecious or partially dioecious, including tree of heaven, common buckthorn (*Rhamnus cathartica*), and Oriental bittersweet (*Celastrus orbiculatus*). The native broadleaf dioecious *Amaranthus* species Palmer amaranth (*A. palmeri*) and common waterhemp (*A. rudis*) have produced herbicide-resistant biotypes (Steckel 2007). Hydrilla (*Hydrilla verticillata*), an invasive non-native aquatic which has both monoecious and dioecious biotypes, has developed herbicide (fluridone) resistance in three of its dioecious phenotypes, all associated with mutations of the *pds* gene (Arias et al. 2005), which indicates a potential capacity for genetic adaptation as well as relationship between dioecy and the propensity for development of herbicide resistance. Dioecy in plants is positively associated with polyploidy, but it remains unclear if polyploidy is the evolutionary cause or consequence (or more likely both) of dioecy (Ashman et al. 2013). Hybridization also leads to production of polyploids (Soltis and Soltis 2009).

It has been hypothesized that hybridization improves invasion success through the generation of novel phenotypes or increased genetic variation (Ellstrand and Schierenbeck 2000; Parepa et al. 2014; Schierenbeck and Ellstrand 2009). Hybridization may occur between two invasive species, sometimes resulting in a hybrid with superior characteristics compared to the parents, or it may occur with a closely related native species, resulting in the loss of native species alleles. Morrow honeysuckle (*Lonicera morrowii*) and Tatarian honeysuckle (*L. tatarica*) as well as Japanese knotweed (*Fallopia japonica*) (typically male-sterile in its invasive range) (Tiébré et al. 2007) and giant knotweed (*F. sachalinense*) are examples of non-native species within the same genus that hybridize. The hybrid for the *Lonicera* species is *L.* × *bella* which does not show any known advantage over the parent species, while the hybrid for the *Fallopia* species, *F.* × *bohemica*, can backcross with its parents and exhibits higher vegetative regeneration than its parents (Bimova et al. 2003) as well as novel secondary metabolites (Piola et al. 2013). Hybridization between several species within the *Spartina* genus has resulted in allopolyploid (having two or more complete sets of chromosomes derived from different species) genomes or hybrid swarms thought to contribute to the success of several invasive *Spartina* species (Ayres et al. 2004). Smooth cordgrass (*Spartina alterniflora*), which is native to the US East Coast but non-native on the West Coast, has hybridized with the native western species California cordgrass (*S. foliosa*), producing hybrids with superior fitness that have spread rapidly through California marshes (Ainouche et al. 2009; Ayres et al. 2008). In contrast, hybridization between Oriental bittersweet and the native American bittersweet (*C. scandens*) has led to significant declines of American bittersweet due to unidirectional pollen flow from the non-native plant and poor seed set of the hybrid, essentially wasting female reproductive effort and eventually alleles of the native (Zaya et al. 2015). Invasive hybrids can invade from their native range, as exemplified by diffuse knapweed × spotted knapweed (*Centaurea diffusa* × *C. stoebe*) (Blair and Hufbauer 2010). Hybrids can also result when species introduced separately from allopatric or only minimally overlapping native ranges hybridize unaided once in close proximity in the invaded range, as is the case for hybrid toadflax (yellow or common toadflax × Dalmatian toadflax) (*Linaria vulgaris* × *L. dalmatica*) (Ward et al. 2009), Bohemian knotweed (*Fallopia japonica* × *F. sachalinense*, also known as *F.* × *bohemica*) (Walls 2010), and saltcedar or tamarisk (*Tamarix chinensis* × *T. ramosissima*) (Gaskin and Schaal 2002).

Potential hybridization between non-native and native species that are similar in appearance, such as Oriental bittersweet and American bittersweet and Japanese angelica (*Aralia elata*) and devil's walking stick (*A. spinosa*), can be easy to overlook (Sarver et al. 2008). An expansion of what was thought to be the native devil's walking stick north of its traditional range has instead been confirmed as the non-native Japanese angelica, which can only reliably be distinguished from devil's walking stick by its inflorescence (Moore et al. 2009). This is another example where an invasive species may be outcompeting other native vegetation. Not only could the formation of new hybrids go undetected, managers could inadvertently remove native species or not treat invasive species because of misidentifications (Verloove 2010). The development of genetic barcodes, possibly involving the ITS2 region for morphologically similar species, may facilitate more reliable identifications in the future (Hollingsworth et al. 2011; Yao et al. 2010).

Though information on seed banks and germination requirements is lacking on many common invasive plants and native plants used for restoration of invaded sites, seed banks of various invasive plants have been estimated to range from none to a year (e.g., tree of heaven (Kostel-Hughes and Young 1998) and Amur honeysuckle (*Lonicera maackii*) (Luken and Goessling 1995)), between 3 and 7 years (e.g., Japanese stiltgrass (Barden 1987) and mile-a-minute weed (*Persicaria perfoliata*) (Hough-Goldstein et al. 2015)), to as long as 20 or more years (e.g., multiflora rose (*Rosa multiflora*) (Kay et al. 1995; Luginbuhl et al. 1999)). The cost implications in restoration involving invasive plants with long-lived (anything more than a year) seed banks are obvious. Plants possessing long-lived seed banks are often associated with more specific germination requirements; therefore, seed bank longevity of most invasive plants is likely to vary with site conditions (Huebner 2011; Kostel-Hughes and Young 1998). Likewise, successful restoration of a site requires knowledge of the site and any existing native seed banks as well as the ability to predict germination rates and future regeneration of the native species that are being reintroduced. Unfortunately, an unintended outcome associated with using commercially produced native seed mixes for restoration purposes has been the inadvertent selection for particular genotypes which can negatively impact the genetic integrity of those species targeted for restoration (Dyer et al. 2016).

Not all discrete populations of a given invasive plant spread at equal rates, and understanding these differences will lead to more strategic management. For example, 33% of garlic mustard (*Alliaria petiolata*) populations have growth rates <1 (Evans et al. 2012). Given the relatively high variation in population growth rates, each invasive plant species and attendant strategy for control need to be considered separately. Ongoing and future detailed population studies of invasive plants that evaluate growth rates under varying biotic and abiotic conditions (topography, climate, soils, other species present) may show that some invasive plants die out on their own under certain conditions. For instance, older populations of Japanese stiltgrass show a decline in survival perhaps due to changes in the root fungal community (Cunard and Lankau 2017). In such cases, it might be more cost-effective to allow declining populations of these plants to go unmanaged, except for monitoring, so that active management efforts can be directed at populations that are expanding or spreading.

Over time, garlic mustard populations were found to select for conspecifics that released lower levels of allelopathic compounds (Lankau et al. 2009). Allelopathic suppression can be amplified in the invaded range, thus affecting overall plant community diversity (Ledger et al. 2015). Though field effects of noted allelopathic invasive plants (including tree of heaven, multiflora rose, Oriental bittersweet, and Japanese stiltgrass) on native species have not been documented (Pisula and Meiners 2010), allelopathic suppression of native species by Bohemian knotweed (the hybrid of Japanese and giant knotweed) was confirmed in field-based assessments in Europe (Murrell et al. 2011). Field-based allelopathic suppression by *Fallopia* spp. was determined to be only a contributing factor in the overall impact of *Fallopia* spp. on other plants in the United States (Siemens and Blossey 2007). The time since invasive plant introduction influences release rates, activity, and persistence of allelopathic compounds in the soil and also determines how frequently this novel weapon becomes consistently repressed among naturalized invasive plant populations of touch-me-not (*Impatiens glandulifera*) (Gruntman et al. 2016).

Evidence for the development of coevolutionary tolerance to allelopathic compounds produced by invasive plants in native plants and soil microbes has been reported for garlic mustard (Lankau 2010, 2012) and tree of heaven (Lawrence et al. 1991). Increasing tolerances may be the result of ameliorating effects of resident soil microbial communities; conversely, native species dependent on soil microbes can suffer negative indirect impacts when these microbes are affected by allelopathy (Cipollini et al. 2012). Evaluating evolving tolerances to allelochemicals under different site conditions might identify those soil microbial communities that are more likely to be protective against allelochemicals versus microbial communities that are more likely to be harmed by allelopathic compounds. Such responses are also likely to vary with the environment (soil type, topography, and climate) and thus be unpredictable and hard to incorporate into management plans.

Changes in soil chemistry can reflect the decomposition of invasive plant leaf litter containing highly concentrated nutrients. For example, tree of heaven is associated with high species richness beneath its canopy (Masaka et al. 2013), and this can be attributed to a facilitative effect of its high-nutrient, rapidly decomposing litter (Gomez-Aparicio and Canham 2008). Nitrogen-fixing species such as Russian and autumn olive (*Elaeagnus angustifolia* and *E. umbellata*) alter soil properties on marginal sites through the action of symbiotic actinorhizal associations that increase soil nutrients (DeCant 2008; Funk et al. 1979). One invasive species, Japanese stiltgrass, may indirectly facilitate the success of another invasive species, garlic mustard, by suppressing other plant species and thus increasing light availability (Flory and Bauer 2014). Increases in soil nutrients from invasive plants might facilitate their regeneration as well as native species leading to co-existence and changes in overall species composition in the community, instead of competitive exclusion of one or more native species.

Plant invasions alter species-area relationships such that larger invaded areas tend to have higher richness values than occur in uninvaded sites of similar size (Powell et al. 2013). This is due to a disproportionately greater impact of invasive plants on the abundance of common rather than rare native species. This finding supports the need to better understand how invasive plant species change plant community composition, rather than focusing on the extinction of native plants attributable to invasive plants (Gurevitch and Padilla 2004). The impacts of changing plant community compositions due to invasive plants are also evident at other trophic levels. For example, the abundance of native insectivorous and nectarivorous bird species decreased where the urban streetscape is dominated by non-native plants (White et al. 2005). Also, native insects have been shown to be in lower abundance on non-native plants than native plants (Zuefle et al. 2008).

Several invasive plants are known to have pathogens that impact their fitness. These include *Verticillium nonalfalfae* on tree of heaven and *Bipolaris* spp. on Japanese stiltgrass. Other pathogens do not appear to result in high mortality but help control the host plant species; these include powdery mildew (*Erysiphe cruciferarum*) on garlic mustard, rose rosette disease on multiflora rose, and soybean rust (*Phakopsora pachyrhizi*) on kudzu (*Pueraria lobata*) (Flory and Clay 2013). In addition, many invasive plants show signs of herbivory caused by several species of insects or other invertebrates. For example, the Ailanthus webworm can cause extensive defoliation of the invasive tree of heaven. However, many invasive plants demonstrate greater tolerance toward generalist herbivores than associated native plant species (Jogesh et al. 2008). A better understanding of the collective role of pathogen and predator species accumulation on invasive plants may assist managers in predicting if and when certain invasive plants could become less of a threat. It has been suggested that many of these "volunteer" pathogens might be suitable for use as biocontrol agents. However, the potential impact of these organisms on native species needs to be determined before they can be considered for release.

The claim that invasive plants are successful invaders because they are more plastic may only be true in the initial stages of an invasion after which selection for optimal phenotypes is likely (Palacio-López and Gianoli 2011). Thus, attempting to define general traits that determine invasiveness is not likely to be productive, because such traits will vary with the environment and the stage of invasion. Competing and/or facilitative co-occurring plants, herbivores, pathogens, and symbionts may or may not co-migrate in response to a changing environment (Van der Putten et al. 2010). Models that incorporate these interactions will better predict future invasive plant distributions in response to global change. For example, tree of heaven has shown signs of evolving since its invasion, evidenced by the fact that its current range has expanded beyond the climatic range predicted from its native distribution (Albright et al. 2010). Range expansion is likely to be common among invasive plant species. Understanding the length of time required for range expansion to develop, what species interactions may be linked to this expansion, and the ability of an invasive plant to evolve into a new range in response to a changing climate will improve our success in managing current and future plant invasions.

Some invasive plants facilitate the presence of other invasive plants. The term "invasional meltdown" has been used to describe sites that are composed of invasive species that facilitate each other's sustained presence at an increasing rate of establishment (Green et al. 2011; Rodriguez-Echeverria 2010; Simberloff 2006). Such complex interactions across the same and different trophic levels may make it exceptionally difficult to restore plant communities that are predominantly composed of invasive species. There may be cases where manipulation of one invasive species, which serves as an analogous "keystone" species, may improve the likelihood of restoring native species because removal of the former may have a domino effect on other invasive plants.

Outcomes of these varied considerations are additional inputs required for establishing priorities in invasive species management (Box 7.2).

7.4.5 Invasive Terrestrial Vertebrates

While a system-wide approach is often most desirable for managing invasive species (Mack et al. 2000), the most impactful invasive vertebrates represent diverse taxa that each require specific approaches, thus making it difficult to render generalizations about their management and control. Strategies often must be tailored to specific attributes of species' natural history or behavior. Some of the most threatening invasive terrestrial vertebrates are presented in Table 7.4.

The natural history of most vertebrate species has been well-understood for many years. Some invasive vertebrate groups were intentionally introduced and have been well-established for many years (e.g., horses (*Equus caballus*), cats (*Felis catus*), and swine (*Sus scrofa*)); however, feral populations cause substantial economic and ecological damage. Most recent work has been directed at either understanding impacts or testing control technologies such as evaluating various chemical or fertility control drugs. Recent advances in understanding the ecology of certain species and in improving the technological capacity of traps have enhanced the efficacy of trapping as a control tool. For example, Sparklin et al. (2007) reported evidence that matriarchal social groups of feral swine ("sounders") are territorial (i.e., they defend a home range against other sounders). This means that more effective control may be achieved on large tracts by trapping whole sounders while systematically moving across the area (Sparklin et al. 2007). The mechanics of whole-sounder trapping have recently been facilitated by technological advances in trap designs. Historically, feral swine were captured in small corral traps with a gate that closed when a pig feeding on bait in the trap hit a tripwire or similar trigger mechanism. Any pigs not yet inside the enclosure when the gate closed were not only not captured, they also became educated as to the danger of the trap and were much more difficult to capture in future encounters. Traps now available from several manufacturers employ motion-sensing cameras and cellular technology to either send photographs (via text message) or livestream video of pig activity in the trap. Trappers then can activate the trap with a wireless command whenever they choose to do so. Thus, the trapper controls activation of the trap, not the pigs. Additional research is needed to confirm both the degree of territoriality in feral swine under variable resource conditions and the effectiveness of whole-sounder trapping.

Table 7.4 Significant invasive vertebrates and management approaches[a] in the United States

Common name	Scientific name	Management approaches[b]
Rats	*Rattus* spp.	PC, TS, CC
House mouse	*Mus musculus*	PC, TS, CC
Nutria	*Myocastor coypus*	TS, CC
Feral cat	*Felis catus*	PC, TS, R
Feral horse	*Equus caballus*	PC, TS, R
Feral swine	*Sus scrofa*	PC, TS, R
Rock pigeon	*Columba livia*	PC, TS, R
European starling	*Sturnus vulgaris*	PC, TS, R
House sparrow	*Passer domesticus*	PC, TS
Nile monitor	*Varanus niloticus*	TS
Burmese python	*Python molurus*	TS
Brown treesnake	*Boiga irregularis*	TS, CC

[a]Management approaches listed are not "recommended"; rather, they are a summary of approaches that have been studied and may also be used in some operational invasive species management programs
[b]*PC* physical control, *TS* trap/shoot, *CC* chemical control, *R* reproduction control

7.4.6 Invasive Aquatic Animals

Invasive species scenarios can unfold rapidly in aquatic systems through a variety of transmission pathways and can result in ecosystem-altering effects (Penaluna et al. 2017). Aquatic invasive species are in a very high rate of flux, with many taxa in apparent early stages of invasion, many being highly managed to prevent their attaining a foothold in regional waters, and others having been more established and expanding their distribution. Some of the most damaging invasive aquatic organisms are listed in Table 7.5.

Research is progressing to understand ecosystem responses to aquatic invasive species, including food web alterations, and the ecological potential and ramifications for biotic homogenization. Warmwater aquatic invaders appear to have become more established in many ecosystems (Sanderson et al. 2009); therefore with projections of climate change, the distributions of many warmwater species are expected to increase in certain regions (e.g., with latitude and elevation) (Perry et al. 2005). Current modeling efforts are directed at identifying cool-water refuges for native aquatic species that are unlikely to be subject to such invasions (USDA Forest Service 2017) and examining climate niches of aquatic invasive species known to have adverse effects on native systems. These activities should aid managers in establishing monitoring priorities (e.g., Olson et al. 2013).

7.4.7 Key Findings

- Many non-native species established in the United States are not economic pests in their native range, and, initially, little is known about their biology, ecology, host interac-

Table 7.5 Significant invasive aquatic organisms and management approaches[a] in the United States

Common name	Scientific name	Management approaches[b]
Asian carp	*Hypophthalmichthys nobilis* (bighead carp); *Mylopharyngodon piceus* (black carp); *Ctenopharyngodon idella* (grass carp); *Hypophthalmichthys molitrix* (silver carp)	RC, PC
Spiny waterflea	*Bythotrephes longimanus*	RC
Sea lamprey	*Petromyzon marinus*	RC, CC
Zebra mussel	*Dreissena polymorpha*	RC, CuC
Chinese mitten crab	*Eriocheir sinensis*	RC, PC
New Zealand mud snail	*Potamopyrgus antipodarum*	RC, PC, CuC, CC, BC

[a]Management approaches listed are not "recommended"; rather, they are a summary of approaches that have been studied and may also be used in some operational invasive species management programs
[b]*RC* regulatory control, *PC* physical control, *CuC* cultural control, *CC* chemical control, *BC* biological control

tions, and ecological and economic impacts nor management.

- There has been considerable progress in understanding the life cycle, genetics, host range, dispersal, semiochemical communication, and host interactions of many significant invasive insects, diseases, plants, vertebrates, and aquatic organisms.
- There have been recent advances in developing molecular techniques that are needed to identify new invasive species, characterize their lineages and distributions, and determine their country of origin.
- Significant advances have been made in analytical chemistry for identifying semiochemical attractants for insects that can be used for detection and monitoring.
- Sophisticated methods have been developed and used for determining rates of spread of invasive pests including the use of insect flight mills, aerial survey for inoculum and disease spread, and dispersal models for invasive pests.
- Progress has been made in understanding environmental and spatial factors that influence the risk of invasion.
- Significant advances have been made in understanding mechanisms of plant and animal resistance to diseases.
- Progress has been made in understanding plant mating and breeding systems, including how dioecy and hybridization influence genetic variation and resistance.
- Knowledge has been gained on the impact of allelopathic compounds produced by invasive plants and how they influence native plants and soils.

7.4.8 Key Information Needs

- Understanding interactions between native and invasive species, elucidating impacts of invasives on the abundance of native species and the ecosystem affected, and ascertaining if coexistence can be facilitated and is an evolutionary stable strategy
- Additional information on the accumulation and impacts of native and introduced natural enemies on invasive insects and plants
- Greater understanding of the etiology and epidemiology of emerging disease complexes
- Information on polyploidy, dioecy, and mating systems of plants necessary to manipulate reproduction, inbreeding, timing of control, and herbicide resistance
- Greater understanding of how environmental factors and soil microbial communities influence allelopathic compounds released from invasive plants and how these compounds impact native plants
- Greater understanding of how characteristics of population ecology and behavior of vertebrate invasive species can be utilized to improve their control

7.5 Approaches to Management of Invasive Species

Approaches to manage invasive species and specific control tactics used (Fig. 7.1) may vary by the general type of invading organism, whether in terrestrial or aquatic ecosystems (see Table 7.6). Some management approaches are effective for several taxa in different ecosystems, while others are practical only for certain kinds of invasive species. For instance, outreach and education are critical components of management programs for all invasive species taxa. Similarly, the application of pesticides is used across all invasive species taxa. Regulatory control is used for invasive insects, pathogens, terrestrial wildlife, and aquatic organisms, but is not as effective for control of invasive terrestrial or aquatic plants. Biological control using natural enemies is used operationally primarily for both invasive insects and plants. Vaccination control is only practical for protecting wildlife from invasive pathogens and is not used operationally for management of other invasive taxa. Similarly, reproduction control is used primarily for invasive insects or wildlife and has limited practicality for managing invasive pathogens or plants. Developing hosts resistant to invasive pests is a management approach used against invasive insects and pathogens, but is not practical for managing plants or wildlife that invade ecosystems rather than specific host trees. Table 7.6 provides a general overview of management approaches that are operationally used for each invasive taxonomic group. A more detailed discussion of management approaches and

Table 7.6 Operationally used invasive species management approaches by invader type[a]

Invader type	Regulatory	Outreach and education	Physical	Cultural	Chemical or pesticidal	Biological	Vaccination	Host resistance	Reproduction	IPM[b]
Insects	X	X	X	X	X	X		X	X	X
Pathogens of trees	X	X	X	X	X			X		X
Pathogens of wildlife	X	X	X	X	X		X			
Terrestrial plants		X	X	X	X	X				X
Aquatic plants		X			X					X
Terrestrial wildlife	X	X	X		X		X		X	
Aquatic organisms	X	X	X	X	X		X		X	

[a]Management approaches listed are not "recommended"; rather, the table presents a summary of management approaches that are currently used
[b]IPM = integrated pest management

examples of how they are used for specific invasive species or taxonomic groups is provided in subsequent sections.

Selection of the optimal combination of management approaches and tactics depends on the management goals and objectives, economic and ecological priorities, the invading species and its impacts, available tools, and ecosystem interactions.

7.5.1 Regulatory Control

Regulatory control is a society-based strategy designed to exclude or monitor pathways that are available for introducing an invasive species into a suitable habitat or ecosystem. State and Federal seed certification programs provide verification that seeds are free of pests such as insects, plant pathogens, and even unwanted plant seed (AOSCA 2016). Plant quarantines established by the Federal Government, States, or other countries restrict the movement of items that may result in movement of unwanted invasive species within or among states, or into the United States (see Chap. 6). Best management practices (see Sect. 7.5.4 below) developed by users or commodity groups often become regulatory controls (USDA NAL 2016).

Implementing regulations is one of the first and most effective measures available to reduce further spread and impacts of established populations of invasive pests (Liebhold et al. 1992; Vander Zanden and Olden 2008). Several significant invasive insects are regulated by Federal or State quarantines that restrict movement of the insects and any potentially infested host material including firewood, nursery stock, logs, chips, and cuttings. Other examples of regulatory control measures include inspection and removal of gypsy moth egg masses from outdoor household articles transported out of regulated areas (USDA APHIS 2010) and removal and destruction by chipping, grinding, or burning of trees infested by emerald ash borer (Poland and McCullough 2006), hemlock woolly adelgid (NHDFL 2015), and Asian longhorned beetle (Meng et al. 2015). Research is being conducted on new regulatory treatments for infested wood, including vacuum or microwave treatment (see Chap. 6). Adaptations of regulatory policies to cover soil potentially infested with fire ants, root-knot nematode, and other pests such as invasive earthworms have been researched and are being proposed to prevent their introduction (Callaham, Jr. et al. 2006). The Pennsylvania Department of Agriculture established an internal quarantine restricting the movement of many commodities to evaluate options for managing spotted lanternfly following discovery of the insect.

Regulatory approaches are perhaps more commonly used for most recent introductions of invasive species (e.g., sudden oak death, rapid `ōhi'a death (*Ceratocystis* A, *Ceratocystis* B), and spotted lanternfly) rather than for long-established species (e.g., Dutch elm disease, white pine blister rust) that are widely distributed.

Many invasive plants are so widespread that regulatory control is impractical. Consequently, only 4 of the 79 common invasive plants in the United States listed in Table 7.3 are also listed as Federal noxious weeds (USDA NRCS 2016a). According to the Federal Noxious Weed Act, if species included on the Federal Noxious Weeds List are discovered, active management and adequately funded management programs must be implemented against them (FNWA 1974). Individual States vary greatly in how they categorize invasive plants and how they define those categories. Many States prohibit planting or selling the species, but do not require their active management. They may refer to these species as invasive rather than noxious. However, some States also use the term "noxious," but use of the term only prohibits seed of these species from contaminating other seed supplies and does not require their removal or management. California, Massachusetts, Oregon, and Washington have more comprehensive lists of noxious weeds than those occurring in most of the other States (USDA NRCS 2016b), but this does not correlate with the occurrence of more invasive plant problems in other States. Indeed, Maryland has one of the shortest lists of noxious weeds (USDA NRCS 2016b), yet has a very active invasive species council that provides a much longer list of invasive species of concern (with no regulations associated with them); a similar situation exists in several other States. Consistency among the States on how they define and code regulations would assist scientists in evaluating the effectiveness of such regulations.

Range expansion of feral swine has occurred because swine are transported illegally and intentionally released in new areas in order to establish populations suitable for hunting (Mayer and Brisbin 1991). In response, several States have established or are considering tighter regulations for controlling the transport of swine as well as educating hog hunters.

Pet trade and quarantine regulations may aid in preventive management of animal diseases such as amphibian chytridiomycosis, a disease caused by a fungal pathogen (Liu et al. 2013; Stokstad 2014). The US Fish and Wildlife Service published an interim rule declaring 201 non-native salamander species as injurious wildlife under the US Lacey Act (Federal Register Docket No. FWS-HQ-FAC-2015-000; GPO 2017; USFWS 2016). A petition to include all live amphibians in trade as injurious unless certified as free of the causative fungal pathogen is also under review.

Some aquatic invasive species have status listings as nuisance or injurious species because they pose adverse effects on native ecosystems or local economies; some of these species are being targeted for significant regulatory control efforts (see Penaluna et al. 2017). The USDA highlights some aquatic nuisance species (USDA NAL 2017), and the US Fish and Wildlife Service identifies several aquatic species as injurious and covered under the Lacey Act (GPO 2017; USFWS 2017b). Aquatic invasive species trigger cross-jurisdiction considerations due to the occurrence of common waterways. Significant resources are expended to control some aquatic invasive species. For example, in the Pacific Northwest, invasive species councils are well-established in British Columbia, Canada, and the US States of Alaska, Oregon, and Washington, and each council addresses both pathways of spread and control of infestations. In Washington State, the 2015 Report to the Legislature (WDFW 2015) reported results from 2011 to 2013, which included (1) >27,000 boat inspections, with decontamination of 83 boats that contained aquatic invasive species, of which 19 boats had zebra or quagga mussels (*Dreissena bugensis*), and (2) 6 new infestations of New Zealand mud snails (*Potamopyrgus antipodarum*). Despite region-wide efforts, some species are recognized as requiring continuous management, whereas for others, effectiveness of control methods is poor. As a result, some invasive species seem to be fully established.

Purposeful introductions of non-native aquatic species has been a norm throughout the world, because these actions have served to promote ecosystem services of local societies, such as food provisioning, recreation, pest control, and general well-being. For example, the introductions of the European brown trout (*Salmo trutta*) into South America (Soto et al. 2006) and New Zealand (Townsend 1996) and the eastern mosquitofish (*Gambusia holbrooki*) into Australia (Hamer et al. 2002) were intentional. However, we now know that these introductions have subsequently caused major reductions in native fauna, including fish, amphibians, and invertebrates (Townsend 1996; Wissinger et al. 2006). Similar adverse effects on native amphibians and other ecosystem components have been exacerbated by fish-stocking practices common throughout the United States (see reviews by Dunham et al. 2004; Kats and Ferrer 2003). Throughout the Western United States, forested lakes at higher elevations that were devoid of fish have been intentionally stocked with game fishes such as the brook trout (*Salvelinus fontinalis*), which is native to Eastern North America (Macneale et al. 2010).

In the United States, States have key jurisdiction over fish-stocking practices, and their practices have come under increasing scrutiny because State stocking of fish occurs in wildlands and affects native species including some that are threatened or endangered (T&E). For example, in California, a court-ordered moratorium was placed on non-native fish stocking of numerous water bodies due to risks to native species including those listed T&E species (CDFW 2010). Additionally, recovery efforts for the California golden trout (*Oncorhynchus mykiss aguabonita*) in the Kern Plateau of the Sierra Nevada range have included not only a moratorium on non-native fish stocking but also additional efforts to remove non-native trout (which hybridize with or prey on golden trout) by using chemical treatments and installing barriers to prevent upstream migration into golden trout habitats (Pister 2008). Stocking practices continue in other areas, however, due to the strong desire to provide positive fishing experiences. Additionally, modifying fish harvest regulations designed to increase predation pressure on invasive aquatic species such as the Asian carp species (Cyprinidae) (Hein et al. 2006; Tsehaye et al. 2013) has been effective at reducing the abundance of some invasive aquatics.

7.5.2 Outreach and Education

Besides legal or regulatory approaches, human behavior can also be influenced to manage invasive species through efforts involving outreach and education. These efforts are conducted to increase awareness of quarantines and regulations, reduce introduction and spread of invasive species, aid in identifying and reporting new detections, facilitate rapid response, and enhance support and successful implementation of control tactics. Education and outreach are components of management responses at all stages of the continuum from strategies to approaches and to tactics (Fig. 7.1). This extension of information contributes to the success of programs for prediction and prevention of invasive species, early detection, rapid response, management, and restoration. Informing and engaging the public on invasive species issues is one of the first opportunities to disrupt the progression of invasions by preventing the introduction and spread of non-native species (see Chap. 6). In managing established populations of invasive species, education and outreach tools are used across all taxonomic groups. Outreach and education can be accomplished in a myriad of ways using various media outlets to increase understanding of controversial issues that may surround approaches to manage invasive species and to increase public support for management programs and enhance their success. These engagement and communication efforts are critical to the success of cooperative management programs. While the overall management approach and optimal combination of tactics are specific to particular invasive species and ecosystems, public outreach and education are important components of management across all categories of invasive species.

Invasive species management programs can be controversial and in some cases have been delayed or halted because of opposition from organized groups (Warner and Kinslow 2013). For example, criticism and petitions from community members over aerial applications of pheromones for the eradication of light brown apple moth (*Epiphyas postvittana*) in California led to the eventual suspension of the aerial spray program and an overall change in management tactics (Ben-Haim et al. 2013). Public support can be critical to the success of management projects, and understanding the underlying attitudes of the public can help in development of outreach education activities. The level of support for control and eradication programs was generally higher among people who had prior knowledge of control and eradication projects and members of conservation organizations, indicating the important role of awareness and education in increasing public support for invasive species management projects (Bremner and Park 2007). The Pennsylvania Department of Agriculture implemented an extensive and effective outreach program to inform affected citizens and businesses on available detection, identification, and control methods of spotted lanternfly. This program engages property owners in egg scraping, tree banding, trap tree establishment, host removal to kill all life stages, and reporting any findings (PDA 2019).

Public outreach in practice contributes to the success of regulatory control programs by enhancing awareness of quarantines and compliance with regulations (Peterson and Diss-Torrance 2012; Warner and Kinslow 2013). For example, major campaigns have been implemented that included radio and television advertisements, billboards, bumper stickers, and social media to stop the movement of firewood that may harbor invasive insects and pathogens (Poland and McCullough 2010). Pest alerts, brochures, identification cards, "wanted" posters, doorknockers, fliers, and identification kits containing pest and damage specimens have been distributed with the intent to educate the public to be on the lookout for and to report major insect pests such as the Asian longhorned beetle and emerald ash borer (Haack et al. 2002).

Cooperative Weed Management Areas (CWMA), Partnerships for Regional Invasive Species Management (PRISM), and Invasive Plant Partnerships (IPP), all hereafter referred to as CWMAs, have become common in many States, with some States, like New York, dividing the entire State area into CWMAs. Some CWMAs are actually CWPMAs (Cooperative Weed and Pest Management Areas) and incorporate all invasive pests and not just plants as a focus. All of these CWMAs are partnerships of Federal, State, and local government agencies, tribes, individuals, and interested groups who manage invasive species in a defined area (MIPN 2016). The goal within these management areas is to engage all private and public landowners across a common landscape or watershed to enhance cooperation and communication and facilitate more effective management of invasive species (CISM 2016). Outreach and education are essential activities that can promote effective cooperation and communication of science-based approaches that are included in CWMA management plans.

Invasive vertebrates are perhaps unique in that they include species which, though highly impactful ecologically, enjoy considerable support among the public (Witmer et al. 2007) which often objects to their lethal control, no matter how humane the euthanasia may be or how destructive the damage is (Simberloff 2014). For example, feral horses and feral cats can be removed relatively easily as compared to many other species, but objections based on ethical, emotional, cultural, and historical grounds require that alternative approaches are used such as fertility control (Kirkpatrick et al. 1982). Unfortunately, while fertility control may limit population growth, it often permits existing problems to persist. Control of vertebrate invasives is further complicated by the fact that the public generally does not distinguish between native and non-native animals, particularly if they fail to experience immediate or direct impact from damage (Witmer et al. 2007). Since most introductions of non-native vertebrates are a result of anthropogenic activity, educational and regulatory approaches to prevent such introductions are critical. For example, many reptiles and amphibians are now established in Florida as a result of accidental or intentional releases of pets (Reed 2005). While increased regulation of the pet trade may help in preventing future releases, education of the public is also important. The Florida Fish and Wildlife Conservation Commission conducts an annual Python Challenge (FFWCC 2016), a competition in which members of the public compete to capture (and remove) the most and largest Burmese pythons (*Python molurus bivittatus*). However, the objective of the challenge is not focused on population control but rather to direct attention to the issue of invasive pythons and consequently enhance public awareness and participation (Dorcas et al. 2017).

Public outreach and education are also important in the management of invasive aquatic animals. This may include posting signs at vulnerable areas, providing brochures that explain the issues, initiating programs in schools to educate youth, launching workshops at environmental centers to inform the public, and encouraging professional aquatic biologists to spread the word through face-to-face encounters with the public. For example, the "Don't Turn it Loose" brochure, produced by Partners in Amphibian and Reptile Conservation, is available online and has been distributed nationwide (PARC 2016). These efforts are linked to cultural changes involving past practices that previously were not known to cause damage to native ecosystems. In addition, continuing education programs and online training courses are available to assist professionals in learning about newly developed approaches (USDA NRCS 2016c). Online outreach tools are continually being developed and incorporate rapid response plans, which are critical in preventing non-native aquatic species introductions.

Humans are responsible for much of the overland movement of aquatic animals among waterways (Buchan and Padilla 1999). Consequently, management activities intended to stop the movement of invasive aquatic animals must incorporate efforts to influence and manage human behavior. Research in the field of social science is investigating the influence that outreach, education, and law have on risky behaviors associated with the spread of invasive aquatic animals. High-risk behaviors include releasing unwanted pet fish or unused live bait fish into the wild (Drake and Mandrak 2014) or failure to decontaminate boats and gear when moving from one body of water to another (Puth and Post 2005; Rothlisberger et al. 2010). Multiple studies indicate that, to date, efforts have been insufficient to reduce high-risk behaviors to acceptable levels (Drake et al. 2015; Kilian et al. 2012; Nathan et al. 2014; Prinbeck et al. 2011).

Live bait used for fishing has been linked to introductions of both non-native species and diseases (e.g., Ranavirus on salamanders in the bait trade; Picco and Collins 2008). Similarly, with the increase in residential water gardens and non-native pets, a variety of aquatic invasive species can be spread inadvertently (Keller and Lodge 2007). The pet trade along with schools (biological laboratory classes) have also been sources for introduction of non-native aquatic species (Larson and Olden 2008). Possibly a combination of such practices has led to the introduction of nine non-native freshwater turtles to Hawaii, where effects on native biota are increasing as aquatic communities are changing. In conclusion, there has been increased awareness of the adverse effects associated with releases of non-native species. Hopefully, education and outreach campaigns will be effective in forestalling these practices in the future.

7.5.3 Physical Control

Physical control is defined in this chapter as either mechanical methods of physically removing invasive species (e.g., hand-pulling small invasive plant infestations before flowering, mowing plants, trapping and removing or shooting terrestrial invasive animals) or physically precluding them from areas. Erecting barriers to prevent invasive species access to a protected stand, ecosystem, or water course is another example of controlling spread.

Controlling insects by physically removing them or using barriers to prevent their movement is difficult and often impractical due to the small size and cryptic nature of many species, large population numbers, and the ability to disperse rapidly over great distances by flight. At the individual tree or local level, burlap bands, sticky bands, glues, oils, or grease applied to the stems of trees may be used as physical barriers to prevent movement of caterpillars such as gypsy moth (Thorpe and Ridgway 1994) that crawl up from the ground to infest the tree canopy. Invasive insects such as Asian longhorned beetle may also be controlled by physically removing and chipping or burning infested trees (Meng et al. 2015). Physical control is the primary method employed to control spotted lanternfly. The trap tree/host removal method shows that populations are significantly reduced in the areas where tree removal and trap tree treatments have been conducted (Parra et al. 2017).

Several invasive plants are controlled by combining a physical control (fire, cutting, girdling, or mowing) with a chemical application. In many cases, cutting or top-"killing" (e.g., after a burn) the plant often leads to prolific sprouting or root suckering. However, a few invasive plants can be controlled on a local level by physical control alone. Examples include garlic mustard (a biennial herb) via hand-pulling before seed production (Chapman et al. 2012) and Japanese stiltgrass (a shade-tolerant annual grass; Shelton 2012) and tall oatgrass (*Arrhenatherum elatius*, a perennial grass) (Wilson and Clark 2001) via intensive (very close to the ground) mowing prior to seed production. Conversely, whereas mowing has little effect on yellow bluestem (*Bothriochloa ischaemum*, a perennial grass of Texas prairies), a growing-season fire significantly reduces its abundance (Simmons et al. 2007).

Exclusion of invasive animals, particularly ungulates including swine and horses, via fencing has been used to mitigate damage and protect sensitive resources. Likewise, rodent damage can be minimized by using a combination of exclusion, sanitation, and habitat manipulation (Witmer et al. 2007). The use of sound devices to frighten wildlife tends to be effective only for limited periods because animals habituate to the stimulus (Bomford and O'Brien 1990). However Mahjoub et al. (2015) reported that a nonlinear ultrasonic parametric array effectively created a sonic net that repelled European starlings (*Sturnus vulgaris*). Repellents such as taste-aversive agents have been explored with varied success, mainly for use against ungulates including feral swine and horses. However, the benefits of such physical approaches are localized spatially and temporally and may serve to defer or shift the damage.

Lethal control, exemplified by trapping or shooting, is the most frequent approach used to manage many vertebrate and aquatic invertebrate invasive species. Use of baited wire-minnow traps was more effective at reducing the most reproductive crayfish than use of predatory fish (Hein et al. 2006). Even among species such as rodents on which chemical control is used, trapping can be an important supplemental tool; however, in most cases, it has only achieved short-term population control because surviving animals become trap shy.

Shooting, either aerially from a helicopter or from the ground over baited traps, is an approach frequently used for feral swine, while shooting over bait is an effective method used to control nutria (*Myocastor coypus*) (LeBlanc 1994). Newer technologies such as night-vision and thermal optics have enhanced the efficiency of such operations (McCann and Garcelon 2008).

Some species have been controlled successfully using the Judas animal approach (McIlroy and Gifford 1997), wherein a radio-marked individual of the target species is released into a control area and subsequently tracked, directing managers to other individuals which may then be euthanized. Campbell et al. (2005) described an improved sterilization procedure for both male and female Judas goats (*Capra hircus*) that allowed for preservation of normal sexual behavior (but not function) and hence drive to locate conspecifics. Similarly, preliminary evidence suggests that Judas pythons may enhance direct capture and control of Burmese pythons in Florida (Dorcas and Willson 2011; Dorcas et al. 2017).

Physical control measures have been implemented for control of aquatic invasive fauna and flora, and in some situations, citizen involvement can increase the scope of efforts. Common methods used for manual removal of fish include angling and netting. For bullfrogs (*Lithobates catesbeianus*), shooting, spears/gigs, bow and arrow, clubs, traps, and hand capture have been used, and removal of egg clutches can be effective in depleting populations over time. For aquatic plant control, manual removal can be effective; however, it can be costly in terms of time and effort involved. Furthermore, the resultant fragmentation of plants may also increase their spread.

7.5.4 Cultural Control

Cultural control refers to activities conducted by humans during the culture or management of the resource of concern with the intent to minimize the likelihood of establishment, spread, or build-up of invasive species. As used in this chapter, cultural control may include steps taken to avoid an invasive species (e.g., not planting susceptible trees on a site known to have a particular invasive insect or plant pathogen of concern) or manipulations of growth designed to maintain overall health and vigor of the resource being protected, e.g., forest trees, grassland, rangeland, wildlife, wetlands, and bodies of water.

Silvicultural thinning, increasing stand diversity, improving tree vigor, and urban tree care have been shown to reduce tree vulnerability to gypsy moth (Gottschalk 1993). Hemlock woolly adelgid density was found to be highest on seedlings grown in dense shade and decreased with increasing light; therefore, silvicultural treatments that increase light exposure might reduce hemlock woolly adelgid abundance (Brantley et al. 2017). Thinning mixed hemlock-hardwood stands may also improve tree vigor and growth of hemlock (*Tsuga* spp.) trees, thereby altering their foliar chemistry to make them less vulnerable to infestations by hemlock woolly adelgid (Fajvan 2008; Piatek et al. 2016). Sirex woodwasp infests only stressed or weakened pines, so removal of stressed trees, pre-commercial thinnings, and other silvicultural treatments that improve tree vigor can reduce the incidence of damage (Dodds et al. 2007). Healthy hemlock trees are able to withstand higher infestations of hemlock woolly adelgid than trees with low vigor (McClure 1995); therefore, mulching and irrigation can be used to help improve tree health (Ward et al. 2004). Submerging emerald ash borer-infested black ash (*Fraxinus nigra*) logs in running water for at least 3 months kills the larvae inside, preserves the wood for use in Native American basket-making, and may help prevent spread of the beetle when logs are transported to tribal lands for basket-making ceremonies (Poland et al. 2015).

Cultural approaches are critical to management of disease epidemics (e.g., Dutch elm disease, white pine blister rust, oak wilt) and are also important in management of new invasive diseases (e.g., sudden oak death). For example, effectiveness of using a vibratory plow to sever connected roots and thus prevent belowground spread of the oak wilt fungus was evaluated over a period of 6 years in 25 mixed hardwood infection centers. Results indicate that spread was stopped in 84% of the treated centers for 4–6 years (Juzwik et al. 2010). Two sanitation options were also evaluated to reduce the potential for aboveground transmission of the pathogen by insects. Annual removal of wilted red oaks (*Quercus* spp.) within the outermost root-cutting line during the same year of plowing would have resulted in 64% fewer removals than a strategy that required felling of all red oaks (healthy or diseased) inside the outermost line (Juzwik et al. 2010).

Recent advances have been made in understanding and implementing habitat modification to control mosquito populations that vector West Nile virus. Cultural controls include sanitation involving removal of tires or other sources of standing water that serve as breeding sites for larvae and the use of pumps, culverts, and networks of shallow ditches for seasonal water flow management in marshes and wetlands that allow ecosystem function but reduce mosquito reproductive habitat (Floore 2006). Removing invasive trees and shrubs mitigates degradation of habitat and reduces breeding habitat for mosquitoes and is also a component of a program to conserve the greater sage-grouse in Montana and Wyoming (Walker et al. 2007). In comparing effects of native versus non-native shrubs on the ecology of a common vector of West Nile virus, *Culex pipiens*, leaf detritus of the invasive shrubs Amur honeysuckle and autumn olive were linked to higher adult mosquito emergence rates, while leaf litter from native blackberry (*Rubus allegheniensis*) functioned as an ecological trap because it was found to be correlated with high rate of oviposition but low adult emergence rates (Gardner et al. 2015). Trapping protocols were evaluated to determine if presence and abundance of West Nile virus-vectoring mosquitos could explain the 2003 die-off of American white pelican chicks (*Pelecanus erythrorhynchos*) at Medicine Lake National Wildlife Refuge in northern Montana. Results indicate that significantly more West Nile virus-infected mosquitos were associated with shelterbelts comprised of mixed dense stands of invasive Russian olive and caragana (*Caragana arborescens*) than in marshy or grassland habitats (Friesen and Johnson 2013).

Cultural control of white nose syndrome of bats includes modifying bat hibernacula environments to eliminate the disease pathogen and increase bat survival (modifying temperature and humidity, providing alternate sources of food and water, and treating hibernacula with chemical or biological control agents), conserving genetic diversity of bats to increase development of immunity and resistance, and reducing human-assisted dispersal of the disease-causing fungus by decontaminating clothing and equipment for anyone planning to enter areas where bats hibernate (e.g., US Fish and Wildlife Service National White-Nose Syndrome Decontamination Protocol – Version 04.12.2016; WNS DT 2016). Culling heavily diseased individual bats or populations was also proposed, but disease models suggest this approach may not be effective (Hallam and McCracken 2010).

Management of amphibian chytridiomycosis may involve several activities which include environmental manipulation, controlling amphibian introductions, and deploying ex situ conservation efforts to reduce the disease-causing bacteria in the environment and on hosts or to increase population buffering capacity (Scheele et al. 2014). Temperature control has also been tested to control the fungus causing amphibian chytridiomycosis. Four of five studies found that increasing water temperature eliminated infection from amphibians (Sutherland et al. 2015).

Regeneration of many native plant species in several US forest types, including most eastern forests, is accomplished by increasing the quantity of light reaching the forest understory and reducing competition from other species and often occurs following a harvest, fire, or both. Such disturbances, depending on their severity or frequency, increase the likelihood of invasion by non-native plants (Haeussler et al. 2002; Nelson et al. 2008). However, a disturbance can be so severe in some community types that the affected site is resource-limited and is more likely to be colonized by native species than by non-native plants (Hebel et al. 2009). Defining light and competition levels (and corresponding harvesting and burn frequencies) that will promote regeneration of native species but deter invasion of non-native species is needed for all forest types.

Invasion by non-native plants may occur primarily from unsustainable land management practices that have resulted in a seed bank depleted of native seed and the loss of native plant species because they are the most merchantable or most preferred by herbivores such as deer (Cervidae) or cattle(*Bos taurus*) (Beauchamp et al. 2013; DiTommaso et al. 2014). Several studies have demonstrated that most non-native plant species are not preferred forage. Though this may change with time (e.g., Japanese honeysuckle is now a preferred deer food), controlling deer alone (e.g., via fencing) may reduce current and future invasions by non-native plants. It may require decades to realize an effect, but evidence shows that non-native species decline in abundance (DiTommaso et al. 2014; Kalisz et al. 2014) and depleted native species may recover (Tanentzap et al. 2009). Likewise, overgrazing by cattle and sheep (*Ovis aries*) may be prevented by rotating sites used for pasture and by fencing these pastures. Abandoned grazing areas that were overgrazed often become an epicenter for new plant invasions. Depending on the condition of the site and the type of grazer, simply removing the animals may not prevent invasions. For example, adding goats has proven to be effective at controlling several invasive plants, and removing goats has been detrimental to some systems (Zavaleta et al. 2001). Conversely, removal of feral sheep and cattle from Santa Cruz Island (beginning in 1981, with full eradication thought to have been achieved by 1997) initially resulted in increases in exotic fennel (*Foeniculum vulgare*) and yellow star-thistle and only a slight increase in one of the native species (Klinger et al. 1994, 2002). However, passive recovery 28 years after the removal of the feral sheep from Santa Cruz Island shows a transition from non-native plants to native woody vegetation (Beltran et al. 2014).

Changes in logging severity, fire intensity or frequency, or grazing pressure should incorporate best management practices practices aimed at preventing the introduction of invasive plant propagules. Plant propagules may be reintroduced via contaminated equipment used in a previous harvest or burn (Bryson and Carter 2004; Westbrooks 1998), transport of seed in animal dung via animal rotations from contaminated pastures (Bartuszevige and Endress 2008), and use of contaminated hay as forage (Bryson and Carter 2004; Westbrooks 1998) or gravel for road cover (Christen and Matlack 2009; Mortensen et al. 2009; Westbrooks 1998). Adoption of best management practices (that are not regulations) by private landowners may be enhanced by providing economic incentives. Matta et al. (2009), using a multinominal logit model, conducted a landowner survey which estimated that most private forest landowners would require an incentive of $95.54 per ha per year to voluntarily participate in a program using best management practices that were not required at that time (2009).

For invasive aquatic plants, divers can remove some early infestations of submerged plants by hand or with hand tools. Smothering or shading with mats or bottom barrier materials can be used to control smaller patches of invasive aquatic plants such as yellow floating heart (*Nymphoides peltata*) (DiTomaso et al. 2013). Mechanical removal of aquatic invasive plants (see Haller 2014) can be achieved by deploying boats with skimmers to remove surface-growing plants such as hyacinth (*Hyacinthus* spp.) and salvinia (*Salvinia molesta*), boats with lawnmower-like blades to mow or harvest plants, and/or rotovators (large aquatic rototiller) and dredges. In a comprehensive review by Sutherland and others (2015) as part of the Conservation Evidence project that summarizes information from 156 conservation journals, removal of two invasive aquatic plants (swamp stonecrop (*Crassula helmsii*) and reed canary grass (*Phalaris arundinacea*)) was found to increase abundance of native amphibians.

Habitat manipulation is a frequently used approach to forestall the adverse effects of aquatic invasive species on native species. Draining wetlands or reducing water levels is one approach used for both plants and animals (Hine et al. 2017; Hussner et al. 2017). Reducing wetland levels before summer or prior to extreme winter conditions can expose unwanted plants (e.g., Eurasian watermilfoil (*Myriophyllum spicatum*), Brazilian elodea (*Egeria densa*)) to freezing or drying conditions that can kill them (Haller 2014).

7.5.5 Chemical or Pesticidal Control

Chemical or pesticidal control of invasive species involves the use of natural or synthetic chemicals or microbial agents to prevent infestation, eliminate populations, reduce damage and impacts, or slow the spread by significantly reducing the population.

The use of insecticides is a very effective means of controlling many invasive insects if the insect can be effectively brought into contact with the applied material. Research is conducted to evaluate pesticide efficacy, delivery method(s), translocation within hosts, fate in the environment, and impacts on other species. Systemic insecticides applied by trunk or soil injection or basal bark sprays can provide control of emerald ash borer (McCullough et al. 2011) and hemlock woolly adelgid (Cowles and Cheah 2002; Whitmore 2014) in urban or high-value landscape trees but are not practical for large-scale management of forest stands. Similarly, horticultural oil or insecticidal soap sprays have been found to be effective in reducing hemlock woolly adelgid populations on accessible trees but are not practical at the forest landscape level (Cowles and Cheah 2002; McClure 1995). Systemic insecticides are generally ineffective in controlling ambrosia beetles such as shot hole borer and redbay ambrosia beetle; however, prophylactic spraying of bark may help prevent attacks on individual trees (Peña et al. 2011). Ground application of soaps, oils, and contact or systemic insecticides are effective for controlling balsam woolly adelgid (*Adelgespiceae*) (Ragenovich and Mitchell 2006). Rapid testing of insecticides to control spotted lanternfly resulted in the registration of several products for spotted lanternfly control in Pennsylvania (PSE 2019).

The microbial insecticide *Bacillus thuringiensis* var. *kurstaki* (*Btk*) and the gypsy moth nucleopolyhedrosis virus product Gypchek are used to manage gypsy moth populations (Elkinton and Liebhold 1990; USDA 2012). Aerial sprays of Bt_k are used to eradicate isolated populations and control gypsy moth over large suburban or rural areas. Gypchek is highly specific to gypsy moth but can only be produced by infecting larvae; therefore, it is not mass produced commercially and is only available through the Forest Service for limited applications (Podgwaite 1999). Both chemical and microbial insecticides are also used to control mosquitos that vector West Nile virus or Zika virus.

Use of pesticides against invasive pathogens of trees and other plants is not common in forests and wildlands. Rather, they are used on a small spatial scale within higher-value landscapes where economics or other values justify its use (e.g., Dutch elm disease, oak wilt, sudden oak death). For example, the systemic fungicide potassium phosphite has been widely used in California on high-value landscape trees as a bark spray (with or without a bark penetrant) applied to lower trunks of *Quercus* species to suppress sudden oak death development in trees newly infected with *P. ramorum* or to prevent infection of "at-risk," healthy trees (UCB FPM Lab 2017). In a current long-term study, potassium phosphite is being evaluated for its potential to protect tanoaks from *P. ramorum* in forest settings (Phytosphere Research 2013). This treatment has also demonstrated efficacy in protecting avocado, pineapple (*Ananas comosus*), and cocoa (*Theobroma cacao* L.) crops as well as jarrah (*Eucalyptus marginata*) (Pegg et al. 1990).

Chemical control is commonly used for management of invasive plants infesting small areas of refuges or other protected areas. Conducting research to identify the safest and most effective herbicides to control existing and new invasive plants is an ongoing need as are standardized protocols for systematically surveying and testing for herbicide resistance. Yellow star-thistle has developed resistance to four auxin inhibitors, including triclopyr (Miller et al. 2001), and hydrilla has developed resistance to fluridone, which inhibits carotenoid biosynthesis (Michel et al. 2004). Invasive plants that produce numerous seeds or spores and have long-distance dispersal are most likely to develop herbicide resistance after repeated applications. Most invasive plants have these characteristics. Likewise, the acetolactate synthase (ALS) inhibitor herbicides, which are commonly used in invasive plant management in natural areas and include imazapyr, imazapic, and metsulfuron-methyl, are the most likely to develop resistance in natural areas because they have shown the highest resistance development in agricultural settings (Hutchinson et al. 2007). With increasing use of herbicides in natural areas to combat invasive plants and the repeated use of the same herbicides, many of which are ALS inhibitors, the development of resistance is likely but could be prevented with the use of proper protocols. Such protocols may include developing multiple herbicides in different herbicide families that can be used in rotation to prevent herbicide resistance (Hutchinson et al. 2007). However, the rotational use of multiple herbicides on public lands requires approval under National Environmental Policy Act (NEPA) guidelines. Control of invasive plants may require applications in consecutive years. However, the impacts of herbicide on nontarget species may also increase with the frequency of herbicide application, especially if they are applied at intervals of less than 4–5 years (Crone et al. 2009; Huebner et al. 2010). The acceptable number of herbicide applications requires a delicate balance between reducing the abundance of the non-native species and ensuring that treatments do not eliminate native species. Ideally, applied research should be combined with basic ecological assessments such as competition and demographic studies to define optimal application rates and timing of treatments.

Rodenticides have been used extensively to control invasive rats (*Rattus* spp.) and mice (*Mus musculus*), particularly on islands where extirpation is achievable. Anticoagulants such as brodifacoum are the most widely used option for treatment. Howald et al. (2007) reviewed the literature on attempts to eradicate invasive rodents on islands worldwide and reported that 332 of 387 attempts were successful. Acetaminophen baits have been shown to be effective for controlling brown treesnakes (*Boiga irregularis*; Savarie et al. 2001), although methods for optimal delivery continue to be investigated (Lardner et al. 2013).

Although no registered toxicants are currently available for use against feral swine in the United States, development is ongoing, and US Environmental Protection Agency (EPA) registration is being sought (Snow et al. 2016). Sodium nitrite, developed and licensed for use in Australia and New Zealand, works through binding hemoglobin and causing death from methemoglobinemia, which causes rapid depletion of oxygen to the brain and vital organs. Nitrite toxicosis is considered to be humane (Cowled et al. 2008; IMVS 2010; Shapiro et al. 2015), and the risk of secondary toxicosis to nontarget species is slight (Lapidge et al. 2012). Sodium nitrite baits are delivered using specialized feeding stations that are designed to minimize or prevent access by nontarget species (Campbell et al. 2013; Lapidge et al. 2012). However, additional research is needed to further evaluate and minimize nontarget effects.

Sylvatic plague is a rodent-associated, flea-borne disease of animals caused by the gram-negative bacterium *Yersinia pestis* that can be transmitted to humans. The disease affects nonurban wildlife including the endangered black-footed ferret (*Mustela nigripes*), an obligate predator of highly plague-susceptible prairie dogs (*Cynomys* spp.) (Jachowski et al. 2011). Management of plague-vectoring fleas has emerged as a significant factor in the conservation of endangered species. Reintroduction efforts for black-footed ferret are dependent on developing tools to control plague vectors that affect prairie dogs. Initial treatments focused on applications of insecticidal deltamethrin dust (DeltaDust® – Bayer

Environmental Science, Montvale, NJ) and targeted fleas in prairie dog burrows (Biggins et al. 2010; Bodenchuk et al. 2013; Dinsmore 2013). Insecticidal treatment of prairie dog burrows affected the food supply of a ground-nesting insectivorous bird, the mountain plover (*Charadrius montanus*), which preferentially nests on prairie dog colonies. Treatments effectively lowered nest survival because adults spent more time away from nests searching for prey or were forced to switch to lower-quality insect prey. Jones et al. (2012) determined that dusting not only allowed black-tailed prairie dogs (*Cynomys ludovicianus*) to persist in plague-affected areas during epizootics but also generated refugia of genetic diversity in treated colonies. Furthermore, the increased survival of resident and immigrant individuals created a more robust base population for reestablishing old colonies or starting new colonies.

Although numerous control techniques are available for use in aquatic ecosystems, once an invasive animal is established, eradication is rarely attained and meaningful control is often achieved only at great expense (Johnson et al. 2009a, b). For example, control of the sea lamprey (*Petromyzon marinus*) in the Laurentian Great Lakes using the lampricide TFM (3-trifluoromethyl-4-nitrophenol) costs more than $20 million each year and is an ongoing annual expense that is required to keep sea lamprey populations at levels sufficient to minimize their predation on valuable sportfish (Hansen and Jones 2008). Research is progressing to develop innovative control techniques for controlling multiple high-impact aquatic invasive animals, including zebra and quagga mussels (Meehan et al. 2014) and Asian carp species (Zielinski and Sorensen 2016); methods include the application of pesticides such as rotenone and antimycin to kill invasive fish (Sato et al. 2010).

Numerous pesticides are registered for use against unwanted aquatic species. A total of 45 chemicals were identified as piscicides and are listed in a US Geological Survey report (USGS 2017). The most effective pesticides available for removing invasive fish include antimycin and rotenone, though these products may not fully remove adequate numbers of unwanted individuals. For example, rotenone was used at Diamond Lake in Oregon in 2006 to remove the invasive tui chub (*Gila bicolor*) which had been adversely altering the lake ecosystem (Finlayson et al. 2014).

Applications of herbicides and algacides are well-established treatments used to control aquatic invasive plants. Fourteen herbicides are registered for use in US aquatic systems (Netherland 2014). Foliar spray treatments are useful for free-floating plants such as water lettuce (*Pistia stratiotes*) or salvinia; however, systemic herbicides can be more selectively administered and can be more effective. In Washington, invasive Japanese knotweed has been controlled along watercourses using herbicidal mixtures, and application methods include stem injections or foliar spray incorporating glyphosate, imazapyr, and vegetable oil (Claeson and Bisson 2013). Altering salinity has been used in some circumstances. Reducing available nutrients is an approach available in the broader portfolio of methods used to control algae (Lembi 2014).

There is concern about the potential for the inadvertent spread of aquatic invasive species and diseases when water is drawn for wildfire management or other uses (Olson et al. 2013), transferred from fish hatcheries for stocking of nonnative fishes, or used for transportation of people and supplies. Treatment of water with ammonia compounds or bleach has been instituted in some regions to forestall such disease transmission (Olson et al. 2013; USDA FS 2016).

7.5.6 Biological Control

Biological control, or biocontrol, is essentially using living organisms to reduce the numbers of pest organisms, the goal being to achieve sustainable and targeted management of the pest or invasive species. Biological control agents frequently involve insects (e.g., predators or parasitoids), but selected microorganisms such as fungi, bacteria, and viruses may also be utilized. Biological control is one of the more successful methods available for achieving long-lasting, widespread, and environmentally safe management of invasive species. Biological control is sustainable, selective, and cost-effective, and its use may successfully avoid the ecological and economic collateral damage often associated with pesticides (Rinella et al. 2009; Suckling and Sforza 2014; Van Driesche and Hoddle 2016). Developing biological control programs requires a significant investment in basic and applied research which includes exploring for natural enemies, developing rearing and release methods for selected biological control agents, evaluating impacts on nontarget species, monitoring the establishment of the released agent, and assessing control of the pest and the level of protection provided to host plants (Van Driesche et al. 2008).

Biological control has been used for many invasive forest insects (Van Driesche and Reardon 2013). For example, foreign exploration, collection, mass rearing, and release of insect and microbial biological control agents to manage the gypsy moth and brown-tail moth (*Euproctis chrysorrhoea*) began in the early 1900s and continued into the 1970s. More recently, three species of wasps were discovered parasitizing emerald ash borer in its native range in China, and following extensive host-specificity testing, these species were approved in 2007 for release in the United States. They are currently being mass reared, released, and evaluated in long-term studies to determine their impact on emerald ash borer populations and ash tree health (Bauer et al. 2015). A fourth parasitic wasp from Russia was also approved for release beginning in 2015 (USDA APHIS 2015) and is currently being reared and released.

Biological control agents have also been evaluated for their efficacy against the Asian longhorned beetle; however, host-specificity screening indicates that several of these species that were recovered from the pest's native range have a broad host range and may have an impact on nontarget species in North America (Meng et al. 2015). Beetle predators of the hemlock woolly adelgid have been released widely to achieve biological control, and their establishment and efficacy are currently being evaluated (Havill et al. 2014). Similarly, beetle predators from Europe have been introduced for biological control of balsam woolly adelgid (MacQuarrie et al. 2016). Augmentative biological control of sirex woodwasp using a nematode, *Deladenus siricidicola*, has been very successful in confining sirex woodwasp infestations to small localized areas in Australia (Carnegie et al. 2005); however, in North America, differences in species and strains of nematodes and their associated fungal symbionts have been shown to affect their virulence and efficacy as biological control agents. Considerable research is currently underway to better understand these complex interactions (Morris et al. 2012). Entomopathogens, including nuclear polyhedrosis virus (NPV) and the fungus *Entomophaga maimaiga*, can cause significant mortality in gypsy moth populations. Climatic factors that favor the development of fungal epizootics, and methods to release infected gypsy moth larval cadavers, have been investigated (Siegert et al. 2012; Smitley et al. 1995) with intent to optimize the use of this pathogen for biological control of gypsy moth populations. *Ooencyrtus kuvanae* (Howard) (Hymenoptera: Encyrtidae), an egg parasite of spotted lanternfly, was discovered in Pennsylvania in 2016 (Liu and Mottern 2017). The parasite was introduced in 1908 for gypsy moth control and is known to attack multiple host species. An unidentified species of native Dryinidae (solitary wasp) parasitized spotted lanternfly nymphs in Pennsylvania, and an unidentified fungus that was found infecting nymphs is being identified (Parra et al. 2017).

Major success stories in biological control point to the great potential this tool holds for controlling invasive plants at a landscape scale (Seastedt 2015; Van Driesche et al. 2002). The use of biological control agents against invasive plants includes some fungi such as rusts (Hasan and Ayres 1990) and herbivorous insects (McFadyen 1998). However, biological control agents sometimes fail to impact weed populations even where they become established. The worldwide success rate of projects using biological control attempts against invasive plants is estimated at 20–30% (Crawley 1989; Raghu et al. 2006; Van Driesche et al. 2010; Van Klinken and Edwards 2002), as compared to 62% of projects that achieved complete control of target invasive arthropods (Van Driesche et al. 2010). More recently, Cock et al. (2016) found the overall success rate for complete biological control of invasive insects was 10% and has been declining since the 1970s, while the number of introductions of classical biological control agents has decreased. Ineffective biological control agents, even if host-specific, can persist and may cause unwanted ecological effects (Pearson and Callaway 2003, 2005, 2006, 2008; Pearson and Fletcher 2008; Ortega et al. 2004; but see Van Driesche and Hoddle 2016). The practical application of biological control is universally challenged by difficulties in understanding (e.g., quantifying and/or verifying) single and interacting factors that influence its success or failure and predicting the efficacy of individual agents (Carson et al. 2008).

Host specificity of potential biological control agents may influence the success of biological control and potential for adverse impacts on nontarget organisms. The ability to test for and predict host specificity of herbivores has improved greatly in recent decades (Sheppard et al. 2005). However, our ability to predict efficacy of agents has not followed suit, and consequently it's currently not possible to predict with confidence if a biocontrol agent will reduce populations of the target pest once released. Several avenues of research can advance the science of biocontrol and improve its efficacy. These include (1) choosing an appropriate target plant, (2) selecting the best biocontrol agent using genetics and chemical ecology, and (3) understanding and exploiting climate change effects.

Selecting an appropriate target plant is important because recent work suggests that some plant species are more amenable to biocontrol than others. The enemy release hypothesis (Elton 1958) underlies the theory of biocontrol and proposes that non-native species transplanted outside of their native range thrive because they leave most or all of their natural enemies behind (Keane and Crawley 2002; Müller-Schärer and Schaffner 2008). Thus, biological control is most appropriate for use against plants that are "released" from their natural enemies in the invaded range, though these conditions are not usually fully demonstrated because of limited studies in the native range (Hierro et al. 2005). Other recent work indicates that plant traits could be used to predict plant species most amenable to biological control. For example, "easy targets" for biocontrol are those species that are not known to become overly abundant or negatively impactful in their native range, i.e., asexual species and/or species that occur in aquatic or wetland habitats (Paynter et al. 2012). Naturally, some invasive plants are more likely to be successfully controlled using biological control, but uncertainty exists in identifying susceptible (and unsusceptible) target plants.

Identifying, prior to release, which agents are most likely to be successful in reducing the abundance of invasives is critical for successful biological control (McFadyen 1998). Recent advances in genetics and chemical ecology have shown promise in improving agent selection. Molecular approaches have significantly contributed to the resolution of taxonomic issues associated with both target plants and candidate biological control agents (Gaskin et al. 2011; Goolsby et al. 2006). Genetic diagnostics have also allowed us to pinpoint the geographic origin of invasive plants (Gammon and Kesseli 2010; Gaskin et al. 2013a, 2013b; Tarin et al. 2013; Williams et al. 2005) and to reconstruct routes of invasion (Buckley and Catford 2016; Estoup and Guillemaud 2010; Le Roux and Wieczorek 2009). This knowledge is essential to properly test the enemy release hypothesis and can guide the search for effective biocontrol agents. This can be especially critical when local adaptation results in herbivores and especially pathogens that have become highly host-specific to certain populations or genotypes. Examples of these include Brazilian peppertree (*Schinus terebinthifolius*) (Cuda et al. 2012; Diaz et al. 2015; Manrique et al. 2014), rush skeletonweed (*Chondrilla juncea*) (Campanella et al. 2009), and invasive knotweeds (Grevstad et al. 2013).

Plant-insect chemical ecology has only recently been applied to weed biological control (aside from host-specificity testing, which is based largely on plant chemistry). Chemistry plays a central role in determining ecological outcomes between plants and insects and should provide information that can be used to better predict those candidate agents that are most likely to be effective (Wheeler and Schaffner 2013). For example, hybridization may function as an extreme example of hypothesized evolution of increased competitive ability, whereby plants introduced into new areas in the absence of natural enemies evolve reduced allocation to costly chemical defenses, which then allows them to increase allocation to growth and/or reproduction (Blossey and Nötzold 1995). In hybrid plants, heterosis (hybrid vigor) resulting in increased allocation to growth and reproduction can be associated with novel phytochemistry, which produces confusing signals for biocontrol agents coevolved with either of the hybrids' parental species (Hubbard 2016).

Evidence suggests that some types of herbivores will be positively affected by climate change, whereas others will be negatively affected (Robinson et al. 2012; Runyon et al. 2012). Similarly, climate change will likely affect biological control responses due to phenological differences in responses of hosts and biological control agents to changes in temperature (Reeves et al. 2015). These changes could potentially be exploited in biocontrol, for example, by focusing on agents that respond most positively to climate change (see Chap. 4).

Research is underway to investigate the use of biological control against invasive diseases of terrestrial vertebrates. Cornelison et al. (2014) found that the ubiquitous soil-associated gram-positive bacterium *Rhodococcus rhodochrous* strain DAP 96253 demonstrated potential for biological control of the white nose syndrome in bats by inhibiting conidial growth of the fungus in infected tissues.

Although highly desirable, biological control has not been widely used in managing invasive diseases of trees. The mycoparasitic species complexes of *Clonostachys* and *Trichoderma* have been shown to be effective against *Crinipellis roreri*, a fungal disease of cocoa in Ecuador (Evans et al. 2003). Infecting the chestnut blight fungus (*Cryphonectria parasitica*) with hypoviruses (namely, *Cryphonectria hypovirus* 1 (CHV-1), CHV-2, CHV-3, and CHV-4) has proven effective against chestnut blight in some locations in Europe and in Michigan (MacDonald and Fulbright 1991), but has failed almost completely in the Eastern United States possibly due to vegetative incompatibility among host individuals that prevents the virus from spreading (Milgroom and Cortesi 2004). Biological control of *Heterobasidion* root disease (*Heterobasidion annosum*) has been achieved by exclusion of sugar resources on freshly cut stumps by a native decay fungus, *Phlebiopsis gigantea* (BioForest Technologies 2016).

Biological control agents being considered for various invasive aquatic plants include mollusks, fungi, carp, and invertebrates such as moths, thrips, mites, and chironomid midges (Van Driesche and Bellows 1996). Beetles are the dominant insects being used for biological control of aquatic plants (Cuda 2014; Cuda et al. 2014). Aquatic predators have been considered as biological control agents, but their lack of prey specificity may restrict their utility. In 2016, the predaceous sterile hybrid tiger trout (female brown trout × male brook trout (*Salmo trutta* × *Salvelinus fontinalis*)) was evaluated as a control measure for reinvasion by tui chub and golden shiner (*Notemigonus crysoleucas*) at Diamond Lake, OR (Carroll and Miller 2016).

7.5.7 Vaccination

Vaccines involve the treatment (e.g., via oral or direct injection) of host organisms with killed microorganisms or attenuated strains of the invasive disease organism (e.g., bacterium, virus) to render the potential host immune or only mildly susceptible to infection and development of disease.

Recent efforts to conserve black-footed ferret populations have focused on developing a sylvatic plague vaccine and delivery system for prairie dogs (Abbott et al. 2012; Rocke et al. 2010). A vaccine, which is in the final phases of field testing for animal safety and efficacy, uses recombinant raccoon poxvirus (RCN) to vector proteins of *Y. pestis*, F1 and V, already approved for use in a human-injectable plague vaccine (Rocke et al. 2014). The vaccine is delivered in a palatable (peanut butter-flavored) bait matrix incorporating rhodamine B, a biomarker used to track uptake (Fernandez and Rocke 2011; Tripp et al. 2014).

Vaccines and immunomodulators are also being used to increase resistance of bat populations to white nose syndrome (Lilley et al. 2017) and to treat island scrub-jays (*Aphelocoma insularis*) that occur only on Santa Cruz Island and are susceptible to infection by the mosquito-vectored West Nile virus (Boyce et al. 2011).

Vaccination and application of antifungal or probiotic agents were categorized as having low or moderate effectiveness against amphibian chytridiomycosis (Grant et al. 2016). Booroolong frogs (*Litoria booroolongensis*) infected with an isolate of the disease-causing fungus, treated with itraconazole, a triazole fungicide, to clear infection, and then re-exposed to the fungus did not acquire immunity from the initial exposure, suggesting that a vaccine is unlikely to be effective (Cashins et al. 2013).

Aquatic invasive species that are emerging infectious diseases are candidates for control using vaccination. To date, most attempts to use vaccination for control have been directed against farmed fish and widespread diseases (Gudding et al. 1999) that are not considered aquatic invasive species. Research is continuing in this area.

7.5.8 Host Resistance

Resistance is a result of genetic traits of the potential host species (e.g., tree, terrestrial animal) that render it mostly "immune" to the invasive species, results in tolerance of attack or infection by the invasive species, or is manifested as a morphological barrier to infestation by invasive species infestation (Fritz and Simms 1992).

Host resistance, if it exists, can be propagated in populations across the landscape, can be an effective long-term defense against invasive pests, and can serve as a tool for restoring impacted landscapes (see Chap. 8 for more details). Although all North American ash species encountered by emerald ash borer to date may be infested, relative preferences and susceptibility vary among species and appear to be related to differences in volatiles, nutrition, and defense compounds (Chen and Poland 2010; Chen et al. 2011; Cipollini et al. 2011; Pureswaran and Poland 2009; Whitehill et al. 2012). Within an ash species, some individual "lingering" ash trees persist in stands where all of the surrounding ash trees have succumbed to emerald ash borer within 5–6 years. Asian ash species typically have higher levels of resistance than North American species, and this trait may have a role in developing a resistance breeding program for our native ash species. Traditional and hybrid breeding programs have been utilized to select, screen, and develop ash cultivars that exhibit increased resistance to emerald ash borer (Koch et al. 2012) (see Chap. 8). Foliar chemistry has been linked to hemlock infestation and susceptibility to hemlock woolly adelgid (Pontius et al. 2006), and the relative resistance of North American hemlock species as compared to Chinese hybrids has been evaluated (Montgomery et al. 2009).

Identification of resistance and selection or breeding for resistant species, phenotypes, or genotypes generally is a most common strategy for use against widespread, long-established invasive pathogens whose resultant diseases have broad geographic distributions. Within genetically diverse populations of trees, there can be a small number of individuals that exhibit some level of resistance to invasive diseases (see Chap. 8 for more details). Silvicultural strategies aimed at decreasing the proportion of susceptible individuals in a stand, and, therefore, increasing the proportion of resistant individuals, can, in some cases, be an effective tool for use in disease management. Single tree selection and removal of American beech (*Fagus grandifolia*) trees actively infected with beech bark disease resulted in an 11.5% increase in disease-free (apparently resistant) basal area 50 years after treatment (Leak 2006). The basal area of trees with *Neonectria* infection decreased from 67% in the untreated stands to 27% in the treated stands, indicating that removing susceptible trees may have also decreased the level of fungal inoculum (Leak 2006). Beech bark disease is caused by an introduced scale insect (*Cryptococcus fagisuga*) that provides a pathway for entry for the bark canker pathogens (*Neonectria ditissima* and *N. faginata*). Proteomic investigation of scale-resistant and scale-susceptible trees in eight geographically isolated stands led to the discovery that different protein profiles occurred in diseased and healthy trees (Mason et al. 2013). Further study of these proteins is underway with the goal of developing biomarkers that will aid managers in identifying and retaining resistant trees and removing susceptible trees as a preemptive measure to minimize the impacts of beech bark disease.

Increasing resistance of native species to infection by an emerging invasive infectious disease is a relevant topic for continuing research. For example, it was discovered that a complex microbiota appears to be interacting on amphibian skin and that some species have a controlling effect on these disease microbiota. Amphibian skin harbors symbiotic resident bacteria that possess antifungal properties that are being examined for their potential to combat the amphibian chytridiomycosis fungus. The probiotic bacteria *Janthinobacterium lividum* is thought to provide some resistance to the chytridiomycosis fungus and is being tested for that purpose in susceptible frogs (Bletz et al. 2013).

7.5.9 Reproduction Control

Reproduction control involves using a natural or synthesized chemical or genetic manipulation to impede or prevent mating or development of offspring in the invasive species population. Reproduction control includes tactics such as mating disruption of invasive insect species using pheromones or controlling fertility in terrestrial invasive vertebrates.

Mating disruption is used to eradicate or slow the spread of sparse gypsy moth and control brown apple moth populations by applying female mating pheromone to saturate the environment and thus interfere with male location of females (Leonardt et al. 1996; Soopaya et al. 2015; USDA 2012). Sterile insect release of irradiated males has been attempted for control of gypsy moth (USDA 2012) and light brown apple moth (Stringer et al. 2013), but is challenging due to reduced reproductive fitness of irradiated insects and is not practical for most invasive forest insects which are difficult to mass rear. Recent novel approaches for control of invasive fish involve the intentional release of genetically modified fish that are designed to disrupt reproduction of target invasive fish species (Kapuscinski and Sharpe 2014). This involves manipulating the chromosomes to skew sex ratios, or using recombinant DNA techniques to insert damaging genes into the genome of target invasive fish to disrupt the reproductive cycle, or a combination of both (Thresher et al. 2014).

Immunocontraceptives have been evaluated for population control in several species of invasive vertebrate pests (Fagerstone et al. 2010). OvoControl® P is an oral contraceptive approved by the EPA for use on rock pigeons (*Columba livia*). Vaccines such as gonadotropin-releasing hormone (GnRH) and porcine zona pellucida (PZP) are used in mammals such as feral horses and feral swine. A single dose of GnRH vaccine can render an animal infertile for 1–5 years (Killian et al. 2008; Massei et al. 2008). However, while such approaches enjoy greater public support than using toxicants and other lethal methods, their use has been limited due to the high costs associated with delivery as compared to other methods and the relatively low effectiveness and length of time required to achieve population reduction (Massei et al. 2011). This technology might be most appropriate on islands or other areas where immigration and emigration are limited.

7.5.10 Integrated Pest Management Programs

Integrated pest management (IPM) is the optimization of several pest control methods in an economically and ecologically sound manner. In natural ecosystems, it is an environmentally based strategy that focuses on attaining long-term efficacy by deploying a combination of tactics in a compatible manner to maintain pest damage below an economic threshold, while protecting against hazards to humans, animals, plants, and the environment. IPM may involve use of several techniques such as biological control, habitat manipulation, cultural practices, pesticides, and resistant varieties incorporated into a unified program. IPM requires clear articulation of management goals, knowledge of the pest and its impacts on the ecosystem, technology to monitor the presence and abundance of the pest, guidance on when management is worthwhile, a suite of complementary tools and strategies to affect the abundance and/or reduce impact of the pest, and the methodology to evaluate the success of interventions.

IPM programs have been developed for several invasive forest species and many agricultural pests. The gypsy moth Slow the Spread program is currently the largest and most successful IPM program in the United States for managing the spread of an invasive forest pest (Sharov et al. 2002) and is recognized as a model approach for managing invasive species. A grid of pheromone-baited traps is deployed to detect the presence of adult male gypsy moths just ahead of the advancing front of the generally infested area. Analysis of the pattern of moth captures is then used to identify areas that require treatment. The network of traps identifies new infestations of gypsy moth that are well below population densities that cause defoliation; therefore management has the option to apply specific treatments aimed to eliminate or reduce sparse populations, such as trapping, mating disruption, and application of microbial pesticides such as *Btk* and gypsy moth nucleopolyhedrosis virus. The overall success of the program can be attributed to the integrated and coordinated involvement by the Forest Service, USDA Animal and Plant Health Inspection Service (APHIS), State plant pest regulatory officials, and State foresters. The project is managed at the landscape level and focuses use of standardized protocols for data collection and analysis, decision-making, and allocation of funds across all States and agencies that participate in the project (Sharov et al. 2002).

The Hemlock Woolly Adelgid Initiative also utilizes an integrated and coordinated approach to manage this invasive pest, which is currently established in 19 eastern States (Ferguson et al. 2013). Research identified the geographic region or country of origin of eastern US hemlock woolly adelgid populations, and then foreign explorations were conducted to locate candidate natural enemies that could be considered for release to control local hemlock woolly adelgid populations. The program utilizes insecticides to protect hemlocks while biological control agents become established and is investigating levels of pest resistance among hemlock species in the infested area (Onken and Keena

2008). SLow Ash Mortality (SLAM) is an integrated program for managing the emerald ash borer (Mercader et al. 2015; Poland and McCullough 2010). This program uses a combination of detection and monitoring traps, tree removal, systemic insecticides, biological control, and behavioral modifications with clusters of girdled trap trees to achieve a greater level of control. In a large-scale multiagency pilot study of the program in an area over 350 km^2, both girdled trees and insecticide treatments reduced emerald ash borer densities and protected ash trees in areas surrounding the treatments. Model results indicated that emerald ash borer spread rates were reduced from areas with girdled trees. Trees treated with the systemic insecticide also reduced larval abundance in subsequent years (Mercader et al. 2015, 2016). The Pennsylvania Department of Agriculture has developed and implemented a pest management strategy to suppress the spotted lanternfly population focusing on the core of the infested area and working outward using the trap tree/host removal method as well as pesticide applications. This approach has not been fully validated, but preliminary results show that lanternfly populations are significantly reduced in the areas where trap tree/tree removal treatments have been completed (PDA 2019). Success of large-scale IPM programs requires support from residents and landowners in the affected area and can be attained by fully informing the public about the program goals, methods used, and anticipated results. Success also requires a commitment to a sustained level of resources over time. Research is also needed to evaluate and model the success of these programs.

Successful management of individual species of invasive plants requires that knowledge exists about which treatments are effective against particular species; however, many sites are invaded by multiple species of invasive plants or are surrounded by adjacent populations of non-native plants in the landscape. Success in controlling one invasive species often facilitates invasion by another invasive plant. Consequently, it's necessary to be able to predict how multiple species are likely to respond to the removal of one or more invasive plant species (Kuebbing et al. 2013).

It may not be economically feasible to restore some plant communities that incur invasional meltdowns (as discussed in Sect. 7.4.1), and even removal of some invasive plants could cause more damage than good. For example, the removal of the non-native saltcedar (*Tamarix* spp.) may initially impact the federally endangered southwestern willow flycatcher (*Empidonax traillii*), because in degraded and invaded habitats, saltcedar can serve as important habitat for the flycatcher, though some *Tamarix* stands are unsuitable (Schlaepfer et al. 2011; Sogge et al. 2008; York et al. 2011; USFWS 1997). Other examples of non-native species with potential new conservation value are non-native plant species used to reclaim coal mine grasslands which serve as habitat for Henslow's sparrow (*Ammodramus henslowii*) in Indiana (Bajema et al. 2009), melaleuca (*Melaleuca quinquenervia*) which provides habitat for snail kite (*Rostrhamus sociabilis*) in the Everglades (Chen 2001), and European legume gorse (*Ulex europaeus*) which protects the endangered New Zealand weta (very large stenopelmatid orthopterans) from predators (Gibbs 1998) and serves as a nurse plant for native forest regeneration if grazing is stopped (Sullivan et al. 2007). These examples do not negate the well-documented harm these non-native invasive plants can do in other settings. Indeed, they may both provide a service and cost. For instance, melaleuca also decreases the primary food source for the kite (Chen 2001), and regeneration with European legume gorse as a nurse plant results in a successional trajectory toward lower species richness (Sullivan et al. 2007). Some plant communities that experience meltdown, especially those occurring in urban areas, are often labeled as novel communities. These communities are so different from the original after invasion that recovery is deemed unlikely. More importantly, the communities now appear to serve an ecosystem service (benefits provided by ecosystems including food and water, regulating climate and disease, providing nutrient cycling, crop pollination, or recreational values), because few if any other species could grow in some of these sites. In some situations, removal of non-native species could harm native species that are now dependent on services or resources provided by the non-native species (e.g., native birds using non-native shrubs as nest sites, native pollinators using non-native plants to forage for pollen or nectar) (Schlaepfer et al. 2011). The willingness to allow the existence of some non-native communities, and accept coexistence between non-native and native species, is termed "conciliation biology" (Carroll 2011). Ecological and economic costs associated with conciliation biology can be estimated, and these data would provide additional input for prioritizing invasive species management efforts (Box 7.2).

Ecosystems and invasive plants may be best managed as part of a landscape mosaic composed of dynamic land uses (Chabrerie et al. 2007; Vila and Ibanez 2011). These uses may help move invasive plants (corridors and disturbed patches), while others may serve as barriers (actively cultivated agricultural land and large uninvaded forest patches). It is also a social landscape in which some landowners who choose not to control or prevent the occurrence of an invasive species may serve as the source of invasion for other landowners. In order to successfully manage invasive plants, knowledge of the landscape spatial composition, landscape ecology, management and design, and coordinated control and prevention efforts must be shared among the various affected landowners (Epanchin-Niell et al. 2010). This is a goal of most CWMAs, but it is not clear how successful such organizations have been. Without documentation of success, it is not possible to evaluate the value of applying economic incentives to further successful management (Hershdorfer et al. 2007).

A combination of harvesting, fire, herbicide application, strategic grazing, deer control, and/or biocontrol may be the key to ensure that sustainable forests, grasslands, rangelands, and wetlands avoid large damaging invasions by non-native plants. It may be possible to reduce the negative impacts on nontarget species attributed to repeat applications of herbicides, or prescribed burning, by rotating their application. Biocontrol treatments, when available, may be the best initial step to decrease plant population abundance where large non-native plant populations occur. In situations where invasive plants are less abundant or not widely distributed, repeat applications of herbicide may not be needed. More detailed information on the efficacy of each biocontrol agent (percent reduction in population size and spatial patterns of establishment) in different environments will help to define management objectives. For instance, garlic mustard populations are most impacted (63% reduction in population size) by a root-mining weevil (*Ceutorhynchus scrobicollis*) because it attacks both the rosette and flowering stages; adding a stem-attacking weevil (*C. alliariae*) reduced the overall population by 88% (Evans et al. 2012). Likewise, the strategy for releasing a biocontrol agent may be dependent on the distance between patches or populations of the invasive plant, as well as the micro-physiography of each patch (Pratt et al. 2003). The application of multiple biocontrol agents may be needed in sites that contain several abundant invasive plants (invasion meltdown sites). It's important to understand how biocontrol agents may interact with each other, as well as with any existing native insects and/or pathogens (some of which may also impact the non-native species with time) within the target site.

Land managers may choose a proactive approach if economic assessments demonstrate that employing sustainable forestry to promote native species recovery and deter invasions is more economically viable than treating invasions after the disturbance occurs. Such assessments should incorporate the true cost of invasion, including opportunity costs associated with the impact of invasions on future loss of forest regeneration (Holmes et al. 2009). Likewise, there needs to be an assessment of the measures used to prevent invasives associated with best management practices (BMPs) and the cost of using them, as compared to the cost of invasion when they are not utilized. Such economic assessments will provide landowners with more tangible evidence to support why investing in BMPs and lower-impact harvesting regimes is cost-effective.

IPM has been implemented for some tree diseases (e.g., sudden oak death, white pine blister rust, oak wilt). For example, an IPM program for management of white pine blister rust has been developed (Schoettle and Sniezko 2007) that consists of manipulating the forest composition, improving host vigor, using rust-resistant planting stock, reducing pest populations, and diversifying age structure. For oak wilt disease, the greatest success in management has occurred when early diagnosis is followed by creative and integrative use of control tools tailored for local sites (Juzwik et al. 2011). Similarly, appropriate site-specific strategies are the basis for management of sudden oak death in California (Swiecki and Bernhardt 2011). Many aquatic invasive species have established populations in situations that require sustained management. IPM programs may be implemented using a variety of approaches over time to control the non-native species. To manage fish populations, such programs may include barriers, manipulating water levels, targeted overharvest, stocking of predators, sterilants, toxic baits, selective piscicides, attractants and repellants, immunocontraceptive agents, viruses, chromosomal manipulations, gynogenesis, and transgenics (Fausch et al. 2009; USGS 2017).

7.5.11 Key Findings

- Considerable research has been conducted to develop and evaluate management of invasive species under each of the major management approaches including regulatory control, education and outreach, physical control, cultural control, chemical control, vaccination, biological control, reproduction control, host resistance, and IPM programs.
- Federal and State quarantines regulate movement of many significant invasive pests. Recent research addresses efficacy of current regulations and has led to new regulations for treating solid wood packing material.
- Public outreach and education promote awareness and support of regulations and control actions for invasive vertebrates and aquatic animals, including trapping and shooting. Research evaluates the efficacy of various outreach activities on influencing human behavior and compliance with regulations.
- Efficacy of physical control is being evaluated for use against invasive vertebrates including fencing, sound devices to frighten animals, lethal control in the form of trapping and shooting, angling, netting, water skimmers, and hand removal of egg clutches for invasive aquatic organisms.
- Cultural control practices including mulching, irrigation, mechanical root cutting, sanitation, harvesting, prescribed fire, and silvicultural manipulations have been developed and implemented for management of invasive species.
- Research on pesticides includes evaluation of efficacy of insecticides, identification of the safest and most effective herbicides, use of rodenticides, and efficacy of toxicants for feral swine and rotenone for invasive fish.

- Biological control research has led to the identification of natural enemies of hemlock woolly adelgid, gypsy moth, and emerald ash borer in their native ranges; development and evaluation of rearing, release, and recovery methods; biological control of invasive weeds with insects; use of soil bacteria for biological control of white nose syndrome of bats; and evaluation of predators for control of invasive aquatic organisms.
- Vaccination and immunomodulation are being evaluated to control vertebrate diseases including sylvatic plague, white nose syndrome, and amphibian chytridiomycosis.
- Considerable research has been conducted on development of host resistance for many tree diseases including chestnut blight, Dutch elm disease, and beech bark disease as well as development of resistant ash against emerald ash borer, possible host resistance in hemlock against hemlock woolly adelgid, and frogs that are resistant to amphibian chytridiomycosis.
- Research on reproduction control has led to mating disruption for management of gypsy moth and immunocontraceptives for feral horses and swine.
- IPM programs incorporate multiple techniques such as biological control, habitat manipulation, modification of cultural practices, use of pesticides, and use of resistant varieties that are consolidated into a unified program. Integrated ecosystem- or landscape-level programs are being developed and evaluated for hemlock woolly adelgid, emerald ash borer, gypsy moth, invasive plants, and aquatic organisms.

7.5.12 Key Information Needs

Additional research is needed on a number of issues, including the following:

- Assessment of the effectiveness of legislative control and different outreach methods related to managing human behavior and for informing the development of practical and effective strategies for employing outreach, education, laws, and other social incentives and deterrents to slow substantially the human-mediated spread of invasive species
- Posttreatment monitoring and evaluation of invasive species management responses and efficacy
- Development of improved pesticide and toxicant application delivery methods, rates, and frequencies to effectively control invasive species but minimally affect nontarget native species and the ecosystem
- Better understanding of plant-plant interactions, system-specific plant-insect chemical ecology, and cross-trophic level interactions
- Knowledge of how climate change will affect different biological control applications, chemical pesticide efficacy, and cultural control treatments
- Better integration of methods for combining toxicant and reproductive controls for invasive vertebrates that are more society-friendly
- Better species-specific methods for assessing invasive species density/abundance in order to more effectively evaluate the relative success of control programs
- Improved decision-support tools that take into account ecological and economic factors to assist managers in prescribing management approaches, designing integrated pest management strategies, and determining conciliatory strategies when an invasive species cannot be stopped
- Development of standardized protocols for systematically surveying and testing for pesticide resistance
- Development of improved rearing, release, and recovery methods for introduced natural enemies and evaluation of their interactions with each other and native species for biological control of major invasive insects and plants
- Better assessment of efficacy of integrated pest management programs and adaptation of implementation guidelines

7.6 Recent Advances in Development of Tools for Invasive Species Management

Significant advances have been made in the past decade in developing tools for managing invasive species. One of the first steps in containing and managing an invasive species is to accurately identify the damaging agent and determine its distribution. The longer that species introductions go undetected or unidentified, the more difficult it becomes to control the introduced population (Simberloff 2003). Development of new and improved monitoring and detection technologies including traps, lures, and molecular tools has enhanced both early detection and our ability to monitor and manage invasive species. New tools are also being developed for suppressing and managing established populations of invasive species, while existing technologies must be evaluated and modified for use against new invasive species. New management tools may include discovery of new biological control agents or host resistance traits that are specific to the particular invasive species. Considerable progress has been made in developing new data management and decision systems for use in pest management.

7.6.1 Advances in Surveys and Traps for Monitoring and Early Detection of Invasive Species

Semiochemical attractants have been identified, and traps and lures have been developed for early detection of many insect species, including gypsy moth (Sharov et al. 2002), polyphagous and Kuroshio shot hole borers (*Euwallacea* spp.) (Dodge et al. 2017), and sirex woodwasp (Cooperband et al. 2012), and are being improved for other species such as emerald ash borer (Crook and Mastro 2010; Ryall et al. 2012), Asian longhorned beetle (Nehme et al. 2010, 2014), and redbay ambrosia beetle (Kuhns et al. 2014). Other technologies being evaluated include using acoustic detection for Asian longhorned beetle (Mankin et al. 2008), biosurveillance of emerald ash borer utilizing solitary ground nesting predaceous wasps (Careless et al. 2014), and using sniffer dogs to locate and identify Asian longhorned beetle (Errico 2013). The reliance on simple visual surveys to detect spotted lanternfly hinders suppression efforts. Research is needed to determine pheromone behavioral cues and trapping methods for nymphs and adults.

Intensive sampling methods that were used primarily for detection of rare plant species have been used to detect invasive plant species in early stages of invasion (Huebner 2007; Moore et al. 2011). Aerial photography combined with multispectral imagery (includes visible and near-infrared frequencies), hyperspectral sensors, or satellite imagery (e.g., LiDAR, Landsat) has been used successfully to detect invasive plant populations (Huang and Asner 2009) with distinct physical structures (Asner et al. 2008; Gavier-Pizarro et al. 2012), fruit characteristics (Rebbeck et al. 2015), or leaf phenology (many invasive species leaf out earlier in the spring and remain in leaf longer in the fall; Resasco et al. 2007). Despite the usefulness of the data obtained using aerial and remote-sensing technologies, they are limited by labor and equipment costs, safety issues, and a combination of these factors. Furthermore, this technology has no utility for detecting plants that don't possess unique characteristics.

Selection of the best tools for detecting invasive animals is also limited by cost and labor. Jarrad et al. (2011) suggest combining detection of mammals, amphibians, and reptiles for inclusion in a comprehensive surveillance program for invasive terrestrial vertebrates. This approach begins with risk analyses to identify preferred habitats of high-risk invaders, choosing survey areas that match and would therefore have a higher probability of hosting the invader. At those sites, a combination of species-appropriate traps and direct biological surveys for the target invasive species could be carried out (Jarrad et al. 2011).

7.6.2 Advances in Molecular Tools for Detection

Recent advances in molecular diagnostic tools are improving detection and identification of invasive insects, plant pathogens, plants, animal diseases, and aquatic organisms. DNA barcodes are being developed to aid in rapid identification of new invasive species (Ball and Armstrong 2006; Hollingsworth et al. 2011). A major advantage of DNA-based identification is that it provides the ability to identify morphologically indistinct immature stages of insects such as eggs and larvae, as well as damaged specimens, which would have been difficult to identify using conventional methods. Molecular methods are also useful in distinguishing between morphologically similar species that occur in a cryptic species complex (Cooperband et al. 2016; Lopez et al. 2014). Detection technology for plant pathogens has evolved rapidly over the past decade, progressing from molecular-based polymerase chain reaction methods that require 1–2 days to complete to newer techniques that are more rapid. The newer methodology does not require DNA extraction or extensive training to complete, uses portable equipment that can be utilized in the field, and is more specific than immunologically based methods. A recent example of this evolution is the development of diagnostics for *P. ramorum*-infected tissues. Application of sorption real-time assay and/or loop-mediated isothermal amplification (LAMP) assay provides sensitive and specific detection of this pathogen in 30 and 45 min, respectively (Tomlinson et al. 2007). An on-site device has been developed that can identify pathogens within 1 h and does not require specialized equipment (Tomlinson et al. 2010). Additionally, a species-specific assay for *P. ramorum* that uses recombinase polymerase amplification was developed that produces rapid results (as little as 15 min), uses portable equipment, and does not require DNA extraction or extensive training (Miles et al. 2015). Most recently, surface-enhanced Raman scattering (SERS) has been used for label-free and species-specific detection of *P. ramorum* in infected rhododendron (*Rhododendron* spp.) leaves (Yuksel et al. 2015). Accurate detection of the laurel wilt pathogen (*R. lauricola*) has been difficult, due in part to the occurrence of related fungi in the same affected plant. Because *R. lauricola* samples contain relatively low concentrations of diseased tissue, currently available diagnostic methods don't possess the sensitivity to reliably detect the fungus in woody tissue. However, using primers to amplify two taxon-specific simple-sequence repeat (SSR) loci with 0.1 ng detection limit for *R. lauricola* has improved the sensitivity of these tests. This method is now routinely used in diagnostic clinics and by researchers at the University of Florida to identify isolates obtained from suspect host tissues (Dreaden et al. 2014).

During the past decade, there has been a dramatic increase in the technical ability to conduct surveillance for invasive aquatic animals, producing new technology that is rapid, inexpensive, and highly sensitive (Trebitz et al. 2017). Much of this progress can be attributed to the development of environmental DNA (eDNA) methods that allow researchers to screen for the presence of very rare aquatic species, using technologies which can detect just a few cells in a water sample (Jerde et al. 2011; Wilcox et al. 2013) (see Chap. 9).

Recently developed innovations in molecular diagnostics have produced accurate and conclusive confirmation that *P. destructans* is the causative agent of white nose syndrome of bats (Shuey et al. 2014). Long-wave ultraviolet (UV) light, which produces a distinctive orange-yellow florescence in response to microscopic skin lesions present on the wings of infected bats, is a reliable nondestructive diagnostic method that can be used to detect white nose syndrome without disturbing hibernating bats (USGS 2014). More details on eDNA tools are covered in Chap. 10.

7.6.3 Tools for Suppression of Invasive Species

Tools have also been developed to suppress established invasive pest populations. Some examples include (1) improved delivery techniques and new pesticide chemistries for chemical control of several invasive species; (2) development of rearing, release, and recovery methods for natural enemies used for biological control; and (3) traditional and transgenic breeding tools for developing resistant hosts. An improved formulation of the insecticide emamectin benzoate, TREE-äge®, was developed and tested (Herms and McCullough 2014; McCullough et al. 2011) along with new injection tools to improve its delivery into trees for protection against emerald ash borer (Doccola et al. 2015). Rearing and release methods have been developed for parasitoids of emerald ash borer (Duan et al. 2012, 2013; Gould et al. 2011) and predators of hemlock woolly adelgid (Havill et al. 2011), and trapping methods are being tested to recover released natural enemies (Abell et al. 2015). Transgenically developed resistant elm and American chestnuts have been developed and are currently being evaluated for their resistance to disease (Newhouse et al. 2007, 2014).

Recent advances in three-dimensional printing technology and computer applications have facilitated the development of highly technical tools for use in management of invasive pests. New material processes for bioreplication have furthered the development of nanoreplication of beetles that possess an accurate nanostructure of the exoskeleton and physical properties that include iridescent color reflection (Domingue et al. 2014a). Nano-fabricated and three-dimensional printed emerald ash borer decoy females, electrically charged with high DC voltage, have been developed to attract and electrocute males (Domingue et al. 2014b). A recent application that tracks and monitors releases of parasitoids for biological control of emerald ash borer is now available for use on smart phone devices (Mapbiocontrol 2016).

7.6.4 Current Research that May Lead to the Development of New Management Tools

Development of a standardized DNA barcoding system for plants has been more challenging than those developed for animals, fungi, and insects, which use a portion of the cytochrome oxidase 1 (CO1) mitochondrial gene. Unfortunately, the low rate of nucleotide substitution in plant mitochondrial genomes precludes use of CO1 as a universal plant barcode. Most researchers now concur that multiple markers will be required to adequately discriminate among plant species. Currently, most plant DNA barcodes can only identify to species group and not species (Hollingsworth et al. 2011).

A new plant mating-disruption tool may include use of cytoplasmic male sterility (CMS). CMS has been observed in over 150 plant species and is a maternally inherited condition in which a plant is unable to produce functional pollen. It is associated with chimeric mitochondria and may be induced by interspecific crosses (Schnable and Wise 1998). It's been theorized that the use of CMS to control invasive plants is potentially effective against non-selfing species (dioecious and self-incompatible species) and should be further evaluated. CMS could spread in some populations despite causing severe reduction in the invasive species' fitness, resulting in rapid population extinction (Hodgins et al. 2009).

It has been suggested for several years that unmanned aerial vehicles (drones) could be deployed to locate invasive plants in remote areas (Jay et al. 2009; Pajares 2015a, b); deposit biocontrol agents, herbicides, or ignition sources; and monitor the efficacy of control measures; however, current regulations and safety guidelines have limited their use.

The opportunity for and direct costs of invasion control for invasive plants needs to be incorporated into existing forest growth and yield models or individual-based forest yield models, such as the Forest Service's Forest Vegetation Simulator (FVS) model.

By combining structured demography (integral projection models) with spatial spread models in discrete time, detailed projections of population growth and spread, as well as sensitivities and elasticities associated with both growth and spread, can be determined. Such models may allow us to predict the most and least sensitive stages of growth and spread of invasive plants and determine how variation at the different growth stages contributes to the spread of that plant (Jongejans et al. 2011).

Invasion simulations of a single invasive species into a food web indicate that food webs with the most species but the least connections among those species are the most likely to be invaded (Romanuk et al. 2009). Additional ecological network simulations that involve multiple invasive species introductions are needed to understand how invasional meltdowns occur. Do they become part of an existing stable ecological network or do they form a new highly connected network?

Invasions are primarily human-mediated, and consequently any success at managing invasive plants will require including a human dimension. Large global datasets that include information on population growth and environmental change may be integrated with economic costs of invasions to develop decision models. Terra Populus, developed by the Minnesota Population Center at the University of Minnesota, may be one such model. Terra Populus combines census and survey data (from the Integrated Public Use Microdata Series) with data on agricultural acreage and yields (Global Landscapes Initiative) and data from the Global Land Cover 2000 and WorldClim datasets. Other large data sets including the Global Invasive Species Database (ISSG 2017), Forest Service Forest Inventory and Analysis invasive plant data, USDA Natural Resources Conservation Service PLANTS database, and/or the Global Compendium of Weeds (HEAR 2017) may also be integrated in decision models incorporating human dimensions, population growth, and environmental change.

7.6.5 Data System Design

Advancements in computer hardware and software have expedited more sophisticated designs for data management and data systems needed in pest management programs. System design is the process of defining the architecture, components, modules, interfaces, and data for a system to achieve desired objectives. Data system designs include requirements for input and output, data storage and processing, and system control. Reliable data on pests and pest management are necessary for building reliable models, performing accurate analyses, developing effective policies, and making good management decisions. Data systems are needed for database management, integration of quantitative and spatial data, analytical algorithms, and decision-support tools. Data input and delivery may occur in real time through a network, and output and decisions may be delivered through web servers. Advancement of new technologies in remote sensing and spatially linked data loggers will expedite development of more sophisticated database and data processing systems.

7.6.6 Key Findings

- Major advances have been made in providing molecular tools for identification and detection of invasive pests.
- Improved chemistries and delivery tools have been developed and evaluated for pesticides used to control invasive pests.

7.6.7 Key Information Needs

- Development of new rearing, release, and recovery methods for natural enemies used in biological control programs, along with tools and models for evaluating their efficacy
- Traditional and transgenic breeding tools for developing resistant hosts
- Development of new high-technology tools such as three-dimensional printing and nanoreplication of insect decoy traps, mobile applications for tracking biological control releases, remote sensing using unmanned aerial vehicles (drones), and invasion simulation and dispersal models
- Improved detection tools for fast and accurate identification of new invasive species and broad-scale monitoring of invasive tree pathogens
- DNA barcoding to distinguish look-alike native and non-native invasive congeners and to identify species more reliably
- Models that include competing and/or facilitative co-occurring plants, herbivores, pathogens, and symbionts and that predict how they may or may not co-migrate in response to a changing environment
- Further investigation into the use of new management tools including novel genetic manipulations and cytoplasmic male sterility of plants
- Invasive species databases that link with global population, land use, and global climate change global databases

Literature Cited

Abbott RD, Osorio JE, Bunck CM, Rocke TE (2012) Sylvatic plague vaccine: a new tool for conservation of threatened and endangered species? EcoHealth 9:243–250

Abell K, Poland T, Cosse A, Bauer L (2015) Trapping techniques for emerald ash borer and its introduced parasitoids. In: Van Driesche RG, Reardon RC (eds) Biology and control of emerald ash borer, FHTET-2014-09. USDA Forest Service, Forest Health Technology Enterprise Team, Morgantown, pp 113–129. http://www.fs.fed.us/foresthealth/technology/pdfs/FHTET-2014-09_Biology_Control_EAB.pdf

Ainouche ML, Fortune PM, Salmon A et al (2009) Hybridization, polyploidy and invasion: lessons from *Spartina* (Poaceae). Biol Invasions 11:1159–1173

Albright TP, Chen H, Chen L, Guo Q (2010) The ecological niche and reciprocal prediction of the disjunct distribution of an invasive species: the example of *Ailanthus altissima*. Biol Invasions 12:2413–2427

Ambourn A, Juzwik J, Moon RD (2005) Seasonal dispersal of the oak wilt fungus by *Colopterus truncatus* and *Carpophilus sayi* in Minnesota. Plant Dis 89:1067–1076

Ambourn AK, Juzwik J, Eggers J (2006) Flight periodicities, phoresy rates, and colonization characteristics of *Pseudopityophthorus minutissimus* in oak wilt centers. For Sci 52:243–250

Anderson LW (2005) California's reaction to *Caulerpa taxifolia*: a model for invasive species rapid response. Biol Invasions 7(6):1003–1016

Arias RS, Netherland MD, Scheffler BE et al (2005) Molecular evolution of herbicide resistance to phytoene desaturase inhibitors in *Hydrilla verticillata* and its potential use to generate herbicide-resistant crops. Pest Manag Sci 61:258–268

Ashman TL, Kwok A, Husband BC (2013) Revisiting the dioecy-polyploidy association: alternate pathways and research opportunities. Cytogenet Genome Res 140:241–255

Asner GP, Knapp DE, Kennedy-Bowdoin T et al (2008) Invasive species detection in Hawaiian rainforests using airborne imaging spectroscopy and LiDAR. Remote Sens Environ 112:1942–1955

Association of Official Seed Certifying Agencies (AOSCA) (2016) Seed certification. http://www.aosca.org/page/Seed_Certification.aspx. 6 December 2016.

Aukema JE, Leung B, Kovacs K et al (2011) Economic impacts of non-native forest insects in the continental United States. PLoS One 6(9):e24587

Ayres DR, Smith DL, Zaremba K et al (2004) Spread of exotic cordgrasses and hybrids (*Spartina* sp.) in the tidal marshes of San Francisco Bay, California, USA. Biol Invasions 6:221–231

Ayres DR, Grotkopp E, Zaremba K et al (2008) Hybridization between invasive *Spartina densiflora* (Poaceae) and native *S. foliosa* in San Francisco Bay, California, USA. Am J Bot 95(6):713–719

Bajema RA, DeVault TL, Scott PE, Lima SL (2009) Reclaimed coal mine grasslands and their significance for Henslow's sparrows in the American Midwest. Auk 118:422–431

Ball SL, Armstrong KF (2006) DNA barcodes for insect pest identification: a test case with tussock moths (Lepidoptera: Lymantriidae). Can J For Res 36:337–350

Barden LS (1987) Invasion of *Microstegium vimineum* (Poaceae), an exotic, annual, shade-tolerant, C4 grass, into a North Carolina floodplain. Am Midl Nat 118:40–45

Barney JN, Tekiela DR, Dollete ES, Tomasek BJ (2013) What is the "real" impact of invasive plant species? Front Ecol Environ 11(6):322–329

Bartuszevige AM, Endress BA (2008) Do ungulates facilitate native and exotic plant spread? Seed dispersal by cattle, elk and deer in northeastern Oregon. J Arid Environ 72:904–913

Bataille A, Cashins SD, Grogan L et al (2015) Susceptibility of amphibians to chytridiomycosis is associated with MHC class II conformation. Proc R Soc Lond B Biol Sci 282:20143127

Bauer LS, Duan JJ, Gould JG, Van Driesche RG (2015) Progress in the classical biological control of *Agrilus planipennis* Fairmaire (Coleoptera: Buprestidae) in North America. Can Entomol 147:300–317

Beauchamp VB, Ghuznavi N, Koontz SM, Roberts RP (2013) Edges, exotics and deer: the seed bank of a suburban secondary successional temperate deciduous forest. Appl Veg Sci 16:571–584

Beisner BE, Haydon DT, Cuddington K (2003) Alternative stable states in ecology. Front Ecol Environ 1(7):376–382

Beltran RS, Kreidler N, Van Vuren DH et al (2014) Passive recovery of vegetation after herbivore eradication on Santa Cruz Island, California. Restor Ecol 22(6):790–797

Ben-Haim Y, Osteen CD, Moffitt LJ (2013) Policy dilemma of innovation: an info-gap approach. Ecol Econ 85:130–138

Berger L, Speare R, Daszak P et al (1998) Chytridiomycosis causes amphibian mortality associated with population declines in the rain forests of Australia and Central America. Proc Natl Acad Sci 95:9031–9036

Biggins DE, Godbey JL, Gage KL et al (2010) Vector control improves survival of three species of prairie dogs (*Cynomys*) in areas considered enzootic for plague. Vector Borne Zoonotic Dis 10(1):17–26

Bimova K, Mandak B, Pysek P (2003) Experimental study of vegetative regeneration in four invasive *Reynoutria* taxa (Polygonaceae). Plant Ecol 166:1–11

BioForest Technologies (2016) Healthy forests: RotstopC for heterobasidion root disease. http://www.bioforest.ca/index.cfm?fuseaction=content&menuid=11&pageid=1030

Blair AC, Hufbauer RA (2010) Hybridization and invasion: one of North America's most devastating invasive plants shows evidence for a history of interspecific hybridization. Evol Appl 3:40–51

Blehert DS, Meteyer CU (2011) Bat white nose syndrome in North America. In: Fungal diseases: an emerging threat to human, animal, and plant health: workshop summary. National Academies Press, Washington, DC, 167 p

Blehert D, Hicks AC, Behr M et al (2009) Bat white-nose syndrome: an emerging fungal pathogen? Science 323(5911):227

Bletz MC, Loudon AH, Becker MH et al (2013) Mitigating amphibian chytridiomycosis with bioaugmentation: characteristics of effective probiotics and strategies for their selection and use. Ecol Lett 16(6):807–820

Blossey B, Nötzold R (1995) Evolution of increased competitive ability in invasive nonindigenous plants: a hypothesis. J Ecol 83:887–889

Bodenchuk M, Halstead TD, Yeary MA (2013) Prairie dog management and conservation benefits. In: Armstrong JB, Gallagher GR (eds) Proceedings of the 15th wildlife damage management conference. http://digitalcommons.unl.edu/icwdm_wdmconfproc/163/

Boissin E, Hurley B, Wingfield MJ et al (2012) Retracing the routes of introduction of invasive species: the case of the *Sirex noctilio* woodwasp. Mol Ecol 21:5728–5744

Bomford M, O'Brien PH (1990) Deterrents in animal damage control: a review of device tests and effectiveness. Wildl Soc Bull 18:411–422

Boyce WM, Vickers W, Morrison SA et al (2011) Surveillance for West Nile virus and vaccination of free-ranging island scrub-jays (*Aphelocoma insularis*) on Santa Cruz Island, California. Vector Borne Zoonotic Dis 11(8):1063–1068

Brantley ST, Mayfield AE, Jetton RM et al (2017) Elevated light levels reduce hemlock woolly adelgid infestation and improve carbon balance of infested eastern hemlock seedlings. For Ecol Manag 385:150–160

Bray AM, Bauer LS, Poland TM et al (2011) Genetic analysis of emerald ash borer (*Agrilus planipennis* Fairmaire) populations in Asia and North America. Biol Invasions 13:2869–2887

Bremner A, Park K (2007) Public attitudes to the management of invasive non-native species in Scotland. Biol Conserv 139:306–314

Brownlee-Bouboulis SA, Reeder DM (2013) White-nose syndrome-affected little brown myotis (*Myotis lucifugus*) increase grooming and other active behaviors during arousals from hibernation. J Wildl Dis 49(4):850–859

Bruzzone OA, Villacide JM, Bernstein C, Corley JC (2009) Flight variability in the woodwasp *Sirex noctilio* (Hymenoptera: Siricidae): an analysis of flight data using wavelets. J Exp Biol 212(5):731–737

Bryson CT, Carter R (2004) Biology of pathways for invasive weeds. Weed Technol 18(sp1):1216–1220

Buchan LA, Padilla DK (1999) Estimating the probability of long-distance overland dispersal of invading aquatic species. Ecol Appl 9(1):254–265

Buckley YM, Catford J (2016) Does the biogeographic origin of species matter? Ecological effects of native and non-native species and use of origin to guide management. J Ecol 104:4–17

California Department of Fish and Wildlife (CDFW) (2010) CDFW hatchery operations: final CDFW environmental impact report (EIR)/Environmental impact statement (EIS). https://wildlife.ca.gov/Fishing/Hatcheries/EIR

California Oak Mortality Task Force (COMTF) (2016) Sudden oak death: clonal lineages and distribution. http://www.suddenoakdeath.org/research/clonal-lineages-and-distribution/

Callaham MA Jr, Gonzalez G, Hale CM et al (2006) Policy and management responses to earthworm invasions in North America. Biol Invasions 8:1317–1329

Campanella DM, McEvoy PB, Mundt CC (2009) Interaction effects of two biological control organisms on resistant and susceptible weed biotypes of *Chondrilla juncea* in western North America. Biol Control 50:50–59

Campbell KJ, Baxter GS, Murray PJ et al (2005) Increasing the efficiency of Judas goats by sterilization and pregnancy termination. Wildl Res 32:737–743

Campbell TA, Foster JA, Bodenchuk MJ et al (2013) Effectiveness and target-specificity of a novel design of food dispenser to deliver a toxin to feral swine in the United States. Int J Pest Manag 59:197–204

Careless P, Marshall SA, Gill BD (2014) The use of *Cerceris fumipennis* (Hymenoptera: Crabronidae) for surveying and monitoring emerald ash borer (Coleoptera: Buprestidae) infestations in eastern North America. Can Entomol 146:90–105

Carnegie AJ, Eldridge RH, Waterson DG (2005) History and management of sirex wood wasp in pine plantation in New South Wales, Australia. N Z J For Sci 35(1):3–24

Carroll SP (2011) Conciliation biology: the eco-evolutionary management of permanently invaded biotic systems. Evol Appl 4:184–199

Carroll T, Miller R (2016) Diamond lake: a new approach to restoration. Newsletter from Oregon Lake Association. September 2016, pp 3–4

Carson WP, Hovick SM, Baumert AJ et al (2008) Evaluating the post-release efficacy of invasive plant biocontrol by insects: A comprehensive approach. Arthropod Plant Interact 2:77–86

Carter M, Smith M, Harrison R (2010) Genetic analyses of the Asian longhorned beetle (Coleoptera, Cerambycidae, *Anoplophora glabripennis*), in North America, Europe and Asia. Biol Invasions 12:1165–1182

Cashins SD, Grogan LF, McFadden M et al (2013) Prior infection does not improve survival against the amphibian disease chytridiomycosis. PLoS One 8:e56747

Center for Invasive Species and Ecosystem Health (CISEH) (2016) Invasive and exotic species of North America. Center for Invasive Species and Ecosystem Health. http://www.invasive.org/

Center for Invasive Species Management (CISM) (2016) Developing a CWMA/CISMA. http://weedcenter.org/cwma/index.html

Chabrerie O, Roulier F, Hoeblich H et al (2007) Defining patch mosaic functional types to predict invasion patterns in a forest landscape. Ecol Appl 17(2):464–481

Chambers JC, Bradley BA, Brown CS et al (2014) Resilience to stress and disturbance, and resistance to *Bromus tectorum* L. invasion in cold desert shrublands of western North America. Ecosystems 17(2):360–375

Chapman JI, Cantino PD, McCarthy BC (2012) Seed production in garlic mustard (*Alliaria petiolata*) prevented by some methods of manual removal. Nat Areas J 32(3):305–315

Chen LY (2001) Cost savings from properly managing endangered species habitats. Nat Areas J 21:197–203

Chen Y, Poland TM (2010) Nutritional and defensive chemistry of three North American ash species: possible roles in host performance and preference by emerald ash borer adults. Great Lakes Entomologist 43:20–33

Chen Y, Whitehill JGA, Bonello P, Poland TM (2011) Differential response in foliar chemistry of three ash species to emerald ash borer adult feeding. J Chem Ecol 37:29–39

Cheplick GP (2005) Biomass partitioning and reproductive allocation in the invasive, cleistogamous grass *Microstegium vimineum*: influence of the light environment. J Torrey Bot Soc 132(2):214–224

Chornesky EA, Bartuska AM, Aplet GH et al (2005) Science priorities for reducing the threat of invasive species to sustainable forestry. Bioscience 55(4):335–348

Christen DC, Matlack GR (2009) The habitat and conduit functions of roads in the spread of three invasive plant species. Biol Invasions 11:453–465

Cipollini D (2015) White fringetree as a novel larval host for emerald ash borer. J Econ Entomol 108(1):370–375

Cipollini D, Wang Q, Whitehill JGA et al (2011) Distinguishing defense characteristics in the phloem of ash species resistant and susceptible to emerald ash borer. J Chem Ecol 37:450–459

Cipollini D, Rigsby CM, Barto EK (2012) Microbes as targets and mediators of allelopathy in plants. J Chem Ecol 38:714–727

Claeson SM, Bisson PA (2013) Passive reestablishment of riparian vegetation following removal of invasive knotweed (*Polygonum*). Invasive Plant Sci Manag 6:208–218

Cock MJ, Murphy ST, Kairo MT et al (2016) Trends in the classical biological control of insect pests by insects: an update of the BIOCAT database. BioControl 61(4):349–363

Coleman TW, Seybold SJ (2011) Collection history and comparison of the interactions of the goldspotted oak borer, *Agrilus auroguttatus* Schaeffer (Coleoptera: Buprestidae), with host oaks in southern California and southeastern Arizona, U.S.A. Coleopt Bull 65(2):93–108

Cooperband MF, Boroczky K, Hartness A et al (2012) Male-produced pheromone in the European woodwasp, *Sirex noctilio*. J Chem Ecol 38:52–62

Cooperband MF, Stouthamer R, Carrillo D et al (2016) Biology of two members of the *Euwallacea fornicatus* species complex (Coleoptera: Curculionidae: Scolytinae), recently invasive in the USA, reared on an ambrosia beetle artificial diet. Agric For Entomol 18(3):223–237

Cornelison CT, Gabriel KT, Barlament C, Crow SA Jr (2014) Inhibition of *Pseudogymnoascus destructans* growth from conidia and mycelial extension by bacterially produced volatile organic compounds. Mycopathologia 177:1–10

Cowled BA, Elsworth P, Lapidge SJ (2008) Additional toxins for feral pig (*Sus scrofa*) control: identifying and testing Achilles' heel. Wildl Res 35:651–662

Cowles RS, Cheah CAS-J (2002) Foliar sprays for control of hemlock woolly adelgid, 2001. Arthropod Manag Tests 27:G48

Crawley MJ (1989) Insect herbivores and plant population dynamics. Annu Rev Entomol 34:531–564

Crone EE, Marler M, Pearson DE (2009) Non-target effects of broadleaf herbicide on a native perennial forb: a demographic framework for assessing and minimizing impacts. J Appl Ecol 46:673–682

Crook DJ, Mastro VC (2010) Chemical ecology of the emerald ash borer *Agrilus planipennis*. J Chem Ecol 36:101–112

Cryan PM, Uphoff Meteyer C, Boyles JG, Blehert DS (2010) Wing pathology of white-nose syndrome in bats suggests life-threatening disruption of physiology. BMC Biol 20(108):135–142

Cuda J (2014) Chapter 8: Introduction to biological control of aquatic weeds. In: Gettys LA, Haller WT, Petty DG (eds) Biology and control of aquatic plants: a best management practices handbook, 3rd edn. Aquatic Ecosystem Restoration Foundation, Marietta, pp 51–58

Cuda JP, Christ LR, Manrique V et al (2012) Role of molecular genetics in identifying 'fine tuned' natural enemies of the invasive

Brazilian peppertree, *Schinus terebinthifolius*: a review. BioControl 57:227–233

Cuda JP, Shearer JF, Weeks ENI et al (2014) Compatibility of an insect, a fungus, and a herbicide for integrated pest management of dioecious hydrilla. J Aquat Plant Manag 54:20–25

Culver CS, Kuris AM (2000) The apparent eradication of a locally established introduced marine pest. Biol Invasions 2(3):245–253

Cunard CE, Lankau RA (2017) Declining survival across invasion history for *Microstegium vimineum*. PLoS One 12(8):e0183107

Davidson JH (2009) North America's great carbon ocean: protecting prairie grasslands keeps carbon in the soil and slows the pace of climate change. In: Saving land, vol 29(1). Land Trust Alliance, Washington, DC

DeCant JP (2008) Russian olive, *Elaeagnus angustifolia*, alters patterns in soil nitrogen pools along the Rio Grande River, New Mexico, USA. Wetlands 28:896–904

Diaz R, Dickey AM, Shatters RG Jr et al (2015) New species diversity revealed from molecular and morphological characterization of gall-inducing *Calophya* spp. (Hemiptera: Calophyidae) from Brazilian peppertree. Fla Entomol 98:776–779

Dinsmore SJ (2013) Mountain plover responses to deltamethrin treatments on prairie dog colonies in Montana. Ecotoxicology 22(2):415–424

DiTomaso JM, Kyser GB, Oneto SR et al (2013) Weed control in natural areas in the Western United States. University of California Weed Research and Information Center, Davis, 544 p

DiTommaso A, Morris SH, Parker JD et al (2014) Deer browsing delays succession by altering aboveground vegetation and belowground seed banks. PLoS One 9(3):e91155

Doccola JJ, Herms DA, Miller FD et al (2015) Innovation under pressure: new tools and methods in systemic injection. In: Buck J, Parra G, Lance D, compilers et al (eds) Proceedings of the emerald ash borer national research and technology development meeting, 15–16 October, 2014, FHTET 2015-07. USDA Forest Service, Wooster, pp 14–17

Dodds KI, Cooke RR, Gilmore DW (2007) Silvicultural options to reduce pine susceptibility to attack by a newly detected invasive species, *Sirex noctilio*. North J Appl For 24:165–167

Dodge C, Coolidge J, Cooperband M et al (2017) Quercivorol as a lure for the polyphagous and Kuroshio shot hole borers, *Euwallacea* spp. nr. *fornicatus* (Coleoptera: Scolytinae), vectors of Fusarium dieback. PeerJ 5:e3032v1

Domingue MJ, Lakhtakia A, Pulsifer DP et al (2014a) Bioreplicated visual features of nanofabricated buprestid beetle decoys evoke stereotypical male mating flights. Proc Natl Acad Sci 111:14106–14,111

Domingue MJ, Pulsifer DP, Lakhtakia A et al (2014b) Nanofabricated decoys used in an electrocution trap for emerald ash borer and other Agrilus spp. In: McManus KA, Gottschalk KW, Compilers (eds) Proceedings of the USDA interagency research forum on invasive species 2014, FHTET 2014-01, vol 11, pp 11–12

Dorcas ME, Willson JD (2011) Invasive pythons in the United States: ecology of an introduced predator. University of Georgia Press, Athens, 156 p

Dorcas ME, Pittman SE, Willson JD (2017) Burmese pythons. In: Pitt W, Beasley JC, Witmer GW (eds) Ecology and management of terrestrial vertebrate invasive species in the United States. CRC Press, Boca Raton, 403 p

Drake DAR, Mandrak NE (2014) Ecological risk of live bait fisheries: a new angle on selective fishing. Fisheries 39:201–211

Drake DAR, Mercader R, Dobson T, Mandrak NE (2015) Can we predict risky human behaviour involving invasive species? A case study of the release of fishes to the wild. Biol Invasions 17:309–326

Dreaden TJ, Davis JM, Harmon CL et al (2014) Development of multilocus PCR assays for *Raffaelea lauricola*, causal agent of laurel wilt disease. Plant Dis 98:379–383

Duan JJ, Bauer LS, Hansen JA et al (2012) An improved method for monitoring parasitism and establishment of *Oobius agrili* (Hymenoptera: Encyrtidae), an egg parasitoid introduced for biological control of the emerald ash borer (Coleoptera: Buprestidae) in North America. Biol Control 60:255–261

Duan JJ, Bauer LS, Abell KJ et al (2013) Establishment and abundance of *Tetrastichus planipennisi* (Hymenoptera: Eulophidae) in Michigan: potential for success in classical biocontrol of the invasive emerald ash borer (Coleoptera: Buprestidae). J Econ Entomol 106:1145–1154

Dunham JB, Pilliod DS, Young M (2004) Assessing the consequences of nonnative trout in headwater ecosystems. Fisheries 29:18–26

Dyer AR, Knapp EE, Rice KJ (2016) Unintentional selection and genetic changes in native perennial grass populations during commercial seed production. Ecol Restor 34(1):39–48

Elkinton JS, Liebhold AM (1990) Population dynamics of gypsy moth in North America. Annu Rev Entomol 35:571–596

Ellstrand NC, Schierenbeck KA (2000) Hybridization as a stimulus for the evolution of invasiveness in plants? Proc Natl Acad Sci U S A 97(13):7043–7050

Elton CS (1958) The ecology of invasions by animals and plants. Methuen, London, 181 p

Epanchin-Niell RS, Hufford MB, Aslan CE et al (2010) Controlling invasive species in complex social landscapes. Front Ecol Environ 8(4):210–216

Errico M (2013) Asian longhorned beetle detector dog pilot project. In: McManus KA, Gottschalk KW (eds) Proceedings, 23rd U.S. Department of Agriculture interagency research forum on invasive species 2012, Gen. Tech. Rep. NRS-P-114. U.S. Department of Agriculture, Forest Service, Northern Research Station, Newtown Square, p 18

Eskalen A, Stouthamer R, Lynch SC et al (2013) Host range of fusarium dieback and its ambrosia beetle (Coleoptera: Scolytinae) vector in Southern California. Plant Dis 97:938–951

Estoup A, Guillemaud T (2010) Reconstructing routes of invasion using genetic data: why, how, and so what? Mol Ecol 19:4113–4130

Evans HC, Holmes KA, Thomas SE (2003) Endophytes and mycoparasites associated with an indigenous forest tree, *Theobroma gileri*, in Ecuador and a preliminary assessment of their potential as biocontrol agents of cocoa diseases. Mycol Prog 2(2):149–160

Evans JA, Davis AS, Raghu S et al (2012) The importance of space, time, and stochasticity to the demography and management of *Alliaria petiolate*. Ecol Appl 22(5):1497–1511

Fagerstone KA, Miller LA, Killian G, Yoder CA (2010) Review of issues concerning the use of reproductive inhibitors, with particular emphasis on resolving human-wildlife conflicts in North America. Integrative Zoology 1:15–30

Fajvan MA (2008) The role of silvicultural thinning in eastern forests threatened by hemlock woolly adelgid (*Adelges tsugae*). In: Deal RL, tech (eds) Integrated restoration of forested ecosystems to achieve multiresource benefits. Proceedings of the 2007 national silviculture workshop. 2007 May 7–10, Ketchikan, AK. Gen. Tech. Rep. PNW-733. U.S. Department of Agriculture, Forest Service, Pacific Northwest Research Station, Portland, pp 247–256

Fauci AS, Morens DM (2016) Zika virus in the Americas — Yet another arbovirus threat. N Engl J Med 374:601–604

Faush KD, Rieman BE, Dunham JB et al (2009) Invasion versus isolation: trade-offs in managing native salmonids with barriers to upstream movement. Conserv Biol 23:859–870

Federal Noxious Weed Act (FNWA) (1974) Federal noxious weed act. Pub.L. 93–629, 88 Stat. 2148

Ferguson TL, Buccowich M, Schneeberger N (2013) Hemlock woolly adelgid initiative protecting the hemlock resource in the East. http://www.na.fs.fed.us/ra/specialinitiatives/hwa/2013/hemlock-woolly-adelgid-130131.pdf. 8 June 2016

Fernandez JRR, Rocke TE (2011) Use of Rhodamine B as a biomarker for oral plague vaccination of prairie dogs. J Wildl Dis 47(3):765–768

Finlayson BJ, Eilers JM, Huchko HA (2014) Fate and behavior of rotenone in Diamond Lake, Oregon, USA following invasive tui chub eradication. Environ Toxicol Chem 33(7):1650–1655

Floore TG (2006) Mosquito larval control practices: past and present. J Am Mosq Control Assoc 22(3):527–533

Florida Fish and Wildlife Conservation Commission (FFWCC) (2016) Python challenge: python removal competition. http://pythonchallenge.org/participate/python-removal-competition/

Flory SL, Bauer JT (2014) Experimental evidence for indirect facilitation among invasive plants. J Ecol 102:12–18

Flory SL, Clay K (2013) Pathogen accumulation and long-term dynamics of plant invasions. J Ecol 101:607–613

Fraedrich SW, Harrington TC, Rabaglia RJ et al (2008) A fungal symbiont of the redbay ambrosia beetle causes a lethal wilt in redbay and other *Lauraceae* in the southeastern United States. Plant Dis 92:215–224

Friesen KM, Johnson GD (2013) Mosquito and West Nile virus surveillance in northeast Montana, U.S.A., 2005 and 2006. Med Vet Entomol 28:85–93

Fritz RS, Simms EL (1992) Plant resistance to herbivores and pathogens: ecology, evolution, and genetics. University of Chicago Press, Chicago, 600 p

Funk DT, Schlesinger RC, Ponder F Jr (1979) Autumn-olive as a nurse plant for black walnut. Botanical Gazette. 140 Supplement: S110-S114.

Gammon MA, Kesseli R (2010) Haplotypes of *Fallopia* introduced to the US. Biol Invasions 12:421–427

Gardner AM, Allan BF, Frisbie LA, Muturi EJ (2015) Asymmetric effects of native and exotic invasive shrubs on ecology of the West Nile virus vector *Culex pipiens* (Diptera: Culicidae). Parasit Vectors 8:329–337

Gargas A, Trest MT, Christensen M et al (2009) *Geomyces destructans* sp. nov. associated with bat white-nose syndrome. Mycotaxon 108:147–154

Garnica DP, Pinzón AM, Quesada-Ocampo LM et al (2006) Survey and analysis of microsatellites from transcript sequences in *Phytophthora* species: frequency, distribution, and potential as markers for the genus. BMC Genomics 7:245

Gaskin JF, Schaal BA (2002) Hybrid *Tamarix* widespread in U.S. invasion and undetected in native Asian range. Proc Natl Acad Sci U S A 99:11256–11,259

Gaskin JF, Bon MC, Cock MJ et al (2011) Applying molecular-based approaches to classical biological control of weeds. Biol Control 58:1–21

Gaskin JF, Schwarzländer M, Kinter CL et al (2013a) Propagule pressure, genetic structure, and geographic origins of *Chondrilla juncea* (Asteraceae): an apomictic invader on three continents. Am J Bot 100:1871–1882

Gaskin JF, Schwarzländer M, Hinz HL et al (2013b) Genetic identity and diversity of perennial pepperweed (*Lepidium latifolium*) in its native and invaded ranges. Invasive Plant Sci Manag 6:268–280

Gavier-Pizarro GI, Kuemmerle T, Hoyos LE et al (2012) Monitoring the invasion of an exotic tree (*Ligustrum lucidum*) from 1983 to 2006 with Landsat TM/ETM + satellite data and Support Vector Machines in Cordoba, Argentina. Remote Sens Environ 122:134–145

Gibbs GW (1998) Why are some weta (Orthoptera: Stenopelmatidea) vulnerable yet others are common? J Insect Conserv 2:161–166

Gomez-Aparicio L, Canham CD (2008) Neighbourhood analyses of the allelopathic effects of the invasive tree *Ailanthus altissima* in temperate forests. J Ecol 96:447–458

Goolsby JA, Van Klinken RD, Palmer WA (2006) Maximising the contribution of native-range studies towards the identification and prioritisation of weed biocontrol agents. Aust J Entomol 45:276–286

Gottschalk K (1993) Silvicultural guidelines for forest stands threatened by the gypsy moth, General Technical Report No. NE-171. U.S. Department of Agriculture, Forest Service, Northeastern Forest Experiment Station, Radnor. ii + 50 p

Gould J, Ayer T, Fraser I (2011) Effects of rearing conditions on reproduction of *Spathius agrili* (Hymenoptera: Braconidae), a parasitoid of the emerald ash borer (Coleoptera: Buprestidae). J Econ Entomol 104:379–387

Government Printing Office (GPO) (2017) 50 CFR 16 injurious wildlife. https://www.gpo.gov/fdsys/granule/CFR-2006-title50-vol1/CFR-2006-title50-vol1-part16

Grant EHC, Muths EL, Katz RA et al (2016) Salamander chytrid fungus (*Batrachochytrium salamandrivorans*) in the United States—Developing research, monitoring, and management strategies. U.S. Geological Survey Open-File Report 2015-1233. https://doi.org/10.3133/ofr20151233

Green PT, O'Dowd DJ, Abbott KL et al (2011) Invasional meltdown: invader-invader mutualism facilitates a secondary invasion. Ecology 92(9):1758–1768

Grevstad F, Shaw R, Bourchier R et al (2013) Efficacy and host specificity compared between two populations of the psyllid *Aphalara itadori*, candidates for biological control of invasive knotweeds in North America. Biol Control 65:53–62

Gruntman M, Zieger S, Tielbörger K (2016) Invasive success and the evolution of enhanced weaponry. Oikos 125:59–65

Gudding R, Lillehaug A, Evensen Ø (1999) Recent developments in fish vaccinology. Vet Immunol Immunopathol 72(1–2):203–212

Guggisberg A, Welk E, Sforza R et al (2012) Invasion history of North American Canada thistle, *Cirsium arvense*. J Biogeography 39:1919–1931

Gurevitch J, Padilla DK (2004) Are invasive species a major cause of extinctions? Trends Ecol Evol 19(9):470–474

Haack RA, Jendek E, Liu H et al (2002) The emerald ash borer: a new exotic pest in North America. Newsl Michigan Entomol Soc 47(3 & 4):1–5

Haeussler S, Bedford L, Leduc A et al (2002) Silvicultural disturbance severity and plant communities of the Southern Canadian Boreal Forest. Silva Fennica 36(1):307–327

Hallam TG, McCracken GF (2010) Management of the panzootic white-nose syndrome through culling of bats. Conserv Biol 25:189–194

Haller W (2014) Chapter 7: Mechanical control of aquatic weeds. In: Gettys LA, Haller WT, Petty DG (eds) Biology and control of aquatic plants: a best management practices handbook, 3rd edn. Aquatic Ecosystem Restoration Foundation, Marietta, Georgia, pp 43–50

Hamer AJ, Land SJ, Mahony MJ (2002) The role of introduced mosquitofish (*Gambusia holbrooki*) in excluding the native green and golden bell frog (*Litoria aurea*) from original habitats in south-eastern Australia. Oecologia 132:445–452

Hansen GJA, Jones ML (2008) A rapid assessment approach to prioritizing streams for control of Great Lakes sea lampreys (*Petromyzon marinus*): a case study in adaptive management. Can J Fish Aquat Sci 65:2471–2484

Hansen E, Peterson E, Hulbert J (2014) Source or sink? The role of soil and water borne inoculum in the dispersal of *Phytophthora ramorum* in Oregon tanoak forests. For Ecol Manag 322:48–57

Hanula JL, Mayfield AE III, Fraedrich SW, Rabaglia RJ (2008) Biology and host associations of redbay ambrosia beetle (Coleoptera: Curculionidae: Scolytinae), exotic vector of laurel wilt killing redbay trees in the Southeastern United States. J Econ Entomol 101(4):1276–1286

Harrington TC, Fraedrich SW, Aghayeva DN (2008) *Raffaelea lauricola*, a new ambrosia beetle symbiont and pathogen on the *Lauraceae*. Mycotaxon 104:399–404

Harrington TC, Aghayeva DN, Fraedrich SW (2010) New combinations in *Raffaelea*, *Ambrosiella* and *Hyalorhinocladiella*, and four new species from the redbay ambrosia beetle, *Xyleborus glabratus*. Mycotaxon 111:337–361

Hasan S, Ayres PG (1990) The control of weeds through fungi; principles and prospects. New Phytol 115(2):201–222

Havill NP, Montgomery ME, Yu G et al (2006) Mitochondrial DNA from hemlock woolly adelgid (Hemiptera: Adelgidae) suggests cryptic speciation and pinpoints the source of the introduction to eastern North America. Ann Entomol Soc Am 99(2):195–203

Havill NP, Davis G, Klein J et al (2011) Hemlock woolly adelgid biological control: molecular methods to distinguish *Laricobius nigrinus, L. rubidus*, and their hybrids. In: McManus KA, Gottschalk KW (eds) Proceedings. 21st U.S. Department of Agriculture interagency research forum on invasive species, 12–15 January 2010, Annapolis, MD. Gen. Tech. Rep. NRS-P-75. U.S. Department of Agriculture, Forest Service, Northern Research Station, Newtown Square, pp 25–28

Havill NP, Vieira LC, Salom SM (2014) Biology and control of hemlock woolly adelgid, FHTET-2014-05. U.S. Department of Agriculture, Forest Service, Forest Health Technology Enterprise Team, Morgantown, 21 p

Hawaiian Ecosystem at Risk (HEAR) (2017) Global compendium of weeds. http://hear.org/gcw/

Hebel CL, Smith JE, Cromack K Jr (2009) Invasive plant species and soil microbial response to wildfire burn severity in the Cascade Range of Oregon. Appl Soil Ecol 42:150–159

Hein CL, Roth BM, Ives AR, Vander Zanden MJ (2006) Fish predation and trapping for rusty crayfish (*Orconectes rusticus*) control: a whole-lake experiment. Can J Fish Aquat Sci 63:383–393

Herms DA, McCullough DG (2014) Emerald ash borer invasion of North America: history, biology, ecology, impact and management. Annu Rev Entomol 59:13–30

Hershdorfer ME, Fernandea-Gimenez ME, Howery LD (2007) Key attributes influence the performance of local weed management programs in the Southwest United States. Rangel Ecol Manag 60:225–234

Hierro JL, Maron JL, Callaway RM (2005) A biogeographical approach to plant invasions: the importance of studying exotics in their introduced and native range. J Ecol 93:5–15

Hine CS, Hagy HM, Horath MM et al (2017) Response of aquatic vegetation communities and other wetland cover types to floodplain restoration at Emiquon Preserve. Hydrobiologia 804(1):59–71

Hobbs RJ (2007) Setting effective and realistic restoration goals: key directions for research. Restor Ecol 15(2):354–357

Hobbs RJ, Arico S, Aronson J et al (2006) Novel ecosystems: theoretical and management aspects of the new ecological world order. Glob Ecol Biogeogr 15(1):1–7

Hodgins KA, Rieseberg L, Otto SP (2009) Genetic control of invasive plant species using selfish genetic elements. Evol Appl. ISSN 1752-4571:555–569

Hollingsworth PM, Graham SW, Little DP (2011) Choosing and using a plant DNA barcode. PLoS One 6(5):1–13

Holmes TP, Aukema JE, Von Holle B et al (2009) Economic impacts of invasive species in forests. Ann N Y Acad Sci 1162:18–38

Hough-Goldstein J, Lake E, Reardon R, Wu Y (2015) Biology and biological control of mile-a-minute weed, FHTET-2008-10 Revised July 2015. U. S. Department of Agriculture, Forest Service, Forest Health Technology Enterprise Team, Morgantown, 75 p

Howald G, Donlan CJ, Galvan JP et al (2007) Invasive rodent eradication on islands. Conserv Biol 21:1258–1268

Hu J, Angeli S, Schuetz S et al (2009) Ecology and management of exotic and endemic Asian longhorned beetle *Anoplophora glabripennis*. Agric For Entomol 11:359–375

Huang C-Y, Asner GP (2009) Applications of remote sensing to alien invasive plant studies. Sensors 9:4869–4889

Hubbard CE (2016) Olfactory preference and reproductive isolation of two *Mecinus* species (Coleoptera: Curculionidae): implications for biological control of Dalmatian, yellow and hybrid populations of toadflax, *Linaria* species. M.S. Thesis, Montana State University. Bozeman, MT

Huebner CD (2007) Detection and monitoring of invasive exotic plants: a comparison of four sampling methods. Northeast Nat 14(2):183–206

Huebner CD (2011) Seed mass, viability, and germination of Japanese stiltgrass (*Microstegium vimineum*) under variable light and moisture conditions. Invasive Plant Sci Manag 4(3):274–283

Huebner CD, Gottschalk KW, Miller GW, Brose PH (2010) Restoration of three forest herbs in the Liliaceae family by manipulating deer herbivory and overstorey and understorey vegetation. Plant Ecolog Divers 3(3):259–272

Hussner A, Stiers I, Verhofstad MJJM et al (2017) Management and control methods of invasive alien freshwater aquatic plants: a review. Aquat Bot 136:112–137

Hutchinson JT, MacDonald GE, Langeland KA (2007) The potential for herbicide resistance in non-native plants in Florida's natural areas. Nat Areas J 27(3):258–263

Institute of Medical and Veterinary Science (IMVS) (2010) Assessing the humaneness and efficacy of a new feral pig bait in domestic pigs. Report for the Australian Government Department of the Environment, Water, Heritage and the Arts. Canberra, Australia. http://www.environment.gov.au/system/files/resources/091b0583-f35c-40b3-a530-f2e0c307a20c/files/pigs-imvs-report.pdf

Invasive Species Specialist Group (ISSG) (2017) Tools: global invasive species database. www.issg.org

Ivors K, Garbelotto M, Vries IDE et al (2006) Microsatellite markers identify three lineages of *Phytophthora ramorum* in US nurseries, yet single lineages in US forest and European nursery populations. Mol Ecol 15:1493–1505

Jachowski DS, Gitzen RA, Grenier MG et al (2011) The importance of thinking big: large-scale prey conservation drives black-footed ferret reintroduction success. Biol Conserv 144:1560–1566

James TY, Toledo LF, Rödder D et al (2015) Disentangling host, pathogen, and environmental determinants of a recently emerged wildlife disease: lessons from the first 15 years of amphibian chytridiomycosis research. Ecology and Evolution 5:4079–4097

Jarrad FC, Barrett S, Murray J et al (2011) Ecological aspects of biosecurity surveillance design for the detection of multiple invasive animal species. Biol Invasions 13:803–818

Jay S, Lawrence R, Repasky K, Keith C (2009) Invasive species mapping using low cost hyperspectral imagery. ASPRS Annual Conference, March 9–13, 2009. Baltimore, MD. 9 p

Jerde CL, Mahon AR, Chadderton WL (2011) "Sight-unseen" detection of rare aquatic species using environmental DNA. Conserv Lett 4:150–157

Jogesh T, Carpenter D, Cappuccino N (2008) Herbivory on invasive exotic plants and their non-invasive relatives. Biol Invasions 10:797–804

Johnson BM, Arlinghaus R, Martinez PJ (2009a) Are we doing all we can to stem the tide of illegal fish stocking? Fisheries 34(8):389–394

Johnson PTJ, Olden JD, Solomon CT, Vander Zanden MJ (2009b) Interactions among invaders: community and ecosystem effects of multiple invasive species in an experimental aquatic system. Oecologia 159:161–170

Jones PH, Biggins DE, Eads DA et al (2012) Deltamethrin flea-control preserves genetic variability of black-tailed prairie dogs during a plague outbreak. Conserv Genet 13:183–195

Jongejans E, Shea K, Skarpaas O et al (2011) Importance of individual and environmental variation for invasive species spread: a spatial integral projection model. Ecology 92(1):86–97

Juzwik J, Skalbeck TC, Neuman MF (2004) Sap beetle species (Coleoptera: Nitidulidae) visiting fresh wounds on healthy oaks during spring in Minnesota. For Sci 50:757–764

Juzwik J, O'Brien J, Evenson Castillo P, Mahal G (2010) Controlling spread of the oak wilt pathogen (*Ceratocystis fagacearum*) in a Minnesota urban forest park reserve. Arboricult Urban For 36:171–178

Juzwik J, Appel DN, MacDonald WE, Burks S (2011) Challenges and successes in managing oak wilt in the United States. Plant Dis 95:888–900

Kalisz S, Spigler RB, Horvitz CC (2014) In a long-term experimental demography study, excluding ungulates reversed invader's explosive population growth rate and restored natives. Proc Natl Acad Sci 111(12):4501–4506

Kapuscinski AR, Sharpe LM (2014) Introduction: genetic biocontrol of invasive fish species. Biol Invasions 16(6):1197–1200

Kats LB, Ferrer RP (2003) Alien predators and amphibian declines: review of two decades of science and the transition to conservation. Divers Distrib 9:99–110

Kay SH, Lewis WM, Langel KA (1995) Integrated management of multiflora rose in North Carolina. North Carolina Agricultural Extension Service. Publication No. AG-536. 17 p

Keane RM, Crawley MJ (2002) Exotic plant invasions and the enemy release hypothesis. Trends Ecol Evol 17:164–170

Keena MA (2002) *Anoplophora glabripennis* (Coleoptera: Cerambycidae) fecundity and longevity under laboratory conditions: comparison of populations from New York and Illinois on *Acer Saccharum*. Environ Entomol 31(3):490–498

Keena MA, Cote M-J, Grinberg PS, Wallner WE (2008) World distribution of female flight and genetic variation in Lymantria dispar (Lepidoptera: Lymantriidae). Environ Entomol 37(3):636–649

Keller RB, Lodge DM (2007) Species invasions from commerce in live aquatic organisms: problems and possible solutions. Bioscience 57(5):428–436

Kilian JV, Klauda RJ, Widman S et al (2012) An assessment of a bait industry and angler behavior as a vector of invasive species. Biol Invasions 14:1469–1481

Killian G, Thain D, Diehl NK et al (2008) Four-year contraception rates of mares treated with single-injection GnRH vaccines and intrauterine devices. Wildl Res 35:531–539

Kirkpatrick J, Turner JW Jr, Perkins A (1982) Reversible chemical fertility control in feral horses. Equine Veterinary Sci. July/August: 114–118

Klinger RC, Schuyler PT, Sterner JD (1994) Vegetation response to the removal of feral sheep from Santa Cruz Island. In: Halvorson WL, Maender GJ (eds) The fourth California Islands symposium: update on the status of resources. Santa Barbara Museum of Natural History, Santa Barbara, pp 341–350

Klinger RC, Schuyler P, Sterner JD (2002) The response of herbaceous vegetation and endemic plant species to the removal of feral sheep from Santa Cruz Island, California. In: Vetch RR, Clout MN (eds) Turning the tide: the eradication of invasive species. IUCN SSC Invasive Species Specialist Group, Gland, pp 141–154

Koch JL, Carey DW, Knight KS et al (2012) Breeding strategies for the development of emerald ash borer-resistant North American ash. In: Sniezko RA, Yanchuk AD, Kliejunas JT et al (eds) tech. coords. Proceedings of the 4th international workshop on genetics host-parasite interactions in forestry: disease and insect resistance in forest trees, Gen. Tech. Rep. PSW-GTR-240. U.S. Department of Agriculture, Forest Service, Pacific Southwest Research Station, Albany, pp 235–239

Kostel-Hughes F, Young TP (1998) The soil seed bank and its relationship to the aboveground vegetation in deciduous forests in New York City. Urban Ecosyst 2:43–59

Kuebbing SE, Nuñez MA, Simberloff D (2013) Current mismatch between research and conservation efforts: the need to study co-occurring invasive plant species. Biol Conserv 160:121–129

Kuhns EH, Tribuiani Y, Martini X et al (2014) Volatiles from the symbiotic fungus *Raffaelea lauricola* are synergistic with manuka lures for increased capture of the redbay ambrosia beetle *Xyleborus glabratus*. Agric For Entomol 16(1):87–94

LaBonte NR, Ostry ME, Ross-Davis A, Woeste KE (2015) Estimating heritability of disease resistance and factors that contribute to long-term survival in butternut (*Juglans cinerea* L.). Tree Genet Genomes 11:63

Lankau RA (2010) Soil microbial communities alter allelopathic competition between *Alliaria petiolate* and a native species. Biol Invasions 12:2059–2068

Lankau RA (2012) Coevolution between invasive and native plants driven by chemical competition and soil biota. Proc Natl Acad Sci 109(28):11240–11,245

Lankau RA, Nuzzo V, Spyreas G, Davis AS (2009) Evolutionary limits ameliorate the negative impact of an invasive plant. Proc Natl Acad Sci 106(36):15362–15,367

Lapidge S, Wishart J, Staples L et al (2012) Development of a feral swine toxic bait (Hog-Gone®) and bait hopper (Hog-Hopper™) in Australia and the USA. In: Proceedings of the 14th wildlife damage management conference, paper 1158. USDA National Wildlife Research Center, Fort Collins

Lardner B, Adams AAY, Savidge JA et al (2013) Effectiveness of bait tubes for brown treesnake control on Guam. Wildl Soc Bull 37:664–673

Larson ER, Olden JD (2008) Do schools and golf courses represent emerging pathways for crayfish invasions? Aquat Invasions 3(4):465–468

Lawrence JG, Colwell A, Sexton OJ (1991) The ecological impact of allelopathy in *Ailanthus altissima* (Simaroubaceae). Am J Bot 78(7):948–958

Le Roux J, Wieczorek AM (2009) Molecular systematics and population genetics of biological invasions: towards a better understanding of invasive species management. Ann Appl Biol 154:1–17

Leak WB (2006) Fifty-year impacts of the beech bark disease in the Bartlett Experimental Forest, New Hampshire. North J Appl For 23:141–143

LeBlanc DJ (1994) Nutria. In: Hyngstrom SE, Timm RM, Larson GE (eds) The handbook: prevention and control of wildlife damage. Paper 16. U.S. Department of Agriculture, National Wildlife Research Center, Fort Collins, pp B71–B80

Ledger KJ, Pal RW, Murphy P et al (2015) Impact of an invader on species diversity is stronger in the non-native range than in the native range. Plant Ecol 216:1285–1295

Lembi C (2014) Chapter 13: The biology and management of algae. In: Gettys LA, Haller WT, Petty DG (eds) Biology and control of aquatic plants: a best management practices handbook, 3rd edn. Aquatic Ecosystem Restoration Foundation, Marietta, pp 97–104

Leonardt BA, Mastro VC, Leonard DS et al (1996) Control of low-density gypsy moth (Lepidoptera: Lymantriidae) populations by mating disruption with pheromone. J Chem Ecol 22:1255–1272

Leopardi S, Blake D, Puechmaille SJ (2015) White nose syndrome fungus introduced from Europe to North America. Curr Biol 25:217–219

Liebhold AM, Halverson JA, Elmes GA (1992) Gypsy moth invasion in North America: a quantitative analysis. J Biogeogr 19(5):513–520

Lilley TM, Prokkola JM, Johnson JS et al (2017) Immune responses in hibernating little brown myotis (*Myotis lucifugus*) with white-nose syndrome. Proc R Soc B 284(1848):20162232

Liu H-P, Mottern J (2017) An old remedy for a new problem? Identification of *Ooencyrtus kuvanae* (Hymenoptera: Encyrtidae), an egg parasitoid of *Lycorma delicatula* (Hemiptera: Fulgoridae) in North America. J Insect Sci 18:1–6

Liu X, Rohr JR, Li Y (2013) Climate, vegetation, introduced hosts and trade shape a global wildlife pandemic. Proc R Soc B Biol Sci 280:20122506

Lopez VM, Hoddle MS (2014) Effects of body size, diet, and mating on the fecundity and longevity of the goldspotted oak borer (Coleoptera: Buprestidae). Ann Entomol Soc Am 107(2):539–548

Lopez VM, Rugman-Jones PF, Coleman TW et al (2014) Population genetics of goldspotted oak borer, *Agrilus auroguttatus* Schaeffer (Coleoptera: Buprestidae): Investigating the origin of an invasive pest of native oaks in California. Biol Invasions 16(11):2393–2402

Lopez VM, Hoddle MS, Francese JA et al (2017) Assessing flight potential of the invasive Asian longhorned beetle (Coleoptera: Cerambycidae) with computerized flight mills. J Econ Entomol 110(3):1070–1077

Lorch JM, Meteyer CU, Behr MJ et al (2011) Experimental infection of bats with *Geomyces destructans* causes white-nose syndrome. Nature 480:376–378

Luginbuhl JM, Harvey TE, Green JT Jr et al (1999) Use of goats as biological agents for the renovation of pastures in the Appalachian region of the United States. Agrofor Syst 44:241–252

Luken JO, Goessling N (1995) Seedling distribution and potential persistence of the exotic shrub *Lonicera maackii* in fragmented forests. Am Midl Nat 133(1):124–130

MacDonald WL, Fulbright DW (1991) Biological control of chestnut blight: use and limitations of transmissible hypovirulence. Plant Dis 75(7):656–661

MacDougall AS, Turkington R (2005) Are invasive species the drivers or passengers of change in degraded ecosystems? Ecology 86:42–55

Mack RN, Simberloff D, Lonsdale WM et al (2000) Biotic invasions: causes, epidemiology, global consequences, and control. Ecol Appl 10:689–710

Macneale KH, Sanderson BL, Courbois JYP, Kiffney PM (2010) Effects of non-native brook trout (*Salvelinus fontinalis*) on threatened juvenile Chinook salmon (*Oncorhynchus tshawytscha*) in an Idaho stream. Ecol Freshw Fish 19:139–152

MacQuarrie CJ, Lyons DB, Seehausen ML, Smith SM (2016) A history of biological control in Canadian forests, 1882–2014. Can Entomol 148(S1):S239–S269

Magnoli SM, Kleinhesselink AR, Cushman JH (2013) Responses to invasion and invader removal differ between native and exotic plant groups in a coastal dune. Oecologia 173(4):1521–1530

Maher SP, Kramer AM, Tomlin Pulliam J et al (2012) Spread of white-nose syndrome on a network regulated by geography and climate. Nat Commun 3:1306

Mahjoub G, Hinders MK, Swaddle JP (2015) Using a "sonic net" to deter pest bird species: excluding European starlings from food sources by disrupting their acoustic communication. Wildl Soc Bull 39:326–333

Mankin RW, Smith MT, Tropp JM et al (2008) Detection of *Anoplophora glabripennis* (Coleoptera: Cerambycidae) larvae in different host trees and tissues by automated analyses of sound-impulse frequency and temporal patterns. J Econ Entomol 101(3):838–849

Manrique V, Diaz R, Erazo L et al (2014) Comparison of two populations of *Pseudophilothrips ichini* (Thysanoptera: Phlaeothripidae) as candidates for biological control of the invasive weed *Schinus terebinthifolia* (Sapindales: Anacardiaceae). Biocontrol Sci Tech 24:518–535

MapBiocontrol. 2016. Emerald ash borer release guidelines and emerald ash borer and hemlock woolly adelgid release and recovery databases. http://www.mapbiocontrol.org/

Maron JL, Marler M (2008) Field-based competitive impacts between invaders and natives at varying resource supply. J Ecol 96:1187–1197

Martel A, Spitzen-van der Sluijs A, Blooi M et al (2013) *Batrachochytrium salamandrivorans* sp. nov. causes lethal chytridiomycosis in amphibians. Proc Natl Acad Sci 110:15325–15329

Masaka K, Yamada K, Sato H et al (2013) Understory plant richness and native tree invasion in exotic *Robinia pseudoacacia* stands in Hokkaido, Japan. For Sci 59(5):589–597

Mason ME, Koch JL, Krasowski M, Loo J (2013) Comparisons of protein profiles of beech bark disease resistant and susceptible American beech (*Fagus grandifolia*). Proteome Sci 11:2

Massei G, Cowan DP, Coats J et al (2008) Effect of GnRH vaccine GonaCon on the fertility, physiology and behavior of wild boar. Wildl Res 35:540–547

Massei G, Roy S, Bunting R (2011) Too many hogs? A review of methods to mitigate impact by wild boar and feral hogs. Human-Wildlife Interactions 5:79–99

Matta JR, Alavalapati JRR, Mercer DE (2009) Incentives for biodiversity conservation beyond the best management practices: are forestland owners interested? Land Econ 85(1):132–143

Mayer JJ, Brisbin LL (1991) Wild pigs of the United States: their history, morphology, and current status. University of Georgia Press, Athens, 336 p

Mayfield AE, MacKenzie M, Cannon PG et al (2013) Suitability of California bay laurel and other species as hosts for the non-native redbay ambrosia beetle and granulate ambrosia beetle. Agric For Entomol 15:227–235

McCann BE, Garcelon DK (2008) Eradication of feral pigs from Pinnacles National Monument. J Wildl Manag 72:1287–1295

McClure MS (1995) Managing hemlock woolly adelgid in ornamental landscapes. Bulletin 925. Connecticut Agricultural Experiment Station. 7 p

McCullough DG, Poland TM, Anulewicz AC et al (2011) Evaluation of *Agrilus planipennis* control provided by emamectin benzoate and two neonicotinoid insecticides, one and two seasons after treatment. J Econ Entomol 104:1599–1612

McFadyen REC (1998) Biological control of weeds. Annu Rev Entomol 43:369–393

McIlroy JC, Gifford EJ (1997) The 'Judas' pig technique: a method that could enhance control programmes against feral pigs, *Sus scrofa*. Wildl Res 24:483–491

Meehan S, Gruber B, Lucy FE (2014) Zebra mussel control using Zequanox® in an Irish waterway. Manag Biol Invasions 5(3):279–286

Meinesz A, Belsher T, Thibaut T et al (2001) The introduced green alga *Caulerpa taxifolia* continues to spread in the Mediterranean. Biol Invasions 3:201–210

Meng PS, Hoover K, Keena MA (2015) Asian longhorned beetle (Coleoptera: Cerambycidae), an introduced pest of maple and other hardwood trees in North America and Europe. J Integrated Pest Manag 6(1):4–16

Mercader RJ, McCullough DG, Storer AJ et al (2015) Evaluation of the potential use of a systemic insecticide and girdled trees in area wide management of the emerald ash borer. For Ecol Manag 350:70–80

Mercader RJ, McCullough DG, Storer AJ et al (2016) Estimating local spread of recently established emerald ash borer, *Agrilus planipennis,* infestations and the potential to influence it with a systemic insecticide and girdled ash trees. For Ecol Manag 366:87–97

Michel A, Arias RS, Sheffler BE et al (2004) Somatic mutation-mediated evolution of herbicide resistance in the nonindigenous invasive plant hydrilla (*Hydrilla verticillata*). Mol Ecol 13:3229–3237

Midwest Invasive Plant Network (MIPN) (2016) Cooperative weed management area resource page. http://www.mipn.org/cwma-resources/

Miles TD, Martin FN, Coffey MD (2015) Development of rapid isothermal amplification assays for detection of *Phytophthora* in plant tissue. Phytopathology 105:265–278

Milgroom MG, Cortesi P (2004) Biological control of chestnut blight with hypovirulence: a critical analysis. Annu Rev Phytopathol 42:311–338

Miller TW, Shinn SL, Thill DC (2001) Cross-resistance in a chemical control of auxinic herbicide-resistant yellow starthistle (*Centaurea solstitialis*). Weed Technol 15:293–299

Montgomery ME, Bentz SE, Olsen RT (2009) Evaluation of hemlock (*Tsuga*) species and hybrids for resistance to *Adelges tsugae* (Hemiptera: Adelgidae) using artificial infestation. J Econ Entomol 102:1247–1254

Moore G, Glenn SD, Ma J (2009) Distribution of the native *Aralia spinosa* and non-native *Aralia elata* (Araliaceae) in the Northeastern United States. Rhodora 111(946):145–154

Moore JL, Hauser CE, Bear JL et al (2011) Estimating detection-effort curves for plants using search experiments. Ecol Appl 21(2):601–607

Morin RS, Liebhold AM, Gottschalk KW (2009) Anisotropic spread of hemlock woolly adelgid in the eastern United States. Biol Invasions 11:2341–2350

Morris EE, Hajek AE (2014) Eat or be eaten: fungus and nematode switch off as predator and prey. Fungal Ecol 11:114–121

Morris EE, Jimenez A, Long SJ et al (2012) Variability in growth of *Deladenus siricidicola* on strains of the white rot fungus *Amylostereum areolatum*. BioControl 57:677–686

Mortensen DA, Rauschert ESJ, Nord AN, Jones BP (2009) Forest roads facilitate the spread of invasive plants. Invasive Plant Sci Manag 2(3):191–199

Müller-Schärer H, Schaffner U (2008) Classical biological control: exploiting enemy escape to manage plant invasions. Biol Invasions 10:859–874

Murrell C, Gerber E, Krebs C et al (2011) Invasive knotweed affects native plants through allelopathy. Am J Bot 98:38–43

Nathan LM, Simmons M, Wegleitner BJ et al (2014) Quantifying environmental DNA signals for aquatic invasive species across multiple detection platforms. Environ Sci Technol 48:12800–12,806

Nehme ME, Keena MA, Zhang A et al (2010) Evaluating the use of male-produced pheromone components and plant volatiles in two trap designs to monitor *Anoplophora glabripennis*. Environ Entomol 39:169–176

Nehme ME, Trotter RT, Keena MA et al (2014) Development and evaluation of a trapping system for *Anoplophora glabripennis* in the United States. Environ Entomol 43:1034–1044

Nelson CR, Halpern CB, Agee JK (2008) Thinning and burning result in low-level invasion by nonnative plants but neutral effects on natives. Ecol Appl 18(3):762–770

Nemeth N, Young G, Ndaluka C et al (2009) Persistent West Nile virus infection in the house sparrow (*Passer domesticus*). Arch Virol 154:783–789

Netherland M (2014) Chapter 11: Chemical control of aquatic weeds. In: Gettys LA, Haller WT, Petty DG (eds) Biology and control of aquatic plants: a best management practices handbook, 3rd edn. Aquatic Ecosystem Restoration Foundation, Marietta, pp 71–88

New Hampshire Division of Forests and Lands (NHDFL). 2015. Action plan to restrict the spread and manage hemlock woolly adelgid within the state of New Hampshire. https://www.nhdfl.org/library/pdf/Forest%20Health/HWA%20action%20plan%2015.pdf

Newhouse AE, Schrodt F, Liang H et al (2007) Transgenic American elm shows reduced Dutch elm disease symptoms and normal mycorrhizal colonization. Plant Cell Rep 26:977–987

Newhouse AE, Polin LD, McGuigan KA et al (2014) Transgenic American chestnuts show enhanced blight resistance and transmit the trait to T1 progeny. Plant Sci 228:88–97

Nichols JD, Eaton MJ, Martin J (2014) Chapter 2: Thresholds for conservation and management: structured decision-making as a conceptual framework. In: Guntensperger GR (ed) Application of threshold concepts in natural resource decision-making. Springer Science and Business Media, New York, pp 9–28

Olson DH, Aanensen DM, Ronnenberg KL et al (2013) Mapping the global emergence of *Batrachochytrium dendrobatidis*, the amphibian chytrid fungus. PLoS One 8(2):e56802

Onken B, Keena M (2008) Hemlock woolly adelgid initiative: progress and future direction. In: Onken B, Reardon R, comps (eds) Fourth Symposium on hemlock woolly adelgid in the eastern United States, 2008 February 12–14; Hartford, CT. FHTET 2008-01. U.S. Forest Service, Forest Health Technology Enterprise Team, Morgantown, pp 214–220

Ortega YK, Pearson DE (2010) Effects of picloram application on community dominants vary with initial levels of spotted knapweed (*Centaurea stoebe*) invasion. Invasive Plant Sci Manag 3(1):70–80

Ortega YK, Pearson DE, McKelvey KS (2004) Effects of biological control agents and exotic plant invasion on deer mouse populations. Ecol Appl 14:241–253

Ortega YK, McKelvey KS, Six DL (2006) Invasion of an exotic forb impacts reproductive success and site fidelity of a migratory songbird. Oecologia 149(2):340–351

Pajares G (2015a) Overview and current status of remote sensing applications based on unmanned aerial vehicles (UAVs). Photogramm Eng Remote Sens 81(4):281–329

Pajares G (2015b) Overview and current status of remote sensing applications based on unmanned aerial vehicles (UAVs). Photogramm Eng Remote Sens 81(4):281–330

Palacio-López K, Gianoli E (2011) Invasive plants do not display greater phenotypic plasticity than their native or non-invasive counterparts: a meta-analysis. Oikos 9:1393–1401

Pannell JR (2015) Evolution of the mating system in colonizing plants. Mol Ecol 24:2018–2037

Parepa M, Fischer M, Krebs C, Bossdorf O (2014) Hybridization increases invasive knotweed success. Evol Appl 7:413–420

Parra G, Moylett H, Bulluck R (2017) Technical working group summary report spotted lanternfly, *Lycorma delicatula* (White, 1845). USDA-APHIS-PPQ-CPHST. 4700 River Road, Riverdale, MD 20737. 42 p

Partners in Amphibian and Reptile Conservation (PARC) (2016) Don't turn it loose. http://www.parcplace.org/publications/213-parc-dont-turn-it-loose.html

Paynter Q, Overton JMC, Hill RL et al (2012) Plant traits predict the success of weed biocontrol. J Appl Ecol 49:1140–1148

Pearson DE, Callaway RM (2003) Indirect effects of host-specific biological control agents. Trends Ecol Evol 18(9):456–461

Pearson DE, Callaway RM (2005) Indirect nontarget effects of host-specific biological control agents: implications for biological control. Biol Control 35:288–298

Pearson DE, Callaway RM (2006) Biological control agents elevate hantavirus by subsidizing mice. Ecol Lett 9:443–450

Pearson DE, Callaway RM (2008) Weed biocontrol insects reduce native plant recruitment through second-order apparent competition. Ecol Appl 18:1489–1500

Pearson DE, Fletcher RJ Jr (2008) Mitigating exotic impacts: restoring native deer mouse populations elevated by an exotic food subsidy. Ecol Appl 18(2):321–334

Pearson DE, Ortega YK, Runyon J, Butler J (2016a) Secondary invasion: the bane of weed management. Biol Conserv 197:8–17

Pearson DE, Ortega YK, Eren O, Hierro JL (2016b) Quantifying "apparent" impact and distinguishing impact from invasiveness in multispecies plant invasions. Ecol Appl 26:162–173

Pegg KG, Whiley AW, Hargreaves PA (1990) Phosphonic (phosphorous) acid treatments control *Phytophthora* diseases in avocado and pineapple. Australas Plant Pathol 19(4):122–124

Peña JE, Crane JH, Capinera JL et al (2011) Chemical control of the redbay ambrosia beetle, *Xyleborus glabratus*, and other Scolytinae (Coleoptera: Curculionidae). Fla Entomol 94(4):882–896

Penaluna BE, Olson DH, Flitcroft RL et al (2017) Aquatic biodiversity in forests: a weak link in ecosystem resilience. Biodivers Conserv 26:3125–3155

Pennsylvania Department of Agriculture (PDA) (2019) Spotted lanternfly. Department of Agriculture (PDA), Pennsylvania. Last accessed 7 Feb 2019. https://www.agriculture.pa.gov/Plants_Land_Water/PlantIndustry/Entomology/spotted_lanternfly/Pages/default.aspx

Pennsylvania State Extension (PSE) (2019) Updated insecticide recommendations for spotted lanternfly on grape. PennState Extension. Last accessed 7 Feb 2019. https://extension.psu.edu/updated-insecticide-recommendations-for-spotted-lanternfly-on-grape

Pérez-Ramírez E, Llorente F, Jiménez-Clavero MA (2014) Experimental infections of wild birds with West Nile virus. Viruses 6(2):752–781

Perry AL, Low PJ, Ellis JR, Reynolds JD (2005) Climate change and distribution shifts in marine fishes. Science 308(5730):1912–1915

Petanidou T, Godfree RC, Song DS et al (2012) Self-compatibility and plant invasiveness: comparing species in native and invasive ranges. Perspect Plant Ecol Evol Syst 14:3–12

Petersen CE, Lovich RE, Phillips CA et al (2016) Prevalence and seasonality of the amphibian chytrid fungus *Batrachochytrium dendrobatidis* along widely separated longitudes across the United States. EcoHealth 13(2):368–382

Peterson K, Diss-Torrance A (2012) Motivation for compliance with environmental regulations related to forest health. J Environ Manag 112:104–119

Phytophthora ramorum Multilocus Genotyping Project (PRMGP) (2016) *Phytophthora ramorum* Multilocus Genotyping Database. http://people.oregonstate.edu/~grunwaln/phytophthora.php

Phytosphere Research (2013) Management of *Phytophthora ramorum* in Tanoak and Oak Stands. http://phytosphere.com/publications/2013%20June%20Phytosphere%20Final%20Report%2011-JV-076.pdf

Piatek KB, Fajvan MA, Turcotte RM (2016) Thinning effects on foliar elements in eastern hemlock: implications for managing the spread of the hemlock woolly adelgid. Can J For Res 47(1):81–88

Picco AM, Collins JP (2008) Amphibian commerce as a likely source of pathogen pollution. Conserv Biol 22:1582–1589

Pickett ST, Cadenasso ML (2002) The ecosystem as a multidimensional concept: meaning, model, and metaphor. Ecosystems 5(1):1–10

Pimentel D, Zuniga R, Morrison D (2005) Update on the environmental and economic costs associated with alien-invasive species in the United States. Ecol Econ 52(3):273–288

Piola F, Bellvert F, Meiffren G et al (2013) Invasive *Fallopia* × *bohemica* interspecific hybrids display different patterns in secondary metabolites. Ecoscience 20(3):230–239

Piotrowski JS, Annis SL, Longcore JE (2004) Physiology of *Batrachochytrium dendrobatidis*, a chytrid pathogen of amphibians. Mycologia 96:9–15

Pister EP (2008) Restoration of the California golden trout in the South Fork Kern River, Kern Plateau, Tulare County, California, 1966–2004, with reference to Golden Trout Creek. California Department of Fish and Game Administrative Report 2008-1. 122 p

Pisula NL, Meiners SJ (2010) Relative allelopathic potential of invasive plant species in a young disturbed woodland. J Torrey Bot Soc 137(1):81–87

Podgwaite JD (1999) Gypchek: biological insecticide for the gypsy moth. J For 97(3):16–19

Poland TM, McCullough DG (2006) Emerald ash borer: invasion of the urban forest and the threat to North America's ash resource. J For 104(3):118–124

Poland TM, McCullough DG (2010) SLAM: a multi-agency pilot project to SL.ow A.sh M.ortality caused by emerald ash borer in outlier sites. Newsl Michigan Entomol Soc 55(1 & 2):4–8

Poland TM, Ciaramitaro TM, Emery M et al (2015) Submergence of black ash logs to control emerald ash borer and preserve wood for American Indian basketmaking. Agric For Entomol 17:412–420

Pontius JA, Hallett RA, Jenkins JC (2006) Foliar chemistry linked to infestation and susceptibility to hemlock woolly adelgid (Homoptera: Adelgidae). Environ Entomol 35:112–120

Powell KI, Chase JM, Knight TM (2013) Invasive plants have scale-dependent effects on diversity by altering species-area relationships. Science 339:316–318

Prasad AM, Iverson P et al (2010) Modeling the invasive emerald ash borer risk of spread using a spatially explicit cellular model. Landsc Ecol 25:353–369

Pratt PD, Slone DH, Rayamajhi MB (2003) Geographic distribution and dispersal rate of *Oxyops vitiosa* (Coleoptera: Curculionidae), a biological control agent of the invasive tree *Melaleuca quinquenervia* in South Florida. Environ Entomol 32(2):397–406

Prinbeck G, Lach D, Chan S (2011) Exploring stakeholders' attitudes and beliefs regarding behaviors that prevent the spread of invasive species. Environ Educ Res 17:341–352

Prospero S, Hansen EM, Grünwald NJ, Winton LM (2007) Population dynamics of the sudden oak death pathogen *Phytophthora ramorum* in Oregon from 2001 to 2004. Mol Ecol 16:2958–2973

Pureswaran DS, Poland TM (2009) Host selection and feeding preference of *Agrilus planipennis* (Coleoptera: Buprestidae) on ash (*Fraxinus* spp.). Environ Entomol 38:757–765

Puth LM, Post DM (2005) Studying invasion: have we missed the boat? Ecol Lett 8:715–721

Ragenovich IR, Mitchell RG (2006) Balsam woolly adelgid. Forest insect and disease leaflet No. 118, USDA Forest Service, R6-NR-FID-PR-04-2006. 12 p.

Raghu S, Wilson JR, Dhileepan K (2006) Refining the process of agent selection through understanding plant demography and plant response to herbivory. Aust J Entomol 45:308–316

Rebbeck J, Kloss A, Bowden M et al (2015) Aerial detection of seed-bearing female *Ailanthus altissima*: a cost-effective method to map an invasive tree in forested landscapes. For Sci 61(6):1068–1078

Reed RN (2005) An ecological risk assessment of nonnative boas and pythons as potentially invasive species in the United States. Risk Anal 25(3):753–766

Reeves JL, Blumenthal DM, Kray JA, Derner JD (2015) Increased seed consumption by biological control weevil tempers positive CO_2 effect on invasive plant (*Centaurea diffusa*) fitness. Biol Control 84:36–43

Reisen WK (2013) Ecology of West Nile Virus in North America. Viruses 5(9):2079–2105

Rejmanek M, Pitcairn MJ (2002) When is eradication of exotic plant pests a realistic goal? In: Veith CR, Clout MN (eds) Turning the tide: the eradication of invasive species. IUCN SSC Invasive Species Spec. Group, Cambridge, pp 249–253

Renner SS (2014) The relative and absolute frequencies of angiosperm sexual systems: dioecy, monoecy, gynodioecy, and an updated online database. Am J Bot 101:1588–1596

Resasco J, Hale AN, Henry MC, Gorchov DL (2007) Detecting an invasive shrub in a deciduous forest understory using late-fall Landsat sensor imagery. Int J Remote Sens 28(16):3739–3745

Ricciardi A, Hoopes MF, Marchetti MP, Lockwood JL (2013) Progress toward understanding the ecological impacts of nonnative species. Ecol Monogr 83(3):263–282

Richgels KL, Russell RE, Adams MJ et al (2016) Spatial variation in risk and consequence of *Batrachochytrium salamandrivorans* introduction in the USA. Open Sci 3:150616

Rinella MJ, Maxwell BD, Fay PK et al (2009) Control effort exacerbates invasive-species problem. Ecol Appl 19:155–162

Robinson EA, Ryan GD, Newman JA (2012) A meta-analytical review of the effects of elevated CO2 on plant-arthropod interactions highlights the importance of interacting environmental and biological variables. New Phytol 194:321–336

Rocke TE, Pussini N, Smith SR et al (2010) Consumption of baits containing raccoon pox-based plague vaccines protects black-tailed prairie dogs (*Cynomys ludovicianus*). Vector Borne Zoonotic Dis 10(1):53–58

Rocke TE, Kingstad-Bakke B, Berlier W, Osorio JE (2014) A recombinant raccoon poxvirus vaccine expressing both *Yersinia pestis* F1 and truncated V antigens protects animals against lethal plague. Vaccine 2(4):772–784

Rödder D, Kielgast J, Bielby J et al (2009) Global amphibian extinction risk assessment for the panzootic chytrid fungus. Diversity 1:52–66

Rodriguez-Echeverria S (2010) Rhizobial hitchhikers from down under: invasional meltdown in a plant-bacteria mutualism? J Biogeogr 37:1611–1622

Romanuk TN, Zhou Y, Brose U et al (2009) Predicting invasion success in complex ecological networks. Philos Trans R Soc B 3654:1743–1754

Rothlisberger JD, Chadderton WL, McNulty J, Lodge DM (2010) Aquatic invasive species transport via trailered boats: What is being moved, who is moving it, and what can be done. Fisheries 35(3):121–132

Runyon JB, Butler JL, Friggens MM et al (2012) Chapter 7: Invasive species and climate change. In: Finch DM (ed) Climate change in grasslands, shrublands, and deserts of the interior American West: a review and needs assessment, Gen. Tech. Rep. RMRS-GTR-285. U.S. Department of Agriculture, Forest Service, Rocky Mountain Research Station, Fort Collins, pp 97–115

Ryall KL, Silk PJ, Mayo P et al (2012) Attraction of *Agrilus planipennis* (Coleoptera: Buprestidae) to a volatile pheromone: effects of release rate, host volatile, and trap placement. Environ Entomol 41:648–656

Sanderson BL, Barnas KA, Wargo-Rub AM (2009) Nonindigenous species of the Pacific Northwest: an overlooked risk to endangered salmon? Bioscience 59:245–256

Sarver MJ, Treher A, Wilson L et al (2008) Mistaken identity? Invasive plants and their native look-alikes, an identification guide for the Mid-Atlantic. Delaware Department of Agriculture and USDA Natural Resources Conservation Service, Dover, 62 p

Sato M, Kawaguchi Y, Nakajima J, Mukai T (2010) A review of the research on introduced freshwater fishes: new perspectives, the need for research, and management implications. Landsc Ecol Eng 6:99–108

Savage AE, Zamudio KR (2011) MHC genotypes associate with resistance to a frog-killing fungus. Proc Natl Acad Sci 108:16705–16,710

Savarie PJ, Shivik JA, White GC et al (2001) Use of acetaminophen for large-scale control of brown treesnakes. J Wildl Manag 65:356–365

Scheele BC, Hunter DA, Grogan LF et al (2014) Interventions for reducing extinction risk in chytridiomycosis-threatened amphibians. Conserv Biol 28:1195–1205

Schierenbeck KA, Ellstrand NC (2009) Hybridization and the evolution of invasiveness in plants and other organisms. Biol Invasions 11:1093–1105

Schlaepfer MA, Sax DF, Olden JD (2011) The potential conservation value of non-native species. Conserv Biol 25:428–437

Schnable PS, Wise RP (1998) The molecular basis of cytoplasmic male sterility and fertility restoration. Trends Plant Sci 3:175–180

Schoettle A, Sniezko R (2007) Proactive intervention to sustain high-elevation pine ecosystems threatened by white pine blister rust. J For Res 12:327–336

Schweitzer C, Clark SL, Gottschalk KW et al (2014) Proactive restoration: planning, implementation, and early results of silvicultural strategies for increasing resilience against gypsy moth infestation in upland oak forests on the Daniel Boone National Forest, Kentucky. J For 112(5):401–411

Seastedt TR (2015) Biological control of invasive plant species: a reassessment for the Anthropocene. New Phytol 205:490–502

Shapiro L, Eason C, Bunt C et al (2015) Efficacy of encapsulated sodium nitrite as a new tool for feral pig management. J Pest Sci 89(2):489–495

Sharov AA, Leonard D, Liebhold AM et al (2002) "Slow The Spread": a national program to contain the gypsy moth. J For 100(5):30–36

Shelton AL (2012) Mowing any time after midsummer can manage Japanese stiltgrass. Invasive Plant Sci Manag 5(2):209–216

Sheppard AW, van Klinken RD, Heard TA (2005) Scientific advances in the analysis of direct risks of weed biological control agents to nontarget plants. Biol Control 35:215–226

Shuey MM, Drees KP, Lindner DL et al (2014) Highly sensitive quantitative PCR for the detection and differentiation of *Pseudogymnoascus destructans* and other *Pseudogymnoascus* species. Appl Environ Microbiol 80:1726–1731

Siegert NW, McCullough DG, Wheeler MM, Hajek AE (2012) Evaluation of potential versus realized primary infection of gypsy moth (Lepidoptera: Lymantriidae) by *Entomophaga maimaiga* (Zygomycetes: Entomophthorales). Environ Entomol 41:1115–1124

Siegert NW, McCullough DG, Liebhold AM, Telewski FW (2014) Dendrochronological reconstruction of the epicentre and early spread of emerald ash borer in North America. Divers Distrib 20:847–858

Siemens TJ, Blossey B (2007) An evaluation of mechanisms preventing growth and survival of two native species in invasive Bohemian knotweed (*Fallopia* × *bohemica*, Polygonaceae). Am J Bot 94:776–783

Simberloff D (2003) How much information on population biology is needed to manage introduced species? Conserv Biol 17:83–92

Simberloff D (2006) Invasional meltdown 6 years late: important phenomenon, unfortunate metaphor, or both? Ecol Lett 9:912–919

Simberloff D (2014) Biological invasions: What's worth fighting and what can be won? Ecol Eng 65:112–121

Simmons MT, Windhager S, Power P et al (2007) Selective and non-selective control of invasive plants: the short-term effects of growing-season prescribed fire, herbicide, and mowing in two Texas prairies. Restor Ecol 15(4):662–669

Slippers B, Hurley BP, Wingfield MJ (2015) *Sirex* woodwasp: a model for evolving management paradigms of invasive forest pests. Annu Rev Entomol 60:601–619

Smitley DR, Bauer LS, Hajek AE et al (1995) Introduction and establishment of *Entomophaga maimaiga*, a fungal pathogen of gypsy moth (Lepidoptera: Lymantriidae) in Michigan. Environ Entomol 24(6):1685–1695

Snow NP, Halseth JM, Lavelle MJ et al (2016) Bait preference of free-ranging feral swine for delivery of novel toxicant. PLoS One 11(1):e0146712

Sogge MK, Sferra SJ, Paxton EH (2008) Tamarix as habitat for birds: implications for riparian restoration in the Southwestern United States. Restor Ecol 16(1):146–154

Soltis PS, Soltis DE (2009) The role of hybridization in plant speciation. Annu Rev Plant Biol 60:561–588

Soopaya R, Woods B, Lacey I et al (2015) Feasibility of mating disruption for agricultural pest eradication in an urban environment: light brown apple moth (Lepidoptera: Tortricidae) in Perth. J Econ Entomol 108(4):1930–1935

Soto D, Arismendi I, González J et al (2006) Southern Chile, trout and salmon country: invasion patterns and threats for native species. Rev Chil Hist Nat 79:97–117

Sparklin BD, Mitchell MS, Hanson LB et al (2007) Territoriality of feral pigs in a highly persecuted population on Fort Benning, Georgia. J Wildl Manag 73:497–502

Steckel LE (2007) The dioecious *Amaranthus* spp.: here to stay. Weed Technol 21:567–570

Stokstad E (2014) The coming salamander plague. Science 346(6209):530–531

Strayer DL (2009) Twenty years of zebra mussels: lessons from the mollusk that made headlines. Front Ecol Environ 7(3):135–141

Stringer LD, Sullivan NJ, Sullivan TE et al (2013) Attractiveness and competitiveness of irradiated light brown apple moths. Entomol Exp Appl 148(3):203–212

Suckling DM, Sforza RFH (2014) What magnitude are observed non-target impacts from weed biocontrol? PLoS One 9(1):e84847

Sullivan JJ, Williams PA, Timmins SM (2007) Secondary forest succession differs through naturalized gorse and native kanuka near Wellington and Nelson. N Z J Ecol 31:22–38

Sutherland WJ, Dicks LV, Ockendon N, Smith RK (2015) What works in conservation. Open Book Publishers, Cambridge, 328 p

Swiecki, T.J.; Bernhardt, E.A. 2011. Managing sudden oak death in California forests and woodlands: before, during and after *Phytophthora ramorum* invasion. In: Standiford, R.B.; Purcell, K.L., tech. cords. Proceedings of the seventh California oak symposium. Gen. Tech. Rep. PSW-GTR-242. Albany: U.S. Department of Agriculture, Forest Service, Pacific Southwest Research Station: 563–572.

Tanentzap AJ, Burrows LE, Lee WG et al (2009) Landscape-level vegetation recovery from herbivory: progress after four decades of invasive red deer control. J Appl Ecol 46:1064–1072

Tarin D, Pepper AE, Goolsby JA et al (2013) Microsatellites uncover multiple introductions of clonal giant reed (*Arundo donax*). Invasive Plant Sci Manag 6:328–338

Taylor RAJ, Bauer LS, Poland TM, Windell K (2010) Flight performance of *Agrilus planipennis* (Coleoptera: Buprestidae) on a flight mill and in free flight. J Insect Behav 23:128–148

Thompson MJ (1996) Winter foraging response of elk to spotted knapweed removal. Northwest Sci 70:10–19

Thorpe KW, Ridgway RL (1994) Effects of trunk barriers on larval gypsy moth (Lepidoptera: Lymantriidae) density in isolated and contiguous oak trees. Environ Entomol 23:832–836

Thresher RE, Hayes K, Bax NJ et al (2014) Genetic control of invasive fish: technological options and its role in integrated pest management. Biol Invasions 16:1201–1216

Tiébré M-S, Vanderhoeven S, Saad L, Mahy G (2007) Hybridization and sexual reproduction in the invasive alien *Fallopia* (Polygonaceae) complex in Belgium. Ann Bot 99(1):193–203

Tluczek AR, McCullough DG, Poland TM (2011) Influence of host stress on emerald ash borer (*Agrilus planipennis* Fairmaire) (Coleoptera: Buprestidae) adult density, development, and distribution in *Fraxinus pennsylvanica* trees. Environ Entomol 40:357–366

Tobin PC, Blackburn LM (2007) Slow the spread: a national program to manage the gypsy moth, Gen. Tech. Rep. NRS-6. U.S. Department of Agriculture, Forest Service, Northern Research Station, Newtown Square, 109 p

Tobin PC, Kean JM, Suckling DM et al (2013) Determinants of successful arthropod eradication programs. Biol Invasions 16:401–414

Tomlinson JA, Dickinson MJ, Boonham N (2007) Faster, simpler, more-specific methods for improved molecular detection of *Phytophthora ramorum* in the field. Appl Environ Microbiol 73:4040–4047

Tomlinson JA, Dickinson MJ, Boonham N (2010) Rapid detection of *Phytophthora ramorum* and *P. kernoviae* by two-minute DNA extraction followed by isothermal amplification and amplicon detection by generic lateral flow device. Phytopathology 100:143–149

Townsend CR (1996) Invasion biology and ecological impacts of brown trout *Salmo trutta* in New Zealand. Biol Conserv 78:13–22

Trebitz AS, Hoffman JC, Darling JA et al (2017) Early detection monitoring for aquatic non-indigenous species: optimizing surveillance, incorporating advanced technologies, and identifying research needs. J Environ Manag 202:299–310

Tripp DW, Rocke TE, Streich SP et al (2014) Season and application rates affect vaccine bait consumption by prairie dogs in Colorado and Utah, USA. J Wildl Dis 50(2):224–234

Trotter RT, Hull-Sanders HM (2015) Quantifying dispersal of the Asian longhorned beetle (*Anoplophora glabripennis*, Coleoptera) with incomplete data and behavioral knowledge. Biol Invasions 17(12):3359–3369

Tsehaye I, Catalano M, Sass G et al (2013) Prospects for fishery-induced collapse. Fisheries 38:445–454

University of California Berkeley Forest Pathology and Mycology Lab (UCB FPM Lab) (2017). Chemical treatment for sudden oak death.

Urban DL, O'Neill RV, Shugart HH (1987) Landscape ecology. Bioscience 37(2):119–127

U.S. Department of Agriculture (USDA) (2012). Supplemental environmental impact statement. Gypsy moth management in the United States: a cooperative approach. Forest Service and APHIS. NA-MB-01-12. August 2012

U.S. Department of Agriculture Animal and Plant Health Inspection Service (USDA APHIS) (2015) Field release of the parasitoid *Spathius galinae* for the biological control of the emerald ash borer (*Agrilus planipennis*) in the contiguous United States. Environmental Assessment 2015. USDA APHIS Marketing and Regulatory Programs, Pest Permitting, Plant Protection and Quarantine, Riverdale. 33 p

U.S. Department of Agriculture Animal and Plant Health Inspection Service (USDA-APHIS) (2010) Gypsy moth program manual. https://www.aphis.usda.gov/import_export/plants/manuals/domestic/downloads/gypsy_moth.pdf

U.S. Department of Agriculture Forest Service (USDA Forest Service). 2016. Aquatic invasive species – fire operations guidance background: aquatic invasive species. http://www.fs.usda.gov/detail/r4/landmanagement/resourcemanagement/?cid=fsbdev3_016113

U.S. Department of Agriculture Forest Service (USDA FS) (2017) Air water and aquatics environment program. http://www.fs.fed.us/rm/boise/AWAE/projects/NorWeST.html

U.S. Department of Agriculture National Agricultural Library (USDA NAL) (2016) National Invasive Species Information Center. Managers' toolkit: prevention: best management practices. https://www.invasivespeciesinfo.gov/toolkit/preventionbmp.shtml

U.S. Department of Agriculture National Agricultural Library (USDA NAL) (2017) National invasive species information center. Aquatic species. http://www.invasivespeciesinfo.gov/aquatics/main.shtml

U.S. Department of Agriculture Natural Resources Conservation Service (USDA NRCS) (2016a) Introduced, invasive, and noxious plants. https://plants.usda.gov/java/noxiousDriver

U.S. Department of Agriculture Natural Resources Conservation Service (USDA NRCS) (2016b) Introduced, invasive, and noxious plants: federal and state noxious weeds. https://plants.usda.gov/java/noxComposite

U.S. Department of Agriculture Natural Resources Conservation Service (USDA NRCS) (2016c) Managers tool kit: outreach tools. https://www.invasivespeciesinfo.gov/toolkit/outreach.shtml

U.S. Fish and Wildlife Service (US FWS) (1997) Endangered and threatened wildlife and plants; Final determination of critical habitat for the Southwestern Willow Flycatcher. Fed Regist 62(140):39129–39,147

U.S. Fish and Wildlife Service (US FWS) (2016) Injurious wildlife species; listing salamanders due to risk of Salamander Chytrid Fungus. Federal Register. Docket No. FWS-HQ-FAC-2015-000, January 13, 2016

U.S. Fish and Wildlife Service (US FWS) (2017a) Volunteers and Invasive plants: learning and lending a hand. https://www.fws.gov/invasives/volunteersTrainingModule/

U.S. Fish and Wildlife Service (US FWS) (2017b) Fish and Aquatic Conservation. http://www.fws.gov/fisheries/ANS/index.html

U.S. Geological Survey (USGS) (2014) Ultra-violet light works as screening tool for bats with white-nose syndrome. https://www.whitenosesyndrome.org/sites/default/files/files/wns_uv_light_release_final_05292014.pdf

U.S. Geological Survey (USGS) (2017) Integrated management techniques to control non-native fishes. http://www.umesc.usgs.gov/reports_publications/integrated_techniques.html

U.S. Geological Survey (USGS) (2018) Nonindigenous aquatic species: hydrologic unit codes. https://nas.er.usgs.gov/hucs.aspx

Van der Putten WH, Macel M, Visser ME (2010) Predicting species distributions and abundance responses to climate change: why it is essential to include biotic interactions across trophic levels. Philos Trans R Soc B 365:2025–2034

Van Driesche R, Bellows T (1996) Biological Control. Springer, New York, 539 p

Van Driesche R, Hoddle M (2016) Non-target effects of insect biocontrol agents and trends in host specificity since 1985. CAB Rev 11(4):1–66. https://doi.org/10.1079/PAVSNNR201611044

Van Driesche R, Reardon R (2013) The use of classical biological control to preserve forests in North America, FHTET-2013-02. U. S. Department of Agriculture, Forest Service, Forest Health Technology Enterprise Team, Morgantown, 433 p

Van Driesche R, Blossey B, Hoddle M et al (2002) Biological control of invasive plants in the Eastern United States. FHTET-2002-04, U.S. Department of Agriculture, Forest Service, Forest Health Technology Enterprise Team, Morgantown. 434 p. https://www.fs.fed.us/foresthealth/technology/pdfs/BiocontrolsOfInvasivePlants02_04.pdf

Van Driesche R, Hoddle M, Center TD (2008) Control of pests and weeds by natural enemies. In: An introduction to biological control. Blackwell Publishing, Oxford. x + 473 p. ISBN 978-14051-4571

Van Driesche RG, Carruthers RI, Center T et al (2010) Classical biological control for the protection of natural ecosystems. Biol Control 54:S2–S33

Van Klinken RD, Edwards OR (2002) Is host-specificity of weed biological control agents likely to evolve rapidly following establishment? Ecol Lett 5:590–596

VanDalen KK, Hall JS, Clark L et al (2013) West Nile virus infection in American robins: new insights on dose response. PLoS One 8(7):e68537. https://doi.org/10.1371/journal.pone.0068537

Vander Zanden MJ, Olden JD (2008) A management framework for preventing the secondary spread of aquatic invasive species. Can J Fish Aquat Sci 65:1512–1522

Verant ML, Meteyer CU, Speakman JR et al (2014) White-nose syndrome initiates a cascade of physiologic disturbances in the hibernating bat host. BMC Physiol 14:10

Verloove F (2010) Invaders in disguise. Conservation risks derived from misidentifications of invasive plants. Manag Biol Invasions 1:1–5

Vila M, Ibanez I (2011) Plant invasions in the landscape. Landsc Ecol 26:461–472

Voyles J, Young S, Berger L et al (2009) Pathogenesis of chytridiomycosis, a cause of catastrophic amphibian declines. Science 326:582–585

Walker BL, Naugle DE, Doherty KE, Cornish TE (2007) West Nile virus and greater sage-grouse: estimating infection rate in a wild bird population. Avian Dis 51:691–696

Walls RL (2010) Hybridization and plasticity contribute to divergence among coastal and wetland populations of invasive hybrid Japanese knotweed *s.l.* (*Fallopia* spp.). Estuar Coasts 33:902–918

Ward JS, Cheah CAS-J, Montgomery ME, Onken BP, Cowles RS (2004) Eastern hemlock forests: guidelines to minimize the impacts of hemlock woolly adelgid, Technical Paper NA-TP-03-04. U.S. Department of Agriculture, Forest Service, Northeastern Area State and Private Forestry, Morgantown, 27 p

Ward SM, Fleischmann CE, Turner MF, Sing SE (2009) Hybridization between invasive populations of dalmatian toadflax (*Linaria dalmatica*) and yellow toadflax (*Linaria vulgaris*). Invasive Plant Sci Manag 2:369–378

Warnecke L, Turner JM, Bollinger TK et al (2013) Pathophysiology of white-nose syndrome in bats: a mechanistic model linking wing damage to mortality. Biol Lett 9:20130177

Warner KD, Kinslow F (2013) Manipulating risk communication: value predispositions shape public understandings of invasive species science in Hawaii. Public Underst Sci 22(2):203–218

Washington Department of Fish and Wildlife (WDFW) (2015) Washington state aquatic invasive species prevention and enforcement program: 2015 report to the legislature. http://wdfw.wa.gov/publications/01697/

Westbrooks RG (1998). Invasive plants: changing the landscape of America. Federal Interagency Committee for the Management of Noxious and Exotic Weeds. All U.S. Government Documents (Utah Regional Depository). Paper 490, 109 p. http://digitalcommons.usu.edu/govdocs/490

Wheeler GS, Schaffner U (2013) Improved understanding of weed biological control safety and impact with chemical ecology: a review. Invasive Plant Sci Manag 6:16–29

Wheeler SS, Langevin SA, Brault AC et al (2012) Detection of persistent West Nile virus RNA in experimentally and naturally infected avian hosts. Am J Trop Med Hyg 87(3):559–564

White Nose Syndrome Decontamination Team (WNS DT) (2016) National white-nose syndrome decontamination protocol – Version 04.12.2016. https://www.whitenosesyndrome.org/sites/default/files/files/national_wns_decon_protocol_04.12.2016.pdf

White JG, Antos MJ, Fitzsimons JA, Palmer GC (2005) Non-uniform bird assemblages in urban environments: the influence of streetscape vegetation. Landsc Urban Plan 71:123–135

Whitehill JGA, Opiyo SO, Koch JL et al (2012) Interspecific comparison of constitutive ash phloem phenolic chemistry reveals compounds unique to Manchurian ash, a species resistant to emerald ash borer. J Chem Ecol 38:499–511

Whitmore M (2014) The hemlock woolly adelgid, a pest we can manage. Branching Out Newsl 21(5):2 p

Wilcox TM, McKelvey KS, Young MK et al (2013) Robust detection of rare species using environmental DNA: the importance of primer specificity. PLoS One 8(3):e59520

Wilcox A, Warnecke L, Turner JM et al (2014) Behaviour of hibernating little brown bats experimentally inoculated with the pathogen that causes white-nose syndrome. Anim Behav 88:157e164

Williams DW, Li G, Gao R (2004) Tracking movements of individual *Anoplophora glabripennis* (Coleoptera: Cerambycidae) adults: application of harmonic radar. Environ Entomol 33:644–649

Williams DA, Overholt WA, Cuda JP et al (2005) Chloroplast and microsatellite DNA diversities reveal the introduction history of Brazilian peppertree (*Schinus terebinthifolius*) in Florida. Mol Ecol 14:3643–3656

Willis CKR, Wilcox A (2014) Hormones and hibernation: possible links between hormone systems, winter energy balance and white-nose syndrome in bats. Horm Behav 66:66–73

Wilson MV, Clark DL (2001) Controlling invasive *Arrhenatherum elatius* and promoting native prairie grasses through mowing. Appl Veg Sci 4:129–138

Wissinger SA, McIntosh AR, Greigh HS (2006) Impacts of introduced brown and rainbow trout on benthic invertebrate communities in shallow New Zealand lakes. Freshw Biol 51(11):2009–2028

Witmer GW, Burke PW, Pitt WC, Avery ML (2007) Management of invasive vertebrates in the United States: an overview. In: Witmer GW, Pitt WC, Fagerstone KA (eds) Managing vertebrate invasive species. Proceedings of an international symposium. USDA/APHIS Wildlife Services, National Wildlife Research Center, Fort Collins, pp 127–137

Yao H, Song J, Liu C et al (2010) Use of ITS2 region as the universal DNA barcode for plants and animals. PLoS One 5:1–9

York P, Evangelista P, Kumar S et al (2011) A habitat overlap analysis derived from maxent for tamarisk and south-western willow flycatcher. Front Earth Sci 5(2):120–129

Yuksel S, Schwenkbier L, Pollok S et al (2015) Label-free detection of *Phytophthora ramorum* using surface-enhanced raman spectroscopy. Analyst 140(21):7254–7262

Zavaleta ES, Hobbs RJ, Mooney HA (2001) Viewing invasive species removal in a whole-ecosystem context. Trends Ecol Evol 16(8):454–459

Zaya DN, Leicht-Young SA, Pavlovic NB et al (2015) Genetic characterization of hybridization between native and invasive bittersweet vines (*Celastrus* spp.). Biol Invasions 17:2975–2988

Zielinski DP, Sorensen PW (2016) Bubble curtain deflection screen diverts the movement of both Asian and common carp. N Am J Fish Manag 36(2):267–276

Zuefle ME, Brown WP, Tallamy DW (2008) Effects of non-native plants on the native insect community of Delaware. Biol Invasions 10:1159–1169

8 Restoration of Landscapes and Habitats Affected by Established Invasive Species

Jennifer Koch, Dean E. Pearson, Cynthia D. Huebner, Michael K. Young, and Richard A. Sniezko

8.1 Introduction

Preventing invasions of aquatic and terrestrial habitats is the preferred and most cost-effective approach to address the invasive species threat (see Chap. 6). However, when prevention efforts fail, invasive species can become widespread and deeply embedded in native ecosystems, causing severe impacts (see Chaps. 2, 3, and 4). In such cases, invader control (see Chap. 7) accompanied by restoration facilitates recovery of native species and prevents reinvasion (e.g., Pearson et al. 2016). Here, we summarize the current state of restoration science and highlight critical information gaps that must be overcome to advance ecosystem restoration in terrestrial and aquatic systems affected by invasive plants, insects, diseases, and vertebrates.

Restoration objectives and associated management strategies vary by the type of invader, the extent of its impacts, characteristics of the affected site, and the value of the affected system in terms of its biological diversity, uniqueness, or the ecosystem services it provides to humans. Accordingly, objectives may range from classic ecosystem restoration strategies intended to fully restore a system to its pre-invasion state (ecological restoration) to more pragmatic strategies such as redirecting invasion trajectories toward desirable ecosystem services (functional restoration), despite deviations from historic composition and function, as in the case of "novel ecosystems" that have been severely transformed by multiple invaders (Forest Service Manual 2016; Hobbs et al. 2009). The decision to expend time and resources to attempt to fully restore a particular system (versus lesser restoration goals) is determined by assessing ecological, economic, and societal values of the recipient ecosystem; susceptibility to reinvasion by the same or other invaders; availability of effective restoration tools or tactics; and the defined management objectives (see example Chap. 7, Box 7.1). Several inputs are required for developing effective management and restoration strategies for affected ecosystems (see Chap. 7, Box 7.2). Generally, invasive species must be controlled to some degree in order for ecosystem restoration to be successful (see Chap. 7). Control efforts may occur prior to or in conjunction with initiation of the restoration process or may require implementation on a recurring basis for continued suppression of the invasive population, since invasive species are rarely eradicated (Myers et al. 2000). In general, ecosystem restoration efforts should emphasize ecosystem function, resilience, and resistance to future invasions in order to ensure long-term successes (Forest Service Manual 2016).

The degree and type of restoration effort depend on which native system components are affected and to what extent. Passive restoration occurs when native systems naturally recover following suppression or removal of the invasive species. For example, passive recovery may occur following control of terrestrial invertebrate or vertebrate invaders that harm native fauna allowing recovery and the subsequent return of vagile animals to formerly invaded areas, reestablishing functional populations once the invader is removed. However, this is commonly not the case for plants which

J. Koch (✉)
U.S. Department of Agriculture, Forest Service, Northern Research Station, Delaware, OH, USA
e-mail: jennifer.koch@usda.gov

D. E. Pearson
U.S. Department of Agriculture, Forest Service, Rocky Mountain Research Station, Missoula, MT, USA

Ecology and Evolution, Division of Biological Sciences, University of Montana, Missoula, MT, USA

C. D. Huebner
U.S. Department of Agriculture, Forest Service, Northern Research Station, Morgantown, WV, USA

M. K. Young
U.S. Department of Agriculture, Forest Service, Rocky Mountain Research Station and National Genomics Center for Wildlife and Fish Conservation, Missoula, MT, USA

R. A. Sniezko
U.S. Department of Agriculture, Forest Service, Dorena Genetic Resource Center, Cottage Grove, OR, USA

T. M. Poland et al. (eds.), *Invasive Species in Forests and Rangelands of the United States*,
https://doi.org/10.1007/978-3-030-45367-1_8

often fail to reestablish following control of target pests for a variety of reasons (Pearson et al. 2016) including seed limitation (Seabloom et al. 2003) and various legacy effects (Magnoli et al. 2013). Hence, restoration is often an active process requiring an array of management strategies including reintroduction of propagules and/or nutrient or soil treatments to facilitate plant recovery. In aquatic systems, restocking native fish populations may be necessary following the use of chemicals like rotenone that kills both native and invasive fish in isolated water bodies.

Successful restoration requires an understanding of what makes an ecosystem resistant to invasion, what levels of genetic diversity enhance restoration efforts, and how to accelerate the search for and development of genetic resistance to insects and pathogens. Summarizing past activities and learning from past successes and failures is critical in directing future efforts, and such information can help enlist public and cooperator support for future restoration projects. Below, we discuss the nuances associated with restoration efforts in a variety of terrestrial and aquatic communities.

8.2 Restoration of Forests and Grasslands Affected by Invasive Plants

Understanding how a plant community has been degraded is key to its restoration. Systems may gradually degrade in a predictable linear fashion with change in the environmental conditions, producing a change in plant species composition and function that is proportional to the environmental change. Or, systems may be seemingly resilient to environmental changes, until a critical threshold is reached, at which point an unpredicted and rapid change occurs. In such cases, if one understands the trigger or driver of the sudden change, removal of the driver may reverse the change. Alternatively, some systems may respond to the same environmental change by reaching multiple stable states, termed hysteresis (Suding and Hobbs 2009). The restoration pathway could be very different from that which led to the degraded state (Beisner et al. 2003), making this possibly the most challenging type of restoration. Recent literature reviews examining the efficacy of weed management (all types, including herbicide, mechanical, and biocontrol, with herbicide being the most common) indicate that weed control alone often does not promote full system recovery (Kettenring and Adams 2011; Pearson et al. 2016; Reid et al. 2009). Lack of success may be caused by failing to address the type of system degradation. For example, deer (Cervidae) browse may have been at such a duration and extent that removal of deer and invasive plants does not result in full recovery of native species diversity (Royo et al. 2010). Similarly, removing invasive aquatic plants from a riparian system is only a temporary solution, if underlying nutrient loading of the system is not addressed (Lotze et al. 2006).

Active restoration may be required following invasive plant control and may include (1) initial reintroduction of the desired species or (2) use of a predictable successional trajectory to eventually reach the final desired species composition. The more degraded the site, the more likely option two will be necessary (Chazdon 2008). It is important to first assess the need for active restoration by evaluating the native seedbank's likelihood that passive recovery (spontaneous succession) may occur after the invader is removed (Holl and Aide 2011; Tropek et al. 2010).

For grasslands dominated by annual grasses in the Western United States, management and restoration emphasize resilience to both wildfire and species invasions. In these semiarid ecosystems, resilience and invasion resistance are strongly associated with soil moisture and temperature regimes (Chambers et al. 2014, 2017). Coates et al. (2016) found that areas with low resistance and resilience to cheatgrass (*Bromus tectorum*) could be targeted for using wildfire suppression efforts to avoid domination by cheatgrass due to fire.

An important obstacle to actively restoring native plant communities is secondary invasion by non-target invasive weeds following control of the target invader (Dickens et al. 2016; Pearson et al. 2016). System recovery is more likely to occur if the risk of secondary invaders is anticipated and managed in conjunction with control of the primary target weed. Restoration may also require recovery of the initial soil conditions because some plant invaders may alter original soil conditions by adding allelopathic compounds (Grove et al. 2012; Lankau et al. 2014; Suseela et al. 2016) or changing soil pH and/or nutrient levels (Castro-Diez et al. 2012; Liao et al. 2008), soil microbial composition (Middleton and Bever 2012), or other ecosystem properties in ways that either inhibit native plant recovery or facilitate invasion by exotic plants (Kardol and Wardle 2010; Suding et al. 2004; Symstad 2004; Yelenik et al. 2004). Amending degraded soil is especially important if changes in soil condition triggered a threshold response.

An overabundance of herbivores (DiTommaso et al. 2014; see also next section on vertebrates) and non-sustainable forestry practices (Haeussler et al. 2002; Puettmann 2011) are examples of drivers that have led to depletion of native seed banks, making some sites less resilient to disturbance and more susceptible to rapid secondary invasions. Consequently, gaps created in these types of systems by such target invaders are not able to recover passively upon removal of the invader and should be filled as quickly as possible with native plants or transitional non-invasive exotic species. Establishing a rich native plant community can increase a system's resistance to reinvasion (Maron and Marler 2007; Pokorney et al. 2005), with the caveat that richer plant communities are also often the most vulnerable to invasion (Huebner and Tobin 2006; Stohlgren et al. 2003). Thus, increasing native species richness alone may not prevent invasion at resource-rich sites (Davis et al. 2000). A better strategy may be to actively plant aggressive, early successional native species that are more likely to outcompete non-native invasive plants (Funk et al. 2008), especially if present in high enough numbers. Once these early successional native species are established, mid- to late-successional native species can be added, possibly allowing a predictable successional trajectory to then take its course. Succession ecology provides a useful framework for approaches that rapidly fill the "invader gap" and move the system toward a native stable-state community (Cox and Anderson 2004; Sheley et al. 2006). However, movement toward a native stable-state community is only possible if existing undesirable drivers are addressed. In highly degraded sites, arrested novel communities may need to be accepted, but such communities, theoretically, could be managed to maximize native species richness and abundance and to deliver particular ecosystem services (Hobbs et al. 2009; Suding et al. 2004; Tognetti et al. 2010).

Failure to reseed native species is often a major obstacle to restoration (Firn et al. 2010; Hulet et al. 2010), especially for mid- to late-successional seres (Prach and Walker 2011). Plants from local and diverse seed sources are most likely to be successfully established in a restoration site. Seed source distance from the restoration site is an important consideration, and it is equally important to base seed source over an environmental gradient. Doing so increases diversity and may enable the species to adapt more quickly in a degraded environment (Johnson et al. 2010) and adapt to changes in climate (Havens et al. 2015). One method for ensuring native species survival over a relatively broad environmental gradient is to develop provisional seed zones, because they incorporate species-specific information concerning fitness in different environments and could be of great value when selecting species for restoration (Bower et al. 2014). Provisional seed zones have been developed for a few species, but these need further testing to assess their effectiveness (Bower et al. 2014; Johnson et al. 2013) even if they are deemed locally suitable (Tischew et al. 2011). An unintended outcome of using commercially produced native seed mixes obtained from different parts of an environmental gradient for restoration has been the introduction of non-site-specific genotypes which can negatively alter the local genetic integrity of species (Dyer et al. 2016). This increases genetic heterozygosity at the site but may compromise local fitness.

Even in cases where restoration of the desired plant species composition appears successful, much remains unknown regarding the reestablishment of ecosystem processes (nutrient cycling and interactions with other trophic levels). Indeed, for restored systems that have been evaluated (which are few), current restoration practices have failed to recover the original level of ecosystem functions (Moreno-Mateos et al. 2012; Yelenik and Levine 2010).

Globally, there are growing environmental markets that pay for ecosystem services (PES) to fund restoration efforts (e.g., Florida Everglades, Costa Rican and Madagascan reforestation efforts, and China's conversion of sloping cropland to forest and pasture). A problem associated with PES is the question of long-term sustainability related to dependency on available funds. Depending on the source of the funds, restoration via PES may also skew the restoration approach toward certain services while neglecting others. If not adequately regulated, this approach could lead to promotion of restoration projects that fall short of the full suite of ecosystem functions that the site once provided (Bullock et al. 2011; Palmer and Filaso 2009).

8.2.1 Key Information Needs

1. Improvements in seeding technology (e.g., Madsen et al. 2014) and other revegetation approaches to promote system recovery following invasive plant control
2. Identification of protocols to prevent loss of genetic diversity while sustaining cost-effective production of native seeds for use in restoration
3. Understanding and incorporating ecosystem process and function as recovery targets into all restoration activities
4. Developing guidelines differentiating degraded systems that (1) are a linear and proportional response to an environmental change, (2) respond to environmental change only after a threshold is reached and associated with a trigger, and (3) can take several pathways leading to multiple stable states, many of which are distinctly unlike the system's original state
5. Evaluation of pay for ecosystem services (PES) as a means of funding more restoration projects, including the need for regulating such payments so that ecosystem services and plant compositions associated with the undisturbed site take priority
6. Formulation of an economic model that helps determine when costs of restoration are too great to warrant an attempt at restoration

8.3 Genetic Considerations for the Restoration of Forests Affected by Invasive Insects and Disease

Once an invasive insect or pathogen has been successfully established, high mortality of the host tree species populations can occur rapidly. In rare cases, this results in extirpation or extinction of a species. A striking example was the elimination of the American chestnut (*Castanea dentata*) as a dominant species in eastern hardwood forests by chestnut blight, an invasive fungal disease from Asia caused by *Cryphonectria parasitica* (Paillet 2002). Chestnut blight arrived within a matter of decades after ink disease (caused by the invasive pathogen *Phytophthora cinnamomi*) had caused widespread death of American chestnut in lower elevations throughout the Southeastern United States (Anagnostakis 2002). More recently, emerald ash borer (*Agrilus planipennis*) threatens potential extinction of the North American ash (*Fraxinus* spp.) resource (see www.iucnredlist.org). Long-term monitoring revealed mortality rates of 99–100% of green (*Fraxinus pennsylvanica*) and white ash (*F. americana*) trees greater than 10 cm in diameter at breast height, within 5–7 years of infestation in stands throughout southwestern Michigan and northwestern Ohio (Gandhi et al. 2014; Knight et al. 2013). Dozens of forest tree species throughout the United States are currently under threat of widespread mortality due to invasive insects and pathogens (Table 8.1). Genetic conservation efforts are underway for some of these species (Knight et al. 2010; Mangold 2011; Sniezko et al. 2011).

Although management sometimes reduces the abundance of these destructive invasive species, complete eradication is seldom accomplished. Focus then turns to identifying sources of resistance and enriching genetic resistance in native populations of the affected host tree species (Waring and O'Hara 2005). In natural ecosystems, genetically diverse populations of tree species, through a long-term co-evolutionary history with insects or pathogens, develop an array of resistance alleles that may involve the interactions of multiple genes or a smaller number of genes of large effect (Budde et al. 2016; Ennos 2015). When a non-native insect or pathogen is introduced, existing plant resistance mechanisms developed for defense against native pests and pathogens may offer some benefit. Allelic variation in genes directed against existing native threats may provide some protection against invasive insects and pathogens, but the degree of effectiveness can vary. These evolutionary variants are likely to be uncommon because, in the absence of the invasive species, they confer little or no selective advantage to trees that possess them. As a result of the huge amount of standing genetic variation found in the large effective population sizes of most obligate outcrossing forest tree species, response to an invasive insect or pathogen can range from entire populations that die quickly, to individual trees that remain symptom-free, to those that are less affected and therefore able to survive longer, to species that are no more susceptible to the invader than its original host was (Budde et al. 2016; Ennos 2015, Sniezko et al. 2014; Telford et al. 2015). Genetic variants that confer a level of resistance to invasive species may have evolved in response to different selection pressures, served a different function before the introduction of the invasive species, or form part of the gene network directed against a related species.

As susceptible trees succumb to an invasive insect or pathogen, gaps are created that provide opportunities for regeneration to occur under the continued high selection pressure from the invasive species. This can result in dramatic shifts in allele frequencies that favor survival or confer resistance. This process of natural selection can, over many generations, lead to passive restoration of species and ecosystems that have been impacted by invasive insects and diseases. The generation time for tree species can take decades; thus, the slow process of natural selection leading to the development of resistance may not take place within an acceptable time frame for forest managers. Successful natural regeneration of a self-sustaining population is dependent on several factors. First and most importantly, there must be a level of genetic resistance among the surviving members of the population. The type and complexity of resistance is also a factor; for example, single gene resistance can sometimes be rapidly overcome by pathogens, while resistance conferred by many genes is generally more durable (Sniezko et al. 2014). The frequency and geographic distribution of resistant individuals plays an important role as well. If resistance is present at low frequency in the population, as is often the case without a shared co-evolutionary history, the surviving trees may be scattered over large areas, limiting their opportunities for reproduction. Tree species that have prolific seed production along with long-distance pollen and seed dispersal would be favored to overcome such obstacles. Not surprisingly, evidence indicating the possible occurrence of natural selection for resistance has been reported in natural populations of two such species, American elm (*Ulmus americana*), which has been heavily affected by Dutch elm disease (*Ophiostoma ulmi* and *Ophiostoma novo-ulmi*), and European ash (*Fraxinus excelsior*), which is threatened by ash dieback disease (caused by the fungal pathogen *Hymenoscyphus fraxineus*). Elm seedlings from regional seed collections, conducted in 1959 and again in 1981, were inoculated to assess tolerance to Dutch elm disease (Smalley and Guries 1993). This study revealed an increase in the frequency of tolerance in seed collected from trees that had experienced a longer exposure time to Dutch elm disease (Smalley and Guries 1993). Similarly, a progeny test of 320 open-pollinated European ash families from seed collected from Lithuania and seven additional European countries across a southern gradient found that families from regions with the longest history of ash dieback disease had the highest survival rates and lowest incidence of disease damage (Pliūra et al. 2011). These examples indicate that potential for resistance exists and, if warranted, natural selection could be accelerated or supplemented through breeding programs.

In some cases, mortality caused by invasive insects and diseases is so extensive it severely reduces genetic diversity

Table 8.1 Forest tree breeding programs for development of resistance to invasive insects and diseases in the United States

Host tree species		Invasive species			
Common name	Scientific name	Common name	Scientific name	Organization(s)	Status of breeding program
American x Chinese chestnut hybrids	*Castanea dentata X Castanea mollissima*	Chestnut blight	*Cryphonectria parasitica*	The American Chestnut Foundation, Connecticut Agricultural Experiment Station, USDA Forest Service NRS	Breeding program for resistance
American x Chinese chestnut hybrids	*Castanea dentata X Castanea mollissima*	Ink rot disease	*Phytophthora cinnamomi*	The American Chestnut Foundation, Clemson University	Resistance detected in genetic/provenance trials
Butternut	*Juglans cinerea*	Butternut canker	*Sirococcus clavigignenti-juglandacearum*	USDA Forest Service ORSO and NRS; University of Tennessee	Evidence for genetic variation in resistance in seedling or clone screens
Butternut x Japanese walnut hybrids	*Juglans cinerea X Juglans ailantifolia*	Butternut canker	*Sirococcus clavigignenti-juglandacearum*	USDA Forest Service NRS, University of Tennessee	Evidence for genetic variation in resistance in seedling or clone screens
Tanoak	*Notholithocarpus densiflorus*	Sudden oak death	*Phytophthora ramorum*	USDA Forest Service DGRC, University of California at Berkeley, Oregon State University	Efforts to identify resistance initiated
Whitebark pine	*Pinus albicaulis*	White pine blister rust	*Cronartium ribicola*	USDA Forest Service DGRC, RMRS, and CDA	Resistant planting stock deployed
Rocky mountain bristlecone pine	*Pinus aristata*	White pine blister rust	*Cronartium ribicola*	USDA Forest Service DGRC, PSWRS, RMRS	Evidence for genetic variation in resistance in seedling or clone screens
Foxtail pine	*Pinus balfouriana*	White pine blister rust	*Cronartium ribicola*	USDA Forest Service DGRC, R5 Placerville	Efforts to identify resistance initiated
Limber pine	*Pinus flexilis*	White pine blister rust	*Cronartium ribicola*	USDA Forest Service DGRC, PSWRS, RMRS	Evidence for genetic variation in resistance in seedling or clone screens
Sugar pine	*Pinus lambertiana*	White pine blister rust	*Cronartium ribicola*	USDA Forest Service DGRC and R5 Placerville	Resistant planting stock deployed
Great Basin bristlecone pine	*Pinus longaeva*	White pine blister rust	*Cronartium ribicola*	USDA Forest Service DGRC, PSWRS and RMRS	Evidence for genetic variation in resistance in seedling or clone screens
Western white pine	*Pinus monticola*	White pine blister rust	*Cronartium ribicola*	USDA Forest Service DGRC, CDA, R5 Placerville	Resistant planting stock deployed
Southwestern white pine	*Pinus strobiformis*	White pine blister rust	*Cronartium ribicola*	USDA Forest Service DGRC, PSWRS, CDA	Evidence for genetic variation in resistance in seedling or clone screens
Eastern white pine	*Pinus strobus*	White pine blister rust	*Cronartium ribicola*	USDA Forest Service ORSO and DGRC; University of Minnesota	Resistant planting stock deployed
Port-Orford-cedar	*Chamaecyparis lawsoniana*	Root rot	*Phytophthora lateralis*	USDA Forest Service DGRC	Resistant planting stock deployed
Port-Orford-cedar	*Chamaecyparis lawsoniana*	Stigmina foliage blight	*Pseudocercospora thujina*	USDA Forest Service DGRC	Resistance detected in genetic/provenance trials
American elm	*Ulmus americana*	Dutch elm disease	*Ophiostoma novo-ulmi*	USDA Forest Service NRS, University of Minnesota	Evidence for genetic variation in resistance in seedling or clone screens

(continued)

Table 8.1 (continued)

Host tree species		Invasive species			
Common name	Scientific name	Common name	Scientific name	Organization(s)	Status of breeding program
Oak species	*Quercus* spp.	Sudden oak death	*Phytophthora ramorum*	USDA Forest Service PSWRS	Evidence for genetic variation in resistance in seedling or clone screens
Green ash	*Fraxinus pennsylvanica*	Emerald ash borer	*Agrilus planipennis*	USDA Forest Service NRS	Evidence for genetic variation in resistance in seedling or clone screens
White ash	*Fraxinus americana*	Emerald ash borer	*Agrilus planipennis*	USDA Forest Service NRS	Evidence for genetic variation in resistance in seedling or clone screens
Black ash x Manchurian ash hybrids	*Fraxinus nigra X Fraxinus mandshurica*	Emerald ash borer	*Agrilus planipennis*	USDA Forest Service NRS	Evidence for genetic variation in resistance in seedling or clone screens
American beech	*Fagus grandifolia*	Beech scale (beech bark disease)	*Cryptococcus fagisuga*	USDA Forest Service NRS	Resistant planting stock deployed
Fir species	*Abies* spp.	Balsam woolly adelgid	*Adelges piceae*	North Carolina State University	Evidence for genetic variation in resistance in seedling or clone screens
Eastern hemlock	*Tsuga canadensis*	Hemlock woolly adelgid	*Adelges tsugae*	University of Rhode Island, North Carolina State University	Evidence for genetic variation in resistance in seedling or clone screens
Carolina hemlock hybrids	*Tsuga caroliniana X Tsuga* spp.	Hemlock woolly adelgid	*Adelges tsugae*	US National Arboretum	Evidence for genetic variation in resistance in seedling or clone screens
Redbay	*Persea borbonia*	Laurel wilt	*Raffaelea lauricola*	University of Florida	Evidence for genetic variation in resistance in seedling or clone screens
Koa	*Acacia koa*	Koa wilt	*Fusarium oxysporum*	Hawaii Agriculture Research Center, USDA Forest Service DGRC	Resistant planting stock deployed
Ōhi'a	*Metrosideros polymorpha*	Rapid 'Ōhi'a Death	*Ceratocystis huliohia Ceratocystis lukuohia*	University of Hawaii, USDA ARS, USDA Forest Service PWRS, DGRC and Region 5 Forest Health Protection	Efforts to identify resistance initiated

Abbreviations: *NRS* Northern Research Station, *ORSO* Oconto River Seed Orchard, *DGRC* Dorena Genetic Resource Center, *RMRS* Rocky Mountain Research Station, *CDA* Couer D'Alene Nursery, USFS, *PSWRS* Pacific Southwest Research Station, *R5 Placerville* USDA Forest Service Region 5 Genetics program at Placerville nursery

of affected populations. This can lead to inbreeding depression and a catastrophic loss of essential allelic variants, resulting in a loss of fitness in the next generation. The few surviving trees may be resistant to the disease or insect, but their progeny may be more vulnerable to secondary threats (native pests, other invasive insects, pathogens, or abiotic stresses such as drought and fire) because the resistance alleles have been lost. Even in situations where recovery is possible, the species' distribution or density is likely to have been drastically reduced. The remaining genetic diversity may be insufficient to allow adaptation to new threats, including adaptation to a potentially changing climate. Despite indications of natural selection in American elm for Dutch elm disease resistance and in European ash for resistance to ash dieback, both are facing new threats. Recent outbreaks of elm yellows (*Candidatus Phytoplasma ulmi*) have been documented in the United States, and although emerald ash borer is native to Far Eastern Russia, it has now become established in Western Russia (Moscow) where it is poised to spread throughout Europe (Herath et al. 2010;

Orlova-Bienkowskaja 2014; Sherald 2009). The loss of genetic diversity in these species has likely made them even more vulnerable to new threats, as will be the case for North American ash trees that survive emerald ash borer, should ash dieback ever invade the United States. Black ash (*Fraxinus nigra*) is reported to be highly susceptible to ash dieback, and green ash is reported to be moderately susceptible (Drenkhan and Hanso 2010; Gross and Sieber 2016).

Silviculture can be used to favor and support regeneration of forest species impacted by invasive insects and diseases with the goal of retaining or increasing genetic diversity and increasing population-level resistance (Waring and O'Hara 2005). Natural selection can be accelerated by careful selection and removal of diseased or infested trees, with the goal of eliminating susceptible alleles from the gene pool while retaining genetic diversity. A test comparing seedling progeny from parent trees from an unmanaged stand of American beech (*Fagus grandifolia*) with seedling progeny from parent trees in a stand where single tree selection and removal of trees with symptoms of beech bark disease had been performed 12 years earlier indicated that there was a 50% increase in the number of beech bark disease-resistant seedlings from the managed stand compared to seedlings from the unmanaged stand (Koch 2010; Koch et al. 2010). Despite these promising results, residual stands ravaged by invasive insects and diseases frequently consist of a small number of widely dispersed surviving individuals that are unable to recover naturally. In these situations, the process of stand recovery can be accelerated through breeding remaining resistant individuals to retain genetic diversity and adaptive capacity and to harness and even improve upon the combinations of genes that proved advantageous to their survival.

8.3.1 Breeding Resistance to Invasive Forest Pathogens

The most enduring and successful efforts at breeding for resistance against an invasive pathogen attacking forest trees have focused on the fungus *Cronartium ribicola*, the cause of white pine blister rust, which has been in the United States since 1898 (Kinloch 2003; Sniezko et al. 2014). All nine species of white pine (*Pinus* spp.) native to the United States are susceptible to blister rust, and all except *P. longaeva* (Great Basin bristlecone pine) have been infected in their native range (Sniezko et al. 2011; Tomback and Achuff 2010). Infection levels as high as 70 to 100% have been reported in the northern range of whitebark (*P. albicaulis*) and limber pine (*P. flexilis*) (Kinloch 2003). Three species are vulnerable to extinction because of the negative effect of blister rust on regeneration (Tomback and Achuff 2010). Complete resistance conferred by a single dominant major gene has been identified in four of the white pine species (Kinloch 2000; Kinloch and Dupper 2002), and several types of partial or quantitative (controlled by multiple genes) resistance have been identified (Sniezko et al. 2014). The best approach and current focus for obtaining durable resistance is to breed for both complete and partial resistance in production populations (Sniezko et al. 2014). Assessment of genetic resistance is underway in all nine species, and more extensive efforts for breeding resistance are ongoing for several of the species (summarized in Table 8.1; see Fig. 8.1). Resistant seedlings are currently being deployed (as part of artificial regeneration strategies) for four of the species to date (Waring and Goodrich 2012), and substantial progress has been reported (see Box 8.1). Pathogens such a *C. ribicola* have the capacity for rapid evolution as they can complete many generations over the lifetime of their host tree. Field trials are monitoring the durability and stability of tree resistance over time, on a range of sites with varying risk of rust infection. Western white pine (*P. monticola*) and sugar pine (*P. lambertiana*) results are encouraging (Kinloch et al. 2012; Sniezko et al. 2012a, 2014). Efforts are underway to further increase the level of resistance in several of these species.

Box 8.1 On the Road to Success: Tangible Evidence of the Impact of Host Resistance

According to the International Union for Conservation of Nature's Red List of Threatened Species, the species status of Port-Orford-cedar (*Chamaecyparis lawsoniana*), listed as "vulnerable" in 2000, has been downgraded to "near threatened" as of 2013, with anticipation that it will be listed as a species of "least concern" within 10 years if current conservation actions, including planting resistant seedlings (see Fig. 8.2), are successful and maintained (Farjon 2013). The interagency, inter-regional genetic resistance program in Port-Orford-cedar is based at the USDA Forest Service Dorena Genetic Resource Center in the Pacific Northwest Region and has produced one of the most quickly implemented and effective resistance programs in forest trees.

In 2011, in another example, the US Fish and Wildlife Service determined that "the whitebark pine warrants protection under the Endangered Species Act" and was under imminent threat of high magnitude, assigning it a listing priority number (LPN) of 2 in 2011. In 2015, the LPN was reduced to eight (U.S. Fish and Wildlife Service 2015), a decision that was attributed, in part, to the identification and propagation of genetically resistant trees. This program is coordinated by USDA Forest Service efforts (see Table 8.1).

Fig. 8.1 Whitebark pine seedlings that have been inoculated with white pine blister rust at the USDA Forest Service Dorena Genetic Resource Center. Resistant seedlings are those that have remained green and healthy (Photo by Richard Sniezko, USDA Forest Service, Dorena Genetic Resource Center)

Fig. 8.2 Restoration of Port-Orford-cedar. Volunteers and USDA Forest Service personnel plant 900 Port-Orford-cedar seedlings along Grayback Creek in southern Oregon to help restore a riparian area used by salmon and steelhead trout. The seedlings will be monitored to determine if resistance to root disease caused by *Phytophthora lateralis* is durable. Community groups involved with the planting include the Middle Rogue Steelheaders Trout Unlimited Chapter, Illinois Valley Watershed Council, The Nature Conservancy, Forestry Action Committee, and Southern Oregon Fly Fishers (Photo by Scott Kopak, USDA Forest Service Umpqua National Forest)

The hybrid (interspecies) breeding approach exploits the original host tree resistance to the invasive insect or pathogen. Frequently, the original host species is related to the newly invaded tree species allowing hybridization. Butternut (*Juglans cinerea*), a native North American species that has experienced severe decline because of butternut canker (*Sirococcus clavigignenti-juglandacearum*), hybridizes naturally with Japanese walnut (*Juglans ailantifolia*) (Hoban et al. 2009). The Japanese walnut, which is resistant to butternut canker, was widely planted in the United States since it was introduced in the nineteenth century and has since naturalized in many areas (Hoban et al. 2012). As a result, hybrids of Japanese walnut and butternut, including advanced-generation hybrids, occur in natural populations and retain the pathogen resistance from the Japanese walnut lineage (Broders et al. 2015; Hoban et al. 2009). The lack of evidence for heritable resistance within native butternut populations, combined with the low number of surviving pure butternut (the species is locally extinct throughout much of the former range), strongly suggests that it will be necessary to use hybrid breeding, with Japanese walnut as a source of resistance alleles, to achieve levels of resistance adequate for butternut restoration (LaBonte et al. 2015).

In examples like butternut, hybrid breeding may be the only option. Hybrid breeding programs are challenging because selection for resistance is not always sufficient and the program must also ensure that the ecosystem function and adaptive traits of the native species are retained. The chestnut blight resistance breeding program of The American Chestnut Foundation is among the most notable and long-standing hybrid breeding programs in the United States. Although there is no evidence of complete, single gene resistance to chestnut blight in either the American chestnut or other Asian species of chestnut, quantitative (multi-gene) resistance is present in the Chinese chestnut (*Castanea mollissima*). Very little information is available in the literature related to the presence of or testing for quantitative resistance in the American chestnut, possibly due to the early focus placed on the hybrid breeding approach (Budde et al. 2016). The American Chestnut Foundation program produced hybrids between American chestnut and Chinese chestnut and selected for blight resistance using several generations of backcrossing and intercrossing to retain desirable American traits (Hebard 2012). More recently, the program has also selected American chestnut for resistance to ink disease. Complete resistance to this pathogen has been identified in one of the hybrid breeding lines (Jeffers et al. 2012; Zhebentyayeva et al. 2013), but additional testing will be necessary to confirm this result. Significant resources have been directed to support using advanced-generation hybrids from The American Chestnut Foundation's program to develop and optimize artificial regeneration procedures for chestnut restoration (Clark et al. 2014). Although the first widespread tests of advanced-generation hybrids demonstrated promising levels of early seedling establishment, susceptibility to both ink disease and chestnut blight remains a significant issue (Clark et al. 2015; Pinchot et al. 2014). The American Chestnut Foundation program continues to pursue increasing the resistance to both pathogens.

Dutch elm disease, caused by the invasive pathogens *Ophiostoma ulmi* and *Ophiostoma novo-ulmi*, was first identified in Ohio in 1930 killing a tremendous number of American elm trees that were highly valued street trees for their beautiful arching canopies. Early efforts to identify resistance were focused mostly on developing cultivars by selection, propagation, and testing of large surviving elm trees or by screening large numbers of seedlings (Smalley and Guries 1993; Townsend and Douglass 2001). The cultivars that were developed were mostly used as landscape and nursery trees. Efforts to restore American elm in naturally forested areas focused initially on deploying five of these previously developed nursery cultivars and later were expanded to include progeny from a cross between two of the five cultivars (Slavicek et al. 2005). The long-term performance of these plantings may be problematic, as genetic diversity is limited and three of these cultivars are susceptible to elm yellows (Sinclair et al. 2001; Smalley and Guries 1993). Recently, efforts have been initiated to enhance adaptive capacity and expand genetic diversity through breeding (Slavicek and Knight 2013).

One of the most significant successes attributed to a forest tree resistance breeding is development of populations of Port-Orford-cedar (*Chamaecyparis lawsoniana*) that have genetic resistance to a root disease caused by *Phytophthora lateralis* (Sniezko et al. 2012b) (see Box 8.1). This program identified both single gene and multi-genic sources of resistance, delineated breeding zones, and established seed orchards. The program is now producing seed from several breeding zones (Sniezko et al. 2012b). Reforestation and restoration efforts using Port-Orford-cedar seed resistant to *P. lateralis* are now underway. Field trials to monitor the efficacy and durability of resistance have been established. The program's success will be decided over time as seed used in field trials is evaluated for restoration and reforestation efforts.

Laurel wilt is among the newest invasive diseases affecting trees in the United States. The disease is vectored by the redbay ambrosia beetle (*Xyleborus glabratus*), which introduces its fungal symbiont (*Raffaelea lauricola*) into the sapwood of host trees. The beetle was first detected in 2002 and, along with its fungal symbiont, was associated with extensive mortality of redbay (*Persea borbonia*) in 2003. The disease spreads rapidly within stands, and mortality levels greater than 90% have been reported within just a few years (Hughes et al. 2015). Since then, laurel wilt has been confirmed in eight southeastern states. Eight additional forest and landscape tree or shrub species (including two that are considered threatened or endangered) have been confirmed to be infected, and five additional species are vulnerable, based on results from artificial inoculation (Hughes et al. 2015). Researchers at the University of Florida identified and clonally propagated redbay trees that remained asymptomatic in heavily diseased natural areas (Hughes and Smith 2014). Field trials demonstrated tolerance in a few selected genotypes, and additional genotypes are being propagated for testing.

8.3.2 Breeding Resistance to Invasive Insects

Although laurel wilt is caused by an invasive pathogen vectored by an invasive insect, some invasive pathogens are vectored by native insects. The original vector for Dutch elm disease was the native elm bark beetle (*Hylurgopinus rufipes*), but over time as non-native insect species entered the United States, the list of vectors of Dutch elm disease has grown (Jacobi et al. 2007; Lee et al. 2009). Control of these insect vectors, whether native or invasive, can be critical for managing the diseases they vector. Genetic traits of the host can play a pivotal role in control of insects associated with disease. For example, the correlation between early flushing and reduced susceptibility to Dutch elm disease in both American and European elm may be due to a mismatch between the period of maximum host susceptibility and timing of beetle emergence in the spring, thus allowing the trees to escape infection (Ghelardini and Santini 2009). In the case of beech bark disease, breeding has focused on selecting for resistance to the beech scale insect (*Cryptococcus fagisuga*). Although this scale insect is not a vector of the fungal species (*Neonectria ditissima*, *N. faginata*) that cause the disease, its feeding activity creates many entry points for the fungal spores (Ehrlich 1934). In the absence of scale infestation, species of *Neonectria* alone have not been associated with the high tree mortality levels characteristic of beech bark disease. Effective techniques to screen for resistance to the scale insect in the field, in potted seedlings, or in grafts have been developed (Koch and Carey 2014), as have efficient methods to propagate resistant beech trees (Carey et al. 2013). Genetic studies have confirmed that resistance to the scale insect is heritable, and regional seed orchards of grafted resistant American beech trees are being established (Koch and Heyd 2013; Koch et al. 2010). Unfortunately, the American beech now appears to be threatened by another potentially invasive pest causing what is currently referred to as beech leaf disease (Pogacnik and Macy 2016). Research is being conducted to follow up on the identification of a foliar nematode as a possible causal agent of this disease. The nematode appears to be closely related to the recently reported *Litylenchus crenatae*, found in *Fagus crenata*, a beech species native to Japan (Carta 2018; Kanzaki et al. 2019). In addition to confirmation of the causal agent of this disease, continued monitoring will be needed to fully understand what the long-term, landscape-scale impacts will be, but current information is pointing to what could quite possibly be a devastating and wide-ranging impact to American beech, particularly given the stress it is already enduring due to beech bark disease.

Other invasive insect species cause significant damage and mortality directly as a consequence of completing their life cycle on or within specific tree hosts. Emerald ash borer and hemlock woolly adelgid (*Adelges tsugae*) are currently among the most significant threats to forests in the Eastern United States. Efforts are underway to identify and breed for resistance to these invasive species (Table 8.1). Genetic variation in the susceptibility of green ash to emerald ash borer was first observed in natural stands where, despite long-term emerald ash borer infestation resulting in the death of over 95% of the ash, occasional surviving trees were identified

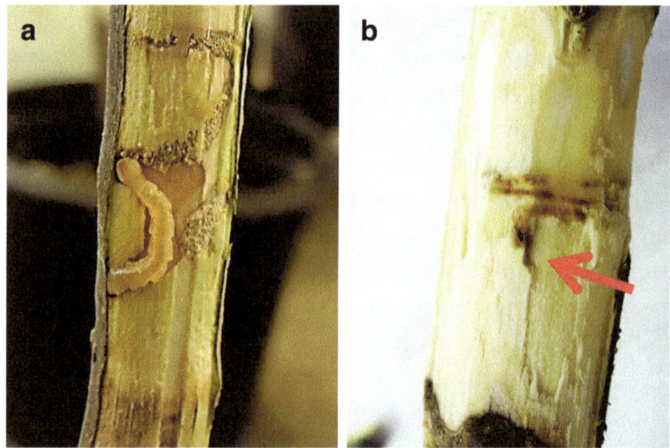

Fig. 8.3 Results of emerald ash borer egg bioassay. The larva in *panel A* has successfully developed to a late instar on this susceptible green ash tree, while the larva in *panel B* has been killed by defense responses mounted by the tree, indicating resistance (Photo by Jennifer Koch, USDA Forest Service, Northern Research Station)

(Knight et al. 2012a). Bioassay experiments performed on grafted replicates of some of these green ash trees have confirmed that the increased level of resistance observed in select trees is due to multiple types of host defense responses, including mortality of early instar larvae, larvae with significantly lower weights, and reduced adult feeding on foliage (Koch et al. 2015) (Fig. 8.3). Although additional field testing and genetic studies are needed, these results indicate that there is potential for developing of a successful emerald ash borer resistance breeding program.

A citizen science approach was used to identify eastern hemlock trees (*Tsuga canadensis*) that survived in areas heavily infested by hemlock woolly adelgid (Ingwell and Preisser 2010). Rooted cuttings of some surviving trees were compared to susceptible controls and hemlock woolly adelgid-resistant western hemlock (*Tsuga heterophylla*) using an artificial infestation technique. Results showed that, although there was no difference between the initial establishment of hemlock woolly adelgid sistens (winter generation) on resistant versus susceptible eastern hemlock cuttings, significantly lower adult sistens were produced on the resistant genotype (Ingwell and Preisser 2010). Current efforts are focused on optimizing the inoculation technique to develop an efficient screen for resistance as a first step toward developing a breeding program (Powers et al. 2015).

8.3.3 Deployment of Resistant Planting Stock or Alternative Non-host Species

The development of a resistance breeding program is only one facet involved in the process of restoration. In some cases, there may be no detectable host resistance within the species, and interspecies hybridization may not be successful. In other cases, there may be evidence for resistance, but the threat to sensitive habitats warrants implementation of restoration activities before resistant planting stock is available. In such cases, proactive planting of alternative non-host species may be considered. For example, American elm cultivars and a full-sibling elm seedling family, along with oak (*Quercus* spp.) and sycamore (*Platanus occidentalis*) seedlings, were used in plantings to restore ash-dominated riparian forests at risk of high ecological impact due to near synchronous emerald ash borer-induced mortality of ash (Knight et al. 2012b; Nesbit et al. 2015). Similar approaches are being considered for northern black ash forests threatened by emerald ash borer. The destruction of black ash has the potential to cause a transition to non-forest conditions (Palik et al. 2012) and, subsequently, a significant rise in the water table (Slesak et al. 2014). Extensive analysis of emerald ash borer risk, potential impacts of climate change, and co-occurring species has been performed to identify potential candidate species for replacing black ash (Iverson et al. 2015).

Whether resistant hosts are developed through breeding, or non-host trees are used as an alternative to restore forest ecosystems affected by invasive species, the same basic steps are required. These include (1) seed production and collection; (2) mass propagation of planting stock in a nursery or greenhouse; (3) site preparation, which can vary significantly across sites; and 4) post-planting maintenance during the establishment phase, which may include occasional watering, weed removal, and herbivore control. Each step requires a significant investment in labor, with needs for specific skill sets, tools, facilities, and infrastructure (Campbell and Schlarbaum 2014). Applied nucleation is a promising strategy for cost-effective restoration of forests affected by invasive insects and diseases. This approach utilizes natural successional processes and is based on planting trees in irregularly spaced clusters, which serve as the focal point of propagule establishment and provide islands of habitat that attract birds and mammals, which then aid in seed dispersal. Recent reports indicate that this approach reduces the need for management activities (such as watering, fertilization, and invasive plant removal) and, consequently, is a lower-cost alternative to a regularly spaced plantation design for facilitation of forest recovery and restoration (Corbin and Holl 2012; Corbin et al. 2016).

8.3.4 Key Information Needs: Toward Developing Capacity for Solutions

Recent literature reviews suggest that host resistance, forest genetics, and tree improvement may be the most effective approaches for retaining and restoring resilience in our forests under the continuous threat of invasive insects and diseases (Budde et al. 2016; Ennos 2015; Telford et al. 2015; Wheeler et al. 2015). Resistance breeding programs can provide real solutions to invasive insects and diseases within reasonable time periods, especially when supported by the necessary infrastructure and skills. Such programs have a long record of success in crops, fruit and nut trees, and forest trees. A resistance breeding approach does not require the use of genetically modified organisms (GMOs) produced through transgenic or gene editing technologies, although established breeding programs can, under certain circumstances, successfully incorporate these techniques to develop strategies to accelerate breeding (see Chap. 9). GMOs or even gene-edited plants are not always widely accepted by the public, and despite assertions that GMOs provide sustainable solutions more quickly or more cost-effectively than a well-designed and executed breeding program, there is no evidence available to support this because there are currently no GMO forest trees that have been approved for planting in natural forests. Use of genetic engineering to produce a plant with resistance can have application in plantation and crop forestry, but a tree genetically modified for resistance represents a single genotype and lacks the genetic diversity to achieve the resilience needed for ecosystem restoration. For utilization in ecosystem restoration, a GMO would have to be integrated with a breeding program to incorporate genetic diversity and adaptive capacity, an approach that The American Chestnut Foundation intends to pursue, should they successfully obtain regulatory approval (Steiner et al. 2017). Therefore, the use of GMOs would not negate or reduce the need for traditional breeding programs. Unfortunately, the demand for forest genetics and tree improvement programs to address invasive insect and disease problems comes at a time when the infrastructure and expertise required for such programs has been declining for two decades (Campbell and Schlarbaum 2014; Wheeler et al. 2015). In addition to basic infrastructure, expertise in forest tree breeding and forest tree genetics is essential for implementing cost-effective breeding programs. Tree breeding requires the development and the disciplined execution of long-term plans based on (1) a solid understanding of the desired phenotype and the genetic basis for the phenotype (Groβkinsky et al. 2015; Zivy et al. 2015), (2) screening and development of breeding populations, (3) maintenance of local adaptation and genetic diversity, and (4) testing and deployment of the resistant trees. Even in a situation where resistance breeding strategies have succeeded in developing resistant trees, the lack of necessary infrastructure (seed orchards to produce improved seed, nurseries to mass produce planting stock) for carrying out restoration may prevent effective deployment of such resistance. Effective long-term planning also requires (1) strategies for germplasm conservation; (2) an understanding of the long-term interaction of the invasive pathogen or insect on the standing genetic variation of the host species across the landscape (e.g., would natural selection and natural regeneration be sufficient for recovery?); (3) validation of resistance and assessment of durability through long-term monitoring of field trials across different sites and environments; (4) assessment of the different types of resistance and the distribution and frequency of such resistance within a species; (5) an understanding of the dynamics of the disease triangle (host, pest/pathogen, and environment) across the native range of the host, including the environments in which most people interact with trees (streets, parks, urban woodlands, arboreta, tree plantation, and highly fragmented natural forests); and (6) development of updated seed zones, taking into consideration predicted environmental changes due to climate change.

8.4 Restoration of Forests and Grasslands Affected by Invasive Vertebrates

The incursion of invasive vertebrates in terrestrial ecosystems can impact both native plants and wildlife (see Chap. 2). From a restoration perspective, the first step in restoring systems affected by invasive vertebrates is to suppress populations of the target invader to a level sufficient to reduce their impacts, thus allowing restoration to proceed (see Chap. 7). However, frequently restoration activities must begin in concert with control of the invasive vertebrate, and, in some cases, it may be necessary to suppress several invasive vertebrates and/or plants simultaneously or in a specific sequence in order to avoid releasing other invaders from suppression (Bergstrom et al. 2009; Chapuis et al. 2004; Morrison 2007; Zavaleta et al. 2001). For example, on sub-Antarctic Macquarie Island, extirpation of feral cat (*Felis catus*) populations resulted in an increase in exotic rabbit (*Oryctolagus cuniculus*) populations which devastated certain native vegetation communities and released exotic plants like annual bluegrass (*Poa annua*) (Bergstrom et al. 2009). Control of invasive vertebrate predators can also threaten native vertebrates if strategies are not carefully executed, such as the control of feral pigs (*Sus scrofa*) on the Channel Islands that resulted in increased predation by golden eagles (*Aquila chrysaetos*) on the endangered island fox (*Urocyon littoralis*) (Morrison 2007; Roemer et al. 2001). Invasive vertebrate herbivores, such as feral hogs, goats, and horses, can directly harm native plants and cause disturbances that can facilitate plant invasions (Campbell and Donlan 2005). The resulting plant invaders may frequently persist or they may increase, along with other invasive plant species, following control of invasive vertebrate herbivores (Chapuis et al. 2004; Morrison 2007). Hence, weed control (see Chap. 7) and vegetation restoration practices described elsewhere in this chapter (Sect. 8.2) should be implemented in conjunction with control of invasive vertebrate herbivores. In more extreme cases involving vertebrates that act as ecosystem engineers (see Chap. 2), such as nutria (*Myocastor coypus*), which disrupt waterways (Carter and Leonard 2002), physical restoration of the hydrology of the system may be required before vegetation and other system components can recover.

Reductions in invasive vertebrate populations may facilitate natural recovery of vagile native fauna. For example, reducing populations of invasive cavity nesting birds such as starlings (*Sturnus vulgaris*) provides cavities that can be utilized by native cavity nesters (e.g., Smith 2005), and reductions of invasive predator populations permit native prey populations to recover as in New Zealand where rat (*Rattus* spp.) control is critical for bird conservation (Moorhouse et al. 2003). However, in cases where the affected native species is rare or threatened, it may be necessary to reintroduce the species or subsidize their populations initially, applying traditional wildlife management approaches (Mills 2012). The presence of multiple invasive vertebrates can greatly complicate management and restoration because invaders can develop strong interactions with native vertebrates and other invasive vertebrates (Zavaleta et al. 2001). In such cases, control of any one target invader can initiate complex outcomes with extreme unintended side effects (Bergstrom et al. 2009; Zavaleta et al. 2001). In the case of multiple invasive species, particularly at multiple trophic levels, it may be necessary to implement integrated management strategies that address each invader separately to protect against intended outcomes (Zavaleta et al. 2001).

8.4.1 Key Information Needs

Restoration of native communities affected by exotic vertebrates requires better understanding of the impacts that vertebrate invaders have on different native taxa so that control efforts can be planned in ways that mitigate unintended consequences associated with releasing secondary invaders (Bergstrom et al. 2009; Zavaleta et al. 2001); in addition, follow-up restoration efforts can be carefully directed to recover system components that may not recover without assistance. While it is recognized that impacts of vertebrate pests on plant communities may require active restoration efforts (see above), research is needed to document the extent to which other taxa may or may not recover naturally following activities to control vertebrate pests.

8.5 Restoration of Aquatic Habitats Invaded by Aquatic Species

In many aquatic ecosystems, restoration of native-only communities is not feasible. The few and often non-selective, broad-spectrum strategies available for species removal—

generally mechanical removal by netting or electrofishing or chemical removal using piscicides—preclude their widespread application. Furthermore, the frequent and often long-range movements that are characteristic of some non-native species make them resistant to eradication. Even the process of removing species from relatively simple habitats can require near-impossible efforts (Pacas and Taylor 2015), and long-term success is contingent on preventing reinvasion. In flowing water systems, this generally consists of using artificial barrier to prevent movement, a tactic that is accompanied by its own suite of problems (Fausch et al. 2009). In some cases, the objective has been to control but not eradicate non-native species, and native species have occasionally rebounded (Propst et al. 2015). However, "control fatigue" sometimes sets in if there is an erosion of funds or management focus, resulting in the rapid reestablishment of non-native species populations (Meyer et al. 2006). A further impediment to applying control measures is the substantial societal resistance to removal of non-native species that are prized for recreation, food, or outdoor-related economic activity. Fresh water for municipal, agricultural, and industrial use constitutes a fundamental ecosystem service. Diverse uses of fresh water provided by many federal lands add enormous complexity to management efforts to reduce non-native species in favor of native ones.

Within the subset of habitats where non-native species removal is feasible and undertaken, opportunities exist for restoring native species. Recolonization by native species may be rapid and begin immediately following removal of non-native species (i.e., within days or weeks). In fluvial aquatic ecosystems, passive restoration of native animal species following removal of non-native species typically relies on downstream drift of organisms originating from untreated, upstream reaches. This is a standard strategy for addressing aquatic insects in streams and is often used as a measure of ecosystem recovery (e.g., Kjærstad et al. 2015). Upstream migration by native taxa from areas downstream of treated reaches is rarely feasible; these locations are often the sources of the invading non-native taxa, and the installation of structures that prevent recolonization by invasive species similarly constrains native taxa (Fausch et al. 2009). In standing water, amphibians have successfully recolonized mountain lakes following removal of introduced fish (Knapp et al. 2007). However, for many native taxa, passive recolonization is not feasible due to their limited mobility, few local sources of potential re-colonists, or lack of access to restored habitats (Knapp and Sarnelle 2008).

Despite the often poor probability of recolonizing native aquatic species, active restoration continues to be a standard practice. This is not unexpected, because management of aquatic species, particularly fishes, has a 150-year tradition of fostering the introduction and spread of species of value to society. Some hatchery facilities are dedicated solely to the production of rare native species for use in conservation programs (e.g., Dexter National Fish Hatchery, NM), whereas others are responsible for maintaining local stocks of conservation value. The establishment of broodstocks of native species that are used for restoration has evolved substantially, particularly with respect to maintaining genetic diversity both in the hatchery and in recipient populations (Fisch et al. 2015). This continues to be an ongoing challenge, because even a single generation of hatchery domestication results in some degree of artificial selection (Christie et al. 2016). Introductions of native species may involve any or all age classes, from sexually mature adults to fertilized embryos, and programs can rely on translocations of wild individuals instead of hatchery products. Also, local populations of native species are sometimes salvaged from an area before treatment to remove non-native species and transferred off-site to nearby water bodies or hatcheries until treatment is completed. As an aside, many hatchery facilities now propagate non-native fish that are sterile; these fish are widely introduced to promote recreational fishing, and though they may have ecosystem effects, these effects are expected to be minimal because these fish are unable to establish reproducing populations or cause hybridization.

The decision involving where to practice aquatic species restoration is rather arbitrary and is often dictated by convenience or opportunity. Efforts to prioritize conservation actions have been underway for decades, including sophisticated approaches that (1) weigh a host of variables to identify sites that are critical to the survival of a species or are representative of rare or at-risk habitats (Groves 2003), (2) rank the value and vulnerability of individual populations, stocks, or species (McElhany et al. 2000), or (3) integrate the two (Fausch et al. 2009). According to Wenger et al. (2011), aquatic habitats are highly dynamic and climate change is likely to drive future changes, including the distribution of native and non-native species. The development of massive bio-geo databases, coupled with tools to perform synthetic analyses, now permits a further step: the site-specific, probabilistic assessment of occupancy by native species in light of both climate change and non-native species invasions (e.g., for bull trout (*Salvelinus confluentus*) and cutthroat trout (*Oncorhynchus clarkii lewisi*) in the Northwestern United States) (Isaak et al. 2015). Many taxa, however, must await assembly of existing data, deployment of new species surveillance techniques (e.g., eDNA sampling) (McKelvey et al. 2016; Thomsen et al. 2012), and, for the myriad lesser known species, a better understanding of their ecology.

8.5.1 Key Information Needs

The preceding text touched on the critical areas of research that are needed or underway. Barring an unlikely consensus in the way societies view the introduction and redistribution of non-native aquatic species—from an unavoidable consequence of global commerce, a positive addition to biodiversity, or an ecological menace—the issues associated with non-native species invasions are likely to continue. Moreover, their ecological idiosyncrasies and ability to rapidly adapt to new environments may necessitate employing individualized control measures against the non-native species, perhaps while managing targeted native species to ensure they persist. Consequently, and perhaps regrettably, the need and scope for new tactics, strategies, and understanding is likely to continue to grow.

Disclaimer Text The findings and conclusions in this publication are those of the authors and should not be construed to represent any official USDA or U.S. Government determination or policy.

Literature Cited

Anagnostakis SL (2002) The effect of multiple importations of pests and pathogens on a native tree. Biol Invasions 3:245–254

Beisner BE, Haydon DT, Cuddington K (2003) Alternative stable states in ecology. Front Ecol Environ 1:376–382

Bergstrom DM, Lucieer A, Kiefer K et al (2009) Indirect effects of invasive species removal devastate World Heritage Island. J Appl Ecol 46:73–81

Bower AD, St. Clair JB, Erickson V (2014) Generalized provisional seed zones for native plants. Ecol Appl 24:913–919

Broders K, Boraks A, Barbison L et al (2015) Recent insights into the pandemic disease butternut canker caused by the invasive pathogen *Ophiognomonia clavigignenti-juglandacearum*. For Pathol 45:1–8

Budde KB, Nielsen LR, Ravn HP, Kjaer ED (2016) The natural evolutionary potential of tree populations to cope with newly introduced pests and pathogens-lessons learned from forest health catastrophes in recent decades. Tree Genet 2:19–29

Bullock JM, Aronson J, Newton AC et al (2011) Restoration of ecosystem services and biodiversity: conflicts and opportunities. Trends Ecol Evol 26(10):541–549

Campbell K, Donlan CJ (2005) Feral goat eradications on islands. Conserv Biol 19:1362–1374

Campbell FT, Schlarbaum SE (2014) Fading forests III. American Forests: what choice will we make? The Nature Conservancy and The University of Tennessee, 155 p., online only. http://www.nature.org/ourinitiatives/habitats/forests/fading-forests-3-complete-report.pdf

Carey DW, Mason ME, Bloese P, Koch JL (2013) Hot callusing for propagation of American beech by grafting. HortScience 48:620–624

Carta L (2018 Personal correspondence with unpublished report dated August 31, 2018 to Jennifer Koch. On file with Koch at U.S. Department of Agriculture, Forest Service, Northern Research Station, 359 Main Rd., Delaware, OH 43015

Carter J, Leonard BP (2002) A review of the literature on the worldwide distribution, spread of, and efforts to eradicate the coypu (*Myocastor coypus*). Wildl Soc Bull 30:162–175

Castro-Diez P, Fierro-Brunnenmeister N, Gonzalez-Munoz N, Gallardo A (2012) Effects of exotic and native tree leaf litter on soil properties of two contrasting sites in the Iberian Peninsula. Plant Soil 350:179–191

Chambers JC, Bradley BA, Brown CS et al (2014) Resilience to stress and disturbance, and resistance to Bromus tectorum L. invasion in cold desert shrublands of western North America. Ecosystems 17:360–375

Chambers JC, Maestas JD, Pyke DA et al (2017) Using resilience and resistance concepts to manage persistent threats to sagebrush ecosystems and Greater Sage-Grouse. Rangel Ecol Manag 70:149–164

Chapuis JL, Frenot Y, Lebouvier M (2004) Recovery of native plant communities after eradication of rabbits from the subantarctic Kerguelen Islands, and influence of climate change. Biol Conserv 117:167–179

Chazdon RL (2008) Beyond deforestation: restoring forests and ecosystem services on degraded lands. Science 320:1458–1460

Christie MR, Marine ML, Fox SE et al (2016) A single generation of domestication heritably alters the expression of hundreds of genes. Nat Commun 7: 10676, online only. https://doi.org/10.1038/ncomms10676

Clark SL, Schlarbaum SE, Pinchot CC et al (2014) Reintroduction of American chestnut in the National Forest System. J For 112:502–512

Clark SL, Schlarbaum SE, Saxton AM, Hebard FV (2015) Establishment of American chestnuts (*Castanea dentata*) bred for blight (*Cryphonectria parasitica*) resistance: influence of breeding and nursery grading. New For 47:243–270

Coates PS, Riccaa MA, Prochazkaa BG et al (2016) Wildfire, climate, and invasive grass interactions negatively impact an indicator species by reshaping sagebrush ecosystems. Proc Natl Acad Sci 113:12745–12750

Corbin JD, Holl KD (2012) Applied nucleation as a forest restoration strategy. For Ecol Manag 265:37–46

Corbin JD, Robinson GR, Hafkemeyer LM, Handel SN (2016) A long-term evaluation of applied nucleation as a strategy to facilitate forest restoration. Ecol Appl 26:104–114

Cox RD, Anderson VJ (2004) Increasing native diversity of cheatgrass-dominated rangeland through assisted succession. J Range Manag 57:203–210

Davis MA, Grime JP, Thompson K (2000) Fluctuating resources in plant communities: a general theory of invisibility. J Ecol 88:528–534

Dickens SJM, Mangla S, Preston KL, Suding KN (2016) Embracing variability: environmental dependence and plant community context in ecological restoration. Restor Ecol 24:119–127

DiTommaso A, Morris SH, Parker JD et al (2014) Deer browsing delays succession by altering aboveground vegetation and belowground seed banks. PLoS One 9:1–9

Drenkhan R, Hanso M (2010) New host species for *Chalara fraxinea*. New Dis Rep 22:16

Dyer AR, Knapp EE, Rice KJ (2016) Unintentional selection and genetic changes in native perennial grass populations during commercial seed production. Ecol Restor 34:39–48

Ehrlich J (1934) The beech bark disease: a *nectria* disease of *Fagus*, following *cryptococcus fagi* (Baer). Can J Res 10:593–692

Ennos RA (2015) Resilience of forests to pathogens: an evolutionary ecology perspective. Forestry 88:41–52

Farjon A (2013) *Chamaecyparis lawsoniana*. The IUCN Red List of Threatened Species 2013: e.T34004A2840024. https://doi.org/10.2305/IUCN.UK.2013-1.RLTS.T34004A2840024.en

Fausch KD, Rieman BE, Dunham JB et al (2009) Invasion versus isolation: trade-offs in managing native salmonids with barriers to upstream movement. Conserv Biol 23:859–870

Firn J, House APN, Buckley YM (2010) Alternative states models provide an effective framework for invasive species control and restoration of native communities. J Appl Ecol 47:96–105

Fisch KM, Kozfkay CC, Ivy JA et al (2015) Fish hatchery genetic management techniques: integrating theory with implementation. N Am J Aquac 77:343–357

Forest Service Manual (2016) Chapter 2020 ecosystem restoration. Federal Register April 27, 2016 81(81): 24785–24794. https://www.gpo.gov/fdsys/pkg/FR-2016-04-27/pdf/2016-09750.pdf

Funk JL, Cleland EE, Suding KN, Zavaleta ES (2008) Restoration through reassembly: plant traits and invasion resistance. Trends Ecol Evol 23:695–703

Gandhi KJK, Smith A, Hartzler DM, Herms DA (2014) Indirect effects of emerald ash borer-induced ash mortality and canopy gap formation on epigaeic beetles. Environ Ecol 43:546–555

Ghelardini L, Santini A (2009) Avoidance by early flushing: a new perspective on Dutch elm disease. J Biogeosci For 2:143–153

Gross A, Sieber TN (2016) Virulence of *Hymenoscyphus albidus* and native and introduced *Hymenoscyphus fraxineus* on *Fraxinus excelsior* and *Fraxinus pennsylvanica*. Plant Pathol 65:655–663

Grove S, Haubensak KA, Parker IM (2012) Direct and indirect effects of allelopathy in the sol legacy of an exotic plant invasion. Plant Ecol 213:1869–1882

Groves CR (2003) Drafting a conservation blueprint: a practitioner's guide to planning for biodiversity. Island Press, Washington, DC, 404 p

Großkinsky DK, Pieruschka R, Svensgaard J et al (2015) Phenotyping in the fields: dissecting the traits of quantitative genetics and digital farming. New Phytol 207:950–952

Haeussler S, Bedford L, Leduc A et al (2002) Silvicultural disturbance severity and plant communities of the Southern Canadian boreal forest. Silva Fennica 36:307–327

Havens K, Vitt P, Still S, Kramer AT et al (2015) Seed sourcing for restoration in an era of climate change. Nat Areas J 35:122–133

Hebard FV (2012) The American chestnut foundation breeding program. In: Sniezko RA, Yanchuk AD (eds) Proceedings, fourth international workshop on the genetics of host-parasite interactions in forestry: disease and insect resistance in forest trees, General Technical Reports PSW-GTR-240. Forest Service, U.S. Department of Agriculture, Pacific Southwest Research Station, Albany, pp 221–234

Herath P, Hoover GA, Andelini E, Moorman GW (2010) Detection of elm yellows phytoplasma in elms and insects using real-time PCR. Plant Dis 94:1355–1360

Hoban SM, McCleary TS, Schlarbaum SE, Romero-Severson J (2009) Geographically extensive hybridization between the forest trees American butternut and Japanese walnut. Biol Lett 5:324–327

Hoban SM, McCleary TS, Schlarbaum SE et al (2012) Human impacted landscapes facilitate hybridization between a native and an introduced tree. Evol Appl 5(7):720–731

Hobbs RJ, Higgs E, Harris JA (2009) Novel ecosystems: implications for conservation and restoration. Trends Ecol Evol 24:599–605

Holl KD, Aide TM (2011) When and where to actively restore ecosystems? For Ecol Manag 261:1558–1563

Huebner CD, Tobin PC (2006) Invisibility of mature and 15-year-old deciduous forests by exotic plants. Plant Ecol 186:57–68

Hughes MA, Smith JA (2014) Vegetative propagation of putatively laurel wilt-resistant redbay (*Persea borbonia)*. Native Plants J 15:42–50

Hughes MA, Smith JA, Ploetz RC et al (2015) Recovery plan for laurel wilt on redbay and other forest species caused by *Raffaelea lauricola* and disseminated by *Xyleborus glabratus*. Plant Health Progr 16:174–210

Hulet A, Roundy BA, Jessop B (2010) Crested wheatgrass control and native plant establishment in Utah. Rangeland Ecol Manag 63:450–460

Ingwell LL, Preisser EL (2010) Using citizen science programs to identify host resistance in pest-invaded forests. Conserv Biol 25:182–188

Isaak DJ, Young MK, Nagel DE et al (2015) The cold-water climate shield: delineating refugia for preserving salmonid fishes through the twenty-first century. Glob Chang Biol 21:2540–2053

Iverson L, Knight KS, Prasad A et al (2015) Potential species replacements for black ash (*Fraxinus nigra*) at the confluence of two threats: emerald ash borer and a changing climate. Ecosystems 19:248–270

Jacobi WR, Koski RD, Harrington TC, Witcosky JJ (2007) Association of *Ophiostoma novo-ulmi* with *Scolytus schevyrewi* (Scolytidae) in Colorado. Plant Dis 91:245–247

Jeffers SN, Meadows IN, James JB, Sisco PH (2012) Resistance to *Phytophthora cinnamomi* among seedlings from backcross families of hybrid American chestnut. In: Sniezko RA, Yanchuck AD, Kliejunas JT (eds) Proceedings, fourth international workshop on the genetics of host-parasite interactions in forestry: Disease and insect resistance in forest trees, General Technical Reports PSW-GTR-240. Forest Service, U.S. Department of Agriculture, Pacific Southwest Research Station, Albany, pp 194–195

Johnson R, Stritch L, Olwell P et al (2010) What are the best seed sources for ecosystem restoration on BLM and USFS lands? Native Plants 11:117–131

Johnson RC, Hellier BC, Vance-Borland KW (2013) Genecology and seed zones for tapertip onion in the US Great Basin. Botany 91:686–694

Kanzaki N, Ichihara Y, Aikawa T et al (2019) *Litylenchus crenatae* n. sp. (Tylenchomorpha: Anguinidae) a leaf gall nematode parasitizing *Fagus crenata* Blume. Nematology 21:5–22

Kardol P, Wardle DA (2010) . How understanding aboveground-belowground linkages can assist restoration ecology. Trends Ecol Evol 25:670–679

Kettenring KM, Adams CR (2011) Lessons learned from invasive plant control experiments: a systematic review and meta-analysis. J Appl Ecol 48:970–979

Kinloch BB Jr (2000) Genetic interactions in the white pine/blister rust pathosystem. Proceedings, disease resistance genetics in forest trees, North Carolina Arboretum, August 8–9, 2000, Asheville, pp 1–2

Kinloch BB Jr (2003) White pine blister rust in North America: past and prognosis. Phytopathology 93:1044–1047

Kinloch BB Jr, Dupper GE (2002) Genetic specificity in the white pine-blister rust pathosystem. Phytopathology 92:278–280

Kinloch BB Jr, Burton D, Davis DA et al (2012) Strong partial resistance to white pine blister rust in sugar pine. In: Sniezko RA, Yanchuk AD, Kliejunas JT (eds) Proceedings, fourth international workshop on the genetics of host-parasite interactions in forestry: Disease and insect resistance in forest trees, General Technical Reports PSW-GTR 240. U.S. Department of Agriculture, Forest Service, Pacific Southwest Research Station, Albany, pp 80–91

Kjærstad G, Arnekleiv JV, Speed JD (2015) Effects of three consecutive rotenone treatments on the benthic macroinvertebrate fauna of the River Ogna, central Norway. River Res Appl 32(4):572–582

Knapp RA, Sarnelle O (2008) Recovery after local extinction: factors affecting re-establishment of alpine lake zooplankton. Ecol Appl 18:1850–1859

Knapp RA, Boiano DM, Vredenburg VT (2007) Removal of nonnative fish results in population expansion of a declining amphibian (mountain yellow-legged frog, *Rana muscosa*). Biol Conserv 135:11–20

Knight KS, Karrfalt RP, Mason ME (2010) Methods for collecting ash (Fraxinus spp) seeds, General Technical Reports NRS-55. U.S. Department of Agriculture, Forest Service, Northern Research Station, Newtown Square, 14 p

Knight KS, Herms D, Plumb R et al (2012a) Dynamics of surviving ash (*Fraxinus* spp.) populations in areas long infested by emerald ash borer (*Agrilus planipennis*). In: Sniezko RA, Yanchuk AD (eds) Proceedings, fourth international workshop on the genetics of host-parasite interactions in forestry: disease and insect resistance in forest trees, General Technical Reports PSW-GTR-240, Albany, U.S. Department of Agriculture, Forest Service, Pacific Southwest Research Station, p 9

Knight KS, Slavicek JM, Kappler R et al (2012b) Using Dutch elm disease-tolerant elm to restore floodplains impacted by emerald ash borer. In: Sniezko RA, Yanchuk AD (eds) Proceedings, fourth international workshop on the genetics of host-parasite interactions in forestry: disease and insect resistance in forest trees, General Technical Reports PSW-GTR-240. U.S. Department of Agriculture, Forest Service, Pacific Southwest Research Station, Albany, pp 317–323

Knight KS, Brown JP, Long RP (2013) Factors affecting the survival of ash (*Fraxinus* spp.) trees infested by emerald ash borer (*Agrilus planipennis*). Biol Invasions 15:371–383

Koch JL (2010) Beech Bark Disease: the oldest "new" threat to American beech in the United States. Outlooks Pest Manag 21:64–68

Koch JL, Carey DW (2014) A technique to screen American beech for resistance to the beech scale insect (*Cryptococcus fagisuga* Lind.). J Visualized Exp. e51515

Koch JL, Heyd RL (2013) Battling beech bark disease: establishment of beech seed orchards in Michigan. Newslett Michigan Entomol Soc 58(1&2):11–14

Koch JL, Carey DW, Mason ME, Nelson CD (2010) Assessment of beech scale resistance in full- and half-sibling American beech families. Can J For Res 40:265–272

Koch JL, Carey DW, Mason ME et al (2015) Intraspecific variation in *Fraxinus pennsylvanica* responses to emerald ash borer (*Agrilus planipennis*). New For 46:995–1011

LaBonte NR, Ostry ME, Ross-Davis A, Woeste KE (2015) Estimating heritability of disease resistance and factors that contribute to long-term survival in butternut (*Juglans cinerea* L.). Tree Genet Genomes 11(3):63

Lankau RA, Bauer JT, Anderson MR, Anderson RC (2014) Long-term legacies and partial recovery of mycorrhizal communities after invasive plant removal. Biol Invasions 16:1979–1990

Lee JC, Aguayo I, Aslin R et al (2009) Co-occurrence of the invasive banded and European elm bark beetles (*Coleoptera scolytidae*) in North America. Ann Entomol Soc Am 102:426–436

Liao C, Peng R, Luo Y et al (2008) Altered ecosystem carbon and nitrogen cycles by plant invasion: a meta-analysis. New Phytol 177:706–714

Lotze HK, Lenihan HS, Bourque BJ et al (2006) Depletion, degradation, and recovery potential of estuaries and coastal seas. Science 312(5781):1806–1809

Madsen MD, Davies KW, Mummey DL, Svejcar TJ (2014) Improving restoration of exotic annual grass-invaded rangelands through activated carbon seed enhancement technologies. Rangeland Ecol Manag 67:61–67

Magnoli SM, Kleinhesselink AR, Cushman JH (2013) Responses to invasion and invader removal differ between native and exotic plant groups in a coastal dune. Oecologia 173(4):1521–1530

Mangold RD (2011) The U.S. Forest Service's renewed focus on gene conservation of five-needle pine species. In: Keane RE, Tomback DF, Murray MP, Smith CM (eds) The future of high-elevation, five-needle white pines in Western North America: Proceedings of the High Five Symposium. 28–30 June 2010; Missoula, MT, Proceedings RMRS-P-63. U.S. Department of Agriculture, Forest Service, Rocky Mountain Research Station, Fort Collins, p 151. http://www.fs.fed.us/rm/pubs/rmrs_p063.html

Maron J, Marler M (2007) Native plant diversity resists invasion at both low and high resource levels. Ecology 88:2651–2661

McElhany P, Ruckelshaus MH, Ford MJ et al (2000) Viable salmonid populations and the recovery of evolutionarily significant units, Technical memorandum NMFSNWFSC- 42. National Oceanic and Atmospheric Administration, Seattle

McKelvey KS, Young MK, Knotek WL et al (2016) Sampling large geographic areas for rare species using environmental DNA: a study of bull trout Salvelinus confluentus occupancy in western Montana. J Fish Biol 88:1215–1222

Meyer KA, Lamansky JA Jr, Schill DJ (2006) Evaluation of an unsuccessful brook trout electrofishing removal project in a small Rocky Mountain stream. N Am J Fish Manag 26:849–860

Middleton EL, Bever JD (2012) Inoculation with a native soil community advances succession in a grassland restoration. Restor Ecol 20:218–226

Mills LS. (2012) Conservation of wildlife populations: demography, genetics, and management. Wiley. ISBN 978-470-67150-4

Moorhouse R, Greene T, Dilks P et al (2003) Control of introduced mammalian predators improves kaka *Nestor meridionalis* breeding success: reversing the decline of a threatened New Zealand parrot. Biol Conserv 110:33–44

Moreno-Mateos D, Power ME, Comin FA, Yockteng R (2012) Structural and functional loss in restored wetland ecosystems. PLoS Biol 10:1–8

Morrison SA (2007) Reducing risk and enhancing efficiency in non-native vertebrate removal efforts on islands: a 25 year multi-taxa retrospective from Santa Cruz Island, California. Managing Vertebr Invasive Species 31:398–401

Myers JH, Simberloff D, Kuris AM, Carey JR (2000) Eradication revisited: dealing with exotic species. Trends Ecol Evol 15:316–320

Nesbit D, Kreutzweiser D, Sibley P, Scarr T (2015) Ecological risks posed by emerald ash borer to riparian forest habitats: a review and problem formulation with management implications. For Ecol Manag 258:165–173

Orlova-Bienkowskaja MJ (2014) Ashes in Europe are in danger: the invasive range of *Agrilus planipennis* in European Russia is expanding. Biol Invasions 16:1345–1349

Pacas C, Taylor MK (2015) Nonchemical eradication of an introduced trout from a headwater complex in Banff National Park, Canada. N Am J Fish Manag 35:748–758

Paillet FL (2002) Chestnut: history and ecology of a transformed species. J Biogeogr 29:1517–1530

Palik BJ, Ostry ME, Venette RC, Abdela E (2012) Tree regeneration in black ash (*Fraxinus nigra*) stands exhibiting crown dieback in Minnesota. For Ecol Manag 269:26–30

Palmer MA, Filoso S (2009) Restoration of ecosystem services for environmental markets. Science 325:575–576

Pearson DE, Ortega YK, Runyon J, Butler J (2016) Secondary invasion: the bane of weed management. Biol Conserv 197:8–17

Pinchot CC, Schlarbaum SE, Clark SL et al (2014) Impact of silvicultural treatment on chestnut seedling growth and survival. Acta Hortic 1019:191–197

Pliūra A, Lygis V, Suchockas V, Bartkevičius E (2011) Performance of twenty-four European *Fraxinus excelsior* populations in three Lithuanian progeny trials with a special emphasis on resistance to *Chalara fraxinea*. Balt For 17:17–34

Pogacnik J, Macy T (2016) Beech leaf disease. Ohio Dept. Natural Resources Forest Health Pest Alert. http://forestry.ohiodnr.gov/portals/forestry/pdfs/BLDAlert.pdf

Pokorney ML, Sheley RL, Zabinski CA et al (2005) Plant functional group diversity as a mechanism for invasion resistance. Restor Ecol 13:448–459

Powers ZL, Mayfield AE, Frampton J, Jetton RM (2015) Comparison of suspended branch and direct infestation techniques for artificially infesting hemlock seedlings with the hemlock woolly adelgid for resistance screening. Forests 6:2066–2081

Prach K, Walker LR (2011) Four opportunities for studies of ecological succession. Trends Ecol Evol 26:119–123

Propst DL, Gido KB, Whitney JE et al (2015) Efficacy of mechanically removing nonnative predators from a desert stream. River Res Appl 31:692–703

Puettmann KJ (2011) Silvicultural challenges and options in the context of global change: "Simple" fixes and opportunities for new management approaches. J For 109:321–331

Reid AM, Morin L, Downey PO et al (2009) Does invasive plant management aid the restoration of natural ecosystems? Biol Conserv 142:2342–2349

Roemer GW, Coonan TJ, Garcelon DK et al (2001) Feral pigs facilitate hyperpredation by golden eagles and indirectly cause the decline of the island fox. Anim Conser Forum 4:307–318

Royo AA, Stout SL, deCalesta DS, Pierson TG (2010) Restoring forest herb communities through landscape-level deer herd reductions: is recovery limited by legacy effects? Biol Conserv 143:2425–2434

Seabloom EW, Borer ET, Boucher VL et al (2003) Competition, seed limitation, disturbance, and reestablishment of California native annual forbs. Ecol Appl 13(3):575–592

Sheley RL, Mangold JM, Anderson JL (2006) Potential for successional theory to guide restoration of invasive-plant-dominated rangeland. Ecol Monogr 76:365–379

Sherald JL (2009) Elms of the monumental core: history and management plan. NPS/NCR/NRR-2009/001. National Park Service, Center for Urban Ecology, Washington, D.C. 92 pp. https://www.nps.gov/nationalmallplan/Documents/Studies/ ElmsoftheMonuCore_HistandMgmtPlan_122009.pdf

Sinclair WA, Townsend AM, Sherald JL (2001) Elm yellows phytoplasma lethal to Dutch elm disease-resistant *Ulmus americana* cultivars. Disease Notes 85:560

Slavicek JM, Knight KS (2013) Development of methods for the restoration of the American elm in forested landscapes. In: Sniezko RA, Yanchuk AD, Kliejunas JT et al (eds) Proceedings, Fourth international workshop on the genetics of host-parasite interactions in forestry: Disease and insect resistance in forest trees, General Technical Reports PSW-GTR-240. Pacific Southwest Research Station, Forest Service, U.S. Department of Agriculture, Albany, pp 337–341

Slavicek JM, Boose A, Balser D, Cavender N (2005) Restoration of the American elm in forested landscapes. In: Gottschalk KW (ed) Proceedings, 16th U.S. Department of Agriculture interagency research forum on gypsy moth and other invasive species 2005, General Technical Reports NE-337. U.S. Department of Agriculture, Forest Service, Northeastern Research Station, Newtown Square, p 74

Slesak RA, Lenhart CF, Brooks KN et al (2014) Water table response to harvesting and simulated emerald ash borer mortality in black ash wetlands in Minnesota, USA. Can J For Res 44:961–968

Smalley EB, Guries RP (1993) Breeding elms for resistance to Dutch elm disease. Annu Rev Phytopathol 31:325–352

Smith KW (2005) Has the reduction in nest-site competition from starlings *Sturnus vulgaris* been a factor in the recent increase of great spotted woodpecker *Dendrocopos major* numbers in Britain? Bird Study 52:307–313

Sniezko RA, Mahalovich MF, Schoettle AW, Vogler DR (2011) Past and current investigation of the genetics of resistance to *Cronartium ribicola* in high-elevation five-needle pines. In: Keane RE, Tomback DF, Murray MP (eds) Proceedings, high five symposium 2010; Missoula, MT, RMRS-P-63. U.S. Department of Agriculture, Forest Service, Rocky Mountain Research Station, Fort Collins, pp 246–264

Sniezko RA, Schoettle A, Dunlap J et al (2012a) Ex Situ gene conservation in high elevation white pine species in the United States-a beginning. In: Keane RE, Tomback DF, Murray MP, Smith CM (eds) The future of high-elevation, five-needle white pines in Western North America: Proceedings of the High Five Symposium. 28–30 June 2010; Missoula, MT, Proceedings RMRS-P-63. U.S. Department of Agriculture, Forest Service, Rocky Mountain Research Station, Fort Collins, pp 147–149

Sniezko RA, Hamlin J, Hansen EM (2012b) Operational program to develop *Phytophthora lateralis*-resistant populations of Port-Orford-cedar (*Chamaecyparis lawsoniana*). In: Sniezko RA, Yanchuk AD, Kliejunas JT, Palmieri KM, Alexander JM, Frankel SJ, Tech. coords. 2012. Proceedings of the fourth international workshop on the genetics of host-parasite interactions in forestry: Disease and insect resistance in forest trees. General Technical Reports PSW-GTR-240. Pacific Southwest Research Station, Forest Service, U.S. Department of Agriculture, Albany, pp 65–79

Sniezko RA, Smith J, Liu J, Hamelin RC (2014) Genetic resistance to fusiform rust in southern pines and white pine blister rust in white pines-a contrasting tale of two rust pathosystems-current status and future prospects. Forests 5:2050–2083

Steiner KC, Westbrook JW, Hebard FV et al (2017) Rescue of American chestnut with extraspecific genes following its destruction by a naturalized pathogen. New For 48:317–336

Stohlgren TJ, Barnett DT, Kartesz JT (2003) The rich get richer: patterns of plant invasions in the United States. Front Ecol Environ 1:11–14

Suding KN, Hobbs RJ (2009) Threshold models in restoration and conservation: a developing framework. Trends Ecol Evol 24:271–279

Suding KN, Gross KL, Houseman GR (2004) Alternative states and positive feedbacks in restoration ecology. Trends Ecol Evol 19(1):46–53

Suseela V, Alpert P, Nakatsu CH, Armstrong A (2016) Plant-soil interactions regulate the identity of soil carbon in invaded ecosystems: implication for legacy effects. Funct Ecol 30:1227–1238

Symstad AJ (2004) Secondary invasion following the reduction of *Coronilla varia* (Crownvetch) in Sand Prairie. Am Midl Nat 152:183–189

Telford A, Cavers S, Ennos RA, Cottrell JE (2015) Can we protect forests by harnessing variation in resistance to pests and pathogens? Forestry 88:3–12

Thomsen PF, Kielgast J, Iversen LL et al (2012) Monitoring endangered freshwater biodiversity using environmental DNA. Mol Ecol 21:2565–2573

Tischew S, Youtie B, Kirmer A, Shaw N (2011) Farming for restoration: building bridges for native seeds. Ecol Restor 29:219–222

Tognetti PM, Chaneton EJ, Omacini M et al (2010) Exotic vs. native plant dominance over 20 years of old-field succession on set aside farmlands in Argentina. Biol Conserv 143:2494–2503

Tomback DF, Achuff P (2010) Blister rust and western forest biodiversity: ecology, values, and outlook for white pines. For Pathol 40:186–225

Townsend AM, Douglass LW (2001) Variation among American elm clones in long-term dieback, growth, and survival following *Ophiostoma* inoculation. J Environ Hortic 19:100–103

Tropek R, Kadlec T, Karesova P et al (2010) Spontaneous succession in limestone quarries as an effective restoration tool for endangered arthropods and plants. J Appl Ecol 47:139–147

U.S. Fish and Wildlife Service (2015) Endangered and threatened wildlife and plants; review of native species that are candidates for listing as endangered or threatened; annual notice of findings on resubmitted petitions; annual description of progress on listing actions. Federal register. A Notice by the Fish and Wildlife Service on 12/24/2015. https://federalregister.gov/a/2015-32284

Waring KM, Goodrich BA (2012) Artificial regeneration of five-needle pines of western North America: a survey of current practices and future needs. Tree Planter's Notes 55:55–71

Waring KM, O'Hara KL (2005) Silvicultural strategies in forest ecosystems affected by introduced pests. For Ecol Manag 209:27–41

Wenger SJ, Isaak DJ, Luce CH et al (2011) Flow regime, temperature, and biotic interactions drive differential declines of trout species under climate change. Proc Natl Acad Sci 108:14175–14180

Wheeler NC, Steiner KC, Schlarbaum SE, Neale DB (2015) The evolution of forest genetics and tree improvement research in the United States. J For 113:500–510

Yelenik SG, Levine JM (2010) Native shrub reestablishment in exotic annual grasslands: do ecosystem processes recover? Ecol Appl 20:716–727

Yelenik SG, Stock WD, Richardson DM (2004) Ecosystem level impacts of invasive *Acacia saligna* in the south African fynbos. Restor Ecol 12:44–51

Zavaleta ES, Hobbs RJ, Mooney HA (2001) Viewing invasive species removal in a whole-ecosystem context. Trends Ecol Evol 16:454–459

Zhebentyayeva T, Chandra A, Abbott AG et al (2013) Genetic and genomic resources for mapping resistance to *Phytophthora cinnamomi* in chestnut. Acta Hortic 1019:263–270

Zivy M, Wienkoop S, Renaut J et al (2015) The quest for tolerant varieties: the importance of integrating "omics" techniques to phenotyping. Front Plant Sci 6:448

Sectoral Impacts of Invasive Species in the United States and Approaches to Management

9

Anne S. Marsh, Deborah C. Hayes, Patrice N. Klein, Nicole Zimmerman, Alison Dalsimer, Douglas A. Burkett, Cynthia D. Huebner, Robert Rabaglia, Laura A. Meyerson, Bonnie L. Harper-Lore, Jamie L. Davidson, Marla R. Emery, Travis Warziniack, Rebecca Flitcroft, Becky K. Kerns, and Vanessa M. Lopez

9.1 Introduction

Invasive species have a major effect on many sectors of the U.S. economy and on the well-being of its citizens. Their presence impacts animal and human health, military readiness, urban vegetation and infrastructure, water, energy and transportations systems, and indigenous peoples in the United States (Table 9.1). They alter bio-physical systems and cultural practices and require significant public and private expenditure for control. This chapter provides examples of the impacts to human systems and explains mechanisms of invasive species' establishment and spread within sectors of the U.S. economy. The chapter is not intended to be comprehensive but rather to provide insight into the range and severity of impacts. Examples provide context for ongoing Federal programs and initiatives; support State and private efforts to prevent the introduction and spread of invasive species; and to eradicate and control established invasive species.

Invasive species are a problem for humans as well as the ecosystems they inhabit as economic impact analyses have indicated (Pimentel 2011; Pimentel et al. 2005). Approximately 50,000 non-native species have been introduced into the United States, including plants, animals, and microbes (Pimentel et al. 2005). Many of these species were intentionally introduced to control erosion or as agricultural

A. S. Marsh (✉) · D. C. Hayes
U.S. Department of Agriculture, Forest Service, Sustainable Forest Management Research, Washington, DC, USA
e-mail: anne.marsh@usda.gov

P. N. Klein
U.S. Department of Agriculture, Forest Service, Research and Development, Landscape and Ecosystem Services Research, Washington, DC, USA

N. Zimmerman
Ecological Society of America, Washington, DC, USA

A. Dalsimer
U.S. Department of Defense, Mark Center, Alexandria, VA, USA

D. A. Burkett
U.S. Department of Defense, Armed Forces Pest Management Board, Silver Spring, MD, USA

C. D. Huebner
U.S. Department of Agriculture, Forest Service, Northern Research Station, Morgantown, WV, USA

R. Rabaglia
U.S. Department of Agriculture, Forest Service, State & Private Forestry, Forest Health Protection, Washington Office, Washington, DC, USA

L. A. Meyerson
University of Rhode Island, Kingston, RI, USA

B. L. Harper-Lore
U.S. Department of Transportation, Federal Highway Administration, Washington, DC, USA

J. L. Davidson
U.S. Department of Agriculture, Forest Service, Allegheny National Forest, Warren, PA, USA

M. R. Emery
U.S. Department of Agriculture, Forest Service, Northern Research Station, Burlington, VT, USA

T. Warziniack
U.S. Department of Agriculture, Forest Service, Rocky Mountain Research Station, Fort Collins, CO, USA

R. Flitcroft · B. K. Kerns
U.S. Department of Agriculture, Forest Service, Pacific Northwest Research Station, Corvallis, OR, USA

V. M. Lopez
U.S. Department of Agriculture, Forest Service, State & Private Forestry, Forest Health Protection, Washington Office, Washington, DC, USA

T. M. Poland et al. (eds.), *Invasive Species in Forests and Rangelands of the United States*,
https://doi.org/10.1007/978-3-030-45367-1_9

Table 9.1 Key impacts of invasive species on the humans and the systems that support them[a]

Sector	Impact
Animal and human health	Spread of infectious disease; dermatitis; respiratory allergies; wounds and bites; loss of native species
Military readiness	Diminished military force health; decreased aircraft safety; decreased safety on training grounds; hampered movement of equipment, vehicles, and supplies; equipment and infrastructure damage
Urban areas	Tree damage and mortality; infrastructure damage; loss of native species
Water resources and systems	Decreased water quality and quantity; flooding; water infrastructure damage; diminished recreational opportunities
Energy systems	Damage to electrical equipment, utility lines, and poles; damage to hydropower facilities
Transportation systems	Reduced visibility; flooding; increased fire risk; damage to roadways and railroad tracks and ties
Forest and grassland products	Decreased forage; cattle poisoning; disease transmission to cattle; loss of timber and non-timber forest products
Indigenous peoples	Loss of native foods, medicines, ceremonial materials, and species with cultural value

[a]With proper control and management measures, many impacts can be avoided or abated

crops, livestock, landscaping, game species, or pets. Others were accidental hitchhikers on imported products or came by natural catastrophic disturbances, such as hurricanes. Once established in the United States, many of these species became invasive (~6500 species) and have caused significant unintended economic losses, threatened human and animal health and safety, and disrupted human activities. Some of the consequences include: reductions in agriculture, forestry, and livestock production; damage to infrastructure; alteration of water and nutrient cycles; changes in the availability of clean water; and introduction and spread of new or re-emerging diseases. In addition, invasive species can alter disturbance regimes and threaten biodiversity through atypical fire, predation, competition, and disease transmission, often in concert with other invasive species, to cause cascading impacts throughout entire ecosystems (Doherty et al. 2016).

Humans, and their support systems, may be particularly vulnerable to impacts from invasive species. Invasive species are frequently introduced to urban areas or economic centers, where they may influence humans because of their close proximity and relative abundance. Human behavior in urban areas tends to encourage the increase and spread of invasive species, making humans and their support systems even more vulnerable to their impacts. These behaviors include landscaping, clearing native vegetation, and creating disturbed environments in which few species other than invasives can survive.

Costs are incurred in the public and private sectors because of damage to landscapes, animals, infrastructure, and humans, and through management efforts to prevent and mitigate this damage and ensure human safety. Unfortunately, in many cases, these costs have not been quantified. Economic cost data are often not available, and when available, there may be no system in place to summarize or analyze costs. Costs may also be masked within other expenses as a component of larger management initiatives such as habitat restoration. Additionally, there is no direct way to evaluate costs associated with impacts on aesthetics or cultural practices, so estimates of indirect costs or willingness to pay are used. At this point in time it is not possible, even with the best available data, to provide a full accounting of costs related to invasive species (see Chap. 14).

9.2 Invasive Species and Animal and Human Health

Animal and human health have been impacted by a wide range of invasive species, including vertebrate animals (mammals, birds, reptiles, fish), invertebrate animals (insects), microbial organisms (fungus, bacteria, viruses), and plant species. These species can affect animals and humans directly by spreading disease or acting as chemical or biological irritants, or indirectly by serving as vectors for invasive or non-invasive organisms that affect health. Invasive species alter the provision of goods and clean water, and may influence safety, as discussed in other sections. Health-related impacts from invasive species may be challenging to prevent when there are lag times between introduction, establishment, and early detection.

9.2.1 Impacts of Invasive Mammals on Animal and Human Health

Invasive mammals in the United States include dogs, cats, horses, pigs, and other domesticated animals that were introduced for livestock and pets but have become feral, as well as wild species that were accidentally introduced. The European or black rat (*Rattus rattus*), Norway or brown rat (*Rattus norvegicus*), and house mouse (*Mus musculus*), for example, were unintentionally introduced through trans-Atlantic shipping and other international trade routes (Pimentel et al. 2005).

Rats are known to destroy stored food and grain, damage crops, and prey on poultry, with estimated economic losses of over $19 billion per year to the agricultural sector (Pimentel et al. 2000). Moreover, invasive rodents are vectors of serious diseases that infect humans and animals such as salmonellosis, leptospirosis, plague, and murine typhus (Pimentel et al. 2005). Some of these bacterial diseases are directly spread through contamination of feed, food, and water systems; others, like plague, are spread by fleas carried by rodents. The rodent fleas can infect domestic pets which reside in homes, and subsequently transmit disease to humans (Ettestad 2017; Weniger et al. 1984). In addition to their capacity to transmit disease, invasive mammals (rats and feral cats, dogs, and pigs) are considered to be a major factor in the extinction of 87 bird, 45 mammal, and 10 reptile species worldwide. These invasive predators also threaten 596 species listed as either vulnerable, endangered, critically endangered, or possibly extinct (Doherty et al. 2016).

Feral and other free-roaming domestic cats (*Felis catus*), whose populations have been estimated at over 30 million and rising, prey on native birds, small mammals, amphibians, and reptiles (Pimentel et al. 2005). It is estimated that feral cats kill at least 240 million birds each year in the United States (Jessup 2004; Pimentel et al. 2000). The total damage to the U.S. bird population is estimated to be at least $17 billion per year when lost recreational costs (hunting, bird watching, etc.) are included (Pimentel et al. 2000). The free-ranging behavior of feral cats in shared habitats with native wildlife also predisposes them to exposure to common infections associated with numerous parasites and other transmissible infectious diseases (Jessup 2004). Many of these diseases are zoonotic or capable of infecting humans and domestic animals.

Feral cats can harbor the bacterium *Pasteurella multocida*, which can be transmitted by cat bites and cause virulent disease in wild birds (avian cholera) and domestic poultry (fowl cholera), and severe infections in humans. Feral cats also are the host for *Toxoplasma gondii*, a protozoan parasite causing infection in humans and animals in the United States (Dubey 2010; Dubey and Jones 2008). Additionally, Toxoplasma-infected feces can enter coastal waters and cause toxoplasmosis in marine mammals (Barberi et al. 2016; Harris et al. 2002; Miller et al. 2008; U.S. Department of Commerce National Oceanic and Atmospheric Administration 2016).

Feral swine (*Sus scrofa*) have become established across the United States and have been reported in at least 35 States. Their origins date back to the late 1400s to early 1500s when European explorers and settlers brought their domestic swine to the Americas to serve as a source of food. Free-ranging feral swine populations were established in the United States as a result of open livestock management practices and opportunistic escapes from shooting preserves or commercial farm operations. Subsequent intentional releases of Eurasian wild boar for sport hunting occurred in the 1900s in some parts of the country. Wild boar and domestic swine easily inter-bred, forming additional hybridized populations of feral swine. It is currently estimated that there are over six million feral swine in the United States (Fall et al. 2011).

Feral swine are known to carry at least 30 viral and bacterial diseases such as swine brucellosis (*Brucella suis*), tuberculosis, leptospirosis, hepatitis (hepatitis E virus), pseudorabies, and influenza, and nearly 40 parasitic diseases, including toxoplasmosis and trichinosis, which can be transmitted to humans, pets, livestock, and wildlife (Fall et al. 2011; Meng et al. 2009; Witmer et al. 2003). *Brucella suis* infections were reported in hunters who contracted the disease from feral swine in Florida (Meng et al. 2009). Pederson et al. (2012) found feral swine exposure to brucellosis in 13 States, with exposure rates ranging from 1% to 14% of the feral swine population. Swine brucellosis and pseudorabies pose a significant threat to the commercial swine industry, affecting the health and productivity of individual animals resulting in infertility, abortions, lameness in adults, and high mortality in young pigs. The industry spends millions of dollars each year in surveillance and prevention measures dedicated to eradicating disease, supporting animal well-being, and reducing economic impacts attributed to animal mortality (USDA APHIS 2016; Witmer et al. 2003). Feral swine also present a threat for introducing or spreading foreign animal diseases in the United States such as foot-and-mouth disease, classical swine fever, and African swine fever. These foreign agricultural diseases can be introduced either through illegal importation of wild boar from countries where those diseases are prevalent, or by facilitating spread when outbreaks occur in commercial livestock. Swine-adapted influenza viruses also circulate in some feral swine populations that occur in proximity to commercial swine operations. These viruses can be easily transmitted between swine populations and can infect people, as documented in historic and more recent "swine flu" epidemics in the United States and other countries (Miller et al. 2017; Olsen 2002).

Escherichia coli O157:H7 is a virulent strain of bacteria causing severe intestinal and hemorrhagic disease in people (Griffin and Tauxe 1991). Although cattle have been considered a primary reservoir of this bacterium (Hancock et al. 1994), it has been isolated from feral swine feces. In 2006, people in 26 States and Canada became ill from *E. coli* O157:H7 after consuming contaminated, bagged baby spinach harvested from feral swine fecal-contaminated agricultural fields in central California (Jay et al. 2007).

9.2.2 Impacts of Invasive Birds on Animal and Human Health

Non-native birds introduced to the United States include species such as rock pigeons (*Columba livia*), English/house sparrows (*Passer domesticus*), and European starlings (*Sturnus vulgaris*). Pigeons were intentionally introduced with poultry and livestock in the 1600s. House sparrows and starlings were introduced in the 1800s, house sparrows to control canker worm and starlings as ornamental birds. Since introduction, these species have become established throughout the United States and thrive in association with human activities, particularly in urban areas. These birds harbor and transmit more than 50 diseases to humans and other animals, and are common reservoirs for bacterial diseases such as salmonellosis, colibacillosis (*E. coli*), and chlamydiosis; fungal diseases (histoplasmosis); parasitic diseases (toxoplasmosis); and viral diseases including influenza and West Nile Virus (WNV) encephalitis. For example, Kauffman and LeJeune

(2011) report that starlings can carry and transmit *E.coli* O157:H7 to cattle. Humans that consume dairy and beef products from infected cattle may subsequently become infected with the bacteria (Hancock et al. 1994; Kauffman and LeJeune 2011). House sparrows are known to be an important host for WNV, facilitating spread of this introduced vector-borne viral disease (Hayes et al. 2005).

9.2.3 Impacts of Insect Vectors and Vector-borne Diseases on Humans and Animals

Vector-borne diseases are transmitted to animals and humans by insect vectors such as muscid flies, mosquitos, ticks, and fleas, typically through bites that introduce infectious pathogens. These vectors can carry pathogenic viruses, bacteria, and parasites and transfer these infectious agents among susceptible hosts. In the United States, there are at least 14 notifiable vector-borne diseases that are threats to national public health and that can cause significant illness and mortality (Beard and Eisen 2016). Several of these, such as Lyme disease and Rocky Mountain spotted fever, have spread to new areas over the past decade, due, in part, to the expanded geographical ranges of their vectors (Beard and Eisen 2016).

Many of these vectors are invasive species. For example, approximately 15 invasive mosquitoes now inhabit Florida, most of which arrived in the past decade. Two new invasive arrivals from Latin America and the Caribbean, *Culex panocossa* and *Aedeomyia squamipennis*, were just detected near Everglades National Park in late 2016 (Blosser and Burkett-Cadena 2017). These mosquitoes are suspected vectors of Venezuela equine encephalitis virus, which can cause fatal infections in humans, horses, and other animals. In Hawaii, a State with no native mosquito species, the introduced *Culex quinquefasciatus* mosquito spread quickly, vectoring avian malarias that have devastated Hawaii's native bird populations (van Riper et al. 1986).

Other vector-borne diseases have recently been reported in the United States, including WNV, dengue, chikungunya, and Zika virus. These diseases have been introduced to the United States either through infected travelers carrying the disease or the inadvertent movement of invasive vectors inside international cargo containers. Once in the United States, competent vectors may spread diseases further by infecting susceptible hosts. For example, the yellow fever mosquito, *Aedes aegypti*, which originated in Africa and has been reported in Florida, along the Gulf coast, and the Washington, D.C. region, can carry and transmit more recently arrived pathogens such as those that cause dengue, chikungunya, and Zika virus.

9.2.4 Impacts of Microbial Pathogens on Animals

White-nose syndrome (WNS) is a lethal fungal disease prevalent in North American bats and caused by an introduced non-native fungus, *Pseudogymnoascus destructans* (*Pd*). WNS was first reported in upstate New York in 2006, and has since spread to 30 U.S. States and five Canadian provinces. The *Pd* fungus is a cryophilic (cold-loving) species originating in Europe and Asia. It thrives in cave environments and is thought to have been introduced into the United States through international recreational caving activities. Models suggest that WNS will eventually spread across the entire United States (Maher et al. 2012) since cave climatological data indicate sufficient microclimates exist for growth of the fungus throughout North America (Perry 2013).

WNS has been found in nine species of hibernating bats in North America, with mass mortality and significant population declines primarily in the little brown bat (*Myotis lucifugus*), northern long-eared bat (*Myotis septentrionalis*), Indiana bat (*Myotis sodalis*), and tri-colored bat (*Perimyotis subflavus*) (Turner et al. 2011). In 2012, the U.S. Fish and Wildlife Service and collaborating scientists estimated that up to six million bats had died from WNS since 2006 (Reeder et al. 2012; Turner et al. 2011). Due to severe population declines due to WNS, the northern long-eared bat was listed in 2016 as endangered. Bats are important in sustaining healthy forest and agricultural ecosystems, and provide many ecosystem services benefits to humans, including crop pollination and insect control (Boyles et al. 2011; López-Hoffman et al. 2014). Tourism associated with bat-viewing has been estimated to contribute $6.5 million annually to some local economies (Bagstad and Wiederholt 2013).

Another invasive and infectious fungal disease, chytridiomycosis, has been spreading globally for many decades causing worldwide declines and extirpations of amphibian populations. Chytridiomycosis is caused by the chytrid fungus, *Batrachochytrium dendrobatidis* (*Bd*), which has been linked to the commercial exportation of African clawed frogs (*Xenopus laevis*) for use in research, human pregnancy testing, and the pet trade. More recent evidence suggests that cycling weather patterns (flooding, hurricanes, the El Niño Southern Oscillation) also may have contributed to spread of this virulent disease (Daszak et al. 2000; Kolby and Daszak 2016). A closely related pathogenic fungus, *Batrachochytrium salamandrivorans* (*Bsal*), has emerged recently in Europe and is causing widespread mortality in salamanders. Although not yet reported in the United States, major concerns exist for its potential introduction into North America, and the risk of causing catastrophic losses to native U.S. salamander species, including many species likely susceptible to *Bsal* (Grant et al. 2016; Martel et al. 2013; Yap et al. 2015).

9.2.5 Impacts of Invasive Plants on Animal and Human Health

Invasive plants can affect animal and human health by providing habitat for vectors of disease. For example, Japanese barberry (*Berberis thunbergii*) was introduced into the United States in the late 1800s as an ornamental plant and promoted as a replacement for the common barberry (*Berberis vulgaris*). Now present in over 30 States and four Canadian provinces, it forms dense stands or thickets favorable to blacklegged/deer tick (*Ixodes scapularis*) populations (Elias et al. 2006; Williams and Ward 2010). Deer ticks are a key vector for Lyme disease in humans and animals, caused by the bacterial spirochete *Borrelia burgdorferi* (Beard and Eisen 2016).

Other invasive plants cause harm to humans through direct contact, causing irritation to the skin or other tissues. For example, giant hogweed (*Heracleum mantegazzianum*) was imported from the western Caucasus area into Europe and North America as an ornamental plant due to its immense size. However its milky sap can cause severe dermatitis, much like poison oak (*Toxicodendron diversilobum*) and poison ivy (*Toxicodendron radicans*) (Jones and Russell 1968). The sap of tree of heaven (*Ailanthus altissima*) can cause myocarditis (an inflammation of the heart muscle) in humans (Bisognano et al. 2005). In addition, the architecture of some invasive plants can cause harm, as large thorns may rip skin or clothing. An infestation of yellow star-thistle (*Centaurea solstitialis*) may effectively eliminate an area from any use or activity, such as grazing, recreational activities (such as hiking or birding), and military purposes (see Sect. 9.3.3).

9.2.6 Next Steps in Management and Mitigation of Invasive Species for Animal and Human Health

Global travel and trade, international movement of livestock, translocation of wildlife and pets, modern agricultural practices, and changing climate patterns have led to the introduction of non-native invasive vertebrate and invertebrate animal, plant, and microbial species, and the emergence of novel infectious diseases across North America and other continents. Expansion of the geographic range of invasive species and introduced diseases arises from changes in the inter-relationships between the native host, invasive pathogen, and the local environment or ecosystem. Intervention in one or more of these factors can disrupt or prevent the disease cycle. The same integrated approach can apply in developing control strategies to disrupt the life cycles of invasive species. Multi-disciplinary investigations are needed into the ecology, disease transmission and pathogenesis, and population biology of invasive species and introduced diseases to identify underlying causes and ways to prevent spread (Daszak et al. 2000).

In addition, collaboration across Federal, State, local, and tribal governments; stakeholders; and the private sector can help minimize and mitigate the spread of invasive species. Research is needed to develop new technologies and methods that enhance early detection and rapid response to eradicate or control invasive species and restore native species and ecosystems. Increased efforts to promote public education, address regulatory gaps, and coordinate with national and foreign governments to prevent introduction, minimize movements, and address impacts of invasive species may also be effective. Since many health impacts begin with local invasions, either intentional or inadvertent, actions by individual citizens and local communities to prevent, mitigate, and eradicate invasive species may be particularly effective.

9.3 Invasive Species and Military Readiness

Invasive species impact military operations directly and indirectly through effects on combat readiness; human health, as discussed above; and recovery efforts for threatened and endangered species. Military personnel conduct crucial training, testing, and operational activities in air, land, and aquatic landscapes that must be managed and maintained for mission readiness. The Department of Defense (DoD) manages approximately 25 million acres of land encompassing over 420 large installations, of which 342 have natural resources significant enough to require management under the Sikes Act (DoD 2017). Access to DoD lands is often restricted, resulting in undisturbed areas that harbor rare and sometimes unique habitats. In fact, DoD lands have higher levels of biological diversity and more sensitive species per acre than other Federal lands (NatureServe 2015). Invasive species can have significant detrimental effects on these sensitive habitats. In addition to the impacts on natural resources, invasive species can affect testing, operational, and training activities; military personnel health, security, and installation infrastructure; and global movement of personnel and equipment that can and do result in land use restrictions and added expenses for DoD.

9.3.1 Operational Impacts Due to Invasive Species

9.3.1.1 Impacts on Military Personnel Health

Safety and health of military personnel and operations can be significantly impacted by invasive species in a variety of ways. Invasive species can cause injury, transmit disease-causing pathogens, and serve as a refugia for vectors, hosts, and pathogens. They may interfere with safety and security by obscuring unexploded ordnance, serving as fuel for wildfires, and impeding lines-of-site monitoring for security forces personnel. For instance, military personnel and dogs have been attacked by invasive feral swine. Soldiers have also been injured with burns and temporary blindness caused by contact with giant hogweed during training operations. Red imported fire ants (*Solenopsis invicta*) and recently little fire ants (*Wasmannia auropunctata*) in the Pacific can cause intense burning and swelling through their mass biting/stinging, and may also seriously injure, blind, or kill sensitive animals. Invasive plants such as honeysuckles (*Lonicera* spp.) and barberries (*Berberis* spp.) serve as refugia for infected ticks that transmit diseases to military personnel (Dalsimer 2002).

The DoD is very concerned about introducing disease pathogens and their reservoirs, as well as arthropods that may vector diseases (see Sects. 9.2.3 and 9.2.4). Introductions of invasive diseases can impact local and deployed forces, and require significant surveillance, testing, and control measures at affected installations (DoD AFPMB 2016). The 2016 Zika outbreak in the New World interfered with missions, and resulted in country clearance burdens requiring aircraft disinsection (spraying aircraft for insects) to eliminate potential mosquito vectors.

9.3.1.2 Impacts on Personnel and Operational Safety

Invasive species and wildlife management is essential around training ranges and airfields to protect aircrews and aircraft and ensure safe military flight operations as these can attract birds and other wildlife that pose bird/wildlife aircraft strike hazards (DoD 2017). In addition to impacting flight operations, invasives also can damage runways, infrastructure, and surrounding areas. Managing for invasives in and around airfields and training ranges is a DoD priority.

Invasive plants, such as the common reed (*Phragmites* spp.) in wetland habitats at several military installations along the Chesapeake Bay, not only affect wildlife but can obscure clear lines-of-sight, thus threatening base security. The DoD actively manages *Phragmities* through targeted herbicide spraying and reseeding with native plants to limit common reed growth and expansion (DoD 2011).

Invasive insects that cause tree mortality also create safety hazards from falling trees until dead trees can be removed. The invasive coconut rhinoceros beetle (*Oryctes rhinoceros*) bores holes into crowns of coconut and other palm trees (Arecaceae), often resulting in widespread tree mortality and interference with training operations in Hawaii and Guam. In Hawaii, the DoD is collaborating with the U.S. Department of Agriculture, Animal and Plant Health Inspection Service (USDA APHIS), State of Hawaii Department of Agriculture, and the University of Hawaii to eradicate the coconut rhinoceros beetle from Hawaii through an integrated program that includes population monitoring, removal of breeding sites, and public outreach and education campaigns (HDOA 2016).

9.3.2 Environmental Impacts of Invasive Species on Defense

9.3.2.1 Impacts to Threatened, Endangered, and At-risk Species

Approximately 60 listed species and over 70 species at risk of needing listing protection occur only on DoD lands, which harbor a higher density of sensitive species than on any other federally managed lands (NatureServe 2015). DoD strives to prevent and control the introduction, spread, and establishment of invasive species which can have devastating effects on sensitive species. For instance, DoD manages invasive feral swine through monitoring, hunting, trapping, and public outreach and education to help prevent the swine from destroying nesting habitats, eating endangered plants, and damaging infrastructure and training lands. Another example is at Marine Corps Base Hawaii where the military conducts amphibious vehicle training in mudflats invaded by pickleweed (*Salicornia* spp.). The activity not only breaks up and destroys the pickleweed, but helps protect the endangered native Hawaiian stilt (*Himantopus mexicanus knudseni*), which nests in the mudflats, by creating additional nesting habitat (U.S. Marine Corps 2001).

9.3.2.2 Impacts on Wildfire

Invasive grasses, including fountain grass (*Pennisetum setaceum*) in Hawaii and cheatgrass (*Bromus tectorum*) in the Western United States, have become the dominant ground cover on many military installations. Species composition is dramatically altered following invasion by these grasses which creates wildfire-friendly conditions in ecosystems that are not wildfire adapted (Coates et al. 2016).

Uncontrolled wildfires threaten the safety of military personnel and their families. Wildfires pose a risk to infrastructure and degrade testing and training lands, and may spread beyond military boundaries to local communities. These fires are expensive to control and mitigate. In addition, smoke and atmospheric particulate matter negatively impact air quality and can inhibit training activities and military flight operations (DoD 2017).

The DoD is actively engaged in managing both invasive species and wildfire. Military natural resource managers use prescribed fire to reduce or clear excess vegetation including cheatgrass, halogeton (*Halogeton* spp.), musk thistle (*Carduus nutans*), salt cedar (*Tamarix* spp.), tumbleweed (e.g., *Kali* spp.), and Sahara mustard (*Brassica tournefortii*) to support recovery of fire-tolerant native species such as longleaf pine (*Pinus palustris*) (Sustainable Fort Bragg 2017).

9.3.3 Economic Impacts on Defense

9.3.3.1 Mitigation Costs

Invasive species surveillance, prevention, management, and control are integrated in a variety of DoD programs including Operations, Training, Pest Management, Natural Resources, and Morale and Welfare. Collectively, DoD invests significant resources into managing and mitigating invasive species. For instance, Marine Corps Base Camp Pendleton in southern California spent approximately $1.2 million over a 5-year period to control invasive species that could have severe impacts on infrastructure, native ecosystems, and training activities (DoD 2017). The Navy also invests considerable funds by implementing voluntary adoption and application of U.S. Coast Guard guidelines for controlling ship ballast water to prevent the introduction of invasive aquatic organisms into U.S. waters. Similarly, the fee-for-service military agricultural pre-clearance inspection program (Defense Transportation Regulation 2018a, b) helps to prevent accidental introduction of invasive species, and facilitates the safe return of military personnel, vehicles, equipment, and cargo from locations around the world.

To avoid the inadvertent introduction of invasives when moving equipment, vehicles, and supplies for military and humanitarian relief activities, DoD implements costly management practices including utilizing cleaning facilities and equipment, and having plant and pest quarantine officers inspect military equipment and personnel (DoD AFPMB 2017). Inspections and cleanup operations are performed in partnership with USDA APHIS, and are intended to prevent the transport of non-native animal and plant pests. In addition, DoD has phytosanitary regulations (e.g., DoD 4149.01-M-1) for wood packing materials (WPM) to prevent introduction of pests through transport of uninspected WPMs used in shipping (DoD 2017).

9.3.3.2 Lost Capability Costs

Invasive species also have indirect economic impacts through lost capability, including limiting field maneuvers and training exercises. For instance, several years ago, parachute training had to be halted at Fort Hunter Liggett, CA due to a severe infestation of yellow star-thistle in a parachute drop zone, which would have injured soldiers, snagged and torn their parachutes, and clogged vehicle air filters. The Fort used an integrated pest management strategy incorporating hand pulling, mowing, herbicide treatment, and biological control that significantly reduced yellow star-thistle density on about 10,000 acres. Costs for control and alternative protocols to continue mission activities were substantial (DoD 2011). Whenever invasive species interfere with operational activities, costly workarounds must be implemented to ensure that military personnel can sustain mission readiness.

9.3.4 DoD Invasive Species Management

9.3.4.1 Policy and Guidance

The DoD's policy and guidance related to invasive species management focuses on Pest Management, Operations and Maintenance, and Natural Resources programs. DoD provides policy and guidance to (1) prevent the introduction of invasive species; (2) detect and respond rapidly to, and control populations of such species using integrated pest management (IPM) techniques; (3) monitor invasive species populations accurately and reliably; (4) restore native species and habitat conditions in ecosystems that have been invaded; (5) conduct research on invasive species, develop technologies to prevent introduction, and provide the latest IPM techniques for their control; and (6) promote public education on invasive species (DoD 2017). The Military Services also comply with Executive Order 13751, Safeguarding the Nation from the Impacts of Invasive Species (No. E.O. 2016).

DoD natural resources managers control invasive species by implementing integrated pest management plans, which focus on identifying and monitoring pests, pathogens, and insects; setting action thresholds; and preventing and controlling pests. Integrated Natural Resources Management Plans (INRMPs) are used to guide how military installations will manage their natural resources, including control of invasive species. Installations are required to identify, prioritize, monitor, and control invasive and noxious species and feral animals whenever feasible (DoD 2018). In addition, DoD requires that native species be used, where feasible, to restore habitats impacted by invasive species. Installations endeavor to conserve and protect water resources, use locally adapted native plants, avoid using invasive species, and minimize the use of pesticides and supplemental watering.

9.3.4.2 Managing Invasive Species Through Biosecurity Plans

Increasingly, the DoD is using biosecurity plans as a means of addressing mission impacts associated with invasive species often in conjunction with Endangered Species Recovery efforts. For instance, the Department of the Navy developed a regionally vetted plan called the Regional Biosecurity Plan (RBP) for Micronesia and Hawaii to guide relocation of U.S. Marine Corps personnel to Guam, and invested nearly $4 million to proactively address regional concerns about the risks of invasive species to terrestrial, freshwater, and marine environments. The RBP was produced in cooperation with the U.S. Geological Survey, Smithsonian Environmental Research Center, National Invasive Species Council, University of Guam, and USDA APHIS and assesses invasive species risk pathways with a focus on threats from the coconut rhinoceros beetle, brown tree snake (*Boiga irregularis*), and other serious invasives in the Pacific (University of Guam and the Secretariat of the Pacific Community 2014). The Department of the Navy has also implemented an invasive species-focused biosecurity plan for Naval assets on San Clemente Island in California (DoD 2017), Hawaii, and other locations.

9.4 Invasive Species Effects on Urban Plant Communities, Infrastructure, and Safety

Urban areas are the epicenter of many, if not most, non-native plant, insect, and pathogen invasions (Alston and Richardson 2006; Smith et al. 2006). Urbanization and the import of plant species for landscaping are directly linked. Most imported plants are non-native and some may become invasive, potentially spreading into surrounding rural and natural areas (Duguay et al. 2007; Guo et al. 2006; McDonald et al. 2008).

Invasive plants may change the community structure of the natural vegetation in urban areas such that native species are less dominant (Lambrinos 2000). They may alter nutrient cycling or add novel allelopathic compounds to the soil, resulting in disturbed conditions that favor further invasion and deter re-establishment of native species (Gomez-Aparicio and Canham 2008a; Murrell et al. 2011; Stinson et al. 2006). Such changes in ecosystem (Ehrenfeld 2003; Gomez-Aparicio and Canham 2008b) and disturbance (Mack and D'Antonio 1998) processes may occur anywhere along the urban-rural interface. Urbanization also produces disturbed open areas subject to invasion (Alston and Richardson 2006), heat islands (Botkin and Beveridge 1997; Oke 1995; Pickett et al. 2001), impervious surfaces, and toxic soils (Cannon and Horton 2009; Gill et al. 2007; Leishman and Thomson 2005) that only a few, often non-native, plants can tolerate.

Urban residents value the decorative nature of many non-native plant species, and the sale of these plants and landscaping costs associated with their planting and maintenance can be a key component of urban and suburban economies, including housing valuation and landscaping, and other businesses associated with horticulture and supplying plants (Des Rosiers et al. 2007; Drew et al. 2010; With 2002). According to the U.S. Census of Horticulture Specialties, there were over 23,000 horticulture operators in 2014 that conducted about $13.8 billion in sales, and approximately 12,400 of these operators were in the floriculture sector, which conducted $5.9 billion in sales (USDA National Agricultural Statistics Service 2014). Of course, not all horticulture sales are of non-native plants, and not all non-native plant sales pose a risk of invasion. Unfortunately, there are substantial difficulties associated with identifying which non-native plants will become invasive and managing a highly disaggregated industry. Controlling risk associated with commercial horticulture and nurseries, therefore, requires balancing the risk of invasion with the benefits people derive from non-native plants (Knowler and Barbier 2005).

Invasive insects and diseases may be inadvertently imported with landscaping plants or other imported goods, damaging both native and non-native plant communities. Invasive plants, insects, and animals can harm the built environment, incurring public and private costs for repair and replacement, and presenting safety risks to urban residents.

9.4.1 Invasive Species and Urban Plant Community Composition

When regionally distinct native communities are replaced with locally expanding, non-native communities with no barriers to expansion, there may be homogenization of species (Collins et al. 2002; McKinney and Lockwood 1999; Olden and Rooney 2006). The result is that one urban area often shares many of the same plant species with other urban areas. Although these urban areas may share many of the same non-native plants, they may still have a rich flora of different, albeit rare, native species that could be maintained (Kowarkik 2011; Schwartz et al. 2006).

Possibly, homogenization in the United States may not be fully realized since the oldest cities are only a few centuries old (Clements and Moore 2003; Lososova et al. 2012). A study of native and non-native flora of Boston; New York; Philadelphia; Washington, D.C.; Detroit; Chicago; Minneapolis; and St. Louis found that those cities only shared 7.5% of their non-native species. They also only shared 11.6% of their native species, with distinct East Coast and Midwest city clusters, suggesting geography still plays an important role in defining species compositions in these urban areas (Clements and Moore 2003). A study of 110 cities worldwide found that the majority of urban plant species were still native; however, their densities had declined substantially to 25% of the overall density of all plant species.

This change in density is correlated most strongly with land cover and city age (Aronson et al. 2014).

Phylogenetic diversity, or differences in evolutionary origin, of non-native plants is lower than that of native plants, both within single cities and across cities; phylogenetic diversity is even lower for non-native species with residence times of multiple centuries. Non-native species in the United States are significantly overrepresented by six families: Boraginaceae, Brassicaceae, Fabaceae, Caryophyllaceae, Chenopodiaceae, and Solanaceae (Ricotta et al. 2009). Functionally related species often share a common origin and phylogenetic history, suggesting that limited phylogenetic diversity equates to limited functional diversity (Darwin 1859). Thus, low functional diversity may limit the range of ecosystem services that non-native species in urban areas can provide.

Globally, pollinator declines are attributed to habitat loss associated with urbanization. Nonetheless, there is empirical support suggesting that non-native entomophilous plants (reliant on insects for pollination) are easily integrated into existing native plant-pollinator networks, acting as additional or in some cases the only sources of pollen and nectar. Thus, non-native plants (including entomophilous invasive plants) may serve to keep pollinator populations viable (Potts et al. 2010; Stout and Morales 2009). Unfortunately, this pollination "buffer" may only benefit generalist pollinators. Many native species interact with specialist pollinators; consequently, displacement or reduction in abundance of these native plants by non-native species also leads to a subsequent reduction in specialist pollinators (Traveset and Richardson 2006).

9.4.2 Effects of Invasive Insects and Disease on Urban Plant Health

Previous chapters have described the effects of invasive insects and disease on forest and grassland communities, but they can also affect urban areas. Urban forests and trees provide critical benefits such as helping filter air and water, controlling storm water, conserving energy, and providing animal habitat and shade. They also reduce noise, provide places to recreate, support mental health, and add economic value to our communities. According to Nowak et al. (2002), the total value for urban forests in the contiguous United States is about $2.4 trillion. They place the value of New York City's urban forest near $5.2 billion. These estimates are based on compensatory values, the value homeowners would have to be paid to be as well off should one of their trees be removed. These values largely reflect the cost of replacing the tree (purchasing a new tree and having it planted). A broader look at values associated with urban trees is given in Roy et al. (2012), which reviews 115 urban tree studies. Many of the papers reviewed discuss the role of trees in making the urban environment more pleasant, in increasing property values, in providing shade and wind protection, and in lowering energy costs. Many also address the costs associated with trees, including prevalence of pollen allergies, release of volatile organic compounds, and reducing solar access. One study in Minnesota found that a 10% increase in tree cover within 100 m of a home increases the home's sale price by about $1400 (Sanders et al. 2010). Donovan and Butry (2009) found that tree cover on the west and south sides of houses in their sample resulted in a 5% reduction in summertime electricity use.

With overarching branches that create beautiful shaded canopies, elm trees (*Ulmus* spp.) are well suited as street trees, and they have been planted in rows along streets and walkways of large cities and small towns across North America. The uniformity of these monocultures is not only aesthetically pleasing but efficient from the standpoint of nursery production, planting, and maintenance. However, these monocultures are susceptible to insects or pathogens that can invade and quickly move through the area. Dutch elm disease, caused by the invasive fungi *Ophiostoma ulmi* and *Ophiostomoma novo-ulmi*, has spread rapidly through urban areas and produced severe damage to vegetation. Not only is the invasive pathogen able to quickly colonize the native host, but it is able to efficiently move to neighboring trees because it is vectored by the non-native European elm bark beetle (*Scolytus multistriatus*), whose populations exploded with the abundance of dying host material. When the Dutch elm disease fungus was first introduced into North America around 1930, there were approximately 77 million elms in cities and towns across North America; however, by the 1970s, only about 34 million elms survived. The costs of removal of many dead elms and of protection of the remaining trees has had a significant economic impact on municipalities and homeowners.

A historical analysis of the economic impacts of non-native forest insects by Aukema et al. (2011) found that homeowners and local governments bear the largest costs from invasive insects. Local government expenditures for management, and residential property value losses were the two highest cost categories. Additionally, they found that woodborers were consistently the most economically damaging insect feeding guild. Nowak et al. (2002) found that in select cities, the potential tree losses from the Asian longhorned beetle (*Anoplophora glabripennis*) ranged from 12% to 61% of city tree populations with an estimated value of $72 million to $2.3 billion per city. As of the end of 2016, Asian longhorned beetle infestations were being managed by USDA in New York, Massachusetts, and Ohio. In an effort to eradicate this non-native woodborer in Massachusetts and Ohio, more than 40,000 host trees have been removed from these two States since 2008 and 2011, respectively (USDA APHIS 2017).

Perhaps the invasive insect with the most significant impact in urban areas is one of the most recently established species, the emerald ash borer (EAB) (*Agrilus planipennis*). This woodborer, native to Asia, was first found in the Detroit, MI–Windsor, Ontario area in 2002 but was likely introduced five to 10 years prior to then (see Chap. 7 for details on EAB biology and management). As with Dutch elm disease, EAB moved rapidly along city streets and caused extensive mortality where ash trees (*Fraxinus* spp.), especially green and white ash (*Fraxinus pennsylvanica* and *Fraxinus americana*) and their cultivars, were commonly planted in monocultures as elm replacements in the aftermath of Dutch elm disease. It is estimated that 100 million ash trees in both rural and urban landscapes have been killed by EAB, but perhaps the greatest economic impact has been realized in urban forests. A 2009 estimate of the potential cost of EAB treatment, removal, and replacement in urban communities over a 10-year period was $10.7 million (Kovacs et al. 2010). Simulations in this study were based on EAB expanding to 25 States between 2009–2019 and affecting 17 million ash trees. As of the spring of 2020, EAB was found in 35 States (USDA APHIS 2020).

The management of EAB in North America highlights some of the challenges faced when dealing with a new invasive. Very little was known about EAB prior to 2002, even in the Asian literature. Research in North America basically had to start from scratch to understand EAB biology, life cycle, host relationships, and population biology. One of the greatest challenges was to develop tools to effectively survey for this insect. Although regulations were quickly established to limit the movement of EAB, it became readily apparent that infestations had already spread beyond the known infested area. As these isolated infestations developed, the beetles again were unknowingly spread to additional locations. In many cases it was determined that EAB was in an area for 3–5 years before being detected.

Across North America, other invasive insects and pathogens have had economic and social impacts in the urban environment. Although some of these invasive species, such as chestnut blight (*Cryphonectria parasitica*) and gypsy moth (*Lymantria dispar dispar*) have had greater impacts in natural forests, they also have impacted urban and suburban forests (see Sect. 9.8). In the Western United States, the walnut twig beetle/thousand cankers disease complex (*Pityophthorus juglandis* and *Geosmithia morbida*) and the shothole borer (*Euwallacea* spp.)/*Fusarium* species complex have recently caused significant local impacts on urban forests. In the Eastern United States, bacterial leaf scorch (*Xylella fastidiosa*) has reportedly caused gradual mortality of a number of different species of urban street and park trees (see Chap. 2).

9.4.3 Impacts on Urban Safety and the Built Environment

Unless removed, dead and dying trees damaged by invasive species can be extremely hazardous in the urban environment. Limbs and even whole trees can fall on roads, houses, cars, and power lines, destroying property and threatening human safety. Dead trees and invasive grasses, such as cheatgrass and buffelgrass (*Pennisetum ciliare*) (Balch et al. 2013), may elevate the risk of wildfire, particularly in drought years, affecting open spaces in western cities such as Denver and Boulder, CO. The National Storm Damage Center estimates that trees cause about $1 billion in property damage each year attributed to severe weather.

Invasive species may also cause harm to the built environment (ISAC 2016). English ivy (*Hedera helix*) can root in the grouting between masonry, destroying the structural integrity of brick structures over time, while African land snails (*Achatina fulica*) can eat the concrete stucco of buildings (ISAC 2016). The Formosan termite (*Coptotermes formosanus*) has also caused significant damage to structures in the United States. It can enter buildings from the ground through cracks, joints, and utility conduits, and can even maintain colonies on flat rooftops (Su and Scheffrahan 2016); once established it consumes cellulose in wood. Prevention, control, and repair of structural damage is estimated to cost consumers over $1 billion per year (Lax and Osbrink 2003).

Invasive birds and mammals may cause additional damage to structures. Birds can leave droppings, crack windows, bore into wood, and build nests that may be a nuisance. Control costs for rock pigeons are estimated at $2 billion per year in the United States (Pimentel et al. 2005). Norway rats can live below building foundations and inhabit walls, cellars, and lower floors of buildings; their gnawing and burrowing can damage the wood and insulation of the buildings and undermine the foundations (Timm 1994).

9.4.4 Management Options

Urban areas are not as conducive to management tools often used in rural and natural areas, such as fire, pesticides, poison baits, addition of top predators, and fencing to reduce herbivory. Eliminating pests and managing for native species will entail increased engagement of community members (Dearborn and Kark 2009), which may be facilitated through public education and volunteer engagement programs, including those that engage community members in restoration and citizen science. Local arboreta and public and private botanical gardens can often be of assistance, providing both education and local expertise to the community.

Urban areas could potentially avoid homogenization or a loss of native species densities through planting of native species (preferably, historically common to the area or locally suitable), reducing the presence of invasive species, and applying a landscape-level approach to urban planning. A landscape approach might involve establishing interconnecting gardens dominated by native and/or non-invasive non-native plants via corridors, such as utility rights-of-way, road medians, subdivision entry displays, and walking paths. This scaling up from the garden to the neighborhood, and finally to a city-wide landscape of connected gardens, parks, and planted corridors, requires cooperation among a range of stakeholders across all sectors of society (home owners, city planners and parks departments, and housing developers). Such collaboration is often only possible when economic incentives (e.g., resulting increase in property value) or measurable ecosystem services (e.g., improved community health and well-being) (Goddard et al. 2009; Ignatieva et al. 2011) are provided.

Researchers have begun developing provisional seed zones for native species (Bower et al. 2014). Having seed zones established for native species, especially those that are nested within an ecoregion and include more detailed species-specific fitness (growth and reproductive) responses over an environmental gradient, will enable urban planners and individual city dwellers to select native species that are genetically and ecologically suitable to their site. More importantly, they could select species that can survive within a relatively broad environmental gradient and would therefore be less vulnerable to changes in temperature or precipitation, and to invasive species (Havens et al. 2015).

Urban areas can limit impacts of invasive plants if the remnant ecosystems are managed sustainably. Managers can actively remove invasive species and avoid practices that create openings suitable for invasive species, such as harvest and prescribed fire, or involve equipment that might aid the spread of invasive species. Successful regeneration of native species after removal of invasive species has had mixed success (Kettenring and Adams 2011) and may require active control of other factors, such as excessive deer herbivory (Huebner et al. 2010), toxic runoff from adjacent impervious surfaces (Gill et al. 2007; Leishman and Thomson 2005), new insects and pathogens introduced with non-native landscape plants (Liebhold et al. 2012), soil compaction, urban air pollution and nitrogen/sulfur and ozone deposition (Nowak 2010; Ziska et al. 2004), and urban heat (Pickett et al. 2001). Many people are supportive of planting native species preferentially over better-known non-natives as long as they are aware of the native species, can purchase them, and can control where they plant them (Doody et al. 2010). Thus, educating the public about native species that are available for them to plant instead of non-native species is just as important as educating them about the potential negative impacts of non-native plants. The City of Portland, OR, for example, publishes a plant list of native plants that homeowners and landscapers are encouraged to use, and a list of nuisance plants that people are encouraged not to plant and that are prohibited on city grounds and projects (Bureau of Planning and Sustainability 2016). In some instances, cities may opt to quantify costs associated with invasive species in an effort to incentivize businesses not to sell potentially profitable invasive plants.

As non-native plant imports are the major source of inadvertent introductions of invasive insects and pathogens (Liebhold et al. 2012), use of native species in urban green spaces will also decrease the abundance of invasive insects and pathogens in urban areas. Once established, the impact of insects and pathogens can be substantial, and management actions required to reduce such impacts will be required. In urban areas, the management of Dutch elm disease usually is accomplished in one of three ways: reduction of the bark beetle vector populations, prophylactic protection of elms with fungicides, or the long-term use of breeding for host resistance. Large-scale treatment of elms with insecticide spraying produced significant ecological impacts, and for the most part has been abandoned (Karnosky 1979). The use of bark beetle trap trees along with pheromone-baited traps has shown promise (Lanier 1989) but has not been utilized widely.

Perhaps what has proven to be the most effective strategy for most municipalities is the diligent surveillance and removal of infected trees or branches. Scouting to identify new infections, followed by timely pruning of infected trees, or injections of fungicides to stop spread within a tree, have proven to be very cost-effective. Sanitation, or the rapid removal of beetle- and disease-infested trees, is also an integral component of effective management. Cities such as Washington, D.C. and Winnipeg, Canada still have extensive elm populations due to their utilization of integrated management programs. Although it is expensive for municipalities to sustain effective management programs (Winnipeg spends approximately $3 million annually), it is still less costly than the city-wide removal of large, dead trees (Pines 2009).

9.5 Impacts of Invasive Species on Water Resources and Systems

Globally, human use of freshwater surpasses long-term supplies by approximately 25%; this is possible due to water transfers between watersheds and extraction of groundwater supplies (Catford 2017). Invasive plants, vertebrates, and invertebrates can alter water supplies by altering water courses via accretion or clogging of waterways, or increasing evapotranspiration rates; invasive species also impede human use of water systems through eutrophication, alteration of sediments, fouling, and impeding access. Therefore, a key concern attributed to invasive species in the United States is the potential reduction of available and usable water resources. Concern about these impacts is increasing because predicted global climate change is anticipated to alter patterns of discharge and temperature, potentially enhancing habitat for invasive species rather than for native species, further exacerbating the economic effects of invasive species management and remediation.

9.5.1 Impacts of Invasive Species on Water Quantity

As the human population continues to increase in the United States and globally, water demands for irrigation, drinking, and other household uses will continue to rise. This is particularly challenging in areas that face frequent droughts and water shortages, such as California, an increasingly populated state. The July 2016 estimated population of California was 39.4 million and is projected to grow by 0.76% annually, or 6.5 million additional people by 2036 (California Department of Finance 2017). This will place even higher demands on water infrastructure.

Invasive plants are a particular concern because they can reduce freshwater availability through high rates of transpiration. For example, saltcedar (*Tamarix* spp.), originally introduced to North America as nursery stock in the 1800s (DiTomaso 1998), is now common in the Southwestern and interior Northwestern United States (Kerns et al. 2009). Although there have been studies that document high rates of water use by native plants (Cohn 2005), some have found that saltcedar uses significantly more water than native species (~1.4–3.0 billion m^3 annually), at an estimated cost of $26.3–$67.8 million dollars in water that could be used for agriculture, drinking water, or hydropower (Pejchar and Mooney 2009; Zavaleta 2000). A large saltcedar tree can uptake 760 L of water in a single day, leading to desiccation of streams and springs, potentially lowering the water table and limiting water supplies to native species (DiTomaso 1998 and references therein). While saltcedar was originally marketed as an ornamental shrub, it was also used for windbreaks and to stabilize eroding stream banks. In places, the dense and stable root architecture of salt cedar exceeds that of native riparian species, leading to immobilized channels with increased sediment deposition. This process gradually restricts channel width and increases flow rates, which enhances flooding and promotes dispersal of invasive species further away from the stream bank (DiTomaso 1998).

Tall-statured grasses such as the giant reed (*Arundo donax*) and the common reed have significantly altered wetlands in North America (Zedler and Kercher 2010). Both species readily spread via clonal rhizome fragments, grow in near monocultures (outcompeting native plants), use the C3 photosynthetic pathway, which makes them less efficient in their use of water than C4 grasses, and produce highly flammable aboveground biomass, thus making them a fire hazard (Bell 1997; Meyerson et al. 2000). Giant reed colonizes arid regions such as southern California, where its high rate of water consumption (nearly 9 mm of water per day in one study (i.e., Watts and Moore 2011) and 1700 mm/year in another (i.e., Iverson 1998)) further reduces water availability and exacerbates drought conditions. The common reed is among the most widely spread and best-studied species globally, and is highly invasive in both freshwater and brackish wetlands (Meyerson et al. 2016). This invasive wetland plant has transformed wetlands into uplands because of its high productivity and production of slowly decomposing biomass, which acts as a significant barrier to water accessible for recreation. Martin and Blossey (2013) estimated that over a 4-year period in the United States, nearly $20 million dollars were spent on common reed control (often with pesticides) with limited success at eradication.

In Hawaii, invasive species may exacerbate already significant reductions in freshwater supply. Computer models that simulate changes in water yields on the Big Island of Hawai'i found that stream flows could be reduced by 50% due to climate change, and project an additional 2% loss due to invasive species such as strawberry guava (*Psidium cattleianum*) and invasive ginger (*Hedychium gardnerianum*) (Ayron et al. 2016).

9.5.2 Impacts of Invasive Species on Infrastructure and Water Quality

Invasive aquatic invertebrates may be particularly problematic for human water infrastructure. Invasive Dreissenid mussels, which include the zebra mussel (*Dreissena polymorpha*) and the quagga mussel (*Dreissena bugensis*), are well-known for their harm to native mussels and water quality and costly effects on infrastructure, fish hatcheries, navigational locks, and recreation. Zebra mussels are found in at least 30 U.S. States and in over 600 lakes and reservoirs, including the Great Lakes (Benson 2017). Zebra mussel populations can become so dense that they effectively smother aquatic plants and organisms on the bottom of lakes and encrust water intake pipes. This may significantly impact cooling systems of power plants, turbines used in hydropower generation, and water treatment facilities with costs of hundreds of millions of dollars across all sectors (Connelly et al. 2007; Strayer 2009).

In addition, aquatic invasive species can affect ambient water quality. Zebra mussels and other introduced invasive bivalves (e.g., the quagga mussel and Asiatic clam (*Corbicula fluminea*)) are highly productive filter feeders. When their populations are sufficiently large, they can significantly reduce the concentration of phytoplankton in the systems that they colonize. This may increase water clarity but can lead to a collapse in the pelagic community as seen in the Hudson River in 1991 (Strayer 2010). Selective filter feeding by zebra mussels has also promoted blooms of the toxin-producing cyanobacterium *Microcystis aeruginosa*, which can poison wildlife, domestic animals, and humans if ingested (Butler et al. 2009; Falconer 2005; Vanderploeg et al. 2001)

9.5.3 Impacts of Invasive Species on Recreation and Public Safety

Many invasive species, such as common reed or Japanese knotweed (*Fallopia japonica*), can impede water views and public access to recreational water bodies. Invaders can also pose public hazards associated with build-up of flammable biomass. Such species aggressively colonize shorelines of lakes, rivers, and coastal areas and require repeated and costly management efforts for their control. In some cases, the dense vegetation associated with invasive species can provide cover for illegal or illicit activities. Zebra mussels and introduced bivalves colonize lake and river bottoms. Their sharp shells can cut bare feet making water shoes a necessity and barefoot swimming a thing of the past.

Many States have implemented regulations that require cleaning the hulls of boats and other equipment to prevent the unintended spread of invasive plants and animals from lake to lake and river to river. For example, Minnesota, Missouri, Alaska, Vermont, and Maryland either prohibit or discourage anglers from wearing "porous soled" boots (often with felt bottoms) because these types of fishing boots can transport many invasive species: zebra mussel larvae, rock snot algae (*Didymosphenia geminata*), water fleas (*Bythotrephes longimanus* (*cederstroemi*), *Cercopagis pengoi*), viral hemorrhagic septicemia (*Oncorhynchus 2 novirhabdovirus*) which affects fish, whirling disease (*Myxobolus cerebralis*), faucet snails (*Bithynia tentaculata*), and New Zealand mud snails (*Potamopyrgus antipodarum*) (MN DNR 2017). Traditional cleaning methods (e.g., bleach) do not adequately kill the hitchhikers on these boots, therefore switching to alternative types of non-felt bottomed footwear is preferred. While necessary, such measures place an unwanted burden on recreational fishermen that may discourage their compliance.

Some non-native fishes are deliberately stocked with intent to provide recreational opportunities for anglers. For example, in the West, recreational fisheries have been stocked with smallmouth (*Micropterus dolomieu*) and largemouth bass (*M. salmoides*). There are recreational trophy fisheries for Atlantic salmon (*Salmo salar*) in Oregon, while fisheries for rainbow trout (*Oncorhynchus mykiss*) exist in many locations in Eastern North America. In 2015, the American Sportfishing Association estimated that freshwater sport fisheries support a $115-billion annual industry (ASA 2015). When introduced non-native sport fish become invasive, the strong economic value of non-native sport fishes may complicate efforts to restore or protect native fishes (Cambray 2003). Some non-native sport fishes are strong predators and may dominate food sources or habitat thereby restructuring native ecosystems (Cambray 2003).

9.5.4 Management Approaches for Water Resources and Systems

It will continue to be a challenge to manage water resources colonized by invasive species, particularly with the emergence of climatic variability and change. Because invasive species interact with both the biotic and abiotic environment, detailed knowledge of the life history of invaders and information on the site being invaded is needed to maximize management efforts. In addition, the interaction of novel groupings of species and the advent of climate change are creating new ecosystems (Hobbs et al. 2009), often through destruction and re-structuring of existing native systems (Simberloff and Von Holle 1999). Consequently, it will be a challenge to develop effective strategies for managing water resources. Managers of aquatic systems are likely to encounter new invaders that require treatments (Strayer 2010) which may strain budgets. Long-term monitoring of invaded and restored sites, and real-time reporting of invasive species sightings in publicly available datasets, could facilitate detecting new introductions (or decreases in invader populations). Such information would assist managers in targeting high-priority areas for control, and potentially improve management efficiency by allowing managers to target more than one invasive species at a time (Catford 2017). Continued efforts by Federal, State, and local agencies; non-governmental organizations; and other private entities to educate the public about the negative effects of invasive species on water resources, and to employ outreach efforts to engage the public, are critical in combatting and reporting aquatic invasive species. Additional details for early detection, management approaches, restoration, and rehabilitation for aquatic invasives and aquatic ecosystems are also discussed in Chaps. 6, 7 and 8.

9.6 Invasive Species and Energy Systems

Terrestrial and aquatic energy systems can be conduits for connectivity and dispersal of invasive species. Terrestrial energy systems are often extensions of urban ecosystems and connect rural areas through utility corridors. In aquatic systems, reservoirs or long slow-water river reaches associated with hydropower dams and energy generation often serve as prime habitat for invasive predatory warm-water fishes (that ultimately may have strong recreational value) and affect native fish survival and the trophic structure of river reaches. In both terrestrial and aquatic environments, invasive species can directly damage energy infrastructure, contribute to hazards that can damage infrastructure, and incur costs for removal.

9.6.1 Impacts of Invasive Species and Energy Systems

Several species have caused direct damage to electrical equipment. Invasive tawny crazy ants (*Nylanderia fulva*) infest electrical equipment and cause short circuits, equipment failures, and obstructions in switching mechanisms. It is estimated that tawny crazy ants cause $146.5 million in damage to electrical equipment (ISAC 2016). The monk parakeet (*Myiopsitta monachus*), which is native to South America, causes short circuits and electrical fires because of its nesting behavior. The damages and power outages attributed to the stick-built nests lead to an increase in operation and maintenance costs (Booy et al. 2017). Brown tree snakes were accidentally introduced to the island of Guam after World War II. They cause power outages by actively crawling up utility poles, climbing power lines, and entering transformers. It is estimated that a single island-wide power outage caused by the snake can cost the power utility company more than $6 million per year (Pimentel et al. 2005).

Invasive aquatic mollusks (i.e., zebra mussel, quagga mussel, Asian clam (*Potamocorbula amurensis*)) can cause significant expense if/when they contaminate hydropower facilities. Because these species deeply encrust exposed pipes, openings, or surfaces, they can clog water intakes and pipelines (Rosaen et al. 2012). These invasives can also infiltrate interior portions of power plants and dams, greatly affecting operations and increasing maintenance costs (ISAC 2016). The Anderson Economic Group cites mollusk infestation as the reason behind the closure of the Detroit Edison (later named DTE Energy) Power Generation Plant in Monroe, MI; customers were without power for 3 days (Park and Hushak 1999).

When invasive species damage or kill trees, these trees and associated limbs can pose a threat to utility lines and poles. Utility companies face increasing costs to maintain power lines in areas where trees are infested or diseased (Aukema et al. 2011). Power companies not only need to repair damaged wires and poles after treefall, but need to clear corridors of trees to prevent potential damage to lines. The highly invasive vine kudzu (*Pueraria montana* var. *lobata*) can smother poles and lines, increasing fire hazards and weighing down lines; it is estimated that kudzu costs power companies $1.5 million per year (Blaustein 2001).

Occasionally, energy systems play a role in the conveyance of non-native species, for example, utility corridors may function as a focal point for invasion of adjacent natural ecosystems by non-native plant species (Zink et al. 1995). Activities associated with oil and gas development can facilitate the spread of invasive species through large-scale land clearing, movement of people and equipment, and transport of soil. Development activities promote the transfer of people and equipment from distant locations and thus facilitate the transport of invasive species to and from sites. Recently disturbed land lends itself to easy movement of invasive animals such as feral swine and rats (IOGP et al. 2010) and establishment of invasive plants that thrive in the disturbed soils and shade-free environment. Much like fire breaks, these corridors can promote the spread of invasive species into wildland areas (Keeley 2006).

In aquatic settings, hydropower dams used for energy generation cause significant changes to free-flowing rivers. Reservoirs created by dams and changes in river gradient due to altered flow patterns may alter thermal regimes and flow pathways (Poff et al. 1997). Additionally, reservoirs become important destinations for anglers, boaters, and others seeking recreation. Because boats are a primary vector in the unintentional introduction of aquatic invasive species, reservoirs may be among the initial areas contaminated in any particular region (Rothlisberger et al. 2010). Lastly, intentional introduction of invasive sport fishes may alter trophic systems and increase pressure on native fishes that are also contending with significant habitat changes associated with hydropower dams.

9.6.2 Management Options for Energy Systems

Management of invasive species that affect energy systems varies in terrestrial and aquatic ecosystems and includes a range of treatment options described in greater detail in Chap. 7. Invasive plants are often controlled in energy corridors through mowing regimes, removal of vines, and use of herbicides.

In areas impacted by zebra mussels, hydropower companies maintain water intakes by clearing mussels from intake pipes. Anti-fouling paint and hypochlorite systems can be used to inhibit further colonization. In the Great Lakes region, hydropower companies spend millions of dollars annually to mitigate the mussel problem and deliver lake water to power generation facilities (Rosaen et al. 2012). The Bonneville Power Administration estimated that it would cost approximately $1.8 million to install hypochlorite systems and apply anti-fouling paint to 26 hydropower dams in Idaho to address the zebra mussel problem (Phillips et al. 2005).

There has been some discussion of using the harvested biomass of some of the established fast-growing invasive plant species as feedstock for biofuels (Jakubowski et al. 2010; Young et al. 2011). However, numerous challenges remain including the limited number of biorefineries; the costs of biomass harvest, transportation, and storage; environmental damage during harvest; and the potential to further spread invasive species during transport (Quinn et al. 2014).

9.7 Invasive Species and Transportation Systems

U.S. transportation systems include roads, railways, airways, and water transport. They not only crisscross the Nation but connect all private and public lands. As such, they can be important corridors of invasive species. Invasive species can impact transportation infrastructure, the safety of travelers, and property values. Costs associated with control of invasive species are extensive but are not fully tracked. For example, States do not itemize the cost to control invasive plants in roadside construction or maintenance.

This section covers road and rail transportation systems but does not cover air systems. Impacts associated with water transportation are addressed in Sects. 9.5 and 9.6.

9.7.1 Road and Railways as Conduits of Invasive Species Spread

Roadsides are prone to the invasion of non-native plant (Gelbard and Belnap 2003; Mortensen et al. 2009; Rentch et al. 2005) and animal (Hulme 2009; King et al. 2009) species as roads and road maintenance are vectors for their dispersion. For example, spread rates of Japanese stiltgrass (*Microstegium vimineum*) are notably higher along roads than in forests and wetlands (Mortensen et al. 2009). Fire ants have been spread over long distances and on different roads because the soil used for roadside fills came from the same ant-infested soil depot (King et al. 2009; Stiles and Jones 1998). Invasive cane toads (*Rhinella marina*) in Australia travel more quickly using roads than the adjacent vegetated sites (Brown et al. 2006), a particular concern as cane toads have also been introduced to Hawaii and the Caribbean. Several studies also show that the likelihood of an invasive plant being present in abundance increases significantly with proximity to a road (Flory and Clay 2009; Huebner and Tobin 2006). Roads provide a constant source of disturbance in the form of open, bare areas with high light levels that make them susceptible to rapid colonization by opportunistic species, including invasive species (Parendes and Jones 2000).

Invasions are particularly common in roads or railways that are paved, have a wide verge, or are in open areas as opposed to forested areas (Hanson and Clevenger 2005; Joly et al. 2011; Kalwij et al. 2008; Neher et al. 2013). Paved roads tend to have alkaline soils and are also characterized by higher metal concentrations, conditions that may be better tolerated by some invasive plants than native plants (Barbosa et al. 2010; Neher et al. 2013). Off-road vehicle trails are also an important conduit for invasive species into more remote forested sites (Rooney 2005). Some invasive plants, such as the common reed, can be found on older roadsides that have a history of exposure to de-icing salts, ditch digging, and nitrogen input from nearby agricultural fields (Jodoin et al. 2008). Invasive earthworms may also be found on older roads with reduced forest cover and on agricultural soils (Cameron and Bayne 2009; Hulme 2009).

During the process of road-building or upgrading a highway, top soils are scraped, stockpiled, and moved throughout the construction site. When the soils are moved, existing invasive seed and vegetative parts as well as invasive insects, earthworms, and pathogens may also move. Sand and gravel materials imported for building the road may also contain invasive species (Godefroid and Koedam 2004; Johnston and Johnston 2004). Top soils or mulch imported for special plantings may not be screened or treated before transport, and thus may harbor additional invasive propagules (Kruse et al. 2004).

In some areas, invasive species such as sweetclover (*Melilotus officinalis*), smooth brome (*Bromus inermis*), perennial rye (*Lolium perenne*), bird's-foot trefoil (*Lotus corniculatus*), crownvetch (*Securigera varia*), and reed canary grass (*Phalaris arundinacea*) have been purposefully introduced post-construction to control erosion because many native warm season or perennial plant species are slow to establish. However, some research indicates that highly effective seed mixes could be provided for these corridors if greater efforts were directed at selecting the right native seed mixes (Seastedt et al. 2008; Tinsley et al. 2006).

Vehicles can carry invasive seed and produce air turbulence that spreads seeds (Ansong and Pickering 2013; Taylor et al. 2012; Von der Lippe and Kowarik 2007). A review of 13 studies conducted across the globe involving seed dispersal associated with cars concluded that 626 species of seed were collected on vehicles (Ansong and Pickering 2013).

9.7.2 Impacts of Invasive Species on Roads and Railways

Once established on roadsides, invasive plant species have been reported to reduce visibility and block roads and railways because of falling trees or accumulation of leaf litter. The United Kingdom has estimated such interference has cost at least £30 million ($39 million) annually (Williams et al. 2010). Existing or future colonization of non-native invasive plants can reduce the biodiversity of flora and the presence of associated butterflies and moths in roadside verges (Valtonen et al. 2006). Nutria (*Myocastor coypus*), a large invasive semi-aquatic rodent, can burrow into flood levees, undermining their structural integrity and leaving low-lying road and railways vulnerable to water damage (Leblanc 1994).

Invasive species can also enhance natural disturbances that can damage transportation infrastructure. For instance, cheatgrass increases the frequency of wildfires (Whisenant 1989), which may damage roads and railway systems by warping roadways and tracks, producing toxic gas, and destroying railway ties (Bonati et al. 2012; Rutz et al. 2014; Schartel et al. 2010).

9.7.3 Management Options for Roads and Railways

Management options are available to limit the introduction, establishment, and spread of invasive plant species along roadways. Within the National Highway System, each State is responsible after initial construction or upgrade for maintenance of all interstate, State, county, and local roads. States are eligible for Federal financing to prevent and control noxious weeds (invasive plants that cross the threshold of harm to agriculture, environment, or human health) on rights-of-way or roadsides; however, few States have fully used funding for weed control because State maintenance departments must compete against funding for pavement and bridges within their State's budgets.

Control of invasive plants prior to construction or upgrades can limit regrowth of invasive plants on the site later. Although the Federal Highway Administration recommends that all States inventory their roadside vegetation (including existing species, planted species, and invasive plants) and has developed a GPS-inventory protocol, few States keep good baseline records. As a result, States may not be aware of the presence of invasive plants at a given site or if their control efforts were successful on that site.

To limit the transfer of invasive weeds and seed, mulches applied for erosion control and sand and gravel pits used as sources for materials can be certified as weed-feed. In addition, roadside crews can use seed mixes for erosion control and roadside landscaping that are free of invasive species. The use of native grassland species requires less mowing and less use of chemicals to control weeds. Plantings of native grasses may actually limit the invasion of new invasive species (Pokorny et al. 2005; Sheley and James 2010). The Federal Highway Administration has also supported the development of a native erosion control sod to fulfill the needs of National Pollutant Discharge Elimination System erosion control permits.

Spread of invasive plants can be controlled after roadside maintenance. Portable washing machines are available so crews can wash equipment on site after mowing and construction operations and before equipment is moved offsite (Fleming 2008). If timed correctly, mowing can be an effective control method for some invasive plants. For example, if Japanese stiltgrass is mowed low right after it flowers in late August and early September, but before it fruits in late September or early October, this annual will fail to reproduce seed that year (Shelton 2012). Unfortunately, roadsides are often inhabited by a wide variety of species with different flowering and fruiting times. In addition, with many roads to mow, coordinating the timing of mowing may not be a realistic option. Attempts at reducing the spread of invasive species along roads are also most effective if all landowners along the corridor coordinate their efforts. However, this can be a challenge (Epanchin-Niell et al. 2010).

By spot spraying sites for invasive species rather than blanket spraying herbicides, or by using biocontrol alternatives, crews can limit disturbance that might allow for the colonization of invasive plants. Site disturbance can also be reduced by limiting grazing on difficult-to-access steep slopes and wetlands. The use of yearly prescribed fires on road shoulders may be effective to control invasive plants in some grassland areas with low traffic. By alternating herbicides, managers may limit increases in chemical resistance in weeds.

Knowledge of best management practices varies among States and even among State maintenance districts, and therefore training may be useful. Botanical expertise within State Departments of Transportation (DOT) can help states identify site-specific native seed mixes preferred for use in projects. Roadside managers with sufficient expertise and capacity can become frontline reporters of new invasive plants. In Minnesota, a DOT maintenance worker identified and reported oriental bittersweet (*Celastrus orbiculatus*) on a roadside, providing Minnesota DOT the opportunity to mobilize a response to limit its spread. GPS inventory methods for examining transportation corridors can help minimize the costs associated with the inventory of invasive species.

State partnerships with public and private lands across the right-of-way fence line can also be established to minimize invasions. For example, in 2009 the Minnesota DOT entered into a Memorandum of Understanding with the Fond du Lac Band of Lake Superior Chippewa in order to address concerns over use of herbicides and tree encroachment. Both parties agreed on the need to control invasive species.

9.8 Invasive Species and Forest and Grassland Products

As invasive species have become established in many U.S. forests and rangelands, they may interact with introduced livestock as well as the forest species we have come to rely on for timber and non-timber forest products. They can impact the health, productivity, and survival of these provisioning species and the management expenses associated with the production of food, feed, fuel, and fiber (DiTomaso 2000; French 2017; Kenis et al. 2017).

Invasive species are of great concern in most agricultural production; however, production systems such as croplands, orchards, and enclosed contained commercial animal operations are beyond the scope of this assessment. However, for cattle grazing on pastures or rangelands in Texas, the invasive cattle ticks, *Rhipicephalus annulatus* and *R. microplus*, carry the parasite *Boophilus microplus*, which causes tick fever (bovine babesiosis). Millions of dollars have been spent to keep the ticks from entering the United States from Mexico, and quarantine zones have been implemented along the border of Texas. In 2010, Texas A&M University estimated the costs of treatment of cattle ticks at $122,983,110 (Anderson et al. 2010). The ticks can also use native and imported ungulates as hosts, thus potentially spreading throughout native rangelands. White-tailed deer (*Odocoileus virginianus*) have been the most important alternative hosts. A January 2017 survey has shown the ticks are again advancing and extending their range throughout Texas (Texas Animal Health Commission 2017).

Invasive plant species can also decrease the quantity and quality of forage for livestock (DiTomaso 2000). Some have low palatability or may be poisonous, or may change soil or hydrologic regimes. Knapweeds and star-thistle (*Centaurea* spp.) and leafy spurge (*Euphorbia esula*) may reduce grazing capacity by as much as 50% (Olsen 1999). In California, the occurrence of yellow star-thistle on private land is estimated to cost cattle ranchers $7.65 million dollars annually in lost forage (Eagle et al. 2007). Leafy spurge in Montana, North Dakota, South Dakota, and Wyoming has been estimated to decrease grazing capacity by approximately 90,000 cattle (Leistritz et al. 2004).

Alteration of fire regimes by invasive annual grasses such as cheatgrass, medusahead (*Taeniatherum caput-medusae*), and ventenata (*Ventenata dubia*) is a particular threat to western rangeland products. Invasion by grass species following fire can promote strong feedbacks, a process frequently referred to as the "grass-fire cycle" (Chap. 2, Box 2.1, this volume). It is widely recognized that there is an emergent risk of rangeland degradation and reduction in forage because of invasive-dominated grasslands that burn readily (Chambers et al. 2014). New approaches, such as using late-season cattle grazing, are now being evaluated to reduce cheatgrass and other annual invasives (Schmelzer et al. 2014). Due to the poor forage quality, ranchers must provide protein supplements to cattle grazing on these grasses (Schmelzer et al. 2014).

Since invasive species can impact forest species composition and productivity, and the capacity of native species to regenerate after disturbance, they can impact the supply of both timber and non-timber forest products. Invasive invertebrates, pathogens, and plants have a direct impact on forest products in the United States, although there is limited information available on the extent of economic damages to the forest products industry (Kenis et al. 2017).

Emerald ash borer, which affects all species of ash trees in the Eastern United States (Herms and McCullough 2014), is estimated to cause $60 million in timber value loss (Aukema et al. 2011). White ash is the most commercially important of the ash species; white ash logs are processed into diverse products including tool handles, baseball bats, railroad ties, flooring, cabinets, furniture, and cargo crates (MacFarlane and Meyer 2005). Additionally, mortality of black ash (*Fraxinus nigra*) has been of particular concern to Native Americans and other artisans who use black ash for basket-making (Diamond and Emery 2011).

The ability of Asian longhorned beetle to attack multiple genera of apparently healthy hardwood trees is estimated to cause $41 billion in potential losses to forest products, commercial fruit, maple syrup, nursery, and tourist industries (Fleming et al. 2002; Meng et al. 2015; USDA APHIS PPQ 2007). This woodborer is of great concern in North America because of the economic importance of maple (*Acer* spp.), one of its primary host genera. Maple is a prominent component of the northern hardwood forests that range from south-eastern Canada to central New England west to the Great Lakes region, and is valued for its colorful fall foliage and maple syrup products (Dodds and Orwig 2011).

Gypsy moth larvae cause damage to broadleaved trees such as oak (*Quercus* spp.) and aspen (*Populus* spp.) by feeding on foliage. Defoliation has resulted in declines in the radial increment of oak trees (Muzika and Liebold 1999), and successive years of defoliation may cause extensive tree mortality. Aukema et al. (2011) estimate annual timber losses due to gyspy moth at $4.6 million.

The walnut twig beetle, a species native to the Southwestern United States but invasive in the East, is vector to a fungus (*Geosmithia mobida*) that causes thousand cankers disease in walnut trees (*Juglans* spp.) (Utley et al. 2013). Black walnut (*J. nigra*) is prized for its resistance to decay, and its uses in wood products include furniture, specially items, gunstocks, and paneling (Voulgaridis and Vassiliou 2005). Black walnuts are valued as food and medicine, and are of commercial value. Hammons Products estimates that it harvests over 25 million pounds of black walnuts, mostly from wild trees (Chamberlain et al. 1998).

The invasive fungi *Ceratocystis fimbriata* and *Fusarium oxysporum* have been of particular concern in Hawaii. On the Big Island, *C. fimbriata* has damaged native ʻōhiʻa (*Metrosideros polymorpha*), which is not only a culturally important endemic tree, but also important for construction and specialty wood products, and whose flowers are used to make lei. Leaves of infected trees turn yellow and brown, and tree mortality may occur within a few weeks; 100% stand mortality can occur in 2–3 years (Friday et al. 2015). *F. oxysporum* has been responsible for wilt disease in koa trees (*Acacia koa*) (Gardner 1980), which has made regeneration of koa a challenge. Koa is one of Hawaii's largest native trees, valued for its wood, which is used to make furniture, veneer, and crafts (Skolmen 1974). The wood is one of the most expensive in the world, and due to its tonal properties is used to make instruments such as the ukulele and acoustic guitar (Shafto and McDaniel 2009); dyes are also made from its bark (Krauss 1974; Krohn-Ching 2016).

In some areas, invasive plants may outcompete native tree species, making it difficult or more costly for landowners to sustain healthy forests that ultimately can be harvested for forested products. Kudzu is a significant challenge for forest managers because it grows rapidly, and can wind itself around trees, shading native foliage and inhibiting photosynthesis (Forseth and Innis 2004). Affected trees may die within 2–3 years (Britton et al. 2002; Mitich 2000). Lost forest productivity due to kudzu is estimated at $100–500 million per year (Forseth and Innis 2004). In Hawaii, invasive grass-fire complexes have replaced much native dry forest (Bruegmann 1996). These forests are a source of many forest products including those derived from koa and ʻōhiʻa .

9.8.1 Management Options to Limit Damage to Forest and Grassland Products

Management options to limit the introduction, establishment, and spread of invasive species and to restore impacted ecosystems have been addressed in Chaps. 6, 7, and 8 and, consequently, are only briefly addressed in this section.

On rangelands, management of invasive plants can include a range of tactics including mechanical control, management of grazing, prescribed fire, revegetation programs, biological and chemical control, and integrated approaches (DiTomaso 2000). Mechanical control may include tilling, mowing, hand-pulling, and use of heavy machinery for invasive woody plants. Shifts in grazing strategies may be used to change the intensity and type of foraging and disturbance to soils (Olsen 1999). Prescribed fire can be effective but must be carefully timed prior to when seed production of invasive species occurs, in order to prevent their further spread. Timing of herbicide applications is also critical and may vary depending on the herbicide used (DiTomaso 2000).

For invasive insects such as emerald ash borer, goldspotted oak borer (*Agrilus auroguttatus*), and Asian longhorned beetle, significant effort has been placed on limiting transport of host material, including firewood, logs, and nursery stock (Haack 2006; Haack et al. 2010a; Koch et al. 2012a; Seybold et al. 2010). In 2009, USDA APHIS formed the National Firewood Task Force to develop a coordinated response to address outreach and voluntary and regulatory aspects of firewood movement at a national level (USDA APHIS PPQ 2010). Quarantines have been used to limit the spread of several invasive invertebrates. Several States have restricted the movement of walnut from counties with thousand cankers disease (USDA Forest Service et al. 2011). The Asian longhorned beetle has been under quarantine by State and Federal officials since its initial detection in the United States in 1996; eradication programs have been put in place in detected areas (Haack et al. 1997; Haack et al. 2010b; Lopez et al. 2017). Heating and fumigation may allow for the use and marketing of some infected wood (Audley et al. 2015; Wang 2012).

The spread of gypsy moths has been slowed by using pheromone traps to monitor low-level populations and then aerially applying pheromone flakes to disrupt mating (Sharov et al. 2002). To suppress or eradicate populations of gypsy moth, managers have often used aerial applications of *Bacillus thuringiensis* var. *kurstaki*, a bacterial pesticide (Kauffman et al. 2017). A naturally occurring nucleopolyhedrosis virus (NPV) is also being used in aerial and ground applications to suppress gypsy moth populations. The NPV product Gypchek was first registered as a general use pesticide with the U.S. Environmental Protection Agency in 1978 and, until recently, was produced through a contract with the USDA Forest Service (Reardon et al. 2012).

Research has also focused on developing more resilient forest stock. For example, efforts are underway to breed EAB-resistant ash. This work has involved examining the genome of Asian ash species that are less prone to infection, as well as the genome of isolated native ash populations which have survived in infected areas (Knight et al. 2012; Koch et al. 2012b; Whitehill et al. 2012). Scientists in Hawaii have worked to identify wilt-resistant koa seed for restoration and reforestation. With technical assistance from the USDA Forest Service, the Hawaii Agriculture Research Center has identified resistant parent trees, established seed orchards, and identified provisional seed zone guidelines (Dudley et al. 2017).

9.9 Invasive Species and Indigenous Peoples

Native cultures and economies are grounded in land and natural resources (Pretty 2002). Any species that adversely impacts the water, land, plants, and animals on which indigenous peoples rely has negative consequences for these communities and cultures. Native peoples also have millennia of experience adapting to social and ecological change, which includes non-native species (Crosby 1986). Understanding indigenous experiences and traditional ecological knowledge is essential to developing invasive species management programs that honor the U.S. trust responsibility to Native peoples, and also is of benefit to others.

As of 2016, there were 566 federally recognized American Indian and Alaska Native tribes or communities in the United States, and an additional 60 tribes have been recognized through formal processes of U.S. State governments (National Conference of State Legislatures 2017). Tribes and Native communities hold over two million acres of land (National Congress of American Indians 2015) and have retained rights to hunt, fish, and gather on millions of additional acres. Treaty and case law require the Federal government to work with tribes and Native communities as sovereign nations for the protection of their lands and the resources on which they depend (National Congress of American Indians 2015). This doctrine extends to identifying, monitoring, and managing invasive species. Tribal cultural perspectives are covered extensively in Chap. 12.

9.9.1 Impacts of Invasive Species on Indigenous Peoples

Tribal natural resource professionals indicate that invasive species are having adverse impacts on the lands and communities they serve (for in-depth discussion see Chap. 12). Native perspectives on these species and what should be done about them are informed by recent and historical experience. In cases where recently arrived biota result in the rapid reorganization of landscapes, and/or replacement of culturally important species in two human generations or less, invasive species may be culturally impoverishing (Pfeiffer and Voeks 2008), with effects that include loss of access to cultural keystone species for food, medicinal, ceremonial, and other purposes (Garibaldi and Turner 2004) and interruptions of place-based traditions that literally ground indigenous identity (Pretty 2002). In other cases, invasives have enriched Native peoples' diets and pharmacopoeias, as exemplified by the weedy northern European species, English plantain (*Plantago major* L.) (Crosby 1986).

Indigenous communities are mindful of the social and political histories of many biological invasions. For example, the Dawes Act of 1887 resulted in privatization and ownership of land by non-Indians inside the boundaries of the Crow Indian Reservation, setting the stage for the contemporary invasion of Russian olive (*Elaeagnus angustifolia* L.) (Pretty Paint-Small 2013). Further, indigenous communities have suffered negative impacts from some programs designed to address invasive species, including use of toxic chemicals and biocontrol agents (Norgaard 2007; Pfeiffer and Voeks 2008).

The introduction of new species as a result of climate variability and change (see Chap. 4) increases demands on overcommitted Native and tribal natural resources departments, and creates both opportunities and challenges for collaboration with non-Native institutions. Each new invasive species identified will create a need for these staffs (with moderate resources) to work with community members, tribal government, and external land management agencies, in addition to their normal activities on the land. While there will be opportunities for tribes and Native communities to form learning partnerships with non-Native institutions, in the absence of established relationships of mutual respect and protocols for collaboration, the incidence of contested responses to invasive species may increase.

The incidence of rapid reorganization of landscapes, and replacement of culturally important native species, likely will increase with changing climate. As noted above, indigenous peoples have extensive experience adapting to new circumstances, but the pace of change can outstrip the capacity to adapt successfully, without significant cultural and economic disruption. In some cases, the consequence may be wide ranging for the health and well-being of Native and tribal communities, including loss of access to foods, medicines, crafts, and ceremonial materials, as well as the means for pursuing culturally appropriate sources of income.

9.9.2 Management Options

Future responses of indigenous communities to invasive species will reflect their cultures and economies. A non-native species is not necessarily regarded as undesirable to Native cultures. This judgment may hinge on whether a species is seen to compromise social, cultural, and economic well-being. For example, the Eastern Band of Cherokee Indians defines an invasive species as one that causes net harm to the tribe's economic or cultural resources. Where a species is determined to be harmful, some commonly used control and eradication methods may be unacceptable, and consequently indigenous communities may object to the invasive species management programs used or recommended by other land management agencies. One such case is the response of the

Kashia Band of Pomo Indians to sudden oak death due to the pathogen *Phytophthora ramorum*. Although sudden oak death threatens a cultural keystone species (chishkale, or tanoak (*Notholithocarpus densiflorus*)), its primary inoculum source also is a culturally important species (pepperwood (*Umbellularia californica*)). Removal of pepperwood is one approach to managing sudden oak death. The Kashia decline to choose one culturally important tree species over another (Alexander et al. 2017).

Human health implications of pesticides and herbicides are paramount concerns for indigenous peoples, who may rely particularly heavily on food, medicine, and other materials harvested directly from the environment by community members. The Shoalwater Bay Tribe's experience with the tidal zone invasive plant spartina (*Spartina alterniflora* Loisel.) offers an example. Initially, the tribe declined to issue permits for herbicide applications to remove spartina on reservation land. When tribal members became increasingly aware of spartina's negative impacts on fishing and the State of Washington adopted an alternative chemical for its eradication, the Shoalwater Tribe partnered with the State's Spartina Technical Working Group. The tribe ultimately consented to targeted applications of the alternative herbicide on its lands, conducted by its own Natural Resources Department.

9.10 Key Findings and Gaps in Information and Technology

9.10.1 Key Findings

- Global travel and trade, international movement of livestock, translocation of wildlife and pets, modern agricultural practices, and changing climate patterns have led to the introduction of invasive vertebrate and invertebrate animals, plants, and microbial species, and the emergence of novel infectious diseases across North America and other continents.
- Expansion of the geographic range of invasive species and introduced diseases arises from changes in the interrelationships between the native host, invasive pathogen, and the local environment or ecosystem. Intervention in one or more of these parameters can disrupt or prevent the disease cycle.
- Invasive species pose numerous site-specific challenges to the U.S. military, many of which significantly impact mission critical activities. DoD implements a wide array of approaches to combat the spread of, and threats from, invasive species, often with the support of partners (public and private).
- Invasive species can damage infrastructure in urban and rural areas (e.g., buildings, utility lines, electrical systems), and pose a risk to human health and safety. Invasive species may increase the risk of wildfire in the wildland-urban interface and number of hazardous trees in a community.
- Proper maintenance of urban plant communities at residential to city-wide scales may deter the spread of existing invaders. Creating green corridors can help reduce physical properties conducive to invasive species. Increasing the quantity of interconnected green infrastructure in urban areas may eventually make it easier to re-introduce native species.
- Invasive species can affect the amount of water available to communities, as well as the quality of untreated water. Control of invasive mussels is a serious maintenance problem for boat owners, and for power plants and other facilities that have water intake.
- Roads and railways are conduits for invasive species. Treatment of invaded areas, following nationwide policies, can help slow the spread of invasive species and limit invasion of nearby natural areas as well as rural communities.
- If not properly managed, invasive species can diminish the value of grazing lands, decrease the productivity of timberlands, and reduce production of non-timber forest products. Lost value and expenditures for management present major economic challenges for ranchers and forest managers.
- Invasive species have profound impacts on indigenous cultures and economies, compromising food supplies, traditional health systems, and spiritual practices at the heart of indigenous identity. Valuable insights for controlling invasive species may be acquired from Tribes' and Native communities' traditional ecological knowledge and land management practices (see Chap. 12).
- Better coordination of invasive species activities in all sectors of the United States may yield novel approaches, gains in efficiency, and increased capacity to respond to future invasions.

9.10.2 Key Information and Technology Needs

- Better understand the ecology, disease transmission, pathogenesis, and population biology of invasive species that impact human health, and the environmental variables that affect their distribution and spread.
- Develop new technologies and methods for early detection and monitoring of invasive insects and pathogens in forests, rangelands, urban areas, and transport systems, and enhance the availability of real-time data.
- Develop new technologies and methods for reducing the spread of invasive species and restoring native species along the urban-rural-natural area interface.

- Identify the trade-offs associated with controlling invasive species (e.g., costs of control versus eradication, use of pesticides, damage to native species, and costs of not controlling the invasion).
- Better understand the interaction of climate change and invasive species, and associated sectoral impacts.
- Conduct research on genetic resilience to invasive pests and pathogens.
- Develop additional regional and species-specific information on the economic costs of invasive species (due to damage, management costs, and lost revenue) on infrastructure, military resources, forest and grassland products, urban vegetation, recreation, and tribal resources.
- Develop economic estimates of the costs of invasive species that factor in valuation of ecosystem services lost (e.g., pollination, soil carbon storage) with the decline or disappearance of native species.
- Identify how to best prioritize roads for treatment of invasive plants and restoration of native species.
- Develop a better understanding of gaps and barriers to interagency collaboration to prevent importation of high-risk animals into the United States, and to oversee imported animals.
- Better understand how economic incentives might be used to encourage planting of native species in urban environments with intent to limit the establishment and spread of invasive species.
- Identify the effects of indigenous land management practices on invasive species.
- Identify best practices for socio-cultural adaptation planning and management in order to address the impacts of invasive species in Native communities.

Acknowledgments The authors would like to thank those who provided input and reviewed this chapter. Special thanks to Derrick Golla, Valerie Leone, Kimberly Alles, and Booz Allen Hamilton for their document coordination and edits to the section on military readiness.

Disclaimer Text The findings and conclusions in this publication are those of the authors and should not be construed to represent any official USDA or U.S. Government determination or policy.

Literature Cited

Alexander JM, Frankel SJ, Hapner N et al (2017) Working across cultures to protect Native American natural and cultural resources from invasive species in California. J For 115:473–479

Alston KP, Richardson DM (2006) The roles of habitat features, disturbance, and distance from putative source populations in structuring alien plant invasions at the urban/wildland interface on the cape peninsula, South Africa. Biol Conserv 132:183–198

American Sportfishing Association (ASA) (2015) Economic contributions of recreational fishing: U.S. Congressional Districts. American Sportfishing Association. Southwick Associates. http://asafishing.org/uploads/Congressional_District_Sportfishing_Impacts_2015_Report.pdf

Anderson DP, Hagerman AD, Teel PD et al (2010) Economic impact of expanded fever tick range. Agricultural and Food Policy Center, Texas A&M University System, College Station, 28 p

Ansong M, Pickering C (2013) Are weeds hitchhiking a ride on your car? A systematic review of seed dispersal on cars. PLoS One 8:e80275. https://doi.org/10.1371/journal.pone.0080275

Aronson MF, La Sorte FA, Nilon CH et al (2014) A global analysis of the impacts of urbanization on bird and plant diversity reveals key anthropogenic drivers. Proc R Soc B 281:20133330. https://doi.org/10.1098/rspb.2013.3330

Audley J, Taylor A, Klingman WE et al (2015) Insecticide dip treatments to prevent walnut twig beetle colonization of black walnut logs. For Prod J 66(34):235–240

Aukema JE, Leung B, Kovacs K et al (2011) Economic impacts of non-native forest insects in the continental United States. PLoS One 9:e24587

Ayron M, Bruland GL, MacKenzie RA, Giardina CP (2016) Soil and hydrological responses to wild pig (*Sus scofa*) exclusion from native and strawberry guava (*Psidium cattleianum*)-invaded tropical montane wet forests. Geoderma 279:53–60

Bagstad KJ, Wiederholt R (2013) Tourism values for Mexican free-tailed bat viewing. Hum Dimens Wildl Int J 18:307–311

Balch JK, Bradley BA, D'Antonio CM, Gomez-Dans J (2013) Introduced annual grass increases regional fire activity across the arid western USA (1980–2009). Glob Chang Biol 19:173–183

Barberi MM, Kashinsky L, Littnan CL (2016) Protozoal-related mortalities in endangered Hawaiian monk seals (*Neomonachus schauinslandi*). Dis Aquat Org 121:85–95

Barbosa NPU, Fernandes GW, Carneiro MAA, Junior LAC (2010) Distribution of non-native invasive species and soil properties in proximity to paved roads and unpaved roads in a quartzitic mountainous grassland of southeastern Brazil (rupestrian fields). Biol Invasions 12:3745–3755

Beard CB, Eisen RJ (2016) Chapter 5 vector-borne diseases. In: The impacts of climate change on human health in the United States: a scientific assessment. U.S. Global Change Research Program, Washington, DC, pp 129–156

Bell GP (1997) Ecology and management of *Arundo donax*, and approaches to riparian habitat restoration in southern California. In: Brock JH, Wade M, Pyšek P, Green D (eds) Plant invasions. Studies from North America and Europe. Backhuys Publishers, Leiden, pp 103–113

Benson AJ (2017) The exotic Zebra Mussel. US Fish and Wildlife Service. https://www.fws.gov/midwest/endangered/clams/zebra.html

Bisognano JD, McGrody KS, Spence AM (2005) Myocarditis from the Chinese sumac tree. Ann Intern Med 143(2):159

Blaustein R (2001) Kudzu's invasion into southern United States life and culture. In: The great reschuffling: human dimensions of invasive species. IUCN, Gland, pp 55–62

Blosser EM, Burkett-Cadena ND (2017) Culex (Melanoconion) panocossa from peninsular Florida, USA. Acta Trop 157:59–63

Bonati A, Merusi F, Polacco G et al (2012) Ignitability and thermal stability of asphalt binders and mastics for flexible pavements in highway tunnels. Constr Build Mater 37:660–668

Booy O, Cornwell L, Parrott D et al (2017) Impact of biological invasions on infrastructure. In: Vilà M, Hulme PE (eds) Impact of biological invasions on ecosystem services. Springer International Publishing, Cham, pp 240–241

Botkin DB, Beveridge CE (1997) Cities as environments. Urban Ecosyst 1:3–19

Bower AD, St. Clair JB, Erickson V (2014) Generalized provisional seed zones for native plants. Ecol Appl 24:913–919

Boyles JG, Cryan PM, McCracken GF, Kunz TH (2011) Economic importance of bats in agriculture. Science 332(6025):41–42

Britton KO, Orr D, Sun J (2002) Chapter 25, Kudzu. In: Van Driesche R, Blossey B, Hoddle M et al (eds) Biological control of invasive plants in the eastern United States. USDA Forest Service Publication FHTET-2002-04, Morgantown

Brown GP, Phillips BL, Webb JK, Shine R (2006) Toad on the road: use of roads as dispersal corridors by cane toads (*Bufo marinus*) at an invasion front in tropical Australia. Biol Conserv 133:88–94

Bruegmann MM (1996) Hawaii's dry forests. Endanger Species Bull 11:26–27

Bureau of Planning and Sustainability (2016) Portland plant list. https://www.portlandoregon.gov/citycode/article/322280

Butler N, Carlisle JC, Linville R, Washburn B (2009) Microcystins: a brief overview of their toxicity and effects, with special reference to fish, wildlife, and livestock. California Environmental Protection Agency, Office of Environmental Health Hazard Assessment, 17 p. https://oehha.ca.gov/media/downloads/ecotoxicology/document/microcystin031209.pdf

California Department of Finance (2017) http://www.dof.ca.gov/Forecasting/Demographics/Projections/

Cambray JA (2003) Impact on indigenous species biodiversity caused by the globalization of alien recreational freshwater fisheries. Hydrobiologia 500:217–230

Cameron EK, Bayne EM (2009) Road age and its importance in earthworm invasion of northern boreal forests. J Appl Ecol 46:28–36

Cannon WF, Horton JD (2009) Soil geochemical signature of urbanization and industrialization – Chicago, Illinois, USA. Appl Geochem 24:1590–1601

Catford JA (2017) Hydrological impacts of biological invasions. In: Vilà M, Hulme PE (eds) Impact of biological invasions on ecosystem services. Springer International Publishing, Switzerland, Cham, pp 63–80

Chamberlain J, Bush R, Hammett AL (1998) Non-timber forest products: the other forest products. For Prod J 48(10):10–19

Chambers JC, Bradley BA, Brown CS et al (2014) Resilience to stress and disturbance, and resistance to *Bromus tectorum* L. invasion in cold desert shrublands of western North America. Ecosystems 17:360–375

Clements S, Moore G (2003) Patterns of species richness in eight Northeastern United States cities. Urban Habitats 1(1): ISSN 1541-7115. Http://www.urbanhabitats.org

Coates PS, Ricca MA, Prochazka BG et al (2016) Wildfire, climate, and invasive grass interactions negatively impact an indicator species by reshaping sagebrush ecosystems. Proc Natl Acad Sci 113:12745–12750

Cohn JP (2005) Tiff over tamarisk: can a nuisance be nice, too? Bioscience 55:648–654

Collins MD, Vazquez DP, Sanders NJ (2002) Species area curves, homogenization and the loss of global diversity. Evol Ecol Res 4:457–464

Connelly NA, O'Neill CR Jr, Knuth BA, Brown TL (2007) Economic impacts of zebra mussels on drinking water treatment and electric power generation facilities. Environ Manag 40:105–112

Crosby AW (1986) Ecological imperialism: the biological expansion of Europe, 900–1900. Cambridge University Press, Cambridge

Dalsimer A (2002) Why DOD is interested in invasive species. Fed Facil Environ J 13:41–54

Darwin C (1859) The origin of species by means of natural selection. Murray, London

Daszak P, Cunningham AA, Hyatt AD (2000) Emerging infections diseases of wildlife – threats to biodiversity and human heath. Science 287:443–449

Dearborn DC, Kark S (2009) Motivations for conserving urban biodiversity. Conserv Biol 24(2):432–440

Defense Transportation Regulation (2018a) Department of Defense customs and border clearance policies and procedures. Chapter 505: agricultural cleaning and inspection requirements

Defense Transportation Regulation (2018b) Department of Defense customs and border clearance policies and procedures. Chapter 506 DoD Preclearance program customs and agriculture inspections

Department of Defense (DoD) (2011) Commander's guide on invasive species. http://www.dodnaturalresources.net/files/Commanders_Guide_on_Invasive_Species_10.11.pdf

Department of Defense (DoD) (2017) Department of Defense: invasive species challenges and solutions. http://www.dodinvasives.org/Invasives_and_Military_Feb2017.pdf

Department of Defense (DoD) (2018) Department of Defense Instruction Number 4715.03. Natural resource conservation program. https://www.esd.whs.mil/Portals/54/Documents/DD/issuances/dodi/471503p.pdf?ver=2018-11-13-125207-877

Department of Defense Armed Forces Pest Management Board (DoD AFPMB) (2016) DoD guidance for the surveillance, control and testing of *Ae. aegytpti*, *Ae. albopictus* and *Ae. polynesiensis*. http://www.acq.osd.mil/eie/afpmb/docs/zika/Strategy_Control_Zika_Virus.pdf

Department of Defense Armed Forces Pest Management Board (DoD AFPMB) (2017) Technical guide No. 31, guide for agricultural preparation of military gear and equipment for redeployment. https://www.acq.osd.mil/eie/afpmb/docs/techguides/tg31.pdf

Des Rosiers F, Theriault M, Kestens Y, Villeneuve P (2007) Landscaping attributes and property buyers' profiles: their joint effect on house prices. Hous Stud 22:945–964

Diamond AK, Emery MR (2011) Black ash (*Fraxinus nigra* Marsh.): local ecological knowledge of site characteristics and morphology associated with basket–grade specimens in New England (USA) 1. Econ Bot 65:422–426

DiTomaso JM (1998) Impact, biology, and ecology of Saltcedar (*Tamarix* spp.) in the Southwestern United States. Weed Technol 12:326–336

DiTomaso JM (2000) Invasive weeds in rangelands: species, impacts and management. Weed Sci 48:255–265

Dodds KJ, Orwig DA (2011) An invasive urban forest pest invades natural environments – Asian longhorned beetle in northeastern US hardwood forests. Can J For Res 41:1729–1742

Doherty TS, Glen AS, Dickman CR et al (2016) Invasive predators and global biodiversity loss. Proc Natl Acad Sci U S A 113:11261–11265

Donovan GH, Butry DT (2009) The value of shade: estimating the effect of urban trees on summertime electricity use. Energ Buildings 41:662–668

Doody BJ, Sullivan JJ, Meurk CD et al (2010) Urban realities: the contribution of residential gardens to the conservation of urban forest remnants. Biodivers Conserv 19:1385–1400

Drew J, Anderson N, Andow D (2010) Conundrums of a complex vector for invasive species control: a detailed examination of the horticultural industry. Biol Invasions 12:2837–2851

Dubey JP (2010) Toxoplasmosis of animals and humans, 2nd edn. CRC Press, Boca Raton, 313 p

Dubey JP, Jones JL (2008) Toxoplasma gondii infection in humans and animals in the United States. Int J Parasitol 38:1257–1278

Dudley N, Jones T, James R et al (2017) Applied genetic conservation of Hawaiian Acacia koa: an eco-regional approach. In: Sniezko RA, Man G, Hipkins V, Woeste K, Gwaze D, Kliejunas JT, McTeague BA, tech. cords (2017) Gene conservation of tree species—banking on the future. Proceedings of a workshop. General Technical Report PNW-GTR-963. U.S. Department of Agriculture, Forest Service, Pacific Northwest Research Station, Portland, OR, pp 78–91

Duguay S, Eigenbrod F, Fahrig L (2007) Effects of surrounding urbanization on non-native flora in small forest patches. Landsc Ecol 22:589–599

Eagle AJ, Eiswerth ME, Johnson WS et al (2007) Costs and losses imposed on California ranchers by yellow star thistle. Rangel Ecol Manag 60:369–377

Ehrenfeld JG (2003) Effects of exotic plant invasions on soil nutrient cycling processes. Ecosystems 6:503–523

Elias SP, Lubelczyz CB, Rand PW et al (2006) Deer Browse resistant exotic-invasive understory: an indicator of elevated human risk of exposure to *Ixodes scapularis* (Acari: Ixodidae) in Southern Coastal Maine Woodlands. J Med Entomol 43:1142–1152

Epanchin-Niell RS, Hufford MB, Aslan CE et al (2010) Controlling invasive species in complex social landscapes. Front Ecol Environ 8:210–216

Ettestad P (2017) Overview of plague. Merck Manual, Veterinary Manual. In: http://www.merckvetmanual.com/generalized-conditions/plague/overview-of-plague

Falconer IR (2005) Is there a human health hazard from microcystins in the drinking water supply? Acta Hydrochim Hydrobiol 33:64–71

Fall MW, Avery ML, Witmer G (2011) Chapter sixteen: rodents and other vertebrate invaders in the United States. In: Pimentel D (ed) Biological invasions: economic and environmental costs of alien plant, animal, and microbe species, 2nd edn. CRC Press, Boca Raton, pp 381–410

Fleming J (2008) Comparison of vehicle washing system. U.S. Forest service, prepared for Army Corps of Engineers and Engineer Research and Development Center. Strategic Environmental Research and Development Program Project SI-1545. http://www.dtic.mil/dtic/tr/fulltext/u2/a512807.pdf

Fleming MR, Hoover K, Janowiak JJ et al (2002) Microwave irradiation of wood packing material to destroy the Asian longhorned beetle. For Prod J 52(11/12):1–7

Flory SL, Clay K (2009) Effects of roads and forest successional age on experimental plant invasions. Biol Conserv 142:2531–2537

Forseth IN, Innis AF (2004) Kudzu (*Pueraria montana*): history, physiology, and ecology combine to make a major ecosystem threat. Crit Rev Plant Sci 23:401–413

French NP (2017) Impacts of non-native species on livestock. In: Vilà M, Hulme PE (eds) Impact of biological invasions on ecosystem services. Springer International Publishing, Cham, pp 139–154

Friday J, Keith LM, Hughes F (2015) Ceratocystis Wilt of 'Ohi'a. Extension Publications. PD-107. College of Tropical Agriculture and Human Resources, University of Hawai'i at Manoa

Gardner DE (1980) Acacia koa seedling wilt caused by *Fusarium oxysporum*. Phytopathology 70:594–597

Garibaldi A, Turner NJ (2004) Cultural keystone species: implications for ecological conservation and restoration. Ecol Soc 9(3):Art 1

Gelbard JL, Belnap J (2003) Roads as conduits for exotic plant invasions in a s semiarid landscape. Conserv Biol 17:420–432

Gill SE, Handley JF, Ennos AR, Pauleit S (2007) Adapting cities for climate change: the role of the green infrastructure. Built Environ 33:115–1333

Goddard MA, Dougill AJ, Benton TG (2009) Scaling up from gardens: biodiversity conservation in urban environments. Trends Ecol Evol 25:90–98

Godefroid S, Koedam N (2004) The impact of forest paths upon adjacent vegetation: effects of the path surfacing material on the species composition and soil compaction. Biol Conserv 119:405–419

Gomez-Aparicio L, Canham CD (2008a) Neighborhood analyses of the allelopathic effects of the invasive tree *Ailanthus altissima* in temperate forests. J Ecol 96:447–458

Gomez-Aparicio L, Canham CD (2008b) Neighborhood models of the effects of invasive tree species on ecosystem processes. Ecol Monogr 78:69–86

Grant EHC, Muths E, White CL et al (2016) Salamander chytrid fungus (*Batrachochytrium salamandrivorans*) in the United States—Developing research, monitoring, and management strategies. U.S. Geological Survey Open-File Report 2015–1233, 16 p. https://doi.org/10.3133/ofr20151233

Griffin PM, Tauxe RV (1991) The epidemiology of infections caused by *Escherichia coli* O157: H7, other enterohemorrhagic *E. coli*, and the associated Hemolytic Uremic Syndrome. Epidemiol Rev 13:60–98

Guo Q, Qian H, Ricklefs RE, Xi W (2006) Distributions of exotic plants in eastern Asia and North America. Ecol Lett 9:827–834

Haack RA (2006) Exotic bark- and wood-boring Coleoptera in the United States: recent establishments and interceptions. Can J For Res 36:269–288

Haack RA, Law KR, Mastro VC et al (1997) New York's battle with the Asian long-horned beetle. J For 95:11–15

Haack RA, Petrice TR, Wiedenhoeft AC (2010a) Incidence of bark-and wood-boring insects in firewood: a survey at Michigan's Mackinac bridge. J Econ Entomol 103:1682–1692

Haack RA, Hérard F, Sun J, Turgeon JJ (2010b) Managing invasive populations of Asian longhorned beetle and citrus longhorned beetle: a worldwide perspective. Annu Rev Entomol 55:521–546

Hancock DD, Besser TE, Kinsel ML et al (1994) The prevalence of *Escherichia coli* O157.H7 in dairy and beef cattle in Washington State. Epidemiol Infect 113:199–207

Hansen MJ, Clevenger AP (2005) The influence of disturbance and habitat on the presence of non-native plant species along transport corridors. Biol Conserv 125:249–259

Harris E, Ames JA, Packham AE, Conrad PA (2002) Coastal freshwater runoff is a risk factor for *Toxoplasma gondii* infection of southern sea otters (*Enhydra lutris nereis*). Int J Parasitol 32:997–1006

Havens K, Vitt P, Kramer AT et al (2015) Seed sourcing for restoration in an era of climate change. Nat Areas J 35:122–133

Hawaii Department of Agriculture (HDOA) (2016) Coconut Rhinoceros Beetle Program annual report/Farm Bill Survey – July 1, 2015 – June 30, 2016

Hayes EB, Komar N, Nasci RS et al (2005) Epidemiology and transmission dynamics of West Nile Virus disease. Emerg Infect Dis 11:1167–1173

Herms DA, McCullough DG (2014) Emerald ash borer invasion of North America: history, biology, ecology, impacts, and management. Annu Rev Entomol 59:13–30

Hobbs RJ, Higgs E, Harris JA (2009) Novel ecosystems: implications for conservation and restoration. Trends Ecol Evol 24:599–605

Huebner CD, Tobin PC (2006) Invasibility of mature and 15-year-old deciduous forests by exotic plants. Plant Ecol 186:57–68

Huebner CD, Gottschalk KW, Miller GW, Brose PH (2010) Restoration of three forest herbs in the Liliaceae family by manipulating deer herbivory and overstorey and understorey vegetation. Plant Ecolog Divers 3:259–272

Hulme PE (2009) Trade, transport and trouble: managing invasive species pathways in an era of globalization. J Appl Ecol 46:10–18

Ignatieva M, Stewart GH, Meurk C (2011) Planning and design of ecological networks in urban areas. Landsc Ecol 7:17–25

International Association of Oil and Gas Producers (IOGP), International Petroleum Industry Environmental Conservation Association (IPIECA), Little D (2010) Alien invasive species and the oil and gas industry. IOGP Report Number 436. http://www.ipieca.org/resources/good-practice/alien-invasive-species-and-the-oil-and-gas-industry/

Invasive Species Advisory Committee (ISAC) (2016) White paper: invasive species impacts on infrastructure. U.S. Department of the Interior, Washington, DC

Iverson M (1998) Effects of *Arundo donax* on water resources. CalEPPC News. California Exotic Pest Plant Council, Trabuco Canyon, p 10

Jakubowski AR, Casler MD, Jackson RD (2010) The benefits of harvesting wetland invaders for cellulosic biofuel: an ecosystem services perspective. Restor Ecol 18:789–795

Jay MT, Cooley M, Mandrell RE et al (2007) *Escherchia coli* O157:H7 in feral swine near spinach fields and cattle in Central California coast. Emerg Infect Dis 13:1908–1911

Jessup DA (2004) The welfare of feral cats and wildlife. J Am Vet Med Assoc 225:1377–1383

Jodoin YD, Lovoie P, Villeneuve M et al (2008) Highways as corridors and habitats for the invasive common reed *Phragmites australis* in Quebec, Canada. J Appl Ecol 45:459–466

Johnston FM, Johnston SW (2004) Impacts of road disturbance on soil properties and on exotic plant occurrence in subalpine areas of the Australian Alps. Arct Antarct Alp Res 36:201–207

Joly M, Bertrand P, Gbangou RY et al (2011) Paving the way for invasive species: road type and the spread of common ragweed (*Ambrosia artemisiifolia*). Environ Manag 48:514–522

Jones JG, Russell DG (1968) Giant hogweed dermatis. Practitioner 200(1199):704–706

Kalwij JM, Milton SJ, McGeoch MA (2008) Road verges as invasion corridors? A spatial hierarchical test in an arid ecosystem. Landsc Ecol 23:439–451

Karnosky D (1979) Dutch elm disease: a review of the history, environmental implications, control, and research needs. Environ Conserv 6:311–322

Kauffman MD, LeJeune J (2011) European starlings (*Sturnus vulgaris*) challenged with *Escherichia coli* O157 can carry and transmit the human pathogen to cattle. Lett Appl Microbiol 53:596–601

Kauffman BW, Clatterbuck WK, Liebhold AM, Coyle DR (2017) Gypsy moth in the southeastern U.S.: biology, ecology, and forest management strategies. SREF-FH-008. Southern Regional Extension Forestry, Athens, 10 p

Keeley JE (2006) Fire management impacts on invasive plants in the western United States. Conserv Biol 20:375–384

Kenis M, Roques A, Santini A, Liebhold AM (2017) Impact of non-native invertebrates and pathogens on market forest tree resources. In: Vilà M, Hulme PE (eds) Impact of biological invasions on ecosystem services. Springer International Publishing, Cham, pp 103–117

Kerns BK, Naylor BJ, Buonopane M et al (2009) Modeling tamarisk (*Tamarix* spp.) habitat and climate change effects in the northwestern United States. Invasive Plant Sci Manag 2:200–215

Kettenring KM, Adams CR (2011) Lessons learned from invasive plant control experiments: a systematic review and meta–analysis. J Appl Ecol 48:970–979

King JR, Tschinkel WR, Ross KG (2009) A case study of human exacerbation of the invasive species problem: transport and establishment of polygyne fire ants in Tallahassee, Florida, USA. Biol Invasions 11:373–377

Knight KS, Herms D, Plumb R et al (2012) Dynamics of surviving ash (*Fraxinus* spp.) populations in areas long infested by emerald ash borer (*Agrilus planipennis*). In: Sniezko RA, Yanchuk AD, Kliejunas JT, Palmieri KM, Alexander JM, Frankel SJ tech. coords (2012) Proceedings of the fourth international workshop on the genetics of host-parasite interactions in forestry: Disease and insect resistance in forest trees. General Technical Report PSW-GTR-240. Pacific Southwest Research Station, Forest Service, U.S. Department of Agriculture, Albany, CA, pp 143–152

Knowler D, Barbier E (2005) Importing exotic plants and the risk of invasion: are market-based instruments adequate? Ecol Econ 52:341–354

Koch FH, Yemshanov D, Magarey RD, Smith WD (2012a) Dispersal of invasive forest insects via recreational firewood: a quantitative analysis. J Econ Entomol 105:438–450

Koch JL, Carey DW, Knight KS et al (2012b) Breeding strategies for the development of emerald ash borer-resistant North American ash. In: Sniezko RA, Yanchuk AD, Kliejunas JT, Palmieri KM, Alexander JM, Frankel SJ (eds) Proceedings of the 4th international workshop on the genetics of host–parasite interactions in forestry: disease and insect resistance in forest trees. USDA Forest Service, Pacific Southwest Research Station, Albany, pp 235–239

Kolby JE, Daszak P (2016) The emerging amphibian fungal disease, chytridiomycosis: a key example of the global phenomenon of wildlife emerging infectious diseases. Microbiol Spectr 4(3):EI10-0004-2015. https://doi.org/10.1128/microbiolspec.EI10-0004-2015

Kovacs KF, Haight RG, McCullough DG et al (2010) Cost of potential emerald ash borer damage in US communities, 2009–2019. Ecol Econ 69:569–578

Kowarik I (2011) Novel urban ecosystems, biodiversity, and conservation. Environ Pollut 159:1974–1983

Krauss BH (1974) Ethnobotany of Hawaii. Department of Botany, University of Hawaii, Honolulu, pp 127–130

Krohn-Ching VF (2016) Hawaii dye plants and dye recipes. University of Hawaii Press, Honolulu, 136p

Kruse R, Bend E, Bierzychudek P (2004) Native plant regeneration and introduction of non-natives following post-fire rehabilitation with straw mulch and barley seeding. For Ecol Manag 196:299–310

Lambrinos JG (2000) The impact of invasive alien grass *Cortaderia jubata* (*Lemoine*) Stapf. On an endangered Mediterranean-type shrubland in California. Divers Distrib 6:217–231

Lanier GN (1989) Trap trees for control of Dutch elm disease. J Arboric 15:105–111

Lax AR, Osbrink WL (2003) United States Department of Agriculture-Agriculture Research Service research on targeted management of the Formosan subterranean termite *Coptotermes formosanus* Shiraki (Isoptera: Rhinotermitidae). Pest Manag Sci 59:788–800

LeBlanc DJ (1994) Nutria. In: Hygnstrom SE, Timm RM, Larson GE (eds) Prevention and control of wildlife damage. University of Nebraska, Lincoln

Leishman MR, Thomson VP (2005) Experimental evidence for the effects of additional water, nutrients and physical disturbance on invasive plants in low fertility Hawkesbury sandstone soils, Sydney. Aust J Ecol 93:38–49

Leistritz FL, Bangsund DA, Hodur NM (2004) Assessing the economic impact of invasive weeds: the case of leafy spurge (*Euphorbia esula*). Weed Technol 18:1392–1395

Liebhold A, Brokerhoff EG, Garrett LJ et al (2012) Live plant imports: the major pathway for forest insect and pathogen invasions of the US. Front Ecol Environ 10(3):135–143

Lopez VM, Hoddle MS, Francese JA et al (2017) Assessing flight potential of the invasive Asian longhorned beetle (Coleoptera: Cerambycidae) with computerized flight mills. J Econ Entomol 110:1070–1077

López-Hoffman L, Wiederholt R, Semmens D et al (2014) Market forces and technological substitutes cause fluctuations in the value of bat-control services for cotton. PLoS One 9(2):e87912

Lososova Z, Chyrtry M, Tichy L et al (2012) Biotic homogenization of Central Europe urban floras depends on residence time of alien species and habitat type. Biol Conserv 145:179–184

MacFarlane DW, Meyer SP (2005) Characteristics and distribution of potential ash tree hosts for emerald ash borer. For Ecol Manag 213:15–24

Mack MC, D'Antonio CM (1998) Impacts of biological invasions on disturbance regimes. Trends Ecol Evol 13:195–198

Maher SP, Kramer AM, Drake JM et al (2012) Spread of white-nose syndrome on a network regulated by geography and climate. Nat Commun 3:1306

Martel A, Spitzen-van der Sluijsb A, Pasmans F et al (2013) *Batrachochytrium salamandrivorans* sp. nov. causes lethal chytridiomycosis in amphibians. PNAS 110:15325–15329

Martin L, Blossey B (2013) The runaway weed: costs and failures of *Phragmites australis* management in the USA. Estuar Coasts 36:626–632

McDonald RI, Kareiva P, Forman RIT (2008) The implications of current and future urbanization for global protected areas and biodiversity conservation. Biol Conserv 141:1695–1703

McKinney ML, Lockwood JL (1999) Biotic homogenization: a few winners replacing many losers in the next mass extinction. Trends Ecol Evol 14:450–453

Meng XJ, Lindsay DS, Sriranganthan N (2009) Wild boars as sources for infectious diseases in livestock and humans. Philos Trans R Soc 364:2697–2707

Meng PS, Hoover K, Keena MA (2015) Asian longhorned beetle (Coleoptera: Cerambycidae), an introduced pest of maple and other hardwood trees in North America and Europe. J Integr Pest Manag 6(1):4. https://doi.org/10.1093/jipm/pmv003

Meyerson LA, Saltonstall K, Windham L et al (2000) A comparison of *Phragmites australis* in freshwater and brackish marsh environments in North America. Wetl Ecol Manag 8:89–103

Meyerson LA, Cronin JT, Pyšek P (2016) Phragmites as a model organism for plant invasions. Biol Invasions 18:2421–2431

Miller MA, Miller WA, Conrad PA et al (2008) Type X *Toxoplasma gondii* in a wild mussel and terrestrial carnivores from coastal California: new linkages between terrestrial mammals, runoff and toxoplasmosis of sea otters. Int J Parasitol 38:1319–1328

Miller RS, Sweeney SJ, Slootmaker C et al (2017) Cross-species transmission potential between wild pigs, livestock, poultry, wildlife, and humans: implications for disease risk management in North America. Sci Rep 7:1–14

Minnesota Department of Natural Resources (MN DNR) (2017) Fly-anglers and wader users: felt-soled wader issues and alternatives. http://www.dnr.state.mn.us/invasives/felt.html

Mitich LW (2000) Intriguing world of weeds. Kudzu *Pueraria lobata* (Willd.) Ohwi. Weed Technol 14:231–235

Mortensen DA, Rauschert ESJ, Nord AN, Jones BP (2009) Forest roads facilitate the spread of invasive plants. Invasive Plant Sci Manag 2:191–199

Murrell CE, Gerber C, Krebs M et al (2011) Invasive knotweed affects native plants through allelopathy. Am J Bot 98:38–43

Muzika RM, Liebhold AM (1999) Changes in radial increment of host and nonhost tree species with gypsy moth defoliation. Can J For Res 29:1365–1373

National Conference of State Legislatures (2017) Federal and state recognized tribes. National Conference of State Legislatures. http://www.ncsl.org/research/state-tribal-institute/list-of-federal-and-state-recognized-tribes.aspx#State

National Congress of American Indians (2015) An introduction to Indian Nations in the United States. http://www.ncai.org/resources/ncai_publications/tribal-nations-and-the-united-states-an-introduction

NatureServe (2015) Species at-risk on Department of Defense Lands: 2014 updated analysis, report, and maps. http://www.natureserve.org/conservation-tools/projects/species-risk-dod-installations

Neher DA, Asmussen D, Lovell ST (2013) Roads in northern hardwood forests affect adjacent plant communities and soil chemistry in proportion to the maintained roadside area. Sci Total Environ 449:320–327

No, E.O. (2016) No. Executive Order. 13751. Safeguarding the nation from the impacts of invasive species. *Federal Register* 81: 88609–88614

Norgaard KM (2007) The politics of invasive weed management: gender, race, and risk perception in rural California. Rural Sociol 72:450–477

Nowak DJ (2010) Urban biodiversity and climate change. In: Muller N, Werner P, Kelcey JG (eds) Urban biodiversity and design. Wiley-Blackwell Publishing, Hoboken, pp 101–117

Nowak DJ, Crane DE, Dwyer JF (2002) Compensatory values of urban trees in the U.S. J Arboric 28:194–199

Oke TR (1995) The heat island of the urban boundary layer: characteristics, causes, and effects. In: Cermak JE et al (eds) Wind climate in cities. Kluwer Academic Publishers, Amsterdam, pp 81–107

Olden JD, Rooney TP (2006) On defining and quantifying biotic homogenization. Glob Ecol Biogeogr 15:113–120

Olsen BE (1999) Grazing and weeds. In: Shely RL, Petroff JK (eds) Biology and management of noxious weeds. Oregon State University Press, Corvalis, pp 85–96

Olsen CW (2002) The emergence of novel swine influenza viruses in North America. Virus Res 85:199–210

Parendes LA, Jones JA (2000) Role of light availability and dispersal in exotic plant invasion along roads and streams in the H.J. Andrews Experimental Forest, Oregon. Conserv Biol 14:64–75

Park J, Hushak LJ (1999) Zebra mussel control costs in surface water using facilities. Technical summary Ohio State University, OHSU-TS-028. Ohio Sea Grant College Program, The Ohio State University, Columbus

Pederson K, Bevins SN, Schmidt BS et al (2012) Apparent prevalence of swine brucellosis in feral swine in the United States. Hum–Wildl Interact 6:38–47

Pejchar L, Mooney HA (2009) Invasive species, ecosystem services and human well-being. Trends Ecol Evol 24:497–504

Perry R (2013) White-nose syndrome in bats: an overview of current knowledge for land managers. . General technical report SRS-GTR-184. U.S. Department of Agriculture Forest Service, Southern Research Station, Asheville, 9 p

Pfeiffer JM, Voeks RA (2008) Biological invasions and biocultural diversity: linking ecological and cultural systems. Environ Conserv 35:281–293

Phillips S, Darland T, Systma M (2005) Potential economic impacts of Zebra mussels on the hydropower facilities in the Columbia River basin. Pacific States Marine Fisheries Commission, Portland. http://www.aquaticnuisance.org/wordpress/wp-content/uploads/2009/01/Phillips-Darland-and-Sytsma.pdf

Pickett STA, Cadenasso MI, Grove JM et al (2001) Urban ecological systems: linking terrestrial ecological, physical and socioeconomic components of metropolitan areas. Annu Rev Ecol Syst 32:127–157

Pimentel D (2011) Environmental and economic costs associated with alien invasive species in the United States. In: Pimentel D (ed) Biological invasions: economic and environmental costs of alien plant, animal, and microbe species. CRC Press, Boca Raton, pp 411–430

Pimentel D, Lach L, Zuniga R, Morrison D (2000) Environmental and economic costs of nonindigenous species in the United States. Bioscience 50:53–65

Pimentel D, Zuniga R, Morrison D (2005) Updates on the environmental and economics costs associated with alien-invasive species in the United States. Environ Econ 52:273–288

Pines IL (2009) Forest pests in Manitoba 2008: Dutch elm disease management program in Manitoba. Proceedings of the forest pest management forum, 2–4 December, 2008, Gatineau, PQ (pp 88–94). Natural Resources Canada, Ottawa, ON. http://www.glfc.forestry.ca/VLF/ForumPresentations/pinesmanitoba_08e.pdf

Poff NL, Allan JD, Bain MB et al (1997) The natural flow regime. Bioscience 47:769–784

Pokorny ML, Sheley RL, Zabinski CA et al (2005) Plant functional group diversity as a mechanism for invasion resistance. Restor Ecol 13:448–459

Potts SG, Biesmeijer JC, Kremen C et al (2010) Global pollinator declines: trends, impacts and drivers. Trends Ecol Evol 25:345–353

Pretty J (2002) Landscapes lost and found. In: Pretty J (ed) Agriculture: reconnecting people, land and nature. Earthscan Publishing, London, pp 10–26

Pretty Paint-Small V (2013) Linking culture, ecology, and policy: the invasion of Russian Olive (*Elaeagnus angustifolia* L.) on the Crow Indian Reservation, South-Central Montana, USA. Doctoral dissertation, Bioagricultural Sciences and Pest Management, Colorado State University, Fort Collins, Colorado

Quinn LD, Endres AB, Voigt TB (2014) Why not harvest existing invaders for bioethanol? Biol Invasions 16:1559. https://doi.org/10.1007/s10530-013-0591-z

Reardon R, Podgwaite J, Zerillo R (2012) Gypchek-environmentally safe viral insecticide for gypsy moth control. FHTET-2012–01. U.S. Department of Agriculture, Forest Service, Forest Health Technology Enterprise Team, Morgantown, 49 p

Reeder DM, Frank CL, Turner GG et al (2012) Frequent arousal from hibernation linked to severity of infection and mortality in bats with white-nose syndrome. PLoS One 7(6):e38920

Rentch JS, Fortney RH, Stephenson SL et al (2005) Vegetation-site relationships of roadside plant communities in West Virginia, USA. J Appl Ecol 42:129–138

Ricotta C, La Sorte FA, Pysek P et al (2009) Pyloecology of urban alien floras. J Ecol 97:1243–1251

Rooney TP (2005) Distribution of ecologically-invasive plants along off-road vehicle trails in the Chequmegon National Forest, Wisconsin. Mich Bot 44:178–182

Rosaen AL, Grover EA, Spencer CW, Anderson PL (2012) The costs of aquatic invasive species to the Great Lakes States. Anderson Economic Group

Rothlisberger JD, Chadderton WL, McNulty J, Lodge DM (2010) Aquatic invasive species transport via trailered boats: what is being moved, who is moving it, and what can be done. Fisheries 35:121–132

Roy S, Byrne J, Pickering C (2012) A systematic quantitative review of urban tree benefits, costs, and assessment methods across cities in different climatic zones. Urban For Urban Green 11:351–363

Rutz FR, Riley TM, Foster P (2014) Fire and reconstruction at Lobato bridge in New Mexico. J Preserv Technol 45:37–42

Sanders H, Polaska S, Haight RG (2010) The value of urban tree cover: a hedonic property price model in Ramsey and Dakota counties, Minnesota, USA. Ecol Econ 69:1646–1656

Schartel B, Bahr H, Braun U, Recknagel C (2010) Fire risks of burning asphalt. Fire Mater 34:333–340

Schmelzer L, Perryman B, Bruce B et al (2014) Case study: reducing cheatgrass (*Bromus tectorum* L.) fuel loads using fall cattle grazing. Prof Anim Sci 30:270–278

Schwartz MW, Thorne JH, Viers JH (2006) Biotic homogenization of the California flora in urban and urbanizing regions. Biol Conserv 127:282–291

Seastedt TR, Hobbs RJ, Suding KN (2008) Management of novel ecosystems: are novel approaches required? Front Ecol Environ 6:547–553

Seybold SJ, Coleman TW, Flint ML (2010) Burn it, chip it, or tarp it, but just don't move it: managing oak firewood infested with the goldspotted oak borer, *Agrilus coxalis auroguttatus*. In Sudden Oak Death Fourth Science Symposium, p 365

Shafto TD, McDaniel L (2009) Contemporary Hawai'i woodworkers: the wood, the art, the aloha. Contemporatry Publications, Mountain View, 223p

Sharov AA, Leonard DS, Liebhold AM et al (2002) Slow the spread: a national program to contain the gypsy moth. J For 100:30–35

Sheley RL, James J (2010) Resistance of native plant functional groups to invasion by medusahead. Invasive Plant Sci Manag 3:294–300

Shelton AL (2012) Mowing any time after midsummer can manage Japanese stiltgrass. Invasive Plant Sci Manag 5:209–216

Simberloff D, Von Holle B (1999) Positive interactions of nonindigenous species: invasional meltdown? Biol Invasions 1:21–32

Skolmen RG (1974) Some woods of Hawaii, properties and uses of 16 commercial species. USDA Forest Service general technical report PSW-8. PSW Forest and Range Experimental Station, Berkeley, p 30

Smith RM, Thompson K, Hodgson JG et al (2006) Urban domestic gardens (IX): composition and richness of the vascular plant flora, and implications for native biodiversity. Biol Conserv 129:312–322

Stiles JH, Jones RH (1998) Distribution of the red imported fire ant, *Solenopsis invicta*, in road and powerline habitats. Landsc Ecol 335:335–346

Stinson KA, Campbell SA, Powell JR et al (2006) Invasive plant suppresses the growth of native tree seedlings by disrupting belowground mutualisms. PLoS Biol 4(5):e104

Stout J, Morales CL (2009) Ecological impacts of invasive alien species on bees. Apidologie 40:388–409

Strayer D (2009) Twenty years of zebra mussels: lessons from the mollusk that made headlines. Front Ecol Environ 7:135–141

Strayer D (2010) Alien species in fresh waters: ecological effects, interactions with other stressors, and prospects for the future. Freshw Biol 55:152–174

Su N, Scheffrahan RH (2016) Formosan subterranean termite. Featured Creatures, University of Florida Institute of Food and Agricultural Sciences. Publication Number: EENY-121. http://entnemdept.ufl.edu/creatures/urban/termites/formosan_termite.htm

Sustainable Fort Bragg (2017) Burning towards success

Taylor K, Brummer T, Taper ML et al (2012) Human-mediated long-distance dispersal: an empirical evaluation of seed dispersal by vehicles. Divers Distrib 18:942–951

Texas Animal Health Commission (2017) Fever tick situation report of January 30, 2017. Fact sheet. Texas

Timm RM (1994) Norway rats. The handbook: prevention and control of wildlife damage. 5. http://digitalcommons.unl.edu/icwdmhandbook/5

Tinsley MJ, Simmons MT, Windhager S (2006) The establishment success of native versus non-native herbaceous seed mixes on a revegetated roadside in Central Texas. Ecol Eng 26:231–240

Traveset A, Richardson DM (2006) Biological invasions as disruptors of plant reproductive mutualisms. Trends Ecol Evol 21:208–216

Turner GG, Reeder DM, Coleman JTH (2011) A five-year assessment of mortality and geographic spread of White-nose Syndrome in North American bats and a look to the future. Bat Res News 52:13–27

U.S. Deparatment of Agriculture, National Agricultural Statistics Service (2014) Census of agriculture, 2014 census of horticultural specialties. https://www.nass.usda.gov/Publications/AgCensus/2012/Online_Resources/Census_of_Horticulture_Specialties/

U.S. Department of Agriculture, Animal and Plant Health Inspection Service (USDA APHIS) (2016) Feral Swine – risks to people and domestic animals. https://www.aphis.usda.gov/aphis/ourfocus/wildlifedamage/operational-activities/feral-swine/feral-swine-damage/feral-swine-risks-pets-people

U.S. Department of Agriculture, Animal and Plant Health Inspection Service (USDA APHIS) (2017) August is tree check month: public's help needed to stop the invasive Asian Longhorned Beetle in its tracks. https://www.aphis.usda.gov/aphis/newsroom/news/sa_by_date/sa-2017/alb-tree-check-month

U.S. Department of Agriculture, Animal and Plant Health Inspection Service (USDA APHIS) (2020). Emerald Ash Borer. https://www.aphis.usda.gov/aphis/ourfocus/planthealth/plant-pest-and-disease-programs/pests-and-diseases/emerald-ash-borer

U.S. Department of Agriculture, Animal and Plant Health Inspection Service, Plant Protection and Quarantine (USDA APHIS PPQ) (2007) Asian Longhorned Beetle: questions and answers

U.S. Department of Agriculture Animal and Plant Health Inspection Service, Plant Protection and Quarantine (USDA APHIS PPQ) (2010) National Firewood Task Force recommendations, USDA-APHIS-PPQ Fact Sheet. https://www.aphis.usda.gov/aphis/ourfocus/planthealth/import-information/firewood

U.S. Department of Agriculture, Forest Service, Animal and Plant Health Inspection Service, National Association of State Foresters, National Plant Board (2011) National response framework for Thousand Cankers Disease (TCD) on Walnut. https://www.aphis.usda.gov/plant_health/plant_pest_info/tcd/downloads/NationalResponseFramework.pdf

U.S. Department of Commerce National Oceanic and Atmospheric Administration (2016) Main Hawaiian Islands Monk seal management plan. U.S. Department of Commerce National Oceanic Atmospheric Administration, National Marine Fisheries Service, Pacific Islands Regional Office, Honolulu, 72 p

U.S. Marine Corps (2001) Marine Corps Base Hawaii, supporting readiness and global projection. http://www.mcbhawaii.marines.mil/News/News-Article-Display/Article/538881/aavs-improve-bird-habitat/

University of Guam and the Secretariat of the Pacific Community (2014) Regional biosecurity plan for Micronesia and Hawaii. http://guaminsects.net/anr/sites/default/files/RBPC.pdf

Utley C, Nguyen T, Roubtsova T et al (2013) Susceptibility of walnut and hickory species to *Geosmithia morbida*. Plant Dis 97:601–607

Valtonen A, Jantunen J, Saarinen K (2006) Flora and lepidoptera fauna adversely affected by invasive *Lupinus polyphyllus* along road verges. Biol Conserv 133:389–396

van Riper C, van Riper SG, Goff ML, Laird M (1986) The epizootiology and ecological significance of malaria in Hawaiian land birds. Ecol Monogr 56:327–344

Vanderploeg HA, Liebig JR, Carmichael WW et al (2001) Zebra mussel (*Dreissena polymorpha*) selective filtration promoted toxic Microcystis blooms in Saginaw Bay (Lake Huron) and Lake Erie. Can J Fish Aquat Sci 58:1208–1221

Von der Lippe M, Kowarik I (2007) Long-distance dispersal of plants by vehicles as a driver of plant invasions. Conserv Biol 21:986–996

Voulgaridis V, Vassiliou VG (2005) The walnut wood and its utilization to high value products. Acta Hortic 705:69–81

Wang X (2012) Heat treatment of wood for invasive forest pests. In: Bradshaw BK, Ross RJ, Wang X, Wiemann MC (eds) Wood utilization options for urban trees infested by invasive species. University of Minnesota, Duluth

Watts DA, Moore GW (2011) Water-use dynamics of an invasive reed, *Arundo donax*, from leaf to stand. Wetlands 31:725–734

Weniger BG, Warren J, Forseth V et al (1984) Human bubonic plague transmitted by a domestic cat scratch. J Am Med Assoc 251:927–928

Whisenant SG (1989) Changing fire frequencies on Idaho's Snake River Plains: ecological and management implications. Symposium on Cheatgrass Invasion, Shrub Die-Off and Other Aspects of Shrub Biology and Management. Las Vegas, NV. April 5–7, pp 4–10

Whitehill JGA, Opiyo SO, Koch JL et al (2012) Interspecific comparison of constitutive ash phloem phenolic chemistry reveals compounds unique to Manchurian ash, a species resistant to emerald ash borer. J Chem Ecol 38:499–511

Williams SC, Ward JS (2010) Effects of Japanese Barbarry (Ranunculales: Beridacaeae) removal and resulting microclimate changes on *Ixodes scapularius* (Acari: Ixodidae) abundances in Connecticut, USA. Environ Entomol 39:1911–1921

Williams F, Eschen R, Harris A et al (2010) The economic cost of invasive non-native species of Great Britain. CABI, Surrey, pp 1–199

With KA (2002) The landscape ecology of invasive spread. Conserv Biol 16:1192–1203

Witmer GW, Sanders RB, Taft AC (2003) Feral swine – are they a disease threat to livestock in the United States? In: Fagerstone KA, Witmer GW (eds) Proceedings of the 10th wildlife damage management conference. USDA National Wildlife Research Center – Staff Publications, Hot Springs

Yap TA, Koo MS, Ambrose RF et al (2015) Averting a North American biodiversity crisis: a newly described pathogen poses a major threat to salamanders via trade. Science 349(6247):481–482

Young SL, Gopalakrishnan G, Keshwani DR (2011) Invasive plant species as potential bioenergy producers and carbon contributors. J Soil Water Conserv 66:45A–50A

Zavaleta E (2000) The economic value of controlling an invasive shrub. Ambio 29:462–467

Zedler JB, Kercher S (2010) Causes and consequences of invasive plants in wetlands: opportunities, opportunists, and outcomes. Crit Rev Plant Sci 23:431–452

Zink TA, Allen MF, Heindl-Tenhunen B, Allen EB (1995) The effect of a disturbance corridor on an ecological reserve. Restor Ecol 3:304–310

Ziska LH, Bunce JA, Goins EW (2004) Characterization of an urban-rural CO2/temperature gradient and associated changes in initial plant productivity during secondary succession. Oecologia 139:454–458

Inventory and Monitoring of Invasive Species

10

Sonja Oswalt, Chris Oswalt, Alycia Crall, Robert Rabaglia, Michael K. Schwartz, and Becky K. Kerns

10.1 The Need for Inventory and Monitoring

Inventory and monitoring programs and high-quality distribution data are needed to manage invasive species and to develop quantitative ecological, economic, and social impact assessments. Strong inventory and monitoring programs provide valuable insights that help direct efforts to prevent and contain invasive species. Understanding where to focus management efforts, in addition to educational outreach, is key to developing economically efficient and ecologically relevant programs. As noted in Chap. 2, additional research is needed to address large-scale impacts and co-occurring impacts of multiple invaders. Inventory and monitoring efforts help to inform where this research can be most effectively applied by identifying areas where invasives are most abundant, most likely to spread, or most likely to be contained (Byers et al. 2002; Myers et al. 2000). Furthermore, as climate change continues to pose new and uncertain impacts on ecosystems, monitoring at the edge of invasive species' ranges will become more important to management efforts (Hellmann et al. 2008).

Inventory and monitoring of invasive species in the United States currently occur at a variety of scales (Fig. 10.1); for example, programs include national efforts like the Forest Inventory and Analysis (FIA) program of the USDA Forest Service, which samples many invasive plants systematically across the entire country, State-level Natural Heritage Programs that provide detailed information on invasive organisms at small scales, and multi-scale citizen science efforts. Understanding where invasive species are encroaching on native ecosystems can facilitate more focused management plans and activities, identify research needs, and suggest targeted approaches to the prevention of establishment and spread. This chapter explores how invasive species are being monitored at multiple scales, across multiple taxa, and in a wide variety of ecological systems by reviewing some recognized invasive species programs in the United States. While we cannot adequately cover all the inventory and monitoring programs in the country, we provide several prominent examples (Table 10.1) at each scale [note: many of the programs described cover multiple scales] to illustrate the utility of inventory in decision making and research planning.

10.2 National-Scale Inventories of Invasive Species

National and sub-national programs provide a mechanism for identifying invasive species "hot spots" and detecting potential areas where spread is likely to occur. In addition, these programs afford monitoring of well-known invasives to track impact and spread. Federal agencies and bureaus that are

S. Oswalt (✉)
U.S. Department of Agriculture, Forest Service, Forest Inventory and Analysis Program, Washington, DC, USA

Department of Forestry, Wildlife, & Fisheries, University of Tennessee, Knoxville, TN, USA
e-mail: sonja.n.oswalt@usda.gov

C. Oswalt
U.S. Department of Agriculture, Forest Service, Southern Research Station Forest Inventory and Analysis, Knoxville, TN, USA

A. Crall
National Ecological Observatory Network, Boulder, CO, USA

R. Rabaglia
U.S. Department of Agriculture, Forest Service, State & Private Forestry, Forest Health Protection, Washington, DC, USA

M. K. Schwartz
U.S. Department of Agriculture, Forest Service, National Genomics Center for Wildlife and Fish Conservation, Missoula, MT, USA

B. K. Kerns
U.S. Department of Agriculture, Forest Service, Pacific Northwest Research Station, Corvallis, OR, USA

T. M. Poland et al. (eds.), *Invasive Species in Forests and Rangelands of the United States*,
https://doi.org/10.1007/978-3-030-45367-1_10

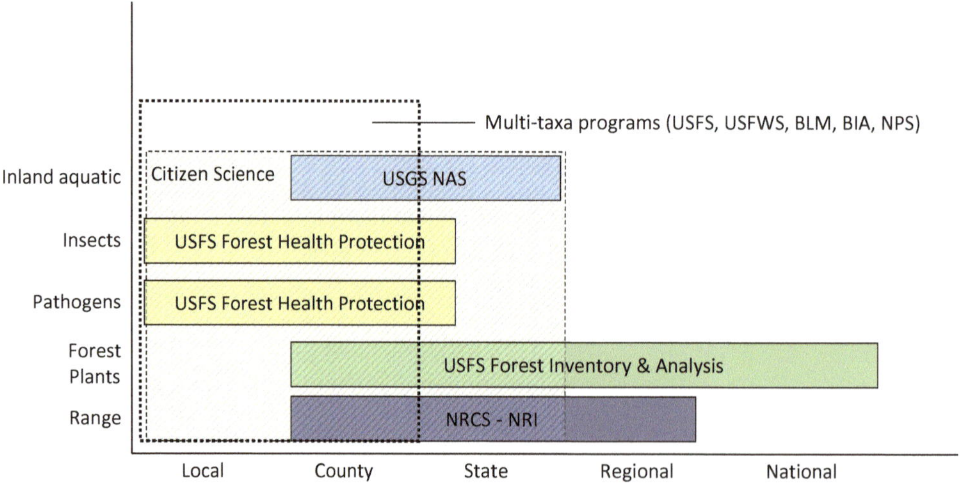

Fig. 10.1 The operating space of select large inventory and monitoring programs in the United States. Scale (*on the x-axis*) represents the scope of regular, systematic sampling. *USFS* USDA Forest Service, *USFWS* US Fish and Wildlife Service, *BLM* Bureau of Land Management, *BIA* Bureau of Indian Affairs, *NPS* National Park Service, *USGS* *US* Geological Survey, *NAS* Nonindigenous Aquatic Species, *NRCS* Natural Resources Conservation Service, *NRI* National Resources Inventory

overseen by the members of the National Invasive Species Council (NISC) inventory and monitor certain plant and animal populations and conduct research (Table 10.1). Information gained from these efforts is used to guide invasive species early detection and rapid response efforts that are broadly coordinated on a national scale by frameworks developed by NISC. Many of the broad-scale national efforts discussed here are considered "surveillance monitoring" programs (Nichols and Williams 2006) and represent strategic inventories. Tactical inventories or monitoring is equally important and often implemented on much smaller scales. In a scientific context, one can think of strategic inventories or monitoring as a hypothesis generator and tactical inventories as those monitoring systems used to guide specific tests or science programs. From a management perspective, the same dichotomy can be thought of as a system to generate broad policies, with the latter as a system to guide localized actions or management. Both are imperative for a robust national system to monitor invasive species across large landscapes.

Examples of national-scale inventory and monitoring programs include the FIA program, which includes some invasive plant monitoring; the USDA Natural Resources Conservation Service National Resources Inventory (NRI), which tracks invasive plants on grasslands and rangelands in the United States; the Forest Health Protection (FHP) program of the Forest Service, which provides a broad-scale example of the strength of survey and monitoring for invasive insects and disease; the U.S. Geological Survey (USGS) Nonindigenous Aquatic Species (NAS) program, which documents invasive aquatic invertebrates, vertebrates, and aquatic plants; and the USDA PLANTS database for native and invasive species which documents herbaria records of plant occurrence nationwide.

10.2.1 Forest Inventory and Analysis and Invasive Plants

The FIA program has existed since the mid-1930s as the country's pre-eminent forest inventory program. While much of the history of the program centered on summarizing tree data to better understand timber availability and resource capacity, invasive species have been a component of the program for the better part of a decade. With over 350,000 plots across the Nation on public and private land, FIA supplies a mechanism for identifying and tracking broad-scale patterns for some invasive plants that plague forested landscapes. The FIA program collects data on plots distributed in a random, systematic fashion on both private and public land across the United States. The plot design consists of four 0.067-ha fixed-radius subplots arranged in a "tri-areal" configuration (Bechtold and Patterson 2005). Several environmental conditions, tree-level variables, and abiotic measures are recorded on each subplot. Detailed explanations of the FIA plot design and sampling phases are given in Bechtold and Patterson (2005). Details specific to invasive plants are also covered in Gray (2008)

Historically, data on invasive plants on FIA plots were collected using different methodologies in each of the program's various regions, which led to difficulties if one wished to compare patterns across the national landscape or among or between regions. Currently, FIA is working to collect invasive plant data using a more uniform methodology that facilitates broad-scale comparisons. However, not all areas have implemented these protocols. For all invasive plant inventory and monitoring programs, the sheer number of plant species present is one of the many challenges to monitoring. Therefore, lists of priority plants are frequently developed, although each region maintains its own lists of

Table 10.1 Examples of Federal invasive species monitoring programs

Department/ Organization	Agency/Program	Implementation program	Scale of invasive species I&M	Taxonomic focus areas	Short description	More information
U.S. Department of Agriculture (NISC chair	Forest Service	National Forest System	Local, Regional	Terrestrial and aquatic plants, animals, insects, disease	Multi-taxa invasive sampling occurs on individual National Forests that can be aggregated for large-scale analyses when procedures are comparable.	http://www.fs.fed.us/ invasivespecies/framework.shtml
		Forest Inventory and Analysis	Regional, National	Terrestrial plants, insects, disease	A systematic national sample of various forest attributes on public and private forest land. Includes an invasive plant component as well as insect and disease damage detection.	
		Forest Health Protection	Regional, National	Terrestrial plants, insects, disease	Utilizes data from ground plots and surveys, aerial surveys, and other biotic and abiotic data sources and develops novel analytical approaches. Data are often aggregated from multiple localized surveys or monitoring efforts to address national analyses.	
		State and Private Forestry	Local through grants and cooperatives			
	Natural Resources Conservation Service	National Resources Inventory	National non-federal	Terrestrial plants	A nationally consistent sample of non-federal lands across the country. Line point intercept data are utilized in summaries of non-native invasive herbaceous species, native invasive woody species.	
U.S. Department of the Interior (NISC chair)	Fish and Wildlife Service (ANS Task Force Co-Chair)	Aquatic Nuisance Species Program	Local, Regional	Aquatic plants and animals	A targeted program wherein specific threatening invasive species are monitored in localized efforts (e.g., zebra mussels in the Lake States).	https://www.fws.gov/fisheries/ans/
		National Wildlife Refuge System	Local, Regional	Terrestrial and aquatic plants and animals	Multi-taxa invasive sampling occurs across the Wildlife Refuge System that can be aggregated for large-scale analyses when procedures are comparable.	https://www.fws.gov/invasives/ nwrs.html
	National Park Service	Inventory and Monitoring Vital Signs Program	Local, Regional	Ecosystem/ park dependent	Multi-taxa invasive sampling occurs on individual national parks that can be aggregated for large-scale analyses when procedures are comparable. Most national parks also maintain aggressive eradication programs that are supported by ad-hoc invasive species sampling.	https://www.nps.gov/im/ vital-signs.htm
	Bureau of Land Management	AIM-Monitoring Program	Local, Regional	Terrestrial plants and animals	Multi-taxa invasive sampling occurs on individual Bureau of Land Management parcels that can be aggregated for large-scale analyses.	https://landscape.blm.gov/ geoportal/catalog/AIM/AIM.page
	United States Geological Survey	Nonindigenous Aquatic Species	Local, Regional, National	Aquatic plants, invertebrates, and vertebrates	The program provides timely, reliable data about the presence and distribution of nonindigenous aquatic species aggregated from multiple sources and monitoring efforts including small-scale research projects.	https://nas.er.usgs.gov/
	Bureau of Indian Affairs	Agriculture and Rangeland Development Inventory Program	Local, Regional	Invasive plants and insects	Multi-taxa invasive sampling occurs on individual Bureau of Indian Affairs-managed lands that can be aggregated for large-scale analyses when procedures are comparable.	http://www.bia.gov/WhoWeAre/ BIA/OTS/NaturalResources/ AgrRngeDev/index.htm

(continued)

Table 10.1 (continued)

Department/ Organization	Agency/Program	Implementation program	Scale of invasive species I&M	Taxonomic focus areas	Short description	More information
US Department of Commerce (NISC chair)	National Oceanic and Atmospheric Administration (NOAA) (ANS Task Force Co-Chair)	National Marine Fisheries Service, Restoration Center -Invasive Species Program				
		Great Lakes Environmental Research Laboratory				https://www.glerl.noaa.gov//
		Southeast Fisheries Science Center				https://www.fisheries.noaa.gov/about/southeast-fisheries-science-center
Natureserve	State Governments	Individual State Heritage Programs in all 50 states	Local, State, Regional	Terrestrial and aquatic plants, animals, insects	Multi-taxa invasive species sampling as a component of natural heritage programs in each state. States individually determine the extent of inventory and monitoring of invasive species.	http://www.natureserve.org/about-us/our-history
Citizen Science	Citizen Science Association		Local, Regional		International membership-based organization advancing the field; manages a community listserv.	www.citizenscience.org
	Citizen Science: Theory and Practice		Local, Regional		Open-access, peer-reviewed journal for the field.	http://theoryandpractice.citizenscienceassociation.org
	Early Detection and Distribution Mapping System		Local, Regional		Web-based mapping system with smartphone applications for documenting invasive species distribution.	https://www.eddmaps.org/
	SciStarter		Local, Regional		Database of existing citizen science projects searchable by category.	www.SciStarter.com
	Federal Crowdsourcing and Citizen Science Toolkit		Local, Regional		Toolkit developed for federal agencies on how to design and maintain citizen science projects.	https://www.citizenscience.gov/toolkit/#

invasive forest plants. These lists may be developed in conjunction with Forest Service regional experts or State experts on the species of concern within their forest systems. Sometimes the lists are developed using State noxious weed lists, which tend to focus more narrowly on nuisance species in agricultural settings. In addition, given that the physiography, climate, and geography of each region differs, the species that become truly noxious tend to differ as well.

Another issue common to all invasive plant programs is that the reliable identification of species in the field requires considerable botanical expertise needed to distinguish them from the numerous and varied native species. Some species can be very distinctive, but many invasive plants belong to families that require advanced skills to reliably key out and identify (e.g., Compositae or Gramineae). Focused inventories or lists, as noted above, of a single or a few selected species, however, can effectively use non-specialist personnel by providing some training and thorough guides for plant identification. These guides can usually incorporate more tips on identification and distinguishing among similar species than what are generally available in formal keys of published flora.

As noted above, having consistent FIA data on an individual invasive plant may be problematic. Thus, when viewing forest invasives nationally, it may be most useful to consider invasives by lifeform, using maps to identify hot spots of invasion and areas where invasive plants appear to be progressing (i.e., invasion "fronts"). In addition, use of the FIA inventory and monitoring plots may be difficult for programs whose primary goal is early detection of new invasions. With a standard grid density of one plot per 2430 ha and a sample area of 0.067 ha, FIA plots sample only one-36,000th of the landscape (Gray 2008).

Despite these challenges, researchers have been able to produce the first national map of invasive plants in U.S. forests using FIA data (Oswalt and Oswalt 2012; Oswalt et al. 2015), showcasing how national-scale data can inform research at smaller scales (Fig. 10.2). Regional analyses have also been conducted for some areas. Gray (2008) noted that one or more non-native species were recorded on 63% of all sampled plots in Oregon and Washington. The discrepancy between results from Gray (2008) and the national map presented in Fig. 10.2

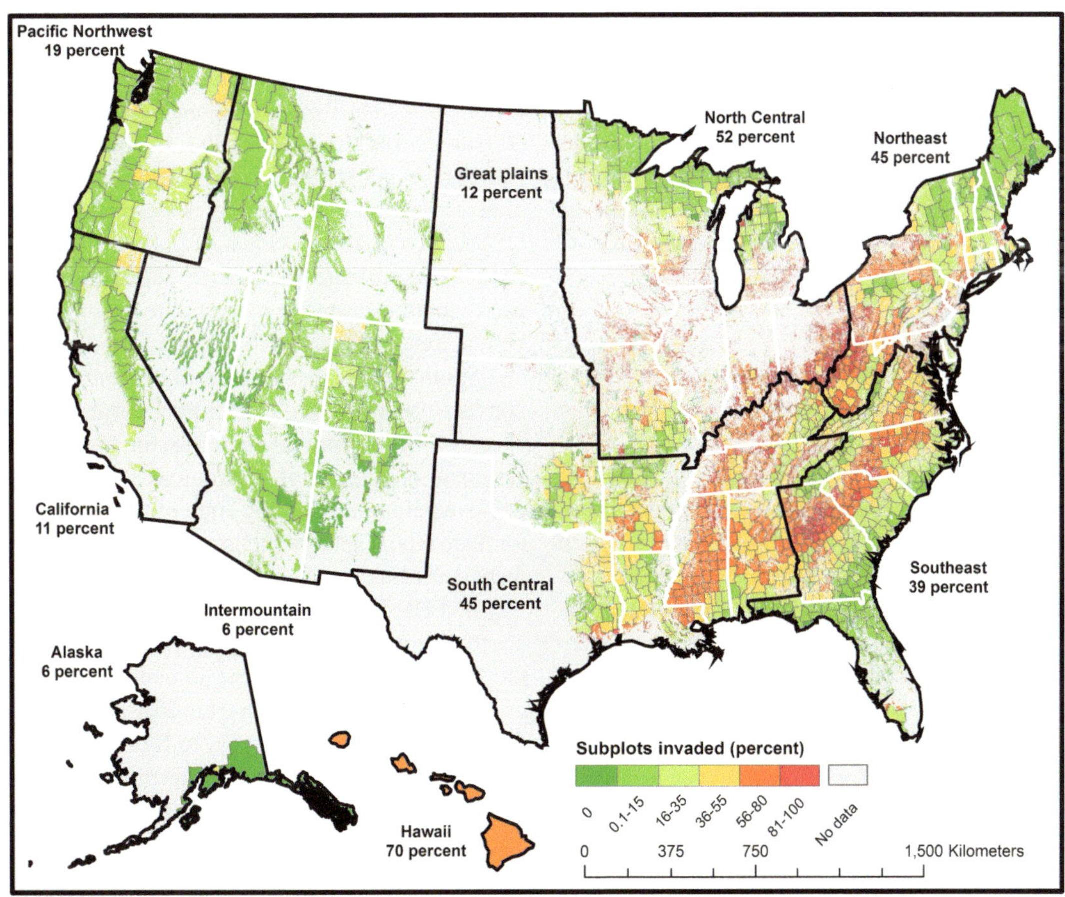

Fig. 10.2 Percent of forested subplots invaded by one or more monitored plant species as collected by the USDA Forest Service Forest Inventory and Analysis program

is most likely due to analysis of different lists of which species are considered non-native versus invasive or "monitored." Similarly, FIA invasive plant data have been used to highlight the continued northward expansion of Chinese tallowtree (*Triadica sebifera*), an invasive in the southern U.S. Coastal Plain. Reports based on FIA data documenting tallowtree expansion were picked up by local and regional news organizations and received extensive coverage. This increased exposure may increase private landowner awareness and, potentially, influence public response and behavior.

Recently, invasive plant data collected by the FIA program and compiled nationally has led to significant advancements in understanding biological invasions and working with large national datasets. Potter et al. (2016) outlined an approach using spatial neighborhood analysis to identify geographic areas with high and low invasive plant diversity across the United States. Invasive plant data collected by FIA have also been used to illustrate the improved understanding of biotic resistance to invasions gained by accounting for sub-regional variability in ecological processes (Iannone et al. 2016). When aggregated with trait-based data, this national dataset has also promoted the documentation of sub-continental patterns of invasive species across the United States (Iannone et al. 2015a). In concert, recent utilization of FIA invasive plant data has helped advance the relatively new field of macroecology (Fei et al. 2015; Iannone et al. 2015b). However, issues related to consistency among the FIA programs, including which species are identified or recorded as invasive, should be considered when using this large national database. Increased consistency among the FIA programs would help to eliminate these issues.

10.2.2 Natural Resources Conservation Service National Resources Inventory

Invasive plants in rangeland systems can differ greatly from invasive plants in forests, although there is often overlap in drier forest ecosystems. The NRI, the Bureau of Land Management (BLM), and the Forest Service all conduct invasive inventory and monitoring programs in range systems in the United States at varying scales. The NRI utilizes geographic information systems (GIS), remotely sensed imagery, soil surveys, and climate models combined with line-point intercept data collected on the ground across 17 Western States from North Dakota to Texas and westward to estimate invasive plant species distribution (Breidt and Fuller 1999; Nusser and Goebel 1997). Recent analyses of NRI data have demonstrated that 48% of rangeland area and 50% of total rangeland plant cover are impacted by invasive plants (Herrick et al. 2010).

10.2.3 USDA Forest Service Forest Health Protection Program

Detection and monitoring are critical components of effective invasive insect and disease management. Risk assessments, pathway analysis, and early detection (Chap. 6) are critical to cost-effective eradication and management actions. Once invasive insects and pathogens are established, continuing surveys to delimit their spread and monitor associated impacts are important.

The FHP program of the Forest Service provides a broad-scale example of the strength of survey and monitoring for invasive insects and disease. Forest Health Protection works closely with a network of forest health specialists in States, tribes, universities, and other Federal agencies to conduct broad-scale surveillance and reporting of insect and disease infestations. Each year, FHP and partners conduct systematic aerial and ground-based pest detection surveys on 400–500 million acres of forest lands across the United States. Although these surveys may not be targeted specifically at invasive species, any damage or infestation detected is checked for causal agent which may be an invasive species.

Forest Health Protection also coordinates several invasive insect- and pathogen-specific surveys to detect or delimit newly introduced or established species (e.g., detection surveys for bark and ambrosia beetles, sudden oak death (*Phytophthora ramorum*), walnut twig beetle (*Pityophthorus juglandis*), and hemlock woolly adelgid (*Adelges tsugae*)). The Slow the Spread program is a cooperative program conducted by the Forest Service FHP, the USDA Animal and Plant Health Inspection Service (APHIS), and State partners to detect, monitor, and treat gypsy moth (*Lymantria dispar*) infestations throughout the United States (Chap. 7). In the uninfested areas of the West and South, APHIS works with States and other partners to conduct systematic surveys for the European and Asian subspecies of gypsy moth. The basis for the Slow the Spread Program is a grid of pheromone-baited traps that are deployed along the leading edge of the infestation to monitor male moth captures and is used to help make future treatment decisions.

The FHP Forest Health Monitoring (FHM) program, in coordination with FIA, collects data on the condition of the Nation's forests and the impacts of invasive species over time. Combining data from FIA plots and pest survey data, FHM conducts analyses to report on tree health and other ecosystem components. Forest Health Monitoring disseminates survey and monitoring data in annual pest conditions reports and through an online database (https://www.fs.fed.us/foresthealth/protecting-forest/forest-health-monitoring/).

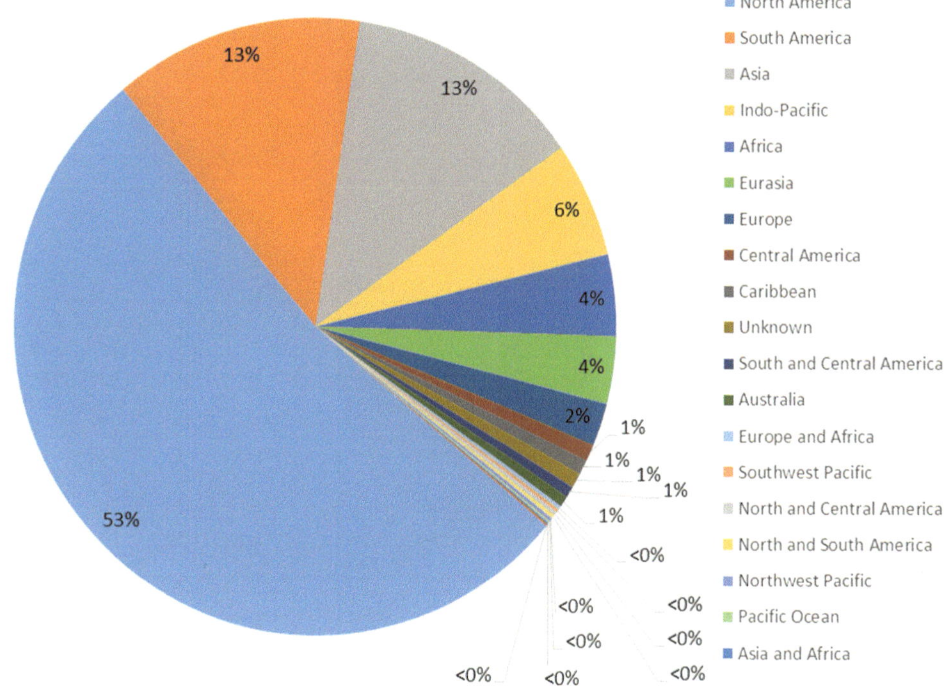

Fig. 10.3 Origin of introduction for monitored aquatic invasives per U.S. Geological Survey data

10.2.4 U.S. Geological Survey Nonindigenous Aquatic Species Program

While most aquatic species are monitored at local or regional scales, there is an effort to compile information from small-scale monitoring programs adequate to meet national information needs. The USGS NAS Database maintains a nationwide data repository and reporting, tracking, and alert system for more than 1100 aquatic invertebrates, aquatic plants, freshwater animals, and marine fish that has been operational for over 30 years. The NAS database is spatially referenced and provides data that can be queried, species profiles, spatial data sets, distribution maps, and general information related to invasives. Information is obtained from literature, museums, agency biologists, and online reporting by the general public. The NAS database is used by State, Federal, and non-governmental agencies for a wide variety of research and management needs, including learning about new introductions to their localities or regions in real time. The NAS database also accounts for the movement of invasives from other countries into the United States by tracking imports from foreign countries (Fuller and Cannister 2013).

According to the NAS database, the majority of introduced aquatic species (both invasive and noninvasive) from outside North America has been imported from South America or Asia (Fig. 10.3). While the majority of introduced aquatics in the United States consists of intentionally stocked game fish, many unintended introductions to aquatic systems are aquarium and bait releases, a portion of which become invasive (Fig. 10.4). That type of inventory information provides pathways for improving educational messages and helps identify potential actions that might prevent unintentional introduction from the pet trade.

10.3 Regional Inventory and Monitoring Programs

Regional inventory and monitoring programs focus on identifying invasives that are a threat to a specific ecosystem of concern and that may span multiple States or other political boundaries but are sub-national in scale. For example, bodies of water or mountain ranges often span multiple political boundaries, but the management and monitoring of invasive species that inhabit these areas require a coordinated approach. While the above-mentioned national programs help support regional-scale analyses and management actions, intensive monitoring for specific species at scales useful for regional analysis may be too costly, inefficient, or ecologically nonsensical at a national scale. While regional programs may be implemented at the Federal or State level, oftentimes regional monitoring efforts are collaborative in nature and involve partners from a variety of technical fields and organizations. These cooperative projects enable implementation of programs across a span of political boundaries and provide platforms for storing data, formulating action plans, and disseminating educational information. Examples of regional monitoring programs implemented by national or sub-national governmental agencies include those oper-

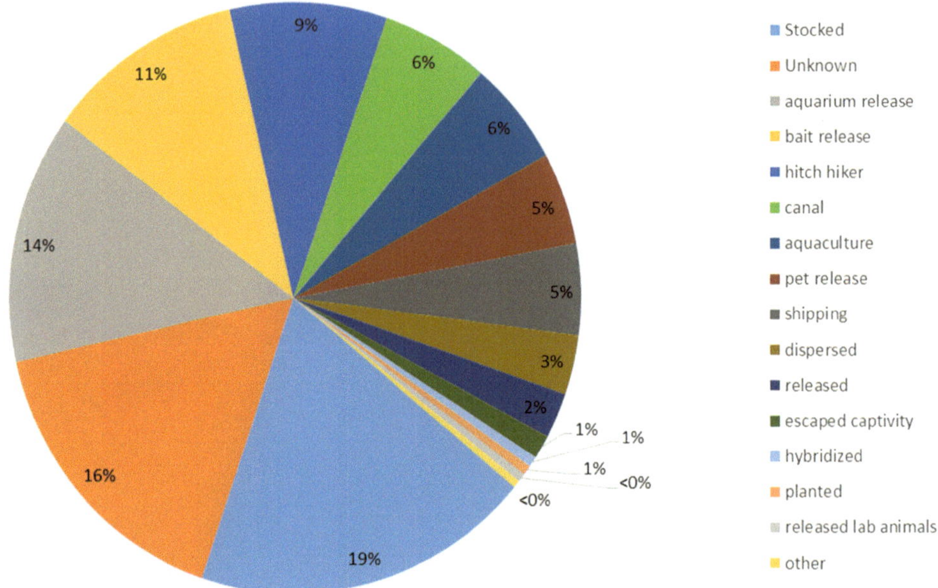

Fig. 10.4 Identified aquatic invasive introduction pathways per U.S. Geological Survey data

ated by the National Oceanic and Atmospheric Administration (NOAA) Great Lakes Environmental Research Laboratory, which inventories the Great Lakes for species like the highly reproductive zebra mussel (*Dreissena polymorpha*), and the NOAA Southeast Fisheries Science Center, which investigates invasive species in the Southeastern United States. Examples of multi-organizational inventory and monitoring cooperatives, partnerships, and working groups are widespread and varied and cannot be adequately documented here. However, one example is the St. Lawrence Eastern Lake Ontario Partnership for Regional Invasive Species Management (SLELO PRISM), which formed in 2011 to provide region-wide coordination for invasive species monitoring and management across the terrestrial and aquatic ecosystems within five counties around Lake Ontario.

As mentioned earlier, the NAS also contains regionally collected information. A significant amount of effort has been spent in recent years monitoring Asian carp (*Cyprinus* spp.) in the upper Mississippi River drainage and the Great Lakes, and though this sampling effort is specific to carp, it may detect other species. Some of the most sampled-for species are the zebra mussel and quagga mussel (*Dreissena bugensis*). Because of the habitat-changing effects they have, there are concerted efforts nationwide to detect these species in waters where they do not occur. Now that they have largely spread through much of the Mississippi Basin, detection efforts are largely concentrated in the West. Plankton samples are checked for veligers, and surfaces are checked for attached adults. Several natural resource agencies are currently working together to ensure that the Columbia River Basin is adequately monitored. These groups pool their anticipated sampling locations yearly in an effort to avoid duplication or data gaps.

The Aquatic Nuisance Species (ANS) Program of the U.S. Fish and Wildlife Service (USFWS) is another example of regional targeted monitoring. The ANS program is focused on species-specific monitoring in targeted ecosystems (e.g., zebra mussels in the Great Lakes). The ANS program has been a significant partner in the Great Lakes Restoration Initiative in which efforts are underway to use new and emerging monitoring technologies such as environmental DNA (see Chap. 8) to track invasive aquatic density in and around the Great Lakes. The USFWS also tracks invasive species on 545 national wildlife refuges, encompassing approximately 96 million acres across the United States. While monitoring programs are generally focused on singular or small groups of refuges and targeted toward localized management efforts, data can be aggregated to address larger scale issues and analyses.

Consistent with monitoring that occurs within the National Wildlife Refuge System of the USFWS, agencies such as the National Park Service (NPS), BLM, Bureau of Indian Affairs (BIA), and the Forest Service National Forest System (NFS) all monitor invasive species locally across multiple taxa. The NPS Inventory and Monitoring Vital Signs Monitoring program is currently collecting data on invasive plant and animal species in 238 park units. At least 229 park units are collecting data for invasive plant species, and 29 park units are collecting data on invasive animal species (see https://www.nps.gov/subjects/invasive/index.htm). The BLM's Assessment, Inventory, and Monitoring (AIM) Strategy for Integrated Renewable Resources Management was developed in 2011 with the goal of developing consistent monitoring procedures to be utilized across all BLM-stewarded land. The BLM AIM program provides for consistent monitoring data of invasive species on a localized level (individual BLM land units) that can facilitate easily combining data for larger scale analyses. Through the BIA Agriculture and Rangeland Development Inventory Program, small-scale monitoring occurs to address localized infestations with a strong emphasis on managing the monitored infestations. Much like the agency programs immediately above, the Forest Service tracks invasive species on each individual national forest. Attempts are made to collect data consistently to facilitate aggregation of data across all regions of the NFS.

10.4 Citizen Science as a Tool for Inventory and Monitoring of Invasive Species

Citizen science, involvement of the public in scientific research, has become a popular tool for expanding the reach of inventory and monitoring for many invasive species (Crowl et al. 2008; Dickinson and Bonney 2012; Dickinson et al. 2010) to support management and public policy (McKinley et al. 2015) (Fig. 10.5). In 2010, a study found 249 such programs in the United States alone (Crall et al. 2010). This number continues to grow as the availability of new technologies allows anyone with Internet access or a smartphone to submit data to a growing number of online data repositories (Newman et al. 2012; Wallace and Bargeron 2014). In addition, as resources for professional monitoring programs decline, citizen science approaches increase the number of observers in the field to identify and report species of concern (Ashcroft et al. 2012; Bodilis et al. 2014; Bryce et al. 2011; Simpson et al. 2009) (see Table 10.2). Several studies have demonstrated how citizen science efforts have enhanced knowledge on the distribution of existing species beyond that gleaned from professional efforts (Ashcroft et al. 2012; Delaney et al. 2008; Goldstein et al. 2014).

With the growth of citizen science in the past decade, there have been concerns raised about the ability of these programs to generate high-quality data (Nature Editorial 2015). Indeed, some protocols using volunteers have resulted in issues such as the misidentification of certain taxonomic groups (Bloniarz and Ryan 1996; Brandon et al. 2003; Crall et al. 2011) and lower detection rates for small pest infestations (Fitzpatrick et al. 2009). However, other studies have shown these programs can generate useful data of high quality when researchers develop protocols for volunteers that emphasize data quality (Boudreau and Yan 2004; Crall et al. 2011; Delaney et al. 2008; Newman et al. 2010). Specifically, studies have shown that volunteer experience (Bloniarz and Ryan 1996), participant confidence in identification (Crall et al. 2011), and volunteer certification (Dickinson et al. 2010) result in improvements in accuracy when implemented as part of a protocol. Examination of data quality among data collectors with varying expertise demonstrated that the quality of data should be a concern of data generated from volunteer efforts (Crall et al. 2011). The variability in findings across studies demonstrates the need to use best practices when developing new protocols and then iteratively testing and refining them to assess and, if needed, adjust for any limitations in data quality.

Technology will continue to play a vital role in increasing public engagement while ensuring data quality (Bonney et al. 2009; Newman et al. 2012; Wallace and Bargeron

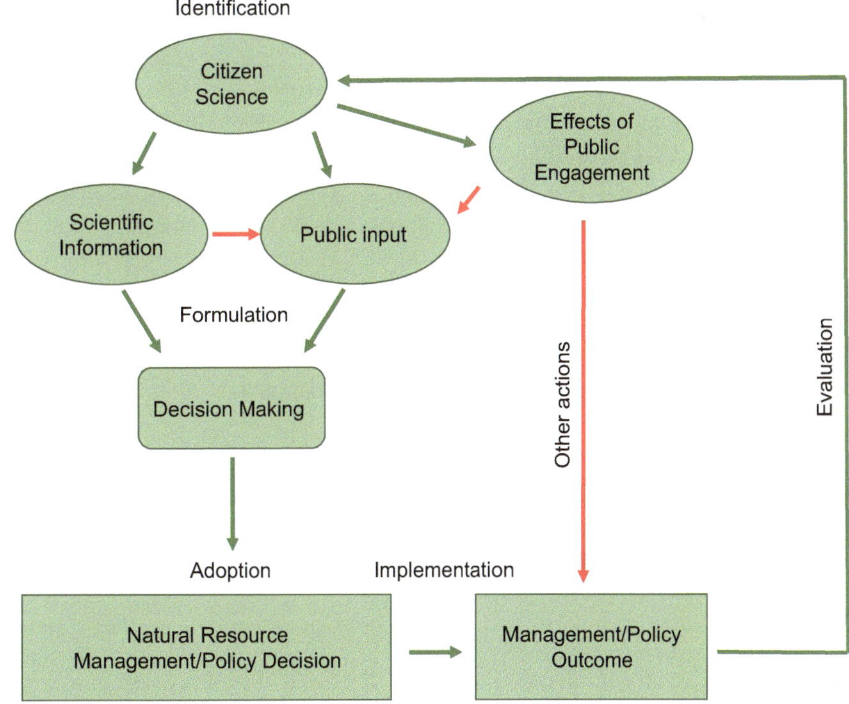

Fig. 10.5 Pathways that citizen science can take to influence natural resource management by (1) generating scientific information and (2) facilitating direct (*green arrows*) and indirect (*red arrows*) public input and engagement. (Figure reproduced with permission from the Ecological Society of America (see McKinley et al. (2015)))

Table 10.2 General resources available for designing an effective citizen science program

Resource	Description	Website
Citizen Science Association	International membership-based organization advancing the field; manages a community listserv	https://www.citizenscience.org
Citizen Science: Theory and Practice	Open-access, peer-reviewed journal for the field	https://theoryandpractice.citizenscienceassociation.org
Early Detection and Distribution Mapping System	Web-based mapping system with smartphone applications for documenting invasive species distribution	https://www.eddmaps.org
SciStarter	Database of existing citizen science projects searchable by category	https://scistarter.com
Federal Crowdsourcing and Citizen Science Toolkit	Toolkit developed for Federal agencies on how to design and maintain citizen science projects	https://www.citizenscience.gov/toolkit

2014). Additional details on the availability of tools and technology are provided in Chap. 11. The project iNaturalist uses a crowd-sourced approach where the online community must agree with the identification of a species made by an observer. Based on the level of community agreement with the observation, the observation will be tagged as "casual" or "research" grade. Other projects, like eBird, have embedded code that automatically performs data quality checks on new sightings entered into the system (Bonter and Cooper 2012). Photo vouchers uploaded to regional websites that include a network of trained taxonomists facilitate rapid verification and response for this and other projects (Crall et al. 2012; Simpson et al. 2009).

As researchers and managers begin to adopt citizen science more broadly, efforts should be initiated to better integrate the data collection efforts of both professionals and volunteers. Through data synergy, researchers and natural resource managers can fill gaps in species distributions to support the development of habitat suitability models that can be used to predict future spread (Ashcroft et al. 2012; Crall et al. 2015; Crowl et al. 2008; Lodge et al. 2006). Less rigorous monitoring approaches can also be used to inform more rigorous approaches (Aslan and Rejmanek 2010). For example, Crall et al. (2010) propose an iterative sampling design that integrates monitoring efforts of these two groups guided by habitat suitability models. It distinguishes the role of the professional from the volunteer and provides recommendations to build on the benefits and weaknesses of each approach. Testing of this approach in the field indicated more species could be found with less sampling effort (Crall et al. 2015). Protocols utilizing and building upon both volunteer and professional data collection in the field have been proposed and tested but not broadly implemented.

Since a majority of volunteer data is collected opportunistically (Crall et al. 2010), sampling design and other monitoring procedures should be considered when designing as well as integrating data from these multiple groups. Appropriate metadata can be used to provide the necessary information to help identify only those data appropriate to address certain research questions (Dickinson et al. 2010).

10.5 Integration Across Scales

Monitoring at multiple geographic scales delivers unique strengths but may also impart some misperceptions. Researchers must recognize that monitoring and inventory systems at large scales are useful for different management goals and storytelling than are monitoring and inventory programs at local levels. For example, at a local scale, an infestation by an invasive species may severely impact a resource, but that impact may be limited to a particular waterbody, forest, or neighborhood. Thus, to the local community, that invasive species may be a serious threat to local biodiversity or economies. In contrast, the same invasive species may be relatively limited at a regional or national scale. This "problem of scale" can become particularly challenging when scientists studying impacts at the stand level (for forests) or similar report devastating effects to the ecosystem in question, while impacts to the overall landscape remain relatively limited. Engaging scientists, landowners, managers, and politicians in discussions involving the impacts of species at various scales is important in developing invasive species prevention and management strategies.

One method of using large-scale inventory and monitoring data is that of "hot spot" detection and/or prioritizing where to direct resources at the broad scale. For example, in a map of invasive plant occurrence on southern U.S. forest land, it is immediately apparent where highly infested areas occur. Noting where these "hot spots" exist can help direct resources to the edges of those hotspots in surrounding counties to help potentially reduce the spread of species beyond their current boundaries.

10.6 Key Findings, Gaps, and Challenges in Inventory and Monitoring

10.6.1 Key Findings

- Inventory and monitoring of invasive species across multiple scales and taxa are integral to understanding where to focus management time, effort, and financial resources.

- National and sub-national programs are considered surveillance monitoring programs and can be used to identify invasive species "hot-spots," to monitor spread, and to conduct other broad-scale studies.
- Early detection of the spread of species is the most cost-effective method for controlling invasives, yet most national programs are better suited for monitoring spread rather than early detection.
- Assessment of inventory data has revealed that invasives impact a broad portion of the terrestrial and aquatic landscape.
- Tracking invasive species will become increasingly important in light of a changing climate because invasive species are generally well suited to rapid establishment in novel systems.
- The use of citizen science in monitoring is becoming an increasingly accepted and utilized method for collecting information.

10.6.2 Key Information Needs

Despite the need for inventory and monitoring, and the wide variety of programs and multiple scales tasked with tracking invasive species, there are still gaps in knowledge and challenges related to inventory and monitoring.

- **Paucity of Spatially Referenced Standardized Data** Although national programs exist, the scale of spatial data is still broad, and generating simple distribution data for many species remains a challenge, especially recently identified species not yet on monitoring lists and species in non-forest systems and aquatic habitats. There is also a significant need to standardize inventory approaches among agencies and partners. Many organizational datasets and citizen science datasets containing invasive species data remain isolated. Opportunities for improvement include working to create consistency within and among programs regionally and nationally, integrating isolated monitoring programs, and developing additional methodologies for monitoring certain taxa (e.g., insects, disease, and genetics). There is also a need to integrate professional and volunteer datasets for monitoring and early detection/rapid response across political boundaries.
- **Longitudinal Datasets** Many inventory programs are conducted at temporal scales that do not allow assessment of drivers of invasive species spread. These types of datasets are needed to understand how invasive species may respond to climate change and other environmental stressors or disturbances.
- **Adoption and Reliability of Citizen Science and Project Design** There are still concerns among managers and researchers regarding the quality of citizen science data. These concerns need to be adequately addressed to broaden the adoption of citizen science as a tool to monitor and manage invasive species. Existing and future research on project successes should be used to support the establishment of new and iterative citizen science projects.
- **New Survey and Monitoring Techniques for Non-native Insects and Pathogens** New tools and technologies for inventory and monitoring need to be developed and deployed to improve inventory and monitoring of those species that are challenging to survey.

Disclaimer Text The findings and conclusions in this publication are those of the authors and should not be construed to represent any official USDA or U.S. Government determination or policy.

Literature Cited

Ashcroft MB, Gollan JR, Batley M (2012) Combining citizen science, bioclimatic envelope models and observed habitat preferences to determine the distribution of an inconspicuous, recently detected introduced bee (*Halictus smaragdulus* Vachal Hymenoptera: Halictidae) in Australia. Biol Invasions 14:515–527

Aslan CE, Rejmanek M (2010) Avian use of introduced plants: ornithologist records illuminate interspecific associations and research needs. Ecol Appl 20:1005–1020

Bechtold WA, Patterson PL (eds) (2005) The enhanced forest inventory and analysis program – national sampling design and estimation procedures, Gen. Tech. Rep. SRS-80. U.S. Department of Agriculture, Forest Service, Southern Research Station, Asheville, 85 p

Bloniarz DV, Ryan HDP (1996) The use of volunteer initiatives in conducting urban forest resource inventories. J Arboric 22:75–82

Bodilis P, Louisy P, Draman M et al (2014) Can citizen science survey non-indigenous fish species in the eastern Mediterranean Sea? Environ Manag 53:172–180

Bonney R, Cooper CB, Dickinson J et al (2009) Citizen science: a developing tool for expanding science knowledge and scientific literacy. Bioscience 59:977–984

Bonter DN, Cooper CB (2012) Data validation in citizen science: a case study from Project FeederWatch. Front Ecol Environ 10:305–307

Boudreau SA, Yan ND (2004) Auditing the accuracy of a volunteer-based surveillance program for an aquatic invader *Bythotrephes*. Environ Monit Assess 91:17–26

Brandon A, Spyreas G, Molano-Flores B et al (2003) Can volunteers provide reliable data for forest vegetation surveys? Nat Areas J 23:254–261

Breidt FJ, Fuller WA (1999) Design of supplemented panel surveys with application to the National Resources Inventory. J Agric Biol Environ Stat 4:391–403

Bryce R, Oliver MK, Davies L et al (2011) Turning back the tide of American mink invasion at an unprecedented scale through community participation and adaptive management. Biol Conserv 144:575–583

Byers JE, Reichard S, Randall JM et al (2002) Directing research to reduce the impacts of nonindigenous species. Conserv Biol 16:630–640

Crall AW, Newman GJ, Jarnevich C et al (2010) Improving and integrating data on invasive species collected by citizen scientists. Biol Invasions 12:3419–3428

Crall AW, Newman G, Waller DM et al (2011) Assessing citizen science data quality: an invasive species case study. Conserv Lett 4:433–442

Crall AW, Renz M, Panke BJ et al (2012) Developing cost-effective early detection networks for regional invasions. Biol Invasions 14:2461–2469

Crall AW, Jarnevich CS, Young NE et al (2015) Citizen science contributes to our knowledge of invasive plant species distributions. Biol Invasions 17:2415–2427

Crowl TA, Crist TO, Parmenter RR et al (2008) The spread of invasive species and infectious disease as drivers of ecosystem change. Front Ecol Environ 6:238–246

Delaney DG, Sperling CD, Adams CS, Leung B (2008) Marine invasive species: validation of citizen science and implications for national monitoring networks. Biol Invasions 10:117–128

Dickinson JL, Bonney R (eds) (2012) Citizen science: public participation in environmental research. Comstock Publishing Associates, Ithaca, 304 p

Dickinson JL, Zuckerberg B, Bonter DN (2010) Citizen science as an ecological research tool: challenges and benefits. Annu Rev Ecol Evol Syst 41:149–172

Fei S, Iannone BV III, Oswalt CM et al (2015) Understanding macroscale invasion patterns and processes with FIA data. In: Stanton SM, Christensen GA (eds) Pushing boundaries: new directions in inventory techniques and applications. Forest Inventory and Analysis (FIA) symposium 2015, December 8–10, Portland, OR, Gen. Tech. Rep. PNW-GTR-931. U.S. Department of Agriculture, Forest Service, Pacific Northwest Research Station, Portland, pp 92–93

Fitzpatrick MC, Preisser EL, Ellison AM, Elkinton JS (2009) Observer bias and the detection of low-density populations. Ecol Appl 19:1673–1679

Fuller P, Cannister M (2013) USGS nonindigenous aquatic species database with a focus on the introduced fishes of the lower Tennessee and Cumberland drainages. In: Johansen R, Estes LD, Hamilton SW, Barrass AN (eds) On the natural history of lower Tennessee and Cumberland River Valleys. Center for Excellence for Field Biology, Austin Peay State University, Clarksville, pp 29–42

Goldstein EA, Lawton C, Sheehy E, Butler F (2014) Locating species range frontiers: a cost and efficiency comparison of citizen science and hair-tube survey methods for use in tracking an invasive squirrel. Wildl Res 41:64–75

Gray A (2008) Monitoring and assessment of regional impacts from nonnative invasive plants in forests of the Pacific Coast, United States. In: Kohli RK, Jose S, Singh HP, Batish DR (eds) Invasive plants and forest ecosystems. CRC Press, Boca Raton, pp 217–235

Hellmann JJ, Byers JE, Bierwagen BG, Dukes JS (2008) Five potential consequences of climate change for invasive species. Conserv Biol 22:534–543

Herrick JE, Lessard VC, Spaeth KE et al (2010) National ecosystem assessments supported by scientific and local knowledge. Front Ecol Environ 8:403–408

Iannone BV III, Oswalt CM, Liebhold AM et al (2015a) Region-specific patterns and drivers of macroscale forest plant invasions. Divers Distrib 21:1181–1192

Iannone BV III, Potter KM, Guo Q et al (2015b) Biological invasion hotspots: a trait-based perspective reveals new sub-continental patterns. Ecography 38:1–9

Iannone BV III, Potter KM, Hamil KD et al (2016) Evidence of biotic resistance to invasions in forests of the Eastern USA. Landsc Ecol 31:85–99

Lodge DM, Williams S, MacIsaac HJ et al (2006) Biological invasions: recommendations for US policy and management. Ecol Appl 16:2035–2054

McKinley DC, Miller-Rushing AJ, Ballard HL et al (2015) Investing in citizen science can improve natural resource management and environmental protection. Issues Ecol 19:1–27

Myers JH, Simberloff D, Kuris AM, Carey JR (2000) Eradication revisited: dealing with exotic species. Trends Ecol Evol 15:316–320

Nature Editorial (2015) Rise of the citizen scientist. Nature 524:265

Newman G, Crall A, Laituri M et al (2010) Teaching citizen science skills online: implications for invasive species training programs. Appl Environ Educ Commun 9:276–286

Newman G, Wiggins A, Crall A et al (2012) The future of citizen science: emerging technologies and shifting paradigms. Front Ecol Environ 10:298–304

Nichols JD, Williams BK (2006) Monitoring for conservation. Trends Ecol Evol 21:668–673

Nusser SM, Goebel JJ (1997) The national resources inventory: a long-term multi-resource monitoring programme. Environ Ecol Stat 4:181–204

Oswalt SN, Oswalt CM (2012) The national picture of nonnative plants in the United States according to FIA data. In: Morin RS, Liknes GC (eds) Moving from status to trends: forest inventory and analysis (FIA) symposium 2012; 2012 December 4–6; Baltimore, MD, Gen. Tech. Rep. NRS-P-105. U.S. Department of Agriculture, Forest Service, Northern Research Station. [CD-ROM], Newtown Square, pp 262–267

Oswalt CM, Fei S, Guo Q et al (2015) A subcontinental view of forest plant invasions. NeoBiota 24:49–54

Potter KM, Koch FH, Oswalt CM, Iannone BV (2016) Data, data everywhere: detecting spatial patterns in fine-scale ecological information collected across a continent. Landsc Ecol 31:67–84

Simpson A, Jarnevich C, Madson J et al (2009) Invasive species information networks: collaboration at multiple scales for prevention, early detection, and rapid response to invasive alien species. Biodiversity 10:5–13

Wallace RD, Bargeron CT (2014) Identifying invasive species in real time: early detection and distribution mapping system (EDDMapS) and other mapping tools. In: Ziska LH, Dukes JS (eds) Invasive species and global climate change. CABI, Wallingford, pp 219–231

11 Tools and Technologies for Quantifying Spread and Impacts of Invasive Species

Matt Reeves, Inés Ibáñez, Dana Blumenthal, Gang Chen, Qinfeng Guo, Catherine Jarnevich, Jennifer Koch, Frank Sapio, Michael K. Schwartz, Ross K. Meentemeyer, Bruce K. Wylie, and Stephen Boyte

11.1 The Need for Tools and Technologies

The need for tools and technologies for understanding and quantifying invasive species has never been greater. Rates of infestation vary on the species or organism being examined across the United States, and notable examples can be found. For example, from 2001 to 2003 alone, ash (*Fraxinus* spp.) mortality progressed at a rate of 12.97 km year $^{-1}$ (Siegert et al. 2014), and cheatgrass (*Bromus tectorum*) is expected to increase dominance on 14% of Great Basin rangelands (Boyte et al. 2016). The magnitude and scope of problems that invasive species present suggest novel approaches for detection and management are needed, especially those that enable more cost-effective solutions. The advantages of using technologically advanced approaches and tools are numerous, and the quality and quantity of available information can be significantly enhanced by their use. They can also play a key role in development of decision-support systems; they are meant to be integrated with other systems, such as inventory and monitoring, because often the tools are applied after a species of interest has been detected and a threat has been identified. In addition, the inventory systems mentioned in Chap. 10 are regularly used in calibrating and validating models and decision-support systems. For forested areas, Forest Inventory and Analysis (FIA) data are most commonly used (e.g., Václavík et al. 2015) given the long history of the program. In non-forested systems, national inventory datasets have not been around as long (see Chap. 10), but use of these data to calibrate and validate spatial models is growing. These inventory datasets include the National Resources Inventory (NRI) (e.g., Duniway et al. 2012) and the Assessment Inventory and Monitoring program (AIM) (e.g., McCord et al. 2017). Similarly, use of the Nonindigenous

M. Reeves (✉)
U.S. Department of Agriculture, Forest Service, Rocky Mountain Research Station, Human Dimensions Program, Department of Forestry and Environmental Resources Campus, Missoula, MT, USA
e-mail: matt.c.reeves@usda.gov

I. Ibáñez
School of Environment and Sustainability, University of Michigan, Ann Arbor, MI, USA

D. Blumenthal
U.S. Department of Agriculture, Agricultural Research Service, Rangeland Resources & Systems Research Unit, Fort Collins, CO, USA

G. Chen
Department of Geography and Earth Sciences, University of North Carolina at Charlotte, Charlotte, NC, USA

Q. Guo
U.S. Department of Agriculture, Forest Service, Southern Research Station, Eastern Forest Environmental Threat Assessment Center, Research Triangle Park, NC, USA

C. Jarnevich
U.S. Deparatment of Interior, U.S. Geological Survey, Fort Collins Science Center, Fort Collins, CO, USA

J. Koch
U.S. Department of Agriculture, Forest Service, Northern Research Station, Delaware, OH, USA

F. Sapio
Forest Health Assessment and Applied Sciences Team, Forest Health Protection, U.S. Department of Agriculture, Forest Service, Fort Collins, CO, USA

M. K. Schwartz
U.S. Department of Agriculture, Forest Service, National Genomics Center for Wildlife and Fish Conservation, Missoula, MT, USA

R. K. Meentemeyer
Center for Geospatial Analytics, North Carolina State University, Raleigh, NC, USA

B. K. Wylie
U.S. Department of Interior, U.S. Geological Survey, EROS Science Division, Sioux Falls, SD, USA

S. Boyte
U.S. Department of the Interior, U.S. Geological Survey, Sioux Falls, SD, USA

T. M. Poland et al. (eds.), *Invasive Species in Forests and Rangelands of the United States*,
https://doi.org/10.1007/978-3-030-45367-1_11

Aquatic Species (NAS) database is growing as well (e.g., Evangelista et al. 2017). The consistent protocols employed by these programs prove valuable for developing better tools, but the data they afford are generally limited for some tools because the sampling intensity is too low.

The rapidly accelerating impact of invasive species suggests that development and implementation of geospatial tools and technologies will need to be expanded, given the high cost and impracticality of wide-area in situ reconnaissance. Although improvements are definitely needed, with increasing remote sensing data availability and significant computer processing capacity, geospatial tools and models are now sufficiently useful to be applied in an operational mode.

While tools and technologies offer significant promise, they should not be automatically applied to all situations. Ground surveys provide the most fundamental data for species management, and nearly all imagery studies suggest that georeferenced field data will not be replaced by technology anytime soon (Underwood et al. 2013). Though the need for consistent and comprehensive data describing the extent and location of invasive species has never been greater, interagency collaboration toward this end continues to be deficient. This is significant since the effective tools and technologies invariably require access to plot data for calibration and validation. In turn, fieldwork can become more efficient through the use of tools that identify high-priority locations for management or those that are likely to experience new invasions or outbreaks in the future (Underwood et al. 2013). An accounting of these tools and technologies is therefore essential to aid in prioritization, to determine which tools merit further development, and to identify gaps where further research is needed.

Recently, Underwood et al. (2013) provided excellent overviews of remote sensing and geospatial tools, focusing on more specific examples than are discussed in this chapter. We sought to build on their previous efforts and add some non-spatially explicit technologies by focusing more attention on different types of tools needed for evaluating invasive species. As a result, our assessment is broad and introduces new ideas and concepts that represent the state of the science. This chapter provides a synthesis of tools, technologies, and techniques that are available across a number of disciplines, to quantify, estimate, and characterize presence, spread, and impacts of invasive species. Many tools and technologies specific to management activities (e.g., development of new trapping methods) are covered in Chap. 7. Here we focus on five areas including detection and mapping, predicting establishment and spread, decision-support systems, genetic tools for restoration, and key findings, information needs, and opportunities. The systems presented are not a comprehensive list but do offer a state-of-the-science assessment of prominent tools and techniques.

11.2 Detection and Mapping of Species Occurrence or Spread

In this section, we refer to invasive species as a subject, which can include floral and faunal species. Some of the ideas, tools, and techniques can apply to both floral and faunal population, but, in general, we focus our efforts on vegetation. Understanding and reducing the spread and impact of invasive species begin with detection (Lodge et al. 2006). Detection involves surveying, reporting, and verifying the presence of a non-native species. This is a critical process because it is the basis for initiating a rapid and timely response before an invasive species spreads so widely that eradication is no longer feasible (U.S. Department of Interior 2016). There are many approaches used to detect invasive species before they become established, but the most successful approaches are relatively inexpensive and have the potential for wide application. Thus, approaches like citizen science to detect colonization of new habitats by invasive species can be especially effective since they meet the criteria of being inexpensive and widely applicable (Crall et al. 2011; Delaney et al. 2008; Hawthorne et al. 2015). Citizen science is particularly relevant when the target species is conspicuous and easily identified (Darwall and Dulvy 1996), which is rarely the case for insects and pathogens. Yet, there are circumstances when those involved in citizen science lack the more advanced training or technology that is required to detect and identify invasive species (Fore et al. 2001). The subject of citizen science is discussed in Chap. 10 and in this chapter (including Geospatial Participatory Modeling). Tools for early detection included in this chapter include remote sensing, traditional geospatial modeling, analysis of inventory databases, and molecular detection techniques. Of these, remote sensing is probably the most widely used method for gathering data, repeatedly over large areas, at a relatively low cost.

11.2.1 Remote Sensing

A diverse suite of sensors spanning numerous temporal, spatial, and spectral resolutions has been evaluated for use in detection strategies. However, all remote sensing systems have some inherent limitations that constrain their applicability to specific situations, species, and temporal and spatial domains. Some of these limitations can be at least partially addressed through greater collaboration and sharing of resources across disciplines, organizations, and agencies. If image processing, data warehousing, and spectral analysis algorithms were partitioned among different organizations, great economies of scale could be realized, and this process would foster greater interagency communication, data collection, sharing, and cooperation.

Some species are particularly unsuited for identification with remote sensing instruments because they are too small, mobile, or simply look like non-target species. For example, consider a remote sensing early detection system designed to find new outbreaks of kudzu (*Pueraria lobata*). Because kudzu is normally a sub-canopy species, it will be practically impossible to identify using most remote sensing platforms. Likewise, detection of new outbreaks of knapweeds (*Centaurea* spp.) at low densities will be equally as difficult given the similarities of spectral qualities with other species. Successful strategies for early detection with remote sensing instruments exploit characteristics of a target invasive species that are separable from background vegetation, by color, spectral response, phenology, or inference. Thus, for the purpose of this chapter, we limit our discussion to remote sensing studies, techniques, and platforms that have demonstrated some utility for effective early detection.

Remote sensing is only useful for detection if outbreaks of invasive species populations cause changes in spectral response from airborne or space-borne instruments (Asner et al. 2008a). Current remote detection and mapping of the ecological impacts of invasive species typically rely on the measurements of the disparities in spectral, structural, and temporal characteristics. These disparities, however, resulting from symptoms of invasion such as dead tree canopies caused by sudden oak death (caused by *Phytophthora ramorum*), often manifest several seasons after initial infection (Haas et al. 2016). This delay in expression of symptoms is problematic for early detection regardless of the characteristics of the sensors being used. In addition, detection of invasive vegetation using remotely sensed data can be difficult because non-native plants are often obscured by, or commingled with, natural vegetation, making them difficult to identify using relatively moderate spatial/spectral resolution images.

Spectral Considerations Changes in the spectral response of a vegetation canopy may be used to identify an invasion. Because such phenomena can be directly captured in the visible portion of the electromagnetic spectrum (wavelengths from approximately 400–700 nm), many remote sensing systems are equipped to detect the noticeable discoloration. Typical systems that have proven effective to detect foliage-level spectral variation include those that employ coarse-spatial resolution for broad-scale detection, for example, 1-km MODIS (Moderate Resolution Imaging Spectroradiometer) (Coops et al. 2009) and 30-m Landsat (Skakun et al. 2003), and high-spatial resolution for monitoring fine-scale patchy distributions of tree mortality, for example, aerial photography (Kelly and Meentemeyer 2002) and 2.5-m QuickBird (Wulder et al. 2008). Limiting detection strategies to only visible bands is problematic if, for example, an affected forest remains at the pre-visual green mortality stage, where tree foliage contains slightly reduced chlorophylls and water content. To address this challenge, researchers seek to discover a stronger relationship between plant physiological stress and spectral reflectance from the near infrared (wavelengths from approximately 700–1300 nm) and the short wave infrared (wavelengths from approximately 1300–2500 nm) spectral ranges (Knipling 1970; Laurent et al. 2005). Most of today's remote sensors have the capacity to record near infrared radiation, while the short wave infrared bands are often available from the sensors with medium to coarse spatial resolution (e.g., Thematic Mapper (TM) and Landsat 8 Operational Land Imager (OLI)).

At times, however, more narrow spectral channels offered on hyperspectral platforms offer improved identification of the subtle spectral discrepancies between healthy and damaged vegetation. Fine spectral resolution data acquired through hyperspectral imaging provide a viable solution by using dozens to hundreds of narrow and contiguous spectral bands. Successful applications of hyperspectral sensors have primarily emerged since 2000 (e.g., Asner et al. 2008b; Chen et al. 2015; Cheng et al. 2010; Coops et al. 2003; McNeil et al. 2007; Noujdina and Ustin 2008; Pu et al. 2008). Unfortunately, the majority of the applications have been limited to relatively small areas because hyperspectral sensors, to date, are mostly mounted on airborne platforms, such as NASA's AVIRIS (Airborne Visible/Infrared Imaging Spectrometer), Canada's CASI (Compact Airborne Spectrographic Imager), and Australia's HyMap (hyperspectral mapper). A few space-borne hyperspectral systems (e.g., NASA's Hyperion) are available; however, their applications have been restricted due to limited spatial coverage and high spectral noise. NASA's next generation HyspIRI (Hyperspectral Infrared Imager) is expected to deliver consistent, global coverage imagery using contiguous 10-nm spectral bands from the visible, short wave infrared to the thermal infrared spectral range.

Spatial Resolution Considerations Satellite technology has been used to study issues of environmental concern over large geographical areas (Rose et al. 2015), but high-spatial resolution satellite and airborne imagery, such as sub-meter aerial photos, are more suitable for detecting fine-scale disturbances, where the infested native species are within small, discrete patches (Meddens et al. 2011). Applying high spatial resolution data can be advantageous for capturing spatial details for monitoring plant structural dynamics; however, limited data availability, high acquisition costs, and reduced processing efficiency often become major obstacles in real-world applications.

Despite these issues, perhaps the most straightforward approach for detecting invasive species is using visual inspection of high spatial resolution images. Using this simple technique, it is possible to pinpoint certain species based on their unique spatial patterns, phenological characteristics (Huang and Asner 2009), or color of inflorescence. Species with colorful, diagnostic flowers, such as leafy spurge (*Euphorbia esula*) and saltcedar (*Tamarix chinensis*), have been identified using visible wavelengths (400–700 nm) in aerial photographs taken during the flowering seasons (Everitt et al. 1995, 1996). The USDA National Agriculture Imagery Program (NAIP) has archived color (RGB) and color infrared (CIR) aerial photographs with resolutions ranging from a few centimeters (in the case of aerial videography) to ~2 m. Each State has its own acquisition schedule and choice of imagery, but the timing of data acquisition is crucial because the data may only be useful if collected when the targeted non-native plant is distinct from its background and neighboring areas (e.g., flowering).

Techniques and sensors also exist that enable exploitation of unique spatial patterns. For example, 4-m multispectral IKONOS imagery has been used to identify Melaleuca (*Melaleuca quinquenervia*) in South Florida because the spatial pattern of Melaleuca is highly aggregated (Huang and Asner 2009). However, even with such a diagnostic spatial pattern, 4-m spatial resolution was insufficient to identify this tree at lower densities (Fuller 2005). Likewise, smaller statured species, such as grasses and forbs, often require even higher resolution imagery, again exemplifying the need to design specific protocols for each target species. For example, QuickBird (2.4 m), another multispectral satellite system, has been used to estimate (accuracy assessment ≥65%) the presence of multiple non-native plants including purple loosestrife (*Lythrum salicaria*), common reed (*Phragmites australis*), and water chestnut (*Trapa natans*) in the Hudson River National Estuarine Research Reserve (Laba et al. 2008), and to delineate (accuracy assessment ≥86%) giant cane (*Arundo donax*) in Texas (Everitt et al. 2005). Again, despite success with relatively high resolution from satellite platforms, some situations require still greater resolution necessitating use of aerial platforms.

Very-high-resolution imagery taken from light aircraft or unmanned aerial vehicles (UAVs) can be used to find and even measure invasive species that blend in with other vegetation. While not yet practical for regional mapping, it has been used effectively for finding invasive species, measuring changes in invasive species abundance over time, and understanding environmental correlates of invasive species success (Blumenthal et al. 2012; Calviño-Cancela et al. 2014; Wan et al. 2014; Zaman et al. 2011).

Digital images taken from slow-flying lightweight aircraft have attained resolution (ground sample distances) of as little as 1–2 mm per pixel (Booth and Cox 2008). The advantage of such methods is that species not visible with coarser resolution methods can be identified and measured (Fig. 11.1). Herbaceous invasive species that have been effectively measured through visual analysis of such images include Dalmatian toadflax (*Linaria dalmatica*), leafy spurge, and cheatgrass (*Bromus tectorum*) (Blumenthal et al. 2007; Booth et al. 2010; Mealor et al. 2012). The disadvantage of using such very-high-resolution methods is that they currently offer narrow fields of view and thus low spatial coverage (17.5–48.5 m/image in the above examples). Consequently, they favor subsampling rather than fully mapping invasive species. For example, 2049 images used to study Dalmatian toadflax in mixed-grass prairie covered only 2.4% of the 4.1-km^2 sampling area (Blumenthal et al. 2007).

Fig. 11.1 Digital aerial image of mixed-grass prairie containing multiple patches of Dalmatian toadflax. *Upper inset* shows an individual toadflax plant with six stems. *Lower inset* shows a dense patch of toadflax. *Both insets* also contain prairie sagewort (*Artemisia frigida* Willd.), with lighter gray-green foliage

For many invasive species, moderately lower resolution images are sufficient to distinguish them from native vegetation, facilitating complete spatial coverage. For example, 25-cm spatial resolution visible and near-infrared images taken from a UAV were sufficient to identify patches of common reed (Zaman et al. 2011) across a 31-km^2 area over several hours of flight time. Similarly, 20–40-cm spatial resolution hyperspectral imagery has been effective for mapping broad Melaleuca patches across the Florida Everglades. Even without hyperspectral imagery, 2.4-cm ground resolution was sufficient to distinguish Brazilian pepper (*Schinus terebinthifolius*) from similar woody vegetation in Florida (Pearlstine et al. 2005).

As technology continues to improve, the trade-off between resolution and spatial coverage should be eliminated. This prospect could be realized with a combination of improved sensors flown at greater altitudes and faster processing and storage speeds, allowing more images to be taken per unit time (Anderson and Gaston 2013; Calviño-Cancela et al. 2014). Visual examination of images can be accurate and quick when determining presence/absence, which is often all that is needed to detect new infestations (Blumenthal et al. 2007). Visual measurements of species cover, however, can take considerably longer; consequently, developing techniques for automating image processing will be key to measuring invasive species cover over larger areas. These can rely on spectral characteristics, texture, shape, and phenology, and sometimes involve machine learning (Bradley 2014; Pearlstine et al. 2005; Wan et al. 2014). Given images with sufficient spatial and spectral resolution, such methods can often achieve >95% accuracy in identifying invasive species (Calviño-Cancela et al. 2014; McCormick 1999; Wan et al. 2014; Zaman et al. 2011).

Regardless of the spatial resolution, detecting invasive species below forest canopies, especially when forests are dense, is practically impossible. There is, however, promising research suggesting that by using LiDAR (light detection and ranging), it is possible to penetrate the forest canopy and characterize forest 3D structure (e.g., biomass and leaf area index) with no apparent sign of saturation (Zhao et al. 2011). Over the past two decades, LiDAR attracted considerable attention and is already established as one of the standard remote sensing tools for mapping forest biophysical parameters such as tree height, crown size, basal area, timber volume, and canopy fuel (Asner et al. 2008a; Chen and Hay 2011; Lim et al. 2003; Zhao et al. 2011). Recent studies suggest that even if the structure of the native plants has yet to be significantly altered, LiDAR could still be used to detect understory invasive plant species (Singh et al. 2015). Most LiDAR systems are mounted on airborne platforms, and, currently, there is no space-borne LiDAR specifically designed for studying forest ecosystems; however, the Geoscience Laser Altimeter System (GLAS) instrument (footprint: 70 m in diameter; point spacing: 170 m along the track) onboard the NASA Ice, Cloud, and land Elevation satellite (ICESat) was utilized (data available from 2003 to 2010) to map forest height and carbon variability at the regional to global scales (Saatchi et al. 2011).

Temporal Resolution Considerations In contrast to the very-high-resolution applications introduced above, moderate to low spatial resolution imagery usually covers the largest areas in the greatest temporal detail but at reduced spatial resolution. The relatively coarse spatial resolution of high temporal resolution imagery generally limits its early detection to invasive species that cover large, relatively homogenous areas, or that delineate strong canopy differences between infested and non-infested sites. Typically, annual or bi-annual data are used because the intra-annual variability in native species (e.g., broadleaf trees) may also be induced by seasonal changes in climatic variables. To date, remotely sensed time-series data have been applied successfully to assess the impacts of invasive diseases/insects (e.g., Townsend et al. 2012; Wulder et al. 2008) and invasive plants (Bradley and Mustard 2005). It should be noted that time-series data are collected at different dates and are possibly affected by various atmospheric conditions. Thus, conducting an effective radiometric correction becomes essential for extracting 'real' changes in forests (Song et al. 2001).

Although most of the remote sensing systems offer repeated monitoring capability, medium- to coarse-spatial resolution sensors are more frequently used in time-series analysis owing to the short revisit intervals (e.g., MODIS: half a day; Landsat: 16 days); this provides flexibility to collect high-quality data, especially in the cloud-prone tropical regions (Asner 2001). Using medium- to coarse-spatial resolution sensors can further mitigate the joint effects of sensor/sun angles and tree 3D structure, which typically cause high spectral variation in high spatial resolution imagery (Chen et al. 2011). Another solution for analyzing high-spatial resolution time-series data is employing Geographic Object-Based Image Analysis (GEOBIA), using image objects (groups of pixels) rather than individual pixels as the basic study units (Chen et al. 2012). These groups of pixels reduce within-class variability while maximizing between class variance. For example, Chen et al. (2015) applied this approach to high-spatial, high-spectral MASTER (MODIS/ASTER Airborne Simulator) airborne images for assessing the severity of wildfire burn in forests affected by sudden oak death. Additionally, the USDA Forest Service Region 1 (Northern Region) Existing Vegetation Database (VMap) (www.fs.usda.gov/goto/r1/VMap) applied this approach to imagery from NAIP at ~1 m to detect a host of forest stand attributes and outbreaks of insects and disease across the Northern Region. Regardless of the approach used to evaluate the presence of invasive species, the timing of data acquisition (time of overpass) is one of the most critical aspects for consideration and can be a significant determinant in the success of the analysis. For example, if imagery is acquired during a senescent period, the efficacy of the analysis will be reduced, especially when the invasive species (e.g., leafy spurge) can be easily differentiated from background and other species when flowering. Likewise, in forested systems, the early-stage establishment of invasive species can influence trees' photosynthetic capacity by altering leaf nitrogen content and chlorophyll pigment (Kattge et al. 2009). This rapid change in leaf optical properties indicates a need to carefully consider the timing of acquisition and repeat frequency.

Application on Invasive Annual Grasses Some of the most successful uses of high temporal resolution imagery are those leveraging distinct phenological cycles of cool season invasive annual grasses, such as cheatgrass, red brome (*Bromus rubens*), and medusahead (*Taeniatherum caput-medusae*) (Clinton et al. 2010). These invasive grasses are among the most serious invasive species present in the Western United States. These invasive annuals displace native species and accelerate fire cycles. A variety of satellites at different spatial, temporal, and spectral resolutions have been used to detect and quantify the presence of invasive annual grasses.

The most common platforms for quantifying presence or abundance of invasive annual grasses include Landsat TM and ETM+ (Enhanced Thematic Mapper Plus), Advanced Very High Resolution Radiometer (AVHRR), and MODIS. Peterson (2003) used Landsat ETM+ to quantify percent cover of cheatgrass over roughly 12.9 million ha in the Great Basin to a root mean square error (RMSE) of 9%. In the same region, Bradley and Mustard (2006) used time series of Landsat TM and ETM+ and AVHRR data to classify areas infested with cheatgrass. The interannual amplified response to rainfall was quite distinct from native shrub/bunch grass, which enabled a 71% classification accuracy of detecting cheatgrass. Other examples of successfully detecting invasive annual grasses include Peterson (2007) (Owyhee uplands; Landsat TM), Singh and Glenn (2009) (Southern Idaho; Landsat 7 ETM+), and Boyte et al. (2015) (Great Basin and vicinity; MODIS). All these successful studies leverage the distinct phenological pattern via time-series analysis portrayed by invasive annual grasses and focus, most often, on the presence/absence of invasive annual grasses.

In terms of ecological effects and the resistance to control, it is helpful to understand the relative abundance of annual grasses in addition to their presence/absence. Such information enables both ongoing regional-scale analysis and management while still facilitating local patch-based application. The US Geological Survey (USGS) has developed two methods that detect and map invasive annual grasses in the Western United States. One method created a time series of cheatgrass percent-cover maps (Fig. 11.2) by keying in on dynamic phenological characteristics of cheatgrass that differ from other vegetation types, using a fine-scale temporal resolution satellite product at 250-m spatial resolution. The other method created a snapshot of annual

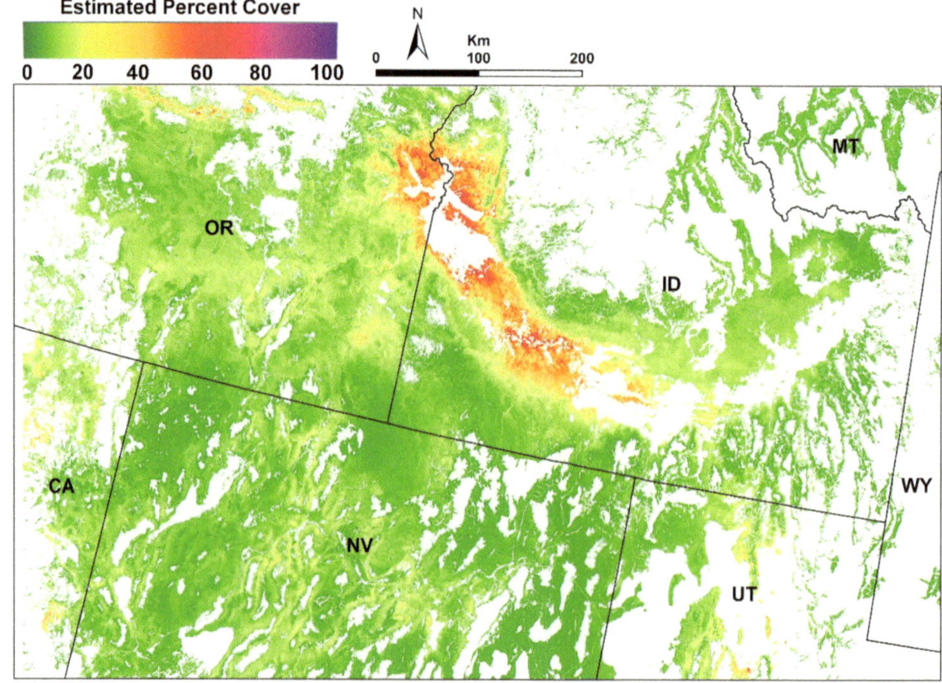

Fig. 11.2 Mean cheatgrass percent cover (2000–2013). The map represents the mean cheatgrass percent-cover value for each pixel during 14 years. The mapping model was developed using regression-tree software driven by annual eMODIS NDVI at 250 m and biogeophysical data. Values ranged from 0 to 86 with an overall mean of 9%. The mask (*white areas*) hides 2001 National Land Cover Database classifications other than shrub or grassland/herbaceous and elevations higher than 2000 m

grass abundance by transforming scaled-up field data using multiple spectral bands from two satellite products with different resolutions to separate the response of invasive annual grasses from spectra of other vegetation types at 30-m spatial resolution. To improve the time series of information for mapping invasive annual grasses, high temporal resolution imagery at 250-m resolution from the enhanced Moderate Resolution Imaging Spectroradiometer (eMODIS) (Brown et al. 2015) can be used in conjunction with higher spatial resolution training data.

The weekly composites from the eMODIS enable characterization of cheatgrass abundance by analyzing the phenology of cheatgrass. This is because the species commences spring growth (increases in plant greenness are measured by normalized difference vegetation index (NDVI) values), sets seed, senesces, and dies within a period of few to several weeks. The eMODIS NDVI is adept at capturing cheatgrass green-up because the data product represents near-daily acquisitions of satellite images that are composited into the best available pixel for each 7-day period (Jenkerson et al. 2010) (Fig. 11.3).

The USGS developed model parameters and algorithms using two years of data on invasive annual grasses (Nevada Natural Heritage Program 2015) stratified by percent-cover cohorts at more than 35,000 points resulting in 250-m predicted cheatgrass percent-cover time series (2000–2013). A substantial advantage of focusing on cheatgrass phenology using eMODIS NDVI and regression-tree models is that both an historical time series of cheatgrass percent-cover maps and annual near-real-time cheatgrass percent-cover maps can be developed. The same eMODIS data, used in conjunction with other data sources for detecting invasive annual grasses, are used for evaluating the status of current forest health and identifying the presence and extent of pest outbreaks across the United States (e.g., Chastain et al. 2015a).

Application on Forest Pests It's important to distinguish between "pests" and invasive species as they are defined in this assessment. The term "pest" is used here because it is commonly used throughout many State and Federal government programs, and thus cannot be easily removed from the discussion. A pest is an organism out of place, though it may not cause a disturbance. A pest can be a plant, an insect, or a pathogen in our context of management. Often pests are invasive, sometimes they are native, and often they are exotic, but this distinction is without value because some problems are native to the country but are new to certain areas and causing pestilence, economic, and ecological harm. A good example of this condition arises from the goldspotted oak borer (*Agrilus auroguttatus*), which is native to North America and the Pacific Southwest region of the United States. However, this pest is killing and weakening numerous tree species in southern California and Mexico.

Forest pests influence millions of hectares of both private and federally owned forest land. As noted in Chap. 10, forest health surveys are conducted annually to detect and evaluate the scope and impact of forest pest activity and are a major component of the Forest Service Forest Health Protection (FHP) program's strategy to minimize the impact of both native and exotic invasive pests. FHP surveys for distur-

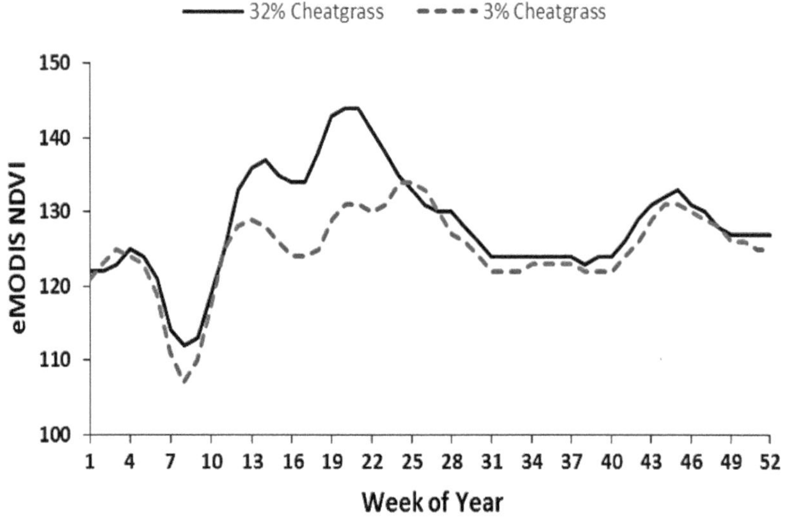

Fig. 11.3 The chart illustrates differences between eMODIS NDVI profiles in adjacent pixels, one estimated with moderate cheatgrass percent cover and the other with very low cheatgrass percent cover. The spike in the profiles starting about week 9 and ending about week 14 represents the period of short-lived cheatgrass growth. The second spike in the profiles starting about week 17 and ending about week 21 could represent (1) the emergence from an extended wet period where high soil moisture deflated NDVI values, (2) a second wave of cheatgrass growth, or (3) other annual species' green-up

bances and pests, and the Forest Service's Forest Health Assessment and Applied Sciences Team (FHAAST) (formerly the Forest Health Technology Enterprise Team (FHTET)) facilitates these surveys by offering a range of data collection, data analysis, and forest pest information reporting technologies. The FHAAST is unique in that it is one of few operational pest detection programs that employs remote sensing technology in forests across the United States. The process model (Fig. 11.4) describes how FHP and its partners in all 50 States identify, survey, and report on forest pest activity.

Forest pests, including those considered to be invasive, are detected annually through a variety of means. The surveys are organized around a reporting construct called a pest event. Pest events are characterized as annual pest activity for a given organism or damage-causing agent, where the pest activity is homogeneous in nature over a given geography. Pest events can be quite large, sometimes covering large portions of entire States, or they may be multistate in nature, exemplified by Fig. 11.5, which depicts the proportion of forested area damaged by gypsy moth (*Lymantria dispar*) from 2011 to 2015. A recent lodgepole pine (*Pinus contorta*) mortality event triggered by the mountain pine beetle (*Dendroctonus ponderosae*) covered most of Colorado and parts of Utah and extended well into Wyoming. While that pest event is largely concluded, the mortality caused by mountain pine beetle continues to surface in the northern part of the lodgepole pine range. This continuing tree mortality would likely be described as a separate event. Pest events also often include damage caused by extreme weather events, where trees are killed, broken, or uprooted. Forest fires are not typically mapped by FHP surveys (active fire mapping at the national level is led by the Forest Service through the Monitoring Trends in Burn Severity (MTBS) program, www.mtbs.gov), though many cooperating State partners assume that responsibility.

FHP and its partners issue pest reports in one of two ways. Often times, pest events are not specifically surveyed, and even though their location and intensity are known locally, they are not mapped with geospatial data and transmitted to the national office (FHAAST). FHP refers to these observations as unstructured pest reports. Frequently, pest events of this kind are difficult to survey or have a technological or a diagnostic limitation, or there is a lack of agency commitment to survey based upon local need. These events are described with software called the pest event reporter (PER). This web-based tool is specifically designed to address pest conditions where actual geospatial data may not exist or are inconsistently acquired nationally.

Secondly, broad-scale impact pests achieve a level of importance indicating that a forest pest survey is necessary. These surveys are often interagency in nature and target a specific invasive species. Pests such as the mountain pine beetle or the emerald ash borer (*Agrilus planipennis*) are

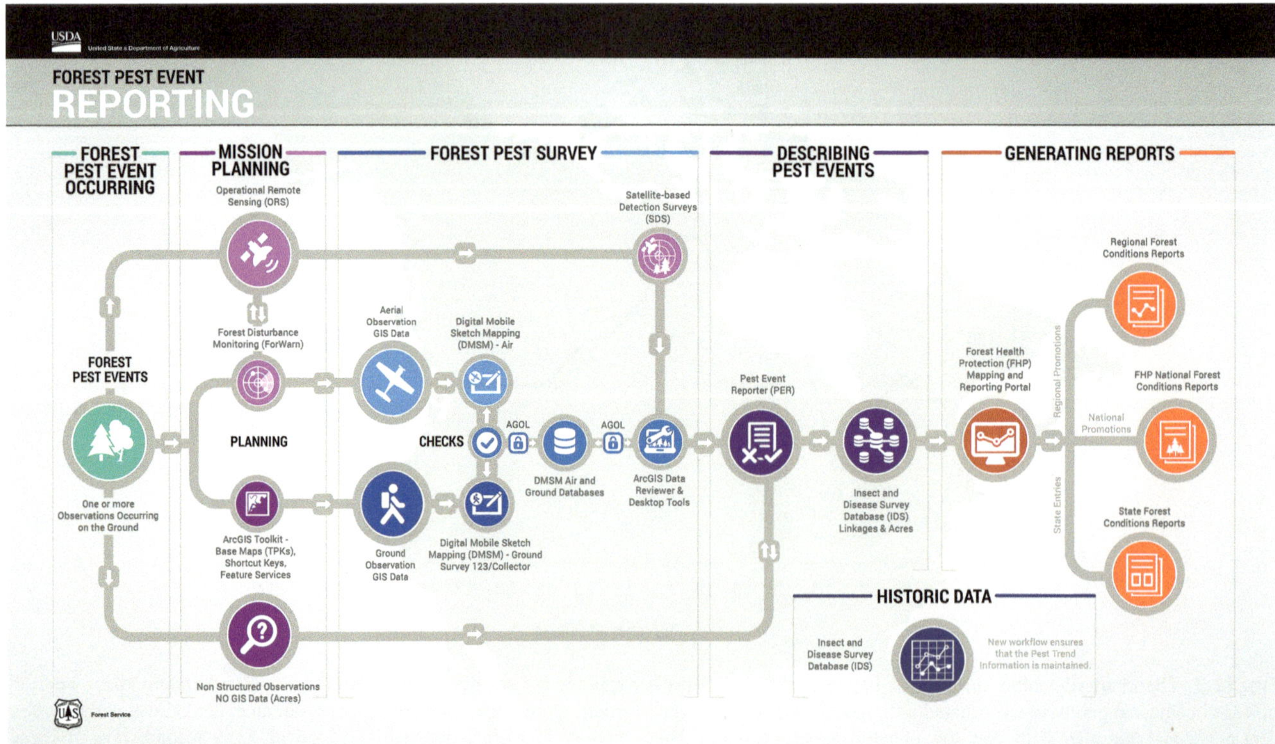

Fig. 11.4 FHP forest disturbance reporting process

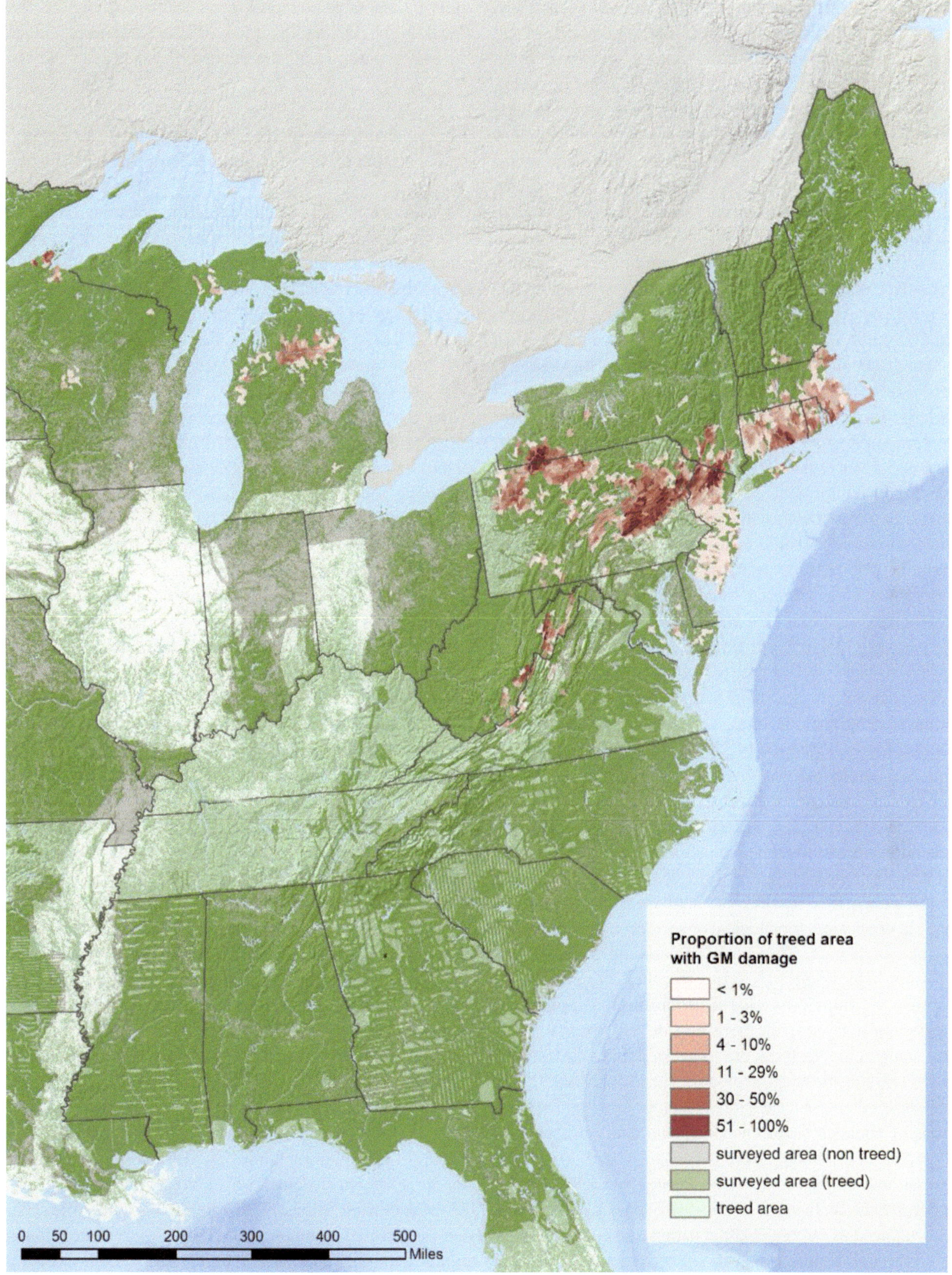

Fig. 11.5 Insect and Disease Survey by Subwatersheds (6th Level Hydrologic Unit Codes (HUCs)) proportion of treed area with damage from gypsy moth from 2011 to 2015 (Map date: June 2016)

recent examples of pests that warranted a multiagency survey response. Mission planning is critical for these pests and occurs on an annual basis. Initial queues of outbreaks often come from the public, overview surveys, or remote sensing initiatives such as the Forest Disturbance Monitor (FDM) (Chastain et al. 2015a). At present, forest disturbance monitoring remotely from satellite is only intended to trigger a subsequent aviation or ground-based survey, where causal agents, intensity, and impact can be adequately estimated. These efforts represent the mission planning phase, which focuses on enabling surveys that incorporate appropriate reference data, maps, and scope of the event(s) to be surveyed.

Forest pest surveys can be conducted in multiple ways. Aerial Detection Surveys (ADS) make up most of the area evaluated annually (202,343 ha). Structured ground surveys, where observers are equipped with pen-based tablets and specialized software, are expected to increase the area where detection surveys are conducted and, to some degree, displace aerial surveys. Surveillance using satellite remote sensing is also improving through programs like Operational Remote Sensing (ORS) (Chastain et al. 2015b). ORS was expected to surveil 30 million ha in 2016 and targeted 60 million ha in 2017. Efforts to increase the number of observations or the area surveyed have been facilitated through recent successes using massive parallel computer processing, such as Google Earth Engine, coupled with a robust imagery stack of satellite data (Chastain et al. 2015b).

Pest observation data, in the form of point-, polygon-, and cell-based geographic features, are captured through a range of technologies supported by the FHAAST. Recent efforts focus on Android-based tablet software, where field observations are collected and reported to FHAAST annually for correction, summation, and sharing through a variety of web applications (https://www.fs.fed.us/foresthealth/publications/fhaast/index.shtml). Without additional adornment, these data are just geographic features and attributes which lack any temporal dimension or biological or socioeconomic reference to scope or impact. Through a process called Pest Event Reporting, and using web software called the Pest Event Reporter (https://www.fs.fed.us/foresthealth/applied-sciences/mapping-reporting/data-app-development.shtml), these geographic features are organized into biologically based events at the local, regional, and national level. Pest outbreaks or their associated forest damage are organized into groups of counties based on geographic data or other non-geographic information. These events are vetted at both the regional and national level by designated specialists within the Forest Service. In the case of invasive species, these events are recognized as important at the national level and are publicized through a variety of web reporting applications. The FHP Mapping and Reporting Portal provides forest health information annually to internal and external partners through a variety of web applications. This information is increasingly incorporated into local forest health reports, State forest health highlights, and a variety of internal reports including the FHP report on annual pest conditions.

11.2.2 Environmental DNA Sampling for Early Detection of Invasive Species

One technological advance that has the potential to be highly effective in the early detection of invasive species is environmental DNA sampling (eDNA). Environmental DNA is the trace DNA in samples of water, soil, or air from shed or eliminated parts (or the whole) of multiple organisms (Bohmann et al. 2014). Sampling eDNA has been used effectively to detect the presence of endangered (Goldberg et al. 2011; McKelvey et al. 2016) and invasive (Mahon et al. 2013; Wilcox et al. 2013, 2016) species at low densities. While most often used to detect invasive fish and amphibians, eDNA can also be used to detect a suite of other vertebrates including semi-aquatic mammals, terrestrial mammals, and reptiles (Padgett-Stewart et al. 2016; Piaggio et al. 2014; Rodgers and Mock 2015; Schwartz et al. 2017). Most eDNA protocols involve filtering water from streams, rivers, ponds, or oceans to collect DNA for detection of invasive species. However, other approaches involve sampling the soil, honey from beehives, carrion flies, or leeches, all of which may contain traces of target invasive species (reviewed in Bohmann et al. 2014). The key to using eDNA successfully is in the development of the laboratory methods that are sensitive to detection of trace amounts of targeted DNA, and in the development of field protocols where the probability of detection and behavior of eDNA under different field conditions is accounted for (Jane et al. 2015; Pilliod et al. 2014).

The advantages of sampling eDNA for invasives are the ease at which field sampling can be conducted, and improved sensitivity as compared to traditional sampling approaches used for detecting rare, invasive amphibians and fish (Dejean et al. 2012; Wilcox et al. 2016). Furthermore, no specialized expertise is required to conduct the field sampling, making it ideal for pairing with citizen science. The diagnosis for the invasive species is determined when the sample is brought into the molecular genetics laboratory. In the laboratory, quantitative polymerase chain reaction (qPCR) or droplet digital PCR (ddPCR) can be used to detect one or a few target species with extreme sensitivity. Wilcox et al. (2016) found that samples containing an average of two DNA copies led to positive detections of target invasive species in 72–86% of trials using qPCR. Alternatively, multiple species can be detected simultaneously through metabarcoding and related genomics approaches, which involve high-throughput sequencing of all DNA in the sample. Overall, the combination of sensitivity, reliability, and efficiency afforded by eDNA sampling enhances the potential to expand the detection and monitoring of invasive species across broad geographic ranges.

11.3 Predicting Occurrence and Spread

Prediction in relation to early detection and intervention, pathway analysis, and risk assessments is covered in Chap. 6. This section is more focused on species occurrence and spread models. Predicting the potential occurrence, suitable habitat, and spread of invasive species encompasses both static and dynamic approaches. To date, static approaches to species distribution models are more common than dynamic approaches, with parameters based on expert-defined rules or statistical estimation of relationships between species occurrence and environmental variables. Logistic regression, for example, is commonly used to model disease occurrence (i.e., presence/absence) at a site (Meentemeyer et al. 2012). Dynamic models allow phenomena to change through time to represent population dynamics and fluctuating behavior of an epidemic. Since they are process-based, dynamic models have the advantage of being applicable for projecting future epidemics. This is a critical advantage over static models since the process driving the invasion (e.g., colonization, establishment, or spread) can be identified. Process information is critical for management (Simberloff 2009), as approaches to management differ depending on the phase of infestation. For example, the goal of early detection and rapid response (EDRR) programs is to avoid colonization of invasive species, while the prevention of favorable conditions (e.g., disturbances) interrupts the spread and dominance of invasive species. Necessary precursors to forecast invasions at each phase are (1) the availability of field-based observations from which the spatial distribution can be estimated and (2) the demographic data from which population dynamics can be inferred. Spatial distribution data can aid identification of other locations where the species is likely to thrive and become problematic. For some species, distribution data from a species' native ranges have been collected, but more comprehensive data are needed from native ranges, including life history, genetic, and abundance. Data on demographic transitions, propagule to juvenile, juvenile to reproductive adult, and dispersal mechanisms are also essential to identify both bottlenecks and opportunities in the invasion process. Moreover, integrating distributional data (from both native and invaded ranges) with available demographic data could greatly enhance the results of current modeling efforts (Guo et al. 2009; Ibáñez et al. 2009).

High-quality distribution data in exotic ranges (e.g., county-level or FIA program data for forest ecosystems in the United States) are needed, but corresponding distribution data from the native regions are equally important (Guo 2006; Ibáñez et al. 2009). This is based on the hypothesis (which has been confirmed in many cases) that species with large native ranges are likely to have larger ranges in exotic regions (Guo et al. 2006) (Fig. 11.6) and on the prediction that invasive species may expand their climatic ranges (Broennimann et al. 2007; Ibáñez et al. 2009) (Fig. 11.7).

Once a species has been introduced into a new region, data related to the dispersal of propagules are essential to assess the potential for its spread into new areas. Assessment of the colonization potential depends on the ability of the introduced species to establish without human intervention and to survive and initiate self-sustaining populations. Proliferation of the invasive species to the point where it has a detrimental effect on the native community is dependent on its growth and reproductive capacity (Ricklefs et al. 2008). Thus, demographic data that include a species' reproduction, survival, growth, and dispersal, and its response to disturbances and resources, will be critical to model the invasion process (Foxcroft et al. 2011; Gurevitch et al. 2011).

The success of some highly invasive species can be attributed to the rapid changes in their genetics ("rapid evolution") after invading new habitats (Ellstrand 2009; Whitney and Gabler 2008). Currently, the Forest Service is building a database with 29 major life history/genetic trait categories for over 4000 introduced plant species in the United States. Once this work is completed, two ranking systems for all the species can be developed based on (1) current distributions and (2) traits that can indicate future spread (some species with invasive traits now have limited distribution but could spread rather quickly in the future with or without climate change). Albright et al. (2010) demonstrate an example of this predictive capability where data collected abroad on the native range of tree of heaven

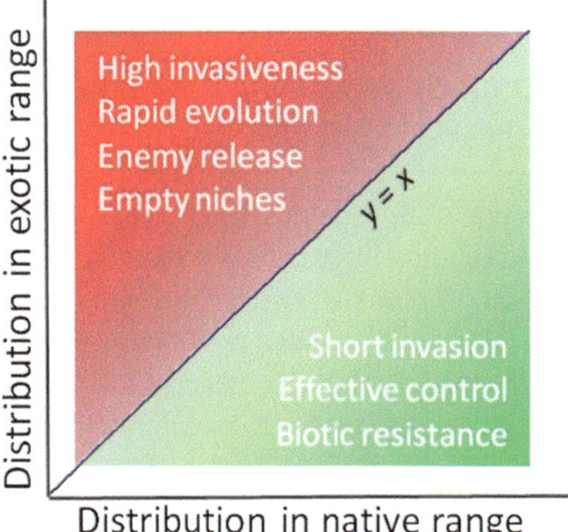

Fig. 11.6 Diagram showing that native distribution can be used to predict present and future exotic distribution and spread in conjunction with species traits (i.e., life history and genetic elasticity). The position of a particular species in the graph may be determined by multiple factors as discussed in this chapter (e.g., time, rapid evolution, interspecies facilitation, mutualisms). (Modified from Guo (2006))

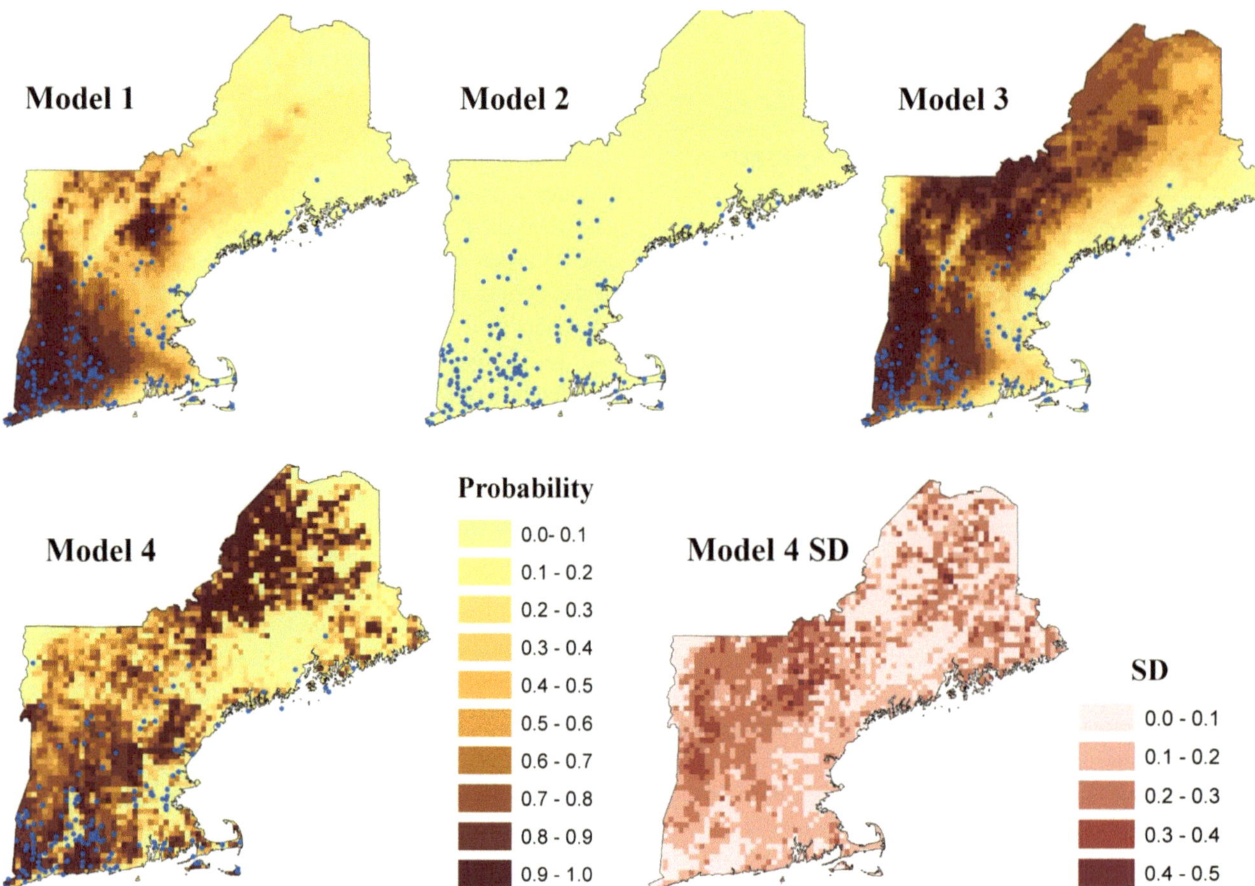

Fig. 11.7 Distribution and climate data models used to predict invasion. *Model 1*: distribution and climate data from invaded range used to predict invasion in invaded range of Japanese barberry (*Berberis thunbergii*). *Model 2*: distribution and climate data from native range used to predict invasion in invaded range. *Model 3*: distribution and climate data from native and invaded ranges used to predict invasion in invaded range. *Model 4*: model ensemble and standard deviation (SD) about the mean

(*Ailanthus altissima*) can be used to estimate potential distribution in the invaded range (Fig. 11.8).

11.3.1 Integrating Available Data into Forecasting Models

Integrated models are analytical approaches that link local demographic/genetic information with coarse-resolution models of environmental suitability into a predictive assessment of species invasion (Ibáñez et al. 2014). Integrated models combine knowledge of the processes that take place during invasions, based mainly on life history/demographic data (e.g., genetics, reproduction, growth, survival, dispersal), along with available distribution data (native and/or invaded ranges). These models are highly flexible and easily adaptable to analyze whatever types of data (e.g., demographic, distributional, observational, experimental) are available. Statistical techniques like hierarchical/multilevel models are highly suitable for use in such integration (Clark and Gelfand 2006; Clark et al. 2010). Hierarchical models allow the combination of data collected at different temporal and spatial scales. They can also include latent processes that are not directly observed, for example, establishment and reproduction, that are then modeled as a function of the available demographic data. Another advantage of such models is that they can quantify the uncertainty associated with each model component that can then be tracked when invasion forecasts for different scenarios are generated. They can quantify the varying effects of certain variables along gradients (e.g., the effect of disturbance along climatic gradients), and can identify and quantify both bottlenecks and windows of opportunity for invasion. These models can produce forecasts of potential impact of invasions at any particular site, from a local to regional scale, and for different scenarios of environmental conditions. This information could be then employed to guide monitoring efforts for early detection of plant invasions. In summary, results from integrated models have the potential for enhancing model realism, explanatory insight, and predictive capability (Ibáñez et al. 2014).

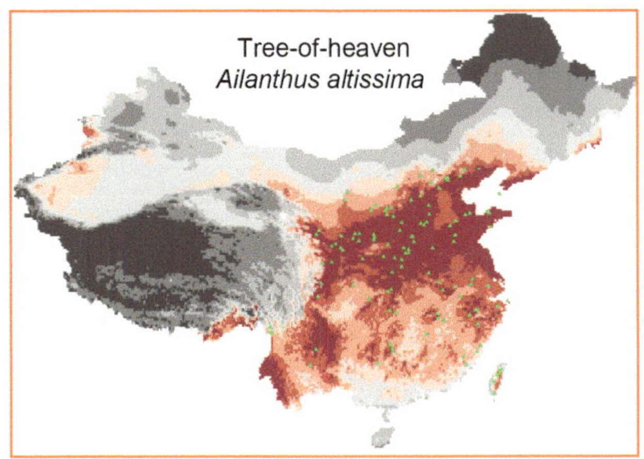

$+ \text{maxtmp} - \text{maxtmp}^2 + \text{precip} + \text{precip}^2 + \text{mintmp} - \text{mintmp}^2 + \text{agurb} - \text{agurb}^2$

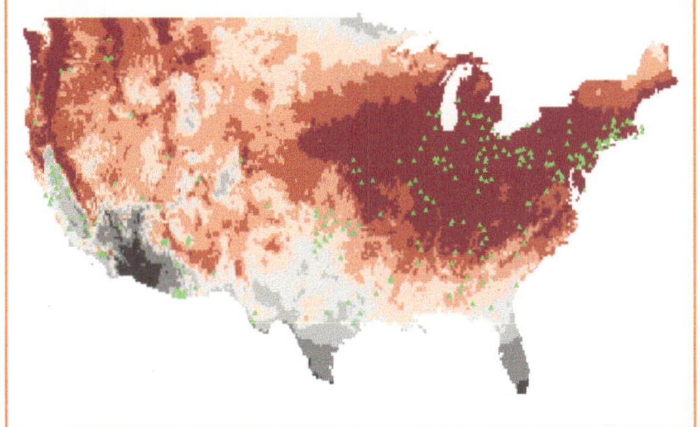

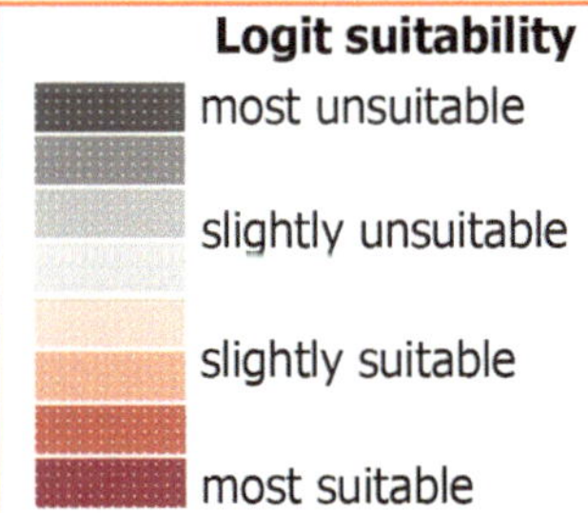

Fig. 11.8 Relative suitability for tree of heaven estimated by a simple Bayesian model based on distribution information from the United States and prior parameters derived from China. In the maps, *point symbols* represent occurrence records and areas with different suitabilities (i.e., modeled logit values using posterior parameter estimates). Both suitable and unsuitable areas are divided into quartiles of relative probability (i.e., suitability). (Courtesy of Thomas P. Albright and reproduced from Albright and others (2010))

11.4 Decision-Support Systems

A decision-support system is a program that analyzes data and presents results in a format that supports decision making for management, operations, and planning activities. Though these systems vary greatly in complexity, they generally utilize multiple sources of data. The models underlying the data analyses can be developed specifically to address management questions by involving decision makers and managers in their development. These tools can predict the dynamics of invasions, including the area invaded through time, and also guide selection of the management activities, monitoring, and treatments that are most likely to result in suitable outcomes (Provencher et al. 2016). Model-based decision-support systems include optimization models and simulation models. Optimization models are computationally difficult, usually require a small spatial extent, and consider minimal alternative actions from which the optimal solution is determined. Conversely, simulation models can be used to address larger areas, and can provide a comparison of a large number of proposed activities. Both types of models have been used to evaluate management activities for buffelgrass (*Pennisetum ciliare*) in southern Arizona but focused on different aspects of response (Büyüktahtakin et al. 2014). The optimization model attempted to minimize damage over time for an 800-ha area, subject to budget and labor constraints, and focused on where to treat. The simulation models (Frid et al. 2013a; Jarnevich et al. 2015) focused on comparing allocation of resources between inventory and treatment activities and levels of activities (resources involved).

Even when sophisticated models of biological invasions are available (Cunniffe et al. 2016), informing and mobilizing stakeholders to use them for making timely decisions is still a challenge (see Chap. 12). Participatory approaches (Reed 2008; Voinov and Bousquet 2010) have been designed to facilitate stakeholder engagement in research and management of invasive species. Perera et al. (2006) suggest that involving stakeholders (Fig. 11.9) throughout the modeling process maximizes information transfer, helps generate buy-in, and creates advocates for their inclusion in complex circumstances. This conceptual framework known as participatory modeling (Voinov and Bousquet 2010)—may move participants from passive or didactic learning to experiential learning through immersion in what Colella (2000) called the "computational sandbox," that is, simulations with realism adequate to temporarily suspend disbelief and constitute a shared experience.

A new modeling tool called Tangible Landscape (Fig. 11.10), which is being developed at the Center for Geospatial Analytics at North Carolina State University, gives stakeholders the ability to visualize place- and time-dependent management scenarios with real-time feedback (Petrasova et al. 2015). Using simple, tangible gestures on a physical, 3D representation of landscape data (Petrasova et al. 2015), the tool uses simultaneous 3D laser scanning and liquid crystal display (LCD) projection to connect a computational model to a physical 3D model. Tangible Landscape will soon be used to help stakeholders develop collaborative solutions for managing sudden oak death in California. Although these recent efforts in invasive species modeling offer advantages over earlier models, at increasingly larger scales of analysis, landscape heterogeneity and diverse patterns of ownership still present significant chal-

Fig. 11.9 Stakeholder engagement in invasive species research and management. (**a**) Field crew chemically treats the stumps of recently cut California bay laurel (*Umbellularia californica*) trees to prevent re-sprouting of this reservoir host species; (**b**) disease prevention in some areas involves complete removal of host trees located within 15 feet of susceptible oak species; (**c**) a stakeholder defines a disease management scenario by placing intervention markers on the 3D Tangible Landscape; (**d**) markers are laser scanned and resulting intervention areas provide real-time input for exploring simulated scenarios of management actions at particular place and time. (Photos courtesy of Ross Meentemeyer)

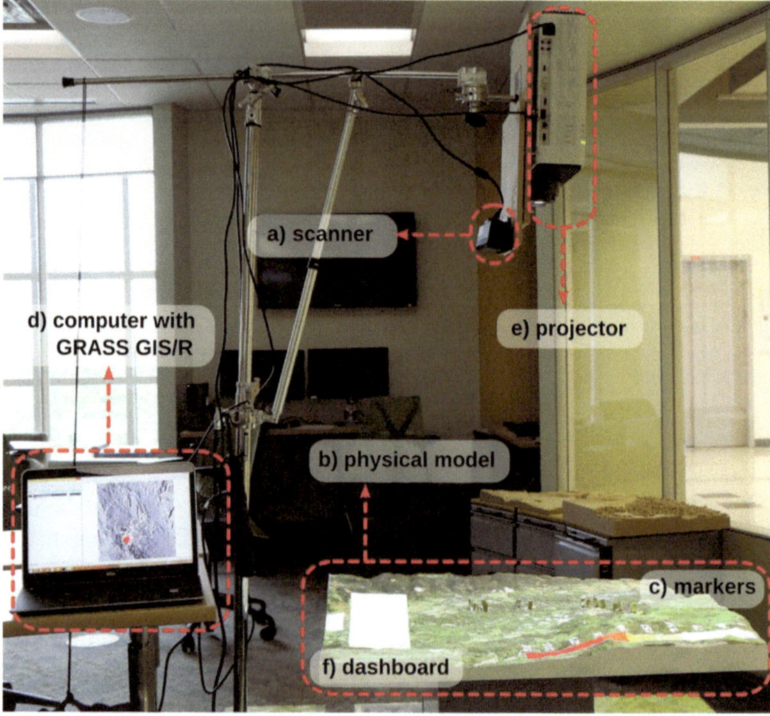

Fig. 11.10 Overview of the Tangible Landscape system. Photos courtesy of Ross Meentemeyer

lenges. In addition, development and utilization of models is further complicated by the lack of consistency among organizations and agencies in their priorities, data availability, modeling expertise, and ecological classifications.

Although organizations have disparate ideas, programs, and directives for managing invasive species, there is a clear need to improve consistency in data collection, ecological classification, and modeling. The national framework for early detection and rapid response is a positive approach toward consolidating thoughts and actions among agencies (https://www.fws.gov/ficmnew/FICMNEW_EDRR_FINAL.pdf). Another example of publicly accessible citizen science data is the Early Detection and Distribution Mapping System (https://www.eddmaps.org). EDDMapS is the most easily accessible and comprehensive spatially explicit database available that describes the extent and magnitude of biological invasions.

Invasive species present a national challenge most efficiently addressed through consolidated, all lands approaches, which necessitates developing more consistent databases for describing ecological processes. Though numerous challenges exist toward realizing this idyllic scenario, in 2013, all major land management agencies participated in the development of the Interagency Ecological Site Handbook for Rangelands. Ecological sites provide a kind of decision-support and land-classification system that describes the ecological potential and ecosystem dynamics of land areas (http://www.ars.usda.gov/Research/docs.htm?docid=18502). Ecological Site Descriptions provide narratives of each site and most often include conceptual state and transition models (STMs), which are box-and-arrow diagrams that depict vegetation communities (states depicted by boxes) and shifts between them (transitions depicted by arrows) (Bestelmeyer et al. 2004; Westoby et al. 1989) (Fig. 11.11). States can be defined based on vegetation cover types, structural stages, and ages, while transitions are processes or thresholds that shift the vegetation between states, including natural processes (e.g., dispersal, succession, fire) and

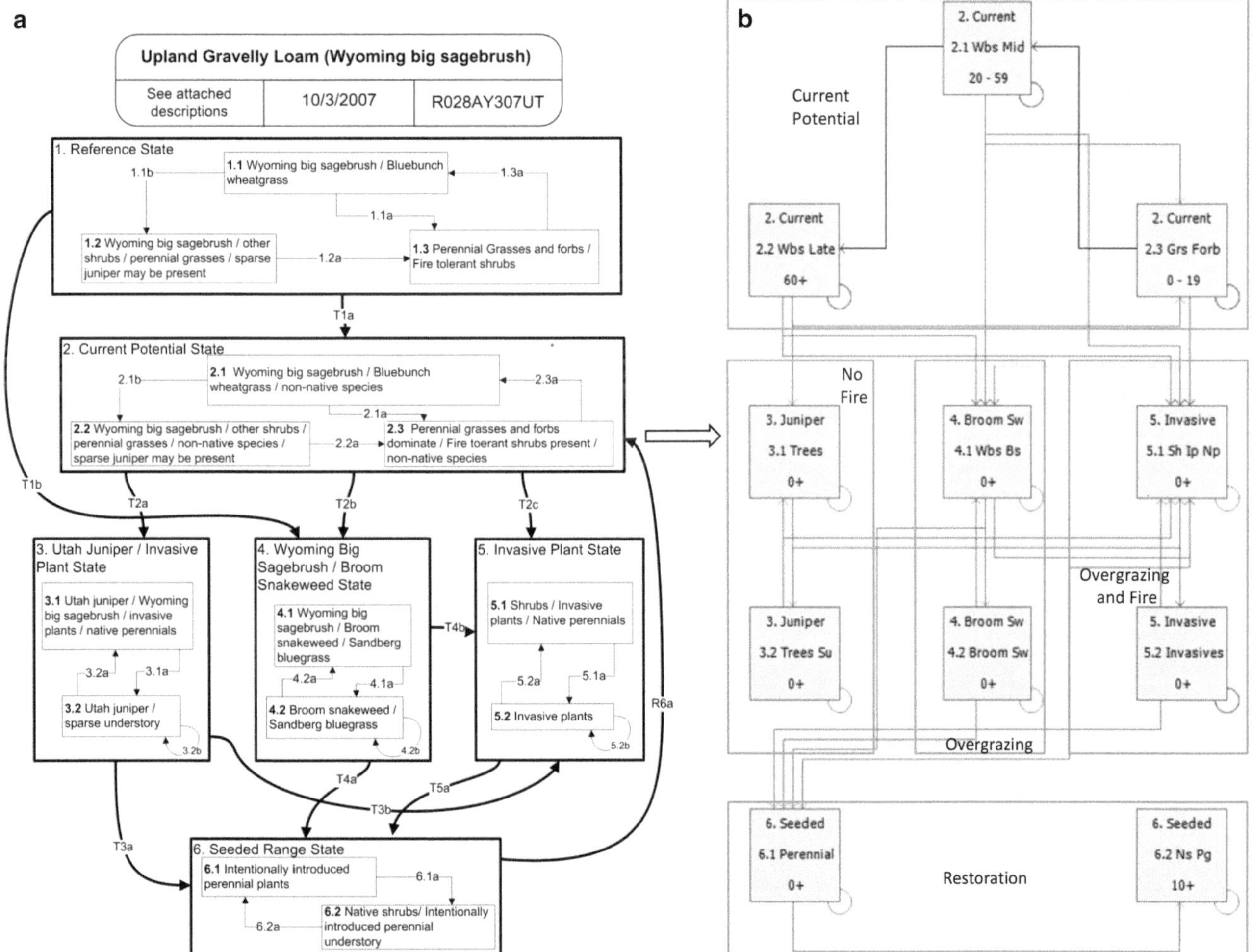

Fig. 11.11 (**a**) Example of how a state and transition model from the Upland Gravelly Loam Ecological Site becomes digitized, and (**b**) acts as input to a simulation system such as ST-SIM. This site is highly invasible (very low resistance) by cheatgrass. Note that the Reference State with no non-native species is unlikely on this site and is therefore omitted from this example digitized model

anthropogenic processes (e.g., herbicide spraying, grazing, development). These conceptual STMs can be subsequently digitized and converted to state and transition simulation models (STSM) enabling stochastic simulation of ecological processes including invasive species dynamics. State and transition simulation modeling conducted in a quantitative manner is an analytical framework for consolidating resource management issues under different scenarios (Provencher et al. 2016). Additionally, using STSMs to explore the impacts of different assumptions can quantify uncertainty and help guide future research activities, identifying key information that is needed to answer management questions. For example, Jarnevich et al. (2015) developed an STSM for buffelgrass in Ironwood Forest National Monument, AZ, with the states representing cover class categories, and the transitions including growth (increase in abundance), detection of patches through surveys, and decrease in abundance associated with treatments to control buffelgrass (Fig. 11.12). In this buffelgrass example, Jarnevich et al. (2015) compared invasion over time related to the degree of management occurring in the monument across all lands within the ownership matrix included in the monument (Fig. 11.13). STSMs have been used to aid decisions for several invasive plant species by assessing whether to prioritize small or large patches for treatment (Frid et al. 2013b; Frid and Wilmshurst 2009) and how to allocate resources between inventory and treatment activities (Frid et al. 2013a; Jarnevich et al. 2015), and for assessing varying amounts and types of treatment (Frid et al. 2013b; Jarnevich et al. 2015). In addition, Provencher et al. (2016) used STSM to investigate management issues and uncertainty associated with exotic annual *Bromus* species.

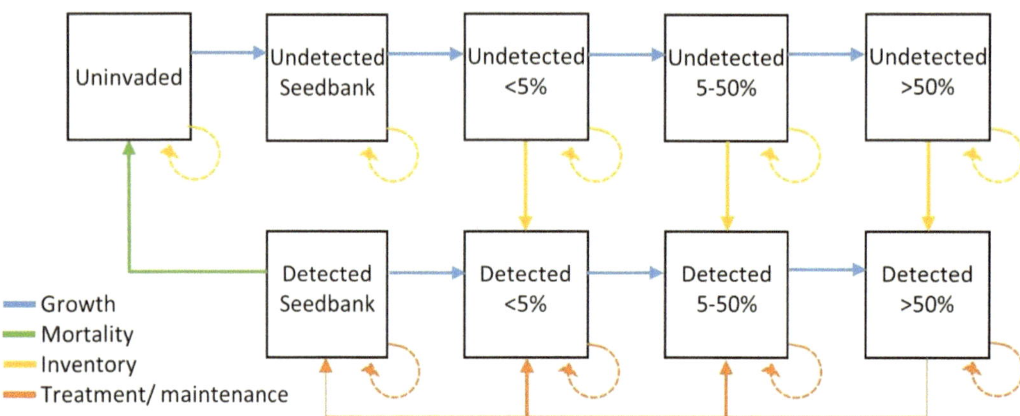

Fig. 11.12 State and transition simulation model for buffelgrass in Ironwood Forest National Monument, AZ. Shown here is the conceptual model where each *box* represents the state with regards to buffelgrass cover (uninvaded, seedbank, <5% cover, 5–50% cover, or >50% cover; *right to left*) and detection (undetected or detected; *top to bottom*). The *color-coded arrows* represent different types of transitions including growth (invasion, establishment, spread), detection (failure and success), and management (treatment and maintenance failure and success). *Solid lines* represent success; *dotted lines* represent failure

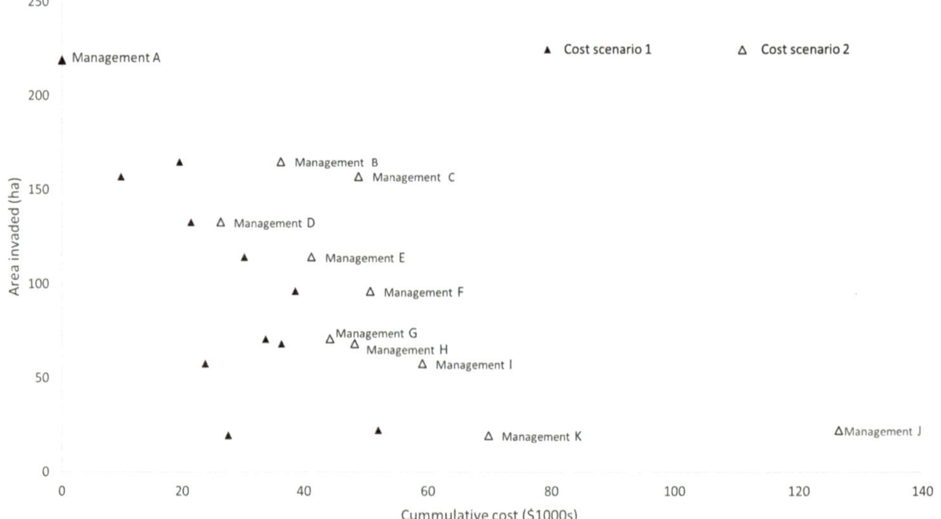

Fig. 11.13 Simulation model results for 11 "what if" management scenarios (*A to K*) showing area invaded after 20 simulated years on the *y-axis* and cumulative cost of management activities (varying amounts of intensive field surveys, vehicle-mounted herbicide spraying, backpack herbicide spraying, volunteer hand pulling, contract hand pulling, helicopter herbicide spraying, and follow-up maintenance of buffelgrass) over a 20-year period on the *x-axis*. (Figures adapted from Jarnevich and others (2015))

The increase in availability of STMs associated with ecological sites, combined with new spatially explicit simulation models, such as those developed with the ecological modeling software ST-SIM (http://www.apexrms.com/), provides an emerging toolset capable of estimating the effects of management and climate on biological invasions across the United States (Daniel et al. 2016). Although Ecological Sites are works in progress, they represent a potentially cost-effective framework that is applicable on an interagency basis for local to regional simulation of how invasive species respond to disturbances.

11.5 Genetic Tools to Accelerate Restoration and Facilitate Management

Advances in technology have provided a plethora of genetic-based tools that can be used to facilitate management of genetic resources and/or advance restoration efforts by accelerating breeding programs aimed at developing planting stock with resistance to invasive insects and pathogens. Genetic markers, pieces of DNA that can be used to distinguish an individual, population, or species, can facilitate management of genetic resources. Such markers can also be used to identify locations in the genome that contain genes for desired traits such as resistance. Breeding can be accelerated by using genetic markers to select for desired traits instead of labor-intensive, costly, long-term classic phenotyping methods (evaluating traits and performance over time), a process known as indirect selection. Genes that have been identified as having a role in flowering can be used to accelerate breeding by inducing early flowering so that breeding can be done without waiting years or decades for a superior tree to reach sexual maturity so that it can serve as a parent. Genetic engineering is a tool that can introduce genes, such as those to induce early flowering or even those that would confer resistance, into a plant to produce transgenic trees that express these desired traits. Advances in technology that allow rapid sequencing of the DNA that incorporates the entire genome of a tree, referred to as next-generation sequencing or high-throughput sequencing, have advanced our abilities to develop genetic markers and to identify genes and even differences in the ways these genes are expressed (called transcriptomics) that may play a role in resistance. In this section, we provide a brief overview of genetic tools, their potential uses in restoration and management, and their current limitations.

Common examples of genetic markers include SSRs (simple sequence repeats) and SNPs (single nucleotide polymorphism). These important tools can improve efficiencies of conventional breeding programs through a variety of applications, including the evaluation of genetic diversity in breeding populations; the confirmation and tracking of identity, parentage, and relatedness; and the assessment of pollen flow/contamination in seed orchards (Neale and Kremer 2011; Porth and El-Kassaby 2014). Breeding can be a long-term process in forest trees, with some species requiring a decade or more to reach reproductive status. Technical advances in sequencing DNA have significantly reduced the costs of obtaining thousands of markers that are dispersed throughout the genome. The ability to achieve such dense genome coverage can provide information on genetic variation relevant to a desired phenotype through the development of indirect selection techniques, including marker assisted selection (MAS) and genomic selection (GS). These two tools can potentially streamline the conventional breeding process by allowing the breeder to use markers to "pre-select" trees at a young age, or to select parent trees directly from natural stands. Pre-selection will help by minimizing the number of trees whose phenotypes will need to be carefully confirmed over a range of time and environments.

Markers linked to traits for use in traditional MAS are identified through the development of a genetic linkage map which relies on analyzing patterns of segregation of markers from parents to progeny to identify the number of linkage groups and to place markers in an orderly fashion on each group. Markers that are closer together are more likely to be inherited together in the progeny. Many different genes may contribute to expression of the desired trait. The process of identifying markers associated with the region of the genome that contains each of these genes, called a quantitative trait locus (QTL), is referred to as QTL analysis. Despite the enormous investment in resources that have been expended to identify markers, very few operational breeding programs for either crops or trees use MAS (Isik 2014; Muranty et al. 2014; Neale and Kremer 2011; Xu and Crouch 2008). Several factors have contributed to this, but often markers associated with QTLs are identified in studies using a relatively small number of progeny, and consequently they may not be useful when tested in other families because the markers are not located sufficiently close to the genes responsible for the trait of interest (Nilausen et al. 2016). In addition, traits like resistance are often complex, so it can be difficult to detect all the different loci involved. Development of an operational MAS system is more likely to be successful when conducted in conjunction with an existing breeding program that has access to carefully phenotyped progeny and parents.

Breeding programs that do successfully employ MAS in trees, for example, those associated with domesticated fruit and nut producing species, are typically tracking a single locus or a small number of loci with very large effect, including major gene resistance (Ru et al. 2015; Sathuvalli et al. 2011). A common use of MAS in crops and fruit trees is pyramiding multiple major effect resistance alleles, which is difficult using traditional phenotyping methods (Muranty et al. 2014; Ru et al. 2015). In forest tree species, a potentially useful application of MAS would be for within family selection in dealing with rapidly evolving pathogens, such as white pine blister rust, and when the goal is to pyramid major gene resistance with quantitative resistance to increase durability of resistance (Sniezko et al. 2011).

Another type of indirect selection known as genomic selection (GS) does not require identification of specific marker-trait associations of individual QTLs. Instead, GS relies on phenotyping and using such a high density of markers to genotype a large enough sample of the breeding population (called the training population) that the majority of loci that contribute to a quantitative trait are closely located to one or more markers. The effects of all markers are then estimated simultaneously (unlike MAS) and used to predict genomic breeding values in a test population without needing phenotypic data (Jannink et al. 2010; ReSende et al. 2012). Although GS is routinely used in animal breeding programs (Hayes and Goddard 2010), and has been used successfully in crop breeding programs (Lorenzana and Bernardo 2009), the strategy has only been tested in forest trees in simulation studies (Grattapaglia and Resende 2011; Iwata et al. 2011) and in preliminary trials (Resende et al. 2012; Zapata-Valenzuela et al. 2013). The studies in forest trees show promise but the results should be interpreted with caution because studies were conducted in small populations; additional 'proof of concept' studies need to be performed using larger populations (Plomion et al. 2016). Successful application of GS to forest tree breeding will ultimately require correlating thousands or even hundreds of thousands of markers with a desired phenotype in a sufficiently large training population in order to develop models to select the best performing trees in the breeding population. Both training and breeding populations need to have undergone at least some breeding and been carefully phenotyped for resistance, a genetic resource that likely will only be available in the most advanced forest tree breeding programs (Isik 2014; Jannink et al. 2010; Zapata-Valenzuela et al. 2013). Analysis of the economic feasibility of incorporating GS is also needed before it will be accepted for widespread use in forest tree breeding programs (Plomion et al. 2016).

Another tool that can accelerate breeding is the use of genetic engineering to induce early flowering. Transgenic approaches to manipulate flowering have been developed in many woody plants and, because the effect of early flowering caused by the introduction of the transgene is dominant, it is only required in one parent and can be selected against in the progeny so that the final selected genotype is not transgenic (Van Nocker and Gardiner 2014). New technology, using a virus as a vector to introduce genes that control flowering, has been used successfully to produce early flowering apple and pear trees (Yamagishi et al. 2016). A simple heat treatment can be used to eliminate the viral vector from resultant seedlings. Successful application of this technology in forest trees offers the potential to reduce breeding cycles from several decades to mere months.

Although it is commonly asserted that the use of genetic engineering to insert a gene to develop a transgenic plant that conveys resistance to an insect or disease is a quicker, less expensive alternative to traditional breeding, this is not necessarily the case. It takes an estimated timeframe of 7–24 years (average 13.1 years) to discover, develop, and obtain regulatory authorization to distribute seed for a crop plant carrying a new transgenic trait, and an average cost of $136 million (McDougall 2011). Despite this investment, resistance based on a single transgene can sometimes be overcome or inactivated or have unintended fitness costs (Finnegan and McElroy 1994; Gurr and Rushton 2005; Tabashnik and Carriere 2017; Tian et al. 2003). In the case of forest tree species, it is not possible yet to do an accurate cost assessment because a transgenic forest tree has not yet obtained authorization to be released into natural forests. Development and deployment of a transgenic forest tree will likely require a similar investment of time and money. In addition, once a transgenic tree is produced, it will be necessary to incorporate genetic diversity adequate to maintain the adaptive capacity to multiple stresses and environments through traditional breeding (Steiner et al. 2016). Transgenic technology alone cannot replace a breeding program, rather successful deployment of transgenics (assuming regulatory approval) for restoration purposes will depend on the existence of a breeding program, the possible exception being industrial forestry plantations or horticultural cultivars, where it may be appropriate to deploy material with limited genetic diversity. The same would be true if the new genome editing technologies, such as CRISPR-Cas9, were used to develop a resistant tree (Puchta 2016). Unlike the development of a resistant plant through transgenics, genome editing directly alters specific genes in the genome. This requires extensive basic knowledge about the number and function of genes involved in the trait being altered (i.e., susceptible altered to become resistant) that is not always available and can take years or even decades of research to acquire. Genetic engineering through transgenics or genome editing is a valuable tool that can facilitate the study and confirmation of gene function. Once the necessary basic knowledge is accumulated, development of a resistant tree through transgenics or genome editing may be possible but should only be pursued in conjunction with a breeding program if the ultimate goal is restoration using resistant planting stock.

Transcriptomics, the process of sequencing expressed genes to evaluate the level of gene expression, is a genomic tool used to study patterns of differential gene expression in forest trees, often with the goal of identifying candidate genes involved in resistance to insects and diseases, for potential downstream utility in developing markers for implementation in breeding programs, or for use in developing transgenic trees with resistance (Bai et al. 2011; Barakat et al. 2009). However, a recent review article by Feder and Walser (2005) concluded that using transcriptomics to compare gene expression patterns in plants exposed to a stress (abiotic or biotic), or to identify the genes responsible for a specific phenotype or response to the stress, such as resistance, is rarely successful. This can be explained because the measured differences in expression levels of genes do not accurately predict abundance or activity of the resulting proteins that are produced from these expressed genes (Feder and Walser 2005). The authors recommended employing a careful cost/benefit analysis prior to employing such a strategy.

Many of the genomic tools and technologies that we have presented are in their infancy and have not yet been widely implemented; additionally, thorough review of their usefulness and success over the long term is not always available. An analysis of the benefits derived from utilizing many of these techniques, relative to the significant financial investment required to develop them, is lacking. Ultimately, technological advances will undoubtedly reduce the cost of many of these tools, and continued research will lead to improved rates of success, but it's also important to remember that, although these technologies have the potential to accelerate restoration, (breeding), they are not essential for implementing a successful breeding program. The vast majority of breeding programs (see Table 8.1 in Chap. 8) have achieved success without the use of genomic tools. It is of equal importance to note that an active breeding program is a prerequisite for the successful development and implementation of these genetic tools. Investment in both technology/tool development and traditional breeding programs is essential for providing the appropriate genetic tools that can facilitate the accelerated and cost-effective production of resistant planting stock needed to manage and restore forests impacted by invasive insects and diseases.

11.6 Key Findings, Information Needs, and Opportunities

Advances in biotechnology have produced genetic tools with the potential to accelerate the production of resistant planting stock for restoration. However, these tools cannot replace the need for traditional breeding programs, and, in fact, they depend upon such programs for their efficient and effective development and implementation or deployment. Forest managers and tree breeders will need to weigh public opinion as well as the costs and benefits of such tools before adopting them. Continued technical developments may help decrease costs in the future.

Improvements in computational resources and model complexity have enhanced the ability of managers and scientists to evaluate the extent and magnitude of invasive species. This situation is exemplified by FHP and other national systems that provide essential monitoring of forest conditions. In addition, these improvements have led to greater use of models to forecast potential future conditions. Overall availability of data and modeling is better in forested ecosystems relative to other ecosystems, suggesting that there is a significant opportunity and need for improved modeling, data collection, and mapping in non-forested systems, including aquatic habitats.

The most significant gap consistently identified across the numerous disciplines discussed in this chapter is the relative paucity of spatially referenced data describing the extent and magnitude of invasive species, especially on non-forested lands. A consistent, comprehensive, easily accessible database provides the underpinning for identifying new outbreaks, calibrating and validating models, improving remote sensing analyses, and monitoring management effectiveness. US land management agencies do not currently perform consistent annual inventory and monitoring of non-forested landscapes; however, the situation is slowly changing. For example, the FIA program has established protocols for non-forested landscapes very similar to the forested data collection, but less than 1% of the FIA plots represent non-forested landscapes (All Conditions Inventory). In 2011, the Bureau of Land Management began data collection for the Assessment, Inventory, and Monitoring (AIM) project. To date, data have been collected on about 5000 "national" and 4000 "project" plots. National and field office–collected AIM plots use the same protocols and can be used to report the same AIM core indicators regarding terrestrial ecosystems. National plots are designed, however, to yield a statistically valid sample that can be used to derive inferences across the landscape, like FIA. The situation is considerably better on privately owned landscapes where the USDA Natural Resources Conservation Service supports and maintains the National Resources Inventory (USDA 2015, ~20,000 plots), which includes non-forested plots. Non-forested lands of the coterminous United States occupy roughly 268 million ha (Reeves and Mitchell 2011), of which approximately 166 million ha are privately owned (USDA 2015), indicating that there is approximately 1 plot every 83 km^2 of privately owned land that is non-forested (assuming an even distribution). Given that many of these plots are unevenly distributed, many vegetation types and regions of the country are considerably unrepresented, and therefore many new invasions may go unnoticed. On 3 October 2016, the search words of "FIA invasive species" with Google yielded ~41,000 results. This suggests that regular, consistent, and comprehensive data collection could be invaluable. Considering the large number of publications, reports, and analyses supported by FIA, it follows that a program designed to acquire data (or simply an expansion of existing programs) on the roughly 268 million ha of non-forested land (Reeves and Mitchell 2011) could be equally invaluable.

Although an improved strategy for combatting invasive species would include a consolidated, interagency, publicly available, spatially explicit database, other noteworthy gaps were identified during the conduct of this assessment. These gaps can be classified in one of three categories including data limitations, institutional limitations, or a combination of these. With respect to data limitations (data gathering, dissemination, and analysis), these gaps are generally summarized as:

- Lack of consistency among stakeholders with image data processing, cataloging, and distribution
- Lack of consistency among stakeholders with data collection protocols (e.g., many habitats go unsampled each year) and data storage and distribution

With respect to institutional limitations, gaps are generally summarized as:

- Lack of interagency communication and cooperation
- Lack of shared vision and priorities
- Lack of consistency in data collection protocols, data storage, distribution, and sharing (including both georeferenced plot data and image data)

The existence of these gaps does not automatically suggest that more programs are needed, but it does indicate that there is a need for greater sharing, communication, and collaboration. If all stakeholders and management agencies could define priorities nationally, and then regionally define strategies and tactics, this could foster more efficient data acquisition and analysis and ultimately enhance the control of invasive species.

Disclaimer Text The findings and conclusions in this publication are those of the authors and should not be construed to represent any official USDA or U.S. Government determination or policy.

Literature Cited

Albright TP, Chen H, Chen LJ et al (2010) The ecological niche and reciprocal prediction of the disjunct distribution of an invasive species: the example of *Ailanthus altissima*. Biol Invasions 12:2413–2427

Anderson K, Gaston KJ (2013) Lightweight unmanned aerial vehicles will revolutionize spatial ecology. Front Ecol Environ 11:138–146

Asner GP (2001) Cloud cover in Landsat observations of the Brazilian Amazon. Int J Remote Sens 22:3855–3862

Asner GP, Hughes RF, Vitousek PM et al (2008a) Invasive plants transform the three-dimensional structure of rain forests. Proc Natl Acad Sci U S A 105:4519–4523

Asner GP, Jones MO, Martin RE et al (2008b) Remote sensing of native and invasive species in Hawaiian forests. Remote Sens Environ 112:1912–1926

Bai X, Rivera-Vega L, Mamidala P et al (2011) Transcriptomic signatures of ash (*Fraxinus* spp.) phloem. PLoS One 6:1–12

Barakat A, DiLoreto DS, Zhang Y et al (2009) Comparison of transcriptomes of American chestnut (*Castanea dentata*) and Chinese chestnut (*Castanea mollissima*) in response to the chestnut blight infection. BMC Plant Biol 9:51

Bestelmeyer BT, Herrick JE, Brown JR et al (2004) Land management in the American Southwest: a state-and-transition approach to ecosystem complexity. Environ Manag 34:38–51

Blumenthal D, Booth DT, Cox SE et al (2007) Large-scale aerial images capture details of invasive plant populations. Rangel Ecol Manag 60:523–528

Blumenthal DM, Norton AP, Cox SE et al (2012) *Linaria dalmatica* invades south-facing slopes and less grazed areas in grazing-tolerant mixed-grass prairie. Biol Invasions 14:395–404

Bohmann K, Evans A, Gilbert MTP et al (2014) Environmental DNA for wildlife biology and biodiversity monitoring. Trends Ecol Evol 29:358–367

Booth DT, Cox SE (2008) Image-based monitoring to measure ecological change in rangeland. Front Ecol Environ 6:185–190

Booth DT, Cox SE, Teel D (2010) Aerial assessment of leafy spurge (*Euphorbia esula* L.) on Idaho's deep fire burn. Nativ Plants J 11:327–340

Boyte SP, Wylie BK, Major DJ (2015) Mapping and monitoring cheatgrass dieoff in rangelands of the northern Great Basin, USA. Rangel Ecol Manag 68:18–28

Boyte SP, Wylie BK, Major DJ (2016) Cheatgrass percent cover change: comparing recent estimates to climate change driven predictions in the northern Great Basin. Rangel Ecol Manag 69:265–279

Bradley BA (2014) Remote detection of invasive plants: a review of spectral, textural and phenological approaches. Biol Invasions 16:1411–1425

Bradley BA, Mustard JF (2005) Identifying land cover variability distinct from land cover change: Cheatgrass in the Great Basin. Remote Sens Environ 94:204–213

Bradley BA, Mustard JF (2006) Characterizing the landscape dynamics of an invasive plant and risk of invasion of invasion using remote sensing. Ecol Appl 16:1132–1147

Broennimann O, Treier UA, Müller-Schärer H et al (2007) Evidence of climatic niche shift during biological invasion. Ecol Lett 10:701–709

Brown JF, Howard D, Wylie B et al (2015) Application-ready expedited MODIS data for operational land surface monitoring of vegetation condition. Remote Sens 7:16226–16240

Büyüktahtakin İE, Feng Z, Olsson AD et al (2014) Invasive species control optimization as a dynamic spatial process: an application to buffelgrass (*Pennisetum ciliare*) in Arizona. Invasive Plant Sci Manag 7:132–146

Calviño-Cancela M, Méndez-Rial R, Reguera-Salgado J et al (2014) Alien plant monitoring with ultralight airborne imaging spectroscopy. PLoS One 9:e102381

Chastain RA, Fisk H, Ellenwood JR et al (2015a) Near-real time delivery of MODIS-based information on forest disturbances. In: Lippitt CD, Stow DA, Coulter LL (eds) Time sensitive remote sensing. Springer, New York, pp 147–164

Chastain RA, Housman I, Finco M (2015b) Forest health protection operational remote sensing: algorithm development and pilot study area assessments, RSAC-10107-RPT1. U.S. Department of Agriculture, Forest Service, Remote Sensing Applications Center, Salt Lake City, 34 p

Chen G, Hay GJ (2011) A support vector regression approach to estimate forest biophysical parameters at the object level using airborne

LiDAR transects and QuickBird data. Photogramm Eng Remote Sens 77:733–741
Chen G, Hay GJ, Castilla G et al (2011) A multiscale geographic object-based image analysis to estimate lidar-measured forest canopy height using QuickBird imagery. Int J Geogr Inf Sci 25:877–893
Chen G, Hay GJ, Carvalho LMT, Wulder MA (2012) Object-based change detection. Int J Remote Sens 33:4434–4457
Chen G, Metz MR, Rizzo DM et al (2015) Object-based assessment of burn severity in diseased forests using high-spatial and high-spectral resolution MASTER airborne imagery. ISPRS J Photogramm Remote Sens 102:38–47
Cheng T, Rivard B, Sánchez-Azofeifa GA et al (2010) Continuous wavelet analysis for the detection of green attack damage due to mountain pine beetle infestation. Remote Sens Environ 114:899–910
Clark JS, Gelfand AEE (2006) Hierarchical modelling for the environmental sciences statistical methods and applications. Oxford University Press, Oxford, 205 p
Clark JS, Bell D, Chu C et al (2010) High-dimensional coexistence based on individual variation: a synthesis of evidence. Ecol Monogr 80:569–608
Clinton NE, Potter C, Crabtree B et al (2010) Remote sensing-based time-series analysis of cheatgrass (*Bromus tectorum* L.) phenology. J Environ Qual 39:955–963
Colella V (2000) Participatory simulations: building collaborative understanding through immersive dynamic modeling. J Learn Sci 9:471–500
Coops N, Stanford M, Old K et al (2003) Assessment of dothistroma needle blight of *Pinus radiata* using airborne hyperspectral imagery. Ecol Epidemiol 93:1524–1532
Coops NC, Wulder MA, Iwanicka D (2009) Large area monitoring with a MODIS-based Disturbance Index (DI) sensitive to annual and seasonal variations. Remote Sens Environ 113:1250–1261
Crall AW, Newman GJ, Stohlgren TJ et al (2011) Assessing citizen science data quality: an invasive species case study. Conserv Lett 4:433–442
Cunniffe NJ, Cobb RC, Meentemeyer RK et al (2016) Modeling when, where, and how to manage a forest epidemic, motivated by sudden oak death in California. Proc Natl Acad Sci U S A 113:5640–5645
Daniel C, Frid L, Sleeter B et al (2016) State-and-transition simulation models: a framework for forecasting landscape change. Methods Ecol Evol 7:1413–1423
Darwall WRT, Dulvy NK (1996) An evaluation of the suitability of non-specialist volunteer researchers for coral reef fish surveys. Mafia Island, Tanzania: a case study. Biol Conserv 78:223–231
Dejean T, Valentini A, Miquel C et al (2012) Improved detection of an alien invasive species through environmental DNA barcoding: the example of the American bullfrog *Lithobates catesbeianus*. J Appl Ecol 49:953–959
Delaney DG, Sperling CD, Adams CS et al (2008) Marine invasive species: validation of citizen science and implications for national monitoring networks. Biol Invasions 10:117–128
Duniway MC, Karl JW, Schrader S et al (2012) Rangeland and pasture monitoring: an approach to interpretation of high-resolution imagery focused on observer calibration for repeatability. Environ Monit Assess 184:3789–3804
Ellstrand NC (2009) Evolution of invasiveness in plants following hybridization. Biol Invasions 11:1089–1091
Evangelista PH, Young NE, Schofield PJ et al (2017) Modeling suitable habitat of invasive red lionfish *Pterois volitans* (Linnaeus, 1758) in North and South America's coastal waters. Aquat Invasions 11:313–326
Everitt JH, Anderson GL, Escobar DE et al (1995) Use of remote sensing for detecting and mapping leafy spurge (*Euphorbia esula*). Weed Technol 9:599–609
Everitt JH, Escobar DE, Alaniz MA et al (1996) Using spatial information technologies to map Chinese tamarisk (*Tamarix chinensis*) infestations. Weed Sci 44:194–201
Everitt JH, Yang C, Deloach CJ (2005) Remote sensing of giant reed with QuickBird satellite imagery. J Aquat Plant Manag 43:81–85
Feder ME, Walser JC (2005) The biological limitations of transcriptomes in elucidating stress and stress responses. J Evol Biol 18:901–910
Finnegan J, McElroy D (1994) Transgene inactivation: plants fight back! Biotechnology 12:883–888
Fore LS, Paulsen K, O'Laughlin K (2001) Assessing the performance of volunteers in monitoring streams. Freshw Biol 46:109–123
Foxcroft LC, Pickett STA, Cadenasso ML (2011) Expanding the conceptual frameworks of plant invasion ecology. Perspect Plant Ecol Evol Syst 13:89–100
Frid L, Wilmshurst JF (2009) Decision analysis to evaluate control strategies for crested wheatgrass (*Agropyron cristatum*) in Grasslands National Park of Canada. Invasive Plant Sci Manag 2:324–336
Frid L, Hanna D, Korb N et al (2013a) Evaluating alternative weed management strategies for three Montana landscapes. Invasive Plant Sci Manag 6:48–59
Frid L, Holcombe T, Morisette JT et al (2013b) Using state and transition modeling to account for imperfect knowledge in invasive species management. Invasive Plant Sci Manag 6:36–47
Fuller D (2005) Remote detection of invasive Melaleuca trees (*Melaleuca quinquenervia*) in South Florida with multispectral IKONOS imagery. Int J Remote Sens 26:1057–1063
Goldberg CS, Pilliod DS, Arkle RS et al (2011) Molecular detection of vertebrates in stream water: a demonstration using Rocky Mountain tailed frogs and Idaho giant salamanders. PLoS One 6:e22746
Grattapaglia D, Resende MDV (2011) Genomic selection in forest tree breeding. Tree Genet Genomes 7:241–255
Guo Q (2006) Intercontinental biotic invasions: what can we learn from native populations and habitats? Biol Invasions 8:1451–1459
Guo Q, Qian H, Ricklefs RE et al (2006) Distributions of exotic plants in eastern Asia and North America. Ecol Lett 9:827–834
Guo QF, Falcone J, Brownsmith J (2009) Building the database for introduced plants in the United States. In: McManus KA, Gottschalk KW (eds) Proceedings of 20th U.S. department of agriculture interagency research forum on invasive species, Gen. Tech. Rep. NRS-P-51. USDA, Forest Service, Annapolis, 73 p
Gurevitch J, Fox GA, Wardle GM et al (2011) Emergent insights from the synthesis of conceptual frameworks for biological invasions. Ecol Lett 14:407–418
Gurr SJ, Rushton PJ (2005) Engineering plants with increased disease resistance: what are we going to express? Trends Biotechnol 23:275–282
Haas SE, Cushman JH, Dillon WW et al (2016) Effects of individual, community, and landscape drivers on the dynamics of a wildland forest epidemic. Ecology 97:649–660
Hawthorne TL, Elmore V, Strong A et al (2015) Mapping non-native invasive species and accessibility in an urban forest: a case study of participatory mapping and citizen science in Atlanta, Georgia. Appl Geogr 56:187–198
Hayes B, Goddard M (2010) Genome wide association and genomic selection in animal breeding. Genome 53:876–883
Huang C, Asner GP (2009) Applications of remote sensing to alien invasive plant studies. Sensors 9:4869–4889
Ibáñez I, Silander JA, Wilson AM et al (2009) Multi-variate forecasts of potential distribution of invasive plant species. Ecol Appl 19:359–375
Ibáñez I, Diez JM, Miller LP et al (2014) Integrated assessment of biological invasions. Ecol Appl 24:25–37
Isik F (2014) Genomic selection in forest tree breeding: the concept and an outlook for the future. New For 45:379–401
Iwata H, Hayashi T, Tsumura Y (2011) Prospects for genomic selection in conifer breeding: a simulation study of *Cryptomeria japonica*. Tree Genet Genomes 7:747–758
Jane SF, Wilcox TM, McKelvey KS et al (2015) Distance, flow and PCR inhibition: eDNA dynamics in two headwater streams. Mol Ecol Resour 15:216–227

Jannink JL, Lorenz AJ, Iwata H (2010) Genomic selection in plant breeding: from theory to practice. Brief Funct Genomics 9:166–177

Jarnevich CS, Holcombe TR, Cullinane Thomas C et al (2015) Simulating long-term effectiveness and efficiency of management scenarios for an invasive grass. AIMS Environ Sci 2:427–447

Jenkerson CB, Maiersperger TK, Schmidt GL (2010) eMODIS – a user-friendly data source, U.S. Geological Survey Open-File Report 2010-1055, Reston

Kattge J, Knorr W, Raddatz T, Wirth C (2009) Quantifying photosynthetic capacity and its relationship to leaf nitrogen content for global-scale terrestrial biosphere models. Glob Chang Biol 15:976–991

Kelly M, Meentemeyer RK (2002) Landscape dynamics of the spread of sudden oak death. Photogramm Eng Remote Sens 68:1001–1009

Knipling EB (1970) Physical and physiological basis for the reflectance of visible and near-infrared radiation from vegetation. Remote Sens Environ 1:155–159

Laba M, Downs R, Smith S et al (2008) Mapping invasive wetland plants in the Hudson River National Estuarine Research Reserve using QuickBird satellite imagery. Remote Sens Environ 112:286–300

Laurent EJ, Shi H, Gatziolis D et al (2005) Using the spatial and spectral precision of satellite imagery to predict wildlife occurrence patterns. Remote Sens Environ 97:249–262

Lim K, Treitz P, Wulder M et al (2003) LiDAR remote sensing of forest structure. Prog Phys Geogr 27:88–106

Lodge DM, Williams S, MacIsaac HJ et al (2006) Biological invasions: recommendations for US policy and management. Ecol Appl 16:2035–2054

Lorenzana RE, Bernardo R (2009) Accuracy of genotypic value predictions for marker-based selection in biparental plant populations. Theor Appl Genet 120:151–161

Mahon AR, Jerde CL, Galaska M et al (2013) Validation of eDNA surveillance sensitivity for detection of Asian carps in controlled and field experiments. PLoS One 8:e58316

McCord SE, Buenemann M, Karl JW et al (2017) Integrating remotely sensed imagery and existing multiscale field data to derive rangeland indicators: application of Bayesian additive regression trees. Rangel Ecol Manag 70:644–655

McCormick CM (1999) Mapping exotic vegetation in the Everglades from large-scale aerial photographs. Photogramm Eng Remote Sens 65:179–184

McDougall Phillips, consultants (2011) The cost and time involved in the discovery, development and authorization of a new plant biotechnology derived trait. A Consultancy Study for Crop Life International. 24 p. https://croplife.org/wp-content/uploads/2014/04/Getting-a-Biotech-Crop-to-Market-Phillips-McDougall-Study.pdf.

McKelvey KS, Young MK, Knotek WL et al (2016) Sampling large geographic areas for rare species using environmental DNA: a study of bull trout (*Salvelinus confluentus*) occupancy in western Montana. J Fish Biol 88:1215–1222

McNeil BE, de Beurs KM, Eshleman KN et al (2007) Maintenance of ecosystem nitrogen limitation by ephemeral forest disturbance: an assessment using MODIS, Hyperion, and Landsat ETM+. Geophys Res Lett 34:1–5

Mealor BA, Cox S, Booth DT (2012) Postfire downy brome (*Bromus tectorum*) invasion at high elevations in Wyoming. Invasive Plant Sci Manag 5:427–435

Meddens AJH, Hicke JA, Vierling LA (2011) Evaluating the potential of multispectral imagery to map multiple stages of tree mortality. Remote Sens Environ 115:1632–1642

Meentemeyer RK, Haas SE, Václavík T (2012) Landscape epidemiology of emerging infectious diseases in natural and human-altered ecosystems. Annu Rev Phytopathol 50:379–402

Muranty H, Jorge V, Bastien C et al (2014) Potential for marker-assisted selection for forest tree breeding: lessons from 20 years of MAS in crops. Tree Genet Genomes 10:1491–1510

Neale DB, Kremer A (2011) Forest tree genomics: growing resources and applications. Nat Rev Genet 12:111–122

Nevada Natural Heritage Program, State of Nevada (2015). Available at http://heritage.nv.gov/gis. Last Accessed 1 Sept 2015

Nilausen C, Gelinas N, Bull G (2016) Perceived acceptability of implementing marker-assisted selection in the forests of British Columbia. Forests 7:1–9

Noujdina NV, Ustin SL (2008) Mapping downy brome (*Bromus tectorum*) using multidate AVIRIS data. Weed Sci 56:173–179

Padgett-Stewart TM, Wilcox TM, Carim KJ et al (2016) An eDNA assay for river otter detection: a tool for surveying a semi-aquatic mammal. Conserv Genet Resour 8:5–7

Pearlstine L, Portier KM, Smith SE (2005) Textural discrimination of an invasive plant, *Schinus terebinthifolius*, from low altitude aerial digital imagery. Photogramm Eng Remote Sens 71:289–298

Perera A, Buse A, Crow T (2006) Knowledge transfer in forest landscape ecology: a primer. In: Perera A, Buse L, Crow T (eds) Forest landscape ecology. Transferring knowledge into practice. Springer, New York, pp 1–18

Peterson EB (2003) Mapping percent-cover of the invasive species *Bromus tectorum* (Cheatgrass) over a large portion of Nevada from satellite imagery. U.S. Fish and Wildlife Service, Nevada State Office, Reno, Carson City, 40 p

Peterson EB (2007) A map of annual grasses in the Owyhee Uplands, Spring 2006, derived from multitemporal Landsat 5 TM imagery. U.S. Department of Interior, Bureau of Land Management, Nevada State Office, Reno, Carson City, 32 p

Petrasova A, Harmon B, Petras V et al (2015) Tangible modeling with open source GIS. Springer, Cham, 202 p

Piaggio AJ, Engeman RM, Hopken MW et al (2014) Detecting an elusive invasive species: a diagnostic PCR to detect Burmese python in Florida waters and an assessment of persistence of environmental DNA. Mol Ecol Resour 14:374–380

Pilliod DS, Goldberg CS, Arkle RS et al (2014) Factors influencing detection of eDNA from a stream-dwelling amphibian. Mol Ecol Resour 14:109–116

Plomion C, Bastien C, Bogeat-Triboulot MB et al (2016) Forest tree genomics: 10 achievements from the past 10 years and future prospects. Ann For Sci 73:77–103

Porth I, El-Kassaby YA (2014) Assessment of the genetic diversity in forest tree populations using molecular markers. Diversity 6:283–295

Provencher L, Frid L, Czembor C, Morisette JT (2016) State-and-transition models: conceptual versus simulation perspectives, usefulness and breadth of use, and land management applications. In: Germino M, Chambers J, Brown C (eds) Exotic brome-grasses in arid and semiarid ecosystems of the western US. Springer, Cham, pp 371–407

Pu R, Kelly M, Anderson GL et al (2008) Using CASI hyperspectral imagery to detect mortality and vegetation stress associated with a new hardwood forest disease. Photogramm Eng Remote Sens 74:65–75

Puchta H (2016) Using CRISPR/Cas in three dimensions: towards synthetic plant genomes, transcriptomes and epigenomes. Plant J 87:5–15

Reed MS (2008) Stakeholder participation for environmental management: a literature review. Biol Conserv 141:2417–2431

Reeves MC, Mitchell JE (2011) Extent of coterminous US rangelands: quantifying implications of differing agency perspectives. Rangel Ecol Manag 64:1–12

Resende MDV, Resende MFR, Sansaloni CP et al (2012) Genomic selection for growth and wood quality in Eucalyptus: capturing the missing heritability and accelerating breeding for complex traits in forest trees. New Phytol 194:116–128

Ricklefs RE, Guo Q, Qian H (2008) Growth form and distribution of introduced plants in their native and non-native ranges in Eastern Asia and North America. Divers Distrib 14:381–386

Rodgers TW, Mock KE (2015) Drinking water as a source of environmental DNA for the detection of terrestrial wildlife species. Conserv Genet Resour 7:693–696

Rose RA, Byler D, Eastman JR et al (2015) Ten ways remote sensing can contribute to conservation. Conserv Biol 29:350–359

Ru S, Main D, Evans K et al (2015) Current applications, challenges, and perspectives of marker-assisted seedling selection in Rosaceae tree fruit breeding. Tree Genet Genomes 11:8

Saatchi SS, Harris NL, Brown S (2011) Benchmark map of forest carbon stocks in tropical regions across three continents. Proc Natl Acad Sci U S A 108:9899–9904

Sathuvalli VR, Mehlenbacher SA, Smith DC (2011) DNA markers linked to eastern filbert blight resistance from a hazelnut selection from the Republic of Georgia. J Am Soc Hortic Sci 136:350–357

Schwartz MK, Penaluna BE, Wilcox TM (2017) Not just for fisheries biologists anymore: environmental DNA sampling makes strides in wildlife. Wildl Prof 11:47–51

Siegert NW, McCullough DG, Liebhold AM, Telewski FW (2014) Dendrochronological reconstruction of the epicenter and early spread of emerald ash borer in North America. Divers Distrib 20:847–858

Simberloff D (2009) The role of propagule pressure in biological invasions. Annu Rev Ecol Evol Syst 40:81–102

Singh N, Glenn NF (2009) Multitemporal spectral analysis for cheatgrass (*Bromus tectorum*) classification. Int J Remote Sens 30:3441–3462

Singh KK, Davis AJ, Meentemeyer RK (2015) Detecting understory plant invasion in urban forests using LiDAR. Int J Appl Earth Obs Geoinf 38:267–279

Skakun RS, Wulder MA, Franklin SE (2003) Sensitivity of the thematic mapper enhanced wetness difference index to detect mountain pine beetle red-attack damage. Remote Sens Environ 86:433–443

Sniezko RA, Mahalovich MF, Schoettle AW (2011) Past and current investigations of the genetic resistance to *Cronartium ribicola* in high-elevation five-needle pines. In: Keane RE, Tomback DF, Murray MP, Smith CM (eds) The future of high-elevation, five-needle white pines in western North America: proceedings of the high five symposium, Proceedings RMRS-P-63. U.S. Department of Agriculture, Forest Service, Rocky Mountain Research Station, Fort Collins, pp 246–264

Song C, Woodcock CE, Seto KC et al (2001) Classification and change detection using Landsat TM data: when and how to correct atmospheric effects? Remote Sens Environ 75:230–244

Steiner KC, Westbrook JW, Hebard FV et al (2016) Rescue of American chestnut with extraspecific genes following its destruction by a naturalized pathogen. New For 48:1–2

Tabashnik BE, Carriere Y (2017) Surge in insect resistance to transgenic crops and prospects for sustainability. Nat Biotechnol 35:926–935

Tian D, Traw MB, Chen JQ et al (2003) Fitness costs of R-gene-mediated resistance in *Arabidopsis thaliana*. Nature 423:74–77

Townsend PA, Singh A, Foster JR et al (2012) A general Landsat model to predict canopy defoliation in broadleaf deciduous forests. Remote Sens Environ 119:255–265

U.S. Department of Agriculture (2015) Summary report: 2012 national resources inventory. Natural Resources Conservation Service, Washington, DC, and Center for Survey Statistics and Methodology, Iowa State University, Ames, Iowa

U.S. Department of the Interior (2016) Safeguarding America's lands and waters from invasive species: a national framework for early detection and rapid response, Washington DC, 55 p

Underwood MC, Allan DH, James FQ (2013) Geospatial tools for identifying and managing invasive plants. In: Jose S, Singh HP, Batish DR, Kohli RK (eds) Invasive plant ecology. CRC Press, Boca Raton, pp 175–202

Václavík TÅ, Kanaskie A, Hansen EM et al (2015) Predicting potential and actual distribution of sudden oak death in Oregon: prioritizing landscape contexts for early detection and eradication of disease outbreaks. For Ecol Manag 260:1026–1035

Van Nocker S, Gardiner SE (2014) Breeding better cultivars, faster: applications of new technologies for the rapid deployment of superior horticultural tree crops. Hortic Res 1:14022

Voinov A, Bousquet F (2010) Modelling with stakeholders. Environ Model Softw 25:1268–1281

Wan H, Wang Q, Jiang D et al (2014) Monitoring the invasion of *Spartina alterniflora* using very high resolution unmanned aerial vehicle imagery in Beihai, Guangxi (China). Sci World J 24:1–9

Westoby M, Walker B, Noy-Meir I (1989) Opportunistic management for rangelands not at equilibrium. J Range Manag 42:266–274

Whitney KD, Gabler CA (2008) Rapid evolution in introduced species, 'invasive traits' and recipient communities: challenges for predicting invasive potential. Divers Distrib 14:569–580

Wilcox TM, McKelvey KS, Young MK et al (2013) Robust detection of rare species using environmental DNA: the importance of primer specificity. PLoS One 8:e59520

Wilcox TM, McKelvey KS, Young MK et al (2016) Understanding environmental DNA detection probabilities: a case study using a stream-dwelling char *Salvelinus fontinalis*. Biol Conserv 194:209–216

Wulder MA, White JC, Coops NC et al (2008) Multi-temporal analysis of high spatial resolution imagery for disturbance monitoring. Remote Sens Environ 112:2729–2740

Xu Y, Crouch JH (2008) Marker-assisted selection in plant breeding: from publications to practice. Crop Sci 48:391–407

Yamagishi N, Li C, Yoshikawa N (2016) Promotion of flowering by apple latent spherical virus vector and virus elimination at high temperature allow accelerated breeding of apple and pear. Front Plant Sci 7:1–10

Zaman B, Jensen AM, McKee M (2011) Use of high-resolution multispectral imagery acquired with an autonomous unmanned aerial vehicle to quantify the spread of an invasive wetlands species. In: Geoscience and Remote Sensing Symposium (IGARSS), 2011 IEEE international. IEEE, pp 803–806

Zapata-Valenzuela J, Whetten RW, Neale DB et al (2013) Genomic estimated breeding values using genomic relationship matrices in a cloned population of loblolly pine. Genes Genomes Genet 3:909–916

Zhao K, Popescu S, Meng X et al (2011) Characterizing forest canopy structure with LiDAR composite metrics and machine learning. Remote Sens Environ 115:1978–1996

12 Social and Cultural Dynamics of Non-native Invasive Species

John Schelhas, Janice Alexander, Mark Brunson, Tommy Cabe, Alycia Crall, Michael J. Dockry, Marla R. Emery, Susan J. Frankel, Nina Hapner, Caleb R. Hickman, Rebecca Jordan, Michael J. LaVoie, Zhao Ma, Joe Starinchak, and Jelena Vukomanovic

12.1 Introduction

Invasive species and their management represent a complex issue spanning social and ecological systems. Invasive species present existing and potential threats to the nature of ecosystems and the products and services that people receive from them. Humans can both cause and address problems through their complex interactions with ecosystems. Yet, public awareness of invasive species and their impact is highly uneven, and public support for management and control of invasive species can be variable. Public perceptions often differ markedly from the perspectives of concerned scientists, and perceptions and support for management are influenced by a wide range of social and ecological values. In this chapter, we present a broad survey of social science research across a diversity of ecosystems and stakeholders in order to provide a foundation for understanding the social and cultural dimensions of invasive species and plan more effective management approaches. This chapter also addresses tribal perspectives on invasive species, including traditional ecological knowledge, unique cultural dimensions for tribes, and issues critical to engaging tribes as partners and leaders in invasive species management. Recognizing that natural resource managers often seek to change people's perceptions and behaviors, we present and discuss some promising approaches that are being used to engage human communities in ways that empower and enlist stakeholders as partners in management.

J. Schelhas (✉)
U.S. Department of Agriculture, Forest Service, Southern Research Station, Athens, GA, USA
e-mail: john.schelhas@usda.gov

J. Alexander
University of California Cooperative Extension, Marin County, Novato, CA, USA

M. Brunson
Department of Environment and Society, Utah State University, Logan, UT, USA

T. Cabe · C. R. Hickman · M. J. LaVoie
Natural Resource Department, Eastern Band of Cherokee Indians, Cherokee, NC, USA

A. Crall
National Ecological Observatory Network, Boulder, CO, USA

M. J. Dockry
Department of Forest Resources, University of Minnesota, Saint Paul, MN, USA

M. R. Emery
U.S. Department of Agriculture, Forest Service, Northern Research Station, Burlington, VT, USA

S. J. Frankel
U.S. Department of Agriculture, Forest Service, Pacific Southwest Research Station, Albany, CA, USA

N. Hapner
Kashia Department of Environmental Planning, Kashia Band of Pomo Indians, Santa Rosa, CA, USA

R. Jordan
Department of Community Sustainability, Michigan State University, East Lansing, MI, USA

Z. Ma
Department of Forestry and Natural Resources, Purdue University, West Lafayette, IN, USA

J. Starinchak
U.S. Department of the Interior, U.S. Fish & Wildlife Service – Headquarters, Fish and Aquatic Conservation Program, Falls Church, VA, USA

J. Vukomanovic
Department of Parks, Recreation, and Tourism Management, Faculty Fellow in the Center for Geospatial Analytics, North Carolina State University, Raleigh, NC, USA

T. M. Poland et al. (eds.), *Invasive Species in Forests and Rangelands of the United States*,
https://doi.org/10.1007/978-3-030-45367-1_12

Humans are a fundamental component of invasive species issues (McNeely 2011). People have long transported species across biogeographic boundaries, both accidentally and intentionally, and this has increased with globalization of the economy and society (McNeely 2011). Humans modify landscapes in ways that precipitate, facilitate, and exacerbate invasions (Rotherham and Lambert 2011). As a result, non-native species, some of which are invasive, are deeply woven into the fabric of modern life (McNeely 2011). Yet, public awareness and knowledge of invasive species remain low even where they are a significant ecological threat (see, for example, Dodds et al. 2014). Only a few invasive species with significant economic and cultural impacts have garnered broad levels of public concern and widespread management attention across different ownership types at the landscape level (Keller et al. 2015; McNeely 2001, 2011; Mooney and Hobbs 2000). Management actions themselves can cause public reactions and objections for a variety of reasons ranging from lack of public acceptance of chemical control methods, to animal rights issues, objections to costs, and cultural preferences for invasive species themselves (McNeely 2011).

Despite this strong human connection, invasive species are not often studied by social scientists, and the existing studies, which generally have examined the issue through particular disciplinary lenses, have produced a fragmented body of knowledge. Anthropologists, historians, and others have analyzed narratives and discourses about non-native species and the ways in which they have affected public opinion in general. Some researchers, using methods of social psychology, have explored the relationships among individuals' knowledge, attitudes, and personal behaviors related to invasive species. Studies of the impacts of educational programs have examined the effectiveness of efforts to raise awareness, increase knowledge, and motivate behavior change. Sociologists have offered insights into social norms regarding invasive species, and have addressed the collective action and institutional challenges that are required for communities and society to address invasive species management. Policy scientists help us understand how invasive species laws and regulations are formed and implemented by governmental bodies, and why environmental and industry groups respond as they do to invasive species management proposals. Ultimately, sustained and interdisciplinary efforts are required to generate the necessary social science understanding to address this issue.

12.2 Understanding the Human Dimensions

A growing community of researchers has recognized that managing invasive species is as much a social issue involving various human factors as it is an ecological or technical issue (Bremner and Park 2007; Epanchin-Niell et al. 2010; Gobster 2011; Kueffer 2010; Reaser 2001). Invasive species impose huge conservation or economic costs on society (Pimentel et al. 2005; Wilcove et al. 1998). However, when citizens consider the full range of environmental risks, invasive species often do not rank very high. Slimak and Dietz (2006) surveyed members of the public as well as selected U.S. environmental professionals, asking them to rank 24 ecological risk items from climate change to hazardous wastes to sport hunting and fishing. Among the lay public, invasive species ranked 19th, just behind overgrazing and ahead of damming rivers but well below the greatest perceived risks: hazardous waste sites and persistent organic pesticides. Professional risk assessors ranked invasive species ninth, suggesting that beliefs about the threat of invasive species are highly influenced by knowledge and experience. Yet, while general public awareness and concern may be low, there clearly are locations where public interest and knowledge are greater. For example, a survey in Hawai'i found that 96% of respondents were aware of invasion by the coqui frog (*Eleutherodactylus* spp.), and 82% held negative views toward the frogs (Kalnicky 2012).

Geographically, more research has been conducted in Australia, New Zealand, and parts of Europe to assess public attitudes toward invasive species and potential management options using interviews, focus groups, and surveys (e.g., Bardsley and Edwards-Jones 2006; Barr et al. 2002; Bremner and Park 2007; Coates 2015; Fraser 2001; Fischer and van der Wal 2007; Fitzgerald et al. 2007; García-Llorente et al. 2008; Johnston and Marks 1997; Manchester and Bullock 2000; Meech 2005; Nimmo and Miller 2007; Selge et al. 2011; Shine 2015a, b; Veitch and Clout 2001). In contrast, fewer studies have been conducted in the United States to examine public perceptions and behaviors toward invasive species.

12.2.1 Broad Issues and Narratives

People view and relate to the general issue of invasive species in diverse and complicated ways that reflect their underlying values across a range of environmental and social issues. These underlying values, and the narratives in which they are situated, play an important role in shaping perceptions, attitudes, and responses to specific invasive species and their management. Understanding these general issues and the ways they are often discussed, which differ significantly from the ways scientists talk about invasive species, is important for scientists, managers, and policymakers, and can help them avoid major pitfalls, understand why stakeholders may hold different ideas and desires about invasive species and their management, identify mutually acceptable solutions, and determine how to encourage stakeholders to get more involved in control and prevention.

General Attitudes and Stakeholders Relatively few members of the public are likely to see non-native species as inherently problematic. As McNeely (2011) noted, human dietary needs worldwide are largely met by species introduced from elsewhere, and maintaining food production often requires the introduction of non-native species. Many non-native species, including some that are invasive or have deleterious ecosystem impacts, are beloved by people. The ring-necked pheasant (*Phasianus colchicus*), of Asian origin, is a popular upland game bird in the United States and has been adopted as the state bird of South Dakota (Coates 2006). Honeysuckle (*Lonicera* spp.), introduced as an ornamental, for erosion control, and to improve habitat for birds, is valued by gardeners and has been incorporated into Southern culture to the point where it is not widely recognized as a non-native species (Geier 2015; Luken and Thieret 1996; McNeely 2011). Kudzu (*Pueraria* spp.) presents a similar, if more complicated, story. Introduced as an ornamental to shade porches and courtyards, it was later used as inexpensive livestock forage on overgrazed pastures, and then promoted and distributed throughout the South for erosion control before being classified as a weed (Blaustein 2001). Over time, it seeped into Southern culture as "the vine that ate the South," appearing in band names, logos, festivals, crafts (baskets), and poems at the same time that it triggered economic impacts costing millions of dollars (Blaustein 2001).

People have introduced plants and animals for food, economic gain, aesthetics, and to remind them of the past, and until recently, intentional introductions were generally viewed positively (Borowy 2011). Widespread environmental concern related to introduced species only surfaced in the United States in the 1990s, aside from a few particularly prolific and damaging invasive species such as the chestnut blight (caused by *Cryphonectria parasitica*), sea lamprey (*Petromyzon marinus*), kudzu, and saltcedar (*Tamarix* spp.) (Simberloff 2011). A further complication is that any individual invasive species may affect people and stakeholder groups differently. McNeely (2011) notes that many introductions are beneficial to most people, others benefit some individuals or interest groups while harming others, and a few, generally disease organisms and forest or agricultural pests, are clearly harmful to everyone. When costs and benefits are unevenly distributed across stakeholder groups and over the short and long term, interest groups can be expected to view invasive species issues differently, sometimes to be in conflict, and even to change their positions over time (McNeely 2011). There are numerous examples of stakeholder conflicts over invasive species in the United States. In Chicago, prairie restoration involved removal of large non-native trees that were preferred by some people over prairie, use of herbicides and fire that were seen as risky, and removal of non-native shrubby boundaries that were valued by some for screening and wildlife (Gobster 2011). In San Francisco, dominant native coastal scrub and dune ecosystems are open and treeless, but fire suppression and afforestation have made non-native eucalyptus (*Eucalyptus* spp.) trees familiar components of the landscape (Coates 2006; Gobster 2011).

Public awareness of invasive species impacts is often limited, and at times public opposition can prove to be an obstacle to invasive species management (Gherardi 2011; Keller et al. 2015; McNeely 2001; Reaser 2001). Groups opposing eradication or control of invasive species, by engaging in protests and lawsuits, can have considerable power to prevent or delay control efforts (McNeely 2001; Sandiford et al. 2015; Simberloff 2011). Examples where opposition has been significant include eucalyptus removals in California, removal of "Australian pine" (*Casuarina* spp.) on the Florida coast, release of a biological control agent for strawberry guava (*Psidium cattleianum*) in Hawai'i, removal of wild parrots in San Francisco, and hunting of wild pigs (*Sus scrofa*) in Hawai'i (Simberloff 2011). Understanding the human dimensions of invasive species is vital for building political and community support to implement policies, laws, and regulations (McNeely 2001).

Lurking beneath many of these issues are questions and debates about what is and what is not native and, ultimately, different views of the fundamental relationship between humans and ecosystems. Particular introduced species have arrived at different times, but it is common to view as non-native only those introduced after Columbus initiated the colonization of the Americas. A number of authors argue that there is no unambiguous point at which an established non-native species is considered native, and thus these distinctions are inherently arbitrary (Chew 2011; Coates 2006; Rotherham and Lambert 2011; Smout 2011). Opponents to control efforts sometimes reference this ambiguity in their narratives, presenting complex challenges to managers. Wild horses and burros (*Equus* spp.) represent a prime example of this. Large non-native mammals are attractive to many people, and horses and burros have particular cultural salience in the Western United States. Furthermore, in spite of ecological and economic costs involved, opponents to their removal have sought to change the underlying basis of the discussion by arguing that wild horses and burros fill an ecological niche once occupied by equine ancestors, some of which may have been eliminated by early humans, and that their introduction should therefore be considered re-wilding (Donlan et al. 2005; Pimentel et al. 2005).

Cross-Over Between Social and Environmental Thought and Language It has been common in U.S. environmental thought and management to draw a sharp line between untrammeled wilderness and human-modified nature (Cronon 1996; Turner 2012), although social scientists today are likely to recognize the social construction of views of nature (Javelle et al. 2011; Smout 2011). To say that nature is socially constructed does not mean that there is no underlying ecological reality, but rather that our interpretations of what is "natural," desired states of nature, and management goals are often the result of socio-cultural rather than scientific thinking (Borowy 2011; Coates 2006; Smout 2011). These debates are not limited to the social sciences and humanities. Some biological scientists have argued that notions of pure natural ecosystems have led to exaggeration of the seriousness and breadth of problems caused by introduced species (Allison 2011; Simberloff 2011). Other biologists argue that it is unrealistic to eliminate invasive species and that the focus should be on managing invasive species within novel ecosystems (Hobbs et al. 2013). While a recent trend in ecology may appear to sidestep these questions by merging nature and people into the concept of social-ecological systems (see, for example, Gunderson and Holling 2001; Liu et al. 2016), the notion of "natural ecosystems" is deeply embedded in popular and scientific cultures and difficult to move beyond.

Ecological and social narratives often encode fundamental value differences that affect public acceptance of natural resource management. For example, there may be fundamental differences between worldviews, such as between animal rights perspectives, where the intrinsic right to exist is seen as resting at the level of the individual animal, and ecological approaches to invasive species management, where concern focuses at the level of populations of species within ecosystems (McNeely 2001, 2011). Narratives may also reflect intentional boundary blurring and use of invasive species issues as a means to an unrelated end. Sandiford et al. (2015) give examples where debates over invasive carp (*Cyprinus* spp.) became attached to unrelated agendas, as when a power company framed its proposed increase of wastewater discharge as a way to combat an invasive. Shine (2015b) discusses how an outspoken leader of a naturalist group leveraged invasive species media visibility into a much broader agenda and election to a high public office. Conservation and natural resource management are undertaken by human society and mixing of social and ecological thinking is common.

A common crossover between social and ecological narratives occurs when words and ideas that have strong human associations are used in talking about invasive species. People's linguistic frameworks contain many assumptions, unarticulated values, implications, and consequences (McNeely 2001), and they are also subject to multiple interpretations and discursive uses. Animals and plants have long been used symbolically by humans (Bloch 1998; Lévi-Strauss 1966). Language use can be a fundamental driver of disagreement in invasive species discussions because it often implies cultural oppositions, such as native/alien, pure/contaminated, harmless/harmful, original/degraded, and diversity/homogeneity. Also, ideals such as ecological integrity and authenticity are frequently associated with opposition to modern forces of social and economic globalization (McNeely 2001).

Two issues, labeling and xenophobia, have been particularly salient drivers of conflict in public discourse about invasive species. Name-calling—a form of labeling broadly defined as using words with common negative associations in social worlds to talk about ecological issues—is rampant in invasive species management and public messaging around the world and is often divisive. Invasive species may be described in ways that are divisive, derogatory, and cause conflicts (Chew 2009; Coates 2006, 2011, 2015; Larson 2005; Sandiford et al. 2015; Verbrugge et al. 2016). Word choice, or labeling more generally, when discussing introduced species can drive larger narratives. Sandiford et al. (2015) notes that when introductions have been promoted, usually by the government or private sector, there is often a historical pattern of rhetoric that begins with an overly optimistic assessment of the species prior to importation and transforms through metaphorical reinvention into a negative narrative through a process in which traits once considered virtues—for example, high reproductive rates and hardiness—become threatening qualities. In the aquatic realm, introduced species such as carp and tilapia have been promoted as the "cheapest food for the greatest number of people" or "poor man's fish," later to be disdained by anglers as "trash fish" (Sandiford et al. 2015). Sometimes a narrative seeks to change a previously undesirable species into a valuable addition to the ecosystem. For example, grass carp (*Ctenopharyngodon idella*) have been renamed "white amur" and, tapping into societal fears about herbicides and other chemicals, promoted as a benign "green" tool for cleaning not just fish farm ponds, but thousands of weed-infested lakes, reservoirs, and ditches (Sandiford et al. 2015).

At the extreme, invasive species are referenced using culturally loaded and divisive terms (Coates 2006). This tendency occurs worldwide, perhaps linked to fundamental social and psychological tendencies of humans to distinguish between ingroups and outgroups (Giles and Giles 2012).

Occurrences of inflammatory and derogatory language in reference to invasive species be examples of rhetorical convenience, limited to contrarians, or more common in the past, as Coates (2006) and Simberloff (2011) have maintained. However, when public debates over control of invasive species are characterized as divisive (Coates 2006), the overall cause of invasive species management is likely to be harmed. There has been insufficient research to know the extent to which language issues and nativism have affected public values in general or even in specific cases of invasive species management (Coates 2006; Simberloff 2011). But, language containing negative social associations can be powerful in both positive and negative ways, and there is little doubt that words and issue framing can significantly influence public opinion (Gobster 2005).

Avoiding Pitfalls and Making Progress This section has reviewed general social and cultural issues and common narratives involving invasive species. The ways in which these issues emerge and play out in individual cases are complex, and subsequent sections of this chapter will examine more specific research. Clearly, the language and imagery used when talking about invasive species matters (McNeely 2011; Smout 2011). There are dangers in ecological purism (Smout 2011), and it is important not to forget that people love and have deep ties to many invasive species (Rothernham and Lambert 2011). Some ecological writing has involved barely disguised racism and xenophobia, and such negative social connotations can distance science and management from the public (Rotherham and Lambert 2011). Pooley (2011) advises, based on past issues, that we should pause and reflect on the values and trends that inform our current thinking and messages about invasive species and their management and how these might differ from the ways that stakeholders think and talk about them. At a minimum, work to date underscores the importance of scientists and managers becoming aware that some individuals and groups often hold different viewpoints and values toward invasive species. In the same ways that our language has evolved to avoid marginalizing people by gender, race, and physical and mental abilities, we must also sharpen the language that we use to discuss invasive species in order to avoid what stakeholders may see as value-laden and potentially offensive terms.

In spite of examples of social opposition to and debate about invasive species control, evidence suggests that when a non-native species becomes highly invasive, destabilizing ecosystems and causing economic harm, control may become more socially acceptable. McNeely (2011) notes that killer bees (or Africanized Honey Bees, a hybrid between *Apis mellifera* and *A. mellifera scutellata*), water hyacinth (*Eichhornia crassipes*), kudzu, spruce budworm (*Choristoneura fumiferana, Choristineura orae*), various pathogens, and agricultural weeds have been sufficiently damaging to generate strong consensus that they should be controlled, bringing together interest groups that might otherwise be in opposition. Gobster (2005) suggests focusing on the more positive term of ecological restoration rather invasive species control. Rotherham and Lambert (2011) argue for focusing on problem species rather than invasive species in general, recognizing that people value and even celebrate some invasive plants and animals, and accepting that management often involves subjective decisions that require open discussions and stakeholder debates.

12.2.2 Research on People and Invasive Species in Ecosystems

Research on the human dimensions of invasive species is typically carried out in particular ecosystem types. There has been some general research on invasive species that focused on public protected areas. Sharp et al. (2011) surveyed visitors to a national park in Georgia, and their results show that visitor support for invasive species control is associated with their knowledge, perceived threats, age, education, previous experience of visiting national parks, and environmental value orientations. Seidl and Klepeis (2011) interviewed and surveyed residents around the Adirondack State Park in New York, and found a positive attitude toward invasive earthworms (in North America belonging primarily to two families, the Lumbricidae and the Megascolecidae) and their ecological effects, which lead to local residents' casual disposal or use of them. Schlueter and Schneider (2016) surveyed visitors to a State park in Minnesota about emerald ash borer (*Agrilus planipennis*) management approaches, and found that wood regulations, sanitation cutting, and progressive thinning were the most acceptable actions, while complete harvest, chemical treatment, and doing nothing were unacceptable. Light-handed approaches, such as biological control and doing nothing, were most acceptable in natural areas, indicating that landscape context influences social acceptability (Schlueter and Schneider 2016). However, significant bodies of research exist for forest, rangeland, and aquatic ecosystems, and here we present reviews of this literature by ecosystem type. While each of these ecosystem types has unique aspects and issues related to invasive species, they share common issues related to individual behavior and collective action.

Forest Ecosystems In the United States, relatively little has been done to examine the human dimensions of invasive species in forest ecosystems, specifically the role of private forest landowners in preventing and controlling invasive species, even though they control 56% of the forest land (Butler 2008; Steele et al. 2006). Anecdotal evidence has suggested that forest landowners are not always aware of invasive species infestations, and are not actively preventing new invasions or controlling existing infestations. Forestry professionals, based on their own experience interacting with non-industrial private forest landowners, have observed that as an invasive plant species becomes abundant on the landscape, landowner observation of the species and awareness of the associated problems seem to increase, which could potentially motivate landowners to take actions to manage invasions locally (Carlson 2014; Fig. 12.1). Although little research has empirically examined the relationship between invasion intensity and landowner awareness, anecdotal evidence has also suggested that landowners often fail to recognize and/or act upon such invasions until they are at or near the point where eradication is highly unlikely. Furthermore, less is known regarding how to move the point when landowner awareness typically begins down the curve to an earlier point when eradication or control is still ecologically and financially feasible. Thus, effective invasive species prevention and control require not only an accurate understanding of invasion risks on the landscape, but also a comprehensive assessment of the awareness, attitudes, and behaviors of forest landowners, as well as a better understanding of how they perceive and respond to invasions and invasion risks; how and why they engage in collaborative management; and policy implications.

There have been few studies that explicitly examine the human dimensions of invasive insects in forest ecosystems in the United States. Cartwell (2007) mentions the importance of engaging the public in eradicating Asian longhorned beetles (*Anoplophora glabripennis*) and emerald ash borer by increasing public reports of sightings. Several studies were found to assess landowner awareness of invasive plants and their adoption of prevention and control practices. Steele et al. (2006) found that the issue of invasive plants was moderately salient among private forest landowners in West Virginia, but that many landowners could only identify a limited number of invasive plant species. A majority of landowners who recognized certain invasive plants as undesirable had used mechanical methods to remove them, while placing less emphasis on applying herbicides or establishing desirable plants. In a subsequent study, Steele et al. (2008) found that private forest landowners who farmed on their property, held recreation objectives, and lived in the local area were more likely than their counterparts to have received outreach information about invasive plants; however, a majority of these landowners had not heard or received such information. More recently, Fischer and Charnley (2012) surveyed private forest owners in Oregon's ponderosa pine (*Pinus ponderosa*) zone and found that they had widely different perceptions of invasion risks, ranging from a lack of concern, to the opinion that non-native plant invasions have discrete causes and controllable consequences, to the opinion that invasions have gone out of control. All three studies discussed the importance of raising landowner awareness and the need for communicating invasive plant information in a way that resonates with landowners and that is consistent with their management objectives.

Thus far, there has been limited research on how forest landowners perceive and make decisions about invasive species management, specifically minimizing invasion risks and adopting effective strategies to control and eradicate already established invasions. In addition to this knowledge gap, research has shown that only a small segment of forest landowners is committed to active management (Kittredge 2004). This suggests that there is a significant need and justification for using social science research to inform effective engagement of various stakeholders, especially forest landowners, in controlling invasive species in forest ecosystems.

Because invasive species generally occur at the landscape level and easily cross property boundaries, their management requires coordinated and collective action. For example, as forest land in the United States becomes increasingly fragmented and parcelized, any one public or private entity may assume responsibility for only a small portion of the total damage caused by invasive species, per-

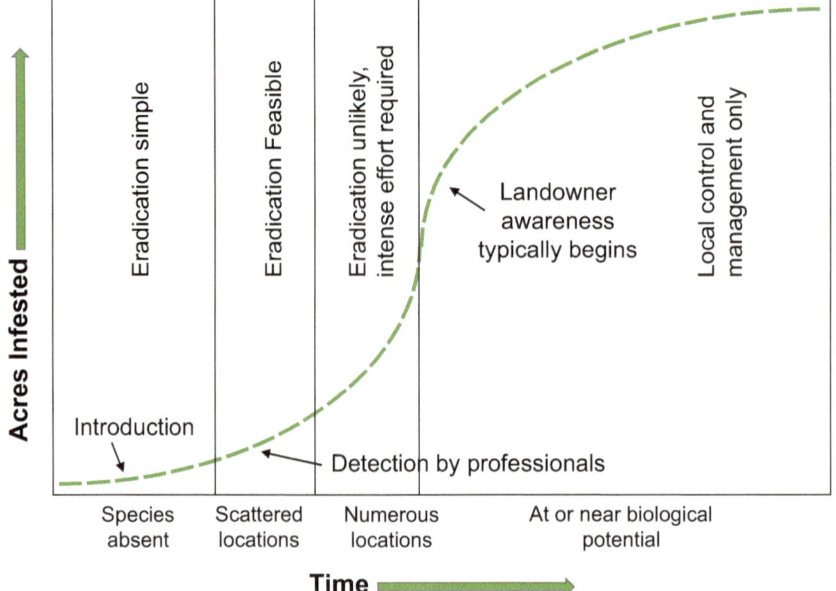

Fig. 12.1 Four stages of invasive plant infestation and possible treatment at each stage. For a given invasive plant species, ease of treatment declines and cost of treatment increases *from left to right*. (Adapted from Hobbs and Humphries (1995) and Indiana Division of Forestry (2010))

ceive that the management of invasive species—particularly wildlife species—is the responsibility of others, or not feel motivated or adequate to tackle an invasive species problem because the extent of the problem goes beyond their own properties. However, an entity opting not to control invasions will increase control costs for neighboring private and public entities by allowing their land to act as a source for invader propagules (Epanchin-Niell et al. 2010; Simberloff et al. 2005). Thus, invasive species threaten a public good—the health of forest ecosystems—which makes invasive species management a problem that requires collective action. Invasive species management as a collective action problem has been more commonly discussed in grassland and rangeland systems than in forest ecosystems (e.g., Epanchin-Niell et al. 2010; Yung and Belsky 2007). Niemiec et al. (2016) investigated landscape-scale invasive species control in Hawai'i and noted the importance of reciprocity in promoting a community good and the presence of social norms as motivators of invasive species control, and suggested that these are important complements to approaches that focus on individuals like education campaigns and subsidies.

In the forestry context, a significant body of literature has been developed to examine landowner cooperation in forest management in general. For example, Kittredge (2005) reviewed and analyzed private forest landowner cooperation in temperate nations with developed economies, including the United States. He emphasized the need and potential for enhanced landowner cooperation to increase individual ownership benefits, as well as the greater public benefits that may result from better managed forest landscapes. Despite the importance of landowner cooperation, previous research has shown a relatively low level of interest in cooperation among private forest owners (e.g., Erickson et al. 2002; Jacobson 2002; Jacobson et al. 2000; Rickenbach and Jahnke 2006). Researchers have identified various factors that influence landowner willingness to cooperate with one other. These include environmental values, stewardship ethics, concerns about maintaining control and property rights, ability to see immediate outcomes from cooperation, using neighboring properties for non-consumptive recreation activities, trust, existing social networks among landowners, time constraint, demographics, and various dimensions of the institutional environment that supports landowner cooperation (e.g., Blinn et al. 2007; Finley et al. 2006; Rickenbach and Kittredge 2009; Rickenbach and Reed 2002; Rickenbach et al. 2011; Vokoun et al. 2010; Wagner et al. 2007; Wolf and Hufnagl-Eichiner 2007).

In contrast to the literature on cross-boundary cooperation in the context of timber harvesting and other forest management activities, little has been done to apply this collective action approach to manage invasive species in forest ecosystems. There is much to be learned about forest landowner attitudes and willingness to prevent and control invasive species that travel and spread across property boundaries collectively. There may also be opportunities to innovate on the areas of outreach and policy to facilitate the formation of a collective invasive plant management norm within larger forest landowner communities.

Invasive Species and Rangelands Invasive species, particularly plants, have impacted rangelands at increasing rates, causing significant economic and ecological impacts (Vasquez et al. 2010). Human roles in the rangeland invasive species issues include transporting and introducing invasive species and disturbing and fragmenting ecosystems in ways that facilitate their invasion, and thus coordinated management actions are required (Vasquez et al. 2010). Rangeland communities may have a greater level of awareness of invasive species than those in other ecosystem types. Tidwell (2005) surveyed residents in the Southwest United States about their beliefs regarding invasive forbs and their management and found that 94% of respondents were concerned about invasive species. A 2010 survey in the Great Basin found that more than 83% of respondents believed annual invasive grasses pose a threat to healthy rangelands (Gordon et al. 2014).

Management of rangeland invasive species requires both awareness and acceptable control methods. Ambivalence about invasive species control can be common, because it may involve methods considered potentially risky to humans. Norgaard (2007) examined a controversy over control of spotted knapweed (*Centaurea maculosa*; syn. *Centaurea stoebe* L. ssp. *micranthos* (Gugler) Hayek) in northern California and found that natural resource managers considered herbicide application to be safe while community members did not. Similarly, Shindler et al. (2011) reported that residents of rural parts of the Great Basin were more than twice as likely as urban respondents to believe herbicide application is an appropriate management tool on public lands. Tidwell (2005) found that chemical control of invasive plants was judged to be more acceptable for use on multiple-use or agricultural lands than on protected or residential lands. However, he did not find similar differences in support for other forms of management.

Invasive species program managers often seek to heighten awareness and concern about invasive plants with the intention of promoting willingness to support or engage in control activities. It is therefore useful to learn what influences changes in attitudes toward invasive species. However, findings have been somewhat disappointing. Gordon et al. (2014) measured attitudes toward invasive species control methods in 2006 and again in 2010, and found that attitude change was less affected by changes in awareness than by changes in trust of the government agencies that would be applying those methods.

Efforts to detect, manage, and control invasive species typically involve a combination of top-down measures (e.g., laws and ordinances) and grassroots volunteer action. Legal measures directed against invasive species are reviewed in Chap. 15 and are not repeated here. However, it is worth noting that the absence of legal support for invasive species control can pose a barrier to management effectiveness. Kelley et al. (2013) found that the fact that cheatgrass (*Bromus tectorum*) was not designated as a noxious weed in Wyoming was seen as a barrier to management by 34% of ranchers and 39% of natural resource professionals. As one manager stated in a focus group that Kelley et al. conducted, "One interesting thing with regard to cheatgrass for me is that my budget is such that, I get money to treat noxious weeds and as of yet, cheatgrass isn't a noxious weed." The same study found that 77% of ranchers felt other weeds were a higher priority than cheatgrass.

Because volunteers are an important tool in the battle against invasives, Tidwell and Brunson (2008) queried respondents about their willingness to volunteer for weed management activities including control, monitoring, education, and restoration. Ten percent of respondents said they had engaged previously in weed-related volunteer activities, and 43% expressed their willingness to participate. Among those willing individuals, more were interested in directly participating in control activities (57%) or monitoring (55%) than education (39%) or restoration (38%).

Collective action is important, because individual actions are not sufficient to control invasive species (Epanchin-Niell and Wilen 2014; Yung et al. 2015). The nature of collective action needed has received some attention in the rangeland literature. Graham (2013), based on interviews with landholders and agency personnel, suggests there are three ways that communities can encourage greater weed control: sharing information, providing support, and applying pressure on other landowners. Analyzing the case of yellow star-thistle (*Centaurea solstitialis*) in the Sierra Nevada foothills of California, Epanchin-Niell et al. (2010) suggest that no single management regime will control invasive species and suggest multi-level approaches. These levels include bottom-up—public and private landowners, middle-level—cooperative weed management areas and weed districts, and top-down—regulatory and financial support from local, State, and Federal governments (Epanchin-Niell et al. 2010). Researchers also have emphasized the need for socioecological research and data in these cooperative and landscape-level management programs (Epanchin-Niell et al. 2010), bringing together collaborative and scientific efforts (Miller and Schelhas 2008; Schelhas et al. 2012). Weed districts and Cooperative Weed Management Areas have been particularly effective platforms for these efforts in the Western United States (Forcella and Harvey 1988; Schelhas et al. 2012).

A number of collaborative groups have arisen in recent years to address invasive plant management. These volunteer groups (e.g., Arizona's Sonoran Desert Weedwackers) typically involve multiple public agencies as well as citizen groups that collaboratively engage volunteers in mapping, monitoring, and controlling invasive plant infestations. Some groups focus on all species in a specific locale while others are organized around particular taxa, but nearly all are geographically limited and supported by a combination of public and private funds. Such groups are increasingly popular because they can accomplish tasks that are not supported by limited tax revenues, can inform management decisions with stakeholder perspectives, and can breach communications barriers between groups of people (e.g., ranchers and environmentalists) to achieve goals of common interest. Fernandez-Gimenez et al. (2004) identified more than 100 collaboratives working in Arizona's rangelands alone, although only a small subset of those focused on invasive plants.

Hershdorfer et al. (2007) surveyed coordinators of 53 local weed programs in Arizona, Colorado, New Mexico, and Utah to determine how attributes of the programs were linked to performance of control, education, monitoring, and integrated weed management. They found that programs that used volunteers did more monitoring but less direct control than those that relied entirely on paid employees. Contrary to the researchers' expectations, more regulatory action did not translate to better control. In fact, groups that had regulatory authority but generally refrained from punitive enforcement treated more infestations, partly because staffs are typically small and enforcement takes time, and partly because a gentler approach with private landowners seems to yield better results.

Invasive Species in Aquatic Ecosystems There have been very few general studies of aquatic invasive species in the United States. Weber and Ringold (2015) studied people's preferences for river and stream features in an arid landscape in Arizona and found that there was concern about both invasive plants and animals. Responses were generally linked to positive concern or threats to native species (Weber and Ringold 2015). There are a number of invasive species studies that focused on aquatic vertebrate and invertebrate species. For example, Limburg et al. (2010) surveyed four homeowner communities near Lake Ontario. They found that most respondents noticed and valued improved water clarity, which in fact reflected the loss of ecological functions due to invasive zebra and quagga mussels (*Dreissena polymorpha* and *D. bugensis*) lowering phytoplankton biomass and pelagic production in the lake. Luizza et al. (2016) modeled potential spread of Alaska's first freshwater invasive plant (*Elodea* spp.) with climate change, and combined these results with subsistence use by Alaska Natives and concerns related to Chinook salmon (*Oncorhynchus tshawytscha*) and whitefish (*Coregonus nelsonii*). This highlighted the positive contribution of community involvement to risk assessment and incorporating stakeholder concerns into management.

Aquatic invasive species often arrive unintentionally through shipping and recreation vectors, although they can also be associated with pet and landscape trades or through deliberate introductions (Cambray 2003; Pyšek and Richardson 2010). Recreational transport of aquatic invasives represents a significant risk. Cambray (2003) emphasized the significance of deliberate and accidental introductions of sport fish as a problem that is increasing due to globalization. Anderson et al. (2015), while noting the existence of limited literature on recreation and aquatic invasive species, note that there are several publications on transport by recreational boaters between rivers and lakes and by yachts in marine environments (Rothlisberger et al. 2010; Thresher 1999; Willette et al. 2014). Studies indicate that the diversity of aquatic non-native species, including plants, algae, and invertebrates is higher where recreational boating or yachting took place than at control sites, with vectors including hulls of boats, ballast and bilge water, and anglers (Anderson et al. 2015). Waterkeyn et al. (2010) showed that aquatic invertebrates could be dispersed among wetlands at very local scales by footwear and vehicles. Pradhananga et al. (2015), studying boaters in Illinois, found nature- versus human-oriented values had predicted environmental concern but had little impact on behavior; behavioral intentions were most influenced by habit and concern about aquatic invasive species. Overall management recommendations include raising awareness and bio-security measures for tires, boots, boats, and other equipment (Anderson et al. 2015; Pradhananga et al. 2015). Sharp et al. (2016) found that recreational boaters understood the importance of managing aquatic invasive species and supported inspections and regulations. The Cornell Human Dimensions Research Unit has published a series of integrated reports on aquatic invasive species in the Great Lakes region that examines human dimensions across a range of vectors, stakeholders, and issues. This research addresses vectors, such as anglers, boaters, and other recreationists (Lauber et al. 2015a), the role of bait dealers and boating facilities, and angler and boating organizations (Connelly et al. 2014a, b, c; Heck et al. 2013; Lauber et al. 2014), aquarium and plant trades (Lauber et al. 2015b), and factors affecting communication success and outreach capacity in recreational communities (Lauber et al. 2015a).

Intentional fish introductions can have negative impacts. McNeely (2011) notes the harmful effects of introduced trout (*Oncorhynchus* and *Salvelinus* spp.) on amphibian populations. Varble and Secchi (2013) analyzed the results of the first national survey on the attitudes of U.S. fish consumers toward invasive Asian carp, and suggested harvesting Asian carp for human consumption as a potentially promising strategy for controlling this invasive species. However, Nuñez et al. (2012) cautioned policymakers, practitioners, and researchers that controlling invasive species through human consumption should be carefully examined to avoid creating a market that engenders pressure to maintain that problematic species. Sandiford (2015) discusses the long history of rhetoric, both positive and negative, around introductions of different species of carp in the Mississippi River Basin. Various species of carp, which have been promoted by fish farmers and for weed control in ponds, have caused impacts on recreational fishing and other aquatic species. Moreover, Carlson and VonDracek (2014) acknowledged a "dearth of sociological research on Asian carps represents a barrier to predictive management." They further stated that human dimensions research in the prevention and control of Asian carp can contribute to understanding public attitudes, enhancing stakeholder engagement, fostering harmony between agencies and stakeholders, and gaining social knowledge for effective management.

12.3 Tribal Perspectives and Engagement

Native peoples of North America have millennia of experience adapting to social and ecological change. Among these changes, indigenous communities in the United States and its territories have been responding to the presence of non-native species, some of them satisfying the definition of invasives, at least since the beginning of the Columbian Exchange in the late fifteenth century (Crosby 1972). This experience and traditional ecological knowledge are reflected in indigenous attitudes toward and approaches to invasive species, and are essential to the development of invasive species management programs that honor the U.S. trust responsibility to Native peoples.

There are over 560 federally recognized tribes in the United States (Bureau of Indian Affairs 2016). Each tribe has distinct cultures, histories, and lands. Additionally, within each tribe, members hold multiple perspectives, attitudes, beliefs, and relationships to the natural environment. While tribal governments may take many forms, they are responsible for managing tribal natural resources. The U.S. Government has a trust responsibility to ensure proper management of tribal resources as well as Federal lands. The Federal trust responsibility is codified in treaties, the

U.S. Constitution, case law, Presidential Executive Orders, United States Department of Agriculture (USDA) Regulations, and the USDA Forest Service Handbook. The Federal trust responsibility includes the legal responsibility to consult with individual tribal nations on a government-to-government basis on programs and actions that may impact or are important to federally recognized tribes. This includes a requirement to consult with tribes on planning and actions related to invasive species management and control (see Donoghue et al. 2010; Haskew 1999).

Tribal governments are modern institutions. One of their many responsibilities is to manage, conserve, and protect tribal lands. Tribes approach this in various ways. Some tribes maintain large natural resource, forestry, environmental, and fish and wildlife departments which operate with Federal and tribal funding, while other tribes have smaller departments directly supported by the Bureau of Indian Affairs staff. Most, if not all, of these tribal institutions are concerned about the impacts invasive species are having or could have on tribal ecosystems, tribal resources, tribal enterprises, tribal communities, the Federal Government's trust responsibility, and tribal sovereignty.

Tribal natural resource management staff and tribal community members have indicated that invasive species are one of the most important issues facing tribal natural resources, especially in conjunction with climate change and integrated forest management (see, for example, Gordon et al. 2013; Sustainable Development Institute 2012). While there is no one national group that works exclusively with tribal invasive species management, tribes have formed partnerships with local, State, and Federal institutions to manage invasive species. There are also intertribal organizations that coordinate and share information about invasive species at national, State, and local levels. These partnerships strive to manage invasive species on lands and waters with reserved treaty rights, as well as lands that have the potential to impact tribal resources. Often, non-tribal institutions learn valuable perspectives from tribal partners concerning management, control, and social and cultural impacts.

Because invasive species impact tribal communities on environmental, social, spiritual, and economic levels, tribes throughout the country are actively working on invasive species management. For example, the Menominee Nation in Wisconsin is a leader in sustainable forest management and has established their own forest health department. They are actively working on control of invasive species including garlic mustard (*Alliaria petiolata*), oak wilt (*Bretziella fagacearum*), beech bark disease (*Neonectria* spp.), and emerald ash borer. These, and other invasive species management projects, are important projects for the Menominee Nation and are related to maintaining high quality saw timber for the tribal saw mill, a diversity of species within the tribal forest, and culturally important species for traditional and contemporary use. Another example is the Shoalwater Bay Tribe in Washington State, which is working on several invasive species control projects including control for the aquatic invasive plant Spartina (*Spartina alterniflora*). This plant impacts native plant species, hydrology, bird habitat, and fish communities, which in turn affects many aspects of tribal life including subsistence fishing, recreation, and spiritual practices. Tribes in the Midwest and East are working on projects related to the ecological and cultural impacts of emerald ash borer. Tribes in the Pacific Northwest, including the Confederated Tribes of Warm Springs, the Yakama Nation, and the Colville Confederated Tribes, are working on white pine blister rust (*Cronartium ribicola*) in collaboration with the Forest Service.

Invasive species can have important cultural impacts and meanings for Native people. While recognizing there is no single Native culture but, rather, many sovereign nations with distinct cultures, there are common teachings about the roles, responsibilities, and relationships between human beings and the rest of the biotic and abiotic world (hereafter, "Creation"). These teachings are grounded in the spiritual traditions and lived experience that form the basis for traditional ecological knowledge (Emery et al. 2014). This traditional ecological knowledge teaches that all elements of Creation are relatives and each has roles and responsibilities. The right relationship between humans and Creation is one of mutual respect and caring. The relationship between people and Creation becomes out of balance when humans cease to honor their responsibilities to care for and behave in a respectful way toward their non-human relatives, which includes active stewardship and respectful use. When this happens, the plants and animals that provide for humans may cease to be present, pushed out by or replaced by other aggressive species. In some cases, these species may have the capacity and responsibility to repair damage done by human's poor treatment of the land.

Recent and historical experiences also inform Native perspectives on invasive species and what should be done about them. In their review of 70 case studies of the sociocultural implications of invasive species around the world, Pfeiffer and Voeks (2008) note that the impacts of invasive flora and fauna on indigenous communities are far from uniform. In some cases, especially where recently arrived biota result in the rapid reorganization of landscapes and/or replacement of culturally important native species within two human generations or less, invasive species may be culturally impoverishing. Effects can include loss of access to cultural keystone species for food, medicinal, ceremonial, and other purposes (Garibaldi and Turner 2004) and interruption of place-based traditions that literally ground indigenous identity (Pretty 2002). Such impacts are especially acute for indigenous groups already struggling to revitalize their cultures.

In other cases, invasive species have enriched Native peoples' diets and pharmacopoeias, particularly where a species originating elsewhere has been present for 100 or more years, providing time for its absorption into individual and group practices (Pfeiffer and Voeks 2008). For example, the weedy northern European species English plantain (*Plantago major*) was widely adopted into North American indigenous healing practices (Crosby 1986).

Forced relocation and voluntary moves also are common in North American indigenous communities. In such cases, highly cosmopolitan species that were used in the home territory and also are present in the new location may assist displaced Native peoples in maintaining cultural practices. Indeed, it has been noted that disturbed habitats that create the conditions necessary for the establishment of invasive species tend to be readily accessible and rich in plant species with medicinally useful secondary compounds (Voeks 2004).

It has been noted that biological invasions and responses to them have social and political histories, as well as biotic roots (Crosby 1986; Robbins 2004), a process with which indigenous peoples around the world have abundant experience. For example, Pretty Paint-Small (2013) notes that the Dawes Act of 1887, which resulted in privatization and ownership of land by non-Indians inside the boundaries of reservations throughout the Western United States, set the stage for the contemporary invasion of Russian-olive (*Elaeagnus angustifolia* L.) on the Crow Indian Reservation in Montana. The cessation of traditional land management practices due to loss of sovereignty and control over ancestral territories likely is a component in other instances (see, for example, Long et al. 2016; Ortiz 2008a).

Indigenous communities also have suffered negative impacts from programs designed to address invasive species, including the use of toxic chemicals and escaped biocontrol agents (Pfeiffer and Voeks 2008). When chemical applications affect culturally important foods, entire communities may suffer, with children and elders at particular risk. Likewise, cultural practices such as those involved in processing basket materials may present increased risks of exposure (Norgaard 2007).

There are numerous examples of collaborative efforts between indigenous communities and government agencies, which are designed to replicate traditional resource management practices in order to control invasive species and support biocultural diversity (Pfeiffer and Voeks 2008). Yet there is relatively little published scientific literature on tribes and invasive species, in spite of the fact that many tribes are faced with invasive species issues. In the following discussions, two tribes, the Eastern Band of Cherokee Indians and the Kashia Band of Pomo Indians, present their perspectives on and experiences with invasive species.

Cherokee Perspective of Invasive Species[1]

Through art, subsistence, and culture, the Cherokee people have relied on the great biodiversity of the Southern Appalachians for thousands of years. However, this biodiversity is threatened by invasive species. The Eastern Band of the Cherokee, therefore, considers invasive species one of the greatest threats to cultural and economic stability (EBCI 2013). Because some non-native species have taken the place of native organisms as cultural resources throughout American society, appropriately defining an invasive species as more than exotic is paramount, especially when the term invasive is often confused with nuisance or exotic species.

In accordance with Cherokee priorities and modern scientific designations, we believe that invasive species are most appropriately defined as species that causes net harm to our economic or cultural resources. We are still in the early stages of species census and evaluation of impacts, but we have categorized several organisms that we plan to manage as potential invasive species. We plan to list plants and animals that can cause either direct or indirect harm to our resources. For example, the hemlock woolly adelgid (*Adelges tsugae*) has nearly eliminated culturally and ecologically important eastern hemlocks (*Tsuga canadensis*) from our area, and the newly arriving didymo algae (*Didymosphenia geminata*) can choke out stream bottoms that in turn influence trout food, which would be detrimental to our fishing industry.

With our definition of a "net harm to economic or cultural resources," some situations will require evaluation. We do not consider a non-native designation as equivalent to invasive. There are many examples of non-native or exotic species that are used for horticultural and agricultural purposes that become naturalized, leaving a net positive benefit to people, ecosystems, or communities (Brown and Sax 2004; Colautti and MacIsaac 2004). For example, Japanese honeysuckle (*Lonicera japonica*) is deemed an invasive species by many Federal and State agencies, but the Cherokee people have adopted this East Asia native as culturally important for making baskets. Therefore, we require an economic and cultural impact assessment before we would deem this species invasive. As our census efforts progress, we will consider each species with a Federal or State invasive listing on a case-by-case basis for our own listing process. Feral hogs (*Sus scrofa*) represent another example of a species that can have both positive and negative impacts but is almost universally listed as invasive among State and Federal agencies. People in our region have long used these animals as livestock as well as game. Groups have intentionally released hogs to hunt, which creates a difficult dynamic when manag-

[1] This perspective was written by Caleb R. Hickman, Michael J. LaVoie, and Tommy Cabe of the Eastern Band of Cherokee Indians Natural Resources program.

ing for eradication versus for a sustainable resource. Based on Federal and State designations as invasive, and a level of damage on our lands, we have implemented feral hog management by creating an open hunting season and supporting research to understand hog and disease movement.

Some species with potential to be invasive might have a higher net economic benefit to people. Trout are native to the Southern Appalachians and a culturally important organism to the Cherokee. Native to Cherokee lands, brook trout (*Salvelinus fontinalis*) populations declined over the past century due to unregulated harvesting and habitat changes. In order to restore this harvesting connection, brook trout were replaced in many areas by a non-native game fish, rainbow trout (*Oncorhynchus mykiss*). Although rainbow trout are considered invasive in certain situations (i.e., western States when hybridizing with native trout), we do not have sufficient information to deem them invasive on tribal lands. Despite a lack of designation, we are committed to careful stocking so that we only manage areas where naturalized rainbow trout exist and preserve reaches with only native brook trout.

Compared to neighboring State and Federal agencies, we might have a different designation for a species based solely on its impact on culturally important organisms. For example, we are describing coyotes (*Canis latrans*) as invasive because of their potential to reduce populations of culturally important white-tailed deer (*Odocoileus virginianus*), which occur in small densities on tribal lands. Even though they are native to the United States, coyotes are new to our area and their predation pressure can cause decreases in deer populations. We consider deer culturally significant because they represent one of the Cherokee clans and serve as a focal point in stories and history. To assess coyote impacts on deer, we are currently leveraging a bounty system to understand their movement and diet.

Our management process consists of these particular areas: assess, monitor, mitigate, and manage. First, we need to determine if a species is invasive by targeting those listed as invasive by State and Federal designation or determining if they produce a net cost to Cherokee economic or cultural resources. In addition to biological surveys, our assessments will include rigorous scientific review and professional research from our staff and experts in the field. If we deem a species invasive, we monitor for prevalence and attempt to decrease impacts by creating management plans with expert involvement. Our actions for invasive species will be within larger management plans. We are currently finishing both a Wildlife Action Plan and Forest Management Plan that will outline strategies to deal with invasive species. Most of the daily challenges will be addressed through eradication treatments of the invasive species, education of people, and enforcement when introductions are a factor. We rely on our tribal municipal code to deal with legal actions and limitations imposed by our enforcement. We feel our program is poised to exercise our sovereignty goals of protecting our natural resources against invasive species.

The Kashia Band of Pomo Indians Respond to Sudden Oak Death[2]

> We'll be gathering acorns, and they'll teach us how to sing the songs that are appropriate for gathering, and why we're singing those songs. We'll learn language and stories… It's what we call the University of Kashaya. It's our school. *Phytophthora ramorum* is threatening that. (Reno Franklin (Kashaya Pomo) 2007)

In the Kashaya Pomo language, tanoak (*Notholithocarpus densiflorus*) is *chishkale*, meaning beautiful tree (Gifford 1967), so in the late 1990s when unprecedented tanoak mortality was recognized on tribal lands near Stewarts Point and on surrounding traditional gathering areas in western Sonoma County, tribal members became quite distressed (Bowcutt 2013). The tanoaks were dying from sudden oak death, caused by *Phytophthora ramorum,* an invasive, exotic, microscopic pathogen, new to science. The pathogen was introduced to the United States on ornamental nursery plants (Mascheretti et al. 2009); once established, its spores spread through forests by wind-blown rain (Rizzo et al. 2005). Tribal environmental staff, elders, and leaders discussed the issue internally and reached out to plant pathologists for assistance to develop management plans, and to share their concerns, they hosted outreach and education workshops for tribal and non-tribal neighbors.

Thousands of trees died near the 40-acre Stewarts Point Rancheria (Ortiz 2008b). Reno Franklin, former Kashia chairperson, described the loss in 2007 as, "We still continue to pass on our ceremonies, our traditions, our prayers, and our songs, and some of those songs and prayers and ceremonies are centered around tanoak and these acorns. We still have roundhouse ceremonies that celebrate and give thanks for what we're taking from those tanoak trees in the form of acorns. You could have a tanoak that's maybe three or four hundred years-old where five or six generations of Kashaya families go. We've got families whose entire gathering areas have been wiped out. And it's hard to take seven generations of a family and remove something like that, and then try and fill that void" (Reno Franklin, quoted in Ortiz et al. 2008b).

As Franklin explains, the ecological, cultural, and community impacts of this invasive, quarantined pathogen pose a serious threat to the tribe including loss of highly valued acorn-producing trees (tanoak and oak (*Quercus* spp.)) and disruption of traditions. The disease's primary source of inoculum, California bay laurel or pepperwood (*Umbellularia californica*), is also culturally important to the Kashia and

[2]This perspective was written jointly by Susan J. Frankel, USDA Forest Service, Pacific Southwest Research Station; Janice Alexander, University of California Cooperative Extension, Marin County; and Nina Hapner, Kashia Band of Pomo Indians.

other California North Coast tribes. The berries are collected for food, and other plant parts are used in ceremony and for medicinal and household purposes. While *P. ramorum* infection is not lethal to pepperwood, transporting infected leaves could contribute to disease spread.

Former Kashia Chairman Eric Wilder, when describing the impacts of sudden oak death asked, "What happens when you take that element from your people that is a significant ceremony, and a practice of your people that happened for thousands and thousands of years, and it's suddenly gone? In our traditional belief, when we go out and we gather these acorns and anything from the land, the Creator has put that here for us.... This is a sacred ceremony that we do.... According to the teachings of our people from thousands and thousands of years, if you don't respect the creation, and we don't follow those rules that we were given to gather, this is the kind of thing that will happen.... In the traditional people's view...creation's showing us what happens when you don't respect it..., so we feel like we're...responsible for what's happening, too...." (Ortiz 2008a).

Different management approaches are needed to meet tribal needs. The Kashia's relationship to tanoaks and pepperwood causes the tribe to be reluctant to use pest management practices commonly used on lands of other ownerships. Recommendations to control sudden oak death include removal of pepperwood trees to protect oaks, and thereby eliminate the inoculum reservoir for spores that spread to highly susceptible oaks (Swiecki and Bernhardt 2013). Favoring oaks over pepperwood is a preference the Kashia do not agree to because both trees are utilized for food and ceremony.

Because Kashia collect and consume acorns, they are also concerned that a systemic pesticide used to prevent *P. ramorum* infection (Lee et al. 2011) may contaminate acorns. A preliminary study of the efficacy of phosphonate to protect tanoak was conducted on Kashia tribal lands, after Kashia staff conducted extensive education and outreach with the tribal community concerning the risks of sudden oak death and the use of phosphonate. Only with agreement from the Kashia community was the application allowed (N. Hapner, personal observation). The chemical composition of tanoak acorns was analyzed (Meyers et al. 2006), but there was insufficient information to determine toxicity. Despite the desire to protect tanoak trees, the Kashia Band of Pomo Indians is reluctant to treat trees because of concerns about pesticide exposure to the acorns (N. Hapner, personal observation).

12.4 Promising Approaches for Changing Awareness, Attitudes, and Behaviors

The preceding review of the social and cultural literature highlights the extent to which public values, opinions, and behaviors related to invasive species and their management are deeply embedded in larger societal and cultural processes. Incorporating the human dimensions into invasive species programs means that managers endeavor to consider a broad range of relevant social science research and to engage with the public and stakeholder groups to incorporate their diverse perspectives, develop broadly supported priorities, and identify strategies that can promote change. Verbrugge et al. (2013) show how people's basic understandings of nature and the relationship between humans and nature influence their perceptions of invasive species and their management, and highlight the importance of early stakeholder participation and risk communication. McLeod et al. (2015) find that providing information has been the primary strategy implemented to change attitudes and behaviors toward invasive species, in spite of the fact that research on human behavior has demonstrated that knowledge transfer alone rarely brings about change. Recognizing that there are many behavioral change models in the social sciences, McLeod et al. (2015) developed a tool employing multiple theories to identify key leverage points and apply them through a range of intervention strategies. Notably, they are able to link a broad range of policy and management approaches to these intervention strategies. Several other studies (Dalrymple et al. 2013; Howell et al. 2015) draw on social networks and diffusion of innovation theories to target behavioral change efforts at opinion leaders, who serve as important and respected sources of information, in these cases targeting vendors of fishing supplies to reach the broader recreational fishing population.

While care must be taken to understand stakeholder perspectives and respect cultural differences, education will continue to be an important part of invasive plant management strategies both to increase participation in management efforts (Marler et al. 2005) and to influence policy (Hershdorfer et al. 2007). Forms of public outreach vary, from relatively low-cost options such as printed materials and electronic resources to direct engagement of citizens in activities (DiTomaso 2000). Marler et al. (2005) report using a suite of educational and citizen engagement efforts in Missoula, MT, that included stewardship opportunities (Adopt-a-Switchback on a popular trail; a Prairie Keepers program that organized activities that included weed pulls, seed collecting, and K-12 education); an annual "weed fair" education project that drew significant attention in the community; and a Grow Native project that engaged junior high school students in restoration activities. Strategies that directly engage learners, either through active participation or involvement in discussion, have been shown to work better than unidirectional or rote-learning approaches used for increasing knowledge (DiEnno and Hilton 2005).

The lack of evidence that short-term training increases long-term participation in invasive species control (Crall et al. 2013; Jordan et al. 2011) suggests that more careful attention must be paid to audience, curriculum design, and desired behavior change. Two curricula were created as part of the Ecologically-Based Invasive Plant Management (EBIPM) program, instituted by the USDA Agricultural Research Service to provide science-based solutions to annual grass invasions (Crall et al. 2013). The university curriculum (Kartchner 2013) provides an example of how an education program can be structured when direct engagement is not always feasible. EBIPM offers a decision-making framework for landowners and managers; therefore, the curriculum covers each of the decision steps in separate modules that include synoptic reading, case studies, in-class and field activities, review questions, additional resources, and a PowerPoint presentation. The modular curriculum was developed with collaborators, including plant ecologists, weed scientists, social scientists, economists, range managers, and media developers.

Public engagement in invasive species monitoring and management can both complement and amplify the work of natural resource professionals, and will be essential if larger cultural changes in the understanding of invasive species and implementation of widespread management actions are to occur. There are a number of promising new public science and engagement techniques that have potential to simultaneously address a suite of human dimension needs for invasive species management, including changing attitudes, engaging stakeholders, instilling a landscape-level perspective, generating a common vision to motivate cross-boundary cooperation, changing behaviors, and complementing and expanding the work of public agencies. Here we review experiences with several of these techniques, while emphasizing the need for them to be accompanied by two-way communication and learning between scientists and the public. Citizen science can improve our scientific understanding of invasive species issues while facilitating attitude and, perhaps, behavior change. Geospatial Participatory Modeling helps engage people to increase their understanding of invasive species issues at landscape and regional scales, envision alternative futures, and establish the conditions for collective action. Social marketing applies marketing principals and strategies to social and environmental issues, and has shown particular promise for developing high profile campaigns with multiple partners to promote behavior changes to limit the spread of invasive species, for example, dispersal by recreationists or the pet trade. These are just a sampling of possibilities, but broad thinking informed by public input, social science research, ecological science, and consideration of the full range of intervention possibilities are the paths most likely to develop successful programs to address invasive species.

Citizen Science Citizen science, which involves members of the public in scientific research, has the potential to drive social change with respect to environmental issues. Through knowledge and skills training to engage the public in scientific activities, citizen science provides opportunities to facilitate change through improvements in participants' science literacy, knowledge of an issue, attitudes surrounding an issue, and behavior to address that issue (Bonney et al. 2009, 2015; Brossard et al. 2005; Evans et al. 2005; Jordan et al. 2012; Shirk et al. 2012; Trumbull et al. 2000). In addition, it is often suggested that public involvement in research induces social change by building social capital, enhancing community capacity, and promoting trust among various stakeholders (Bonney et al. 2015; Jordan et al. 2012; Kountoupes and Oberhauser 2008; Overdevest et al. 2004; Shirk et al. 2012). However, few studies have empirically supported this (Bonney et al. 2015).

This holds true for citizen science projects that involve invasive species. Jordan et al. (2011) examined knowledge gain and behavior change among participants following participation in a project called "Spotting the Weedy Invasives." As part of the training program, instruction included invasive plant species ecology and implementation of the project protocol. Participation resulted in increased knowledge of invasive plant species, improved skills in recognizing invasive plants, and increased awareness of invasive plant impacts. However, participation did not improve understanding of the scientific process or result in changes in behavior. The authors suggest modifications to the training program that better align project design with its goals and participant motivations to reach desired outcomes.

Crall et al. (2013) examined changes in participants' attitudes, behavior, and science literacy following their participation in a citizen science project that focused on invasive plant species. The day-long training included presentations covering an introduction to invasive species, global positioning systems (GPS), sampling design, and the project's vegetation monitoring protocol. A field component included identification of plant species, marking and navigating with a GPS, and implementing the protocol. Although the study found no changes in general science literacy or attitudes following participation, it did note improvements in science literacy and knowledge using context-specific measures. In addition, participants expressed their intention to engage in more pro-environmental activities following the training which included volunteering for environmental organizations, attending community events, removing invasive species, and educating others about them.

Despite the potential, very little research has focused on outcomes from citizen science invasive species initiatives. Outcomes can be considered to be scientific (such as ecological datasets or publications based on these data) and social (such as abilities, skills, and knowledge). Historically, the study of citizen science programmatic outcomes was focused on verifying data quality and individual benefits.

More recently, however, the field has begun to widen its scope and address larger scale social changes that may result from citizen science programs including the potential for long-term impacts that involve attaining conservation outcomes as well as human well-being (Jordan et al. 2011; Shirk et al. 2012). For example, Jordan et al. (2016) report on preliminary data that suggest that collaboratively structured citizen science can produce social outcomes such as increased involvement in natural resource stewardship.

In a recent review of citizen science programs, Conrad and Hilchey (2011) found that some programs increase in environmental engagement, scientific literacy, and social capital. Direct evidence for benefits to the ecosystem, however, was not well-documented, perhaps because data collected from citizen science are often not shared through the management phase. Furthermore, citizen science impacts on conservation behavior, as opposed to behavioral intentions, have not been well-measured (Gray et al. 2017 is an exception). Perhaps conservation scientists and resource managers, by adopting a socio-ecological or adaptive management perspective, can not only gather essential data with respect to invasive species, but, through carefully structured training and education, also use learning as a management tool that increases conservation behavior and improves decision making (Jordan et al. 2016). A recent review suggests programs consider project design, metrics to measure outcomes, ways to engage new audiences, and new directions for research (Bonney et al. 2015). As the field continues to expand, more evidence will be available on best practices for generating desired outcomes through citizen science.

Geospatial Participatory Modeling Despite the availability of bigger data and better models, many efforts to manage invasive species have not been as effective as we have hoped. For complex systems with multi-scale interactions across ecological, social, and economic domains, even the best applied research will not yield solutions without the addition of sustained and meaningful stakeholder participation. Decisions that comprehensively involve stakeholders in the management of invasive species—from data collection to policy—are more likely to be viewed as legitimate, more likely to be accepted, and more likely to succeed (Groffman et al. 2010; Reed 2008). Yet, most public science projects fail to gain traction in shaping collaborative solutions, because either they do not follow best practices for participatory research or they use abstract or aspatial representations of data and models that fail to engage stakeholders. Advancements in geospatial analytics are helping generate more data and better models, raising the question of how to use geospatial technologies effectively to make a difference.

Geospatial Participatory Modeling (GPM) provides an opportunity to improve the connection between communities and the environment and offers three ways to involve stakeholders in research better. Dynamic, adaptive geospatial models enable multiple stakeholders to visualize and explore the roles of (1) place; (2) spatial interaction; and (3) multi-scale processes through all steps of a research process. GPM is an umbrella term that combines (1) adaptive modeling of complex multi-scale/multi-domain processes; (2) geospatial tools and technologies to conceptualize and visualize modeled processes; and (3) the principles and best practices of participatory research, where stakeholders are meaningfully involved throughout the research process. Examples of specific models that could be incorporated into a GPM approach include process-based models such as those that model the spread of invasive insects (Fitzpatrick et al. 2012) and plant pathogens (Cunniffe et al. 2016; Meentemeyer et al. 2011).

By offering a spatial context through maps and interactive spatial media, GPM can evoke and establish stakeholders' sense of place and spatial awareness (Brown and Raymond 2007; Silbernagel et al. 2015). This geospatial framing offers participants insight into how the spaces around them (home, neighborhood, landscape) might be impacted by personal or policy decisions. The first-person positionality and spatial orientation possible with geospatial media can highlight causal relationships between users' behavior and environmental outcomes for locations familiar to or used by participants. Illuminating this causality can be highly persuasive for changing behavior or management practices; stakeholders are more likely to become involved when they recognize that the places they care about are being affected. For example, when citizen scientists were able to contribute data from their own backyards, there was a boost to monitoring efforts in normally under-sampled urban ecosystems (Meentemeyer et al. 2015). Geospatial models and representations can help the world to move away from abstract ideas and vague representations and to bring those problems home. Whether it is pests impacting agricultural commodities and farmer livelihoods, such as the Asian citrus psyllid (*Diaphorina citri*), or the death of species with cultural or spiritual values, such as the oak trees impacted by sudden oak death, making it spatial makes it personal.

Biological invasions and management outcomes are rarely confined to one place. The connections between places can be very difficult to visualize or understand without thinking spatially. Tools from geospatial analytics allow us to examine spatial interactions between invasion and affected people and places, which can catalyze an understanding of the connectedness of our world. For example, geospatial information regarding (1) where; (2) when; and (3) how severe a problem or threat is can shape discussions about management tradeoffs, offering stakeholders improved opportunities to represent their interests. Epanchin-Niell et al. (2010) reported that 75% of interviewed ranchers stated that they were negatively affected, in terms of reduced cattle forage, by the invasion of the rangeland weed yellow starthistle from neighboring land parcels. One-quarter of those interviewed also reported that due to the cost associated with continual reinvasion, if their neighbors did not treat for the invasive species, they would reduce their own investment in control efforts. The importance of spatial interactions is clear—what happens on one stakeholder's property directly affects another stakeholder's livelihood.

A GPM approach could also accommodate localized stakeholder knowledge and understanding to promote sustainability. Stakeholders have unique and often deep knowledge of both the environment and their community and can offer tremendous insight on the use and management of local resources. Geospatial models can contextualize spatial interactions—what happens here affects there—and stakeholders can contribute iteratively to alternative future scenarios by evaluating options and eliminating non-starters.

Geospatial information can also help stakeholders to better understand multi-scale processes and to know where critical geographic boundaries lie. Complex problems are better assessed, and solutions are more sustainable, when stakeholders consider dynamic cross-scale linkages (Cash et al. 2006). For example, multi-scale scenario exercises can highlight cross-scale interactions that manifest or have strong impacts at one scale but not at others (Biggs et al. 2007). Heavy impacts at a local scale, such as a localized outbreak, may be lost when considering a regional assessment or perspective. Conversely, widespread, but low-level, invasions may not be recognized as a threat at a local level but may be seen as a problem when we scale up and realize that a whole region is impacted. With reference to sudden oak death, Cunniffe et al. (2016) showed that it is no longer feasible to eradicate and probably impossible to significantly slow the geographical spread of this disease. However, countless trees can still be protected locally with careful forest management in high priority landscapes, such as national parks and places of cultural heritage. GPM has the ability to incorporate these scenarios within a spatially explicit framework, making it easier to understand mismatches between scales at which decisions are made and scales at which ecological processes occur.

Socio-ecological systems function at multiple scales; however, stakeholders typically engage and manage within a local or jurisdictional boundary unrelated to the boundaries of biophysical and ecological processes. Rarely is there one scale where optimal, equitable solutions exist for multiple stakeholders. Nevertheless, there is an urge to simplify issues of scale in order to control and manage these complex systems more easily (Cash et al. 2006). Local actions often compound to create environmental and social tradeoffs. For example, land owners may choose to divert resources elsewhere rather than control for invasive species. They would be making a decision based on the perceived damage to their land. However, this may contribute to increased invasion at a landscape scale and an additional cost to others because the untreated parcel now serves as a propagule source (Epanchin-Niell et al. 2010). In order to understand and manage complex natural resource issues effectively, it is critical to clarify potential effects at multiple scales. Geospatial analytics has become an invaluable tool to visualize geographic boundaries and to understand and contextualize multi-scale processes. GPM offers a method for allowing stakeholders to see themselves in a connected world, with considerations ranging from site-specific to global perspectives.

Stakeholder involvement in the research process will be vital for developing lasting sustainability solutions, and GPM offers three ways to improve stakeholder engagement. Contextualizing "place" in a problem strongly motivates people to explore how an issue affects them; making it spatial makes it personal. Visualizing "spatial interaction" catalyzes new understandings of the connectedness of our world; people learn that what happens here affects there! Defining "spatial scale" helps visualize geographical boundaries of a problem, including knowledge of where policy and funding mechanisms operate at multiple and overlapping levels. We must move beyond specialized computational environments (and so-called "decision-support tools") that continue to inhibit discussion and co-learning of complex problems between professionals and the public. Technical solutions alone cannot provide sustainable futures for environmental management, rather we need integrated approaches with new tools for envisioning the future and evaluating tradeoffs that arise from multiple social, economic, and environmental drivers.

Social Marketing Social marketing, an approach derived from the applied social sciences, has considerable potential for changing awareness, attitudes, and behaviors of targeted audiences. It provides a method for improving our relationship with the environment and promoting the adoption of a conservation ethnic and sustainable behaviors. Social marketing is a discipline that is grounded in education and psychology, and can be combined with other applied social science disciplines, like branding and storytelling, to promote environmental behaviors.

Public engagement in many natural resource management functions can both complement and amplify the work of conservation professionals. While new science and management techniques are addressing invasive species issues, it is also critical to influence citizen behavior to prevent the spread and introduction of invasive species. Kotler and Zaltman (1971) introduced social marketing as a method to influence behaviors for good—those behaviors that improve health, prevent injuries, protect the environment, and contribute to communities. Social marketing does this by applying marketing principles and practices to bring about positive social change, to improve society and the environment, or to enhance the health and/or social status of individuals within society. In the 1980s, social marketing was used by agencies such as the World Bank to address personal hygiene and sanitation and by the Centers for Disease Control to influence the health behaviors of individuals or the behavior of policymakers.

There are seven steps involved in social marketing: selecting behaviors; uncovering barriers and benefits to the behaviors; researching the target audience's knowledge, attitudes, and related behaviors; developing communication strategies that address these barriers and benefits; combining branding and storytelling; piloting this strategy; and implementing it broadly when it is cost-effective (Goodman 2008; Jiwa 2015; McKenzie-Mohr and Smith 1999). This process can and has been used to foster a wide range of sustainable behaviors, ranging from individual health behaviors like smoking cessation and family planning, to pro-environmental behaviors like recycling, litter prevention, and use of public transportation.

Social marketing provides a voluntary approach that simplifies a very complex resource management issue, makes the issue relevant to a targeted audience, and empowers individuals to become part of the solution. In 2002, the U.S. Fish and Wildlife Service (FWS) created the "Stop Aquatic Hitchhikers!" campaign to elevate the aquatic invasive species issue and empower people with cleaning behaviors that are designed to prevent the introduction and spread of these harmful organisms into other waters. By leveraging social marketing with branding and storytelling, the campaign has created an international community of grassroots organizations that support, engage in, and promote behaviors that inhibit the spread of invasive species. Branding has introduced the need for a consistent message, and storytelling enables people to understand a very complex natural resource issue (Ries and Trout 2001). Branding the issue and the behaviors created an action step that empowers people beyond raising awareness about different non-native invasive species. All 50 State fish and wildlife agencies have joined the campaign, as well as all of the Canadian provinces and the countries of New Zealand, Scotland, England, and Ireland.

The campaign is particularly effective because the grassroots branding strategy enables the campaign to transcend State borders and promote a unified message. The same empowering brand is seen in Florida, Alaska, Maine, and California. After 13 years, the campaign is currently undergoing a brand refresh process and website update. "Stop Aquatic Hitchhikers!" has led to the emergence of different regional initiatives over the years, and the new face of the campaign will include a multi-initiative strategy that will leverage the market-defining impacts of the slogan. Currently, the campaign has been written up as a case study example of effective grassroots branding in a social marketing textbook written by Kotler and Lee (2011), producing a return on investment of 5:1, leveraging $5 from external partners for every $1 of Federal money spent. Over 2000 grassroots organizations have joined the campaign and are using the marketing collateral to promote the same message.

Due to the success of "Stop Aquatic Hitchhikers!," the pet and aquarium industry approached the FWS to develop a similar type of campaign. The high cultural importance of individual rights and identities in the modern United States has led to a considerable increase in the purchase of exotic pets and plants. Unfortunately, many of these species are impulse purchases and consumers may not realize how large they will grow and what other issues might arise. Ultimately, for a variety of reasons, these people may not be able to properly care for these species, and they may end up releasing them into the environment, thinking this is beneficial for the pet. If the pet survives, it has the potential to wreak havoc on the environment; a perfect example is the introduction of Burmese pythons (*Python bivittatus*) in south Florida and the impacts these species are having on the Everglades ecosystem. In partnership, the FWS and pet industry created "Habitattitude," a social marketing campaign designed to promote the environmentally friendly surrender of pets and discourage the spread of aquatic plants.

In addition to these two campaigns, the State of Minnesota's Department of Natural Resources has created the "PlayCleanGo" campaign to target those who recreate on the land. Seeds and plant fragments from non-native, invasive terrestrial plants have the potential to "hitchhike" on hiking boots, tires of off-road vehicles, horses, and other mobile vectors and thereby introduce and spread these harmful species to other areas. While this campaign was initiated by the state of Minnesota, it is seeing a similar growth trajectory as "Stop Aquatic Hitchhikers!" and the people who created the campaign are currently exploring ways to transfer the management of the campaign to a national organization to enhance growing interest.

12.5 Using Human Dimensions Research to Inform Invasive Species Policies

At the global and national levels, the World Conservation Union and Convention on Biological Diversity as well as the National Invasive Species Council in the United States have developed guiding principles for the prevention and control of invasive species. However, their guiding principles are largely about what governments should or should not do and thus fall short of utilizing human dimensions research to inform policies that can motivate and direct actions to minimize the spread of invasive species on the ground (Reaser 2001). Researchers have emphasized the need to incorporate more human dimensions research into invasive species policy development. Warren (2001) argued that it is important to incorporate the management of invasive species within a system of legislation, public policy, and resource management that reflects public interest and is informed by values, cultures, and other human dimensions considerations. Carlson and Vondracek (2014) state that even though some of the current invasive species management approaches in the United States are progressive and anticipatory, they are deficient in human dimensions, and there is a need for predictive models, management paradigms, and human dimensions research to design ecologically effective, economically feasible, and socially acceptable management policies and strategies. This general need for using human dimensions research to inform invasive species policies is further exacerbated by the fact that the management of invasive species is a multi-scalar, cross-boundary problem that requires various stakeholders at different levels to work collaboratively (Stokes et al. 2006).

In the United States, significant public policy efforts have been made to improve the ability of government agencies and the general public to detect, report, and verify suspected new invasive species and to assess and respond to verified new infestations. The Federal Interagency Committee for the Management of Noxious and Exotic Weeds established the National Early Detection and Rapid Response (EDRR) System for Invasive Plants to foster interagency cooperation and public-private partnerships needed to address new and emerging invasive plant species in agricultural, forest, and other ecosystems. Several regional networks were established to coordinate EDRR efforts operating across State lines (e.g., Great Lakes Early Detection Network, Mid-Atlantic Early Detection Network). A number of States also have their own EDRR system incorporated within their State invasive species management plan. Undoubtedly, investment and coordination by Federal, regional, and State officials are important for invasive species prevention and control (Leung et al. 2012; Lodge et al. 2006). These existing public policy efforts strongly rely on the idea that "the best offense is a good defense" (Mehta et al. 2007) because prevention and early detection of invasive species are considered to be more effective than eradication and control (Hobbs and Humphries 1995; Mehta et al. 2007). However, these public policy efforts have been mostly focused on public lands, while relatively little is known about private landowners' ability and willingness to prevent invasions and detect early infestations. Human dimensions research could provide important insights to address this need.

In addition, communication between government agencies and the general public with respect to invasive species prevention and control has generally focused on discernible terrestrial and aquatic wildlife species (e.g., Burmese pythons, feral pigs, Asian carp, and zebra mussels). Insufficient attention has been directed at communication between government agencies and private landowners about invasive species in a forestry setting. Understanding how private landowners perceive and respond to invasive species will be critical for informing effective outreach and communication strategies targeting these people who are at the forefront in efforts to control invasive species.

Finally, limited effort has been made to assess the extent to which previous and current public policy efforts effectively address local needs and concerns and motivate individual citizens to engage in invasive species management actions on their own. So far, few studies have evaluated the various public outreach efforts that aim to increase public awareness and willingness to report sightings of, eradicate, and/or control invasive species (Fritts 2007; Hawley 2007; Martin 2007; Reaser and Meyers 2007). The use of human dimensions research to evaluate a broader range of invasive species policies and programs will provide important insights that can be used in the development of future policies and programs to incentivize the public (including private landowners) to engage in invasive species management actions individually or collectively. It can also help policymakers and resource managers to anticipate and minimize conflicts over invasive species management rooted in diverse stakeholder values (Buckley and Han 2014; Estévez et al. 2015; Gobster 2011; Larson et al. 2011).

12.6 Conclusions: Key Findings and Information Needs

12.6.1 Key Findings

Social and cultural research is of fundamental importance in addressing the issue of non-native invasive species. Invasive species can threaten many of the fundamental ecosystem values and services on which society depends. The spread of invasive species is largely through human actions, including intentional introductions, accidental introduction through global movements of products and people, and human disturbances that facilitate their introduction and spread. In spite of the threats that invasive species impose on ecosystems and human well-being, public awareness of their presence and impact is generally low except in a few cases that involve highly problematic species. Broad awareness of an invasive species is generally only achieved once a species is widespread and well-established, but unfortunately this is also the point at which control is most difficult. Invasive species are generally viewed quite differently by the public than by scientists. People have complicated relationships with invasive species, with some being viewed very positively. Public views are further divided among stakeholder groups, who, depending on their relationship with a particular invasive species, may differ in their attitudes and associated interests.

Human dimensions research addresses a wide variety of topics and plays a critical role in informing scientists and managers about the larger social and cultural contexts in which people relate to invasive species. It also provides knowledge of awareness, attitudes and values, behaviors, and management preferences in relation to specific invasive species issues. Public support for management and control of invasive species is variable and often influenced by other values, such as the ways that people think about ecosystems and nature, and by the specific control measure being used. Language employed to call attention to invasive species and support management actions should be chosen carefully, because there is substantial evidence that language that is divisive or offensive to some people can create reactions that hinder efforts to carry out invasive species management activities. Differences in public opinion and interest related to invasive species can create conflict over control and management actions, and stifle efforts to promote widespread behavior changes.

Viewing invasive species from social and cultural perspectives highlights the importance of public dialogue that involves both listening to and educating the public to develop sufficient common understanding and concern to support needed management and policy actions. Collective action to address invasive species across ownership boundaries and at the landscape level is known to be important, yet more research is needed to learn the key motivating factors and steps necessary to promote collective action. Public engagement in invasive species monitoring and management can both complement and amplify the work of natural resource professionals, and will be essential if larger cultural changes in understanding invasive species issues and implementing management actions at landscape levels are to occur. Innovative human dimension techniques often simultaneously address various human dimension issues, including attitude change, stakeholder engagement, instilling a landscape-level perspective, generating a common vision to motivate cross-boundary cooperation, behavior change, and complementing and expanding the efforts of public agencies. Some of the promising new public science and engagement techniques being used for invasive species include citizen science, Geospatial Participatory Modeling, and social marketing.

Relatively little attention has been directed at determining how racial and ethnic diversity in the United States affects invasive species and their management. Structural issues, such as resource rights and environmental justice, and cultural differences, such as the ways that different species and ecosystems are used and are valued, are critical in managing many invasive species. Yet, relatively little is known about these issues. This chapter has highlighted the perspectives of tribes and invasive species. Tribes bring unique cultural perspectives and traditional ecological knowledge to invasive species management, and their close ties with the land and ecosystems can amplify its importance. Because of tribal sovereignty, the Federal Government relates to tribes on a government-to-government basis, yet overlapping trust responsibilities and common interests have resulted in numerous collaborative efforts to address invasive species issues.

12.6.2 Key Information Needs

Social science research conducted in the United States has been both limited and uneven in addressing general awareness, attitudes, and behaviors toward invasive species and how these are situated in larger social and cultural contexts. There is an obvious need for broad research in these areas. More specific research is also needed on particular invasive species, in the full range of ecological contexts and using a diversity of social science approaches and methods. Currently, research results have been insufficient to support strong decisions and actions by managers and policymakers. More research is needed on all aspects of the human dimensions of forest invasive species in public ownerships, such as national and State parks and forests. For private forest owners, we need to know: (1) how they perceive invasive species problems, particularly how they perceive invasions and associated impacts, both on an individual and landscape scale; and (2) what types of information, assistance, and resources will be most useful for helping landowners detect and manage invasions that have occurred and the potential risk of future invasions. There is also a need for a more explicit focus on the role of scale in landowner perceptions of invasive species and invasion risks, concerns about invasive species and invasion risk, and willingness to take actions.

Public awareness of invasive species in grasslands and aquatic ecosystems may be greater than it is in forest ecosystems, yet there is relatively little published research available for both. There is a clear need for more research on (1) attitudes and behaviors of individuals; (2) mechanisms to generate public concern regarding invasive species and support for their management; (3) collective action responses at the landscape level on both public and private lands; and (4) how laws and polices interact with other human dimensions issues, and how can their effectiveness be improved.

We know that cooperation across land ownerships is of fundamental importance, but we need more research on collective action practice in order to know: (1) what factors determine the likelihood of landowner cooperation and the effectiveness of their cooperation; (2) whether there are tradeoffs between the increased likelihood of landowner cooperation (potentially by reducing the scale at which landowners cooperate with one another) and decreased effectiveness of landowner cooperation (e.g., as few landowners cooperate, whether the ability of the group to prevent and control invasive species and affect landscape outcomes could be compromised); and (3) at what scale landowners should cooperate with one another in order to realize invasive species management at a landscape scale.

There is a need for interdisciplinary research to better understand the interactions between biological and social complexities and uncertainties, in order to more effectively manage invasive species and reduce the associated social conflicts among stakeholders (Kokotovich and Andow 2017). An accurate understanding of current and future invasion risks is critical to achieving effective invasive species management and to enhance strategic planning and policymaking at the regional level (Leung et al. 2012; Lodge et al. 2006). Assessing and predicting invasion risks require a holistic understanding of various interacting components of invasions (Catford et al. 2009; Richardson et al. 2000). However, current invasion risk assessments tend to focus on the biological characteristics of invaders, environmental factors that make a recipient system more or less likely to be invaded, and a number of biological, ecological, and (in a few cases) land-use drivers. Considerable research has shown that socioeconomic drivers on both local and regional scales can influence the distribution, abundance, and species richness of invasive plants and animals (Chhabra et al. 2006; Vilà and Ibáñez 2011; With 2002). Therefore, to better predict invasion risks, modeling efforts need to incorporate changing ecological and landscape characteristics, as well as socioeconomic conditions over time. Only by incorporating human dimensions data on landowner willingness-to-manage invasive species into invasion risk models will we be able to achieve a more realistic understanding of future invasion risks.

There is relatively little published literature on tribes and invasive species. There are a number of successful examples of collaboration between tribes and biological scientists to address invasive species, particularly involving tree pests and diseases. There is a need to conduct collaborative research with tribes in order to better document the cultural, traditional ecological knowledge, and sovereignty and other policy issues that are often key factors in invasive species management.

While promising new approaches are being developed to increase public awareness and actions related to invasive species, it is critical that new research focuses on the outcomes and effectiveness of these approaches. We need to know how participation in citizen science projects and geospatial modeling and exposure to story maps and social marketing change knowledge of invasive species, attitudes, and behavior for both the public and scientists. We also know very little about the broader ecological, community, and social impacts of these new approaches.

Disclaimer Text The findings and conclusions in this publication are those of the authors and should not be construed to represent any official USDA or U.S. Government determination or policy.

Literature Cited

Allison SK (2011) The paradox of invasive species in ecological restoration: do restorationists worry about them too much? In: Rotherham ID, Lambert RA (eds) Invasive and introduced plants and animals: human perceptions, attitudes and approaches to management. Earthscan, London, pp 265–275

Anderson LG, Rocliffe S, Haddaway NR et al (2015) The role of tourism and recreation in the spread of non-native species: a systematic review and meta-analysis. PLoS One 10:e0140833. https://doi.org/10.1371/journal.pone.0140833

Bardsley D, Edward-Jones G (2006) Stakeholders' perceptions of the impacts of invasive exotic plant species in the Mediterranean region. GeoJournal 65:199–210

Barr JJF, Lurz PWW, Shirley MDF et al (2002) Evaluation of immunocontraception as a publicly acceptable form of vertebrate pest species control: the introduced grey squirrel in Britain as an example. Environ Manag 30:342–351

Biggs R, Raudsepp-Hearne C, Atkinson-Palombo C et al (2007) Linking futures across scales: a dialog on multiscale scenarios. Ecol Soc 12:17

Blaustein RJ (2001) Kudzu's invasion into southern United States life and culture. In: McNeely JA (ed) The great reshuffling: human dimensions of invasive alien species. IUCN, Gland, pp 55–62

Blinn CR, Jakes PJ, Sakai M (2007) Forest landowner cooperatives in the United States: a local focus for engaging landowners. J For 105:245–251

Bloch M (1998) Why trees, too, are good to think with: towards an anthropology of the meaning of life. In: Rival L (ed) The social life of trees: anthropological perspectives on tree symbolism. Berg, Oxford, pp 39–55

Bonney R, Ballard H, Jordan R et al (2009) Public participation in scientific research: defining the field and assessing its potential for informal science education. A CAISE inquiry group report. Center for Advancement of Informal Education (CAISE), Washington, DC

Bonney R, Phillips TB, Ballard HL et al (2015) Can citizen science enhance public understanding of science? Public Underst Sci 25:2–16

Borowy I (2011) The other side of bio-invasion: the example of acclimatization in Germany. In: Rotherham ID, Lambert RA (eds) Invasive and introduced plants and animals: human perceptions, attitudes and approaches to management. Earthscan, London, pp 153–166

Bowcutt F (2013) Tanoak landscapes: tending a native American nut tree. Madrono 60:64–86

Bremner A, Park K (2007) Public attitudes to the management of invasive non-native species in Scotland. Biol Conserv 139:306–314

Brossard D, Lewenstein B, Bonney R (2005) Scientific knowledge and attitude change: the impact of a citizen science project. Int J Sci Educ 27:1099–1121

Brown G, Raymond C (2007) The relationship between place attachment and landscape values: toward mapping place attachment. Appl Geogr 27:89–111

Brown JH, Sax DF (2004) An essay on some topics concerning invasive species. Austral Ecol 29:530–536

Buckley YM, Han Y (2014) Managing the side effects of invasion control. Science 344:975–976

Bureau of Indian Affairs (2016) Notice: Indian entities recognized and eligible to receive services from the United States Bureau of Indian Affairs. Fed Regist 81(86):26826–26832. Washington, DC

Butler BJ (2008) Family forest owners of the United States, 2006. General technical report NRS-27. U.S. Department of Agriculture, Forest Service, northern Research Station, Newtown Square, 72 p

Cambray JA (2003) Impact on indigenous species biodiversity caused by the globalization of alien recreational freshwater fisheries. In: Martens K (ed) Aquatic biodiversity. Springer, Dordrecht, pp 217–230

Carlson DE (2014) Personal correspondence, email letter dated April 16, 2014 to Zhao Ma. On file with Ma at Department of Forestry and Natural Resources, Purdue University, West Lafayette, IN 47907

Carlson AK, Vondracek B (2014) Synthesis of ecology and human dimensions for predictive management of Bighead and Silver Carp in the United States. Rev Fish Sci Aquacult 22:284–300

Cartwell CG (2007) Efforts to eradicate these pests. In: Cartwell CG (ed) Invasive forest pests. Nova Science Publishers, New York, pp 17–34

Cash D, Adger W, Berkes F et al (2006) Scale and cross-scale dynamics: governance and information in a multilevel world. Ecol Soc 11:8

Catford JA, Jansson R, Nilsson C (2009) Reducing redundancy in invasion ecology by integrating hypotheses into a single theoretical framework. Divers Distrib 15:22–40

Chew M (2009) The monstering of tamarisk: how scientists made a plant into a problem. J Hist Biol 42:131–266

Chew MK (2011) Anekeitaxonomy: botany, place, and belonging. In: Rotherham ID, Lambert RA (eds) Invasive and introduced plants and animals: human perceptions, attitudes and approaches to management. Earthscan, London, pp 137–151

Chhabra A, Geist H, Houghton RA et al (2006) Multiple impacts of land-use/cover change. In: Lambin EF, Geist H (eds) Land-use and land-cover change. Springer, Berlin/Heidelberg, pp 71–116

Coates P (2006) American perceptions of immigrant and invasive species: strangers on the land. University of California Press, Berkeley, 266 p

Coates P (2011) Over here: American animals in Britain. In: Rotherham ID, Lambert RA (eds) Invasive and introduced plants and animals: human perceptions, attitudes and approaches to management. Earthscan, London, pp 39–54

Coates P (2015) A tale of two squirrels: a British case study of the sociocultural dimensions of debates over invasive species. In: Keller RP, Cadotte MW, Sandiford G (eds) Invasive species in a globalized world: ecological, social, and legal perspectives on policy. University of Chicago Press, Chicago, pp 44–71

Colautti RI, MacIsaac HJ (2004) A neutral terminology to define 'invasive' species. Divers Distrib 10:135–141

Connelly NA, Biedron I, Lauber TB (2014a) Roles of boating facilities, bait dealers, and angler and boating organizations in preventing the spread of aquatic invasive species in the Lake Ontario Basin. HDRU publication no. 14–12. Department of Natural Resources, College of Agriculture and Life Sciences, Cornell University, Ithaca, 44 p

Connelly NA, Lauber TB, Stedman RC (2014b) Reducing the spread of aquatic invasive species and fish pathogens in the Great Lakes: the role of bait dealers. HDRU publication no. 14–8. Department of Natural Resources, College of Agriculture and Life Sciences, Cornell University, Ithaca, 41 p

Connelly NA, Lauber TB, Stedman RC (2014c) Reducing the spread of aquatic invasive species and fish pathogens in the Great Lakes: the role of anglers. HDRU publication no. 14–7. Department of Natural Resources, College of Agriculture and Life Sciences, Cornell University, Ithaca, 36 p

Conrad CC, Hilchey KG (2011) A review of citizen science and community-based environmental monitoring: issues and opportunities. Environ Monit Assess 176:273–291

Crall AW, Jordan R, Holfelder K et al (2013) The impacts of an invasive species citizen science training program on participant attitudes, behavior, and science literacy. Public Underst Sci 22:745–764

Cronon W (1996) The trouble with wilderness; or, getting back to the wrong nature. In: Cronon W (ed) Uncommon ground: rethinking the human place in nature. W.W. Norton, New York, pp 69–90

Crosby A (1972) The Columbian exchange: biological and cultural consequences of 1492. Greenwood Press, Westport, 283 p

Crosby A (1986) Ecological imperialism: the biological expansion of Europe 900–1900. Cambridge University Press, Cambridge, 388 p

Cuniffe NJ, Cobb RC, Meentemeyer RK et al (2016) Modeling when, where, and how to manage a forest epidemic: sudden oak death in California. Proc Natl Acad Sci U S A 113:5640–5645

Dalrymple KE, Shaw BR, Brossard D (2013) Following the leader: using opinion leaders in environmental strategic communication. Soc Nat Resour 26:1438–1453

DiEnno CM, Hilton SC (2005) High school students' knowledge, attitudes, and levels of enjoyment of an environmental education unit on nonnative plants. J Environ Educ 37:13–25

DiTomaso JM (2000) Invasive weeds in rangelands: species, impacts, and management. Weed Sci 48:255–265

Dodds NMW, Miller MH, Lamm AJ (2014) Floridians' perceptions of invasive species. Publication AEC524. University of Florida/ Agricultural Education and Communication Department, Institute of Food and Agriculture, Gainesville. http://edis.ifas.ufl.edu/wc186

Donlan J, Greene HW, Berger J et al (2005) Re-wilding North America. Nature 436:913–914

Donoghue EM, Thompson SA, Bliss JC (2010) Tribal-federal collaboration in resource management. J Ecol Anthropol 14:22

Eastern Band of the Cherokee Indians (2013) Integrative resource management plan. Eastern Band of the Cherokee Indians, Cherokee

Emery MR, Wrobel A, Hansen MH et al (2014) Using traditional ecological knowledge as a basis for targeted forest inventories: paper birch (*Betula papyrifera*) in the U.S. Great Lakes Region. J For 112:207–214

Epanchin-Niell RS, Hufford MB, Aslan CE et al (2010) Controlling invasive species in complex social landscapes. Front Ecol Environ 8:210–216

Erickson DL, Ryan RL, De Young R (2002) Woodlots in the rural landscapes: landowner motivations and management attitudes in a Michigan (USA) case study. Landsc Urban Plan 58:101–112

Espanchin-Niell RS, Wilen JE (2014) Individual and cooperative management of invasive species in human-mediated landscapes. Am J Agric Econ 97:180–198

Estévez RA, Anderson CB, Pizarro JC et al (2015) Clarifying values, risk perceptions, and attitudes to resolve or avoid social conflicts in invasive species management. Conserv Biol 29:19–30

Evans C, Abrams E, Reitsma R et al (2005) The Neighborhood Nestwatch program: participant outcomes of a citizen-science ecological research project. Conserv Biol 19:589–594

Fernandez-Gimenez ME, LeFebre S, Conley A et al (2004) Collaborative stewardship of Arizona's rangelands: making a difference in resource management. Rangelands 26:24–30

Finley AO, Kittredge DB, Stevens TH et al (2006) Interest in cross-boundary cooperation: identification of distinct types of private forest owners. For Sci 52:10–22

Fischer AP, Charnley S (2012) Private forest owners and invasive plants: risk perception and management. Invasive Plant Sci Manag 5:375–389

Fischer A, van der Wal R (2007) Invasive plant suppresses charismatic seabird—the construction of attitudes towards biodiversity management options. Biol Conserv 135:256–267

Fitzgerald G, Fitzgerald N, Davidson C (2007) Public attitudes towards invasive animals and their impacts. University of Canberra/Invasive Animals Cooperative Research Centre, Canberra, 57 p

Fitzpatrick MC, Preisser EL, Porter A et al (2012) Modeling range dynamics in heterogeneous landscapes: invasion of the hemlock woolly adelgid in eastern North America. Ecol Appl 22:472–486

Forcella F, Harvey SJ (1988) Patterns of weed migration in Northwestern U.S.A. Weed Sci 36:194–201

Fraser W (2001) Introduced wildlife in New Zealand: a survey of general public views. Landcare research science series no. 23. Manaaki Whenua Press, Canterbury, 45 p

Fritts E (2007) Wildlife and people at risk: a plan for keeping rats out of Alaska. Alaska Department of Fish and Game, Juneau, 190 p

Garcia-Llorente M, Martin-Lopez B, Gonzalez JA et al (2008) Social perceptions of the impacts and benefits of invasive alien species: implications for management. Biol Conserv 141:2969–2983

Garibaldi A, Turner NJ (2004) Cultural keystone species: implications for ecological conservation and restoration. Ecol Soc 9:Article 1

Geier S (2015) A lesson in honeysuckle. Southern Things Issue 4, Fall 2015. http://southernthings.web.unc.edu/honeysuckle/. 18 Apr 2016

Gherardi F (2011) Public perception of invasive alien species in Mediterranean Europe. In: Rotherham ID, Lambert RA (eds) Invasive and introduced plants and animals: human perceptions, attitudes and approaches to management. Earthscan, London, pp 185–200

Gifford EW (1967) Ethnographic notes on the southwestern Pomo, University of California anthropological records 25. Kraus Reprint, Millwood, pp 1–48

Giles H, Giles J (2012) Ingroups and outgroups. In: Kurylo A (ed) Inter/cultural communication: representation and construction of culture. Sage, Thousand Oaks, pp 141–162

Gobster PH (2005) Invasive species as ecological threat: is restoration an alternative to fear-based resource management? Ecol Restor 23:261–270

Gobster PH (2011) Factors affecting people's responses to invasive species management. In: Rotherham ID, Lambert RA (eds) Invasive and introduced plants and animals: human perceptions, attitudes and approaches to management. Earthscan, London, pp 249–263

Goodman A (2008) Storytelling as best practice: how stories strengthen your organization, engage your audience, and advance your mission, 4th edn. Goodman, A, Los Angeles, 46 p

Gordon J, Sessions J, Bailey J et al (2013) Assessment of Indian forests and forest management in the United States final report—the third Indian Forest Management Assessment Team for the Intertribal Timber Council (White R (ed)). Intertribal Timber Council, Portland. Volumes I and II

Gordon R, Brunson MW, Shindler B (2014) Acceptance, acceptability and trust for sagebrush restoration options in the Great Basin: a longitudinal response. Rangel Ecol Manag 67:573–583

Graham S (2013) Three cooperative pathways to solving a collective weed management problem. Aust J Environ Manag 20:116–129

Gray S, Mellor D, Jordan R et al (2017) Combining participatory modelling and citizen science to support volunteer conservation action. Biol Conserv 208:76–86

Groffman PM, Stylinski C, Nisbet MC et al (2010) Restarting the conversation: challenges at the interface between ecology and society. Front Ecol Environ 8:284–291

Gunderson LH, Holling CS (2001) Panarchy: understanding transformation in human and natural systems. Island Press, Covello, 536 p

Haskew DC (1999) Federal consultation with Indian tribes: the foundation of enlightened policy decisions, or another badge of shame? Am Indian Law Rev 24:21–74

Hawley N (2007) Custom trucks, radio snake jingles, and temporary tattoos: an overview of a successful public awareness campaign related to brown tree snakes in the Commonwealth of the Northern Mariana Islands. In: Managing vertebrate invasive species: proceedings of an international symposium. U.S. Department of Agriculture, APHIS, National Wildlife Research Center, Fort Collins, pp 53–56

Heck N, Lauber TB, Stedman RC (2013) Pathogens and invasive species in the Great Lakes: understanding manager responses targeting bait dealers and anglers. HDRU publication no. 13-9. Department of Natural Resources, College of Agriculture and Life Sciences, Cornell University, Ithaca, 39 p

Hershdorfer ME, Fernandez-Gimenez ME, Howery LD (2007) Key attributes influence the performance of local weed management programs in the Southwest United States. Rangel Ecol Manag 60:225–234

Hobbs RJ, Humphries SE (1995) An integrated approach to the ecology and management of plant invasions. Conserv Biol 9:761–770

Hobbs RJ, Higgs ES, Hall CM (2013) Introduction: why novel ecosystems? In: Hobbs RJ, Higgs ES, Hall CM (eds) Novel ecosystems: intervening in the new ecological world order. Wiley-Blackwell, Oxford, pp 3–8

Howell AP, Shaw BR, Alvarez G (2015) Bait shop owners as opinion leaders: a test of the theory of planned behavior to predict pro-environmental outreach behaviors and intentions. Environ Behav 47:1107–1126

Indiana Division of Forestry (2010) Indiana statewide forest assessment. Indiana Department of Natural Resources Division of Forestry, Indianapolis

Jacobson MG (2002) Ecosystem management in the Southeast United States: interest of forest landowners in joint management across ownership. Small Scale For 1:71–92

Jacobson MG, Abt RC, Carter DR (2000) Attitudes toward joint forest planning among private landowners. J Sustain For 11:95–112

Javelle A, Kalaora B, Decocq G (2011) A view from continental Europe: the case if *Prunus serotina* in France in comparison with other invasives. In: Rotherham ID, Lambert RA (eds) Invasive and introduced plants and animals: human perceptions, attitudes and approaches to management. Earthscan, London, pp 277–2288

Jiwa B (2015) Meaningful: the story of ideas that fly. Perceptive Press, Melbourne, 176 p

Johnston MJ, Marks CA (1997) Attitudinal survey of vertebrate pest management in Victoria. Vertebrate pest research unit report series number 3. Department of Natural Resources and Environment, Frankston, 35 p

Jordan RC, Gray SA, Howe DV et al (2011) Knowledge gain and behavior change in citizen-science programs. Conserv Biol 25:1148–1154

Jordan R, Ballard HL, Phillips TB (2012) Key issues and new approaches for evaluating citizen-science learning outcomes. Front Ecol Environ 10:307–309

Jordan R, Gray S, Sorensen A et al (2016) Studying citizen science through adaptive management and learning feedbacks as mechanisms for improving conservation. Conserv Biol 30:487–495

Kalnicky EA (2012) A coupled and natural systems approach to understanding an invasive frog, *Eleutherodactylus coqui*, in Hawai'i. PhD dissertation in Ecology, Utah State University, Logan, 235 p. https://digitalcommons.usu.edu/etd/1412

Kartchner H (2013) Development of ecologically-based invasive plant management curriculum for university audiences. MS in Human dimensions of ecosystem science and management, Utah State University, Logan, 99 p. https://digitalcommons.usu.edu/gradreports/278

Keller RP, Cadotte MW, Sandiford G (eds) (2015) Invasive species in a globalized world: ecological, social, and legal perspectives on policy. University of Chicago Press, Chicago, 416 p

Kelley WK, Fernandez-Gimenez ME, Brown CS (2013) Managing downy brome (*Bromus tectorum*) in the Central Rockies: land manager perspectives. Invasive Plant Sci Manag 6:521–535

Kittredge DB (2004) Extension/outreach implications for America's family forest owners. J For 102:15–18

Kittredge DB (2005) The cooperation of private forest owners on scales larger than their individual properties. Forest Policy Econ 7:671–688

Kokotovich AE, Andow DA (2017) Exploring tensions and conflicts in invasive species management: the case of Asian carp. Environ Sci Policy 69:105–112

Kotler P, Lee N (2011) Social marketing—influencing behaviors for good. Sage, Washington, DC, 520 p

Kotler P, Zaltman G (1971) Social marketing: an approach to planned social change. J Mark 35:3–12

Kountoupes D, Oberhauser K (2008) Citizen science and youth audiences: educational outcomes of the Monarch Larva monitoring project. J Commun Engage Scholarsh 1:10–20

Kueffer C (2010) Transdisciplinary research is needed to predict plant invasion in an era of global change. Trends Ecol Evol 25:620–621

Larson BMH (2005) The war of the roses: demilitarizing invasion biology. Front Ecol Environ 3:495–500

Larson DL, Phillips-Mao L, Quiram G et al (2011) A framework for sustainable invasive species management: environmental, social and economic objectives. J Environ Manag 92:14–22

Lauber TB, Connelly NA, Stedman RC (2014) Aquatic invasive species outreach to boaters and anglers in the Lake Ontario basin. HDRU publication no. 14-5. Department of Natural Resources, College of Agriculture and Life Sciences, Cornell University, Ithaca, 59 p

Lauber TB, Connelly NA, Stedman RC (2015a) Assessing capacity for aquatic invasive species outreach in recreational communities. HDRU publication no. 15-7. Department of Natural Resources, College of Agriculture and Life Sciences, Cornell University, Ithaca, 58 p

Lauber TB, Connelly NA, Stedman RC (2015b) Perspectives of New York farmers, aquarium owners, and water gardeners on invasive species. HDRU publication no. 15-9. Department of Natural Resources, College of Agriculture and Life Sciences, Cornell University, Ithaca, 23 p

Lee C, Valachovic Y, Garbelotto M (2011) Protecting trees from sudden oak death before infection. UCANR publication 8426, February 2011, 14 p

Leung B, Roura-Pascual N, Bacher S et al (2012) TEASIng apart alien species risk assessments: a framework for best practices. Ecol Lett 15:1475–1493

Lévi-Strauss C (1966) The savage mind. University of Chicago Press, Chicago, 310 p

Limburg KE, Luzadis VA, Ramsey M et al (2010) The good, the bad, and the algae: perceiving ecosystem services and disservices generated by zebra and quagga mussels. J Great Lakes Res 36:86–92

Liu J, Hull V, Yang W et al (2016) Pandas and people: coupling human and natural systems for sustainability. Oxford University Press, Oxford, 304 p

Lodge DM, Williams S, MacIsaac HJ et al (2006) Biological invasions: recommendations for US policy and management. Ecol Appl 16:2035–2054

Long JW, Anderson MK, Quinn-Davidson L et al (2016) Restoring California black oak ecosystems to promote tribal values and wildlife. General technical report PSW-GTR-252. US Department of Agriculture, Forest Service, Pacific Southwest Research Station, Albany

Luizza MW, Evangelista PH, Jarnevich CS et al (2016) Integrating subsistence practice and species distribution modeling: assessing invasive elodea's potential impact on Native Alaskan subsistence of Chinook salmon and whitefish. Environ Manag 58:144–163

Luken JO, Thieret JW (1996) Amur honeysuckle, its fall from grace. Bioscience 46:18–24

Manchester SJ, Bullock JM (2000) The impacts of non-native species on UK biodiversity and the effectiveness of control. J Appl Ecol 37:845–864

Marler MJ, Supplee K, Wessner M et al (2005) Changing attitudes about grassland conservation in Missoula, Montana—"Weed Capital of the West". Ecol Restor 23:29–34

Martin C (2007) Promoting awareness, knowledge and good intentions. In: Managing vertebrate invasive species: proceedings of an international symposium. U.S. Department of Agriculture, APHIS, WS, National Wildlife Research Center, Fort Collins, pp 57–62

Mascheretti S, Croucher P, Kozanitas M et al (2009) Genetic epidemiology of the sudden oak death pathogen *Phytophthora ramorum* in California. Mol Ecol 18:4577–4590

McKenzie-Mohr D, Smith W (1999) Fostering sustainable behavior: an introduction to community-based social marketing. Education for sustainability series. New Society Publishers, Gabriola Island, 176 p

McLeod LJ, Hine DW, Please PM, Driver AB (2015) Applying behavioral theories to invasive animal management: towards an integrated framework. J Environ Manag 161:63–71

McNeely JA (2001) An introduction to human dimensions of invasive alien species. In: McNeely JA (ed) The great reshuffling: human dimensions of invasive alien species. IUCN, Gland, pp 5–20

McNeely JA (2011) Xenophobia or conservation: some human dimensions. In: Rotherham ID, Lambert RA (eds) Invasive and introduced plants and animals: human perceptions, attitudes and approaches to management. Earthscan, London, pp 19–36

Meech H (2005) Eradicating non-native mammals from islands: facts and perceptions. ECOS—Br Assoc Nat Conservat 26:72–80

Meentemeyer RK, Cunniffe NJ, Cook et al (2011) Epidemiological modeling of invasion in heterogeneous landscapes: spread of sudden oak death in California (1990–2030). Ecosphere 2:art17

Meentemeyer RK, Dorning MA, Vogler JB et al (2015) Citizen science helps predict risk of emerging infectious disease. Front Ecol Environ 13:189–194

Mehta SV, Haight RG, Homans FR et al (2007) Optimal detection and control strategies for invasive species management. Ecol Econ 61:237–245

Meyers KJ, Swiecki TJ, Mitchell AE (2006) Understanding the native Californian diet: identification of condensed and hydrolyzable tannins in tanoak acorns (*Lithocarpus densiflorus*). J Agric Food Chem 54:7686–7691

Miller JH, Schelhas J (2008) Adaptive collaborative restoration: a key concept in invasive plant management. In: Kohli RK, Jose S, Singh H, Batish D (eds) Invasive plants and forest ecosystems. CRC Press, Boca Raton, pp 251–265

Mooney HA, Hobbs RJ (2000) Invasive species in a changing world. Island Press, Washington, DC, 384 p

Niemiec RM, Ardoin NM, Wharton CR, Asner GP (2016) Motivating residents to combat invasive species on private lands: social norms and community reciprocity. Ecol Soc 21:30

Nimmo D, Miller K (2007) Ecological and human dimensions of management of feral horses in Australia: a review. Wildl Res 34:408–417

Norgaard KM (2007) The politics of invasive weed management: gender, race, and risk perception in northern California. Rural Sociol 72:450–477

Nuñez MA, Kuebbing S, Dimarco RD et al (2012) Invasive species: to eat or not to eat, that is the question. Conserv Lett 5:334–341

Ortiz BR (2008a) Contemporary California Indians, oaks, and sudden oak death (*Phytophthora ramorum*). In: Merenlender A, McCreary D, Purcell KL (eds) Proceedings of the sixth California oak symposium: today's challenges, tomorrow's opportunities. General technical report PSW-GTR-217. US Department of Agriculture, Forest Service, Pacific Southwest Research Station, Albany, pp 39–56

Ortiz BR (2008b) Contemporary California Indian uses for food of species affected by *Phytophthora ramorum*. In: Frankel SJ, Kliejunas JT, Palmieri KM (eds) Proceedings of the sudden oak death third science symposium. General technical report PSW-GTR-214. Pacific Southwest Research Station, Forest Service, U.S. Department of Agriculture, Albany, pp 419–425

Overdevest C, Orr CH, Stepenuck K (2004) Volunteer stream monitoring and local participation in natural resource issues. Hum Ecol Rev 11:177–185

Pfeiffer JM, Voeks RA (2008) Biological invasions and biocultural diversity: linking ecological and cultural systems. Environ Conserv 35:281–293

Pimentel D, Zuniga R, Morrison D (2005) Update on the environmental and economic costs associated with alien-invasive species in the United States. Ecol Econ 52:273–288

Pooley S (2011) Fire and loathing in the Fynbos: notions of indigenous and alien vegetation in South Africa's Western Cape. In: Rotherham ID, Lambert RA (eds) Invasive and introduced plants and animals: human perceptions, attitudes and approaches to management. Earthscan, London, pp 327–341

Pradhananga A, Davenport MA, Seekamp E et al (2015) Preventing spread of aquatic invasive species: boater concerns, habits, and future behaviors. Hum Dimens Wildl 20:381–393

Pretty J (2002) Landscapes lost and found. In: Pretty J (ed) Agri-culture: reconnecting people, land and nature. Earthscan Publications, London, pp 10–26

Pretty Paint-Small V (2013) Linking culture, ecology and policy: the invasion of Russian-olive (*Elaeagnus angustifolia* L.) on the Crow Indian Reservation, South-Central Montana, USA. Doctoral dissertation, University of Colorado, Ft. Collins, 94 p

Pyšek P, Richardson DM (2010) Invasive species, environmental change and management, and health. Annu Rev Environ Resour 35:25–55

Reaser JK (2001) Invasive alien species prevention and control: the art and science of managing people. In: McNeely JA (ed) The great reshuffling: human dimensions of invasive alien species. IUCN, Gland/Cambridge, pp 89–104

Reaser J, Meyers N (2007) Habitattitude: getting a backbone about the pet release pathway. In: Witmer GW, Pitt WC, Fagerstone KA (eds) Managing vertebrate invasive species: proceedings of an international symposium. U.S. Department of Agriculture, APHIS, WS, National Wildlife Research Center, Fort Collins, pp 63–71

Reed MS (2008) Stakeholder participation for environmental management: a literature review. Biol Conserv 141:2417–2431

Richardson DM, Pysek P, Rejmanek M et al (2000) Naturalization and invasion of alien plants: concepts and definitions. Divers Distrib 6:93–107

Rickenbach M, Janhke AD (2006) Wisconsin private sector foresters' involvement in nonindustrial private forestland cross-boundary forestry practices. North J Appl For 23:100–105

Rickenbach M, Kittredge DB (2009) Time and distance: comparing motivations among forest landowners in New England. Small Scale For 8:95–108

Rickenbach MG, Reed AS (2002) Cross-boundary cooperation in a watershed context: the sentiments of private forest landowners. Environ Manag 30:584–594

Rickenbach M, Schulte LA, Kittredge DB et al (2011) Cross-boundary cooperation: a mechanism for sustaining ecosystem services from private lands. J Soil Water Conserv 66:91A–96A

Ries A, Trout J (2001) Positioning: the battle for your mind. McGraw-Hill, New York, 224 p

Rizzo DM, Garbelotto M, Hansen E (2005) *Phytophthora ramorum*: integrative research and management of an emerging pathogen in California and Oregon forests. Annu Rev Phytopathol 43:309–335

Robbins P (2004) Comparing invasive networks: cultural and political biographies of invasive species. Geogr Rev 94:139–156

Rotherham ID, Lambert RA (2011) Good science, good history and pragmatism. In: Rotherham ID, Lambert RA (eds) Invasive and introduced plants and animals: human perceptions, attitudes and approaches to management. Earthscan, London, pp 355–366

Rothlisberger JD, Chadderton WL, McNulty J et al (2010) Aquatic invasive species transport via trailered boats: what is being moved, who is moving it, and what can be done. Fisheries 35:121–132

Sandiford G, Keller RP, Cadotte M (2015) Final thoughts: nature and human nature. In: Keller RP, Cadotte MW, Sandiford G (eds) Invasive species in a globalized world: ecological, social, and legal perspectives on policy. University of Chicago Press, Chicago, pp 381–394

Schelhas J, Miller JH, Chambers J (2012) Non-native plants and adaptive collaborative approaches to ecosystem restoration. In: Stanturf J, Madsen P, Lamb D (eds) A goal-oriented approach to forest landscape restoration. Springer, New York, pp 63–186

Schlueter AC, Schneider IE (2016) Visitor acceptance and confidence in emerald ash borer management approaches. For Sci 62:316–322

Seidl DE, Klepeis P (2011) Human dimensions of earthworm invasion in the Adirondack State Park. Hum Ecol 39:641–665

Selge S, Fischer A, van der Wal R (2011) Public and professional views on invasive non-native species—a qualitative social scientific investigation. Biol Conserv 144:3089–3097

Sharp RL, Larson LR, Green GT (2011) Factors influencing public preferences for invasive alien species management. Biol Conserv 144:2097–2104

Sharp RL, Cleckner LB, DePillo S (2016) The impact of on-site educational outreach on recreational users' perceptions of aquatic invasive species and their management. Environ Educ Res 23:1200–1210

Shindler B, Gordon R, Brunson MW (2011) Public perceptions of sagebrush ecosystem management in the Great Basin. Rangel Ecol Manag 64:335–343

Shine C (2015a) Developing invasive species policy for a major free trade bloc: challenges and progress in the European Union. In: Keller RP, Cadotte MW, Sandiford G (eds) Invasive species in a globalized world: ecological, social, and legal perspectives on policy. University of Chicago Press, Chicago, pp 303–326

Shine R (2015b) The ecological, evolutionary, and social impact of invasive cane toads in Australia. In: Keller RP, Cadotte MW, Sandiford G (eds) Invasive species in a globalized world: ecological, social, and legal perspectives on policy. University of Chicago Press, Chicago, pp 303–326

Shirk JL, Ballard HL, Wilderman CC et al (2012) Public participation in scientific research: a framework for deliberate design. Ecol Soc 17:29

Silbernagel J, Host G, Hagley C et al (2015) Linking place-based science to people through spatial narratives of coastal stewardship. J Coast Conserv 19:181–198

Simberloff D (2011) The rise of modern invasion biology and American attitudes towards introduced species. In: Rotherham ID, Lambert RA (eds) Invasive and introduced plants and animals: human perceptions, attitudes and approaches to management. Earthscan, London, pp 121–135

Simberloff D, Parker IM, Windle PN (2005) Introduced species policy, management, and future research needs. Front Ecol Environ 3:12–20

Slimak MW, Dietz T (2006) Personal values, beliefs, and ecological risk perception. Risk Anal 26:1689–1705

Smout C (2011) How the concept of alien species emerged and developed in 20th-century Britain. In: Rotherham ID, Lambert RA (eds) Invasive and introduced plants and animals: human perceptions, attitudes and approaches to management. Earthscan, London, pp 55–66

Steele J, Chandran RS, Grafton WN et al (2006) Awareness and management of invasive plants among West Virginia woodland owners. J For 104:248–253

Steele J, McGill DW, Chandran RS et al (2008) Landowner characteristics associated with receiving information about invasive plants and implications for outreach providers. J Ext 46:6FEA7

Stokes KE, O'Neill KP, Montgomery WI et al (2006) The importance of stakeholder engagement in invasive species management: a cross-jurisdictional perspective in Ireland. Biodivers Conserv 15:2829–2852

Sustainable Development Institute (2012) Shifting seasons great lakes forest industry, products, and resources summit—summit report and findings (Fellman D, Edler J, Dockry MJ (eds)). College of Menominee Nation Sustainable Development Institute, Keshena

Swiecki TJ, Bernhardt EA (2013) A reference manual for managing sudden oak death in California, USDA forest service general technical report PSW-GTR-242. US Department of Agriculture, Forest Service, Pacific Southwest Research Station, Albany, 129 p

Thresher RE (1999) Diversity, impacts and options for managing invasive marine species in Australian waters. Aust J Environ Manag 6:137–148

Tidwell LS (2005) Information sources, willingness to volunteer, and attitudes toward invasive plants in the Southwestern United States. MS thesis in Forestry, Utah State University, Logan, 131 p

Tidwell LS, Brunson MW (2008) Volunteering to manage rangeland weeds: results of a citizen survey in the southwestern United States. Rangelands 30:19–24

Trumbull DJ, Bonney R, Bascom D et al (2000) Thinking scientifically during participation in a citizen-science project. Sci Educ 84:265–275

Turner JM (2012) The promise of wilderness: American environmental politics since 1964. University of Washington Press, Seattle, 576 p

Varble S, Secchi S (2013) Human consumption as an invasive species management strategy. A preliminary assessment of the marketing potential of invasive Asian carp in the US. Appetite 65:58–67

Vasquez EA, James JJ, Monaco TA, Cummings DC (2010) Invasive plants on rangelands: a global threat. Rangelands 32:3–5

Veitch CR, Clout MN (2001) Human dimensions in the management of invasive species in New Zealand. In: McNeely JA (ed) The great reshuffling: human dimensions of invasive alien species. IUCN Biodiversity Policy Coordination Division, Gland/Cambridge, pp 63–74

Verbrugge LNH, Van den Born RJG, Lenders HJR (2013) Exploring public perception of non-native species from a visions of nature perspective. Environ Manag 52:1562–1573

Verbrugge LNH, Leuven RSEW, Zwart HAE (2016) Metaphors in invasion biology: implications for risk assessment and management of non-native species. Ethics Policy Environ 19:273–284

Vilà M, Ibáñez I (2011) Plant invasions in the landscape. Landsc Ecol 26:461–472

Voeks RA (2004) Disturbance pharmacopoeias: medicine and myth from the humid tropics. Ann Assoc Am Geogr 94:868–888

Vokoun M, Amacher GS, Sullivan J et al (2010) Examining incentives for adjacent non-industrial private forest landowners to cooperate. Forest Policy Econ 12:104–110

Wagner MW, Kreuter UP, Kaiser RA et al (2007) Collective action and social capital of wildlife management associations. J Wildl Manag 71:1729–1738

Warren P (2001) Dealing with the human dimensions of invasive alien species within New Zealand's biosecurity system. In: McNeely JA (ed) The great reshuffling: human dimensions of invasive alien species. IUCN, Gland/Cambridge, pp 105–111

Waterkeyn A, Vanschoenwinkel B, Elsen S et al (2010) Unintentional dispersal of aquatic invertebrates via footwear and motor vehicles in a Mediterranean wetland area. Aquat Conserv Mar Freshwat Ecosyst 20:580–587

Weber MA, Ringold PL (2015) Priority river metrics for residents of an urbanized arid watershed. Landsc Urban Plan 133:37–52

Wilcove DS, Rothstein D, Dubow J et al (1998) Quantifying threats to imperiled species in the United States. Bioscience 48:607–615

Willette DA, Chalifour J, Debrot AOD et al (2014) Continued expansion of the trans-Atlantic invasive marine angiosperm *Halophila stipulacea* in the Eastern Caribbean. Aquat Bot 112:98–102

With KA (2002) The landscape ecology of invasive spread. Conserv Biol 16:1192–1203

Wolf SA, Hufnagl-Eichiner S (2007) External resources and development of forest landowner collaboratives. Soc Nat Resour 20:675–688

Yung L, Belsky JM (2007) Private property rights and community goods: negotiating landowner cooperation amid changing ownership on the Rocky Mountain Front. Soc Nat Resour 20:689–703

Yung L, Chandler J, Haverhals M (2015) Effective weed management, collective action, and landownership change in Western Montana. Invasive Plant Sci Manag 8:193–202

13 The Role of International Cooperation in Invasive Species Research

Andrew M. Liebhold, Faith T. Campbell, Doria R. Gordon, Qinfeng Guo, Nathan Havill, Bradley Kinder, Richard MacKenzie, David R. Lance, Dean E. Pearson, Sharlene E. Sing, Travis Warziniack, Robert C. Venette, and Denys Yemshanov

13.1 Introduction

The root cause of the biological invasion problem is globalization, which has facilitated the planet-wide breakdown of biogeographic barriers to species migration (Mooney and Hobbs 2000). In order to understand and manage the problem, coordination on a global scale is essential, and international cooperation among affected countries as well as with countries of pest origin must therefore play a critical role in virtually all aspects of research on biological invasions (Chornesky et al. 2005; McNeely et al. 2001; Perrings et al. 2010; Wingfield et al. 2015). Here we discuss key aspects of research on biological invasions, where international collaboration and coordination are important, and what infrastructures play a role in this work.

The study of invasive species in both their native and introduced ranges is critical to mitigating the invasion problem. The translocation of organisms beyond their native ranges can, in some cases, simply extend the range of species that are already pests, and in other cases it can create new pests. It is widely hypothesized that such translocations result in novel ecological interactions, which may cause these introduced (non-native) species to become more abundant and/or modify their ecosystem impacts in their new range (e.g., Broennimann et al. 2007; Torchin et al. 2003). Based on this assumption, several mechanisms have been proposed to explain why introduced species sometimes become serious pests in their new ranges (Colautti et al. 2004; Mitchell et al. 2006). Remarkably few studies have actually quantified the abundance and impact of invading species in both native and introduced ranges, to test the

A. M. Liebhold (✉)
U.S. Department of Agriculture, Forest Service, Northern Research Station, Morgantown, WV, USA
e-mail: andrew.liebhold@usda.gov

F. T. Campbell
Center for Invasive Species Prevention, Springfield, VA, USA

D. R. Gordon
Environmental Defense Fund, Washington, DC, USA

Q. Guo
U.S. Department of Agriculture, Forest Service, Southern Research Station, Eastern Forest Environmental Threat Assessment Center, Research Triangle Park, NC, USA

N. Havill
U.S. Department of Agriculture, Forest Service, Northern Research Station, Hamden, CT, USA

B. Kinder
U.S. Department of Agriculture, Forest Service, Washington Office, Washington, DC, USA

R. MacKenzie
U.S. Department of Agriculture, Forest Service, Pacific Southwest Research Station, Institute of Pacific Islands Forestry, Hilo, HI, USA

D. R. Lance
U.S. Department of Agriculture, Animal and Plant Health Inspection Service, Plant Protection and Quarantine, Center for Plant Health Science and Technology, Buzzards Bay, MA, USA

D. E. Pearson
U.S. Department of Agriculture, Forest Service, Rocky Mountain Research Station, Missoula, MT, USA

Ecology and Evolution, Division of Biological Sciences, University of Montana, Missoula, MT, USA

S. E. Sing
U.S. Department of Agriculture, Forest Service, Rocky Mountain Research Station, Bozeman, MT, USA

T. Warziniack
U.S. Department of Agriculture, Forest Service, Rocky Mountain Research Station, Fort Collins, CO, USA

R. C. Venette
U.S. Department of Agriculture, Forest Service, Northern Research Station, St. Paul, MN, USA

D. Yemshanov
Natural Resources Canada, Canadian Forest Service, Great Lakes Forestry Centre, Sault Ste. Marie, ON, Canada

T. M. Poland et al. (eds.), *Invasive Species in Forests and Rangelands of the United States*,
https://doi.org/10.1007/978-3-030-45367-1_13

occurrence of assumed novel ecological interactions (but see Firn et al. 2011; Parker et al. 2013). Nonetheless, there are well-documented examples of non-native plants or animals that became serious pests as a result of their invasions altering hydrology, nutrient availability, fire regimes, etc. in ways that extensively change the recipient environment (see Chap. 2). Many vertebrate predator species not known to be problem species in their native ranges have become devastating pests after being introduced to island ecosystems that historically lacked predators (Blackburn et al. 2004). Extreme population growth and impacts on susceptible host trees, exhibited by many non-native herbivorous insects and tree pathogens, can be attributed in some cases to their lack of top-down control by natural enemies, or, in other systems, to a lack of coevolved resistance in host trees (Bonello et al. 2006; Colautti et al. 2004; Keane and Crawley 2002). In many cases, the transformations that create pests are much more subtle and can only be understood via international collaborations by studying the invader's population dynamics and community interactions in both its native and introduced ranges (Hierro et al. 2005).

In order to address the question of why non-native species transform into invasive pests when introduced into a new geographic range, it is critical to understand the specific changes in their ecology and community interactions that allow them to become pests. Acquiring that knowledge requires understanding how ecological interactions differ in a species' native and introduced ranges. For example, the enemy release hypothesis postulates that non-native species become problematic in their newly invaded ranges because they are released from population controls afforded by their specialist natural enemies (Keane and Crawley 2002). If this hypothesis is valid, then introducing natural enemies from the pest's native range could potentially reestablish control over its populations in the newly invaded range. This concept provides the theoretical basis for classical biological control, i.e., the introduction of natural enemies from the pest's native range to control its populations in the invaded range (van Driesche and Bellows 1996), a widely applied management tool for invasive pests which in many cases has proven to be highly successful (Huffaker and Kennett 1959; McFadyen 1998; van Driesche et al. 2008) (see Chap. 7). Similarly, when the pestilence of non-native insects or pathogens can be attributed to a lack of coevolved host tree resistance, this suggests that benefits may be gained through breeding for tree resistance (Sniezko 2006) (see Chaps. 7 and 8).

Whether translocations create new invasive pests or simply extend populations of pests, study of the invasive pest in both its native and newly invaded range can be crucial to the development of effective management strategies (e.g., McEvoy and Coombs 1999). Accordingly, for a non-native organism which becomes an invasive pest due to its translocation to a new range, comparing its population biology in the native and invaded ranges can identify which factor(s) drives population release in the newly invaded range, information that may be critical in formulating management strategies. For organisms that are known pest species in both ranges, studies of their populations in the native range before they are sufficiently abundant to study in the new range can be critical to containing and controlling the invader before it becomes widespread. Furthermore, when eradication is being employed, it is impractical to work with target organisms in the field in the introduced range; aside from working in a quarantine facility, the only alternative for acquiring needed research involves conducting biological studies in the native range or elsewhere within the introduced range. Consequently, international collaboration among researchers and managers across the native and introduced ranges is crucial.

All biological invasions can be partitioned into three major phases: arrival, establishment, and spread. Management strategies corresponding to each of these phases are prevention of species arrival, eradication (purposefully driving a species to extinction over a specific area) to prevent establishment, and containment to prevent or slow spread (see Chap. 6). Once non-native species become widely established and eradication is no longer possible, other management options, such as biological control or breeding resistant trees, may be appropriate (see Chaps. 7 and 8).

13.2 Prevention

A key step in the prevention of destructive invasions is to identify potentially damaging species in native habitats in their regions of origin and use this information to designate import quarantines. However, obtaining information about potential invaders, both in their native ranges and invaded ranges outside of the United States, may be difficult. Thus, international collaboration and data sharing are critical to implementing effective biosecurity strategies.

Available Data Information that is available through the public domain has vastly increased our ability to access and share specialized information on invasive species. The International Association for the Plant Protection Sciences disseminates largely informal reports on plant pests and their management in both their native and invaded ranges through a dedicated website, Global Plant Protection News (https://iapps2010.me). The Centre for Agriculture and Bioscience International's (CABI) Invasive Species Compendium (https://www.cabi.org/isc) is an encyclopedic resource with detailed datasheets that have been sourced from experts and peer-reviewed literature and includes images and range maps. The North American Plant Protection Organization's

(NAPPO) Phytosanitary Alert System (PAS) https://www.pestalerts.org) provides up-to-date information on developing plant pest problems that are just beginning or are likely to become significant in North America. The mission of the PAS is to facilitate awareness, detection, prevention, and management of new or potential exotic pest species through official pest reports and emerging pest alerts. Official pest reports are provided by the respective national plant protection organizations of Canada, the United States, and Mexico and serve as the official communication from the country of origin to comply with the International Plant Protection Convention's (IPPC) Standard on Pest Reporting (ISPM-17). Pest reports typically contain useful information on an invasive species: (1) detection or eradication in a specified location, (2) updated regulated areas, and (3) reports on new establishments or expansion of quarantine areas. Emerging pest alerts are not official NAPPO communiqués but instead are relevant (but unvetted) communications obtained from public sources that are posted on the PAS. Alerts are intended to function as an early warning tool for emerging plant pests that are not yet present in North America. The National Agricultural Pest Information System's Pest Tracker (http://pest.ceris.purdue.edu/index.php) provides information and distribution maps for invasive species already established in the United States, organized by taxonomic category as follows: bacteria, fungi, gastropods, insects, mites, nematodes, phytoplasma, plants, viruses, and others. This resource could be useful for identifying invasions occurring in new areas of the United States or forecasting potential local invasion sites based on their proximity to established populations and habitat similarities.

The US Department of Agriculture Animal and Plant Health Inspection Service (USDA APHIS) has regulatory responsibility for protecting U.S. agricultural and natural resource interests. The plant health branch of the APHIS, Plant Protection and Quarantine (PPQ), has a mandate to safeguard U.S. agriculture and natural resources against the entry, establishment, and spread of economically and environmentally significant pests and to facilitate the safe trade of agricultural products. The Center for Plant Health Science and Technology (CPHST) depends on international data and collaborations to provide information, tools, and technology for scientific support of PPQ's regulatory decisions and operations. Within the CPHST, the Plant Epidemiology and Risk Analysis Laboratory (PERAL) assesses pest risks incurred through the import and export of plant products. PERAL participates in a range of risk assessment activities under the auspices of various working groups and units. The New Pest Advisory Group (NPAG) assesses non-native plant pests that are new or not yet present in the United States but may pose a risk to US agriculture or the environment. The Exotic Pest Information Collection and Analysis (EPICA) project provides early warning of emerging non-native plant pest threats before they reach the United States, through weekly e-mailed reports based on open-source plant health information analyzed and communicated within a PPQ-relevant context. The Global Pest and Disease Database is a secure database of scientific information about potentially invasive pests of concern to U.S. agriculture and is maintained and operated by the National Science Foundation's Center for Integrated Pest Management, with oversight and input on the content and direction of the database provided by the CPHST. The predictive weed screening model has been developed by PERAL to identify plants that are likely to become weedy or invasive in the United States, based on species biological traits; impacts to agricultural, natural, and anthropogenic systems; and history of invasiveness elsewhere in the world (see Chap. 6).

Risk Assessment For prevention purposes, international data collection and collaboration are particularly critical to understand whether species have become invasive anywhere outside their native range. This single variable, whether or not a species is already reported as invasive elsewhere, is among the most consistent predictors of the risk of species establishment in a new location (e.g., Bomford et al. 2009; Panetta 1993; Samways 1999). However, the surge in international trade means that increasing numbers of species with no history of prior translocation will also be introduced (Perrings et al. 2010). Risk assessment methods will, therefore, be increasingly important (see Chap. 6).

The International Pest Risk Research Group (IPRRG) is a science network that enables sharing of data and expertise for prevention of invasive species. The IPRRG is an interdisciplinary group with a diverse international composition that facilitates collaborative research and information sharing to improve plant pest risk modeling and mapping methods. Pest risk in this context refers to the likelihood that an alien species will invade and cause harm within an endangered area. Pest risk models and maps are powerful tools to support decision-making in international trade, domestic quarantines, biosecurity surveillance, or pest-incursion responses. Research in this area helps provide a thorough description of the potential harmful impacts of invasive pests across space and time and ultimately to provide a more rigorous estimation of risk and more useful information for decision-makers. This research is inherently multidisciplinary, so the IPRRG includes more than 100 ecologists, economists, modelers, and practicing risk analysts from around the world. The group also works to communicate the research findings of its members to policymakers and other biosecurity, production, and natural resource sector stakeholders through peer-reviewed publications, technical presentations, and training sessions.

Research on plant pests is a dynamic field that has been rapidly advancing in the last two decades. With many new concepts and advanced methodologies being developed, it becomes increasingly difficult for practitioners and policymakers to identify and use the most appropriate risk assessment methods for a particular pest organism. Research produced by IPRRG members has helped identify the most important activities and risk assessment methodologies to prevent the introduction of invasive alien plant pests (Venette et al. 2010). The IPRRG considers some of the most pressing research needs, ensuring inclusion of better measures and representations of uncertainty, reliable metrics of impact, incorporation of climate change (into pest risk forecasts), and clarification of how human activities might affect the course of invasions. Initial solutions to some of these needs were addressed by the group in a special issue of the international journal NeoBiota (Kriticos and Venette 2013) and a textbook (Venette 2015).

In contrast with species-specific risk assessment for plant pests, a more generalized weed risk assessment system has now been tested in many regions and countries, with productive data sharing (Chap. 6). The tool, first developed in Australia (Pheloung et al. 1999), has demonstrated similar accuracy in tropical and temperate climates and island and continental geographies (Gordon et al. 2008) and forms the basis of the weed risk assessment system developed by the USDA APHIS PPQ (Koop et al. 2012). Additional research has demonstrated that weed risk assessment systems developed for one geographical area are relevant in new locations with similar climates (Chong et al. 2011). This result illustrates the benefits of online archiving of weed risk assessment datasheets (Chap. 6). An international collaboration has also developed specific guidance for use of weed risk assessments to increase consistency and utility of these tools (Gordon et al. 2010).

Another risk assessment approach for identifying species abroad that hold potential to damage forest trees in the United States or elsewhere is the use of sentinel plantings (Britton et al. 2010; Roques et al. 2015). Under this concept, either new plantings of non-native plants or existing plantings in arboreta are monitored for damage by insects or diseases in overseas locations. Such monitoring in foreign locations serves to identify potentially invasive insects and diseases capable of severely damaging native tree species should they be introduced. This information then can be used to inform the implementation of quarantines and other practices designed to prevent these pest species from establishing in the native range of the sentinel species. This is based upon the tendency of many of the worst invasive pests to cause extensive damage because native tree species lack resistance to introduced insects and diseases with which they have no prior evolutionary exposure (Ploetz et al. 2013). The International Plant Sentinel Network (http://www.plantsentinel.org) was established to provide an early warning system for new and emerging pest and pathogen risks monitored via sentinel plants. Scientific evidence of known quarantine organisms and potential new risks collected from member gardens (e.g., botanical gardens, arboreta) helps NPPOs prioritize plant health activities, thereby safeguarding susceptible plant species worldwide.

Quarantine The World Trade Organization's Agreement on the Application of Sanitary and Phytosanitary Standards (SPS) identifies the IPPC as the international body responsible for setting standards for phytosanitary treatments and other plant quarantine activities (MacLeod et al. 2010). At the heart of these standards is the premise that all phytosanitary rules imposed by countries must be scientifically based. The quest for scientifically based quarantine practices thus provides challenges for researchers to develop methods for both identifying and mitigating risks.

The International Forestry Quarantine Research Group (IFQRG) is a scientific advisory body composed of scientists from around the world (including the USDA Forest Service) that provides the IPPC with key information for setting policy and identifying research priorities on quarantine and other prevention activities. Over the last two decades, Forest Service scientists working in support of the IFQRG have played key roles in developing ISPM-15, an international standard for phytosanitary measures ratified by the IPPC that specifies mandatory treatment of solid wood packing material used to ship products between countries. Its main purpose is to provide a mandatory, harmonized phytosanitary treatment to prevent the international transport and spread of plant diseases and pest insects. Forest Service scientists played a key role in the original specification of the ISPM-15 treatment and in evaluation of its operational impact (Brockerhoff et al. 2014; Haack and Petrice 2009; Haack et al. 2014).

Additional Prevention Approaches Another area where international scientific collaboration plays a key role in preventing the introduction of invasive species is through identifying invasion pathways. By analyzing patterns of historical establishments around the world and categorization of organisms intercepted by port inspectors, scientists have made progress in identifying important pathways for insect, disease, vertebrate, and plant invasions, which include solid packing material, imported live plants, air passengers, and containerized cargo (Brockerhoff et al. 2016; Liebhold et al. 2006, 2012, 2016a; Smith et al. 2009). Identification of invasion pathways provides critical information necessary for implementing phytosanitary measures. Pathway identification has also made it possible to identify when and where to strategically apply prophylactic treatments such as fumigation, insecticide applications, disinfection, cold, heat, and irradiation or physical pest removal offshore, as part of a systems approach to phytosanitation (Hennessey et al. 2014).

The implementation of quarantine practices outside of North America can be highly effective in preventing the establishment of new invasions in the United States. One reason for this is the "bridgehead effect," in which a species may initially invade one part of the world where it becomes abundant and thus a source for accidental transport elsewhere (Garnas et al. 2016). Implementing globally harmonized quarantine measures is ultimately a more effective approach to mitigating economic impacts than implementing activities made after a pest is detected in the United States (Perrings et al. 2010). Realizing this approach, however, may necessitate investing in capacity building of quarantine and research programs in other countries, particularly those with developing economies. The Food and Agriculture Organization (FAO) of the United Nations conducts extensive capacity building through their Regional Forest Invasive Species Networks (http://www.fao.org/forestry/pests/en).

The USDA APHIS' International Technical and Regulatory Capacity Building Center specializes in program coordination, including technical and regulatory capacity building efforts with US and foreign government counterparts. The Center focuses on training and technology transfer needed to support sanitary and phytosanitary (SPS) issues related to safeguarding of US agriculture from foreign plant pests and animal diseases as well as expanding US international trade.

The USDA Foreign Agricultural Service also supports international SPS capacity building. Scientists in the United States are invested in overseas research capacity building in the fields of taxonomy and species identification, surveillance, and phytosanitary practices that will facilitate pest-free imports to the United States. These ongoing efforts to share scientific expertise abroad ultimately benefit the United States by reducing the probability that invasive species enter invasion pathways.

International education on the nature of invasive species and their impacts is an important measure to limit the global spread of species. This is especially true for plant, vertebrate, and aquatic species which often are intentionally introduced. Many purposeful introductions are initiated with good intentions, often by educated but ill-informed professionals who are not fully aware of the potential repercussions of introducing species to a new habitat. For example, many international nongovernmental organizations (NGOs), in their quest to seek alternative food resources for developing countries, stock aquaculture ponds with non-native tilapia or other fish. Once these fish escape, they invade adjacent aquatic habitats and often have deleterious impacts on native fish and invertebrates (e.g., Peterson et al. 2005). More recently, NGOs and countries have attempted to combat mosquito-borne diseases (e.g., Zika virus, dengue fever) by introducing non-native predators such as mosquito fish (*Gambusia affinis*) or guppies (*Poecilia reticulata*) (Azevedo-Santos et al. 2016). Some of these introductions may have resulted in negative impacts to native aquatic habitats, and yet their effectiveness has often never been determined.

A key area of research that requires international cooperation is genetic characterization of pest species across their invaded and native ranges. Such analyses can provide critical information about the chronology and source of historical invasions (e.g., for plant pests: Boissin et al. 2012; Dutech et al. 2012; Havill et al. 2016; for invasive plants: Gaskin et al. 2013; Tarin et al. 2013; Williams et al. 2005). Since it is usually not clear exactly where an introduced pest originated, genetic analysis can often pinpoint the source locality and better determine the pathway of introduction. This information is particularly useful when conducting foreign exploration for new biological control agents and for identifying natural enemies that are well adapted to the specific strain of pest that was introduced. Genetic characterization can also provide critical information for identifying cryptic strains, siblings, or hybrids of species (e.g., Gwiazdowski et al. 2011; Toševski et al. 2011; Ward et al. 2009). For example, hemlock woolly adelgid (*Adelges tsugae*), an introduced insect pest of hemlock (*Tsuga* spp.) trees in the Eastern United States, was known to occur throughout East Asia and Western North America; however, the source of the introduction was unknown. Genetic studies showed that there are up to eight genetically divergent groups in Asia, that the source of the introduction to the Eastern United States was Southern Japan, and that the strain in Western North America is native to that region and probably arrived there just prior to the last glacial period (Havill et al. 2016). Consequently, the search for the most effective natural enemies of hemlock woolly adelgid is now focused in Japan and Western North America (Havill et al. 2014). Completing studies of this nature requires international collaborations between scientists in the invaded countries and scientists in source countries, who can provide critical local taxonomic expertise and logistical support for collecting samples.

13.3 Surveillance/Eradication

Surveillance for populations of newly arrived non-native species is crucial to their early detection, and eradication is most likely to be successful when target populations are still small (Liebhold et al. 2016b; Rejmánek and Pitcairn 2002) (see Chap. 6).

Most surveillance programs for non-native insects utilize attractant traps baited with pheromones or host attractants. In order to develop such lures, it is often necessary to conduct field tests abroad in the species' native range. For example, studies of pheromone communication used in surveillance trapping can most effectively be carried out in world regions where these species are already established. International collaboration is thus key to developing effective surveillance programs. There are numerous examples of Forest Service scientists collaborating with overseas scientists to test semiochemical attractants abroad (e.g., Fan et al. 2010; Meng et al. 2014). One of the challenges arising in programs to eradicate newly discovered insect populations is the lack of information on efficacious treatment methods. Frequently, little information is available on how to control these species even in their native ranges. Consequently, it may be necessary to evaluate treatment methods in the organism's native range. For example, Forest Service scientists collaborated with APHIS PPQ staff and Chinese scientists to test the efficacy of different insecticide treatments for Asian long-horned beetle (*Anoplophora glabripennis*) populations in China so that these methods could be applied for eradication of incipient Asian long-horned beetle populations in the United States (Poland et al. 2006; Wang et al. 2003) (Fig. 13.1).

In recent years, there has been considerable progress in understanding the population biology of invading organisms and applying that knowledge to develop more effective strategies for eradicating invasive populations. As in other settings, sharing of experiences and knowledge about eradication efforts among countries can provide valuable information. One example of this approach was formation of a working group at the National Center for Ecological Analysis and Synthesis (University of California, Santa Barbara) entitled "Applying population ecology to strategies for eradicating invasive forest insects." This group consisted of an international team of scientists that investigated more effective strategies for detecting and eradicating invading

Fig. 13.1 Dr. Leah Bauer (USDA Forest Service) (left) and Xiaoyi Wang (Chinese Academy of Forestry) examining parasitized emerald ash borer larvae in ash trees in Liaoning Province, China, 2013 (Photo by Jian Duan, ARS)

forest insect populations (Liebhold et al. 2016b). One product of this work was the assembly and analysis of a global database of insect eradication programs that yielded new insight into the determinants of successful eradication programs (Kean et al. 2016; Tobin et al. 2014).

Once non-native species become established, they enter the spread phase, during which they expand their range into previously unoccupied habitats. Predictions of future spread are valuable to many different activities, including forest management, where knowledge of when an invading species is likely to invade new regions may be crucial. An example of how international collaboration has provided critical knowledge about spread is provided by the *Sirex* woodwasp (*Sirex noctilio*). This species, native to Europe, invaded most pine-growing regions of the Southern Hemisphere several decades ago but only recently became established in the Northeastern United States. An international team of insect ecologists assembled and analyzed data on the historical spread of this insect in various regions of the Southern Hemisphere and used this valuable information to predict the future spread of *Sirex* woodwasp in North America (Lantschner et al. 2013).

In the cases of plant invasions, international collaboration is needed particularly with reciprocally exchanged species. Plant ecologists in Eastern Asia, Europe, and North America are focusing on exchanges of species among the three continents as a direct result of rapidly growing travel and trade, the implications of these exchanges in facilitating the introduction and spread of invasive species, and the environmental and economic impacts associated with these invasions. Because Eastern Asia and North America share a wide range of similar environments, species native to one region increasingly find suitable habitat for establishment in the other (Heberling et al. 2017). Some species have become invasive, disrupting ecosystems and food webs, threatening native species, causing economic losses, and occasionally jeopardizing the health of wildlife, domestic animals, and human populations (e.g., Gordon 1998).

13.4 Mitigation

Biological control is an important component of invasive species research programs. A key aspect of biological control research is the search for natural enemies of invasive pests in their native range followed by intensive studies and analysis of their population ecology and host specificity, to ascertain their potential utility in the safe regulation of targeted pests and mitigation of their impacts at the population level.

The emerald ash borer (EAB) (*Agrilus planipennis*) biological control effort is an excellent example of a successful overseas exploration for natural enemies of an invasive pest.

A team comprised of USDA Forest Service, Agricultural Research Service (ARS), and APHIS scientists worked with collaborators in China and Russia (Fig. 13.2) to search out populations of EAB in its native range from which parasitoids were reared and eventually shipped to quarantine facilities in the United States. There, the team performed host specificity experiments and ultimately released one egg and four larval parasitoids which have established in parts of the EAB's range (Bauer et al. 2015; Duan et al. 2012, 2015a; Liu et al. 2003). While these introductions are not anticipated to stop the current EAB outbreak that is sweeping across the Eastern and Central United States, there is a good chance that these agents will eventually contribute to the regulation of post-epidemic populations at levels low enough to allow a population of host ash (*Fraxinus* spp.) trees to persist and reproduce (Duan et al. 2015b).

Classical biological control of invasive toadflaxes (*Linaria* spp.) has relied on collaboration among US (Forest Service, Colorado and Montana State Universities), Canadian (Agriculture and Agri-Food Canada), Swiss, and Serbian (CABI) researchers. Initial reports based on greenhouse- and garden-based host specificity tests conducted overseas indicated that most biocontrol agents were equally effective against yellow toadflax (*L. vulgaris*) and Dalmatian toadflax (*L. dalmatica*). These proved to be inaccurate under North American field conditions. Subsequent molecular diagnostics revealed that releases (and redistributions) of the toadflax stem mining weevil (*Mecinus janthinus*) (Coleoptera: Curculionidae) included a cryptic sister species (*M. janthiniformis*) and that the two weevil species were highly host specific (Toševski et al. 2011). Climate matching challenges and the confirmed hybridization of yellow and Dalmatian toadflax (Ward et al. 2009) have necessitated the continuing search for candidate agents.

Several organizations provide support for biocontrol research collaboration among countries. Voluntary membership in the International Organization for Biological Control (IOBC) facilitates collaboration on a global scale among foreign and domestic scientists working toward biological control solutions. The ARS Overseas Biological Control Laboratories provide research infrastructure and networks critical for conducting foreign exploration for natural enemies and for coordinating host specificity testing. The CABI and the Biotechnology and Biological Control Agency are international nonprofit organizations that play a key role in both the development of knowledge and its application for managing invasions of both plants and plant pests.

Collaborative research on biological control must be cognizant of, and responsive to, international treaties, a challenge not incurred by many domestic research programs.

One such treaty, the Convention on Biological Diversity (CBD), is an international agreement with three main objectives: (1) the conservation of biological diversity, (2) the sustainable use of the components of biological diversity, and (3) the fair and equitable sharing of the benefits arising out of the utilization of genetic resources (https://www.cbd.int/intro/default.shtml). The United States supports the CBD though it has never officially ratified the treaty. In 2010, the CBD Nagoya Protocol was negotiated to address the third objective. It was adopted in 2014 to recognize and protect sovereign rights over the national genetic resources of participating countries. The Nagoya Protocol broadly describes the access and benefit sharing (ABS) of national genetic resources, and practical considerations for ABS of specific genetic resources are stipulated in agreements established between provider and recipient nations. Development and implementation of ABS legislation and regulation is the responsibility of each participating nation, so there is a high

Fig. 13.2 Dr. Therese Poland (USDA Forest Service) (upper right) and collaborators from the Chinese Academy of Forestry and Baiyin City Forest Protection Bureau dissecting trees injected with systemic insecticides and examining Asian long-horned beetle mortality 4 months after treatment in Gansu Province, China, October 2000 (Photo by Leah Bauer, USDA Forest Service)

probability that interpretation of Nagoya Protocol Article 8 "Special Considerations" will vary considerably. ABS in certain cases may involve payments equivalent to royalties from the nation receiving genetic resources to the nation providing the genetic resources. The relevance of ABS to classical biological control depends on the providing nation's determination, under Article 8, that access to biological control organisms supports research benefiting agriculture and food security and as such should be subject to less complicated and costly ABS conditions due to its noncommercial nature (Gourlay et al. 2013; van Lenteren et al. 2011). Guidance on this issue was provided to the FAO in a report prepared by the IOBC (Cock et al. 2009).

Planting genetically resistant tree varieties is another method forest managers may adopt for the long-term management of invasive insect and pathogen species (Sniezko 2006). For example, one approach to developing American chestnuts (*Castanea dentata*) resistant to the chestnut blight fungus (*Cryphonectria parasitica*) involves hybridization of American chestnut with Chinese chestnut (*C. mollissima*) and repeated backcrossing to American chestnut (Diskin et al. 2006). A key component of such resistance development programs is having access to a wide global selection of plant germplasm. International collaboration thus may play an important role in facilitating such resistance programs.

13.5 Overarching Efforts

In the United States, the Forest Service International Programs (IP) plays a key role in facilitating international cooperation in research on biological invasions. Their work involves providing logistical assistance to Forest Service scientists working abroad as well as to foreign scientists visiting the United States. The IP also supports technical cooperation projects that fund key aspects of invasion research as well as overseas capacity building. Furthermore, the IP engages in building partnerships with foreign governments and research organizations to help facilitate international research.

The Forest Service has established several bilateral agreements with other countries with objectives that include invasion and biosecurity research issues. One example of this is the United States-New Zealand Joint Commission on Science and Technology Cooperation which promotes collaborative research between the two countries on several topics including forest biosecurity.

In North America, cross-border collaboration among the United States, Canada, and Mexico promotes research on topics such as the analysis of spread and prediction of impacts. Invasions do not stop at borders, so the sharing of data and analyses across borders is invaluable. Information on cross-border movement of invasive organisms also helps better coordinate national surveillance and control programs and ultimately improves the efficacy of regulatory and containment measures directed against high-impact species.

Several organizations coordinate sharing of data and research on invasive species among the United States, Canada, Mexico, and, in some cases, Caribbean nations. These include consortia of the NGOs, including the North American Invasive Species Network and Weeds Across Borders. Other organizations are made up of government agencies, including the US Aquatic Nuisance Species Task Force (its Western and Northeast Regional Panels work with Canada); the North American Forest Commission Insects, Diseases, and Invasive Plants Working Group; and the binational Great Lakes Water Quality Agreement, which works to assess and improve technologies for control, eradication, and detection of invasive aquatic species and to assess these species' ecosystem impacts. The Cooperative Program in Research and Technology for the Northern Region (known as PROCINORTE) is a network of national agricultural research bodies in Mexico, the United States, and Canada. This network supports agricultural trade through sound science and knowledge sharing on issues of regional relevance.

As is the case with most scientific work, research on biological invasions is often constrained by the availability of data. In particular, there is a strong need for sharing of data on invasive species among scientists in different parts of the world. The availability of such international databases can greatly facilitate work on risk analysis, control, and other aspects of invasive species management. Several efforts have been made to develop global and regional databases on invasive species. These include the Global Invasive Species Information Network (Simpson 2004; Simpson et al. 2006) and the International Union for Conservation of Nature Invasive Species Specialist Group's (IUCN ISSG) Global Invasive Species Database (Pagad et al. 2015). International efforts such as the Global Eradication and Response Database (Kean et al. 2016) and Pherobase (El-Sayed 2014) synthesize scientific information that is used as a resource in developing surveillance and eradication programs.

International scientific societies also comprise key infrastructures that facilitate international collaboration among scientists engaged in research on invasive species. One of the most active societies is the International Union of Forest Research Organizations (IUFRO), which hosts several international research working parties (WP) and a task force that focus on the biological invasion problem. These include WP 2.02.15, "Breeding and genetic resources of five-needle pines"; WP 4.04.07, "Risk analysis"; WP 5.03.06, "Wood protection for quarantine, food packaging and trade in wood"; WP 7.03.12, "Alien invasive species and international trade"; WP 7.03.13, "Biological control of forest insects and pathogens"; WP 8.02.04, "Ecology of alien invasives"; and the

Task Force on Biological Invasions in Forests. The IUFRO is the preeminent international body for scientific networking on forest research and, as such, plays a crucial role in the exchange of information among scientists. The IUFRO also hosts the FORENT and FORPATH Internet listservs which facilitate communication among forest entomologists and forest pathologists, respectively, on a variety of topics.

Other organizations and working groups such as the Organisation for Economic Co-operation and Development Co-operative Research Programme and the IUCN ISSG operate programs that facilitate the synthesis of scientific knowledge for use by policymakers and stakeholders. The ISSG hosts the Aliens-L Internet listserv that facilitates communication among scientists studying biological invasions as well as with policymakers and managers.

13.6 Key Findings

- International collaboration plays a key role in research on prevention of future invasions. Understanding species in their native ranges and how they enter invasion pathways is critical to the analysis of risk, used to guide quarantine measures.
- Scientific research conducted overseas is essential for the development of more effective surveillance and eradication methods. Testing of survey tools and treatments is often only practically conducted in foreign regions.
- History of invasions elsewhere and risk assessments conducted overseas can inform assessment of the probability of invasion in the United States, potentially enhancing prevention and eradication efforts.
- The implementation of classical biological control, and sometimes breeding of genetic resistance, requires studying the invasive species overseas and is dependent on effective coordination with foreign scientists.
- A variety of international organizations facilitate collaborative research and the exchange of data.
- International collaboration includes coordination of research among North American countries as well as among different continents and sharing of data across borders.

Disclaimer Text The findings and conclusions in this publication are those of the authors and should not be construed to represent any official USDA or U.S. Government determination or policy.

Literature Cited

Azevedo-Santos VM, Vitule JR, Pelicice FM et al (2016) Nonnative fish to control Aedes mosquitoes: a controversial, harmful tool. Bioscience 67:84–90

Bauer LS, Duan JJ, Gould JR, Van Driesche RG (2015) Progress in the classical biological control of *Agrilus planipennis* Fairmaire (Coleoptera: Buprestidae) in North America. Can Entomol 147:300–317

Blackburn TM, Cassey P, Duncan RP et al (2004) Avian extinction and mammalian introductions on oceanic islands. Science 305:1955–1958

Boissin E, Hurley B, Wingfield MJ et al (2012) Retracing the routes of introduction of invasive species: the case of the *Sirex noctilio* woodwasp. Mol Ecol 21:5728–5744

Bomford M, Kraus F, Barry SC, Lawrence E (2009) Predicting establishment success for alien reptiles and amphibians: a role for climate matching. Biol Invasions 11:713–724

Bonello P, Gordon TR, Herms DA et al (2006) Nature and ecological implications of pathogen-induced systemic resistance in conifers: a novel hypothesis. Physiol Mol Plant Pathol 68:95–104

Britton KO, White P, Kramer A, Hudler G (2010) A new approach to stopping the spread of invasive insects and pathogens: early detection and rapid response via a global network of sentinel plantings. N Z J For Sci 40:109–114

Brockerhoff EG, Kimberley M, Liebhold AM et al (2014) Predicting how altering propagule pressure changes establishment rates of biological invaders across species pools. Ecology 95:594–601

Brockerhoff EG, Bulman LS, Liebhold AM, Monge JJ (2016) Role of sea containers in unintentional movement of invasive contaminating pests (so-called "hitchhikers"), and opportunities for mitigation measures. Report to IPPC Commission on Phytosanitary Measures

Broennimann O, Treier UA, Müller-Schärer H et al (2007) Evidence of climatic niche shift during biological invasion. Ecol Lett 10:701–709

Chong KY, Corlett RT, Yeo DCJ, Tan HTW (2011) Towards a global database of weed risk assessments: a test of transferability for the tropics. Biol Invasions 13:1571–1577

Chornesky EA, Bartuska AM, Aplet GH, Britton KO, Cummings-Carlson J, Davis FW, Eskow J, Gordon DR, Gottschalk KW, Haack RA, Hansen AJ (2005) Science priorities for reducing the threat of invasive species to sustainable forestry. Bioscience 55:335–348

Cock MJW, van Lenteren JC, Brodeur J et al (2009) The use and exchange of biological control agents for food and agriculture. FAO Background Study No. 47, FAO Rome. 88 pp

Colautti RI, Ricciardi A, Grigorovich IA, MacIsaac HJ (2004) Is invasion success explained by the enemy release hypothesis? Ecol Lett 7:721–733

Diskin M, Steiner KC, Hebard FV (2006) Recovery of American chestnut characteristics following hybridization and backcross breeding to restore blight-ravaged *Castanea dentata*. For Ecol Manag 223:439–447

Duan JJ, Yuchenko G, Fuester RW (2012) Occurrence of emerald ash borer (Coleoptera: Buprestidae) and biotic factors affecting its immature stages in the Russian Far East. Environ Entomol 41:245–254

Duan JJ, Gould JR, Fuester RW (2015a) Evaluation of the host specificity of *Spathius galinae* (Hymenoptera: Braconidae), a larval parasitoid of the emerald ash borer (Coleoptera: Buprestidae) in Northeast Asia. Biol Control 89:91–97

Duan JJ, Bauer LS, Abell KJ et al (2015b) Population dynamics of an invasive forest insect and associated natural enemies in the aftermath of invasion: implications for biological control. J Appl Ecol 52:1246–1254

Dutech C, Barres B, Bridier J et al (2012) The chestnut blight fungus world tour: successive introduction events from diverse origins in an invasive plant fungal pathogen. Mol Ecol 21:3931–3946

El-Sayed AM (2014) The Pherobase: database of pheromones and semiochemicals. http://www.pherobase.com

Fan JT, Miller DR, Zhang LW, Sun JH (2010) Effects of bark beetle pheromones on the attraction of *Monochamus alternatus* to pine volatiles. Insect Sci 17:553–556

Firn J, Moore JL, MacDougall AS et al (2011) Abundance of introduced species at home predicts abundance away in herbaceous communities. Ecol Lett 14:274–281

Garnas JR, Auger-Rozenberg MA, Roques A et al (2016) Complex patterns of global spread in invasive insects: eco-evolutionary and management consequences. Biol Invasions 18:935–952

Gaskin JF, Schwarzländer M, Kinter CL et al (2013) Propagule pressure, genetic structure, and geographic origins of *Chondrilla juncea* (Asteraceae): an apomictic invader of three continents. Am J Bot 100:1871–1882

Gordon DR (1998) Effects of invasive, non-indigenous plant species on ecosystem processes: lessons from Florida. Ecol Appl 8:975–989

Gordon DR, Onderdonk DA, Fox AM, Stocker RK (2008) Consistent accuracy of the Australian Weed Risk Assessment system across varied geographies. Divers Distrib 14:234–242

Gordon DR, Mitterdorfer B, Pheloung PC et al (2010) Guidance for addressing the Australian Weed Risk Assessment questions. Plant Prot Q 25:56–74

Gourlay AH, Shaw R, Cock MJW (2013) Workshop report: the Nagoya protocol on access to genetic resources under the convention on biological diversity. pp 493–495 In: Wu Y, Johnson T, Sing SE, Raghu S, Wheeler G, Pratt P, Warner K, Center T, Goolsby J, Reardon R (eds) Proceedings of the XIII international symposium on biological control of weeds, September 11–16, 2011, Waikoloa, HI. FHTET-2012-07

Gwiazdowski RA, Vea IM, Andersen JC, Normark BB (2011) Discovery of cryptic species among North American pine-feeding Chionaspis scale insects (Hemiptera: Diaspididae). Biol J Linn Soc 104:47–62

Haack RA, Petrice TR (2009) Bark-and wood-borer colonization of logs and lumber after heat treatment to ISPM 15 specifications: the role of residual bark. J Econ Entomol 102:1075–1084

Haack RA, Britton KO, Brockerhoff EG et al (2014) Effectiveness of the International Phytosanitary Standard ISPM No. 15 on reducing wood borer infestation rates in wood packaging material entering the United States. PLoS One 9(5):e96611

Havill NP, Vieira LC, Salom SM (2014) Biology and Control of Hemlock Woolly Adelgid, vol FHTET-2014-05. USDA Forest Service, Forest Health Technology Enterprise Team, Morgantown

Havill NP, Shiyake S, Galloway A et al (2016) Ancient and modern colonization of North America by hemlock woolly adelgid, *Adelges tsugae* (Hemiptera: Adelgidae), an invasive insect from East Asia. Mol Ecol 25:2065–2080

Heberling JM, Jo I, Kozhevnikov A, Lee H, Fridley JD (2017) Biotic interchange in the Anthropocene: strong asymmetry in East Asian and eastern North American plant invasions. Glob Ecol Biogeogr 26:447–458

Hennessey MK, Jeffers L, Nendick D et al (2014) Chapter 10: Phytosanitary treatments. In: Gordh G, McKirdy S (eds) The handbook of plant biosecurity. Springer, Dordrecht, pp 269–308

Hierro JL, Maron JL, Callaway RM (2005) A biogeographical approach to plant invasions: the importance of studying exotics in their introduced and native range. J Ecol 93:5–15

Huffaker CB, Kennett CE (1959) A ten-year study of vegetational changes associated with biological control of Klamath weed. J Range Manag 12:69–82

Kean JM, Suckling DM, Sullivan NJ et al (2016) Global eradication and response database. http://b3.net.nz/gerda. Accessed 20 Mar 2016

Keane RM, Crawley MJ (2002) Exotic plant invasions and the enemy release hypothesis. Trends Ecol Evol 17:164–170

Koop A, Fowler L, Newton L, Caton B (2012) Development and validation of a weed screening tool for the United States. Biol Invasions 14:273–294

Kriticos DJ, Venette RC (eds) (2013) Advancing risk assessment models to address climate change, economics and uncertainty. NeoBiota 18:1–7

Lantschner MV, Villacide JM, Garnas JR et al (2013) Temperature explains variable spread rates of the invasive woodwasp *Sirex noctilio* in the Southern Hemisphere. Biol Invasions 16:29–339

Liebhold AM, Work TT, McCullough DG, Cavey JF (2006) Airline baggage as a pathway for alien insect species invading the United States. Am Entomol 52:48–54

Liebhold AM, Brockerhoff EG, Garrett LJ et al (2012) Live plant imports: the major pathway for forest insect and pathogen invasions of the US. Front Ecol Environ 10:135–143

Liebhold AM, Yamanaka T, Roques A et al (2016a) Global compositional variation among native and non-native regional insect assemblages emphasizes the importance of pathways. Biol Invasions 18:893–899

Liebhold AM, Berec L, Brockerhoff EG et al (2016b) Eradication of invading insect populations: from concepts to applications. Annu Rev Entomol 61:335–352

Liu HP, Bauer LS, Gao RT et al (2003) Exploratory survey for the emerald ash borer, *Agrilus planipennis* (Coleoptera: Buprestidae), and its natural enemies in China. Great Lakes Entomol 36:191–204

MacLeod A, Pautasso M, Jeger MJ, Haines-Young R (2010) Evolution of the international regulation of plant pests and challenges for future plant health. Food Secur 2:49–70

McEvoy PB, Coombs EM (1999) Biological control of plant invaders: regional patterns, field experiments, and structured population models. Ecol Appl 9:387–401

McFadyen RE (1998) Biological control of weeds. Annu Rev Entomol 43:369–393

McNeely JA, Mooney HA, Neville LE et al (2001) Global strategy on invasive alien species. Global Invasive Species Programme (GISP), Gland/Cambridge. 63 pp

Meng PS, Trotter RT, Keena MA et al (2014) Effects of pheromone and plant volatile release rates and ratios on trapping Anoplophora glabripennis (Coleoptera: Cerambycidae) in China. Environ Entomol 43:1379–1388

Mitchell CE, Agrawal AA, Bever JD et al (2006) Biotic interactions and plant invasions. Ecol Lett 9:726–740

Mooney HA, Hobbs RJ (2000) Invasive species in a changing world. Island Press, Washington, DC

Pagad S, Genovesi P, Carnevali L et al (2015) IUCN SSC Invasive Species Specialist Group: invasive alien species information management supporting practitioners, policy makers and decision takers. Manage Biol Invasions 6:127–135

Panetta FD (1993) A system for assessing proposed plant introductions for weed potential. Plant Prot Q 8:10–14

Parker JD, Torchin ME, Hufbauer RA et al (2013) Do invasive species perform better in their new ranges? Ecology 94:985–994

Perrings C, Burgiel S, Lonsdale M et al (2010) International cooperation in the solution to trade-related invasive species risks. Ann N Y Acad Sci 1195:198–212

Peterson MS, Slack WT, Woodley CM (2005) The occurrence of non-indigenous Nile tilapia, *Oreochromis niloticus* (Linnaeus) in coastal Mississippi, USA: ties to aquaculture and thermal effluent. Wetlands 25:112–121

Pheloung PC, Williams PA, Halloy SR (1999) A weed risk assessment model for use as a biosecurity tool evaluating plant introductions. J Environ Manag 57:239–251

Ploetz RC, Hulcr J, Wingfield MJ, De Beer ZW (2013) Destructive tree diseases associated with ambrosia and bark beetles: black swan events in tree pathology? Plant Dis 97:856–872

Poland TM, Haack RA, Petrice TR et al (2006) Field evaluations of systemic insecticides for control of *Anoplophora glabripennis* (Coleoptera: Cerambycidae) in China. J Econ Entomol 99:383–392

Rejmánek M, Pitcairn MJ (2002) When is eradication of exotic plants a realistic goal? In: Veitch CR, Clout MN (eds) Turning the tide: the eradication of invasive species. IUCN SSC Invasive Species Specialist Group, IUCN [World Conservation Union], Gland/Cambridge, pp 249–253

Roques A, Fan JT, Courtial B, Zhang YZ et al (2015) Planting sentinel European trees in eastern Asia as a novel method to identify potential insect pest invaders. PloS one 10(5):e0120864

Samways MJ (1999) Managing insect invasions by watching other countries. In: Sandlund OT, Schei PJ, Viken A (eds) Invasive species and biodiversity management. Kluwer Academic Publishers, Dordrecht, pp 295–304

Simpson A (2004) The global invasive species information network: what's in it for you? Bioscience 54:613–614

Simpson A, Sellers E, Grosse A, Xie Y (2006) Essential elements of online information networks on invasive alien species. Biol Invasions 8:1579–1587

Smith KF, Behrens M, Schloegel LM et al (2009) Reducing the risks of the wildlife trade. Science 324:594–595

Sniezko RA (2006) Resistance breeding against nonnative pathogens in forest trees—current successes in North America. Can J Plant Pathol 28(S1):S270–S279

Tarin D, Pepper AE, Goolsby JA et al (2013) Microsatellites uncover multiple introductions of clonal giant reed (*Arundo donax*). Invasive Plant Sci Manag 6:328–338

Tobin PC, Kean JM, Suckling DM et al (2014) Determinants of successful arthropod eradication programs. Biol Invasions 16:401–414

Torchin ME, Lafferty KD, Dobson AP et al (2003) Introduced species and their missing parasites. Nature 421:628–630

Toševski I, Caldara R, Jović J et al (2011) Morphological, molecular and biological evidence reveal two cryptic species in Mecinus janthinus Germar (Coleoptera, Curculionidae), a successful biological control agent of Dalmatian toadflax, Linaria dalmatica (Lamiales, Plantaginaceae). Syst Entomol 36:741–753

Van Driesche RG, Bellows TS Jr (eds) (1996) Biological control. Chapman & Hall, New York

Van Driesche R, Hoddle M, Center T (2008) Control of pests and weeds by natural enemies: an Introduction to biological control. Wiley, New York

van Lenteren JC, Cock MJW, Brodeur J et al (2011) Will the Convention on Biological Diversity put an end to biological control? Revista Brasileira de Entomologia 55:1–5

Venette RC (ed) (2015) Pest risk modelling and mapping for invasive alien species. CAB International, Wallingford

Venette RC, Kriticos DJ, Magarey RD et al (2010) Pest risk maps for invasive alien species: a roadmap for improvement. Bioscience 60:349–362

Wang B, Gao R, McLane WH et al (2003) Evaluation of insecticides for controlling the Asian longhorned beetle, *Anoplophora glabripennis* – a synthesis presentation. In: Fosbroke SLC, Gottschalk KW (eds) Proceedings, U.S. Department of Agriculture interagency research forum on gypsy moth and other invasive species 2002; 2002 January 15–18; Annapolis, MD. Gen. Tech. Rep. NE-300. U.S. Department of Agriculture, Forest Service, Northeastern Research Station, Newtown Square, pp 97–99

Ward SM, Fleischmann CE, Turner MF, Sing SE (2009) Hybridization between invasive populations of Dalmatian toadflax (*Linaria dalmatica*) and yellow toadflax (*Linaria vulgaris*). Invasive Plant Sci Technol 2:369–378

Williams DA, Overholt WA, Cuda JP, Hughes CR (2005) Chloroplast and microsatellite DNA diversities reveal the introduction history of Brazilian peppertree (*Schinus terebinthifolius*) in Florida. Mol Ecol 14:3643–3656

Wingfield MJ, Brockerhoff EG, Wingfield BD, Slippers B (2015) Planted forest health: the need for a global strategy. Science 349:832–836

Economics of Invasive Species

14

Travis Warziniack, Robert G. Haight, Denys Yemshanov, Jenny L. Apriesnig, Thomas P. Holmes, Amanda M. Countryman, John D. Rothlisberger, and Christopher Haberland

14.1 Introduction

While the subset of introduced species that become invasive is small, the damages caused by that subset and the costs of controlling them can be substantial. This chapter takes an in-depth look at the economic damages non-native species cause, methods economists often use to measure those damages, and tools used to assess invasive species policies. Ecological damages are covered in other chapters of this book. To put the problem in perspective, Federal agencies reported spending more than half a billion dollars per year in 1999 and 2000 for activities related to invasive species ($513.9 million in 1999 and $631.5 million in 2000 (U.S. GAO 2000)). Approximately half of these expenses were spent on prevention. Several states also spend considerable resources on managing non-native species; for example, Florida spent $127.6 million on invasive species activities in 2000 (U.S. GAO 2000), and the Great Lakes states spend about $20 million each year to control sea lamprey (*Petromyzon marinus*) (Kinnunen 2015). Costs to government may not be the same as actual damages, which generally fall disproportionately on a few economic sectors and households. For example, the impact of the 2002 outbreak of West Nile virus exceeded $4 million in damages to the equine industries in Colorado and Nebraska alone (USDA APHIS 2003) and more than $20 million in public health damages in Louisiana (Zohrabian et al. 2004). Zebra mussels (*Dreissena polymorpha*) cause $300–$500 million annually in damages to power plants, water systems, and industrial water intakes in the Great Lakes region (Great Lakes Commission 2012) and are expected to cause $64 million annually in damages should they or quagga mussels (*Dreissena bugensis*) spread to the Columbia River basin (Warziniack et al. 2011).

Studies on economic impacts from invasive species vary in their rigor and usefulness for informing policy decisions. This chapter discusses economic impacts and methods used to calculate them, how to distinguish impact studies that were done well from those that were done poorly, and appropriate use of values calculated in impact studies. The chapter also discusses key contributions of economics to invasive species science and provides a quick overview of behavioral and economic responses to invasive species risk.

The chapter is organized according to four main themes. The first section focuses on introduction and establishment of species into an area. Economic research on the introduction of species has focused on people's understanding of invasion risk and potential impacts and how they respond to that risk, human-mediated vectors of introductions, and development of trade and regulatory policies that prevent the movement of invasive species into uninvaded areas. The sec-

T. Warziniack (✉)
U.S. Department of Agriculture, Forest Service, Rocky Mountain Research Station, Fort Collins, CO, USA
e-mail: travis.w.warziniack@usda.gov

R. G. Haight
U.S. Department of Agriculture, Forest Service, Northern Research Station, St Paul, MN, USA

D. Yemshanov
Natural Resources Canada, Canadian Forest Service, Great Lakes Forestry Centre, Sault Ste. Marie, ON, Canada

J. L. Apriesnig
Julie Ann Wrigley Global Institute of Sustainability, Arizona State University, Tempe, AZ, USA

T. P. Holmes
U.S. Department of Agriculture, Forest Service, Southern Research Station, Research Triangle Park, NC, USA

A. M. Countryman
Department of Agricultural and Resource Economics, Colorado State University, Fort Collins, CO, USA

J. D. Rothlisberger
U.S. Department of Agriculture, Forest Service, Washington Office, Washington, DC, USA

C. Haberland
U.S. Department of Agriculture, Forest Service, National Center for Natural Resource Economics, Washington, DC, USA

T. M. Poland et al. (eds.), *Invasive Species in Forests and Rangelands of the United States*,
https://doi.org/10.1007/978-3-030-45367-1_14

ond section provides an overview of methods used to measure damages and costs related to invasive species. There exists a rather mature literature on market damages from invasive species, a maturing literature on non-market damages from invasive species, and a very young literature linking the production of ecosystem services with their market and non-market values. These estimates are essential to formulate a realistic examination of policy, as discussed in the third section on optimal policies and strategies for preventing and controlling invasive species. New models on optimal policy link introduction and establishment through the use of geographical models that depict invasion and its negative impacts in temporal and spatial domains. Such models indicate that establishment in one area makes introduction into neighboring areas more likely, and are often used to estimate the cost-effectiveness of control or slow-the-spread measures that are applied, subject to geographic constraints on policy and various environmental variables. The chapter concludes with a discussion of future research needs and a table (Table 14.1) summarizing damages from invasive species found in the literature.

Table 14.1 Representative studies estimating potential impacts and sources of risk of invasive species, broken down roughly by stage of invasion

Pre-invasion modeling/risk assessment			
Species	Impact estimate	Background	Author, date
Asian carp	$78.5 million (total committed)	Fed. gov't committed funds for prevention to date	U.S. Fish and Wildlife Service (2012)
Vampire bats (*Desmodus rotundus*)	$7–9.2 million, annually	Input-output model (impacts of bat-introduced rabies if vampire bats introduced rabies to South Texas)	Anderson et al. (2014)
Feral swine (*Sus scrofa*)	$12 million (from modeled 45-day foot-and-mouth outbreak in Missouri)	Bioeconomic model (potential disease spread and IMPLAN)	Cozzens et al. (2010)
Brown tree snakes (*Boiga irregularis*)	$593 million–$2.14 billion, annually	Input-output model (medical, power, tourism impacts)	Shwiff et al. (2010)
Emerald ash borer (*Agrilus planipennis*) in Eastern United States	$10.7 billion (25-State area), 2009–2019	Cost of potential emerald ash borer damage in US communities; estimated treatment removal and replacement costs	Kovacs et al. (2010)
Sudden oak death caused by non-native pathogen *Phytophthora ramorum* in California	$142 million, 2010–2020	Cost of potential damage to coastal live oak in California communities; estimated treatment, removal, and replacement costs plus residential property value loss	Kovacs et al. (2011)
455 non-native phytophagous forest insects	$2.53 billion, 2011–2021	Estimated Federal gov't, local gov't, and household expenditures; residential property value loss; and landowner timber loss, via expert elicitation with Bayesian estimation	Aukema et al. (2011)
Miconia (*Miconia calvescens*)	$4.7 million–2.36 billion from 50 years to 200 years after introduction	Optimal control estimates of impacts of potential spread in Hawaii	Kaiser (2006)
Red streaked leafhopper (*Balclutha rubrostriata*)	$132 million–1.7 billion, 2015–2040 in Louisiana	IMPLAN input-output model (lost economic output)	Piper and Liu (2014)
Invasive plants	$34.7 billion, annually, in the United States	Bioeconomic forecasting of invasive species by ecological syndrome	Schmidt et al. (2012)
Non-indigenous aquatic species	$138 million, annually (2007 US dollars)	Structured expert judgment and cost-benefit forecast	Rothlisberger et al. (2012)
Zebra mussel (*Dreissena polymorpha*) and quagga mussel (*D. bugensis*)	$59–67 million in Columbia River basin, annually	Bioeconomic model (estimating damages and relative sources of risk)	Warziniack et al. (2011)
Hemlock woolly adelgid (*Adelges tsugae*)	$24.6 million, 2007–2011 (7500 km^2 in New England)	Potential loss in residential property value (estimated by value transfer)	Li et al. (2014)
Impacts of Established Species			
Zebra mussel (*Dreissena polymorpha*)	$300–$500 million, annually		Great Lakes Commission (2012)
Burmese python (*Python bivittatus*)	$1.4 million (Key Largo woodrat); $101.2 million (wood stork recovery)	1999–2009, State agency expenditures on recovery for these species to date	U.S. Fish and Wildlife Service (2012)

(continued)

Table 14.1 (continued)

Species	Impact estimate	Background	Author, date
Pre-invasion modeling/risk assessment			
Borers (emerald ash borer, *Agrilus planipennis*)	$760 million (household tree removal); $830 million (property value loss); $130 million (timber loss)	Net present value of damage costs based on estimated tree impacts	Aukema et al. (2011)
Sap feeders (hemlock woolly adelgid, *Adelges tsugae*)	$130 million (household tree removal); $260 million (property value loss); $4.2 million (timber loss)	Net present value of damage costs based on estimated tree impacts	Aukema et al. (2011)
Foliage feeders (gypsy moth, *Lymantria dispar dispar*)	$160 million (household tree removal); $410 million (property value loss); $18 million (timber loss)	Net present value of damage costs based on estimated tree impacts	Aukema et al. (2011)
Noxious weeds	$83.5 million/year (lost income)	Input-output model of Oregon	The Research Group (2014)
Southern pine beetle (*Dendroctonus frontalis*)	$3.7–78.2 million (1982 dollars, Texas)	Price and welfare effects of catastrophic forest damage from southern pine beetle (supply and demand model)	Holmes (1991)
West Nile virus	$2.98 million (California, 2005)	Direct cost estimates of medical costs, productivity loss, vector control	Barber et al. (2010)
Emerald ash borer (*Agrilus planipennis*)	$70 million, annually (Ohio)	Computable general equilibrium (welfare impact)	McDermott et al. (2013)
Leafy spurge (*Euphorbia esula*)	$129.5 million, annually (Montana, North Dakota, South Dakota, Wyoming)	Input-output model	Leitch et al. (1996)
Knapweed (*Centaurea* spp.)	$42 million, annually (1994 dollars)	Input-output model	Hirsch and Leitch (1996)
Saltcedar (*Tamarix* spp.)	$7.33–16 billion, 55 years (1998 dollars)	Water provision replacement cost, farm budget residual method, hydropower replacement cost, river recreation contingent valuation/willingness to pay, avoided flood damages, wildlife contingent valuation/willingness to pay	Zavaleta (2000)
Tropical soda apple (*Solanum viarum*)	$8.1 million, annually (2006 dollars)	IMPLAN model	Salaudeen et al. (2013)
Gypsy moth (*Lymantria dispar dispar*)	$1175/acre	Hedonic price method	Payne et al. (1973)
Cost of Control			
Sea lamprey (*Petromyzon marinus*)	$20 million, annually		Great Lakes Commission (2012)
Nutria (*Myocastor coypus*)	$2 million (2005)	US Fish and Wildlife Service eradication in Maryland	U.S. Fish and Wildlife Service (2012)
Birds	$185–238 million (immediate); $21–29 million (long run)	Partial equilibrium model estimating impacts to total surplus of fruit tree growers partaking in bird management	Elser et al. (2016)
Borers (emerald ash borer, *Agrilus planipennis*)	$92 million (Federal); $1.7 billion (local)	Net present value of control costs based on estimated tree impacts	Aukema et al. (2011)
Sap feeders (hemlock wooly adelgid, *Adelges tsugae*)	$14 million (Federal); $170 million (local)	Net present value of control costs based on estimated tree impacts	Aukema et al. (2011)
Sudden oak death pathogen (*Phytophthora ramorum*)	$0.1 million (local)	Net present value of control costs based on estimated tree impacts	Aukema et al. (2011)

Note: many studies cover multiple stages of invasion
Southern pine beetle is a domestic species but has experienced population explosions and associated damages well beyond its normal range

14.2 Introduction of Invasive Species, Risk Perceptions, and Human Vectors

More than 450 species of non-native forest insects and at least 16 pathogens have been established and detected in the United States since 1860, with approximately 2.5 established non-native forest insects detected per year between 1860 and 2006 (Aukema et al. 2010). Intentional introductions attributed to nurseries, botanical gardens, and private plant enthusiasts are responsible for most introductions of terrestrial plant species into the United States (Reichard and White 2001). Unintentionally introduced species, often called "hitchhikers," arrive on trade and transportation vectors. For example, herbaceous invasive species are often introduced through crop seed contamination (Baker 1986; Mack 1991); aquatic species are often introduced via biofouling and ship ballast water (Baker 1986; Drake and Lodge 2007; Keller et al. 2011); wood-boring insects are often introduced with wood packaging materials and by movement of fuel wood (Barlow et al. 2014; IPPC 2002; Jacobi et al. 2011; Koch et al. 2012; Liebhold and Tobin 2008; McNeely et al. 2001); and non-native plant insects and diseases often arrive on live plant imports (Liebhold et al. 2012). Nearly 70% of damaging forest insects and pathogens established in the United States between 1860 and 2006 most likely entered on imported live plants (Liebhold et al. 2012).

Human-mediated transport facilitates the spread of non-native species populations at rates and distances well beyond what would occur naturally (Blakeslee et al. 2010). Patterns of historical trade and settlement (Brawley et al. 2009), marine trade, road transportation (Bain et al. 2010; Kaluza et al. 2010; Yemshanov et al. 2013), and recent economic and demographic changes (Pysek et al. 2010) have all been linked to the distribution of invasive species. In recent decades, long-distance transport of raw commodities of both domestic and international trade has grown as a key driver of species spread (Aukema et al. 2010; Bain et al. 2010; Pysek et al. 2010; Warziniack et al. 2013), a trend that is expected to continue as the proportional growth of trade volume exceeds rates of economic growth (UNCTAD 2007; WTO 2008).

Non-native insects, pathogens, and other organisms are often inadvertently transported to novel territories in shipping containers and commercial transports where they may become established as ecologically and/or economically harmful invasive species (Hulme et al. 2008; Hulme 2009; Kaluza et al. 2010; Lounibos 2002; Tatem et al. 2006; Westphal et al. 2008). While the rate of accumulation of forest pests has been relatively constant since 1860 (Aukema et al. 2010), changes in trade and phytosanitary practices have likely altered the relative importance of particular pathways. For example, Aukema et al. (2010) found that establishment of wood-borers increased faster than any other insect guild since the 1980s; they attributed this increase to the increased volume of containerized freight and accompanying wood packaging material. The magnitude of economic factors that influence trade flows (and the potential introductions of non-native species) is projected to increase (Pysek et al. 2010). Hopefully, analyses of evolving world trade networks can facilitate the development of new approaches for preventing the movement of non-native species (Banks et al. 2015). Recent analyses have shown the increasing importance of countries such as China and South Korea as world trade hubs (Fagiolo et al. 2010), a trend that is consistent with increasing detections of wood-boring insects originating from Asia. Although it is difficult to demonstrate the effectiveness of international trade policies on the rate of accumulation of forest pests, Lovett et al. (2016) find that rates of introduction to the United States from China of wood-boring species decreased after policies were put in place that require phytosanitary treatment of wood packing material.

Several modeling approaches have been developed that take into account local and long-distance dispersal (due to factors such as transportation networks). For example, gravity models and random utility models have been used to predict invasions when human-mediated dispersal is important (Bossenbroek et al. 2009; Chivers and Leung 2012). Each takes into account distance as well as the attractiveness of alternative locations, and therefore can incorporate differential traffic to each site and its consequences on patterns of spread.

Eliminating all risks of invasive species to a region is usually not possible without significantly affecting local economies, so economic research often focuses on the "right" amount of risk, or the "optimal amount of invaders." Risk of introduction is assessed in relation to the appeal of owning exotic species (e.g., exotic house and landscaping plants, aquarium plants and fish, and exotic pets), the role of trade in economic growth, and gains from trade (Fraser and Cook 2008; Knowler and Barbier 2005; Warziniack et al. 2013). When considering protective regulations, agencies face the possibility of making Type I (false positive) and Type II (false negative) errors, which can lead to either over-regulation or under-regulation, respectively. The challenges associated with quantifying the costs of Type II errors, in combination with political influences (Simberloff 2005), may cause biosecurity agencies to focus on minimizing the costs associated with Type I errors (e.g., management costs) while neglecting the potential for economic damage (Davidson et al. 2015).

By controlling the vectors of introduction or influencing the composition of goods produced in a region, managers can affect exposure (Tu et al. 2008), and thus should an invasion occur, people can adapt to environmental changes (Settle and Shogren 2004). Not only does the environment respond to human activity, but human activity also responds to environmental conditions (Finnoff et al. 2005; Merel and Carter 2008; Shogren 2000). Damage estimates should be sensitive to the fact that people can adjust their behavior both pre and post invasion. For example, Finnoff et al. (2010) proposed an endogenous risk framework in which probability of a species' presence in the transportation network depends on prevention choices. Should an invasive species enter the transportation network, managers can try to either eradicate the species or control it to reduce severity of damages. Should all efforts prove either ineffective or too costly, society can limit damages through adaptation.

14.3 Establishment of Species in an Area and Measurement of Damages

The presence of harmful non-native organisms causes damages to economically valuable host resources and negatively affects the state of native ecosystems and economically important crops. Assessing economic risks entails a valuation of economic consequences and impacts from an introduction and spread of non-native organisms. The severity of economic damages may justify the establishment of quarantine and other regulatory actions aimed at containing the invading populations or, if containment is no longer possible, at slowing the rate of spread. This section discusses the state of the science for valuation methods used in those decisions.

Partial Budgeting, Replacement Costs, and Costs of Control Partial budgeting helps evaluate the economic consequences of small adjustments in production (such as agricultural crop production) and is based on the principle that a small change in production may reduce some costs and revenues while adding other costs and revenues (Soliman et al. 2010). Partial budgeting methods generally focus on the net decrease or increase in income resulting from a change in production. The method requires a relatively modest amount of data and personnel time (Holland 2007); accordingly, it has been widely used to assess the economic impacts of agricultural and forest pests (FAO IPPC 2004; Macleod et al. 2003). While the analyses can be scaled up to the national level (see Breukers et al. 2008; Macleod et al. 2003), the method cannot measure multi-sectorial impacts because it relies on fixed budgets with defined prices to describe the economic activities of the firm or enterprise (Soliman et al. 2010).

Using replacement and control costs to estimate the economic impact of an invasive species is also a relatively straightforward and easy way to interpret and measure damages. For example, to estimate the economic impact of emerald ash borer (EAB), *Agrilus planipennis*, Kovacs et al. (2010) estimated the discounted cost of treatment, removal, and replacement of landscape ash (*Fraxinus* spp.) trees on developed land within communities in a 25-State study area centered on Detroit using simulations of EAB spread and infestation over the next decade (2009–2019). An estimated 38 million ash trees exist on this land base. The simulations predicted an expanding EAB infestation that will likely encompass most of the 25 States and warrant treatment, removal, and replacement of more than 17 million ash trees with a discounted cost equal to $10.7 billion. Note that replacement and control costs address only one side of the cost-benefit analysis; they do not determine whether or not those costs are worth incurring.

Single-Industry Impacts (Partial Equilibrium Models) Partial equilibrium modeling represents another common assessment technique, especially useful when an invasion is expected to change the producers' surplus or consumers' demand value (Mas-Colell et al. 1995). The methodology evaluates the welfare effects on participants in a market that is affected by an introduction of a harmful non-native species. The approach defines relationships for supply and demand for the commodity of interest (such as agricultural or forest commodities that may be negatively affected by the introduction of the invader) to determine the final combination of prices and quantities that leads to a market equilibrium (Mas-Colell et al. 1995).

As shown in Box 14.1 on partial equilibrium impacts, such models estimate the aggregate impact of a non-native species by measuring differences in equilibrium price and quantity and changes in welfare before and after the introduction. Introduction of harmful organisms may lead to an increase in the production costs and a decrease in the quantity (or quality) of a susceptible host resource (such as valuable crops or forest tree species), which also affects the supply curve and the equilibrium price. Changes in welfare are estimated from the aggregated changes in producers' and consumers' welfare (Just et al. 1982). Partial equilibrium models have been used widely as policy assessment tools in agriculture, forestry, and trade (Cook 2008; Elobeld and Beghin 2006; Holmes 1991; Kaye-Blake et al. 2008; Qaim and Traxler 2005; Schmitz et al. 2008), for risk assessments of quarantine pests (Arthur 2006; Breukers et al. 2008; Surkov et al. 2009), and to evaluate changes in exports and access to markets (Cook 2008; Elliston et al. 2005; Julia et al. 2007).

Box 14.1: Partial Equilibrium Model of Invasive Species Impacts

Consider the following example, adapted from Arthur (2006), that looks at the trade-off between gains from trade to Australian apple consumers and damages from an invasive apple blight. Without trade, the domestic production of apples in Australia is Q_0, and Australian consumers pay P_0. Pre-trade welfare is measured by the sum of consumer surplus *(triangle XP_0Y, the area below the demand curve D but above the price*) and producer surplus (ZP_0Y, *the area above the initial supply curve S_0 but below the price*). Opening the market to trade allows consumers to buy apples at world price P_W, which increases consumption to Q_1 and decreases domestic production to Q_2. Producer surplus falls to P_WTZ, but consumer surplus increases to XP_WU. *The shaded triangle marked GAINS* represents the increase in welfare from trade. Trade, however, also brings potential damages from apple blight. Prevention measures and crop damages increase costs of production to Australian growers, causing the domestic supply curve to shift to S_{INV} and domestic production to fall to Q_{INV}. Consumer surplus is not affected, but producer surplus shrinks. *The shaded area marked LOSS* shows the loss to Australian apple growers from the invasion. The total impact of trade, accounting for losses of invasion, is GAINS – LOSS, which could be either positive or negative.

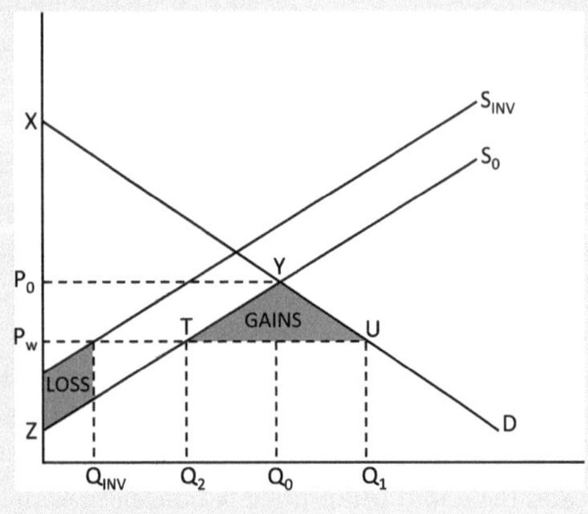

Economy-Wide Impacts (Input-Output and Computable General Equilibrium Models) Input-output and computable general equilibrium (CGE) models are used when impacts from invasive species are likely to affect multiple sectors of the economy, when indirect effects are likely due to impacts on factors of production, or when income effects are likely to be large. Input-output analysis elucidates the interdependencies of sectors in an economy and predicts an economy-wide impact of changes within a particular sector (Leontief 1986). Input-output analysis also requires a description of the monetary flows of inputs and outputs among the productive sectors of an economy (Miller and Blair 1985). Changes in product demands in a sector generate effects on the economy as a whole and cause direct changes in the purchasing policies of the affected sector. The suppliers of the affected sector must change their purchasing policies in order to satisfy the changed demands, and so on. Input-output analysis can estimate the impact of an invasive species on an economy by adjusting the final demand in the affected sector (such as agriculture or forestry) in response to the expected changes in demand (such as decrease in the production of agricultural commodities or reduction in exports) (Elliston et al. 2005; Julia et al. 2007). Overall, the approach helps measure short-term impacts across broad sectors of the economy.

CGE models are composed of sets of equations that specify demand, production, and interactions between domestic production and imports, prices, and other equilibrium conditions. CGE models are similar in flavor to input-output models, but they place more emphasis on the behavioral equations that underlie the economic system and allow for price adjustments. CGE helps assess economy-wide impacts across sectors and regions, and considers long-term consequences. CGE models are also appropriate when assessing impacts of trade restrictions due to invasive species policy and when agents in the economy can substitute away from sectors of the economy impacted by the invasion (Warziniack et al. 2011). Using the invasive emerald ash borer as an example, McDermott et al. (2013) developed a CGE model for the State of Ohio and estimated annual damages from the beetle to be about $70 million. The majority of this damage ($57 million) is incurred by the parks and recreation sectors, households, and State government. The parks and recreation sectors must add the costs of removing infested ash trees to their primary production costs. Households must reduce their disposable income by the cost of ash removal while these expenditures flow to the garden sector as an increase in demand for their services. The government must make revenue adjustments to account for ash removal with those expenditures moving to the garden sector.

Impacts on Non-market Values and Ecosystem Services Some more difficult-to-assess risks include impacts on social infrastructure, recreational use (such as fishing), existence values of native species threatened by invasive species, aesthetics, and factors associated with human health (such as water quality). The value of damages and impacts on these ecosystem services is more difficult to measure because these services are not traded in markets and therefore do not have observable prices. Thus, economists estimate the value of changes in non-market ecosystem services by leveraging the information conveyed by individuals' observable decisions. Information obtained from observable decisions in hypothetical markets created by the analyst is known as stated preference data. In contrast, revealed preference data are obtained from observable choices concerning "consumption" of non-market ecosystem services such as where to recreate, how votes on ballot referenda might influence non-market ecosystem services, and how people behave in markets for a weak complement to the non-market ecosystem service. In such cases, the choices and trade-offs people make reflect their willingness to pay to access or obtain ecosystem services.

Stated and revealed preference methods used to estimate the value of non-market ecosystem services are reviewed by Ninan (2014) and Binder et al. (2016). Applications to estimate people's willingness to pay for forest insect control programs that reduce insect-related damage to non-market forest ecosystem services are reviewed in Rosenberger et al. (2012). New and promising areas of research extending non-market valuation methods to the suite of ecosystem services provided by natural areas are discussed in Charles and Dukes (2007) and Boyd et al. (2013). Two key examples of revealed preference techniques include hedonic pricing models and travel costs methods. Hedonic price studies look at the effect of an invasion on the value or market price of a closely related good—most often housing prices. Travel cost methods use expenditures people incur to visit a location (most often a recreation site) as a proxy for willingness to pay for that visit. Olden and Tamayo (2014), for example, used a hedonic model to measure damages from Eurasian milfoil (*Myriophyllum spicatum*) in King County, WA. They found the presence of Eurasian milfoil decreases the value of homes near invaded lakes by $94,000, or about 14%. Using similar methods, Horsch and Lewis (2009) found Eurasian milfoil decreases home values by about 13% on invaded lakes in Wisconsin. Nunes and van den Bergh (2004) used travel cost methods to measure the impact of algal blooms caused by non-native species along Dutch beaches. They found management actions required to reduce algal blooms would be worth about 225 million euros to area residents and visitors.

Information Needs of Various Types of Impact Studies The assessment of economic impacts from non-native invasive species often initiates from qualitative estimates based on expert judgments (Brunel et al. 2009; Sansford 2002; Soliman et al. 2010). Expert judgments are used because of very low costs and availability of expert knowledge, but often lack transparency and rigor (Sansford 2002). From there, the application of partial budgeting, partial equilibrium, input-output analysis, and CGE is often dictated by the goal of the study, the methodology used, and the level of detail available (see Dixon and Parmenter 1996; Holland 2007; Miller and Blair 1985). Partial budgeting is better suited to estimate immediate impacts of invasive species introductions, whereas partial equilibrium models can provide insights on the changes in the production volumes and effects on commodity prices that may be affected by the introductions. Partial equilibrium models can also include many sectors so that the spillover effects between sectors can be analyzed. This method, however, requires defining the structure of the affected markets and the level of homogeneity for products from exogenous markets, and may require large amounts of data (Baker et al. 2009; Rich et al. 2005). If nationwide economic impacts or multi-sectorial effects are expected, then input-output analysis or CGE would be an appropriate choice because they each recognize the feedback loops that exist within the economy and address behavioral complexities that many of the other methods cannot deal with. However, input-output analysis and CGE also require a large amount of data and computational expertise.

14.4 Optimal Policies and Strategies

Biological invasions usually proceed in stages where each stage is associated with one or more management actions and a vector of economic costs and damages (Fig. 14.1). Economic analysis proceeds by seeking efficient strategies either within a stage (partial analysis) or across stages (global analysis). This section identifies research on prevention and control strategies, as well as factors such as risk and uncertainty that make designing an optimal policy extremely difficult.

Preventing Arrival and Introduction Prevention policies focus on trade vectors and optimal inspection rates and must balance costs of policy, risk of introduction, and gains from trade (Chen et al. 2018; Leung et al. 2002; McAusland and Costello 2004). International trade is the major pathway for the introduction of non-native forest pests (Liebhold et al. 1995), and the importation of live plants is the most probable pathway of introduction for most damaging forest insects and pathogens established in the United States (Liebhold et al. 2012). Wood packing materials are the most common pathway of introduction for wood-boring forest insects, and the rapid acceleration in the use of these materials over the past decade is an increasing concern (Aukema et al. 2010; Strutt et al. 2013).

A general economic strategy for preventing the introduction of invasive species is to internalize the costs of biological invasions using tariffs in combination with improved port inspections (Perrings et al. 2005). Economic optimization suggests that the importing country should set the tariff equal to the sum of expected damages from contaminated units not detected during inspections plus the costs of inspections (McAusland and Costello 2004). When it is possible to estimate the probability of a successful invasion, each biosecurity

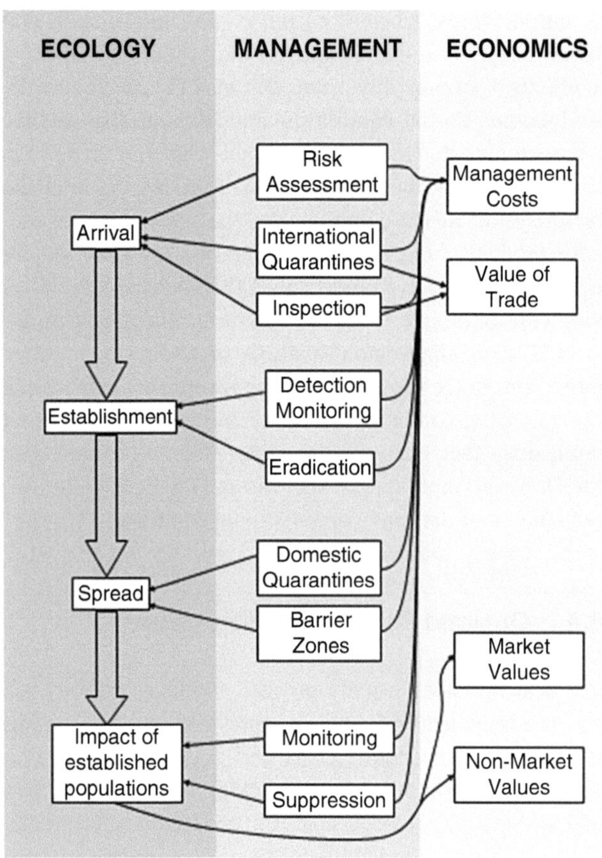

Fig. 14.1 Stages of invasion and associated management activities (from Holmes et al. 2014). Ecological sciences generally focus on species' arrival, establishment, spread, and ecological impacts, which are affected by management actions. Management actions determine economic outcomes, and economic values should inform management decisions

facility should optimally set the marginal cost of undertaking preventive measures equal to marginal expected benefits (damages avoided), taking into account the probability that a species might invade through a different facility (Horan et al. 2002).

Inspection of shipments at ports of entry is one approach to reduce the introduction of invasive pests. For example, inspection of live plant imports is a prominent component of the system used by the US Department of Agriculture, Animal and Plant Health Inspection Service (USDA APHIS) Plant Protection and Quarantine program to protect US agriculture and natural resources from unwanted and damaging pests (Venette et al. 2002). Inspectors examine a number of selected plant units from each incoming shipment. If a regulated pest or pathogen is detected in the sample units, inspectors may require that the shipment be treated, returned, or destroyed. Inspection strategies have been developed to allocate a fixed inspection budget among shipments to minimize the expected number or cost of accepted infested shipments (Springborn 2014; Surkov et al. 2009). Alternatively, the inspection budget can be allocated among shipments to minimize the expected number or cost of infested plant units in accepted shipments (slippage) (Chen et al. 2018; Yamamura et al. 2016). Sampling strategies that minimize expected slippage instruct inspectors to focus on larger shipments with higher plant infestation rates, while strategies that minimize the number of accepted infested shipments allocate sampling effort to shipments with higher infestation rates with less regard to shipment size. For live plant import inspections, optimization, based on the number of accepted infested plants, is most relevant because the number of introductions of a pest into the environment is a key predictor of establishment.

Most of the analysis on trade focuses on international imports and trade with foreign partners. However, of the *100 of the World's Worst Invasive Alien Species* listed in the Global Invasive Species Database (Lowe et al. 2000), 86 species have already been introduced to the United States or are increasing their range within the United States, 7 species are indigenous or non-threatening to other areas of the United States, and only 7 species have not been introduced. Surprisingly, few studies acknowledge how this change in perspective has affected optimal policies and methods for measuring impacts. Warziniack et al. (2013), for example, demonstrated that correcting the externality with a tax on the risky vector is virtually impossible when hitchhiking species are linked to tourism and the incidence of private and recreational vehicles coming into the area.

Surveillance and Eradication to Prevent Establishment The probability of successful establishment depends on the frequency and size of arrivals (propagule pressure), spatial habitat suitability, and temporal environmental fluctuations (Leung et al. 2004; Von Holle and Simberloff 2005), all of which are highly uncertain. Most preventative strategies are based on reducing propagule pressure, which is a measure of the expected number of individuals (e.g., the number of fecund adults of the species of interest) reaching an uninvaded location and is commonly expressed in terms of the rate, probability, or likelihood of arrival (Johnston et al. 2009; Simberloff 2009). However, if new species are repeatedly introduced through similar or novel invasion pathways, Allee effects and stochastic population dynamics are much less likely to cause initial populations to go extinct, thereby increasing the likelihood that isolated populations become established.[1]

[1]An Allee effect, as defined by Drake and Kramer (2011), "is a positive association between absolute average individual fitness and population size over some finite interval." In some cases, Allee effects imply a minimum population size necessary for a species to become established.

Surveillance systems designed to detect newly established species that evade port inspections are critical to reducing the potential for ecological and economic damage (Lodge et al. 2006). Cost-effective surveillance systems for newly established populations balance the intensity and cost of surveillance (which increase with the level of effort) with the costs of damage and eradication of newly detected populations (which may be less if detected early) (Epanchin-Niell and Hastings 2010). Economic models that account for this trade-off have assumed the pest location is unknown (Mehta et al. 2007); rates of pest establishment, spread, and damage vary across locales (Epanchin-Niell et al. 2012, 2014); small invasive populations establish ahead of an advancing front (Homans and Horie 2011); or that the likelihood of detection increases with the size of an infestation (Bogich et al. 2008). Research efforts have also focused on the properties of optimal one-time surveillance across multiple sites when species' presence is uncertain prior to detection, accounting for heterogeneity in species occurrence, probability, and detectability across sites (Hauser and McCarthy 2009). Other models of one-time surveillance have investigated the impact of uncertainty regarding the extent (rather than simply the presence) of an infestation (Horie et al. 2013) and to maximize the coverage of the locations from where an invasive species is likely to spread to the uninvaded area (Yemshanov et al. 2015).

Economic models of long-term surveillance programs with constant surveillance effort have been developed using optimization algorithms and indicate that greater surveillance effort is warranted in locations that have higher establishment rates, higher damage and eradication costs, or lower sampling costs (Epanchin-Niell et al. 2012, 2014). In applying their model to the design of an optimal surveillance program for gypsy moth (*Lymantria dispar dispar*) detection and eradication in California, Epanchin-Niell et al. (2012) found that California's 2010 county-level trapping densities correspond closely to the optimal trapping policy derived from the model; however, reductions in trapping densities in just 3 counties (out of 58 counties) might reduce long-term costs of surveillance and eradication in California by up to 30%. Using optimal control theory to calculate time-dependent surveillance policies that minimize the total cost of sampling, eradication, and damage by an invasive, Holden et al. (2016) developed rules of thumb to determine when intense initial sampling, followed by a sharp decrease in sampling effort, is more cost-effective than strategies that are constant through time. For invaders with high rates of establishment from an outside source, constant effort surveillance strategies are cost-effective. However, when reintroductions are infrequent, an intense early search for the invader can drastically reduce costs, depending on initial pest prevalence and the economic benefit-to-cost ratio of sampling.

Active research is currently underway to develop optimal surveillance and eradication policies when there is uncertainty about invasion dynamics and detectability. This line of research recognizes that surveillance may not provide accurate information, and therefore researchers have used partially observable Markov decision processes to address optimal invasive species surveillance (Regan et al. 2007), monitoring, and control strategies (Haight and Polasky 2010). More generally, partially observable decision models have been used to allocate management resources for networks of cryptic diseases, pests, and threatened species (Chadès et al. 2011).

Optimal Control to Slow the Spread When a non-native species becomes established, various strategies can be used to reduce the expansion of its range, including initiating a domestic quarantine to reduce the chances of accidental movement of organisms to uninfested areas, detecting and eradicating isolated colonies, or applying control treatments to slow or stop the spread of the core population. Research has focused on developing optimal control strategies for slowing or eradicating populations and addressing questions such as when, where, and how much control should be applied (see Epanchin-Niell and Hastings 2010 for a review).

Invasive species control models generally include pest population dynamics and a stated objective of minimizing the sum of discounted control costs and invasion damages over time. The most basic models of invasive species dynamics focus on the numbers of individuals or the area of infestation and ignore spatial description (Eiswerth and Johnson 2002; Saphores and Shogren 2005; Sharov and Liebhold 1998). A general principle emerging from this research is that, if the invasive species stock is initially greater than its optimal equilibrium level, the highest level of management effort should be applied initially, and then should decline over time until the steady state is reached (Eiswerth and Johnson 2002). When the goal is to control the population front, the optimal strategy changes from eradication to slowing the spread to doing nothing; as the initial area occupied by the species increases, the negative impact of the pest per unit area decreases or the discount rate increases. Preventing population spread is not viewed as an optimal strategy unless natural barriers to population spread exist (e.g., Sharov and Liebhold 1998). These basic population models have been extended to account for uncertainty in invasion growth. The optimal control strategy is obtained using discrete-time stochastic dynamic programming (Eiswerth and van Kooten 2002; Olson and Roy 2002) or a real options framework in continuous time (Marten and Moore 2011; Saphores and Shogren 2005).

Policy is also complicated by politics—the political scales of policies rarely match the ecological scales of invasions. Effective management depends in part on coordination across jurisdictions, heterogeneous landscapes, heterogeneous populations (Epanchin-Niell et al. 2012), and international borders (Gren et al. 2010; Knowler and Barbier 2005; Tu et al. 2005). When bio-invasions occur at landscape scales and with multiple landowners, each landowner's control decisions can impact their neighbors' decisions by affecting invasion spread across boundaries (Epanchin-Niell et al. 2010; Epanchin-Niell and Wilen 2015; Wilen 2007). When landholders make control decisions based only on damages occurring on their own land, an externality occurs because those landholders taking action confer uncompensated benefits to those in advance of the invading front (Wilen 2007). As a result, managers may under-control from a systemwide perspective, leading to increased invasion of the landscape (Wilen 2007). Decision makers responsible for controlling a bio-invasion can internalize this diffusion externality and increase total net benefits across ownerships (e.g., Bhat and Huffaker 2007; Feder and Regev 1975; Richards et al. 2010; Sims et al. 2010).

Recently, spatially explicit models of invasive species dynamics and control have been developed for invasive plants (Blackwood et al. 2010; Büyüktahtakın et al. 2015), reptiles (Kaiser and Burnett 2010), insects (Sims et al. 2010; Kovacs et al. 2014), and generic pests (Epanchin-Niell and Wilen 2012; Hof 1998). All of the aforementioned models define the landscape as a set of discrete patches or map cells, define control activities for each patch, and predict the growth and dispersal of the invasive species among patches as a function of the selected controls. Spatial-dynamic models use a variety of models of pest population dynamics, including pest occupancy (Epanchin-Niell and Wilen 2012), pest population size (Blackwood et al. 2010; Büyüktahtakın et al. 2015; Hof 1998; Kaiser and Burnett 2010), and host and pest population sizes (Kovacs et al. 2014; Sims et al. 2010), depending on the characteristics of the ecological system.

Although these spatial-dynamic models are complicated to solve, they can provide pragmatic guidance to forest managers. For example, Kovacs et al. (2014) developed a spatial-dynamic model for the optimal control of EAB in the Twin Cities metropolitan area of Minnesota. They focused on managing valuable host trees by applying preventative insecticide treatment or pre-emptively removing infested trees to slow EAB spread. The model incorporates spatial variation in the ownership and benefits of host trees, the costs of management, and the budgets of municipal jurisdictions. The authors developed and evaluated centralized strategies for 17 jurisdictions surrounding the infestation. The central planner determines the quantities of trees in public ownership to treat and remove over time to maximize benefits associated with net costs of managing surviving trees across public and private ownerships, subject to constraints on municipal budgets, management activities, and access to private trees. The results suggest that centralizing the budget across jurisdictions, rather than increasing any one municipal budget, does more to increase total net benefits. Further, strategies incorporating insecticide treatments are superior to those with pre-emptive removal because they reduce the quantity of susceptible trees at lower cost and protect the benefits of healthy trees. Finally, increasing the accessibility of private trees to public management substantially slows EAB spread and improves total net benefits. The change from local to centralized control increased the percentage of healthy trees remaining on the landscape by 18% and more than doubled the total net benefits.

Much of the literature on invasive species management in multi-ownership landscapes examines two polar cases characterizing control choices. In the first case, myopic landowners choose their own control without considering the impact of their choices on the probability that other landowners' at-risk lands will become infested. In the second case, a social planner is assumed to control actions for landowners at a landscape level, thus internalizing all externalities by choosing controls that maximize social welfare. Epanchin-Niell and Wilen (2015) point out that this dichotomy fails to capture the often-observed case in which landowners cooperate with other managers to control the spread of an invasive species by engaging in an invasive species cooperative control district. The classical way to boost cooperation has been addressed in the economics literature is through a bargaining mechanism such as a transfer payment. For example, Cobourn et al. (2016) developed a theory of cooperation for invasive species management using an axiomatic Nash bargaining game assuming that the threat exists for an invasive species to spread from an infested to an uninfested municipality. Without bargaining, the infested municipality chooses control efforts that maximize its own benefits and likely invests too little from a social perspective because its choice of control influences the probability that the invasive species will spread to the uninfested municipality. Therein lies the potential for bargaining: the uninfested municipality has an incentive to bargain with the infested municipality to share in the latter's control costs, using a transfer payment in exchange for applying a higher level of control that would effectively reduce the probability that the invasive species will spread. Cobourn et al. (2016) calibrate their bargaining model to represent the emerald ash borer invasion in the Twin Cities of Minneapolis and St. Paul, MN. Their results suggest that bargaining improves the public benefits across communities relative to the case without bargaining. Further, bargaining may achieve the social planner's optimal level of control when the uninfested municipality possesses a substantial advantage in terms of relative bargaining power. Short-term bargaining agreements are unlikely to succeed, suggesting that there may be a role for a higher government involvement to facilitate long-term bargaining agreements.

In many situations, land managers specify a desired outcome in terms of ecosystem attributes, such as species composition, vegetation structure, pest population size, or likelihood of pest occurrence, striving for a management strategy that achieves these attributes at least cost. A manager may have many mutually exclusive least-cost projects to select for investment given a limited budget. Conservation priorities are generally made with an eye solely on benefits of management actions, largely ignoring costs (Brooks et al. 2006; Groves et al. 2002). A more thorough (and efficient) method would use return-on-investment (ROI) as a decision criterion for prioritizing projects, making use of both benefit and cost data. As discussed in Polasky (2008), a number of studies have shown that, for a wide range of conservation objectives, more variability exists in costs of land management options than exists in the ecological benefits. Bode et al. (2008), for example, used seven different taxonomic measures of biodiversity to allocate funding among 34 of the world's terrestrial biodiversity hotspots with the objective of minimizing total species loss. They found the optimal decision was far less dependent on the measure of biodiversity than it was on cost of conservation. Similar studies using ROI as a decision criterion have been conducted for ecological restoration in Hawaii (Goldstein et al. 2008), temperate forests in North America and Mediterranean ecoregions (Murdoch et al. 2007), and Mediterranean forests, woodlands, and scrub (Underwood et al. 2008). The ROI criterion could be used as a decision criterion for prioritizing invasive species management projects.

Optimal Control and Risk Across Invasion Stages Economic models that focus on a single stage of the invasion process cannot provide globally optimal solutions because they ignore potential trade-offs among defensive actions across the stages of an invasion. Optimal allocation among prevention and control depends on the nature of prevention and control cost curves and the decision maker's preferences over risky events. Research has shown that, under some conditions, invasive species can be managed most cost-effectively using greater investments in prevention relative to control because damages can be catastrophic (Leung et al. 2002). Other research has shown that, if decision makers are risk averse and if control options are thought to be more certain than prevention, then control may be preferred to prevention (Finnoff et al. 2007). Recent innovations in the analysis of trade-offs among invasion stages include the development of spatial models of prevention, detection, and control (Sanchirico et al. 2010). Such interdependencies between prevention and control are highlighted in Burnett et al. (2006), with examples from Hawaii for both a current invader (miconia, *Miconia calvescens*) and a potential invader (brown tree snake, *Boiga irregularis*). The primary lesson is that focusing on a subset of transmission pathways, on only one or two controls, or on a single region ignores important interactions that are critical in identifying cost-effective policy recommendations.

Investments in prevention should also be made recognizing the ability to detect the invader in the environment and the ability to control or eradicate it should it become established (Haight and Polasky 2010; Mehta et al. 2007). Studies such as Homans and Horie (2011) and Epanchin-Niell et al. (2014) consider optimal control decisions post detection when determining optimal levels of investment in surveillance.

Special methods have also been introduced to develop optimal policies to manage invasive species in the presence of uncertainty (Eiswerth and Johnson 2002; Haight and Polasky 2010; Hester and Cacho 2012; Horie et al. 2013; Hyytiainen et al. 2013; Olson and Roy 2002). Adding the dimensions of risk and uncertainty requires that decision makers consider their perceptions of risk, such as risk aversion (Olson and Roy 2005). Often, risk of invasion is treated in optimal management policies as exogenous (Leung et al. 2002; Ranjan et al. 2008), with fewer attempts to represent the risk of invasion as endogenous (Finnoff and Shogren 2004).

14.5 Gaps and Future Research Needs

Invasive species science has largely remained in the domain of natural sciences. Greater research in economics and other social sciences could help to better integrate governance and management policy, address the objectives of multiple stakeholders, account for risk perceptions, and promote bargaining and cooperative behavior among decision makers. To date, there has not been a comprehensive investigation of impacts from terrestrial and aquatic invasive species, including the full value of ecosystem services lost. Acquiring such data is necessary for conducting cost-benefit analyses. This omission prevents policymakers from establishing a meaningful list of priorities and realistic management strategies. Table 14.1 gives a few representative damage estimates from the literature, focusing primarily on local and regional studies. Estimates of damages from invasive species at national (or even global) scales usually combine values from several studies or generalize across diverse landscapes, invaders, and impact methods. While the large impact numbers such studies generate are popular with policymakers and scientists looking to emphasize the importance of their research problem, they violate some of the most basic rules of economic analysis and generally do more harm than good to the science.

Uncertainty about invasive species remains a serious challenge in the development of effective control and management policies, and will require special analytic and modeling tools to factor uncertainty into optimal management policies. Another important issue that remains to be addressed is the practical validation of optimal management policies that have been developed. While many countries have introduced strategies to reduce the rates of non-native species introductions, such as sanitary and phytosanitary policies that regulate the movement of pest-associated commodities, more efforts will be required to assess the practical utility and transaction costs of implementing those measures.

14.6 Key Findings

- Changes in trade and phytosanitary practices have altered the relative risk of species introductions and the importance of particular pathways. Introductions of woodborers increased faster than any other insect guild since the 1980s due to the increased volume of containerized freight and accompanying wood packaging material (Aukema et al. 2010).
- For live plant import inspections, optimizing based on the number of accepted infested plants is most relevant because the number of introductions of a pest into the environment is a key predictor of establishment. This optimization results in strategies that allocate limited sampling resources to larger shipments that have higher infestation rates (Chen et al. 2018).
- For invaders with high rates of establishment from an outside source, constant effort surveillance strategies are cost-effective (Epanchin-Niell et al. 2012, 2014). However, when reintroductions are infrequent, an intense early search for the invader can drastically reduce costs, depending on initial pest prevalence and the economic benefit-to-cost ratio of sampling (Holden et al. 2016).
- When landholders make control decisions based only on damages occurring on their own land, an externality occurs because controllers confer uncompensated benefits to those in advance of the invading front (Epanchin-Niell and Wilen 2015). This externality creates an incentive for landowners to cooperate in the cross-boundary control of invading populations. One mechanism for furthering cooperation is bargaining for a transfer payment from an invasion-free landowner to fund increased control by an invaded landowner (Cobourn et al. 2016).

14.7 Key Information Needs

- Expanding applications of non-market valuation methods to address the suite of ecosystem services provided by natural areas that are at risk because of invasive species would facilitate more comprehensive cost-benefit analyses.
- Developing models that account for cooperative management among landowners and across policy jurisdictions would help to address behavioral interactions among landowners and jurisdictions.
- Conservation priorities are often determined solely on the basis of the benefits of proposed management actions, not including costs. A more thorough analysis, such as ROI, would employ both benefit and cost data, when decision criterion for prioritizing projects is desired.
- Economic models that focus on a single stage of the invasion process cannot provide globally optimal solutions because they ignore potential trade-offs among defensive actions across the stages of an invasion. Ideally, economic models should consider such trade-offs to provide a fuller accounting of invasion economic effects.
- Effective control and management depends on model improvement to account for uncertainty surrounding impacts and the probability of introductions.
- Additional studies are needed to investigate the spread of invasive species through domestic trade, and how policies may differ between foreign and domestic sources of risk.

Disclaimer Text The findings and conclusions in this publication are those of the authors and should not be construed to represent any official USDA or U.S. Government determination or policy.

Literature Cited

Anderson AM, Shwiff SS, Shwiff SA (2014) Economic impact of the potential spread of vampire bats into South Texas. USDA National Wildlife Research Center – Staff Publications. 1762

Arthur M (2006) An economic analysis of quarantine: the economics of Australian's ban on New Zealand apple imports. In: Proceedings of the New Zealand agricultural and resource economics society annual conference, Nelson. 2425

Aukema JE, McCullough DG, Von Holle B et al (2010) Historical accumulation of nonindigenous forest pests in the continental United States. Bioscience 60(11):886–897

Aukema JE, Leung B, Kovacs K et al (2011) Economic impacts of non-native forest insects in the continental United States. PLoS One 6(9):e24587

Bain MB, Cornwell ER, Hope KM et al (2010) Distribution of an invasive aquatic pathogen (viral hemorrhagic septicemia virus) in the Great Lakes and its relationship to shipping. PLoS One 5:e10156

Baker HG (1986) Patterns of plant invasion in North America. In: Ecology of biological invasions of North America and Hawaii. Springer, New York, pp 44–57

Baker RHA, Battisti A, Bremmer J et al (2009) PRATIQUE: a research project to enhance pest risk analysis techniques in the European Union. EPPO Bull 39:87–93
Banks NC, Paini DR, Bayliss KL, Hodda M (2015) The role of global trade and transport network topology in human-mediated dispersal of alien species. Ecol Lett 18:188–199
Barber LM, Schleier JJ, Peterson RKD (2010) Economic cost analysis of West Nile virus outbreak, Sacramento County, California, USA, 2005. Emerg Infect Dis 16(3):480–486
Barlow L-A, Cecile J, Bauch CT, Anand M (2014) Modelling interactions between forest pest invasions and human decisions regarding firewood transport restrictions. PLoS One 9:e90511
Bhat MG, Huffaker RG (2007) Management of a transboundary wildlife population: a self-enforcing cooperative agreement with renegotiation and variable transfer payments. J Environ Econ Manag 53(1):54–67
Binder S, Haight RG, Polasky S et al (2016) Assessment and valuation of forest ecosystem services: state of the science review, Gen. Tech. Rep. NRS-170. U.S. Department of Agriculture Forest Service, Northern Research Station, pp 1–47
Blackwood J, Hastings A, Costello C (2010) Cost-effective management of invasive species using linear-quadratic control. Ecol Econ 69:519–527
Blakeslee AMH, McKenzie CH, Darling JA et al (2010) A hitchhiker's guide to the Maritimes: anthropogenic transport facilitates long-distance dispersal of an invasive marine crab to Newfoundland. Divers Distrib 16:879–891
Bode M, Wilson KA, Brooks TM et al (2008) Cost-effective global conservation spending is robust to taxonomic group. Proc Natl Acad Sci 105(17):6498–6501
Bogich TL, Liebhold AM, Shea K (2008) To sample or eradicate? A cost minimization model for monitoring and managing an invasive species. Journal of Applied Ecology 45:1134–1142
Bossenbroek JM, Finnoff DC, Shogren JF, Warziniack TW (2009) Advances in ecological and economic analyses of invasive species: dreissenid mussels as a case study. In: Bioeconomics of invasive species: integrating ecology, economics, policy, and management. Oxford University Press, New York, pp 244–265
Boyd IL, Freer-Smith PH, Gilligan CA, Godfray HCJ (2013) The consequences of tree pests and diseases for ecosystem services. Science 342:1235773
Brawley SH, Coyer JA, Blakeslee AMH et al (2009) Historical invasions of the intertidal zone of Atlantic North America associated with distinctive patterns of trade and emigration. Proc Natl Acad Sci 106:8239–8244
Breukers A, Mourits M, van der Werf W, Lansink OA (2008) Costs and benefits of controlling quarantine diseases: a bio-economic modeling approach. Agric Econ 38:137–149
Brooks TM, Mittermeier RA, da Fonseca GAB et al (2006) Global biodiversity conservation priorities. Science 313(5783):58–61
Brunel S, Petter F, Fernandez-Galiano E, Smith I (2009) Approach of the European and Mediterranean plant protection organization to the evaluation and management of risks presented by invasive alien plants. In: Inderjit (ed) Management of invasive weeds. Springer, Dordrecht, pp 319–343
Burnett K, Kaiser B, Pitafi BA, Roumasset J (2006) Prevention, eradication, and containment of invasive species: illustrations from Hawaii. Agric Resour Econ Rev 35(1):63
Büyüktahtakın IE, Kibis EY, Cobuloglu HI et al (2015) An age-structured bio-economic model of invasive species management: insights and strategies for optimal control. Biol Invasions 17:2545–2563
Chadès I, Martin TG, Nicol S et al (2011) General rules for managing and surveying networks of pests, diseases, and endangered species. Proc Natl Acad U S A 108(20):8323–8328
Charles H, Dukes JS (2007) Impacts of invasive species on ecosystem services. In: Biological invasions. Springer, Berlin/Heidelberg, pp 217–237
Chen C, Epanchin-Niell RS, Haight RG (2018) Optimal inspection of imports to prevent invasive pest introduction. Risk Anal 38(3):603–619
Chivers C, Leung B (2012) Predicting invasions: alternative models of human-mediated dispersal and interactions between dispersal network structure and Allee effects. J Appl Ecol 49:1113–1123
Cobourn KM, Amacher GS, Haight RG (2016) Cooperative management of invasive species: a dynamic Nash bargaining approach. Environ Resour Econ:1–28
Cook DC (2008) Benefit cost analysis of an import access request. Food Policy 33(3):277–285
Cozzens T, Gebhardt K, Shwiff S et al (2010) Modeling the economic impact of feral swine-transmitted foot-and-mouth disease: a case study from Missouri. In: Timm RM, Fagerston KA (eds) Vertebrate pest conference. https://www.aphis.usda.gov/wildlife_damage/nwrc/publications/10pubs/cozzens101.pdf
Davidson AD, Hewitt CL, Kashian DR (2015) Understanding acceptable level of risk: incorporating the economic cost of undermanaging invasive species. PLoS One 10(11):e0141958
Dixon PB, Parmenter BR (1996) Computable general equilibrium modeling for policy analysis and forecasting. In: Amman HM, Kendrick DA, Rust J (eds) Handbook of computational economics, vol I. Elsevier Science B.V., Amsterdam
Drake JM, Kramer AM (2011) Allee effects. Nat Educ Knowl 3(10):2. http://www.nature.com/scitable/knowledge/library/allee-effects-19699394
Drake JM, Lodge DM (2007) Hull fouling is a risk factor for intercontinental species exchange in aquatic ecosystems. Aquat Invasions 2(2):121–131
Eiswerth ME, Johnson WS (2002) Managing nonnative invasive species: insights from dynamic analysis. Environ Resour Econ 23(3):319–342
Eiswerth ME, van Kooten GC (2002) Uncertainty, economics, and the spread of an invasive plant species. Am J Agric Econ 84(5):1317–1322
Elliston L, Hinde R, Yainshet A (2005) Plant disease incursion management. In: International workshop on multi-agent systems and agent-based simulation. Springer, Berlin/Heidelberg, pp 225–235
Elser JL, Anderson A, Lindell CA, Dalsted N, Bernasek A, Shwiff SA (2016) Economic Impacts of bird damage and management in U.S. Sweet Cherry production. Crop Production. Elsivier. http://dx.doi.org/10.1016/j.croppro.2016.01.014
Elobeld A, Beghin J (2006) Multilateral trade and agricultural policy reforms in sugar markets. J Agric Econ 57(1):23–48
Epanchin-Niell RS, Hastings A (2010) Controlling established invaders: integrating economics and spread dynamics to determine optimal management. Ecol Lett 13(4):528–541
Epanchin-Niell RS, Wilen JE (2012) Optimal spatial control of biological invasions. J Environ Econ Manag 63:260–270
Epanchin-Niell RS, Wilen JE (2015) Individual and cooperative management of invasive species in human-mediated landscapes. Am J Agric Econ 97:180–198
Epanchin-Niell RS, Hufford MB, Aslan CE et al (2010) Controlling invasive species in complex social landscapes. Front Ecol Environ 8:210–216
Epanchin-Niell RS, Haight RG, Berec L et al (2012) Optimal surveillance and eradication of invasive species in heterogeneous landscapes. Ecol Lett 15(8):803–812
Epanchin-Niell R, Brockerhoff E, Kean J, Turner J (2014) Designing cost-efficient surveillance for early detection and control of multiple biological invaders. Ecol Appl 24:1258–1274
Fagiolo G, Reyes R, Schiavo S (2010) The evolution of the world trade web: a weighted-network analysis. J Evol Econ 20:479–514
FAO IPPC (2004) Pest risk analysis for quarantine pests including analysis of environmental risks. International standards for phytosanitary measures, Publication No. 11. Rev. 1. FAO, Rome

Feder G, Regev U (1975) Biological interactions and environmental effects in the economics of pest control. J Environ Econ Manag 2:75–91

Finnoff D, Shogren JF (2004) Endogenous risk as a tool for nonindigenous species management. Weed Control 18:1261–1265

Finnoff D, Shogren JF, Leung B, Lodge D (2005) The importance of bioeconomic feedback in invasive species management. Ecol Econ 52(3):367–381

Finnoff D, Shogren JF, Leung B, Lodge D (2007) Take a risk: preferring prevention over control of biological invaders. Ecol Econ 62(2):216–222

Finnoff D, McIntosh C, Shogren JF et al (2010) Invasive species and endogenous risk. Ann Rev Resour Econ 2(1):77–100

Fraser R, Cook D (2008) Trade and invasive species risk mitigation: reconciling WTO compliance with maximising the gains from trade. Food Policy 33(2):176–184

Goldstein JH, Pejchar L, Daily GC (2008) Using return-on-investment to guide restoration: a case study from Hawaii. Conserv Lett 1(5):236–243

Great Lakes Commission, & St. Lawrence Cities Initiative (2012) Restoring the natural divide: separating the Great Lakes and Mississippi River basins in the Chicago Area waterway system. http://projects.glc.org/caws//pdf/CAWS-PublicSummary-mediumres.pdf

Gren I-M, Thierfelder T, Berglund H (2010) Country characteristics and non-indigenous species. Environ Dev Econ 18:51–70

Groves CR, Jensen DB, Valutis LL et al (2002) Planning for biodiversity conservation: putting conservation science into practice a seven-step framework for developing regional plans to conserve biological diversity, based upon principles of conservation biology and ecology, is being used extensively by the nature conservancy to identify priority areas for conservation. Bioscience 52(6):499–512

Haight RG, Polasky S (2010) Optimal control of an invasive species with imperfect information about the level of infestation. Resour Energy Econ 32(4):519–533

Hauser C, McCarthy M (2009) Streamlining 'search and destroy': cost-effective surveillance for invasive species management. Ecol Lett 12:683–692

Hester S, Cacho O (2012) Optimization of search strategies in managing biological invasions: a simulation approach. Hum Ecol Risk Assess Int J 18:181–199

Hirsch SA, Leitch JA (1996) The impact of knapweed on Montana's economy agricultural economics report February, 355. http://mtweed.org/wp-content/uploads/impact-of-knapweed-mt-economy.pdf

Hof J (1998) Optimizing spatial and dynamic population-based control strategies for invading forest pests. Nat Resour Model 11(3):197–216

Holden MH, Nyrop JP, Ellner SP (2016) The economic benefit of time-varying surveillance effort for invasive species management. J Appl Ecol 53(3):712–721

Holland J (2007) Tools for institutional, political, and social analysis of policy reform. A source book for development practitioners. The World Bank/Oxford University Press, Washington, DC

Holmes TP (1991) Price and welfare effects of catastrophic forest damage from southern pine beetle epidemics. For Sci 37(2):500–516

Holmes TP, Aukema J, Englin J et al (2014) Economic analysis of biological invasions in forests. In: Kant S, Alavalapati J (eds) Handbook of forest resource economics. Routledge, New York, p 560

Homans F, Horie T (2011) Optimal detection strategies for an established invasive pest. Ecol Econ 70:1129–1138

Horan RD, Perrings C, Lupi F, Bulte E (2002) Biological pollution prevention strategies under ignorance: the case of invasive species. Am J Agric Econ 84(5):1303–1310

Horie T, Haight RG, Homans FR, Venette R (2013) Optimal strategies for the surveillance and control of forest pathogens. Ecol Econ 86:78–85

Horsch EJ, Lewis DJ (2009) The effects of aquatic invasive species on property values: evidence from a quasi-experiment. Land Econ 85(3):391–409

Hulme PE (2009) Trade, transport and trouble: managing invasive species pathways in an era of globalization. J Appl Ecol 46:10–18

Hulme PE, Bacher S, Kenis M et al (2008) Grasping at the routes of biological invasions: a framework for integrating pathways into policy. J Appl Ecol 45:403–414

Hyytiäinen K, Lehtniemi M, Niemi JK, Tikka K (2013) An optimization framework for addressing aquatic invasive species. Ecol Econ 91:69–79

International Plant Protection Convention (IPPC) (2002) Guidelines for regulating wood packaging material in international trade. Secretariat of the international plant protection convention. FAO, Rome. https://www.ippc.int/servlet/BinaryDownloaderServlet/16259_ISPM_15_English.pdf?filename=1055161712885_ISPM15_e.pdf&refID=16259

Jacobi WR, Goodrich BA, Cleaver CM (2011) Firewood transport by national and state park campers: a risk for native or exotic tree pest movement. Arboricult Urban For 37:126–138

Johnston EL, Piola RF, Clark GF (2009) The role of propagule pressure in invasion success. In: Rilov G, Crooks JA (eds) Biological invasions in marine ecosystems. Ecological studies (Analysis and synthesis), vol 204. Springer, Berlin/Heidelberg

Juliá R, Holland DW, Guenthner J (eds) (2007) Assessing the economic impact of invasive species: the case of yellow starthistle (*Centaurea solsitialis* L.) in the rangelands of Idaho, USA. J Environ Manag 85:876–882

Just RE, Rausser GC, Zilberman D (1982) Modeling equity and efficiency in agricultural production systems. Growth Equity Agric Dev:120–138

Kaiser BA (2006) Economic impacts of non-indigenous species: Miconia and the Hawaiian economy. Euphytica 148(1–2):135–150

Kaiser BA, Burnett KM (2010) Spatial economic analysis of early detection and rapid response strategies for an invasive species. Resour Energy Econ 32:566–585

Kaluza P, Kolzsch A, Gastner MT, Blasius B (2010) The complex network of global cargo ship movements. J R Soc Interface 7:1093–1103

Kaye-Blake WH, Saunders CM, Cagatay S (2008) Genetic modification technology and producer returns: the impacts of productivity, preferences, and technology uptake. Rev Agric Econ 30(4):692–710

Keller RP, Drake JM, Drew MB, Lodge DM (2011) Linking environmental conditions and ship movements to estimate invasive species transport across the global shipping network. Divers Distrib 17(1):93–102

Kinnunen RE (2015) Sea lamprey control in the Great Lakes, Michigan State University Extension Newsletter. http://msue.anr.msu.edu/news/sea_lamprey_control_in_the_great_lakes

Knowler D, Barbier E (2005) Importing exotic plants and the risk of invasion: are market-based instruments adequate? Ecol Econ 52(3):341–354

Koch FH, Yemshanov D, Magarey RD, Smith WD (2012) Dispersal of invasive forest insects via recreational firewood: a quantitative analysis. J Econ Entomol 105:438–450

Kovacs KF, Haight RG, McCullough DG et al (2010) Cost of potential emerald ash borer damage in US communities, 2009–2019. Ecol Econ 69:569–578

Kovacs K, Vaclavik T, Haight RC et al (2011) Prediction the economic costs and property value losses attributed to sudden oak death damage in California (2010–2020). J Environ Manag 92:1291–1302

Kovacs KF, Haight RG, Mercader RJ, McCullough DG (2014) A bioeconomic analysis of an emerald ash borer invasion of an urban forest with multiple jurisdictions. Resour Energy Econ 36(1):270–289

Leitch JA, Leistritz LF, Bangsund DA (1996) Economic effect of leafy spurge in the Upper Great Plains: methods, models, and results. Impact Assess 14(4):419–433

Leontief W (1986) Input–output economics. Oxford University Press, New York

Leung B, Lodge DM, Finnoff D et al (2002) An ounce of prevention or a pound of cure: bioeconomic risk analysis of invasive species. Proc R Soc Lond B Biol Sci 269(1508):2407–2413

Leung B, Drake JM, Lodge DM (2004) Predicting invasions: propagule pressure and the gravity of Allee effects. Ecology 85:1651–1660

Li X, Preisser EL, Boyle KJ et al (2014) Potential social and economic impacts of the hemlock woolly adelgid in southern New England. Southeastern Nat 13(sp6):130–146

Liebhold AM, Tobin PC (2008) Population ecology of insect invasions and their management. Annu Rev Entomol 53:387–408

Liebhold AM, Macdonald WL, Bergdahl D, Mastro VC (1995) Invasion by exotic forest pests: a threat to forest ecosystems. In: Forest science monograph. Society of American Foresters, Bethesda, p 30

Liebhold AM, Brockerhoff EG, Garrett LJ et al (2012) Live plant imports: the major pathway for forest insect and pathogen invasions of the US. Front Ecol Environ 10(3):135–143

Lodge DM, Williams S, MacIsaac HJ et al (2006) Biological invasions: recommendations for U.S. policy and management. Ecol Appl 16:2035–2054

Lounibos LP (2002) Invasions by insect vectors of human disease. Annu Rev Entomol 47:233–266

Lovett G, Weiss M, Liebhold A et al (2016) Non-native forest insects and pathogens in the US: impacts and policy options. Ecol Appl 26(5):1437–1455

Lowe S, Browne M, Boudjelas S, De Poorter M (2000) 100 of the world's worst invasive alien species: a selection from the global invasive species database. https://www.iucn.org/content/100-worlds-worst-invasive-alien-species-selection-global-invasive-species-database

Mack RN (1991) The commercial seed trade: an early disperser of weeds in the United States. Econ Bot 45(2):257–273

Macleod A, Head J, Gaunt A (2003) The assessment of the potential economic impact of *Thrips palmi* on horticulture in England and the significance of a successful eradication campaign. Crop Prot 23:601–610

Marten AL, Moore CC (2011) An options based bioeconomic model for biological and chemical control of invasive species. Ecol Econ 70:2050–2061

Mas-Colell A, Whinston MD, Green JR (1995) Microeconomic theory. Oxford University Press, New York

McAusland C, Costello C (2004) Avoiding invasives: trade-related policies for controlling unintentional exotic species introductions. J Environ Econ Manag 48(2):954–977

McDermott SM, Finnoff DC, Shogren JF (2013) The welfare impacts of an invasive species: endogenous vs. exogenous price models. Ecol Econ 85:43–49

McNeely JA, Mooney HA, Neville LE et al (2001) A global strategy on invasive alien species. World Conservation Union (IUCN), Gland/Cambridge

Mehta SV, Haight RG, Homans FR et al (2007) Optimal detection and control strategies for invasive species management. Ecol Econ 61(2):237–245

Mérel PR, Carter CA (2008) A second look at managing import risk from invasive species. J Environ Econ Manag 56(3):286–290

Miller R, Blair P (1985) Input output analysis: foundations and extensions. Prentice-Hall, Englewood Cliffs

Murdoch W, Polasky S, Wilson KA et al (2007) Maximizing return on investment in conservation. Biol Conserv 139(3):375–388

Ninan KN (ed) (2014) Valuing ecosystem services: methodological issues and case studies. Edward Elgar, Cheltenham

Nunes PA, van den Bergh JC (2004) Can people value protection against invasive marine species? Evidence from a joint TC–CV survey in the Netherlands. Environ Resour Econ 28(4):517–532

Olden JD, Tamayo M (2014) Incentivizing the public to support invasive species management: Eurasian milfoil reduces lakefront property values. PLoS One 9(10):e110458

Olson LJ, Roy S (2002) The economics of controlling a stochastic biological invasion. Am J Agric Econ 84:1311–1316

Olson LJ, Roy S (2005) On prevention and control of an uncertain biological invasion. Rev Agric Econ 27:491–497

Payne BR, White WB, McCay RE, McNichols RR (1973) Economic analysis of the gypsy moth problem in the northeast: II. Applied to residential property, Res. Pap. NE-285. US Department of Agriculture, Forest Service, Northeastern Forest Experiment Station, Upper Darby, 6 p

Perrings C, Dehnen-Schmutz K, Touza J, Williamson M (2005) How to manage biological invasions under globalization. Trends Ecol Evol 20:212–215

Piper B, Liu L (2014) Predicting the total economic impacts of invasive species: the case of the red streaked leafhopper. SSRN Electron J. https://doi.org/10.2139/ssrn.2536457

Polasky S (2008) Why conservation planning needs socioeconomic data. Proc Natl Acad Sci 105(18):6505–6506

Pysek P, Jarošik V, Hulme PE et al (2010) Disentangling the role of environmental and human pressures on biological invasions across Europe. Proc Natl Acad Sci 107:12157–12162

Qaim M, Traxler G (2005) Roundup ready soybeans in Argentina: farm level and aggregate welfare effects. Agric Econ 32(1):73–86

Ranjan R, Marshall E, Shortle J (2008) Optimal renewable resource management in the presence of endogenous risk of invasion. J Environ Manag 89(4):273–283

Regan TJ, McCarthy MA, Baxter PWJ et al (2007) Optimal eradication: when to stop looking for an invasive plant. Ecol Lett 9(7):759–766

Reichard SH, White P (2001) Horticultural introductions of invasive plant species: a North American perspective in the great reshuffling. In: McNeeley JA (ed) Human dimensions of invasive species. IUCN, The World Conservation Union, Gland/Cambridge

Rich KM, Miller GY, Winter-Nilson A (2005) A review of economic tools for the assessment of animal disease outbreaks. Paris Sci Tech Rev Off Int Epizooties 24:833–845

Richards TJ, Ellsworth P, Tronstad R, Naranjo S (2010) Market-based instruments for the optimal control of invasive insect species: *B. tabaci* in Arizona. J Agric Resour Econ 35:349–367

Rosenberger RS, Bell LA, Champ PA, Smith EL (2012) Nonmarket economic values of forest insect pests: an updated literature review, Gen. Tech. Rep. RMRS 275. U.S. Department of Agriculture Forest Service, Fort Collins, 46 p

Rothlisberger JD, Finnoff DC, Cooke RM, Lodge DM (2012) Ship-borne nonindigenous species diminish Great Lakes ecosystem services. Ecosystems 15(3):1–15

Salaudeen T, Thomas M, Harding D, Hight SD (2013) Economic impact of tropical soda apple (*Solanum viarum*) on Florida cattle production. Weed Technol 27(2):389–394

Sanchirico JN, Albers HJ, Fischer C, Coleman C (2010) Spatial management of invasive species: pathways and policy options. Environ Resour Econ 45:517–535

Sansford C (2002) Quantitative versus qualitative: pest risk analysis in the UK and Europe including the European and Mediterranean Plant Protection (EPPO) system. NAPPO International Symposium on Pest Risk Analysis, Puerto Vallarta

Saphores J, Shogren J (2005) Managing exotic pests under uncertainty: optimal control actions and bioeconomic investigations. Ecol Econ 52:327–339

Schmidt JP, Springborn M, Drake JM (2012) Bioeconomic forecasting of invasive species by ecological syndrome. Ecosphere 3(5):1–19

Schmitz TG, Giese CR, Shultz CJ (2008) Welfare implications of EU enlargement under the CAP. Can J Agric Econ 56(4):555–562

Settle C, Shogren JF (2004) Hyperbolic discounting and time inconsistency in a native–exotic species conflict. Resour Energy Econ 26(2):255–274

Sharov AA, Liebhold AM (1998) Bioeconomics of managing the spread of exotic pest species with barrier zones. Ecol Appl 8(3):833–845

Shogren JF (2000) Risk reduction strategies against the 'explosive invader'. In: The economics of biological invasions. E. Elgar, Cheltenham, pp 56–69

Shwiff SA, Gebhardt K, Kirkpatrick KN, Shwiff SS (2010) Potential economic damage from introduction of brown tree snakes *Boiga irregularis* (Reptilia: Colubridae), to the Islands of Hawai'i. Pac Sci 1:1–10

Simberloff D (2005) The politics of assessing risk for biological invasions: the USA as a case study. Trends Ecol Evol 20(5):216–222

Simberloff D (2009) The role of propagule pressure in biological invasions. Annu Rev Ecol Evol Syst 40:81–102

Sims C, Aadland D, Finnoff D (2010) A dynamic bioeconomic analysis of mountain pine beetle epidemics. J Econ Dyn Control 34(12):2407–2419

Soliman T, Mourits MCM, Oude Lansink AGJM, van der Werf W (2010) Economic impact assessment in pest risk analysis. Crop Prot 29:517–524

Springborn MR (2014) Risk aversion and adaptive management: insights from a multi-armed bandit model of invasive species risk. J Environ Econ Manag 68(2):226–242

Strutt A, Turner JA, Haack RA, Olson L (2013) Evaluating the impacts of an international phytosanitary standard for wood packaging material: global and United States trade implications. Forest Policy Econ 27:54–64

Surkov IV, Oude Lansink AGJM, van der Werf W (2009) The optimal amount and allocation of sampling effort for plant health inspection. Eur Rev Agric Econ 36:295–320

Tatem AJ, Rogers DJ, Hay SI (2006) Global transport networks and infectious disease spread. Adv Parasitol 62:293–343

The Research Group, LLC (2014) Economic impact from selected noxious weeds in Oregon. Prepared for Oregon Department of Agricultural Noxious Weed Control Program

Tu A, Beghin J, Gozlan E (2005) Tariff escalation and invasive species risk; Working Paper 05-WP 407. Center for Agricultural and Rural Development: Iowa State University, Ames

Tu A, Beghin J, Gozlan E (2008) Tariff escalation and invasive species damages. Ecol Econ 67:619–629

U.S. Fish and Wildlife Service (2012) The cost of invasive species. https://www.fws.gov/verobeach/PythonPDF/CostofInvasivesFactSheet.pdf

Underwood EC, Shaw RM, Wilson KA et al (2008) Protecting biodiversity when money matters: maximizing return on investment. PLoS One 3(1):e1515

United Nations Conference on Trade and Development (UNCTAD) (2007) Review of maritime transport. UNCTAD, Geneva

United States General Accounting Office (GAO) (2000) Invasive species: federal and selected state funding to address harmful, nonnative species. http://www.gao.gov/products/RCED-00-219

USDA Animal and Plant Health Inspection Service (USDA APHIS) (2003) Economic impact of West Nile virus on the Colorado and Nebraska equine industries: 2002. Safeguarding American Agriculture, Info Sheet. https://www.aphis.usda.gov/animal_health/nahms/equine/downloads/wnv2002_CO_NB.pdf

Venette RC, Moon RD, Hutchison WD (2002) Strategies and statistics of sampling for rare individuals. Annu Rev Entomol 47(1):143–174

Von Holle B, Simberloff D (2005) Ecological resistance to biological invasion overwhelmed by propagule pressure. Ecology 86:3212–3218

Warziniack T, Finnoff D, Bossenbroek J et al (2011) Stepping stones for biological invasion: a bioeconomic model of transferable risk. Environ Resour Econ 50(4):605–627

Warziniack TW, Finnoff D, Shogren JF (2013) Public economics of hitchhiking species and tourism-based risk to ecosystem services. Resour Energy Econ 35(3):277–294

Westphal MI, Browne M, MacKinnon K, Noble I (2008) The link between international trade and the global distribution of invasive alien species. Biol Invasions 10:391–398

Wilen J (2007) Economics of spatial-dynamic processes. Am J Agric Econ 89:1134–1144

World Trade Organization (WTO) (2008) World trade report 2008. Trade in a globalizing world. WTO Secretariat. http://www.wto.org/english/res_e/booksp_e/anrep_e/world_trade_report08_e.pdf

Yamamura K, Katsumata H, Yoshioka J et al (2016) Sampling inspection to prevent the invasion of alien pests: statistical theory of import plant quarantine systems in Japan. Popul Ecol 58:63–80

Yemshanov D, Koch FH, Ducey MJ et al (2013) Exploring critical uncertainties in pathway assessment of human-assisted introductions of alien forest species in Canada. J Environ Manag 129:173–182

Yemshanov D, Haight RG, Koch FH et al (2015) Optimal allocation of invasive species surveillance with the maximum expected coverage concept. Divers Distrib 21:1349–1359

Zavaleta E (2000) Valuing ecosystem services lost to Tamarix invasion in the United States. In: Mooney HA, Hobbs RJ (eds) Invasive species in a changing world. Island, Washington, DC, pp 261–300

Zohrabian A, Meltzer MI, Ratard R et al (2004) West Nile virus economic impact, Louisiana, 2002. Emerg Infect Dis 10(10):1736–1744

Legislation and Policy

15

Faith T. Campbell, Hilda Diaz-Soltero, and Deborah C. Hayes

15.1 Introduction

In the United States, biological invaders are managed by all Federal agencies that have responsibility for natural resources, as well as the States, territories, and occasionally regional entities. Federal agencies' invasive species programs are implemented under the mandates and guidance provided by dozens of laws, which include statutes enacted by the Congress, Executive Orders issued by the President, and regulations adopted by the relevant agencies. Although there are numerous laws implemented by the States or occasionally regional entities, this chapter will focus on Federal legislation and regulations that guide work on all public and private forests, rangelands, and grasslands in the United States. There are three categories of laws: (1) laws to prevent introduction or initial spread; (2) laws for management or control of invasive species; and (3) more generally defined land management laws which serve as an umbrella for invasive species activities.

Some of the most important Federal laws were enacted to prevent introduction and interstate spread of known or potential invaders. Some laws regulate modes of transport as well as the organisms themselves. Usually these powers do not reside in land- or water-managing agencies. Determination of the legal authority and responsible agency is based primarily on the anticipated impact of the nonnative invasive organism. Thus, the US Department of Agriculture (USDA) Animal and Plant Health Inspection Service (APHIS) is authorized to deal with a range of taxonomic groups considered to be "plant pests," as well as diseases of livestock and poultry. The Department of the Interior (DOI) US Fish and Wildlife Service (USFWS) regulates the introduction and spread of vertebrate animals and some invertebrates that the Secretary of the Interior has determined to be "injurious" to human beings; to the interests of agriculture, horticulture, forestry; or to wildlife or the wildlife resources of the United States.[1]

A small group of laws authorize Federal agencies to control invasive species where they have invaded. These laws often specifically direct either USDA or DOI as the lead for control or management; however, they apply to all Federal land management agencies. In addition to authorizing or mandating control of invasive species on Federal lands, there is often language that encourages cooperation between Federal agencies and States and private landowners. The law may even require an infrastructure for cooperation, such as appointing a coordinator within each program.

Other laws authorize and occasionally mandate more general programs implemented by land- and water-managing agencies. Most of these laws incorporate invasive species management into the agency's broader management goals for protecting wildlife (sometimes limited to specific taxa), habitats, natural resources, historic or cultural sites, etc. A few laws provide for the application of user fees to manage invasive species, or allow private entities operating under permit to be charged for the cost of managing invasive species.

This chapter will summarize the statutes which provide authority to regulate the introduction and spread of invasive species, including:

[1]Many of the older laws use older terms such as plant pests, noxious weeds, nuisance species, etc. rather than "invasive species."

F. T. Campbell (✉)
Center for Invasive Species Prevention, Springfield, VA, USA

H. Diaz-Soltero
U.S. Department of Agriculture, Animal and Plant Health Inspection Service, San Juan, PR, USA

D. C. Hayes
U.S. Department of Agriculture, Forest Service, Sustainable Forest Management Research, Washington, DC, USA

T. M. Poland et al. (eds.), *Invasive Species in Forests and Rangelands of the United States*,
https://doi.org/10.1007/978-3-030-45367-1_15

- The Plant Protection Act of 2000 [7 U.S.C. §7701, *et seq.* (2000)]
- The "injurious wildlife" portions of the Lacey Act [18 U.S.C. § 42(a)(1)]
- The Nonindigenous Aquatic Nuisance Prevention and Control Act (NANPCA) of 1990; as reauthorized by the National Invasive Species Act (NISA) of 1996 [P.L. 104–332; 16 U.S.C. § 4701, note]

Authorities specific to the USDA Forest Service will also be summarized or listed in this chapter.

In all cases, references will be provided to the full statute so that the specific legal authority may be analyzed in greater depth by those who are interested.

15.2 Laws Regulating Introduction and Interstate Spread of Invasive and Potentially Invasive Species

15.2.1 Laws Regulating Movement of Invasive Plants and Plant Pests

Regulatory authority over the introduction and spread of legally designated invasive plants (called "noxious weeds") as well as insects, nematodes, plant pathogens, and other invertebrates or micro-organisms that threaten native plants (called "plant pests") is exercised by USDA APHIS under the authority of the Federal Noxious Weed Act and the Plant Protection Act of 2000.

Federal Noxious Weed Act of 1974 The Federal Noxious Weed Act (FNWA) has been superseded by the Plant Protection Act of 2000 (7 U.S.C. § 7701, *et seq.*) except for the amendment adopted in 2004, Noxious Weed Control and Eradication Act of 2004 (P.L. 108–412), which remains in effect.[2] The amended statute provides, among other things, that USDA APHIS may prohibit or restrict the importation, entry, exportation, or movement in interstate commerce of any noxious weed, article, or means of conveyance, if the Secretary of Agriculture determines that the prohibition or restriction is necessary to prevent the introduction into the United States or the dissemination of the noxious weed within the United States. The term "noxious weed" is defined as any plant or plant product that can directly or indirectly injure or cause damage to crops (including nursery stock and plant products), livestock, poultry, or other interests of agriculture, irrigation, navigation, the natural resources of the United States, and the public. One hundred and twelve species are listed as Federal noxious weeds. The list is posted at https://plants.usda.gov/java/noxious or https://www.aphis.usda.gov/plant_health/plant_pest_info/weeds/downloads/weedlist.pdf. Citizens may petition the Secretary to add or delete species from the list.

Federal Seed Act of 1939 (7 U.S.C. § 1551, *et seq.*) The Federal Seed Act authorizes USDA APHIS to regulate interstate and foreign commerce in seeds and addresses "noxious weed seeds" that may be present in agriculture or vegetable seed.

Plant Protection Act (7 U.S.C. § 7701, *et seq.*) The Plant Protection Act (https://www.aphis.usda.gov/plant_health/plant_pest_info/weeds/downloads/PPAText.pdf), adopted in June 2000, provides the legal foundation for USDA APHIS to regulate the importation, exportation, and interstate movement of plant pests, designated noxious weeds, biological control organisms, and plants or other articles which could transport these pests or weeds. Regulation of pests and invasive plants within individual States is carried out by that State except in cases when the Secretary of Agriculture declares an extraordinary emergency.

Organisms covered by the statute, in addition to noxious weeds (discussed above), are plant pests, defined as any living stage of any of the following that can directly or indirectly injure, cause damage to, or cause disease to any plant or plant product, including protozoans, nonhuman animals, parasitic plants, bacterium, fungus, virus, and infectious agents.

The Plant Protection Act prohibits importation or movement in interstate commerce (including by mail) of any plant pest unless otherwise authorized under the statute. The statute provides a petition process for adding or deleting organisms from the plant pest regulations. Policies and regulations must be based on sound science, transparent, and accessible.

The Secretary of Agriculture is authorized to prohibit or restrict the importation or movement in interstate commerce of any plant, plant product, biological control organism, noxious weed, article, or means of conveyance, if the Secretary determines that the prohibition or restriction is necessary to prevent the introduction into the United States or the dissemination of a plant pest or noxious weed within the United States.

The Secretary is also authorized to hold, seize, quarantine, treat, or apply other remedial measures to destroy or otherwise dispose of any plant, plant pest, noxious weed, biological control organism, plant product, article, or means of conveyance that—

1. is moving into or through the United States or interstate, or has moved into or through the United States or interstate, and—
   (A) the Secretary has reason to believe is a plant pest or noxious weed or is infested with a plant pest or noxious weed at the time of the movement; or
   (B) is or has been otherwise in violation of this title;

[2]Numerous plant species considered to be invasive by various State agencies or other authorities are not listed under the FNWA, so they are not subject to its provisions.

2. has not been maintained in compliance with a post-entry quarantine requirement; or
3. is the progeny of any plant, biological control organism, plant product, plant pest, or noxious weed that is moving into or through the United States or interstate, or has moved into the United States or interstate, in violation of this title.

The Secretary may order the owner of any plant, biological control organism, plant product, plant pest, noxious weed, article, or means of conveyance subject to action under subsection (a) of the Plant Protection Act, or the owner's agent, to treat or apply other remedial measures to destroy or otherwise dispose of the plant, biological control organism, plant product, plant pest, noxious weed, article, or means of conveyance, without cost to the Federal government.

USDA APHIS inspects imports of living plants—usually at one of its 16 Plant Inspection Stations. All other imports—including fruits, vegetables, and grains for consumption and miscellaneous merchandise in packaging made from wood—are inspected by the Bureau of Customs and Border Protection (CBP), a division of the Department of Homeland Security. The CBP inspections follow protocols and rules established by USDA APHIS.

Descriptions of USDA APHIS programs targeting major plant pests, including those that attack forest trees, are posted on the USDA APHIS website at https://www.aphis.usda.gov/aphis/ourfocus/planthealth.

15.2.2 Statutes Regulating Movement of Invasive Animals

Regulatory authority over the introduction and spread of invasive or potentially invasive animals and their diseases is divided among several agencies, including the USFWS, Coast Guard, Environmental Protection Agency, US Army Corps of Engineers (Army COE), and US Department of Commerce National Oceanic and Atmospheric Administration (NOAA).

Lacey Act, Title 18 Available at https://www.fws.gov/le/pdffiles/Lacey.pdf. The injurious wildlife provisions of the Lacey Act (18 U.S.C. § 42(a)(1))[3] regulate importation and interstate movement of potentially harmful species belonging to certain specified taxa of wildlife. The Act was enacted in 1900 and has been subsequently amended. The Lacey Act is implemented by the USFWS.

The organisms covered by the Lacey Act include wild mammals, wild birds, fishes, reptiles, amphibians, mollusks, and crustaceans. Some animals have been listed under the statute by law, including mongooses, fruit bats, brown tree snake (*Boiga irregularis*), zebra mussel (*Dreissena polymorpha*), and bighead carp (*Hypophthalmichthys nobilis*). Other species have been listed by regulation. As of May 2016, 611 species of wildlife are listed as injurious, including 301 fish, 201 amphibians, 92 mammals, 9 reptiles, 4 birds, 3 crabs, and 1 mollusk. Of the fish, 170 are salmonids that can serve as hosts of pathogens. All of the amphibians listed are salamanders that can serve as hosts of a pathogen.

The injurious wildlife provisions of the Lacey Act (18 U.S.C. § 42(a)(1)) state:

> The importation into the U.S., any territory of the U.S., the District of Columbia, the Commonwealth of Puerto Rico, or any possession of the U.S., or any shipment between the continental U.S., the District of Columbia, Hawaii, the Commonwealth of Puerto Rico, or any possession of the U.S., of [listed species] is hereby prohibited. All such prohibited mammals, birds, fish (including mollusks and crustacea), amphibians, and reptiles, and the eggs or offspring therefrom, shall be promptly exported or destroyed at the expense of the importer or consignee.

The USFWS' authority to regulate movements of listed species among the continental States has been reversed by the decision in *United States Association of Reptile Keepers, Inc. v. Jewell,* 103 F. Supp. 3d 133 (U.S.D.C., D.C. 2015). The decision by the District Court for the District of Columbia was upheld by the DC Circuit Court of Appeals.

Listing decisions are subject to the regulations at Title 50, Code of Federal Regulations (C.F.R.), Part 16. Species are added to the list of injurious wildlife to prevent their introduction, establishment, and spread in the United States, and to prevent harm they may cause to humans and the interests of agriculture, horticulture, forestry, or wildlife of the United States.

Since the statute does not allow for the listing of invertebrates other than mollusks and crustaceans, the pathogens and parasites that can be carried by wildlife cannot be listed by themselves. However, the USFWS has listed injurious wildlife because of their potential to introduce and spread harmful pathogens and parasites in the United States. Thus, USFWS has listed all members of the salmon family as injurious because of certain pathogens they may carry that can infect native and farmed salmonids. The USFWS also listed 201 species of salamanders because they might carry a fungus lethal to many salamander species.

Species are usually listed as injurious only as alive, but a listing may include dead individuals if the factor(s) that made the species injurious still exist(s) in a dead specimen. Gametes, viable eggs, and hybrids may also be included in a listing.

Possession of a listed species within State boundaries, and movement within the State, is the responsibility of each State and is not regulated by an injurious wildlife listing

[3]Other sections of the Lacey Act (16 U.S.C. §§ 3371–3378) pertain to trafficking in protected species of wildlife and plants.

unless the organism has previously been permitted for importation or interstate transport. Similarly, export is not regulated by an injurious wildlife listing, provided the shipment is exported directly from a designated port without crossing State lines. Under the Lacey Act, USFWS may grant permits for the importation or interstate transportation of injurious wildlife or their offspring or eggs for zoological, educational, medical, or scientific purposes.

Nonindigenous Aquatic Nuisance Prevention and Control Act (NANPCA) of 1990; as reauthorized by the National Invasive Species Act (NISA) of 1996 P.L. 101–646 (104 Stat. 4761), as amended by the National Invasive Species Act (NISA) of 1996 (P.L. 104–332; 110 Stat. 4073) The text of NISA is available at https://www.govinfo.gov/content/pkg/PLAW-104publ332/html/PLAW-104publ332.htm.

The USFWS, US Coast Guard, Environmental Protection Agency, Army COE, and NOAA share responsibilities to develop a program of prevention, monitoring, control, and study to prevent introduction of and to control the spread of introduced aquatic nuisance species and the brown tree snake. The statutes also created the Aquatic Nuisance Species Task Force (ANSTF), which is co-chaired by the Undersecretary of Commerce for Oceans and Atmosphere and the USFWS Director. USDA is among the Task Force members. The Army COE is directed to establish an Aquatic Nuisance Species Research Program for developing control technologies to address aquatic invasive species (other than invasive aquatic plants). The Secretary of the Department of Homeland Security is required to ensure, to the maximum extent practicable, that aquatic nuisance species are not discharged into waters of the United States from vessels. Regulations require vessels to carry out management practices necessary to reduce the probability of unintentional discharges resulting from ship operations other than ballast water discharge (e.g., hull fouling). Jurisdiction under NISA (and NANPCA earlier) applies to the US territorial sea or within 12 nautical miles.

15.2.3 Laws Regulating Species in Order to Prevent Introduction or Spread of Wildlife Diseases

Animal Health Protection Act (P.L. 107–171, 7 U.S.C. § 8301, *et. seq.*) The Animal Health Protection Act (AHPA) authorizes USDA APHIS to prohibit the importation or entry of any animal or animal product that is likely to disseminate any foreign pest or disease of livestock (including farmed deer) or poultry within the United States. The AHPA prohibits the use of any means of conveyance in connection with the importation or entry of livestock if the means of conveyance has not been maintained in a clean and sanitary condition. The Act also provides that USDA APHIS should maintain, in all regions of the United States, a sufficient number of Federal veterinarians and accredited State veterinarians who are well trained in recognition and diagnosis of exotic and endemic animal diseases. Under this statute, USDA APHIS has provided expertise to diagnose diseases in wild birds because the diseases could threaten domestic poultry.

Lacey Act, Title 18 The Lacey Act, Title 18, also regulates animals as a potential vector of disease. Refer to discussion above under Sect. 15.2.2.

15.3 Laws Authorizing Federal Agencies to Control Invasive Species

This section covers laws pertaining to specific Federal agencies to control invasive species.

Some of these laws go beyond just invasive species to give the agency a broad mandate within the original agency mission.

Animal Damage Control Act of 2000 (7 U.S.C. §§ 426-426c, as amended) The Animal Damage Control Act gives USDA APHIS the authority to control wildlife damage on Federal, State, or private land. The purpose is to protect field crops, vegetables, fruits, nuts, horticultural crops, and commercial forests; freshwater aquaculture ponds and marine species cultivation areas; livestock on public and private range and in feedlots; public and private buildings and facilities; civilian and military aircraft; and public human health. In addition, USDA APHIS has broad authority for investigation, demonstrations, and control of mammalian predators, rodents, and birds. An amendment adopted in 1991 (P.L. 101–190 (Div. A, title III, Sec. 348, Dec. 5, 1991)) directs the Secretary of Agriculture to take actions to prevent introduction of brown tree snakes from Guam into other areas of the United States.

Federal Noxious Weed Act §15 of the Federal Noxious Weed Act of 1974 (7 U.S.C. § 2801, note; 7 U.S.C. § 2814) Adopted in 1990, Section 15 titled "Management of Undesirable Plants on Federal Lands" authorizes the Secretary of Agriculture to cooperate with other Federal and State agencies and others in carrying out operations or measures to eradicate, suppress, control, prevent, or retard the spread of any noxious weed. Each Federal agency shall:

1. designate an office or person adequately trained to develop and coordinate an undesirable plants management program for control of undesirable plants on Federal lands under the agency's jurisdiction;
2. establish and adequately fund an undesirable plants management program through the agency's budgetary process;

3. complete and implement cooperative agreements with State agencies regarding the management of undesirable plant species on Federal lands; and
4. establish integrated management systems to control or contain undesirable plant species targeted under cooperative agreements.

Public Law 109–59—Safe, Accountable, Flexible, Efficient Transportation Equity Act: A Legacy for Users or "SAFETEA-LU" 23 U.S. Code § 145 Text can be found at https://www.gpo.gov/fdsys/pkg/PLAW-109publ59/pdf/PLAW-109publ59.pdf. For the first time, SAFETEA-LU provided eligibility for States to spend Federal-aid dollars on noxious weed programs. This is not mandatory spending; the choice is strictly up to the individual States.

Sec. 6006. Environmental Restoration and Pollution Abatement; control of noxious weeds and aquatic noxious weeds and establishment of native species.

(a) Modification to NHS/STP for Environmental Restoration, Pollution Abatement, Control of Noxious Weeds and Aquatic Noxious Weeds.—
   1. Modifications to national highway system.—Section 103(b)(6) of title 23, United States Code, is amended by adding at the end the following:
   "(Q) Environmental restoration and pollution abatement in accordance with section 328.
   (R) Control of noxious weeds and aquatic noxious weeds and establishment of native species in accordance with section 329."
   2. Modifications to surface transportation program.—Section 133(b) of title 23, is amended by striking paragraph (14) and inserting the following:
   "(14) Environmental restoration and pollution abatement in accordance with section 328.
   (15) Control of noxious weeds and aquatic noxious weeds and establishment of native species in accordance with section 329."

Part 329—Eligibility for control of noxious weeds and aquatic noxious weeds and establishment of native species.

"(b)Included Activities.—The establishment and management under subsection (a)(1) and (a)(2) may include—
   1. right-of-way surveys to determine management requirements to control Federal or State noxious weeds as defined in the Plant Protection Act (7 U.S.C. 7701 *et seq.*) or State law, and brush or tree species, whether native or nonnative, that may be considered by State or local transportation authorities to be a threat with respect to the safety or maintenance of transportation systems;
   2. establishment of plants, whether native or nonnative with a preference for native to the maximum extent possible, for the purposes defined in subsection (a)(1);
   3. control or elimination of plants as defined in subsection (a)(2);
   4. elimination of plants to create fuel breaks for the prevention and control of wildfires; and
   5. training.".

15.4 Laws Authorizing Invasive Species Management Programs as Part of Resource Management Agencies' Overall Programs

15.4.1 General Land Management Authorities

These authorities deal with general land management and, for the most part, apply to all natural resource agencies. However, within the specific aim of the authorities, management of invasive species may be necessary to achieve the stated goals. This list of authorities is not exhaustive of all the general land management or other authorities which can be used.

Statutes pertinent to all resource management agencies

- Endangered Species Act (ESA) of 1973 (16 U.S.C. §§1531 *et seq.*)
- Surface Mining Control and Reclamation Act of 1977 (30 U.S.C. 1201, 1201 (note), 1236, 1272, 1305). §515
- North American Wetland Conservation Act 1989 (16 U.S.C. 4401 (note), 4401–4413, 16 U.S.C. 669b (note)). §9 [U.S.C. 4408]
- National Historic Preservation Act of 1966 [16 U.S.C. §§470 *et seq.*]
- Sikes Act (Fish and Wildlife Conservation) of September 15, 1960 (16 U.S.C. 670g-670l, 670o, P.L. 86–797), as amended. §201
- Wilderness Act of 1964 (16 U.S.C. §§1131 *et seq.*)
- Clean Water Act, section 404 (333 U.S.C. §1344) on wetlands determination by Army COE and USFWS

15.4.2 Statutes Authorizing Invasive Species Management Programs Implemented Specifically by the USDA Forest Service

15.4.2.1 Statutes Providing for Use of User Fees or Billing Permittees to Cover Costs of Managing Invasive Species

Granger-Thye Act of April 24, 1950 (16 U.S.C. §§ 572) This statute authorizes the Secretary of Agriculture to cooperate with adjacent public and private landowners in performing work done for the "administration, protection, improvement, reforestation, and such other kinds of work the USDA Forest Service is authorized to do on lands of the United States." This law thus allows for cooperative abatement activities pertaining to invasive species on intermingled Federal and non-Federal land.

Cooperative Forestry Assistance Act (16 U.S.C. 2101, *et seq.*) This statute allows the USDA Forest Service to provide assistance to States for the management, protection, and development of nonindustrial private forest lands. Similarly, there is an urban and community forestry assistance program.

15.4.2.2 Statutes Authorizing USDA Forest Service Management

Included in this category are numerous statutes that broadly authorize invasive species management as a component of the agency's resource management activities in forests and grasslands of the United States:

- Organic Administration Act of 1897 (16 U.S.C. §§ 473, *et seq.*). This law authorizes the Secretary to take such actions and regulate the occupancy and use of the National Forests to protect the forests thereon from "destruction by fire and depredations." 16 U.S.C. § 551.
- Forest and Rangeland Renewable Resources Research Act of 1978; especially §3(a)(3); 16 U.S.C. 1642. Section 3(a)(3) authorizes research for protecting renewable resources from "fires, insects, diseases, noxious plants, animals…."
- International Forestry Cooperation Act of 1990 (16 U.S.C. § 4501). Among many things, this statute authorizes the Secretary of Agriculture to provide international assistance for the "prevention and control of insects, diseases, and other damaging agents." 16 U.S.C. § 4501.
- Healthy Forests Restoration Act of 2003 (H.R. 1904) [16 U.S.C. 6501–6502, 6511–18, 6541–42, 6571–78]. The Restoration Act amended statutory processes for hazardous fuel reduction projects on certain types of at-risk National Forest System (NFS) and Bureau of Land Management (BLM) lands and provides other authorities and direction to help reduce hazardous fuel and restore healthy forest and rangeland conditions on lands of all ownerships. The act also establishes a program to combat infestations by forest-damaging insects and associated diseases (although the species prompting principal concern, the mountain pine beetle, is not an invasive species).
- Wyden Amendment (P.L. 109–54, Section 434). The Wyden amendment authorizes the Secretaries of Agriculture and Interior to enter into cooperative agreements to benefit resources within watersheds on certain lands under their respective jurisdictions. These agreements may support or conduct invasive species management activities on aquatic and terrestrial areas owned by local and State governments, tribes, other Federal agencies, private individuals, or organizations to benefit and protect public lands, the NFS, and other resources within a watershed at risk from invasive species.

15.4.2.3 Regulations Governing USDA Forest Service Management of Invasive Species

According to Section 2901.02 of the USDA Forest Service Manual, the Forest Service's authority to manage invasive species on NFS lands was delegated from the Secretary of Agriculture to the Under Secretary for Natural Resources and Environment at Title 7, C.F.R., Section 2.20 (7 C.F.R.§ 2.20). This authority has been delegated in turn from the Under Secretary to the Chief of the Forest Service at Title 7, C.F.R. Section 2.60 (7 C.F.R. § 2.60). Title 36, C.F.R. (including Parts 221, 222, 228, 241, 251, 261, 290, 292, 293, 296, and 297), provides additional authorities to manage and regulate invasive species across the NFS (national forests and national grasslands), including establishing requirements and prohibitions to prevent and control aquatic and terrestrial invasive species. Forest Service regulations at 36 C.F.R. § 222.8 acknowledge the Agency's obligation to work cooperatively in identifying invasive species (including noxious weeds) problems and initiating control programs in aquatic and terrestrial areas of the NFS.

15.4.2.4 Other Regulations and Policies Governing Specific Aspects of USDA Forest Service Invasive Species Management Programs

Policy on Noxious Weed Management Departmental Regulation 9500–10 (DR 9500–10) (January 18, 1990) Established USDA policy to manage and coordinate noxious weed activities among USDA agencies in order to improve the quality and ecological conditions of crops and rangeland in the United States.

Policy on the Management of Wildlife, Fish, and Plant Habitat Departmental Regulation 9500–4 (DR 9500–4) Guides the management of Wildlife, Fish, and Plant Habitat on public lands.

Gypsy Moth Policy (USDA) of 1990 Departmental Regulation 5600–001 (DR 5600–001) This regulation established the Departmental Gypsy Moth Program. It assigns responsibilities to USDA agencies and defines agency roles to avoid duplication and provide maximum coordination of USDA activities dealing with gypsy moth.

Departmental Regulation 9500–4 (USDA policy on wildlife, fish, and plant habitat management on NFS lands and waters) This regulation provides that the Department will promote the concept and use of integrated pest management (IPM) practices in carrying out its responsibilities for pest control and will seek to alleviate damage by plant and animal pests to farm crops, livestock, poultry, forage, forest and urban trees, wildlife, and their habitats. Departmental agencies, through management and research programs, will develop or assist in developing new techniques and methodologies for the prevention of damage to agricultural or forestry production. The agencies also will strive to reduce potential depredations through improved management of USDA programs. Pest control techniques and considerations will be incorporated into appropriate management and education programs.

Native Plant Materials Policy (FSM 2070) of Feb 13, 2008 This policy provides USDA Forest Service manual direction on the use of native plant materials in revegetation, rehabilitation, and restoration of both aquatic and terrestrial ecosystems across the NFS.

Pesticide Use Management and Coordination Policy (FSM Chapter 2150) (March 19, 2013) This provides agency policy and guidance on the use of pesticides as part of an IPM approach. Additional guidance is provided in the Pesticide Use Management Handbook (FSH 2109).

In December 2011, the USDA Forest Service finalized a new title in the Forest Service Manual, FSM 2900—Invasive Species Management. The goal of the directive in FSM 2900 is to integrate invasive species prevention, early detection and rapid response, control, restoration, mitigation, cooperation, education, and awareness activities across forests and grasslands in the NFS resource management programs, Forest Plans, project-level planning activities, and other NFS operations.

The Manual states that all NFS invasive species management activities will be conducted within the following strategic objectives:

- **Prevention** Take proactive approaches to manage all aquatic and terrestrial areas of the NFS in a manner to protect native species and ecosystems from the introduction, establishment, and spread of invasive species.
- **Early Detection and Rapid Response (EDRR)** Inventory and survey susceptible aquatic and terrestrial areas of the NFS to quickly detect invasive species infestations, and subsequently implement immediate and specific actions to eradicate those infestations.
- **Control and Management** Conduct integrated invasive species management activities on priority aquatic and terrestrial areas of the NFS consistent with guidance from the National Invasive Species Council. (This statement includes the USDA guidance to use IPM.)
- **Restoration** Proactively manage aquatic and terrestrial areas of the NFS to increase the ability of those areas to be self-sustaining and resistant (resilience) to the establishment of invasive species.
- **Organizational Collaboration** Cooperate with other Federal agencies, State agencies, local governments, tribes, academic institutions, and private sector on invasive species issues.

The management of aquatic and terrestrial invasive species (including vertebrates, invertebrates, plants, and pathogens) will be based on an integrated pest management approach, throughout the NFS.

15.4.3 Principal Statutes Governing Other Land-Managing Agencies

Federal Lands Policy and Management Act of 1976 (FLPMA, 43 U.S.C. § 1701 *et seq.*) FLPMA directs the BLM to "take any action necessary to prevent unnecessary or undue degradation of the public lands," to "manage the public lands under principles of multiple use and sustained yield," and to include protection of "wildlife and fish."

Public Rangelands Improvement Act (43 U.S.C. § 1901 *et seq.*) This Act directs BLM to improve the range conditions of the public rangelands. BLM activities under the Act include reducing invasive species.

Taylor Grazing Act (43 U.S.C. § 315, *et seq.*) The Taylor Grazing act directs BLM to "make provision for the protection, administration, regulation, and improvement of [Federal] grazing districts" (43 U.S.C. § 315a); "to preserve the land and its resources from destruction or unnecessary injury" *Id.*; "to provide for the orderly use, improvement, and development of the range" *Id.*; and "to perform such work as may be necessary amply to protect and rehabilitate the areas subject to the provisions of this Act" *Id.* The Act also authorizes BLM to "cooperat[e] with … official State agencies engaged in conservation or propagation of wildlife interested in the use of the grazing districts" *Id.* at § 315 h.

National Park Service Organic Act (54 U.S.C. § 100,101, *et seq.*) The Organic Act established the National Park Service (NPS). It requires the agency to promote and regulate the areas known as national parks to conserve the scenery and the natural and historic objects and the wildlife therein and to provide for the enjoyment of the same in such manner and by such means as will leave them unimpaired for the enjoyment of future generations. This law generally gives the NPS the authority to regulate the movement of invasive species. Section 15.3 expressly authorizes the destruction of detrimental animals, which includes invasive species.

National Wildlife Refuge Administration Act (16 U.S.C. §§668dd-ee) This Act establishes the National Wildlife Refuge (NWR) System and requires USFWS to administer lands to provide for the conservation of fish, wildlife, plants, and their habitats and to ensure that biological integrity and diversity are maintained. Under implementing regulations promulgated at 50 C.F.R. § 27.52, USFWS prohibits introduction of plants and animals taken elsewhere to any NWR except as authorized.

15.5 Executive Orders

Executive Order 13112 Executive Order (EO) 13112 (available at https://www.invasivespeciesinfo.gov/executive-orders-invasive-species) directs Federal agencies to (1) identify actions that may affect status of an invasive species; (2a) prevent introduction of such species; (2b) detect and control such species; (2c) monitor population of such species; (2d) provide for restoration of native species; (2e) conduct research on invasive species and develop technologies to prevent introduction of such species; (2f) promote public education of such species; and (3) not authorize, fund, or carry out actions likely to cause the introduction or spread of invasive species in the United States or elsewhere, unless the benefits of the action clearly outweigh the harm and the agencies take steps to minimize the harm.

Executive Order 13112 defines invasive species as "an alien species whose introduction does or is likely to cause economic or environmental harm or harm to human health." To understand the full meaning of the term "invasive species" within the EO, the definition of "Introduction" needs to be inserted. "Introduction" means the intentional or unintentional escape, release, dissemination, or placement of a species into an ecosystem *as a result of human activity*. Thus, the actual working definition to use for invasive species is "an alien species whose introduction (the intentional or unintentional escape, release, dissemination, or placement of a species into an ecosystem as a result of human activity) does or is likely to cause economic or environmental harm or harm to human health." The definition of what is meant by harm is further elaborated in a white paper issued by the Invasive Species Advisory Council. The paper is available at https://www.doi.gov/sites/doi.opengov.ibmcloud.com/files/uploads/ISAC%20Definititions%20White%20Paper%20%20-%20FINAL%20VERSION.pdf.

Executive Order 13112 also establishes the National Invasive Species Council (NISC). NISC is co-chaired by the Secretaries of Agriculture, Commerce, and the Interior. Its membership encompasses 13 member departments and their constituent agencies. The Council's work is supported by a small staff. NISC responsibilities include promoting action at local, State, tribal, and ecosystem levels; identifying recommendations for international cooperation; facilitating a coordinated network to document, evaluate, and monitor invasive species' effects; developing a web-based information network on invasive species; and developing guidance on invasive species for Federal agencies to use in implementing the National Environmental Policy Act.

Executive Order 13112 also established the Invasive Species Advisory Council. Members are appointed by the Secretary of Interior; the Council operates under transparency and other provisions of the Federal Advisory Committee Act (FACA) (Pub. L. 92–463, § 1, Oct. 6, 1972, 86 Stat. 770).

Future Invasive Species Research Challenges and Opportunities

16

Becky K. Kerns, Therese M. Poland, Robert C. Venette, Toral Patel-Weynand, Deborah M. Finch, Allen Rowley, Deborah C. Hayes, and Mike Ielmini

16.1 Introduction

Introductions of new invasive species continue with ever-expanding global trade, and the increase in invasive species does not show any sign of saturation for most taxonomic groups. However, some non-native species may arrive in an area and fail to become established or have no measurable impact. Thus, the protection of valued ecosystems depends on reliable knowledge about the species that are or might be invading, the ecosystems that are being invaded, the potential magnitude of different types of harm, the availability and effectiveness of management responses, and the desired state of ecosystems under protection. Primary research is a vital means of gathering that knowledge.

There has been considerable progress in research that focuses on understanding the broad range of impacts invasive species have in their introduced environments. This includes their interactions with ecosystem processes, climate change and disturbance, social and cultural dynamics, and economics. There have also been advancements in management tools and techniques for predicting and preventing their establishment, detecting and managing established populations, and restoring impacted ecosystems. The previous chapters provide an assessment of our current state of knowledge about invasive species, their effects on ecosystem processes, and their management with a particular emphasis on lessons learned that apply to multiple species. Each chapter identifies critical knowledge gaps and areas for future research that are specific to a particular topic area. The intent of this chapter is to highlight common and significant research challenges and opportunities. Our goal is not an exhaustive review of the material of each chapter, but to highlight four broad research challenges that, if tackled, could make far-reaching contributions to our understanding of, and response to, biological invasions.

16.2 Improve Our Understanding and Quantification of the Ecological Impacts of Invasive Species and the Mechanisms of Invasion

Invasive species impact terrestrial and aquatic ecosystems, both directly and indirectly, by numerous means such as killing resident species, lowering the abundance of resident species, reducing biotic diversity, or altering nutrient cycles. As noted in Chaps. 2 and 3, most of the research quantifying the

B. K. Kerns (✉)
U.S. Department of Agriculture, Forest Service, Pacific Northwest Research Station, Corvallis, OR, USA
e-mail: becky.kerns@usda.gov

T. M. Poland
U.S. Department of Agriculture, Forest Service, Northern Research Station, Lansing, MI, USA

R. C. Venette
U.S. Department of Agriculture, Forest Service, Northern Research Station, St. Paul, MN, USA

T. Patel-Weynand · D. C. Hayes
U.S. Department of Agriculture, Forest Service, Sustainable Forest Management Research, Washington, DC, USA

D. M. Finch
U.S. Department of Agriculture, Forest Service, Rocky Mountain Research Station, Albuquerque, NM, USA

A. Rowley
U.S. Department of Agriculture, Forest Service, National Forest System, Washington Office, Washington, DC, USA

M. Ielmini
U.S. Department of Agriculture, Forest Service, Washington, DC, USA

T. M. Poland et al. (eds.), *Invasive Species in Forests and Rangelands of the United States*, https://doi.org/10.1007/978-3-030-45367-1_16

impacts of invasive species has been species-specific, short-term, qualitative, locally scaled, and focused on ecological and biological systems. Quantitative impact assessments, long-term impact studies, and models that characterize impacts are needed across larger, regional scales, with an increased focus on multiple species and a much wider range of taxa at the population and community levels. Quantifying impacts is essential to justify the need to be proactive in addressing invasive species issues, to quantify costs, and to promote collaborative agencywide efforts, and is critical to help decision makers weigh options so cost-efficient and effective programs can be developed, particularly under highly constrained budgetary environments. Understanding complex interactions, ecosystem processes, and impacts that cascade across trophic levels is a major research challenge. For example, invasive species may directly or indirectly affect other trophic levels in food webs, and subsequently alter carbon and nutrient cycling, but how this happens is not well known. Examining impacts to other disturbances, such as the alteration of fire regimes from invasive grasses, allows more comprehensive quantification of the extensive deleterious impacts that invasive species can have across the landscape. In addition, more research is needed to assess the regional impacts of invasive species on important ecosystem services, such as providing ample clean air and water.

The ability to estimate the potential impacts of invasion would be facilitated by a better understanding of the mechanisms of invasion, as well as ecosystem invasibility and resilience. Many studies have examined what makes an ecosystem prone to invasion, including which species are likely to be invasive, and the role of propagule pressure. There is still no clear paradigm, however, that can explain the invasion process (Chap. 5). For example, how much is the impact of invasion dependent on the abundance or density of an invading species? It is extremely challenging to design research to develop a deeper understanding of the interactions among propagule pressure, functional traits of invasive species, abiotic and biotic resistance, and the role of disturbance. New knowledge is needed on the ecological differences between invasive species and non-native species that do not become invasive, including the development of models to predict the likelihood of invasion into specific ecosystems, based on species functional traits and interactions with the resident ecosystem.

Finally, inventory and monitoring programs and high-quality distribution data are necessary for managing invasive species and for developing quantitative ecological, economic, and social impact assessments (Chap. 9). A wide variety of programs track invasive species at multiple scales. Nevertheless, there are still gaps in our knowledge. For instance, the USDA Forest Service Forest Health Protection program monitors forest conditions, and there are insect and disease risk maps for forested landscapes. Yet there is a great need for improved modeling, data collection, and mapping of non-forested systems, aquatic habitats, and invasive plants. The most consistently identified gap across numerous disciplines is the relative paucity of spatially referenced data describing the extent and magnitude of invasive species. The use of citizen science or crowd source data can help address this issue, but data quality must be improved to broaden its applicability as a tool. There is also a great deal of work that needs to be done to standardize inventory, survey, and monitoring protocols among agencies and partners.

16.3 Develop Better Knowledge and Models of How Climatic Variability and Change May Impact Invasive Species

Environmental changes, and changes in precipitation and temperature, will alter all aspects of invasion, from introduction pathways to species establishment, spread, and impacts across multiple sectors (Chap. 4). Understanding how invasive species may respond to disturbances will allow the development of impact assessments, as well as prevention, management, and restoration tactics that consider the future. Responses and approaches may change across time and space as environmental conditions change. To begin to understand the effects of environmental changes and invasive species impacts in the future, it will be necessary to identify multiple interacting climatic and other drivers of invasive species establishment and spread. Most studies have examined the impacts of climatic variability and change on individual invasive species in simplified systems based on one physical component of climate change (e.g., temperature, soil moisture, elevated carbon dioxide (CO_2)); these components are rarely studied in conjunction with biological interactions (e.g., competition) and other stressors and disturbances. For example, increased CO_2 may increase plant biomass or productivity, but how competition may factor into this isn't well known. Realistic studies in natural settings over larger landscapes and across multiple environmental gradients could provide insights into the potential future of how invasive species distribution or function may change.

The potential rate at which species may adapt to future climate conditions through evolutionary processes should be considered (Chap. 4). For example, knowing the genetic structure of invading populations is most likely necessary to identify how climate change can drive or inhibit their spread. Genomic studies of relatedness and genetic variations among invading populations are needed for developing robust models of spread in a changing climate. Research is also needed to determine how and to what extent climate change acts as a selection process on invasive species. New knowledge on the interactions between climate change and genetic processes, such as hybridization and polyploidization, will be critical for explaining how invasive species adapt to climate change.

While predictive modeling frameworks that incorporate climatic change are highly useful, they may require the development and application of spatially explicit models of invasive species distribution and spread at relevant spatial scales. This endeavor involves constructing species-specific models capable of projecting potential future spread under a range of climate scenarios. Many current species distribution models can predict areas of potential introduction (climate matching or suitability) and potential changes in future climatic habitats, but they rely on past conditions which often makes them less successful in predicting what might happen in the future, especially as conditions become more novel. The ability to project future invasive species distributions in response to climate change can be enhanced when multiple models using different approaches (e.g., empirically based or process based) are used in conjunction with multiple climate change scenarios. As with other aspects of invasive species research, high-quality baseline species distribution data on invasive species are needed for developing quantitative predictive models.

There is also evidence that climate change may alter the efficacy of management tactics for control and restoration of invaded ecosystems, including the use of herbicides or mechanical methods to manage invasive plants. This will require a better understanding of how climate change will impact the interactions between invasive organisms and their chemical and biocontrol agents, as well as the effectiveness and resilience to agents. Research is also needed for improving other management and restoration strategies under varying climates following invasions.

16.4 Develop Additional Cost-Effective Tools, Techniques, and Approaches for Invasive Species Detection, Prevention, and Control As Well As Approaches for Restoration of Invaded Ecosystems

Four broad strategies are used to protect natural and other resources from invasive species: prediction and prevention, early detection and rapid response, management and mitigation, and rehabilitation and restoration (Chaps. 6, 7 and 8). In general, preventative measures that are targeted at the highest risk species (i.e., those species that are most likely to invade and cause severe harm) are considered the most cost-effective. Rapid responses (i.e., eradication or containment) to newly invading populations are considered the next best alternative to exclude potentially damaging invasive species from occupying their full potential range. Reactive efforts to manage and mitigate invasive species and their impacts can be costly, and "success" can be a matter of perspective. However, the public may not be aware of the benefits of prevention and, understandably, may be focused on those species that are well established and causing harm. For those pernicious species, the goal is not necessarily to eliminate all individuals but to reduce their densities to manageable levels.

Research to distinguish non-native species that might cause harm from those that are likely to be benign and to identify pathways by which harmful species might be introduced would help to predict and prevent high-risk species from invading. For potentially harmful species, regulators and resource managers need to know where these species are likely to occur and cause harm over time; this information is integrated in spatially explicit invasive species risk assessments. As noted above, incorporating the effects of future climate and human behaviors into the assessments could provide valuable new insights. New approaches are needed to reduce and reflect uncertainties in those assessments in ways that support decision making. Extensive empirical observations of invasive species in both their native and non-native ranges would be extremely useful to rigorously test new theories and models and identify opportunities for substantive improvements in prevention.

Preventative efforts are seldom perfect, making rapid response strategies necessary to protect natural resources from new invading species. These strategies would benefit from an improved understanding of the dynamics of low-density populations, along with new techniques and technologies to detect and eliminate invading species, and bioeconomic analyses to determine when such efforts might be cost-effective. Researchers should evaluate the role citizen scientists might play in national surveillance systems for the early detection of invasive species. When a target is found, managers need eradication or containment options that are specific to the target and leave resident species unaffected. Managers would benefit from new frameworks to incorporate social issues related to eradication into their decision-making processes.

For well-established invasive species, natural resource managers need tools for analyzing risk and uncertainty in order to prioritize when, where, and how intensively management efforts should occur. Unfortunately, many public and private land managers often operate in "triage mode" when it comes to managing invasive species. They will require new, effective tools and approaches; these management alternatives may stem from advances in genetic technologies, remote sensing, chemical ecology, biological control, or other fields. Their decisions should account for ecosystem uniqueness, potential invader impacts, management goals, available tools, and chances of success. The idiosyncratic nature of each decision requires information from ongoing lines of research and will inspire new avenues of research. Some questions may be very specific, relating to the etiology and epidemiology of an emerging disease or the impact of microbial communities on allelopathic compounds from invasive plants, for example. Others are seemingly more general, such as how the severity of impact changes as the abundance of an invading species declines.

Regardless of the target restoration state (historic or other defined future condition), research suggests that restoration will be more successful if ecosystem function, resilience, and resistance to future invasions are emphasized and monitored to ensure long-term success. However, a well-developed knowledge base and infrastructure to accomplish this goal for all invasive species do not exist. For example, a greater understanding of the resilience and resistance of different ecosystems is needed, as are improvements in plant seeding technology, native plant seed availability, and other revegetation approaches to advance system recovery following degradation or invasive plant control. Guidance and protocols need to be developed that include taking predicted environmental changes into consideration while keeping costs down. Breeding programs are needed to develop resistant hosts that can provide actual solutions to invasive insects and diseases within reasonable time periods, but there is currently a lack of infrastructure and expertise necessary to support such programs. For aquatic invasive species, with their ecological idiosyncrasies and potential for rapid adaptation to new environments, the need for new tactics, strategies, and understanding at a broad scope is likely.

Even when existing science is available for guidance on restoration and rehabilitation, costs to implement many approaches can be operationally prohibitive, especially if landscape-level restoration is needed. Economic models (discussed below) can help to determine the costs and benefits of implementing restoration activities versus accepting the degraded site as a novel community and utilizing any ecosystem services it may provide. However, there is a dearth of landscape-scale operational, economic, and feasible restoration approaches for many invasive species.

16.5 Broaden the Understanding of the Cultural, Social, and Economic Impacts of Invasive Species Presently and Over Time

Invasive species can create important economic and social impacts across many diverse segments and sectors of society. Indeed, understanding economic and social impacts is critical for raising awareness among decision makers and the public of how invasive species may detrimentally impact multiple sectors of society, and why actions to treat invasive species may be proposed. The lack of comprehensive investigations into impacts from invasive species, including the full value of ecosystem services lost, can be a barrier to policymakers and others in forming meaningful lists of priorities and realistic management strategies. Across most aquatic and terrestrial invasive species taxa, decision makers lack quantified economic impacts (both direct and indirect) of invasive species at the local, regional, national, and international levels. As noted in Chap. 9, there is increased recognition of impacts to other resources and systems, such as animal and human health, urban ecosystems, public safety, infrastructure and transportation, and indigenous peoples. Most damage estimates from invasive species at the national (or even global) scale merely add up values from several studies or generalize across diverse landscapes, invasive species, and impact methods. Such studies estimate very high impact values but may violate some of the most basic rules of economic analysis and can do more harm than good (Chap. 14).

In many respects, the challenges from, and responses to, invasive species are social constructs. "Harm" from invasive species reflects a reduction in social, cultural, and economic value that people ascribe to resident individuals, populations, or ecosystems. Similarly, responses to invasive species reflect the degree of willingness to protect or restore threatened entities and to preserve their associated current or future value. Of course, those values must be balanced against other social values such as freedom of movement or economic development. There has been little social science research conducted in the United States on general awareness, attitudes, and behaviors toward invasive species or the entities they threaten. More information is needed about how and why public and private land managers perceive invasive species as problems, and what types of information, assistance, and resources would be most helpful in managing current invasions and the risk of future invasions.

Economics provides a formal framework for exploring tradeoffs among different management alternatives for invasive species. Those tradeoffs are not inherently fiscal, but currency is a useful metric to measure many but not all values. Increasingly, managers and policymakers are considering the efforts to control invasive species as an investment. Research to determine how to measure and forecast future returns on such investments, perhaps as a basis for prioritizing management activities, is needed. Of course, progress in this area will depend on more general advancements in the non-market valuation of ecosystem services. In addition, economics may provide useful approaches for incorporating ecological and management uncertainties into decision-support tools.

Socioeconomic research also provides a lens into human behavior and, in this case, may provide insights about drivers at local, regional, or international scales that can influence the distribution, abundance, and diversity of invasive plants, pathogens, and animals. This may require new methods to incorporate socioeconomic conditions and human dimensions data into pest risk models and to characterize how future human behaviors and related patterns might change pest risks. In particular, more research is needed on the effectiveness of actions specifically intended to change behaviors (e.g., laws and regulations) to lower the spread of invasive species.

Socioeconomics plays a critical role in the success of invasive species management. For example, cooperation across landowners is essential for successful areawide management of invasive species. A greater understanding of factors that determine the likelihood and effectiveness of landowner cooperation, and whether there are tradeoffs between likelihood and effectiveness of landowner cooperation, is needed. Information is also needed about the scale at which landowners should cooperate with one another to be able to affect invasive species management at a landscape scale. Landscape-scale management can engender conflict, particularly among stakeholder groups with different perceptions of the values at risk from an invasive species. Greater interdisciplinary research between biological and social scientists could unveil new approaches for addressing the complexities and uncertainties in invasive species management and reduce the potential for associated social conflicts among stakeholders.

Invasive species can have unique cultural impacts for Native people (Chap. 12). Tribes throughout the country are actively working on invasive species management, often in partnership with other agencies, to protect biocultural diversity. A number of successful collaborations exist between Native tribes and biological scientists to address invasive species, particularly tree pests and diseases, yet more research is needed by or with tribes to understand how cultural values, traditional knowledge, political sovereignty, and other policy issues affect invasive species management.

16.6 Conclusion

The assessment consolidates the state of science for invasive species from a comprehensive array of topics pertinent to invasive species. The information from this assessment is targeted to be useful for multiple stakeholders and decision- and policymakers to more effectively manage invasive species. While considerable progress has been made in understanding multiple aspects of invasions, we highlight four broad research challenges and gaps: (1) improved understanding and quantification of the ecological, economic, and social impacts of invasive species and the mechanisms of invasion; (2) development of better knowledge and models of how climate change may impact invasive species; (3) more cost-effective early intervention, eradication, and control methods and approaches for managing existing invasive species populations; and (4) a better understanding of the economic, social, and cultural dynamics of invasive species. High-quality standardized spatial data for invasive species distribution and abundance across the landscape, change detection (monitoring, longitudinal data sets), cutting-edge tools and technologies, and international cooperation will be needed to support progress for all of the challenges (Chaps. 10, 11 and 13).

The challenges identified here subsume numerous other questions that are described in the preceding chapters. The lists of research needs should not belie the significant advancements that are being made in invasive species research which are also captured in the preceding chapters. The knowledge gained has unquestionably improved our ability to protect lands and waters from the harmful effects of invasive species, yet so much more needs to be done.

Progress to meet these challenges will require close collaboration among professional and citizen scientists, land managers, decision makers, the international community, and other stakeholders. There is an overarching need to develop meaningful measures of progress toward achieving these outcomes. Progress cannot be measured merely by the number of research products (professional presentations, peer-reviewed publications, patents, etc.) produced, but ultimately by the degree to which people, cultures, and natural resources are protected from the harmful effects of invasive species.

Disclaimer Text The findings and conclusions in this publication are those of the authors and should not be construed to represent any official USDA or U.S. Government determination or policy.

Correction to: Invasive Species in Forests and Rangelands of the United States

Therese M. Poland, Toral Patel-Weynand, Deborah M. Finch, Chelcy Ford Miniat, Deborah C. Hayes, and Vanessa M. Lopez

Correction to:
T. M. Poland et al. (eds.), *Invasive Species in Forests and Rangelands of the United States*, https://doi.org/10.1007/978-3-030-45367-1

The original version of this book was revised as the editor has provided belated corrections for Page xiv, Page 2, Page 19, Page 32, Page 141, Page 159, Page 191, Page 192, Page 194, Page 213, Page 325, Page 343, Page 366, Page 368, and Page 423.

Editor has also requested to add "Disclaimer Text" at the end of each chapter. The book corrections has now been incorporated and approved by the Editor.

The updated version of this book can be found at
https://doi.org/10.1007/978-3-030-45367-1

T. M. Poland et al. (eds.), *Invasive Species in Forests and Rangelands of the United States*,
https://doi.org/10.1007/978-3-030-45367-1_17

Appendix: Regional Summaries

The following summaries include information for nine regions of the United States. These summaries provide specific material regarding major issues and invasive species for that region. The regions defined in this report were based on the US Global Climate Change Research Program Report: Climate Change Impacts in the United States, available at http://nca2014.globalchange.gov/downloads.

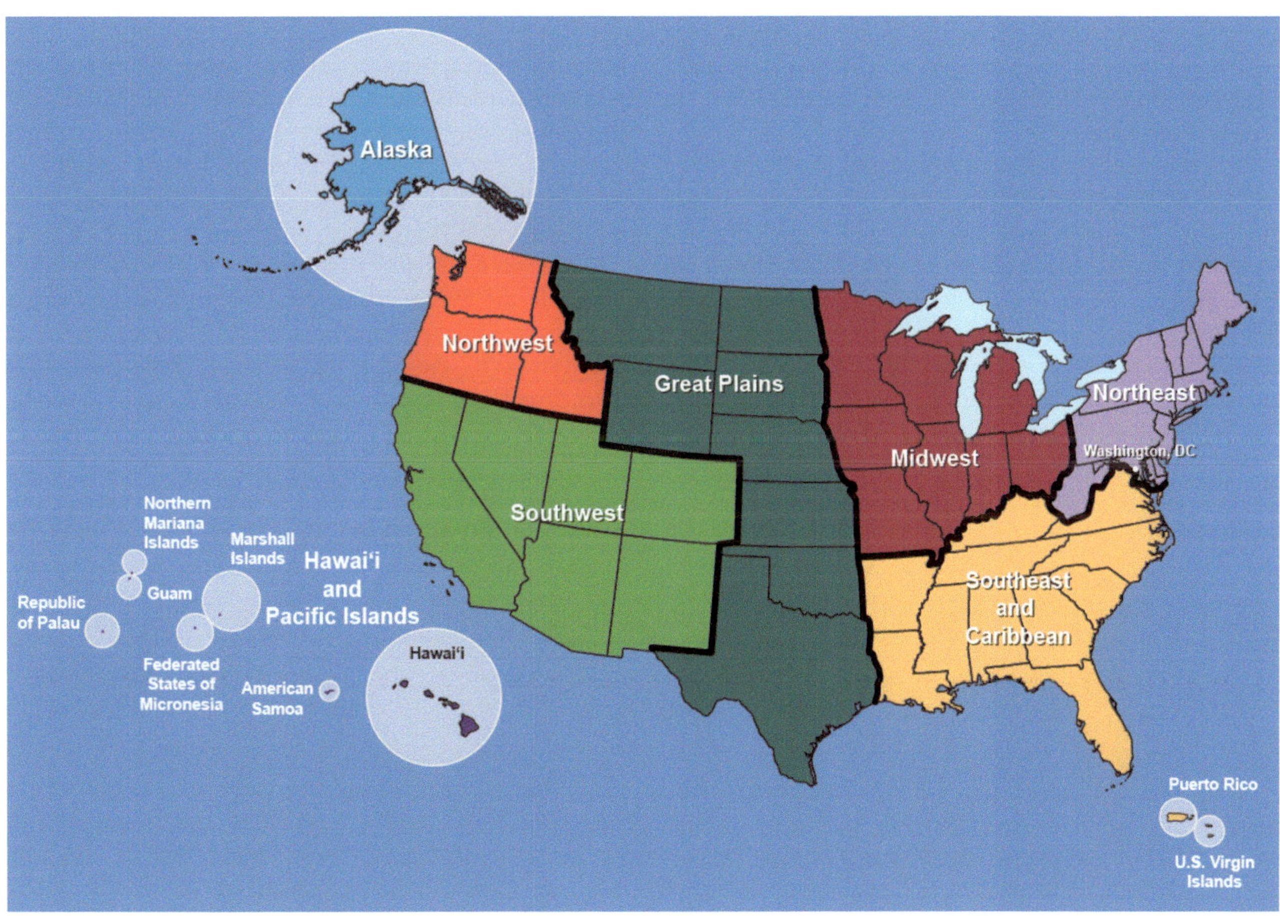

T. M. Poland et al. (eds.), *Invasive Species in Forests and Rangelands of the United States*,
https://doi.org/10.1007/978-3-030-45367-1

Alaska Region

Tricia L. Wurtz and Bethany K. Schulz

Introduction

Alaska has fewer invasive species and is less impacted by invasive species than most places on Earth. Until recently, Alaska has been protected by its cold climate and by its comparative lack of roads and other development. However, these barriers are eroding. Warming climate trends and longer shoulder seasons have reduced the climate filter that so far may have prevented some invasive species from establishing in the State (Figs. A1.1 and A1.2; Carlson et al. 2015; Jarnevich et al. 2014; Sanderson et al. 2012; Wolken et al. 2011). More extensive wildland fire combined with increasing activity in mining, oil and gas extraction, and wilderness tourism are extending the network of travel corridors and altered landscapes that are vulnerable to the establishment and spread of invasive species (Cortes-Burns et al. 2008; Spellman et al. 2014). With 10,680 km of coastline and at least 2670 named islands, Alaska is also vulnerable to invaders in the nearshore marine environment.

Invasive species have been introduced to Alaska via a variety of pathways. Plants have been intentionally introduced for agricultural and commercial purposes, and non-native animals have been introduced for subsistence and sport hunting. Many introductions of invasive species to Alaska, however, have been unintentional. Such pathways include contaminants in agricultural or forestry products; the movement of contaminated road vehicles, boats, and aircraft; and the disposal of live animals and plants from aquaria. Problem species span many taxa: both terrestrial and aquatic plants, mammals, birds, fish, insects, earthworms, and a marine tunicate.

Pathways

Animal Introductions Alaska has a long history of animal introductions, especially on its islands (Bailey 1993; Paul 2009). The first deliberate releases of Arctic foxes (*Vulpes lagopus*) for fox ranching on several Aleutian Islands occurred in 1750 (Black 1984); the first known accidental release of Norway rats (*Rattus norvegicus*) was prior to 1780 via a shipwreck (Brechbill 1977). The rats spread so prolifically that the island where the shipwreck occurred later came to be known as Rat Island (Ebbert and Byrd 2002). It was the site of an intensive, successful rat eradication effort in 2008 for the purpose of restoring seabird nesting habitat (Croll et al. 2016; Dunham 2012; Fritts 2007; USFWS 2007). Cattle (*Bos taurus*), Sitka black-tailed deer (*Odocoileus hemionus sitkensis*), and elk (*Cervus canadensis*) have all been released on islands in Alaska and have proven difficult to manage or remove when their populations thrived beyond intention or when management objectives changed (Ebbert and Byrd 2002).

Alaska encompasses a very large geographic area, and species that occur naturally in one part of the State may behave invasively and problematically when introduced to another. Two prime examples are northern pike (*Esox lucius*) and Alaska blackfish (*Dallia pectoralis*), both of which are native to parts of Alaska north or west of the Alaska Range. Both have been illegally transplanted to Southcentral Alaska, starting in the 1950s, resulting in a number of established populations. The diets of introduced blackfish are similar in composition to the diets of native juvenile salmonids (*Oncorhynchus* spp.) and stickleback (*Gasterosteus aculeatus* and *Pungitius pungitius*), pointing to the potential for competition for prey between blackfish and native fish species (Eidam et al. 2016). Northern pike are highly predatory; where introduced, they have greatly reduced the

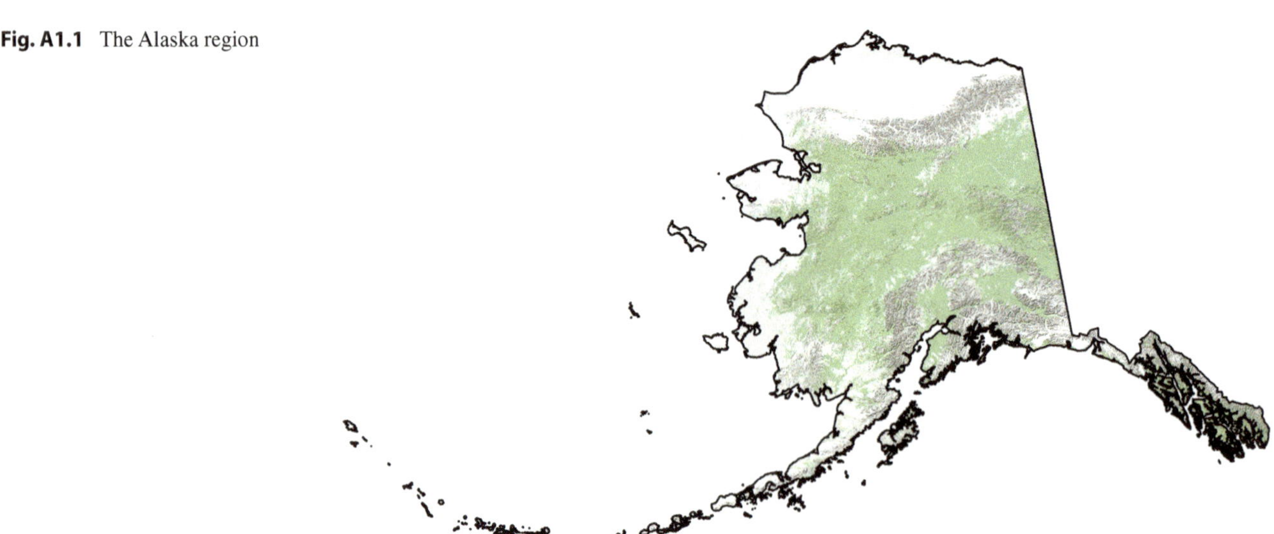

Fig. A1.1 The Alaska region

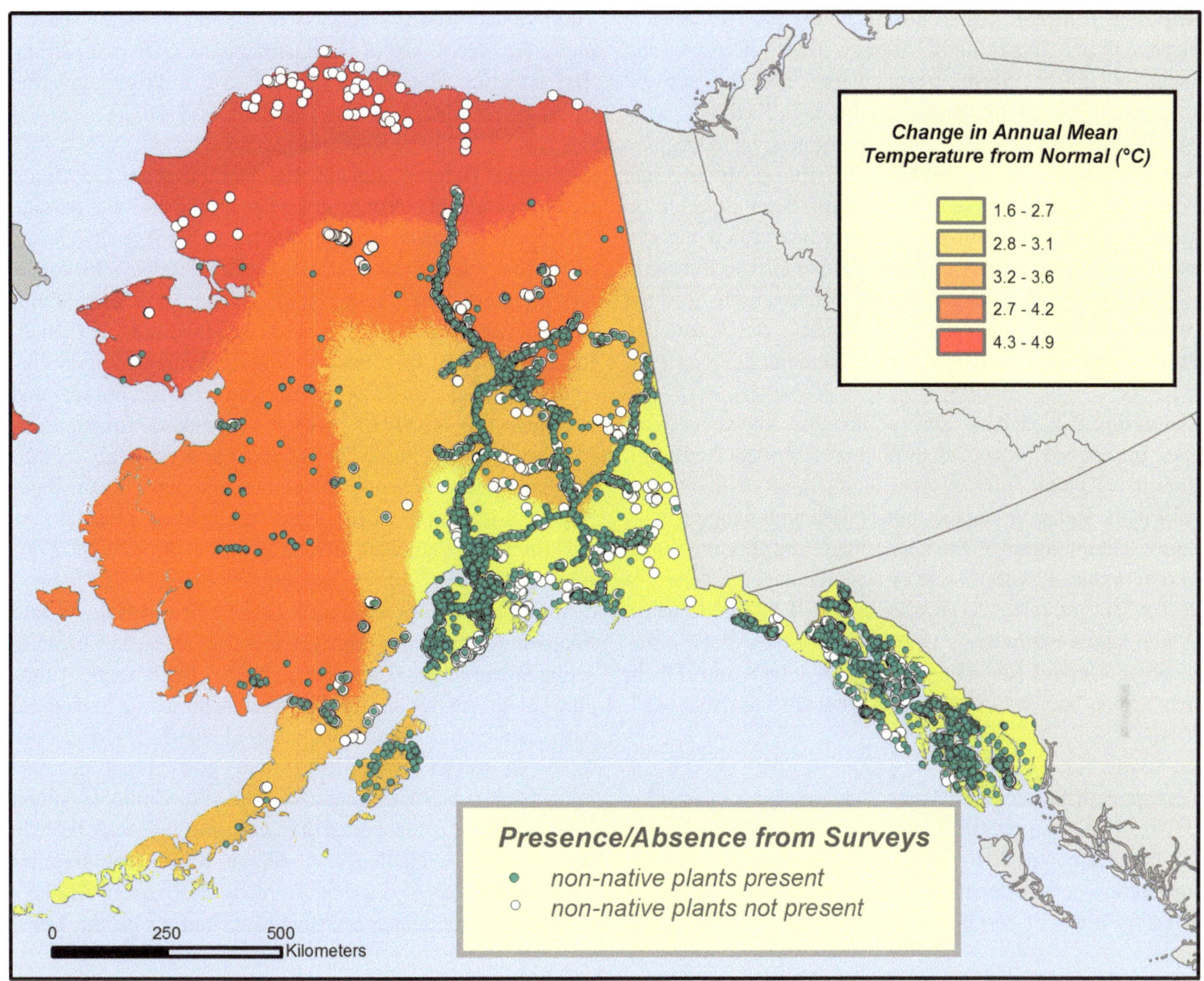

Fig. A1.2 Current known distribution of non-native plant species in Alaska, shown with the projected change in annual mean temperature from the reference period 1970–1999 to the future period 2030–2059. The projected climate is the average of five climate model projections (CMIP5) for a higher emissions scenario (RCP 8.5). Plant data are from the Alaska Exotic Plant Information Clearinghouse database (https://accs.uaa.alaska.edu/invasive-species/non-native-plants/); climate projections are from the US Department of the Interior Alaska Climate Science Center. (Data courtesy of Scenarios Network for Alaska and Arctic Planning (SNAP 2016))

presence of native fish species. Rutz (1999) identified five species of Pacific salmon juveniles in the stomachs of harvested pike in the Susitna River drainage, with coho salmon (*O. kisutch*) juveniles in 59% of non-empty stomachs. The Alaska Department of Fish and Game considers northern pike to be the highest-priority invasive threat in Southcentral Alaska (ADFG 2002).

Agriculture and Horticulture Several of Alaska's most aggressive invasive plant species were initially introduced through agriculture, either for forage (e.g., reed canary grass (*Phalaris arundinacea*), bird vetch (*Vicia cracca*)) or to improve soils (e.g., white sweetclover (*Melilotus albus*)), or accidentally through seed mix impurities (e.g., creeping thistle (*Cirsium arvense*)). Some of these agricultural species were later repurposed for erosion control during road construction, mineral exploration, and mine reclamation. This enlarged the area occupied by the introduced species, providing a greater source for further spread. Horticultural activities have greatly increased the variety of species being introduced, again both intentionally—such as Japanese knotweed (*Fallopia japonica*), Siberian peashrub (*Caragana arborescens*), European bird cherry (*Prunus padus*), Maltese cross (*Silene chalcedonica*), and orange hawkweed (*Hieracium aurantiacum*)—and inadvertently such as perennial sowthistle (*Sonchus arvensis*) in the soil of containerized imported ornamental plants (Conn et al. 2008a). Imported hay and straw have been shown to carry a variety of viable weed seeds (Conn et al. 2010).

Shipping In recent years, introductions have increased in frequency and opportunity through trade (Carlson and Shephard 2007). Shrink-wrapped bundles of firewood originating in Washington State and sold in Fairbanks were found to harbor five species of living insects, including two exotic bark beetles (FS-R10-FHP 2012). European and Asian gypsy moths (*Lymantria dispar*) have been detected in Alaska (either captured in insect traps or intercepted in ports) seven times since 1985 (FS-R10-FHP 2012). The European gypsy moth egg masses were likely carried north from the lower 48 States on recreational vehicles; the Asian egg masses were found on cargo ships from Asia.

Recreational Activities Several taxa are known to have been introduced or spread through recreational activities. Non-native plants are common along most of the hiking trails in the Kenai Mountains, but are rare within natural vegetation communities of the area, suggesting that the plants were introduced by trail users or trail maintenance activities (Bella 2011; Develice 2003; Ware et al. 2012). Three species of non-native earthworms (Lumbricidae) have been introduced on Alaska's Kenai Peninsula, with at least one of them arriving via bait abandonment by anglers (Saltmarsh et al. 2016).

Transport of Aquaculture Gear A marine tunicate invader, "D-vex" (*Didemnum vexillum*), was discovered in Sitka's Whiting Harbor in 2010. D-vex is known to foul shellfish aquaculture gear, hamper scallop movement, and overgrow extensive areas of benthic habitat. This find represented a 1000-km northward extension of the range of this species along the West Coast of North America (Cohen et al. 2011). The attention it garnered has led to a multi-organization effort to test control options (McCann et al. 2013).

Aquarium Release Aquatic plants of the genus *Elodea* are native to much of North America but are not native to Alaska. The genus is widely used in aquaria, and in fact elodea was likely first introduced to Alaska's wild waterways by aquarium dumping. Since the initial find of a small amount of elodea in Eyak Lake in 1982, it has been found in more than 20 locations around the State, with several infestations occurring in urban lakes and waterways in Anchorage and Fairbanks (FS-R10-FHP 2016). It was released from aquaria (one infestation is immediately behind an elementary school) and has spread via boats, floatplanes, and downstream water flow (FS-R10-FHP 2016). Once established in a lake or slow-moving waterway, elodea grows aggressively (Figs. A1.3 and A1.4), with the potential to degrade fish habitat, displace native flora and fauna, impede boat travel and safe floatplane operation, decrease water flow rates, and increase sedimentation rates (Luizza et al. 2016). The amount of suitable habitat for this species in Alaska is projected to increase with the warming climate (Luizza et al. 2016). If elodea continues to spread in Alaska, it may pose direct negative impacts on subsistence practices related to Chinook salmon (*O. tshawytscha*) and whitefish (*Coregonus nelsonii*), thereby posing a major challenge to Alaska Native communities (Luizza et al. 2016). Elodea is already challenging Alaska's natural resource managers. Small infestations on the Kenai

Fig. A1.3 Dense stands of the invasive aquatic plant elodea in a waterway near Fairbanks, AK, and close-up view of elodea (*inset*). (USDA Forest Service photos)

Fig. A1.4 Elodea washed up on the shore of Eyak Lake, in spring 2016. (USDA Forest Service photo)

Peninsula and the Anchorage Bowl have been successfully treated with aquatic herbicides (Morton et al. 2014), but getting control of the State's larger and more challenging infestations (some in remote locations, some in flowing water) will take significant funding and cross-agency cooperation.

In 1982, a school teacher on Chichagof Island in Southeast Alaska purchased frog eggs from a biological supply company for a classroom project and later released about two dozen newly metamorphosed juvenile red-legged frogs (*Rana aurora*) into a small pond there. Red-legged frogs are not native to Alaska. By 2006, the frogs had spread to occupy over 6000 ha of wetland and forested habitats, completely displacing the native amphibian species (Lerum and Piehl 2007; Rozell 2009).

Ecological Effects

Information on the ecological effects of invasive species in Alaska is scant, partly because, beyond the historical mammal releases on islands, invaders are only beginning to move into natural ecosystems from areas disturbed by humans (Oswalt et al. 2015; Rose and Hermanutz 2004). Between 1999 and 2004, an outbreak of larch sawfly (*Pristiphora erichsonii*), an insect native to Europe, impacted an estimated 240,000 ha of interior Alaska, killing roughly 80% of the larch trees (*Larix laricina*) in the affected area (Burnside et al. 2010). In 2010, the green alder sawfly (*Monsoma pulveratum*), an insect native to Europe and North Africa, was found to be widespread in Southcentral Alaska; in some areas it completely defoliated large patches of alder (*Alnus* spp.) (FS-R10-FHP 2011; Kruse et al. 2010). How this sawfly was originally introduced to Alaska is unknown. The European bird cherry, a popular landscape tree, produces fruits that are readily consumed and spread by wild birds. Near monocultures of European bird cherry have formed in some forested urban parklands of Anchorage, including along two urban streams that support runs of wild salmon. Roon (2011) studied the invertebrates present on bird cherry foliage, the biomass of insects falling to the stream below, and the consumption of those insects by juvenile salmon. Riparian bird cherry trees had significantly less invertebrate biomass on their foliage and lower stream input of insects compared to native deciduous trees (Roon et al. 2016). Reduced terrestrial prey subsidies to streams are likely to have negative consequences for salmon as European bird cherry continues to spread (Roon et al. 2016).

White sweetclover was originally introduced to Alaska to increase nitrogen and organic matter content in agricultural soils. Later, sweetclover was planted for roadside stabilization and for reclamation projects associated with oil and gas exploration, pipeline construction, and mining. Alaskan beekeepers have sown it to enhance foraging opportunity for honeybees. From roadsides and reclamation project sites, sweetclover has spread onto the floodplains of at least four major rivers, where its seeds are easily carried downstream. Sweetclover spread from the town of Telegraph Creek in British Columbia down the Stikine River to the Stikine-LeConte Wilderness Area in Southeast Alaska (Conn et al. 2008b, 2011). On early successional floodplain sites, dense stands of sweetclover create novel shade environments; such areas had 50% greater mortality of native seedlings than areas without, suggesting that sweetclover infestations have the potential to change plant community composition (Spellman and Wurtz 2011). A study of reproductive interactions between sweetclover and native cranberry (*Vaccinium vitis-idaea*), blueberry (*V. uliginosum*), and Labrador tea (*Rhododendron groenlandicum*) found a complex relationship, with negative, neutral, and positive effects on pollination and fruit set of native species (Spellman et al. 2013, 2015). Habitat suitability models have projected increasing white sweetclover habitat in Alaska as a consequence of climate change (Jarnevich et al. 2014).

Regulation

Few regulatory measures related to invasive species have been enacted in Alaska (Environmental Law Institute 2002). A Prohibited and Restricted Noxious Weed List was established in 1987, initially focused on species that are likely to be introduced as contaminants in crop seed. Because of this focus, the list fails to include many of Alaska's most aggressive invasive plants. State laws prohibit stocking any fish into waters of Alaska without a permit and releasing unwanted pets into the wild. In 2012, the State banned the use of felt-soled waders to reduce the risk of introducing non-native aquatic pests and in 2014 enacted a quarantine that prohibits the importation, sale, and distribution of five aquatic plant species, including elodea. Although these laws and regulations have been enacted, the State's ability to enforce them is very limited. The State has also initiated voluntary weed-free gravel and forage certification programs.

The Alaska Committee for Noxious and Invasive Plant Management (CNIPM), established in 2000, is an effective network with more than 40 Federal, State, local, and private member organizations. Its strategic plan includes objectives in coordination, education and outreach, prevention, inventory and monitoring, control and management, and research (Alaska CNIPM 2016). Its members use a weed-ranking system (Carlson et al. 2008) and a statewide database of known infestations of non-native plants (Alaska Center for Conservation Science 2016). CNIPM identifies research opportunities to address existing knowledge gaps, determine potential impacts of a changing climate, and improve methods for managing invasive pests (Alaska CNIPM 2016). Monthly conference calls and an annual workshop bring people together to exchange information. CNIPM facilitates communications between regulatory agencies to remove roadblocks to timely treatments and between funding sources and small community organizations that support activities such as local weed surveys, weed pulls, and a vehicle-washing station at the ferry terminal of one island community. In 2016, the focus of this organization broadened from plants to all taxa of invasive species, becoming the Alaska Committee for Noxious and Invasive Pests Management (CNIPM). In 2017 CNIPM changed its name to Committee for Noxious and Invasive Pests Management to reflect the ongoing collaboration across taxa. In 2018 CNIPM changed its name once again to the Alaska Invasive Species Partnership (AISP).

The likelihood of invasive species reaching new parts of Alaska is increasing with resource development and climate change, as demonstrated in modeling exercises (Bella 2011; Carlson et al. 2015; Jarnevich et al. 2014). New gas lines and mines are being developed; new and far-ranging roads are proposed or are under construction. The increasing extent and severity of wildfire means that more land area will be disturbed, much of it vulnerable to the introduction and spread of invasive plants (Cortes-Burns et al. 2008; Spellman et al. 2014). Fortunately, many Alaskans are observant and tuned in to their environment; they're natural citizen scientists (Sigman et al. 2015). In several well-documented cases, initial reports of new invaders have been made by citizens. In 2010, the village environmental officer in the tiny Southeast Alaska community of Kake reported the State's first (and so far, only) infestation of giant hogweed (*Heracleum mantegazzianum*); she recognized the species, Alaska's only known infestation of a Federally listed noxious weed, from a photo included in an outreach booklet. Citizens are monitoring for European green crabs (*Carcinus maenas*) and invasive tunicates in Ketchikan, Sitka, Juneau, Seward, Kodiak, Tatitlek, and Cordova (ADFG 2016). Outreach to both urban and rural communities is an ongoing effort and has led to greater awareness by land managers and the public. Alaskans are taking the lessons learned in other parts of the country to heart and putting those lessons into action.

Literature Cited

Alaska Center for Conservation Science (2016) Alaska exotic plants information clearinghouse (AKEPIC) [Database]. http://aknhp.uaa.alaska.edu/maps-js/integrated-map/akepic.php

Alaska Committee for Noxious and Invasive Plant Management (2016) Strategic plan. https://www.uaf.edu/ces/pests/cnipm/strategic-plans/CNIPM-Strategic-Plan_2016.pdf

Alaska Department of Fish and Game (ADFG) (2002) Alaska aquatic nuisance species management plan. Juneau, Alaska. RIR 5J02–10. 103 p. http://www.adfg.alaska.gov/FedAidPDFs/RIR.5J.2002.10.pdf

Alaska Department of Fish and Game (ADFG) (2016) Green crab monitoring by Alaskan citizen scientists. http://www.adfg.alaska.gov/index.cfm?adfg=invasiveprofiles.europeangreencrab_getinvolved

Bailey EP (1993) Introduction of foxes to Alaskan islands: history, effects on avifauna, and eradication. Resource publication 193. U.S. Dept. of the Interior, Fish and Wildlife Service, Washington, DC. 53 p

Black LT (1984) Atka: an Ethnohistory of the western Aleutians. Limestone Press, Kingston, 219 p

Bella EM (2011) Invasion prediction on Alaska trails: distribution, habitat, and trail use. Invasive Plant Sci Manag 4:296–305

Brechbill RA (1977) Status of the Norway rat. In: Merritt ML, Fuller RG (eds) The environment of Amchitka Island, Alaska. Technical Information Center, Energy Research and Development Administration, TID-26712. Springfield, pp 261–267

Burnside R, Schultz M, Lisuzzo N, Kruse J (2010) Chapter 14. Assessing mortality and regeneration of larch (*Larix laricina*) after a 1999–2004 landscape level outbreak of the larch sawfly (*Pristiphora erichsonii*) in Alaska. In: Potter KM, Conkling BL (eds) Forest health monitoring: National status, trends, and analysis. USDA Forestry Service. General Technical Report. SRS-176, pp 143–150

Carlson ML, Shephard M (2007) Is the spread of non-native plants in Alaska accelerating? In: Meeting the challenge: invasive plants in Pacific Northwest ecosystems. General Technical Report PNW GTR-694. U.S. Department of Agriculture, Forest Service, Pacific Northwest Research Station, Portland, pp 117–133

Carlson ML, Lapina IV, Shephard M et al (2008) Invasiveness ranking system for non-native plants of Alaska. R10-TP-143. U.S. Department of Agriculture, Forest Service, Alaska Region, Anchorage. 21 p

Carlson ML, Aisu M, Trammell EJ, Nawrocki T (2015) Biotic change agents: invasive species. In: Trammell EJ, Carlson ML, Fresco N, Gotthardt T, McTeague ML, Vadapalli D (eds) North Slope rapid ecoregional assessment. Prepared for the Bureau of Land Management, U.S. Department of the Interior. Anchorage, Alaska, pp 109–130

Cohen CS, McCann L, Davis T et al (2011) Discovery and significance of the colonial tunicate *Didemnum vexillum* in Alaska. Aquat Invasions 6:263–271

Conn J, Stockdale C, Morgan JC (2008a) Characterizing pathways of invasive plant spread to Alaska: I. Propagules from container-grown ornamentals. Invasive Plant Sci Manag 1:331–336

Conn J, Beattie K, Shephard M et al (2008b) Alaska *Melilotus* invasions: distribution, origin, and susceptibility of plant communities. Arct Antarct Alp Res 40(2):298–308

Conn J, Stockdale C, Werdin-Pfisterer N, Morgan J (2010) Characterizing pathways of invasive plant spread to Alaska: II. Propagules from imported hay and straw. Invasive Plant Sci Manag 3:276–285

Conn J, Werdin-Pfisterer N, Beattie K, Densmore R (2011) Ecology of invasive *Melilotus albus* on Alaskan glacial river floodplains. Arct Antarct Alp Res 43:343–354

Cortés-Burns H, Lapina I, Klein S et al (2008) Invasive plant species monitoring and control—areas impacted by 2004 and 2005 fires in interior Alaska: a survey of Alaska BLM lands along the Dalton, Steese, and Taylor highways. Prepared for the Bureau of Land Management—Alaska State Office. Alaska Natural Heritage Program, University of Alaska Anchorage, Anchorage. 162 p

Croll DA, Newton KM, McKown M et al (2016) Passive recovery of an island bird community after rodent eradication. Biol Invasions 18:703–715

DeVelice R (2003) Non-native plant inventory: Kenai Trails. USDA Forest Service, Chugach National Forest, R10-TP-124. 24 p

Dunham M (2012) Cleared of rodents, Rat Island gets renamed. Alaska Dispatch News. May 31, 2012. Available from: http://www.adn.com/article/20120531/cleared-rodents-rat-island-gets-renamed

Ebbert SE, Byrd GV (2002) Eradications of invasive species to restore natural biological diversity on Alaska maritime National Wildlife Refuge. In: Veitch CR, Clout MN (eds) Turning the tide: the eradication of invasive species. IUCN SSC Invasive Species Specialist Group. IUCN, Gland/Cambridge, pp 102–109

Eidam DM, von Hippel FA, Carlson ML et al (2016) Trophic ecology of introduced populations of Alaska blackfish (*Dallia pectoralis*) in the cook Inlet Basin, Alaska. Environ Biol Fish 99:557–569

Environmental Law Institute (2002) Halting the invasion: state tools for invasive species management. http://www.eli.org/sites/default/files/eli-pubs/d12-06.pdf

Fritts EI (2007) Wildlife and people at risk: a plan to keep rats out of Alaska. Alaska Department of Fish and Game. Juneau Alaska. 190 p. http://www.adfg.alaska.gov/static/species/nonnative/invasive/pdfs/invasive_rodent_plan.pdf

FS-R10-FHP (2011) Forest health conditions in Alaska 2010. Anchorage, Alaska. USDA Forest Service Alaska Region. R10-PR-23

FS-R10-FHP (2012) Forest health conditions in Alaska 2011. Anchorage, Alaska. USDA Forest Service Alaska Region. R10-PR-25

FS-R10-FHP (2016) Forest health conditions in Alaska 2015. Anchorage, Alaska. USDA Forest Service Alaska Region. R10-PR-038

Jarnevich CS, Holcombe TR, Bella EM et al (2014) Cross-scale assessment of potential habitat shifts in a rapidly changing climate. Invasive Plant Sci Manag 7:491–502

Kruse JJ, Smith DR, Schiff NM (2010) *Monsoma pulveratum* (Retzius) (Hymenoptera: Tenthredinidae: Allantinae), A palaearctic sawfly defoliator of alder in Alaska and new to the United States. Proc Entomol Soc Wash 112(2):332–335

Lerum L, Piehl R (2007) Southeast Alaska, Chichagof Island redlegged frog population status. Progress report. U.S. Department of Agriculture, Forest Service], Admiralty Island National Monument, Juneau, 7 p

Luizza MW, Evangelista PH, Jarnevich CS et al (2016) Integrating subsistence practice and species distribution modeling: assessing invasive elodea's potential impact on native Alaskan subsistence of Chinook salmon and whitefish. Environ Manag 58(1):144–163. https://doi.org/10.1007/s00267-016-0692-4

McCann LD, Holzer KK, Davidson IC et al (2013) Promoting invasive species control and eradication in the seas: options for managing the tunicate invader *Didemnum vexillum* in Sitka. Alaska Mar Pollut Bull 77:165–171

Morton JM, Blackburn BN, Bella E and others (2014) Integrated pest management plan for eradicating elodea from the Kenai Peninsula. Prepared for the Kenai Peninsula Cooperative Weed Management Area. http://www.kenaiweeds.org/pdfs/Integrated%20Pest%20Management%20Plan%20for%20Eradicating%20Elodea_ver3%203.pdf

Oswalt CM, Fei S, Guo Q et al (2015) A subcontinental view of forest plant invasions. NeoBiota 24:49–54. https://doi.org/10.3897/neobiota.24.8378

Paul TW (2009) Game transplants in Alaska. Technical bulletin No. 4, 2nd edn. Alaska Department of Fish and Game. Juneau. 150 pp

Roon DA (2011) Ecological effects of invasive European bird cherry (*Prunus padus*) on salmonid food webs in Anchorage, Alaska streams. M.Sc. thesis, University of Alaska Fairbanks, Fairbanks [Unknown pagination]

Roon DA, Wipfli MS, Wurtz TL, Blanchard AL (2016) Invasive European bird cherry (*Prunus padus*) reduces terrestrial prey subsidies to urban Alaskan salmon streams. Can J Fish Aquat Sci 73(11):1679–1690

Rose M, Hermanutz L (2004) Are boreal ecosystems susceptible to alien plant invasion? Evidence from protected areas. Oecologia 139:467–477

Rozell N (2009) Invasion of the redlegged frogs. Available from https://www.gi.alaska.edu/alaska-science-forum/invasion-redlegged-frogs

Rutz DS (1999) Movements, food, availability and stomach contents of northern pike in selected river drainages, 1996–1997. Division of Sport Fish, Alaska Department of Fish and Game. Fishery Data Series Report No. 99–5. Anchorage. http://www.adfg.alaska.gov/FedAidPDFs/fds99-05.pdf

Saltmarsh DM, Bowser ML, Morton JM et al (2016) Distribution and abundance of exotic earthworms within a boreal forest system in southcentral Alaska. NeoBiota 28:67–86. https://doi.org/10.3897/neobiota.28.5503

Sanderson LA, McLaughlin JA, Antunes PM (2012) The last great forest: a review of the status of invasive species in the North American boreal forest. Forestry 85:329–340

Scenarios Network for Alaska and Arctic Planning (SNAP), University of Alaska (2016) Historical monthly temperature and precipitation 2km CRU TS 3.1, Projected monthly temperature and precipitation 2km CMIP5/AR5, Projected derived temperature products 2km CMIP 5/AR5. Retrieved 11/2016 from http://www.snap.uaf.edu/data.php

Sigman M, Behé C, Dublin R et al (2015) Community-based monitoring of Alaska's coastal and ocean environment: best practices for linking Alaska citizens with science. Alaska Sea Grant, SG-ED-78. Anchorage, AK. 37 p

Spellman BT, Wurtz TL (2011) Invasive sweetclover (*Melilotus albus*) impacts seedling recruitment along floodplains in Alaska. Biol Invasions 13:1779–1790

Spellman KV, Mulder CPH, Carlson ML (2013) Effects of white sweetclover (*Melilotus albus*) invasion on the pollination and berry production of *Vaccinium* sp. in Alaska [Abstract]. In: Proceedings of the ecological society of america annual meeting, Washington, DC

Spellman KV, Mulder CPH, Hollingsworth TN (2014) Susceptibility of burned black spruce (*Picea mariana*) forests to non-native plant invasions in interior Alaska. Biol Invasions 16(9):1879–1895. https://doi.org/10.1007/s10530-013-0633-6

Spellman KV, Schneller LC, Mulder CPH, Carlson ML (2015) Effects of non-native *Melilotus albus* on pollination and reproduction in two boreal shrubs. Oecologia 179:495–507. https://doi.org/10.1007/s00442-015-3364-9

US Fish and Wildlife Service (USFWS) (2007) Restoring wildlife habitat on Rat Island, Alaska Maritime National Wildlife Refuge, Aleutian Islands Unit: Environmental Assessment. Fish and Wildlife Service, Homer, AK. Alaska Maritime National Wildlife Refuge. 152 p

Ware C, Bergstrom DM, Muller E, Alsos IG (2012) Humans introduce viable seeds to the Arctic on footwear. Biol Invasions 12:567–577

Wolken JM et al (2011) Evidence and implications of recent and projected climate change in Alaska's forest ecosystems. Ecosphere 2(11):124. https://doi.org/10.1890/ES11-00288.1

Hawaii and US-Affiliated Pacific Islands

Susan Cordell

Introduction

Hawaii and the US-affiliated Pacific Islands (see Figs. A2.1 and A2.2 for associated map) have high levels of endemic native biodiversity, largely as a function of varied ecosystems and isolation. For example, Hawaii is the most isolated archipelago on Earth where 90% of its 10,000 native species are endemic. This geographic area represents most major forest ecosystem types and includes thousands of oceanic islands, elevational clines from coastal to alpine in temperate and tropical ecosystems, species and communities that occur nowhere else in the world, the tallest mountains on Earth, and a broad range of governments and cultures.

Island ecosystems are particularly vulnerable and susceptible to environmental change from both natural disturbances (wildfire, flood, hurricane, typhoon, and drought) and anthropogenic disturbances (invasive species, deforestation, pollution, and urbanization). Climate change is further exacerbating existing ecological challenges and affecting biological diversity across all biological scales in terrestrial and aquatic ecosystems. This is a global challenge faced around the world, but it is particularly acute in Hawaii and the US-affiliated Pacific Islands, where the loss of biodiversity from extirpations and extinctions is already consequential. An estimated 270 plant and animal species have gone extinct over the past 200 years in Hawaii, with an additional 440 species threatened or endangered (Wagner et al. 1999).

Despite the recognized prominence of invasive species in most of the terrestrial ecosystems in Hawaii and the US-affiliated Pacific Islands, it is unclear if this is a result of more invasive species being introduced or if human disturbance allows greater establishment. In general, Hawaii is considered ground zero for invasive species (Dawson et al. 2017), and with many other ecosystems likely to be impacted, Hawaii can serve as a model system through innovative approaches to both reduce the impacts of invasive species and enhance the resilience of native species assemblages.

Hawaii Overview

Hawaii is considered the most isolated archipelago on Earth with an estimated historic rate of one new species established every 100,000 years (Fosberg 1948). Ultimately, the approximate 1100 native flowering plants now in Hawaii originated from approximately 270 to 280 successful colonization events. Further, those species that not only arrived but successfully colonized harbored traits of long distance travel, endurance, and adaptive flexibility (Wagner et al. 1999). As a result, Hawaii's native flora, fauna, and ecosystems are considered disharmonic with taxa either under- or overrepresented relative to the founder biomes. Currently, there are more endangered species per square mile on these islands than any other place on the planet, and most of these species—and the ecosystems in which they live—are found nowhere else in the world (see https://ecos.fws.gov/ecp0/reports/species-listed-by-state-report?state=HI for more information). Hawaii is home to nearly one third of all Federally listed threatened and endangered species and almost half of all listed plants. The total number of listed plant species in Hawaii has increased by 40% over the last three decades, and over 100 of these have fewer than 20 known individuals (Loope 1998) (also see https://ecos.fws.gov/ecp/species-reports for current statistics).

Invasive species have caused significant ecological and economic damage in Hawaii (see https://dlnr.hawaii.gov/hisc/info/). It is estimated that Hawaii spends $50 million

Fig. A2.1 Hawaii and Guam

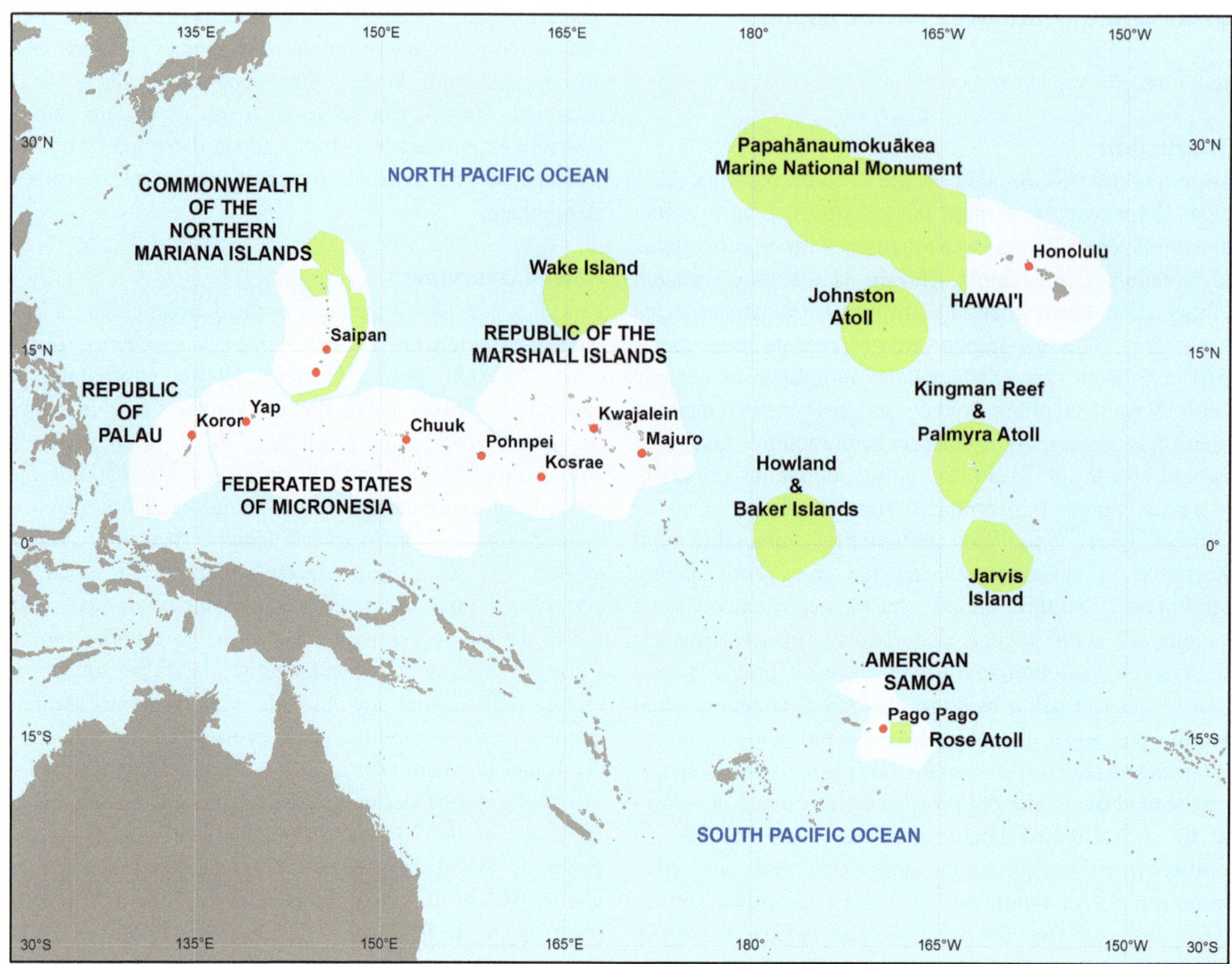

Fig. A2.2 Map of the Hawaii Archipelago and the US-affiliated Pacific Islands

annually to manage invasive species and that costs associated with present and potential invasive species could top $180 million per year. Some important invasive species impacting Hawaii's native and agricultural ecosystems include wild boar (*Sus scrofa*), coconut rhinoceros beetle (*Oryctes rhinoceros*), and the pathogen causing rapid 'ōhi'a death (*Ceratocystis lukuohia and huliohia*). Please see Table A2.1, Fig. A2.3, and https://dlnr.hawaii.gov/hisc/info/ for a list of top invaders. The impacts of invasive species are vast in the Hawaii and Pacific region, especially when considering the diversity of taxa that has established in these susceptible island ecosystems. For example, a newly identified invasive pathogen that causes rapid 'ōhi'a death, *Ceratocystis lukuohia*, has killed hundreds of thousands of 'ōhi'a (*Metrosideros polymorpha*) across more than 50,000 acres of forests and residential areas in east and south districts of Hawaii Island. 'Ōhi'a is a keystone native tree species which occupies 80% of Hawaiian forests across different elevations and habitats and exists in many forms. The tree mortality caused by rapid 'ōhi'a death is anticipated to have widespread impacts to native wildlife and Hawaiian culture.

Protecting the region from the introduction and potential impacts of invasive species is an integral component of land management, and preventing the establishment and spread of invasive plants has continued to be a high priority because these species have been shown to affect all trophic levels of an ecological community (Loope and Mueller-Dombois 1989; Smith 1985). Therefore, the emphasis of this regional study is focused on the impacts, consequences, and mechanisms associated with invasive plants. For reviews on the environmental and economic impacts of other invasive taxa in Hawaii, please see Arcilla et al. (2015), Chynoweth et al. (2013), Day and Winston (2016), Doherty et al. (2016), Marchetti and Engstrom (2016), Pitt et al. (2017), and Spatz et al. (2017).

Invasive Plants

In today's world, Hawaii's combined native and non-native flora is close to 5000 species. Approximately 22 taxa are introduced per year with 869 non-native species becoming established in the last 200 years (Reichard and White 2001; Wagner et al. 1999). In a report by Schmidt and Drake (2011), 7866 ornamental species were brought into the islands from 1840 to 1999. Of

Table A2.1 A list of the most impactful invasive species in Hawaii by taxon

Terrestrial plants	Albizia (*Falcataria moluccana*)
Banana poka (*Passiflora tarminiana*)	Tibouchina (*Tibouchina herbacea*)
Fountain grass (*Cenchrus setaceus*)	Glory bush (*Tibouchina urvilleana*)
Gorse (Ulex europaeus)	Hiptage (*Hiptage benghalensis*)
Ivy gourd (*Coccinia grandis*)	Kāhili ginger (*Hedychium gardnerianum*)
Miconia (*Miconia calvescens*)	Prickly blackberry (*Rubus argutus*)
Strawberry guava (*Psidium cattleianum*)	Devil weed (*Chromolaena odorata*)
Fireweed (*Senecio madagascariensis*)	**Marine aquatic plants**
Brown licorice (*Kappaphycus striatus*)	Gorilla ogo (*Gracilaria salicornia*)
Leather mudweed (*Avrainvillea amadelpha*)	[no common name] (*Cladophora sericea*)
Smothering seaweed (*Kappaphycus* and *Eucheuma* spp.)	Hookweed (*Hypnea musciformis*)
Spiny prickly seaweed (*Acanthophora spicifera*)	
Terrestrial vertebrate animals	Axis deer (*Axis axis*)
Coqui frog (*Eleutherodactylus coqui*)	Domestic goat (*Capra hircus*)
Giant marine toad (*Bufo marinus*)	Feral cats (*Felis catus*)
Jackson's chameleon (*Chamaeleo jacksonii*)	Mongoose (*Herpestes javanicus*)
Red-vented bulbul (*Pycnonotus cafer*)	Red-whiskered bulbul (*Pycnonotus jocosus*)
Veiled chameleon (*Chamaeleo calyptratus*)	Wild boar (*Sus scrofa*)
Terrestrial invertebrate species	Apple snail (*Pomacea canaliculata*)
Black twig borer (*Xylosandrus compactus*)	Coconut rhinoceros beetle (*Oryctes rhinoceros*)
Coffee berry borer (*Hypothenemus hampei*)	Erythrina gall wasp (*Quadrastichus erythrinae*)
Giant African snail (*Achatina fulica*)	Little fire ant (*Wasmannia auropunctata*)
Nettle caterpillar (*Darna pallivitta*)	Red imported fire ant (*Solenopsis invicta*)
Rosy wolfsnail (*Euglandina rosea*)	Varroa mite (*Varroa destructor*)
Mosquito (*Culex quinquefasciatus*)	Oriental fruit fly (*Bactrocera dorsalis*)
Freshwater aquatic animals	Apple snail (*Pomacea canaliculata*)
Armored catfish (*Hypostomus watwata*)	Jewel cichlid (*Hemichromis elongatus*)
Freshwater bivalve (*Corbicula fluminea*)	
Infectious organisms	Banana bunchy top virus
Coconut heart rot (*Phytophthora katsurae*)	Papaya ringspot virus
Rat lungworm (*Angiostrongylus cantonensis*)	Rapid 'ōhi'a death (*Ceratocystis fimbriata*)
Marine aquatic animals	Australian mullet (*Osteomugil engeli*)
Blueline snapper or ta'ape (*Lutjanus kasmira*)	Peacock grouper or roi (*Cephalopholis argus*)
Snowflake coral (*Carijoa riisei*)	Orange sponge (*Mycale armata*)

these, 420 were considered naturalized, and 141 were listed as weeds with 39 of those listed as noxious. A more recent report by Dawson et al. (2017) which captured native and non-native naturalized flora and fauna in Hawaii lists 1586 native plant species and 1488 non-native plant species (Table A2.2). This list includes dicots, monocots, gymnosperms, and ferns.

The welcome mat for non-native plant species in Hawaii was laid out as the first humans arrived from Polynesia more than 1500 years ago. These Polynesian plant introductions—better known as "canoe plants"—were somewhat benign compared with what was to come as the exponential rise in transport to and from the islands provided numerous opportunities for alien colonizers (Loope et al. 1988). The combination of advances in agricultural technology, increases in population density, and the introduction of alien animals (ungulates, rodents, etc.) led to a multipronged induced decline of Hawaii's native biota (Denslow 2003; Loope et al. 1988). Further, in the 1930s, State and Federal foresters deemed that many of the native tree species in Hawaii lacked "utility" which resulted in the planting of 1026 non-native species into forest reserves, many of which became invasive (Woodcock 2003). Correspondingly, the seed banks of most Hawaiian systems—even those that are native-dominated—are now saturated with non-native species, indicating that these forests will likely be heavily impacted and influenced by non-native and perhaps invasive species when disturbed (Cordell et al. 2009; Drake 1998; Nonner 2005). The extreme habitat complexity of the islands due to large gradients of elevation, productivity, and climate (ranging from dry desert to subalpine climates) has offered comfortable homes for a broad suite of temperate and tropical species. Almost every traveler can find a familiar plant from home while visiting Hawaii.

Why Are Invasive Plant Species So Successful in Hawaii?

Attempts to define unified themes and/or hypotheses about the success of invasive species are as difficult in Hawaii as it is

Fig. A2.3 A pictorial view of some of Hawaii's most problematic invasive species. See Table A2.1 for scientific names. (**a**) albizia; (**b**) rapid 'ōhi'a death; (**c**) strawberry guava; (**d**) fountain grass; (**e**) rhinoceros beetle; (**f**) mongoose; (**g**) wild boar; (**h**) clidemia

Table A2.2 Estimates of the number of introduced and native species in Hawaii grouped by taxon (data are from Zeigler (2002) and Dawson et al. 2017)

Taxon	Total	Introduced and naturalized	Native
Plants	3074	1488	1586
Insects	7982	2592	5390
Spiders	205	77	128
Snails	804	34	770
Fishes	51	45	6
Amphibians	7	7	0
Reptiles	25	25	0
Birds	158	54	104
Mammals	22	20	2

anywhere (Catford et al. 2009). While classic success pathways related to plant invasion are evident and perhaps even predominant in Hawaii (Denslow 2003), Daehler (2003) argues that "outside of super-invaders, increased resource availability and altered disturbance regimes associated with human activities often differentially increase the performance of invaders over natives." Others add that, once disturbed, other classic hypotheses such as superior traits (reproductive output, growth rates) and competitive abilities (resource acquisition) of invasive species over natives as well as the enemy release hypothesis (Keane and Crawley 2002; Kellner et al. 2011), non-native herbivores, and loss of frugivores and pollinators (Carlquist 1980) directly interact and shape a trajectory from native to invaded communities. In a recent paper, where Hawaii leads the globe as the number one hotspot for non-native species, Dawson et al. (2017) show that invasion success correlates with a region's wealth, population density, and climate. They further argue that wealthier islands are more vulnerable because these areas have more points of entry.

Impacts and Effects of Invasive Species

Disturbance Regimes At the ecosystem level, invasive species in Hawaii have introduced a number of severe positive feedback level disturbance regimes that have overwhelmed the natural resource management community. In tropical wet habitats, fast-growing and often nitrogen (N)-fixing invasive tree species have substantially increased nutrient cycling which facilitates further invasion (Hughes and Denslow 2005; Ostertag et al. 2009). Many times, the native species adapted to these ecosystems are often outcompeted and slowly eliminated from the mix. In separate but similar studies, Allison et al. (2004) and Hughes and Denslow (2005) convey evidence of noticeable increases in litter nutrients, decomposition rates, and nutrient availability following invasion in lowland wet forests. Both argue that these shifts facilitate further invasion and alternative stable states.

In tropical dry and subalpine systems, fire-promoting tropical grasses from Africa have introduced a positive-feedback grass-fire cycle resulting in the almost complete loss of native forest. In Hawaii, for example, fires were generally infrequent and limited in size prior to human-induced changes in native ecosystems (Loope and Mueller-Dombois 1989). Over the past century, however, wildfire frequency and size have increased dramatically as a result of invasion by fire-promoting alien grasses (Hughes et al. 1991; Smith and Tunison 1992; Trauernicht et al. 2015). These grasses increase fine fuel loads and alter fuel structure in ways that increase the likelihood of fire ignition and spread. Furthermore, fire effects and post-fire environmental conditions promote recruitment of non-native grasses and inhibit recruitment of native woody species. These changes in community structure and composition result in fuel and microclimate conditions that increase the likelihood of subsequent fire (Freifelder et al. 1998). In this way, non-native grass invasion initiates a grass fire cycle that converts native forest to non-native-dominated grassland (D'Antonio and Vitousek 1992). This cycle is now considered the primary agent of forest-to-grassland conversion in dry and mesic plant communities in Hawaii and elsewhere in the tropics (Mack and D'Antonio 1998). While the total area burned annually in Hawaii may be small relative to that of fire-prone areas of the US mainland, the proportion of land burned in Hawaii is greater (0.48% area burned in Hawaii relative to 0.30% in the continental United States) indicating that the potential damage to natural resources posed by fires in Hawaii is profound (Trauernicht et al. 2015). Over 90% of the original Hawaiian dry forests have been destroyed (Bruegmann 1996; Mehrhoff 1993), and over 25% of the officially listed endangered plant taxa in the Hawaiian flora are from dry forest or dry-scrub ecosystems (Sakai and Wagner, unpublished data).

Carbon and Nutrient Cycling In Hawaii, changes in carbon and nutrient dynamics as a result of invasive species are ecosystem-dependent especially when they possess plant functional traits not represented in the native flora (Vitousek 1986). For example, introduced grass species in Hawaii result in the transformation of a forest to a grassland yielding a massive loss of aboveground carbon and large impacts on nutrient cycling (Mack 2003). In a tropical dry ecosystem in Hawaii, total aboveground biomass was 93% lower within a grass-converted site relative to forested plots (Litton et al. 2006). In wet ecosystems, the highly invasive firetree (*Morella faya*) also reduced carbon stocks, although not through a state transition as described above, but rather through shading out light-dependent native understory and midstory species (Asner et al. 2008). In a lowland wet forest experiment in which all alien species were removed, productivity as measured by litterfall decreased by 40% when compared with the invaded site (Ostertag et al. 2009). Further, in N-limited Hawaiian substrates, invasive N-fixing tree species have the greatest impact on ecosystem function through strong increases in carbon and nutrient cycling and carbon storage (Allison and Vitousek 2004; Hughes and Denslow 2005; Hughes et al. 2014; Mascaro et al. 2012). Even the invasive coqui frog (*Eleutherodactylus coqui*) was shown to increase invasive plant growth and increase litter decomposition rates by reducing herbivores and increasing new leaf production in a nutrient-poor site thereby conferring a competitive advantage to invasive plants in an ecosystem where native species have evolved in nutrient-poor conditions (Sin et al. 2008). Other studies have linked these altered biogeochemical cycles to facilitation of additional problematic invasive species (Hughes and Denslow 2005; Vitousek 1986), ultimately shifting these systems to entirely novel ecosystems (Mascaro et al. 2012). Nutrient additions from invasive species via litterfall inputs are also positively correlated with non-native earthworm (*Eisenia* spp.) density leading to increases in feral pig activity (Aplet et al. 1991; Zou 1993) which in turn disperses non-native fruits. This positive-feedback cycle links invasive species both by altering the nutrient cycle and facilitating the density and spread of additional invaders.

Competition for Resources Invasive species in Hawaii often outcompete native species because they have superior traits and use resources more efficiently (Baruch and Goldstein 1999). In addition, some invaders have the ability to utilize limiting resources by maximizing performance during times of abundance and avoiding stressful conditions during times of scarcity (Stratton and Goldstein 2001; Vitousek 1986). Pattison et al. (1998) showed definitively that invasive species in a wet tropical forest in Hawaii had higher growth rates than natives, particularly when compared in low-light environments. In a similar system, Cordell et al. (2016) indicated a parallel trend in wet forest seedlings where introduced and invasive pioneer species from Central America and Asia grow significantly faster than the highly conservative and slow-growing dominant native forest species. To further the impact, many Hawaiian species in these systems are shade-intolerant so are quickly displaced by shade-tolerant invaders once there is canopy closure (McDaniel and Ostertag 2010; Schulten et al. 2014). High-resolution remote sensing imagery confirms this transition by showing alteration of the three-dimensional structure of a Hawaiian rainforest following invasion through the loss of midstory and understory native species (Asner et al. 2008). Hawaiian forests are typically characterized by more open canopies than their continental tropical counterparts, and most native Hawaiian forest species require high-light environments for germination and survival (Burton and Mueller-Dombois 1984; Drake 1993; Drake and Mueller-Dombois 1993). Unfortunately, the high-light conditions in Hawaiian lowland wet forests are also conducive to the establishment and growth of invasive species. For example, Pattison et al. (1998) found that invasive species in Hawaii were able to capture and utilize light more efficiently than the natives, and their photosynthesis rates were significantly higher in high-light environments.

In tropical dry systems in Hawaii—and even in wet systems to a certain extent—water is a predominant limiting resource. Rainfall in these systems is episodic and somewhat aseasonal, and the flora that evolved and adapted to these conditions tend to have conservative growth strategies. As a result, drought-tolerant and drought-adapted invasive species that exhibit plasticity in morphological and physiological traits when resources are abundant or scarce can quickly outcompete native species. In the leeward (dry) sides of the islands, drought-tolerant grasses from Africa have been highly successful based on their ability to quickly utilize water resources when they are abundant and retreat to dormancy during times of drought (Williams et al. 1995). Experimental invasive grass removal in a tropical dry system on Hawaii Island resulted in a 40% increase in growth rates of native tree species relative to the grass-invaded plots (Cordell and Sandquist 2008). Oxygen isotope data from this study further revealed that the grasses were able to access water from low-rainfall (<10 mm) events more effectively than native species through their shallow root systems. In contrast, invasive mesquite trees (*Prosopis* spp.) in a similar system are able to dominate by accessing groundwater resources though deep-rooted taproots (Miyazawa et al. 2015). In wet systems periodic droughts are also common (Michaud et al. 2015). In an invasive species removal experiment in a lowland wet forest in Hawaii where non-native and invasive species are known to utilize more water resources than natives, Cavaleri et al. (2014) revealed an increase in water use by native species, and stand-level water use within the removal plots was half that of the invaded plots.

Climate Change and Plant Invasion Recent advances in climate modeling in Hawaii using the Coupled Model Intercomparison Project Phase 5 (Timm et al. 2015) indicate warmer-wetter scenarios for the windward slopes of the high islands and warmer-drier conditions for the remainder of the State. Other expected changes include seasonal and interannual variability accentuation (Lauer et al. 2013). These modeling efforts have been used to further predict the fate of Hawaii's plant and animal populations, including invasive species. Dramatic declines in native birds are predicted as warmer temperatures increase the prevalence of avian malaria in suitable bird habitat (Liao et al. 2015), and extinction is projected for some native plant populations when habitat area as defined by climatic envelopes is impinged by sea level rise on the lower end and elevation at the upper end (Fortini et al. 2013). An invasibility metric defined by Vorsino et al. (2014) used current knowledge of invasions from species distribution models and expected climate-driven reductions of native ecosystems to project plant invasion onto a 2100 Hawaii regional climate change scenario. While most of the area occupied by invasive species increased in size, the area of two highly invasive species occupying upper elevation wet forest decreased dramatically.

Management All land management agencies in Hawaii are tasked with managing invasive species because of their detrimental effects on native biodiversity. Numerous strategies have been attempted with efforts primarily invested in chemical and manual control (including exclusion via fences) methodologies; however, many of these efforts are challenging due in part to Hawaii's often inaccessible steep and rugged terrain. Most argue that an integrated pest management approach is required, including programs to exclude and screen new species, strategies to reduce or eradicate species, public education, and a biological control program for the most problematic species (Medeiros et al. 2013). The Hawaii Pacific Weed Risk Assessment (Daehler et al. 2004), a cost-efficient tool that uses a modified version of the Australia and New Zealand tool, has been useful for educating the public and the horticulture industry about the potential effects of species introductions. Beginning in the early 2000s, the State of Hawaii invested in interdepartmental island-specific Invasive Species Councils "for the special purpose of providing policy level direction, coordination, and planning among State departments, Federal agencies, and international and local initiatives for the control and eradication of harmful invasive species infestations throughout the State and for preventing the introduction of other invasive species that may be potentially harmful."

Other Approaches The use of forest plantations and restoration to reduce problematic invaders and promote native biodiversity has been an effective strategy in many parts of the tropics (Holl et al. 2011; Lugo 1997). Unfortunately this is not the case in Hawaii, where forest plantations have been shown to facilitate other weedy and invasive species, rather than increasing native species recruitment (Ostertag et al.

2008). However, forest restoration in dry systems in Hawaii does hold promise as a tool to shade out light-dependent invasive grasses (Cordell et al. 2004; McDaniel and Ostertag 2010), including attempts to reduce the highly invasive gorse (*Ulex europaeus*) using the native koa tree (*Acacia koa*). Girdling and thinning of highly invasive tree species in Hawaii Volcanoes National Park were both effective treatments in reducing new invasion and facilitating native species recruitment (Loh and Daehler 2007), while a new approach that uses functional traits to select species combinations of native and non-native (but noninvasive) to reduce invasion in lowland wet ecosystems in Hawaii holds promise as an effective tool (Cordell et al. 2016; Ostertag et al. 2015).

US-Affiliated Pacific Islands Overview

The state of the science of the impact of invasive plant species in the US-affiliated Pacific Islands is sparse largely due to the remote setting and lack of resources such as trained specialists. These areas include American Samoa, Guam, the Marshall Islands, the Northern Mariana Islands, Palau, and the Federated States of Micronesia (see Fig. A2.2). Moreover, the pronounced diversity of types of island or groups of islands (atolls to volcanic high mountains, isolated to archipelagos) and variation in land-use history makes it challenging to set priorities or find unifying themes. In the late 1990s, the islands of Micronesia and American Samoa were extensively surveyed for invasive plant species at the request of the Pacific Islands Forestry Committee, Council of Western State Foresters (Space and Falanruw 1999). The survey resulted in comprehensive lists of invasive plant species throughout the region grouped in the following five categories:

1. Species that are invasive elsewhere in similar ecosystems but were not listed in the literature during the time of this survey as being present in Micronesia (82 species)
2. Species that are invasive elsewhere and are also invasive in Micronesia (13 species)
3. Species that are not known to be particularly invasive elsewhere but are invasive in Micronesia (3 species)
4. Species that are invasive or weedy elsewhere and are common or weedy but not yet invasive in Micronesia (117 species)
5. Native species that exhibit aggressive behavior following disturbance (16 species)

As a result of these surveys, local forestry departments now prioritize their activities to keep the worst invasives at bay and educate communities about the economic and biological impacts of invasive species. In 2009, Denslow et al. surveyed Pacific Islands extensively to address the roles of biogeographic, environmental, and socioeconomic impacts on the distribution and spread of exotic species. Across the Pacific Islands, size, elevation, and the presence of an airport with a paved runway all correlated with exotic species richness, whereas analysis by country revealed that exotic species richness was associated with size, gross domestic product, and population density (Denslow et al. 2009). A broader survey across tropical oceanic islands (Kueffer et al. 2010) found that the best predictor of invasive species across all islands sampled was human development and habitat diversity. Further, the number of dominant invaders decreased with increasing age of the island. When comparing flora across islands, they conclude that invasive species found in an island group are also almost always present in other groups where they may not yet be considered invasive. This finding clarifies the need for early detection monitoring programs on oceanic islands (Kueffer et al. 2010). In 2017, Pysek et al. looked at many factors associated with naturalized non-native flora of the world. In their analysis, they included information from Hawaii, Guam, American Samoa, and the Northern Mariana Islands. What really stands out in their report is that islands, and in particular tropical islands, have a much higher percentage of alien flora than other systems. For example 66% of Guam's, 48% of Hawaii's, 41% of the Northern Mariana Islands', and 33% of American Samoa's total flora are non-native, whereas 23% of California's and 30% of Florida's flora are represented by non-native species.

Aside from island-wide surveys, weed risk assessments, and comparative studies, very little work has been done in this region on the impacts and consequences of invasive species. However, promising results from two studies indicate that invasive species removal in American Samoa (Hughes et al. 2012) and the cessation of anthropogenic disturbance regimes (primarily fire) in Palau (Costion et al. 2012) have resulted in the recovery of native forest.

Conclusions

The historical and continued degradation of Hawaiian and Pacific Island ecosystems has opened the door to species invasions that have, in turn, transformed entire ecosystems and altered historic disturbance regimes. Together with the unique character of Hawaii and the US-affiliated Pacific Islands' plant and animal species (isolation, biogeography, endemic flora), invasive species are leading many of these ecosystems to a crossroad in which all native systems are no longer sustainable. In essence, Hawaii is ground zero for invasive species. Scientists estimate that more than 10,000 non-native plant and animal species have been brought to Hawaii with more than 100 of those causing extreme damage to Hawaii's native ecosystems. If we are to save these ecosystems and the associated native species from extinction, innovative approaches are needed to both reduce impacts of invasive species and enhance the resilience of native species assemblages.

Literature Cited

Allison SD, Vitousek PM (2004) Rapid nutrient cycling in leaf litter from invasive plants in Hawaii. Oecologia 141:612–619

Aplet GH, Anderson SJ, Stone CP (1991) Association between feral pig disturbance and the composition of some alien plant assemblages in Hawaii volcanoes National Park. Vegetatio 95(1):55–62. https://doi.org/10.1007/bf00124953

Arcilla N, Choi C, Ozaki K, Lepczyk CA (2015) Invasive species and Pacific island bird conservation: a selective review of recent research featuring case studies of Swinhoe's storm petrel and the Okinawa and Guam rail. J Ornithol 156(1):199–207

Asner GP, Hughes RF, Vitousek PM et al (2008) Invasive plants transform the three-dimensional structure of rain forests. Proc Natl Acad Sci 105(11):4519–4523. https://doi.org/10.1073/pnas.0710811105

Baruch Z, Goldstein G (1999) Leaf construction cost, nutrient concentration, and net CO_2 assimilation of native and invasive species in Hawaii. Oecologia 121(2):183–192

Bruegmann MM (1996) Hawaii's dry forests. Endangered Spec Bull 21(1):26–27

Burton PJ, Mueller-Dombois D (1984) Response of *Metrosideros polymorpha* seedlings to experimental canopy opening. Ecology 65(3):779–791

Carlquist S (1980) Hawaii: a natural history, 2nd edn. Pacific Tropical Botanical Garden, Lawai, 468 pp

Catford JA, Jansson R, Nilsson C (2009) Reducing redundancy in invasion ecology by integrating hypotheses into a single theoretical framework. Divers Distrib 15(1):22–40

Cavaleri MA, Ostertag R, Cordell S, Sack L (2014) Native trees show conservative water use relative to invasive trees: results from a removal experiment in a Hawaiian wet forest. Conserv Physiol 2(1):cou016

Chynoweth MW, Litton CM, Lepczyk CA et al (2013) Biology and impacts of Pacific Island invasive species. 9. *Capra hircus*, the feral goat (Mammalia: Bovidae). Pac Sci 67(2):141–156

Cordell S, Sandquist DR (2008) The impact of an invasive African bunchgrass (*Pennisetum setaceum*) on water availability and productivity of canopy trees within a tropical dry forest in Hawaii. Funct Ecol 22(6):1008–1017

Cordell S, Sandquist DR, Litton CM et al (2004) An invasive grass has significant impacts on tropical dry forest ecosystems in Hawaii: the role of science in landscape level resource management and native forest restoration in West Hawaii. In: 16th international conference, society for ecological restoration, Victoria, Canada, pp 1–4

Cordell S, Ostertag R, Rowe B et al (2009) Evaluating barriers to native seedling establishment in an invaded Hawaiian lowland wet forest. Biol Conserv 142(12):2997–3004. https://doi.org/10.1016/j.biocon.2009.07.033

Cordell S, Ostertag R, Michaud J, Warman L (2016) Quandaries of a decade-long restoration experiment trying to reduce invasive species: beat them, join them, give up, or start over? Restor Ecol 24:139–144

Costion CM, Liston J, Kitalong AH et al (2012) Using the ancient past for establishing current threat in poorly inventoried regions. Biol Conserv 147(1):153–162. https://doi.org/10.1016/j.biocon.2011.12.026

D'Antonio CM, Vitousek PM (1992) Biological invasions by exotic grasses, the grass/fire cycle, and global change. Annu Rev Ecol Syst 23:63–87

Daehler CC (2003) Performance comparisons of co-occurring native and alien invasive plants: implications for conservation and restoration. Annu Rev Ecol Evol Syst 34(1):183–211

Daehler CC, Denslow JS, Ansari S, Kuo H-C (2004) A risk-assessment system for screening out invasive pest plants from Hawaii and other Pacific Islands. Conserv Biol 18(2):360–368

Dawson W, Moser D, Kleunen M et al (2017) Global hotspots and correlates of alien species richness across taxonomic groups. Nat Ecol Evol 1:0186

Day MD, Winston RL (2016) Biological control of weeds in the 22 Pacific island countries and territories: current status and future prospects. NeoBiota 30:167–192

Denslow JS (2003) Weeds in paradise: thoughts on the invasibility of tropical islands. Ann Mo Bot Gard 90(1):119–127. https://doi.org/10.2307/3298531

Denslow JS, Space JC, Thomas PA (2009) Invasive exotic plants in the tropical pacific islands: patterns of diversity. Biotropica 41(2):162–170. https://doi.org/10.1111/j.1744-7429.2008.00469.x

Doherty TS, Glen AS, Nimmo DG et al (2016) Invasive predators and global biodiversity loss. Proc Natl Acad Sci 113(40):11261–11265

Drake DR (1993) Germination requirements of *Metrosideros polymorpha*, the dominant tree of Hawaiian lava flows and rain forests. Biotropica 25(4):461–467

Drake DR (1998) Relationships among the seed rain, seed bank and vegetation of a Hawaiian forest. J Veg Sci 9:103–112

Drake DR, Mueller-Dombois D (1993) Population development of rain forest trees in a chronosequence of Hawaiian lava flows. Ecology 74:1012–1019

Fortini L, Price J, Jacobi J et al (2013) A landscape-based assessment of climate change vulnerability for all native Hawaiian plants: University of Hawaii. https://pubs.er.usgs.gov/publication/70111902. Accessed 19 Feb 2019

Fosberg FR (1948) Immigrant plants in the Hawaiian islands. The University of Hawaii, Honolulu, Hawaii, pp 3–17

Freifelder RR, Vitousek PM, D'Antonio CM (1998) Microclimate change and effect on fire following forest-grass conversion in seasonally dry tropical woodland. Biotropica 30(2):286–297

Holl KD, Zahawi RA, Cole RJ et al (2011) Planting seedlings in tree islands versus plantations as a large-scale tropical forest restoration strategy. Restor Ecol 19(4):470–479

Hughes RF, Denslow JS (2005) Invasion by a N_2-fixing tree alters function and structure in wet lowland forests of Hawaii. Ecol Appl 15(5):1615–1628

Hughes F, Vitousek PM, Tunison T (1991) Alien grass invasion and fire in the seasonal submontane zone of Hawai'i. Ecology 72(2):743–747. https://doi.org/10.2307/2937215

Hughes RF, Uowolo AL, Togia TP (2012) Recovery of native forest after removal of an invasive tree, *Falcataria moluccana*, in American Samoa. Biol Invasions 14(7):1393–1413. https://doi.org/10.1007/s10530-011-0164-y

Hughes RF, Asner GP, Mascaro J et al (2014) Carbon storage landscapes of lowland Hawaii: the role of native and invasive species through space and time. Ecol Appl 24(4):716–731. https://doi.org/10.1890/12-2253.1

Keane RM, Crawley MJ (2002) Exotic plant invasions and the enemy release hypothesis. Trends Ecol Evol 17(4):164–170

Kellner JR, Asner GP, Kinney KM et al (2011) Remote analysis of biological invasion and the impact of enemy release. Ecol Appl 21(6):2094–2104. https://doi.org/10.1890/10-0859.1

Kueffer C, Daehler CC, Torres Santana CW et al (2010) A global comparison of plant invasions on oceanic islands. Perspect Plant Ecol Evol Syst 12(2):145–161. https://doi.org/10.1016/j.ppees.2009.06.002

Lauer A, Zhang C, Elison-Timm O et al (2013) Downscaling of climate change in the Hawaii region using CMIP5 results: on the choice of the forcing fields. J Clim 26(24):10006–10030. https://doi.org/10.1175/JCLI-D-13-00126.1

Liao W, Elison Timm O, Zhang C et al (2015) Will a warmer and wetter future cause extinction of native Hawaiian forest birds? Glob Chang Biol 21(12):4342–4352. https://doi.org/10.1111/gcb.13005

Litton CM, Sandquist DR, Cordell S (2006) Effects of non-native grass invasion on aboveground carbon pools and tree population structure in a tropical dry forest of Hawaii. For Ecol Manag 231(1–3):105–113. https://doi.org/10.1016/j.foreco.2006.05.008

Loh RK, Daehler CC (2007) Influence of invasive tree kill rates on native and invasive plant establishment in a Hawaiian forest. Restor Ecol 15(2):199–211. https://doi.org/10.1111/j.1526-100X.2007.00204.x

Loope LL (1998) Hawaii and Pacific islands. In: Status and trends of the nation's biological resources, vol 1, pp 747–774

Loope LL, Mueller-Dombois D (1989) Characteristics of invaded islands with special reference to Hawaii. In: Drake, Di Castri, Groves, Kruger, Mooney, Rejmanek, Williamson (eds) Biological invasions: a global perspective. Wiley, New York, pp 257–280

Loope LL, Hamann O, Stone CP (1988) Comparative conservation biology of oceanic archipelagoes: Hawaii and the Galapagos. Bioscience 38(4):272–282. https://doi.org/10.2307/1310851

Lugo AE (1997) The apparent paradox of reestablishing species richness on degraded lands with tree monocultures. For Ecol Manag 99(1–2):9–19

Mack MC, D'Antonio CM (1998) Impacts of biological invasions on disturbance regimes. Trees 13(5):195–198

Mack RN (2003) Phylogenetic constraint, absent life forms, and preadapted alien plants: a prescription for biological invasions. Int J Plant Sci 164:185–196

Marchetti MP, Engstrom T (2016) The conservation paradox of endangered and invasive species. Conserv Biol 30(2):434–437

Mascaro J, Hughes RF, Schnitzer SA (2012) Novel forests maintain ecosystem processes after the decline of native tree species. Ecol Monogr 82(2):221–228

McDaniel S, Ostertag R (2010) Strategic light manipulation as a restoration strategy to reduce alien grasses and encourage native regeneration in Hawaiian Mesic forests. Appl Veg Sci 13(3):280–290. https://doi.org/10.1111/j.1654-109X.2009.01074.x

Medeiros A, Loope L, Wu Y et al (2013) Weeds of Hawaii's lands devoted to watershed protection and biodiversity conservation: role of biological control as the missing piece in an integrated pest management strategy. In: Proceedings of the XIII international symposium on biological control of weeds, Waikoloa, Hawaii, USA, 11–16 September, 2011. USDA Forest Service, Pacific Southwest Research Station, Institute of Pacific Islands Forestry, pp 206–210

Mehrhoff L (1993) Rare plants in Hawaii: a status report. Plant Conserv 7:1–2

Michaud J, Cordell S, Cole TC, Ostertag R (2015) Drought in an invaded Hawaiian lowland wet forest 1. Pac Sci 69(3):367–383

Miyazawa Y, Dudley BD, Hughes RF et al (2015) Non-native tree in a dry coastal area in Hawai'i has high transpiration but restricts water use despite phreatophytic trait. Ecohydrology 9(7):1166–1176

Nonner E (2005) Seed bank dynamics and germination ecology of fountain grass (*Pennisetum setaceum*). MS master's thesis, University of Hawaii, Honolulu. 59 p

Ostertag R, Giardina CP, Cordell S (2008) Understory colonization of *Eucalyptus* plantations in Hawaii in relation to light and nutrient levels. Restor Ecol 16(3):475–485

Ostertag R, Cordell S, Michaud J et al (2009) Ecosystem and restoration consequences of invasive woody species removal in Hawaiian lowland wet forest. Ecosystems 12:503–515

Ostertag R, Warman L, Cordell S, Vitousek PM (2015) Using plant functional traits to restore Hawaiian rainforest. J Appl Ecol 52:805–809

Pattison RR, Goldstein G, Ares A (1998) Growth, biomass allocation and photosynthesis of invasive and native Hawaiian rainforest species. Oecologia 117:449–459

Pitt WC, Beasley J, Witmer GW (2017) Ecology and management of terrestrial vertebrate invasive species in the United States. CRC Press, Boca Raton, 415 p

Pyšek P, Pergl J, Essl F, Lenzner B, Dawson W, Kreft H, Weigelt P et al (2017) Naturalized alien flora of the world. Preslia 89(3):203–274

Reichard SH, White P (2001) Horticulture as a pathway of invasive plant introductions in the United States: Most invasive plants have been introduced for horticultural use by nurseries, botanical gardens, and individuals. Bioscience 51(2):103–113. https://doi.org/10.1641/0006-3568(2001)051[0103:haapoi]2.0.co;2

Schmidt JP, Drake JM (2011) Time since introduction, seed mass, and genome size predict successful invaders among the cultivated vascular plants of Hawaii. PLoS One 6(3):e17391. https://doi.org/10.1371/journal.pone.0017391

Schulten JR, Cole TC, Cordell S et al (2014) Persistence of native trees in an invaded Hawaiian lowland wet forest: experimental evaluation of light and water constraints. Pac Sci 68(2):267–285. https://doi.org/10.2984/68.2.7

Sin H, Beard KH, Pitt WC (2008) An invasive frog, *Eleutherodactylus coqui*, increases new leaf production and leaf litter decomposition rates through nutrient cycling in Hawaii. Biol Invasions 10(3):335–345. https://doi.org/10.1007/s10530-007-9133-x

Smith CW (1985) Impact of alien plants on Hawaii's native biota. In: Stone CP, Scott JM (eds) Hawaii terrestrial ecosystems: preservation and management, Cooperative National Park Resources Studies Unit. University of Hawaii, Honolulu, pp 180–250

Smith CW, Tunison JT (1992) Fire and alien plants in Hawaii: research and management implications for native ecosystems. In: Stone CP, Smith CW, Tunison JT (eds) Alien plant invasions in native ecosystems of Hawaii: management and research, Cooperative National Park Resources Studies Unit. University of Hawaii, Honolulu, pp 394–408

Space J, Falanruw M (1999) Observations on invasive plant species in Micronesia. (Report prepared for the meeting of the Pacific Islands Committee, Council of Western State Foresters, Majuro, Republic of the Marshall Islands)

Spatz DR, Zilliacus KM, Holmes ND et al (2017) Globally threatened vertebrates on islands with invasive species. Sci Adv 3(10):e1603080. https://doi.org/10.1126/sciadv.1603080

Stratton LC, Goldstein G (2001) Carbon uptake, growth and resource-use efficiency in one invasive and six native Hawaiian dry forest tree species. Tree Physiol 21:1327–1334

Timm OE, Giambelluca TW, Diaz HF (2015) Statistical downscaling of rainfall changes in Hawai'i based on the CMIP5 global model projections. J Geophys Res-Atmos 120(1):92–112. https://doi.org/10.1002/2014JD022059

Trauernicht C, Pickett E, Giardina CP et al (2015) The contemporary scale and context of wildfire in Hawai'i 1. Pac Sci 69(4):427–444

Vitousek PM (1986) Biological invasions and ecosystem properties: can species make a difference? In: Mooney HA, Drake JA (eds) Ecology of biological invasions of North America and Hawai'i. Springer, New York, pp 163–176

Vorsino AE, Fortini LB, Amidon FA et al (2014) Modeling Hawaiian ecosystem degradation due to invasive plants under current and future climates. PLoS One 9(5):e95427. https://doi.org/10.1371/journal.pone.0095427

Wagner WL, Herbst DR, Sohmer SH (1999) Manual of the flowering plants of Hawaii. Bishop Museum Special Publication 97. 2 vols. University of Hawaii Press and Bishop Museum Press, Honolulu. 1919 p

Williams DG, Mack RN, Black RA (1995) Ecophysiology of introduced *Pennisetum setaceum* on Hawaii: the role of phenotypic plasticity. Ecology 76(5):1569–1580

Woodcock D (2003) To restore the watersheds: early twentieth-century tree planting in Hawai 'i. Ann Assoc Am Geogr 93(3):624–635

Zou X (1993) Species effects on earthworm density in tropical tree plantations in Hawaii. Biol Fertil Soils 15(1):35–38. https://doi.org/10.1007/bf00336285

Northwest Region

Steven J. Seybold, Susan J. Frankel, Deanna H. Olson, Rebecca Flitcroft, Becky K. Kerns, Jana C. Lee, Karen L. Ripley, and A. Steve Munson

Introduction

The Northwest region (Idaho, Oregon, and Washington; Figs. A3.1 and A3.2) contains major coastal and inland ports (Coos Bay, Lewiston, Pasco, Portland, Seattle-Tacoma, and The Dalles), waterways (Puget Sound, Columbia River Basin, and Willamette Valley), and major

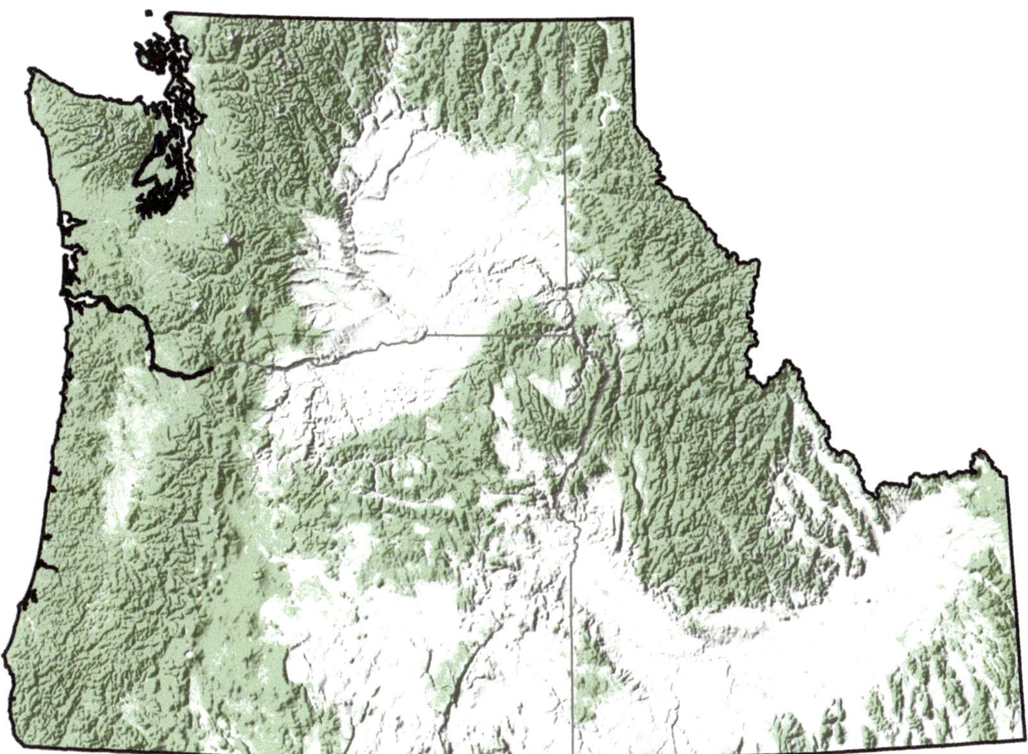

Fig. A3.1 The Northwest Region

Fig. A3.2 The Northwest region of the United States includes Idaho, Oregon, and Washington. (Figure courtesy of Daniel Ryerson, USDA Forest Service Southwestern Region, Forest Health Protection)

highway and rail arteries that provide pathways for invasive plants, pathogens, insects, and vertebrates. Two main pathways in this region for the introduction of invasive forest and horticultural pathogens and insects are shipping infrastructure, especially solid wood packing material from shipping cargo (USDA Forest Service and APHIS 2000), and the live plant trade (Liebhold et al. 2012). Temperate coastal climates, diverse and abundant native vegetation, extensive trade patterns with Pacific Rim nations, and the border with Canada and its major west coast port of Vancouver have facilitated the repeated introductions of major forest pests such as the European (also known as North American) gypsy moth (*Lymantria dispar dispar*), the Asian gypsy moth (*Lymantria dispar asiatica/japonica*), and the sudden oak death pathogen (*Phytophthora ramorum*). The region has a significant horticultural industry, extensive areas of mesic and dryland agriculture, and abundant urban and native forests whose trees can serve as adventive hosts or alternate hosts for invasive insects and pathogens as well as disturbed landscapes for invasive plants. Forested lands are regionally vital to the forest industry and are a recreational base for millions in the Northwest. Invasive species have the potential to inflict severe economic hardship on individuals, local governments, and the businesses involved in the forest, horticultural, agricultural, and tourism industries.

The Northwest region has a long history of invasive forest insects and diseases; for example, white pine blister rust, caused by *Cronartium ribicola*, was first introduced on the West Coast and was "Pest #1" on the historical "Quarantine 37" of the Plant Quarantine Act of 1912, and balsam woolly adelgid (*Adelges piceae*) was first noted in this region in the 1920s (Annand 1928; Keen 1952). More modern regulations such as the International Standards for Phytosanitary Measures No. 15 (ISPM 15), an international treaty signed by 200 nations, play a major role in protecting this region from invasive species that may arrive in solid wood packing material (FAO 2009; Strutt et al. 2013). Also, examples of the more profound impacts of climate change and invasion biology are developing in high-elevation ecosystems of this region where five-needle white pines (*Pinus* spp.) and subalpine fir (*Abies lasiocarpa*) trees are experiencing unprecedented rates of mortality (Shoal and Aubry 2006). In these instances the mountain pine beetle, *Dendroctonus ponderosae*, has been a native invader of high-elevation ecotones (Bentz et al. 2010; Logan and Powell 2001; Logan et al. 2003), whereas *C. ribicola* (on pines) and *A. piceae* (on subalpine fir) are significant nonnative invaders with expanding elevational ranges (Hrinkevich et al. 2016).

Regional State government departments, interagency collaborative groups such as State Invasive Species Councils, regional USDA Forest Service personnel, and other entities have identified nearly 190 species and species groups as regional invasive or nuisance species of key concern. Altogether, the taxa identified as detected or potential invasive species fall into five categories: 78 plants; 11 plant pathogens and parasites; 93 insect species; 10 aquatic invertebrates (noninsect); and 23 vertebrates (Tables A3.1, A3.2, A3.3, A3.4 and A3.5, Bautista 2017; Flitcroft et al. 2016; Invasive Species of Idaho 2017; OIE 2018; OISC 2015; Prather et al. 2016; WISP 2009). Because of the large number of organisms (most notably plants) treated in this summary (Tables A3.1, A3.2, A3.3, A3.4 and A3.5), scientific names are only provided in the text for the most prominent species. This overview will focus on a subset of the species that are established and have caused resource damage or that have been introduced repeatedly into the Northwest region.

Plants

Aquatic invasive plants of this region (Table A3.1) include multiple species of submersed aquatic plants (i.e., *Elodea* spp., *Hydrilla* spp., milfoils, and swollen bladderwort), emergent plants (reeds, *Spartina* spp., loosestrife, rushes, giant salvinia, reed canary grass, foxtail barley, yellow flag iris, and water primrose), and floating plants (parrotfeather, curly-leaf pondweed, water chestnut, yellow floating heart, West Indian spongeplant, and dotted duckweed). Two nonplants, a cyanobacterium (toxic blue-green "alga") and a diatom (rock snot or didymo), are also included in this survey (Table A3.1). Once established, many of these taxa can cause significant impairment of water quality and navigation (by growing as dense mats) (see Chap. 2, Sect. 2.5, for additional discussion of these impacts). Furthermore, established populations may be spread by waterfowl as they move from one location to another, or by human vectors (e.g., boats and fishing gear/tackle). Marine invasive plants are not included in the survey.

Several priority aquatic plants can achieve high densities, leading to ecological problems or nuisance issues for people. Emergent plants can dominate wetland and floodplain areas (i.e., reed canary grass, water primrose, and purple loosestrife), outcompeting or displacing native species. Some invasive aquatic plants were brought initially to the region by the aquarium trade (*Elodea* spp.) or for ornamental use or seeding of wet areas for livestock or waterfowl (reed canary grass (*Phalaris arundinacea*)). Once established, cut fragments of the plants, broken by boat propellers or wildlife, can disperse for colonization. Some species such as *Elodea canadensis* have broad ecological tolerance, which make expansion and invasion a concern. Non-native and native species of water primroses (*Ludwigia* spp.) comprise a plant group whose growth has become problematic in recent years. One species, *L. grandiflora*, native to the Eastern United States, Central America, and South America, is now present in Oregon and Washington (CABI 2018). Physical and chemical alteration of the environment by *L. grandiflora* can cause severe damage to local ecosystems and biodiversity. Dense stands of this plant can reduce floodwater retention, cause hyper-sedimentation and silting, and block slow-moving waterways. The plant also gives off allelopathic elements that impact water quality throughout the year and make it detrimental to vulnerable native flora. Because it can shade out other submersed vegetation, it is generally considered a threat to biodiversity in its introduced range (CABI 2018).

Non-plant toxic "algae" are a health concern for native vertebrates and humans because they create powerful toxins known to kill fish, ducks, geese, marine mammals, and other

Table A3.1 Priority non-native invasive plants of the Northwest region

Scientific name	Common name	ID	OR	WA	R6[a]
Aquatic plants					
Lagarosiphon major	African waterweed or African elodea		x		x
Phragmites australis	Common reed	x	x	x	
Potamogeton crispus	Curly-leaf pondweed	x			x
Butomus umbellatus	Flowering rush	x	x		x
Salvinia molesta	Giant salvinia	x	x		
Hydrilla verticillata	Hydrilla, water thyme	x	x	x	x
Myriophyllum spp. including *M. spicatum, M. aquaticum*	Milfoils: Eurasian, parrotfeather	x		x	x
Lythrum salicaria, Lysimachia vulgaris	Purple loosestrife, garden yellow loosestrife	x		x	x
Phalaris arundinacea; P. arundinacea var. *picta*	Reed canary grass; ribbon grass				x
Didymosphenia geminata	Rock snot (Didymo)[b]		x		x
Chondrilla juncea	Rush skeletonweed	x		x	x
Spartina spp. including *S. alterniflora, S. densiflora*	Spartina (cordgrass)		x	x	
Prymnesium parvum, Cylindrospermopsis raciborskii	Toxic algae (golden, toxic cyanobacteria)[b]		x		
Trapa natans	Water chestnut (European)	x	x	x	
Ludwigia grandiflora	Water primrose[c]		x	x	x
Eichhornia crassipes	Water hyacinth	x			
Egeria densa, Elodea nuttallii, E. canadensis, E. canadensis x *E. nuttallii* hybrid	Brazilian elodea, western waterweed (elodea)	x		x	
Iris pseudacorus	Yellow flag iris	x			x
Nymphoides peltata	Yellow floating heart	x	x		x
Riparian-terrestrial plants					
Peganum harmala	African rue		x		
Alyssum corsicum, A. murale, Berteroa incana	Alyssums	x	x		x
Hyoscyamus niger	Black henbane	x			
Solanum rostratum	Buffalo bur	x			
Buddleja davidii; B. globosa	Butterfly bush			x	
Alhagi maurorum	Camelthorn		x		
Bromus tectorum	Cheatgrass				x
Tussilago farfara	Coltsfoot (European)		x		
Anchusa officinalis, A. arvensis, Echium vulgare	Common bugloss, small bugloss, viper's bugloss	x			x
Crupina vulgaris	Common crupina	x		x	x
Isatis tinctoria	Dyer's woad	x			
Linaria dalmatica, L. vulgaris	Toadflax (Dalmatian, yellow)	x		x	x
Hedera helix	English ivy				x
Ammophila arenaria	European beachgrass				x
Brachypodium sylvaticum	False brome				x
Cabomba caroliniana	Fanwort	x			
Azolla pinnata	Feathered mosquito fern	x			
Convolvulus arvensis	Field bindweed	x			
Alliaria petiolata	Garlic mustard			x	x
Geranium robertianum, G. lucidum	Geranium				x
Heracleum mantegazzianum	Giant hogweed	x	x	x	
Aegilops triuncialis, A. ovata, A. cylindrica	Goatgrasses (barbed, ovate, jointed)	x	x		x
Galega officinalis	Goatsrue		x		
Ulex europaeus	Gorse				x
Hieracium piloselloides, H. pratense, H. pilosella, H. aurantiacum, H. floribundum, H. caespitosum, H. lachenalii	Hawkweeds	x	x	x	x
Rubus armeniacus	Himalayan blackberry			x	x
Cynoglossum officinale	Houndstongue	x			x
Cuscuta japonica	Japanese dodder		x		
Sorghum halepense	Johnsongrass	x			
Centaurea virgata; C. vulgaris; C. jacea x *nigra; C. stoebe*	Knapweeds	x	x	x	x
Fallopia japonica var. *japonica; Polygonum bohemicum*	Knotweeds (Japanese, bohemian)	x		x	x
Kochia scoparia ssp. *scoparia*	Kochia			x	

(continued)

Table A3.1 (continued)

Scientific name	Common name	ID	OR	WA	R6[a]
Pueraria montana var. *lobata*	Kudzu		x	x	
Euphorbia esula	Leafy spurge	x		x	x
Nardus stricta	Matgrass	x	x		
Salvia aethiopis	Mediterranean sage	x			
Taeniatherum caput-medusae	Medusahead				x
Milium vernale	Millium	x			
Ventenata dubia	North Africa grass				x
Euphorbia oblongata	Oblong spurge		x		
Clematis vitalba	Old man's beard				x
Leucanthemum vulgare	Oxeye daisy	x			
Echium plantagineum	Paterson's curse		x		
Lepidium latifolium	Perennial pepperweed	x			
Sonchus arvensis	Perennial sowthistle	x			
Conium maculatum	Poison hemlock	x			
Impatiens glandulifera	Policeman's helmet	x			x
Tribulus terrestris	Puncturevine	x			x
Cyperus rotundus	Purple nutsedge		x		
Cytisus scoparius	Scotch broom	x		x	x
Solanum elaeagnifolium	Silverleaf nightshade		x		
Centaurea iberica, C. calcitrapa, C. solstitialis	Starthistles	x	x	x	
Potentilla recta	Sulfur cinquefoil				x
Zygophyllum fabago	Syrian bean-caper	x	x		
Tamarix spp.	Tamarix (saltcedar)	x		x	x
Senecio jacobaea	Tansy ragwort	x		x	x
Carduus nutans, C. acanthoides; Carthamus baeticus, C. lanatus; Onopordum tauricum, O. acanthium	Thistles	x	x	x	x
Bryonia alba	White bryony	x	x		
Lepidium draba	Whitetop/hoary cress	x			x

[a]R6 = USDA Forest Service Region 6 (OR, WA) priority invasive species
[b]Non-plants (diatom and cyanobacterium, respectively)
[c]Other species of *Ludwigia* (e.g., *L. palustris*) are native to the Northwest region but have become pests

Table A3.2 Priority non-native invasive plant pathogens and other parasites of the Northwest region

Scientific name	Common name	ID	OR	WA	R6[a]	Other[b]	Plant or animal host
Phytophthora lateralis	Port-Orford-cedar root disease		x		x		*Chamaecyparis lawsoniana, Taxus brevifolia*
Phytophthora ramorum	Sudden oak death	x	x		x		*Quercus* spp., *Notholithocarpus* spp., many others
Batrachochytrium dendrobatidis	Amphibian chytrid fungus (*Bd*)					x	
B. salamandrivorans	Amphibian chytrid fungus (*Bsal*)					x	
Ranavirus	Ranavirus					x	
Chronic wasting disease prion	Chronic wasting disease (CWD)		x				
Pseudogymnoascus destructans	White-nose syndrome				x		
Cronartium ribicola	White pine blister rust		x	x	x		*Pinus* spp.
Melampsora larici-populina	Eurasian poplar leaf rust		x	x			*Populus* spp.
Lachnellula willkommii	European larch canker	x					
Puccinia graminis	Black stem rust	x					

[a]R6 = US Forest Service Region 6 (OR, WA) priority invasive species
[b]Other = World Organization for Animal Health (Office International des Epizootics) priorities

Table A3.3 Non-native invasive insects of the Northwest Region include beetles, flies, aphids/adelgids, thrips, wasps, and moths

		Occurrence			Origin	Introduction		Plant or animal host
Scientific name	Common name	ID	OR	WA		Accidental	Intentional	
Coleoptera								
Agrilus cuprescens	Rose stem girdler			x	Europe	•		*Rosa* spp., *Rubus* spp.
Agriotes lineatus	Lined click beetle			x	Europe	•		Grass and plant roots, Potatoes, strawberries
Agriotes obscurus	Dusky click beetle			x	Europe	•		Grass and plant roots Potatoes, strawberries
Amphimallon majale	European chafer		x	x	Europe	•		Grass roots
Anobium punctatum	Furniture beetle			x	Europe	•		Wood products
Apion fuscirostre					Europe		•	Gorse (*Ulex europaeus*)
Apion ulicis	Gorse weevil	x	x	x	Europe		•	Gorse
Brachypterolus pulicarius	Toadflax flower-feeding beetle	x	x	x	Europe		•	Dalmatian and yellow toadflax, *Linaria* spp.
Chrysolina hyperici		x	x	x	Europe via Australia		•	Klamathweed (*Hypericum perforatum*)
Chrysolina quadrigemina	Klamathweed beetle	x	x	x	Europe via Australia		•	Klamathweed (*Hypericum perforatum*)
Cryptorhynchus lapathi	Poplar-and-willow borer	x	x	x	Europe	•		*Salix* spp. and *Populus* spp.
Cyclorhipidion bodoanum			x	x	Asia	•		Broad-leaved trees (woodborer)
Hylastes opacus			x	x	Europe	•		*Pinus*
Ips paraconfusus	California fivespined ips		O	x	California, Oregon	•		*Pinus* (invasive in Washington, native to Oregon)
Laricobius erichsonii			x	x	Europe		•	*Adelges piceae*
Larinus minutus	Lesser knapweed flower weevil	x	x	x	Europe		•	Knapweeds, *Centaurea* sp.
Larinus obtusus	Blunt knapweed flower weevil	x	x	x	Europe		•	Knapweeds, *Centaurea* sp.
Lasioderma serricorne	Cigarette beetle	x	x	x	Europe (cosmopolitan)	•		*Plant products*
Lilioceris lilii	Scarlet lily beetle			x	Eurasia	•		*Fritillaria* spp. and *Lilium* spp. leaves
Lyctus brunneus	Old World lyctus beetle	x	x	x	Europe (cosmopolitan)	•		Wood products
Lyctus linearis	European lyctus beetle	x	x	x	Europe (cosmopolitan)	•		Wood products
Mecinus janthiniformis	Dalmatian toadflax stem-mining weevil	x	x	x	Southeastern Europe		•	Dalmatian toadflax, *Linaria dalmatica*
Mecinus janthinus	Yellow toadflax stem-mining weevil	x	x	x	Eurasia		•	Yellow toadflax, *Linaria vulgaris*
Melanotus cete			x	x	Japan	•		Unknown
Micromalthus debilis	Telephone pole beetle		x	x	Eastern North America	•		Decayed wood products
Mogulones crucifer	Houndstongue root weevil	x		x	Europe	•		Houndstongue, *Cynoglossum officinale*
Nacerdes melanura	Wharf borer	x	x	x	Europe	•		Wood products
Orchestes alni	European elm flea weevil	x	x	x	Europe	•		*Ulmus* spp.
Otiorhynchus ovatus	Strawberry root weevil	x	x	x	Europe	•		Seedling conifers
Otiorhynchus rugosostriatus	Rough strawberry root weevil	x	x		Europe	•		Seedling conifers
Otiorhynchus sulcatus	Black vine weevil	x	x	x	Europe	•		Seedling conifers
Oulema melanopus	Cereal leaf beetle	x	x	x	Eurasia	•		Cereal crops, wild grasses
Phymatodes testaceus	Tanbark borer			x	Europe, Japan, Middle East, North Africa	•		Wood products/bark

(continued)

Table A3.3 (continued)

		Occurrence			Origin	Introduction		Plant or animal host
Scientific name	Common name	ID	OR	WA		Accidental	Intentional	
Pityophthorus juglandis	Walnut twig beetle	x	x	x	Arizona/New Mexico, Mexico	•		*Juglans* spp., *Pterocarya* spp.
Popillia japonica	Japanese beetle	x	x	x	Japan	•		Broad-leaved trees and many ornamentals
Pullus impexus			x		Europe		•	*Adelges piceae*
Pyrrhalta viburni	Viburnum leaf beetle			x	Eurasia	•		*Viburnum* spp.
Rhinusa antirrhini	Toadflax seed-galling weevil	x	x	x	Eurasia, Mediterranean	•		Dalmatian and yellow toadflax, *Linaria* spp.
Rhinusa neta	Toadflax seed-feeding weevil			x	Europe	•		Dalmatian and yellow toadflax, *Linaria* spp.
Saperda populnea		x	x	x	Europe	•		*Populus* spp.
Scolytus multistriatus	Smaller European elm bark beetle	x	x	x	Europe	•		*Ulmus* spp.
Scolytus rugulosus	Shot hole borer	x	x	x	Europe	•		Broad-leaved trees, Rosaceae
Scolytus schevyrewi	Banded elm bark beetle	x	x	x	Asia	•		*Ulmus* spp.
Stegobium paniceum	Drugstore beetle	x	x	x	Cosmopolitan	•		Stored products
Tenebroides mauritanicus	Cadelle	x	x	x	Cosmopolitan	•		Stored products
Trypodendron domesticum	European hardwood ambrosia beetle			x	Europe	•		Broad-leaved trees
Xanthogaleruca luteola	Elm leaf beetle	x	x	x	Europe	•		*Ulmus* spp.
Xestobium rufovillosum	Deathwatch beetle		x		Europe	•		Wood products
Xyleborinus alni = (attenuates)			x	x	Europe/Asia	•		Broad-leaved trees
Xyleborinus saxeseni	Fruit tree pin-hole borer	x	x	x	Europe	•		Broad-leaved trees and conifers
Xyleborus dispar	European shot hole borer	x	x	x	Europe	•		Broad-leaved trees
Xyleborus pfeili			x		Cosmopolitan	•		Broad-leaved trees
Xylosandrus crassiusculus			x		Africa/Asia	•		Broad-leaved trees
Xylosandrus germanus			x	x	Asia	•		Broad-leaved trees
Xyloterinus politus				x	Eastern North America	•		Broad-leaved trees, rare in conifers
Diptera								
Aphidoletes thompsoni			x	x	Europe		•	*Adelges piceae*
Compsilura concinnata		?	x	x	Europe		•	*Lymantria dispar* and other Lepidoptera
Cremifania nigrocellulata			x		Europe		•	*Adelges piceae*
Delia platura	Seedcorn maggot	x	x	x	Europe	•		Seedling conifers
Drosophila suzukii	Spotted wing Drosophila	x	x	x	Asia	•		Fruits/berries
Leucopis obscura			x	?	Europe		•	*Adelges piceae*
Phytomyza ilicis	Holly leafminer		x		Europe	•		*Ilex* spp.
Rhagoletis completa	Walnut husk fly	x	x	x	Eastern USA	•		*Juglans* spp., *Prunus persica*
Hemiptera								
Adelges abietis	Eastern spruce gall aphid	x	x	x	Europe	•		*Picea* spp.

(continued)

Table A3.3 (continued)

		Occurrence			Origin	Introduction		Plant or animal host
Scientific name	Common name	ID	OR	WA		Accidental	Intentional	
Adelges nüsslini (=nordmaneanae)		x	x	x	Europe	•		*Picea* spp., *Abies* spp.
Adelges piceae	Balsam woolly adelgid	x	x	x	Europe	•		*Abies* spp.
Adelges strobilobius (=laricis)	Larch woolly aphid	x	x	x	Europe	•		*Picea* spp., *Larix* spp.
Arocatus melanocephalus	Elm seed bug	x	x	x	Europe	•		*Ulmus* spp.
Asterolecanium minus	Oak pit scale			x	Europe	•		*Quercus* spp.
Cinara tujafilina		x	x	x	Europe	•		Cupressini
Carulaspis juniperi	Juniper scale	x	x	x		•		
Dialeurodes chittendeni	Rhododendron whitefly				Asia	•		*Rhododendron* spp.
Elatobium abietinum	Spruce aphid		x	x	Europe	•		*Picea* spp.
Eriosoma ulmi	European elm leafcurl aphid				Europe	•		*Ulmus* spp.
Eucallipterus tiliae	Linden aphid	x	x	x		•		*Tilia* spp.
Euceraphis punctipennis	European birch aphid				Europe	•		*Betula* spp.
Eulecanium cerasorum	Calico scale	x	x	x		•		
Gossyparia spuria	European elm scale				Europe	•		*Ulmus* spp.
Halyomorpha halys	Brown marmorated stink bug	x	x	x	Asia	•		Fruit/vegetable crops
Icerya purchasi	Cottony cushion scale	x				•		
Lecanium corni	European fruit lecanium				Europe	•		Broad-leaved trees
Lepidosaphes ulmi	Oystershell scale	x	x	x	Europe	•		Broad-leaved trees
Metopoplax ditomoides			x	x	Europe	•		
Nezara viridula				x		•		
Periphyllus californiensis					Asia	•		*Acer* spp.
Periphyllus lyropictus	Norway maple aphid	x	x	x	Europe	•		*Acer platanoides* and other *Acer* spp.
Periphyllus testudinacea					Europe	•		*Acer* spp.
Physokermes piceae	Spruce bud scale	x	x	x	Europe	•		*Picea* spp.
Pineus strobi	Pine bark aphid				Europe	•		*Pinus* spp.
Quadraspidiotus perniciosus	San Jose scale	x	x	x	Asia	•		Broad-leaved trees
Raglius alboacuminatus			x	x	Europe and the Mediterranean Basin	•		Mint (Lamiaceae) seeds
Rhyparochromus vulgaris	Dirt-colored seed bug			x	Europe	•		
Schizolachnus pineti					Europe	•		*Pinus* spp.
Stephanitis pyrioides	Azalea lace bug		x	x	Japan	•		*Azalea* spp., *Rhododendron* spp.
Thysanoptera								
Taeniothrips inconsequens	Pear thrips		x	x				
Hymenoptera								
Agathis pumila		x	x	x	Europe		•	*Coleophora laricella*
Apanteles solitarius					Europe		•	*Stilpnotia salicis*
Caliroa cerasi	Pear sawfly	x	x	x		•		

(continued)

Table A3.3 (continued)

		Occurrence			Origin	Introduction		Plant or animal host
Scientific name	Common name	ID	OR	WA		Accidental	Intentional	
Chrysocharis laricinellae		x	x	x	Europe		•	*Coleophora laricella*
Cladius grandis				x	Eurasia	•		*Alnus* spp., *Populus* spp., *Salix* spp.
Cladius gregarius				x	Eastern North America	•		*Populus* spp.
Craesus alniastri				x	Europe	•		*Alnus* spp.
Diprion similis	Introduced pine sawfly			x	Europe	•		*Pinus* spp.
Eupareophora parca				x	Eastern North America	•		*Carya* spp., *Fraxinus* spp.
Eriocampa ovata	Alder woolly sawfly				Europe	•		*Alnus* spp.
Fenusella nana				x	Europe	•		*Betula* spp.
Fenusa pusilla	Birch leafminer	x	x	x	Europe	•		*Betula* spp.
Fenusa ulmi				x	Europe	•		*Ulmus* spp.
Gilpinia hercyniae	European spruce sawfly			x	Eurasia	•		*Picea* spp.
Halidamia affinis				x	Europe	•		*Gallium* spp.
Heterarthrus nemoratus				x	Eurasia	•		*Betula* spp.
Heterarthrus vagans				x	Eurasia	•		*Alnus* spp.
Kaliofenusa ulmi	Elm leafminer		x	x	Europe	•		*Ulmus* spp.
Macrophya punctumalbum				x	Europe	•		*Ligustrum* spp., *Syringa* spp., *Fraxinus* spp.
Mesoleius tenthredinis					Europe		•	*Pristiphora erichsonii*
Metallus lanceolatus				x	Europe	•		*Geum macrophyllum*
Meteorus versicolor					Europe		•	*Stilpnotia salicis*
Monophadnus pallescens				x	Europe	•		*Ranunculus* spp.
Monostegia abdominalis				x	Europe	•		*Glaux* spp., *Lysimachia* spp., *Anagallis* spp.
Monsoma pulveratum	Green alder sawfly	x	x	x	Europe, Asia Minor and North Africa	•		*Alnus* spp.
Nematus lipovskyi				x	Eastern North America	•		*Rhododendron* spp.
Neodiprion sertifer	European pine sawfly			x	Europe	•		*Pinus* spp.
Polistes dominulus	European paper wasp			x	Europe	•		Omnivore
Pristiphora erichsonii	Larch sawfly	x	x	x	Europe	•		*Larix* spp.
Pristiphora geniculata	Mountain ash sawfly			x	Europe	•		*Crataegus* spp., *Sorbus* spp.
Pristiphora rufipes				x	Central Europe	•		*Aquilegia* spp.
Profenusa thomsoni	Amber-marked birch leafminer			x	Eurasia	•		*Betula* spp.
Trichiocampus viminalis	Poplar sawfly	x	x	x	Europe	•		*Populus* spp., *Salix* spp.
Vespula germanica	German yellowjacket	x			Europe	•		Omnivore
Xiphydria prolongata	Willow wood wasp		x	x	Europe	•		Broad-leaved trees
Lepidoptera								
Aethes rutilana	Pale juniper webworm				Europe	•		*Juniperus* spp.
Anarsia lineatella	Peach twig borer	x	x	x		•		*Prunus* spp.
Archips rosanus					Europe	•		Broad-leaved trees

(continued)

Table A3.3 (continued)

Scientific name	Common name	Occurrence ID	OR	WA	Origin	Introduction Accidental	Intentional	Plant or animal host
Calophasia lunula	Toadflax defoliating moth	x	x	x	Eurasia		•	Dalmatian and yellow toadflax, *Linaria* spp.
Caloptilia negundella	Boxelder leafroller				Europe	•		*Acer* spp.
Caloptilia syringella	Lilac leafminer				Europe	•		Lilac and *Fraxinus* spp.
Cnephasia longana	Omnivorous leaftier				Europe	•		*Pseudotsuga* spp. and broad-leaved trees
Coleophora laricella	Larch casebearer	x	x	x	Europe	•		*Larix* spp.
Coleophora serratella	Birch casebearer		x	x	Europe	•		*Betula* spp.
Dichomeris marginella	Juniper webworm	x	x		Europe	•		*Juniperus* spp.
Enarmonia formosana	Cherry bark tortrix		x	x	Eurasia	•		*Crataegus* spp., *Malus* spp., *Prunus* spp., *Pyrus* spp., *Sorbus* spp.
Epinotia nanana	European spruce needleminer		x	x	Europe	•		*Picea* spp.
Homadaula anisocentra	Mimosa webworm				Unknown	•		*Albizia* spp., *Gleditsia* spp.
Leucoma salicis	Satin moth	x	x	x	Europe	•		*Populus* spp., *Salix* spp.
Leucoptera spartifoliella					Europe		•	Scotch broom, *Cytisus scoparius*
Lymantria dispar dispar[a]	North American gypsy moth				Europe	•		Broad-leaved trees
Lymantria dispar asiatica/japonica[a]	Asian gypsy moth				Asia	•		Broad-leaved trees and conifers
Ocnerostoma piniariellum					Europe	•		*Pinus* spp.
Operophtera brumata	Winter moth		x	x	Europe	•		Broad-leaved trees and agricultural crops
Pandemis cerasana					Europe	•		Broad-leaved trees
Rhyacionia buoliana	European pine shoot moth	x	x	x	Europe	•		*Pinus* spp.
Spilonota ocellana	Eyespotted bud moth		x		Europe	•		*Quercus* spp. and other broad-leaved trees
Synanthedon myopaeformis	Apple clearwing moth			x	Europe, Mediterranean Basin	•		*Crataegus* spp., *Malus* spp., *Prunus* spp., *Pyrus* spp.
Synanthedon scitula	Eastern dogwood borer			x	Eastern North America	•		Broad-leaved trees esp. *Cornus* spp., *Carya* spp., *Malus* spp.
Tyria jacobaeae	Cinnabar moth		x	x	Europe		•	Tansy ragwort, *Senecio jacobaea*

[a]Detected repeatedly in the Northwest region, but eradicated. [O] Native range within the region

This list was compiled primarily from Furniss and Carolin (1977) with additions from Acheampong et al. (2016), Bai et al. (2002), Bellows et al. (1998), Childs and Swanson (2003), Doerr et al. (2008), Foote et al. (1993), Gerberg (1957), Hatch (1953, 1962), Hayes and Ragenovich (2001), Hitchcock et al. (2002), Idaho State Department of Agriculture (2013), Ivie (2002), Johnson (1998), Kruse et al. (2010), LaBonte et al. (2005), LaGasa (2006), LaGasa and Murray (2007), Lee et al. (2009, 2011), Looney et al. (2012, 2016a,2016b), Mudge et al. (2001), Murray et al. (2012, 2013), Nugent (2005), Phillips (2002), Rabaglia et al. (2006, 2019), Rosetta (2013), Sabrosky and Reardon (1976), Seybold and Downing (2009), Sing et al. (2016), Vernon et al. (2001), USDA (1986, 2019), Washington Invasive Species Council (2012), White (1982), and Winston et al. (2014a, b)

US Department of Agriculture 2009. National Agriculture Library. Species Profiles. Accessed online at http://www.invasivespeciesinfo.gov/plants/main.shtml

Other terrestrial invertebrates found in Oregon and Washington include *Amynthas agrestis* (Crazy snake/Asian jumping earthworm), giant African snail (*Achatina fulica*), vineyard snail (*Cernuella virgata*), white garden snail (*Theba pisana*), heath snail (*Xerolenta obvia*)

Table A3.4 Priority non-native invasive aquatic invertebrates (noninsect) of the Northwest region

Scientific name	Common name	ID	OR	WA	R6[a]
Potamocorbula amurensis	Asian clam	x	x		x
Radix auricularia	Big-eared radix				x
Eriocheir sinensis	Chinese mitten crab		x	x	
Cipangopaludina chinensis, C. japonica	Chinese mystery snail, Japanese mystery snail	x			x
Orconectes spp., *O. virilis; Procambarus* spp.	Crayfish (red swamp, rusty, ringed, virile, marbled, signal, red claw, yabby, marron)	x		x	x
Carcinus maenas	European green crab			x	
Potamopyrgus antipodarum	New Zealand mud snail	x		x	x
Philine auriformis	New Zealand sea slug		x		
Bythotrephes longimanus [*cederstroemi*], *Cercopagis pengoi*	Water fleas	x	x		
Dreissena polymorpha, D. rostriformis bugensis	Zebra/quagga mussels	x	x	x	x

[a]R6 = USDA Forest Service Region 6 (OR, WA) priority invasive species

Table A3.5 Priority non-native invasive vertebrates of the Northwest region

Scientific name	Common name	ID	OR	WA	R6[a]
Aquatic vertebrates					
Lithobates catesbeianus (*Rana catesbeiana*)	American bullfrog	x		x	x
Amia calva	Bowfin	x			
Hypophthalmichthys spp., *Mylopharyngodon piceus*	Carp (Asian, black, big head, diploid grass, silver)	x	x	x	
Salmo salar	Atlantic salmon		x	x	
Didemnum vexillum	Didemnum tunicate			x	
Chelydra serpentina serpentina	Eastern snapping turtle	x	x		
Lepisosteidae spp.	Gar	x			
Neogobius melanostomus, Rhinogobius brunneus, Tridentiger bifasciatus	Goby	x	x		
Notemigonus crysoleucas	Golden shiner		x		
Acipenser medirostris	Green sturgeon	x			
Leuciscus idus	Ide	x			
Esox spp.	Muskellunge/northern pike		x		
Serrasalmus spp., *Rosseveltiella* spp., *Pygocentrus* spp.	Piranhas	x			
Trachemys scripta elegans	Red-eared slider	x			
Taricha granulosa	Rough-skinned newt	x			
Scardinius erythrophthalmus	Rudd	x			
Gymnocephalus cernuus	Ruffe	x	x		
Channa spp.	Snakehead	x	x	x	
Dorosoma petenense	Threadfin shad (yellowtails)		x		
Clarias spp.	Walking catfish	x			
Terrestrial vertebrates					
Sus scrofa	Feral swine		x	x	x
Cygnus olor	Mute swan		x		
Myocastor coypus	Nutria	x	x	x	x

[a]R6 = USDA Forest Service Region 6 (OR, WA) priority invasive species

wildlife (Edwards 1999). Toxic algal blooms are enhanced by high water temperatures and fertilizer runoff. The diatom rock snot (didymo) is actually native to the Pacific Northwest, but in the mid-1980s, it became more prolific in its distribution and began to impact recreational activity.

Three invasive terrestrial plants—Himalayan blackberry (*Rubus armeniacus*), Japanese knotweed (*Fallopia japonica*), and giant hogweed (*Heracleum mantegazzianum*)—are problematic in both upland and riparian environments in the Marine West Coast Forest ecoregion (Fig. A3.3a). These very abundant species were introduced intentionally as ornamentals or crops. They tend to shade out smaller native plants, reducing plant diversity and limiting habitat and food sources for both birds and native wildlife. Native to Western Europe, Himalayan blackberry is an evergreen woody vine whose canes have large stiff prickles and form dense thickets. Its growth form approximates a shrub in terms of how animals use it and its height in the environment. The plant is very common in the Northwest region, providing prolific berries that are collected recreationally

Fig. A3.3 Invasive plants in the Northwest region include (**a**) false brome (*Brachypodium* spp.), on the Willamette National Forest, OR; (**b**) roadside infestation of spotted knapweed (*Centaurea stoebe*) on the Deschutes National Forest, OR; (**c**) orange hawkweed (*Hieracium aurantiacum*) on the Wallowa-Whitman National Forest, OR; (**d**) ribbon grass, a striped horticultural variety of reed canary grass (*Phalaris arundinacea*), on an islet in the Metolius River, Deschutes National Forest, OR

and commercially (Stannard 2014). The thickets, mounds, or banks can completely and permanently exclude other plant species and pose a potential fire hazard. Japanese knotweed, native originally to Asia, but introduced to the Netherlands in 1829 (The Knotweed Company Ltd. 2018), and giant hogweed, native to the Caucasus region of Eurasia, can grow as tall as 15–20 ft. and spread rapidly. Japanese knotweed is known globally as one of the world's most destructive invasive species because its large underground root system can damage structures, walls, and architectural sites, as well as reduce channel capacity. Giant hogweed is considered a public health hazard because it causes a phototoxic reaction when animal skin is exposed to sap and ultraviolet radiation.

Many of the invasive plants in the Northwest region were imported initially as ornamentals to either the area, or, more generally, North America. Some examples are English ivy (native to the United Kingdom), old man's beard (native to the United Kingdom), orange hawkweed (native to Europe), yellow archangel (native to Europe/Western Asia), garlic mustard (native to Europe and Asia), Scotch broom (native to Europe and North Africa), purple loosestrife (native to Europe and Asia), and saltcedar (*Tamarix* spp., native to Europe and Asia) (Table A3.1). Saltcedars (see Southwest region summary) are riparian shrubs or small trees that are aggressively invasive. Populations of these plants are prevalent in the warm and dry riparian corridors of the Blue Mountains and Cold Basins Ecoregions of eastern Oregon and Washington along the Owyhee, Snake, and John Day Rivers, and in the Columbia Plateau Ecoregion (Fig. A3.2) (Thorson et al. 2003). These riparian trees are known to decrease stream flows, lower biodiversity, and create salinization issues. Purple loosestrife is a wetland forb that can rapidly establish and replace native vegetation with a dense, homogeneous stand that reduces local biodiversity, endangers rare species, and provides little value to wildlife. English ivy and old man's beard are both vigorous creepers that not only cover terrestrial surfaces but will climb high into trees and smother competing vegetation. Garlic mustard was initially introduced to the East Coast of North America as a medicinal herb, but it has spread through forest understories and competes with native species. Orange hawkweed and yellow archangel are also understory shade plants that spread vigorously and smother competing native plants.

Many of the remaining invasive upland plants (Table A3.1) are weed species that can become dominant in meadow, range, or forest habitats, often creating monocultures. Most of these species were introduced accidentally, although some were introduced as ornamentals (i.e., Dalmatian toadflax). They can severely degrade the quality of range habitats for cattle and other domestic and native ungulates (i.e., cheatgrass, knapweeds). Although they are not toxic specifically to livestock as are some other species (i.e., some knapweeds, tansy ragwort), they can cause mechanical damage to the animals. In addition, invasive annual grasses like cheatgrass, medusahead, and ventenata are considered highly detrimental in the interior Western United States because they have the potential to alter wildfire regimes in some areas (see Chap. 2, Box 2.1; and the Southwest region summary for additional discussion of these issues). These species are problematic largely in the Blue Mountains and Cold Basins Ecoregions of the Northwest region (Fig. A3.2). Western juniper (*Juniperus occidentalis*) is an example of a native species that is often managed as an invasive species due to emergent concerns of its ecological and economic impact in the interior ecoregions, including degrading habitat for the threatened greater sage-grouse (*Centrocercus urophasianus*). Juniper management may be problematic, however, because removal can encourage exotic weedy grasses like cheatgrass to invade cut areas.

Plant Pathogens

In the Pacific Northwest, invasive pathogens (fungi, water molds, bacteria, nematodes, and viruses) are a significant forestry problem for wood production as well as for urban and rural landscaping (Table A3.2). Two of the most damaging invasive species are in the genus *Phytophthora*, "water molds" in the kingdom Straminipila (formerly Chromista), which includes aquatic organisms such as diatoms and kelp (Dick 2001). These fungi-like microbes thrive in wet conditions, so hydric or mesic native ecosystems west of the Cascade Mountains and well-irrigated nurseries of the Northwest region are conducive for their growth and reproduction. Notably, all three of the destructive pathogens described below were introduced on imported nursery stock, illustrating the significance of the nursery pathway (Liebhold et al. 2012) in the Northwest region.

Sudden Oak Death (*Phytophthora ramorum*) First discovered in this region in coastal forests of southwest Oregon in 2001, *Phytophthora ramorum* causes sudden oak death and other diseases. It is lethal to tanoak (*Notholithocarpus densiflorus*) and threatens this species throughout its range in Oregon (Kanaskie et al. 2017) (see Chap. 2, Box 2.5; Chap. 6, Sect. 6.4.2; Chap. 7, Sect. 7.4.2; and the Southwest regional summary for additional discussion of this pathogen). The pathogen also infects *Rhododendron*, *Viburnum*, and other important plant species in Pacific Northwest horticultural nurseries so the Federal and State quarantines affect the nursery industry throughout Oregon and Washington as well as forestry interests in part of Oregon's Curry County. The pathogen was inadvertently introduced to Oregon forests via infested nursery stock that originated in either California or Oregon (Kamvar et al. 2015).

From 2001 to 2012, a Federal and State interagency team attempted to eradicate the pathogen from Oregon, supported by the Oregon quarantine that required destruction of infected and nearby uninfested host plants. Although eradication treatments eliminated this disease from many infested sites, the pathogen continued to spread slowly. The wildland quarantine area has expanded from 22 km^2 (9 mi^2) in 2001 to 1333 km^2 (515 mi^2) in 2017, affecting over 30% of Curry County. Hundreds of thousands of tanoaks have died from *P. ramorum* in southwest Oregon (Kanaskie et al. 2013).

In 2015, a second significant introduction of the pathogen was detected in this region—the first find of the *P. ramorum* EU1 lineage (European Union Lineage One) in a US forest. The detection (on a tanoak tree) was approximately 1.6 km (1 mi) north of a small private nursery (now closed) near the Pistol River which, once again, underscores the importance of the nursery pathway for long-distance invasive pathogen movement (Grünwald et al. 2016). In Europe, the EU1 lineage of *P. ramorum* damages and kills several conifer species of significance to the Northwest region, including larch (*Larix* spp.), Douglas fir (*Pseudotsuga menziesii*), western hemlock (*Tsuga heterophylla*), and grand fir (*Abies grandis*) (Webber et al. 2010). In 2016, the EU1 strain was detected in an Oregon rural forest for a second time, less than 1.6 km (1 mi) south of the original 2015 detection (California Oak Mortality Task Force 2016).

Current management of *P. ramorum* focuses on early detection and rapid response. To support the overall goal of containment, new infections outside of the generally infested area are eradicated. With detection of new infections on the rise, management strategies will need to include a general, integrated pest management approach to minimize the impact to urban and rural forests susceptible to *P. ramorum* in the Northwest region.

Port-Orford-Cedar Root Disease (*Phytophthora lateralis*) Port-Orford-cedar or Lawson's cypress

(*Chamaecyparis lawsoniana*) is a large, beautiful conifer, endemic to southwestern Oregon and northwestern California (Zobel 1990). It is planted frequently in urban areas of the Pacific Northwest. However, the introduction and spread of the exotic fungus-like pathogen *Phytophthora lateralis* has caused high mortality levels on high-risk sites in old-growth Port-Orford-cedar forests (Fig. A3.4) and in ornamental landscapes (Hansen 2011) (see the Southwest regional summary for additional discussion of this pathogen).

The pathogen was introduced on infected nursery stock near Seattle, WA, in the 1920s and moved southward on horticultural plantings until it reached native southwest Oregon forest stands in 1952 (Hansen 2011). Port-Orford-cedar and Pacific yew (*Taxus brevifolia*) are the only native North American tree species known to be susceptible to *P. lateralis* (DeNitto and Kliejunas 1991), but the pathogen can persist in soil for more than 5 years (Hansen and Hamm 1996). The pathogen moves via transport of infected nursery plants, infested soil, and contaminated runoff water, and disease spread is correlated with proximity to roads and rivers (Hansen et al. 2000).

Several management practices are recommended to minimize the impact of Port-Orford-cedar root disease. The best management strategy involves a combination of appropriate techniques tailored for use in a specific site or landscape. These include planting resistant trees, closing forest roads during the wet season, limiting activities involving heavy equipment (e.g., timber harvesting, road maintenance) to the summer dry season, washing vehicles before they enter uninfested areas, paving road surfaces, and using only pathogen-free water for dust abatement and firefighting (Betlejewski et al. 2011; Hansen et al. 2000).

White Pine Blister Rust (*Cronartium ribicola*) In the Northwest region, white pine blister rust (*Cronartium ribicola*) threatens, damages, or kills western white pine (*Pinus monticola*), sugar pine (*P. lambertiana*), and other high-elevation white (five-needle) pines, i.e., whitebark (*P. albicaulis*) and limber (*P. flexilis*). The future of whitebark pine in the Pacific Northwest is a serious concern due to *C. ribicola* infection and the effects of colonization by mountain pine beetle (*Dendroctonus ponderosae*), wildfire, climate change, and other factors (Aubry et al. 2008).

Cronartium ribicola is one of the most damaging invasive pathogens in US forests and parks (Benedict 1981; Boyce 1938; Vitousek et al. 1996) (see also Chap. 7, Sect. 7.3.2; and the Southwest regional summary for additional discussion of this pathogen). The rust, which is native to Asia, was introduced to Western North America around 1910, on nursery stock from France imported into Vancouver, British Columbia, Canada (Liebhold et al. 2012). The 1912 Plant Quarantine Act was prompted by its introduction, with US Quarantine No. 1 prohibiting import of five-needle pines (Maloy 1997).

In the Northwest region, surveys for *C. ribicola* began around Puget Sound, WA, in 1917, with the first detection in Washington reported on black currant (*Ribes* spp.) and western white pine in 1921 (Detwiler 1922, as cited by Geils et al. 2010). Over the next several decades, the pathogen had spread on that host throughout much of Washington and

Fig. A3.4 Port-Orford-cedar killed by *Phytophthora lateralis* on the Gold Beach Ranger District, Rogue River-Siskiyou National Forest. (Figure courtesy of Ellen Goheen, USDA Forest Service Northwestern Region, Forest Health Protection)

Oregon, but was also detected on whitebark pine (Geils et al. 2010).

This rust fungus is an obligate, biotrophic (requires a live host) pathogen with a complex life cycle that requires an alternate host, primarily currants, along with the white pine host for the disease to occur (Geils et al. 2010). The Civilian Conservation Corps undertook extensive efforts to control the disease by removing *Ribes* species in the 1930s (Benedict 1981), but contemporary management favors resistance breeding programs (Kegley et al. 2012; Schoettle et al. 2012), as well as pruning young stands to minimize infections in the lower crowns and protecting larger trees from other mortality agents, such as fire and mountain pine beetle (Goheen and Goheen 2014) (see also Chap. 7, Table 7.2).

Insects

A late twentieth-century summary of extra-continental forest insects known to have been established in Western North America suggested that 75 species had been introduced and that 17 of these were purposeful introductions—releases as biological agents for the control of insects or noxious plants (Furniss and Carolin 1977). Many of these historical introduced species, as well as some new species, have established populations in the Northwest region (Table A3.3) (Furniss and Carolin 1977; Hayes and Ragenovich 2001; LaBonte et al. 2005; Mudge et al. 2001; Rabaglia et al. 2019; Seybold and Downing 2009). A large number of additional purposeful introductions have been made in the Northwest region for the biological control of urban and wildland forest pests (Bellows et al. 1998) or invasive plants (Sing et al. 2016; Winston et al. 2014a, b). Though not invasive species in the strict sense, we include these purposefully introduced taxa (Table A3.3) to maintain continuity with the original summary by Furniss and Carolin (1977). The European (North American) gypsy moth (*Lymantria dispar dispar*) was included in the historical summary (Furniss and Carolin 1977), but the Asian gypsy moth (*Lymantria dispar asiatica/japonica*) has been detected recently in the vicinity of the coastal ports of the Northwest region (see below). Furthermore, this region has sustained new invasions by pest insects that threaten valuable horticultural crops. Several insect species that already damage or threaten to damage forest trees and horticultural crops in the Northwest region are highlighted below.

Balsam Woolly Adelgid (*Adelges piceae*) A tiny (about 1-mm-long) sap-sucking insect, the balsam woolly adelgid (*Adelges piceae*) (Fig. A3.5), is probably the most prominent invasive pest of forest trees in the Northwest region (Ragenovich and Mitchell 2006) (see Southeast and Caribbean regional summary for further discussion of this species). It first appeared on the West Coast in the late 1920s (Annand 1928; Keen 1952). In this region, it infests primarily subalpine fir, Pacific silver fir (*Abies amabilis*), and grand fir (*A. grandis*) (Ragenovich and Mitchell 2006). Other susceptible hosts include noble fir (*A. procera*), Shasta fir (*A. magnifica*), and white fir (*A. concolor*). Subalpine fir and Pacific silver fir are impacted to a greater degree in

Fig. A3.5 Life stages of the balsam woolly adelgid (*Adelges piceae*). (Figure courtesy of USDA Forest Service Northern Region, Forest Health Protection)

mountainous areas, whereas grand fir is impacted to a greater degree in lowland valleys. The symptoms of attack by balsam woolly adelgid, especially on young fir trees, include buds failing to open and twigs becoming enlarged at the nodes and around the buds. Feeding by the insect results in stem and twig injury whereby the adelgid injects a substance into the inner bark, resulting in abnormal cell division and differentiation in the inner bark and newly formed wood. In the 1950s, a major balsam woolly adelgid outbreak killed or "seriously weakened" over 3,539,606 m^3 (1.5 billion board feet) of mature Pacific silver fir trees across 161,874 ha (400,000 ac) in southwestern Washington (Johnson et al. 1963). The life cycle of this species (and related adelgids) is unusual (Havill and Foottit 2007; Havill et al. 2011). North American populations are composed entirely of females; thus reproduction is parthenogenetic (i.e., without mating and fertilization). Adults are dark purple to black, nearly spherical, and wingless (Fig. A3.5). They produce a thick mass of a white waxy wool-like material that covers the body and protects the adult and her eggs (Fig. A3.5).

The range of balsam woolly adelgid has been expanding in the eastern portion of the Northwest region (Fig. A3.6) (Gast et al. 1990; Hrinkevich et al. 2016; Lowrey and Davis 2017). In apparent synergy with climate change effects, the invasive threat here is to high-elevation stands of subalpine fir, which play an important ecological function in regulation of snow melt and wildlife habitat and have a modest, yet measureable, timber value (Alexander 1987; Steele et al. 1981). Although 20 predaceous insects were introduced to the West for the biological control of balsam woolly adelgid (Table A3.3) (Bellows et al. 1998; Mitchell and Wright 1967), only 5 species were considered established in 1998 (*Aphidoletes thompsoni*, *Pullus impexus*, *Laricobius erichsonii*, *Cremifania nigrocellulata*, and *Leucopis obscura*), and Bellows et al. (1998) characterized the effectiveness of the biological con-

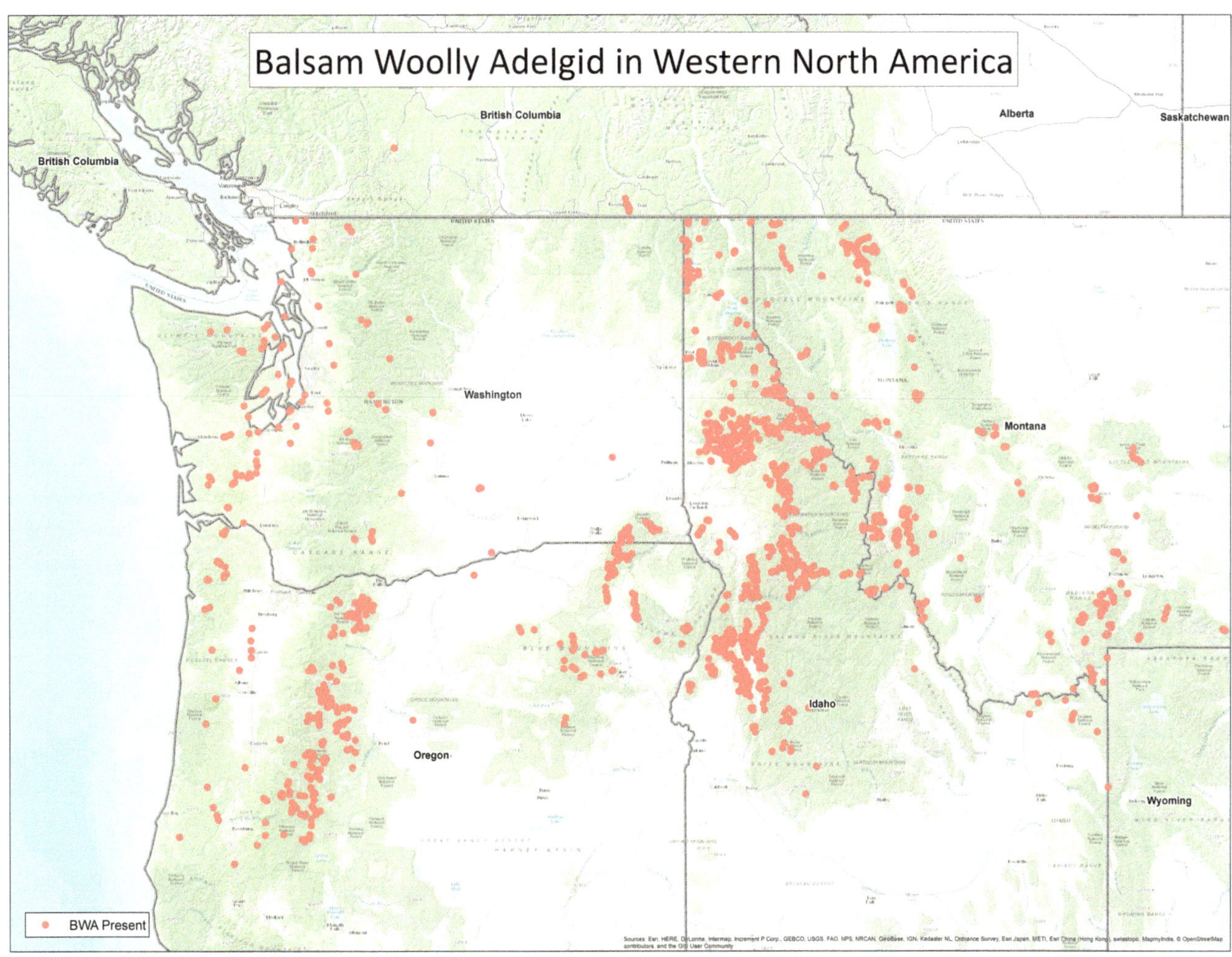

Fig. A3.6 Distribution of the balsam woolly adelgid (*Adelges piceae*) in the Northwest region of the United States and Canada as of spring 2016 based on georeferenced collection records and positive identification of specimens. (Figure courtesy of Amy Gannon, Montana Department of Natural Resources and Conservation, Missoula, MT)

trol as "variable and somewhat limited." It is unknown whether any of these species (released in Oregon, Washington, and British Columbia) may have spread into the expanding range of balsam woolly adelgid in the eastern portion of the Northwest region (Lowrey and Davis 2017).

Hemlock Woolly Adelgid (*Adelges tsugae*) Hemlock woolly adelgid (*Adelges tsugae*) is an economically and ecologically important invasive pest of hemlock trees (*Tsuga* spp.) in eastern North America (McClure et al. 1996) that has only had minor impacts to ornamentals and other plantings in the Northwest region (Furniss and Carolin 1977) (see the Southeast and Caribbean regional summary for further discussion of this species). Western hosts include western hemlock, *T. heterophylla*, and mountain hemlock (*T. mertensiana*), though the adelgid is less commonly encountered on the latter host (Havill et al. 2016). Recent molecular analyses (microsatellite and mitochondrial DNA sequences) have revealed that populations of hemlock woolly adelgid in the Northwest region are a consequence of an ancient colonization event from an ancestor whose host was an Asian hemlock (*T. sieboldii*) (Havill et al. 2016). Colonization of Western North America was estimated to have occurred prior to the last glacial period by adelgids directly ancestral to those in southern Japan, perhaps carried by birds. Havill et al. (2011) report that the earliest North American specimens were collected in 1907 from South Bend, WA (US National Collection of Insects, Beltsville, MD). Other early records from the West include a report of damage to western hemlocks in Vancouver, British Columbia (Chrystal 1916), and specimens collected in Oregon and California that were used to formally describe hemlock woolly adelgid as a new species (Annand 1924). Populations of adelgids in the Northwest Region have served as important sources of natural enemies for the biological control program for hemlock woolly adelgid in the Eastern United States (McClure 2001; Reardon et al. 2004). Two non-native species in this program, a beetle, *Laricobius erichsonii*, and a fly, *Leucopis obscura*, were introduced originally into the Northwest region to control to control balsam woolly adelgid (Furniss and Carolin 1977), but they have not been found on hemlock woolly adelgid (Kohler et al. 2008). However, a native species of beetle, *Laricobius nigrinus*, and two native species of flies, *Leucopis argenticollis* and *Leucopis piniperda*, have been recognized as predators of hemlock woolly adelgid in Idaho (*L. nigrinus* only), Oregon, and Washington and were moved subsequently by specialists to the Eastern United States for biological control (Kohler et al. 2008, 2016; Mausel et al. 2011).

An ensemble of other sapsucking insects has invaded the Northwest region and has, at times, threatened the health of native trees (Table A3.3). These include larch woolly aphid, eastern spruce gall aphid, spruce aphid, European elm scale, European birch aphid, and pine bark aphid.

Larch Casebearer (*Coleophora laricella*) Larch casebearer (*Coleophora laricella*) is a small moth whose larvae feed in the needles of western larch (*Larix occidentalis*), in the Northwest region. The species was introduced from Europe first to the Eastern United States in 1886 (Bellows et al. 1998) and detected in western larch near St. Maries, ID, in 1957 (Tunnock and Ryan 1985). In the 1970s, it caused significant damage to western larch in the Blue Mountains (OR and WA). Its invaded range in this region is now coincident with that of western larch, which is present in all three States of the Northwest region, as well as in Montana and southern Canada (Tunnock and Ryan 1985). Beginning in the 1960s, ten parasitoid wasps were introduced into the region from eastern North America, Europe, and Japan (Bellows et al. 1998); two of the parasitoids, *Agathis pumila* and *Chrysocharis laricinellae*, established populations in the Blue Mountains and appeared to significantly lower the population density of larch casebearer to the point that "the species is no longer considered a pest in the West" (Bellows et al. 1998; Ryan 1997). However, Hayes and Ragenovich (2001) suggested that the level of control of larch casebearer in eastern Oregon relaxed in the late 1990s such that insect population densities and western larch defoliation increased. A subsequent survey in this area for both larch casebearer and the two principal parasitoids revealed that both the defoliator and the introduced natural enemies were widespread, that parasitism rates ranged from 1.8% to 53.4%, and that moth population density was negatively correlated with percentage parasitism by *A. pumila* (Shaw and Oester 2010).

Gypsy Moths (*Lymantria* **spp.**) Larvae of two non-native invasive moths are a constant and recurring threat for introduction and establishment in the Northwest region. These defoliators are the European (North American) gypsy moth (*Lymantria dispar dispar*) and the Asian gypsy moth (*Lymantria asiatica/japonica*) (see Chap. 2, Sect. 2.3; Chap. 7, Sect. 7.4.1; and the Southwest and Southeast and Caribbean regional summaries for additional discussion of this insect complex). Potential sources of European gypsy moth include populations from northeastern and North central North America (introduced and established first in Medford, MA, in 1869) and ancestral locations in Europe; sources of Asian gypsy moth include Siberia, the Russian Far East, Korea, China, and Japan. The Canadian province of British Columbia has also experienced frequent introductions and attempted eradications of these moths (Myers et al. 2000). These ecologically and behaviorally distinct subspecies were distinguished initially by mitochondrial and nuclear DNA sequencing techniques, including microsatellite DNA analysis (Bogdanowicz et al. 1993, 1997, 2000). Other workers have used more sophisticated approaches and methodology, including real-time multiplex PCR (Djoumad et al. 2017; Islam et al. 2015; Stewart et al. 2016). With its broader host range (that includes conifers) and the flight capacity of females, Asian gypsy moth is considered a greater threat to the forests of the Western United States. Other potentially invasive moth defoliators that threaten forest trees in the Northwest region are the nun moth (*L. monacha*), the pink gypsy moth (*L. mathura*), and the Siberian moth (*Dendrolimus superans sibiricus*) (Hayes and Ragenovich 2001). Three other *Lymantria* species in Japan are also recognized as threats to Northwest region forests, *L. albescens*, *L. umbrosa*, and *L. postalba*, which are all listed as quarantine pests by the North Atlantic Plant Protection Organization (NAPPO).

Increased trade with the Russian Far East and other parts of Asia has greatly increased the frequency of encounters that regulatory entomologists face with Asian gypsy moth and other lymantriid moths. In 1990 and 1991, an introduced population of Asian gypsy moth was detected in pheromone-baited traps primarily around Tacoma, WA (Bogdanowicz et al. 1993), which resulted in a 47,146-ha (116,500-ac) eradication program. In a similar scenario in 2015, trap captures of ten Asian gypsy moth males in Washington (primarily in the southern half of the Puget Sound area) and two Asian gypsy moth males near Portland, OR, revealed that an incipient population of Asian gypsy moth with gravid adult females may have occurred on the U.S. mainland. Potential pathways included egg masses or pupae attached to ship cargo from Asia and then moved inland as the cargo was offloaded or dispersal of young larvae or adults from cargo on board or from the ship superstructure to sites on shore. These pathways are likely as Asian gypsy moth egg masses are intercepted annually on ships and cargo in Washington and Oregon ports. One male Asian gypsy moth has been trapped as far inland as northern Idaho (Kootenai County), with speculation that the insect was transported on a shipping container along a nearby rail line (Lech and Livingston 2004; Pederson et al. 2004). To mitigate introductions, USDA entomologists have long pursued a cooperative assistance program with their foreign counterparts in and around ports in the Russian Far East, Japan, Korea, and China to detect and manage these potential invaders (Freyman 2015; Humble et al. 2013; USDA 1993, 2016). To further reduce interceptions of Asian gypsy moth and other lymantriids in North America, NAPPO developed Regional Standards for Phytosanitary Measures (RSPM 33: Guidelines for Regulating the Movement of Vessels from Areas Infested with the Asian Gypsy Moth).

In response to the 2015 detection of Asian gypsy moth, a technical working group proposed a combination of insecticide treatment and delimitation trapping in the areas where moths were detected. Three- and six-mile (4.8- and 9.7-km) radius zones were established around each of six locations. A microbially based material, *Bacillus thuringiensis* var. *kurstaki* (Btk), was applied aerially three times within the designated treatment areas of more than 4047 ha (10,000 acres) in Washington and 2833 ha (7000 ac) in Oregon. Approximately 11,000 delimitation traps were placed in Washington, and approximately 3100 traps were placed in Oregon. The estimated cost for the 2016 Asian gypsy moth eradication program was $5 million. Detections of Asian gypsy moth in the Northwest region are continually being made against a backdrop of detections of European gypsy moth, presumably from introductions from eastern North America. In 2015, a population of European gypsy moth found in the Seattle area raised the concern that this population might mix with the contemporaneously detected Asian gypsy moth populations.

Sawfly Defoliators (Hymenoptera, Symphyta) A major survey for native and non-native invasive sawflies has revealed a large number of new defoliators in Washington and Oregon (Table A3.3) (Looney et al. 2012, 2016a, b). At least 20 species were found when field trap catches or museum collections were evaluated, and these were primarily free feeding or leaf-mining forms on hardwoods (alder, birch, elm, etc.), shrubs, and herbaceous vegetation. Many of the species were originally from Europe, but had first invaded eastern North America before they were introduced into the Northwest region (Looney et al. 2016b). Some key forest pests of note are the green alder sawfly (*Monsoma pulveratum*), which has also been damaging alder in Alaska (Kruse et al. 2010), the introduced and European pine sawflies (*Diprion similis* and *Neodiprion sertifer*, respectively), the European spruce sawfly (*Gilpinia hercyniae*), and various leafminers on birch (*Fenusa pusilla*, *Fenusella nana*, and *Profenusa thomsoni*). The ultimate impact of these new species to the region—mostly collected since the mid-2000s—is unknown, but the local biology, feeding behavior, and interactions within this assemblage of invading defoliators bear future scrutiny.

Spotted Wing Drosophila (*Drosophila suzukii*) The spotted wing drosophila (SWD) (*Drosophila suzukii*) was first found in August 2008 in the US mainland in California (Hauser 2011). Infestations were reported soon thereafter in 2009 in Oregon and Washington (Lee et al. 2011) and in 2012 in Idaho. While SWD is a primary economic concern in blackberry, blueberry, cherry, and raspberry crops, this insect species has a wide host range that includes many wild and ornamental hosts, which enables it to persist in woodland areas (Kenis et al. 2016; Lee et al. 2015). In a 2-year study at 35 farms, SWD were captured earlier in farms that had more woodland area surrounding the farm (Pelton et al. 2016).

These woodland habitats often contained wild blackberry and wild cherry in mid-summer and other fall-bearing fruit that may provide SWD with a late-season host after the crop has been harvested. A study that marked wild blackberry borders revealed that SWD moved from there and into the raspberry crop in the summer (Klick et al. 2015). Within agricultural areas, SWD is primarily managed with insecticides (Bruck et al. 2011). Classical biological control agents for SWD from Asia are being evaluated and are still under quarantine. Endemic parasitoids, such as the pupal parasitoid *Pachycrepoideus vindemiae* and the larval parasitoid *Leptopilina heterotoma*, have been found to attack sentinel SWD in mixed farm, raspberry, blueberry, and riparian habitats of Oregon (Miller et al. 2015). Although parasitism counts were low, conserving or augmenting these biological control agents may lead to greater pest suppression in unmanaged areas.

Brown Marmorated Stink Bug (*Halyomorpha halys*) The brown marmorated stink bug (BMSB) (*Halyomorpha halys*) was first found in Portland, OR, in 2004 and in Vancouver, WA, in 2010 (ODA 2010; Wiman and Lowenstein 2017). As with the invasion process in the East Coast, BMSB was first a nuisance pest reported by homeowners and then appeared in agricultural crops and spread to every county in the Willamette Valley of Oregon, Hood River, and southern Oregon. By 2013, economic losses from BMSB were reported in the northern Willamette Valley (Wiman and Lowenstein 2017). Although BMSB is a primary economic concern to stone fruit, vegetables, and field crop and nut tree growers (Rice et al. 2014), it also has a wide host range that includes many trees: elm, hawthorn, holly, linden, maples, and tree of heaven (Bergman et al. 2014; Leskey and Nielsen 2018). Many ornamental hosts are partial hosts meaning that one or more of BMSB's life stages can develop on the host but not all stages (Bergman et al. 2014, 2016), and these partial hosts may facilitate the movement of BMSB across a landscape. In agricultural areas, BMSB are often treated with insecticides although some may have limited efficacy (Leskey et al. 2012; Leskey and Nielsen 2018). The classical biological control agent, *Trissolcus japonicus*, is being studied for potential release. Genetic analyses of the *T. japonicus* found in the field in Maryland and Oregon-Washington revealed that these populations were not related to the populations in quarantine (stopBMSB.org; Mortenson 2016). In Oregon, the Oregon Department of Agriculture has approved releases of the *T. japonicus* reared from field sources, and efforts are underway in Washington to make similar releases.

Noninsect Invertebrates

Aquatic invasive invertebrates (Table A3.4) include several mollusks (Asian clam, big-eared radix, Chinese mystery snail, New Zealand mud snails, New Zealand sea slug, and zebra/quagga mussels) and crustaceans (crayfish, crabs, and water fleas). Many of the invasive mollusks of concern in the Northwest spread rapidly and can attain large population sizes that displace native species. These taxa can accrue prey resources rapidly, affecting foundation levels of food webs (algae, phytoplankton) in aquatic systems. Along with abundant populations come abundant waste products that can affect environmental systems. In some systems, for example, the tissues or waste products of zebra mussels may accumulate contaminants to 300,000 times the level available in the environment, with subsequent effects on their environment, including contaminating their predators (Snyder et al. 1997) (see Chap. 2, Sect. 2.6, and Box 2.6 for additional discussion of the impacts of zebra mussels). Another concern is that large numbers of mollusks can foul human structures. Introductions of some species are likely tied to inadvertent human transmission, such as in ship ballast water or in boats or fishing gear (i.e., zebra/quagga mussels, spiny water flea (*Bythotrephes longimanus*), and green crabs). The deliberate introduction and consequent escape of some species are also associated with food and medical markets (i.e., Chinese mystery snails, crayfish, and mitten crabs). Nearly all terrestrial invertebrates are insects (see above and Table A3.3).

Vertebrates

Two frogs (American bullfrog, red-legged frog (Ranidae)), a tunicate (*Didemnum vexillum*), a turtle, and 11 fishes (Table A3.5) have established populations in the Northwest region as a consequence of human activities. For example, American bullfrogs are native to the Eastern United States, but were brought to the West to establish food farms and out of nostalgia for their calls. Bullfrogs are carriers of the amphibian chytrid fungus, *Batrachochytrium dendrobatidis* (Bd) (Table A3.2), but do not always exhibit disease symptoms and hence may serve as a reservoir species of the pathogen. Invasive fishes include a mix of species introduced for human food, as bait for recreational fisheries, or from the aquarium or ornamental industry (Table A3.5). They are of concern primarily because of their ecological effects on native ecosystems. Atlantic salmon are native to the North Atlantic Ocean, where they are anadromous, occurring in the ocean and returning to spawn in rivers. Farms in Washington and British Columbia are thought to be the origin of the species found elsewhere in the Northwest, as these fish stray from "natal" streams—even as far as Alaska. Competition with native salmonids, pollution from the farms, and the potential for farm-raised animals to carry pathogens to native stocks are all ecological impacts of these invaders. Certain species of gobies (Table A3.5) are of Asian origin and occur in fresh and brackish water. They are thought to have been introduced to the Northwest region by ballast water and may compete with or prey upon native species. Golden shiners are from the Eastern United States and are pond-cultured fishes that are also used as bait. Golden shiners may displace native species. The carpet sea squirt or ascidian (*Didemnum vexillum*) is a colonial tunicate in the phylum Chordata (hence its inclusion under vertebrates). It seems to be native to Japan, but it has been detected along the Washington coast since 2009 and in two Oregon bays since 2010. It is a fouling organism in marine and estuarine systems, growing rapidly and covering vast surfaces as mats, displacing native biota and encrusting dock pilings and aquatic equipment. It can be introduced in ballast water, or it may hitchhike on the hulls of boats, or with commercial shellfish stock or equipment.

The category of terrestrial invasive vertebrates in the Northwest region contains only six species (Table A3.5), but they can have extensive effects, ecologically and socioeconomically. Three species are strongly associated with wetland and riparian environments (feral swine, mute swan, nutria). Feral swine are escaped domestic pigs whose rooting behavior can have several effects: waterway habitat degradation, provision of an invasion pathway for non-native plants, and damage to agricultural crops and lands. The mute swan was introduced from New York for scenic enjoyment. These aggressive, large (11.3–13.6 kg, 25–30 lb) birds may consume significant quantities of aquatic plants, competing with native birds for food and habitat supplies. Nutria were initially brought to the Pacific Northwest as part of fur farming in the 1920s. Escaped and released animals (after the collapse of this element of the fur industry) subsequently spread throughout the region. They burrow into the banks of streams and agricultural canals, destabilizing natural stream systems and human agricultural infrastructure. See the Pacific Southwest regional summary for additional discussion of nutria. In addition to these three species, European starling and rock doves are strong competitors for nest space and food sources with native birds. The Norway rat is common in urban settings in the Northwest region and is closely affiliated with human structures. These animals may cause extensive damage to human structures and are known carriers of pathogens that may affect human health.

Acknowledgments We greatly appreciate preliminary reviews of this summary by Blakey Lockman (USDA Forest Service Pacific Northwest Region Forest Health Protection, State and Private Forestry, Portland, OR) and Ellen Goheen (USDA Forest Service Pacific Northwest Region Forest Health Protection, State and Private Forestry, Central Point, OR). Connie J. Mehmel (USDA Forest Service Pacific Northwest Region Forest Health Protection, State and Private Forestry, Wenatchee, WA) reviewed the list of invasive insect species; Gina Davis (USDA Forest Service Northern Region Forest Health Protection, Coeur d' Alene, ID) and Laura Lowrey (USDA Forest Service Intermountain Region Forest Health Protection, Boise, ID) provided background and content on the balsam woolly adelgid; and Daniel J. Larkin (University of Minnesota) reviewed the section on aquatic plants. Lori J. Nelson (USDA Forest Service Pacific Southwest Research Station) assisted with manuscript preparation. Shawna Bautista (USDA Forest Service Pacific Northwest Region) assisted with compilation of Northwest Region invasive species. We also appreciate the critical comments from four anonymous reviewers on an earlier version of the manuscript.

Literature Cited

Acheampong S, Strong WB, Schwartz MD et al (2016) First Canadian records for two invasive seed-feeding bugs, *Arocatus melanocephalus* (Fabricius, 1798) and *Raglius alboacuminatus* (Goeze, 1778), and a range extension for a third species, *Rhyparochromus vulgaris* (schilling, 1829) (Hemiptera: Heteroptera). J Entomol Soc Br Columbia 113:74–78

Alexander RR (1987) Ecology, silviculture and management of the Engelmann spruce-subalpine fir type in the central and southern Rocky Mountains. Agricultural handbook 659. U.S. Department of Agriculture, Forest Service, Washington, DC, 144 p

Annand PN (1924) A new species of *Adelges* (Hemiptera, Phyloxeridae). Pan Pac Entomol 1:79–82

Annand PN (1928) A contribution toward a monograph of the Adelginae (Phylloxeridae) of North America. University Series in Biological Sciences, vol 6, no. 1. Stanford University Publication, Palo Alto. 146 p

Aubry C, Goheen D, Shoal R, Ohlson T, Lorenz T et al (2008) Whitebark pine restoration strategy for the Pacific northwest region 2009–2013. USDA Forest Service, Pacific Northwest Region, Portland, 96 p

Bai BB, Worth RA, Johnson KJR, Brown G (2002) Distribution and phenology of the cereal leaf beetle, *Oulema melanopus* (L.) (Coleoptera: Chrysomelidae) in Oregon. Proceedings 61st Annual Pacific Northwest Insect Management Conference, Portland. 4 p. https://research.libraries.wsu.edu:8443/xmlui/bitstream/handle/2376/2253/2002%20Oregon%20CLB%20Survey%20-%20Bai%20et%20al.pdf?sequence=1&isAllowed=y

Bautista S (2017) Personal communication on April 19, 2017 to Rebecca Flitcroft, U.S. Department of Agriculture, Forest Service, Pacific Northwest Research Station, 3200 SW Jefferson Way, Corvallis, OR 97330

Bellows TS, Meisenbacher C, Reardon RC (1998) Biological control of arthropod forest pests of the western United States: a review and recommendations. Publication FHTET-96-21. US Department of Agriculture, Forest Service, Forest Health Technology Enterprise Team, Morgantown, 121 p

Benedict WV (1981) History of white pine blister rust control: a personal account. FS-355. U.S. Department of Agriculture, Forest Service, Washington, DC, 47 p

Bentz BJ, Régnière J, Fettig CJ et al (2010) Climate change and bark beetles of the western US and Canada: direct and indirect effects. Bioscience 60:602–613

Bergmann EJ, Bernhard KM, Bernon G et al (2014) Host plants of the brown marmorated stink bug in the U.S. StopBMSB.org. USDA-NIFA SCRI Coordinated Agricultural Project Northeastern IPM Center. www.stopbmsb.org/where-is-bmsb/host-plants/

Bergmann EJ, Venugopal PD, Martinson HM et al (2016) Host plant use by the invasive *Halyomorpha halys* (Stal) on woody ornamental trees and shrubs. PLoS One 11:e0149975

Betlejewski F, Goheen DJ, Angwin PA, Sniezko RA (2011) Port-Orford-cedar root disease. Forest Insect & Disease Leaflet 131 (revised). US Department of Agriculture, Forest Service, Portland. 12 p

Bogdanowicz SM, Mastro VC, Prasher DC, Harrison RG (1997) Microsatellite DNA variation among Asian and North American gypsy moths (Lepidoptera: Lymantriidae). Ann Entomol Soc Am 90:768–775

Bogdanowicz SM, Schaefer PW, Harrison RG (2000) Mitochondrial DNA variation among worldwide populations of gypsy moths, *Lymantria dispar*. Mol Phylogenet Evol 15:487–495

Bogdanowicz SM, Wallner WE, Bell J et al (1993) Asian gypsy moths (Lepidoptera: Lymantriidae) in North America: evidence from molecular data. Ann Entomol Soc Am 86:710–715

Boyce JS (1938) Forest pathology, 1st edn. McGraw-Hill, New York/London, 600 p

Bruck DJ, Bolda M, Tanigoshi L et al (2011) Laboratory and field comparisons of insecticides to reduce infestation of *Drosophila suzukii* in berry crops. Pest Manag Sci 67:1375–1385

CABI Invasive Species Compendium (2018) *Ludwigia grandiflora* (water primrose). https://www.cabi.org/isc/datasheet/109148#3C51E072-69A3-4DD8-9920-0E5C0E57CE27

California Oak Mortality Task Force (2016) November monthly report. 4 p. http://www.suddenoakdeath.org/wp-content/uploads/2016/11/COMTF-Report-November-2016.pdf

Childs RD, Swanson DC (2003) The winter moth, (*Operophtera brumata* (L.). University of Massachusetts Extension Report, Amherst. 2 p. http://www.massaudubon.org/content/download/7334/133282/file/winter_moth.pdf

Chrystal RN (1916) The life-history of *Chermes cooleyi* Gillette, in Stanley Park, Vancouver, B.C. 46th annual report of the entomological society of Ontario 1915, pp 123–130

DeNitto GA, Kliejunas JT (1991) First report of *Phytophthora lateralis* on Pacific yew. Plant Dis 75:968

Detwiler SB (1922) Blister rust appears in the Puget Sound region. Am For 28:97–98

Dick MW (2001) Straminipilous fungi: systematics of the Peronosporomycetes including accounts of the marine Straminipilous protists, the Plasmodiophorids and similar organisms. Kluwer Academic Publishers, Dordrecht, 660 p

Djoumad A, Nisole A, Zahiri R et al (2017) Comparative analysis of mitochondrial genomes of geographic variants of the gypsy moth, *Lymantria dispar*, reveals a previously undescribed genotypic entity. Sci Rep 7:1–12. https://doi.org/10.1038/s41598-017-14530-6

Doerr MD, Brunner JF, Smith TJ (2008) Biology and management of bark beetles (Coleoptera: Curculionidae) in Washington cherry orchards. J Entomol Soc BC 105:69–81

Edwards N (1999) Anatoxin. University of Sussex at Brighton. Updated 1 September 1999

Flitcroft R, Hansen B, Capurso J, Christiansen K (2016) Coverage of aquatic invasive risk assessment in USFS Region 6. Report to U.S. Forest Service Region 6. US Department of Agriculture, Forest Service, Pacific Northwest Research Station, Corvallis. Available at: https://www.fs.fed.us/pnw/lwm/aem/people/flitcroft.html

FAO (Food and Agriculture Organization) (2009) International standards for phytosanitary measures: revision of ISPM No. 15, Regulation of wood packaging material in international trade. Publication No. 15. Food and Agriculture Organization of the United Nations, Rome. https://en.wikipedia.org/wiki/ISPM_15

Foote RH, Blanc FL, Norrbom AL (1993) Handbook of the fruit flies (Diptera: Tephritidae) of America north of Mexico. Cornell University Press, Ithaca, 571 p

Freyman T (2015) The monitoring of Asian gypsy moth, pink gypsy moth and nun moth at the ports of Vladivostok (Russky Island), Nakhodka, Votochny, Slavyanka, Olga, Vanino, Plastun, Pos'et, Zarubino, Kozmino, Korsakov in 2015. OFO-PR-15-01. Federal Service for Veterinary and Phytosanitary Surveillance, Federal State Budgetary Institution, All-Russia Plant Quarantine Service, Ogden. 51 p

Furniss RL, Carolin VM (1977) Western forest insects. Miscellaneous Publication 1339. U.S. Department of Agriculture, Forest Service, Washington, DC. 654 p

Gast S, Beckman DP, Livingston RL, Gustin J (1990) Distribution of the balsam woolly adelgid in Idaho. Report 90–5. USDA Forest Service Northern Region, Missoula. 8 p

Geils BW, Hummer KE, Hunt RS (2010) White pines, *Ribes*, and blister rust: a review and synthesis. For Pathol 40(3–4):147–185

Gerberg EJ (1957) A revision of the New World species of powder-post beetles belonging to the family Lyctidae. United States Department of Agriculture, Technical Bulletin No.1157, 55 p

Goheen EM, Goheen DJ (2014) Status of sugar and western white pines on federal forest lands in southwest Oregon: Inventory query and natural stand survey results. SWOFIDSC-14-01. US Department of Agriculture, Forest Service, Pacific Northwest Region, Portland. 71 p. https://www.fs.usda.gov/Internet/FSE_DOCUMENTS/stelprdb5447311.pdf

Grünwald NJ, Larsen MM, Kamvar ZN et al (2016) First report of the EU1 clonal lineage of *Phytophthora ramorum* on tanoak in an Oregon forest. Plant Dis 100:1024

Hansen EM (2011) *Phytophthora lateralis,* Species profile. For Phytophthoras. https://doi.org/10.5399/osu/fp.1.1.1816. http://journals.library.oregonstate.edu/ForestPhytophthora/article/view/1816/1811

Hansen EM, Goheen DJ, Jules ES, Ullian B (2000) Managing port-Orford-cedar and the introduced pathogen *Phytophthora lateralis*. Plant Dis 84:4–14

Hansen EM, Hamm PB (1996) Survival of *Phytophthora lateralis* in infected roots of Port Orford cedar. Plant Dis 80:1075–1078

Hatch MH (1953) The beetles of the Pacific Northwest. Part 1: introduction and Adephaga. University of Washington Publications in Biology 16(1). University of Washington Press, Seattle. 340 p

Hatch MH (1962) The beetles of the Pacific Northwest. Part 3: Pselaphidae and Diversicornia. University of Washington Publications in Biology 16(3). University of Washington Press, Seattle. 503 p

Hauser M (2011) A historic account of the invasion of *Drosophila suzukii* (Matsumura) (Diptera: Drosophilidae) in the continental United States, with remarks on their identification. Pest Manag Sci 67:1352–1357

Havill NP, Foottit RG (2007) Biology and evolution of Adelgidae. Annu Rev Entomol 52:325–349. https://doi.org/10.1146/annurev.ento.52.110405.091303

Havill NP, Montgomery ME, Keena M (2011) Hemlock woolly adelgid and its hemlock hosts: a global perspective. In: Onken B, Reardon R (eds) Implementation and status of biological control of the hemlock woolly adelgid. Publication FHTET-2011-04. US Department of Agriculture, Forest Service, Forest Health Technology Enterprise Team, Morgantown, pp 3–14

Havill NP, Shiyake S, Galloway AL et al (2016) Ancient and modern colonization of North America by hemlock woolly adelgid, *Adelges tsugae* (Hemiptera: Adelgidae), an invasive insect from East Asia. Mol Ecol 25:2065–2080. https://doi.org/10.1111/mec.13589

Hayes JL, Ragenovich I (2001) Non-native invasive forest insects of eastern Oregon and Washington. Northwest Sci 75:77–84

Hitchcock ME, Miller S, Pike K, Gould MC (2002) Cereal leaf beetle survey and biocontrol activities in Washington State, 2002. Olympia, Washington: 2002 Entomology Project Report—WSDA Pub 077 (N/8/02), Plant Protection Division, Pest Program Washington State Department of Agriculture. 10 p. https://research.libraries.wsu.edu/xmlui/bitstream/handle/2376/2228/2002%20Washington%20State%20CLB%20Activity%20Report%20-%20Hitchcox%20et%20al.pdf?sequence=6&isAllowed=y

Hrinkevich KH, Progar RA, Shaw DC (2016) Climate risk modelling of balsam woolly adelgid damage severity in subalpine fir stands of western North America. PLoS One. https://doi.org/10.1371/journal.pone.0165094

Humble LM, Mastro V, Munson AS (2013) Asian gypsy moth, it's back! In: McManus KA, Gottschalk KW (eds) Proceedings of the 24th USDA interagency research forum on invasive species; 2013 January 8–11; Annapolis, MD: Publication FHTET-13-01. US Department of Agriculture, Forest Service, Forest Health Technology Enterprise Team, Fort Collins. 124 p

Idaho State Department of Agriculture (2013) Elm seed bug, *Arocatus melanocephalus:* an exotic invasive pest new to the U.S. Fact Sheet, Boise. 4 p. http://extension.oregonstate.edu/malheur/sites/default/files/spring_2013_esb_fact_sheet.pdf

Invasive Species of Idaho (2017) http://invasivespecies.idaho.gov/pests/

Islam MS, Barr NB, Braswell WE et al (2015) A multiplex real-time PCR assay for screening gypsy moths (Lepidoptera: Erebidae) in the United States for evidence of an Asian genotype. J Econ Entomol 108:2450–2457

Ivie MA (2002) 69. Bostrichidae. In: Arnett RH, Frank JH, Thomas MC, Skelley PE (eds) American beetles, volume II: Polyphaga: Scarabaeoidea through Curculionoidea. CRC Press LLC, Boca Raton, pp 233–244

Johnson NE, Mitchell RG, Wright KH (1963) Mortality and damage to Pacific silver fir by the balsam woolly aphid in southwestern Washington. J For 61:854–860

Johnson PJ (1998) *Melanotus cete* Candèze, a second adventitious species of *Melanotus* Eschscholtz in North America (Coleoptera: Elateridae). Pan Pac Entomol 74:118–119

Kamvar ZN, Larsen MM, Kanaskie AM et al (2015) Spatial and temporal analysis of populations of the sudden oak death pathogen in Oregon forests. Phytopathology 105:982–989

Kanaskie A, Hansen E, Goheen EM et al (2013) Detection and eradication of *Phytophthora ramorum* from Oregon forests, 2001–2011. In: Frankel SJ, Kliejunas JT, Palmieri KM, Alexander JM (tech. coords) Proceedings of the sudden oak death fifth science symposium. General Technical Report GTR-PSW-243. U.S. Department of Agriculture, Forest Service, Pacific Southwest Research Station, Albany. 169 p

Kanaskie A, Wiese R, Norlander D et al (2017) Slowing spread of sudden oak death in Oregon forests, 2001–2015. In: Frankel SJ, Harrell KM (tech. coords) Proceedings of the sudden oak death sixth science symposium. General Technical Report GTR-PSW-255. U.S. Department of Agriculture, Forest Service, Pacific Southwest Research Station, Albany. 106 p

Keen FP (1952) Insect enemies of western forests. Miscellaneous Publication 273. US Department of Agriculture, Forest Service, Washington, DC. 280 p

Kegley A, Sniezko RA, Danchok R, Savin DP (2012) Blister rust resistance among 19 families of whitebark pine, *Pinus albicaulis*, from Oregon and Washington—early results from an artificial inoculation trial. In: Sniezko RA, Yanchuk AD, Kliejunas JT, Palmieri KM, Alexander JM, Frankel SJ (tech. coords) Proceedings of the fourth international workshop on the genetics of host-parasite interactions in forestry: disease and insect resistance in forest trees. General Technical Reports. PSW-GTR-240. Pacific Southwest Research Station, Forest Service, U.S. Department of Agriculture, Albany, pp 311–315

Kenis M, Tonina L, Eschen R et al (2016) Non-crop plants used as hosts by *Drosophila suzukii* in Europe. J Pest Sci 89:735–748

Klick J, Yang WQ, Walton VM et al (2015) Distribution and activity of *Drosophila suzukii* in cultivated raspberry and surrounding vegetation. J Appl Entomol 140:37–46

Kohler GR, Stiefel VL, Wallin KF, Ross DW (2008) Predators associated with the hemlock woolly adelgid (Hemiptera: Adelgidae) in the Pacific northwest. Environ Entomol 37:494–504

Kohler GR, Wallin KF, Ross DW (2016) Seasonal phenology and abundance of *Leucopis argenticollis*, *Leucopis piniperda* (Diptera: Chamaemyiidae), *Laricobius nigrinus* (Coleoptera: Deridontidae) and *Adelges tsugae* (Hemiptera: Adelgidae) in the Pacific Northwest USA. Bull Entomol Res 106:546–550. https://doi.org/10.1017/S0007485316000250

Kruse JJ, Zogas K, Hard J, Lisuzzo N (2010) New pest in Alaska and Washington—The green alder sawfly—*Monsoma pulveratum* (Retzius). Pest Alert, R10-PR-022, Anchorage, AK: USDA Forest Service, State and Private Forestry. 2 p

LaBonte JR, Mudge AD, Johnson KJR (2005) Non-indigenous woodboring Coleoptera (Cerambycidae, Curculionidae: Scolytinae) new to Oregon and Washington, 1999–2002: consequences of the intracontinental movement of raw wood products and solid wood packing materials. Proc Entomol Soc Wash 107:554–564

LaGasa E (2006) New pest alert and update: Introduced exotic seed-bugs are new and increasing nuisance problems in areas of western Washington—*Rhyparochromis vulgaris* and *Raglius alboacuminatus*. AGR PUB 805–158, Olympia, WA: 2006 Entomology Pest Alert Pest Program, Plant Protection Division Washington State Department of Agriculture. 1 p. https://agr.wa.gov/PlantsInsects/InsectPests/Exotics/Surveys/seedbugs_06.pdf

LaGasa E, Murray T (2007) Exotic seed-bugs (Lygeoidea: Rhyparochromidae & Oxycarenidae) new to the Pacific Northwest. In: Proceedings 66th annual pacific northwest insect management conference, January 8–9, 2007, pp 5–6. http://www.ipmnet.org/PNWIMC/2007_PNW_Conference_Proceedings.pdf

Lech G, Livingston RL (2004) State of Idaho gypsy moth survey trapping program summary report 2004. Report No. IDL 04–2, Boise

Lee JC, Aguayo I, Aslin R et al (2009) Co-occurrence of the invasive banded and European elm bark beetles (Coleoptera: Scolytidae) in North America. Ann Entomol Soc Am 102:426–436

Lee JC, Bruck DJ, Dreves AJ et al (2011) In focus: spotted wing drosophila, *Drosophila suzukii*, across perspectives. Pest Manag Sci 67:1349–1351

Lee JC, Dreves AJ, Cave AM et al (2015) Infestation of wild and ornamental noncrop fruits by *Drosophila suzukii* (Diptera: Drosophilidae). Ann Entomol Soc Am 108:117–129

Leskey TC, Nielsen AL (2018) Impact of the invasive brown marmorated stink bug in North America and Europe: history, biology, ecology, and management. Annu Rev Entomol 63:599–618

Leskey TC, Lee D-H, Short BD, Wright SE (2012) Impact of insecticides on the invasive *Halyomorpha halys* (Hemiptera: Pentatomidae): analysis of insecticide lethality. J Econ Entomol 105:1726–1735

Liebhold AM, Brockerhoff EG, Garrett LJ et al (2012) Live plant imports: the major pathway for forest insect and pathogen invasions of the U.S. Front Ecol Environ 10:135–143

Logan JA, Powell JA (2001) Ghost forests, global warming, and the mountain pine beetle. Am Entomol 47:160–173

Logan JA, Régnière J, Powell JA (2003) Assessing the impacts of global warming on forest pest dynamics. Front Ecol Environ 1:130–137

Looney C, Sheehan K, Bai B et al (2012) The distribution of a potential new forest pest, *Monsoma pulveratum*

(Hymenoptera: Tenthredinidae), in the Pacific northwest states. Northwest Sci 86:342–345

Looney C, Murray T, Lagasa E, Hellman WE, Passoa SC (2016a) Shadow surveys: how non-target identifications and citizen outreach enhance exotic pest detection. Am Entomol 62:247–254

Looney C, Smith DR, Collman SJ, Langor DW, Peterson MA (2016b) Sawflies (Hymenoptera, Symphyta) newly recorded from Washington state. J Hymenopt Res 49:129–159

Lowrey L, Davis G (2017) Management and research goals for balsam woolly adelgid in the Interior West. USDA Forest Service, Northern Region Forest Health Protection, Unpublished report, Missoula. 9 p

Maloy OC (1997) White pine blister rust control in North America: a case history. Annu Rev Phytopathol 35:87–109

Mausel DL, Van Driesche RG, Elkinton JS (2011) Comparative cold tolerance and climate matching of coastal and inland *Laricobius nigrinus* (Coleoptera: Derodontidae), a biological control agent of hemlock woolly adelgid. Biol Control 58:96–102

McClure MS (2001) Biological control of the hemlock woolly adelgid in the eastern United States. Publication HTET-2000-08. US Department of Agriculture, Forest Service, Forest Health Technology Enterprise Team, Morgantown. 10 p

McClure MS, Salom SM, Shields KS (1996) Hemlock woolly adelgid. Publication FHTET-96-35. US Department of Agriculture, Forest Service, Forest Health Technology Enterprise Team, Morgantown. 18 p

Miller B, Anfora G, Buffington M et al (2015) Seasonal occurrence of resident parasitoids associated with *Drosophila suzukii* in two small fruit production regions of Italy and the USA. Bull Insect 68:255–263

Mitchell RG, Wright KH (1967) Foreign predator introductions for control of the balsam woolly aphid in the Pacific northwest. J Econ Entomol 60:140–147

Mortenson M (2016) Stinkbugs' natural predator has arrived in the Pacific Northwest. Capital Press, July 28, 2016. http://www.capitalpress.com/Oregon/20160728/stinkbugs-natural-predator-has-arrived-in-the-pacific-northwest

Mudge AD, LaBonte JR, Johnson KJR et al (2001) Exotic woodboring Coleoptera (Micromalthidae, Scolytidae) and Hymenoptera (Xiphydriidae) new to Oregon and Washington. Proc Entomol Soc Wash 103:1011–1019

Murray T, LaGasa E, Glass J (2012) Pest alert: Red lily leaf beetle. Home Gardening Series, Washington State University Extension Publication, Pullman. 2 p. http://oregonstate.edu/dept/nurspest/RLLB.pdf

Murray TA, Kohler GR, Wilhite EA (2013) *Ips paraconfusus* Lanier (Coleoptera: Curculionidae): new records of the California fivespined Ips from Washington state and the Columbia River gorge in Oregon. Coleopt Bull 67:28–31

Myers JH, Simberloff D, Kuris AM, Carey JR (2000) Eradication revisited: dealing with exotic species. Trends Ecol Evol 15:316–320

Nugent M (2005) Oregon invasive species action plan. Oregon Department of Fish and Wildlife. Portland. Accessed online at: http://www.oregon.gov/OISC/index.shtml

ODA (Oregon Department of Agriculture) (2010) Pest Alert: Brown marmorated stink bug. https://www.oregon.gov/ODA/shared/Documents/Publications/IPPM/BrownMarmoratedStinkBugPestAlert.pdf. Updated July 2016

OIE (World Organization for Animal Health [Office International des Epizooties]) (2018) OIE-listed diseases, infections and infestations in force in 2018. Available at: http://www.oie.int/animal-health-in-the-world/oie-listed-diseases-2018/

OISC (2015) 100 worst list. Oregon Invasive Species Council. Available at: https://static1.squarespace.com/static/58740d57579fb3b4fa5ce66f/t/5891577c579fb38e735f0f83/1485920126222/OISC_top100_2015.pdf

Pederson LA, Beckman DP, Halsey RL, Stipe LE (2004) Idaho forest insect and disease conditions report. Idaho Department of Lands USDA Forest Service, Northern and Intermountain Regions IDL Report No. 04–3, Boise. 39 p. https://www.fs.usda.gov/Internet/FSE_DOCUMENTS/fsbdev3_015300.pdf

Pelton E, Gratton C, Isaacs R et al (2016) Earlier activity of *Drosophila suzukii* in high woodland landscapes but relative abundance is unaffected. J Pest Sci 89:725–733

Philips TK (2002) 70. Anobiidae. In: Arnett RH, Frank JH, Thomas MC, Skelley PE (eds) American beetles, volume II: Polyphaga: Scarabaeoidea through Curculionoidea. CRC Press LLC, Boca Raton, pp 245–260

Prather T, Robins S, Morishita D (2016) Idaho's Noxious Weeds, 8th edn. University of Idaho Extension, Moscow. Available at: http://www.cals.uidaho.edu/edComm/pdf/BUL/BUL816.pdf

Rabaglia RJ, Cognato AI, Hoebeke ER et al (2019) Early detection and rapid response, a ten-year summary of the USDA Forest Service program of surveillance for non-native bark and ambrosia beetles. Am Entomol 65:29–42

Rabaglia RJ, Dole SA, Cognato AI (2006) Review of American Xyleborina (Coleoptera: Curculionidae: Scolytinae) occurring north of Mexico, with an illustrated key. Ann Entomol Soc Am 99:1034–1056

Ragenovich IR, Mitchell RG (2006) Balsam woolly adelgid. Forest Insect & Disease Leaflet 118 (revised). US Department of Agriculture, Forest Service, Washington, DC. 11 p

Reardon R, Onken B, Cheah C et al (2004) Biological control of the hemlock woolly adelgid. Publication FHTET-2004-04. US Department of Agriculture, Forest Service, Forest Health Technology Enterprise Team, Morgantown. 22 p

Rice KB, Bergh CJ, Bergmann EJ et al (2014) Biology, ecology, and management of brown marmorated stink bug (Hemiptera: Pentatomidae). J Integr Pest Manag 5:1–13

Rosetta R (2013) Azalea lace bug: biology and management in commercial nurseries and landscapes. Oregon State University Extension Service, Publication EM 9066, Corvallis. 6 p. https://catalog.extension.oregonstate.edu/em9066

Ryan RB (1997) Before and after evaluation of biological control of the larch casebearer (Lepidoptera: Coleophoridae) in the Blue Mountains of Oregon and Washington, 1972–1995. Environ Entomol 26:703–715

Sabrosky C, Reardon RC (1976) Tachinid parasites of the gypsy moth, Lymantria dispar, with keys to adults and puparia. Misc Publ Entomol Soc Am 10(2). 126 p

Schoettle AW, Klutsch JG, Sniezko RA (2012) Integrating regeneration, genetic resistance, and timing of intervention for the long-term sustainability of ecosystems challenged by non-native pests—a novel proactive approach. In: Sniezko RA, Yanchuk AD, Kliejunas JT, Palmieri KM, Alexander JM, Frankel SJ (tech. coords) Proceedings of the fourth international workshop on the genetics of host-parasite interactions in forestry: disease and insect resistance in forest trees. General Technical Report. PSW-GTR-240. Pacific Southwest Research Station, Forest Service, U.S. Department of Agriculture, Albany, pp 112–123

Seybold SJ, Downing M (2009) What risk do invasive bark beetles and woodborers pose to forests of the western U.S.? A case study of the Mediterranean pine engraver, *Orthotomicus erosus*. In: Hayes JL, Lundquist JE (compilers) The western bark beetle research group: a unique collaboration with Forest Health Protection. Proceedings of a symposium at the 2007 Society of American Foresters Conference, October 23–28, 2007, Portland, Oregon. General Technical Report, GTR–PNW–784, 134 p. Forest Service, U.S. Department of Agriculture, Portland, pp 111–134

Shaw D, Oester P (2010) Are the introduced parasites of larch casebearer (*Coleophora laricella*), still present in the Blue Mountains, Oregon? USDA Forest Service Western Wildland Environmental Threat Assessment Center, Prineville. WWETAC Project FY10TS74. 1 p. https://www.fs.fed.us/wwetac/old/projects/shaw2.html

Shoal R, Aubry CA (2006) Assessment of whitebark pine health on eight national forests in Oregon and Washington. Pacific Northwest Albicaulis project. US Department of Agriculture, Forest Service, Pacific Northwest Region, Olympia. 22 p

Sing SE, De Clerck-Floate RA, Hansen RW et al (2016) Biology and control of Dalmatian and yellow toadflax. USDA Forest Service, Forest Health Technology Enterprise Team, Morgantown. FHTET-2016-01. 141 p. https://www.fs.usda.gov/treesearch/pubs/52446

Snyder FL, Hilgendorf MB, Garton DW (1997) Zebra mussels in North America: the invasion and its implications. Ohio State University, Ohio Sea Grant, Columbus. 4 p

Stannard ME (2014) Plant guide for Himalayan blackberry (*Rubus armeniacus*). USDA Natural Resources Conservation Service. Plant Materials Center, Pullman. 3 p. https://plants.usda.gov/plantguide/pdf/pg_ruar9.pdf

Steele R, Pfister RD, Ryker RA, Kittams JA (1981) Forest habitat types of central Idaho. General Technical Report INT-114. U.S. Department of Agriculture, Forest Service, Intermountain Forest and Range Experiment Station, Ogden. 138 p

Stewart D, Zahiri R, Djoumad A et al (2016) A multi-species TaqMan PCR assay for the identification of Asian gypsy moths (*Lymantria* spp.) and other invasive lymantriines of biosecurity concern to North America. PLoS One 11:e0160878

Strutt A, Turner JA, Haack RA, Olson L (2013) Evaluating the impacts of an international phytosanitary standard for wood packaging material: global and United States trade implications. Forest Policy Econ 27:54–64

The Knotweed Company Ltd (2018) Japanese knotweed history and biology. http://www.knotweed-removal.co.uk/history-and-biology-of-japanese-knotweed.php

Thorson TD, Bryce SA, Lammers DA et al (2003) Ecoregions of Oregon (color poster with map, descriptive text, summary tables, and photographs). U.S. Geological Survey (map scale 1:1,500,000), Reston. http://people.oregonstate.edu/~muirp/FuelsReductionSWOregon/ToolsResources/EcoregionsOregonLevelIVEPA.pdf

Tunnock S, Ryan RB (1985) Larch casebearer in western larch. Forest Insect & Disease Leaflet 96 (revised). US Department of Agriculture, Forest Service, Washington, DC. 8 p

USDA Agricultural Research Service (1986) Stored-grain insects. Agricultural Handbook No. 500. 57 p

USDA Forest Service and APHIS (1993) Russia and United States Pest Monitoring Program, 2 p

USDA Forest Service and APHIS (2000) Pest risk assessment for importation of solid wood packing materials into the United States, 275 p

USDA APHIS (2016) Pest alert: Asian gypsy moth. APHIS 81–35-027, Issued April 2016, 2 p

USDA Forest Service, Northern Research Station and Forest Health Protection (2019) Alien Forest Pest Explorer—

species map. Database last updated 28 July 2016. https://www.nrs.fs.fed.us/tools/afpe/

Vernon RS, LaGasa EH, Phillip H (2001) Geographic and temporal distribution of *Agriotes obscurus* and *A. lineatus* (Coleoptera: Elateridae) in British Columbia and Washington as determined by pheromone trap surveys. J Entomol Soc Br Columbia 98:257–265

Vitousek PM, Antonio CM, Loope LL, Westbrooks R (1996) Biological invasions as global environmental change. Am Sci 84:468–478

Washington Invasive Species Council (2012) Stop the invasion: Bark boring moths. Fact Sheet, Washington State Recreation and Conservation Office, Washington Invasive Species Council. 2 p. https://www.yumpu.com/en/document/read/12118488/bark-borking-moths-fact-sheet-washington-invasive-species-council

Webber JF, Mullett M, Brasier CM (2010) Dieback and mortality of plantation Japanese larch (*Larix kaempferi*) associated with infection by *Phytophthora ramorum*. New Dis Rep 22:19. https://doi.org/10.5197/j.2044-0588.2010.022.019

White RE (1982) A catalog of the Coleoptera of America north of Mexico. Family: Anobiidae. USDA Agricultural Handbook No. 529–70, 59 p

Wiman NG, Lowenstein DM (2017) Emerging pest: brown marmorated stink bug—a pending threat to Pacific northwest agriculture. In: Hollingsworth CS (ed) Pacific Northwest insect management handbook. Pacific Northwest Extension Publication. Oregon State University, Corvallis. unpaginated. https://pnwhandbooks.org/node/12634/print

Winston R, Randall CB, De Clerck-Floate R et al (2014a) Field guide for the biological control of weeds. University of Idaho Extension. Publication FHTET-2014-08. USDA Forest Service, Forest Health Technology Enterprise Team, Morgantown. 338 p. https://www.ibiocontrol.org/westernweeds.pdf

Winston RL, Schwarzländer M, Hinz HL et al (2014b) Biological control of weeds: a world catalogue of agents and their target weeds, 5th edn. Publication FHTET-2014-04. USDA Forest Service, Forest Health Technology Enterprise Team, Morgantown. 838 p. https://www.ibiocontrol.org/catalog/JulienCatalogueFHTET_2014_04.pdf

WISP (2009) Top priorities. Washington Invasive Species Council. Available at: https://invasivespecies.wa.gov/find-a-priority-species/

Zobel DB (1990) *Chamaecyparis lawsoniana* (A. Murr.) Parl. Port-Orford-Cedar. In: Burns RM, Honkala BH (tech. coords) Silvics of North America: 1. Conifers; 2. Hardwoods. Agriculture Handbook 654. U.S. Department of Agriculture, Forest Service, Washington, DC. 877 p

Southwest Region

Steven J. Seybold, Andrew D. Graves, Susan J. Frankel, Allen White, Carol A. Sutherland, and A. Steve Munson

Introduction

The Southwest region (Arizona, California, Colorado, Nevada, New Mexico, and Utah) (Figs. A4.1 and A4.2) is marked by Mediterranean, montane, and desert climates/ecosystems that provide unique and amenable conditions and habitats for invading plants, pathogens, insects, and vertebrates. Aridity is perhaps the dominant climatic feature framing the forest ecosystems of the Southwest (Peterson 2012). Extreme elevational gradients and the intervening desert landscapes in this region (Fig. A4.2) create pronounced biogeographical boundaries and refugia for endemic species of plants and animals. The southern edge of this region has an extensive, but ecologically contiguous, border with Mexico that facilitates biological invasions. Future climate conditions projected for the southern portion of this region predict a trend of increasing temperature and decreasing precipitation (Cayan et al. 2010; Peterson 2012; Williams et al. 2010). Changing climate will likely place water stress on native trees and other plants, perhaps accelerating the establishment of invasive species (Peterson 2012) and amplifying outbreaks of native pest species (Breshears et al. 2005). These changes may also facilitate the spread of invasive species northward across this international border (e.g., Billings et al. 2014; Moser et al. 2005). The rate of spread of invasive species across this border may be increased by instances of unregulated movement of humans and cargo.

This region also features a wide range of non-native ornamental plants in urban and rural areas, enormously productive and diverse agroecosystems, and huge tracts of public lands with grazing impacts that favor the establishment and spread of invasive plants and pathogens by wild and domestic ungulates and other animals. High property values and residences in and near this region's forests make the impacts of invasive species particularly expensive and difficult to manage, as they often range across varied and numerous land ownerships. From a sociological perspective, the diverse human population of the region provides linkages to many overseas source populations of invasive species, whereas numerous maritime and overland ports-of-entry as well as U.S. and international tourism in response to the attractive natural features and mild winter climate may also enhance the introduction, establishment, and spread of invading organisms.

Plants

Terrestrial invasive plants in the Southwest region include annual, biennial, and perennial species of grasses, forbs,

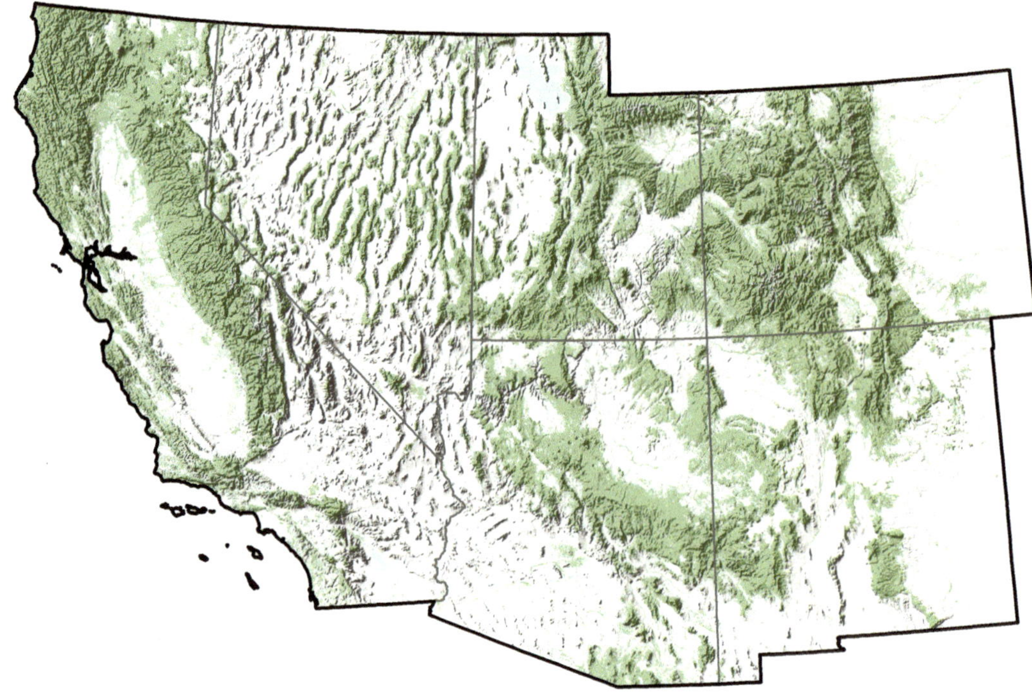

Fig. A4.1 The Southwest region

Fig. A4.2 The Southwest region of the United States includes Arizona, California, Colorado, Nevada, New Mexico, and Utah

shrubs, and trees (Table A4.1). Although some of these plants (e.g., Chinese elm (*Ulmus parvifolia*), Siberian elm (*U. pumila*), and saltcedar (*Tamarix* spp.)) were considered previously as desirable landscaping materials, these and a variety of other species are listed by Southwestern States as noxious weeds, which by statutory regulation require landowners to manage them. However, these species are typically quite difficult to manage once established, and control generally requires repetitions of a variety of separate treatments over a period of years. Availability of financial resources for costly suppression/eradication efforts can also be an impediment. In some cases, only biological control agents (if available)

Table A4.1 The primary invasive plants of the Southwest region include brooms, grasses, knapweeds, thistles, and several trees

Scientific name	Common name	Occurrence						Impacts/potential impacts				
		AZ	CA	CO	NV	NM	UT	Native biodiversity	Land values	Recreation	Fire risk/ ecosystem conversion	Grazing
Acroptilon repens (L.) DC.	Russian knapweed	x	x	x	x	X	x	•	•			•
Aegilops cylindrica Host	Jointed goatgrass	x	x	x	x	X	x	•	•			
Alhagi pseudalhagi (M.Bieb.) Desv. ex B. Keller & Shap.	Camelthorn	x	x	x	x	X	x					•
Bothriochloa bladhii (Retz.) S.T. Blake	Caucasian bluestem					X		•				
Bothriochloa ischaemum (L.) Keng	Yellow bluestem					x		•				
Bromus tectorum L.	Downy brome, cheatgrass	x	x	x	x	x	x	•	•	•	•	•
Cardaria draba (L.) Desv.	Hoary cress, whitetop	x	x	x	x	x	x	•	•			•
Carduus nutans L.	Musk thistle	x	x	x	x	x	x	•				•
Centaurea diffusa Lam.	Diffuse knapweed	x	x	x	x	x	x	•				•
Centaurea melitensis L.	Maltese starthistle				x	x		•	•			•
Centaurea stoebe L. spp. *micranthos* (Gugler) Hayek syn. *C. maculosa* auct. non Lam.	Spotted knapweed	x	x	x	x	x	x	•				•
Centaurea solstitialis L.	Yellow starthistle	x	x	x	x	x	x	•	•	•		•
Centaurea virgata Lam. ssp. *squarrosa* (Willd.) Gugler	Squarrose knapweed	x	x		x		x	•				•
Chondrilla juncea L.	Rush skeletonweed	x	x	x	x		x	•	•			•
Cirsium arvense (L.) Scop.	Canada thistle	x	x	x	x	x	x	•				•
Cynoglossum officinale L.	Houndstongue	x	x	x	x	x	x	•				•
Cytisus scoparius (L.)	Scotch broom		x				x	•			•	
Elaeagnus angustifolia L.	Russian olive	x	x	x	x	x	x	•		•		
Euphorbia esula L.	Leafy spurge	x	x	x	x	x	x	•	•			•
Genista monspessulana (L.) L.A.S. Johnson	French broom		x					•			•	•
Halogeton glomeratus (M. Bieb.) C.A. Mey.	Halogeton	x	x	x	x	x	x					•
Hyoscyamus niger L.	Black henbane			x	x	x	x					•
Hydrilla verticillata (L. f.) Royle	Hydrilla, Florida elodea	x	x					•		•		
Isatis tinctoria L.	Dyer's woad	x	x	x	x	x	x	•				•
Lepidium latifolium L.	Perennial pepperweed	x	x	x	x	x	x	•	•			•
Linaria dalmatica L.	Dalmatian toadflax	x	x	x	x	x	x	•				•
Linaria vulgaris Mill.	Yellow toadflax	x	x	x	x	x	x	•				•
Lythrum salicaria L.	Purple loosestrife	x	x	x	x	x	x	•				

(continued)

Table A4.1 (continued)

Scientific name	Common name	Occurrence AZ	CA	CO	NV	NM	UT	Impacts/potential impacts Native biodiversity	Land values	Recreation	Fire risk/ ecosystem conversion	Grazing
Onopordum acanthium L.	Scotch thistle	x	x	x	x	x	x	•				•
Peganum harmala L.	African rue, Syrian rue	x	x	x	x	x						•
Pennisetum ciliare (L.) Link	Buffelgrass	x	x			x	x	•			•	
Phragmites australis ssp. *australis* (Cavanilles) Trinius ex Steudel	European common reed	x	x	x	x	x	x	•	•	•		
Salvinia molesta Mitchell	Giant salvinia	x	x					•		•		
Sorghum halepense (L.) Pers.	Johnsongrass	x	x	x	x	x	x					•
Taeniatherum caput-medusae (L.) Nevski	Medusahead	x	x		x		x	•			•	•
Tamarix spp. L.	Saltcedar	x	x	x	x	x	x	•			•	

This list was compiled by comparing noxious weed lists of the individual States in the Southwest region (USDA National Resources Conservation Service 2018). Invasive plants that appeared in four or more State lists were included in this table. Distributions for the invasive plants were validated by using EDDMapS (2019). Additional invasive plants that may have occurred in fewer than four regional States were included if local vegetation specialists deemed them to be particularly significant. This list was compiled by Andrew D. Graves and Allen White, USDA Forest Service, Forest Health Protection, Albuquerque, NM, and Richard D. Lee, Ph. D., integrated pest management specialist, U.S. Department of the Interior Bureau of Land Management, Denver, CO

can effectively impact infestations of invasive plant species that have spread extensively over entire landscapes. Several prominent invasive plant species found in the Southwest region are discussed below.

Among the invasive plant species, buffelgrass (*Pennisetum ciliare*) is the single greatest threat to desert ecosystems in the warmer latitudes of the Southwest region (Fig. A4.3a) (USDA FS Southwestern region 2017). Buffelgrass is an invasive bunchgrass from tropical and subtropical arid regions of Africa and Western Asia that was developed in the United States as a drought-tolerant forage grass (Marshall et al. 2012). The perennial species was first planted successfully in Texas in the 1940s for forage and in Arizona in the 1950s to stabilize soils (Marshall et al. 2012); however, it now threatens the Sonoran Desert ecosystem through its expansion into southern Arizona and most of Sonora, Mexico. The northward expansion is currently limited by the relative cold intolerance of buffelgrass; however, regional temperature increases predicted as a consequence of climate change may allow expansion to continue further north and into higher mountain elevations. Although buffelgrass seed is spread long distances by wind, vehicles, and other means, individual patches of buffelgrass can double in place every 2–7 years (Olsson et al. 2012). Buffelgrass can outcompete native desert vegetation for water, nutrients, and sunlight. It also forms dense infestations that allow fires to spread across the landscape on a cyclical basis. The Sonoran Desert evolved without fire, and most native plant species in the desert such as the iconic saguaro cactus (*Carnegiea gigantea*) are fire-intolerant. Consequently, the buffelgrass invasion of the Sonoran Desert is effectively transforming large portions of the desert ecosystem into fire-prone tropical savanna.

Also of concern is musk thistle (*Carduus nutans*), which is a spiny invasive weed in the sunflower family (Asteraceae) with highly branched stems and purplish-red disk flowers (Fig. A4.3b). The flowers "nod" at a 90-degree angle, hence its alternate common name, "nodding" thistle. The species is highly competitive and rapidly invades rangeland, roadsides, and disturbed sites in the Southwest. Musk thistle seed is readily dispersed by wind, water, birds, and other animals. In addition, seed can be carried over long distances by adhering to the surfaces of vehicles and road maintenance equipment. These invasive features make musk thistle difficult to control. A biological control agent, thistle seedhead weevil (*Rhinocyllus conicus* (Coleoptera: Curculionidae)), was imported and released in the United States between 1969 and 1972 to control musk thistle along with other thistles (Louda et al. 1997; Winston et al. 2014a). However, the establishment and expansion of the distribution of the seedhead weevil from early-release sites enabled the weevil to also encounter and attack many native thistles (Louda et al. 1997), including the endangered Sacramento Mountain thistle (*Cirsium vinaceum*) in southern New Mexico. Consequently, interstate shipments of the weevil are no longer permitted by the USDA Animal and Plant Health Inspection Service (APHIS) (Winston et al. 2008).

Fig. A4.3 Invasive plants associated with Arizona and New Mexico include (**a**) buffelgrass (*Pennisetum ciliare*) (photo credit: U.S. Department of the Interior, National Park Service); (**b**) musk thistle (*Carduus nutans*) (photo credit: Norman E. Rees, USDA Agricultural Research Service); (**c**) defoliated saltcedar (*Tamarix* sp.) (photo credit: Debra Allen-Reid, retired, USDA Forest Service, Forest Health Protection, Albuquerque, NM); and (**d**) Russian olive (*Elaeagnus angustifolia*). (Photo credit: J. Scott Peterson, USDA Agricultural Research Service)

One of the most widely dispersed invasive plant species in the Southwest region is saltcedar (*Tamarix* spp.), which occurs frequently as either a shrub or small tree in thick stands along waterways (Fig. A4.3c). Saltcedar can affect riparian systems by altering stream flow (via evapotranspiration) and ecological processes (e.g., soil salinization and microbial activity). However, some detrimental effects attributed to the species such as excessive evapotranspiration may have been overestimated (Glenn and Nagler 2005). Beginning in 2001, four species of the tamarisk leaf beetle (*Diorhabda* spp. (Coleoptera: Chrysomelidae)), from North Africa, the Mediterranean, and Asia were released in several Western US States as a host-specific biocontrol agent (Moran et al. 2009). Both adult tamarisk leaf beetles and larvae damage saltcedar foliage, which can progressively weaken or kill the plant over a period of several years. Since their release, species of *Diorhabda* have spread and are now found in 11 Western US States with pronounced recoveries of the beetle in the major watersheds of eastern Utah, western Colorado, northern Arizona, and New Mexico (Fig. A4.4). Although APHIS no longer issues permits for transporting *Diorhabda* beetle species across state borders (USDA APHIS 2010), beetle spread is expected to continue naturally and possibly by human assistance within individual Western States (Bloodworth et al. 2016). Areas with defoliated or dead saltcedar may eventually allow native plant species such as cottonwood (*Populus* spp.), or willow (*Salix* spp.), to return; however, defoliated saltcedar sites may also be invaded by other weedy plant species, some of which may be even more undesirable and difficult to manage than saltcedar. In addi-

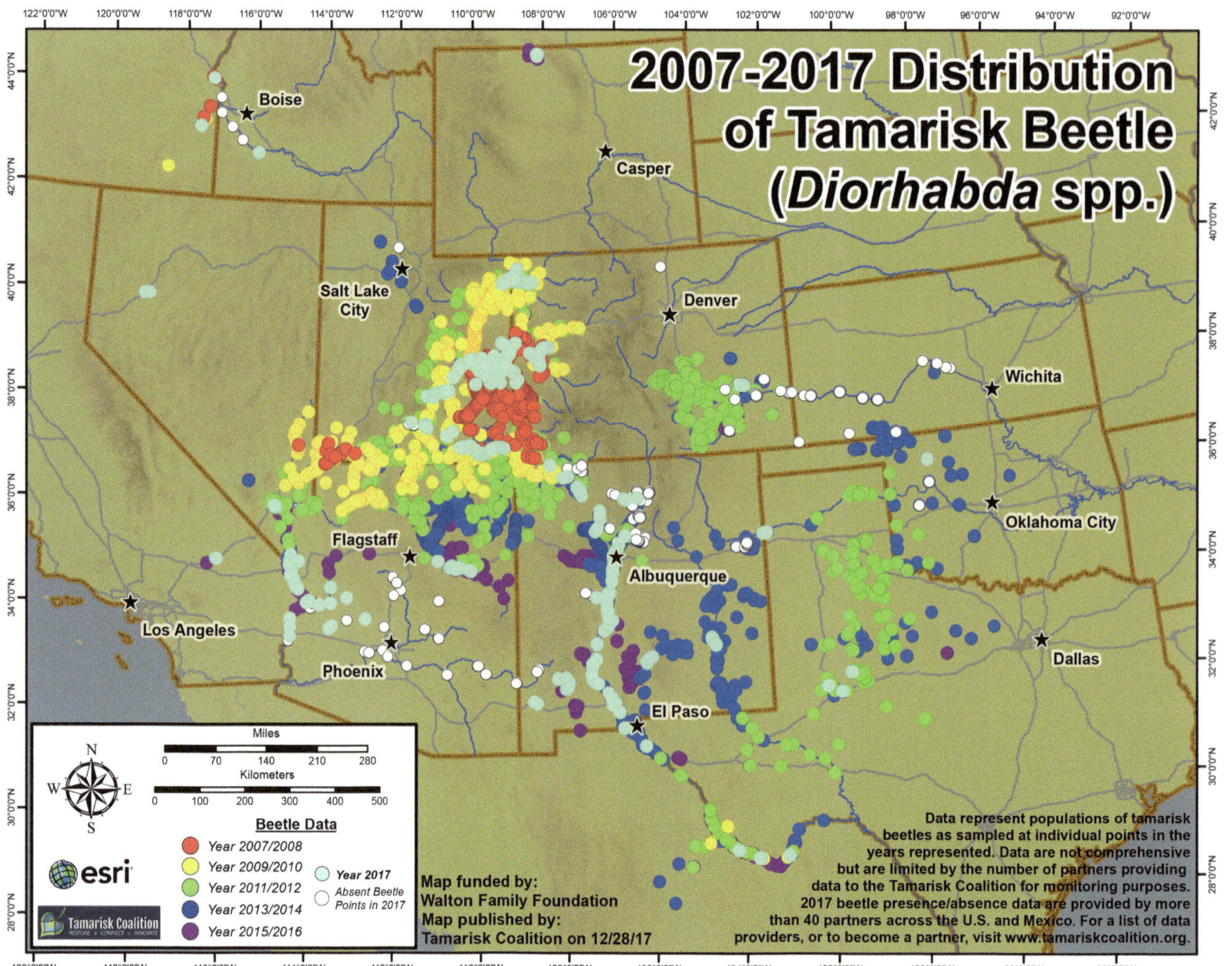

Fig. A4.4 An 11-year record of the distribution of tamarisk beetles (*Diorhabda* spp.), in the Southwest region. Four species of this beetle were introduced in 2001 to provide biological control of invasive saltcedar (*Tamarix* spp.), in the region. Further facilitation of the spread of the beetles was curtailed when it became evident that saltcedar mortality threatened nesting habitat used by the endangered southwestern willow flycatcher (*Empidonax traillii extimus*). In 2016–2017, the Gila (southern Arizona and New Mexico) and Arkansas (southern Kansas) watersheds have been scrutinized intensively for potential newly established populations of *Diorhabda* spp. (Map courtesy of B. Bloodworth (Tamarisk Coalition 2017))

tion, the expanding distribution of *Diorhabda* species threatens nesting habitat used by the endangered southwestern willow flycatcher (*Empidonax traillii extimus*), which nests in saltcedar-dominated plant communities that have replaced native cottonwood and willow species. Large-scale defoliations of tamarisk may cause a temporary loss in flycatcher habitat for at least a decade before the native cottonwoods and willows can return (Paxton et al. 2011).

Over the past century, Russian olive (*Elaeagnus angustifolia*) was planted widely throughout the United States as an ornamental and windbreak tree, but it has since escaped into natural areas of the Southwest region (Fig. A4.3d) (Katz and Shafroth 2003). It is a hardy, fast-growing, deciduous tree with silvery, gray-green leaves, growing to a height of about 10 m. Along with saltcedar, this invasive tree species serves as potential nesting habitat for the southwestern willow flycatcher. Russian olive seed is eaten by birds, and bird droppings with the seed contribute greatly to the tree's spread. The ovoid fruit floats on water and is readily dispersed along waterways. After invasion, Russian olive can become the dominant species in areas due to its adaptability, prolific reproduction, rapid growth rate, and lack of natural enemies. As infestations expand, Russian olive crowds out desirable native riparian trees such as cottonwoods and willows, thereby reducing floral and faunal diversity. Because of its ability to colonize stream banks, the species can alter the natural flooding regime and reduce availability of nutrients and moisture. It is extremely difficult to restore native plant communities once Russian olive has established dense, monotypic stands. The primary tools that land managers have for controlling these trees include repeated bulldozing, root plowing, cut-stump herbicide treatments, or prolonged flooding. These tools may also be used for controlling saltcedar (see above). Based on entomological literature from Asia, there was speculation that Russian olive in the Western United States might serve as a host for the invasive banded elm bark beetle (*Scolytus schevyrewi*) (see below), but there have been no observations or experimental evidence in the United States to substantiate this hypothesis (Lee et al. 2011; Negrón et al. 2005).

In California, the highest-priority noxious weeds include yellow starthistle (*Centaurea solstitialis*), several knapweeds (spotted knapweed (*C. stoebe*), diffuse knapweed (*C. diffusa*), meadow knapweed (*C. debeauxii*)), perennial pepperweed (*Lepidium latifolium*), thistles ((bull thistle (*Cirsium vulgare*), Scotch thistle (*Onopordum acanthium*), musk thistle (*Carduus nutans*), Canada thistle (*Cirsium arvense* and others)), and brooms (Scotch broom (*Cytisus scoparius*) and French broom (*Genista monspessulana*)). More information on these and over 200 California invasive plants is maintained in the California Invasive Plant Council (Cal-IPC) Inventory (CaL-IPC 2006). The northern tier of States in the Southwest region (Colorado, Nevada, and Utah) represent a relatively large landscape, and the most problematic invasive plants vary depending on local conditions (Table A4.1). As a functional group, invasive annual grasses such as cheatgrass (*Bromus tectorum*) and medusahead (*Taeniatherum caput-medusae*) (Table A4.1) impact native plant communities in these northern tier states, as well as in California. They are responsible for landscape-scale conversion of native vegetation to annual grassland in the Great Basin sagebrush steppe and in California grasslands and sage scrub (see Chap. 2, Sects. 2.2.2 and 2.2.3, for more details and discussion of interactions of these regional invasive grasses with fire and pollinators).

Of emerging concern in the Southwest region are two invasive bunchgrass species, yellow bluestem (*Bothriochloa ischaemum*) and Caucasian bluestem (*B. bladhii*). These two bunchgrass species are part of a global complex of invasive bluestem species collectively called Old World bluestems to differentiate them from North American bluestem species. The two bunchgrass species were originally imported from Eurasia and Africa for erosion control and as forage crops (Klataske 2016; Missouri Department of Conservation 2010). These extremely persistent plants form monocultures that can lower biodiversity in native grassland and pastures (Gabbard and Fowler 2007; Klataske 2016). Yellow and Caucasian bluestems have become invasive in pastures and native grasslands of states in the Midwest and the Southern Great Plains (Oklahoma, Texas, and eastern New Mexico). An informal survey conducted by the USDA Forest Service in 2017 indicated that yellow bluestem is present along roadsides on most national forests and national grasslands in the Forest Service's Southwestern Region (Region 3). Caucasian bluestem has also been found on the Coconino National (Agyagos 2018).

Plant Pathogens

In the Southwest region, invasive plant pathogens (Table A4.2) are a significant problem in both urban and rural forests with their impacts dependent on the outbreak location and the attributes of the particular host tree species. Below is a short regional overview of the most damaging invasive pathogens; notably, most of these pathogens have been introduced on imported nursery stock (Liebhold et al. 2012).

White pine blister rust (*Cronartium ribicola*) is killing white (five-needle) pines in all states in the Southwest region including commercial species (i.e., sugar pine (*Pinus lambertiana*) and southwestern white pine (*P. strobiformis*)) and high-elevation pines (limber (*P. flexilis*), whitebark (*P. albicaulis*), foxtail (*P. balfouriana*), western white (*P. monticola*), and bristlecone (*P. aristata* and *P. longaeva*)) (Geils et al. 2010, Fig. A4.5). White pine blister rust is among the most damaging of invasive pathogens in U.S. forests and parks (Benedict 1981; Boyce 1938; Vitousek et al. 1996). The rust, native to Asia, was introduced around 1910 to Western North America near Vancouver, British Columbia, Canada, on nursery stock from France (Liebhold et al. 2012). The 1912 Plant Quarantine Act was a direct result of its introduction, with U.S. Quarantine No. 1 prohibiting importation of five-needle pines (Maloy 1997). The distribution of *C. ribicola* continues to expand southward in California, but the pathogen has not yet moved into and past the Tehachapi Mountains in the Transverse Ranges of southern California (Smith 2018).

A more recent invader, *Phytophthora ramorum*, which is the pathogen associated with sudden oak death, has become the number one cause of tree mortality in California coastal forests (see Chap. 2, Box 2.5; Chap. 6, Sect. 6.4.2; Chap. 7, Sect. 7.4.2; and the Northwest region summary for additional discussion of this pathogen). The pathogen kills tanoak (*Notholithocarpus densiflorus*), coast live oak (*Quercus agrifolia*), and other red oaks. It is also known to infect over 100 other plant species including conifers, herbaceous plants, and ferns. The pathogen is estimated to have been introduced in California sometime around 1980 on rhododendron (*Rhododendron* spp.) nursery stock (Mascheretti et al. 2008) and has since killed millions of trees along the Pacific Coast (California Central Coast north to Curry County, OR). The tree mortality increased dramatically in the late 1990s in the San Francisco Bay Area where over six million people reside, making the management of hazards (dead trees and branches) in residential areas, as well as along roadways and power lines, a chronic issue. The pathogen is also of particular concern to Native American tribes, since they rely on acorns for ceremonies and food

Table A4.2 The primary invasive plant and animal pathogens of the Southwest region include bacteria, fungi, prions, and viruses

Scientific name	Common name	Occurrence AZ	CA	CO	NV	NM	UT	References
Batrachochytrium dendrobatidis	Amphibian chytrid fungus (*Bd*)	X	X	X	X	X	X	http://www.bd-maps.net/
Candidatus Liberibacter asiaticus	Citrus Huanglongbing		X					Kumagai et al. (2013) and McCollum and Baldwin (2017)
Chronic wasting disease prion	Chronic wasting disease (CWD)			X		X	X	https://www.cdc.gov/prions/cwd/occurrence.html
Cronartium ribicola	White pine blister rust	X	X	X	X	X	X	Geils et al. (2010)
Fusarium circinatum	Pitch canker		X					Gordon et al. (2001), Gordon (2016), and Wingfield et al. (2008)
Fusarium euwallaceae	Fusarium decline		X					Eskalen et al. (2013) and Lynch et al. (2016)
Geosmithia morbida	Thousand cankers disease pathogen	X	X	X	X	X	X	Kolařík et al. (2011, 2017) and Tisserat et al. (2011)
Melampsora larici-populina	Eurasian poplar leaf rust		X					Farr and Rossman (2018) and Pinon et al. (1994)
Phytophthora lateralis	Port-Orford-cedar root disease		X					https://www.cabi.org/isc/datasheet/40973
Phytophthora ramorum	Sudden oak death		X					http://www.suddenoakdeath.org
Pseudogymnoascus destructans	White-nose syndrome			X				https://www.whitenosesyndrome.org/mmedia-education/july-2-2018
Puccinia graminis	Black stem rust	x	X	X		X	X	https://www.cabi.org/isc/datasheet/45797; Farr and Rossman 2018
Ranavirus	Ranavirus	X	X	X	X	X	X	https://cwhl.vet.cornell.edu/disease/ranavirus

Fig. A4.5 White pine blister rust is caused by infection of five-needle pines with *Cronartium ribicola*, which is an invasive pathogen that continues to expand its range in the Southwest. The image shows a white pine blister rust canker with aeciospores on southwestern white pine (*Pinus strobiformis*), on the Lincoln National Forest in southeastern New Mexico. (Photo credit: James Jacobs, USDA Forest Service, Forest Health Protection, Albuquerque, NM)

(Long and Goode 2017; Long et al. 2017) and consider the primary host tree species to be sacred (Alexander and Lee 2010).

Of special concern in the Southwest region are invasive pathogens that impact tree species of limited native distribution. Monterey pines (*Pinus radiata*) along the California Central Coast are dying in large numbers from pitch canker, caused by *Fusarium circinatum*, in combination with drought and colonization by bark and twig beetles. Pitch canker was first observed in California in Santa Cruz County in 1986 (Gordon et al. 2001; McCain et al. 1987; Wingfield et al. 2008). Monterey pine mortality has been prevalent in adventive stands along highways and roadsides, but the three largest native forests of Monterey pine in the world are all infested and severely damaged (Wikler et al. 2003). Another invasive species of *Fusarium* (*F. euwallaceae*), in tandem with an invasive ambrosia beetle species complex (see below), threatens native riparian hardwood trees in southern California (Coleman et al. 2013; Eskalen et al. 2013).

Native endemic riparian walnut trees (southern California walnut (*Juglans californica*) and northern California walnut (*J. hindsii*) in California) (Flint et al. 2010; Griffin and Critchfield 1972) and adventive black walnut trees (*J. nigra*) in Colorado and Utah (Tisserat et al. 2011) have been recorded with crown dieback and tree mortality from thousand cankers disease, caused by a newly described fungus, *Geosmithia morbida* (Kolařík et al. 2011, 2017), vectored by the walnut twig beetle (*Pityophthorus juglandis* (Coleoptera: Scolytidae)) (Seybold et al. 2016, 2019). Over 60% (1300 trees) of the black walnut growing in Boulder, CO had to be removed between 2004 and 2010 in response to this disease (Tisserat et al. 2011). Between 2010 and 2014 in survey plots in California, nearly 10% of the trees died in one native stand each of southern California walnut and northern California walnut, whereas levels of infection reached 90–100% for a range of walnut species in a northern California germplasm collection (Hishinuma 2017). The native range of the beetle (Rugman-Jones et al. 2015) and, likely, the fungal pathogen (Zerillo et al. 2014) are sympatric with the northern portion of the distribution of Arizona walnut (*J. major*) (USDA NRCS 2018). In stands of this putative ancestral host of the beetle and pathogen, Graves et al. (2011) reported 6.7% mortality and approximately 50% infection/infestation in national forest survey plots in Arizona and New Mexico.

Nearly the entire native range of Port-Orford-cedar (*Chamaecyparis lawsoniana*) is infested with *Phytophthora lateralis*, an invasive pathogen that causes a lethal root disease (see Northwest regional summary for additional discussion of this pathogen). The tree grows naturally along the Pacific Coast from northern California to southern Oregon but is also a prized ornamental. The pathogen was introduced on nursery stock near Seattle, WA, in the 1920s and spread southward on ornamental plantings until it reached the native stands in 1952. *Phytophthora lateralis* is thought to be native to Asia (Brasier et al. 2010). Once established, the pathogen moves via transport of infested soil on vehicle tires and in infested runoff water. To manage the pathogen, resistant trees have been developed, and, in some instances, local forest roads are closed during the wet season (Hansen et al. 2000).

Insects

Much of the vast Southwest region has not been invaded by non-native forest insects (Table A4.3) (Furniss and Carolin 1977). The incidence of introduction and establishment is far higher in the Eastern United States (Liebhold et al. 2013). However, a recent survey has revealed that at least 22 species of invasive bark and ambrosia beetles (Coleoptera: Scolytidae) have established populations in California (Seybold et al. 2016). As a group, these main stem infesting pests are ranked among the most damaging to the growth and reproduction of trees (Mattson et al. 1988). Many of the spe-

Table A4.3 The primary invasive insects of the Southwest Region include termites, beetles, flies, true bugs and other sucking insects, thrips, ants, wasps, and moths

Scientific name	Common name	Occurrence						References
		AZ	CA	CO	NV	NM	UT	
Isoptera								
Coptotermes formosanus	Formosan subterranean termite		x					Rust et al. (1998)
Cryptotermes brevis[a]	Powderpost termite		x					Evans et al. (2013), Light (1936), and Rust (2004)
Coleoptera								
Agrilus auroguttatus[b]	Goldspotted oak borer	O	x					Coleman and Seybold (2016)
Agrilus cuprescens	Rose stem girdler	x	x	x	x	x	x	Cranshaw et al. (1994)
Agrilus anxius[b]	Bronze birch borer		x	x	x			Carlos et al. (2002), Cranshaw et al. (1994), and Duckles and Švihra (1995)
Agrilus planipennis	Emerald ash borer			x				Colorado State Forest Service (2018)
Agrilus prionurus[b]	Mexican soapberry borer							Billings et al. (2014) and USDA (2019) (in Texas)
Amphimallon majale	European chafer					x		C. Sutherland, personal observation
Anobium punctatum	Furniture beetle		x			x		White (1982)
Anoplophora glabripennis[a]	Asian longhorned beetle		x					Bohne (2007,2008) and Bohne and Rios (2006)
Arhopalus syriacus			x					Seybold et al. (2016)
Brachypterolus pulicarius	Toadflax flower-feeding beetle			x	x		x	Sing et al. (2016)
Chrysolina hyperici							x	Winston et al. (2014a)
Chrysolina quadrigemina	Klamathweed beetle		x	x			x	Winston et al. (2014a)
Chrysophtharta m-fuscum	A chrysomelid leaf beetle of eucalyptus		x					Paine (2016)
Coccinella septempunctata	Seven-spotted lady beetle					x		C. Sutherland, personal observation
Coccotrypes advena	A date palm seed beetle		x					Seybold et al. (2016)
Coccotrypes carpophagus	A date palm seed beetle		x					Rabaglia et al. (2019) and Seybold et al. (2016)
Coccotrypes dactyliperda	A date palm seed beetle	x	x					Seybold et al. (2016)
Coccotrypes rutschuruensis	A date palm seed beetle		x					Atkinson (2018)
Coniatus splendidulus	Saltcedar bud weevil	x	x	x	x	x	x	Bright et al. (2013), C. Sutherland, personal observation, and Winston et al. (2014a)
Cryptorhynchus lapathi	Poplar-and-willow borer			x			x	Anderson (2008), Cranshaw et al. (1994), and USDA (2019)
Curculio caryae[b]	Pecan weevil					x		Mulder et al. (2012) and Sutherland et al. (2017)
Cyclorhipidion bodoanum			x	x				Atkinson (2018), Rabaglia et al. (2019), and Seybold et al. (2016)
Dactylotrypes longicollis	A palm seed beetle		x					Seybold et al. (2016)
Dinapate wrightii[b]	Giant palm borer	x	O		x		x	Ivie (2002) and Olson (1991) (also introduced in Texas)
Dinoderus minutus	A bamboo borer		x					Ivie (2002) and Spilman (1982)
Diorhabda spp. *D. carinata, carinulata, elongata, sublineata*	Tamarisk (saltcedar) leaf beetles	x	x	x	x	x	x	Tamarisk Coalition (2017) and Winston et al. (2014a)
Ernobius mollis	European bark anobiid		x					Seybold and Tupy (1993) and Seybold (2001)
Eustenopus villosus	Yellow starthistle hairy weevil		x			x	x	C. Sutherland, personal observation and Winston et al. (2014a)
Euwallacea sp.	Polyphagous and Kuroshio shot hole borers		x					CFPC (2013), Chen et al. (2017), Seybold et al. (2016), and Umeda et al. (2016)
Exapion fuscirostre	Scotch broom weevil		x					Andreas et al. (2017), EDDMapS (2019), and Winston et al. (2014b)
Exapion ulicis	Gorse seed weevil		x					Andreas et al. (2017(, EDDMapS (2019), and Winston et al. (2014b)
Gonipterus scutellatus	Eucalyptus snout beetle		x					Paine (2016)
Harmonia axyridis	Multicolored Asian lady beetle	x	x	x	x	x	x	C. Sutherland, personal observation and Mizell (2012)

(continued)

Table A4.3 (continued)

Scientific name	Common name	Occurrence						References
		AZ	CA	CO	NV	NM	UT	
Heterobostrychus brunneus			x					Ivie (2002)
Hylastes opacus	European bark beetle		x				x	Atkinson (2018), Rabaglia et al. (2019), and USDA (2019), Also occurs in Texas
Hylastinus obscurus	Clover bark beetle		x				x	Wood and Bright (1992)
Hylurgus ligniperda	Red-haired pine bark beetle		x					Liu et al. (2007) and Seybold et al. (2016)
Hypothenemus californicus			x					Wood and Bright (1992)
Hypothenemus crudiae			x					Rabaglia et al. (2019)
Hypothenemus eruditus			x					Seybold et al. (2016)
Icosium tomentosum			x					Bohne (2007)
Ips calligraphus[b]	Sixspined ips	O	x	O		O	O	Seybold et al. (2016) Invasive in California, native elsewhere in Southwest region
Lasioderma serricorne	Cigarette beetle	x	x	x	x	x	x	CABI (2018), Phillips (2002), USDA (1986), and White (1982)
Lyctus brunneus	Old World lyctus beetle		x					Gerberg (1957), ivie (2002), and Lewis and Seybold (2010)
Lyctus linearis	European powderpost beetle	x	x					Furniss and Carolin (1977), Gerberg (1957), and Ivie (2002)
Mecinus janthinus and *M. janthiniformis*	Toadflax stem-mining weevils		x	x			x	Sing et al. (2016), Toševski et al. (2011), and Willden (2017)
Micromalthus debilis[b]	Telephone pole beetle					x		Philips (2001)
Monarthrum mali[b]	Apple wood stainer		x					Seybold et al. (2016)
Nacerdes melanura	Wharf borer		x					Chamberlin (1953)
Nathrius brevipennis	A walnut twig borer		x					Linsley (1963), Linsley and Chemsak (1997), and Seybold et al. (2016)
Orchestes alni	European elm flea weevil	x		x		x	x	Looney et al. (2012) and USDA (2019)
Orthotomicus erosus	Mediterranean pine engraver	x	x		x			Lee et al. (2005) and Seybold et al. (2016)
Otiorhynchus ovatus	Strawberry root weevil	x	x	x	x	x	x	CABI (2018), Cranshaw et al. (1994), and USDA (2019)
Otiorhynchus rugosostriatus	Rough strawberry root weevil	x	x	x	x	x	x	CABI (2018) and Cranshaw et al. (1994)
Otiorhynchus sulcatus	Black vine weevil	x	x	x	x	x	x	CABI (2018), Cranshaw et al. (1994), and USDA (2019)
Oulema melanopus	Cereal leaf beetle		x				x	Dowell and Pickett (2016)
Phloeotribus liminaris[b]	Peach bark beetle		x					Seybold et al. (2016)
Phloeotribus scarabaeoides	Olive bark beetle		x					Arakelian (2017) and Atkinson (2018)
Phloeosinus armatus	Oriental cypress bark beetle		x					Seybold et al. (2016)
Phoracantha recurva	Yellow *Phoracantha*		x					Hanks et al. (1997), Paine (2016), and Paine et al. (2009)
Phoracantha semipunctata	Eucalyptus longhorned beetle		x					Paine (2016), Paine et al. (2009), and Scriven et al. (1986)
Pityophthorus juglandis[b]	Walnut twig beetle	O	x	x	x	O	x	Seybold et al. (2016)
Popilia japonica	Japanese beetle		x	x		x	x	CABI (2018), Gaimari (2005), and USDA (2019)
Rhinocyllus conicus	Seedhead weevil		x	x	x	x	x	Winston et al. (2014a)
Rhinusa antirrhini	Toadflax seed-galling weevil		x	x				Sing et al. (2016)
Rhynchophorus ferrugineus	Red palm weevil		x					CFPC (2012)
Rhynchophorus palmarum	South American palm weevil	x	x					CFPC (2012) and Hodel et al. (2016)

(continued)

Table A4.3 (continued)

Scientific name	Common name	Occurrence						References
		AZ	CA	CO	NV	NM	UT	
Rhynchophorus vulneratus	Asian palm weevil		x					CFPC (2012) and Hoddle et al. (2017)
Rodolia cardinalis	Cottony cushion scale lady beetle = vedalia beetle		x			x		C. Sutherland, personal observation
Scolytus multistriatus	Smaller European elm bark beetle	x	x	x	x	x	x	Lee et al. (2009) and Seybold et al. (2016)
Scolytus rugulosus	Shot hole borer	x	x	x	x	x	x	Seybold et al. (2016)
Scolytus schevyrewi	Banded elm bark beetle	x	x	x	x	x	x	Lee et al. (2009) and Seybold et al. (2016)
Sinoxylon ceratoniae			x					Furniss and Carolin (1977) and Ivie (2002)
Stegobium paniceum	Drugstore beetle	x	x	x	x	x	x	CABI (2018), Phillips (2002), USDA (1986), and White (1982)
Tenebroides mauritanicus	Cadelle	x	x	x	x	x	x	Furniss and Carolin (1977) and USDA (1986)
Trachymela sloanei	Eucalyptus tortoise beetle		x					Gill (1998) and Paine (2016)
Trichosirocalus horridus	A musk thistle biocontrol weevil			x		x		C. Sutherland, personal observation; Winston et al. (2014a)
Trogoxylon aequale	A powderpost beetle	x						Gerberg (1957) (also occurs in Texas)
Xanthogaleruca luteola	Elm leaf beetle	x	x	x	x	x	x	CABI (2018), Cranshaw et al. (1994), and USDA (2019)
Xestobium rufovillosum	Deathwatch beetle		x					Furniss and Carolin (1977) and Phillips (2002)
Xyleborinus saxeseni	Fruit-tree pinhole borer	x	x	x	x	x	x	Rabaglia et al. (2019) and Seybold et al. (2016)
Xyleborus affinis[b]	Oak-hickory ambrosia beetle		x					Bright 2018; voucher specimen in the California Academy of Sciences, San Francisco
Xyleborus dispar	European shot hole borer		x		x		x	Atkinson (2018), Linsley and MacLeod (1942), and Seybold et al. (2016)
Xyleborus pfeili			x					Seybold et al. (2016)
Xyleborus xylographus[b]	Oak timber beetle		x				x	Atkinson (2018) and Seybold et al. (2016)
Xylosandrus germanus	Black stem borer		x					Rabaglia et al. (2019), Seybold et al. (2016), also occurs in Texas
Diptera								
Aedes aegypti	Yellow fever mosquito		x					CDPH (2019)
Aedes albopictus	Asian tiger mosquito		x					CDPH (2019)
Compsilura concinnata	Gypsy moth parasitoid		x					CABI (2018) and Sabrosky and Reardon (1976)
Delia platura	Seedcorn maggot	x	x	x	x	x	x	CABI (2018)
Drosophila suzukii	Spotted wing drosophila		x	x			x	CABI (2018)
Rhagoletis completa[b]	Walnut husk fly		x	x	x	x	x	Cranshaw et al. (1994), Foote and Blanc (1963), and Foote et al. (1993)
Hemiptera								
Acizzia uncatoides	Acacia psyllid		x					Paine (2016)
Adelges piceae	Balsam woolly adelgid		x					CFPC (2012, 2013, 2017) and USDA (2019)
Aonidiella aurantii	California red scale	x	x					USDA (2019)
Asterolecanium minus	Oak pit scale		x					Koehler and Tamaki (1964) and USDA (2019)
Betulaphis brevipilosa	A European birch aphid		x					Hajek and Dahlsten (1986)
Blastopsylla occidentalis	A eucalyptus psyllid		x					Paine (2016)
Callipterinella calliptera	A European birch aphid		x					Hajek and Dahlsten (1986)
Chionaspis etrusca	Saltcedar armored scale					x		C. Sutherland, personal observation
Cryptoneossa triangular	Lemon gum psyllid		x					Paine (2016)
Ctenarytaina longicauda	A eucalyptus psyllid		x					Paine (2016)
Ctenarytaina spatulata	A eucalyptus psyllid		x					Paine (2016)
Ctenarytaina eucalypti	Blue gum psyllid		x					Paine (2016) and USDA (2019)
Diaphorina citri	Asian citrus psyllid	x	x					Geiger and Woods (2009), Mead and Fasulo (2017), and Milosavljević et al. (2017)
Diuraphis noxia	Russian wheat aphid					x		C. Sutherland, personal observation

(continued)

Table A4.3 (continued)

Scientific name	Common name	Occurrence						References
		AZ	CA	CO	NV	NM	UT	
Elatobium abietinum	Spruce aphid	x	x		x	x	x	Lynch (2004, 2014) and USDA (2019)
Eucallipterus tiliae	Linden aphid		x					USDA (2019)
Eucalytolyma maideni	Spotted gum psyllid		x					Paine (2016)
Euceraphis betulae	European birch aphid		x					Hajek and Dahlsten (1986)
Eulecanium cerasorum	Calico scale		x					USDA (2019)
Glycaspis brimblecombei	Red gum lerp psyllid	x	x					CFPC (2011), Fischer and Woods (2010), Paine (2016), and USDA (2019)
Gossyparia spuria	European elm scale	x	x	x	x	x	x	Cranshaw et al. (1994) and Miller and Miller (1993)
Halyomorpha halys	Brown marmorated stink bug	x	x		x	x	x	CABI (2018)
Homalodisca vitripennis	Glassy-winged sharpshooter		x					Paine (2016)
Icerya purchasi	Cottony cushion scale	x	x			x		C. Sutherland, personal observation; USDA (2019)
Lecanium (Parthenolecanium) corni	European fruit lecanium		x			x	x	CABI (2018)
Lepidosaphes ulmi	Oystershell scale	x	x	x	x	x	x	CABI (2018), Cranshaw et al. (1994), Miller and Davidson (2005), and USDA (2019)
Maconellicoccus hirsutus	Pink hibiscus mealybug		x					USDA (2019)
Melanaphis sacchari	Sugarcane aphid					x		C. Sutherland, personal observation
Melanocallis caryaefoliae[b]	Black pecan aphid	x	x			x		C. Sutherland, personal observation
Meliarhizophagus fraxinifolii	Ash whitefly					x		C. Sutherland, personal observation
Metopoplax ditomoides			x					Wheeler and Hoebeke (2012)
Monellia caryella[b]	Black-margined pecan aphid	x	x			x		C. Sutherland, personal observation
Monelliopsis pecanis[b]	Yellow pecan aphid	x	x			x		C. Sutherland, personal observation
Nezara viridula			x					CABI (2018)
Opsius stactogalus	Saltcedar leafhopper					x		C. Sutherland, personal observation
Parthenolecanium (= Lecanium) corni	European fruit lecanium			x				Cranshaw et al. (1994)
Periphyllus lyropictus	Norway maple aphid		x				x	USDA (2019)
Periphyllus testudinacea	Maple aphid		x	x			x	Furniss and Carolin (1977)
Phyllaphis fagi	Beech woolly aphid		x				x	USDA (2019)
Physokermes piceae	Spruce bud scale		x					USDA (2019)
Pineus strobi	Pine bark aphid			x				Darr et al. (2018)
Quadraspidiotus juglansregiae	Walnut scale		x	x				Cranshaw et al. (1994)
Quadraspidiotus perniciosus	San Jose scale		x	x			x	CABI (2018) and USDA (2019)
Raglius alboacuminatus			x				x	Henry (2004)
Shivaphis celti	Hackberry woolly aphid		x					Lawson and Dreistadt (2005)
Sipha flava	Yellow sugarcane aphid					x		C. Sutherland, personal observation
Sipha maydis	Hedgehog aphid					x		C. Sutherland, personal observation
Siphoninus phillyreae	Ash whitefly					x		C. Sutherland, personal observation
Thysanoptera								
Taeniothrips inconsequens	Pear thrips		x			x	x	USDA (2019)
Hymenoptera								
Anaphes nitens	An egg parasitoid of *G. scutellatus*		x					Paine (2016)
Aprostocetus sp.	A gall-forming eulophid wasp		x					Paine (2016)
Avetianella longoi	An egg parasitoid of *P. semipunctata*		x					Paine (2016)

(continued)

Table A4.3 (continued)

Scientific name	Common name	Occurrence						References
		AZ	CA	CO	NV	NM	UT	
Caliroa cerasi	Pear sawfly		x	x	x	x	x	USDA (2019)
Epichrysocharis burwelli	A gall-forming eulophid wasp		x					Paine (2016)
Eupareophora parca			x					Looney et al. (2016)
Fenusa dohrnii	European alder leafminer			x				Cranshaw et al. (1994)
Fenusa pusilla	Birch leafminer			x				Cranshaw et al. (1994)
Fenusa ulmi	Elm leafminer			x				Cranshaw et al. (1994) and USDA (2019)
Linepithema humile	Argentine ant		x		x			Tsutsui et al. (2001) and Knight (2018)
Nematus ribesii	Imported currantworm			x				Cranshaw et al. (1994)
Ophelimus maskelli.	A gall-forming eulophid wasp		x					Paine (2016)
Polistes dominula	European paper wasp		x					Pilowsky and Starks (2018)
Pristiphora rufipes				x				Cranshaw et al. (1994)
Psix tunetanus	Parasitoid of Pentatomidae	x	x			x	x	Johnson and Masner (1985)
Psyllaephagus bliteus	Parasitoid of *Glycaspis brimblecombei*		x					Paine (2016)
Psyllaephagus parvus	Parasitoid of *Eucalytolyma maideni*		x					Paine (2016)
Psyllaephagus perplexus	Parasitoid of *Cryptoneossa triangular*		x					Paine (2016)
Psyllaephagus pilosus	Parasitoid of *Ctenerytina eucalypti*		x					Paine (2016)
Quadrastichodella nova	Seed-galling eulophid wasp		x					Paine (2016)
Selitrichodes globulus	A gall-forming eulophid wasp		x					Paine (2016)
Solenopsis invicta	Red imported fire ant		x					Greenberg and Kabashima (2013)
Vespula germanica	German yellowjacket		x					Lester and Beggs (2019)
Lepidoptera								
Anarsia lineatella	Peach twig borer	x	x	x	x	x	x	USDA (2019)
Calophasia lunula	Toadflax defoliating moth	x		x				Sing et al. (2016)
Caloptilia negundella	Boxelder leafroller			x				Cranshaw et al. (1994)
Caloptilia syringella	Lilac leafminer			x				Cranshaw et al. (1994)
Epiphyas postvittana	Light brown apple moth		x					Bohne (2008), Brown (2007), CFPC (2012), and Geiger and Woods (2009)
Homadaula anisocentra	Mimosa webworm		x					USDA (2019)
Hyles euphorbiae	Leafy spurge hawk-moth		x	x				Lotts and Naberhaus (2017)
Leucoma salicis	Satin moth		x		x			Phillips (2018) and USDA (2019)
Leucoptera spartifoliella	Scotch broom twig miner		x					Andreas et al. (2017) and Winston et al. (2014b)
Lymantria dispar dispar[a]	North American gypsy moth		x				x	Bohne (2007, 2008)
Lymantria dispar asiatica/japonica[a]	Asian gypsy moth		x					Bohne (2007, 2008) and Bohne and Rios (2006)
Rhyacionia buoliana	European pine shoot moth						x	USDA (2019)
Rhyacionia frustrana[b]	Nantucket pine tip moth		x			x		Brown and Eads (1975) and Cranshaw (1984)
Synanthedon scitula	Eastern dogwood borer			x				Cranshaw et al. (1994)

[a]These species have been introduced repeatedly into the Southwest region, but do not have known established populations

[b]These species are native to other regions of North America, but have invaded the Southwest region (except for *A. auroguttatus, D. wrightii, I. calligraphus*, and *P. juglandis*, which are native originally to certain States in the Southwest region, but have spread to other States within the region. In these instances, the original States are marked with circles in the table)

cies were first detected in heavily urbanized southern California, which appears to be a particularly fertile area for their introduction and establishment due to the proximity of ship traffic and associated cargo, as well as the great diversity of potential hosts in the urban forest. The extra-continental invasive bark and ambrosia beetles in California comprise nearly 30% of the 58 documented invasive species of bark and ambrosia beetles in the United States (Haack 2001, 2006; Haack and Rabaglia 2013; Lee et al. 2007). Since invasive species often reassociate themselves with their hosts of origin (i.e., in this case adventive populations of trees) in the invaded habitat (Mattson et al. 1992; Niemelä and Mattson 1996), it is not surprising that some of the key invasive bark and ambrosia beetles in California are of Mediterranean origin (Mifsud and Knížek 2009). However, there is also a trend toward the introduction and establishment of Asian invasive species in California (e.g., polyphagous and Kuroshio shot hole borers (*Euwallacea* sp.) (Fig. A4.6) and banded elm bark beetle). The latter elm bark beetle has largely replaced the smaller European elm bark beetle (*S. multistriatus*) throughout the Southwest region (Lee et al. 2009). The banded elm bark beetle has also been shown to vector *Ophiostoma novo-ulmi*, the pathogenic agent of Dutch elm disease (Jacobi et al. 2007, 2013). However, unlike urban forests in the Eastern United States and Europe where susceptible elm species were planted frequently, urban forests in many southern cities in the Southwest region are characterized more by Chinese and Siberian elms (*U. parvifolia* and *U. pumila*, respectively) which have been less susceptible to Dutch elm disease (Strobel and Lanier 1981). Among the States of the region, the disease was recorded initially for some years in only California and Colorado (Barger and Hock 1971; Strobel and Lanier 1981). The advent of the banded elm bark beetle, which has a preference for Siberian elm, may change the dynamics of disease transmission to these Asian hosts in the future (Lee et al. 2010).

California has also been a hotbed of invasive species of insects that feed on eucalypts, where nearly 20 insects from 4 feeding guilds, largely derived from Australia, have become established (Paine 2016). The only introduced termite with established populations in the Southwest Region, the Formosan subterranean termite (*Coptotermes formosanus*), began its invasion in southern California as well (Rust et al. 1998). The powderpost termite (*Cryptotermes brevis*) has been detected in California in wood furnishings or structures on occasion, but there are no reports of any established populations in the Southwest region (Evans et al. 2013; Light 1936; Rust 2004; Scheffrahn et al. 2009). Finally, the major invasive ant in the region, the Argentine ant (*Linepithema humile*), was, at the outset, also uniquely established within the region in California (Tsutsui et al. 2001), but has since been detected in both Northern and Southern Nevada (Knight 2018).

Climate change is exerting a broad impact on the population biology of native bark beetles and potentially other subcortical insects in North America (Bentz et al. 2010), with the most obvious effects at higher elevations and latitudes. In the Southwest region, there also appears to be a trend, whereby southern populations of indigenous exotic subcortical species are invading and expanding their populations generally northward. Examples of this include the Mexican pine beetle (*Dendroctonus mexicanus*) (Moser et al. 2005), the Mexican soapberry borer (*Agrilus prionurus*) (Billings et al. 2014), the walnut twig beetle (*P. juglandis*) (Rugman-Jones et al. 2015), the goldspotted oak borer (*Agrilus auroguttatus*) (Coleman and Seybold 2016), and two other newly discovered flatheaded borers in California, *Chrysobothris analis* (Westcott et al. 2015) and *C. costifrons costifrons* (Basham et al. 2015). A variation on this theme has occurred with the South American palm weevil (*Rhynchophorus palmarum*), which was introduced into Tijuana, Baja California, Mexico, and then subsequently expanded its distribution northward into San Diego County, CA (Hodel et al. 2016). Whether or not this trend of "latitudinal creep" will continue to manifest itself in the urban, peri-urban, and wildland forests of the Southwest region remains to be seen.

There have been periodic introductions of two subspecies of gypsy moths (*Lymantria dispar*), in the Southwest region (see Chap. 2, Sect. 2.3; Chap. 7, Sect. 7.4.1; and the Northwest and Southeast and Caribbean regional summaries for additional discussion of this insect complex). Immatures (larvae) of European gypsy moth (*Lymantria d. dispar*), detected but eradicated from Utah, Colorado, and California, and Asian gypsy moth (*L. d. asiatica/japonica*), detected but eradicated from California (Bohne and Rios 2006; Fischer and Woods 2010), are major defoliators of valuable species of street and forest trees (Pogue and Schaefer 2007). Asian gypsy moth occurs in temperate Asia from the Ural Mountains east to China, Korea, and the Russian Far East (north of the Himalayas), whereas European gypsy moth, native originally to Europe, has been established in the eastern portion of the United States since the late 1860s (Liebhold et al. 1989). An extensive detection system is in place throughout most of the developed areas in the Southwest region providing an early detection resource (Bohne 2007; CFPC 2017). When introductions are detected, eradication strategies have been used to avoid population establishment in the Southwest region.

Most of the invasive insect taxa with the highest potential impact for forests of the Southwest region have also been first introduced into southern California. These established species include the goldspotted oak borer (Coleman and Seybold 2016), the polyphagous shot hole borer/Kuroshio shot hole borer complex (Chen et al. 2017; Fig. A4.6), the Mediterranean pine engraver (*Orthotomicus erosus*) (Lee et al. 2005, 2008), and the red-haired pine bark beetle (*Hylurgus ligniperda*) (Liu et al. 2007). An ensemble of massive *Rhynchophorus* weevils (Table A4.3) present a major

Fig. A4.6 The polyphagous and Kuroshio shot hole borers (*Euwallacea whitfordiodendrus and E. kuroshio* (Coleoptera: Scolytidae)) are populations of invasive ambrosia beetles whose invaded ranges are centered in Los Angeles and San Diego Counties, CA, respectively [species nomenclature as per Gomez et al. (2018)]. They originate from Asia and threaten riparian hardwoods in southern California (Coleman et al. 2013). (**a**) Female (*upper*) and male (*lower*) polyphagous shot hole borers (PSHB); (Photo credit: Stacy Hishinuma, University of California, Davis). (**b**) Egg galleries created by female PSHB in the xylem of castor bean (*Ricinus communis* L.), which is itself an invasive plant in southern California. Note the dark staining from symbiotic fungi such as *Fusarium euwallaceae* (Eskalen et al. 2013; Lynch et al. 2016) (Photo credit: Tom W. Coleman, USDA Forest Service, Forest Health Protection, Albuquerque, NM). (**c**) Density of sap stain spots on the bark surface of California sycamore (*Platanus racemosa*) indicates the extent of aggregation as female PSHB colonize the xylem on the main stem of this host. The xylem of exposed roots and branches can also be colonized (Photo credit: T.W. Coleman). (**d**) Landscape-level impact of Kuroshio shot hole borer in the Tijuana River Valley Regional Park, San Diego County, CA. In this river delta, damage is primarily to arroyo and red willow (*Salix* spp.) and castor bean. *Inset* shows bark surface of willow with "toothpick-like" strands of boring dust (frass) expelled from entrance holes by female PSHB. (Photo credit: John Boland, unaffiliated research scientist, and (*inset*) Adrian Poloni, University of California, Davis)

threat to urban Canary Island date palms (*Phoenix canariensis*) and related palms that provide shade and ornament to the urban forests of southern California, Las Vegas, and Phoenix (Hoddle et al. 2017; Hodel et al. 2016). Small introduced populations of the Asian longhorned beetle (*Anoplophora glabripennis*), a woodboring pest of maples (*Acer* spp.) and other hardwoods, have been eradicated from California (Bohne 2007, 2008; Bohne and Rios 2006), but, if established, this pest also presents a high risk to the Southwest region. An example of an invasive species with great potential for urban and riparian forest impact in the Southwest region that has not yet reached California is the emerald ash borer (*Agrilus planipennis*), which has recently established populations in Colorado (Colorado State Forest Service 2018) (see Chap. 2, Box 2.5 and Chap. 7).

Three invasive insect species that bridge agro- and forest ecosystems of the Southwest region are the aforementioned walnut twig beetle, the pecan weevil (*Curculio caryae*), and the light brown apple moth (LBAM) (*Epiphyas postvittana*). The walnut twig beetle and the pecan weevil are both native invasive species with the walnut twig beetle originating from Arizona and New Mexico (Rugman-Jones et al. 2015), whereas the weevil has coevolved with native stands of hickory (*Carya*) species in the Eastern United States (Mulder et al. 2012). The walnut twig beetle has invaded the entire Southwest region; so far, the weevil is only found in New Mexico (Sutherland et al. 2017). The walnut twig beetle cannot develop in pecan (*Carya illinoensis*) (Hefty et al. 2018), but both species may utilize English walnut (*Juglans regia*) as hosts (Hefty et al. 2018; Mulder et al. 2012). Thus, as a duo, the two invaders threaten pecan and walnut nut production in the Southwest region. A complex of aphids from the Eastern United States (Table A4.3) also cause damage to pecan trees in orchards in AZ, CA, and NM. Light brown apple moth, native to Australia, was first found in the San Francisco Bay Area in 2006 (Brown 2007) and has since been detected in 15 coastal and near coastal counties in California (Gutierrez et al. 2010). It is highly polyphagous, but considered primarily as a pest of pome fruits (Rosaceae) and grapes (*Vitis* spp.). For several years, it was regarded as a potential invasive pest of numerous forest and shade trees, but to date has not been recorded causing substantial damage to trees in California (Bohne 2008; CFPC 2012; Geiger and Woods 2009). Using a temperature-driven demographic model and climatic data from 151 locations in California, Gutierrez et al. (2010) predicted that near-coastal regions of California are most favorable for light brown apple moth, northern Central Valley areas were less favorable, and desert regions of Arizona and California are unfavorable.

Vertebrates

A modest number of invasive vertebrates are worth noting in the Southwest region (Table A4.4). A number of ungulate species introduced into the Southwest region have become problematic. Among these are exotic species such as Barbary sheep (*Ammotragus lervia*) and African oryx (*Oryx gazelle*) which were brought over originally from Africa as game animals. The Barbary sheep threatens desert bighorn sheep (*Ovis canadensis Mexicana*), in its native habitat (Novack et al. 2009), whereas expanding oryx populations can damage soil and vegetation resources (Conrod 2004). Of particular concern are invasive feral swine (*Sus*

Table A4.4 The primary invasive vertebrates of the Southwest region include amphibians, fish, mammals, and reptiles

		Occurrence						References
Scientific name	Common name	AZ	CA	CO	NV	NM	UT	
Lithobates catesbeianus (Rana catesbeiana)	American bullfrog	x	x	x	x	x	x	Lever (2003)
Hypophthalmichthys spp. *Ctenopharyngodon* spp., *Mylopharyngodon piceus*	Carp (Asian, black, big head, diploid grass, silver)	x	x	x	x	x	x	NAS (2019)
Salmo salar	Atlantic salmon		x	x	x		x	NAS (2019)
Didemnum vexillum	Carpet sea squirt	x						CABI (2018)
Chelydra serpentina serpentina	Eastern snapping turtle		x					Fuller et al. (2018b)
Neogobius melanostomus, Rhinogobius brunneus, Tridentiger bifasciatus Acanthogobius flavimanus	Goby		x					Nico et al. (2018b)
Notemigonus crysoleucas	Golden shiner	x	x	x	x	x	x	Nico (2018)
Esox spp.	Muskellunge/northern pike	x	x	x	x	x	x	Fuller and Neilson (2018b)
Trachemys scripta elegans	Red-eared slider	x	x			x		Somma et al. (2019)
Scardinius erythrophthalmus	Rudd			x				Nico et al. (2018a)
Channa argus	Northern snakehead		x					Fuller et al. (2018a)
Dorosoma petenense	Threadfin shad (yellowtails)	x	x	x	x	x	x	Fuller and Neilson (2018a)
Clarias spp.	Walking catfish		x		x			Nico et al. (2018c)
Terrestrial vertebrates								
Ammotragus lervia	Barbary sheep	x	x	x		x		CABI (2019)
Cygnus olor	Mute swan		x					CABI (2018)
Myocastor coypus	Nutria		x	x		x	x	CABI (2018)
Oryx gazella	African oryx					x		Morrison (1981)
Sus scrofa	Feral swine	x	x	x	x	x	x	McClure et al. (2018)

scrofa), which include free-roaming European wild boars, former domestic pigs, and hybrids (see Chap. 2, Sect. 2.4 and the Southeast and Caribbean regional summary for additional discussion of this invasive vertebrate). Feral swine rooting in the litter layer and soil for food can cause soil erosion, property damage, destruction of ground nests of birds, and establishment of invasive plants. Feral swine may also transmit diseases to domestic livestock and indirectly to humans by facilitating the spread of ticks and tick-borne pathogens (see Chap. 2, Sect. 2.4) (Sanders et al. 2013). They have been particularly damaging to agricultural properties in California (White et al. 2018). APHIS is currently engaged in limiting westward expansion of feral swine by eradicating local feral swine populations in New Mexico and San Diego County, California. However, feral swine may be protected by local property owners who are interested in maintaining them for hunting. The nutria (*Myocastor coypus*), a large South American rodent, was discovered in California in March 2017 and is now found in five counties to the southeast of the Sacramento-San Joaquin Delta (Sabalow 2018). They feed on native wetland vegetation and burrow into levees and ditch banks, which poses a threat to water management in this vital region. Howard (1953) described an earlier introduction of nutria to Stanislaus County, California (1942–1952), from escapees from several fur farming operations based on animals imported from New Jersey and Louisiana. Howard's description underscores the inherent difficulties associated with eradicating this highly fecund invader. California's Central Valley and parts of Southern Nevada and Western Arizona have been modeled as suitable habitat for nutria (Jarnevich et al. 2017).

Acknowledgments We greatly appreciate reviews of this summary by Jackson Audley (University of California Davis Department of Entomology and Nematology), David Bakke (USDA Forest Service Pacific Southwest Region Forest Health Protection, Vallejo, CA), Tom W. Coleman (USDA Forest Service Southwestern Region Forest Health Protection, Albuquerque, NM, now with Southern Region, Asheville, NC), and Lori J. Nelson (USDA Forest Service Pacific Southwest Research Station, Davis, CA). Both D. Bakke and Warren J. Ririe (USDA Forest Service Intermountain Region, Boise, ID) provided local perspectives on the occurrence and impact of invasive plants in their respective regions. L. Nelson provided information on invasive termites in the Southwest region, and J.B. Knight (Nevada Department of Agriculture, Sparks, NV) reviewed the invasive insects of Nevada. We also appreciate reviews of a previous version of the manuscript by four anonymous reviewers.

Literature Cited

Agyagos J (2018) Personal correspondence, email letter dated January 16, 2018 to Allen White. On file with White at U.S. Department of Agriculture, Forest Service, Southwestern Region, 333 Broadway Blvd SE, Albuquerque NM 8711

Alexander J, Lee CA (2010) Lessons learned from a decade of sudden oak death in California: evaluating local management. Environ Manag 46:315–328

Anderson RS (2008) A review of the genus *Cryptorhynchus* Illiger 1807 in the United States and Canada (Curculionidae: Cryptorhynchinae). Coleopt Bull 62:168–180

Andreas JE, Winston RL, Coombs EM et al (2017) Biology and biological control of Scotch broom and gorse. USDA Forest Service, Forest Health Technology Enterprise Team, Morgantown. FHTET-2017-01. 164 p

Arakelian G (2017) Olive bark beetle (*Phloeotribus scarabaeoides*). Los Angeles County Department of Agricultural Commissioner/Weights and Measures Publication, 1 p. http://file.lacounty.gov/SDSInter/acwm/1025608_Olive_bark_beetle.pdf

Atkinson TH (2018) Regional checklists of bark and ambrosia beetles. http://www.barkbeetles.info/regional_chklist_select.php

Barger JH, Hock WK (1971) Distribution of Dutch elm disease and the smaller European elm bark beetle in the United States as of 1970. Plant Dis Rep 55:271–272

Basham JP, Waayers RJ, Westcott RL (2015) Discovery of *Chrysobothris costifrons costifrons* Waterhouse, 1887 (Coleoptera: Buprestidae) in southern California, U.S.A. Pan-Pacific Entomol 91:200–202

Benedict WV (1981) History of white pine blister rust control: a personal account. FS-355. U.S. Department of Agriculture, Forest Service, Washington, DC. 47 p

Bentz BJ, Régnière J, Fettig CJ et al (2010) Climate change and bark beetles of the western US and Canada: direct and indirect effects. Bioscience 60:602–613

Billings RF, Grosman DM, Pase HA III (2014) Soapberry borer, *Agrilus prionurus* (Coleoptera: Buprestidae): an exotic pest threatens western soapberry in Texas. Southeast Nat 13:105–116

Bloodworth BR, Shafroth PB, Sher AA et al (2016) Tamarisk beetle (*Diorhabda* spp.) in the Colorado River basin: synthesis of an expert panel forum. Scientific and Technical Report No. 1. Colorado Mesa University, Grand Junction. 19 p

Bohne M (2007) California forest pest conditions 2006. California Forest Pest Council, Sacramento. 57 p. https://www.fs.usda.gov/detail/r5/forest-grasslandhealth/?cid=fsbdev3_046704

Bohne M (2008) California forest pest conditions 2007. California Forest Pest Council, Sacramento. 61 p. https://www.fs.usda.gov/detail/r5/forest-grasslandhealth/?cid=fsbdev3_046704

Bohne M, Rios J (2006) California forest pest conditions 2005. California Forest Pest Council, Sacramento. 72 p. https://www.fs.usda.gov/detail/r5/forest-grasslandhealth/?cid=fsbdev3_046704

Boyce JS (1938) Forest pathology, 1st edn. McGraw-Hill, New York/London, 600 p

Brasier CM, Vettraino AM, Chang TT, Vannini A (2010) *Phytophthora lateralis* discovered in an old growth *Chamaecyparis* forest in Taiwan. Plant Pathol 59:595–603

Breshears DD, Cobb NS, Rich PM et al (2005) Regional vegetation die-off in response to global-change type drought. Proc Natl Acad Sci U S A 102:15144–15148

Bright DE, Kondratieff BC, Norton AP (2013) First record of the "splendid tamarisk weevil", *Coniatus splendidulus* (F.) (Coleoptera: Curculionidae: Hyperinae), in Colorado, USA. Coleopt Bull 67:302–303

Bright DE (2018) Personal communication on March 28, 2018 to Steven Seybold, U.S. Department of Agriculture, Forest Service, Pacific Southwest Research Station, 1731 Research Park Drive, Davis, CA 95618

Brown JW (2007) Discovery of light brown apple moth in North America. http://www.tortricidae.com/TORTS_newsletter8%282%29.pdf

Brown LR, Eads CO (1975) Nantucket pine tip moth in Southern California: identity and insecticidal control. J Econ Entomol 68:380–382

CABI (2018) Invasive species compendium. CAB International, Wallingford. www.cabi.org/isc

California Department of Public Health (CDPH) (2019) *Aedes aegypti* and *Aedes albopictus* mosquitoes. https://www.cdph.ca.gov/Programs/CID/DCDC/Pages/Aedes-aegypti-and-Aedes-albopictus-mosquitoes.aspx

California Forest Pest Council (CFPC) (2010–2017) California forest pest conditions 2010–2017. California Forest Pest Council, Sacramento. https://www.fs.usda.gov/detail/r5/forest-grasslandhealth/?cid=fsbdev3_046704

Cal-IPC (2006) California invasive plant inventory. Cal-IPC Publication 2006–02. California Invasive Plant Council, Berkeley. http://www.calflora.org/app/ipl?list_id=px7

Carlos WJ, Johnson W, Skelly J, Knight J (2002) The bronze birch borer. University of Nevada Cooperative Extension, Fact Sheet 02–38, 4 p

Cayan DR, Das T, Pierce TP (2010) Future dryness in the southwest US and the hydrology of the early 21st century drought. Proc Natl Acad Sci U S A 107:21271–21276

Chamberlin WJ (1953) Insects affecting forest products and other materials. Oregon State College Cooperative Association, Corvallis. 159 p

Chen Y, Dallara PL, Nelson LJ et al (2017) Comparative morphometric and chemical analyses of phenotypes of two invasive ambrosia beetles (*Euwallacea* spp.) in the United States of America. Insect Sci 24:647–662. https://doi.org/10.1111/1744-7917.12329

Coleman TW, Seybold SJ (2016) Goldspotted oak borer in California: invasion history, biology, impact, management, and implications for Mediterranean forests worldwide. In: Paine TD, Lieutier F (eds) Insects and diseases of Mediterranean forest systems. Springer, Cham, pp 663–697. https://doi.org/10.1007/978-3-319-24744-1_22

Coleman TW, Eskalen A, Stouthamer R (2013) New pest complex in California: the polyphagous shot hole borer, *Euwallacea* sp., and Fusarium Dieback, *Fusarium euwallaceae*. Pest Alert, R5-PR-032. USDA Forest Service, State and Private Forestry, Vallejo. 5 p

Colorado State Forest Service (2018) Emerald ash borer. https://csfs.colostate.edu/forest-management/emerald-ash-borer/

Conrod B (2004) Highlights: last African oryx removed from White Sands National Monument. Park Sci 22:6

Cranshaw W, Leatherman D, Kondratieff B (1994) Insects that feed on Colorado trees and shrubs. Colorado State University Cooperative Extension Bulletin 506A, Fort Collins. 176 p

Darr M, Salom S, Brooks RK et al (2018) First report of *Pineus strobi* (Hartig, 1839) (Hemiptera: Adelgidae) in western North America. Pan-Pacific Entomol 94:40–42

Dowell RV, Pickett C (2016) The cereal leaf beetle, *Oulema melanopus* Linnaeus, 1758 (Coleoptera: Chrysomelidae) in California: detection and response. Pan-Pacific Entomol 92:165–167

Duckles B, Švihra P (1995) Bronze birch borer discovered in California. HortScript, Aug 1995, No. 16. University of California Cooperative Extension, Novato, 4 p

EDDMapS (2019) Early detection & distribution mapping system. The University of Georgia—Center for Invasive Species and Ecosystem Health. http://www.eddmaps.org/

Eskalen A, Stouthamer R, Lynch SC et al (2013) Host range of *Fusarium* dieback and its ambrosia beetle (Coleoptera: Scolytinae) vector in southern California. Plant Dis 97:938–951

Evans TA, Forschler BT, Grace JK (2013) Biology of invasive termites: a worldwide review. Annu Rev Entomol 58:455–474

Farr DF, Rossman AY (2018) Fungal databases. U.S. National Fungus Collections, Agricultural Research Service, USDA. https://nt.ars-grin.gov/fungaldatabases/

Fischer L, Woods M (2010) California forest pest conditions 2009. California Forest Pest Council, Sacramento. 67 p. https://www.fs.usda.gov/detail/r5/forest-grasslandhealth/?cid=fsbdev3_046704

Flint ML, Graves AD, Seybold SJ (2010) Thousand cankers disease of walnuts spreads in California. CAPCA Advisor Magaz 8:36–39

Foote RH, Blanc FL (1963) The fruit flies or Tephritidae of California. Bulletin of the California Insect Survey, vol. 7. University of California Press, Berkeley. 122 p

Foote RH, Blanc FL, Norrbom AL (1993) Handbook of the fruit flies (Diptera: Tephritidae) of America north of Mexico. Cornell University Press, Ithaca, 571 p

Fuller P, Neilson M (2018a) *Dorosoma petenense* (Günther, 1867): U.S. Geological Survey, nonindigenous aquatic species database, Gainesville. https://nas.er.usgs.gov/queries/factsheet.aspx?SpeciesID=493

Fuller P, Neilson M (2018b) *Esox lucius* Linnaeus, 1758: U.S. Geological Survey, nonindigenous aquatic species database, Gainesville. https://nas.er.usgs.gov/queries/FactSheet.aspx?speciesID=676

Fuller PL, Benson AJ, Nunez G et al (2018a) *Channa argus* (Cantor, 1842): U.S. Geological Survey, nonindigenous aquatic species database, Gainesville. https://nas.er.usgs.gov/queries/factsheet.aspx?speciesid=2265

Fuller P, Foster A, Somma LA (2018b) *Chelydra serpentina serpentina* (Linnaeus, 1758): U.S. Geological Survey, nonindigenous aquatic species database, Gainesville. https://nas.er.usgs.gov/queries/FactSheet.aspx?SpeciesID=1226

Furniss RL, Carolin VM (1977) Western forest insects. Miscellaneous Publication 1339. U.S. Department of Agriculture, Forest Service, Washington, DC. 654 p

Gabbard BL, Fowler NL (2007) Wide ecological amplitude of a diversity-reducing invasive grass. Biol Invasions 9:149–160

Gaimari S (2005) California pest and disease report. California Department of Food & Agriculture, Plant Pest Diagnostics Branch. 78 p

Geiger J, Woods M (2009) California forest pest conditions 2008. California Forest Pest Council, Sacramento. 65 p. https://www.fs.usda.gov/detail/r5/forest-grasslandhealth/?cid=fsbdev3_046704

Geils BW, Hummer KE, Hunt RS (2010) White pines, *Ribes*, and blister rust: a review and synthesis. For Pathol 40:147–185

Gerberg EJ (1957) A revision of the New World species of powder-post beetles belonging to the family Lyctidae. United States Department of Agriculture, Technical Bulletin No.1157. 55 p

Gill RJ (1998) Recently introduced pests of Eucalyptus. Calif Plant Pest Dis Rep 17:21–24

Glenn EP, Nagler PL (2005) Comparative ecophysiology of *Tamarix ramosissima* and native trees in western U.S. riparian zones. J Arid Environ 61:419–446

Gomez DF, Skelton J, Sedonia Steininger M et al (2018) Species delineation within the *Euwallacea fornicatus* (Coleoptera: Curculionidae) complex revealed by morphometric and phylogenetic analyses. Insect Syst Biodivers 2(6):1–11. https://doi.org/10.1093/isd/ixy018

Gordon TR (2016) Pitch canker in California Mediterranean conifer systems. In: Paine TD, Lieutier F (eds) Insects and diseases of Mediterranean forest systems. Springer, Cham, pp 699–710. https://doi.org/10.1007/978-3-319-24744-1_23

Gordon TR, Storer AJ, Wood DL (2001) The pitch canker epidemic in California. Plant Dis 85:1128–1139

Graves AD, Coleman TW, Seybold SJ (2011) Monitoring walnut health and decline in response to thousand cankers disease and infestation by the walnut twig beetle, *Pityophthorus juglandis*, in southern California and New Mexico. Year One Progress Report for USDA Forest Service Forest Health Monitoring Project INT-EM-B-11-03. https://www.fs.fed.us/foresthealth/fhm/em/funded/11/INT-EM-B-11-03.pdf

Greenberg L, Kabashima JN (2013) Red imported fire ant. University of California Statewide Integrated Pest Management Program, Oakland, California, Agriculture and Natural Resources Pest Notes, Publication 7487, Oakland. 4 p. http://ipm.ucanr.edu/PMG/PESTNOTES/pnredimportedfireant.pdf

Griffin JR, Critchfield WB (1972) The distribution of forest trees in California. Research Paper PSW-82 (reprinted with Supplement, 1976). USDA Forest Service, Pacific Southwest Forest and Range Experiment Station, Berkeley. 114 p

Gutierrez AP, Mills NJ, Ponti L (2010) Limits to the potential distribution of light brown apple moth in Arizona-California based on climate suitability and host plant availability. Biol Invasions 12:3319–3331

Haack RA (2001) Intercepted Scolytidae (Coleoptera) at U.S. ports of entry: 1985–2000. Integr Pest Manag Rev 6:253–282

Haack RA (2006) Exotic bark- and wood-boring Coleoptera in the United States: recent establishments and interceptions. Can J For Res 36:269–288

Haack RA, Rabaglia RJ (2013) Exotic bark and ambrosia beetles in the USA: potential and current invaders. In: Peña J (ed) Potential invasive pests of agricultural crops. CAB International, Wallingford, pp 48–74

Hajek AE, Dahlsten DL (1986) Coexistence of three species of leaf-feeding aphids (Homoptera) on *Betula pendula*. Oecologia 68:380–386

Hanks LM, Paine TD, Millar JG, Campbell C (1997) Another tree-killing pest of eucalyptus invades California. Calif Plant Pest Dis Rep 16:19–21

Hansen EM, Goheen DJ, Jules ES, Ullian B (2000) Managing port-Orford-cedar and the introduced pathogen *Phytophthora lateralis*. Plant Dis 84:4–14

Hefty AR, Aukema BH, Venette RC et al (2018) Reproduction and potential range expansion of walnut twig beetle across the Juglandaceae. Biol Invasions 20:2141–2155. https://doi.org/10.1007/s10530-018-1692-5

Henry TJ (2004) *Raglius alboacuminatus* (Goeze) and *Rhyparochromus vulgaris* (schilling) (Lygaeoidea: Rhyparochromidae): two Palearctic bugs newly discovered in North America. Proc Entomol Soc Wash 106:513–522

Hishinuma SM (2017) Quantifying the frequency, severity, and progression of thousand cankers disease (TCD) in various walnut species and regions in California. PhD dissertation, University of California, Davis

Hoddle MS, Hoddle CD, Alzubaidy M et al (2017) The palm weevil *Rhynchophorus vulneratus* is eradicated from Laguna Beach. Calif Agric 71:23–29

Hodel DR, Marika MA, Ohara LM (2016) The south American palm weevil: A new threat to palms in California and the southwest. PalmArbor 3:1–27

Howard WE (1953) Nutria (*Myocastor coypus*) in California. J Mammal 34:512–513

Ivie MA (2002) 69. Bostrichidae. In: Arnett RH, Frank JH, Thomas MC, Skelley PE (eds) American beetles, volume II: Polyphaga: Scarabaeoidea through Curculionoidea. CRC Press LLC, Boca Raton, pp 233–244

Jacobi WR, Koski RD, Harrington TC et al (2007) Association of *Ophiostoma novo-ulmi* with *Scolytus schevyrewi* (Scolytidae) in Colorado. Plant Dis 91:245–247

Jacobi W, Koski R, Negrón JF (2013) Dutch elm disease pathogen transmission by the banded elm bark beetle *Scolytus schevyrewi* (Curculionidae subfamily Scolytinae). For Pathol 43:232–237

Jarnevich CS, Young NE, Sheffels TR et al (2017) Evaluating simplistic methods to understand current distributions and forecast distribution changes under climate change scenarios: an example with coypu (*Myocastor coypus*). NeoBiota 32:107–125

Johnson NF, Masner L (1985) Revision of the genus *Psix* Kozlov & Lê (Hymenoptera: Scelionidae). Syst Entomol 10:33–58

Katz GL, Shafroth PB (2003) Biology, ecology and management of *Elaeagnus angustifolia* L. (Russian olive) in western North America. Wetlands 23:763–777

Klataske R (2016) Invasive old world bluestems spread across the Great Plains. Center for Grassland Studies Newsletter, University of Nebraska, Lincoln. Winter-Spring 22(1). 2 p. https://grassland.unl.edu/documents/CGS%20Winter-Spring%202016%20web.pdf

Knight JB (2018) Personal correspondence, email letter dated August 10, 2018 to Steven Seybold. On file with Seybold at U.S. Department of Agriculture, Forest Service, Pacific Southwest Research Station, 1731 Research Park Drive, Davis, CA 95618

Koehler CS, Tamaki G (1964) Studies on the distribution of the pit scale *Asterolecanium minus* on oak trees. Ann Entomol Soc Am 57:146–150

Kolařík M, Freeland E, Utley C, Tisserat N (2011) *Geosmithia morbida* sp. nov., a new phytopathogenic species living in symbiosis with the walnut twig beetle (*Pityophthorus juglandis*) on *Juglans* in USA. Mycologia 103:325–332

Kolařík M, Hulcr J, Tisserat N et al (2017) *Geosmithia* associated with bark beetles and woodborers in the western USA: taxonomic diversity and vector specificity. Mycologia 109:185–199

Kumagai LB, LeVesque CS, Blomquist CL et al (2013) First report of *Candidatus* Liberibacter asiaticus associated with citrus Huanglongbing in California. Plant Dis 97:283. https://doi.org/10.1094/PDIS-09-12-0845-PDN

Lawson AB, Dreistadt SH (2005) Hackberry woolly aphid. University of California Statewide Integrated Pest Management Program, Oakland, California, Agriculture and Natural Resources Pest Notes, Publication 74111, Oakland. 3 p. https://ucanr.edu/sites/sjcoeh/files/77085.pdf. 5 Mar 2019

Lee JC, Smith SL, Seybold SJ (2005) The Mediterranean pine engraver, *Orthotomicus erosus*. USDA Forest Service, Pest Alert, R5-PR-016. USDA Forest Service, State and Private Forestry, Vallejo. 4 p. http://entomology.ucdavis.edu/files/215004.pdf. or http://www.treesearch.fs.fed.us/pubs/48537

Lee JC, Haack RA, Negrón JF et al (2007) Invasive bark beetles. Forest Insect and Disease Leaflet No. 176. USDA Forest Service, Washington, DC. 12 p. https://www.fs.usda.gov/Internet/FSE_DOCUMENTS/fsbdev2_042986.pdf

Lee JC, Flint ML, Seybold SJ (2008) Suitability of pines and other conifers as hosts for the invasive Mediterranean pine engraver (Coleoptera: Scolytidae) in North America. J Econ Entomol 101:829–837

Lee JC, Aguayo I, Aslin R et al (2009) Co-occurrence of the invasive banded and European elm bark beetles (Coleoptera: Scolytidae) in North America. Ann Entomol Soc Am 102:426–436

Lee JC, Hamud SM, Negrón JF et al (2010) Semiochemical-mediated flight strategies of two invasive elm bark beetles: a potential factor in competitive displacement. Environ Entomol 39:642–652. https://doi.org/10.1603/EN09327

Lee JC, Negrón JF, McElwey SJ et al (2011) Biology of the invasive banded elm bark beetle (Coleoptera: Scolytidae) in the western United States. Ann Entomol Soc Am 104:705–717. https://doi.org/10.1603/AN10150

Lester PJ, Beggs JR (2019) Invasion success and management strategies for social *Vespula* wasps. Annu Rev Entomol 64:51–71

Lever C (2003) Naturalized reptiles and amphibians of the world. Oxford University Press, New York, 344 p

Lewis VR, Seybold SJ (2010) Wood-boring beetles in homes. University of California Statewide Integrated Pest Management Program, Agriculture and Natural Resources Pest Notes, Publication 7418, Oakland, June 2010. 4 p

Liebhold AM, Mastro V, Schaefer PW (1989) Learning from the legacy of Leopold Trouvelot. Bull Entomol Soc Am 35:20–21

Liebhold AM, Brockerhoff EG, Garrett LJ et al (2012) Live plant imports: the major pathway for forest insect and pathogen invasions of the U.S. Front Ecol Environ 10:135–143

Liebhold AM, McCullough DG, Blackburn LM et al (2013) A highly aggregated geographical distribution of forest pest invasions in the USA. Divers Distrib 19:1208–1216

Light SF (1936) A tropical termite in California. Pan-Pacific Entomol 12:125–126

Linsley EG (1963) The Cerambycidae of North America. Part IV. Taxonomy and classification of the subfamily Cerambycinae, tribes Elaphidionini through Rhinotragini, vol 21. University of California Publications in Entomology, University of California Press, Berkeley. 165 p

Linsley EG, Chemsak JA (1997) The Cerambycidae of North America, Part VIII: bibliography, index, and host plant index, vol. 117. University of California Publications in Entomology. University of California Press, Berkeley. 534 p

Linsley EG, MacLeod GF (1942) Ambrosia beetles attacking deciduous fruit trees in California. J Econ Entomol 35:601

Liu D-G, Bohne MJ, Lee JC et al (2007) New introduction in California: the redhaired pine bark beetle, *Hylurgus ligniperda* Fabricius. USDA Forest Service, Pest Alert, R5-PR-07. USDA Forest Service, State and Private Forestry, Vallejo. 3 p. http://entomology.ucdavis.edu/files/215005.pdf. or https://www.fs.fed.us/psw/publications/seybold/psw_2007_seybold001_liu.pdf

Long JW, Goode RW (2017) A mono harvest of California black oak acorns. J For 115:425

Long JW, Goode RW, Gutteriez RJ et al (2017) Managing California black oak for tribal ecocultural restoration. J For 115:426–434

Looney C, Humble LM, Cranshaw W (2012) *Orchestes alni* (L.) (Coleoptera: Curculionidae): new records from western North America with notes on parasitoids. Coleopt Bull 66:63–66

Looney C, Smith DR, Collman SJ et al (2016) Sawflies (Hymenoptera, Symphyta) newly recorded from Washington state. J Hymenopt Res 49:129–159

Lotts K, Naberhaus T (2017) Butterflies and moths of North America. https://www.butterfliesandmoths.org/species/Hyles-euphorbiae

Louda SM, Kendall D, Connor J, Simberloff D (1997) Ecological effects of an insect introduced for the biological control of weeds. Science 277:1088–1090

Lynch AM (2004) Fate and characteristics of *Picea* damaged by *Elatobium abietinum* (Walker) (Homoptera: Aphididae) in the White Mountains of Arizona. Western N Am Nat 64:7–17

Lynch AM (2014) XXIV Spruce aphid (*Elatobium abietinum* Walker) (Hemiptera: Aphididae). In: Van Driesche R, Reardon R (eds) The use of classical biological control to preserve forests in North America. FHTET-2013-2. U.S. Department of Agriculture, Forest Service, Forest Health Technology Enterprise Team, Morgantown, pp 259–270. https://www.fs.fed.us/rm/pubs_journals/2014/rmrs_2014_lynch_a001.pdf

Lynch SC, Twizeyimana M, Sal Mayoquin J et al (2016) Identification, pathogenicity and abundance of *Paracremonium pembeum* sp. nov. and *Graphium euwallaceae* sp. nov.—two newly discovered mycangial associates of the polyphagous shot hole borer (*Euwallacea* sp.) in California. Mycologia 108:313–329

Maloy OC (1997) White pine blister rust control in North America: a case history. Annu Rev Phytopathol 35:87–109

Marshall VM, Lewis MM, Ostendorf B (2012) Buffel grass (*Cenchrus ciliaris*) as an invader and threat to biodiversity in arid environments: a review. J Arid Environ 78:1–12

Mascheretti S, Croucher PJP, Vettraino A et al (2008) Reconstruction of the sudden oak death epidemic in California through microsatellite analysis of the pathogen *Phytophthora ramorum*. Mol Ecol 17:2755–2768

Mattson WJ, Lawrence RK, Haack RA et al (1988) Defensive strategies of woody plants against different insect-feeding guilds in relation to plant ecological strategies and intimacy of association with insects. In: Mattson WJ, Levieux J, Bernard-Dagan C (eds) Mechanisms of woody plant defenses against insects: search for patterns. Springer, New York, pp 3–38

Mattson WJ, Niemelä P, Millers I et al (1992) Immigrant phytophagous insects on woody plants in the United States and Canada: an annotated list. General Technical Report NC-169. USDA Forest Service, North Central Forest Experiment Station, St. Paul. 27 p

McCain AH, Koehler CS, Tjosvold SA (1987) Pitch canker threatens California pines. Calif Agric 41(11–12):22–23

McClure ML, Burdett CL, Farnsworth ML et al (2018) A globally-distributed alien invasive species poses risks to United States imperiled species. Sci Rep 8(1):5331

McCollum G, Baldwin E (2017) Huanglongbing: devastating disease of citrus. Hortic Rev 44:315–361

Mead FW, Fasulo TR (2017) Asian citrus psyllid. University of Florida Publication Number EENY-33. http://entnemdept.ufl.edu/creatures/citrus/acpsyllid.htm

Mifsud D, Knížek M (2009) The bark beetles (Coleoptera: Scolytidae) of the Maltese islands (Central Mediterranean). Bull Entomol Soc Malta 2:25–52

Miller DR, Davidson JA (2005) Armored scale insect pests of trees and shrubs. Cornell University Press, Ithaca, 442 p

Miller DR, Miller GL (1993) Eriococcidae of the eastern United States (Homoptera). Contrib Am Entomol Inst 27:1–91

Milosavljević I, Schall K, Hoddle CD et al (2017) Biocontrol program targets Asian citrus psyllid in California's urban areas. Calif Agric 71:169–177

Missouri Department of Conservation (2010) Old world bluestems. W00046. https://mdc.mo.gov/sites/default/files/downloads/OldWorldBluestems.pdf

Mizell RF (2012) Multicolored Asian lady beetle. University of Florida Publication Number EENY-204. http://entnemdept.ufl.edu/creatures/beneficial/multicolored_asian_lady_beetle.htm

Moran PJ, DeLoach CJ, Dudley TL, Sanabria J (2009) Open field host selection and behavior by tamarisk beetles (*Diorhabda* spp.) (Coleoptera: Chrysomelidae) in biological control of exotic saltcedars (*Tamarix* spp.) and risks to non-target athel (*T. aphylla*) and native *Frankenia* spp. Biol Control 50:243–261

Morrison B (1981) New Mexico's exotic wildlife program: its past, present, and future. Proc Western Assoc State Game Fish Commission 61:88–91

Moser JC, Fitzgibbon BA, Klepzig KD (2005) The Mexican pine beetle, *Dendroctonus mexicanus*: first record in the United States and co-occurrence with the southern pine beetle-*Dendroctonus frontalis* (Coleoptera: Scolytidae or Curculionidae: Scolytinae). Entomol News 116:235–243

Mulder PG, Harris MK, Grantham RA (2012) Biology and management of the pecan weevil (Coleoptera: Curculionidae). J Integr Pest Manag 3:1–9. https://doi.org/10.1603/IPM10027

Negrón JF, Witcosky JJ, Cain RJ et al (2005) The banded elm bark beetle: a new threat to elms in North America. Am Entomol 51:84–94

Nico L (2018) *Notemigonus crysoleucas* (Mitchill, 1814): U.S. Geological Survey, nonindigenous aquatic species database, Gainesville. https://nas.er.usgs.gov/queries/factsheet.aspx?SpeciesID=579

Nico L, Fuller P, Jacobs G et al (2018a) *Scardinius erythrophthalmus* (Linnaeus, 1758): U.S. Geological Survey, nonindigenous aquatic species database, Gainesville. https://nas.er.usgs.gov/queries/factsheet.aspx?SpeciesID=648

Nico L, Fuller P, Neilson M (2018b) *Acanthogobius flavimanus* (Temminck and Schlegel, 1845): U.S. Geological Survey, nonindigenous aquatic species database, Gainesville. https://nas.er.usgs.gov/queries/factsheet.aspx?SpeciesID=707

Nico L, Neilson M, Loftus B (2018c) *Clarias batrachus* (Linnaeus, 1758): U.S. Geological Survey, nonindigenous aquatic species database, Gainesville. https://nas.er.usgs.gov/queries/factsheet.aspx?SpeciesID=486

Niemelä P, Mattson WJ (1996) Invasion of North American forests by European phytophagous insects. Bioscience 46:741–753

Nonindigenous Aquatic Species (NAS) (2019) United States Geological Survey. Wetland and Aquatic Research Center. https://nas.er.usgs.gov/about/default.aspx

Novack A, Fuhrmann K, Dorman-Johnson K, Bartell S (2009) Research report: distribution and abundance of Barbary sheep and other ungulates in Carlsbad caverns National Park. Park Sci 26:92–98

Olson CA (1991) *Dinapate wrighti* Horn, the giant palm borer (Coleoptera: Bostrichidae), reported from Arizona infesting *Phoenix dactylifera* L. Coleopt Bull 45:272–273

Olsson AD, Betancourt JL, Crimmins MA, Marsh SE (2012) Constancy of local spread rates for buffelgrass (*Pennisetum ciliare* L.) in the Arizona upland of the Sonoran Desert. J Arid Environ 87:136–143

Paine TD (2016) Insects colonizing eucalypts in California. In: Paine TD, Lieutier F (eds) Insects and diseases of Mediterranean forest systems. Springer, Cham, pp 711–730. https://doi.org/10.1007/978-3-319-24744-1_24

Paine TD, Dreistadt SH, Millar JG (2009) Eucalyptus longhorned borers. University of California Statewide Integrated Pest Management Program, Oakland, California, Agriculture and Natural Resources Pest Notes, Publication 7425, Oakland, California, September 2009, 4 p. http://ipm.ucanr.edu/PMG/PESTNOTES/pn7425.html

Paxton EH, Theimer TC, Sogge MK (2011) Tamarisk biocontrol using tamarisk beetles: potential consequences for riparian birds in the southwestern United States. Condor 113:255–265

Peterson DL (2012) Southwest. In: Vose JM, Peterson DL, Patel-Weynand T (eds) Effects of climatic variability and change on forest ecosystems: a comprehensive science synthesis for the U.S. forest sector. General Technical Report PNW-GTR-870. United States Department of

Agriculture Forest Service Pacific Northwest Research Station, Portland, pp 219–222

Philips TK (2001) A record of *Micromalthus debilis* (Coleoptera: Micromalthidae) from Central America and a discussion of its distribution. Fla Entomol 84:159–160

Philips TK (2002) 70. Anobiidae. In: Arnett RH, Frank JH, Thomas MC, Skelley PE (eds) American beetles, volume II: Polyphaga: Scarabaeoidea through Curculionoidea. CRC Press LLC, Boca Raton, pp 245–260

Phillips G (2018) White satin moth (*Leucoma salicis*). Nevada Division of Forestry. http://forestry.nv.gov/forestry-resources/forest-health/white-satin-moth-leucoma-salicis/

Pilowsky JA, Starks PT (2018) Displacement and replacement in real time: *Polistes dominula's* impact on *P. fuscatus* in the northeastern US. Biol Invasions 20:1161–1169

Pinon J, Newcombe G, Chastagner GA (1994) Identification of races of *Melampsora larici-populina*, the Eurasian poplar leaf rust fungus, on *Populus* species in California and Washington. Plant Dis 78:101

Pogue MG, Schaefer PW (2007) A review of selected species of *Lymantria* Hübner [1819] including three new species (Lepidoptera: Noctuidae: Lymantriinae). Publication FHTET-2006-07. USDA Forest Service, Forest Health Technology Enterprise Team, Morgantown. 223 p

Rabaglia RJ, Cognato AI, Hoebeke ER et al (2019) Early detection and rapıd response. A ten-year summary of the USDA Forest Service program of surveillance for non-native bark and ambrosia beetles. Am Entomol 65:29–42

Rugman-Jones PF, Seybold SJ, Graves AD et al (2015) Phylogeography of the walnut twig beetle, *Pityophthorus juglandis*, the vector of thousand cankers disease in North American walnut trees. PLoS One 10(2):e0118264. https://doi.org/10.1371/journal.pone.0118264

Rust MK (2004) The effect of urbanization on the distribution of drywood termites of the southwest. In: Proceedings of the 2004 national conference of urban entomology. May 20–22, 2004, Phoenix, AZ, pp 27–31

Rust MK, Reierson DA, Paine EO et al (1998) Ravenous Formosan subterranean termites persist in California. Calif Agric 52:34–37

Sabalow R (2018) Destructive swamp rodents are knocking on the Delta's door. Is it time to panic? The Sacramento Bee, April 17, 2018. https://www.sacbee.com/latest-news/article209144049.html

Sabrosky C, Reardon RC (1976) Tachinid parasites of the gypsy moth, *Lymantria dispar*, with keys to adults and puparia. Misc Publ Entomol Soc Am 10(2):126 p

Sanders DM, Schuster AL, McCardle PW et al (2013) Ixodid ticks associated with feral swine in Texas. J Vector Ecol 38:361–373

Scheffrahn RH, Křeček J, Ripa R, Luppichini P (2009) Endemic origin and vast anthropogenic dispersal of the west Indian drywood termite. Biol Invasions 11:787–799

Scriven GT, Reeves EL, Luck RF (1986) Beetle from Australia threatens eucalyptus. Calif Agric July–August:4–6

Seybold SJ (2001) *Ernobius mollis* (L.) (Coleoptera: Anobiidae): an exotic beetle colonizes Monterey pine, *Pinus radiata* D. Don, in northern California. Pan Pac Entomol 77:51–54

Seybold SJ, Tupy JL (1993) *Ernobius mollis* (L.) (Coleoptera: Anobiidae) established in California. Pan Pac Entomol 69:36–40

Seybold SJ, Klingeman WE III, Hishinuma SM et al (2019) Status and impact of walnut twig beetle in urban forest, orchard, and native forest ecosystems. J For 117:152–163. https://doi.org/10.1093/jofore/fvy081

Seybold SJ, Penrose RL, Graves AD (2016) Invasive bark and ambrosia beetles in California Mediterranean forest ecosystems. In: Paine TD, Lieutier F (eds) Insects and diseases of Mediterranean forest systems. Springer, Cham, pp 583–662. https://doi.org/10.1007/978-3-319-24744-1_21

Sing SE, De Clerck-Floate RA, Hansen RW et al (2016) Biology and control of Dalmatian and yellow toadflax. USDA Forest Service, Forest Health Technology Enterprise Team, Morgantown. FHTET-2016-01. 141 p. https://www.fs.usda.gov/treesearch/pubs/52446

Smith T (2018) Personal correspondence, email letter dated July 17, 2018 to Susan Frankel. On file with Frankel at U.S. Department of Agriculture, Forest Service, Pacific Southwest Research Station, 800 Buchanan Street, Albany, CA 94710

Somma LA, Foster A, Fuller P (2019) *Trachemys scripta elegans* (Weid-Neuwied, 1838): U.S. Geological Survey, Nonindigenous Aquatic Species Database, Gainesville. https://nas.er.usgs.gov/queries/factsheet.aspx?SpeciesID=1261

Spilman TJ (1982) False powder-post beetles of the genus *Dinoderus* in North America (Coleoptera: Bostrichidae). Coleopt Bull 36:193–196

Strobel GA, Lanier GN (1981) Dutch elm disease. Sci Am 245:56–66

Sutherland C, Pierce JB, Lewis B, Heerema R (2017). Pecan weevil: wanted DEAD, not alive. College of Agricultural, Consumer and Environmental Sciences, Cooperative Extension Services Circular 683. Las Cruces. 8 p

Tamarisk Coalition (2017) Yearly distribution (2007–2017) of tamarisk beetle (*Diorhabda* spp.). Annual tamarisk beetle distribution map. http://www.tamariskcoalition.org/sites/default/files/files/2017_Beetle_Map_12_28.pdf

Tisserat N, Cranshaw W, Putnam M et al (2011) Thousand cankers disease is widespread on black walnut, *Juglans nigra*, in the western United States. Plant Health Progress. https://doi.org/10.1094/PHP-2011-0630-01-BR

Toševski I, Caldara R, Jović J et al (2011) Morphological, molecular and biological evidence reveal two cryptic species in *Mecinus janthinus* Germar (Coleoptera: Curculionidae), a successful biological control agent of Dalmatian toadflax, *Linaria dalmatica* (Lamiales, Plantaginaceae). Syst Entomol 36:741–753. https://doi.org/10.1111/j.1365-3113.2011.00593.x

Tsutsui ND, Suarez AV, Holway DA, Case TJ (2001) Relationships among native and introduced populations of the argentine ant (*Linepithema humile*) and the source of introduced populations. Mol Ecol 10:2151–2161

Umeda C, Eskalen A, Paine TD (2016) Polyphagous shot hole borer and Fusarium dieback in California. In: Paine TD, Lieutier F (eds) Insects and diseases of Mediterranean forest systems. Springer, Cham, pp 757–768. https://doi.org/10.1007/978-3-319-24744-1_26

USDA Agricultural Research Service (1986) Stored-grain insects. Agricultural Handbook No. 500. 57 p

USDA Animal Plant Health Inspection Service (APHIS) (2010) USDA APHIS PPQ moratorium for biological control of saltcedar (*Tamarix* species) using the biological control agent *Diorhabda* species (Coleoptera: Chrysomelidae). https://www.usbr.gov/uc/albuq/rm/CBPvegMgmt/saltcedar/pdfs/2010/BeetleMemoUSDA.pdf

USDA Forest Service, Northern Research Station and Forest Health Protection (2019) Alien Forest Pest Explorer—species map. Database last updated 28 July 2016. https://www.nrs.fs.fed.us/tools/afpe/

USDA Forest Service Southwestern Region (2017) Field guide for managing buffelgrass in the Southwest. Report No. TP-R3–16-03. USDA Forest Service Southwestern Region, Albuquerque. 12 p. https://www.fs.usda.gov/Internet/FSE_DOCUMENTS/fseprd563017.pdf

USDA National Resources Conservation Service (2018) Introduced, invasive, and noxious plants. Plants database. https://plants.usda.gov/java/noxiousDriver. 15 Jan 2018

USDA NRCS (2018) The PLANTS Database. National Plant Data Team, Greensboro, NC 27401–4901 USA. https://plants.usda.gov/core/profile?symbol=juma. 30 July 2018

Vitousek PM, Antonio CM, Loope LL, Westbrooks R (1996) Biological invasions as global environmental change. Am Sci 84:468–478

Westcott RL, Haines RD, Hishinuma SM et al (2015) *Chrysobothris analis* LeConte, 1860 (Coleoptera: Buprestidae) discovered in California with notes on *Chrysobothris wintu* Wellso & Manley, 2007. Pan-Pacific Entomol 91:337–341

Wheeler AG Jr, Hoebeke ER (2012) *Metopoplax ditomoides* (costa) (Hemiptera: Lygaeoidea: Oxycarenidae): first Canadian record of a Palearctic seed bug. J Entomol Soc Br Columbia. 109:70–71

White RE (1982) A catalog of the Coleoptera of America north of Mexico. Family: Anobiidae. USDA Agricultural Handbook No. 529–70. 59 p

White MD, Kauffman KM, Lewis JS, Miller RS (2018) Wild pigs breach farm fence through harvest time in southern San Joaquin Valley. Calif Agric 72:122–126

Wikler K, Storer AJ, Newman W et al (2003) The dynamics of an introduced pathogen in a native Monterey pine (*Pinus radiata*) forest. For Ecol Manag 179:209–221

Willden SA (2017) Seasonal development of the biological control agent of Dalmatian toadflax, *Mecinus janthiniformis* (Curculionidae: Coleoptera), in Utah: phenology, overwintering success, and mortality. MS thesis, Utah State University, Logan, Utah, 155 p

Williams AP, Allen CD, Millar CI et al (2010) Forest responses to increasing aridity and warmth in southwestern North America. Proc Natl Acad Sci U S A 107:21289–21294

Wingfield MJ, Hammerbacher A, Ganley RJ et al (2008) Pitch canker caused by *Fusarium circinatum*—a growing threat to pine plantations and forests worldwide. Australas Plant Pathol 37:319–334

Winston R, Hansen R, Schwarzländer M et al (2008) Biology and biological control of exotic true thistles. Publication FHTET-2007-05. USDA Forest Service, Forest Health Technology Enterprise Team, Morgantown. 130 p

Winston R, Randall CB, De Clerck-Floate R et al (2014a) Field guide for the biological control of weeds. University of Idaho Extension. Publication FHTET-2014-08. USDA Forest Service, Forest Health Technology Enterprise Team, Morgantown. 338 p. https://www.ibiocontrol.org/westernweeds.pdf

Winston RL, Schwarzländer M, Hinz HL et al (2014b) Biological control of weeds: a world catalogue of agents and their target weeds, 5th edn. Publication FHTET-2014-04. USDA Forest Service, Forest Health Technology Enterprise Team, Morgantown. 838 p. https://www.ibiocontrol.org/catalog/JulienCatalogueFHTET_2014_04.pdf

Wood SL, Bright DE Jr (1992) A catalog of Scolytidae and Platypodidae (Coleoptera), part 2: taxonomic index, vol 13, Great Basin Naturalist Memoirs. Brigham Young University, Provo. 1553 p

Zerillo MM, Caballero JI, Woeste K et al (2014) Population structure of *Geosmithia morbida*, the causal agent of thousand cankers disease of walnut trees in the United States. PLoS One 9:e112847. https://doi.org/10.1371/journal.pone.0112847

Great Plains Region

Matt Reeves, Jack L. Butler, Michele Schoeneberger, and John C. Kilgo

Introduction

The Great Plains, here encompassing the States of Kansas, Montana, Nebraska, North Dakota, South Dakota, Oklahoma, Texas, and Wyoming (Fig. A5.1), is a diverse landscape consisting of a complex matrix of native, seminative, and non-native grasslands intermixed with riparian and prairie woodlands, shrublands, forests, and intensively cultivated agricultural lands. The composition and abundance of the native vegetation is strongly correlated with a north-south temperature gradient and an east-west precipitation gradient. Increasing pressure for intensive urban, agricultural, and energy development coupled with climate change is threatening maintenance of goods and services in the region. Because of the widespread and complex juxtaposition of privately owned lands with intensive agricultural use intermixed with native vegetation on public lands, invasive plants pose a unique challenge to both private and public land managers. Climate change is likely to enhance pathways for invasive species (see Chap. 4) which increases the risk of some species becoming locally adapted under a changing climate and then dispersed into adjacent lands dominated by native vegetation. Within this context, this regional assessment includes ten invasive plant species (or collections of species), along with examples of invasive animal, pests, and pathogens. Each species, or group of species, was selected for this assessment if the species is not covered extensively in other sections relating to the Great Plains or the species is managerially and ecologically significant. Pests and pathogens are included despite coverage elsewhere in this report since they are germane to the evaluation of invasive species in the Great Plains, especially given the 2016 Technical Report by Bergdahl and Hill (2016).

As a result of these selection criteria, Russian olive (*Elaeagnus angustifolia*), non-native perennial grass assemblages (*Agropyron*, *Bromus*, and *Poa* spp.), buffelgrass (*Pennisetum ciliare*), absinth wormwood (*Artemisia absinthium*), Johnsongrass (*Sorghum halepense*), tumble mustard (*Sisymbrium altissimum*), whitetop (*Lepidium appelianum* Al-Shehbaz), and field (Japanese) brome (*Bromus arvensis*, synonym Japanese brome (*B. japonicus*)) were chosen as examples of problematic invasive species on the Great Plains. Animal species chosen for inclusion are wild horse and burros (*Equus* spp.) and feral pigs (*Sus scrofa*). Invasive pests of trees in the Great Plains included here are emerald ash borer (*Agrilus planipennis*), Balsam woolly adelgid (*Adelges piceae*), European gypsy moth (*Lymantria dispar dispar*), pine wilt (the nematode *Bursaphelenchus xylophilus* which spread via pine sawyer beetles (*Monochamus* spp.)), Dutch elm disease (the fungus *Ophiostoma* spp.), and thousand cankers disease (the fungus *Geosmithia morbida* spread by the walnut twig beetle (*Pityophthorus juglandis*)). Descriptions of pests and pathogens are excerpted from Bergdahl and Hill (2016).

Exotic Perennials

The strong correlation of a north-south temperature gradient and an east-west precipitation gradient with the composition and abundance of plants in the Great Plains means that patterns in the prevalence and distribution of exotic grass species will largely depend upon the photosynthetic pathway of the constituent species. Grasses possessing the C3 photosynthetic pathway (cool-season grasses) are more common and productive in the northern Great Plains, while grasses possessing the C4 photosynthetic pathway (warm-season grasses) are more abundant in the southern Great Plains and eastern tallgrass prairie (Epstein et al. 1997; Terri and Stowe 1976). Where native cool- and warm-season grasses co-occur, they vary in their spatial distribution at the local level with warm-season grasses occupying warmer, open sites, while cool-season grasses tend to occur in cooler, more shaded sites (Barnes et al. 1983; Teeri 1979).

Northern C3-dominated native plant communities of the Great Plains face the threat of invasion by introduced cool-season perennial grasses, particularly smooth brome (*Bromus inermis* Leyss. ssp. *inermis*), Kentucky bluegrass (*Poa pratensis* L.), and crested wheatgrass (*Agropyron cristatum* (L.) Gaertn.) (Christian and Wilson 1999; DeKeyser et al. 2013; Larson et al. 2001). Along those lines, a number of studies, primarily from researchers in Canada, document cases where these three species have escaped cultivation, invaded natural ecosystems, and adversely impacted native species diversity (Christian and Wilson 1999; Fink and Wilson 2011; Hansen 2007; Hansen and Wilson 2006; Henderson and Naeth 2005; Nernberg and Dale 1997; Otfinowski et al. 2007; Vaness and Wilson 2007). In fact, smooth brome was ranked as the eighth most serious invasive alien plant in Canada because of its impact on the abundance and diversity of native prairie species (Catling and Mitrow 2005). Although sparse, research in the United States attributes reductions in native plant diversity (Dillemuth et al. 2009; Frank and McNaugton 1992) and reduced habitat use by native ungulates (Trammell and Butler 1995) to smooth brome. Similar reductions in native plant species diversity have been reported for Kentucky bluegrass (Stohlgren et al. 1998) and crested wheatgrass (Fansler and Mangold 2011; Hulet et al. 2010). Large-scale conversions of native prairie to these exotic perennial grasses can be especially detrimental to prairie specialist butterflies (Swengel and Swengel 2015) and grassland songbirds (Ellis-Felege et al. 2013).

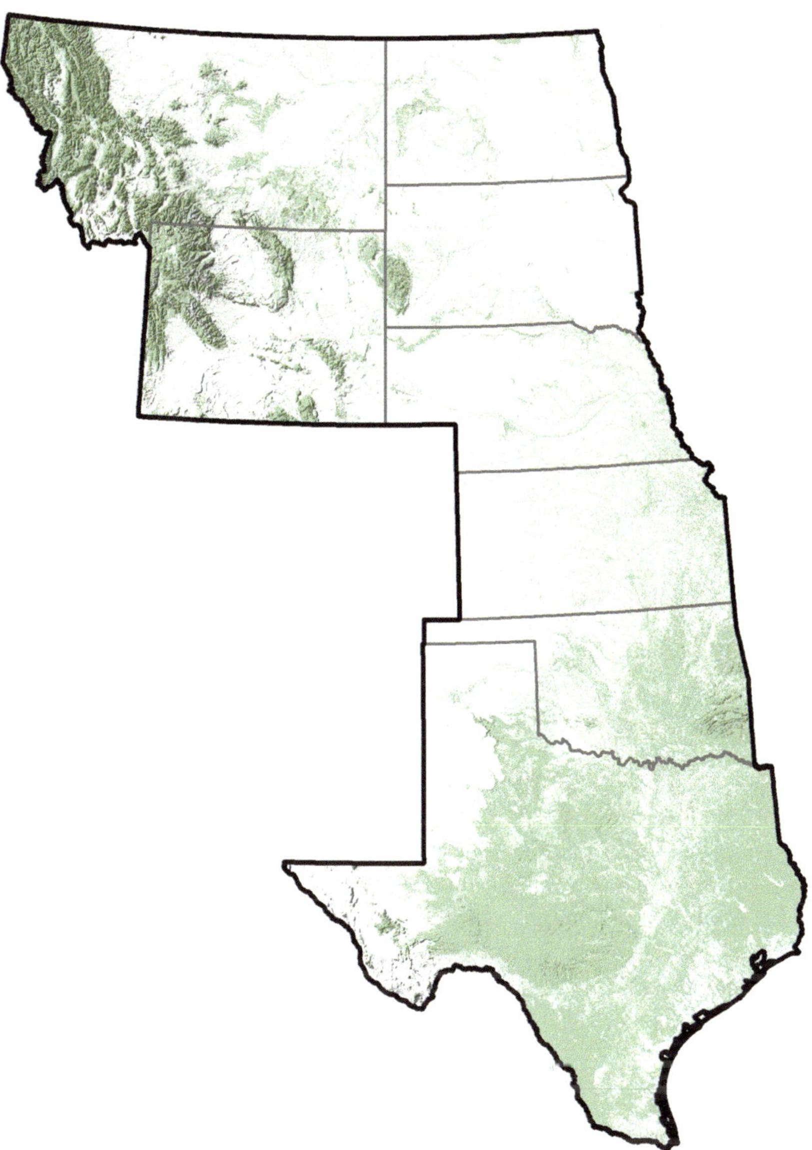

Fig. A5.1 The Great Plains region. (Figure courtesy of Daniel Ryerson and Andy Graves, USDA Forest Service Southwestern Region, Forest Health Protection)

The three cool-season species (smooth brome, crested wheatgrass, and Kentucky bluegrass) and the warm-season species (Johnsongrass) are generally not recognized as invasive, likely because of their perceived forage value. However, introduced exotic forage species are selected for traits that confer persistence under multiple stressors (drought, intensive defoliation, disease, etc.), and they likely often manifested in novel communities with a superior competitive advantage over native species, creating unique challenges in their management (Scasta et al. 2015). The collective evidence strongly indicates that these three non-native, perennial grasses have slowly and inexorably transformed relatively large tracts of non-forested ecosystems, and this transformation has largely gone unnoticed in the United States. The compositional balance of cool- and warm-season native and introduced grasses along the moisture and temperature gradient will undoubtedly be altered by climate change, likely in unknown ways.

Smooth Brome Smooth brome is native to Eurasia (Otfinowski et al. 2007) where it grows along roadsides, riverbanks, and borders of cultivated fields and in pastures (Kennedy 1899). Kennedy (1899) estimated that smooth brome was first introduced into the United States for pasture improvement in 1884 through the California Experiment Station. Initial seeding experiments showed that it was an aggressive rhizomatous species capable of rapidly displacing other plants (Kennedy 1899), a pattern confirmed by more recent experiments (Blankespoor and May 1996; Fink and Wilson 2011), including in areas where it is native (Liu et al. 2008). Once established, smooth brome is capable of higher production than adjacent native grasslands while reducing diversity through reductions in evenness (Fink and Wilson 2011; Otfinowski et al. 2007). Smooth brome readily outcompeted its native neighbors in northern mixed-grass prairie even under drought conditions (Nernberg and Dale 1997; Ulrich and Perkins 2014). Few efforts to control smooth brome have been completely effective (Bahm et al. 2011; Blankenspoor and Larson 1994; Bolwahn-Salesman and Thomsen 2011; Grilz and Romo 1995; Stacy et al. 2005; Willson and Stubbendieck 1996, 1997).

Crested Wheatgrass Crested wheatgrass is a cool-season bunchgrass native to a wide variety of grasslands in Central Europe, the Middle East, Central Asia, Siberia, China, and Mongolia where it is widely recognized as a valuable forage species (Rogler and Lorenz 1983). Crested wheatgrass is a complex of Eurasian species that were first introduced into North America (North Dakota) in 1898 (Dillman 1946). Crested wheatgrass has been widely planted throughout the northern Great Plains since the 1930s (Christian and Wilson 1999). It establishes quickly and is a successful competitor in many grassland ecosystems where it often outproduces and displaces native prairie species (Heidinga and Wilson 2002; Henderson and Naeth 2005). Grasslands dominated by crested wheatgrass contain few native species, especially forbs and grasses with growth forms similar to crested wheatgrass (Christian and Wilson 1999; Henderson and Naeth 2005). Christian and Wilson (1999) also reported that soils dominated by crested wheatgrass had less available nitrogen, total nitrogen, and less total carbon than soils under native prairie in Canada, potentially creating long-term ecosystem impacts. Crested wheatgrass is difficult to control primarily because of a large and persistent seed bank (Fansler and Mangold 2011; Hulet et al. 2010; Wilson and Pärtel 2003).

Kentucky Bluegrass Kentucky bluegrass is undoubtedly one of the most recognized and widespread *Poa* species in Europe, Asia, and North America (DeKeyser et al. 2015). Kentucky bluegrass is strongly rhizomatous, very productive, and highly palatable, making it a popular pasture grass in many ecosystems. Kentucky bluegrass was brought into North America by European traders, explorers, and missionaries in the mid- to late 1600s, largely because of its popularity as a forage grass (Schery 1965). Kentucky bluegrass has greatly expanded its range in North America over the last 100 years and is now a common species in many plant communities where is often considered an invasive species (Dekeyser et al. 2015; Toledo et al. 2014). Lower native species richness and declines in abundance of native warm-season grasses have been attributed to invasion by Kentucky bluegrass (Miles and Knops 2009). In a study on the National Wildlife Refuges in the Dakotas, Kentucky bluegrass accounted for 27–36% of the vegetation and was considered a contributing factor in the decline of the North American prairie (Grant et al. 2009). The current lack of regeneration of native green ash (*Fraxinus pennsylvanica*) woodlands in northern Great Plans grasslands has been attributed to the dense sod formed by the invasion of Kentucky bluegrass that greatly restricts establishment of green ash seedlings (Lesica 2009). The efficacy of using herbicides and fire to control Kentucky bluegrass and restore native species is generally highly variable (Bahm et al. 2011). When Kentucky bluegrass is successfully suppressed, the potential exists for the bare ground created by the reduction of Kentucky bluegrass to produce a secondary invasion by other exotic species (Adkins and Barnes 2013). At the same time, Kentucky bluegrass has also been classified as a major secondary invader following the successful suppression of leafy spurge (*Euphorbia esula*) using classical biological control (Butler and Wacker 2010).

Johnsongrass Johnsongrass (*Sorghum halepense*) is an exotic, perennial warm-season grass hybrid between *S. bicolor* (sorghum or millet) and *S. propinquum* (sorghum). Johnsongrass is a serious problem worldwide, especially in humid warm-temperate and subtropical regions (Follak and Essl 2012). Johnsongrass has long been recognized as aggressive invader of crop systems where heavy infestations can substantially reduce yields (Williams and Hayes 1984). No species of *Sorghum* are native to North America, and Johnsongrass is rapidly becoming a serious invader, adversely impacting the diversity of native prairies in the United States. The plant has several characteristics that are common to some of the most aggressive plant invaders, including a tall growth form and prolific seed production coupled with robust clonal growth through rhizomes. It also produces a defensive cyanogenic glycoside (dhurrin) (Abdul-Wahab and Rice 1967) and an allelopathic molecule (sorgoleone) that is exuded from root hairs (Czarnota et al. 2001). Collectively, these traits play a significant role in the ability of Johnsongrass to displace native species (Abdul-Wahab and Rice 1967; Follak and Essl 2012; Rout et al. 2013). Research is needed on the ecological impact of Johnsongrass in natural ecosystems and possible control strategies.

Buffelgrass Buffelgrass is native to India, Africa, and parts of Asia (Hauser 2008). It was introduced into Texas and Arizona in the 1930s and 1940s for soil stabilization and forage (Hauser 2008). Further establishment has occurred through seeds dispersed from Mexico. In Sonora, it is estimated that over 1000,000 ac of native desert and thornscrub vegetation have been converted to buffelgrass pasture (Burquez et al. 1998, 2002; Franklin et al. 2006). Within the Great Plains, buffelgrass occurs primarily in Texas, with outlying populations in Oklahoma (USDA, NRCS 2008). Within this limited distribution, buffelgrass occurs most often in desert thornscrub, mesquite-dominated shrublands, and cultivated buffelgrass pastures (Hamilton 1980; Hamilton and Scifres 1983; Mayeux and Hamilton 1983). Although the distribution in the Great Plains appears to be constrained by temperature, the ecological effects are significant. Buffelgrass alters plant communities and fire regimes and has been credited with creating "one of the most impressive ecosystem conversions happening in North America" (Nijhuis 2007) and is described as "one of the world's most notorious invaders" (Williams and Baruch 2000). The dramatic effects of buffelgrass on these communities are enhanced by a fire feedback cycle, since buffelgrass is a fire-adapted species (Burquez et al. 2002; Tellman 1997; Van Devender et al. 1997), enabling it to persist and spread following a fire. This is significant because the arid and warm sites that buffelgrass prefers often have extremely long fire intervals and support numerous succulent species that are not fire-adapted. Further, buffelgrass produces much greater fine fuel loads (often exceeding a threefold to fourfold increase in fine fuels (Esque et al. 2007) at the drier end of its invaded range) than native plants in these sites, thereby causing high mortality in native flora and fauna (Esque et al. 2007). Most of the effects of these fires driven by buffelgrass are documented from Sonoran Desert habitats located to the southwest of the Great Plains region. However, in the Chihuahuan Desert of western Texas, the endangered Chisos Mountains hedgehog cactus (*Echinocereus chisosensis*) is very vulnerable to mortality from increased fire frequency and effects from buffelgrass invasion (Hauser 2008). In addition, buffelgrass is competitive and invasive in southern mixed, shortgrass, and semi-desert grasslands of Texas and Oklahoma (Grace and Zouhar 2008; Rice et al. 2008). Since buffelgrass is a fire-adapted species, it is notably difficult to control using managed fire, but success may be enhanced with herbicide treatment or hand-pulling. Another factor increasing the difficulty of chemical control is that buffelgrass has been found to exhibit resistance to three of seven herbicides (Bovey et al. 1986), and older stands tend to tolerate herbicides better than small seedlings (Bovey et al. 1984). Recognizing the potential for very significant changes to southern ecosystems, the need for more research on control techniques has been noted (Hauser 2008).

Absinth Wormwood Absinth wormwood (*Artemisia absinthium*) is a coarse, erect herbaceous or semi-woody, clump-forming perennial that is native to parts of Europe and Asia (Maw et al. 1985; Selleck and Coupland 1961). Absinth wormwood was cultivated on a large scale in Europe for its reported hallucinogenic effects when consumed by humans (Maw et al. 1985) and its use as a folk remedy (see Makrini and Hassam 2016). Maw et al. (1985) further report that it was intentionally introduced into North America as a "medicinal and flavoring plant" in the early 1800s, but was banned in the United States in 1912. An online search in Web of Science using *Artemisia absinthium* in the title produced 170 articles with the vast majority of the papers reporting on the chemical compounds distilled or leached from the plant. Online information from Washington highlights the poisonous nature of the plant and cautions that no part of the plant should be consumed by humans or livestock (King County, WA 2017). Currently, absinth wormwood is naturalized in Canada and is listed as a noxious weed in only three States in the United States (Colorado, North Dakota, and Washington). The plant usually occurs in low densities (Selleck and Coupland 1961), but, because of its poisonous nature and its potential for expansion under climate change and a lack of research on its management, careful monitoring is needed.

Whitetop Three species of whitetop including globe-podded whitetop (*Cardaria pubescens*), lenspod whitetop (*C. chalepensis*), and heart-podded whitetop (*C. draba*) inhabit the Great Plains region. These species probably arrived in North America in the early 1900s (Zouhar 2004). Whitetop has an affinity for waste areas, roadsides, and degraded grasslands but is also attracted to moist environments such as irrigation ditches (Zouhar 2004).

Along the Bighorn River in Wyoming, globe-podded whitetop is often associated with Russian knapweed (*Rhaponticum repens*), Canada thistle (*Cirsium arvense*), and other non-native species co-occurring with saltcedar (*Tamarisk chinensis*), but it is generally rare in native shrublands (Zouhar 2004). Whitetop produces poor forage, crowds out desirable plants, and reduces animal diversity (USDA Forest Service 2014). Whitetop foliage contains glucosinolates, which are toxic to cattle and can impede germination and growth of other species (USDA Forest Service 2014).

Whitetop's extensive root system creates significant control difficulties, and it is a challenge to eradicate large populations once they are established (USDA Forest Service 2014). Treatment with herbicide can be effective, if it is performed during the correct life stage, but fire is not recommended as a solution for managing whitetop infestations. The extensive root system makes *Cardaria* spp. likely to survive even severe fire, but success has been noted using burners at close intervals (Rosenfels and Headley 1944). There have been significant economic and ecological effects of whitetop through reduced crop yields, cost of control, reductions in forage, and reduced quality of some agricultural products (Parsons and Cuthbertson 1992; Scurfield 1962). There is a lot of information available for whitetop control on croplands and heavily impacted lands (Chipping and Bossard 2000; McInnis et al. 2003; Sheley and Stivers 1999) but not for wildlands. Whitetop is considered a "moderate to serious" threat to native plant species in riparian and wetland settings and a "minor" threat in native grasslands (Zouhar 2004).

Russian Olive Russian olive (*Elaeagnus angustifolia*) is a tree or multi-stemmed shrub (5–12 m in height) that is native to Central and Western Asia. It is used worldwide as a nutritional agent or as a natural remedy for a range of illnesses. It was intentionally introduced in North America as a horticultural plant in the early 1900s, to be used for hedge rows and as a shade tree; by the 1940s, it was widely planted in windbreaks throughout the Great Plains (Katz and Shafroth 2003). Russian olive is currently found throughout the United States, where it has become the fourth most dominant woody plant in riparian areas through the Western United States (Friedman et al. 2005). Its rapid spread is potentially attributed to birds consuming the fruits (Edwards et al. 2014). The rapid dominance of Russian olive in riparian settings has generated considerable concern about its impact on natural communities and ecosystems (see review by Collette and Pither 2015). In their review, Collette and Pither (2015) described lower bird species richness and diversity in sites dominated by Russian olive, compared to noninfested sites. They presented evidence of enhanced nitrogen input into streams, likely related to the nitrogen-fixing ability of Russian olive, which could lead to eutrophication. At the same time, there is evidence that Russian olive provides a nesting habitat for the endangered southwestern willow flycatcher (*Empidonax traillii extimus*) and the threatened yellow-billed cuckoo (*Coccyzus americanus*), which creates potential conflicts in the management of the species. Because of its popularity as an ornamental, its ability to invade and dominate riparian areas, and the potential conflicts in its management (adversely affecting communities and ecosystems while providing habitat for endangered and threatened species), additional research is needed to address the ecological implications of the current and future range of Russian olive under climate change (Collette and Pither 2015; Katz and Shafroth 2003).

Exotic Annuals

Field Brome Although cheatgrass or downy brome (*Bromus tectorum*) is, without question, the most notorious and widely recognized annual exotic grass in North America, field brome has life history characteristics similar to cheatgrass (Baskin and Baskin 1981) and shows great potential for being just as invasive (Gasch et al. 2013; Haferkamp et al. 1997; Ogle et al. 2003). Field brome is a native Eurasian winter annual that has long been recognized as major weed of cropland systems worldwide (Sarani et al. 2016). While field brome is not as widely recognized as cheatgrass, it has greatly increased in abundance and distribution in Great Plains prairies (Haferkamp et al. 1997; Harmoney 2007; Ogle et al. 2003). This increase is often attributed to the removal of the interactive effects of fire and grazing, causing increases in litter, which favors the germination and establishment of field brome (Harmoney 2007; Whisenant 1990). It is difficult to assess the specific impacts of field brome on community and ecosystem properties because researchers sometimes lump cheatgrass and field brome together (Gasch et al. 2013; Ogle et al. 2003). Where the two species occur together, Gasch et al. (2013) have reported that annual brome-dominated sites had lower plant community diversity and carbon/nitrogen ratios, higher soil water infiltration rages, and altered soil microbial groups. Ogle et al. (2003) found that removal of both annual bromes resulted in more aboveground and belowground biomass at the end of the growing season. Studies specific to field brome found that, while removal of field brome increased production of associated perennial grasses, total production was reduced, at least for the duration of the study (Haferkam et al. 1997). Efforts to control field brome using fire and grazing suggest that while these treatments, used singly and in combination, may reduce field brome abundance, long-term control strategies are still elusive (Harmoney 2007; Whisenant 1990). Based on field studies conducted on cheatgrass (Blumenthal et al. 2016), long-term control of field brome under climate change may be difficult.

Tumble Mustard Tumble mustard probably came to North America in contaminated seed sources (Kostivkovsky and Young 2000) and is found throughout the continent. Westward expansion of the species was probably enhanced by inadvertent attachment to rail cars (Mitich 1983; Weber and Wittmann 1996). Though widespread, tumble mustard tends to occur most often on degraded sites with very low cover of native perennials and often co-occurs with other invasive annual species (Evans and Young 1970). In addition, tumble mustard is more common in rangeland and agricultural environments than in forested environments above the ponderosa pine (*Pinus ponderosa*) zone. Tumble mustard is a prolific seeder, and it is said that it can produce more than one million seeds per season (Clark and Fletcher 1923; Mitich 1983). Like other invasive annual species, tumble mustard germinates quickly after fire creating a fire-feedback cycle, thus enhancing its ability to regenerate. Tumble mustard is considered the second most invasive alien plant species in the Great Basin (Young and Evans 1972; Young et al. 1970), especially given its more effective seed dispersal mechanisms and earlier germination compared with native herbs (Allen and Knight 1984). With an affinity for degraded lands, tumble mustard is uncommon where there are high proportions of native perennial species and is an indicator of deteriorating land capability (Humphrey 1950). In addition, tumble mustard (and other annuals) can cause significant economic losses through reduction of forage for native and domestic ungulates (Pechanec and Stewart 1949). Other mustard species (*Brassicaceae* spp.) are frequently referenced as having unique tolerance to numerous herbicides, but, like many species, tumble mustard is most susceptible to herbicide application in the rosette stage.

Terrestrial Vertebrates

Historically, the grasslands of the Great Plains region have supported vast numbers of grazing ungulates, most notably the American bison (*Bison bison*). Perhaps it is not surprising then that the most significant animal invasive species affecting ecosystems of this region are ungulates, especially feral swine and feral horses and burros.

Feral swine (hereafter, pigs) in the region are largely restricted at present to the southern plains, particularly Texas and Oklahoma, where they are widespread, although their distribution is expanding continent-wide (Bevins et al. 2014; McClure et al. 2015). Pigs have been released or escaped continually since the arrival of the earliest European explorers in the sixteenth century (Mayer and Brisbin 1991). These animals are not simply grazers but are opportunistic generalists, feeding on plant material of all kinds (roots, stems, foliage, and seeds), fungi, invertebrates, reptiles, amphibians, small mammals, bird eggs, carrion, and refuse. Their rooting and wallowing have an impact on soil stability and chemistry, nutrient cycling, and microbe communities, as well as water quality. Plant community impacts include reduced species diversity, forb cover, leaf litter, and tree regeneration, as well as an increased prevalence of invasive plants (Timmons et al. 2012). In addition to direct impacts on native wildlife through depredation and habitat damage, pigs also compete with native wildlife for important foods (e.g., hard mast).

Feral horses in Western North America are descended from domestic horses of Eurasian and African origin, which were likewise introduced as early as the sixteenth century by European explorers. The number of horses apparently peaked around the mid-nineteenth century, declining thereafter; they were persecuted by grazing interests as competitors to cattle and sheep. Most wild horses and burros now occur on public lands administered by the Bureau of Land Management (BLM) or the USDA Forest Service, and they are protected and managed under provisions of the Wild and Free-Roaming Horses and Burros Act.

This protected status, however, has led to increased populations of wild horses and burros, at levels significantly above management objectives. To promote healthy conditions on the range, the BLM determines the Appropriate Management Level (AML), which is the number of wild horses and burros that can prosper in balance with other public land resources and uses. As of 2016, wild horses and burros exceed AML (which is 26,715) with an estimated population of 67,027, a 15% increase over the 2015 estimate (https://www.blm.gov/programs/wild-horse-and-burro/herd-management). This is consistent with the BLM's finding that wild horse and burro herds double in size about every 4 years. The disturbing trend in the growth of the herd size for these invasive equids has significant implications, given the reductions in rangeland health usually associated with their presence (Beever et al. 2008). Like most invasive species, the management or administration of feral horses and burros carries a significant economic burden; direct costs to the BLM alone topped $75 million in 2015. Depending on a variety of factors that include the abundance of horses in an area, overgrazing and trampling by equids can affect ecosystems through soil erosion and compaction, altered nutrient distribution, and altered plant species composition and abundance (Beever et al. 2008). These impacts, in turn, can affect the diversity and abundance of reptiles and mammals (Beever and Brussard 2004).

Invasive Pests of Trees in the Great Plains

Forests and "trees outside forests" (TOF) represent a relatively small portion of the land cover in the Great Plains. Nonetheless, they have long provided many ecosystem goods and services important to the well-being of humans living in this region (Droze 1977; McKay 1994). Agroforestry, a significant subset of TOFs throughout the Plains, has been used since the 1930s Dust Bowl days to protect soils, crops, livestock, and air and water quality. It is also used today to protect farmsteads, buildings, roads, and communities and to create habitats critical for wildlife, ranging from game species to pollinators (Schoeneberger et al. 2016). These "working trees" in the Great Plains are highly vulnerable to a number of factors (Joyce et al. 2018), including forest insect and disease pests (Bergdahl and Hill 2016; RMR FHP 2010). Exposure to environmental stresses, including the extreme shifts in temperature, moisture, and wind that are pervasive on the Plains, can exacerbate tree susceptibility to these pests (Ball 2016). Further, these severe and erratic weather-related events in the Plains are expected to increase in frequency and intensity in the coming years (Kunkel et al. 2013), further increasing tree vulnerability (Joyce et al. 2017).

While many of these forest pests are native to the region (RMR FHP 2010), there is a growing number of non-native pests threatening many of the key tree species occurring in this region. A few of the most potentially devastating invasive pests of Great Plains tree resources are presented in Table A5.1.

The potential ecological and economic losses related to non-native invasive tree pests have been estimated to be substantial (Lovett et al. 2016; Moser et al. 2009). Lovett et al. (2016) have noted that "non-native forest pests are the only disturbance agent that has effectively eliminated entire

Table A5.1 An overview of key invasive species in the Great Plains

Invasive pest of Great Plains trees	Overview	Current (2016) occurrence in the Great Plains[1/]								
		KS	MT	NE	ND	OK	SD	TX	WY	General comments
Emerald ash borer (EAB)	Highly invasive, exotic insect (*Agrilus planipennis* Fairmaire) introduced from China that attacks and kills all species of North American ash trees	**D**[2/] 2012	**N**	**D** 2016	**N**	**D** 2016	**N**	**D** 2016	**N**	Active monitoring/detection efforts occurring in most of **N** States.
Balsam woolly adelgid	Non-native, invasive insect (*Adelges piceae* Ratzeburg) impacting subalpine and grand fir	**ni**	**D** 2007	**ni**	**N**	**ni**	**ni**	**ni**	**ni**	Important in western Plains States, especially MT.
European gypsy moth (EGM)	*Lymantria dispar dispar* L.	**D***	**D***	**N**	**N**	**N**	**N**	**N**	**N**	Not yet established in any of the Plains States, all of which are suitable habitat for year-long survival of EGM
Pine wilt	Caused by the plant parasitic pinewood nematode (*Bursaphelenchus xylophilus*) via pine sawyer beetles in the genus *Monochamus*	**D** 1979	**N**	**D**	**N**	**D**	**D**	**N**	**N**	Mainly a threat to Scots, Austrian, and other non-native pines used extensively throughout the Plains
Dutch elm disease	Non-native, invasive wilt of elm species caused by the species of the fungus *Ophiostoma*, the most aggressive being *O. novo-ulmi*	**D**	**D**	**D**	**D** 1969	**D**	**D**	**D**	**D**	
Thousand cankers disease (TCD)	Invasive canker disease of black walnut caused by the walnut twig beetle (*Pityophthorus juglandis*) and its fungal associate *Geosmithia morbida*	**N**	**N**	**N**	**N**	**N**	**N**	**N**	**N**	Detected in every State (ID, UT, CO, and NM) bordering the western edges of the Great Plains States

[a]Based on information in the forest health reports up to 2016 from each of the Great Plains States (Montana [MT], Nebraska [NE], North Dakota [ND], Oklahoma [OK], South Dakota [SD], Texas [TX], and Wyoming [WY]) (RMR FHP 2010; Bergdahl and Hill 2016)
[b]**D = detected** (reported date), **N = not yet detected**, **ni = information not found**

tree species or genera from United State forests within decades." A good example of the level and cost of such an invasive tree pest is Dutch elm disease. This disease was responsible for the death and removal of most of the native elms (*Ulmus* spp.) throughout the United States over the past century and is still a disease of concern in the Great Plains (Dunnell and Bergdahl 2016). Thousand cankers disease, which causes widespread mortality of black walnut (*Juglans nigra* L.) and which was only noted in Colorado in 2001, is a major threat to the highly valued black walnut tree throughout the Great Plains (Tisserat and Cranshaw 2016).

Invasive forest pests are a particular concern in the Great Plains because the tree resources suitable to the environment in this region are limited; this greatly reduces tree diversity and, therefore, resilience to such attacks (Bergdahl and Hill 2016). Several of the main tree species long promoted and used in agricultural and community plantings have been removed from recommended planting lists in recent years either due to the high levels of mortality already occurring in the Plains (e.g., Scots (*Pinus sylvestris* L.) and Austrian (*Pinus nigra* L.) pines from pine wilt) or to the high levels of mortality being predicted to occur in the near future (e.g., ash (*Fraxinus* spp.) from the emerald ash borer and black walnut with thousand cankers disease).

Ash is one of the most prevalent species throughout the Plains. It is a significant component of riparian corridors, windbreaks, and community plantings (Rasmussen 2009). In 2008–2009, over four million ash trees were identified in urban settings with an additional 80 million identified in the rural areas just in the four northern Plains States (Schoeneberger et al. 2016). The emerald ash borer (EAB), a highly destructive pest of all North American ash trees, has already been detected in some easternmost areas of Kansas, Nebraska, Oklahoma, and Texas (USDA APHIS 2016a) and is expected to reach most of the Plains States within the next few years. The cost of treatment, removal, and replacement in response to EAB in the Plains could exceed $1 billion per State, along with the additional economic impacts from the loss of ecosystem services important to soil, water, and wildlife resources (Rasmussen 2009).

To better prepare and manage for the EAB and other invasive tree pests in the northern Plains States, State forestry agencies in Kansas, Nebraska, North Dakota, and South Dakota established the Great Plains Tree and Forest Invasives Initiative (GPI) in 2007 (Rasmussen 2009). This effort encompassed a comprehensive assessment of urban and agricultural tree resources across the four States, outreach and monitoring and detection programs, identification of marketing and utilization opportunities, and development of

State and regional planning strategies for EAB readiness and other invasive pests. Results from EAB parasitoid releases in the Northcentral region of the United States (Duan et al. 2017) indicate biocontrol may be a promising option for reducing EAB populations in the Great Plains. Further work is required to determine its success under the more extreme weather conditions and more fragmented ash occurrences experienced in this region.

Many other invasive insect and microbial pests of trees have the potential for significantly impacting Great Plains tree resources in the future. State forest plant health reports in the Plains also include the Asian longhorned beetle (*Anoplophora glabripennis* Motschulsky), which colonizes a wide range of hardwood hosts, and sirex woodwasp (*Sirex noctilio* F.), which has the potential to cause significant mortality of pines. Both are currently established within the United States. In addition, the Asian gypsy moth (*Lymantria dispar asiatica* Vinuskovkij), while not yet established in the United States, represents a major threat to all US tree resources because it can feed on over 100 botanical families (USDA APHIS 2016b). The realities of tree pest invasion in the Plains require more efforts like the GPI to be in place to manage the sustainability of Great Plains tree resources and the ecosystem services important to that region (Schoeneberger et al. 2016).

Literature Cited

Abdul-Wahab A, Rice EL (1967) Plant inhibition by Johnson grass and its possible significance in old-field succession. Bull Torrey Bot Club 94:486–497

Adkins JK, Barnes TG (2013) Herbicide treatment and timing for controlling Kentucky bluegrass (*Poa pratensis*) and tall fescue (*Festuca arundinacea*) in cool season grasslands of Central Kentucky, USA. Nat Areas J 33(1):31–39

Allen EB, Knight DH (1984) The effects of introduced annuals on secondary succession in sagebrush-grassland, Wyoming. Southwest Nat 29(4):407–421

Bahm MA, Barnes TG, Jensen KC (2011) Herbicide and fire effects on smooth brome (*Bromus inermis*) and Kentucky bluegrass (*Poa pratensis*) in invaded prairie remnants. Invasive Plant Sci Manag 4:189–197

Ball J (2016) Chapter 1: Environmental stresses. In: Bergdahl AD, Hill A (tech. coords.) Diseases of trees in the Great Plains. General Technical Report RMRS-GTR-335. U.S. Department of Agriculture, Forest Service, Rocky Mountain Research Station, Fort Collins, pp 4–11

Barnes PW, Tieszen LL, Ode DJ (1983) Distribution, production, and diversity of C3-and C4-dominated communities in a mixed prairie. Can J Bot 61:741–751

Baskin JM, Baskin CC (1981) Photosynthetic pathways indicated by leaf anatomy in 14 summer annuals glades. Photosynthetica 15(2):205–209

Beever EA, Brussard PF (2004) Community- and landscape-level responses of reptiles and small mammals to feral-horse grazing in the Great Basin. J Arid Environ 59:271–297

Beever EA, Tausch RJ, Thogmartin WE (2008) Multi-scale responses of vegetation to removal of horse grazing from Great Basin (USA) mountain ranges. Plant Ecol 196:163–184

Bevins SN, Pedersen K, Lutman et al (2014) Consequences associated with the recent range expansion of nonnative feral swine. Bioscience 64:291–299

Bergdahl AD, Hill A (tech. coords) (2016) Diseases of trees in the Great Plains. General Technical Report RMRS-GTR-335. U.S. Department of Agriculture, Forest Service, Rocky Mountain Research Station, Fort Collins. 229 p

Blankespoor GW, May JM (1996) Alien smooth brome (*Bromus inermis* Leyss.) in a tallgrass prairie remnant: seed bank, seedling establishment, and growth dynamics. Nat Areas J 16:289–294

Blankespoor GW, Larson EA (1994) Response of smooth brome (*Bromus inermis* Leyss.) to burning under varying soil moisture conditions. Am Midl Nat 131:261–272

Blumenthal DM, Kray JA, Ortmans W et al (2016) Cheatgrass is favored by warming but not CO_2 enrichment in a semi-arid grassland. Glob Chang Biol 22(9):3026–3038. https://doi.org/10.1111/gcb.13278

Bolwahn-Salesman J, Thomsen M (2011) Smooth brome (*Bromus inermis*) in tallgrass prairies: a review of control methods and future research directions. Ecol Restor 29:374–381

Bovey RW, Hein H Jr, Meyer RE (1984) Effect of herbicides on the production of common buffelgrass (*Cenchrus ciliaris*). Weed Sci 32(1):8–12

Bovey BW, Meyer RE, Merkle MG, Bashaw EC (1986) Effect of herbicides and handweeding on establishment of kleingrass and buffelgrass. J Range Manag 39(6):547–551

Burquez A, Martinez-Yrzar A, Miller M et al (1998) Mexican grasslands and the changing aridlands of Mexico: an overview and a case study in northwestern Mexico. In: Tellman B, Finch DM, Edminster C, Hamre R (eds) The future of arid grasslands: identifying issues, seeking solutions: Proceedings, 1996 October 9–13, Tucson, AZ. Proceedings RMRS-P-3. U.S. Department of Agriculture, Forest Service, Rocky Mountain Research Station, Fort Collins, pp 21–32

Burquez-Montijo A, Miller ME, Martinez-Yrizar A (2002) Mexican grasslands, thornscrub, and the transformation of the Sonoran Desert by invasive exotic buffelgrass (*Pennisetum ciliare*). In: Tellman, Barbara (ed) Invasive exotic species in the Sonoran region. Arizona-Sonora Desert Museum Studies in Natural History. The University of Arizona Press, The Arizona-Sonora Desert Museum, Tucson, pp 126–146

Butler JL, Wacker SD (2010) Lack of native vegetation recovery following biological control of leafy spurge. Rangel Ecol Manag 63:553–563

Catling PM, Mitrow G (2005) A prioritized list of invasive alien plants in natural habitats in Canada. Can Bot Assoc Bull 38:55–57

Chipping D, Bossard C (2000) *Cardaria chalepensis* (L.) hand-Mazz. And C. draba. In: Bossard CC, Randall JM, Hoshovsky MC (eds) Invasive plants of California's wildlands. University of California Press, Berkeley, pp 80–86

Christian JM, Wilson SD (1999) Long-term ecosystem impacts of an introduced grass in the northern Great Plains. Ecology 80:2397–2407

Clark GH, Fletcher J (1923) Farm weeds of Canada, 2nd edn. Canada Department of Agriculture, Ottawa, 192 p

Collette LKD, Pither J (2015) Russian-olive (*Elaeagnus angustifolia* L.) biology and ecology and its potential to invade North American riparian ecosystems. Invasive Plant Sci Manag 8:1–14

Czarnota MA, Paul RN, Dayan FE et al (2001) Mode of action, localization of production, chemical nature, and activity of sorgoleone: a potent PSII inhibitor in Sorgum spp. root exudates. Weed Technol 15:813–825

DeKeyser ES, Meehan M, Clambey G et al (2013) Cool season invasive grasses in northern Great Plains natural areas. Nat Areas J 33:81–90

DeKeyser ES, Dennhardt LA, Hendrickson J (2015) Kentucky bluegrass (*Poa pratensis*) invasion in the northern Great Plains: a story of rapid dominance in an endangered ecosystem. Invasive Plant Sci Manag 8:255–261

Dillemuth FP, Rietschier EA, Cronin JT (2009) Patch dynamics of a native grass in relation to the spread of invasive smooth brome (*Bromus inermis*). Biol Invasions 11:1381–1391

Dillman AC (1946) The beginnings of crested wheatgrass in North America. J Am Soc Agron 38:237–250

Droze WH (1977) Trees, prairies, and people: a history of tree planting in the plains states. Texas Woman's University Press, Denton, 313 p

Duan JJ, Bauer LS, Van Driesche RG (2017) Emerald ash borer biocontrol in ash saplings: the potential for early stage recovery of North American ash trees. For Ecol Manag 394:64–72

Dunnell KL, Bergdahl AD (2016) Chapter 34: Dutch elm disease. In: Bergdahl AD, Hill A (tech. coords.) Diseases of trees in the Great Plains. General Technical Report RMRS-GTR-335. U.S. Department of Agriculture, Forest Service, Rocky Mountain Research Station, Fort Collins, pp 126–128

Edwards RJ, Clark LC, Beck KG (2014) Russian olive (*Elaeagnus angustifolia*) dispersal by European starlings (*Sturnus vulgaris*). Invasive Plant Sci Manag 7:425–431

Ellis-Felege SN, Dixon CS, Wilson SD (2013) Impacts and management of invasive cool-season grasses in the northern Great Plains: challenges and opportunities for wildlife. Wildl Soc Bull 37:510–516

Epstein HE, Lauenroth WK, Burke IC, Coffin DP (1997) Productivity patterns of C3 and C4 functional types in the US Great Plains. Ecology 78(3):722–731

Esque TC, Schwalbe C, Lissow JA et al (2007) Buffelgrass fuel loads in saguaro National Park, Arizona, increase fire danger and threaten native species. Park Sci 24(2):33–37, 56

Evans RA, Young JA (1970) Plant litter and establishment of alien annual weed species in rangeland communities. Weed Sci 18(6):697–703

Fansler VA, Mangold JM (2011) Restoring native plants to crested wheatgrass stands. Restor Ecol 19:16–23

Farzaei MH, Bahramsoltani R, Abbasabadi Z et al (2015) A comprehensive review on phytochemical and pharmacological aspects of *Elaeagnus angustifolia* L. J Pharm Pharmacol 67:1467–1480

Fink KA, Wilson SD (2011) Bromus inermis invasion of a native grassland: diversity and resource reduction. Botany 89:157–164

Follak S, Essl F (2012) Spread dynamics and agricultural impact of *Sorghum halepense*, an emerging invasive species in Central Europe. Weed Res 53:53–60

Frank DA, McNaughton SJ (1992) The ecology of plants, large mammalian herbivores, and drought in Yellowstone National Park. Ecology 73:2043–2058

Franklin KA, Lyons K, Nagler PL et al (2006) Buffelgrass (*Pennisetum ciliare*) land conversion and productivity in the plains of Sonora, Mexico. Biol Conserv 127(1):62–71

Friedman JM, Auble GT, Shafroth PB et al (2005) Dominance of non-native riparian trees in western USA. Biol Invasions 7:747–751

Gasch CK, Enloe SF, Stahl PD, Williams SE (2013) An aboveground–belowground assessment of ecosystem properties associated with exotic annual brome invasion. Biol Fertil Soils 49(7):919–928

Grant TA, Flanders-Wanner B, Shaffer TL et al (2009) An emerging crisis across northern prairie refuges: prevalence of invasive plants and a plan for adaptive management. Ecol Restor 27:58–65

Grilz PL, Romo JT (1995) Management considerations for controlling smooth brome on fescue prairie. Nat Areas J 15:148–156

Grace JB, Zouhar K (2008) Fire and nonnative invasive plants in the Central bioregion. In: Zouhar K, Smith JK, Sutherland S, Brooks, ML (eds) Wildland fire in ecosystems: fire and nonnative invasive plants. General Technical Report. RMRS-GTR-42-vol. 6. U.S. Department of Agriculture, Forest Service, Rocky Mountain Research Station, Ogden, pp 113–141

Haferkamp M, Heitschmidt K, Karl M (1997) Influence of Japanese brome on western wheatgrass yield. J Range Manag 50:44–50

Hamilton WT (1980) Prescribed burning of improved pastures. In: Hanselka CW (ed) Prescribed range burning in the coastal prairie and eastern Rio Grande Plains of Texas: proceedings of a symposium, 1980 October 16, Kingsville, TX. The Texas A&M University System, Texas Agricultural Extension Service, College Station, pp 114–128

Hamilton WT, Scifres CJ (1983) Buffelgrass (*Cenchrus ciliaris*) responses to tebuthiuron. Weed Sci 31(5):634–638

Hansen MJ (2007) Evaluating management strategies and recovery of an invasive grass (I) using matrix population models. Biol Conserv 140:91–99

Hansen MJ, Wilson SD (2006) Is management of an invasive grass *Agropyron cristatum* contingent on environmental variation? J Appl Ecol 43:269–280

Harmoney KR (2007) Grazing and burning Japanese brome (*Bromus japonicus*) on mixed grass rangelands. Rangel Ecol Manag 60(5):479–486

Hauser AS (2008) *Pennisetum ciliare*. In: Fire effects information system, [Online]. U.S. Department of Agriculture, Forest Service, Rocky Mountain Research Station, Fire Sciences Laboratory (Producer). Available: https://www.feis-crs.org/feis/

Heidinga L, Wilson SD (2002) The impact of an invading alien grass *Agropyron cristatum* on species turnover in native prairie. Divers Distrib 8:249–258

Henderson DC, Naeth MA (2005) Multi-scale impacts of crested wheatgrass invasion in mixed-grass prairie. Biol Invasions 7:639–650

Howard JL (2003) *Sisymbrium altissimum*. In: Fire effects information system, [Online]. U.S. Department of Agriculture, Forest Service, Rocky Mountain Research Station, Fire Sciences Laboratory (Producer). Available: https://www.feis-crs.org/feis/

Hulet A, Roundy BA, Jessop B (2010) Crested wheatgrass control in native plant establishment in Utah. Rangel Ecol Manag 63:450–460

Humphrey RR (1950) Arizona range resources. II. Yavapai County. Bulletin no. 229. University of Arizona, Agricultural Experiment Station, Tucson, 55 p

Joyce LA, Bentrup G, Cheng AS et al (2018) Native and agricultural forests at risk to a changing climate in the Northern Plains. Clim Chang 146:59–74

Katz GL, Shafroth P (2003) Biology, ecology, and management of *Elaeagnus angustifolia* L. (Russian olive) in western North America. Wetlands 4:763–777

Kennedy PB (1899) Smooth bromegrass (*Bromus inermis*). U.S. Department of Agriculture Division of Agrostology. Circular 18

King County WA (2017) Absinth wormwood identification and control. http://www.kingcounty.gov/environment/animals-and-plants/noxious-weeds/weed-identification/absinth-wormwood.aspx

Kostivkovsky V, Young JA (2000) Invasive exotic rangeland weeds: a glimpse at some of their native habitats. Rangelands 22(6):3–6

Kunkel KE, Stevens LE, Stevens SE et al (2013) Regional climate trends and scenarios for the U.S. national climate assessment: part 4. Climate of the U.S. Great Plains. NOAA Technical Report NESDIS 142–4. National Oceanic and Atmospheric Administration, National Environmental Satellite, Data, and Information Service, Washington, DC. 91 p

Larson DL, Anderson PJ, Newton W (2001) Alien plant invasion in mixed-grass prairie: effects of vegetation type and anthropogenic disturbance. Ecol Appl 11:128–141

Lesica P (2009) Can regeneration of green ash (*Fraxinus pennsylvanica*) be restored in declining woodlands in eastern Montana? Rangel Ecol Manag 62:564–571

Liu G, Mao P, Wang Y et al (2008) Effects of adult neighbour and gap size on seedling emergence and early growth of *Bromus inermis* Leyss. Ecol Restor 23:197–205

Lovett GM, Weiss M, Liebhold AM et al (2016) Nonnative forest insects and pathogens in the United States: impacts and policy options. Ecol Appl 0(0):1–19

Makrini NI, Hassam B (2016) *Artemisia absinthium*: burning plant! Pan Afr Med J 23:10

Maw MG, Thomas AG, Stahevitch A (1985) The biology of Canadian weeds. 66 *Artemisia absinthium* L. Can J Plant Sci 65:389–400

Mayer JJ, Brisbin IL (1991) Wild pigs of the United States: their history, morphology, and current status. University of Georgia Press, Athens, 313 pp

Mayeux HS Jr, Hamilton WT (1983) Response of common goldenweed (*Isocoma coronopifolia*) and buffelgrass (*Cenchrus ciliaris*) to fire and soil-applied herbicides. Weed Sci 31(3):355–360

McClure ML, Burdett CL, Farnsworth ML et al (2015) Modeling and mapping the probability of occurrence of invasive wild pigs across the contiguous United States. PLoS One 10(8):e0133771

McInnis ML, Kiemnec GL, Larson LL et al (2003) Heart-podded hoary cress. Rangelands 25(2):18–23

McKay KL (1994) Trails of the past: historical overview of the Flathead National Forest, Montana, 1800–1960. Available at: https://foresthistory.org/wp-content/uploads/2017/01/TRAILS-OF-THE-PAST.pdf

Miles EK, Knops JM (2009) Shifting dominance from native C4 to non-native C3 grasses: relationships to community diversity. Oikos 118:1844–1853

Mitich LW (1983) The intriguing world of weeds. Part XV: Jim Hill mustard. Weeds Today 14(4):5–6

Moser WK, Barnard EL, Billings RF et al (2009) Impacts of nonnative invasive species on US forests and recommendations for policy and management. J For 107(6):320–327

Nernberg D, Dale MRT (1997) Competition of five native prairie grasses with *Bromus inermis* under three moisture regimes. Can J Bot 75:2140–2145

Nijhuis M (2007) Bonfire of the superweeds, [Online]. In: High Country News 39(15). Available: https://www.hcn.org/issues/352

Ogle SM, Reiners WA, Gerow KG (2003) Impacts of exotic annual brome grasses (*Bromus* spp.) on ecosystem properties of northern mixed grass prairie. Am Midl Nat 149(1):46–59

Otfinowski R, Kenkel NC, Catling PM (2007) The biology of Canadian weeds. 134. *Bromus inermis* Leysss. Can J Plant Sci 87:183–198

Parsons WT, Cuthbertson EG (1992) Noxious weeds of Australia. Indata Press, Melbourne, 692 p

Pechanec JF, Stewart G (1949) Grazing spring-fall sheep ranges of southern Idaho. Circular no. 808. U.S. Department of Agriculture, Washington, DC, 34 p

Rasmussen S (2009) Great Plains tree and forest invasives initiative: a multi-State cooperative effort for education, mitigation, and utilization. Unpublished document developed by Kansas Forest Service, Nebraska Forest Service, North Dakota Forest Service, South Dakota Division of Resource Conservation and Forestry, and U.S. Department of Agriculture, Forest Service

Rice PM, McPherson GR, Rew LJ (2008) Fire and nonnative invasive plants in the interior west bioregion. In: Zouhar K, Smith JK, Sutherland S, Brooks ML (eds) Wildland fire in ecosystems: fire and nonnative invasive plants. General technical report RMRS-GTR-42-vol. 6. U.S. Department of Agriculture, Forest Service, Rocky Mountain Research Station, Ogden, pp 141–173

Rocky Mountain Region, Forest Health Protection (RMR FHP) (2010) Field guide to diseases & insects of the Rocky Mountain Region. General Technical Report RMRS-GTR-241. U.S. Department of Agriculture, Forest Service, Rocky Mountain Research Station, Fort Collins. 336 p

Rogler GA, Lorenz RJ (1983) Crested wheatgrass—early history in the United States. J Range Manag 36:91–93

Rosenfels RS, Headley FB (1944) Whitetop eradication. Bulletin No. 170. University of Nevada, Agricultural Experiment Station, Reno. 18 p

Rout ME, Chrzanowski TH, Smith WK et al (2013) Ecological impacts of the invasive grass *Sorghum halepense* on native tallgrass prairie. Biol Invasions 15:327–339

Sarani M, Oveisi M, Rahimian Mashhadi H, Alizadeh H (2016) Modelling interactive effects of herbicide application timing and dose for the control of *Bromus japonicus* in wheat in an arid environment. Weed Res 56(1):50–58

Scasta JD, Engle DM, Fuhlendorf SD et al (2015) Meta-analysis of exotic forages as invasive plants in complex multi-functioning landscapes. Invasive Plant Sci Manag 8:292–306

Schery RW (1965) The migration of a plant. Kentucky bluegrass followed settlers of new world. Nat Hist 74:41–45

Schoeneberger MM, Lister A, Rasmussen S (2016) Chapter 3. Agricultural forests. In: Robertson G, Mason A (eds) Assessing the sustainability of agricultural and urban forests in the United States. FS-1067. USDA-FS, Washington, DC, pp 21–35

Scurfield G (1962) *Cardaria draba* (L.) Desv. J Ecol 50:489–499

Selleck GW, Coupland RT (1961) Studies in the life history of *Artemisia absinthium*. Weeds 9:485–900

Sheley R, Stivers J (1999) Whitetop. In: Sheley RL, Petroff JK (eds) Biology and management of noxious rangeland weeds. Oregon State University Press, Corvallis, pp 401–407

Stacy MD, Perryman BL, Stahl PD et al (2005) Brome control and microbial inoculation effects in reclaimed cool-season grasslands. Rangel Ecol Manag 58:161–166

Stohlegren T, Bull K, Otsuki Y et al (1998) Riparian zones as havens for exotic plant species in the central grasslands. Plant Ecol 138:113–125

Swengel AB, Swengel SR (2015) Grass-skipper (Hesperiinae) trends in Midwestern USA grasslands during 1988–2013. J Insect Conserv 19:279–292

Tellman B (1997) Exotic pest plant introduction in the American southwest. Desert Plants 13(1):3–10

Teeri JA (1979) The climatology of the C4 photosynthetic pathway. In: Solbrig OT, Jain S, Johnson GB, Raven PH (eds) Topics in plant population biology. Columbia University Press, New York, pp 356–374

Terri JA, Stowe LG (1976) Climatic patterns and the distribution of C_4 grasses in North America. Oecologia 23:1–12

Timmons JB, Alldredge B, Rodgers WE, Cathey JC (2012) Feral hogs negatively affect native plant communities. AgriLife Extension Paper SP-467. Texas A&M University, College Station

Tisserat N, Cranshaw W (2016) Chapter 31: Thousand cankers disease. In: Bergdahl AD, Hill A (tech. coords.) Diseases of trees in the Great Plains. General Technical Report RMRS-GTR-335. U.S. Department of Agriculture, Forest Service, Rocky Mountain Research Station, Fort Collins, pp 118–119

Toledo D, Sanderson M, Spaeth K et al (2014) Extent of Kentucky bluegrass on native plant species diversity and ecosystem services in the northern Great Plains of the United States. Invasive Plant Sci Manag 7:543–552

Trammell MA, Butler JL (1995) Effects of exotic plants on habitat utilization by native ungulates. J Wildl Manag 59:808–816

Ulrich E, Perkins L (2014) *Bromus inermis* and *Elymus Canadensis* but not *Poa pratensis* demonstrate strong competitive effects and all benefit from priority. Plant Ecol 215:1269–1275

U.S. Department of Agriculture (USDA), Natural Resources Conservation Service (2008) PLANTS database, [Online]. Available: http://plants.usda.gov/

U.S. Department of Agriculture (USDA) Forest Service (2014) Field guide for managing whitetop in the Southwest. TP-R3-16-20. U.S. Department of Agriculture, Forest Service, Southwest Region, Albuquerque. 7 p

U.S. Department of Agriculture (USDA), Animal Plant Health Inspection Service (USDA APHIS) (2016a) Cooperative emerald ash borer project. On-line map. https://www.aphis.usda.gov/plant_health/plant_pest_info/emerald_ash_b/downloads/MultiState.pdf. 28 Nov 2016

U.S. Department of Agriculture (USDA), Animal Plant Health Inspection Service (USDA APHIS) (2016b) Pest alert: Asian gypsy moth. 2p. https://www.aphis.usda.gov/publications/plant_health/content/printable_version/fs_phasiangm.pdf

U.S. Department of Agriculture (USDA) Forest Service (2017) Working trees. https://nac.unl.edu/Working_Trees/index.htm

U.S. Department of Interior (USDI) Bureau of Land Management (BLM) (2017) Wild horse and burro on-range population estimates. https://www.blm.gov/programs/wild-horse-and-burro/about-the-program/program-data

Van Devender TR, Felger RS, Burquez MA (1997) Exotic plants in the Sonoran Desert region, Arizona and Sonora. In: Kelly M, Wagner E, Warner P (eds) Proceedings, California exotic pest plant council symposium, 1997 October 2–4, Concord, CA, vol 3. California Exotic Pest Plant Council, Berkeley, pp 10–15

Vaness BM, Wilson SD (2007) Impact and management of crested wheatgrass (*Agropyron cristatum*) in the northern Great Plains. Can J Plant Sci 87:1023–1028

Weber WA, Wittmann RC (1996) Colorado flora: eastern slope, 2nd edn. University Press of Colorado, Niwot, 524 p

Whisenant SG (1990) Changing fire frequencies on Idaho's Snake River Plains: ecological and management implications. In: McArthur ED, Romney EM, Smith SD (eds) Proceedings of symposium on cheatgrass invasion, shrub die-off, and other aspects of shrub biology and management, Las Vegas, NV, April 5–7, 1989, pp 5–7

Williams DG, Baruch Z (2000) African grass invasion in the Americas: ecosystem consequences and the role of ecophysiology. Biol Invasions 2:123–140

Williams CS, Hayes RM (1984) Johnsongrass (*Sorghum halepense*) competition in soybeans (Glycine max). Weed Sci 32(4):498–501

Willson GD, Stubbendieck J (1996) Suppression of smooth brome by atrazine, mowing, and fire. Prairie Nat 28:13–20

Willson GD, Stubbendieck J (1997) Fire effects on four growth stages of smooth brome (*Bromus inermis* Leyss.). Nat Areas J 17:306–312

Wilson SD, Pärtel M (2003) Extirpation or coexistence? Management of a persistent introduced grass in prairie restoration. Restor Ecol 11:410–416

Young JA, Evans RA (1972) Germination and establishment of Salsola in relation to seedbed environment. I. Temperature, afterripening, and moisture relations of Salsola seeds as determined by laboratory studies. Agron J 64:214–218

Young JA, Evans RA, Gifford RO, Eckert RE Jr (1970) Germination characteristics of three species of Cruciferae. Weed Sci 18:41–48

Zouhar K (2004) Cardaria spp. In: Fire effects information system, [Online]. U.S. Department of Agriculture, Forest Service, Rocky Mountain Research Station, Fire Sciences Laboratory (Producer). Available: http://www.fs.fed.us/database/feis/plants/forb/carspp3/all.html

Midwest Region

Jennifer Juzwik, Linda Haugen, John Kyhl, Noel F. Schneeberger, John D. Rothlisberger, and Therese M. Poland

Introduction

The Midwest region includes Minnesota, Iowa, Missouri, Wisconsin, Illinois, Michigan, Indiana, and Ohio (Fig. A6.1). Five States border the Great Lakes, in addition to numerous inland lakes and the Missouri and Mississippi River systems. Forty percent of all the water surface area in the continental United States is located within the Midwest. Abundance of water within the region influences trade (shipping ports, river traffic), recreation, agriculture, and ecology. All of these listed factors influence the distribution and impact of invasive species in both terrestrial and aquatic environments.

The diverse and ecologically complex forest ecosystems of the Midwest are dominated by northern and central hardwood forests, bordered by northern boreal forest to the north and prairie ecosystems to the south and west. Forests of the Midwest are productive and valuable, with forest-related businesses ranking in the top 10 for economic importance in every State. The oak-hickory (*Quercus-Carya*) forest type occupies the greatest proportion of the forested area (40%), followed by maple-beech-birch (*Acer-Fagus-Betula*) (15%) and aspen-birch (*Populus-Betula*) (14%). Conifer types, including 9% spruce-fir (*Picea-Abies*) and 6% pine (*Pinus*), are also important, particularly in the Lake States. Bottomland hardwoods rise to importance in this region, with 11% of the area comprising the elm-ash-cottonwood (*Ulmus-Fraxinus-Populus deltoides*) forest type.

The Midwest region also has many large cities and a very high presence of agriculture and industry. Human actions and their interactions with their environment exacerbate the movement and impacts of invasive species. Non-native

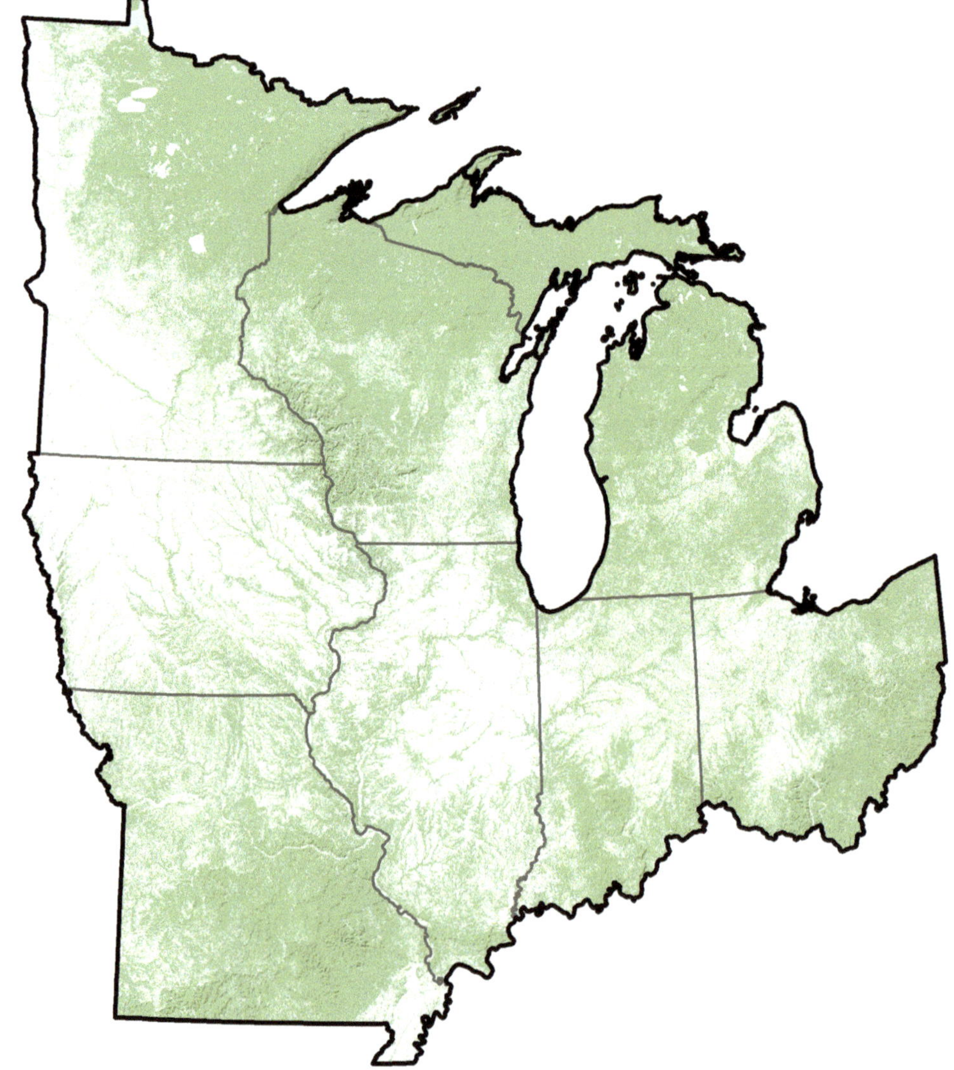

Fig. A6.1 The Midwest region. (Figure courtesy of Daniel Ryerson and Andy Graves, USDA Forest Service Southwestern Region, Forest Health Protection)

invasive species have affected forests and aquatic systems since the time of European settlement, with landscape-level impacts extending into even the most remote areas of the region. We outline selected non-native species below, with focus on current distribution, significant impacts, and current management efforts.

Insect Pests of Trees

Many non-native insect pests occur in the region, and some have caused significant impacts on the region's forests. The focus in this summary is four species that have been of high interest or concern in recent years: gypsy moth (*Lymantria dispar*), hemlock woolly adelgid (HWA) (*Adelges tsugae*), emerald ash borer (EAB) (*Agrilus planipennis*), and Asian longhorned beetle (ALB) (*Anoplophora glabripennis*). Other non-native insects have had impacts that linger in our forests, including larch sawfly (*Pristiphora erichsonii*), larch casebearer (*Coleophora laricella*), Japanese beetle (*Popillia japonica*), birch leafminer (*Profenusa thomsoni*), European pine sawfly (*Neodiprion sertifer*), introduced pine sawfly (*Diprion similis*), and elongate hemlock scale (*Fiorinia externa*).

Gypsy moth caterpillars feed on hundreds of species of trees and shrubs, often causing severe defoliation and contributing to tree decline and mortality. The insect has been the focus of government-sponsored programs for more than 100 years. Currently, gypsy moth is established across Michigan and much of Wisconsin and in portions of Indiana, Illinois, Minnesota, and Ohio. A variety of biological control agents (i.e., parasitoids, predators, and entomopathogens) help regulate gypsy moth populations. In particular, the highly specific insect pathogen *Entomophaga maimaiga* has become widely established in the Midwest and may be contributing to the natural suppression of gypsy moth populations. Management of the insect at the Federal level consists of three distinct strategies (suppression, eradication, and slowing the spread), depending upon where the insect is found (USDA 2012a). Suppression is implemented to reduce adverse effects to trees caused by outbreaks of the insect. Gypsy moth populations in the region remained low between 2007 and 2016, with only Ohio and Wisconsin conducting modest State-led aerial suppression projects on about 44,000 ac (USDA 2017). Eradication is implemented to eliminate colonies of gypsy moth that are detected outside of the currently infested (regulated) area. Between 2007 and 2016, more than 17,000 ac in Indiana, Minnesota, Ohio, and Wisconsin were treated using eradication protocols (USDA 2017). The objective of the Slow the Spread (STS) program, which involves the collaboration of multiple jurisdictions and cooperators, is to slow the natural and short-range human-aided spread of the insect along the leading edge of the area generally infested by the insect. STS is a unique landscape-scale program across a 50-million-ac project area within 11 States from Minnesota to North Carolina. The design and implementation of STS is science-based with the overall strategy founded on research that indicated this was an optimal approach for minimizing spread. Since the start of the program, about 6 million ac have been treated in Iowa, Illinois, Indiana, Minnesota, Ohio, and Wisconsin, mostly employing the application of pheromone flakes to disrupt mating by gypsy moth adults (USDA 2017). Spread rates along the leading edge remained stable in the Midwest region in 2016, while rates across the entire STS project area were low (3.8 km/year).

The hemlock woolly adelgid (HWA) threatens the survival and sustainability of eastern hemlock (*Tsuga canadensis*). Hemlocks are considered a foundation species which define forest structure and control ecosystem dynamics (Havill et al. 2014). The insect, which causes tree decline and mortality, is now present in many eastern States and has recently been confirmed in the Midwest in 13 eastern counties of Ohio and 5 counties in Michigan. The National HWA Initiative, a landscape-scale effort, was established by the USDA Forest Service in 2003 to develop and implement tools to manage HWA and to reduce the adverse effects across the range of eastern hemlock. Current management of HWA in Ohio consists of enhanced survey and monitoring of HWA spread into uninfested areas, as well as the application of systemic insecticides to protect high-value trees in the near term, complemented with the release of biological control agents (predatory beetles) to manage HWA populations in the long term. The HWA predatory beetles *Laricobius nigrinus* and *L. osakensis* have been, and continue to be, released in the infested counties in Ohio. In summer 2015, infestations of HWA were detected in Ottawa and Muskegon Counties in western Michigan. Since then, HWA has also been detected in Allegan, Oceana, and Mason Counties. The State has quarantined the four infested counties and has initiated surveys to delimit the infested area and look for new infestations. Treatments relying heavily on systemic insecticides are being implemented in an attempt to contain local HWA populations. However, it is unlikely that HWA can be eliminated from Lower Michigan. This puts at greater risk more extensive hemlock stands in the Upper Peninsula of Michigan and northern Wisconsin.

Adults of the emerald ash borer (EAB) feed on leaves and larvae tunnel in the phloem. EAB is a significant tree killer that has decimated ash trees across much of the Midwest. Green, white, and black ash (*Fraxinus pennsylvanica*, *F. americana*, and *F. nigra*, respectively) are common and locally abundant. Pumpkin and blue ash (*F. profunda* and *F. quadrangulata*, respectively) are less common but locally important species. All are susceptible to EAB (Klooster et al. 2014). Tree losses from EAB are estimated to be in the hundreds of millions in the Midwest region. A few ash trees have survived in EAB-infested areas which suggests that there may be some resistance or tolerance in the population (e.g., Anulewicz et al. 2007; Knight et al. 2012; Rebek et al. 2008). First discovered in the Detroit metropolitan area in 2002, subsequent detections have occurred in Ohio (2003), Indiana (2004), Illinois (2006), Wisconsin (2008), Minnesota (2009), Iowa (2010), and Missouri (2008). Today, Federal and State quarantines exist in all or parts of every State in the Midwest region. Ash also is a common street and landscape tree in many Midwestern cities. The eventual cost of treatment, removal, and replacement of infested ash trees in communities is estimated to be as high as $10.7 billion over a 10-year period (Kovacs et al. 2010). Commerce and movement of infested nursery stock and wood products such as firewood are major contributors to the spread of the insect. The current management approach focuses on (1) containment of the insect; (2) regulating the movement of potentially infested materials to areas not infested with EAB; (3) survey and monitoring; (4) public outreach; (5) insecticide treatment to protect high value trees; and (6) management of the insect through the release and establishment of (currently) four biological control agents (parasitoids).

Native to China and Korea, the Asian longhorned beetle (ALB) is a wood borer that can penetrate deep into the wood. It poses a serious threat to the Midwest region's forests. At least 13 tree genera, and more than 100 different tree species, are known to be suitable hosts for ALB (USDA 2012b), although the insect mostly prefers maples (*Acer* spp.), poplars (*Populus* spp.), willows (*Salix* spp.), and elms (*Ulmus* spp.). The Midwest region's forests and urban landscapes include a large number of maples, poplars, and willow. The second confirmed detection of ALB in the United States occurred in the Midwest region, in the Chicago metropolitan area in 1998. An aggressive eradication effort was successful, eliminating the insect from that location by 2008. The next ALB detection in the Midwest region occurred in 2011 in Clermont County, OH, which is more rural compared to the Chicago metropolitan area. Current prevention and eradication protocols include (1) detection and monitoring for ALB via intensive surveys; (2) preventing movement of infested material with established quarantines; (3) public outreach and education; (4) removal and destruction of infested and high-risk host trees; and (5) the use of systemic insecticides. The goal is to eradicate the pest from the woodlots and natural forest stands in this Ohio infestation. ALB may spread faster in natural and managed forests than has been observed in urban and suburban environments (Dodds and Orwig 2011; Dodds et al. 2014). Current survey, monitoring, and control tactics developed for urban areas might need to be modified for rural lands.

Pathogens of Trees

Invasive pathogens have caused serious ecological and economic impacts to Midwestern forests. A few of the more significant current problems are highlighted below, in chronological order of recognition or introduction.

White pine blister rust, caused by the fungus *Cronartium ribicola*, was introduced during reforestation efforts in the early 1900s and is currently distributed throughout the range of eastern white pine (*Pinus strobus*). It causes mortality and top dieback, particularly on environmentally conducive sites. It is considered one of the most limiting factors in growing white pine in the region. The disease is managed by appropriate site selection, pathological pruning, and planting of putative resistant nursery stock (Geils et al. 2010).

Dutch elm disease (DED), caused by *Ophiostoma novo-ulmi* and *O. ulmi*, is a vascular wilt disease that has devastated native elms (*Ulmus americana*, *U. rubra*, and *U. thomasii*) across the region since the introduction of the fungi decades ago (*O. ulmi* in the 1930s and O. *novo-ulmi* in the 1970s). Successive waves of mortality can be attributed to ingrowth of susceptible elms and high populations of insect vectors of the DED fungi in affected areas. The vectors known to exist within the region include the native elm bark beetle (*Hylurgopinus rufipes*) and two non-native species, the smaller European elm bark beetle (*Scolytus multistriatus*) and the banded elm bark beetle (*Scolytus schevyrewi*). Management of the disease in urban settings is accomplished by sanitation to control the bark beetle vectors, chemical injections, and use of DED-tolerant cultivars. Operational trials are underway to evaluate the potential use of putative DED-tolerant elms in the restoration of riparian wild areas (Knight et al. 2017).

Oak wilt, caused by *Bretziella fagacearum* (syn. *Ceratocystis fagacearum*), is a devastating disease of red oak species (*Quercus* subsection Lobatae) that was first described in Wisconsin in 1942. It is considered by many experts to be non-native (Juzwik et al. 2008). The disease rapidly kills infected red oaks. It can also kill white oaks (*Quercus* subsection *Quercus*) in the Midwest, but tree death occurs over several to many years. Disease impact is generally more severe in landscapes with abundant red oaks compared to landscapes where white oaks are common. It is currently found in parts of all States in the region. The oak wilt range is expanding along the northern edge of its distribution. Oak wilt is now at epidemic levels in portions of affected States. Oak wilt is managed in urban and wildland environments by disrupting the overland and the belowground portions of the disease cycle to prevent the establishment of new infection centers and the expansion of existing centers. Current approaches to management on forest lands include preventing movement of diseased material, avoiding wounding during high-risk periods, and disruption of connected root systems (Juzwik et al. 2011).

Butternut canker (caused by *Ophiognomonia clavigignenti-juglandacearum*) was first reported on butternut (*Juglans cinerea*) in Wisconsin in 1967. Its origin is unknown, but it is believed to have been introduced to North America (Broders et al. 2014). It is now present throughout the natural range of butternut. The disease has killed up to 90% of the butternut trees in the region and may lead to extirpation of the species (Shultz 2003). Silvicultural approaches for butternut regeneration and selection of resistant trees have been proposed in an effort to promote survival of the species (LaBonte et al. 2015). There are no existing tools for management of the disease at this time.

Beech bark disease (BBD), caused by bark canker fungal species that colonize stylet wound damage of an exotic beech scale (*Cryptococcus fagisuga*), was first detected in the region (Michigan) in 2000 (O'Brien et al. 2001). Since that time, beech mortality has become widespread in parts of Michigan. The disease has also been confirmed in eastern Wisconsin and Ohio. As the disease moves through native forests, it kills a significant proportion of American beech (*Fagus grandifolia*), whose nuts are valuable as wildlife food. Mature beech trees can reach large size and are common in parts of Ohio, Michigan, and eastern Wisconsin. BBD is managed on the advancing front through salvage harvesting with retention of smooth-barked and unaffected trees and preventing the movement of infested materials (McCullough et al. 2005). An operational screening effort is underway to identify and propagate beech resistant to beech scale.

Diseases caused by *Phytophthora* species are an emerging concern throughout the region. White oak mortality in Ohio and Missouri has recently been attributed to *P. cinnamomi*, an exotic root-damaging pathogen (Balci et al. 2010). State and Federal plant regulatory agencies continue to monitor nursery stock for the introduction of *Phytophthora ramorum* which could affect the region's oak and ericaceous plants.

Invasive Plants of Terrestrial and Aquatic Systems

There are many non-native invasive terrestrial and aquatic plants distributed throughout the Midwest region. Many of these terrestrial plant species significantly affect the region's forest ecosystems, displacing native plant species and causing substantial damage. Several of the more important woodland species are highlighted below.

Garlic mustard (*Alliaria petiolata*) is a common invader in all Midwestern States (USDA, NRCS 2018). Brought from Europe as a food plant, this shade-tolerant species is now widely found in settings ranging from intact woodlands to disturbed areas (Kurtz and Hansen 2014). Garlic mustard is a biennial and forms large, nearly monospecific patches through heavy seed production, high seed germination rates, allelopathy, and disruption of mutualistic associations (Stinson et al. 2006). Biological control agents, including stem and root boring *Ceutorhynchus* spp. weevils (Becker et al. 2013), have been studied for nearly 20 years and are currently in the final stages of testing. A variety of tactics are employed to manage garlic mustard, including hand-pulling, removal of flowers before seed set, and herbicide application. Seeds are easily moved by animals, people, equipment, and vehicles, and new introductions are difficult to prevent. It can take years to manage large patches of garlic mustard even using multipronged management approaches.

Japanese barberry (*Berberis thunbergii*) was introduced as an ornamental. This species occurs in all Midwestern States but has a wide distribution in Ohio, Michigan, and Wisconsin (USDA NRCS 2018). It occurs in many habitats (closed canopy forests, open woodlands, wetlands, and fields), forming dense thickets and shading out other plants. It is very shade tolerant and grows under a wide variety of growing conditions. Thorns discourage some herbivores, but rabbits can feed on stems through the winter. Japanese barberry spreads through roots and branches that root when in contact with the soil. Birds and other animals eat the bright red berries and can disperse the seeds long distances. This species is typically managed by cutting, pulling, and herbicide use (Michigan DNR 2012).

Common buckthorn (*Rhamnus cathartica*) was also introduced as an ornamental shrub and is now prevalent in Minnesota, Wisconsin, and Michigan, occurring less frequently in the other Midwestern States (USDA, NRCS 2018). It grows as a shrub or small tree in habitats ranging from open fields to forests, forming dense thickets and crowding out native plants. This species has early leaf out and late leaf senescence and can have a longer growing season than other plants, in some cases by nearly as long as 2 months (Harrington et al. 1989). Common buckthorn is spread by birds that ingest fruit which ripens in the late summer. Control of this species can be difficult and can take years, because the thickets are difficult to work in and often resprout after cutting or pulling. Removal is generally followed by herbicide applications to cut stumps (NRCS 2007).

Exotic honeysuckles (*Lonicera* spp.) are common in forest, edges, wetlands, and disturbed areas, occurring in most counties of all Midwestern States (USDA NRCS 2018). Honeysuckles are shrubs, sometimes reaching 10–15 ft. in height, and produce flowers in spring and early summer that are attractive to bees. Fruits ripen in the fall and are dispersed by birds. Like with buckthorn, control is difficult, generally involving repeated efforts of cutting and stump treatments (Ohio State University Extension 2018).

The tree of heaven (*Ailanthus altissima*) is abundant in Ohio, Indiana, and Illinois and has spotty distributions in most other Midwestern States (USDA NRCS 2018). This fast-growing tree can approach 100 ft. in height and is found in many habitats, ranging from closed canopy forests to open fields and urban areas. Due to allelopathy, high seed production, and aggressive suckering, this species can completely dominate areas in which it grows and is difficult to control with cutting and herbicide stump treatments. Within the last 10–15 years, a soil-borne pathogen (*Verticillium nonalfalfae*) that causes vascular wilt and death in tree of heaven has been found in Ohio, Pennsylvania, and Virginia (Rebbeck et al. 2013). Further research is being conducted on this pathogen and its possible use as a biological control.

Reed canary grass (*Phalaris arundinacea*), phragmites (*Phragmites australis*), and purple loosestrife (*Lythrum salicaria*) are major invasive plants in wetland areas distributed throughout the entire region (USDA NRCS 2018). Biological control with beetles in the genus *Galerucella* has been a success in limiting purple loosestrife (Blossey et al. 2015), while reed canary grass and phragmites are generally managed with consecutive seasonal burns, mechanical removal, and herbicides (Michigan DEQ 2014).

Eurasian watermilfoil *(Myriophyllum spicatum)* is one of several invasive aquatic plants that is distributed widely throughout the region (USDA, NRCS 2018) and which can drastically alter the ecological processes and functioning of aquatic ecosystems. Other invasive aquatic plants in the Midwest include hydrilla (*Hydrilla verticillata*), starry stonewort (*Nitellopsis obtusa*), parrotfeather (*Myriophyllum aquaticum*), and curly-leaf pondweed (*Potamogeton crispus*). Management strategies include harvesting, rotovation, dredging, and aquatic herbicides (Mikulyuk and Nault 2009), but, as with aquatic animals, control of aquatic plants is costly and requires constant effort and investment. Eradication is all but impossible, so preventing new invasions is crucial to avoiding ecological and economic harm.

Invasive Animals of Terrestrial Systems

Invasive vertebrates and noninsect invertebrates threatening terrestrial ecosystems in the Midwest region include feral hogs (*Sus scrofa*) and invasive earthworms. Feral hogs damage native plants and crops and are problematic throughout Missouri, Indiana, Ohio, and Wisconsin. They are managed by trapping and removal, followed by improvement of the degraded habitat. Various species of invasive earthworms have been implicated in the degradation of native plant communities, especially throughout northern Minnesota and Wisconsin (Holdsworth et al. 2007). Best management practices have been developed and implemented to prevent further spread (e.g., Wisconsin Department of Natural Resources 2015).

Invasive Animals and Pathogens of Aquatic Systems

A variety of invasive aquatic animals are recognized as having important negative ecological and economic impacts in the Midwest region. These include fish such as sea lamprey (*Petromyzon marinus*), bighead carp (*Hypophthalmichthys nobilis*), and silver carp (*H. molitrix*); mollusks such as zebra mussel (*Dreissena polymorpha*) and quagga mussel (*D. bugensis*); crustaceans such as rusty crayfish (*Orconectes rusticus*) and spiny water flea (*Bythotrephes longimanus*); and pathogens such as viral hemorrhagic septicemia (VHS). These species and many other invasive aquatic animals in the region have disrupted native food webs and altered ecosystem functioning. In many cases, their impacts have reduced the value of ecosystem services and required the implementation of costly management activities to control invasive species and reduce their impacts. For example, sea lamprey, an invasive parasitic fish that feeds on the blood and body fluids of other fish, played a role in precipitous declines of Great Lakes fish stocks in the mid-twentieth century. Scientists discovered an effective lampricide (TFM, 3-trifluoromethyl-4-nitrophenol) in the late 1950s, and its application, along with several other management techniques, has been used to reduce sea lamprey populations. These control efforts are effective, but cost approximately $20 million each year.

In addition to sea lamprey, which invaded the Great Lakes from the North Atlantic Ocean through man-made canals, many other invasive aquatic animals have been introduced to the Great Lakes by the release of ballast water from transoceanic ships. Ship-borne species include zebra and quagga mussels, spiny and fishhook (*Cercopagis pengoi*) water fleas, round gobies (*Neogobius melanostomus*), and Eurasian ruffe (*Gymnocephalus cernua*). These, and some 50 other non-native aquatic species introduced to the Great Lakes by shipping, are estimated to reduce the value of ecosystem services from wildlife watching, commercial fishing, recreational fishing, and raw water usage by more than $100 million annually (Rothlisberger et al. 2012).

Invasive aquatic species that establish populations in the Great Lakes often spread to the rest of the Midwest and beyond. Zebra mussels, which invaded the Great Lakes in the 1980s, are a well-known biofouling organism. They quickly spread to rivers and inland lakes in the States surrounding the Great Lakes and, more recently, have become established in waterways in the Western United States.

Two invasive crayfish species that have serious impacts in the upper Midwest are native to the Southeast: the rusty crayfish and the red swamp crayfish (*Procambarus clarkii*). These species outcompete and hybridize with native crayfish and prey on native fish, crayfish, and gastropods.

Asian carps, including common carp (*Cyprinus carpio*), bighead carp, black carp (*Mylopharyngodon piceus*), grass carp (*Ctenopharyngodon idella*), and silver carp, are invasive fish that present significant concerns for the region. Asian carp species have had major impacts on native fish populations in the Mississippi River basin. Costly electric barriers to reduce the likelihood of Asian carp movement into the Great Lakes have been installed in the Chicago Ship and Sanitary Canal, a man-made hydrologic connection between the Great Lakes and the Mississippi River basin.

Other invasive fish of concern in the region include round goby and Eurasian ruffe, both of which are voracious benthivorous species with high reproductive rates. The piscivorous northern snakehead fish (*Channa argus*) has also been found in isolated locations in the Midwest region and threatens to become more widespread.

Pathogens that are not native to North America also cause harm to native fish species. Several of the diseases associated with these harmful non-native pathogens include viral hemorrhagic septicemia (VHS), salmonid whirling disease, and bacterial kidney disease. Cost-effective control methods are not yet available for most of the aquatic invasive animals in the Midwest region. Research into more effective and less expensive control methods is ongoing. Current management efforts emphasize spread prevention though campaigns to educate the public about the importance of not intentionally or inadvertently moving species among waterways and best practices for avoiding these movements. Direct intervention efforts such as inspecting and pressure washing recreational boats and trailers to remove invasive species propagules and laws requiring that no water be moved among waterways are also important prevention efforts.

Literature Cited

Anulewicz AC, McCullough DG, Cappaert DL (2007) Emerald ash borer (*Agrilus planipennis*) density and canopy dieback in three North American ash species. Arboricult Urban For 33:338–349

Balci Y, Long RP, Mansfield M et al (2010) Involvement of *Phytophthora* species in white oak (*Quercus alba*) decline in southern Ohio. For Pathol 40:430–442

Becker R, Gerber E, Hinz HL et al (2013) Biology and biological control of garlic mustard. Forest Health Technology Enterprise Team. USDA Forest Service. FHTET-2012-05

Blossey B, Randall C, Schwarzlaender M (2015) Biology and biological control of purple loosestrife, 2nd edn. Forest Health Technology Enterprise Team. USDA Forest Service. FHTET-2015-3

Broders KD, Boland GJ (2011) Reclassification of the butternut canker fungus, *Sirococcus clavigignenti-juglandacearum*, into the genus *Ophiognomonia*. Fungal Biol 115(1):70–79

Broders K, Boraks A, Barbison L et al (2014) Recent insights into the pandemic disease butternut canker caused by the pathogen *Ophiognomonia clavigignenti-juglandacearum*. For Pathol. https://doi.org/10.1111/efp.12161

Dodds KJ, Orwig DA (2011) An invasive urban forest pest invades natural environments Asian longhorned beetle in northeastern US hardwood forests. Can J For Res 41:1729–1742

Dodds KJ, Hull-Sanders HM, Siegert NW et al (2014) Colonization of three maple species by Asian longhorned beetle, *Anoplophora glabripennis*, in two mixed-hardwood forest stands. Insects 5:105–119

Geils BW, Hummer KE, Hunt RS (2010) White pines, *Ribes*, and blister rust: a review and synthesis. For Pathol 40:147–185

Harrington RA, Brown BJ, Reich PB (1989) Ecophysiology of exotic and native shrubs in southern Wisconsin. I. Relationship of leaf characteristics, resource availability, and phenology to seasonal patterns of carbon gain. Oecologia 80:356–367

Havill NP, Vieira LC, Salom SM (2014) Biology and control of hemlock woolly adelgid. USDA Forest Service, Forest Health Technology Enterprise Team. FHTET-2014-05

Holdsworth AR, Frelich LE, Reich PB (2007) Regional extent of an ecosystem engineer: earthworm invasion in northern hardwood forests. Ecol Appl 17:1666–1677

Juzwik J, Harrington TC, MacDonald WL et al (2008) The origin of *Ceratocystis fagacearum*, the oak wilt fungus. Annu Rev Phytopathol 46:13–26

Juzwik J, Appel DN, MacDonald WL et al (2011) Challenges and successes in management of oak wilt in the United States. Plant Dis 95:888–900

Knight KS, Herms D, Plumb R et al (2012) Dynamics of surviving ash (*Fraxinus* spp.) populations in areas long infested by emerald ash borer (*Agrilus planipennis*). In: Proceedings of the 4th international workshop on genetics of host-parasite interactions in forestry. General Technical Report PSW-GTR-240

Knight KS, Haugen LM, Pinchot CC et al (2017) American elm (*Ulmus americana*) in restoration plantings: a review In: Pinchot CC et al (eds) Proceedings of the American elm restoration workshop 2016; 2016 October 25–27; Lewis Center, OH. General Technical Report NRS-P-174. U.S. Department of Agriculture, Forest Service, Northern Research Station, Newtown Square, pp 133–140

Kovacs KF, Haight RG, McCullough DG et al (2010) Cost of potential emerald ash borer damage in U.S. communities, 2009–2019. Ecol Econ 69:569–578

Kurtz CM, Hansen MH (2014) An assessment of garlic mustard in northern U.S. forests. Research Note NRS-199. U.S. Department of Agriculture, Forest Service, Northern Research Station, Newtown Square. 5 p

Labonte NR, Ostry ME, Ross-Davis A et al (2015) Estimating heritability of disease resistance and factors that contribute to long-term survival in butternut (*Juglans cinerea* L.). Tree Genet Genomes 11:63. Available online: https://doi.org/10.1007/s11295-015-0884-8

McCullough DG, Heyd RL, O'Brien JG (2005) Biology and management of beech bark disease. Michigan State University. Extension Bulletin E-2746

Michigan Department of Environmental Quality (DEQ) (2014) A guide to the control and management of invasive phragmites, 3rd edn. http://www.michigan.gov/documents/deq/deq-ogl-ais-guide-PhragBook-Email_212418_7.pdf. Accessed 11 Jan 2018

Michigan Department of Natural Resources (2012) Invasive species best control practices—Japanese barberry. https://mnfi.anr.msu.edu/invasive-species/JapaneseBarberryBCP.pdf. Accessed 11 Jan 2018

Mikulyuk A, Nault ME (2009) Curly-leaf pondweed (*Potamogeton crispus*): a technical review of distribution, ecology, impacts, and management. Wisconsin Department of Natural Resources Bureau of Science Services, PUB-SS-1052 2009, Madison

Natural Resources Conservation Service (NRCS) (2007) Pest management—Invasive plant control buckthorn (common and glossy) conservation practice. Job Sheet MN-797

O'Brien JG, Ostry ME, Mielke ME et al (2001) First report of beech bark disease in Michigan. Plant Dis 85:921

Ohio State University Extension (2018) Controlling non-native invasive plants in Ohio forests: bush honeysuckle factsheet F-68. https://ohioline.osu.edu/factsheet/F-68. Accessed 11 Jan 2018

Rebbeck J, Malone MA, Short DPG et al (2013) First report of Verticillium wilt caused by *Verticillium nonalfalfae* on tree-of-heaven (*Ailanthus altissima*) in Ohio. Plant Dis 97:999

Rebek EJ, Herms DA, Smitley DR (2008) Interspecific variation in resistance to emerald ash borer (Coleoptera: Buprestidae) among North American and Asian ash (*Fraxinus* spp.). Environ Entomol 37:242–246

Rothlisberger JD, Finnoff DC, Cooke RM et al (2012) Ship-borne nonindigenous species diminish Great Lakes ecosystem services. Ecosystems 15:1–15

Schultz J (2003) Conservation assessment for butternut or white walnut (*Juglans cinerea*) L: USDA Forest Service, Eastern Region. Hiawatha National Forest

Stinson K, Campbell S, Powell J et al (2006) Invasive plant suppresses the growth of native tree seedlings by disrupting belowground mutualisms. PLoS Biol 4:e140. https://doi.org/10.1371/journal.pbio.0040140

U.S. Department of Agriculture (USDA) NRCS (2018) The PLANTS database. National Plant Data Team, Greensboro. http://plants.usda.gov. Accessed 11 Jan 2018

U.S. Department of Agriculture (USDA) (2012a) Gypsy moth management in the United States: a cooperative approach. Supplemental environmental impact statement. Forest Service and Animal and Plant Health Inspection Service. NA-MB-01-12, Washington, DC

U.S. Department of Agriculture (USDA) (2012b) Asian longhorned beetle and its host trees. USDA Forest Service, Northeastern Area State and Private Forestry. NA-PR-05-12, Newtown Square

U.S. Department of Agriculture (USDA) (2017) Gypsy moth digest database. https://www.fs.usda.gov/naspf/programs/forest-health-protection/gypsy-moth-digest. USDA Forest Service, Northeastern Area State and Private Forestry, Newtown Square. Accessed 16 Jan 2018

Wisconsin Department of Natural Resources (2015) Jumping worm (*Amynthas* spp.). http://dnr.wi.gov/topic/Invasives/fact/jumpingWorm. Accessed 9 Aug 2018

Northeast Region

Jennifer Juzwik, Linda Haugen, Noel F. Schneeberger, Thomas J. Rawinski, John D. Rothlisberger, and Therese M. Poland

Introduction

The Northeast region is heavily forested with a high diversity of hardwood and conifer forest tree species. Northern hardwoods, including sugar maple (*Acer saccharum*), American beech (*Fagus grandifolia*), yellow (*Betula alleghaniensis*) and paper birch (*B. papyrifera*), and aspen (*Populus tremuloides*) make up 44% of the forests, followed by the oak-hickory (*Quercus-Carya*) type (27%), pine (*Pinus*) types (white-red-jack pine (*P. strobus-P. resinosus-P. banksiana*), loblolly-shortleaf pine (*P. taeda-P. echinata*), and oak-pine) (12%), spruce-fir (*Picea-Abies*) type (11%), and bottomland types (elm/ash/cottonwood (*Ulmus/Fraxinus/Populus deltoides*) and oak/gum/cypress (*Quercus/Liquidambar/Taxodium*)) (5%). Topography, moisture gradient, and disturbance history highly influence where each forest type is found. The Northeast is also water rich, with over 10% of the total area covered by water. Aquatic ecosystems in the region include streams, swamps, lakes and ponds, rivers, and marine and estuarial habitats. In addition, New York has borders on two Great Lakes (Erie and Ontario), while Pennsylvania borders one (Erie).

The Northeast region comprises the New England and Mid-Atlantic States, including Maine, New Hampshire, Vermont, New York, Massachusetts, Connecticut, Delaware, Rhode Island, New Jersey, Maryland, Pennsylvania, and West Virginia (Fig. A7.1), and has a human population density greater than 330 people/mi^2. Many opportunities exist for human-mediated introductions of pests, including international shipping ports, a large urban/rural interface, highly industrialized areas, and high recreational use of forests. This region was colonized by Europeans earlier than most of the rest of the country, and coincidentally has the highest concentrations of invasive forest insects and pathogens in the country (Fig. A7.2). There are many

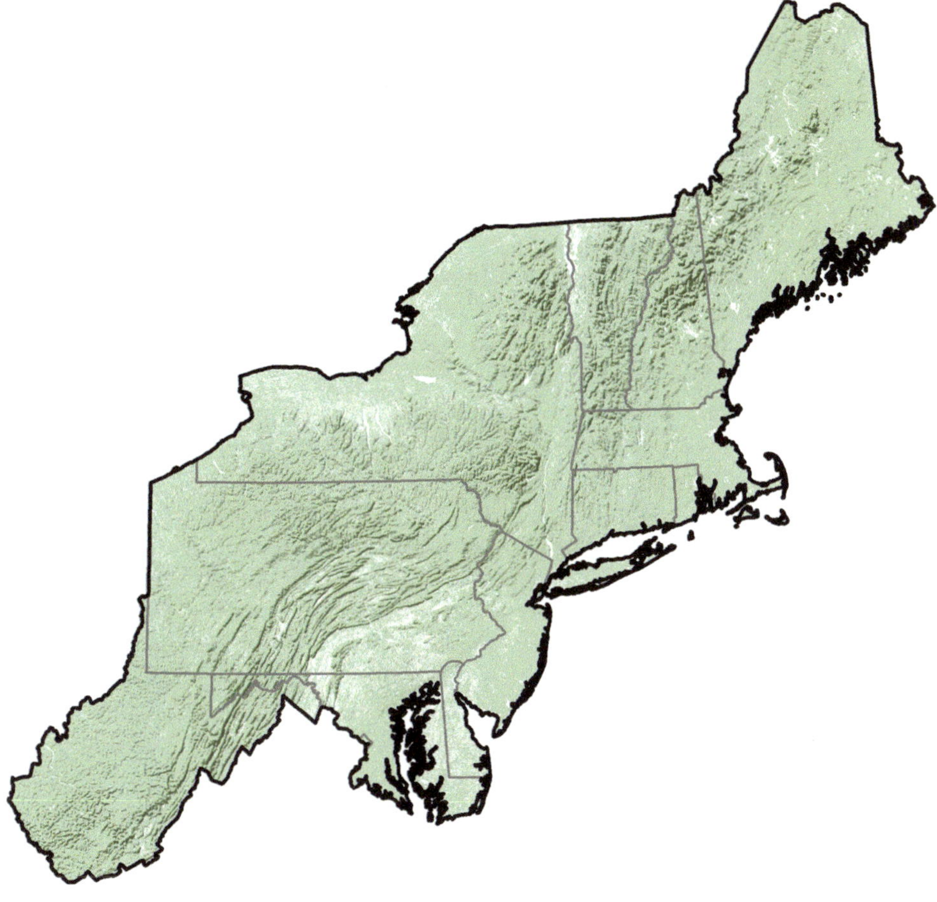

Fig. A7.1 The Northeast region. (Figure courtesy of Daniel Ryerson and Andy Graves, USDA Forest Service Southwestern Region, Forest Health Protection)

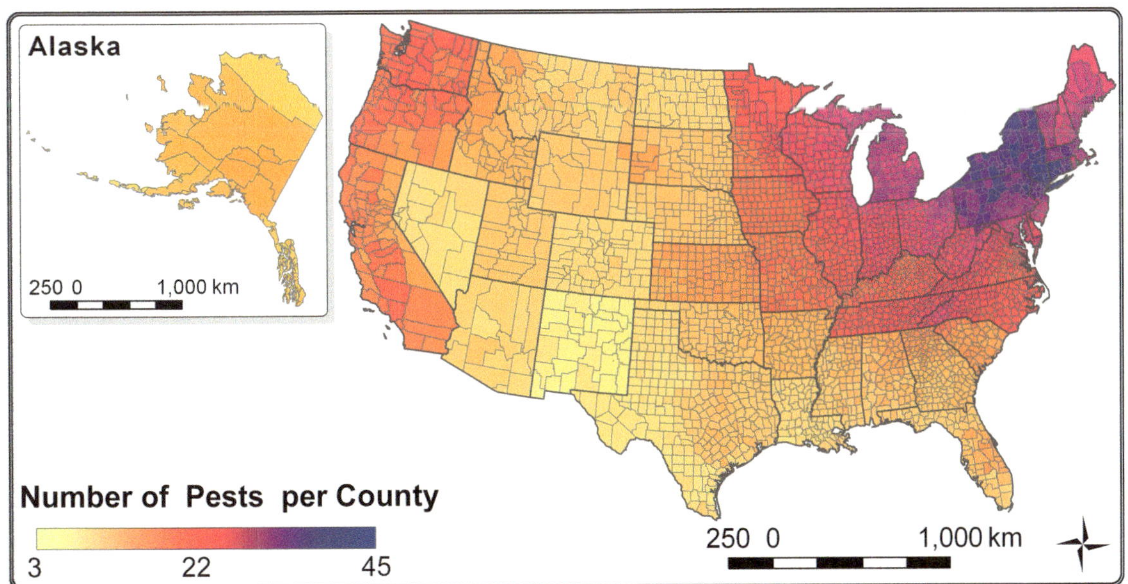

Fig. A7.2 Numbers of damaging invasive forest insects and pathogens per county in the United States. (Source: Liebhold et al. 2013)

significant invasive threats in the region encompassing insects, plant and wildlife pathogens, aquatic animals, and terrestrial and aquatic plants. Recognizing the importance of human interactions in exacerbating pest problems, collaborative organizations such as the forest health subcommittee of the Northeast-Midwest State Foresters Alliance (NMSFA) and New York's Partnerships for Regional Invasive Species Management (PRISMs) are working to identify and prioritize research needs and management/preventive actions (NAASF 2017; CUCE 2017).

Insect Pests of Trees

Gypsy moth (*Lymantria dispar*) feeds on hundreds of species of trees and shrubs and is a serious defoliator of oaks and aspen, often causing severe defoliation, as well as tree decline and mortality in the aftermath of outbreaks. Gypsy moth has caused more than 12 million ac of defoliation in the Northeast region since 2000 (USDA 2017). The insect has been the focus of government-sponsored intervention programs for more than a century, first to eradicate the insect from the United States, and then for its biological control (starting around 1900), and later to manage its adverse effects and slow its spread. Today, gypsy moth resides in all or parts of every State in the Northeast region. Gypsy moth populations are subject to regulation by a variety of biological control agents (i.e., parasitoids, predators, and entomopathogens), but these agents may not prevent periodic outbreaks and subsequent damage from occurring. At the Federal level, management of the insect consists of the integration of three distinct strategies depending on where the insect is found (USDA 2012a). Suppression is implemented to reduce adverse effects to trees caused by outbreaks of the insect. Between 2000 and 2016, State-led aerial treatment projects applied insecticides for gypsy moth control on about 2.3 million ac in Maryland, New Jersey, New York, Pennsylvania, and West Virginia (USDA 2017). With very few exceptions, the entire Northeast region is known to be generally infested by the insect. Therefore, an eradication strategy—that is, elimination of isolated colonies of the insect—is no longer pursued. The third strategy is the Slow the Spread (STS), a unique landscape-scale program in which the objective is to slow the natural and short-range human-aided spread of the insect along the leading edge of the generally infested area. STS is the first of its kind for a forest pest. The design and implementation of STS is science-based, with the overall strategy based on research that indicated this was an optimal approach for minimizing spread. Southern West Virginia is the only portion of the Northeast region located within the STS project area. Since the start of STS, more than 327,000 ac have been treated in West Virginia, mostly (> 90%) employing the application of pheromone flakes to disrupt mating by gypsy moth adults (USDA 2017).

The hemlock woolly adelgid (HWA) (*Adelges tsugae*) threatens the survival and sustainability of eastern hemlocks (*Tsuga canadensis*) in the Northeast region and wildlife species that depend on them. The insect causes tree decline and mortality. Within the region, HWA can be found throughout Connecticut, Delaware, Massachusetts, Maryland, New Jersey, and Rhode Island, with it continuing to spread into uninfested areas in Maine, New Hampshire, New York, Pennsylvania, Vermont, and West Virginia. The National HWA Initiative, a landscape-scale effort, was established by the USDA Forest Service in 2003 to develop and implement tools to manage HWA and reduce the adverse effects across the range of eastern and Carolina (*T. caroliniana*) hemlocks. Current management of HWA in the Northeast region consists of enhanced survey and monitoring of HWA spread into uninfested areas and the application of systemic insecticides to protect high-value trees in the near term complemented with the release of biological control agents (predatory beetles) to manage HWA populations in the long term. Large-scale, State-organized and implemented HWA treatment initiatives on public lands are underway particularly in New Jersey, New York, Maryland, Pennsylvania, and West Virginia. Between 2013 and 2018 more than 75,000 wild-collected and laboratory-reared HWA predatory beetles, *Laricobius nigrinus* and *L. osakensis*, have been released on public lands in nine States in the Northeast region (Massachusetts, Maryland, New Hampshire, New Jersey, New York, Ohio, Pennsylvania, Vermont, and West Virginia). Other areas of focus include region-wide collection and storage of eastern hemlock seed, investigations into plant host resistance and tolerance, and silvicultural treatments to improve hemlock health (Havill et al. 2014).

Adults of the emerald ash borer (EAB) (*Agrilus planipennis*) feed on leaves, and larvae tunnel in the phloem, killing virtually all ash trees in a stand within 5–6 years of infestation. None of the 16 species of ash in North America are immune to EAB; however, some trees do survive in infested areas. Tree losses from EAB are estimated to be in the tens of millions in the Northeast region. The first detection of EAB in the region occurred in 2003 in Maryland. Since then, the insect has been detected in all other States in the Northeast region including Pennsylvania (2007), West Virginia (2007), New York (2009), Massachusetts (2012), Connecticut (2012), New Hampshire (2013), New Jersey (2014), Delaware (2016), Vermont (2018), Maine (2018) and Rhode Island (2018). Commerce and movement of infested nursery stock and wood products such as firewood are major contributors to the spread of the insect. The current management focus is on containment of the insect, regulating the movement of potentially infested materials to areas not infested with EAB, survey and monitoring, public outreach, and management of the insect through the release and establishment of biological control agents (parasitoids).

The Asian longhorned beetle (ALB) (*Anoplophora glabripennis*) threatens urban and forest hardwood trees. At least 13 tree genera and more than 100 different tree species are known to be suitable hosts for ALB (USDA 2012b), although the insect mostly prefers maples, poplars, willows (*Salix* spp.), and elms. Native to China and Korea, ALB was first detected in the United States in New York City in 1996. Other infestations were later found in Jersey City, NJ (2002); Middlesex and Union Counties, NJ (2004); Staten Island,

NY (2007); Worcester, MA (2008); Boston, MA (2010); and most recently Babylon, NY (2013). In the Northeast region, ALB has been successfully eradicated from New Jersey and several areas in New York. Current ALB management in the region includes eradication, intensive surveys, quarantines to regulate the movement of infested material, education and outreach, removal and destruction of infested and high-risk host trees, and the use of systemic insecticides to protect high-value trees in communities. About 70,000 trees have been removed and destroyed, and about 800,000 trees have received insecticide treatments in Massachusetts and New York eradication sites.

Pathogens of Trees

The chestnut blight pathogen, *Cryphonectria parasitica*, was introduced on plant material in the early 1900s and spread rapidly to the natural forest, with devastating impacts on the composition and ecology of the eastern deciduous forest. At the time of settlement, American chestnut (*Castanea dentata*) was a key component of the forest comprising up to 50% or more of some stands. All of the mature American chestnuts in the native range of the host have died due to this pathogen, reducing the ecological position of the species to primarily stump sprouts. Currently, there are no management tools for this disease. Genetic resistance is considered the only hope for restoring this species in the region's forests. Nearly three decades of a backcross breeding program with American chestnut and the blight-resistant Chinese chestnut (*C. mollissima*) by The American Chestnut Foundation yielded the first progeny (BC_3F_3) predicted to have stable blight resistance in 2007 (Clark et al. 2012). A series of field tests using advanced breeding materials were initiated shortly thereafter (Clark et al. 2014). Techniques used to genetically transform chestnut trees that exhibit blight resistance is anticipated to accelerate further traditional breeding work to produce stable resistance trees for restoration work (Zhang et al. 2013).

In similar fashion to chestnut blight, white pine blister rust, caused by *Cronartium ribicola*, was introduced in the early 1900s on nursery stock. Today, it is currently distributed throughout the range of eastern white pine (*Pinus strobus*). It causes mortality and top dieback to white pine, particularly on sites subject to climatic conditions suitable for infections to occur. It also influences the agricultural use of commercial varieties of the alternate *Ribes* host, on which it causes a severe leaf disease. New strains of the pathogen have recently overcome resistance of commercial *Ribes* cultivars. White pine blister rust is considered one of the most limiting factors in growing white pine in the Northeast region. The disease is currently managed by appropriate site selection, pathological pruning, and planting of putatively resistant *Pinus* nursery stock and *Ribes* cultivars (Geils et al. 2010).

The first outbreak of beech bark disease in North America, caused by the interaction of an exotic beech scale (*Cryptococcus fagisuga*) and several canker fungi (*Neonectria* spp.), was observed in Nova Scotia in 1920 and by the 1930s had invaded Maine and other parts of New England. As the disease moves through native forests, it kills a significant proportion of the important mast-producing American beech, leading to loss of wildlife food and predominant tree species. Much of the beech in the region is now part of the forest that is regenerating following beech bark disease invasion (aka the aftermath forest). The disease is currently managed on the advancing front through salvage harvesting with retention of smooth-barked and unaffected trees and preventing movement of infested materials. Management of the disease in the aftermath forest, however, may require multifactor approaches targeting the different biotic agents involved in this complex disease (Cale et al. 2015).

Dutch elm disease (DED) is a vascular wilt disease that has devastated native elms (*U. americana* and *U. rubra*) across the Northeast region since the pathogen *Ophiostoma ulmi* was introduced along with its insect vector, the smaller European elm bark beetle (*Scolytus multistriatus*), on logs before 1935. A second pathogen of unknown origin, *O. novo-ulmi*, emerged later as an even more aggressive component of the disease. The pathogen(s) are also vectored by the native elm bark beetle (*Hylurgopinus rufipes*) and the more recently introduced banded elm bark beetle (*Scolytus schevyrewi*). The disease rapidly destroyed the iconic lines of elm along city streets in the Northeast but also affected the natural floodplain forests in which these trees were a dominant species (Marks 2017). As native elms regenerate on wildland sites, DED causes mortality in temporal wave related to fluctuating populations of the vectoring bark beetles. Management of the disease in urban elms is accomplished by sanitation to control the insect vectors, chemical injections, and use of disease-tolerant cultivars (Haugen 1998). Similar management tools are not available for wildland areas; however, work has begun to enhance resistance by crossing rare, large, surviving American elms with the few DED-resistant American elms (Pinchot et al. 2017).

Butternut canker (caused by *Ophiognomonia clavigignenti-juglandacearum*) was first reported in Wisconsin in 1967 and is believed to have had several introductions to North America (Broders et al. 2014). The disease is now distributed throughout the natural range of butternut (*Juglans cinerea*). The disease kills up to 90% of butternut in affected stands and may lead to extirpation of the species (Shultz 2003). Silvicultural approaches are needed for butternut regeneration, as well as the development of resistance to ensure survival of the species (LaBonte et al. 2015). There are no existing tools for management of the disease.

Oak wilt, caused by *Bretziella fagacearum* (syn. *Ceratocystis fagacearum*), is a devastating disease, particularly of red oak species (*Quercus* subsection *Lobatae*). For decades, the disease has been known to occur within West Virginia, Pennsylvania, and Maryland. It was recently discovered at multiple locations in New York State, making this an emerging problem in the Northeast region. Oak wilt is managed in urban and wildland environments by disrupting the disease cycle to prevent new centers from becoming established and existing centers from expanding. Approaches to disease management on forest lands include preventing movement of diseased material, avoiding wounding or tree harvesting during high-risk periods, and disruption of connected root systems (Juzwik et al. 2011).

Pathogens of Wildlife

White-nose syndrome, caused by *Pseudogymnoascus destructans*, has led to a rapid decline in bat populations since the disease was first detected in New York in 2006 (Frick et al. 2010). The disease is now widespread in the Northeast region. White-nose syndrome is currently managed by restricting access to hibernacula.

Invasive Animals of Aquatic Systems

The aquatic animals that have had the greatest impacts in the Northeast region are sea lamprey (*Petromyzon marinus*) and zebra (*Dreissena polymorpha*) and quagga (*D. bugensis*) mussels. Sea lampreys, which parasitize and kill native and non-native fish sometimes leading to population crashes, are present in the Great Lakes and in several large inland lakes in New York; however, sea lamprey is only considered invasive in the Great Lakes. There are several other invasive fish that are attributed with negative impacts in the region. These include common carp (*Cyprinus carpio*), goldfish (*Carassius auratus*), northern snakehead (*Channa argus*), and oriental weatherfish (*Misgurnus anguillicaudatus*).

Zebra and quagga mussels are present in the Great Lakes as well as in large navigable rivers and many small lakes in the region. They cause economic and ecological harm by fouling infrastructure and personal property and by altering the energy flows in food webs away from pelagic species, which are often valuable sport fish. Other invasive mollusks include Asian clams (*Corbicula fluminea*) and New Zealand mud snail (*Potamopyrgus antipodarum*), which compete with native species and alter nutrient cycling.

Invasive crustaceans in the Northeast include the predatory cladocerans, spiny (*Bythotrephes longimanus*) and fishhook (*Cercopagis pengoi*) water fleas, which compete with juvenile fish for food resources. Chinese mitten crabs (*Eriocheir sinensis*) and Asian shore crabs (*Hemigrapsus sanguineus*) have also been collected in the region and have the potential to alter food webs and damage infrastructure. In the canals of New York State and the Hudson River alone, the economic losses attributable to aquatic invasive species are estimated at $500 million, with impacts affecting commercial and recreational fishing the most (Pimentel 2005). Relative to many other parts of the United States, the Northeast is highly developed and highly populated. The combination of many roads with many water access points and many people traveling those roads and visiting waterways facilitates the human-mediated spread of aquatic invasive species. Similarly, the extensive network of man-made canals in the Northeast has accelerated the spread of introduced species throughout the region.

There are no cost-effective control methods available for most aquatic invasive animals in the Northeast region. Research into more effective and less expensive control methods is ongoing. Current management efforts emphasize spread prevention through campaigns to educate the public about the importance of not intentionally or inadvertently moving species among waterways, and the best practices for avoiding these movements. Direct intervention efforts such as inspecting and pressure washing recreational boats and trailers to remove invasive species propagules and laws banning the movement of species and water among waterways are also important prevention efforts.

Invasive Plants of Aquatic and Terrestrial Systems.

State committees and working groups in the Northeast region have ranked the significance of hundreds of invasive plant species. In New York State, for example, the ranking is based on (1) ecological impact, (2) biological characteristics and dispersal ability, (3) ecological amplitude and distribution, and (4) difficulty of control (Jordan et al. 2012[1]). Of New York's 183 listed invasive plant species, 32 received an invasiveness rank of Very High (Brooklyn Botanical Garden 2013[2]). Of these 32 species, 22 occupy terrestrial habitats: Norway maple (*Acer platanoides*), garlic mustard (*Alliaria petiolata*), Japanese angelica tree (*Aralia elata*), Japanese barberry (*Berberis thunbergii*), slender false brome (*Brachypodium sylvaticum*), Oriental bittersweet (*Celastrus orbiculatus*), black swallow-wort (*Cynanchum louiseae*), pale swallow-wort (*C. rossicum*), autumn olive (*Elaeagnus umbellata*), winged euonymus (*Euonymus alatus*), Japanese knotweed (*Fallopia japonica*), Japanese honeysuckle (*Lonicera japonica*), Amur honeysuckle (*L. maackii*), Morrow's honeysuckle (*L. morrowii*), Japanese

[1] Jordan, M.J.; Moore, G.; Weldy, T.W. 2008 (2012 update). New York State ranking system for evaluating non-native plant species for invasiveness. Unpublished report. On file with: The Nature Conservancy, 250 Lawrence Hill Road, Cold Spring Harbor, NY 11724

[2] Brooklyn Botanic Garden. 2013. Invasiveness assessment scores and ranks for 183 non-native plant species in New York State. Unpublished report. On file with: Brooklyn Botanic Garden, 1000 Washington Avenue, Brooklyn NY 11225

stiltgrass (*Microstegium vimineum*), mile-a-minute weed (*Persicaria perfoliata*), kudzu (*Pueraria montana*), lesser celandine (*Ranunculus ficaria*), common buckthorn (*Rhamnus cathartica*), black locust (*Robinia pseudoacacia*), multiflora rose (*Rosa multiflora*), and wineberry (*Rubus phoenicolasius*).

The remaining 10 of the 32 highly ranked invasive species occupy wetland or aquatic sites: waterthyme (*Hydrilla verticillata*), frogbit (*Hydrocharis morsus-ranae*), Uruguayan primrose-willow (*Ludwigia grandiflora*), floating primrose-willow (*L. peploides*), purple loosestrife (*Lythrum salicaria*), broadleaf watermilfoil (*Myriophyllum heterophyllum*), Eurasian watermilfoil (*M. spicatum*), common reed grass (*Phragmites australis*), gray florist's willow (*Salix atrocinerea*), and water chestnut (*Trapa natans*).

The Federally listed noxious weed giant hogweed (*Heracleum mantegazzianum*), which received an invasiveness rank of High in New York, should also be considered a priority in the Northeast, as it poses a significant threat to human health and safety.

Because New York is somewhat geographically centered in the Northeast, most of the above-listed species occur throughout the region and are considered priority threats. Exceptions would be slender false brome, which has not yet been found in New England States, and black locust, which is actually a native species from Pennsylvania southward. Similarly, glossy buckthorn (*Frangula alnus*) and hardy kiwi (*Actinidia arguta*) are regarded as priority threats in New England States, while arthraxon (*Arthraxon hispidus*) and wavyleaf basketgrass (*Oplismenus undulatifolius*) are priority threats in Mid-Atlantic States.

Educational efforts have been extremely successful in raising awareness of the threats posed by invasive plants. Such threats are not limited to competition for space and resources. American bittersweet (*Celastrus scandens*), for example, is threatened by genetic swamping. It is hybridizing with Oriental bittersweet and no longer breeds true in certain areas (Zaya et al. 2015).

Education has led to action. Many States now have prohibited plant lists, identifying species that may no longer be bought and sold, thereby helping to prevent future spreading through cultivation. Landowners, municipalities, conservation organizations, regional partnerships, and agencies have waged countless battles on invasive plants. Over the years, control efforts have become more strategic and more effective. The emphasis has been on early detection and early control. Practitioners recognize, however, that invasive plants are here to stay. Eradication is virtually impossible. Limited resources are being directed to protect the most important and the most threatened natural resources, using control methods that are often integrated or novel. Similarly, practitioners recognize that infestations are often a symptom, and not the cause, of ecosystem degradation. Invasive plants often thrive in response to anthropogenic perturbations and in forests damaged by overabundant white-tailed deer (*Odocoileus virginianus*). Such forests have lost much of their biotic resistance to exotic plant invasion (Kalisz et al. 2014; Knight et al. 2009). After Rhode Island's largest infestation of Japanese stiltgrass was discovered, the recommendation was to enclose the four-acre area in deer fencing, which restored biotic resistance to the site and all but eliminated the stiltgrass in just 3 years, without the use of any herbicides or any mechanical control measures. Restoring biotic resistance in forest ecosystems and mitigating disturbance impacts hold promise for the effective control of invasive plants in the Northeast and elsewhere.

Literature Cited

Broders K, Boraks A, Barbison L et al (2014) Recent insights into the pandemic disease butternut canker caused by the pathogen *Ophiognomonia clavigignenti-juglandacearum*. For Pathol. https://doi.org/10.1111/efp.12161

Cale JA, Teale SA, Johnston MT et al (2015) New ecological and physiological dimensions of beech bark disease development in aftermath forests. For Ecol Manag 336:99–108

Clark SL, Schlarbaum SE, Saxton AM et al (2012) Lessons from the field: the first tests of restoration American chestnut (*Castanea dentata*) seedlings planted in the Southern Region. In: Butnor JR (ed) Proceedings of the 16th biennial southern silvicultural research conference. General Technical Report-SRS-P-156. U.S. Department of Agriculture, Forest Service, Southern Research Station, Asheville, pp 69–70

Clark SL, Schlarbaum SE, Pinchot CC et al (2014) Reintroduction of American chestnut in the National Forest System. J For 112:502–512

Cornell University Cooperative Extension (CUCE) (2017) New York Invasive Species Information, Partnerships for Regional Invasive Species Management (PRISMs). http://www.nyis.info/index.php?action=prism_partners

Frick WF, Pollock JF, Hicks AC et al (2010) An emerging disease causes regional population collapse of a common North American bat species. Science 329:679–682

Geils BW, Hummer KE, Hunt RS (2010) White pine, *Ribes*, and blister rust: a review and synthesis. For Pathol 40:147–185

Haugen LM (1998) How to identify and manage Dutch elm disease. USDA Forest Service, Northeastern State & Private Forestry NA-PR-07-98

Havill NP, Viera LC, Salom SM (2014) Biology and control of hemlock woolly adelgid. USDA Forest Service, Forest Health Technology Enterprise Team. FHTET-2014-05. 21 p

Juzwik J, Appel DN, MacDonald WL, Burks S (2011) Challenges and successes in management of oak wilt in the United States. Plant Dis 95:888–900

Kalisz S, Spigler RB, Horvitz CC (2014) In a long-term experimental demography study, excluding ungulates reversed invader's explosive population growth rate and restored natives. Proc Natl Acad Sci 111:4501–4506

Knight TM, Dunn JL, Smith LA et al (2009) Deer facilitate invasive plant success in a Pennsylvania forest understory. Nat Areas J 29:110–116

LaBonte NR, Ostry ME, Ross-Davis A et al (2015) Estimating heritability of disease resistance factors that contribute to long-term survival in butternut (*Juglans cinerea* L.). Tree Genet Genomes 11:63. https://doi.org/10.1007/s11295-015-0884-8

Liebhold AM, McCullough DG, Blackburn LM (2013) A highly aggregated geographical distribution of forest pest invasions in the USA. Divers Distrib 19:1208–1216

Marks CO (2017) The ecological role of American elm (*Ulmus americana* L.) in floodplain forests of northeastern North America. In: Pinchot CC et al (eds) Proceedings of the American elm restoration workshop 2016; 2016 October 25–27; Lewis Center, OH. General Technical Report NRS-P-174. U.S. Department of Agriculture, Forest Service, Northern Research Station, Newtown Square, pp 74–98

Northeastern Area Association of State Foresters (NAASF) (2017) Northeastern Area Association of State Foresters webpage. http://www.northeasternforests.org/

Pimentel D (2005) Aquatic nuisance species in the New York State Canal and Hudson River systems and the Great Lakes Basin: an economic and environmental assessment. Environ Manag 35:692–702

Pinchot CC, Flower CE, Knight KS et al (2017) Development of new Dutch elm disease-tolerant selections for restoration of the American elm in urban and forested landscapes. In: Sniezko RA et al (tech. coords) 2017 Gene conservation of tree species—banking on the future. Proceedings of a workshop. General Technical Report PNW-GTR-963. U.S. Department of Agriculture, Forest Service, Pacific Northwest Research Station, Portland, pp 53–63

Shultz J (2003) Conservation assessment for butternut or white walnut (*Juglans cinerea* L.). USDA Forest Service, Eastern Region. Hiawatha National Forest. www.fs.usda.gov/Internet/FSE_DOCUMENTS/fsm91_054130.pdf

U.S. Department of Agriculture (USDA) (2012a) Gypsy moth management in the United States: a cooperative approach. Supplemental environmental impact statement. U.S. Department of Agriculture, Forest Service and Animal and Plant Health Inspection Service. NA-MB-01-12, Washington, DC

U.S. Department of Agriculture (USDA) (2012b) Asian longhorned beetle and its host trees. U.S. Department of Agriculture, Forest Service, Northeastern Area State and Private Forestry. NA-PR-05-12, New Town Square

U.S. Department of Agriculture (USDA) (2017) Gypsy moth digest database. U.S. Department of Agriculture, Forest Service, Northeastern Area State and Private Forestry, Newtown Square. https://www.fs.usda.gov/naspf/programs/forest-health-protection/gypsy-moth-digest

Zaya DN, Leicht-Young SA, Pavlovic NB et al (2015) Genetic characterization of hybridization between native and invasive bittersweet vines (*Celastrus* spp.). Biol Invasions 17:2975–2988

Zhang B, Oakes AD, Newhouse AE et al (2013) A threshold level of oxalate oxidase transgene expression reduces *Cryphonectria parasitica*-induced necrosis in a transgenic American chestnut (*Castanea dentata*) leaf bioassay. Transgenic Res 22:973–982

Southeast and Caribbean

Albert E. Mayfield III, Humfredo Marcano-Vega, and Ariel E. Lugo

Introduction

Wide climatic variations characterize the Southeastern United States (Alabama, Arkansas, Florida, Georgia, Kentucky, Louisiana, Mississippi, North Carolina, Oklahoma, South Carolina, Tennessee, Texas, and Virginia) and Caribbean (Puerto Rico and the US Virgin Islands) region, including tropical, subtropical, warm-temperate, and temperate environments, as well as diverse ecosystems from coastal wetlands and dunes to piedmont savannahs and montane forests (Fig. A8.1). More than 85% of the forest land in the continental Southeast is privately owned, with the region experiencing rapid population growth (particularly around urban centers), as well as increased landscape and ownership fragmentation (Butler and Wear 2013). This population growth and urbanization, along with changing climate, are likely to put stressors on southeastern ecosystems in ways that may increase their invasion by, or decrease their resilience to, non-native invasive species (Duerr and Mistretta 2013; Miller et al. 2013a). The Atlantic, Gulf, and Caribbean Coasts in the region are home to numerous major commercial ports. The large quantity of shipments arriving from international ports daily serves as a constant potential pathway for new invasive pests and/or their propagules into the region.

Invasive species in the Southeast and Caribbean region include a wide variety of taxa and affect both terrestrial and aquatic systems. Wood-boring insect species, such as ambrosia beetles and their microbial associates and

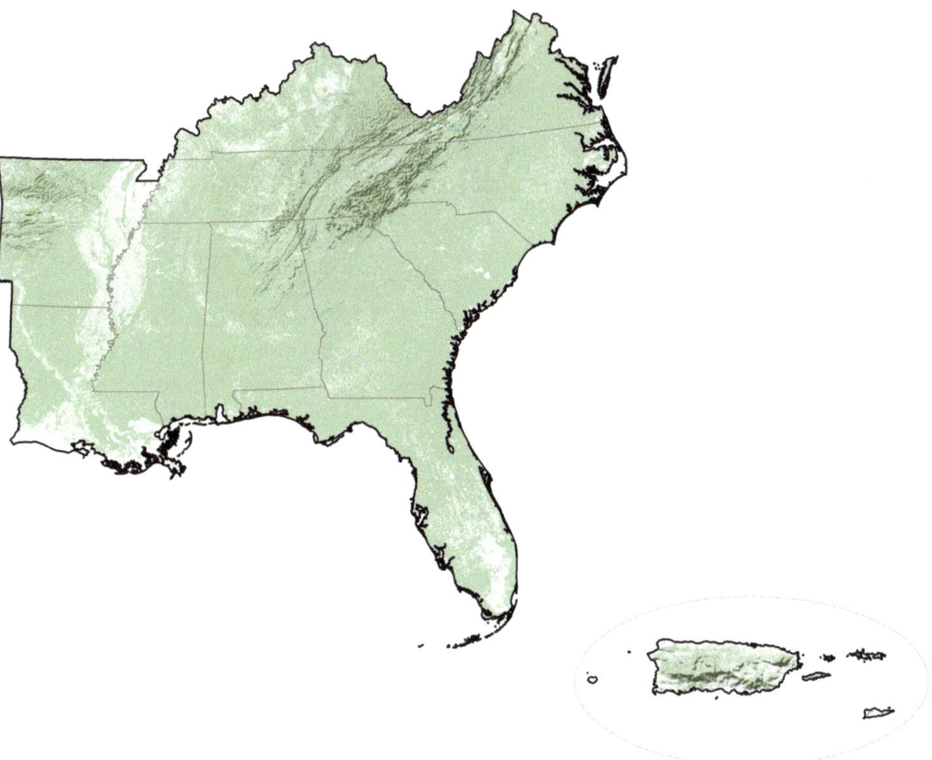

Fig. A8.1 The Southeast and Caribbean Region. (Figure courtesy Daniel Ryerson and Andy Graves, USDA Forest Service Southwestern Region, Forest Health Protection)

symbionts, are easily introduced in solid wood packing material; the warm, humid climate facilitates their establishment in southern forests (Marini et al. 2011). The climate is also very hospitable to the establishment of new invasive plants, which pose a threat not only through their own ecological effects but also via other organisms (e.g., insects and pathogens) they may harbor upon arrival through international plant trade (Miller et al. 2013a). In both the terrestrial and aquatic systems of the Southeast, numerous invasive animal species, including birds, reptiles, amphibians, and fish have, been introduced via the commercial pet trade, for which Florida is a major center of activity (Mazzotti and Harvey 2012; Padilla and Williams 2004; Ruiz-Carus et al. 2006; Russello et al. 2008; Townsend et al. 2003).

Compared to the continental Southeast, the ecosystems of the Caribbean islands of Puerto Rico and the US Virgin Islands, with their diverse sizes and respective levels of anthropogenic activities or land reserve statuses, experience unique environmental and economic effects of non-native invasive species. This chapter will highlight invasive species issues of importance in the region, with separate coverage for the continental Southeast and the Caribbean islands.

Continental Southeast

Invasive Plants

Invasive plants in natural and wildland-urban interface environments of the continental Southeast include over 470 non-native species of trees, shrubs, vines, grasses, and forbs (Miller et al. 2004). Many of these species cause environmental and economic effects that decrease the value of ecosystem services to humans in both terrestrial and aquatic environments. The region has invasive plant collaboration networks organized at local, State, and regional levels, including the Southeast Exotic Pest Plant Council (https://www.se-eppc.org), various cooperative weed management areas, and government agriculture and forestry organizations. These networks function to share information on the identification, control, and management of invasive plant species and to foster collaborative management and research efforts (Miller et al. 2013a). Invasive plants of substantial management importance in the Southeastern United States are numerous; detailed profiles on these plants have recently been covered elsewhere (Miller et al. 2013a, b). Here, key examples representing different plant taxonomic groups are introduced to illustrate the types of impacts and management needs associated with invasive plants in the region.

Cogongrass Cogongrass (*Imperata cylindrica*) is a dense, aggressive, colony-forming grass and is considered one of the top ten worst invasive weeds in the world (MacDonald 2004). In southeastern US forests, it outcompetes native plant species for nutrients, alters decomposition rates, reduces native plant species diversity, and limits light levels near the forest floor to the detriment of endemic flora (Brewer 2008; Brewer and Cralle 2003; Daneshgar and Jose 2009;

Daneshgar et al. 2008; Holly et al. 2009). Cogongrass infestations also increase fine fuel loads, fire intensity, and fire-related tree mortality in pine stands (Lippincott 2000; Platt and Gottschalk 2001) (Fig. A8.2a). Cogongrass forms a dense underground mat of rhizomes with abundant vegetative buds, making propagules difficult to eradicate and easy to spread via heavy equipment. Population genetic studies support historical records of at least two separate introductions of cogongrass into the Southeastern United States from Asia in the early 1900s (Lucardi et al. 2014). In the Southeast region, hundreds of thousands of hectares are infested with cogongrass in Florida, Alabama, and Mississippi (MacDonald 2004), and the species is spreading north and west from these epicenters (Center for Invasive Species and Ecosystem Health 2010). Federal- and State-funded control programs have been aimed at eradicating outlier populations and treating advancing fronts in South Carolina, Georgia, Alabama, Mississippi, Tennessee, Texas (Miller et al. 2004), North Carolina, Florida, and Louisiana. Control relies heavily on repeated applications of herbicides, but a number of cultural management techniques may be integrated to prevent new or treat existing infestations (MacDonald 2004; Miller et al. 2013b). Additional research and technology development on effective control and restoration strategies for cogongrass are ongoing.

Chinese Privet Chinese privet (*Ligustrum sinense*) is a shade-tolerant, evergreen shrub that invades bottomland forests, fencerows, and other habitats. Together with European privet (*Ligustrum vulgare*), it infests more than one million ha throughout the Southeastern United States (Miller et al. 2008). Chinese privet forms dense, mono-species thickets that reduce the abundance and diversity of native herbaceous and woody plants (Greene and Blossey 2012; Hanula et al.

Fig. A8.2 Examples of invasive plant species in the continental Southeastern United States. (**a**) Cogongrass (*Imperata cylindrica*) forms monocultures that outcompete native plant species and increase fire intensity and fire-related tree mortality in southern pine stands. (**b**) Dense, single-species thickets of shade-tolerant Chinese privet (*Ligustrum sinense*) reduce the abundance and diversity of native plants, pollinators, and other taxa in forest understories. (**c**) Common water hyacinth (*Eichhornia crassipes*) has a free-floating, interlocking root system and can rapidly cover the surface of freshwater bodies within a few months of unchecked growth. (**d**) Old World climbing fern (*Lygodium microphyllum*) can cover trees and carry fires into the forest canopy. (Photo credits: (**a**) James R. Meeker, USDA Forest Service, bugwood.org; (**b**) David J. Moorhead, University of Georgia, bugwood.org; (**c**) Graves Lovell, Alabama Department of Conservation and Natural Resources, bugwood.org; (**d**) Tony Pernas, USDI National Park Service, bugwood.org)

2009) (Fig. A8.2b). Experimental removals of Chinese privet in the riparian forests of the southeastern Piedmont have produced positive responses in the abundance, diversity, and/or growth of numerous taxonomic groups, including native herbaceous plants, trees, pollinators, and earthworms (Hudson et al. 2013, 2014; Lobe et al. 2014). Although Chinese privet can be managed using chemical and mechanical treatments, they root sprout readily, and seeds are widely dispersed by birds and other animals, promoting site reinfestations and range expansion. A lace bug, *Leptoypha hospital*, shows promise as a biological control agent, which could provide more efficient long-term control in natural systems (Zhang et al. 2013).

Common Water Hyacinth Common water hyacinth (*Eichhornia crassipes*) is a free-floating, aquatic vascular plant native to South America that has been introduced into freshwater environments in more than 50 countries and five continents (Villamagna and Murphy 2010). It is listed as a noxious weed in four southeastern US States (USDA-NRCS 2017) and is particularly pervasive from eastern Texas through the southern Gulf Region and through the peninsula of Florida (EDDMapS 2016). The plant reproduces both sexually and asexually and forms dense interlocking mats that block light penetration into subsurface water (Fig. A8.2c). Population growth is exponential; a single plant can increase to 500,000 plants within 5 months, covering an acre of water surface and weighing 400 tons (Gettys 2014). Water hyacinth growth inhibits photosynthesis of phytoplankton and submerged vegetation, reduces dissolved oxygen concentrations, alters fish diets through effects on prey species, and severely restricts boating access, navigability, and recreational opportunities (Villamagna and Murphy 2010). Several insect biological control agents have been released to control water hyacinth in the Southeastern United States, and although these have not provided complete control in Florida (Gettys 2014), recent analyses indicate they have been very effective in Louisiana (Nesslage et al. 2016). The species is also managed by intensive chemical or mechanical techniques (Gettys 2014). Water hyacinth tolerates a range of pH, temperature, and nutrient conditions and thrives in certain polluted environments; as such, there is considerable interest in its use for phytoremediation of waters containing excessive nitrogen, heavy metals, and other aquatic pollutants (Alvarado et al. 2008; Fox et al. 2008).

Old World and Japanese Climbing Ferns Introduced as an ornamental plant, Old World climbing fern (*Lygodium microphyllum*) is a twining, vine-like fern that has spread rapidly in natural areas of South and Central Florida since the 1960s. It produces fronds up to 30 m long that spread along the ground, as well as up and over shrubs, mature trees, and other structures (Fig. A8.2d). These fronds form mats of vegetation over 1 m thick that smother, shade, weaken, or kill native vegetation, including native orchids, ferns, and other rare plants found only in unique South Florida ecosystems such as the Everglades. The fronds also act as ladder fuels that carry fire into tree canopies, thereby spreading fire in swamps and other wet areas that would otherwise function as fire boundaries. A partnership known as the Central Florida *Lygodium* Strategy is working to stop the northward spread of Old World climbing fern by monitoring and treating sentinel sites (Langeland et al. 2016; Munger 2005). A related species with greater frost tolerance, Japanese climbing fern (*Lygodium japonicum*) has a wider invasive range, including at least nine southeastern States from Texas to Florida and north to North Carolina. Japanese climbing fern is an economic and regulatory problem for the pine straw industry because the plant is spread via the baling and sale of needles raked from the understory of pine stands (Van Loan 2006). Control of both Old World and Japanese climbing ferns currently relies heavily on herbicides, but a biological control program for the former is being pursued using defoliating and galling insects from the native range (Langeland et al. 2016; Minogue et al. 2009).

Chinese Tallow Chinese tallow (*Triadica sebifera*) is a tree species that produces colorful foliage and white waxy seeds that has been cultivated in China for at least 15 centuries. Introduced to the United States by Benjamin Franklin in 1772, it is valued for numerous practical and economic uses, including seed oil and fat production, medicinal compounds, and biofuel, and as an ornamental. Chinese tallow was promoted commercially in Louisiana and Texas in the early 1900s and has naturalized rapidly throughout the Southeast through repeated introductions, ornamental plantings, and natural spread. Chinese tallow has been increasingly recognized as an invasive species since the 1990s, due to its large seed loads, rapid growth, aggressive response to disturbance, and habitat adaptability. In riparian and floodplain forests, Chinese tallow forms monoculture stands void of other woody vegetation and with less biodiversity. Chemical and mechanical control of Chinese tallow is short-lived and expensive, estimated to cost $200–400 million in Texas, Louisiana, and Mississippi over two decades. There are promising candidate insects for biological control of Chinese tallow, including two from Asia that are under evaluation and one that is naturalized in the invaded range (Duncan et al. 2016; Gao et al. 2016; Wheeler and Ding 2014).

Invasive Forest Insects and Plant Pathogens

There are numerous historic and contemporary examples of invasive insects and pathogens causing substantial tree mortality in the continental Southeast, across a wide range of host plant taxa. Invasive forest insects and pathogens are particularly challenging to manage due to their high rates of reproduction, the ease of spreading them through the transport of wood products and nursery stock, the impracticality (and potential environmental costs) of applying pesticides in forested areas, and their ability to persist in the environment long-term. Management issues and research needs regarding invasive forest insects and disease pathogens are addressed by a diversity of Federal, State and local governments, universities, and private entities throughout the Southeast and are represented by groups such as the Forest Health Committee of the Southern Group of State Foresters and the Southern Forest Insect Work Conference (https://www.sfiwc.org/). Although not comprehensive, below are some of the invasive insect and disease issues of management importance in the continental Southeast region of the United States.

Insects

Hemlock Woolly Adelgid The hemlock woolly adelgid (HWA) (*Adelges tsugae*) is an invasive insect pest threatening the health and sustainability of eastern hemlock (*Tsuga canadensis*) and Carolina hemlock (*Tsuga caroliniana*) in eastern North America. Although hemlock adelgids are native to Western North America, the population of HWA in the Eastern United States was introduced from Japan sometime prior to its initial detection in Virginia in the early 1950s and has spread through more than half the native range of eastern hemlock from Maine to Georgia (Havill et al. 2014). In the Southeastern United States, only a small portion of the hemlock range in Kentucky, Tennessee, and Alabama has not yet become infested (USDA Forest Service 2016). Hemlock mortality and decline caused by HWA has been particularly rapid and pronounced in the Southern Appalachian Mountains (Fig. A8.3c), where winter temperatures are rarely low enough to kill sufficient numbers of adelgids to affect insect population dynamics. Eastern hemlock is a foundation species in riparian areas of the Southern Appalachians, and its elimination by HWA results in numerous short- and long-term ecological effects that include microclimatic changes, altered hydrological and nutrient cycling regimes, and shifts in forest species composition (Vose et al. 2013). Systemic insecticides have been used successfully to temporarily protect trees in parks,

Fig. A8.3 Tree mortality caused by invasive forest insects in the Southern Appalachian Mountains. (**a**) Green ash (*Fraxinus pennsylvanica*) killed by the emerald ash borer (*Agrilus planipennis*) in Virginia. (**b**) Fraser fir (*Abies fraseri*) killed by the balsam woolly adelgid (*Adelges piceae*) in Tennessee. (**c**) Eastern hemlock (*Tsuga canadensis*) killed by the hemlock woolly adelgid (*Adelges tsugae*) in North Carolina. (Photo credits: (**a**) Christopher Asaro, Virginia Department of Forestry, Bugwood.org; (**b**) Ronald F. Billings, Texas A&M Forest Service, Bugwood.org; (**c**) David Casey, USDA Forest Service)

recreational areas, and urbanized landscapes but must be applied on a single-tree basis and are not practical for indefinite use in forests. Several species of predators from the native range of HWA in Asia and the Pacific Northwest have been released and become established in the Southeastern United States; their efficacy is currently being evaluated (Onken and Reardon 2011). Research and management efforts in the areas of gene conservation (i.e., seed collection, banking, and orchard establishment, Jetton et al. 2013), host resistance (Oten et al. 2014), and silviculture and restoration (Jetton 2017; Piatek et al. 2016) are also currently being pursued, but additional research effort is likely needed in each of these areas to viably maintain hemlock in the Eastern United States.

Balsam Woolly Adelgid Like the hemlock woolly adelgid, the balsam woolly adelgid (*Adelges piceae*) kills a tree species with relatively limited regional abundance and distribution but with very high local ecological importance. Since its arrival in the high-elevation spruce-fir forests of the Southern Appalachians in the 1950s, balsam woolly adelgid has killed the vast majority of mature Fraser firs in these unique ecosystems (Fig. A8.3b). Drastic losses of the fir canopy and resulting changes in vegetative composition may pose risks for several rare plant and animal species, some of which are at risk of extinction. Although younger cohorts of Fraser fir have survived the initial balsam woolly adelgid epidemics, it is uncertain whether subsequent outbreaks will gradually reduce the genetic diversity of Fraser fir populations and/or its codominant status (with red spruce (*Picea rubens*)) in these forests. In addition to its ecological impact, balsam woolly adelgid is a major pest of the Fraser fir Christmas tree industry, with an annual cost of more than $1.5 million for insecticide treatments in North Carolina alone. Finding and propagating adelgid-resistant Fraser firs and conserving the genetic diversity of the species will likely be important to the conservation and restoration of these southeastern spruce-fir ecosystems (Potter et al. 2005).

Emerald Ash Borer The emerald ash borer (EAB) (*Agrilus planipennis*) is an invasive phloem-feeding beetle native to Asia that threatens to eliminate the ash (*Fraxinus* spp.) from North America. It was initially detected in Michigan in 2002, and by May 2020, it had spread to at least 36 US States and five Canadian provinces (Canadian Food Inspection Agency 2020; EAB Information Network 2020; Herms and McCullough 2014). In the Southeastern region, EAB has been spreading steadily since 2008 and by May 2020 had been detected in all southeastern States except Florida and Mississippi (EAB Information Network 2020). The EAB is the most destructive forest insect in US history and causes an estimated combined loss of $1.7 billion annually to governments and households and in property devaluation (Aukema et al. 2011). In forests, ash mortality due to EAB can (but does not necessarily) exceed 99% within 6 years (Fig. A8.3a) leaving an "orphaned cohort" of seedlings (Klooster et al. 2014; Knight et al. 2013). These changes alter stand light regimes, nutrient cycles, successional dynamics, susceptibility to invasive plants, and coarse woody debris volume (Gandhi and Herms 2010). There have been extensive investments made to improve early detection traps and lures, but even the best combinations are thought to only detect well-established populations. A unique system of bio-surveillance for EAB has been developed by monitoring nest entry by native *Cerceris* wasps, which prey on EAB and other buprestid species (Careless et al. 2014). Effective systemic insecticides are available for temporary protection of high-value landscape trees (Herms and McCullough 2014). Several introduced biological control agents (parasitoids from the native range in Asia) and native natural enemies are reducing EAB population growth rates and enhancing the survival of young ash trees (Duan et al. 2017). As with hemlock, research and management efforts in the areas of host gene conservation and development of host resistance are being pursued but need continued attention (Herms and McCullough 2014; Poland et al. 2015).

Gypsy Moth The European gypsy moth (*Lymantria dispar*) was introduced to Massachusetts in 1869. Outbreaks of this invasive defoliator in New England, Mid-Atlantic, and Northcentral US forests have been the focus of management and research attention for many decades. With the exception of Virginia and extreme northeastern North Carolina (NC-DACS 2016), the European gypsy moth has not yet established in most of the Southeastern United States, though it is expected to continue advancing into the region (Duerr and Mistretta 2013). The gypsy moth can completely defoliate forest canopies, resulting in a number of short-term ecological effects caused by increased light, reduced transpiration, changes in nitrogen and carbon dynamics, and reduced mast production. Oaks (*Quercus* spp.) and aspens (*Populus* spp.) are preferred gypsy moth hosts, but numerous species may be defoliated during outbreaks, including evergreen species such as pine (*Pinus* spp.) and hemlock (Lovett et al. 2006). The severity of tree mortality due to gypsy moth depends on a number of factors, such as tree species composition and susceptibility, the duration and frequency of outbreaks, tree canopy position, and preexisting tree health (Davidson et al. 1999). The spread of the gypsy moth into the Midwest and Southeast is being slowed through a multifaceted management program (including detection surveys, insecticide treatments, and mating disruption) implemented at the infestation's leading edge (Sharov et al. 2002; Tobin 2008). The extent to which an important fungal pathogen, *Entomophaga maimaiga*, helps suppress gypsy moth outbreaks varies with weather conditions, is difficult to predict, and may change with changing climate and movement of the gypsy moth into the Southeast region (Reilly et al. 2014).

Diseases

Laurel Wilt Laurel wilt is a vascular disease caused by a fungus, *Raffaelea lauricola*, that is carried by the redbay ambrosia beetle (*Xyleborus glabratus)* (Fraedrich et al. 2008; Hughes et al. 2015) (Fig. A8.4). This insect/pathogen complex was likely introduced from Asia in solid wood packing material, and the insect was first detected near Savannah, GA, in 2002 (Fraedrich et al. 2008). Since then, laurel wilt has spread rapidly throughout the southeastern Coastal Plain region and is found from North Carolina to Texas, as well as south into the Everglades (Georgia Forestry Commission 2016; Rodgers et al. 2014). The disease is deadly to redbay (*Persea borbonia*), swamp bay (*Persea palustris*), and silk bay (*Persea humilis*) and has killed millions of stems of these species, with the probability of mortality increasing rapidly as stem diameter increases (Mayfield and Brownie 2013; Shearman et al. 2015). The disease is also killing sassafras (*Sassafras albidum*) and avocado (*Persea americana*) and has been documented in populations of rare plant species including pondberry (*Lindera melissifolia*) and pondspice (*Litsea aestivalis*) (Hughes et al. 2015). Laurel wilt represents a serious potential threat to native California bay laurel (*Umbellularia californica*) populations and to avocado production on the West Coast, as well as to the wide diversity of lauraceous plants elsewhere in the Americas (Kendra et al. 2014; Mayfield et al. 2013). There are no effective management strategies in natural systems as yet. Prophylactic fungicide treatments are short-lived and cost-prohibitive (Mayfield et al. 2008), and current detection lures are not effective for manipulating even low beetle populations (Hanula et al.

Fig. A8.4 Laurel wilt is a vascular disease of woody plants in the family Lauraceae in the Southeastern United States. It is caused by a fungal pathogen, *Raffaelea lauricola*, that is carried into healthy trees by the (**a**) redbay ambrosia beetle (*Xyleborus glabratus*). (**b**) Infected redbay (*Persea borbonia*) trees wilt rapidly and develop a dark discoloration in the outer sapwood. Millions of redbay trees have been killed in (**c**) neighborhoods, (**d**) mixed maritime forests, (**e**) bayhead swamps, and other southeastern United States ecosystems, and populations of other North American species including sassafras (*Sassafras albidum*) and avocado (*Persea americana*) are being negatively affected. (Photo credits: (**a–d**) Albert Mayfield, USDA Forest Service; (**e**) R. Scott Cameron, Advanced Forest Protection, Inc.)

2016). Efforts to find natural resistance to the disease among lingering redbay trees in post-epidemic forests show promise (Hughes et al. 2015), but additional research on management strategies for numerous other hosts is needed if substantial mortality of lauraceous plants elsewhere in the Americas is to be avoided or minimized.

Thousand Cankers Disease Thousand cankers disease of black walnut (*Juglans nigra*) and other *Juglans* spp. is caused by the joint action of the walnut twig beetle (*Pityophthorus juglandis*) and a fungal pathogen, *Geosmithia morbida*, which it carries. Both the insect vector and the pathogen are thought to be native to the Southwestern United States and Northern Mexico, but they have been discovered far outside their historically known ranges in recent decades, causing dieback and mortality of *Juglans* spp. in the Western United States (Rugman-Jones et al. 2015). Thousand cankers disease was first detected on black walnut in the Southeastern United States in 2010–2012, in portions of Tennessee, North Carolina, and Virginia. Black walnut is one of the most economically valuable tree species in the United States, where standing volume is estimated to be worth ~$500 billion. Numerous States have responded to the spread of thousand cankers disease with quarantines and regulations limiting the movement of walnut material (Daniels et al. 2016). The disease has also been detected in Italy, illustrating the risk of spreading the causal organisms internationally via shipment of unprocessed wood. To date, the impact caused by thousand cankers disease in the Eastern United States has not been well quantified, and the degree to which the disease will be a problem in the native range of black walnut is uncertain.

Butternut Canker Butternut (*Juglans cinerea*) has experienced high rates of mortality (70–90%) throughout its range due to butternut canker, a disease caused by a rapidly spreading, invasive fungal pathogen, *Ophiognomonia clavigignenti-juglandacearum* (Broders et al. 2015). Butternut is not a major component of southeastern US forests (less than 0.5%), but it is a locally valued hardwood for carving, furniture, and various uses of the nuts. Dramatic losses due to butternut canker have resulted in butternut being listed as a Species of Special Concern in Kentucky and as Threatened in Tennessee (Duerr and Mistretta 2013). Research on the development of disease-resistant butternut has shown promise using single-tree selection and breeding, but there is a need for continued butternut conservation and genetic characterization of both the host and the pathogen (Michler et al. 2006; Ostry et al. 2003).

Chestnut Blight The chestnut blight fungus (*Cryphonectria parasitica*) along with the ink disease pathogen (*Phytophthora cinnamomi*) functionally eliminated the American chestnut (*Castanea dentata*) from the Appalachian Mountains in the early 1900s (Anagnostakis 1987). Research aimed at reintroducing blight-resistant American chestnut has employed a backcross breeding program with Chinese chestnut (*Castanea mollissima*) (Hebard 2005) and, more recently, the use of transgenic techniques (Maynard et al. 2008; Newhouse et al. 2014). Field tests of chestnut seedlings produced through the backcross breeding program are being evaluated in the Southern Appalachians (Clark et al. 2016).

Dogwood Anthracnose Another invasive pathogen, *Discula destructiva*, has spread from the Northeastern United States into the Southern Appalachians (Hiers and Evans 1997; Jenkins and White 2002) and has caused precipitous declines in the abundance of flowering dogwood (*Cornus florida*), with potential ecological effects that include disruption of the soil calcium cycle, decreased soil fertility, and decreased species richness (Holzmueller et al. 2010). Research suggests that prescribed fire may help to perpetuate dogwood in forests affected by dogwood anthracnose (Holzmueller et al. 2006, 2009).

Invasive Vertebrates

Burmese Python Burmese pythons (*Python molurus bivittatus*), giant constrictors native to Southeast Asia, were recognized as established in southern Florida by 2000. Over the following decade, they increased dramatically in abundance and distribution, including widespread colonization of Everglades National Park (Dorcas et al. 2012). Both indirect and recent experimental evidence now link this invasive apex predator to precipitous declines in populations of numerous mammal species; Burmese pythons also jeopardize threatened and endangered species such as the Key Largo woodrat (*Neotoma floridana smalli*) and wood stork (*Mycteria americana*) (Dorcas et al. 2012; McCleery et al. 2015). Radio tracking in the Everglades reveals that Burmese pythons have large home ranges that may exceed 2000 ha, use sloughs and coastal habitats as core areas, and frequently use tree islands (Hart et al. 2015). Despite their large size, monitoring and estimating Burmese python population sizes are difficult due to the pythons' cryptic nature and aquatic habitats (Piaggio et al. 2014). New detection techniques using environmental DNA (eDNA) may be useful for monitoring Burmese python occurrence and spread (Piaggio et al. 2014). Government-sanctioned bounty hunts and other attempts to kill pythons have helped raise awareness of the issue, but the thousands that have been documented as killed in Florida over the last 16 years likely represent only a small fraction of the population. The Florida Fish and Wildlife Conservation Commission partners with other agencies to implement management strategies that include licensed hunts, use of citizen science reporting, and prevention of new introductions (Florida FWCC 2016), but it is likely that eradicating Burmese pythons from the region is now impossible.

Feral Swine Feral swine, or wild pigs (*Sus scrofa*) (Fig. A8.5a), have repeatedly been introduced around the globe for centuries and are believed to have been introduced into what is now the Southeastern United States by Spanish explorers in the 1500s (Mayer and Brisbin 2008). Historically, US populations of feral swine were limited to the Southeast, Hawaii, and California, but their range expanded dramatically between 1982 and 2012, from 17 to 38 States (Bevins et al. 2014). This range expansion is thought to have been expedited by several factors, including human transport and release, elimination of predators, and increased human alteration of landscapes for agriculture. While some effects of feral swine in the Southeast are viewed as beneficial (e.g., they serve as part of the diet of the endangered Florida panther (*Puma concolor coryi*) and are hunted for sport), they are widely considered a nuisance species due to rooting/feeding damage to crops, native vegetation, and soil-related habitats (Fig. A8.5b) and are associated with declines in numerous rare, threatened, or endangered plant and animal species populations (Bevins et al. 2014). Successful control or reduction of feral swine populations outside of the Southeastern region (e.g., California, Oregon, and Kansas) has involved sustained, integrated use of multiple management strategies such as hunting, trapping, toxicants, and exclusion devices, but substantial proportions of the populations may need to be eliminated annually to restrict negative effects in areas where they have long been established (Bevins et al. 2014; Timmons et al. 2012).

Nutria Nutria (*Myocastor coypus*) are large semiaquatic rodents that are native to South America but that have been introduced worldwide for fur farming and in some cases to control nuisance aquatic vegetation. They were introduced into Louisiana in the 1930s and now occur in all southeastern States with the possible exception of South Carolina (Carter and Leonard 2002). In wetlands, nutria herbivory destroys aboveground vegetation and root mats, retarding marsh development and contributing to their conversion to open water (Evers et al. 1998; Ford and Grace 1998). Nutria also undermine water control structures via burrowing, cause damage to agricultural crops, and harbor parasites of humans, livestock, and other wildlife (Carter and Leonard 2002). Control programs to reduce nutria damage include eradication attempts as well as managing for low populations, such as in Louisiana where nutria are not only considered ecological pests but are also valued for fur and meat by hunters and trappers (Jojola et al. 2005). In 2002, the Louisiana Department of Wildlife and Fisheries began the Coastwide Nutria Control Program, in which participants are provided an incentive payment for each nutria tail delivered. This program has consistently resulted in more than 300,000 nutria harvested annually and is correlated with a reduction in damaged wetland area from 32,400–40,500 ha annually down to 1600–2500 ha annually (Louisiana Department of Wildlife and Fisheries 2016).

Caribbean Islands

The archipelagos of Puerto Rico and the United States Virgin Islands (USVI) are case studies for the level of species introductions and their social and ecological consequences on local, insular, and regional scales that span the whole insular Caribbean. Table A8.1 shows that the number of introduced species in the region is substantial. Introduction of species has increased the total number of flora and fauna species in the Caribbean. Part of the reason is historical, as the islands have experienced a human presence for thousands of years, including over 500 years of European presence in the region.

Fig. A8.5 (**a**) Feral swine (*Sus scrofa*) have been present in the Southeastern United States for hundreds of years, but their range has expanded dramatically in the last three decades largely due to human land-use changes. (**b**) Their rooting and feeding activity can cause substantial soil disturbance and erosion, influencing plant succession and species composition. They are also hunted for sport and comprise part of the diet of large predators such as the Florida panther. (Photo credits: (**a**) Karan A. Rawlins, National Park Service; (**b**) University of Georgia, bugwood.org)

Table A8.1 Estimates of the number of introduced taxa to the Caribbean islands

Taxon	Native (Caribbean)	Introduced (Caribbean)	Native (PR, USVI)	Introduced (PR, USVI)
Families of seed plants	183	25	163	
Genera of seed plants	1474	560	792	
Species of seed plants	10,401	1799	2108	1032
Herbs				302
Trees				271
Shrubs				167
Vines				129
Grasses				119
Succulents or aquatic				44
Crustaceans		2		
Earthworms		1		
Insect		90		
Jellyfishes		1		
Mites		8		
Mollusks		17		
Solifugids		1		
Tunicates		1		
Amphibians		8		
Birds		20		
Fish		37		
Mammals		20		
Reptiles		15		
Fungi		2		
Diseases		2		

The first three rows for the Caribbean and native seed plants in Puerto Rico and the US Virgin Islands (PR, USVI) are from Acevedo-Rodríguez and Strong (2012). Data on introduced plants and their life forms in PR VI are from Rojas-Sandoval and Acevedo Rodríguez (2015). The rest of the data are from Kairo et al. (2003), who caution that the data are incomplete

Empty cells mean the information is not available in the publications cited or is incomplete. However, information for some of the seed plant taxa is available: http://botany.si.edu/antilles/WestIndies/query.cfm or http://www.sil.si.edu/smithsoniancontributions/Botany/. The numbers of Kairo et al. (2003) add to 225 species (121 invertebrates, 100 vertebrates, and 4 fungi and diseases)

The importance of global commerce to the economy of the islands is another reason for so many species of plants and animals to have been introduced to the region (Ewel et al. 2013). More significant from an ecological viewpoint is the fact that a large number of the introduced species have been naturalized in the Caribbean islands. Naturalized introduced species are now part of the natural milieu of insular landscapes, so much so that many of the insular forests are novel forests and many people cannot differentiate between a native endemic and an invasive introduced species. This mixing of species at ecological and social processes leads to proposals that incorporate introduced species into national or municipal symbols.

The geographic theory known as the forest transition model (Rudel 1998) helps explain why so many introduced species naturalize in the Caribbean and why they become part of the natural forest succession of the region. Fig. A8.6 depicts Forest Inventory and Analysis (FIA) data on the historical changes of forest cover in Puerto Rico. It shows a 400-year deforestation process that accelerated in its last 150 years, followed by an equally rapid natural reforestation over the last 60 years or so. The islands of the Caribbean were deforested to accommodate both agricultural activity and fuelwood extraction to power their economies. With the advent of fossil fuels, fuelwood extractions declined and, with the decline of monoculture crops, so did the pressure on the land as people abandoned the land and moved to cities. The process was quicker in Puerto Rico, which today is ahead of other islands in the temporal progression of the forest transition.

Land abandonment left degraded soils and a deforested landscape behind. This gave introduced species an opportunity to compete, establish, and form what are now recognized as novel forests and ecosystems, i.e., forests and ecosystems with novel combinations of plants and animals. The invasive nature of many introduced tree species was an ecologically sound attribute that allowed the restoration of forest conditions and favorable habitats for the regeneration of native species (flora and fauna) that had been limited to isolated refugia during the deforestation period. Most evidence from Puerto Rico and the USVI shows that introduced and native species can cohabitate and complement each other (Lugo et al. 2012). However, current population models of the native Puerto Rico applecactus (*Harrisia portoricensis* Britton) predict that it may become extinct within areas colonized by the non-native guinea grass (*Megathyrsus maximus* (Jacq.) B.K. Simon & S.W.L. Jacobs), suggesting a distinctive circumstance of displacement in Puerto Rico (Rojas-Sandoval et al. 2016a). Here, we highlight examples of introduced species that are considered invasive and problematic to a variety of human activities. Though we are careful to point out whether perceptions are based on objective information or on the geographic origin of the species, the reality of our examples is usually mixed. A species can be problematic under certain social or ecological conditions and not in others and depending on the context of human judgments regarding health, safety, and wealth. Thus, we suggest that, rather than focusing on the non-native species as the source of problems, attention be given to the functions and interactions among species and humans with regard to their intermixed effects on native populations and ecosystems.

Interspecific Interactions with Non-native Invertebrates

Herbivory The invasive South American *Harrisia* cactus mealybug (*Hypogeococcus pungens* Granara de Willink) is severely affecting columnar cacti in Puerto Rico. Native to

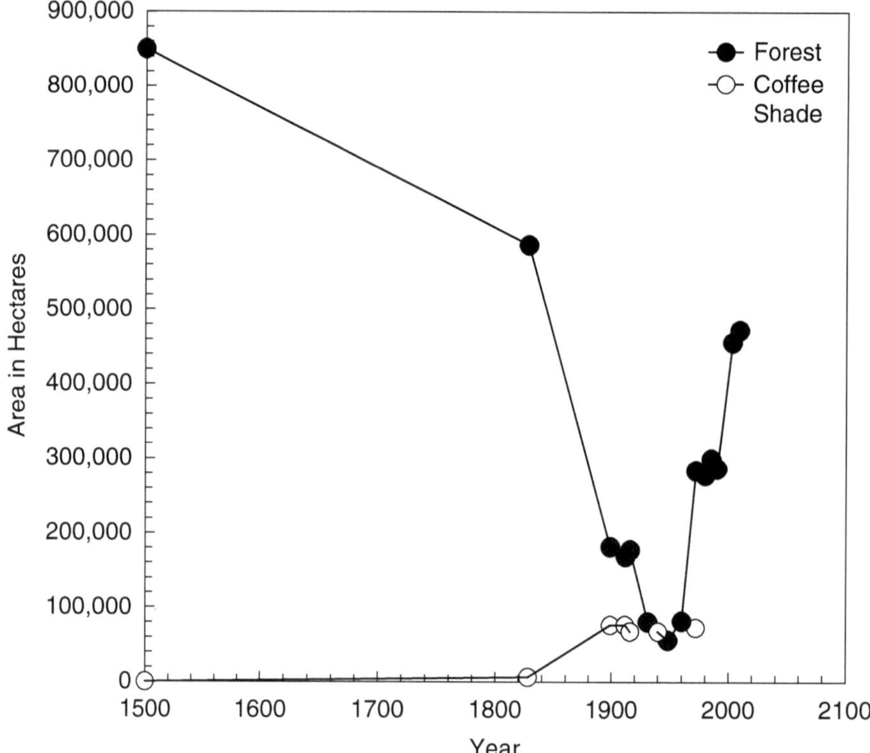

Fig. A8.6 Historical changes of forest cover and coffee shade in Puerto Rico. (Sources of data: Birdsey and Weaver 1982, 1987; Brandeis and Turner 2013a; Brandeis et al. 2007; Franco et al. 1997)

northern Argentina and Chile, westernmost Brazil, Paraguay, and southern Perú, this mealybug was originally intercepted in Puerto Rico on the imported ornamental little hogweed (*Portulaca oleracea* L.) in 2000 and was discovered feeding on the island's cactus during 2005 (Segarra-Carmona et al. 2010). Cacti are important components of dry forests in Puerto Rico and provide food sources for native animals such as endemic bats, birds, moths, and other pollinators (US Department of Agriculture APHIS PPQ 2009). Observed herbivory on cactus hosts includes malformations exposed as gall-like structures, usually associated with arrested growth, prevention of flowering, and the eventual death of some individuals (Zimmerman et al. 2010). The mealybug has substantially extended its distribution and is quickly spreading in the dry forest life zone of the southern coast of Puerto Rico, causing extensive impairment on the tree cactus (*Pilosocereus royenii* (L.) Byles & Rowley). Other members of the Cactaceae family exhibiting severe impairment include the native pitahaya (*Leptocereus quadricostatus* (Bello) Britton & Rose) (Fig. A8.7) and Turks' cap (*Melocactus intortus* (Mill.) Urb.) and the introduced lady of the night cactus (*Cereus hexagonus* (L.) Mill.) found under cultivation. Minor effects have been observed on the native Spanish stenocereus (*Stenocereus fimbriatus* (Lam.) Lourteig) (Segarra-Carmona et al. 2010). The host range of the *Harrisia* cactus mealybug also includes species in families dissimilar to the Cactaceae. Early detection programs and regular inspection of nurseries have been suggested by scholars as a way to limit its spread (Zimmermann et al. 2010). Wind dispersal is considered one of the foremost pathways for the spread of the *Harrisia* cactus mealybug. Thus, its potential expansion from mainland Puerto Rico to other islands of the Puerto Rican Archipelago where endangered cacti are found (i.e., guinea grass) has been highlighted as a concern (Campbell 2015; Zimmerman et al. 2010). The *Harrisia* mealybug is also known to be causing the death of native cacti on the USVI (Campbell 2015).

Chemical control of the *Harrisia* cactus mealybug has been suggested as an appropriate measure specifically for infested plants within nurseries and commercial cultivations. This measure is not feasible for protecting cacti populations in the wild, however, where biological control is recommended instead (Segarra-Carmona et al. 2010). The coccinellid *Decadiomus seini* sp. nov. Segarra & Otero is a new species whose larvae and adults have been discovered feeding on the *Harrisia* cactus mealybug in the dry forests of Puerto Rico (Segarra-Carmona and Otero 2014). In addition, *Gyranusoidea pseudococci* Brèthes, a parasitoid widely distributed in Florida, and the parasitoid wasps of the genus *Leptomastidea* have been influential natural enemies of the *Harrisia* cactus mealybug, thus offering further potential for biological control in Puerto Rico and the Caribbean (US Department of Agriculture APHIS PPQ 2009). Other introduced insects, such as the pink hibiscus mealybug (*Maconellicoccus hirsutus* (Green)), have been successfully controlled in the USVI through the release of nonindigenous wasps in the genus *Leptomastix* (Daley et al. 2012; Meyerdirk

Fig. A8.7 Herbivory of the introduced South American *Harrisia* cactus mealybug (*Hypogeococcus pungens*) on the native pitahaya (*Leptocereus quadricostatus*) induces the proliferation of miniature cacti. Observed herbivory on other native cacti also includes malformations such as gall-like structures; it is usually associated with arrested growth, prevention of flowering, and the eventual death of some individuals. (Photo credit: Humfredo Marcano-Vega, USDA Forest Service Southern Research Station)

et al. 2001). The pink hibiscus mealybug is known to attack diverse crops and ornamental plants, some of which are part of the secondary forests of Puerto Rico and the USVI (Brandeis and Turner 2013a, b; Marcano-Vega et al. 2015; Meyerdirk et al. 2001).

The effects of introduced species such as the *Harrisia* cactus mealybug on landscape elements of the subtropical dry forest (sensu Holdridge; Ewel and Whitmore 1973) are an opportunity to include human ecology principles such as human perceptions (e.g., emotional value) in the study of species invasions within the Caribbean context and abroad (González Bernáldez 1985; Sax et al. 2007). The destructive agave snout weevil (*Scyphophorus acupunctatus* Gyllenhaal), native to the Southwestern United States and Mexico, is present in Puerto Rico and the USVI. Its occurrence has increased the risk of extinction of the USVI endemic and endangered century plant (*Agave eggersiana* Trel.) in St. Croix and of the Puerto Rico Bank endemic corita (*A. missionum* Trel.) in St. John. Chamorro et al. (2016) recently reported the weevil in St. Croix, providing tools for its identification at any stage and recommending a monitoring program throughout the already restricted range of century plant. This initiative is considered to be crucial for prevention because the agave snout weevil's effect is currently described by specialists as virtually impossible to reverse, and landscape management experts have suggested the application of a systemic insecticide around the base of healthy plants (DLC Resources, Inc. 2015). On the other hand, the College of Agricultural Sciences of the University of Puerto Rico at Mayagüez formed the Grupo Antillano de Especies Invasoras (GAEI), which maintains an atlas of invasive insect species. This group shows the *Tabebuia* thrips (*Holopothrips tabebuia* Cabrera & Segarra) as an invasive insect known to be widespread around all mainland Puerto Rico (GAEI 2016; González 2016). Although the genus *Holopothrips* is known to be of neotropical origin, the species' native range is currently uncertain. The *Tabebuia* thrips were found in Puerto Rico in 2006 and are now known to affect the esthetic appearance of leaflets (inward curling forming gall-like structures) and growth of native and introduced tree species in the *Tabebuia* and the *Crescentia* genus in urban and secondary forests (GAEI 2016; González 2016). Currently there is no known mortality of trees as a consequence of the *Tabebuia* thrips, but long-term monitoring is recommended along with the preventive use of systemic insecticides and pruning of affected trees (González 2016). Alternative biocontrol is offered by the insect *Montandoniola moraguesi* Puton (Hemiptera: Anthocoridae), which has been reported as a predator of the *Tabebuia* thrips in Puerto Rico and Florida (GAEI 2016; González 2016).

Interspecific Competition and Mutualism Dispersion of the introduced honeybee (*Apis mellifera* L.) to relatively undisturbed forests at high elevations in northeast Puerto Rico has resulted in novel interactions among species. The observation of swarms of the honeybee indicates that the bees are able to exploit the local forest's floral resources and compete for cavity nesting sites with the endangered Puerto Rican parrot (*Amazona vittata* Boddaert) and other native species (Lugo et al. 2012; Snyder et al. 1987). On the other hand, wild populations of honeybees have also been observed as second in efficiency at depositing pollen on stigmas of flowers of the native terrestrial bromeliad pina cortadora (*Pitcairnia angustifolia* Aiton), resulting in a more complex plant-pollinator relationship than what may be expected from the bromeliad's flower morphology (Fumero-Cabán and Meléndez-Ackerman 2007). Opposing results were found for the native Turk's cap cactus, whose flowers are also visited by honeybees but with negligible effects on the reproductive success of the plant (Fagua and Ackerman 2011). The novel development of partial dependence of native species upon non-natives is developing into a common interspecific interaction in tropical islands (Ewel et al. 2013), but more research is needed to determine the effectiveness and/or consequences of acquired relationships.

Belowground Terrestrial Interactions Additional studies are suggested to better understand the effects of non-native earthworms on tropical belowground ecosystems (González et al. 2006). Studies show how earthworm density decreased while earthworm diversity increased along successional stages of wet forest regeneration after the abandonment of pastures, where *Pontoscolex corethrurus* Müller appeared as the dominant non-native earthworm occurring in all successional stages in Puerto Rico (Zou and González 1997). This earthworm is also found in relatively undisturbed cloud forests, as well as lower altitude tabonuco (*Dacryodes excelsa*) forests in Puerto Rico (Hendrix et al. 1999), but their ecosystem effects remain generally unknown (González et al. 2006). Even so, studies from Puerto Rico show how conversion of forests can increase soil carbon dioxide (CO_2) emissions as litter decomposition and soil respiration can significantly increase under the presence of introduced earthworms such as *P. corethrurus* (Liu and Zou 2002).

Interactions with Non-native Plants

Puerto Rico's secondary forests represent a legacy of anthropogenic activities with new combinations of both native and introduced tree species (Abelleira-Martínez 2010; Fonseca Da Silva 2014; Lugo 2004, 2009). Studies characterizing tree species composition in secondary stands of Puerto Rico highlight the critical function of the introduced African tulip tree (*Spathodea campanulata* P. Beauv.) (Fig. A8.8) as a novel forest facilitator, along with remnant native forests or trees functioning as seed sources for native species recovery within the subtropical moist and wet forest life zones (Abelleira-Martínez 2010; Brandeis and Turner 2013a). The importance of mature native forests as local seed pools is critical because novel forests exhibit fewer endemic species than native forests (Lugo and Helmer 2004). Both novel and native forests, however, share similar structural characteristics (Fonseca Da Silva 2014; Lugo and Helmer 2004). Nevertheless, when comparing novel forest stands with native forests of similar age in analogous Holdridge life zones with similar topographies and elevation, novel forest stands demonstrate lower rates of aboveground biomass, basal area, and soil carbon accumulation (Lugo and Helmer 2004).

In another circumstance, the introduction of Caribbean pine (*Pinus caribaea* Morelet) required the additional introduction of ectomycorrhizal (EM) inoculum to Puerto Rico during the 1950s, for plantations to be successfully established (Briscoe 1959). After reporting the non-native EM fungus *Pisolithus tinctorius* (Pers.) Coker & Couch fruiting under Caribbean pine, populations of both species can be found today outside plantations in disturbed areas (Rivera et al. 2015). It remains uncertain if Caribbean pine will function in the long term as a facilitator or inhibitor of native species during forest regeneration, but some native tree species have been able to reach the canopy layer of small unmanaged plantations after 18 years of growth (Lugo 1992).

Studies show that, after hurricanes, the rapid revitalization of native species in protected old-growth forests within wet areas keeps shade-intolerant introduced species as relatively rare within these areas (Thompson et al. 2007). In contrast, the shade-tolerant introduced rose apple (*Syzygium*

Fig. A8.8 The introduced African tulip tree (*Spathodea campanulata*), with its large orange-red flowers, is widely naturalized in the humid forests of Puerto Rico functioning as a novel forest facilitator along with remnant native forests. (Photo credit: Luis O. Ortiz-López, USDA Forest Service Southern Research Station)

jambos (L.) Alston) has shown the ability to establish populations and persist in mature native forests over a wide range of soil and topographic conditions (Brown et al. 2006; Thompson et al. 2007). But the invasive rust fungus *Puccinia psidii* Winter, known to infect diverse hosts in the Myrtaceae family, has been affecting rose apple in Puerto Rico and acting as an unintentional biological control across the island (Ross-Davis et al. 2013). All the same, the expected increase in frequency of tropical storms and hurricanes in the Caribbean region due to climate change (UNEP 2008) could open opportunities for the colonization of naturalized tree species in old forest reserves (Thompson et al. 2007). We only know of cases of coexistence of native and non-native plant species, but the monitoring of interspecific relationships needs to endure. Continuous studies comparing the ecological performance of native and non-native *Ardisia* (Myrsinaceae) species to identify traits that promote invasiveness in Puerto Rico serve as examples of the importance of monitoring (Muñoz and Ackerman 2011).

Additionally, the introduced autogamous monk orchid (*Oeceoclades maculata* (Lindl.) Lindl), which has shown the capacity to establish in old remnant forests in Puerto Rico, encourages the establishment of long-term studies to determine its influence on native orchids (Cohen and Ackerman 2009). Studies regarding another introduced orchid that has naturalized in Puerto Rico and the USVI—the Philippine ground orchid (*Spathoglottis plicata* Bl.)—have shown that it shares the native florivorous weevil, *Stethobaris polita* Chevrolat, with the native orchid, flor de pasmo (*Bletia patula* Graham) (Recart et al. 2013). A positive relationship has been found between the density of the introduced orchid and the number of weevils and floral damage on the native orchid (Recart et al. 2013). This indirect negative effect on flor de pasmo is expected to continue as distribution models predict that both species of orchids will strongly overlap, having the effect of raising the abundance of the native florivorous weevil as an orchid specialist (Recart et al. 2013).

Furthermore, the non-native red imported fire ant (*Solenopsis invicta* Buren) is attracted to extrafloral nectaries of *S. plicata* and inhibits the native weevil behavior on this non-native orchid. These acquired plant-animal interactions diminish biotic resistance to the continuous spread of *S. plicata* and are proposed as examples of apparent competition between the native and non-native orchid mediated by the native weevil and of invasional meltdowns through facilitation of the non-native ant's invasion by the introduced orchid (Ackerman et al. 2014; Recart et al. 2013). Partial support for the invasional meltdown hypothesis has been provided by showing that red fire ants have a positive effect on the population growth rate of *S. plicata* (Falcón et al. 2016).

In the USVI, the non-native tan tan tree (*Leucaena leucocephala* (Lam.) de Wit) has been perceived as harmful to the local environment because it appears to displace native vegetation while recognizing at the same time that it does not invade closed canopy forests due to its shade intolerance (Daley et al. 2012). An alternative view of tan tan considers the species to be beneficial for Caribbean forest regeneration after chronic anthropogenic disturbance. Tan tan has the ability to regenerate in areas dominated by African grasses and maintained by fires, where native saplings have shown slow growth and high mortality. Tan tan forms mono-dominant stands in the subtropical dry forest life zone in which these conditions prevail (Wolfe and Van Bloem 2012). When anthropogenic disturbance abates, tan tan stands in Puerto Rico evolve toward novel species combinations as native species begin to establish in these sites (Molina Colón et al. 2011). However, as with other novel forests in the different ecological life zones of Puerto Rico and the USVI, a transition to primary forest composition has not been documented (Abelleira Martínez 2010; Aide et al. 2000; Atkinson and Marín-Spiotta 2015; Brandeis et al. 2009; Fonseca Da Silva 2014; Lugo and Helmer 2004; Marcano-Vega et al. 2002, Molina Colón and Lugo 2006). An alternative situation is represented by the cajeput tree (*Melaleuca quinquenervia* (Cav.) S.F. Blake), introduced as an ornamental and forming persisting mono-dominant stands covering small areas (0.06–1.4 ha) within palustrine and herbaceous wetlands in coastal Puerto Rico (Pratt et al. 2005). Sites where the cajeput tree is growing have signs of fire, have been partially drained, and were either farmed in the past or show signs of historical development (Pratt et al. 2005). While the cajeput tree is classified as intolerant of shade (Geary and Woodall 1990), there is no native tree species in Puerto Rico capable of withstanding this combination of flood and fire (Lugo 2004).

Other tree species like neem (*Azadirachta indica* A. Juss.), sweet lime (*Triphasia trifolia* (Burm. f.) P. Wilson), and genip (*Melicoccus bijugatus* Jacq.) have been designated as forest invaders, implying that they have the capacity to establish in forested areas in the USVI without depending on disturbances (Daley et al. 2012). However, forest stands sampled in the USVI show that the presence or dominance of species like sweet lime and genip is the result of secondary forest development after agricultural lands or timber plantations were abandoned (Atkinson and Marín-Spiotta 2015; Brandeis and Oswalt 2007; Brandeis and Turner 2013b). Island-wide forest inventories done in the USVI in 2004, 2009, and 2014 reveal that the number of individuals recorded for sweet lime has increased, while genip appears to be fluctuating (Brandeis and Oswalt 2007; Brandeis and Turner 2013b; Marcano-Vega and Williamson 2017; Miles 2018). Neem has not been found in forested sites in the USVI, and its habitat of establishment has been described as pastureland and cropland (Más and Lugo-Torres 2013). Ongoing studies are needed to determine the interspecific relationships between non-native and native flora and fauna, especially when both native and non-native birds and bats eat their fruits and disperse their seeds (Atkinson and Marín-Spiotta 2015; Daley et al. 2012). Other species introduced as ornamentals in the USVI like the Madagascar rubbervine (*Cryptostegia madagascariensis* Bojer ex Decne.) and the coral vine (*Antigonon leptopus* Hook. & Arn.) (Daley et al. 2012), also present in Puerto Rico (Acevedo-Rodriguez 2005), are similarly in need of field assessments to reveal more specific information regarding their influence on native biodiversity. Madagascar rubbervine has been confirmed as an invasive threatening endemic species in Northeast Brazil (da Silva et al. 2008) and coral vine as having the potential of displacing native species where it invades (CABI 2018), stressing the need for continuous monitoring in Puerto Rico and the USVI.

More than half of the non-native plant species present in Puerto Rico and the USVI have been intentionally introduced for horticulture (Rojas-Sandoval and Acevedo-Rodríguez 2015). Discussions of the wide range of timber and non-timber forest products offered by introduced plant species need to be an essential part of deliberations regarding their management to sensibly balance the conservation of native biodiversity and forest ecosystem services on the insular Caribbean (Benedetti and Negrón Flores 2012; Chamberlain et al. 2018; Davis et al. 2011; Marcano-Vega 2017; Marcano-Vega et al. 2015). Plant families exemplifying the highest numbers of naturalized species considered as invasive (i.e., Poaceae, Fabaceae) in Puerto Rico and the USVI represent intentional introductions of individuals used as nitrogen fixers, fodder, and ornamentals in agroforestry systems and to support cattle activities (Rojas-Sandoval and Acevedo-Rodríguez 2015). On Buck Island, St. Croix, USVI, there have been recommendations for perpetual treatments to remove non-native plant species (Clark et al. 2007), although most of the targeted species, originally cultivated for their offering of various non-timber products (e.g., genip, tamarind (*Tamarindus indica* L.)), have been naturalized for decades and are not known to replace native dry forest species (Atkinson and Marín-Spiotta 2015; Benedetti and Negrón-Flores 2012; Brandeis and Oswalt 2007; Brandeis and Turner 2013b; Little and Wadsworth 1964; Molina Colón et al. 2011). Other targeted species like tan tan have been shown to coexist with native species in novel forest types where tree planting is an option to accelerate natural regeneration or to foster particular species combinations (Atkinson and Marín-Spiotta 2015; Molina-Colón et al. 2011). The tree yellow trumpetbush (*Tecoma stans* (L.) Juss. ex Kunth) is the official flower that has been depicted on the USVI seal since 1991 (Government of the United States Virgin Islands, Office of Management and Budget 2013) and is included in the list for targeted extirpation to provide recruitment areas for native plant species (Clark et al. 2007). As far as we know, yellow trumpetbush is native to the Caribbean including St. Croix (Acevedo-Rodríguez and Strong 2012) and has not been shown to replace native forest plants (Brandeis and Oswalt 2007; Brandeis and Turner 2013b; Little and Wadsworth 1964).

Among non-native plants in aquatic habitats, the seagrass *Halophila stipulacea* (Forssk.) Asch. has spread from the Indian Ocean to the Mediterranean Sea and throughout the Caribbean, most likely assisted by commercial and recreational boat traffic (Willette et al. 2014). Since its arrival, it has intermixed or replaced native seagrasses (Rogers et al. 2014; Willette et al. 2014). *H. stipulacea* has been reported in St. Thomas and St. John, USVI, and Culebra, PR. Additional research has been recommended to ponder what can be considered as the positive or negative effects of *H. stipulacea* on native seagrass ecosystems and if any management action should be pursued (Rogers et al. 2014). Preliminary data from Willette and Ambrose (2012) showed that seagrass beds dominated by *H. stipulacea* had larger individual fish, fewer juvenile fish, and more fish species than seagrass beds dominated by the native manatee grass (*Syringodium filiforme* Kütz). Results also suggest that *H. stipulacea* has had a negative effect on manatee grass by displacement but sustained an equal or higher number of trophic groups than the native seagrass (Willette and Ambrose 2012). Rogers et al. (2014) advocated for more data on herbivory rates, selective feeding, and relative nutritional values of the native and introduced seagrass species to understand the role and effects of *H. stipulacea* beds in habitats for parrotfish, green sea turtles, sea urchins, and other herbivores in the Caribbean. Other suggested research refers to genetic investigations to determine the similarity of *H. stipulacea* in the Caribbean with the seagrass populations in the Indian Ocean and Mediterranean Sea and on the interaction of the introduced seagrass with native Caribbean seagrass (Willette et al. 2014).

Interactions with Non-Native Vertebrates

Mammals Borroto-Páez et al. (2012) published a table of mammals introduced to the West Indies. Populations of introduced rats, mongoose, cats, hogs, goats, donkeys, deer, dogs, and monkeys are generally discussed in the literature as threatening to local native populations of plants and animals. Populations of the brown rat or Norway rat (*Rattus norvegicus* Berk.) and the black rat or ship rat (*Rattus rattus* L.) are the result of human settlement in Puerto Rico and the USVI dating back to the sixteenth and eighteenth centuries (Campbell 1991; Clark et al. 2007; Daley et al. 2012; Joglar 2005). Their threat to the native biodiversity in the Caribbean was established as rat abundance was found to be negatively correlated with significant declines in avian diversity within small islands and cays (0.83–5500 ha) near Puerto Rico and the USVI (Campbell 1991). However, black rats are not regarded as a threat today because they have been integrated into the local food web (Ewel et al. 2013; Willig and Gannon 1996).

The mongoose (*Herpestes auropunctatus* Hodgson) is an example of an intentional introduction in many Caribbean islands. They were brought in for the purpose of controlling rats in sugarcane plantations during the nineteenth century (Seaman and Randall 1962). The adjustments of this species' interspecific relationships (i.e., prey switching) with the native biodiversity produced results counter to the human intentions. Initial accounts portray the mongoose as a severe predator of the black rat but afterward as a heavy predator of native ground-nesting birds, amphibians, and reptiles (Henderson 1991; Pimentel 1955). Moreover, there was an extraordinary recovery of the black rat population after predator-prey adjustments (Seaman and Randall 1962). Henderson (1991) describes at least seven extinctions of West Indian native amphibians and reptiles that were primarily attributable to the introduction of the mongoose, especially on the Lesser Antilles and smaller islands. Numerous accounts of the predation of native fauna (especially amphibians and reptiles) by the mongoose have been registered, as has its emergence as the major vector of rabies and leptospirosis in the area (Borroto-Páez et al. 2012; Pimentel 1955, 2007). Pimentel et al. (2005) estimated that the mongoose causes around $50 million in damages per year in Puerto Rico and Hawaii due to public health effects and the predation of poultry, native birds, reptiles, and amphibians.

On the other hand, the successful translocation and establishment of the critically endangered St. Croix ground lizard (*Ameiva polops* Cope) to Buck Island is an example of a fruitful interagency collaboration, where translocation strategy, life history of the species, habitat quality, and monitoring were identified as indispensable factors for success (Fitzgerald et al. 2015). The strategy included eliminating the mongoose and black rat from Buck Island before translocation.

A study done in Puerto Rico by Guzmán-Colón and Roloff (2014) suggests that the coastal forests of Puerto Rico support a greater abundance of mongoose than wet forests and that the abundance is associated with human activity (e.g., forest trails, garbage). Further studies are planned to confirm initial baseline data and inform mongoose management plans (Guzmán-Colón and Roloff 2014). During the turtle breeding season, the US Fish and Wildlife Service conducts mongoose control near sea turtle nesting sites at the Sandy Point National Wildlife Refuge on St. Croix (Barun et al. 2011), where localized trapping is seen as an alternative providing protection for sensitive insular species facing multiple effects of land-use history. Coblentz and Coblentz (1985) have documented that hawksbill turtles (*Eretmochelys imbricata* L.) nesting in beaches of the USVI, in which they trapped mongooses, experienced no known loss of eggs or young during the period of their study, compared to as much as 23% recorded previously. The Virgin Islands National Park has an ongoing program for controlling mongoose using box traps on the beaches of St. John, USVI (Barun et al. 2011). In Puerto Rico, intense egg predation of the Puerto Rican freshwater turtle (*Trachemys stejnegeri stejnegeri* Schmidt) by the mongoose has been found on the Humacao Nature Reserve of Puerto Rico (León and Joglar 2005), a wetland area that was drained at the beginning of the twentieth century for growing sugarcane and cattle grazing (RNH 2015). Eradication of the mongoose has been suggested to have a direct positive effect on the recruitment of the early life stages of the turtle (Joglar et al. 2007). In addition, the USDA Forest Service and USDA Animal and Plant Health Inspection Service (APHIS) spend around $10,000 per year on mongoose control to protect the critically endangered Puerto Rican parrot (Barun et al. 2011). The Puerto Rican parrot was a common endemic bird into the nineteenth century, with an estimated population of around one million individuals. In 1975, this number declined drastically to 13 birds due to dramatic habitat loss, hunting, and the pet trade (Oberle 2000). All of these instances exemplify present remedies to the product of past human actions.

Mona Island is another example of the effects of introduced mammals in the Archipelago of Puerto Rico. Mona is a 22 mi^2 reserve managed by the Department of Natural and Environmental Resources of Puerto Rico (DNER). It is located approximately 68 km west of the main island and is the single home of the threatened Mona Island iguana (*Cyclura cornuta stejnegeri* Barbour & Noble) (Brandeis et al. 2012; Garcia et al. 2007). Major threats to the iguana come from habitat changes related primarily to litter accumulation of the introduced Australian pine (*Casuarina equisetifolia* L.) at nesting sites and to predation on juveniles by introduced mammals such as feral cats (*Felis silvestris catus* Schreber), as well as on eggs by wild hogs (*Sus scrofa* L.) (Garcia et al. 2007; Wiewandt and García 2000). Wild hog predation on Mona Island iguana eggs has been documented as reaching 100% in dry years (Wiewandt and García 2000). This positions climate change as a potential catalyst of predation rates, as most climate projections for the Caribbean suggest a higher frequency of droughts within the region (UNEP 2008). A survival rate of 0.22 has been recorded during the first 20 weeks of life for the Mona Island iguana hatchlings, which can include effects of potential native predators such as two snakes, *Alsophis* spp. Fitzinger and *Epicrates* spp. Wagler, and two raptors, *Falco* spp. L. (Pérez-Buitrago and Sabat 2007). Introduced feral goats (*Capra hircus aegagrus* Erxleben) are also a threat to the habitat of the Mona Island iguana, coupled with secondary growth areas regenerated after a past disturbance regime that included burning and vegetation clearance (Wiewandt and García 2000). Goats frequently feed on at least four critically endangered plant species in the Cactaceae, Portulacaceae, Orchidaceae, and Rhamnaceae families, some with an unknown reproductive ecology (e.g., Taylor's jujube (*Ziziphus taylorii* (Britton) M.C. Johnst.)) (Meléndez-Ackerman et al. 2008). A study done in Mona Island revealed that the regeneration of understory vegetation following ungulate exclusion was low, suggesting synergistic effects between herbivore-plant interactions and hurricane disturbance on forest conditions (Rojas-Sandoval et al. 2016b). Even so, goats are by now an integral food web component on Mona Island, and no records of goat-driven extinctions have been described five centuries after introduction (Ewel et al. 2013; Lugo et al. 2012). Long-term monitoring efforts on the ecology of natural plant populations and on the effects of ungulate exclusion are encouraged and should be supported (Ewel et al. 2013; Meléndez-Ackerman et al. 2008; Rojas-Sandoval et al. 2016b).

Wild hogs in conjunction with beach erosion are also documented as major threats to the hawksbill sea turtle on Mona Island; fencing beaches to prevent hog predation has corresponded with an increase in the number of hawksbill nests in the last decade (Joglar et al. 2007). The DNER works with other entities to pursue several intervention strategies to manage introduced mammals on Mona Island. These include moving the hunting season to a period not overlapping with the Mona Island iguana nesting and incubation seasons, creating additional nesting sites and fencing them in a way that prevents feral hogs from entering while allowing the free passage of iguanas, opening clearings in Australian pine stands, and fencing remote nest sites (Wiewandt and García 2000). Additional strategies include an ongoing cat eradication program through trapping and hunting, coupled with a headstarting program of young iguanas that consists of keeping hatchlings in captivity until they reach a size less vulnerable to predators such as feral cats and monitoring the headstarted iguanas through radio tracking and direct observation (Garcia et al. 2007).

Traps, fencing, and radio tracking are also planned as intervention strategies in the Virgin Islands National Park where feral species of introduced mammals include hogs, goats, and cats, all of which are a threat to the native biodiversity (Borroto-Páez et al. 2012; Burde and Feldhamer 2005; US Department of the Interior National Park Service Southeast Region 2003). The feral goat population in St. John USVI, an island of approximately 62 km^2, has been estimated at 600–1000 individuals (Burde and Feldhamer 2005). Feral cats are reported as predators or potential predators of numerous reptiles, amphibians, and insects on USVI (Borroto-Páez et al. 2012), but without correlating quantitative data on their effects on native prey populations. Demographic studies of native prey should be encouraged to determine population status and to elucidate key underlying factors explaining dynamics. In Puerto Rico, the predation of bats by feral cats has been described as a factor facilitated by the island's landscape as a mosaic of urban and natural areas (Rodríguez-Durán et al. 2010). In the Virgin Islands National Park, 25 threatened and endangered plant species have been listed as potentially being affected by wild hogs. Pre- and post-reduction hog surveys along with long-term monitoring programs of native vegetation have been included in a sustained reduction plan (US Department of the Interior National Park Service Southeast Region 2003). Other introduced mammals reported in the USVI are the white-tailed deer (*Odocoileus virginianus* Zimmermann), feral donkey (*Equus asinus* L.), and feral dog (*Canis lupus familiaris* L.) (Borroto-Páez et al. 2012). The National Park Service's view of the donkey (Fig. A8.9) as an invasive spe-

Fig. A8.9 Although viewed as an invasive mammal, feral donkeys (*Equus asinus*) represent a valued part of life and history to local St. Johnians. (Photo credit: Humfredo Marcano-Vega, USDA Forest Service Southern Research Station)

cies met with an alternative view from local St. Johnians, who see a species representing a valued part of life and history (Fortwangler 2009). Wild populations of deer and goats are also present in Culebra Island, PR, where the whitetail deer was introduced in 1966, and intervention strategies for controlling these species have been proposed (US Department of the Interior FWS Southeast Region 2012).

Among the non-native mammals in Puerto Rico are wild populations of monkeys, introduced for research purposes (e.g., virus vaccine program, free-ranging ecology). Rough estimates suggest a population of 500 monkeys in southwest Puerto Rico (Jensen et al. 2004). The introduced Asian rhesus monkey (*Macaca mulatta* Zimmermann), the African patas monkey (*Erythrocebus patas* Schreber), and the Central and South American squirrel monkey (*Saimiri sciureus* L.) thrive in Puerto Rico (Grana-Raffucci 2006–2012; Jensen et al. 2004). A survey of rhesus monkeys (n = 57) found that 72% had evidence of B virus infection (Jensen et al. 2004). Nonetheless, risks of transmission to humans are low because an average of only 2–3% of seropositive monkeys shed the virus. An additional public health concern, however, is the potential for monkeys to be exposed to rabies through their contact with the mongoose, which is the major vector of the virus in Puerto Rico (Engeman et al. 2010). The frequency of human encounters and disease exposure is expected to increase as monkeys continue to expand their range and population (Jensen et al. 2004). Populations of rhesus and patas monkeys have extended their range into productive agricultural lands, resulting in substantial crop losses to commercial farmers. Economic losses by commercial farmers have increased from $1.13 to $1.46 million USD from the years 2002 to 2006, and control measures for monkey populations are viewed as highly challenging (Engeman et al. 2010). Research advancements on wildlife contraceptive agents or immunocontraceptives may offer new humane, nonlethal alternatives to manage introduced mammal populations in Puerto Rico and the USVI if deemed necessary (The Humane Society of the United States 2016; US Department of Agriculture APHIS NWRC 2013).

Amphibians and Reptiles Six introduced species of amphibians are described as threatening to native amphibians in Puerto Rico as competitors, predators, and vectors of pathogens and disease (Burrowes and Joglar 2005; Joglar 2005). Coupled with habitat loss, the introduced cane toad (*Rhinella marina* L.) is considered a major factor in explaining the threatened and critically endangered status of the Puerto Rican crested toad (*Peltophryne lemur* Cope) (Joglar et al. 2007). In an effort to increase the Puerto Rican crested toad population, captive-born toadlets are being released, some into artificial ponds. Ongoing research plans include a dietary assessment and characterization of the health threats prevalent in the cane toad individuals cohabiting with Puerto Rican crested toads (Joglar et al. 2007). The fact that *Salmonella* Lignieres has been isolated from the liver of cane toad individuals in Puerto Rico has also been highlighted as a possible public health concern to people on the island (Burrowes and Joglar 2005). The cane toad is also an introduced species in the USVI, where it is estimated to be competing with the native white-lipped frog (*Leptodactylus albilabris* Günther) for limited freshwater resources (Platenberg 2007). The non-native Cuban treefrog (*Osteopilus septentrionalis* Duméril & Bibron) is also believed to be competing with the white-lipped frog in the USVI for ephemeral freshwater, and there is circumstantial evidence of its negative effects on populations of native frogs and lizards (Platenberg 2007).

The green iguana (*Iguana iguana* L.) (Fig. A8.10) was introduced to Puerto Rico in the 1970s and has naturalized and dispersed widely, especially around the main island

Fig. A8.10 The introduced green iguana (*Iguana iguana*) is primarily described as an herbivore, but gut content analysis in Puerto Rico has indicated that it is an opportunistic omnivore (e.g., crabs (*Uca* spp.)). Its naturalization has raised concern for the conservation of endangered and threatened native species and economic losses on nature reserves, gardens, agricultural lands, and airports. Nevertheless, interdisciplinary research shows ways in which the green iguana is providing new opportunities for ecotourism destinations in Puerto Rico. (Photo credit: Humfredo Marcano-Vega, USDA Forest Service Southern Research Station)

(López-Torres et al. 2012). Concerns about the presence of green iguanas in Puerto Rico are mostly related to the habitats of listed endangered or threatened native species, nature reserves, gardens, agricultural lands, and airports (Falcón et al. 2012, 2013; García-Quijano et al. 2011; López-Torres et al. 2012). The reproductive season and nesting sites of the green iguana and the leatherback turtle (*Dermochelys coriacea* Vandelli) overlap. Concerns are mainly due to the digging habits of the iguana and its potential for destroying leatherback eggs, as well as the iguanas' potential for attracting predators that can feed on leatherback eggs (e.g., dogs, mongoose). The digging and herbivory activities of the green iguana in Las Cabezas de San Juan Nature Reserve are resulting in annual investments of up to $10,000 USD for road repair and young tree losses in the plant nursery (López-Torres et al. 2012). The green iguana activities have also affected crops such as yams (*Dioscorea* spp. L.), yautías (*Xanthosoma* spp. Schott), pumpkins (*Cucurbita* spp. L.), and melons (*Cucumis* sp. L.), particularly in southern Puerto Rico (López-Torres et al. 2012). Additionally, although the green iguana is primarily described as an herbivore, gut content analysis in Puerto Rico indicates that it is an opportunistic omnivore (e.g., crabs (*Uca* spp.)), revealing the need for further research regarding community-level function and effects (Govender et al. 2012).

Concerns about economic losses due to green iguana incursions in aircraft traffic areas have also been highlighted for their potential to diminish public trust in the air transport industry. A study over a 2-month period (October–November 2001) in the Luis Muñoz Marín International Airport (SJU) of San Juan, PR, found that operations in airstrip areas had to be stopped on six occasions (Engeman et al. 2005). With an average of 233,000 operations annually at SJU (Engeman et al. 2005), results of the study can be transduced to a rate of 15.4 occasions per 100,000 movements. Events at SJU resulted in the interruption of operations, though they do not seem to have represented actual strikes; rather, they recorded the frequency with which observed green iguanas were judged to be a hazard for a collision with an aircraft (Engeman et al. 2005). In comparison, a rate of 10.8 strikes per 100,000 movements was reported in the United States during the year of the study (US Department of Transportation Federal Aviation Administration and US Department of Agriculture APHIS 2014). Current control measures for the green iguana in Puerto Rico are costing the US Federal Aviation Administration and the Ports Authority $98,000 annually (Falcón et al. 2012).

Ninety-six percent of reports of civil aircraft damage resulting from wildlife strikes in the United States are due to birds. Of a total of 143,800 strikes reported from 1990 to 2013, only 189 (0.1%) involved reptiles, from which 9 refer to strikes with green iguanas (US Department of Transportation Federal Aviation Administration and US Department of Agriculture APHIS 2014). Four of those reports involved a negative effect on a flight, but none gave an account of damage to aircraft or economic losses. This stresses the need for ongoing studies of wildlife hazard assessments with more comparable data to derive implications and potential alternative actions. Considering that it is estimated that most airports account for less than 20% of actual strikes, efforts should be directed to make reporting an essential aspect of contributing to wildlife management (Engeman et al. 2005). Researchers have recommended that a management plan be developed, as the DNER has begun a population control program for the green iguana (López-Torres et al. 2012).

Ethnographic interviews of residents of the San Juan Bay Estuary (SJBE) in Puerto Rico reveal that most people found the systematic killing of green iguanas as a control measure objectionable (García-Quijano et al. 2011). Interest in the green iguana is generating increased participation in local and overseas tourism to coastal and riparian areas, fostering iguana-watching experiences through kayaking and boat tours, as well as iguana-themed souvenirs on the island (García-Quijano et al. 2011). These circumstances provide new opportunities for SJBE residents and for new interest in estuarine ecosystems as ecotourism destinations in Puerto Rico. An adult green iguana census, also by García-Quijano et al. (2011), showed a positive spatial association between human settlements and iguana populations and that iguanas were significantly aggregated toward the edges of mangrove forests. These results illustrate how green iguanas can represent high defoliation rates on localized mangrove stands near forest edges exposed to the effects of human activities. Even so, the majority of the mangrove forests studied were not being affected by the iguanas. The authors emphasized how a conceptual framework for understanding ecosystem function and species introductions needs to incorporate the sociocultural context of host ecosystems (García-Quijano et al. 2011). The green iguana is also present in the USVI, especially around resorts and restaurants on St. Thomas, but neither its origin nor its ecosystem effects have been identified (Platenberg 2007). The expansion of the green iguana across the Caribbean Basin raises conservation concerns, as modeling techniques predict climatic suitability for new invasions in, essentially, all of the insular Caribbean (Falcón et al. 2012). Falcón et al. (2013) offered a review of control efforts and response to management of the green iguana.

Other introduced reptiles in Puerto Rico include the caiman (*Caiman crocodilus* L.) and the red-eared slider turtle (*Trachemys scripta elegans* Wied-Neuwied). The caiman was apparently introduced in the 1950s–1960s and is now naturalized, especially around the northern coast of Puerto Rico (Joglar 2005). There are no ongoing studies on the effects of caimans on the local biodiversity. The introduced red-eared slider turtle is reported to be widespread in the wetlands of Puerto Rico, raising concerns about interspecific competition and hybridization with native turtle populations. Scholars have underscored the need for local agencies on the island to enforce trade laws (Joglar et al. 2007), with research efforts focused on monitoring local populations of the Puerto Rican freshwater turtle through mark-recapture studies and investigation of its reproductive activity and on the identification of predators and non-native turtles within study areas (León and Joglar 2005). The Puerto Rican freshwater turtle population is estimated as being close to threatened due to its restricted distribution and the effects of habitat destruction, human consumption, introduced turtle populations, and (as discussed earlier) intense egg predation by the mongoose (León and Joglar 2005).

Additional introduced species of reptiles and amphibians that are presumed to have an effect in the USVI include the corn snake (*Pantherophis guttatus* L.), the red-footed tortoise (*Geochelone carbonaria* Spix), the house gecko (*Hemidactylus mabouia* Gray), the red-eared slider, the common ground lizard (*Ameiva exsul* Cope), the Puerto Rican coqui (*Eleutherodactylus coqui* Thomas), and the Lesser Antillean whistling frog (*Eleutherodactylus johnstonei* Barbour) (Platenberg 2007). However, their effects on native populations are, for the most part, undetermined, and long-term monitoring programs coupled with habitat protection could be supported to safeguard native herpetofaunal species of conservation concern (Platenberg and Boulon 2006). The corn snake is present in St. Thomas, where it has been highlighted as a potential threat to the endangered USVI tree boa (*Epicrates monensis granti* Stull) (Platenberg 2007).

Birds Members of the Psittacidae (parrots and parakeets), Ploceidae (weaver finches), and Estrildidae (waxbills and allies) families are included among the most numerous introduced bird species in Puerto Rico (Delannoy Juliá 2005). The canary-winged or white-winged parakeet (*Brotogeris versicolurus* Statius Muller) and the monk parakeet (*Myiopsitta monachus* Boddaert) figure among the most successfully established and widespread psittacines in Puerto Rico (Oberle 2000; Raffaele 1989). Studies about interactions of these species with the local biodiversity are not available. Other bird species described as introduced, such as the pin-tailed whydah (*Vidua macroura* Pallas) and the shiny cowbird (*Molothrus bonariensis* Gmelin), are parasitic on other bird species' nests. While the pin-tailed whydah was introduced from Africa, the shiny cowbird is a species originally from South America and Trinidad and Tobago; it expanded its range during the twentieth century, taking advantage of habitat changes due to agricultural activities throughout the Caribbean (Oberle 2000). Shiny cowbirds deposit their eggs in other bird species' nests and exploit parental care, affecting its principal host, the endemic yellow-shouldered blackbird (*Agelaius xanthomus* P. L. Sclater) in Puerto Rico. It is reported that almost all yellow-shouldered blackbird nests have been parasitized by the shiny cowbird since its introduction in 1955 (Delannoy Juliá 2005; Oberle 2000; Raffaele 1989). A trapping program to remove cowbirds from the blackbird's breeding range has been established and is considered a control measure that will probably have to continue indefinitely to avoid the blackbird's eradication from mainland Puerto Rico (Oberle 2000). Although it is documented that cowbirds had reached Mona Island by 1971 (Post and Wiley 1977), they are now absent, and a non-threatened subspecies of the blackbird is present on Mona and Monito Islands (Oberle 2000; Raffaele 1989). The arrival dates of the shiny cowbird in the Lesser Antilles are described as suggesting natural "island hopping" from Trinidad and Tobago (Cruz et al. 1995), while its range movements in North America support the hypothesis of a dispersal pattern unaided by human introductions (Post et al. 1993).

Fishes The Indo-Pacific red lionfish (*Pterois volitans* L.) has spread through the Western Atlantic and Caribbean since 1985, apparently after its release or escape from marine aquaria into Florida coastal waters (Albins and Hixon 2013). As a competitor and generalized predator of native reef fishes, and with its spread into critical nursery ecosystems for native juvenile reef fishes (i.e., mangroves, seagrass beds), the lionfish is considered a potential threat to the recruitment of reef fishes, some of which are the prey of important fishery species or are important food species such as groupers and snappers (Albins and Hixon 2013; Diller et al. 2014; Morris and Akins 2009). A field experiment has shown that the lionfish alone was able to reduce prey fish by 90% after 2 months, but authors indicate that more data are still needed before projecting into the future (Albins and Hixon 2013). Studies in the Bahamas have shown that predation of red lionfish on algae grazers such as parrotfishes, surgeonfishes, and damselfishes has the potential to shift coral reefs to an algal-dominated community, causing declines in biodiversity (Lesser and Slattery 2011). The high growth, reproductive, and consumption rates of the lionfish coupled with its apparent resistance to predation has made biological control a theme of debate (Albins and Hixon 2013; Diller et al. 2014). Yet, recent experimental studies by Diller et al. (2014) have suggested that native predators are capable of consuming healthy, tethered lionfish and that their inexperience preying on lionfish can be overcome by conditioning. In addition, the gourmet food industry is taking advantage of the lionfish's spread, and a cookbook has been written, providing information on ways to catch and prepare the fish (Akins 2012; CBS Interactive 2015).

The nature of Caribbean islands as case studies of the social and ecological consequences of introduced species demands supporting long-term monitoring, education, and management programs at the regional level.

Literature Cited

Abelleira-Martínez OJ (2010) Invasion by native tree species prevents biotic homogenization in novel forests of Puerto Rico. Plant Ecol 211:49–64

Acevedo-Rodriguez P (2005) Vines and climbing plants of Puerto Rico and the Virgin Islands. Contrib US Natl Herb 51:1–483

Acevedo-Rodriguez P, Strong MT (2012) Catalogue of seed plants of the West Indies. Smithsonian contributions to botany 98. Smithsonian Institution Scholarly Press, Washington, DC, 1192 p

Ackerman JD, Falcón W, Molinari J et al (2014) Biotic resistance and invasional meltdown: consequences of acquired interspecific interactions for an invasive orchid, *Spathoglottis plicata* in Puerto Rico. Biol Invasions 16:2435–2447

Aide TM, Zimmerman JK, Pascarella JB et al (2000) Forest regeneration in a chronosequence of tropical abandoned pastures: implications for restoration ecology. Restor Ecol 8:328–338

Akins L (2012) REEF releases Lionfish cookbook. http://www.reef.org/enews/articles/reef-releases-lionfish-cookbook

Albins MA, Hixon MA (2013) Worst case scenario: potential long-term effects of invasive predatory lionfish (*Pterois volitans*) on Atlantic and Caribbean coral-reef communities. Environ Biol Fish 96:1151–1157

Alvarado S, Guédez M, Lué-Merú MP, Nelson G et al (2008) Arsenic removal from waters by bioremediation with the aquatic plants water hyacinth (*Eichhornia crassipes*) and lesser duckweed (*Lemna minor*). Bioresour Technol 99(17):8436–8440

Anagnostakis SL (1987) Chestnut blight: the classical problem of an introduced pathogen. Mycologia 79:23–37

Atkinson EE, Marín-Spiotta E (2015) Land use legacy effects on structure and composition of subtropical dry forests in St. Croix, U.S. Virgin Islands. For Ecol Manag 335:270–280

Aukema JE, Leung B, Kovacs K et al (2011) Economic impacts of non-native forest insects in the continental United States. PLoS One 6(9):e24587

Barun A, Hanson CC, Campbell KJ, Simberloff D (2011) A review of small Indian mongoose management and eradications on islands. In: Veitch CR, Clout MN, Towns DR (eds) Island invasives: eradication and management—Proceedings of the international conference on island invasives. International Union for Conservation of Nature (IUCN) and Centre for Biodiversity and Biosecurity (CBB), Gland/Auckland, pp 17–25

Benedetti M, Negrón-Flores J (2012) 12 árboles amigos: juegos y retos etnobotánicos para Borikén. BotaniCultura, Puerto Rico, 152 p

Bevins SN, Pedersen K, Lutman MW et al (2014) Consequences associated with the recent range expansion of nonnative feral swine. Bioscience 64:291–299

Birdsey RA, Weaver PL (1982) The forest resources of Puerto Rico. resources bulletin SO–85. U.S. Department of Agriculture Forest Service, Southern Forest Experiment Station, New Orleans, 56 p

Birdsey RA, Weaver PL (1987) Forest area trends in Puerto Rico. research note SO-331. U.S. Department of Agriculture Forest Service, Southern Forest Experiment Station, New Orleans, 5 p

Borroto-Páez R, Woods CA, Sergile FE (eds) (2012) Terrestrial mammals of the West Indies: contributions. Florida Museum of Natural History and Wacahoota Press, Gainesville. 481 p

Brandeis TJ, Oswalt SN (2007) The status of U.S. Virgin Islands' forests, 2004. Resources Bulletin SRS–122. U.S. Department of Agriculture Forest Service, Southern Research Station, Asheville. 61 p

Brandeis TJ, Turner JA (2013a) Puerto Rico's forests, 2009. Resources bulletin SRS–191. U.S. Department of Agriculture Forest Service, Southern Research Station, Asheville. 85 p

Brandeis TJ, Turner JA (2013b) U.S. Virgin Islands' forests, 2009. Resources bulletin SRS–196. U.S. Department of Agriculture Forest Service, Southern Research Station, Asheville. 56 p

Brandeis TJ, Helmer EH, Oswalt SN (2007) The status of Puerto Rico's forests, 2003. Resources Bulletin SRS–119. U.S. Department of Agriculture Forest Service, Southern Research Station, Asheville. 72 p

Brandeis TJ, Helmer EH, Marcano-Vega H, Lugo AE (2009) Climate shapes the novel plant communities that form after deforestation in Puerto Rico and the U.S. Virgin Islands. For Ecol Manag 258:1704–1718

Brandeis TJ, Meléndez-Ackerman E, Helmer EH (2012) Forest vegetation cover assessment on Mona Island, Puerto Rico. General Technical Report SRS–165. U.S. Department of Agriculture Forest Service, Southern Research Station, Asheville. 24 p

Brewer S (2008) Declines in plant species richness and endemic plant species in longleaf pine savannas invaded by *Imperata cylindrica*. Biol Invasions 10(8):1257–1264

Brewer JS, Cralle SP (2003) Phosphorus addition reduces invasion of a longleaf pine savanna (Southeastern USA) by a non-indigenous grass (*Imperata cylindrica*). Plant Ecol 167(2):237–245

Briscoe CB (1959) Early results of mycorrhizal inoculation of pine in Puerto Rico. Caribbean For 20:73–77

Broders K, Boraks A, Barbison L et al (2015) Recent insights into the pandemic disease butternut canker caused by the invasive pathogen *Ophiognomonia clavigignenti-juglandacearum*. For Pathol 45:1–8

Brown KA, Scatena FN, Gurevitch J (2006) Effects of an invasive tree on community structure and diversity in a tropical forest in Puerto Rico. For Ecol Manag 226:145–152

Burde JN, Feldhamer GA (2005) Mammals of the national parks. The John Hopkins University Press, Baltimore. 212 p

Burrowes PA, Joglar RL (2005) El estado de los anfibios de Puerto Rico. In: Joglar RL (ed) Biodiversidad de Puerto Rico: vertebrados terrestres y ecosistemas—Serie de historia natural. Editorial del Instituto de Cultura Puertorriqueña, San Juan, pp 60–65

Butler BJ, Wear DN (2013) Forest ownership dynamics of southern forests. Chapter 6. In: Wear DN, Greis JG (eds) The southern Forest futures project: technical report. general technical report SRS-GTR-178. USDA-Forest Service, Southern Research Station, Asheville. 542 p

CABI (2018) *Antigonon leptopus* (coral vine). In: Invasive species compendium. CAB International, Wallingford. https://www.cabi.org/isc/datasheet/112316#05EBAC67-007D-407A-888A-DC279954BCD0

Campbell EW (1991) The effect of introduced roof rats on bird diversity of Antillean cays. J Field Ornithol 62(3):343–348

Campbell F (2015) Cacti under threat—does no one care? Center for invasive species prevention. http://www.nivemnic.us/cacti-under-threat-does-no-one-care/

Canadian Food Inspection Agency (2020) December 7, 2017—Emerald ash borer confirmed in Winnipeg. Emerald Ash Borer—Latest Information. http://www.inspection.gc.ca/plants/plant-pests-invasive-species/insects/emerald-ash-borer/latest-information/eng/1337287614593/1337287715022

Careless P, Marshall SA, Gill BD (2014) The use of *Cerceris fumipennis* (Hymenoptera: Crabronidae) for surveying and monitoring emerald ash borer (Coleoptera: Buprestidae) infestations in eastern North America. Canadian Entomol 146(1):90–105

Carter J, Leonard BP (2002) A review of the literature on the worldwide distribution, spread of, and efforts to eradicate the coypu (*Myocastor coypus*). Wildl Soc Bull 30:162–175

CBS Interactive (2015) Lionfish: from malicious to delicious. http://www.cbsnews.com/news/lionfish-from-malicious-to-delicious/

Center for Invasive Species and Ecosystem Health (2010) Current cogongrass infestation by county in the Southeastern US (online map). http://www.cogongrass.org/distribution/

Chamberlain J, Davis J, Duguid M et al (2018) Non-timber forest products and production. In: National Assessment of non-timber forest products and climate change. General Technical Report, SRS-232, U.S. Department of Agriculture Forest Service, Southern Research Station, Asheville

Chamorro ML, Persson J, Torres-Santana CW et al (2016) Molecular and morphological tools to distinguish *Scyphophorus acupunctatus* Gyllenhal, 1838 (Curculionidae: Dryophthorinae): a new weevil pest of the endangered century plant, *Agave eggersiana* from St. Croix, U.S. Virgin Islands. Proc Entomol Soc Wash 118(2):218–243

Clark DW, Hillis-Starr Z, Furqueron C (2007) National "Rats and weeds and lizards—Oh my!" Eradication of *Rattus rattus* and control of invasive exotic plants on Buck Island, U.S. Virgin Islands. In: Proceedings of the 2007 George Wright Society Conference. http://www.georgewright.org/0719clark.pdf

Clark SL, Schlarbaum SE, Saxton AM, Hebard FV (2016) Establishment of American chestnuts (*Castanea dentata*) bred for blight (*Cryphonectria parasitica*) resistance: influence of breeding and nursery grading. New For 47:243–2708

Coblentz BE, Coblentz BA (1985) Control of the Indian mongoose *Herpestes auropunctatus* on St John, US Virgin Islands. Biol Conserv 33:281–288

Cohen IM, Ackerman JD (2009) *Oeceoclades maculata*, an alien tropical orchid in a Caribbean rainforest. Ann Bot 104:557 563

Cruz A, Manolis TH, Andrews RW (1995) History of shiny cowbird *Molothrus bonariensis* brood parasitism in Trinidad and Tobago. Ibis 137(3):317–321

Daley B, Valiulis J, Slatton R (2012) Exotic invasives—US Virgin Islands—Species effecting forests. Doc. GS-VIDA-1201. St. Croix, US Virgin Islands. 10 p

Daneshgar P, Jose S (2009) Imperata cylindrica, an alien invasive grass, maintains control over nitrogen availability in an establishing pine forest. Plant Soil 320(1–2):209–218

Daneshgar P, Jose S, Collins A, Ramsey C (2008) Cogongrass (*Imperata cylindrica*), an alien invasive grass, reduces survival and productivity of an establishing pine forest. For Sci 54(6):579–587

Daniels DA, Nix KA, Wadl PA, Vito LM et al (2016) Thousand cankers disease complex: a forest health issue that threatens *Juglans* species across the U.S. Forests 7(11):260. https://doi.org/10.3390/f7110260

da Silva JL, Barreto RW, Pereira OL (2008) *Pseudocercospora cryptostegiae-madagascariensis* sp. nov. on *Cryptostegia madagascariensis*, an exotic vine involved in major biological invasions in Northeast Brazil. Mycopathologia 166:87–91

Davidson CB, Gottschalk KW, Johnson JE (1999) Tree mortality following defoliation by the European gypsy moth (*Lymantria dispar* L.) in the United States: a review. For Sci 45:74–84

Davis MA, Chew MK, Hobbs RJ (2011) Don't judge species on their origins. Nature 474:153–154

Delannoy Juliá CA (2005) Aves. In: Joglar RL (ed) Biodiversidad de Puerto Rico: vertebrados terrestres y ecosistemas—Serie de historia natural. Editorial del Instituto de Cultura Puertorriqueña, San Juan, pp 193–237

Diller JL, Frazer TK, Jacoby CA (2014) Coping with the lionfish invasion: evidence that naïve, native predators can learn to help. J Exp Mar Biol Ecol 455:45–49

DLC Resources, Inc (2015) From the case files: Agave snout weevil. http://www.eastmark.com/wp-content/uploads/2015/04/landscape_year-round_agave_snout_weevil.pdf

Dorcas ME, Willson JD, Reed RN et al (2012) Severe mammal declines coincide with proliferation of invasive Burmese pythons in Everglades National Park. Proc Natl Acad Sci 109(7):2418–2422

Duan JJ, Bauer LS, Van Driesche RG (2017) Emerald ash borer biocontrol in ash saplings: the potential for early stage recovery of North American ash trees. For Ecol Manag 394:64–72

Duerr DA, Mistretta PA (2013) Invasive pests—insects and diseases. Chapter 16. In: Wear DN, Greis JG (eds) The Southern Forest Futures Project: technical report. General Technical Report SRS-GTR-178. USDA-Forest Service, Southern Research Station, Asheville. 542 p

Duncan JG, Steininger MS, Wright SA, Wheeler GS (2016) Host range of *Caloptilia triadicae* (Lepidoptera: Gracillariidae): an adventive herbivore of Chinese tallowtree (Malpighiales: Euphorbiaceae). Fla Entomol 99:142–145

EAB Information Network (2020) Initial county EAB detections in North America. Emerald Ash Borer Information Network. http://www.emeraldashborer.info/

EDDMapS (2016) Early detection & distribution mapping system. The University of Georgia—Center for Invasive Species and Ecosystem Health. Available online at http://www.eddmaps.org

Engeman RM, Smith HT, Constantin B (2005) Invasive iguanas as an airstrike hazard at Luis Muñoz Marín international airport, San Juan Puerto Rico. J Aviation/Aerospace Educ Res 14:45–50

Engeman RM, Laborde JE, Constantin BU et al (2010) The economic impacts to commercial farms from invasive monkeys in Puerto Rico. Crop Prot 29:401–405

Evers DE, Sasser CE, Gosselink JG et al (1998) The impact of vertebrate herbivores on wetland vegetation in Atchafalaya Bay. Louisiana Estuar 21:1–13

Ewel JJ, Whitmore JL (1973) The ecological life zones of Puerto Rico and the U.S. Virgin Islands. Research Paper ITF–18. U.S. Department of Agriculture Forest Service, Institute of Tropical Forestry, Río Piedras. 72 p

Ewel JJ, Mascaro J, Kueffer C et al (2013) Islands: where novelty is the norm. In: Hobbs RJ, Higgs ES, Hall CM (eds) Novel ecosystems: intervening in the new ecological world order. Wiley, Oxford, pp 29–44

Fagua JC, Ackerman JD (2011) Consequences of floral visits by ants and invasive honeybees to the hummingbird-pollinated, Caribbean cactus *Melocactus intortus*. Plant Species Biol 26:193–204

Falcón W, Ackerman JD, Daehler CC (2012) March of the green iguana: non-native distribution and predicted geographic range of *Iguana iguana* in the greater Caribbean region. IRCF Reptiles Amphibians 19(3):150–160

Falcón W, Ackerman JD, Recart W, Daehler CC (2013) Biology and impacts of Pacific Island invasive species. 10. *Iguana iguana*, The green iguana (Squamata: Iguanidae). Pac Sci 67(2):157–186

Falcón W, Ackerman JD, Tremblay RL (2016) Quantifying how acquired interactions with native and invasive insects influence population growth rates of a non-indigenous plant. Biol Invasions. https://doi.org/10.1007/s10530-016-1318-8

Fitzgerald LA, Treglia ML, Angeli N et al (2015) Determinants of successful establishment and post-translocation dispersal of a new population of the critically endangered St. Croix ground lizard (*Ameiva polops*). Restor Ecol 23(6):776–786. https://doi.org/10.1111/rec.1224

Florida FWCC (2016) Python: what FWC is doing. Florida Fish and Wildlife Conservation Commission. http://myfwc.com/wildlifehabitats/nonnatives/python/fwc-work/

Fonseca Da Silva J (2014) Species composition, diversity and structure of novel forests of *Castilla elastica* in Puerto Rico. Trop Ecol 55(2):231–244

Ford MA, Grace JB (1998) Effects of vertebrate herbivores on soil processes, plant biomass, litter accumulation and soil elevation changes in a coastal marsh. J Ecol 86(6):974–982

Fortwangler C (2009) A place for the donkey: natives and aliens in the US Virgin Islands. Landsc Res 34(2):205–222

Fox LJ, Struik PC, Appleton BL, Rule JH (2008) Nitrogen phytoremediation by water hyacinth (*Eichhornia crassipes* (Mart.) Solms). Water Air Soil Pollut 194(1–4):199–207

Fraedrich SW, Harrington TC, Rabaglia RJ et al (2008) A fungal symbiont of the redbay ambrosia beetle causes a lethal wilt in redbay and other Lauraceae in the southeastern United States. Plant Dis 92:215–224

Franco PA, Weaver PL, Eggen-McIntosh S (1997) Forest resources of Puerto Rico, 1990. Resources Bulletin SRS-22. U.S. Department of Agriculture, Forest Service, Southern Research Station, Asheville. 45 p

Fumero-Cabán JJ, Meléndez-Ackerman EJ (2007) Relative pollination effectiveness of floral visitors of *Pitcairnia angustifolia* (Bromeliaceae). Am J Bot 94(3):419–424

GAEI (Grupo Antillano de Especies Invasoras) (2016) Atlas de especies invasoras de Puerto Rico. http://atlas.eea.uprm.edu/

Gandhi KJK, Herms DA (2010) Direct and indirect effects of alien insect herbivores on ecological processes and interactions in forests of eastern North America. Biol Invasions 12:389–405

Gao R, Su Z, Yin Y et al (2016) Germplasm, chemical constituents, biological activities, utilization, and control of Chinese tallow (*Triadica sebifera* (L.) small). Biol Invasions 18:809–829

García MA, Pérez-Buitrago N, Álvarez AO, Tolson PJ (2007) Survival, dispersal, and reproduction of headstarted Mona Island iguanas, *Cyclura cornuta stejnegeri*. Appl Herpetol 4:357–363

García-Quijano CG, Carlo TA, Arce-Nazario J (2011) Human ecology of a species introduction: interactions between humans and introduced green iguanas in a Puerto Rican urban estuary. Hum Organ 70(2):164–178

Geary TF, Woodall SL (1990) *Melaleuca quinquenervia* (Cav.) S.T. Blake melaleuca. In: Burns RM, Honkala BH (tech. coords) Silvics of North America. Vol. 2. Hardwoods Agricultural Handbook 654. U.S. Department of Agriculture Forest Service, Washington, DC, pp 461–465

Georgia Forestry Commission (2016) Distribution of counties with Laurel Wilt Disease by year of initial detection (online map, updated 7 Apr 2016). http://www.gatrees.org/forest-management/forest-health/laurel-wilt-disease/

Gettys LA (2014) Water hyacinth: Florida's worst floating weed. University of Florida IFAS Extension, EDIS publication SS-AGR-380, 5 p. http://edis.ifas.ufl.edu/pdffiles/AG/AG38500.pdf

González AL (2016) Trípido del roble—*Holopothrips tabebuia* (Segarra & Cabrera). http://atlas.eea.uprm.edu/content/tr%C3%ADpido-del-roble

González G, Yu Huang C, Zou X, Rodríguez C (2006) Earthworm invasions in the tropics. Biol Invasions 8(6):1247–1256

González Bernáldez F (1985) Invitación a la ecología humana—La adaptación afectiva al entorno. Editorial Tecnos, S.A, Madrid. 159 p

Govender Y, Muñoz MC, Ramírez Camejo LA et al (2012) An isotopic study of diet and muscles of the green iguana (*Iguana iguana*) in Puerto Rico. J Herpetol 46(2):167–170

Government of the United States Virgin Islands, Office of Management and Budget (OMB) (2013) Fiscal Year 2014 Executive Budget for the Government of the U.S. Virgin Islands. St. Thomas, USVI OMB. 641 p. http://caribbeanelections.com/eDocs/budget/vi_budget/vi_executive_budget_2014.pdf

Grana-Raffucci FA (2006–2012) Algunos animales exóticos establecidos en Puerto Rico. Puerto DRNA, navega por el ambiente. Departamento de Recursos Naturales y Ambientales de Puerto Rico. http://drna.pr.gov/historico/oficinas/saux/sap/nse/documentos-e-informes-tecnicos-del-negociado-de-servicios-especializados/algunos-animales-exoticos-establecidos-en-puerto-rico-autor-felix-a-grana-raffucci

Greene B, Blossey B (2012) Lost in the weeds: *Ligustrum sinense* reduces native plant growth and survival. Biol Invasions 14:139–150

Guzmán-Colón D, Roloff GJ (2014) Small Indian mongoose (*Herpestes auropunctatus*) population abundance and effects of habitat features on trapping success in protected areas of eastern Puerto Rico. Caribbean Nat 19:1–12

Hanula JL, Horn S, Taylor JW (2009) Chinese privet (*Ligustrum sinense*) removal and its effect on native plant communities of riparian forests. Invasive Plant Sci Manag 2:292–300

Hanula JL, Mayfield AE, III, Reid LS, Horn S (2016) Influence of trap distance from a source population and multiple traps on captures and attack densities of the redbay ambrosia beetle (Coleoptera: Curculionidae: Scolytinae). J Econ Entomol (in press). https://doi.org/10.1093/jee/tow068

Hart KM, Cherkiss MS, Smith BJ et al (2015) Home range, habitat use, and movement patterns of non-native Burmese pythons in Everglades National Park, Florida, USA. Anim Biotelemetry 3:8. https://doi.org/10.1186/s40317-015-0022-2

Havill NP, Vieira LC, Salom SM (2014) Biology and control of hemlock woolly adelgid. USDA Forest Service, Forest Health Technology Enterprise Team, FHTET-2014-05, Morgantown. 21 pp

Hebard FV (2005) The backcross breeding program of the American chestnut foundation. J Am Chestnut Found 19:55–78

Henderson RW (1991) Consequences of predator introductions and habitat destruction on amphibians and reptiles in the post-Columbus West Indies. Caribb J Sci 28(1):1–10

Hendrix PF, Lachnicht SL, Callaham MA Jr, Zou X (1999) Stable isotopic studies of earthworm feeding ecology in tropical ecosystems of Puerto Rico. Rapid Commun Mass Spectrom 13:1295–1299

Herms DA, McCullough DG (2014) Emerald ash borer invasion of North America: history, biology, ecology, impacts, and management. Annu Rev Entomol 59:13–30

Hiers JK, Evans JP (1997) Effects of anthracnose on dogwood mortality and forest composition of the Cumberland plateau (USA). Conserv Biol 11(6):1430–1435

Holly DC, Ervin GN, Jackson CR, Diehl SV, Kirker GT (2009) Effect of an invasive grass on ambient rates of decomposition and microbial community structure: a search for causality. Biol Invasions 11(8):1855–1868

Holzmueller E, Jose S, Jenkins M et al (2006) Dogwood anthracnose in eastern hardwood forests: what is known and what can be done? J For 104(1):21–26

Holzmueller EJ, Gibson DJ, Suchecki PF (2009) Using prescribed fire to protect flowering dogwood from dogwood anthracnose (Illinois). Ecol Restor 27(4):392–394

Holzmueller EJ, Jose S, Jenkins MA (2010) Ecological consequences of an exotic fungal disease in eastern US hardwood forests. For Ecol Manag 259(8):1347–1353

Hudson JR, Hanula JL, Horn S (2013) Removing Chinese privet from riparian forests still benefits pollinators five years later. Biol Conserv 167:355–362

Hudson JR, Hanula JL, Horn S (2014) Impacts of removing Chinese privet from riparian forests on plant communities and tree growth five years later. For Ecol Manag 324:101–108

Hughes MA, Smith JA, Ploetz RC et al (2015) Recovery plan for laurel wilt on redbay and other forest species caused by *Raffaelea lauricola* and disseminated by *Xyleborus glabratus*. Plant Health Progress. https://doi.org/10.1094/PHP-RP-15-0017

Jenkins MA, White PS (2002) *Cornus florida* L. mortality and understory composition changes in western great smoky mountains national park. J Torrey Bot Soc 129:194–206

Jensen K, Alvarado-Ramy F, González-Martínez J et al (2004) B-virus and free-ranging macaques, Puerto Rico. Emerg Infect Dis 10(3):494–496

Jetton R (2017) Silvicultural and conservation strategies. In: Shepherd W (ed) Proceedings of the 58th southern forest insect work conference, 25–28 July 2017, Melbourne, p 27. https://www.sfiwc.org/proceedings/ (Abstract)

Jetton RM, Whittier WA, Dvorak WS, Rhea J (2013) Conserved *ex situ* genetic resources of eastern and Carolina hemlock: eastern North American conifers threatened by the hemlock woolly adelgid. Tree Planters' Notes 56:59–71

Joglar RL (ed) (2005) Biodiversidad de Puerto Rico: vertebrados terrestres y ecosistemas—Serie de historia natural. Editorial del Instituto de Cultura Puertorriqueña, San Juan. 563 p

Joglar RL, Álvarez AO, Aide TM et al (2007) Conserving the Puerto Rican herpetofauna. Appl Herpetol 4:327–345

Jojola S, Witmer G, Nolte D (2005) Nutria: an invasive rodent pest or valued resource? In: Wildlife Damage Management Conferences—Proceedings. Paper 110. http://digitalcommons.unl.edu/icwdm_wdmconfproc/110. University of Nebraska, Lincoln

Kairo M, Ali B, Cheesman O, Haysom K, Murphy S (2003) Invasive species threats in the Caribbean Region. Report to The Nature Conservancy. 132 p. http://www.bu.edu/scscb/working_groups/resources/Kairo-et-al-2003.pdf

Kendra PE, Montgomery WS, Niogret J et al (2014) North American Lauraceae: Terpenoid emissions, relative attraction and boring preferences of redbay ambrosia beetle, *Xyleborus glabratus* (Coleoptera: Curculionidae: Scolytinae). PLoS One 9(7):e102086

Klooster WS, Herms DA, Knight et al (2014) Ash (*Fraxinus* spp.) mortality, regeneration, and seed bank dynamics in mixed hardwood forests following invasion by emerald ash borer (*Agrilus planipennis*). Biol Invasions 16:859–873

Knight KS, Brown JP, Long RP (2013) Factors affecting the survival of ash (Fraxinus spp.) trees infested by emerald ash borer (*Agrilus planipennis*). Biol Invasions 15:371–383

Langeland KA, Enloe SF, Hutchinson J (2016) Natural area weeds: old world climbing fern. Publication SS-AGR-21, University of Florida IFAS Extension. http://edis.ifas.ufl.edu/ag122

León A, Joglar RL (2005) La jicotea puertorriqueña (*Trachemys stejneri stejneri*). In: Joglar RL (ed) Biodiversidad de Puerto Rico: Vertebrados terrestres y ecosistemas—Serie de historia natural. Editorial del Instituto de Cultura Puertorriqueña, San Juan, pp 181–190

Lesser MP, Slattery M (2011) Phase shift to algal dominated communities at mesophotic depths associated with lionfish (*Pterois volitans*) invasion on a Bahamian coral reef. Biol Invasions 13:1855–1868

Lippincott CL (2000) Effects of *Imperata cylindrica* (L.) Beauv.(Cogongrass) invasion on fire regime in Florida sandhill (USA). Nat Areas J 20(2):140–149

Little EL, Wadsworth EH (1964) Common trees of Puerto Rico and the Virgin Islands. Agricultural handbook 249. U.S. Department of Agriculture Forest Service, Washington, DC. 548 p

Liu ZG, Zou XM (2002) Exotic earthworms accelerate plant litter decomposition in a Puerto Rican pasture and a wet forest. Ecol Appl 12(5):1406–1417

Lobe JW, Callaham MA, Hendrix PF, Hanula JL (2014) Removal of an invasive shrub (Chinese privet: *Ligustrum sinense* Lour) reduces exotic earthworm abundance and promotes recovery of native North American earthworms. Appl Soil Ecol 83:133–139

López-Torres AL, Claudio-Hernández HJ, Rodríguez-Gómez CA et al (2012) Green iguanas (*Iguana iguana*) in Puerto Rico: is it time for management? Biol Invasions 14:35–45

Louisiana Department of Wildlife and Fisheries (2016) Louisiana coastwide nutria control program (online resource). http://nutria.com/site10.php

Lovett GM, Canham CD, Arthur MA et al (2006) Forest ecosystem responses to exotic pests and pathogens in eastern North America. Bioscience 56:395–405

Lucardi RD, Wallace LE, Ervin GN (2014) Invasion success in Cogongrass (*Imperata cylindrica*): a population genetic approach exploring genetic diversity and historical introductions. Invasive Plant Sci Manag 7(1):59–75

Lugo AE (1992) Comparison of tropical tree plantations with secondary forests of similar age. Ecol Monogr 62(1):1–41

Lugo AE (2004) The outcome of alien tree invasions in Puerto Rico. Front Ecol Environ 2(5):265–273

Lugo AE (2009) The emerging era of novel tropical forests. Biotropica 41:589–591

Lugo AE, Helmer E (2004) Emerging forests on abandoned land: Puerto Rico's new forests. For Ecol Manag 190:145–161

Lugo AE, Carlo TA, Wunderle JM Jr (2012) Natural mixing of species: novel plant–animal communities on Caribbean islands. Anim Conserv 15:233–241

MacDonald GE (2004) Cogongrass (*Imperata cylindrica*)—biology, ecology, and management. Crit Rev Plant Sci 23:367–380

Marcano-Vega H (2017) Forests of Puerto Rico, 2014. Resource update FS–121. U.S. Department of Agriculture Forest Service, Southern Research Station, Asheville. 4 p

Marcano-Vega H, Williamson JR (2017) Forests of U.S. Virgin Islands, 2014. Resource update FS–144. U.S. Department of Agriculture Forest Service, Southern Research Station, Asheville. 4 p

Marcano-Vega H, Aide TM, Báez D (2002) Forest regeneration in abandoned coffee plantations and pastures in the cordillera central of Puerto Rico. Plant Ecol 161:75–87

Marcano-Vega H, Brandeis TJ, Turner JA (2015) Los bosques de Puerto Rico, 2009. Resources Bulletin SRS–202. U.S. Department of Agriculture Forest Service, Southern Research Station, Asheville. 115 p

Marini L, Haack RA, Rabaglia R et al (2011) Exploring associations between international trade and environmental factors with establishment patterns of exotic Scolytinae. Biol Invasions 13(10):2275–2288

Más EG, Lugo-Torres Md L (2013) Malezas Comunes en Puerto Rico e Islas Vírgenes Americanas/common weeds in Puerto Rico and U.S. Virgin Islands. University of Puerto Rico, Mayaguez Campus, U.S. Department of Agriculture Natural Resources Conservation Service, Caribbean Area, Mayaguez. 395 p

Mayer JJ, Brisbin IL (2008) Wild pigs in the United States: their history, comparative morphology, and current status. University of Georgia Press, Athens, 313 p

Mayfield AE, Brownie C (2013) The redbay ambrosia beetle (Coleoptera: Curculionidae: Scolytinae) uses stem silhouette diameter as a visual host-finding cue. Environ Entomol 42(4):743–750

Mayfield AE III, Barnard EL, Smith JA et al (2008) Effect of propiconazole on laurel wilt disease development in redbay trees and on the pathogen in vitro. Arboricult Urban For 34(5):317–324

Mayfield AE, MacKenzie M, Cannon PG et al (2013) Suitability of California bay laurel and other species as hosts for the non-native redbay ambrosia beetle and granulate ambrosia beetle. Agric For Entomol 15(3):227–235

Maynard CA, Powell WA, Polin-McGuigan LD et al (2008) Chestnut. In: Kole C, Hall TC (eds) A compendium of transgenic crop plants: forest tree species, Wiley Online Library. Blackwell Publishing, Oxford, pp 169–192

Mazzotti FJ, Harvey RG (2012) The invasion of exotic reptiles and amphibians in Florida. Document WEC320, Wildlife Ecology and Conservation Department, UF/IFAS Extension. http://edis.ifas.ufl.edu/uw365

McCleery RA, Sovie A, Reed RN et al (2015) Marsh rabbit mortalities tie pythons to the precipitous decline of mammals in the Everglades. Proc R Soc Lond B Biol Sci 282(1805):2015.0120

Meléndez-Ackerman EJ, Cortés C, Sustache J et al (2008) Diet of feral goats in Mona Island Reserve, Puerto Rico. Caribb J Sci 44(2):199–205

Meyerdirk DE, Warkentin R, Attavian B et al (2001) Biological control of pink hibiscus mealybug project manual. U.S. Department of Agriculture, Animal and Plant Health Inspection Service, Riverdale. 184 p

Michler CH, Pijut PM, Jacobs DF et al (2006) Improving disease resistance of butternut (*Juglans cinerea*), a threatened fine hardwood: a case for single-tree selection through genetic improvement and deployment. Tree Physiol 26(1):121–128

Miles PD (2018) Forest Inventory EVALIDator web-application Version 1.6.0.03. U.S. Department of Agriculture Forest Service, Northern Research Station, St. Paul. http://apps.fs.fed.us/Evalidator/evalidator.jsp

Miller JH, Chambliss EB, Bargeron CT (2004) Invasive plants of the thirteen southern states. Invasive and exotic species of North America. Online resource at http://www.invasive.org

Miller JH, Chambliss EB, Oswalt CM (2008) Maps of occupation and estimates of acres covered by nonnative invasive plants in southern forests using SRS FIA data. http://www.invasive.org/fiamaps/

Miller JH, Lemke D, Coulston J (2013a) The invasion of southern forests by nonnative plants: current and future occupation, with impacts, management strategies, and mitigation approaches. Chapter 15. In: Wear DN, Greis JG (eds) The southern Forest futures project: technical report. general technical report SRS-GTR-178. USDA-Forest Service, Southern Research Station, Asheville. 542 p

Miller JH, Manning ST, Enloe SF (2013b) A management guide for invasive plants in southern forests. (slightly revised 2013 and 2015). General technical report SRS–131. U.S. Department of Agriculture Forest Service, Southern Research Station, Asheville. 120 p

Minogue PJ, Jones S, Bohn KK, Chevasco ED (2009) Biology and control of Japanese climbing fern (*Lygodium japonicum*). Publication FOR-218, University of Florida IFAS Extension. https://edis.ifas.ufl.edu/fr280

Molina Colón S, Lugo AE (2006) Recovery of a subtropical dry forest after abandonment of different land uses. Biotropica 38(3):354–364

Molina Colón S, Lugo AE, Ramos González OM (2011) Novel dry forests in southwestern Puerto Rico. For Ecol Manag 262:170–177

Morris JA Jr, Akins JL (2009) Feeding ecology of invasive lionfish (*Pterois volitans*) in the Bahamian archipelago. Environ Biol Fish 86:389–398

Munger GT (2005) *Lygodium* spp. In: Fire effects information system, [Online]. U.S. Department of Agriculture, Forest Service, Rocky Mountain Research Station, Fire Sciences Laboratory (Producer). Available: http://www.fs.fed.us/database/feis/plants/fern/lygspp/all.html. 29 December 2016

Muñoz MC, Ackerman JD (2011) Spatial distribution and performance of native and invasive *Ardisia* (Myrsinaceae) species in Puerto Rico: the anatomy of an invasion. Biol Invasions 13:1543–1558

North Carolina—Department of Agriculture and Consumer Services (NC-DACS) (2016) Plant industry—plant protection section gypsy moth quarantined area in North Carolina. http://www.ncagr.gov/plantindustry/Plant/entomology/gypqua.htm

Nesslage GM, Wainger LA, Harms NE, Cofrancesco AF (2016) Quantifying the population response of invasive water hyacinth, *Eichhornia crassipes*. Biol Invasions 18:2107–2115

Newhouse AE, Polin-McGuigan LD, Baier KA et al (2014) Transgenic American chestnuts show enhanced blight resistance and transmit the trait to T1 progeny. Plant Sci 228:88–97

Oberle MW (2000) Puerto Rico's birds in photographs. Editorial Humanitas, San Juan. 129 p

Onken BP, Reardon RC (2011) An overview and outlook for biological control of hemlock woolly adelgid. In: Onken B, Reardon R (eds) Implementation and status of biological control of the hemlock woolly adelgid. USDA Forest Service, Forest Health Technology Enterprise Team, Publication FHTET-2011-04, pp 222–228

Ostry ME, Ellingson B, Seekins D, Ruckheim W (2003) The need for silvicultural practices and collection of butternut germplasm for species conservation. In: Van Sambeek JW, Dawson JO, Ponder F Jr, Loewenstein EF, Fralish JS (eds) Proceedings, 13th central hardwood Forest conference; 2002 April 1–3; Urbana, IL. general technical report NC-234. U.S. Department of Agriculture, Forest Service, North Central Research Station, St. Paul, pp 551–555

Oten KL, Merkle SA, Jetton RM et al (2014) Understanding and developing resistance in hemlocks to the hemlock woolly adelgid. Southeast Nat 13:147–167

Padilla DK, Williams SL (2004) Beyond ballast water: aquarium and ornamental trades as sources of invasive species in aquatic ecosystems. Front Ecol Environ 2(3):131–138

Pérez-Buitrago N, Sabat A (2007) Natal dispersal, home range and habitat use of Mona Island iguana (*Cyclura cornuta stejneri*). Appl Herpetol 4:365–376

Piaggio AJ, Engeman RM, Hopken MW et al (2014) Detecting an elusive invasive species: a diagnostic PCR to detect Burmese python in Florida waters and an assessment of persistence of environmental DNA. Mol Ecol Resour 14(2):374–380

Piatek KB, Fajvan MA, Turcotte RM (2016) Thinning effects on foliar elements in eastern hemlock: implications for managing the spread of the hemlock woolly adelgid. Can J For Res 47:81–88

Pimentel D (1955) Biology of the Indian mongoose in Puerto Rico. J Mammal 36(1):62–68

Pimentel D (2007) Environmental and economic costs of vertebrate species invasions into the United States. Managing vertebrate invasive species. In: Witmer GW, Pitt WC, Fagerstone KA (eds) Managing vertebrate invasive species: proceedings of an international symposium. USDA/APHIS Wildlife Services, National Wildlife Research Center, Fort Collins, pp 2–8. https://www.aphis.usda.gov/wildlife_damage/nwrc/symposia/invasive_symposium/nwrc_TOC_index.shtml

Pimentel D, Zuniga R, Morrison D (2005) Update on the environmental and economic costs associated with alien-invasive species in the United States. Ecol Econ 52:273–288

Platenberg RJ (2007) Impacts of introduced species on an island ecosystem: non-native reptiles and amphibians in the U.S. Virgin Islands. U.S. Department of Agriculture

National Wildlife Research Center Symposia. Managing vertebrate invasive species. Paper 39:168–174

Platenberg RJ, Boulon RH Jr (2006) Conservation status of reptiles and amphibians in the U.S. Virgin Islands. Appl Herpetol 3:215–235

Platt WJ, Gottschalk RM (2001) Effects of exotic grasses on potential fine fuel loads in the groundcover of South Florida slash pine savannas. Int J Wildland Fire 10(2):155–159

Poland TM, Chen Y, Koch J, Pureswaran D (2015) Review of the emerald ash borer (Coleoptera: Buprestidae), life history, mating behaviours, host plant selection, and host resistance. Can Entomol 147:252–262

Post W, Wiley JW (1977) The shiny cowbird in the West Indies. Condor 79(1):119–121

Post W, Cruz A, McNair DB (1993) The North American invasion pattern of the shiny cowbird (El patrón de invasión a Norte América por parte de *Molothrus bonariensis*). J Field Ornithol 64(1):32–41

Potter KM, Frampton J, Sidebottom J (2005) Impacts of balsam woolly adelgid on the southern Appalachian spruce-fir ecosystem and the North Carolina Christmas tree industry. In: Third symposium on Hemlock Woolly Adelgid in the Eastern United States, Asheville, NC. USFS Forest Health Technology Enterprise Team, pp 25–41

Pratt PD, Quevedo V, Bernier L et al (2005) Invasion of Puerto Rican wetlands by the Australian tree *Melaleuca quinquenervia*. Caribb J Sci 41(1):42–54

Raffaelle HA (1989) A guide to the birds of Puerto Rico and the Virgin Islands. Princeton University Press, Oxford. 261 p

Recart W, Ackerman JD, Cuevas AA (2013) There goes the neighborhood: apparent competition between invasive and native orchids mediated by a specialist florivorous weevil. Biol Invasions 15:283–293

Reilly JR, Hajek AE, Liebhold AM, Plymale R (2014) Impact of *Entomophaga maimaiga* (Entomophthorales: Entomophthoraceae) on outbreak gypsy moth populations (Lepidoptera: Erebidae): the role of weather. Environ Entomol 43:632–641

Rivera Y, Kretzer AM, Horton TR (2015) New microsatellite markers for the ectomycorrhizal fungus *Pisolithus tinctorius sensu stricto* reveal the genetic structure of US and Puerto Rican populations. Fungal Ecol 13:1–9

Reserva Natural de Humacao (RNH) (2015) Departamento de Recursos Naturales y Ambientales, Estado Libre Asociado de Puerto Rico. http://www.reservanaturalhumacao.com/index_eng.html

Rodgers L, Derksen A, Pernas T (2014) Expansion and impact of laurel wilt in the Florida Everglades. Fla Entomol 97:1247–1250

Rodríguez-Durán A, Pérez J, Montalbán MA, Sandoval JM (2010) Predation by free-roaming cats on an insular population of bats. Acta Chiropterologica 12(2):359–362

Rogers CS, Willette DA, Miller J (2014) Rapidly spreading seagrass invades the Caribbean with unknown ecological consequences. Front Ecol Environ 12(10):546–547

Rojas-Sandoval J, Acevedo-Rodrıguez P (2015) Naturalization and invasion of alien plants in Puerto Rico and the Virgin Islands. Biol Invasions 17:149–163

Rojas-Sandoval J, Meléndez-Ackerman EJ, Anglés-Alcázar D (2016a) Assessing the impact of grass invasion on the population dynamics of a threatened Caribbean dry forest cactus. Biol Conserv 196:156–164

Rojas-Sandoval J, Meléndez-Ackerman EJ, Fumero-Cabán J et al (2016b) Long-term vegetation dynamics and responses to ungulate exclusion in the dry forest of Mona Island. Caribbean Nat Spec Issue 1:138–156

Ross-Davis AM, Graca RN, Alfena AC et al (2013) Tracking the distribution of *Puccinia psidii* genotypes that cause rust disease on diverse myrtaceous trees and shrubs. In: Chadwick K (Comp) Proceedings of the 61st annual western international forest disease work conference, Alberta, Canada: Western Forestry and Conservation Association, pp 131–137

Rudel TK (1998) Is there a forest transition? Deforestation, reforestation, and development. Rural Sociol 63(4):533–552

Rugman-Jones PF, Seybold SJ, Graves AD, Stouthamer R (2015) Phylogeography of the walnut twig beetle, *Pityophthorus juglandis*, the vector of thousand cankers disease in North American walnut trees. PLoS One 10(2):e0118264. https://doi.org/10.1371/journal.pone.0118264

Ruiz-Carus R, Matheson RE, Roberts DE, Whitfield PE (2006) The western Pacific red lionfish, *Pterois volitans* (Scorpaenidae), in Florida: evidence for reproduction and parasitism in the first exotic marine fish established in state waters. Biol Conserv 128(3):384–390

Russello MA, Avery ML, Wright TF (2008) Genetic evidence links invasive monk parakeet populations in the United States to the international pet trade. BMC Evol Biol 8(1):1

Sax DF, Stachowicz JJ, Brown JH et al (2007) Ecological and evolutionary insights from species invasions. Trends Ecol Evol 66(9):465–471

Seaman GA, Randall JE (1962) The mongoose as a predator in the Virgin Islands. J Mammal 43(4):544–546

Segarra-Carmona AE, Otero M (2014) Four new ladybug species belonging to *Decadiomus* Chapin (Coleoptera: Coccinellidae) from Puerto Rico. Neotrop Entomol 43(6):555–563

Segarra-Carmona AE, Ramírez-Lluch A, Cabrera-Asencio I, Jiménez-López AN (2010) First report of a new invasive mealybug, the Harrisia cactus mealybug *Hypogeococcus pungens* (Hemiptera: Pseudococcidae). J Agric Univ P R 94(1/2):183–187

Sharov AA, Leonard DS, Liebhold AM et al (2002) Slow the spread: a national program to contain the gypsy moth. J For 100:30–35

Shearman TM, Wang GG, Bridges WC (2015) Population dynamics of redbay (*Persea borbonia*) after laurel wilt disease: an assessment based on forest inventory and analysis data. Biol Invasions 17:1371–1382

Snyder NFR, Wiley JW, Kepler CB (1987) The parrots of Luquillo: natural history and conservation of the Puerto Rican parrot. Western Foundation of Vertebrate Zoology, Los Angeles. 384 p

The Humane Society of the United States (2016) Questions and answers about inmunocontraception—a safe way to control populations humanely. http://www.humanesociety.org/issues/wildlife_overpopulation/qa/q_and_a_pzp.html

Thompson J, Lugo AE, Thomlinson J (2007) Land use history, hurricane disturbance, and introduced species survival in a subtropical wet forest in Puerto Rico. Plant Ecol 192:289–301

Timmons J, Higgenbotham B, Lopez R et al (2012) Feral hog population growth, density and harvest in Texas. Texas A&M Institute of Renewable National Resources. Report no. SP-472

Tobin PC (2008) Cost analysis and biological ramifications for implementing the gypsy moth slow the spread program. General technical report NRS-37. U.S. Department of Agriculture, Forest Service, Northern Research Station, Newtown Square. 21 p

Townsend JH, Krysko KL, Enge KM (2003) Introduced iguanas in southern Florida: a history of more than 35 years. Iguana 10(4):111–118

UNEP (United Nations Environment Programme) (2008) Climate change in the Caribbean and the challenge of adaptation. United Nations Environment Programme, Regional Office for Latin America and the Caribbean (ROLAC). Panama City, Panama. 91 p

U.S. Department of Agriculture APHIS (Animal and Plant Health Inspection Service), NWRC (National Wildlife Research Center) (2013) Development of injectable and oral contraceptive technologies and their assessment for wildlife population and disease management. https://www.aphis.usda.gov/wildlife_damage/nwrc/research/reproductive_control/index.shtml

U.S. Department of Agriculture APHIS (Animal and Plant Health Inspection Service), PPQ (Plant Protection and Quarantine) (2009) CPHST (Center for Plant Health Science and Technology) Biological Control Unit 2009 annual report. Raleigh. 53 p

U.S. Department of the Interior FWS (Fish and Wildlife Service) Southeast Region (2012) Plan abarcador de conservación (borrador) y evaluación ambiental para el Refugio Nacional de Vida Silvestre de Culebra, Puerto Rico. Atlanta. 181 p. http://www.fws.gov/southeast/planning/PDFdocuments/Culebra%20Draft%20Spanish/Culebra%20Draft%20CCP&EA%20Espa%C3%B1ol.pdf

U.S. Department of the Interior National Park Service Southeast Region (2003) Sustained reduction plan for non-native wild hogs within Virgin Islands National Park. Virgin Islands National Park Final Environmental Assessment. 94 p. https://www.nps.gov/viis/learn/nature/upload/nps_hogs_3002.pdf

U.S. Department of Transportation Federal Aviation Administration and U.S. Department of Agriculture APHIS (Animal and Plant Health Inspection Service) (2014) Wildlife strikes to civil aircraft in the United States 1990–2013. National Wildlife Strike Database Serial Report Number 20, Washington, DC. 98 p. http://wildlife.faa.gov/downloads/Wildlife-Strike-Report-1990-2013-USDA-FAA.pdf

USDA Forest Service (2016) Hemlock woolly adelgid infestation 2015 distribution map. Northeastern Area, Forest Health Protection. http://www.na.fs.fed.us/fhp/hwa/maps/2015_HWA_Infestation_Map_20160502.pdf

USDA-NRCS (United States Department of Agriculture—Natural Resources Conservation Service) (2017) Introduced, invasive, and noxious plants: state noxious weeds. https://plants.usda.gov/java/noxComposite?stateRpt=yes

Villamagna AM, Murphy BR (2010) Ecological and socio-economic impacts of invasive water hyacinth (*Eichhornia crassipes*): a review. Freshw Biol 55:282–298

Van Loan AN (2006) Japanese climbing fern: the insidious "other" Lygodium. Wildland Weeds Spring 2006:25–27

Vose JM, Wear DN, Mayfield AE, Nelson CD (2013) Hemlock woolly adelgid in the southern Appalachians: control strategies, ecological impacts, and potential management responses. For Ecol Manag 291:209–219

Wheeler GS, Ding J (2014) Is Chinese tallowtree, *Triadica sebifera*, an appropriate target for biological control in the United States? Invasive Plant Sci Manag 7:345–359

Wiewandt TA, García MA (2000) Taxonomic account: Mona Island Iguana. In: Alberts A (ed) West Indian iguanas: status survey and conservation action plan. International Union for Conservation of Nature, West Indian Iguana Specialist Group, Gland/Cambridge, pp 27–31

Willette DA, Ambrose RF (2012) Effects of the invasive seagrass *Halophila stipulacea* on the native seagrass, *Syringodium filiforme*, and associated fish and epibiota communities in the eastern Caribbean. Aquat Bot 103:74–82

Willette DA, Chalifour J, Dolfi Debrot AO et al (2014) Continued expansion of the trans-Atlantic invasive marine angiosperm *Halophila stipulacea* in the eastern Caribbean. Aquat Bot 112:98–102

Willig MR, Gannon MR (1996) Mammals. In: Reagan DP, Waide RB (eds) The food web of a tropical rain forest. The University of Chicago Press, Chicago, pp 399–431

Wolfe BT, Van Bloem SJ (2012) Subtropical dry forest regeneration in grass-invaded areas of Puerto Rico: understanding why *Leucaena leucocephala* dominates and native species fail. For Ecol Manag 267:253–261

Zhang Y, Hanula JL, O'Brien J et al (2013) Evaluation of the impacts of herbivory by lace bugs on Chinese privet (*Ligustrum sinense*) survival and physiology. Biol Control 64:299–304

Zimmermann HG, Pérez M, Sandi Cuen MC et al (2010) The south American mealybug that threatens North American cacti. Cactus Succulent J 82(3):105–107

Zou X, González G (1997) Changes in earthworm density and community structure during secondary succession in abandoned tropical pastures. Soil Biol Biochem 29(314):621–629